AF324920

NEURAL REPAIR AND REGENERATION AFTER SPINAL CORD INJURY AND SPINE TRAUMA

NEURAL REPAIR AND REGENERATION AFTER SPINAL CORD INJURY AND SPINE TRAUMA

Edited by

MICHAEL G. FEHLINGS

Professor of Neurosurgery, Vice Chair Research, Department of Surgery, Halbert Chair in Neural Repair and Regeneration, Co-Chairman Spinal Program, University of Toronto and Staff Neurosurgeon Senior Scientist Krembil Brain Institute, Toronto Western Hospital, University Health Network

BRIAN K. KWON

Canada Research Chair in Spinal Cord Injury and Dvorak Chair in Spine Trauma, Professor of Orthopaedics, University of British Columbia, International Collaboration on Repair Discoveries (ICORD) and Vancouver General Hospital, Vancouver, British Columbia, Canada

ALEXANDER R. VACCARO

The Everett J. and Marion Gordon Professor of Orthopaedic Surgery; Professor of Neurosurgery, Co-Director of the Delaware Valley Spinal Cord Injury Center; Co-Chief Spine Surgery, Co-Director Spine Surgery; Thomas Jefferson University and the Rothman Institute, Philadelphia, PA, United States

F. CUMHUR ONER

University Medical Centre Utrecht, Utrecht, The Netherlands

ACADEMIC PRESS

An imprint of Elsevier

ELSEVIER

Academic Press is an imprint of Elsevier
125 London Wall, London EC2Y 5AS, United Kingdom
525 B Street, Suite 1650, San Diego, CA 92101, United States
50 Hampshire Street, 5th Floor, Cambridge, MA 02139, United States
The Boulevard, Langford Lane, Kidlington, Oxford OX5 1GB, United Kingdom

Copyright © 2022 Elsevier Inc. All rights reserved.

No part of this publication may be reproduced or transmitted in any form or by any means, electronic or mechanical, including photocopying, recording, or any information storage and retrieval system, without permission in writing from the publisher. Details on how to seek permission, further information about the Publisher's permissions policies and our arrangements with organizations such as the Copyright Clearance Center and the Copyright Licensing Agency, can be found at our website: www.elsevier.com/permissions.

This book and the individual contributions contained in it are protected under copyright by the Publisher (other than as may be noted herein).

Notices
Knowledge and best practice in this field are constantly changing. As new research and experience broaden our understanding, changes in research methods, professional practices, or medical treatment may become necessary.

Practitioners and researchers must always rely on their own experience and knowledge in evaluating and using any information, methods, compounds, or experiments described herein. In using such information or methods they should be mindful of their own safety and the safety of others, including parties for whom they have a professional responsibility.

To the fullest extent of the law, neither the Publisher nor the authors, contributors, or editors, assume any liability for any injury and/or damage to persons or property as a matter of products liability, negligence or otherwise, or from any use or operation of any methods, products, instructions, or ideas contained in the material herein.

Library of Congress Cataloging-in-Publication Data
A catalog record for this book is available from the Library of Congress

British Library Cataloguing-in-Publication Data
A catalogue record for this book is available from the British Library

ISBN: 978-0-12-819835-3

For information on all Academic Press publications visit our website
at https://www.elsevier.com/books-and-journals

Publisher: Nikki P. Levy
Acquisitions Editor: Joslyn Chaiprasert-Paguio
Editorial Project Manager: Samantha Allard
Production Project Manager: Niranjan Bhaskaran
Cover Designer: Victoria Pearson

Typeset by TNQ Technologies

This book is dedicated to my family and friends, especially my wife Lauren, who have supported me throughout this wonderful journey of learning and teaching over the decades

—Dr. Vaccaro

This book is dedicated to my wonderful family, especially my wife Darcy, to my patients, colleagues, and students

—Dr. Fehlings

This book is dedicated to my wife Colleen and daughter Caroline, who have sacrificed much in supporting me through a career in pursuit of a better future for spinal cord injured patients

—Dr. Kwon

Contents

1. Anatomy

LAUREEN D. HACHEM, ALI MOGHADDAMJOU, AND
MICHAEL G. FEHLINGS

2. Epidemiology

THORSTEN JENTZSCH, ANOUSHKA SINGH, AND
MICHAEL G. FEHLINGS

3. Classification systems: spine trauma

ARIANA A. REYES, SRIKANTH N. DIVI,
THOMAS J. LEE, DHRUV GOYAL, AND
ALEXANDER R. VACCARO

4. Classification systems: SCI

SUKHVINDER KALSI-RYAN AND
GITA GHOLAMREZAEI

5. Outcome measures

JETAN H. BADHIWALA, CHRISTOPHER D. WITIW,
HETSHREE JOSHI, OMAR KHAN, AND
SUKHVINDER KALSI-RYAN

6. Imaging: spine trauma

PARTHIK D. PATEL, MICHAEL MARKOWITZ, SRIKANTH N. DIVI,
GREGORY D. SCHROEDER, AND ALEXANDER R. VACCARO

7. Advanced imaging for spinal cord injury

MUHAMMAD ALI AKBAR, ALLAN R. MARTIN, DARIO PFYFFER,
DAVID W. CADOTTE, SHEKAR KURPAD, PATRICK FREUND,
AND MICHAEL G. FEHLINGS

8. Intraoperative imaging and image guidance

DAIPAYAN GUHA, ADAM A. DMYTRIW, JAMES D. GUEST, AND
VICTOR X.D. YANG

9. Upper cervical spine and spinal cord injuries

ERIK HAYMAN, ROD J. OSKOUIAN, AND JENS R. CHAPMAN

10. Spine trauma management issues: C-spine

JARED T. WILCOX, MINA AZIZ, RAKAN BOKHARI,
SOLON SCHUR, LIOR ELKAIM, MICHAEL H. WEBER,
AND CARLO SANTAGUIDA

11. Spine trauma management issues: thoracic and lumbar

DAVID BEN-ISRAEL AND W. BRADLEY JACOBS

12. Spine trauma: sacral fractures

CARLO BELLABARBA, HAITAO ZHOU, AND
RICHARD J. BRANSFORD

List of contributors

Bizhan Aarabi Department of Neurosurgery, University of Maryland Medical Center and R Adams Cowley Shock Trauma Center, Baltimore, MD, United States

Christopher S. Ahuja Institute of Medical Science, University of Toronto, Toronto, ON, Canada; Division of Genetics and Development, Krembil Research Institute, University Health Network, Toronto, ON, Canada; Division of Neurosurgery, Department of Surgery, University of Toronto, Toronto, ON, Canada

Muhammad Ali Akbar Division of Neurosurgery, Department of Surgery, University of Toronto, Toronto, ON, Canada; Spine Program, Krembil Neuroscience Centre, Toronto Western Hospital, University Health Network, Toronto, ON, Canada

Mina Aziz Montreal University Health Center, McGill University, Montreal, Quebec, Canada

Jetan H. Badhiwala Division of Neurosurgery, Department of Surgery, University of Toronto, Toronto, ON, Canada

Carlo Bellabarba Department of Orthopaedics and Sports Medicine, Department of Neurological Surgery, University of Washington, Harborview Medical Center, Seattle, WA, United States

David Ben-Israel Division of Neurosurgery, Department of Clinical Neurosciences, University of Calgary, Calgary, AB, Canada

Rakan Bokhari Montreal University Health Center, McGill University, Montreal, Quebec, Canada

Richard J. Bransford Department of Orthopaedics and Sports Medicine, Department of Neurological Surgery, University of Washington, Harborview Medical Center, Seattle, WA, United States

Anthony S. Burns Division of Physical Medicine & Rehabilitation, Department of Medicine, University of Toronto, Toronto, ON, Canada

David W. Cadotte Division of Neurosurgery, Department of Clinical Neurosciences, University of Calgary, Calgary, AB, Canada; Combined Spine Program, Hotchkiss Brain Institute, University of Calgary, Calgary, AB, Canada

Steven Casha Hotchkiss Brain Institute, University of Calgary Spine Program, Department of Clinical Neurosciences, University of Calgary, Calgary, AB, Canada

Brian C.F. Chan KITE Research Institute, University Health Network, Toronto, ON, Canada; Institute of Health Policy, Management and Evaluation, University of Toronto, Toronto, ON, Canada

Vivien K.Y. Chan Division of Neurosurgery, Department of Surgery, University of Toronto, Toronto, ON, Canada

Jens R. Chapman Swedish Neuroscience Institute, Seattle, WA, United States

Newton Cho University of Toronto, Toronto, ON, Canada

B. Catharine Craven Lyndhurst Centre, Toronto Rehabilitation Institute, University Health Network, Toronto, ON, Canada; KITE Research Institute, University Health Network, Toronto, ON, Canada; Department of Medicine, Division of Physical Medicine and Rehabilitation, Institute of Medical Science, University of Toronto, Toronto, ON, Canada

Srikanth N. Divi Department of Orthopaedic Surgery, Rothman Orthopedic Institute, Thomas Jefferson University, Philadelphia, PA, United States; Northwestern University Feinberg School of Medicine, Chicago, IL, United States

Adam A. Dmytriw Massachusetts General Hospital, Harvard Medical School, Boston, MA, United States

Lior Elkaim Montreal University Health Center, McGill University, Montreal, Quebec, Canada

Michael G. Fehlings Institute of Medical Science, University of Toronto, Toronto, ON, Canada; Division of Neurosurgery, Department of Surgery, University of Toronto, Toronto, ON, Canada; Division of Genetics and Development, Krembil Research Institute, University Health Network, Toronto, ON, Canada; Division of Neurosurgery, Toronto Western Hospital, University Health Network, Toronto, ON, Canada; Temerty Faculty of Medicine, University of Toronto, Toronto, ON, Canada; Division of Genetics and Development, Krembil Brain Institute, University Health Network, Toronto, ON, Canada; Spine Program, Krembil Neuroscience Centre, Toronto Western Hospital, University Health Network, Toronto, ON, Canada

Anton Fomenko Section of Neurosurgery, Health Sciences Centre, Winnipeg, MB, Canada

Patrick Freund Spinal Cord Injury Center, Balgrist University Hospital, University of Zurich, Zurich, Switzerland

Julio C. Furlan Lyndhurst Centre, Toronto Rehabilitation Institute, University Health Network, Toronto, ON, Canada; KITE Research Institute, University Health Network, Toronto, ON, Canada; Department of Medicine, Division of Physical Medicine and Rehabilitation, Institute of Medical Science, University of Toronto, Toronto, ON, Canada; Rehabilitation Sciences Institute, University of Toronto, Toronto, ON, Canada; Institute of Health Policy, Management and Evaluation, University of Toronto, Toronto, ON, Canada

Gita Gholamrezaei KITE-UHN, Toronto Rehabilitation Institute, Toronto, ON, Canada

Christina L. Goldstein University of Colorado Health - Memorial Hospital, Colorado Springs, CO, United States

Alwyn Gomez Section of Neurosurgery, Health Sciences Centre, Winnipeg, MB, Canada

Dhruv Goyal Department of Orthopaedic Surgery, Rothman Orthopedic Institute, Thomas Jefferson University, Philadelphia, PA, United States

James D. Guest Department of Neurological Surgery, Miller School of Medicine, University of Miami, Miami, FL, United States; The Miami Project to Cure Paralysis and Neurosurgery, University of Miami Miller School of Medicine, Miami, FL, United States

Daipayan Guha Division of Neurosurgery, McMaster University, Toronto, ON, Canada

Laureen D. Hachem Division of Neurosurgery, Department of Surgery, University of Toronto, Toronto, ON, Canada

Gregory W.J. Hawryluk Section of Neurosurgery, Health Sciences Centre, Winnipeg, MB, Canada

Erik Hayman Department of Neurosurgery, University of South Florida, Tampa, FL, United States

Nader Hejrati Division of Genetics and Development, Krembil Brain Institute, University Health Network, Toronto, ON, Canada

James Hong Krembil Research Institute, Toronto, ON, Canada

W. Bradley Jacobs Division of Neurosurgery, Department of Clinical Neurosciences, University of Calgary, Calgary, AB, Canada

Thorsten Jentzsch Division of Neurosurgery, Department of Surgery, University of Toronto, Toronto, ON, Canada; Division of Neurosurgery, Toronto Western Hospital, University Health Network, Toronto, ON, Canada; Department of Orthopaedics, Balgrist University Hospital, University of Zurich, Zurich, Switzerland

Fan Jiang Toronto Western Hospital, Toronto, Canada; Division of Neurosurgery, Toronto Western Hospital, University Health Network, Toronto, ON, Canada

Hetshree Joshi Division of Neurosurgery, Department of Surgery, University of Toronto, Toronto, ON, Canada

Sukhvinder Kalsi-Ryan KITE-UHN, Toronto Rehabilitation Institute, Department of Physical Therapy, University of Toronto, Toronto, ON, Canada; KITE Research Institute, University Health Network, Toronto, ON, Canada

Omar Khan Division of Neurosurgery, Department of Surgery, University of Toronto, Toronto, ON, Canada

Paul A. Koljonen University of Hong Kong, Hong Kong Special Administrative Region, China

Mark R.N. Kotter Division of Neurosurgery, Department of Clinical Neurosciences, University of Cambridge, Cambridge, United Kingdom

Shekar Kurpad Department of Neurosurgery, Medical College of Wisconsin, Milwaukee, WI, United States

Brian K. Kwon International Collaboration on Repair Discoveries, University of British Columbia, Vancouver, BC, Canada; Vancouver Spine Surgery Institute, Vancouver General Hospital, Department of Orthopaedics, UBC, Vancouver, BC, Canada

Mark J. Lambrechts Rothman Orthopedic Institute at Thomas Jefferson University, Philadelphia, PA, United States

Jeremie Larouche University of Toronto, Toronto, ON, Canada

Thomas J. Lee Department of Orthopaedic Surgery, Rothman Orthopedic Institute, Thomas Jefferson University, Philadelphia, PA, United States

Frank Lyons Mater Misericordiae University Hospital, Dublin, Ireland

Armaan K. Malhotra Division of Neurosurgery, Department of Surgery, University of Toronto, Toronto, ON, Canada

Michael Markowitz Rothman Orthopaedic Institute, Philadelphia, PA, United States

Allan R. Martin Department of Neurological Surgery, University of California, Davis, CA, United States

William Brett McIntyre Institute of Medical Science, University of Toronto, Toronto, ON, Canada

Joseph H. McMordie Department of Neurosurgery, University of Nebraska Medical Center, Omaha, NE, United States

Ali Moghaddamjou Division of Neurosurgery, Department of Surgery, University of Toronto, Toronto, ON, Canada

Andrea J. Mothe Krembil Research Institute, Toronto Western Hospital, University Health Network, Toronto, ON, Canada

Narihito Nagoshi Department of Orthopaedic Surgery, Keio University School of Medicine

F. Cumhur Oner Department of Orthopaedics, University Medical Center Utrecht, Utrecht, the Netherlands

Rod J. Oskouian Swedish Neuroscience Institute, Seattle, WA, United States

Sophie Ostmeier Division of Genetics and Development, Krembil Brain Institute, University Health Network, Toronto, ON, Canada

Parthik D. Patel Rothman Orthopaedic Institute, Philadelphia, PA, United States

Dario Pfyffer Spinal Cord Injury Center, Balgrist University Hospital, University of Zurich, Zurich, Switzerland

Katarzyna Pieczonka Institute of Medical Science, University of Toronto, Toronto, ON, Canada

Noah Poulin University of Cambridge, Cambridge, United Kingdom

Nayaab Punjani Institute of Medical Science, University of Toronto, Toronto, ON, Canada; Division of Genetics and Development, Krembil Research Institute, University Health Network, Toronto, ON, Canada

Ariana A. Reyes Department of Orthopaedic Surgery, Rothman Orthopedic Institute, Thomas Jefferson University, Philadelphia, PA, United States

Zaid Salaheen Temerty Faculty of Medicine, University of Toronto, Toronto, ON, Canada

Carlo Santaguida Montreal University Health Center, McGill University, Montreal, Quebec, Canada

Andrea J. Santamaria The Miami Project to Cure Paralysis, University of Miami Miller School of Medicine, Miami, FL, United States

Pedro M. Saraiva The Miami Project to Cure Paralysis, University of Miami Miller School of Medicine, Miami, FL, United States

Gregory D. Schroeder Rothman Orthopaedic Institute, Philadelphia, PA, United States

Solon Schur Montreal University Health Center, McGill University, Montreal, Quebec, Canada

Vjura Senthilnathan Division of Genetics and Development, Krembil Research Institute, University Health Network, Toronto, ON, Canada; University of Toronto Scarborough, Toronto, ON, Canada

Anoushka Singh Division of Neurosurgery, Toronto Western Hospital, University Health Network, Toronto, ON, Canada

Juan P. Solano Pediatric Critical Care, University of Miami Miller School of Medicine, Miami, FL, United States

Charles H. Tator Professor of Neurosurgery, University of Toronto, and Krembil Brain Institute, Toronto Western Hospital, Founder, ThinkFirst Canada and Parachute Canada

Alexander R. Vaccaro Department of Orthopedic Surgery, Rothman Orthopaedic Institute, Thomas Jefferson University, Philadelphia, PA, United States

Paula Valerie ter Wengel Leiden University Medical Center, Leiden, the Netherlands; HMC, The Hague, the Netherlands; Amsterdam UMC, Amsterdam, the Netherlands

Michael H. Weber Montreal University Health Center, McGill University, Montreal, Quebec, Canada

Jared T. Wilcox Montreal University Health Center, McGill University, Montreal, Quebec, Canada

Jamie R.F. Wilson Department of Neurosurgery, University of Nebraska Medical Center, Omaha, NE, United States

Jefferson R. Wilson St. Michael's Hospital, Toronto, ON, Canada; Division of Neurosurgery, Department of Surgery, University of Toronto, Toronto, ON, Canada

Christopher D. Witiw Division of Neurosurgery, Department of Surgery, University of Toronto, Toronto, ON, Canada; Division of Neurosurgery, St. Michael's Hospital, Toronto, ON, Canada

Ian H.Y. Wong Division of Neurosurgery, Department of Surgery, University of Toronto, Toronto, ON, Canada

Victor X.D. Yang Division of Neurosurgery, University of Toronto, Toronto, ON, Canada

Haitao Zhou Department of Orthopaedics and Sports Medicine, Department of Neurological Surgery, University of Washington, Harborview Medical Center, Seattle, WA, United States

Editor bios

MICHAEL G. FEHLINGS

Dr. Fehlings is the Vice Chair Research for the Department of Surgery at the University of Toronto and a Neurosurgeon at Toronto Western Hospital, University Health Network. Dr. Fehlings is a Professor of Neurosurgery at the University of Toronto, holds the Gerry and Tootsie Halbert Chair in Neural Repair and Regeneration, is a Senior Scientist at the Krembil Research Institute, and a McLaughlin Scholar in Molecular Medicine. In the fall of 2008, Dr. Fehlings was appointed the inaugural Director of the University of Toronto Neuroscience Program (which he held until June 2012) and is currently Co-Director of the University of Toronto Spine Program. Dr. Fehlings combines an active clinical practice in complex spinal surgery with a translationally oriented research program focused on discovering novel treatments to improve functional outcomes following spinal cord injury (SCI). He has published over 1000 peer-reviewed articles chiefly in the area of central nervous system injury and complex spinal surgery. Dr. Fehlings has won numerous international awards for his work and holds a number of prominent international positions.

BRIAN K. KWON

Dr. Brian Kwon is a Professor in the Department of Orthopaedics at the University of British Columbia, the Canada Research Chair in Spinal Cord Injury, and holds the Dvorak Chair in Spine Trauma. He is an attending spine surgeon at Vancouver General Hospital, a level 1 trauma center and regional referral center for SCIs. As a surgeon-scientist and the current Chair of the AO Spine Knowledge Forum in Spinal Cord Injury, he is particularly interested in the bidirectional process of translational research for spinal cord injury, and he leads a research program focused on translation at the International Collaboration on Repair Discoveries. He has worked extensively on establishing biomarkers of human SCI to understand the biology of human injury and to better stratify injury severity and improve the prediction of neurologic outcome. Dr. Kwon has led the development of a novel large animal model of SCI and is utilizing this for both bench-to-bedside and bedside-back-to-bench translational studies.

CUMHUR ONER

Dr. Cumhur Öner obtained his Medical Degree in 1980 in Ankara, Turkey, and did his orthopaedic surgery training in Rotterdam between 1987 and 1993. Since 1993, he has worked as an orthopaedic spine surgeon at the University Medical Centre Utrecht. He obtained his PhD (cum laude) in 1999 from the University of Utrecht. The work for his PhD thesis focused on diagnostic and prognostic parameters in spinal trauma patients. He has been the head of the neuro-orthopaedic spine unit since 2002.

Cumhur Öner has actively participated in the International Spine Trauma Study Group and is currently a member of the steering committee of the AO Spine International Knowledge Forum Trauma. He has been involved in clinical spine research as well as the bone and intervertebral disc RM program at the UMC Utrecht. He was the president of the Dutch Spine Society between 2007 and 2011.

ALEXANDER VACCARO

Dr. Alexander Vaccaro graduated Summa Cum Laude from Boston College in 1983 with a BS in Biology. He received his MD degree from Georgetown University School of Medicine where he was promoted with "Distinction." He earned membership in the Alpha Omega Alpha (AOA) Honor Society and graduated with honors in 1987. Today, Dr. Vaccaro is the Richard H. Rothman Professor and Chairman, Department of Orthopaedic Surgery and Professor of Neurosurgery at Thomas Jefferson University in Philadelphia, Pennsylvania. He was the recipient of the Leon Wiltse award given for excellence in leadership and clinical research for spine care by the North American Spine Society (NASS) and is the past President of Cervical Spine Research Society (CSRS), the American Spinal Injury Association, and the Association for Collaborative Spine Research. Dr. Vaccaro is the President of Rothman Institute, Chairman of the department of Orthopaedic Surgery, Co-Director of the Regional Spinal Cord Injury Center of the Delaware Valley, and Co-Director of Spine Surgery and the Spine Fellowship program at Thomas Jefferson University Hospital where he instructs current fellows and residents in the diagnosis and treatment of various spinal problems and disorders.

The sine qua non encyclopedia for spinal cord neurotrauma, spine trauma, and neurorehabilitation

Charles H. Tator, OC, MD, PhD, FRCSC

Professor of Neurosurgery, University of Toronto, and Krembil Brain Institute, Toronto Western Hospital, Founder, ThinkFirst Canada and Parachute Canada

I have been a clinician and researcher in the spinal cord injury-spine trauma field for more than 50 years. It is my privilege and pleasure to be asked to write the foreword to this absolutely essential, encyclopedic textbook in my field. Furthermore, to be asked by my former student and current colleague, Michael Fehlings, brings me special satisfaction. I write this from the standpoint of looking back on the accomplishments and milestones that I have witnessed. One of my own accomplishments is exemplified by Michael Fehlings in whom I kindled an interest in this field, and in turn he has contributed so richly.

All who enter this field as scientists or clinicians, or both want to make a contribution by doing their best. For example, the patient seen on a Saturday evening who dove into shallow water that day and became quadriplegic or the scientist who wants to push the darkness away that still covers some of the key pathophysiological events of acute SCI or how to counteract the inhibitors preventing rejuvenation of the cord in chronic SCI. I am proud of Michael Fehlings' efforts to innovate, carefully study, teach, and do both clinical research and basic science research

in his lab. There are many others who followed Michael in my lab, and then in his lab who are also authors of important papers in this textbook.

Writing this foreword also reminded me of the heroes I have met who lived the experience of SCI—people like Stewart Yesner who started the International Spinal Research Trust, Chris Reeve who started the Chris and Dana Reeve Foundation, Rick Hansen who started the foundation in his name and then Praxis, and Craig Neilsen who started the Neilsen Foundation, and my patient, colleague, and friend Barbara Turnbull who started the Barbara Turnbull Foundation, the current funder with Brain Canada and the Government of Canada of the world's largest prize for neurotrauma research that bears both of our names. Of course, we must also include Dietrich Mateschitz who started Wings for Life to help his friend's son who had sustained a major SCI.

I have certainly been paid back multiple times over for my own contributions with the payback being the reward of being able to help people who really needed my help. SCI is in the category of "a life and death field" where what

"

you do can save a life, or take a life. Barbara Turnbull could breathe on her own when found by the emergency medical personnel. She was 18 years old, awake on the floor behind the counter of the milk store in which she was working the night shift. She was breathing but unable to move her arms and legs. In the ambulance on the way to the hospital, she stopped breathing and had to be "bagged." With that information, I decided to remove the shattered body of C4 and the bullet lodged extradurally all of which were compressing her spinal cord. My teaching indicated this mass needed to be removed right away, and my reward came three weeks later when she once again breathed without the respirator. My reward also came from Angela Macdonald after I did the same thing for her and rapidly removed the burst fracture of the C5 vertebral body causing her quadriparesis. She recently walked into my office and handed me the manuscript of her new book chronicling some of this and said, "Thank you doc, you saved my life." I never dreamed of such rewards as these!

My rewards have also been from the laboratory. Some of the highlights have been developing the first small animal model of SCI with which we shed some light on the profound vascular changes in the injured spinal cord leading to posttraumatic ischemia and infarction. A recent thrill was the restoration of some bladder function after major SCI in rats treated with an anti-inhibitor antibody.

The Fehlings et al textbook also reminds me of other rewards of working in this field, one of which is the opportunity to interact with health care professionals from so many disciplines who show the same goodwill and intentions to do their best. I have been privileged to work with them as a team all working urgently to get the patient through the storm in the best possible way, and to do the best possible research through a myriad of collaborations for the next breakthrough. Both clinically and in the lab, SCI cannot be done alone and "requires a whole village to raise these babies."

Thus, the personal rewards for me in SCI have been enormous and being asked to write this foreword by Michael Fehlings is one of them. Very likely many of the authors and readers of this textbook will have had or will have similar incredible thrills and rewards from being in this field. Hopefully, these chapters will inspire others to enter this "life and death field" of great importance and do their part to help those in need by their dedicated work in the OR, clinic, and laboratory.

Preface

Spinal cord injury and spine trauma are devastating events that dramatically impact the lives of the individuals affected, their families and friends, as well as the health care system. As spinal specialists, we are often one of the first to see spinal cord injured patients when they arrive at the hospital and are tasked with making critical acute treatment decisions, advising patients and their families, as well as guiding them through the hospital setting toward rehabilitation. Witnessing lives being turned upside down is a truly sobering experience, and while we remain inspired by the efforts made by our patients every day to rehabilitate from the injury, we desperately lack treatment options to aid this process.

Given that no effective treatment options exist to restore function in these patients after the injury, it is clear that there is much work to be done to develop neuroprotective and neuroregenerative tools to help spinal cord injury patients regain function. The past ten years have produced a number of promising research findings that are progressing along the translation pathway from bench to bedside. Enormous strides have been made using neuroprotective molecules and stem cell technologies in preclinical models, while neuromodulation and early surgical decompression are being utilized in the clinic to improve patient outcomes.

This textbook is meant to provide a state of the art update regarding our understanding of spinal cord injury and spine trauma, the best clinically available treatment options, as well as promising neuroprotective and neuroregenerative approaches currently being studied in basic science laboratories. The work discussed within this book results from researchers around the globe who have dedicated their careers to helping those affected by this terrible condition. We hope that you find this book to be of interest and that it inspires new research in the field that will one day lead to a cure. We anticipate that this book will be of interest to clinicians from multiple disciplines who treat spinal cord injury as well as those conducting basic and clinical translational research in neuroscience.

Acknowledgments

The editors gratefully acknowledge the individuals with spinal cord injury (SCI) who are the subjects of the important, cutting-edge research that will ultimately lead to better outcomes for this devastating injury in the future. Also, we acknowledge the clinical and scientific trainees who provide the horsepower to drive the research agenda for SCI and whose youth, enthusiasm, and brilliance portend an optimistic future for the field. And finally, we acknowledge the many funding agencies, foundations, and philanthropists who support the mission of improving the lives of individuals with SCI.

Introduction

Neural Repair and Regeneration after Spinal Cord Injury and Spine Trauma was written to provide the current state-of-the-art on spinal cord injury and spine trauma in one location. The book combines both research and clinical insights into one reference volume geared toward anyone studying or caring for those with a spinal cord injury. This includes clinical spine fellows, residents, neurosurgeons, clinical and basic researchers, orthopaedic surgeons, critical care specialists, medical students, and rehabilitation specialists.

Globally, approximately half a million people suffer a spinal cord injury or spine trauma every year. As a result, the lives of these individuals are altered dramatically and they experience significant impairment, reduced quality of life, and daunting financial costs. Currently, there are no cures for spinal cord injury. Researchers have invested significant resources examining approaches to repair and regenerate the injured spinal cord, including cellular strategies, rehabilitation, surgical techniques, and neuroprotective drugs. This textbook will cover the latest and most promising advances in the management of spine trauma and spinal cord injury.

The book provides an overview of spine anatomy before covering the pathophysiology and epidemiology of spinal cord injury, as well as diagnostics and injury classification systems. Management issues based on the level of injury and specific patient populations are also discussed. From there, the book addresses the use of surgical decompression, immunosuppressants, and rehabilitation to improve patient outcomes. An in-depth examination of the latest spinal cord injury research is then undertaken, touching on promising neuroprotective and neuroregenerative approaches. The book ends with expert commentary on what the future of spinal cord injury research and patient management may look like.

Anatomy

Laureen D. Hachem[1], Ali Moghaddamjou[1], Michael G. Fehlings[1,2]

[1]Division of Neurosurgery, Department of Surgery, University of Toronto, Toronto, ON, Canada;
[2]Division of Neurosurgery, Toronto Western Hospital, University Health Network, Toronto, ON, Canada

Understanding the functional anatomy and biomechanics of the spine is essential in the management of spinal trauma and spinal cord injury. Moreover, insight into the complexity of spinal tracts and level-specific differences in spinal cord anatomy is the foundation of understanding disease processes and developing novel therapies for spinal cord injury. This chapter provides an overview of spine and spinal cord anatomy and highlights relevance to spinal trauma.

Spinal column

The spinal column is composed of several bony and ligamentous structures which surround and protect the spinal cord. As the primary support of the body, trauma to the spinal column can lead to significant deformities and unstable injures, which may compromise the underlying neural elements. Fig. 1.1 shows the anatomical relationship between spinal cord level, vertebral level, and corresponding nerve roots.

Vertebrae

The general structure of spinal vertebrae is largely preserved across spinal levels. Each vertebra is composed of the ventral body and a dorsal arch (made up of the pedicles and laminae). At the junction of the pedicle and lamina arise the superior and inferior articular processes and the transverse processes. The superior and inferior articular processes of adjacent vertebrae form the facet joint. This is surrounded by a capsule which is continuous medially with the ligamentum flavum. At the dorsal aspect of the lamina arises the midline spinous process. The space between pedicles of adjacent vertebrae forms the intervertebral foramen, which houses the exiting nerve root along with associated neurovascular structures. Transforaminal ligaments may be seen particularly in the lumbar levels and may serve to protect structures which traverse the foramen from injury.[1] The intervertebral foramen is bounded by the body and disc anteriorly, facet joint posteriorly, and pedicles superiorly and inferiorly. The lateral recess is an important anatomical space bounded laterally by the pedicle, medially by the thecal sac, anteriorly by the vertebral body and disc, and posteriorly by the facet and ligamentum flavum. This region is a common site of stenosis in lumbar degenerative disc disease.[2]

Variations in vertebral structure throughout the spine allow for the unique biomechanics at each level. The first vertebrae, C1 (the atlas), lacks a vertebral body and instead consists of two

© 2022 Elsevier Inc. All rights reserved.

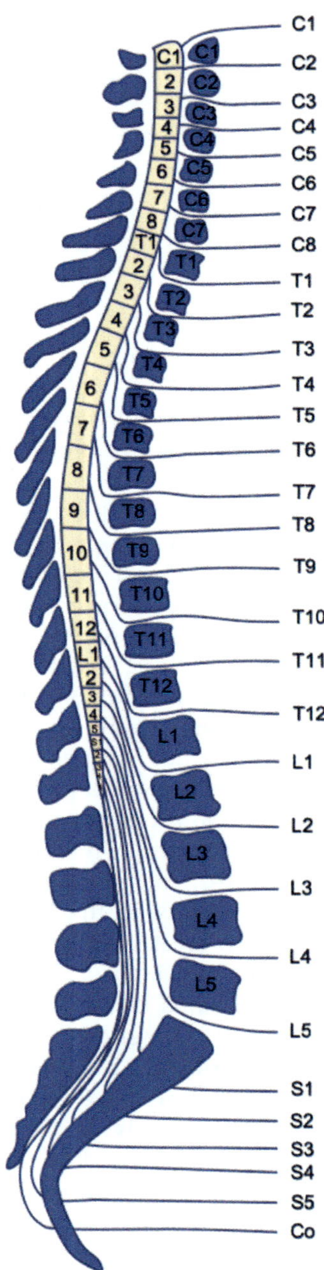

FIGURE 1.1 Relationship between spinal cord level, vertebral level, and nerve roots.

lateral masses bounded by anterior and posterior arches. The facets of C1, which articulate with the occipital condyles, are concave and face medially, thus providing close to 50% of flexion—extension of the neck but limited lateral displacement.[3] C2 (the axis) comprises an odontoid process extending from the vertebral body, which articulates with the posterior aspect of the C1 anterior arch forming the atlanto-axial joint, which provides 50% of rotation of the cervical spine.[3]

The remaining cervical vertebrae (C3—7) are termed the subaxial spine. All cervical vertebrae have transverse foramina, which house the bilateral vertebral arteries at levels C1—6. Between C3—7, the posterolateral lip of each vertebra (termed the uncinate process) articulates with the side of the adjacent rostral vertebrae forming the uncovertebral joint. This joint is a common site of bony spur formation and degeneration.[4] The vertebral body is smallest within the subaxial cervical spine and wider in the lateral dimension with transverse processes projecting out laterally. Typically, spinous processes of C2—6 are short and bifid. The orientation of facet joints at the cervical level allows for free motion in all directions and facilitates the range of motion in the neck.[5] Thoracic vertebrae have heart-shaped bodies with transverse processes directed posteriorly and an articular surface with the corresponding rib. Spinous processes of thoracic vertebrae point inferiorly. At thoracic levels, facet joints allow for lateral flexion and rotation but limited flexion and extension. Lumbar vertebrae have large kidney-shaped bodies with posterior pointing transverse processes and spinous processes projecting horizontally. The orientation of facets in the lumbar region allows for primarily flexion and extension of the spinal column.[5]

Intervertebral discs

Intervertebral discs separate the bodies of vertebrae forming a secondary cartilaginous joint to provide mobility and act as shock absorbers. The central portion of the disc is made up of the nucleus pulposus, a gelatinous substance derived from the embryonic notochord. The nucleus pulposus is composed of 70% water with the remaining component derived from type II collagen fibers and proteoglycans.[6] With age, the water and proteoglycan content of the disc decreases. This biochemical change leads to loss of disc height imparting abnormal loading on adjacent facet joints precipitating degenerative changes.[7] The nucleus pulposus is surrounded by an outer annulus fibrosus composed of multiple sheets of collagen with differing orientations providing increased strength. The superior and inferior aspects of the disc are composed of cartilaginous endplates, which anchor them to the adjacent vertebral bodies. Discs are thicker at the anterior portion in the cervical and lumbar spine contributing to the lordotic curve of the spine at these levels.[8]

Ligaments

Ligaments of the spinal column are essential to maintaining stability. The anterior longitudinal ligament (ALL) runs along the anterior surface of vertebral bodies connecting from the occiput of the skull to the sacrum. Rostrally, the connection between the occiput and C1 is called the anterior occipitoatlantal membrane and between C1 and C2 the anterior atlantoaxial membrane. The posterior longitudinal ligament (PLL) runs along the posterior surface of vertebral bodies and extends rostrally as the tectorial membrane attaching to the basion.

Ligaments of the posterior arch include the ligamentum flavum, which originates midway of the inferior portion of the lamina and inserts onto the superior surface of the caudal lamina. Rostrally it becomes the posterior atlantoaxial membrane and posterior occipitoatlantal membrane. Interspinous ligaments connect the spinous processes of adjacent vertebrae and fuse with the

supraspinous ligaments, which connect the tips of spinous processes. Intertransverse ligaments are thin connections between adjacent transverse processes. Assessment of posterior ligamentous integrity is critical in the evaluation of spinal fractures and a key component in treatment decision-making.[9] Disruption of the posterior ligamentous complex (composed of the supraspinous ligament, interspinous ligament, facet capsule, and ligamentum flavum) indicates a potentially unstable injury, which in many cases may necessitate surgical intervention.[10,11] Fig. 1.2 shows the appearance of spinal ligaments on MRI. Additional ligaments are seen at the craniocervical junction.[12] The apical ligament attaches the dens to the basion with two alar ligaments running lateral to the apical ligament connecting the dens to the occipital condyles. The transverse ligament (TAL) attaches along the anterior arch of the atlas around the odontoid process. Longitudinal bands extend rostral and caudal from the TAL together termed the cruciate ligament.

Ossification of spinal ligaments can often predispose to injuries and fractures in the setting of even low impact traumas. Ossification of the posterior longitudinal ligament (OPLL) is seen in 0.16%—2.4% of the population, most commonly within the cervical spine.[13] Direct ventral compression on the spinal cord leads to symptoms of myelopathy, which are exacerbated in the setting of minor traumas.[14] Diffuse idiopathic skeletal hyperostosis (DISH) is defined as ossification along the anterolateral aspect of at least four contiguous vertebrae.[15] This produces regions of long lever arms putting stress on adjacent vertebrae and thus increasing susceptibility to fractures.[16]

Meninges

The spinal cord is surrounded by the meninges, which is composed of the outer dura mater, intermediate arachnoid, and inner pia. The arachnoid and pia together form the

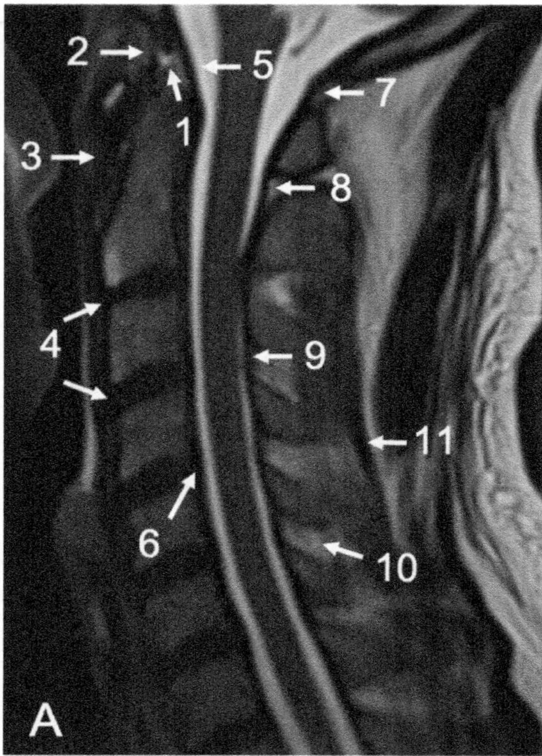

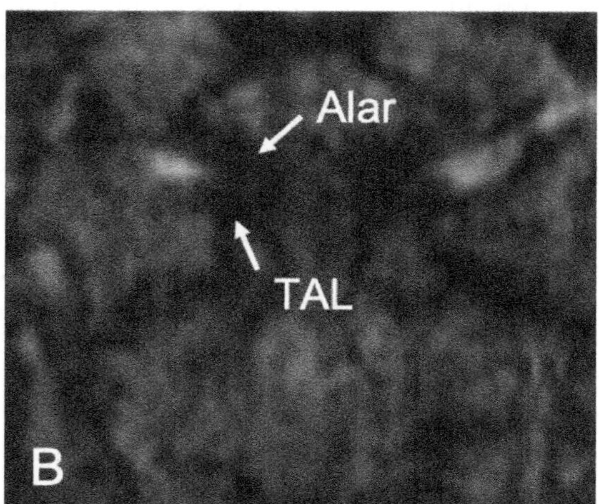

FIGURE 1.2 **Spinal ligaments on MRI.** (A) Sagittal T2-weighted MRI. 1, apical ligament; 2, anterior occipitoatlantal membrane; 3, anterior atlantoaxial membrane; 4, anterior longitudinal ligament; 5, tectorial membrane; 6, posterior longitudinal ligament; 7, posterior occipitoatlantal membrane; 8, posterior atlantoaxial membrane; 9, ligamentum flavum; 10, interspinous ligament; 11, supraspinous ligament. (B) Coronal T2 STIR sequence MRI displaying alar and transverse ligaments.

leptomeninges. The epidural space lies between the vertebral column and the enclosed dura matter. This space contains adipose tissue, venous plexuses along with meningovertebral ligaments that tether the dura to the surrounding supportive ligaments. The subdural space is considered a potential space between the arachnoid and dura matter. The subarachnoid space between the pia and arachnoid contains the cerebrospinal fluid bathing the spinal cord.

Spinal nerves

There are 31 spinal nerves: 8 cervical, 12 thoracic, 5 lumbar, 5 sacral, and 1 coccygeal. C1−7 nerves exit above their respective vertebrae, and all other nerves exit below their corresponding vertebrae. Spinal nerves are made of a dorsal (sensory) and ventral (motor) root, which are in turn composed of multiple rootlets. Both roots are covered by the leptomeninges along with a layer of dura, which continues as the epineurium once it exits the spinal column. Fig. 1.3 outlines the dermatomes and myotomes corresponding to each spinal nerve root.

Dorsal roots relay somatosensory information and are associated with a dorsal root ganglion, which houses the neuronal cell bodies of primary sensory neurons. These pseudo-unipolar neurons relay sensory information from the periphery into the central gray matter. Dorsal rootlets arise from the dorsolateral sulcus in the dorsal root entry zone (DREZ). Ventral roots carry motor information from efferent somatic motor neurons. Between T1 and L2, these fibers are also joined by autonomic preganglionic axons arising from the interomediolateral column of the sympathetic nervous system.

MYOTOMES

C5	Elbow flexors
C6	Wrist extensors
C7	Elbow extensors
C8	Finger flexors
T1	Finger abductors (little finger)
L2	Hip flexors
L3	Knee extensors
L4	Ankle dorsiflexors
L5	Great toe extensors
S1	Ankle plantar flexors

DERMATOMES

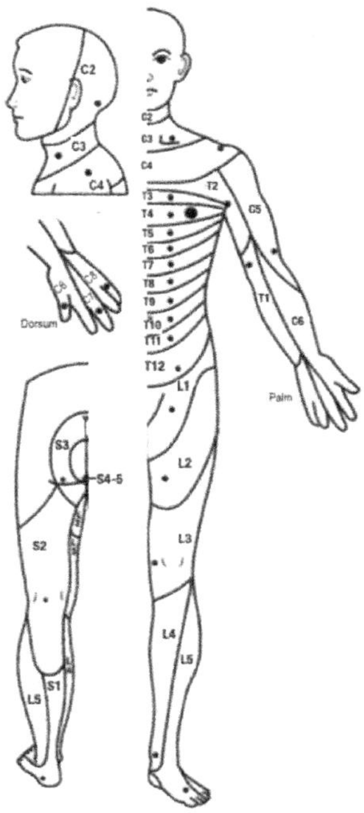

FIGURE 1.3 Myotomes and dermatomes. *Adapted from the International Standards for Neurological Classification of Spinal Cord Injury (ISNCSCI).*

Dorsal and ventral roots combine to form a single spinal nerve. As each spinal nerve exits the intervertebral foramen, it gives off a branch termed the recurrent meningeal branch that enters the vertebral canal and supplies the PLL, portions of the annulus, dura matter, facet joints, and dorsal vertebral periosteum. The continuing spinal nerve then gives off a dorsal and ventral ramus, which each carries mixed motor and sensory information. The ventral rami of spinal nerves supply the anterolateral portions of the trunk and extremities. At the thoracic levels, these form the intercostal nerves, while at other levels, they form various nerve plexuses: cervical plexus (C1—C4), brachial plexus (C5—T1), lumbar plexus (L1—4), and sacral plexus (L4—S4). The dorsal rami mainly supply muscles and dermatomes of the back.

Rami communicantes are additional connections between spinal nerves and the sympathetic chain. White rami communicantes exit at levels T1—L2 and carry preganglionic axons from the spinal cord intermediolateral cell column to ganglia in the sympathetic chain. These fibers either synapse at the ganglion of the respective level or course along the sympathetic chain to distal levels. The postganglionic axons then exit the sympathetic chain and join spinal nerves via gray rami communicantes, which are present at every spinal level.

Gray matter: nuclei and rexed lamina

Gray matter within the spinal cord is located within the central portion of the cord and

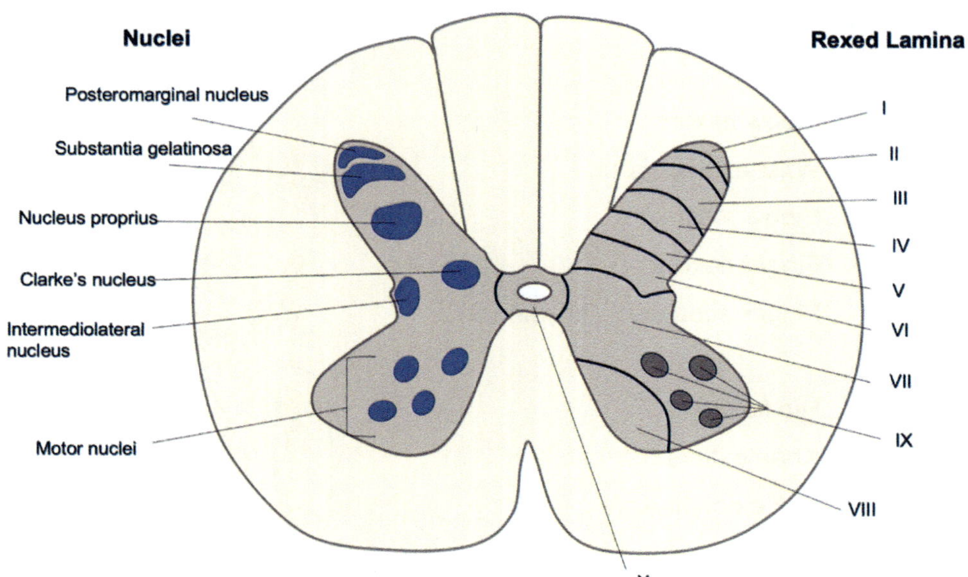

FIGURE 1.4 Spinal cord gray matter organization.

consists of both projection neurons and interneurons. Grossly, the gray matter can be divided into the dorsal horns, which house neurons receiving somatosensory information from the body, and the ventral horn consisting of motor neurons, which project to skeletal muscles. The cervical and lumbosacral enlargements represent increased populations of motor neurons that provide innervation to the upper and lower extremities, respectively. The gray matter of the spinal cord can be further divided into functional nuclei and cellular architecture, which will be discussed in this section (Fig. 1.4).

Rexed lamina

In the early 1950s, Bror Rexed described a system of 10 layers of gray matter (I-X) within the spinal cord, termed the Rexed laminae, corresponding to different portions of the gray matter based on cellular architecture.[17] While often defined as distinct zones, the borders of these laminae are not discrete. Moreover, distinct functional nuclei have been identified within these various laminae.[18]

Lamina I comprises the most dorsal portion of the spinal cord gray matter. Early work in rodents identified four morphological neuronal types, which have subsequently been shown in various other species. Fusiform neurons are the most abundant and typically found in the lateral third of this lamina. Multipolar neurons, pyramidal neurons, and flattened neurons are found in the medial portion, dorsal border, and middle third of lamina I, respectively.[19,20] These neurons receive information from A delta and C fibers carrying pain and temperature sensation from the body along with mechanical stimuli from A-beta fibers. This layer is thought to contribute to nearly half of the spinothalamic outflow to the brain.[21] Moreover, it also receives input from the hypothalamus and brainstem to modulate pain.

Lamina II contains densely packed neurons with rich dendrites and is often termed the substantia gelatinosa. It was initially thought to be composed primarily of stalked (outer portion) and islet (central portion) neurons; however, additional cells have been identified in humans including filamentous and stellate cells.[22] Neurons in this layer serve primarily as interneurons projecting to ventral lamina in the gray matter (III, IV, V). The lateral portion of this lamina receives input from C- and A-delta fibers. Neurons in lamina III and IV have similar functions relaying light touch and proprioception inputs and are modulated by lamina II interneurons.[18]

Lamina V neurons relay nociceptive information to the brain and receive inputs from corticospinal and rubrospinal tracts. Lamina VI contains interneurons involved in spinal reflexes and relays information from muscle spindles to the brain via spinocerebellar tracts. Lamina VII is the largest region of the gray matter composed primarily of interneurons that modulate motoneurons. Lamina VIII contains both interneurons and projection neurons that aid in coordination of motor activity. Lamina IX is composed of groups of motor neurons that mediate skeletal muscle function. Lamina X surrounds the central canal region, termed the gray commissure, and includes axonal decussations of various pathways.

Gray matter nuclei

Populations of neurons may be further subdivided into specific nuclei based on function. The posteromarginal nucleus is found in lamina I of the spinal cord gray matter. Neurons within this nucleus receive information on pain and temperature. The substantia gelatinosa is found in rexed lamina II. It receives input from afferents of the dorsal root ganglia, primarily pain, temperature, and mechanoreception. In addition, it receives input from periaqueductal gray matter, locus coeruleus, gigantocellular reticular nucleus, and nucleus raphe magnus to modulate pain responses. The substantia gelatinosa plays an important role in the modulation of pain responses in the CNS.[23] The nucleus proprius is found in laminae III and IV of the spinal cord

gray matter and relays temperature and mechanical inputs via the spinocerebellar and spinothalamic tracts. Clarke's nucleus is composed of interneurons important for proprioception. This population of neurons is located in lamina VII at T1—L2 spinal levels. This nucleus is the origin of the dorsal spinocerebellar tract, and its fibers project ipsilaterally to the cerebellum. Lamina IX of the gray matter is composed of various groups of specialized motor nuclei innervating skeletal and visceral functions. The phrenic motor nucleus is found at segments C3—7 and innervates muscles of the diaphragm. Onuf's nucleus is found at the lower lumbar and sacral levels and controls micturition. Motor neurons from this nucleus project to perineal muscles and the anal and urethral sphincters.

White matter tracts

The spinal cord white matter is composed of a series of ascending and descending tracts, which relay sensorimotor signals between the brain and peripheral structures (Fig. 1.5).

Descending pathways

Corticospinal tracts: The corticospinal tract is the central motor relay pathway. Layer 5 pyramidal cells in the cortex project through the corona radiata and internal capsule into the cerebral peduncles. Approximately 85% of fibers cross at the pyramidal decussation forming the lateral corticospinal tract. These fibers ultimately synapse onto the anterior horn cells, which project out to the target muscle controlling movement. Classically, there was thought to be a somatotopic representation within the lateral corticospinal tract with upper extremities medial to lower extremities. However, to date, there has been little evidence in humans to support this organization. The remaining 15% of fibers that do not cross at the pyramidal decussation continue down in the anterior corticospinal tract.

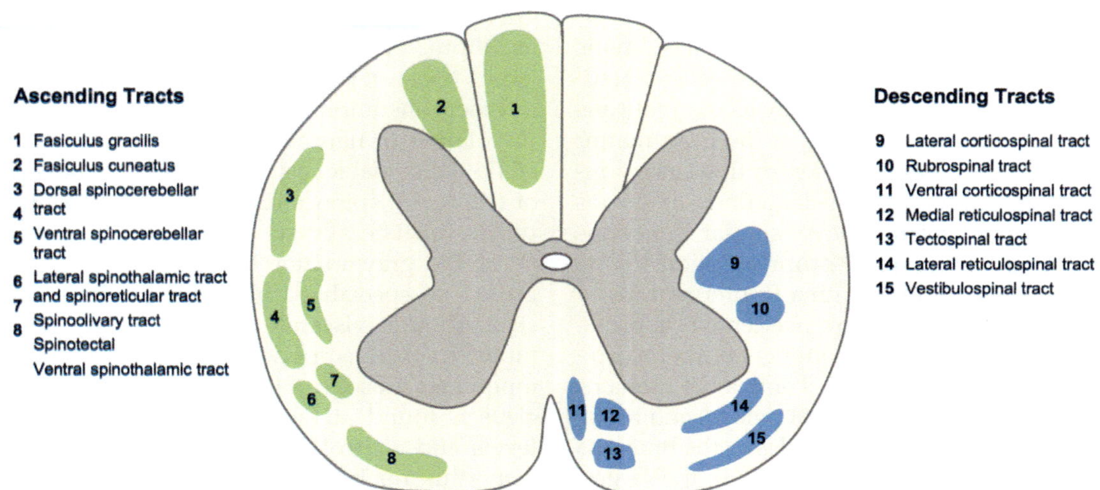

FIGURE 1.5   Spinal cord white matter organization.

Rubrospinal tract: The rubrospinal tract is involved in voluntary movement. Neuronal cell bodies reside in the magnocellular red nucleus of the midbrain with axons that cross over at this level and descend in the lateral funiculus of the spinal cord. Most of these projections terminate in the cervical and lumbar enlargements, and thus, the rubrospinal tract is believed to be important in the modulation of fine motor control. Rubrospinal neurons synapse primarily onto neurons within Rexed laminae V and VI.[24]

Reticulospinal tracts: The reticulospinal tract is involved in the preparation of movements, postural control, and regulation of sensory and autonomic functions. The medial reticulospinal tract originates from neurons in the gigantocellular reticular nucleus and pontine caudal reticular nucleus. These neurons project ipsilaterally dispersed within the ventral and lateral columns and synapse onto laminae VI−IX in the ventral horn increasing muscle tone and voluntary movements. The lateral reticulospinal tract projects from neurons in the gigantocellular reticular nucleus traveling in the ventrolateral funiculus synapsing onto laminae V and VI and reduces muscle tone and voluntary movements.[25]

Vestibulospinal tracts: The vestibulospinal tracts regulate postural extensor activity of the limbs and trunk. The medial vestibulospinal tract originates in the medial vestibular nuclei (with inputs primarily from the semicircular canals). It travels in the ventral funiculus with both ipsilateral and contralateral projections onto laminae VII and VIII to coordinate head position and regulate the vestibulocollic reflex. The lateral vestibulospinal tract runs from the lateral vestibular nucleus (with inputs primarily from the otoliths) in the ipsilateral ventrolateral funiculus and controls extensor tone in the extremities.[24,26,27]

Tectospinal tract: The tectospinal tract is important for the coordination of head and eye movements. Output from the superior colliculus travels in the ventral white matter to upper cervical levels in rexed laminae VI, VII, VIII. The majority of these fibers run contralaterally.

Ascending pathways

Spinothalamic tract: The spinothalamic tract relays pain, temperature, and crude touch from the body. Nociceptive input is sent to layers I, II, and

V of the spinal cord gray matter, whereas visceral input is sent to layers I and V. Second-order neurons within these laminae decussate and ascend rostrally in either the ventral or lateral spinothalamic tract. The ventral spinothalamic tract carries crude touch and pressure sensation, running in the ventral funiculus to join the medial lemniscus at the medulla and pons. The lateral spinothalamic tract transmits pain and temperature sensation and runs in the ventral portion of the lateral funiculus continuing as the spinal lemniscus. Second-order neurons in the spinothalamic tract ultimately synapse onto third-order neurons in the thalamus.

Dorsal columns: The dorsal columns pathway is responsible for proprioception, vibration sense, and fine touch. Primary sensory neurons in the dorsal root ganglion relay information that ascends in the dorsal funiculus. Information from lower extremities arising below T6 ascends more medially in the fasciculus gracilis, and sensory input from upper extremities above T6 ascends in the more lateral fasciculus cuneatus. In the caudal medulla, these neurons synapse onto second-order neurons of the nucleus gracilis and nucleus cuneatus, respectively. The second-order neurons decussate and ascend contralaterally to synapse onto third-order neurons of the thalamus.

Spinocerebellar tracts: Spinocerebellar pathways are composed of a dorsal and ventral component. The dorsal spinocerebellar tract relays afferent proprioceptive input from ipsilateral muscle spindles and Golgi tendons to the cerebellum. First-order neurons enter the dorsal horn and synapse onto second-order neurons within Clarke's nucleus. These second-order neurons then ascend ipsilaterally forming the dorsal spinocerebellar tract within the lateral funiculus entering the cerebellum via the inferior cerebellar peduncle. Inputs from upper thoracic and cervical levels ascend within fasciculus cuneatus and synapse onto second-order neurons within an analogous nucleus in the medulla termed the accessory cuneate nucleus. These neurons then project rostrally forming the cuneocerebellar tract similarly entering the cerebellum via the inferior cerebellar peduncle.

The ventral spinocerebellar tract relays inputs primarily from lower thoracic and lumbar segments. Here, second-order neurons are located within spinal border cells in the lumbar anterior horn (T12—L5), cross at the level of the spinal cord, and ascend in the contralateral lateral funiculus. Fibers enter the cerebellum via the superior cerebellar peduncle and then recross the midline to terminate ipsilateral to their site of origin.

Spinotectal tract: The spinotectal tract is thought to be responsible for nociceptive sensory information particularly with the facilitation of reflexive head movements toward noxious stimuli. The tract ascends through the anterolateral white matter toward the midbrain terminating in the contralateral superior colliculus.

Spinoolivary tract: The spinoolivary tract is responsible for the transmission of unconscious proprioception from tendons and muscles along with cutaneous information from the olivary bodies. Functionally, this tract is important in balance control. The ascending information travels through the dorsal root ganglia synapsing at the second-order neurons in the posterior gray matter. The axons then cross the midline and travel through the intersection between the anterior and lateral white matter columns. The axons synapse at the inferior olivary nuclei at the level of the medulla prior to entering the cerebellum via the inferior cerebellar peduncle.

Vascular supply

Arterial supply

Primary blood supply to the spinal cord is derived from the anterior spinal artery and the paired posterior spinal arteries, which typically arise from the vertebral arteries. The anterior spinal artery runs in the ventral median fissure supplying the anterior two thirds of the spinal cord, and the posterior arteries supply the dorsal columns. Loss of blood supply to these vessels thus produces a classic pattern of deficits whereby anterior spinal artery infarcts lead to leg weakness and loss of pain below the level of injury, and posterior spinal artery infarcts cause loss of proprioception and vibration sense below the level of injury.

At each level, the dorsal ramus of segmental arteries enters the intervertebral foramen as the segmental spinal artery, which in turn gives rise to various branches to supplement blood supply to the cord: (1) branch to the vertebral body and dura, (2) radicular branches (anterior or posterior radicular arteries), which supply the dorsal and ventral nerve roots, and (3) spinal medullary arteries, which augment flow to the anterior and posterior spinal arteries. During the third stage of fetal development, most of the medullary branches involute. However, a large anterior medullary branch remains typically arising from a left posterior intercostal artery between T8 and L2 called the artery of Adamkiewicz. Blood supply from this artery is critical in supplementing cord perfusion. Indeed, injury to this artery can lead to anterior spinal cord infarctions.[28]

Impaired spinal cord perfusion is one of the hallmarks of spinal trauma. Maintaining adequate blood flow to the cord is of utmost importance, and as such, current guidelines recommend maintaining a mean arterial pressure over 85 mmHg for 7 days following injury.[29,30]

Venous drainage

Intrinsic veins (sulcal and radial veins) drain the parenchyma of the spinal cord into the extrinsic venous system. The extrinsic venous system is composed of a network of pial veins, anterior and posterior spinal veins, and radicular veins.[31] The anterior median spinal vein runs adjacent to the anterior spinal artery coursing along the entire length of the cord.[32] The dorsal median spinal vein runs in the posterior median sulcus, and two dorsolateral spinal veins run adjacent to the posterior spinal arteries. Radicular veins drain the anterior and dorsal median veins into the extradural vertebral venous plexus (Batson's plexus). This network is composed of the internal venous plexus (posterior to the vertebral body and anterior to the vertebral arch) and the external venous plexus (anterior to the vertebral body and posterior to the vertebral arch).[33] Basivertebral veins within the vertebral body connect internal and external systems. The veins that comprise Batson's plexus are valveless and thus allow for retrograde flow. This therefore serves a potential route for the spread of urinary infections or metastatic tumors from pelvic organs to the spine.

Autonomic nervous system

The autonomic nervous system (ANS) is a major component in maintaining homeostasis and is responsible for the control of vital functions. The ANS can be divided into the sympathetic and the parasympathetic systems, which have synergistic roles. The general structure of these divisions is made up of a two-neuron pathway composed of a preganglionic and postganglionic neuron. The preganglionic neurons traverse through the ventral roots and synapse with postganglionic neurons near the spinal cord or at the target organ as is the case with the parasympathetic system. The sympathetic preganglionic neurons are located at T1−L2. Surrounding the sympathetic system, the parasympathetic system is located at the cranial nerves and the sacral spine segments.

Cardiovascular function

The autonomic nervous system plays a vital role in the control of the cardiovascular system.

The parasympathetic system decreases cardiac activity by reducing cardiac contractility and heart rate via the vagus nerve. The sympathetic system increases cardiac activity by increasing heart rate and cardiac contractility through T1—T5 thoracic sympathetic output. Spinal cord injury can lead to significant disruptions in the cardiovascular function. Traumatic spinal cord injuries above T6 result in loss of sympathetic tone and unopposed parasympathetic outflow, leading to the phenomenon of neurogenic shock, which is characterized by hypotension and bradycardia. In addition, impaired modulation of the sympathetic activity following trauma can result in autonomic dysreflexia. This phenomenon is characterized by autonomic overactivity in response to noxious stimuli below the neurological level of injury.[34]

Respiratory function

Understanding spinal cord anatomy is essential in the study of respiratory drive. The phrenic nerve originating from C3, C4, and C5 spinal nerves innervates the diaphragm. Accessory muscles including the posterior thoracic muscles (T1—5), pectoralis muscles (C4—T1), intercostal muscles (T2—T11), abdominal muscles (T7—L1), and the trapezius muscle (C2—C6) further aid in respiration. Control of breathing is also mediated through the spinal cord via the phrenic motor neurons, prephrenic interneurons, and white matter tracts. Spinal interneurons play a major role in generating respiratory drive, independent of the medulla.[35–37] Furthermore, there is animal evidence that cervical excitatory interneurons contribute to restoring respiratory function in chronic nontraumatic spinal cord injury.[38]

Conclusion

The anatomy and function of the spinal cord are closely related. The gray matter of the spinal cord organized by the Rexed lamina classification is responsible for the modulation of signals.

The white matter tracts of the cord serve as a relay of ascending and descending information. Through the autonomic nervous system, the spinal cord also plays an important role in maintaining vital respiratory and cardiac functions. Knowledge of the structure of the spinal cord gray matter, autonomic nervous system, and white matter tracts will allow for the localization of pathological lesions in injuries.

References

1. Golub BS, Silverman B. Transforaminal ligaments of the lumbar spine. *J Bone Jt Surg Am Vol* 1969;**51**(5):947—56.
2. Ciric I, Mikhael MA, Tarkington JA, Vick NA. The lateral recess syndrome. A variant of spinal stenosis. *J Neurosurg* 1980;**53**(4):433—43.
3. Bogduk N, Mercer S. Biomechanics of the cervical spine. I: normal kinematics. *Clin Biomech* 2000;**15**(9):633—48.
4. Nagamoto Y, Ishii T, Iwasaki M, et al. Three-dimensional motion of the uncovertebral joint during head rotation. *J Neurosurg Spine* 2012;**17**(4):327—33.
5. Jaumard NV, Welch WC, Winkelstein BA. Spinal facet joint biomechanics and mechanotransduction in normal, injury and degenerative conditions. *J Biomech Eng* 2011;**133**(7):071010.
6. Aladin DM, Cheung KM, Ngan AH, et al. Nanostructure of collagen fibrils in human nucleus pulposus and its correlation with macroscale tissue mechanics. *J Orthop Res* 2010;**28**(4):497—502.
7. Adams MA, Dolan P, Hutton WC, Porter RW. Diurnal changes in spinal mechanics and their clinical significance. *J Bone & Joint Surg Br* 1990;**72**(2):266—70.
8. Been E, Li L, Hunter DJ, Kalichman L. Geometry of the vertebral bodies and the intervertebral discs in lumbar segments adjacent to spondylolysis and spondylolisthesis: pilot study. *Eur Spine J* 2011;**20**(7):1159—65.
9. Vaccaro AR, Lee JY, Schweitzer Jr KM, et al. Assessment of injury to the posterior ligamentous complex in thoracolumbar spine trauma. *Spine J Off J North Am Spine Soc* 2006;**6**(5):524—8.
10. Vaccaro AR, Lehman Jr RA, Hurlbert RJ, et al. A new classification of thoracolumbar injuries: the importance of injury morphology, the integrity of the posterior ligamentous complex, and neurologic status. *Spine* 2005;**30**(20):2325—33.
11. Vaccaro AR, Oner C, Kepler CK, et al. AOSpine thoracolumbar spine injury classification system: fracture description, neurological status, and key modifiers. *Spine* 2013;**38**(23):2028—37.
12. Tubbs RS, Hallock JD, Radcliff V, et al. Ligaments of the craniocervical junction. *J Neurosurg Spine* 2011;**14**(6): 697—709.

13. Smith ZA, Buchanan CC, Raphael D, Khoo LT. Ossification of the posterior longitudinal ligament: pathogenesis, management, and current surgical approaches. A review. *Neurosurg Focus* 2011;**30**(3):E10.

14. Song KJ, Park CI, Kim DY, Jung YR, Lee KB. Is OPLL-induced canal stenosis a risk factor of cord injury in cervical trauma? *Acta Orthop Belg* 2014;**80**(4):567–74.

15. Mader R, Verlaan JJ, Buskila D. Diffuse idiopathic skeletal hyperostosis: clinical features and pathogenic mechanisms. *Nat Rev Rheumatol* 2013;**9**(12):741–50.

16. Taljanovic MS, Hunter TB, Wisneski RJ, et al. Imaging characteristics of diffuse idiopathic skeletal hyperostosis with an emphasis on acute spinal fractures: review. *AJR Am J Roentgenol* 2009;**193**(3 Suppl. l):S10–9. Quiz S20-14.

17. Rexed B. The cytoarchitectonic organization of the spinal cord in the cat. *J Comp Neurol* 1952;**96**(3):414–95.

18. Heise C, Kayalioglu G. Cytoarchitecture of the spinal cord. In: Watson C, Paxinos G, Kayalioglu G, editors. *The spinal cord: a Christopher and Dana Reeve foundation text and atlas*, vol. 1. Amsterdam; Boston: Elsevier/Academic Press; 2009. p. 64–93.

19. Galhardo V, Lima D. Structural characterization of marginal (lamina I) spinal cord neurons in the cat: a Golgi study. *J Comp Neurol* 1999;**414**(3):315–33.

20. Lima D, Coimbra A. A Golgi study of the neuronal population of the marginal zone (lamina I) of the rat spinal cord. *J Comp Neurol* 1986;**244**(1):53–71.

21. Willis WD, Kenshalo Jr DR, Leonard RB. The cells of origin of the primate spinothalamic tract. *J Comp Neurol* 1979;**188**(4):543–73.

22. Gobel S. Golgi studies of the neurons in layer II of the dorsal horn of the medulla (trigeminal nucleus caudalis). *J Comp Neurol* 1978;**180**(2):395–413.

23. Wall PD. The role of substantia gelatinosa as a gate control. *Res Publ Assoc Res Nerv Ment Dis* 1980;**58**:205–31.

24. Watson C, Harvey AR. Projections from the brain to the spinal cord. In: Watson C, Paxinos G, Kayalioglu G, editors. *The spinal cord: a Christopher and Dana Reeve foundation text and atlas*, vol. 1. Amsterdam; Boston: Elsevier/Academic Press; 2009. p. 168–79.

25. Nyberg-Hansen R. Sites and mode of termination of reticulo-spinal fibers in the cat. an experimental study with silver impregnation methods. *J Comp Neurol* 1965;**124**:71–99.

26. Pompeiano O. Spinovestibular relations: anatomical and physiological aspects. *Prog Brain Res* 1972;**37**:263–96.

27. Hayes NL, Rustioni A. Descending projections from brainstem and sensorimotor cortex to spinal enlargements in the cat. Single and double retrograde tracer studies. *Exp Brain Res* 1981;**41**(2):89–107.

28. Phillips D, Dhall SS, Uzelac A, Talbott JF. Gunshot wound causing anterior spinal cord infarction due to injury to the artery of Adamkiewicz. *Spine J Off J North Am Spine Soc* 2016;**16**(9):e603–604.

29. Hawryluk G, Whetstone W, Saigal R, et al. Mean arterial blood pressure correlates with neurological recovery after human spinal cord injury: analysis of high frequency physiologic data. *J Neurotrauma* 2015;**32**(24):1958–67.

30. Squair JW, Bélanger LM, Tsang A, et al. Spinal cord perfusion pressure predicts neurologic recovery in acute spinal cord injury. *Neurology* 2017;**89**(16):1660–7.

31. Krings T. Vascular malformations of the spine and spinal cord*: anatomy, classification, treatment. *Clin Neuroradiol* 2010;**20**(1):5–24.

32. Green K, Reddy V, Hogg JP. Neuroanatomy, spinal cord veins. [Updated 2020 Jul 27]. In: *StatPearls [internet]. Treasure island (FL)*. StatPearls Publishing; January 2021. Available from: https://www.ncbi.nlm.nih.gov/books/NBK542182/.

33. Griessenauer CJ, Raborn J, Foreman P, Shoja MM, Loukas M, Tubbs RS. Venous drainage of the spine and spinal cord: a comprehensive review of its history, embryology, anatomy, physiology, and pathology. *Clin Anat* 2015;**28**(1):75–87.

34. Furlan JC, Fehlings MG. Cardiovascular complications after acute spinal cord injury: pathophysiology, diagnosis, and management. *Neurosurg Focus* 2008;**25**(5):E13.

35. Ghali MG, Marchenko V. Patterns of phrenic nerve discharge after complete high cervical spinal cord injury in the decerebrate rat. *J Neurotrauma* 2016;**33**(12):1115–27.

36. Palisses R, Perségol L, Viala D. Evidence for respiratory interneurones in the C3-C5 cervical spinal cord in the decorticate rabbit. *Exp Brain Res* 1989;**78**(3):624–32.

37. Lane MA, White TE, Coutts MA, et al. Cervical prephrenic interneurons in the normal and lesioned spinal cord of the adult rat. *J Comp Neurol* 2008;**511**(5):692–709.

38. Satkunendrarajah K, Karadimas SK, Laliberte AM, Montandon G, Fehlings MG. Cervical excitatory neurons sustain breathing after spinal cord injury. *Nature* 2018;**562**(7727):419–22.

Epidemiology

Thorsten Jentzsch[1,2,3], *Anoushka Singh*[2], *Michael G. Fehlings*[1,2]

[1]Division of Neurosurgery, Department of Surgery, University of Toronto, Toronto, ON, Canada;
[2]Division of Neurosurgery, Toronto Western Hospital, University Health Network, Toronto, ON,
Canada; [3]Department of Orthopaedics, Balgrist University Hospital, University of Zurich, Zurich,
Switzerland

Overview

Spinal cord injury

Spinal cord injury (SCI) is a devastating, life-changing neurotrauma with a high disease burden.[1] It leads to instant and permanent neurological impairment below the injury level due to disruption of afferent and efferent neural pathways. SCI is often associated with increased morbidity and premature mortality leading to physical, emotional, and social well-being strain for each individual. It also represents a costly diagnosis for the healthcare system with direct and indirect costs in the millions of dollars, particularly if the cervical spine of a young person is affected. Clinical outcome can in part be improved by urgent surgical decompression and rehabilitation.[2,3] Neuroprotective and neuroregenerative interventions continue to be studied to improve motor, sensory, and sphincter (bladder/bowel) functions in acute and chronic stages.[4]

Epidemiology

Epidemiology is derived from the Greek words epi (=upon), demos (=people), and logos (=study) with a literal meaning of studying what is upon people.[5] It is defined as the study of the occurrence and distribution of events or states related to health in specified populations. This includes investigation of the influencing factors and measures to effectively control these health problems.[6] Measures of occurrence include prevalence and incidence. Prevalence describes the existing cases at a designated time and measures disease burden. Point prevalence depicts cases with an outcome at a designated time, while period prevalence defines cases with an outcome over a period of time. The latter is less useful because there is no distinction between existing (prevalent) and new (incident) cases. Incidence describes the frequency of new cases during a time period. Incidence risk depicts the proportion of new cases, while incidence rate defines the frequency of new cases (i.e., new cases

© 2022 Elsevier Inc. All rights reserved.

divided by total person-time at risk). While the incidence risk is useful for rare diseases in a static population, the incidence rate is helpful for common diseases in a dynamic population.[7]

Prevalence, incidence, and costs of spinal cord injury

Understanding the prevalence and incidence of SCI enables healthcare providers to improve prevention and management programs as well as to estimate costs. SCI has a reported prevalence of 906 per million population and an incidence of 40.1 per million population in the United States.[8] This means that over 280,000 people were affected by SCI, with an annual incidence of over 17,000 in 2016.[9] It is important to recognize that these numbers likely represent an underestimate because only patients at the extreme spectrum of disease are usually accounted for. There are several milder clinical conditions, such as degenerative cervical myelopathy, which are usually not taken into consideration in these calculations.[10] The most common causes of traumatic SCI (TSCI) are traffic accidents and falls, while a common cause for nontraumatic SCI (NTSCI) is degenerative cervical myelopathy (DCM).[8] In low- and middle-income developing countries, traffic accidents are the most common cause of SCI with a reported proportion of 35%−53.8% (followed by falls [22.6%−37%]), while falls are the number one cause in high-income developed countries with a proportion of 37.9%−63%.[11] In low- to middle-income countries, low velocity falls are usually associated with carrying heavy loads on the head or crush injuries from collapsing ceilings in younger individuals, while in high-income countries, low velocity falls are usually associated with slipping and tripping in the elderly.[12] In developed countries, there is a demographic shift in TSCI toward the elderly, low velocity falls, and the cervical spine region.

A recent study has shown that the most common cause of new TSCIs in Finland was low velocity falls in 36.2%, while high velocity falls and motor vehicle accidents were the cause in only 25.5% and 19.2%, respectively.[13] Cervical spine SCI increased sixfold from 1970 to 2011, from 59 to 372 cases in Finland,[14] with a projected increase in SCI of the cervical spine from low velocity falls by 50% from 2011 to 2030. With an aging population across the globe, similar findings are reported worldwide.[15]

It is estimated that the SCI costs for society are higher than $4 billion per year.[16] Previous studies have reported varying costs with mean charges during the first year of $523,089, followed by charges of $79,759 per year.[17] These charges can rise even higher if there is a high-level cervical injury with charges during the first year of $1,064,716, followed by $184,891 per year.[18] Therefore, multifactorial strategies for the investigation of better and novel treatment options for SCI remain a very important research topic.[19]

History

Highlights about the history of SCI have been summarized before by New et al.[20] Literature suggests that knowledge about spina bifida, a birth defect with incomplete spinal closure, has been present since the beginning of humanity.[21] Later on, Nicholaas Tulp actually used the term spina bifida in 1641.[22] Hippocrates, who lived 460−377 before common era, assumed that moisture and cold were linked to paraplegia, and it was not until the 16th century that this belief changed.[23] He was also the first to describe SCI due to infection, which may have been tuberculosis.[24] Poliomyelitis was named by Jakob von Heine in 1840 and infected several hundred thousands of patients before an inactivated poliovirus vaccine was developed by Jonas Salk in 1952.[25,26]

Knowledge about SCI due to circulation alterations and inflammation was initiated in 1866 and later amended thanks to an enhanced understanding of the spinal cord's blood supply in the 19th century.[27,28] The first successful intradural and intramedullary spinal cord tumor resections were undertaken by Sir Victor Horsley in 1887 and Christian Fenger in 1890.[29,30] While complication rates were high initially, surgical techniques improved in the second half of the 20th century.[31] Radiotherapeutic options for spinal tumors were described in the 1950s by Wood.[32] Differences between myelomalacia and transverse myelitis were described in 1921.[33] Transverse myelitis was recognized as an inflammatory, autoimmune, and infectious disorder in the middle of the 20th century.[34] Categorization of neurological disorders into trauma, infections, degeneration, demyelination, inflammation, vascular, nutritional, and hereditary was undertaken in the 20th century.[35] Forty-five years ago, Kraus et al. defined SCI as an acute, traumatic spinal cord lesion (including nerve roots) that results in motor and/or sensory deficits.[36] Over the years, the Centers for Disease Control and Prevention refined this definition to a lesion of neural elements in the spinal canal (including not only the spinal cord, but also the cauda equina) and expanded the deficit to bladder and bowel dysfunction.[8]

Pediatrics

The literature on SCI in the pediatric population is more sparse than in adults. This is complicated by the fact that varying terms have been used for nonNTSCI (i.e., spinal cord disease [preferred term], damage, dysfunction, lesion, myelopathy, and paraplegia).[37] Therefore, identification of studies is more difficult, and inclusion criteria often vary limiting the generalizability of results. The International Spinal Cord Injury Data Sets for NTSCI has a basic and extended data set that discriminates congenital—genetic versus (vs.) acquired, which can be expanded to neoplastic, malignant, neural, astrocytoma, and malignant.[38]

It is important to notice that pediatric SCIs have different characteristics than adult SCIs. The incidence of cervical spine injuries was 5%. C2 lesions and violent etiologies were more commonly found in the preteen group (<12 years), while C4—5 lesions were more common in the teen/adult group.[39] The incidence of SCI without radiological anomalies (SCIWORA) has been reported to be as high as 6%.[40] In SCI from sports injuries and child abuse, the incidence may even rise to 75%. This emphasizes the importance of the clinical assessment.[41] A systematic review by Parent et al. pointed out that teenagers are at higher risk for scoliosis but appear to have better neurological recovery than in adults.[42] Apple and associates reported that scoliosis was more common in preteens than adults (23% vs. 5%).[39] Another study showed that 97% of patients that sustained an SCI before their growth spurt developed scoliosis, in contrast to 52% of patients with an SCI after their growth spurt.[43] Wang et al. reviewed 30 patients with SCI. Of 20 patients with complete SCI, 35% died, 35% had no neurological recovery, and 30% improved. Of the patients who improved, 83% became ambulatory. Of 10 patients with incomplete SCI, 80% improved.[44]

Another systematic review identified 862 relevant abstracts and included 25 articles from 14 countries. The age cut-off was 15 years (occasionally 19 years due to paucity of data). It differentiated between SCI and spinal cord disease. The prevalence rates for SCI were highest for the global regions Australia (51.5 cases per million population in 2011), followed by Ireland (12.1 cases per million population in 2015) and Canada (10.1 cases per million population in 2010). The prevalence rates for spinal cord disease were highest in Ireland (19.6 cases per million population in 2015), followed by Australasia (6 per million population in 2010) and Canada (2.5 cases per million population in 2010).

The median SCI incidence rates were highest for the global regions North America (high income, 13.2 million per population per year), followed by Australasia (9.9 per million per year), Asia (east; 5.4 per million population per year), and Western Europe (3.3 per million population per year). The median spinal cord disease incidence rates were highest for Australasia (6.5 per million population per year) and Western Europe (6.2 per million population per year) followed by North America (high income; 2.1 per million population per year). SCI was mostly caused by traffic accidents (46%–74%), followed by falls (12%–35%) and sports/recreation (10% –25%). Spinal cord disease was most commonly caused by tumors (30%–63%) and autoimmune/inflammatory disease (28%–35%).[45]

Studies about the survival time were even more rare and had small sample sizes. It was reported that children <16 years at the time of initial injury had their yearly odds of death increased by one-third compared with elderly patients.[45]

Resources

Data collection

Sophisticated surveillance is needed to precisely identify all SCI cases. This may entail obtaining information not only from acute care institutions (especially emergency, orthopedic, and neurosurgery departments) but also from rehabilitation centers as well as state medical examiners, death certificates, and even surveillance systems in neighboring states, as previously done in Oklahoma and Utah in the United States.[46,47]

In the United States, data from the hospital discharge surveys from the National Center for Health Statistics may also be used.[48] In Canada, for example, Ontario and Alberta have these registries.[49] Furthermore, Alberta operates a Health and Wellness database with medical records of all provincial hospitals. The Offices of the Chief Medical Examiners that investigate unexplained deaths can also be consulted.[50] In provinces where there is only one center for SCI, such as in British Columbia, medical records from that institution can be studied.[51] A review by Singh et al. from 2014 identified nine studies with reports on prevalence and 44 studies commenting on the incidence of acute SCI.[8] Other potential resources can include trauma registries.

Coding

To identify cases of SCI, studies often use ICD codes. ICD-9 codes often include codes 805 ("fracture of the vertebral column without mention of spinal cord lesion"), 806 ("fracture of vertebral column with spinal cord lesion"), and 952 ("spinal cord lesion without spinal bone injury").[8]

As of October 2015, ICD-10 codes are typically used. There is a long list of codes that can be used for SCI. The ones that mention the spinal cord directly are S14.0 ("concussion and edema of cervical spinal cord"), S14.1, S24.0 ("concussion and edema of thoracic spinal cord"), S24.1, S34.0 ("concussions and edema of lumbar spinal cord"), S34.1, T06.0 ("injuries of nerves and spinal cord involving other multiple body regions"), T06.1, T09.3 ("injury of spinal cord, level unspecified"), and T91.3 ("sequelae of injuries, of poisoning and of other consequences of external causes—sequelae of injuries of neck and trunk—sequelae of injury of spinal cord"). Other codes that could potentially be accompanied by SCI are G82 ("paraplegia and tetraplegia"), S12.0 ("fracture of the first vertebrae"), S12.2, S12.7, S13.0 ("traumatic rupture of cervical intervertebral disc"), S13.2, S13.4, S22.0 ("fracture of thoracic vertebrae"), S23.0 ("traumatic rupture of thoracic intervertebral disc"), S23.1, S32.0 ("fracture of lumbar vertebrae"), S33.0 ("traumatic vertebrae of lumbar intervertebral disc"), S33.1, S34.3, and T91.1 ("sequelae of injuries, of poisoning and of other consequences of external causes—sequelae of injuries of neck and trunk—sequelae of fracture of spine").[8]

Prevalence

Global

Prevalence at a certain time point is defined as the proportion of people with a disease in a population. Globally, the prevalence of SCI was estimated between 236 and 1009 per million in 2011.[52] This is similar to the previous estimation of 110–1120 per million in 1995.[53] These numbers are likely to underestimate the real prevalence due to high mortality rates at the scene of injury. They may also not be universally representative since they are from developed countries.

North America

A systematic review by Singh et al. reviewed 5874 articles, of which 48 were included.[8] They reported that the highest prevalence was found in the United States (906 per million),[54] while the lowest was found in France (250 per million).[55] A study by DeVivo et al. calculated a 30-year mean life expectancy after SCI using an estimated incidence rate of 30 cases per million. They estimated the prevalence of 906 per million, which would indicate the need of nine beds per million.[54] An older prevalence study in the United States multiplied the average life duration of 18 years with an incidence of 30 per million to report a prevalence of 525 per million.[56] Another study by Harvey et al. from the United States in 1990 selected area segments in 120 representative primary sampling units to survey households and nursing as well as long-term healthcare facilities to identify patients with SCI. The authors reported that the total prevalence was 721 per million.[57] Canada's SCI prevalence was estimated to be 85,556 persons

in 2012 (considering a population of 34.7 million).[58] The prevalence of SCI has been increasing over the years as described in a study by Griffin et al. in 1985. They counted all patients with residual neurological deficits after SCI in Minnesota to report point prevalence of 197 per million population in 1950 and 473 in 1980.[59,60]

Europe, Asia, and Australia

There are also other countries that have reported on the prevalence of SCI. Iceland stated a prevalence of 526 per million in 2009.[61] Finland used International Statistical Classification of Diseases and Related Health Problems (ICD)-9 and ICD-10 codes from rehabilitation centers, department of Orthopedic Surgery at Helsinki University Central Hospital, local organization of the disabled, local health centers, residential service houses, and announcements in patient magazines (to find cases not included in the previously mentioned sources) to estimate a prevalence of 280 per million in 2005.[62] In Norway, hospital records from eight different hospitals were reviewed, and the prevalence was found to be 365 per million in 2002.[63] In France, as previously mentioned in brief, the prevalence was estimated using a 20-year mean life expectancy after SCI and coined at 250 per million.[55] In Iran, random cluster sampling with 100 addresses as starting points and 25 households as each cluster with actual verification by a nurse was used to estimate a prevalence of 440 per million in 2009.[64] Lastly, Australia used the Spinal Cord Injury Register to estimate a prevalence of 681 per million in 1997.[65]

Low- and middle-income countries

Developing countries comprise more than 80% of the world's population, but information about the epidemiology of SCI in these countries is underrepresented in the literature.[66] It is often pointed out that epidemiological data is lacking (e.g., as in a report from Nepal in 2014[67] or India in 2013[68]). Over 90% of deaths resulting from injury are located in low- and middle-income countries.[68] The number of patients affected by an SCI per year in low- to middle-income countries was estimated as 843,316 (95% CI 393,629–1,292,385) in a recent comprehensive review (considering a population of 6.2 billion).[69]

Traumatic spinal cord injury and nontraumatic spinal cord injury

In Canada, 85,556 people were living with an SCI in 2012. Of those, 51% were TSCI and 49% were NTSCI.[58] Another study found that the prevalence of NTSCI was 1120 per million population in Canada and 2310 per million population in India.[70] The number of publications of NTSC has substantially increased in the past four decades with 1825 publications between 1974 and 1983 to 11,887 between 2004 and 13. This has been accompanied by a trend toward studies with better methodological design including larger sample sizes, multicenter, randomized controlled studies.[71]

Incidence

Incidence is the frequency of new cases during a time period. The United States has the highest reported incidence of 17,700 (54 per million) cases per year, followed by New Zealand (49 per million),[72] Estonia (40 per million),[73] and Japan (39 per million).[74] Within the United States, Alaska had the highest incidence (83 per million), while Alabama had the lowest (29 per million).[8] A study suggested that the United States may be the leader in incidence due to driving behavior (seat belts) and road conditions, as well as violence.[48] Another study suggested that Estonia may show higher numbers because of road safety, partially due to flat landscape, driving speeds, and alcohol consumption.[75] High numbers in Japan were attributed

to a generally increased risk for SCI due to ossification of the posterior longitudinal ligament (OPLL) and congenital stenosis.[76] The lowest rates in Europe were found in Spain (8 per million), Denmark (9 per million),[77] and the Netherlands (12 per million).[78] Divanoglou et al. compared the incidence and causes of SCI between two different cities in Europe. In low- and middle-income developing countries, the incidence of SCI was 25.5 per million per year with ranges from 2.1 to 130.7 per million per year.[66] In India for example, it is estimated that there are around 20,000 cases of SCI per year.[79] The incidence of TSCI in the Middle East and North Africa was 23.2 per year with a general lack of evidence in this region. Fig. 2.1[8] depicts the relative annual incidences of countries, states/provinces, and regions.

A decrease in the incidence of SCI has been observed over time in Ontario and Alberta. While the age-standardized incidence in Ontario was 46 per million in 1994/1995, it decreased to 37 per million in 1998/1999.[49] Similarly, in Alberta, it was 57 per million in 1997/1998 but decreased to 48 per million in 1999/2000.[50] However, there are also reports about increasing incidences of SCI in London, Ontario (Canada) from 21 per million in 1997 to 49 per million in 2000,[80] parts of Minnesota (United States) from 22 per million in 1935–44 to 71 per million in 1975–81,[59] parts of Western Norway from 6 per million in 1952–56 to 26 per million in 1997–2001,[63] and New Zealand from 43 per million in 1979–8 to 49 per million in 1988. In Spain, the incidence has been quite stable with 8 per million in 1972–80, 14 per million in 1981–90, 13 per million in 1991–2000, and 13 per million in 2001–08.[81]

The incidence was found to be higher in Thessaloniki in Greece (34 per million) than in Stockholm in Sweden (20 per million). The most common cause of injury was motor vehicle accidents (51%), followed by falls (37%) in Thessaloniki compared with falls (47%) and motor vehicle accidents (23%) in Stockholm.[82]

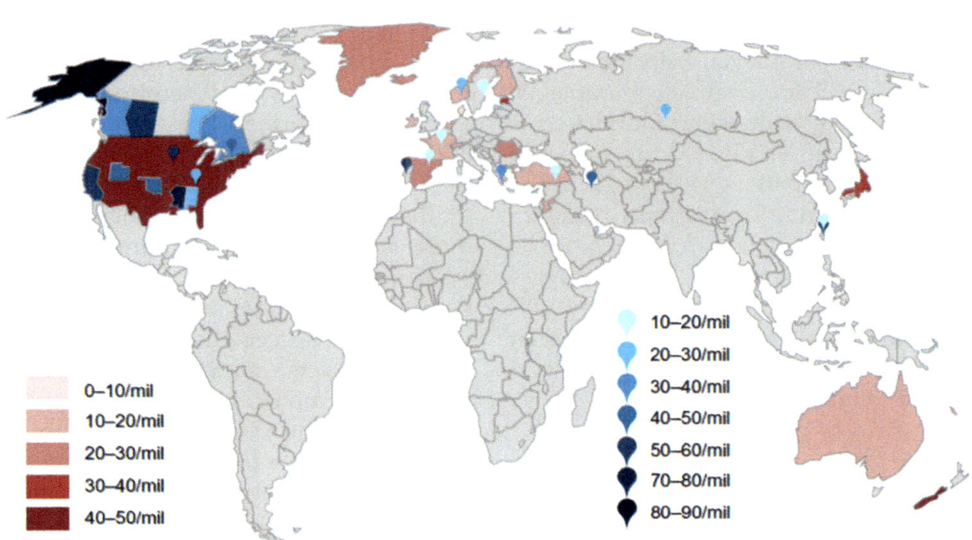

FIGURE 2.1 Relative annual incidences of SCIs in countries, states/provinces, and regions. The red color scheme illustrates incidences of SCI in countries. The blue color scheme highlights incidences in states/provinces and regions. mil, million. *Reproduced from Singh[8] with permission (Dove Medical Press was the original publisher of this work).*

Traumatic spinal cord injury and nontraumatic spinal cord injury

According to a recent review of 102 studies with a metaanalysis of 19 studies, the global incidence of TSCI was 10.5 cases per 100,000 persons. This corresponds to an estimated 768,473 new cases of TSCI per year worldwide. It was higher in low- and middle-income countries than high-income countries (8.7 vs. 13.7 per 100,000 persons). Traffic accidents were more common than falls.[69] In Canada, the incidence of TSCI was 1785 cases per year. The discharge incidence of TSCI was 1389 (41 per million) cases per year.[58]

The incidence rates were highest for North America (76 per million population per year), followed by Australasia (26 per million population per year), Asia Pacific (20 per million population per year), Oceania (9 per million population per year), and Western Europe (6 per million population per year). The discharge incidence of NTSCI was 2286 (68 per million) per year.[58]

Costs

Healthcare costs are generated by various factors during the intensive acute care and chronic management of the disease state as well as complications. Krueger et al. estimated that the overall economic burden per year was 2.7 billion Canadian dollars (CAD) in 2013. Direct costs (1.6 billion CAD) were higher than indirect costs (1.1 billion CAD). The direct costs were more pronounced in tetraplegia (56%—66%) compared with paraplegia (44%—54%). They were mostly driven by attendant care (33%), home modification (12%), healthcare practitioner visits (7%), and hospitalizations (7%). Indirect costs were mainly attributed to loss of productivity, which can be calculated by quality-adjusted life years (QALYs) using the mean annual salary of 47,834 CAD in 2011 and a utility of 0.45, which is typical for patients with SCI. By losing 0.55 QALYs, 26,309 CAD (0.55 × 47,834 CAD) are added to the indirect costs each year after SCI.

If the utility was adapted to the type of SCI, i.e., 0.55 for incomplete paraplegia, 0.45 for incomplete tetraplegia, 0.35 for complete paraplegia, and 0.25 for complete tetraplegia, indirect costs would increase by 18%–23%. Costs are also driven by the age of a patient at the time of the injury. On the one hand, for example, being 25 years instead of 35 years would increase costs by 10%–14%. On the other hand, being 45 years instead of 35 years would decrease costs by 14%–18%. On an individual basis, the lifetime burden is 1.5 million CAD for incomplete paraplegia and twice as high for complete tetraplegia.[83]

Clinical

Classification

SCI and spinal cord disease are classified according to their etiology and neurology using the International Standards for Neurological Classification of Spinal Cord Injury (ISNCSCI). This classification is only valid for patients ≥6 years.[84] Paraplegia indicates the loss of motor and/or sensory function in the legs and pelvis organs, while tetraplegia indicates a loss of motor and/or sensory function in the arms, trunk, legs, and pelvic organs. The majority of cases are incomplete tetraplegia ("tetraparesis"; 34%), followed by complete paraplegia (25%), complete tetraplegia (22%), and incomplete paraplegia ("paraparesis"; 17%).

McKinley et al. reported that 21% of patients present with a clinical syndrome after SCI.[85] Central cord syndrome is the most common form (44%) of incomplete tetraplegia. It is often due to hyperextension injury in patients with cervical spondylosis. It affects the upper extremities more than the lower extremities. Cauda equina syndrome is the second most common form (25%). It is an injury to the lower motor neurons resulting in flaccid paralysis of the limbs and areflexic bladder/bowel. The primarily affected anatomical region is the cervical spine

(44%–62%).[8] In Brown-Séquard syndrome, which is the third most common form (17%), there is a spinal cord hemisection resulting in ipsilateral proprioception/vibration/motor loss and contralateral pain/temperature loss below the lesion (as well as sensation loss at the level of the lesion). Conus medullaris syndrome is quite rare (8%). It can resemble cauda equina syndrome, but it may include upper motor neuron signs. Anterior cord syndrome is very rare (5%) and affects the corticospinal and spinothalamic tracts sparing the dorsal columns resulting in motor/pain/temperature loss with sparing of light touch/position sense. Posterior cord syndrome is the rarest (1%).

The determination of whether there is an incomplete or complete SCI can only be made after the spinal shock has subsided and the bulbocavernosus reflex has returned, which usually occurs around 48 hours after injury. Incomplete SCI is defined by voluntary anal contraction and perianal (S4–5) sensation sparing or muscle contraction below the injury level. Complete SCI (American Spinal Injury Association [ASIA] A) is defined by absent sacral sparing, no perianal sensation, and no muscle contraction below the injury level. ASIA B refers to sensory incomplete (i.e., no motor function), C to motor incomplete (i.e., >50% of key muscles below have <3/5 power), D to motor incomplete (i.e., >50% of key muscles below have ≥3/5 power), and E to normal. The motor level is the lowest key muscle (out of 10) function with at least 3/5 power. The sensory level describes the most caudal intact 2/2 sensation dermatome (out of 28) for pin prick and light touch. The neurological level of injury is determined on each side and represents the lowest level with intact antigravity muscle function and sensation.

Etiology and pathophysiology

There is a bimodal age distribution, whereby TSCI usually affects younger patients and NTSCI or minor trauma SCI affects older patients. The cause of injury is commonly a motor vehicle accident (in around 50% of cases). Up to one-fourth of these injuries may occur due to improper immobilization or transport. The pathophysiology discerns primary, immediate (<2 hours) mechanical injury with neural tissue damage to glial cells and disruption of the blood—spinal cord barrier, which usually functions as a morphological barrier, from secondary injury, which can last from weeks to years (early acute <48 hours, subacute <14 days, intermediate <6 months, and chronic >6 months), with a cascade of pathobiological changes including adjacent tissue damage due to ischemia, lipid peroxidation, homeostasis changes of sodium and calcium, oxidative strain from free radicals, excitotoxicity of the glutamatergic pathway, and programmed cell death or apoptosis.[86]

Diagnosis, treatment, and complications

The gold standard for diagnosis remains magnetic resonance imaging, where intramedullary injury is usually seen as hemorrhage and edema in the acute phase, and posttraumatic cyst, which is predictive of clinical outcome, in the chronic phase.[87] In the future, quantitative neuroimaging techniques may be helpful in quantifying the amount of microstructure disruption in the spinal cord and overcome the limitation of poorly specific conventional MRI.[88]

Most SCIs are treated with surgical decompression and instrumented fusion to achieve the best mechanical conditions for recovery. This may aid in the recovery of one to two levels of function. While one level can recover in around 80% of patients, two levels only recover in 20%, and, unfortunately, full recovery is only made by around 1%. A considerable amount of work has focused on spinal cord regeneration with the aim of avoiding secondary injury by using stem cells, for example.[89] It has

been shown that the use of scaffolds and mesenchymal stem cell transplantation can be effective in animal models in the acute phase of injury.[90]

There are a variety of complications that may follow, such as decubitus ulcers, thromboembolisms, urosepsis, sinus bradycardia, orthostatic hypotension, autonomic dysreflexia, and depression (11%).[91] A retrospective study of 4800 hospitals in the United States found that patients with SCI had higher incidences of drug-related issues than controls between 2007 and 2009.[92] SCI patients display polypharmacy in 74%, and the most commonly used medication is against pain (68%), followed by products against constipation (42%), muscle spasms, and hypertension (42% each), as well as depression (37%).[93] Pressure ulcers can be found in around 10%—30%, with 25% accounting for stage III and IV ulcers. They usually occur around the sacrum. Although the risk remains steady for the first 10 years, it increases 15 years post-SCI. There is also an increasing prevalence in recent years.[94] The incidence rate may be higher in TSCI (71.8%) compared with NTSCI (28%).[95]

Bowel dysfunction occurs in most (68%) patients with SCI. Lower gastrointestinal symptoms, such as constipation (56%—80%) and incontinence (42%—75%), are more prevalent than upper gastrointestinal symptoms, such as bloating (22%) and abdominal distension (31%). It is rated in second place after mobility loss regarding problems after SCI.[96] On a scale from 0 (no problem) to 10 (maximum problem), its mean rating was 5.1, while the one for mobility loss was 6.8 in 115 patients with chronic SCI (median duration of 62 months).[96] Chronic abdominal pain was still present in 33% of responders and inconvenient (6/10), increasing the laxative use from 32% to 52% in a follow-up study after 10 years.[97] A review article concluded that bowel dysfunction in SCI patients requires a comprehensive and individualized approach with lifestyle, toileting routine, stimulation, diet modifications, medications, and surgical procedures.[98] Sexuality is also of concern for patients with SCI. Around 50% of patients reported the ability to have an orgasm after SCI, and the time to orgasm is usually longer.[99]

Etiology

Traumatic spinal cord injury

Traumatic spinal cord injuries are commonly associated with injuries to other body parts, with reported numbers ranging around 33%. Abdominal/pelvic injuries (20%) were more common than extremity (20%), thorax (19%), head (16%), and face/neck injuries (9%).[100] The most common cause of SCI is traffic accidents (42%—63%). The cervical spine is subject to complex biomechanical effects depending on the accident type (e.g., side or rear-end impact). Whiplash injuries may lead to disc herniations, which could potentially cause neurological impairment in later stages of life.[101]

In a few other cities and countries, such as Novosibirsk in Russia (25%),[102] Stockholm in Sweden (23%),[82] and Western Norway (34%—41%),[63] falls are the leading cause of injury. Sports injuries are a rare cause of SCI as seen in Spain (2%)[81] and Greece (4%).[82] These injuries include equestrian sports, where more accidents per hour are sustained than in motorcycling, skiing, and football.[103] While violence is usually a rare cause in Western countries, the highest percentage of SCI was found in parts of Kentucky and Indiana.[104] Violence causes 28% of SCIs in Jordan,[105] and falling weights cause 10% of SCIs in Romania.[106] There are some exceptions to the causes of SCI. Suicide is the cause of SCI in 26% of injuries in Greenland.[107]

War-related SCIs were found in up to 8% of soldiers wounded in a study of the American military personnel serving in Iraq and Afghanistan. It usually affected people with a mean age of 26 years, and the incidence rate ranged from 0.4 to 4.3 per 10,000 person-years,

with the highest incidence rates being found among white men from the Marine Corps. Gunshots and explosions are the most frequent mechanisms. The thoracic spine is usually more commonly injured than the lumbosacral and cervical spine.[108] Other body parts are very commonly injured as well (44%—78%). These injuries are more commonly complete than in the general population (18%—90% vs. 10% —82%). It was shown that 30% of cases with ASIA A converted to a better level within 1 year (ASIA B in 17%, ASIA C in 6%, and ASIA D in 7%). Patients with AIS B or C converted to a better AIS level more often (77%), while few patients deteriorated to a worse level (7%). Patients with ASIA D converted to ASIA E in only 85%, which may be due to a ceiling effect.[109]

Nontraumatic spinal cord injury

Degenerative cervical myelopathy

DCM is defined as acquired cervical spinal canal stenosis, usually due to spondylosis (osteoarthritic degeneration) or OPLL, leading to NTSCI with neurological symptoms and pain. There are several risk factors for accelerated degeneration in the cervical spine. For example, OPLL has been linked to genetic factors related to collagen VI and XI.[110] Collagen IX tryptophan alleles and smoking have also been identified as risk factors.[111] Excessive weight gain in the third to fifth decades of life and diabetes mellitus have also been reported as independent risk factors for the development of OPLL.[112] Furthermore, disc herniation and degeneration with spinal canal stenosis were more common in rugby players than controls, which has been attributed to the fact that rugby scrums may produce up to 1.5 tons on the cervical spine.[113,114] It is estimated that DCM constitutes more than half of the cases of NTSCI in Japan (59%) and the United States (54%), and less than half of the cases of NTSCI in Europe (31%), Australia (22%), and Africa (4%—30%).[70] Spinal canal stenosis affects the cervical spine in around a fourth of cases, and estimates state that around between 250,000 and 500,000 individuals are affected by it in the United States.[115]

DCM is thought to be one of the most common etiologies of spinal cord dysfunction. Although the exact prevalence of the disease is unknown, the estimated prevalence of degenerative cervical myelopathy in the United States is 605 per million, and the prevalence of surgically treated myelopathy in the Netherlands is 1.6 per 100,000 individuals.[10] The reason for this lack of information lies in different underlying degenerative pathologies and the categorization into the broader term of NCSCI. Even in a level I trauma center in the United States, 39% of SCI admissions were NTSCI. Of these, 54% were related to spinal canal stenosis and 26% to tumors. These patients were also more commonly found to have paraplegia and incomplete SCI.[116] Other studies have found hospitalization rates of 7.9 per 100,000 persons per year in the United States and 4.0 per 100,000 persons per year in Taiwan.[117,118] Radiographic criteria for this disease are met in >70% of ≥65 year old patients, while around 25% actually exhibit symptoms.[119] It is becoming more important because of the aging population. The number of elderly persons over 60 years was more than doubled within the past four decades to 982 million in 2017 according to the United Nations.[120] Elderly persons ≥65 years are even projected to increase from 13% to 22% from 2010 to 2050 in the United States.[121] Therefore, there is an expected increase in DCM, which makes diagnostic and treatment advances critical.

The pathobiological cascade involves several aspects. Ischemia has been documented by decreased blood flow in the anterior spinal artery due to mechanical stress that causes hyalinization and wall thickening.[119] Disruption of the blood—spinal cord barrier is associated with increased vascular permeability and inflammation promoting edema and activation

of fractalkine microglia and macrophages.[122] Apoptosis is thought to be mediated by Fas, which is a tumor necrosis factor receptor, and sodium channel activation followed by calcium influx and glutamate excitotoxicity, leading to caspase-mediated cell death.[123] Histopathology usually reveals demyelinating and gliotic features, as well as cavitation, degeneration of the central gray and medial white matter, ascending and descending tract Wallerian degenerative processes, and atrophy of ventral and dorsal horns.[124]

The natural history of degenerative cervical myelopathy is variable. It is thought that 20%—62% of patients experience neurological deterioration after 3—6 years, which can be reliably measured by the modified Japanese Orthopaedic Association (mJOA) scale. It was found that circumferential spinal cord compression constituted a risk factor for neurological deterioration.[125,126] The highest progression rates have been reported in males aged 70—79 years (12.5 cases per 100 patient-years) and females >80 years (9.3 cases per 100 patient-years), although this study is limited by reporting patients at risk of myelopathy instead of patients manifesting clinical symptoms.[127] Of note, a systematic review has shown that nonmyelopathic patients diagnosed with spondylosis and cervical canal stenosis with spinal cord compression progress to clinical signs of myelopathy in 8% over the course of 1 year and 23% after around 3.5 years.[127] One of the largest concerns of untreated degenerative cervical myelopathy is the increased risk of acute SCI after low-energy trauma such as falls. It is estimated that the incidence of this unfortunate event ranges around 2.4—13.9 per 1000 patients with cervical spondylotic myelopathy and 4.1—4.8 per 1000 patients with ossification of the posterior longitudinal ligament.[128] Therefore, surgery is usually recommended for moderate (mJOA 12—14 points) and severe (mJOA 0—11 points) cases as well as progressive mild (mJOA 15—17 points) cases. Surgical treatment is becoming more common. There were 9623 patients admitted for surgery per year in 1993 (around 3.7 per 100,000 population), which increased to 19,212 patients per year in 2002 (around 7.9 per 100,000 population).[117] While decompression usually prevents further progression of disease and neurological deficits, around 7%–11% of patients report neurological deterioration, which may be attributed to molecular and cellular alterations, such as axonal plasticity and ischemia–reperfusion injury, after decompression.[119] The benzothiazole anticonvulsant agent, riluzole, which blocks sodium and glutamate signaling in excitatory neurotransmission and is usually used in amyotrophic lateral sclerosis, has shown some benefit in neuroprotection and functional recovery.[129]

Metastases

Spinal cord compression affects around 5% of cancer patients and 10% of metastatic cancer patients. In adults, the cancer types are often lung, breast, and prostate cancer as well as lymphoma and myeloma being common cancers. The thoracic spine is most commonly affected (in around 70%). If located above the conus medullaris (at around L1), upper motor neuron signs with increased reflexes and tone can usually be found. If situated below the conus, lower motor neuron signs with decreased reflexes and tone as well as present fasciculations can often be observed. Since 35% of patients are affected by metastases at multiple levels, diagnosis is usually made by an MRI of the entire spine.[130]

Multidisciplinary treatment strategies involving neurosurgeons or orthopedic surgeons, oncologists, radiation oncologists, palliative care, and family physicians are usually employed. In acute settings of motor weakness, urgent surgical decompression (mostly with instrumented fusion) is usually the chosen treatment option to prevent further neurological harm and provide the basis for regaining neurological function. Radiation therapy is often added. A single fraction of 8 Gray may be given in slim

prognoses, while complex radiation is usually administered for better prognoses. The guidelines are usually published by the American College of Radiology every 3 years.[131] Dexamethasone is usually given to ameliorate the spinal cord edema and for analgesic purposes. The dose is a 10 mg loading dose and then 4 mg four times per day, followed by a 2-week taper.[132] Once spinal cord compression is diagnosed, the survival time is usually several months, with favorable prognosis in patients who have a slow onset of motor symptoms and/or regain function more successfully. The median survival time is only 1 month once a patient becomes nonambulatory. In these instances, conservative and hospice-based management is often a reasonable option. Palliative care often involves bladder and bowel management, prevention and treatment of decubitus ulcers from immobility, and psychological support with coping strategies.[130]

Gender and age

Gender

Males are more commonly affected than females. One of the highest male:female ratios (5:1) in North America was found in Alaska.[133] This ratio was even higher in Greece (7:1).[82] A smaller ratio was found in Minnesota (3:1).[59] While women usually report more pain, fatigue, skin, and transportation problems, men are usually bothered by health problems, diabetes, and adaptive equipment changes.[134]

Age

Most patients are <30 years old. Although the peak incidence is usually found between 15 and 30 years of age, it was >70 years in Ontario.[135] In males aged 65–74 years, the incidence of TSCI increased from 84 cases per million in 1993 to 131 cases per million in 2012. In patients >65 years, the TSCIs due to falls also increased from 28% in 1997—2000 to 66% in 2010—12. While mortality increased with age, it decreased over time in patients $\geq$85 years from 24.2% in 1993—96 to 20.1%.[136] The incidence of NTSCI increased with age in a study from Australia (38.1 per million in 55—64 year old patients vs. 89.1 per million in >85 year old patients).[137] There, the elderly population is subject to more pronounced disability due to comorbidities, such as pneumonia, which increases from 1.6% in <30 years to 5.4% in >60 years and gastrointestinal hemorrhage.[134,138] Due to inactivity, there often is an increase in body mass index that consequently leads to insulin resistance and diabetes, which ultimately increases mortality for patients with SCI.[139]

Falls are particularly common in the elderly. 19% of patients sustained $\geq$1 injury in the previous year and 10.4% reported a fall with consecutive injury in a study of 759 patients $\geq$65 years.[140] The inactivity due to SCI leads to bone density loss within 1 year. Around one-third of bone mass is lost by 16 months after sustaining the SCI, after which the situation has been shown to reach a steady state.[141] Other studies even report a 50% bone mineral density loss within the first 3 years.[142] This leads to a higher rate of low-energy lower-extremity fractures, for which the rate has been reported to be >30%.[143] Due to overuse of the upper extremities in paraplegia, shoulder pain due to acromioclavicular arthrosis, carpal tunnel syndrome, and metacarpophalangeal joint arthritis of the thumb are also more common.[144,145] Lastly, the incidence of dementia was also found to be increased in 95% of patients with SCI in a study of 5060 gender- and age-matched controls followed over 7 years in Taiwan. The incidence was 1106 per 100,000 person-years.[146]

Organized systems of care

Organized systems of care are a crucial point in the management of SCI. These integrate rapid and safe transport of patients with suspected SCI and excellent acute care, which is linked to timely and comprehensive rehabilitation with effective community service integration. Critical care begins at the scene of the accident, since up to 25% of SCIs develop after the initial injury.[147] A systematic initial management approach is usually based on several factors according to the time period. Within minutes, specific SCI protocols focusing on advanced trauma life support principles to identify potential SCI, spinal immobilization with a rigid cervical collar and a hard backboard with straps to limit motion and further SCI by potentially unstable vertebral segments, vigilance for cardiorespiratory problems, and timely transport[148] with SC perfusion optimization should be sought after.[149] Within hours, acute management should focus on emergent imaging with CT and MRI as well as early spinal cord decompression[2] within 24 hours of the trauma. Within 24 hours, early prevention of complications, such as gastric ulcer prophylaxis, deep vein thrombosis prophylaxis, and early initiation of nutrition, is also considered.[150] Transport is usually safe without worsening of neurological symptoms and can often be done within 24 hours, as reported in 84% of patients, both via ground (41%) and air (59%) including helicopters and fixed-wing aircrafts.[151] Timing has been shown to be of much more importance than the actual method of transportation. Interhospital transport also occurs with the intention to exclude SC in 18%.[152] Designated acute TSCI centers have been associated with improved neurological outcomes, length of hospital stays, and complications, such as pressure ulcers.[153] Upon arrival in the hospital, patients are logrolled, and the backboard and cervical collar are only removed upon clearance with radiographic imaging. Patients are usually assessed with the most widely used ASIA scoring system.[154] Prompt medical and surgical treatment is of great importance for the clinical outcome. SCI-specific rehabilitation in designated centers also helps improve clinical outcome and neurological recovery. In 2011, 35 facilities of acute SCI and rehabilitation centers were identified in Canada.[155] These specialized care facilities are usually in regional proximity to level 1 trauma centers, have a spinal surgery team with on-call coverage, rapid MRI, and operating room accessibility, and offer a wide range of specialists, such as physiotherapists, social workers, and psychologists. These facilities help reduce acute care hospitalizations by almost 50%, and their multidisciplinary care management expertise achieves optimal outcomes through optimizing activities of daily living and empowering to withstand SCI-related illnesses, such as ulcers.[156]

Primary prevention aims to stop the occurrence of an injury, secondary prevention refers to early detection of an injury, and tertiary prevention limits symptoms of injury. Primary SCI prevention is usually done by educating the population targeting knowledge about risky behaviors (e.g., not holding onto a handrail while climbing stairs in elderly) and developing modern equipment, such as air bags.[157] Prevention programs are very cost-effective when looking at literature that states that $ 3 billion could be saved over 2 years by a $ 10 investment per person into prevention programs.

Rehospitalizations

Rehospitalizations are more common in patients with a SCI. Cardenas et al.[158] used the US Department of Education, National Institute on Disability and Rehabilitation Research, Model Spinal Cord Injury Systems database, which includes data from a prospective, multicenter study for outcomes of patients with SCI since 1975, to study rehospitalizations after SCI. Using data from 8668 patients, the mean rehospitalizations were 55% for the first year and around 37% between the 5–20th year of follow-up. The mean days per rehospitalization were between 12 and 14 days. Risk factors for rehabilitation included

lower motor function and payer source. It was shown that a patient with a motor functional independence measurement score of 70 was at 1.25 times higher risk for rehospitalization than a patient with a score ≥10 times higher. Patients with state or federal insurance were also between around 1.45 times more likely to be rehospitalized. The most common reasons for rehospitalization are diseases of the genitourinary system, skin/subcutaneous tissue, and the respiratory system. Patients with tetraplegia (C1–8 ASIA grades A, B, C) were more commonly affected by respiratory issues. The mean length of stay for a rehospitalization was 12 days.

Neurological recovery

It is to be expected that patients affected by an SCI and their families ask about information regarding the recovery and prognosis. Outcome prediction also enables physicians to provide accurate guidance and enable patients with knowledge for informed decisions to provide the best individual treatment, rehabilitation, and follow-up. It is also vital for future research. Lastly, healthcare administrators can plan infrastructure and costs. However, there is a large spectrum of clinical outcomes, and the neurological recovery after SCI cannot be easily predicted due to the heterogeneous nature of the injury and inconsistent studies.

The events that follow SCI are classified as acute, subacute, and chronic.[159] Axonal degeneration is a main contributor of the degenerative pathway to a different degree in each phase. While some studies have tried to target these processes therapeutically with limited success, others have suggested that sprouting induction and axonal regeneration may in fact lead to neuropathic pain.[160,161] A recent systematic review and metaanalysis studied 39 articles and reported that the number of (myelinated) axons, thickness of the myelin sheath, axonal conduction velocity, and internode length progressively decreased over time after the injury.[162] It was also observed that axonal degeneration was increased in the thoracic spine.

Until a recent systematic review and metaanalysis, which provided a detailed summary of the current evidence, knowledge about the prognosis of patients with TSCI was mostly based on studies from North American or European registries.[163] It included studies that reported on follow-up changes in the ASIA impairment scale (AIS) or Frankel or ASIA motor score (AMS) scales. 114 studies with 19,913 patients with AIS/Frankel changes and 6920 patients with AMS changes were included, but the reported quality of evidence was poor. The mean AMS/Frankel improvement was higher for cervical SCI (20.3) than thoracic/lumbar SCI (7.0). Incomplete SCI had better AMS/Frankel improvements (26.6 for cervical and 15.4 for thoracic/lumbar SCI) than complete SCI (8.7 for cervical and 0.9 for thoracic/lumbar SCI). Lower scores in the lower spine may be caused by a ceiling effect since the arms are already neurologically intact. The observed AIS/Frankel conversion rate was highest for grade C (87.3%) and B (73.8%), followed by D (46.5%) and A (19.3%), which was statistically significant for each grade. Incomplete SCI has a relatively good prognosis with 40.4% of AIS/Frankel grade B and 87.3% of grade C improving to grades D or E with improvements of 25.6 and 36.5 points, respectively. Regarding incomplete SCI, the prognosis is usually better for central cord and Brown-Séquard syndrome (regain of about 30 motor points with final motor scores of 76 and 73, respectively) than anterior cord syndrome (regain of 22 points with a final motor score of 38).[164] Other studies reported that around 80% of patients with Brown-Séquard syndrome and 75% of patients with central cord syndrome will have a neurological improvement in the immediate postinjury time period with nearly all and 56%, respectively, regaining hand function and 80% and 59%,

respectively, becoming ambulatory. The anterior cord syndrome has the worst prognosis.[165,166] Patients with cauda equina syndrome often recover some function, but their health-related quality of life does not usually return to baseline (e.g., remaining moderate disability according to Oswestry Disability Index).[167]

Full recovery from complete SCI (AIS/Frankel grade A) and grade B is virtually impossible (0% and 0.3%, respectively), grade C patients regain full recovery in 9.2%, and grade D patients undergo full recovery in 46.5%. Some authors have argued that complete SCI recovery rates have resulted from misclassifying truly incomplete SCI. The initial exam may be influenced by several aspects, such as spinal shock, concomitant injuries (e.g., traumatic brain injury), intoxication, pain, and psychological stress.[163] Nevertheless, patients with complete SCI can even undergo neurological recovery after several months.[168]

Recovery rates were also substantially better if the level of injury was located at the lumbar spine compared with the cervical and thoracolumbar spine as well as the thoracic spine.[163] This is likely due to the cauda equina nerve roots (instead of spinal cord) and the enhanced ability of nerve roots to self-repair and sprout after injury.[169] It was noted that thoracic and penetrating lesions were significantly associated with complete injuries. One explanation may lie in the fact that the thoracic spine is very well shielded by the rib cage, and any harmful trauma to the spine must be very high energy. Another explanation is the small diameter of the spinal canal placing the spinal cord at risk during any injury. The thoracic spinal cord is also in the circulatory watershed area, rendering it more susceptible to injury due to vascular steal phenomena.

Penetrating injuries were 24% less likely to recover than blunt injuries. Since studies with shorter follow-up durations ($\geq$6 months) reported significantly lower recovery rates than studies with long-term follow-up duration (3—5 years), it may be assumed that recovery is a long-term process and follow-up times should have a minimum of 12 months.[163]

Furthermore, motor recovery after TSCI was not associated with the type of treatment (surgery vs. conservative) or country of origin. A potential explanation for the lack of difference is the fact that timing of the surgical decompression could not be considered as a factor and surgical treatment was usually reserved for patients with worsening of neurological deficits and/or persisting spinal cord compression. The fact that neurological recovery did not differ between developed and developing countries may indicate that the initial injury is the most important independent variable and the choice of treatment may not be a strong attributing factor.[163] However, in another systematic review and prospective survey by Fehlings et al., it was shown that $\geq 80\%$ of respondents to the survey $(n = 971)$ preferred surgical decompression of the spinal cord within 24 hours, except in central cord syndrome. They also reported that 73% had the preference to perform surgery within 6 hours for incomplete SCI compared with 46% for complete SCI.[170] Furthermore, large-scale randomized controlled studies are needed to address this issue. A previous review had also concluded that the severity of injury at baseline was the main predictor of neurological recovery and that complete thoracic SCI was linked to worse outcome compared with complete cervical lesions after TSCI.[171]

The selection of an ideal follow-up duration differs between institutions. While some physicians rely on a final neurological examination upon a patients' hospital discharge after around 2—3 months, others perform their follow-up at 6 or 12 months.[163] Since studies have shown conversion rates up to 12 months, it appears reasonable to choose this as the minimum follow-up.[109]

Survival

The severity and level of injury as well as age have a substantial impact on the survival. The mean life expectancy for patients with SCI was estimated to be 30 years.[54] In case of death, the median survival time was 4 years in Norway and 3 years in Estonia. Differences may be explained by socioeconomic differences, less physical activity, and a lack of sufficient injury prevention programs.[75] Patients with NTSCI were found to have a median overall survival of 24 years in a study of 1085 patients with an observation period between 1962 and 2000. The median survival was 22.7 years if the lesion was at the cervical spine, 23.4 years at the thoracic spine, and 27.7 years at the lumbar spine. The level of the injury was evenly distributed throughout the spine (35.4% in the cervical spine, 32.4% in the thoracic spine, and 31.2% in the lumbar spine).[172]

Acknowledgments

We would like to thank all authors from the references for their work and all patients for their participation in research.

References

1. Collaborators GN. Global, regional, and national burden of neurological disorders, 1990-2016: a systematic analysis for the Global Burden of Disease Study 2016. *Lancet Neurol* 2019;**18**:459—80. https://doi.org/10.1016/S1474-4422(18)30499-X.

2. Fehlings MG, et al. A clinical practice guideline for the management of patients with acute spinal cord injury and central cord syndrome: recommendations on the timing ($\leq$24 hours versus >24 hours) of decompressive surgery. *Global Spine J* 2017;**7**:195S—202S. https://doi.org/10.1177/2192568217706367.

3. Seif M, Gandini Wheeler-Kingshott CA, Cohen-Adad J, Flanders AE, Freund P. Guidelines for the conduct of clinical trials in spinal cord injury: neuroimaging biomarkers. *Spinal Cord* 2019;**57**:717—28. https://doi.org/10.1038/s41393-019-0309-x.

4. Cadotte DW, Fehlings MG. Will imaging biomarkers transform spinal cord injury trials? *Lancet Neurol* 2013;**12**:843−4. https://doi.org/10.1016/S1474-4422 (13)70157-1.

5. Nijsten T, Stern RS. How epidemiology has contributed to a better understanding of skin disease. *J Invest Dermatol* 2012;**132**:994−1002. https://doi.org/10.1038/jid.2011.372.

6. Samet JM, Wipfli H, Platz EA, Bhavsar N. American Journal of Epidemiology. *A dictionary of epidemiology.* 5th ed., vol. 170; 2009.

7. Centers for Disease Control and Prevention. *Principles of epidemiology in public health practice: an introduction to applied epidemiology and biostatistics.* 2012. https://www.cdc.gov/csels/dsepd/ss1978/lesson3/section2.html. [Accessed 8 December 2020].

8. Singh A, Tetreault L, Kalsi-Ryan S, Nouri A, Fehlings MG. Global prevalence and incidence of traumatic spinal cord injury. *Clin Epidemiol* 2014;**6**:309−31. https://doi.org/10.2147/CLEP.S68889.

9. National Spinal Cord Injury Statistical Center. Spinal cord injury (SCI) 2016 facts and figures at a glance. *J Spinal Cord Med* 2016;**39**:493−4. https://doi.org/10.1080/10790268.2016.1210925.

10. Nouri A, Tetreault L, Singh A, Karadimas SK, Fehlings MG. Degenerative cervical myelopathy: epidemiology, genetics, and pathogenesis. *Spine* 2015;**40**:E675−93. https://doi.org/10.1097/BRS.0000000000000913.

11. Chiu WT, et al. Review paper: epidemiology of traumatic spinal cord injury: comparisons between developed and developing countries. *Asia Pac J Publ Health* 2010;**22**:9−18. https://doi.org/10.1177/1010539509355470.

12. Lee BB, Cripps RA, Fitzharris M, Wing PC. The global map for traumatic spinal cord injury epidemiology: update 2011, global incidence rate. *Spinal Cord* 2014;**52**:110−6. https://doi.org/10.1038/sc.2012.158.

13. Johansson E, et al. Epidemiology of traumatic spinal cord injury in Finland. *Spinal Cord* 2020. https://doi.org/10.1038/s41393-020-00575-4.

14. Korhonen N, Kannus P, Niemi S, Parkkari J, Sievänen H. Rapid increase in fall-induced cervical spine injuries among older Finnish adults between 1970 and 2011. *Age Ageing* 2014;**43**:567−71. https://doi.org/10.1093/ageing/afu060.

15. Liu H, et al. The changing demographics of traumatic spinal cord injury in Beijing, China: a single-centre report of 2448 cases over 7 years. *Spinal Cord* 2020. https://doi.org/10.1038/s41393-020-00564-7.

16. Sekhon LH, Fehlings MG. Epidemiology, demographics, and pathophysiology of acute spinal cord injury. *Spine* 2001;**26**:S2−12. https://doi.org/10.1097/00007632-200112151-00002.

17. DeVivo M, Chen Y, Mennemeyer S, Deutsch A. Costs of care following spinal cord injury. *Top Spinal Cord Inj Rehabil* 2011;**16**:1−9.

18. Christopher and Dana Reeve Foundation. *Living with paralysis.* 2015. <with_Spinal_Cord_Injury.htm>, http://www.christopherreeve.org/site/c.mtKZKgM-. [Accessed 8 December 2020].

19. Vialle LRG, Fehlings MG, Weidner N. *Spinal cord injury and regeneration. Volume AOSpine Masters Series*, vol. 7; 2016.

20. New PWA. Narrative review of pediatric nontraumatic spinal cord dysfunction. *Top Spinal Cord Inj Rehabil* 2019;**25**:112−20. https://doi.org/10.1310/sci2502-112.

21. Goodrich JT. *Vol. Spina bifida management and outcome 3−17.* Springer Milan; 2008.

22. Tulpius N. *Observationes medicae*, vol. Libri III 231. Elzevirium; 1641.

23. Iacobus H, Houllier J. *De materia chirurgica libri tres.* Mit Holzschnitt-Druckermarke. 1552.

24. Adams F. *The genuine works of hippocrates [translated from the Greek with a preliminary discourse and annotations].* 1849.

25. von Heine J. In: Köhler FH, editor. *Beobachtungen über Lähmungszustände der untern Extremitäten und deren Behandlung*; 1840.

26. Salk JE, et al. Formaldehyde treatment and safety testing of experimental poliomyelitis vaccines. *Am J Public Health Nat Health* 1954;**44**:563−70. https://doi.org/10.2105/ajph.44.5.563.

27. Folliss AGH, Netsky MG. *Progressive necrotic myelopathy.* North- Holland Publishing Company; 1970.

28. Adamkiewicz A. Die Blutegefasse des menschlichen Ruckenmakers: Die Geffasse der Ruckenmarksubstanz. *S Ber Akad Wiss (Wien)* 1881;**III**:85.

29. Gowers WR, Horsley V. Case of tumour of the spinal cord; removal; recovery. *Medico-chirur Trans* 1888;**53**:377−428.

30. Church A, Eisendrath DW. A contribution to spinal cord surgery. *Am J Med Sci* 1892;**103**:403−5.

31. Sciubba DM, Liang D, Kothbauer KF, Noggle JC, Jallo GI. The evolution of intramedullary spinal cord tumor surgery. *Neurosurgery* 2009;**65**:84−91. https://doi.org/10.1227/01.NEU.0000345628.39796.40. Discussion 91-82.

32. Wood EH, Berne AS, Taveras JM. The value of radiation therapy in the management of intrinsic tumors of the spinal cord. *Radiology* 1954;**63**:11–24. https://doi.org/10.1148/63.1.11.

33. Bassoe P, Hassin GB. Myelitis and myelomalacia. A clinicopathalogic study with remarks on the fate of gitter cells. *Arch Neurol Psychiatr* 1921;**6**:32–43.

34. Wilson SAK. *Neurology* 1954;**2**.

35. McHenry LC. *Garrison's history of neurology.* Springfield, IL. 1969.

36. Kraus JF, Franti CE, Riggins RS, Richards D, Borhani NO. Incidence of traumatic spinal cord lesions. *J Chron Dis* 1975;**28**:471–92. https://doi.org/10.1016/0021-9681(75)90057-0.

37. New PW, Delafosse V. What to call spinal cord damage not due to trauma? Implications for literature searching. *J Spinal Cord Med* 2012;**35**:89–95. https://doi.org/10.1179/2045772311Y.0000000053.

38. New PW, Marshall R. International spinal cord injury data sets for non-traumatic spinal cord injury. *Spinal Cord* 2014;**52**:123–32. https://doi.org/10.1038/sc.2012.160.

39. Apple DF, Anson CA, Hunter JD, Bell RB. Spinal cord injury in youth. *Clin Pediatr* 1995;**34**:90–5. https://doi.org/10.1177/000992289503400205.

40. Cirak B, et al. Spinal injuries in children. *J Pediatr Surg* 2004;**39**:607–12. https://doi.org/10.1016/j.jpedsurg.2003.12.011.

41. Brown RL, Brunn MA, Garcia VF. Cervical spine injuries in children: a review of 103 patients treated consecutively at a level 1 pediatric trauma center. *J Pediatr Surg* 2001;**36**:1107–14. https://doi.org/10.1053/jpsu.2001.25665.

42. Parent S, Mac-Thiong JM, Roy-Beaudry M, Sosa JF, Labelle H. Spinal cord injury in the pediatric population: a systematic review of the literature. *J Neurotrauma* 2011;**28**:1515–24. https://doi.org/10.1089/neu.2009.1153.

43. Dearolf WW, et al. Scoliosis in pediatric spinal cord-injured patients. *J Pediatr Orthop* 1990;**10**:214–8.

44. Wang MY, Hoh DJ, Leary SP, Griffith P, McComb JG. High rates of neurological improvement following severe traumatic pediatric spinal cord injury. *Spine* 2004;**29**:1493–7. https://doi.org/10.1097/01.brs.0000129026.03194.0f. Discussion E1266.

45. New PW, et al. Global mapping for the epidemiology of paediatric spinal cord damage: towards a living data repository. *Spinal Cord* 2019;**57**:183–97. https://doi.org/10.1038/s41393-018-0209-5.

46. Price C, Makintubee S, Herndon W, Istre GR. Epidemiology of traumatic spinal cord injury and acute hospitalization and rehabilitation charges for spinal cord injuries in Oklahoma, 1988–1990. *Am J Epidemiol* 1994;**139**:37–47. https://doi.org/10.1093/oxfordjournals.aje.a116933.

47. Thurman DJ, Burnett CL, Jeppson L, Beaudoin DE, Sniezek JE. Surveillance of spinal cord injuries in Utah, USA. *Paraplegia* 1994;**32**:665–9. https://doi.org/10.1038/sc.1994.107.

48. Bracken MB, Freeman DH, Hellenbrand K. Incidence of acute traumatic hospitalized spinal cord injury in the United States, 1970–1977. *Am J Epidemiol* 1981;**113**:615–22. https://doi.org/10.1093/oxfordjournals.aje.a113140.

49. Pickett W, Simpson K, Walker J, Brison RJ. Traumatic spinal cord injury in Ontario, Canada. *J Trauma* 2003;**55**:1070–6. https://doi.org/10.1097/01.TA.0000034228.18541.D1.

50. Dryden DM, et al. The epidemiology of traumatic spinal cord injury in Alberta, Canada. *Can J Neurol Sci* 2003;**30**:113–21. https://doi.org/10.1017/s0317167100053373.

51. Lenehan B, et al. The epidemiology of traumatic spinal cord injury in British Columbia, Canada. *Spine (Phila Pa 1976)* 2012;**37**:321–9. https://doi.org/10.1097/BRS.0b013e31822e5ff8.

52. Cripps RA, et al. A global map for traumatic spinal cord injury epidemiology: towards a living data repository for injury prevention. *Spinal Cord* 2011;**49**:493–501. https://doi.org/10.1038/sc.2010.146.

53. Blumer CE, Quine S. Prevalence of spinal cord injury: an international comparison. *Neuroepidemiology* 1995;**14**:258–68. https://doi.org/10.1159/000109801.

54. DeVivo MJ, Fine PR, Maetz HM, Stover SL. Prevalence of spinal cord injury: a reestimation employing life table techniques. *Arch Neurol* 1980;**37**:707–8. https://doi.org/10.1001/archneur.1980.00500600055011.

55. Minaire P, et al. Life expectancy following spinal cord injury: a ten-years survey in the Rhône-Alpes Region, France, 1969–1980. *Paraplegia* 1983;**21**:11–5. https://doi.org/10.1038/sc.1983.2.

56. Kurtzke JF. Epidemiology of spinal cord injury. *Exp Neurol* 1975;**48**:163–236. https://doi.org/10.1016/0014-4886(75)90175-2.

57. Harvey C, Rothschild BB, Asmann AJ, Stripling T. New estimates of traumatic SCI prevalence: a survey-based approach. *Paraplegia* 1990;**28**:537–44. https://doi.org/10.1038/sc.1990.73.

58. Noonan VK, et al. Incidence and prevalence of spinal cord injury in Canada: a national perspective. *Neuroepidemiology* 2012;**38**:219−26. https://doi.org/10.1159/000336014.

59. Griffin MR, Opitz JL, Kurland LT, Ebersold MJ, O'Fallon WM. Traumatic spinal cord injury in Olmsted County, Minnesota, 1935−1981. *Am J Epidemiol* 1985;**121**:884−95. https://doi.org/10.1093/oxfordjournals.aje.a114058.

60. Griffin MR, O'Fallon WM, Opitz JL, Kurland LT. Mortality, survival and prevalence: traumatic spinal cord injury in Olmsted County, Minnesota, 1935−1981. *J Chron Dis* 1985;**38**:643−53. https://doi.org/10.1016/0021-9681(85)90018-9.

61. Knútsdóttir S, et al. Epidemiology of traumatic spinal cord injuries in Iceland from 1975 to 2009. *Spinal Cord* 2012;**50**:123−6. https://doi.org/10.1038/sc.2011.105.

62. Dahlberg A, Kotila M, Leppänen P, Kautiainen H, Alaranta H. Prevalence of spinal cord injury in Helsinki. *Spinal Cord* 2005;**43**:47−50. https://doi.org/10.1038/sj.sc.3101616.

63. Hagen EM, Eide GE, Rekand T, Gilhus NE, Gronning M. A 50-year follow-up of the incidence of traumatic spinal cord injuries in Western Norway. *Spinal Cord* 2010;**48**:313−8. https://doi.org/10.1038/sc.2009.133.

64. Rahimi-Movaghar V, et al. Prevalence of spinal cord injury in Tehran, Iran. *J Spinal Cord Med* 2009;**32**:428−31. https://doi.org/10.1080/10790268.2009.11754572.

65. O'Connor PJ. Prevalence of spinal cord injury in Australia. *Spinal Cord* 2005;**43**:42−6. https://doi.org/10.1038/sj.sc.3101666.

66. Rahimi-Movaghar V, et al. Epidemiology of traumatic spinal cord injury in developing countries: a systematic review. *Neuroepidemiology* 2013;**41**:65−85. https://doi.org/10.1159/000350710.

67. Shrestha D. Traumatic spinal cord injury in Nepal. *Kathmandu Univ Med J* 2014;**12**:161−2.

68. Øderud T. Surviving spinal cord injury in low income countries. *Afr J Disabil* 2014;**3**:80. https://doi.org/10.4102/ajod.v3i2.80.

69. Kumar R, et al. Traumatic spinal injury: global epidemiology and worldwide volume. *World Neurosurg* 2018;**113**:e345−63. https://doi.org/10.1016/j.wneu.2018.02.033.

70. New PW, Cripps RA, Bonne Lee B. Global maps of non-traumatic spinal cord injury epidemiology: towards a living data repository. *Spinal Cord* 2014;**52**:97−109. https://doi.org/10.1038/sc.2012.165.

71. New PW, et al. Trends, challenges, and opportunities regarding research in non-traumatic spinal cord dysfunction. *Top Spinal Cord Inj Rehabil* 2017;**23**:313−23. https://doi.org/10.1310/sci2304-313.

72. Dixon GS, Danesh JN, Caradoc-Davies TH. Epidemiology of spinal cord injury in New Zealand. *Neuroepidemiology* 1993;**12**:88−95. https://doi.org/10.1159/000110305.

73. Sabre L, et al. High incidence of traumatic spinal cord injury in Estonia. *Spinal Cord* 2012;**50**:755−9. https://doi.org/10.1038/sc.2012.54.

74. Shingu H, Ikata T, Katoh S, Akatsu T. Spinal cord injuries in Japan: a nationwide epidemiological survey in 1990. *Paraplegia* 1994;**32**:3−8. https://doi.org/10.1038/sc.1994.2.

75. Sabre L, Hagen EM, Rekand T, Asser T, Kõrv J. Traumatic spinal cord injury in two European countries: why the differences? *Eur J Neurol* 2013;**20**:293−9. https://doi.org/10.1111/j.1468-1331.2012.03845.x.

76. Shingu H, Ohama M, Ikata T, Katoh S, Akatsu T. A nationwide epidemiological survey of spinal cord injuries in Japan from January 1990 to December 1992. *Paraplegia* 1995;**33**:183−8. https://doi.org/10.1038/sc.1995.42.

77. Biering-Sørensen E, Pedersen V, Clausen S. Epidemiology of spinal cord lesions in Denmark. *Paraplegia* 1990;**28**:105−18. https://doi.org/10.1038/sc.1990.13.

78. van Asbeck FW, Post MW, Pangalila RF. An epidemiological description of spinal cord injuries in the Netherlands in 1994. *Spinal Cord* 2000;**38**:420−4. https://doi.org/10.1038/sj.sc.3101003.

79. Equebal A, Anwer S, Kumar R. The prevalence and impact of age and gender on rehabilitation outcomes in spinal cord injury in India: a retrospective pilot study. *Spinal Cord* 2013;**51**:409−12. https://doi.org/10.1038/sc.2013.5.

80. Pickett GE, Campos-Benitez M, Keller JL, Duggal N. Epidemiology of traumatic spinal cord injury in Canada. *Spine* 2006;**31**:799−805. https://doi.org/10.1097/01.brs.0000207258.80129.03.

81. Van Den Berg M, Castellote JM, Mahillo-Fernandez I, de Pedro-Cuesta J. Incidence of traumatic spinal cord injury in Aragón, Spain (1972−2008). *J Neurotrauma* 2011;**28**:469−77. https://doi.org/10.1089/neu.2010.1608.

82. Divanoglou A, Levi R. Incidence of traumatic spinal cord injury in Thessaloniki, Greece and Stockholm, Sweden: a prospective population-based study. *Spinal Cord* 2009;**47**:796−801. https://doi.org/10.1038/sc.2009.28.

83. Krueger H, Noonan VK, Trenaman LM, Joshi P, Rivers CS. The economic burden of traumatic spinal cord injury in Canada. *Chronic Dis Inj Can* 2013;**33**: 113−22.

84. Kirshblum SC, et al. International standards for neurological classification of spinal cord injury (revised 2011). *J Spinal Cord Med* 2011;**34**:535−46. https://doi.org/10.1179/204577211X13207446293695.

85. McKinley W, Santos K, Meade M, Brooke K. Incidence and outcomes of spinal cord injury clinical syndromes. *J Spinal Cord Med* 2007;**30**:215−24. https://doi.org/10.1080/10790268.2007.11753929.

86. Fehlings MG, Tetzlaff W. Summary statement: repair of the injured spinal cord. *Spine* 2001;**26**:S23. https://doi.org/10.1097/00007632-200112151-00004.

87. Huber E, Lachappelle P, Sutter R, Curt A, Freund P. Are midsagittal tissue bridges predictive of outcome after cervical spinal cord injury? *Ann Neurol* 2017;**81**: 740−8. https://doi.org/10.1002/ana.24932.

88. Kearney H, Miller DH, Ciccarelli O. Spinal cord MRI in multiple sclerosis–diagnostic, prognostic and clinical value. *Nat Rev Neurol* 2015;**11**:327−38. https://doi.org/10.1038/nrneurol.2015.80.

89. Kwon BK, Tetzlaff W. Spinal cord regeneration: from gene to transplants. *Spine* 2001;**26**:S13−22. https://doi.org/10.1097/00007632-200112151-00003.

90. Yousefifard M, et al. A combination of mesenchymal stem cells and scaffolds promotes motor functional recovery in spinal cord injury: a systematic review and meta-analysis. *J Neurosurg Spine* 2019;**32**:269−84. https://doi.org/10.3171/2019.8.SPINE19201.

91. Orthobullets. *Spinal cord injuries*. 2020. https://www.orthobullets.com/spine/2006/spinal-cord-injuries. [Accessed 8 December 2020].

92. Kitzman P, Cecil D, Kolpek JH. The risks of polypharmacy following spinal cord injury. *J Spinal Cord Med* 2017;**40**:147−53. https://doi.org/10.1179/2045772314Y.0000000235.

93. Patel T, Milligan J, Lee J. Medication-related problems in individuals with spinal cord injury in a primary care-based clinic. *J Spinal Cord Med* 2017;**40**:54−61. https://doi.org/10.1179/2045772315Y.0000000055.

94. Dinh A, et al. Management of established pressure ulcer infections in spinal cord injury patients. *Med Maladies Infect* 2019;**49**:9−16. https://doi.org/10.1016/j.medmal.2018.05.004.

95. Taghipoor KD, et al. Factors associated with pressure ulcers in patients with complete or sensory-only preserved spinal cord injury: is there any difference between traumatic and nontraumatic causes? *J Neurosurg Spine* 2009;**11**:438−44. https://doi.org/10.3171/2009.5.SPINE08896.

96. Glickman S, Kamm MA. Bowel dysfunction in spinal-cord-injury patients. *Lancet* 1996;**347**:1651−3. https://doi.org/10.1016/s0140-6736(96)91487-7.

97. Nielsen SD, Faaborg PM, Christensen P, Krogh K, Finnerup NB. Chronic abdominal pain in long-term spinal cord injury: a follow-up study. *Spinal Cord* 2017;**55**:290−3. https://doi.org/10.1038/sc.2016.124.

98. Qi Z, Middleton JW, Malcolm A. Bowel dysfunction in spinal cord injury. *Curr Gastroenterol Rep* 2018;**20**:47. https://doi.org/10.1007/s11894-018-0655-4.

99. Alexander M, Marson L. Orgasm and SCI: what do we know? *Spinal Cord* 2018;**56**:538−47. https://doi.org/10.1038/s41393-017-0020-8.

100. Selassie AW, Varma A, Saunders LL, Welldaregay W. Determinants of in-hospital death after acute spinal cord injury: a population-based study. *Spinal Cord* 2013;**51**:48−54. https://doi.org/10.1038/sc.2012.88.

101. Pettersson K, Hildingsson C, Toolanen G, Fagerlund M, Björnebrink J. Disc pathology after whiplash injury. A prospective magnetic resonance imaging and clinical investigation. *Spine* 1997;**22**:283−7. https://doi.org/10.1097/00007632-199702010-00010. Discussion 288.

102. Silberstein B, Rabinovich S. Epidemiology of spinal cord injuries in Novosibirsk, Russia. *Paraplegia* 1995;**33**:322−5. https://doi.org/10.1038/sc.1995.72.

103. Papachristos A, Edwards E, Dowrick A, Gosling C. A description of the severity of equestrian-related injuries (ERIs) using clinical parameters and patient-reported outcomes. *Injury* 2014;**45**:1484−7. https://doi.org/10.1016/j.injury.2014.04.017.

104. Burke DA, Linden RD, Zhang YP, Maiste AC, Shields CB. Incidence rates and populations at risk for spinal cord injury: a regional study. *Spinal Cord* 2001;**39**:274−8. https://doi.org/10.1038/sj.sc.3101158.

105. Otom AS, Doughan AM, Kawar JS, Hattar EZ. Traumatic spinal cord injuries in Jordan–an epidemiological study. *Spinal Cord* 1997;**35**:253−5. https://doi.org/10.1038/sj.sc.3100402.

106. Soopramanien A. Epidemiology of spinal injuries in Romania. *Paraplegia* 1994;**32**:715−22. https://doi.org/10.1038/sc.1994.116.

107. Pedersen V, Müller PG, Biering-Sørensen F. Traumatic spinal cord injuries in Greenland 1965−1986. *Paraplegia* 1989;**27**:345−9. https://doi.org/10.1038/sc.1989.52.

108. Furlan JC, Gulasingam S, Craven BC. Epidemiology of war-related spinal cord injury among combatants: a systematic review. *Global Spine J* 2019;**9**:545−58. https://doi.org/10.1177/2192568218776914.

109. Spiess MR, et al. Conversion in ASIA impairment scale during the first year after traumatic spinal cord injury. *J Neurotrauma* 2009;**26**:2027−36. https://doi.org/10.1089/neu.2008.0760.

110. Wilson JR, et al. Genetics and heritability of cervical spondylotic myelopathy and ossification of the posterior longitudinal ligament: results of a systematic review. *Spine* 2013;**38**:S123–46. https://doi.org/10.1097/BRS.0b013e3182a7f478.

111. Wang ZC, et al. The role of smoking status and collagen IX polymorphisms in the susceptibility to cervical spondylotic myelopathy. *Genet Mol Res* 2012;**11**:1238–44. https://doi.org/10.4238/2012.May.9.2.

112. Kobashi G, et al. High body mass index after age 20 and diabetes mellitus are independent risk factors for ossification of the posterior longitudinal ligament of the spine in Japanese subjects: a case-control study in multiple hospitals. *Spine* 2004;**29**:1006–10. https://doi.org/10.1097/00007632-200405010-00011.

113. Scher AT. Premature onset of degenerative disease of the cervical spine in rugby players. *S Afr Med J* 1990;**77**:557–8.

114. Berge J, Marque B, Vital JM, Sénégas J, Caillé JM. Age-related changes in the cervical spines of front-line rugby players. *Am J Sports Med* 1999;**27**:422–9. https://doi.org/10.1177/03635465990270040401.

115. Bajwa NS, Toy JO, Young EY, Ahn NU. Establishment of parameters for congenital stenosis of the cervical spine: an anatomic descriptive analysis of 1,066 cadaveric specimens. *Eur Spine J* 2012;**21**:2467–74. https://doi.org/10.1007/s00586-012-2437-2.

116. McKinley WO, Seel RT, Hardman JT. Nontraumatic spinal cord injury: incidence, epidemiology, and functional outcome. *Arch Phys Med Rehabil* 1999;**80**:619–23. https://doi.org/10.1016/s0003-9993(99)90162-4.

117. Lad SP, et al. National trends in spinal fusion for cervical spondylotic myelopathy. *Surg Neurol* 2009;**71**:66–9. https://doi.org/10.1016/j.surneu.2008.02.045. Discussion 69.

118. Wu JC, et al. Epidemiology of cervical spondylotic myelopathy and its risk of causing spinal cord injury: a national cohort study. *Neurosurg Focus* 2013;**35**:E10. https://doi.org/10.3171/2013.4.FOCUS13122.

119. Badhiwala JH, et al. Degenerative cervical myelopathy - update and future directions. *Nat Rev Neurol* 2020;**16**:108–24. https://doi.org/10.1038/s41582-019-0303-0.

120. United Nations, Department of Economic and Social Affairs & Population Division. *World population ageing 2017 - highlights*. ST/ESA/SER.A/397. 2017. p. 1.

121. The World Bank. *DataBank: population estimates and projections*. 2019. http://databank.worldbank.org/data/reports.aspx?source=health-nutrition-and-population-statistics:-population-estimates-and-projections#. [Accessed 8 December 2020].

122. Hong J, et al. Level-specific differences in systemic expression of pro- and anti-inflammatory cytokines and chemokines after spinal cord injury. *Int J Mol Sci* 2018;**19**. https://doi.org/10.3390/ijms19082167.

123. Letellier E, et al. CD95-ligand on peripheral myeloid cells activates Syk kinase to trigger their recruitment to the inflammatory site. *Immunity* 2010;**32**:240–52. https://doi.org/10.1016/j.immuni.2010.01.011.

124. Fehlings MG, Skaf G. A review of the pathophysiology of cervical spondylotic myelopathy with insights for potential novel mechanisms drawn from traumatic spinal cord injury. *Spine* 1998;**23**:2730–7. https://doi.org/10.1097/00007632-199812150-00012.

125. Karadimas SK, Erwin WM, Ely CG, Dettori JR, Fehlings MG. Pathophysiology and natural history of cervical spondylotic myelopathy. *Spine* 2013;**38**:S21–36. https://doi.org/10.1097/BRS.0b013e3182a7f2c3.

126. Rhee J, et al. Nonoperative versus operative management for the treatment degenerative cervical myelopathy: an updated systematic review. *Global Spine J* 2017;**7**:35S–41S. https://doi.org/10.1177/2192568217703083.

127. Wilder FV, Fahlman L, Donnelly R. Radiographic cervical spine osteoarthritis progression rates: a longitudinal assessment. *Rheumatol Int* 2011;**31**:45–8. https://doi.org/10.1007/s00296-009-1216-9.

128. Chen LF, et al. Risk of spinal cord injury in patients with cervical spondylotic myelopathy and ossification of posterior longitudinal ligament: a national cohort study. *Neurosurg Focus* 2016;**40**:E4. https://doi.org/10.3171/2016.3.FOCUS1663.

129. Tetreault LA, Zhu MP, Wilson JR, Karadimas SK, Fehlings MG. The impact of riluzole on neurobehavioral outcomes in preclinical models of traumatic and nontraumatic spinal cord injury: results from a systematic review of the literature. *Global Spine J* 2020;**10**:216–29. https://doi.org/10.1177/2192568219835516.

130. Skoch BM, Sinclair CT. Management of urgent medical conditions at the end of life. *Med Clin* 2020;**104**:525–38. https://doi.org/10.1016/j.mcna.2019.12.006.

131. Lo SS, et al. ACR appropriateness Criteria® metastatic epidural spinal cord compression and recurrent spinal metastasis. *J Palliat Med* 2015;**18**:573–84. https://doi.org/10.1089/jpm.2015.28999.sml.

132. Kumar A, et al. Metastatic spinal cord compression and steroid treatment: a systematic review. *Clin Spine Surg* 2017;**30**:156–63. https://doi.org/10.1097/BSD.0000000000000528.

133. Warren S, Moore M, Johnson MS. Traumatic head and spinal cord injuries in Alaska (1991–1993). *Alaska Med* 1995;**37**:11–9.

134. McColl MA, Charlifue S, Glass C, Lawson N, Savic G. Aging, gender, and spinal cord injury. *Arch Phys Med Rehabil* 2004;**85**:363−7. https://doi.org/10.1016/j.apmr.2003.06.022.

135. Pickett A. Re-engineering clostridial neurotoxins for the treatment of chronic pain: current status and future prospects. *BioDrugs* 2010;**24**:173−82. https://doi.org/10.2165/11534510-000000000-00000.

136. Jain NB, et al. Traumatic spinal cord injury in the United States, 1993−2012. *J Am Med Assoc* 2015;**313**:2236−43. https://doi.org/10.1001/jama.2015.6250.

137. New PW, Sundararajan V. Incidence of non-traumatic spinal cord injury in Victoria, Australia: a population-based study and literature review. *Spinal Cord* 2008;**46**:406−11. https://doi.org/10.1038/sj.sc.3102152.

138. Stolzmann KL, Gagnon DR, Brown R, Tun CG, Garshick E. Longitudinal change in FEV1 and FVC in chronic spinal cord injury. *Am J Respir Crit Care Med* 2008;**177**:781−6. https://doi.org/10.1164/rccm.200709-1332OC.

139. Garshick E, et al. A prospective assessment of mortality in chronic spinal cord injury. *Spinal Cord* 2005;**43**:408−16. https://doi.org/10.1038/sj.sc.3101729.

140. Saunders LL, Krause JS. Injuries and falls in an aging cohort with spinal cord injury: SCI aging study. *Top Spinal Cord Inj Rehabil* 2015;**21**:201−7. https://doi.org/10.1310/sci2103-201.

141. Garland DE, et al. Osteoporosis after spinal cord injury. *J Orthop Res* 1992;**10**:371−8. https://doi.org/10.1002/jor.1100100309.

142. Edwards WB, Schnitzer TJ. Bone imaging and fracture risk after spinal cord injury. *Curr Osteoporos Rep* 2015;**13**:310−7. https://doi.org/10.1007/s11914-015-0288-6.

143. Chan LW, et al. Special considerations in the urological management of the older spinal cord injury patient. *World J Urol* 2018;**36**:1603−11. https://doi.org/10.1007/s00345-018-2326-3.

144. Eriks-Hoogland I, Engisch R, Brinkhof MW, van Drongelen S. Acromioclavicular joint arthrosis in persons with spinal cord injury and able-bodied persons. *Spinal Cord* 2013;**51**:59−63. https://doi.org/10.1038/sc.2012.89.

145. Akbar M, et al. Prevalence of carpal tunnel syndrome and wrist osteoarthritis in long-term paraplegic patients compared with controls. *J Hand Surg Eur* 2014;**39**:132−8. https://doi.org/10.1177/1753193413478550.

146. Huang SW, Wang WT, Chou LC, Liou TH, Lin HW. Risk of dementia in patients with spinal cord injury: a nationwide population-based cohort study. *J Neurotrauma* 2017;**34**:615−22. https://doi.org/10.1089/neu.2016.4525.

147. Shank CD, Walters BC, Hadley MN. Management of acute traumatic spinal cord injuries. *Handb Clin Neurol* 2017;**140**:275−98. https://doi.org/10.1016/B978-0-444-63600-3.00015-5.

148. Ahn H, et al. Pre-hospital care management of a potential spinal cord injured patient: a systematic review of the literature and evidence-based guidelines. *J Neurotrauma* 2011;**28**:1341−61. https://doi.org/10.1089/neu.2009.1168.

149. Yue JK, et al. Update on critical care for acute spinal cord injury in the setting of polytrauma. *Neurosurg Focus* 2017;**43**:E19. https://doi.org/10.3171/2017.7.FOCUS17396.

150. Witiw CD, Fehlings MG. Acute spinal cord injury. *J Spinal Disord Tech* 2015;**28**:202−10. https://doi.org/10.1097/BSD.0000000000000287.

151. Burney RE, Waggoner R, Maynard FM. Stabilization of spinal injury for early transfer. *J Trauma* 1989;**29**:1497−9. https://doi.org/10.1097/00005373-198911000-00008.

152. Flabouris A. Clinical features, patterns of referral and out of hospital transport events for patients with suspected isolated spinal injury. *Injury* 2001;**32**:569−75. https://doi.org/10.1016/s0020-1383(01)00071-7.

153. DeVivo MJ, Go BK, Jackson AB. Overview of the national spinal cord injury statistical center database. *J Spinal Cord Med* 2002;**25**:335−8. https://doi.org/10.1080/10790268.2002.11753637.

154. Savic G, Bergström EM, Frankel HL, Jamous MA, Jones PW. Inter-rater reliability of motor and sensory examinations performed according to American Spinal Injury Association standards. *Spinal Cord* 2007;**45**:444−51. https://doi.org/10.1038/sj.sc.3102044.

155. Parent S, Barchi S, LeBreton M, Casha S, Fehlings MG. The impact of specialized centers of care for spinal cord injury on length of stay, complications, and mortality: a systematic review of the literature. *J Neurotrauma* 2011;**28**:1363−70. https://doi.org/10.1089/neu.2009.1151.

156. Ong B, Wilson JR, Henzel MK. Management of the patient with chronic spinal cord injury. *Med Clin* 2020;**104**:263−78. https://doi.org/10.1016/j.mcna.2019.10.006.

157. Bellon K, et al. Evidence-based practice in primary prevention of spinal cord injury. *Top Spinal Cord Inj Rehabil* 2013;**19**:25−30. https://doi.org/10.1310/sci1901-25.

158. Cardenas DD, Hoffman JM, Kirshblum S, McKinley W. Etiology and incidence of rehospitalization after traumatic spinal cord injury: a multicenter analysis. *Arch Phys Med Rehabil* 2004;**85**:1757−63. https://doi.org/10.1016/j.apmr.2004.03.016.

159. Hosseini M, Yousefifard M, Aziznejad H, Nasirinezhad F. The effect of bone marrow-derived mesenchymal stem cell transplantation on allodynia and hyperalgesia in neuropathic animals: a systematic review with meta-analysis. *Biol Blood Marrow Transpl* 2015;**21**:1537−44. https://doi.org/10.1016/j.bbmt.2015.05.008.

160. Nagoshi N, Fehlings MG. Investigational drugs for the treatment of spinal cord injury: review of preclinical studies and evaluation of clinical trials from Phase I to II. *Expet Opin Invest Drugs* 2015;**24**:645−58. https://doi.org/10.1517/13543784.2015.1009629.

161. Macias MY, et al. Pain with no gain: allodynia following neural stem cell transplantation in spinal cord injury. *Exp Neurol* 2006;**201**:335−48. https://doi.org/10.1016/j.expneurol.2006.04.035.

162. Hassannejad Z, et al. Axonal degeneration and demyelination following traumatic spinal cord injury: a systematic review and meta-analysis. *J Chem Neuroanat* 2019;**97**:9−22. https://doi.org/10.1016/j.jchemneu.2019.01.009.

163. Khorasanizadeh M, et al. Neurological recovery following traumatic spinal cord injury: a systematic review and meta-analysis. *J Neurosurg Spine* 2019:1−17. https://doi.org/10.3171/2018.10.SPINE18802.

164. Pollard ME, Apple DF. Factors associated with improved neurologic outcomes in patients with incomplete tetraplegia. *Spine* 2003;**28**:33−9. https://doi.org/10.1097/00007632-200301010-00009.

165. Bosch A, Stauffer ES, Nickel VL. Incomplete traumatic quadriplegia. A ten-year review. *J Am Med Assoc* 1971;**216**:473−8.

166. Lazaro R, Reina-Guerra S, Quiben M. *Umphred's neurological rehabilitation*. 7 edn. Mosby; 2019.

167. McCarthy MJ, Aylott CE, Grevitt MP, Hegarty J. Cauda equina syndrome: factors affecting long-term functional and sphincteric outcome. *Spine* 2007;**32**:207−16. https://doi.org/10.1097/01.brs.0000251750.20508.84.

168. Donovan WH, Cifu DX, Schotte DE. Neurological and skeletal outcomes in 113 patients with closed injuries to the cervical spinal cord. *Paraplegia* 1992;**30**:533−42. https://doi.org/10.1038/sc.1992.111.

169. Marino RJ, Herbison GJ, Ditunno JF. Peripheral sprouting as a mechanism for recovery in the zone of injury in acute quadriplegia: a single-fiber EMG study. *Muscle Nerve* 1994;**17**:1466−8. https://doi.org/10.1002/mus.880171218.

170. Fehlings MG, Rabin D, Sears W, Cadotte DW, Aarabi B. Current practice in the timing of surgical intervention in spinal cord injury. *Spine* 2010;**35**:S166−73. https://doi.org/10.1097/BRS.0b013e3181f386f6.

171. Wilson JR, Cadotte DW, Fehlings MG. Clinical predictors of neurological outcome, functional status, and survival after traumatic spinal cord injury: a systematic review. *J Neurosurg Spine* 2012;**17**:11−26. https://doi.org/10.3171/2012.4.AOSPINE1245.

172. Ronen J, et al. Survival after nontraumatic spinal cord lesions in Israel. *Arch Phys Med Rehabil* 2004;**85**:1499−502. https://doi.org/10.1016/j.apmr.2003.11.015.

Classification systems: spine trauma

Ariana A. Reyes, Srikanth N. Divi, Thomas J. Lee, Dhruv Goyal, Alexander R. Vaccaro

Department of Orthopaedic Surgery, Rothman Orthopedic Institute, Thomas Jefferson University, Philadelphia, PA, United States

Introduction

The use of classification systems to define spinal trauma dates back to the 1920s.[1,2] Thereafter, various classification systems have been proposed to describe and define distinct injuries in the cervical, thoracolumbar, and sacral regions of the spine. Past and present classification systems have been developed from either mechanistic or anatomic features using different imaging modalities such as radiographs, magnetic resonance imaging (MRI), or computerized topography (CT) to define spinal injuries.

Although numerous classification systems have been proposed, there continues to be a lack of consensus of the ideal classification system to describe injuries in the cervical, thoracolumbar, and sacral regions. Most studies agree that an optimal classification system should be clinically relevant, validated, and accurate to facilitate communication between surgeons.[2–5] Additionally, these classification systems should be comprehensive enough to cover the wide range of injury patterns, neurologic status, and other patient-specific characteristics that may influence treatment and prognostic guidelines.[5–7] The goal of this chapter is to provide an overview of earlier and current classification systems that have been proposed to describe cervical, thoracolumbar, and sacral spine injuries in addition to a brief overview of the epidemiology and anatomy of each spinal region.

Cervical spine trauma classification systems

Incidence of overall cervical spine trauma

Despite limited evidence on the overall prevalence of cervical spine trauma, past reports estimate the overall incidence rate of cervical spine traumas in the general population to be about 12–65 per 100,000.[8–10] The most common etiology of cervical spine fractures is motor vehicle collisions, followed by falls and accidents involving pedestrians.[8–10] Additionally, in an epidemiology study by Passias et al., C2 and C7 were found to be the most common location

© 2022 Elsevier Inc. All rights reserved.

of closed cervical fractures.[10] Furthermore, a study of patients with traumatic spinal cord injury in Canada reported that 56% of spinal cord injuries were found to have an associated cervical fracture.[11]

Anatomy

The cervical spine is divided into the upper (occiput to C2) and subaxial spine (C3–C7). The occiput meets the cervical spine at the craniocervical junction which is formed by the occipito atlantal joints and occipital atlantal membrane.[12,13] Progressing caudally, the upper cervical spine is comprised of the atlas (C1) and axis (C2). The atlas is a ring-shaped bony structure that articulates with the odontoid (dens) process of the axis.[12,13] Additionally, the atlantoaxial junction has three articulation surfaces that are formed by two atlantoaxial facet joints and a central atlantoaxial joint with its stability determined by the integrity of the ligamentous complex.[12,13] The intrinsic ligaments provide the most stability for the upper cervical spine and include the joint capsules, tectorial membrane, cruciate ligament which contains the transverse atlantal ligament (TAL), and alar ligaments.[13] The extrinsic ligaments of this region also contribute to stability and contain the ligamentum nuchae, anterior longitudinal ligament, and the ligamentum flavum.[13] This junction also primarily allows for rotational motion of the cervical spine.[12–14] The subaxial spine is comprised of vertebral bodies with similar features and the orientation of their facet joints allows for flexion and extension motion.[14,15] The subaxial spine is further divided into an anterior and posterior column using the two-column concept.[15–17] The anterior column contains the anterior and posterior longitudinal ligament, anterior and posterior annulus, vertebral bodies, and transverse processes whereas the posterior column includes the ligamentum flavum, interspinous ligament, and facet capsule.[15–17]

Upper cervical spine (occiput to C2)

Occipital condyle

Although relatively uncommon fractures, past reports have estimated the incidence rate of occipital condyle fractures to range from 1% to 4% and occurring mostly after high-energy blunt craniocervical trauma.[18–21] The classification proposed by Anderson and Montesano using CT evaluation is the most commonly used in practice and is described below[22,23]:

Type I: Stable, impacted occipital condyle fracture that may result from compressive forces
Type II: Stable, occipital condyle fractures with a basilar fracture and may result from lateral cranial compressive forces
Type III: Potentially unstable, displaced avulsed fracture involving one or both occipital condyles

Similarly, Tuli et al. described three types of occipital condyle fractures using MRI and also included the presence or absence of an atlanto-occipital dislocation.[23,24]

Type I: Stable, not displaced, or has minimal displacement (<2 mm)
Type IIa: Stable, displaced without atlanto-occipital dislocation
Type IIb: Unstable, displaced with atlanto-occipital dislocation

In addition, Traynelis et al. described injury patterns based on displacement of the injury whereas the authors of the Harborview classification defined a three-stage classification system using MRI to further address craniocervical instability. The Harborview classification system is described below.[6,25,26]

Stage I: Stable or minimally displaced injury with evidence of injury to the craniocervical ligament. Craniocervical alignment is within 2 mm of accepted normal. Distraction is ≤ 2 mm.
Stage II: Similar description to Stage I except with distraction is > 2 mm.
Stage III: Craniocervical malalignment and cervical distraction with traction > 2 m.

Odontoid (dens) fractures

Odontoid fractures are the most common type of fractures of the C2 vertebra (axis).[27−29] This type of fracture is often found in the elderly and may be due to the potential bone loss seen in this population.[30,31] The Anderson and D'Alonzo classification is currently the most recognized classification system for odontoid fractures and divides fractures into three types[28,30,32,33]:

Type I: Oblique fracture of the upper part of the odontoid, uncommon
Type II: Base of the odontoid and does not involve body of C2 (axis)
Type III: Fractures that extend into the body of the C2 (axis)

Additionally, Hadley et al. expanded on the aforementioned classification mentioned to include Type IIa fracture described as a comminuted Type II fracture.[30,34] Furthermore, Grauer et al. proposed a modified classification of the Anderson and D'Alonzo system to further distinguish between Type II and Type III fractures by stratifying Type II fractures to improve treatment guidelines.[35]

Subtype A: Nondisplaced fractures
Subtype B: Displaced transverse or anterior superior to posterior inferior fracture line
Subtype C: Comminuted or anterior inferior to posterior superior fracture line

Although infrequently used, the Roy-Camille classification system alternatively describes fractures based on the direction of the fracture line.[36,37] Type I fractures are those with an anterior oblique fracture line that slopes forward with anterior displacement of the odontoid, Type II fractures are those with a posterior oblique fracture line that slopes backward with posterior displacement of the odontoid, and Type III fractures include horizontal fractures with anterior or posterior displacement of the odontoid.[36,37]

Fractures of the atlas

In general, fractures of the atlas are rare and estimated to be about 2%−13% of cervical spine injuries.[38,39] Injuries can occur after traumatic axial loading, motor vehicle accidents, or falling.[38,39] Additionally, various classification systems have been proposed to describe fractures of the atlas including the Jefferson, Landells, and Gehweiler classifications.[40−43] Discussed below is the commonly used Jefferson classification system that describes five types of fracture patterns[39,41]:

Type I: Fracture through the posterior arch
Type II: Fracture through the anterior arch
Type III: Fracture through the posterior and anterior arch
Type IV: Fracture through the lateral mass
Type V: Fracture through the lateral mass and posterior arch

The Landells classification divides fracture patterns into three types while the Gehweiler system, similarly to the Jefferson classification system, describes five different fracture patterns but further subdivides injuries of Jefferson Type 3 fractures into either stable (3a) or unstable (3b).[40,42,43] Stable injuries are defined by an intact TAL, whereas unstable injures are those with a ruptured TAL.[40,43]

Fractures of the axis (C2)

It is estimated that about 9%—18% of cervical spine fractures involve the C2 (axis) and of those, 11%—25% are from traumatic spondylolisthesis and fractures involving the body of C2 (Hangman's fracture).[27,44—46] These fractures can occur from high-energy trauma in the younger population or falls in the older population.[27,47—50]

Traumatic spondylolisthesis of the axis is described as a fracture through the bilateral pars of C2 and is also known as a Hangman's fracture.[51,52] A classification system for these types of fractures was first described by Effendi, then further modified by Levine and Edwards, which is currently the most widely used and accepted classification system.[53—55]

> **Type I**: Stable, < 3 mm subluxation of C2 on C3, often due to axial loading and hyperextension, rigid collar treatment for 4—6 weeks.
>
> **Type II**: > 4 mm of subluxation and disruption of the posterior longitudinal ligament, with angulation and vertical fracture due to hyperextension—axial load forces related to severe flexion.
>
> **Type IIA**: Unstable, less subluxation, and more angular deformity. Traction is contraindicated.
>
> **Type III**: Unstable, C2, C3 unilateral or bilateral facet joint dislocation due to flexion—compression forces (rare).

Most recently, the AOSpine upper cervical spine fractures classification system was created to further simplify previous systems regarding cervical spine fractures to facilitate communication and guide evaluation of prognosis and management.[6,56] This classification system separates injuries by level from occiput to C3 and also incorporates clinical modifiers and neurologic status (Fig. 3.1).[6,57]

AOSpine upper cervical injury classification system

> Category I: Occipital Condyle and Craniocervical Junction

Type A: Isolated bony injury (condyle)
Type B: Nondisplaced ligamentous injury
Type C: Any injury with displacement of the craniocervical junction
Category II: C1 and C1—C2 Joint Injury
Type A: Isolated bone injury
Type B: Transverse atlantal ligament
Type C: Atlantoaxial joint instability or translation along any plane
Category III: C2 and C2—C3 Joint Injury
Type A: Bony injury only, with no ligamentous, tension band, or discal injury
Type B: Tension band and/or ligamentous injury with or without bony injury
Type C: Any injury that leads to C2 body translation in any directional plane
Case-specific modifiers
M1: Injuries with significant potential for instability
M2: Injuries that have significant risk for nonunion with conservative treatment
M3: Refers to patient-specific demographics and comorbidities
M4: Used to describe signs of vertebral artery injury

Subaxial spine

The subaxial spine is defined as the region of the cervical spine from the C3 to C7 vertebral body. The majority of injuries that occur in the cervical spine occur in this region and more commonly in the C5—C7 levels (Figs. 3.2 and 3.3).[58—60] White et al. first described cervical instability, an important concept in guiding treatment.[15,61] In addition, Holdsworth et al. proposed a two-column classification system and proposed five patterns of trauma that can cause cervical spine injury: flexion, flexion/rotation, extension, compression, and shear.[16,62,63] The Allen classification system was one of the first classification systems proposed and later, Harris et al. similarly proposed a classification system with six mechanisms of injury with the addition of descriptions of rotational forces with flexion and extension injuries.[62,64—66] The subaxial cervical spine injury classification (SLIC), Cervical

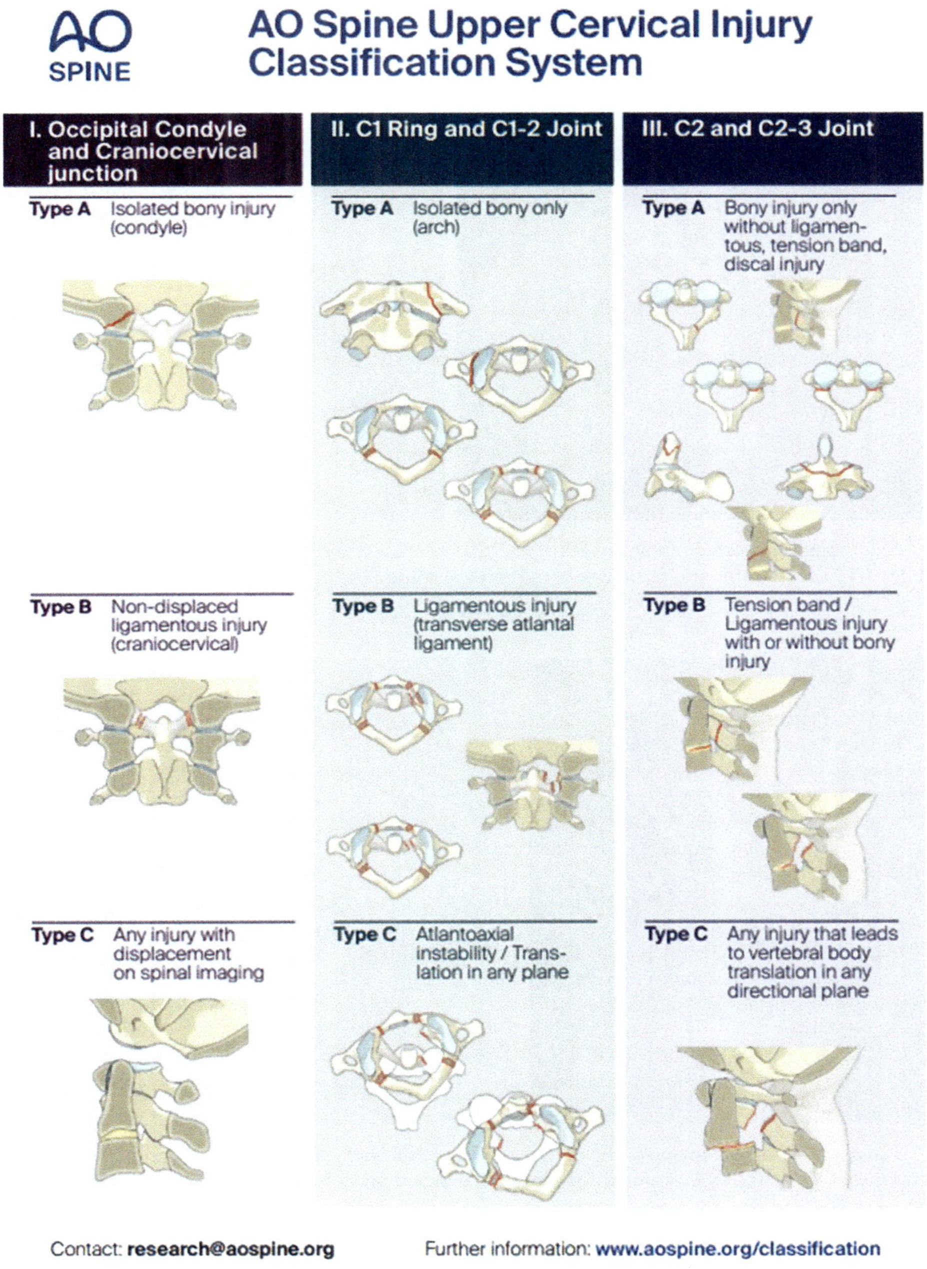

FIGURE 3.1 Upper cervical spine classification system. *Reprinted with permission from AOSpine International © AO Foundation, Switzerland.*

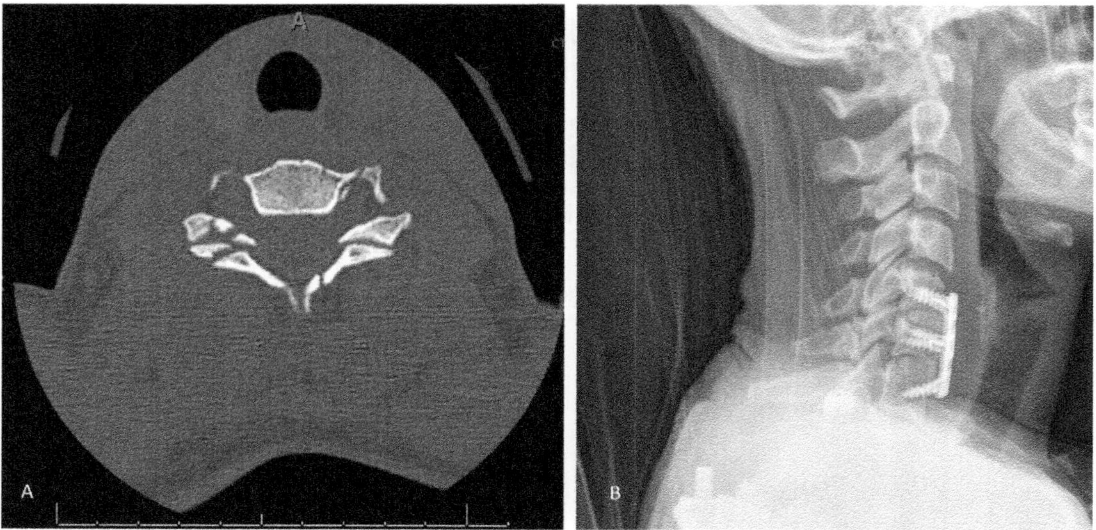

FIGURE 3.2   Axial computed tomographic images of the cervical spine demonstrating C6 floating lateral mass fracture with potential instability (A). Lateral radiographic image of the cervical spine after treatment with anterior cervical discectomy and fusion from C5–C7 (B).

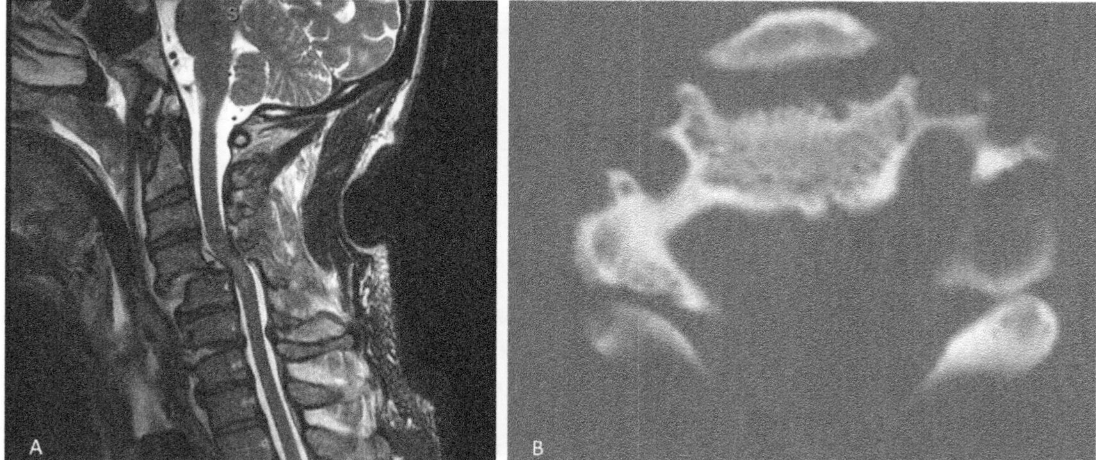

FIGURE 3.3   T2-weighted sagittal image of the cervical spine with flexion-distraction injury and bilateral facet dislocation with spinal cord injury (ASIA B) (A). Computed tomographic image demonstrating bilateral facet dislocation at the level of injury (B).

Spine Injury Severity Score (CSISS), and AO subaxial spine classifications are more recent classification systems proposed[3,6,60,62,65,67].

Allen classification

Allen and Ferguson et al. first described a classification system based on the mechanism of injury and position of the neck during the subaxial cervical spine injury using radiographical findings in the 1980s.[68] Based on 165 patients' radiographs, six mechanistic classification groups were formed.[65,69,70] This classification is mostly used in the research setting given its low interreliability and has limited clinical use.[6,62,71]

This classification system describes six mechanisms of injury and each type of injury is further staged based on severity of injury.[61,62,68] Compressive flexion injury contains five stages that range from "blunting of the anterior–superior vertebral margin" (Stage 1) to displacement into the neural canal and failures of the posterior ligamentous complex (Stage 5).[62,65] Vertical compression injuries contain three stages and the authors describe a "cupping" deformity in the first two stages with more severe injury and bursting of the vertebral body in Stage 3.[62,65] Distractive flexion contains four stages and describes severity of facet subluxation and dislocation with Stage 5 described as severe translation of the vertebral body.[62,65] Compressive

extension injuries contain five stages and describe unilateral or bilateral fractures of the vertebral arch.[62,65] Distractive extension injuries and lateral flexion injuries contain two stages with widening disc space described in the former and compressive fractures and vertebral arch fractures described in the latter.[62,65]

Subaxial cervical spine injury classification

The Spine Trauma Study group and Vaccaro et al. proposed a classification system to simplify existing classification systems.[60] This system classifies subaxial spine injuries based on severity of morphology, integrity of the discoligamentous complex, and degree of spinal cord injury involvement using CT and MRI.[58,60,62] The patient then receives a score to further assist with treatment guidelines (Table 3.1).[58,60,62] In addition, past reports have demonstrated comparable or improved reliability of the SLIC system when compared to the previous Allen and Ferguson and Harris proposed systems.[58,66,71–73]

TABLE 3.1   SLIC classification system.[59,71]

	Description	Points
Injury morphology	No abnormality	0
	Simple compression fracture	1
	Burst fracture	2
	Distraction	3
	Rotation/translation injury	4
Discoligamentous complex	Intact	0
	Indeterminate	1
	Disrupted	2
Spinal cord injury status	Intact	0
	Root injury	1
	Complete cord injury	2
	Incomplete cord injury	3
	Incomplete with ongoing cord compression	4
Treatment	Nonsurgical	≤3
	Indeterminate	4
	Surgical realignment/stabilization with decompression (if warranted)	≥5

Cervical Spine Injury Severity Score

To further expand on the concept of stability in guiding treatment and intervention, Anderson et al. proposed the CSISS to describe stability in the cervical spine for guidance on nonoperative or surgical treatment.[62,74] This classification system proposed a scoring system based on the severity of fracture and ligamentous injury. Additionally, the authors describe four spinal columns: anterior, posterior, right pillar, and left pillar.[67,74] Each column is then given a score from 0 to 5 (5 being most severe) based on bony and ligamentous injury using CT imaging with a maximum score of 20 points.[74] Description of structures within each column is mentioned below[74]:

Anterior column includes the vertebral body, vertebral disc, anterior and posterior longitudinal ligaments, uncinate processes, and transverse processes.
Posterior column includes the spinous processes, laminae, posterior ligamentous complex, ligamentum flavum.
Right and Left pillars include lateral masses, pedicle, transverse processes, superior and inferior articular processes, and the facet capsules.

Past reports have also reported excellent reliability of the CSISS classification system.[66,67,71,74] Moore et al. demonstrated an interobserver intraclass correlation coefficient ranging from 0.75 to 0.98, with an average of 0.88 for five random cases.[67] Similarly, Anderson et al. reported an interobserver reliability of 0.833 and an interobserver reliability of 0.977[74].

AOSpine subaxial injury classification system

The AO subaxial spine injury classification is a more comprehensive classification that was proposed to address gaps of previous classification systems to guide treatment and prognosis.[6,56] This system describes injuries based on morphology, type, and location of injury, in addition to ligamentous involvement.[56] Case-specific modifiers are also used for unique cases to further assist treatment guidelines[3,56,57] (Fig. 3.4).

Type A: compression injuries

Type A0: Minor, nonstructural fracture. Either no bony injury or a minor injury to the spine.
Type A1: Wedge-compression fracture. A compression fracture involving a single endplate without involvement of the posterior wall of the vertebral body.
Type A2: Split fracture. Coronal split or pincer fracture involving both endplate without involvement of the posterior wall of the vertebral body.
Type A3: Incomplete burst fracture. A burst fracture involving a single endplate with involvement of the posterior vertebral wall.
Type A4: Complete burst fracture. A burst fracture of sagittal split involving both endplates.

Type B: tension band injuries

Type B1: Bony posterior tension band injury. Physical separation through fractured bony structures only.
Type B2: Bony capsuloligamentous or ligamentous posterior tension band injury. Complete disruption of the posterior capsuloligamentous or bony capsuloligamentous structures together with a vertebral body, disk, and/or facet injury.
Type B3: Anterior tension band injury. Physical disruption or separation of the anterior structure (bone or disk) with tether of the posterior elements.

Type C: displacement/translation injuries

Translational injury in any axis-displacement or translation of one vertebral body relative to another in any direction.

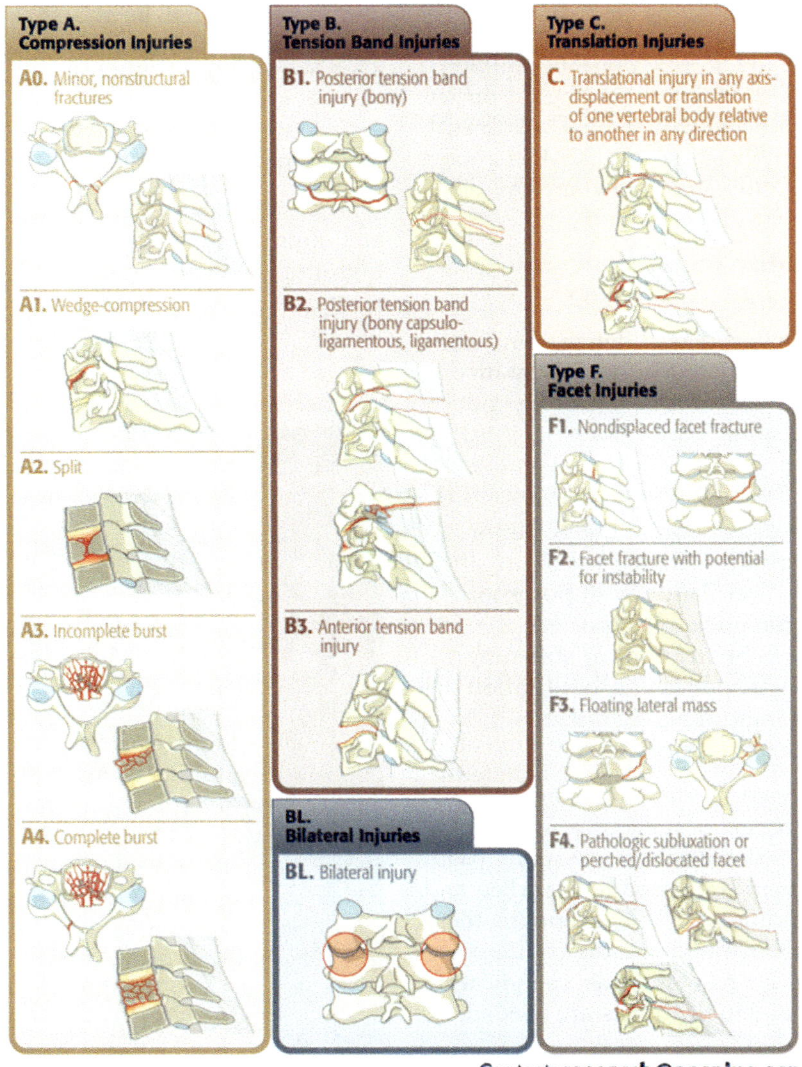

FIGURE 3.4   Subaxial cervical spine classification system. *Reprinted with permission from AOSpine International © AO Foundation, Switzerland.*

Type F: facet injuries

Type F1: Nondisplaced facet fracture. Defined as a fragment <1 cm in height and <40% of later mass.
Type F2: Facet fracture with potential for instability. Defined as a fragment >1 cm and >40% lateral mass or a displaced fragment.
Type F3: Floating lateral mass. Facet fracture causing a lateral segment of the vertebrae to become displaced.
Type F4: Pathologic subluxation or perched/dislocated facet.

Neurological status

N0: Intact
N1: Transient neurological deficit
N2: Signs and symptoms of radiculopathy
N3: Incomplete spinal cord injury
N4: Complete spinal cord injury

Case-specific modifiers

M1: Complete disruption of the posterior capsuloligamentous complex
M2: Severe disc herniation with posterior protrusion of the nucleus pulposus
M3: Metabolic bone disorder or stiffening such as ankylosis spondylitis, ossification of the posterior longitudinal ligament
M4: Signs vertebral artery injury

Past studies have reported average interobserver and intraobserver reliability for the AO subaxial spine classification system with improved agreement compared to the Allen and Ferguson classification.[3,6,56,75] Although these studies have demonstrated reliable communication between surgeons with this classification system, further validation studies are needed.

Thoracolumbar classification systems

Incidence

Previous reports have described an increase in the incidence of lumbosacral spinal trauma.[76,77] Additionally, the most common cause reported for these injuries was motor vehicle collusions with falls coming in second.[78,79] Although there are limited reports on the prevalence of spinal cord injuries associated with thoracolumbar fractures, an epidemiological study from Canada demonstrated that most spinal cord injuries in the thoracic and lumbar regions had an associated fracture (Fig. 3.5).[11]

Anatomy

The thoracolumbar region has unique anatomic features that can make it prone to high-impact energy trauma.[80] The thoracic spine is kyphotic and has more axial rotation while the lumbar spine is lordotic which allows for more

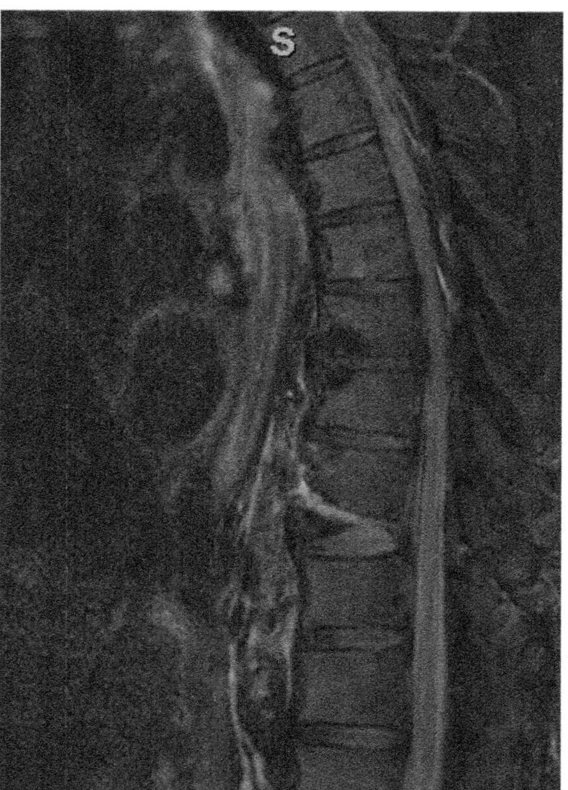

FIGURE 3.5   Sagittal T2-weighted image of T9 extension-distraction fracture. *S*, sagittal slices of the MRI.

flexion and extension movements.[80–82] The thoracolumbar junction is neutral with slight kyphosis and is a common site of injury.[80] The posterior ligamentous complex is an important feature of this region to maintain stability and is comprised of the supraspinous ligament, interspinous ligament, ligamentum flavum, and the facet joints.[16,83–85]

Historical context

Thoracolumbar fractures were first described by Boehler followed by Watson-Jones who then expanded on morphologic descriptions.[1,86,87] In addition, Nicoll et al. introduced the concept of stability in the thoracolumbar spine.[87,88] Kelly and Whitesides then proposed a two-column classification system.[87,89] In the 1970s, Holdsworth further expanded on the two-column system and classified patterns of injury into five categories.[16,87] Denis et al. and later, Mcafee et al. proposed a three-column spine classification system that also addressed stability in the lower cervical and thoracolumbar spine.[90,91] The three columns described include the anterior, middle, and posterior column.[87,90,92] Furthermore, the authors distinguished between minor and major injuries that included five different types of burst fractures (A–E) and described two types of seat-belt injuries using computerized axial tomography.[87,90,93] Ferguson et al. expanded on their cervical classification and proposed a similar system for the thoracolumbar region to address stability.[68,87] McCormack et al. proposed a points-based classification system where points were allocated based on the severity of the collapse of the vertebral body, distance between fragmentation, and degree of correction after surgery.[87,94,95] The authors proposed that a higher score correlated with increased risk for short-segment failed arthrodesis.[87,94,95]

Classification systems

AO-magerl classification

After reviewing 1445 cases with thoracolumbar injuries over a 10-year span, Magerl et al. proposed a classification system based off of morphology and injury patterns using radiographic criteria.[79,87] The Magerl classification system proposes a comprehensive system utilizing the two-column concept.[16,87,96] This classification system relies exclusively on CT findings and describes three types of injuries A–C.[90,92,96]

Type A: vertebral body compression

A1: Impaction fractures
A1.1: Endplate impaction
A1.2: Wedge impaction
A1.3: Vertebral body collapse
A2: Split fractures
A2.1: Frontal split fracture
A2.2: Sagittal split fracture
A2.3: Pincer fracture
A3: Burst fractures
A3.1: Incomplete burst fracture
A3.2: Burst split fracture
A3.3: Complete burst fracture
A3.3.1: Pincer
A3.3.2: Flexion
A3.3.3: Axial

Type B: distraction injuries

B1: Predominantly transligamentous flexion–distraction injury
B1.1: With transverse disc disruption
B1.1.1: Flexion subluxation
B1.1.2: Anterior dislocation
B1.1.3: either of the above with fractures of the articular processes
B1.2: With Type A vertebral body fracture
B2: Predominantly osseous flexion–distraction injury
B2.1: Transverse bicolumn fracture
B2.2: Posterior osseous disruption with transverse disc disruption

B2.2.1: Through the pedicles
B2.2.2: Through the interarticular portions (flexion spondylolysis)
B2.3: With Type A vertebral body fracture
B2.3.1: Through the pedicles
B2.3.2: Through the isthmus
B3: Anterior disruption through the disc
B3.1: Hyperextension-subluxation
B3.2: Hyperextension-spondylolysis
B3.3: Posterior dislocation

Type C: torsion injuries

C1: Rotation—compression injury
C1.1: Impaction
C1.2: Split
C1.3: Burst
C2: Rotation—distraction injury
C2.1: With transligamentous flexion-distraction
C2.2: With transosseous flexion-distraction
C2.3: With hyperextension-distraction
C3: Rotational shear injury

Additionally, past reports have demonstrated moderate reliability with this classification system. Given that the AO-Magerl classification does not account for neurologic status or stability, its role in clinical practice is restricted[4,5,87,97–100].

Thoracolumbar injury classification and severity score

The thoracolumbar injury severity score (TLISS) was developed by the Spine Trauma Study Group using mechanisms to define injury patterns but was later changed to the thoracolumbar injury classification and severity (TLICS) score as morphologic descriptions were incorporated.[4,85] The authors proposed this classification system to address ambiguity and clinical reliability in previous classification systems established.[4,85] TLICS scores use the morphology and location of the spinal injury as well as neurologic involvement to predict treatment and prognosis (Table 3.2).[85,100–104] In addition, the TLICS score is based on points that are given that correlate with the extent of the injury involvement in morphology, posterior ligamentous complex integrity, and neurologic status.[85]

The initial TLISS was reproduced and validated by Harrop et al. and was found to have moderate intrarater reliability and significant

TABLE 3.2   TLICS classification system.[86,91,124]

	Description	Points
Morphology	Compression fracture	1
	Wedge-compression fracture	2
	Burst fracture	3
	Translation/rotation injury	4
	Distraction	
Posterior ligamentous complex	Intact	0
	Suspected injury/indeterminate	2
	Injury	3
Neurological involvement	Intact	0
	Nerve root	2
	Cord/conus medullaris (complete)	2
	Cord/conus medullaris (incomplete)	3
	Cauda equina	3
Treatment	Nonoperative treatment	1–3
	Nonoperative treatment or surgical intervention	4
	Consideration for surgical treatment	≥ 5

accuracy in predicting treatment plan in 90% of cases.[4,105] In a validation study for the TLICS system, the authors reported good interobserver and intraobserver agreement and accuracy in predicting conservative or surgical treatment, but less reliability when assessing the posterior ligamentous complex.[100]

AOSpine thoracolumbar injury classification system

The AOSpine classification for this region was developed using concepts from the TLICS and AO-Magerl classification systems to create a universal system that could have a useful role in both clinical practice and to address previous limitations of being overly comprehensive (AO-Magerl) and low reliability of assessing the PLC complex for classification (TLICS).[3,6,85,96,106] Similar to the aforementioned AOSpine classification systems, this classification also takes into account morphology and neurological status[3,57,103,106,107] (Fig. 3.6).

Type A: compression injuries

Type A0: Minor, nonstructural fractures. Fractures, which do not compromise the structural integrity of the spinal column such as transverse process or spinous process fractures.
Type A1: Wedge-compression fracture. Fracture of a single endplate without involvement of the posterior wall of the vertebral body.
Type A2: Split fracture. Fracture of both endplates without involvement of the posterior wall of the vertebral body.
Type A3: Fracture with any involvement of the posterior wall; only a single endplate fractured. Vertical fracture of the lamina is usually present and does not constitute a tension band failure.
Type A4: Complete burst. Fracture with any involvement of the posterior wall and both endplates. Vertical fracture of the lamina is usually present and does not constitute a tension band failure.

Type B: distraction injuries

Type B1: Transosseous tension band disruption chance fracture. Monosegmental pure osseous failure of the posterior tension band. The classical chance fracture (Fig. 3.7).
Type B2: Posterior tension band disruption. Bony and/or ligamentary failure of the posterior tension band together with a Type A fracture. Type A fracture should be classified separately.

Type C: displacement or dislocation injury

Translational injuries can occur in any direction and can present with a variety of configurations. Injuries can also have an associated type A or B injury.

Case-specific modifiers

M1: Fractures with an indeterminate injury to the tension band
M2: Patient-specific comorbidity

Various studies from both the authors of this classification and independent studies have shown good interobserver and intraobserver reliability and agreement with identifying fracture type with this classification system.[75,107,108] Additionally, the authors of one study found that the AOSpine thoracolumbar classification system had better reliability and reproducibility when compared to TLICS.[108] A recent systematic review demonstrated good reliability of morphologic descriptions when subtypes were excluded, although the authors also reported that further more robust, independent studies are needed to validate this classification system.[109]

Sacral classification systems

Incidence

Sacral fractures remain relatively uncommon with an estimated occurrence rate of 2 per 100,000 people.[110,111] Despite this, the incidence of sacral fractures that were nonosteoporotic was estimated to increase from 0.67 to 2.09 per

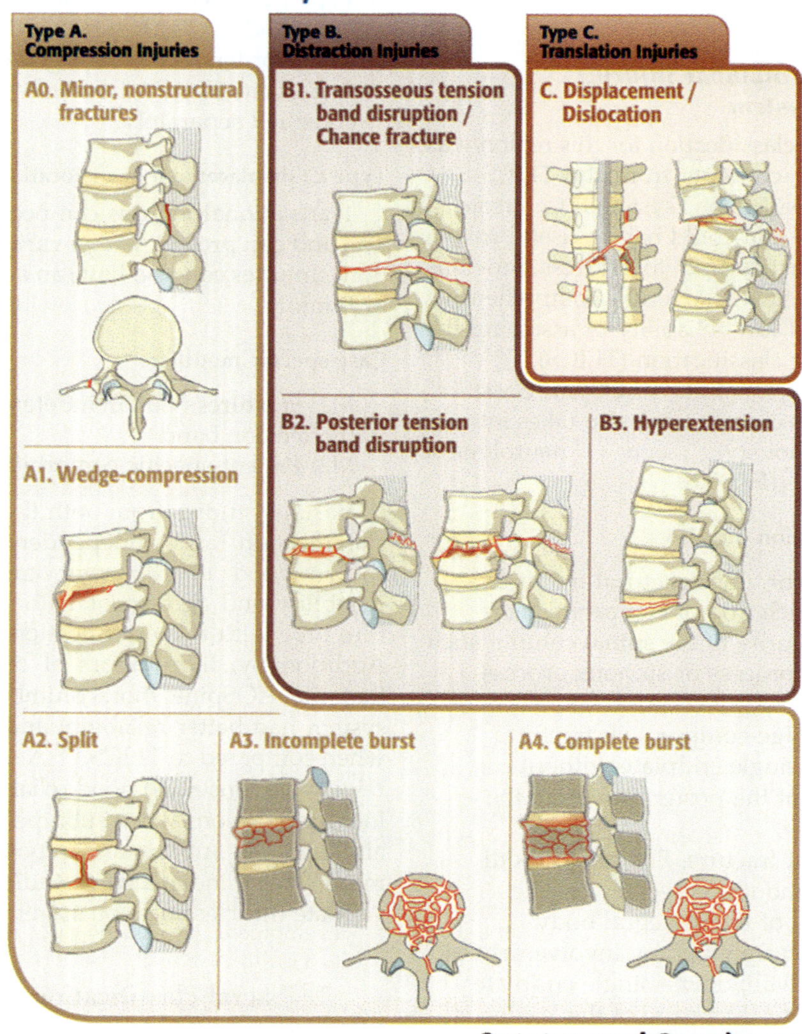

FIGURE 3.6   Thoracolumbar classification system. *Reprinted with permission from AOSpine International © AO Foundation, Switzerland.*

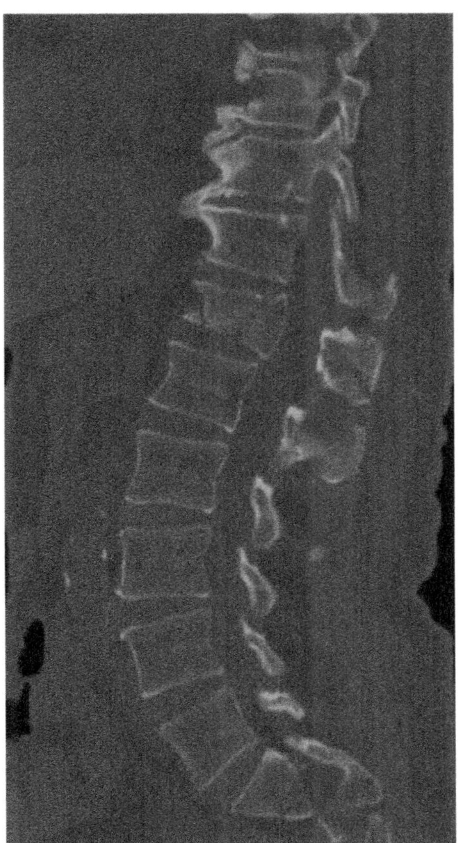

FIGURE 3.7 Chance fracture: sagittal computed tomographic image of L1 flexion-distraction injury (Chance fracture). *Imaging also demonstrates history of congenital six lumbar vertebrae.*

100,000 persons in the time period between 2002 and 2011.[111] These injuries may occur after high-energy trauma or low-energy in the elderly or patients with a history of metabolic or other pathologic bone diseases.[111–114] Past reports estimated that 57% of sacral injuries occur as a result of motor vehicle collusions with a fewer proportion occurring from either falls or crush-related injuries.[112–114] Besides, sacral fractures are often associated with other injuries such as vascular, neurologic, or soft tissue insults which may necessitate surgical intervention.[112,115–119]

Anatomy

The sacrum is the stabilizing structure of the spine and is in the center of the pelvic ring. This structure has a slight kyphotic angulation and is comprised of five fused vertebrae.[112,120] Additionally, this structure is important to help transmit axial loading from the rest of the spine to the lower extremities. Various nerve structures that can be injured with sacral injuries include the cauda equina, filum terminale, sciatica, and sacral plexus.[112,120,121] Likewise, if nerves in this region are injured, the patient can lose bowel or bladder function or have sexual dysfunction.

Classification systems

Denis sacral injury classification system

The Denis sacral injury classification system was first described in 1988 and primarily uses an anatomic approach and continues to be one of the most widely used systems.[6,114] This classification divides the sacrum into three zones:[114]

Zone 1: Lateral to the neuroforamina
Zone 2: Through the neuroforamina
Zone 3: Medial to the neuroforamina

In addition, Denis et al. reported that 21.6% of patients with a sacral injury had an associated neurological involvement.[114] The authors also demonstrated that Zone 1 injuries had the lowest risk of neurologic injury with an L5 radiculopathy being the most common type of resulting neurologic deficit.[114] Zone 2 injuries have an intermediate risk of neurologic involvement with resulting sacral radiculopathy being the most common type, whereas Zone 3 injuries have the highest rate of neurologic involvement with cauda equina syndrome being the most common type of deficit.[114]

Isler classification system

In 1990 Isler et al. proposed a classification to predict the severity of lumbosacral instability based on the relationship of the fracture and the S1 facet.[122] This classification defines three types of fractures including: Type I Isler fractures designated as those lateral to the L5-S1 facet joint, Type II which include fractures that extend into the L5-S1 facet joint, and Type III that describe fractures that are medial to the L5-S1 facet joint and may also involve the spinal canal.[110,122] Surgical intervention in Type I fractures is directed toward addressing pelvic instability while both pelvic and spinal instability are addressed for Types II and III fractures.[122,123] In addition, this classification system is limited by its specificity and application for only vertical fractures involving the foramen and its lack of consideration for associated neurologic and soft tissue injuries.[124]

Modified Roy-Camille

Roy-Camille and colleagues proposed a classification system to further subclassify Zone III Denis fractures that was further modified by Strange-Vognsen et al.[110,112,123,125] The modified Roy-Camille sacral injury classification describes four different morphologies for transverse sacral fractures.[110,125,126] Types I and II fractures are caused by hyperflexion of the sacrum with kyphosis of the fracture with no displacement. Similarly, Type II fractures are caused by hyperflexion of the sacrum but posterior displacement of the superior vertebral body segment may also be seen with these fractures. These fractures are also unstable and have a higher risk of neurologic involvement. In contrast, Type III fractures are caused by hyperextension of the sacrum and have anterior displacement of the superior vertebral body. Type III fractures are also unstable and have the highest risk of neurologic injury. Type IV fractures are caused by axial loading and can have comminution of the S1 and S2 vertebral body segments. Although these fractures rarely occur, they are considered to be unstable and also have a high risk of associated neurologic injury.[125,126]

Lumbosacral injury classification system

More recently, the lumbosacral injury classification system (LSICS) was proposed to address limitations of previous classification systems and to guide clinical treatment decisions.[127] This classification describes injuries based on morphology, the posterior ligamentous complex, and neurologic status with a composite score from 1 to 10 that is then assigned.[127] Injuries with a score of less than 4 can be managed conservatively while surgical intervention is considered for injuries that score greater than 4.[110,127] Table 3.3 describes this scoring system.

AOSpine sacral injury classification system

The AOSpine trauma knowledge forum in combination with AO trauma members developed The AOspine sacral injury classification system which proposed a comprehensive and simplified system to improve communication and account for limitations in previous classification systems such as accounting for posterior pelvic instability.[6,124,128] This classification is divided into three subsections (A–C) that are then further defined through a hierarchical structure.[124,128] In addition, this system describes neurologic status and clinical modifiers for further specificity of injury descriptions to guide clinical decision-making[57,124,128] (Fig. 3.8).

Type A: lower sacrococcygeal injuries

No impact on posterior pelvic or spino-pelvic instability.

Type A1: Coccygeal or compression and ligamentous avulsion fractures.
Type A2: Nondisplaced transverse fractures below the S1 joint. This classification has no implication on stability. Low likelihood of cauda equina injury.
Type A3: Displaced transverse fractures below the S1 joint. Higher likelihood of a neurologic injury than A1 or A2 (displacement). May possibility benefit from reduction and stabilization.

TABLE 3.3 LSICS classification system.[122,126]

	Description	Points
Morphology	Flexion compression:	1
	≤20 degrees kyphosis	2
	>20 degrees kyphosis	2
	Axial compression without sacral canal or neuroforaminal encroachment	3
	With sacral canal or neuroforaminal encroachment	3
	Translation/rotational anterior or posterior translation of upper sacrum	4
	Lumbosacral facet injury or dislocation	
	Vertical translation or instability	
	Blast/shear (severe comminution)	
Posterior ligamentous complex	Intact	0
	Indeterminate	1
	Disrupted	2
Neurologic status	Intact	0
	Paresthesias only	1
	Lower extremity motor deficit	2
	Bowel/bladder dysfunction	3
	Progressive neurologic deficit	4
Treatment	Nonoperative treatment	1–3
	Nonoperative treatment or surgical intervention	4
	Consideration for surgical treatment	≥5

Type B: posterior pelvic injuries

Primary impact is on posterior pelvic stability.

Type B1: Central fracture. Involves spinal canal. Longitudinal injuries only—rare type of Denis Zone III injuries. Low likelihood of neurological injury.

Type B2: Does not involve foramina or spinal canal. Unilateral Denis Zone I injury.

Type B3: Transforaminal fracture. Involves foramina but not spinal canal. Denis Zone II injury.

Type C: spino-pelvic injuries

Spino-pelvic instability.

Type C0: Nondisplaced sacral U-type variant. Commonly seen low-energy insufficiency fracture.

Type C1: Alternative-sacral U-type variant without posterior pelvic instability. Any unilateral B-subtype where ipsilateral superior S1 facet is discontinuous with medial part of sacrum. May impact spino-pelvic stability (Isler).

Type C2: Bilateral complete Type B injuries without transverse fracture. More unstable and higher likelihood of neuro injury than C1.

Type C3: Displaced U-type sacral fracture. Worst combination of instability and likelihood of neuro injury. Displaced transverse sacral fracture and canal compromise (Fig. 3.9).

Neurologic status

Nx: Patient cannot be examined
N0: No neurological deficits found
N1: Transient neurologic injury
N2: Nerve root injury
N3: Cauda equina syndrome

Case-specific modifiers

M1: Significant soft tissue injury
M2: Metabolic bone disease

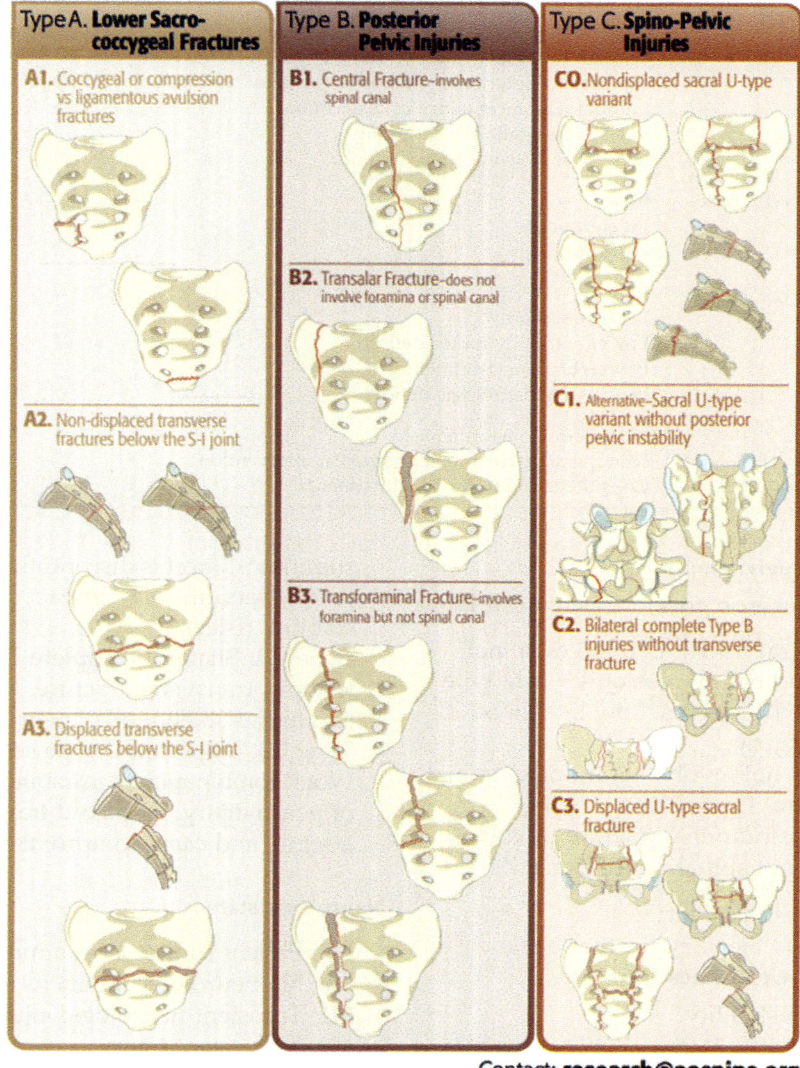

FIGURE 3.8   AOSpine Sacral classification system showing Types A, B, and C injuries with their subclassifications. *Reprinted with permission from AOSpine International © AO Foundation, Switzerland.*

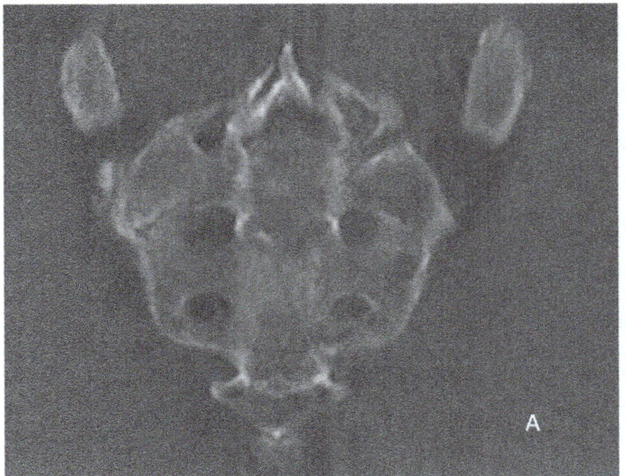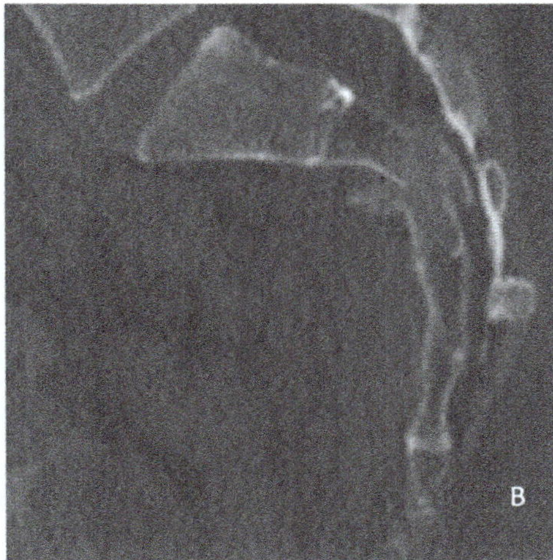

FIGURE 3.9 Sacral Injury. Coronal (A) and sagittal (B) computed tomographic images of transverse (U-shaped) sacral fractures with potential pelvic instability.

M3: High-energy injury and may be associated with anterior pelvic ring injury, acetabular or vascular injury
M4: Altered anatomy of lumbosacral junction (such as from fusion)

Although this classification system was recently developed, further validation studies are needed for this classification system.

Summary

A number of classification systems have been proposed and developed in the past century. An efficient classification system is comprehensive and provides reliable communication between surgeons to guide treatment decisions and prognosis. Although there are commonly used classification systems for each region of the spine, further work is need to validate newer, more comprehensive classification systems to ultimately set a standard classification for cervical, thoracolumbar, and sacral injuries.

Acknowledgment

AOSpine is a clinical division of the AO Foundation—an independent medically guided nonprofit organization. The AOSpine Knowledge Forums are pathology focused working groups acting on behalf of AOSpine in their domain of scientific expertise. Each forum consists of a steering committee of up to 10 international spine experts who meet on a regular basis to discuss research, assess the best evidence for current practices, and formulate clinical trials to advance spine care worldwide. Study support is provided directly through AOSpine's Research department and AO's Clinical Investigation and Documentation unit.

References

1. Böhler L. *Die Marknagelung nach Kuntscher: 3 Band of Bohler's Technik der Knochenbruchbehandlung im Frieden und im Kriege 9*. Maundrich; 1944.
2. Van Middendorp JJ, Audigé L, Hanson B, Chapman JR, Hosman AJF. What should an ideal spinal injury classification system consist of? A methodological review and conceptual proposal for future classifications. *Eur Spine J* 2010;**19**:1238—49.
3. Schnake KJ, Schroeder GD, Vaccaro AR, Oner C. AOSpine classification systems (subaxial, thoracolumbar). *J Orthop Trauma* 2017;Vol. 31:S14—23.
4. Bono CM, Vaccaro AR, Hurlbert RJ, Arnold P, Oner FC, Harrop J, et al. Validating a newly proposed classification system for thoracolumbar spine trauma: looking to the future of the thoracolumbar injury classification and severity score. *J Orthop Trauma* September 2006;**20**(8):567—72.
5. Mirza SK, Mirza AJ, Chapman JR, Anderson PA. Classifications of thoracic and lumbar fractures: rationale and supporting data. *J Am Acad Orthop Surg* 2002;**10**(5):364—77.

6. Divi SN, Schroeder GD, Oner FC, Kandziora F, Schnake KJ, Dvorak MF, et al. AOSpine-spine trauma classification system: the value of modifiers: a narrative review with commentary on evolving descriptive principles [Internet] *Glob Spine J* May 8, 2019;9(1 Suppl):77S–88S. Available from: https://www.ncbi.nlm.nih.gov/pubmed/31157149.

7. Van Middendorp JJ, Slooff WBM, Nellestein WR, Öner FC. Incidence of and risk factors for complications associated with halo-vest immobilization: a prospective, descriptive cohort study of 239 patients. *J Bone Jt Surg Ser A* 2009;91(1):71–9.

8. Fredø HL, Rizvi SAM, Lied B, Rønning P, Helseth E. The epidemiology of traumatic cervical spine fractures: a prospective population study from Norway [Internet] *Scand J Trauma Resusc Emerg Med* August 6, 2012;20:85. Available from: https://www.ncbi.nlm.nih.gov/pmc/articles/PMC3546896/.

9. Hu R, Mustard CA, Burns C. Epidemiology of incident spinal fracture in a complete population [Internet] *Spine (Phila Pa 1976)* 1996;21(4):492–9. Available from: http://www.ncbi.nlm.nih.gov/pubmed/8658254.

10. Passias PG, Poorman GW, Segreto FA, Jalai CM, Horn SR, Bortz CA, et al. Traumatic fractures of the cervical spine: analysis of changes in incidence, cause, concurrent injuries, and complications among 488,262 patients from 2005 to 2013 [Internet] *World Neurosurg* 2018;110. https://doi.org/10.1016/j.wneu.2017.11.011. e427–37. Available from:.

11. Pickett GE, Campos-Benitez M, Keller JL, Duggal N. Epidemiology of traumatic spinal cord injury in Canada. *Spine (Phila Pa 1976)* 2006;31(7):799–805.

12. Morishita Y, Falakassa J, Naito M, Hymanson HJ, Taghavi C, Wang JC. The kinematic relationships of the upper cervical spine. *Spine (Phila Pa 1976)* 2009; 34(24):2642–5.

13. Bransford RJ, Alton TB, Patel AR, Bellabarba C. Upper cervical spine trauma. *J Am Acad Orthop Surg* November 2014;22(11):718–29.

14. Bogduk N, Mercer S. Biomechanics of the cervical spine. I: normal kinematics. *Clin Biomech* 2000;15(9): 633–48.

15. White 3rd AA, Johnson RM, Panjabi MM, Southwick WO. Biomechanical analysis of clinical stability in the cervical spine. *Clin Orthop Relat Res* 1975;109:85–96.

16. Holdsworth F. Fractures, dislocations, and fracture-dislocations of the spine. *J Bone Joint Surg Am* December 1970;52(8):1534–51.

17. Bellabarba AB, Chapman J, Dvorak M, Fehlings M, Kandziora F, Kepler C, et al. Spinal fractures classification system [Internet] *AOSPine* 2015:1–55. Available from: http://www.aospine.org.

18. Karam YR, Traynelis VC, Conditions P. Occipital condyle fractures [Internet] *Neurosurgery* March 1, 2010;66(suppl_3). https://doi.org/10.1227/01.NEU.0000365751.84075.66. A56–9. Available from:.

19. Theodore N, Aarabi B, Dhall SS, Gelb DE, Hurlbert RJ, Rozzelle CJ, et al. Occipital condyle fractures [Internet] *Neurosurgery* March 1, 2013;72(suppl_3):106–13. https://doi.org/10.1227/NEU.0b013e3182775527. Available from:.

20. Link TM, Schuierer G, Hufendiek A, Horch C, Peters PE. Substantial head trauma: value of routine CT examination of the cervicocranium. *Radiology* September 1995;196(3):741–5.

21. Mueller FJ, Fuechtmeier B, Kinner B, Rosskopf M, Neumann C, Nerlich M, et al. Occipital condyle fractures. Prospective follow-up of 31 cases within 5 years at a level 1 trauma centre. *Eur spine J Off Publ Eur Spine Soc Eur Spinal Deform Soc Eur Sect Cerv Spine Res Soc* February 2012;21(2):289–94.

22. Anderson PA, Montesano PX. Morphology and treatment of occipital condyle fractures. *Spine (Phila Pa 1976)* July 1988;13(7):731–6.

23. Krüger A, Oberkircher L, Frangen T, Ruchholtz S, Kühne C, Junge A. Fractures of the occipital condyle clinical spectrum and course in eight patients [Internet] *J Craniovertebral Junction Spine* July 2013; 4(2):49–55. Available from: https://www.ncbi.nlm.nih.gov/pubmed/24744561.

24. Tuli S, Tator CH, Fehlings MG, Mackay M. Occipital condyle fractures. *Neurosurgery* August 1997;41(2): 367–8.

25. Traynelis VC, Marano GD, Dunker RO, Kaufman HH. Traumatic atlanto-occipital dislocation. Case report. *J Neurosurg* December 1986;65(6):863–70.

26. Bellabarba C, Mirza SK, West GA, Mann FA, Dailey AT, Newell DW, et al. Diagnosis and treatment of craniocervical dislocation in a series of 17 consecutive survivors during an 8-year period. *J Neurosurg Spine* 2008;4(6):429–40.

27. Robinson A-LL, Moller A, Robinson Y, Olerud C, Möller A, Robinson Y, et al. C2 fracture subtypes, incidence, and treatment allocation change with age: a retrospective cohort study of 233 consecutive cases. *Biomed Res Int* 2017;2017:8321680.

28. Smith HE, Kerr SM, Fehlings MG, Chapman J, Maltenfort M, Zavlasky J, et al. Trends in epidemiology and management of type II odontoid fractures: 20-year experience at a model system spine injury tertiary referral center. *J Spinal Disord Tech* December 2010; 23(8):501–5.

29. Ryan MD, Henderson JJ. The epidemiology of fractures and fracture-dislocations of the cervical spine. *Injury* 1992;23(1):38–40.

30. Iyer S, Hurlbert RJ, Albert TJ. Management of odontoid fractures in the elderly: a review of the literature and an evidence-based treatment algorithm [Internet] *Neurosurgery* November 18, 2017;**82**(4):419−30. https://doi.org/10.1093/neuros/nyx546. Available from:.

31. Amling M, Hahn M, Wening VJ, Grote HJ, Delling G. The microarchitecture of the axis as the predisposing factor for fracture of the base of the odontoid process. A histomorphometric analysis of twenty-two autopsy specimens. *J Bone Joint Surg Am* December 1994;**76**(12):1840−6.

32. Anderson LD, Alonzo RTD, D'Alonzo RT, Alonzo RTD. Fractures of the odontoid process of the axis. *J Bone Jt Surg* December 1974;**56-A**(8):1663−74.

33. Gonschorek O, Vordemvenne T, Blattert T, Katscher S, Schnake KJ. Trauma SS of the GS for O and. Treatment of odontoid fractures: recommendations of the spine section of the german society for orthopaedics and trauma (DGOU) [Internet] *Glob Spine J* Septemeber 7, 2018;**8**(2_suppl):12S−7S. Available from: https://www.ncbi.nlm.nih.gov/pubmed/30210956.

34. Hadley MN, Browner CM, Liu SS, Sonntag VK. New subtype of acute odontoid fractures (type IIA). *Neurosurgery* January 1988;**22**(1 Pt 1):67−71.

35. Grauer JN, Shafi B, Hilibrand AS, Harrop JS, Kwon BK, Beiner JM, et al. Proposal of a modified, treatment-oriented classification of odontoid fractures. *Spine J* 2005;**5**(2):123−9.

36. Roy-Camille R, Saillant G, Judet T, de Botton G, Michel G. [Factors of severity in the fractures of the odontoid process (author's transl)]. *Rev Chir Orthop Reparatrice Appar Mot* 1980;**66**(3):183−6.

37. Mead 2nd LB, Millhouse PW, Krystal J, Vaccaro AR. C1 fractures: a review of diagnoses, management options, and outcomes [Internet] *Curr Rev Musculoskelet Med* September 2016;**9**(3):255−62. Available from: https://www.ncbi.nlm.nih.gov/pubmed/27357228.

38. Matthiessen C, Robinson Y. Epidemiology of atlas fractures–a national registry-based cohort study of 1,537 cases. *Spine J* November 2015;**15**(11):2332−7.

39. Schleicher P, Pingel A, Kandziora F. Safe management of acute cervical spine injuries [Internet] *EFORT open Rev* May 21, 2018;**3**(5):347−57. Available from: https://www.ncbi.nlm.nih.gov/pubmed/29951274.

40. Kandziora F, Scholz M, Pingel A, Schleicher P, Yildiz U, Kluger P, et al. Treatment of atlas fractures: recommendations of the spine section of the German Society for Orthopaedics and Trauma (DGOU) [Internet] *Glob Spine J* September 1, 2018;**8**(2_suppl):5S−11S. https://doi.org/10.1177/2192568217726304. Available from:.

41. Jefferson G. Fracture of the atlas vertebra. Report of four cases, and a review of those previously recorded [Internet] *BJS* January 1, 1919;**7**(27):407−22. https://doi.org/10.1002/bjs.1800072713. Available from:.

42. Landells CD, Van Peteghem PK. Fractures of the atlas: classification, treatment and morbidity. *Spine (Phila Pa 1976)* May 1988;**13**(5):450−2.

43. Gehweiler JA, Osborne RL, Becker RF. *The radiology of vertebral trauma.* WB Saunders Company; 1980.

44. Greene KA, Dickman CA, Marciano FF, Drabier JB, Hadley MN, Sonntag VK. Acute axis fractures. Analysis of management and outcome in 340 consecutive cases. *Spine (Phila Pa 1976)* August 1997;**22**(16):1843−52.

45. Benzel EC, Hart BL, Ball PA, Baldwin NG, Orrison WW, Espinosa M. Fractures of the C-2 vertebral body. *J Neurosurg* August 1994;**81**(2):206−12.

46. Hadley MN, Dickman CA, Browner CM, Sonntag VK. Acute axis fractures: a review of 229 cases. *J Neurosurg* November 1989;**71**(5 Pt 1):642−7.

47. Malik SA, Murphy M, Connolly P, O'Byrne J. Evaluation of morbidity, mortality and outcome following cervical spine injuries in elderly patients [Internet]. 2008/01/15 *Eur Spine J* April 2008;**17**(4):585−91. Available from: https://www.ncbi.nlm.nih.gov/pubmed/18196293.

48. Robinson A-L, Olerud C, Robinson Y. Epidemiology of C2 fractures in the 21st century: a national registry cohort study of 6,370 patients from 1997 to 2014 [Internet] *Adv Orthop* October 17, 2017;**2017**:6516893. Available from: https://www.ncbi.nlm.nih.gov/pubmed/29181200.

49. Lomoschitz FM, Blackmore CC, Mirza SK, Mann FA. Cervical spine injuries in patients 65 years old and older: epidemiologic analysis regarding the effects of age and injury mechanism on distribution, type, and stability of injuries. *AJR Am J Roentgenol* March 2002;**178**(3):573−7.

50. Hadley MN, Walters BC, Grabb PA, Oyesiku NM, Przybylski GJ, Resnick DK, et al. Management of combination fractures of the atlas and axis in adults. *Neurosurgery* March 2002;**50**(3 Suppl):S140−7.

51. Al-Mahfoudh R, Beagrie C, Woolley E, Zakaria R, Radon M, Clark S, et al. Management of typical and atypical hangman's fractures [Internet]. 2015/09/09 *Glob spine J* May 2016;**6**(3):248−56. Available from: https://www.ncbi.nlm.nih.gov/pubmed/27099816.

52. Schneider RC, Livingston KE, Cave AJ, Hamilton G. "Hangman's fracture" of the cervical spine. *J Neurosurg* February 1965;**22**:141−54.

53. Effendi B, Roy D, Cornish B, Dussault R, Laurin C. Fractures of the ring of the axis. A classification based on the analysis of 131 cases. *J Bone Jt Surg* 1981;**63-B**(3): 319−27.

54. Levine AM, Edwards CC. The management of traumatic spondylolisthesis of the axis. *J Bone Joint Surg Am* February 1985;**67**(2):217−26.

55. Li X-F, Dai L-Y, Lu H, Chen X-D. A systematic review of the management of hangman's fractures [Internet]. 2005/10/19 *Eur Spine J* March 2006;**15**(3):257−69. Available from: https://www.ncbi.nlm.nih.gov/pubmed/16235100.

56. Vaccaro AR, Koerner JD, Radcliff KE, Oner FC, Reinhold M, Schnake KJ, et al. AOSpine subaxial cervical spine injury classification system. *Eur Spine J* July 2016;**25**(7):2173−84.

57. Meinberg EG, Agel J, Roberts CS, Karam MD, Kellam JF. Fracture and dislocation classification compendium-2018. *J Orthop Trauma* January 2018;**32**(Suppl 1):S1−170.

58. Feuchtbaum E, Buchowski J, Zebala L. Subaxial cervical spine trauma [Internet] *Curr Rev Musculoskelet Med* December 2016;**9**(4):496−504. Available from: https://www.ncbi.nlm.nih.gov/pubmed/27864669.

59. Aebi M. Surgical treatment of upper, middle and lower cervical injuries and non-unions by anterior procedures. *Eur spine J Off Publ Eur Spine Soc Eur Spinal Deform Soc Eur Sect Cerv Spine Res Soc* March 2010; **19**(Suppl 1):S33−9.

60. Vaccaro AR, Hulbert RJ, Fisher C, Anderson P, Oner FC, Fehlings M, et al. The subaxial cervical spine injury classification system. *Spine (Phila Pa 1976)* 2007; **32**(21):2365−74.

61. Kwon BK, Vaccaro AR, Grauer JN, Fisher CG, Dvorak MF. Subaxial cervical spine trauma. *J Am Acad Orthop Surg* February 2006;**14**(2):78−89.

62. Aarabi B, Walters BC, Dhall SS, Gelb DE, Hurlbert RJ, Rozzelle CJ, et al. Subaxial cervical spine injury classification systems [Internet] *Neurosurgery* May 12, 2013; **72**(Suppl.2):170−86. Available from: https://academic.oup.com/neurosurgery/article/72/suppl_3/170/2557384.

63. Holdsworth FW. Neurological diagnosis and the indications for treatment of paraplegia and tetraplegia, associated with fractures of the spine. *Manit Med Rev* January 1968;**48**(1):16−8.

64. Harris JHJ, Edeiken-Monroe B, Kopaniky DR. A practical classification of acute cervical spine injuries. *Orthop Clin North Am* January 1986;**17**(1): 15−30.

65. Allen BLJ, Ferguson RL, Lehmann TR, O'Brien RP. A mechanistic classification of closed, indirect fractures and dislocations of the lower cervical spine. *Spine (Phila Pa 1976)* 1982;**7**(1):1−27.

66. Vaccaro AR, Hulbert RJ, Patel AA, Fisher C, Dvorak M, Lehman RAJ, et al. The subaxial cervical spine injury classification system: a novel approach to recognize the importance of morphology, neurology, and integrity of the disco-ligamentous complex [Internet] *Spine (Phila Pa 1976)* May 12, 2007;**32**(21):2365. Available from: https://journals.lww.com/spinejournal/Abstract/2007/10010/The_Subaxial_Cervical_Spine_Injury_Classification.15.aspx.

67. Moore TA, Vaccaro AR, Anderson PA. Classification of lower cervical spine injuries. *Spine (Phila Pa 1976)* May 2006;**31**(11 Suppl):37−43.

68. Ferguson RL, Allen BL. A mechanistic classification of thoracolumbar spine fractures [Internet] *Clin Orthop Relat Res* 1984;**189**:77−88. Available from: http://files/225/FergusonandAllen-1984-Amechanisticclassificationofthoracolumbarspin.pdf.

69. Marcon R, Cristante A, Teixeira W, Narasaki D, Oliveira R, Peter M, et al. Fractures of the cervical spine. *Pract Proced Orthop Trauma Surg A Trainee's Companion* 2015;**20**(5):269−82.

70. Torretti J, Sengupta D. Cervical spine traum. *Indian J Orthop* 2007;**41**(4):255−67.

71. Stone AT, Bransford RJ, Lee MJ, Vilela MD, Bellabarba C, Anderson PA, et al. Reliability of classification systems for subaxial cervical injuries. *Evid Based Spine Care J* December 2010;**1**(3):19−26.

72. Whang PG, Patel AA, Vaccaro AR. The development and evaluation of the subaxial injury classification scoring system for cervical spine trauma [Internet] *Clin Orthop Relat Res* March 21, 2011;**469**(3):723−31. Available from: http://link.springer.com/10.1007/s11999-010-1576-1.

73. Joaquim AF, Lawrence B, Daubs M, Brodke D, Patel AA. Evaluation of the subaxial injury classification system. *J Craniovertebral Junction Spine* July 2011; **2**(2):67−72.

74. Anderson PA, Moore TA, Davis KW, Molinari RW, Resnick DK, Vaccaro AR, et al. Cervical spine injury severity score assessment of reliability. *J Bone Jt Surg Ser A*. 2007;**89**(5):1057−65.

75. Urrutia J, Zamora T, Yurac R, Campos M, Palma J, Mobarec S, et al. An independent inter- and intraobserver agreement evaluation of the AOSpine subaxial cervical spine injury classification system. *Spine (Phila Pa 1976)* March 2017;**42**(5):298−303.

76. Kattail D, Furlan JC, Fehlings MG. Epidemiology and clinical outcomes of acute spine trauma and spinal cord injury: experience from a specialized spine trauma center in Canada in comparison with a large national registry. *J Trauma - Inj Infect Crit Care* 2009; **67**(5):936−42.

77. Doud AN, Weaver AA, Talton JW, Barnard RT, Meredith JW, Stitzel JD, et al. Has the incidence of thoracolumbar spine injuries increased in the United States from 1998 to 2011? *Clin Orthop Relat Res* 2015;**473**(1): 297−304.

78. Katsuura Y, Osborn JM, Cason GW. The epidemiology of thoracolumbar trauma: a meta-analysis [Internet] *J Orthop* 2016;**13**(4):383−8. Available from: http://files/1259/Katsuuraetal.-2016-TheepidemiologyofthoracolumbartraumaAmeta-a.pdf.

79. Hsu JM, Joseph T, Ellis AM. Thoracolumbar fracture in blunt trauma patients: guidelines for diagnosis and imaging. *Injury* June 2003;**34**(6):426−33.

80. Smith HE, Anderson DG, Vaccaro AR, Albert TJ, Hilibrand AS, Harrop JS, et al. Anatomy, biomechanics, and classification of thoracolumbar injuries [Internet] *Semin Spine Surg* 2010;**22**(1):2−7. https://doi.org/10.1053/j.semss.2009.10.001. Available from:.

81. Stagnara P, De Mauroy JC, Dran G, Gonon GP, Costanzo G, Dimnet J, et al. Reciprocal angulation of vertebral bodies in a sagittal plane: approach to references for the evaluation of kyphosis and lordosis. *Spine (Phila Pa 1976)* 1982;**7**(4):335−42.

82. Vialle R, Levassor N, Rillardon L, Templier A, Skalli W, Guigui P. Radiographic analysis of the sagittal alignment and balance of the spine in asymptomatic subjects. *J Bone Jt Surg Ser A* 2005;**87**(2):260−7.

83. Lee JY, Vaccaro AR, Schweitzer KM, Lim MR, Baron EM, Rampersaud R, et al. Assessment of injury to the thoracolumbar posterior ligamentous complex in the setting of normal-appearing plain radiography. *Spine J* 2007;**7**(4):422−7.

84. Terk MR, Hume-Neal M, Fraipont M, Ahmadi J, Colletti PM. Injury of the posterior ligament complex in patients with acute spinal trauma: evaluation by MR imaging. *AJR Am J Roentgenol* June 1997;**168**(6):1481−6.

85. Vaccaro AR, Zeiller SC, Hulbert RJ, Anderson PA, Harris M, Hedlund R, et al. The thoracolumbar injury severity score: a proposed treatment algorithm [Internet] *J Spinal Disord Tech* 2005;**18**(3):209−15. Available from: http://files/207/Vaccaroetal.-2005-Thethoracolumbarinjuryseverityscoreapropose.pdf.

86. Watson-Jones R. The results of postural reduction of fractures of the spine. *JBJS* 1938;**20**(3):567−86.

87. Sethi MK, Schoenfeld AJ, Bono CM, Harris MB. The evolution of thoracolumbar injury classification systems. *Spine J* September 2009;**9**(9):780−8.

88. Nicoll EA. Fractures of the dorso-lumbar spine [Internet] *J Bone Joint Surg Br* 1949;**31B**(3):376−94. Available from: http://www.ncbi.nlm.nih.gov/pubmed/18148776.

89. Whitesides TE. Traumatic kyphosis of the thoracolumbar spine [Internet] *Clin Orthop Relat Res* 1977;**128**:78−92. Available from: http://www.ncbi.nlm.nih.gov/pubmed/340100.

90. Denis F. The three column spine and its significance in the classification of acute thoracolumbar spinal injuries. *Spine (Phila Pa 1976)* 1983;**8**(8):817−31.

91. McAfee P, Yuan H, Fredrickson B, Lubicky JP. The value of computed tomography in thoracolumbar fractures [Internet] *J Bone Jt* 1983;**65A**(4):461−73. Available from: http://citeseerx.ist.psu.edu/viewdoc/download?doi=10.1.1.132.9115&rep=rep1&type=pdf.

92. Rudol G, Gummerson NW. (ii) Thoracolumbar spinal fractures: review of anatomy, biomechanics, classification and treatment [Internet] *Orthop Trauma* 2014;**28**(2):70−8. https://doi.org/10.1016/j.mporth.2014.01.003. Available from:.

93. Denis F. Spinal instability as defined by the three-column spine concept in acute spinal trauma. *Clin Orthop Relat Res* October 1984;**189**:65−76.

94. McCormack T, Karaikovic E, Gaines RW. The load sharing classification of spine fractures [Internet] *Spine (Phila Pa 1976)* 1994;**19**(15):1741−4. Available from: http://www.ncbi.nlm.nih.gov/pubmed/7973969.

95. Avanzi O, Landim E, Meves R, Caffaro MF, de Albuquerque Araujo Luyten F, Faria AA. Thoracolumbar burst fracture: load sharing classification and posterior instrumentation failure [Internet] *Rev Bras Ortop* May 17, 2015;**45**(3):236−40. Available from: https://www.ncbi.nlm.nih.gov/pmc/articles/PMC4799079/.

96. Magerl F, Aebi M, Gertzbein SD, Harms J, Nazarian S. A comprehensive classification of thoracic and lumbar injuries [Internet] *Eur Spine J* 1994;**3**(4):184. Available from: http://files/215/MagerlF.,AebiM.,GertzbeinS.D.,HarmsJ.andS-1994-Acomprehensiveclassificationofthoracicandlum.pdf.

97. Wood KB, Khanna G, Vaccaro AR, Arnold PM, Harris MB, Mehbod AA. Assessment of two thoracolumbar fracture classification systems as used by multiple surgeons [Internet] *JBJS* 2005;**87**(7). Available from: https://journals.lww.com/jbjsjournal/Fulltext/2005/07000/Assessment_of_Two_Thoracolumbar_Fracture.2.aspx.

98. Blauth M, Bastian L, Knop C, Lange U, Tusch G. [Interobserver reliability in the classification of thoracolumbar spinal injuries]. *Orthopade* August 1999;**28**(8):662−81.

99. Oner FC, Ramos LMP, Simmermacher RKJ, Kingma PTD, Diekerhof CH, Dhert WJA, et al. Classification of thoracic and lumbar spine fractures: problems of reproducibility. A study of 53 patients using CT and MRI. *Eur spine J Off Publ Eur Spine Soc Eur Spinal Deform Soc Eur Sect Cerv Spine Res Soc* June 2002;**11**(3):235−45.

100. Lewkonia P, Paolucci EO, Thomas K. Reliability of the thoracolumbar injury classification and severity score and comparison with the denis classification for injury to the thoracic and lumbar spine. *Spine (Phila Pa 1976)* December 2012;**37**(26):2161–7.

101. Koh YWDW, Kim DJ, Koh YWDW. Reliability and validity of thoracolumbar injury classification and severity score (TLICS). *Asian Spine J* 2010;**4**(2):109.

102. Patel AA, Vaccaro AR. Thoracolumbar spine trauma classification. *J Am Acad Orthop Surg* 2010;**18**(2):63–71.

103. Bonfante E, Tenreiro A, Choi J, Supsupin E, Riascos R. Thoracolumbar spine trauma: pearls and pitfalls of the newer classification systems. *Neurographics* 2018;**8**(2):86–96.

104. Vaccaro AR, Lehman RAJ, Hurlbert RJ, Anderson PA, Harris M, Hedlund R, et al. A new classification of thoracolumbar injuries: the importance of injury morphology, the integrity of the posterior ligamentous complex, and neurologic status. *Spine (Phila Pa 1976)* October 2005;**30**(20):2325–33.

105. Harrop JS, Vaccaro AR, Hurlbert RJ, Wilsey JT, Baron EM, Shaffrey CI, et al. Intrarater and interrater reliability and validity in the assessment of the mechanism of injury and integrity of the posterior ligamentous complex: a novel injury severity scoring system for thoracolumbar injuries. Invited submission from the joint sectio. *J Neurosurg Spine* February 2006;**4**(2):118–22.

106. Vaccaro AR, Oner C, Kepler CK, Dvorak M, Schnake K, Bellabarba C, et al. AOSpine thoracolumbar spine injury classification system. *Spine (Phila Pa 1976)* 2013;**38**(23):2028–37.

107. Kepler CK, Vaccaro AR, Koerner JD, Dvorak MF, Kandziora F, Rajasekaran S, et al. Reliability analysis of the AOSpine thoracolumbar spine injury classification system by a worldwide group of naïve spinal surgeons. *Eur Spine J* 2016;**25**(4):1082–6.

108. Kaul R, Chhabra HS, Vaccaro AR, Abel R, Tuli S, Shetty AP, et al. Reliability assessment of AOSpine thoracolumbar spine injury classification system and thoracolumbar injury classification and severity score (TLICS) for thoracolumbar spine injuries: results of a multicentre study. *Eur spine J Off Publ Eur Spine Soc Eur Spinal Deform Soc Eur Sect Cerv Spine Res Soc* May 2017;**26**(5):1470–6.

109. Abedi A, Mokkink LB, Zadegan SA, Paholpak P, Tamai K, Wang JC, et al. Reliability and validity of the AOSpine thoracolumbar injury classification system: a systematic review. *Global Spine J* 2019;**9**:231–42.

110. Beckmann NM, Chinapuvvula NR. Sacral fractures: classification and management [Internet] *Emerg Radiol* December 27, 2017;**24**(6):605–17. Available from: http://link.springer.com/10.1007/s10140-017-1533-3.

111. Bydon M, De la Garza-Ramos R, Macki M, Desai A, Gokaslan AK, Bydon A. Incidence of sacral fractures and in-hospital postoperative complications in the United States: an analysis of 2002–2011 data [Internet] *Spine (Phila Pa 1976)* 2014;**39**(18):E1103–1109. Available from: http://www.ncbi.nlm.nih.gov/pubmed/24875962.

112. Rodrigues-Pinto R, Kurd MF, Schroeder GD, Kepler CK, Krieg JC, Holstein JH, et al. Sacral fractures and associated injuries. *Glob Spine J* 2017;**7**(7):609–16.

113. Mehta S, Auerbach JDJ, Born CT, Chin KR. Sacral fractures [Internet] *J Am Acad Orthop Surg* 2006;**14**(12):656–65. Available from: http://europepmc.org/abstract/MED/17077338.

114. Denis F, Davis S, Comfort T. Sacral fractures: an important problem. Retrospective analysis of 236 cases [Internet] *Clin Orthop Relat Res* 1988;**227**:67–81. Available from: http://europepmc.org/abstract/MED/3338224.

115. Mirghasemi A, Mohamadi A, Ara AM. Completely displaced S-1 / S-2 growth plate fracture in an adolescent: case report and review of literature. *J Orthop Trauma* 2009;**23**(10):16–20.

116. Zelle BA, Gruen GS, Hunt T, Speth SR. Sacral fractures with neurological injury: is early decompression beneficial? *Int Orthop* 2004;**3**:244–51.

117. Lombardi G, Popolo G Del. Clinical outcome of sacral neuromodulation in incomplete spinal cord injured patients suffering from neurogenic lower urinary tract symptoms. *Spinal Cord* 2009;**47**(6):486–91. https://doi.org/10.1038/sc.2008.172. Available from:.

118. Vastenholt JM, Snoek GJ, Buschman HPJ, Van Der Aa HE, Alleman ERJ, Ijzerman MJ, et al. A 7-year follow-up of sacral anterior root stimulation for bladder control in patients with a spinal cord injury: quality of life and users' experiences. *Spinal Cord* 2003;**41**(7):397–402.

119. Nork SE, Jones CB, Harding SP, Mirza SK, Routt MLC. Percutaneous stabilization of U-shaped sacral fractures using iliosacral screws: technique and early results. *J Orthop Trauma* 2001;**15**(4):238–46.

120. Vaccaro A, Kim D, Brodke D, Harris M, Chapman J, Schildhauer T, et al. Diagnosis and management of sacral spine fractures. *Instr Course Lect* 2004;**53**:375–85.

121. Denis F, Davis S, Comfort T. Sacral fractures: an important problem. Retrospective analysis of 236 cases. *Clin Orthop Relat Res* 1988;**227**:67–81.

122. Isler B. Lumbosacral lesions associated with pelvic ring injuries. *J Orthop Trauma* 1990;**4**(1):1–6.

123. Strange-Vognsen H, Lebech A. An unusual type of fracture in the upper sacrum. *J Orthop Trauma* 1991;**5**(2):200–3.

124. Schroeder GD, Kurd MF, Kepler CK, Krieg JC, Wilson JR, Kleweno CP, et al. The development of a universally accepted sacral fracture classification: a survey of AOSpine and AOTrauma members [Internet] *Glob Spine J* April 1, 2016;**6**(1_suppl). https://doi.org/10.1055/s-0036-1582908. s-0036-1582908-s-0036-1582908. Available from:.

125. Roy-Camille R, Saillant G, Gagna G, Mazel C. Transverse fracture of the upper sacrum. Suicidal jumper's fracture. *Spine (Phila Pa 1976)* 1985;**10**(9):838—45.

126. Robles LA. Transverse sacral fractures [Internet] *Spine J* 2009;**9**(1):60—9. https://doi.org/10.1016/j.spinee.2007.08.006. Available from:.

127. Lehman RA, Kang DG, Bellabarba C. A new classification for complex lumbosacral injuries [Internet] *Spine J* 2012;**12**(7):612—28. https://doi.org/10.1016/j.spinee.2012.01.009. Available from:.

128. Bellabarba C, Schroeder GD, Kepler CK, Kurd MF, Kleweno CP, Firoozabadi R, et al. The AOSpine sacral fracture classification [Internet] *Glob Spine J* April 1, 2016;**6**(1_suppl). https://doi.org/10.1055/s-0036-1582696. s-0036-1582696-s-0036-1582696. Available from:.

Classification systems: SCI

Sukhvinder Kalsi-Ryan[1], Gita Gholamrezaei[2]

[1]KITE-UHN, Toronto Rehabilitation Institute, Department of Physical Therapy, University of Toronto, Toronto, ON, Canada; [2]KITE-UHN, Toronto Rehabilitation Institute, Toronto, ON, Canada

Introduction

Humans have been faced with spinal cord injuries (SCIs) since ancient times. Prior to the 1900s individuals with paraplegia simply died due to the injury and related secondary complications such as: pressure ulcers and urinary tract infections (at that time these were considered fatal complications).[1] Traumatic injury to the spinal cord usually occurs as a result of an insult to the spinal vertebrae/column, such as a fracture or vertebral dislocation, which can result in the immediate consequence of nerve damage and the loss of sensory and motor functions. The first documented cases of spinal cord injury date back to Edwin Smith surgical papyrus in Egypt circa 2500 BCE.[2,3]

The earliest efforts in developing treatment to manage SCI and procedures to prevent fatal complications following SCI were made by pioneers during the late 19th and early 20th centuries; at which time the survival rate of paraplegics began to increase. However, tetraplegics continued to have a high mortality rate until the late 1950s. Donald Munro the "father of the treatment of paraplegia" set up the first effective treatment center for individuals with SCI, and Sir Ludwig Guttmann the founder of the modern treatment of spinal injuries instituted an integrated and comprehensive treatment program in an SCI unit at Stoke-Mandeville Hospital. These two individuals are pioneers of modern SCI care.[1]

New discoveries and progress in different branches of medicine also led to the advancement of medical care and management in SCI. An example of this is the discovery of various anesthetics and progress in knowledge of sterilization, which forged a safer path for painless surgeries of the spine. Discovery of the X-ray by William C. Roentgen and the application of radiography in understanding the pathology of SCI had a tremendous impact on survival after SCI.[4,5]

With these advancements, numerous specialized spine units were established around the world, during and after World War II. The National Spinal Injuries Centre at Stoke-Mandeville Hospital in the United Kingdom was one of the first specialized spine centers, founded and directed by Sir Ludwig Guttmann, one of the most influential pioneers in this field.[1] Sir Ludwig Guttmann was also the first president of the International medical society of paraplegia, founded in 1961 with the purpose of studying all problems related to SCI. This society is now the International Spinal Cord Society

© 2022 Elsevier Inc. All rights reserved.

(ISCoS) that facilitates the development of the International Standards of Neurological Classification of SCI (ISNCSCI). The annual scientific meetings for this society occur yearly and every 4 years coincide with the summer paralympics, where experts in the field of SCI gather from all over the world to exchange knowledge and expertise.[5] It was in fact Guttman who incepted the paralympics. This leads to the expansion of medical and scientific interactions, not only in introducing the treatment procedures globally, but also in finding a universal way to discuss the needs in SCI medical care and assessment, such as the need to have a common nomenclature of paraplegia and a neurological classification for a more accurate diagnosis. The documentation and clinical quantification of neurological and structural injury of the spine has been an area of intense work for the past four decades.

The discussion on neurological classification and nomenclature of paraplegia began at the annual scientific meeting of the international medical society of paraplegia in 1969, acknowledging the importance of uniform standards for classification of SCI, and recommending the use of a single method in all SCI centers to define the level of injury.[6] How spinal cord injury is measured, classified, and documented is an essential area of work to assist in defining disease, studying the natural history of SCI and understanding what interventions can improve the status of the injury. Studying natural history of disease is an essential factor when establishing methods to treat disease. With respect to SCI there has been a great deal of measurement development in the area of outcomes.[7,8] With respect to classification systems for SCI, there are three main areas of classification development that will be addressed in this chapter.

Classification systems

In order to establish a standardized model of care for individuals with spinal cord injury,

classification systems were developed not only to facilitate the exchange of knowledge among physicians and medical practitioners with standards and consistent terminology, but also to accurately evaluate, identify, and characterize the injury in finding the optimal treatment protocol. The first categorization of spinal injuries was published by Böhler in 1929. This classification was based on plain radiographic examinations of SCI, done during World War I.[9] Since then numerous classifications have been developed. There are two primary types of classifications for traumatic SCI: neurological and structural. Neurological classification systems are defined based on the severity of the injury and the extent of loss of neurological functions due to the injury. Whereas, the structural classification systems are defined based on the nature and location of the damage. The third type of classification for SCI is related to nontraumatic SCI. These three types of classifications in SCI will be discussed in the following sections.

Neurological classification systems for traumatic SCI

Neurological classifications are used to grade severity of neurological loss after SCI. Here we will address the classifications that are widely accepted and used for assessing and classifying neurological deficit after SCI.

The Frankel classification of SCI

The Frankel Grade was among the first classification systems for SCI and was introduced in 1969 as a 5-point severity scale (Table 4.1). According to Frankel et al.[10] the severity of the injury can be classified as: A = Complete, B= Sensory only, C= Motor useless, D = Motor useful, or E = Recovery.

Although Frankel and his colleagues acknowledged that their classification of the bony lesion and its neurological scale were relatively crude, it was widely appreciated by clinicians and medical

TABLE 4.1 Frankel classification of SCI.

Classification	Grading	Definition
A	Complete	A complete loss of both motor and sensation below the lesion level
B	Sensory only	Presence of some sensation below the lesion level with complete motor paralysis
C	Motor useless	Presence of some motor function below the lesion level by means of movements with no practical application
D	Motor useful	Presence of some useful motor function below the lesion level by means of movements in lower limbs (walking with or without aids)
E	Recovery	A complete recovery with no neurological deficits

practitioners around the world and was used to determine the extent of neurological deficits related to SCI for years. In the 1980s it was the basis for the development of the International Standards for Neurological and Functional Classification of Spinal Cord Injury Patients.[10,11]

According to van Middendorp et al.[12] the Frankel scale had two major limitations. One was the nonspecificity of the classification in which the level of the injury is not incorporated. The second was the built-in subjectivity of the scale in differentiating between "motor useful" and "motor useless" functions.[12] Young W. noted in his online article "Spinal cord injury levels and classifications" that the Frankel classification not only introduced subjectivity into the scale in evaluating the usefulness of the lower limb functions but it also ignored upper extremity (UE) functions in individuals with cervical spinal cord injury.[13]

ASIA neurological standards and impairment scale

The American Spinal Injury Association (ASIA) published the first version of the ASIA Standards for Neurological and Functional Classification of Spinal Cord Injury in 1982 for the National SCI Statistical Center Database.[11,14] The first version of the ASIA standards was developed from discussions and consensus among a group of experts in the field of spinal cord injury with the purpose of developing a more precise definition of the extent of injury and neurologic levels. The purpose was for the American group to gather more reliable and consistent data in the National Database. Based on the committee consensus, key muscle groups and key sensory dermatome locations were chosen for neurological examinations, and the Frankel scale was incorporated into the standards.[14,15]

This neurological classification and measurement system have evolved through the years and undergone several revisions and upgrades. The revisions were made to increase validity and reliability of the measurement system by enhancing the muscle gradings, determining more precise sensory levels introducing anatomical landmarks within the dermatomes, clarification of the differentiation between grade C and D, and also redefining the zone of partial preservation for motor and sensory functions.[15,16]

In 1992, the ASIA revision committee published a major revision of the ASIA standard known as the International Standards for Classification of Spinal Cord Injury (ISCSCI) to minimize its poor intrarater reliability. It was in this major revision that the ASIA Impairment Scale (AIS) replaced the modified Frankel scale turning to an international gold standard for SCI classification.[11,17,18] The ISCoS endorsed the ISCSCI which was renamed to the ISNCSCI in 1996.[19] The latest revision and most widely used revision of the ISNCSCI is the 2011 version along with the 2015 version of the data collection form.[20–24] Available to supplement use has also been the development of online classification systems to assist in classifying cases.[25] However, in April of 2019 a revised version of ISNCSCI was released and uptake/adoption will require 1–2 years.[26]

ISNCSCI is a worldwide accepted standard method for assessment of the severity and level of injuries in individuals with SCI and consists of three sets of neurological examinations. It includes a motor examination based on 10 key muscles (paired points) assessable in the supine position, a sensory examination based on 28 key points on dermatomes (56 locations in total bilaterally) to determine the sensory score using light touch and pin prick (discrimination) sensation, and an anorectal examination. The ISNCSCI also includes a 5-point severity scale known as the AIS, which is the guide for injury classification.[12,18]

Based on the AIS, an individual with no motor or sensory function preserved in the sacral segments S4—S5 is assigned to ASIA A (Complete). An ASIA B (Incomplete), the sensory function but not motor is preserved below the neurological level and includes the sacral segments S4—S5. When motor function is preserved below the neurological level, and more than half of the key muscles below the neurological level have a muscle grade less than 3, the injury can be classified as ASIA C (Incomplete), and when the motor function is preserved below the neurological level and also more than half of the key muscles below the neurological level have a muscle grade of 3 or more, the injury is classified as ASIA D (Incomplete). In ASIA E (Normal), the sensory and motor functions are normal with no detectable neurological deficits (Table 4.2). This is usually assigned in a follow-up assessment where the original SCI minor and the subtle sensory or motor weakness may no longer be detectable.[13,18]

Reliability of the ISNCSCI administered early on (less than 72 hours) postinjury is rather poor. Thus, it has been recommended for clinicians to rely on the results of an ISNCSCI exam done greater than 72 hours postinjury.[27,28] It is much more accurate and does not account for any potential "spinal shock." Spinal shock is a temporary absence or depression of reflexes below the SCI and can skew the ISNCSCI exam results

TABLE 4.2 ASIA impairment scale.

Classification	Type	Definition
A	Complete	No motor or sensory function preserved in the sacral segments S4—S5
B	Incomplete	Sensory but not motor function preserved below the neurological level and includes the sacral segments S4—S5
C	Incomplete	Motor function is preserved below the neurological level, and more than half of key muscles below the neurological level have a muscle grade less than 3
D	Incomplete	Motor function is preserved below the neurological level and the majority of key muscles below the neurological level have a muscle grade of 3 or more
E	Normal	Sensory and motor functions are normal

during the earlier time frames.[29] Thus, confirming that the ISNCSCI does have limitations during the early acute phase of tSCI. Despite the limitations, it is still the most widely and accepted assessment used during the acute, subacute, and chronic phases of tSCI.[30,31]

In light of the ISNCSCI being such a heavily used measure, the existing data in the Model Systems, Rick Hansen Registry and European Multi-Centre SCI Study have been used to create predictive models of recovery, and particularly to understand what the functional outcomes or recovery can be with an early ISNCSCI assessment.[32,33] Predictive models using ISNCSCI usually take an early exam within 4 weeks of injury and predict what the functional outcome may be at 12 months postinjury. There is specific work to predict walking[34] and specific work that discusses prediction based on severity of SCI (AIS).[35,36] These predictive models have been somewhat useful; however, stronger, more accurate predictive models could have a broader application and the reduce burden of completing randomized controlled trials with large samples and long-term follow-up.

The ISNCSCI is the current gold standard in the field globally and is used as international nomenclature to define SCI and severity of disease. All clinicians and scientists that work within the fields of SCI should understand the ISNCSCI classification system and be proficient in administering the assessment as well as how to classify the findings of the test. The standards committee has made it very possible for all clinicians to become proficient in the ISNCSCI through the ASIA learning center and online tools.[37]

Structural classifications of the spine

As mentioned above, injury to the spinal column is the cause of SCI. Damage to the spinal column may or may not cause neurological deficit. Thus, the fracture or column damage is classified separately from the SCI. The purpose of classifying the fracture separate from the SCI is because both may or may not be related, and the decision to manage either may or may not be consistent. Furthermore, fractures are classified separately as, the cause of the fracture may or may not be traumatic. Fractures of the spine can be caused by trauma, or be idiopathic in nature, such as from osteoporosis or oncological causes. Classifying spine fractures can be complex and has evolved as imaging has contributed to field.

The purpose of classifying spine fractures is as important as any other, to create a system that allows the international community to have a common international nomenclature, for treatment indication and outcome. An international language will enable practices such as: consistent management of fractures, ability to combine data across centers, and the opportunity to compare outcomes.

Spine fracture classification was established in the late 1930s and was pioneered by Sir Watson-Jones. His work was followed by a number of orthopedic surgeons, who added to his developing system. At least until the 1980s at which time, classification of spine fractures began incorporating the use of radiology and then computed tomography scans (CT scan). A very thorough report of this evolution is documented by Srinivasan.[38] He thoroughly documents the progression and evolution of spine fracture classification systems, until the work by the AOSpine Group. The AOSpine group initiated the process of developing a comprehensive, collaborative, and measurable spine fracture classification system in 2009. The AOSpine recognized that the global spine community was not using consistent methodology in assessing and managing spine fractures.[38] Based on an iterative process with spine surgeons around the world the AOSpine has not only established/developed four spinal fracture classification systems, Upper Cervical Classification System, Subaxial Classification System, Thoracolumbar Classification System, and the Sacral Classification System.[39–41] The AOSpine has led the process of validation of these tools[42] on a global platform and finally on the website[41] has the classification cards available along with teaching material on use in the way of written instructions and videos. Each system considers the important components of fracture classification, including bony fracture, integrity of ligamentous structures, and displaced fracture. The classification systems maintain these three components with slightly different details that cover the nuances occurring in different regions of the spine.

With respect to tSCI, there remain some limitations in that the health of the neurological tissue is not incorporated within the imaging classifications. The absence of a fracture may not mean that the spinal cord and soft tissue is unaffected. Thus, a clinical index or flow diagram which incorporates imaging (X-ray, CT,

MRI) and physical/clinical exam would be superior to any one classification system that exists.[43]

The AOSpine will now look at the uptake of these classifications, how well they are applied and if their use facilitates a change in practice.

Classifications of nontraumatic SCI (degenerative cervical myelopathy)

Nontraumatic SCI in this chapter refers to degenerative cervical myelopathy (DCM). DCM is considered an SCI, however, does present differently than traumatic SCI at different stages of the disease. Thus, the field is currently moving to separate the measurement methods for these two subgroups of SCI. DCM is a disorder involving chronic compression of the cervical spinal cord and is the most common form of spinal cord impairment in adults.[44] DCM can result from a wide range of pathologies including degenerative disc disease, spondylosis, and hypertrophy or ossification of the spinal ligaments.[45–49]

Cervical cord compression leads to nerve damage over time, resulting in loss of function and reduced quality of life.[46,47,50,51] Individuals with DCM usually present with at least one of the following symptoms: weakness and/or numbness of the UEs, reduced manual dexterity, gait and balance impairment, lower extremity (LE) spasticity, neuropathic pain, and bowel/bladder dysfunction. Although DCM is common, detection can be challenging as impairment can be quite subtle during the mild stage of the disease. Because DCM has a varying presentation due to the severity of progression the classification system used needs to be a multidomain assessment with sensitivity.

To date the ISNCSCI has been used in DCM, such as the CSM protect study,[52] which did show differences between treatment and control groups, but not with respect to the ISNCSCI. The challenge in using the ISNCSCI with DCM is that it is not sensitive enough, and does not

pick up the subtle deficits of DCM. Alternatively, in some countries where the ISNCSCI does not have significant uptake, DCM measures have been applied to traumatic SCI. Despite these practices, the field is transitioning to defining the measurement aspects of traumatic and nontraumatic SCI as different practices.

There are two classification systems that have been frequently used for DCM: the Nurick Scale[53,54] and the modified Japanese Orthopaedic Association Scale (mJOA).[55,56] The Nurick Scale was developed to assess gait impairment in patients with DCM (see Table 4.3). It is a clinician-based measure containing six grades ranging from 0 to 5, with a focus on gait impairment. As the grade increases the disability increases, 0 being no gait deficit and 5 being that the individual is unable to ambulate at all or bed bound. There is one study that evaluated the validity, reliability, and responsiveness of the Nurick Scale.[53,54] However, due to its limited application to gait only, the Nurick Scale is used much less and the mJOA has become the "gold standard" in the field. Another reason why the Nurick Scale lost its popularity is because the field was no longer waiting for DCM patients

TABLE 4.3   Nurick grade.

Grading	Definition
0	Signs and symptoms or root involvement, but without evidence of spinal cord disease
1	Signs of spinal cord disease but difficulty walking
2	Slight difficulty in walking which did not prevent full-time employment
3	Difficulty in walking which prevented full-time employment or ability to do all housework, but which was not so severe as to require someone else's help to walk
4	Able to walk with someone else's help or with the aid of a frame
5	Chair bound or bedridden

Reproduced from Nurick (1972).[53,54]

to develop significant gait deficits before treating them. In fact, a significant group of DCM individuals present with mild or moderate DCM and usually present with mild gait deficits that are not picked up by the Nurick Scale. Neither of these classification systems were rigorously designed and it was only after their uptake some work was done to validate the mJOA.

The mJOA[55,56] is a modified version of the original JOA scale. It is scored on a 0- to 18-point scale with the lowest score representing greater disability (Table 4.4). It is a clinician-based measure that covers items such as upper and lower extremity, motor function, hand sensation, and micturition. The original development study did test for interobserver reliability.[57,58] However, other properties were not tested and reported until 2015. The authors of this chapter define the reliability, validity, and responsiveness of the mJOA. Furthermore, large observational cohort studies such as the CSM-North America and CSM-International studies sponsored by the AOSpine did use the data to define how the mJOA could be used. Defining severity of DCM was aligned with scoring thresholds through these large studies, where mild patients would score between 17 and 15, moderate between 14 and 10, and severe were below 10.[59] These severity thresholds have impacted the field quite significantly and contributed to how the research is conducted today. Not only how research is conducted, but also how clinical management has been defined.[60]

While it is recognized that the mJOA is a useful screening and classifying tool, there is the need to

TABLE 4.4   Modified Japanese orthopaedic assessment.

	Domains
I. Motor dysfunction score of the upper extremities (/5)	0 — Inability to move hands 1 — Inability to eat with a spoon but able to move hands 2 — Inability to button shirt but able to eat with a spoon 3 — Able to button shirt with great difficulty 4 — Able to button shirt with slight difficulty 5 — No dysfunction
II. Motor dysfunction score of the lower extremities (/7)	0 — Complete loss of motor and sensory function 1 — Sensory preservation without ability to move legs 2 — Able to move legs but unable to walk 3 — Able to walk on flat floor with a walking aid (i.e., cane or crutch) 4 — Able to walk up- and/or downstairs with hand rail 5 — Moderate to significant lack of stability but able to walk up- and/or downstairs without hand rail 6 — Mild lack of stability but walk unaided with smooth reciprocation 7 — No dysfunction
III. Sensation (/3)	0 — Complete loss of hand sensation 1 — Severe sensory loss or pain 2 — Mild sensory loss 3 — No sensory loss
IV. Sphincter dysfunction (/3)	0 — Inability to micturate voluntarily 1 — Marked difficulty with micturition 2 — Mild to moderate difficulty with micturition 3 — Normal micturition
Total (/18):	

Benzel et al. (1991).[55]

measure this population with more sensitive assessments for early detection of disease and to assess subtle changes through conservative management and as a result of interventions.[61]

assessments early after injury. We do know clinical assessment alone can be unreliable during the acute stages. Essentially as the field of SCI evolves, the classification tools required to define the injury should as well.

Summary

There are three domains of classification that currently exist for SCI and can be applied within your practice: (1) the neurological classification system for SCI; (2) the fracture classification for spinal injuries; and (3) the classification of DCM (nontraumatic SCI). We have reviewed these three domains and a brief history on these systems. The emphasis of the chapter is to provide you with tools to support your ability to assess and document SCI related to your patients. Classification systems are not typically ideal outcome measures and use of these systems should be coupled with measures that capture the subtleties of disease.

The classification tools presented in this chapter are available online and in the public domain:

- International Standards of Neurological Classification of Spinal Cord Injury: https://asia-spinalinjury.org/.
- AOSpine fracture classification tools: https://aospine.aofoundation.org/clinical-library-and-tools/ao-spine-classification-systems.
- Modified Japanese Orthopaedic Assessment: https://operativeneurosurgery.com/doku.php?id=modified_japanese_orthopaedic_association_scale.

Although, this is the current state of the field, there can be some ongoing evolution of classification systems of SCI, as there remain a few gaps. As imaging becomes much more sophisticated, and easy to administer, efforts to combine imaging with clinical assessment can promote improved and more precise classification of injury. More precise scales would promote improved predictive models as they relate to functional outcomes, as well as more accurate

References

1. Silver JR. *History of the treatment of spinal injuries.* London, England: Kluwer Academic/Plenum Publishers; 2005. p. 111—9.
2. Hughes JT. The Edwin Smith surgical papyrus: an analysis of the first case reports of spinal cord injuries. *Paraplegia* 1988;**26**:71—82.
3. Bennett G. Historical chapter. In: Howorth MB, Petrie JG, editors. *Injuries of the spine.* Baltimore: Williams and Wilkins; 1964. p. 1—19.
4. Wiltse LL. In: Frymoyer JW, editor. *The history of spinal disorders in the adult spine: principles and practice.* New York: Raven Press Ltd; 1991. p. 3—16.
5. Donovan WH. Spinal cord injury-past, present, and future. *J Spinal Cord Med* 2007;**30**:85—100.
6. Michaelis LS. International inquiry on neurological terminology and prognosis in paraplegia and tetraplegia. *Paraplegia* 1969;**7**:1—5.
7. *SCIRE professional, spinal cord injury research evidence, outcome measures.* 2020 [Accessed 20 January 18], https://scireproject.com/outcome-measures/.
8. Shirley Ryan Abilitylab. *Rehabilitation measures database.* 2020 [Accessed 20 January 18], https://www.sralab.org/rehabilitation-measures.
9. Böhler L. The treatment of fractures. In: Steinberg ME, Maudrich W, editors. *A translation of "Technik der Knochenbruchbehandlung im Frieden und im Kriege"*; 1929. Vienna, Austria.
10. Frankel HL, Hancock DO, Hyslop G, et al. The value of postural reduction in the initial management of closed injuries of the spine with paraplegia and tetraplegia. I. Paraplegia *Spinal Cord* 1969;**7**:179—92.
11. ASIA. *Standards for neurological classification of spinal injured patients.* Chicago: ASIA; 1982.
12. van Middendorp JJ, Goss B, Urquhart S, Atresh S, Williams RP, Schuetz M. Diagnosis and prognosis of traumatic spinal cord injury. *Global Spine J* 2011;**1**:1—8.
13. Young W. *Spinal cord injury levels and classifications.* 2019 [Accessed 19 September 21], https://www.sci-info-pages.com/levels-and-classification/.
14. Waring WP, Biering-Sorensen F, Burns S, Donovan W, Graves D, Jha A, Jones L, Kirshblum S, Marino R, Mulcahey MJ, Reeves R, Scelza WM, Schmidt-Read M, Stein A. 2009 review and revisions of the international

standards for the neurological classification of spinal cord injury. *J Spinal Cord Med* 2010;**33**(4):346—52.

15. Cohen ME, Ditunno JF, Donovan WH, Maynard FM. A test of the 1992 international standards for neurological and functional classification of spinal cord injury. *Spinal Cord* 1998;**36**:554—60.

16. Furlan JC, Fehlings MG, Tator CH, Davis AM. Motor and sensory assessment of patients in clinical trials for pharmacological therapy of acute spinal cord injury: psychometric properties of the ASIA Standards. *J Neurotrauma* 2008;**25**:1273—301.

17. Harris P. Editorial: the international standards booklet for neurological and functional classification of spinal cord injury. *Paraplegia* 1994;**32**:69.

18. Roberts TT, Leonard GR, Cepela DJ. Classifications in brief: American Spinal Injury Association (ASIA) impairment scale. *Clin Orthop Relat Res* 2017;**475**(5): 1499—504.

19. ASIA. *International standards for neurological classification of spinal injured patients*. Chicago: ASIA; 1996.

20. Kirshblum SC, Waring W, Biering-Sorensen F, Burns SP, Johansen M, Schmidt-Read M, Donovan W, Graves D, Jha A, Jones L, Mulcahey MJ, Krassioukov A. Reference for the 2011 revision of the international standards for neurological classification of spinal cord injury. *J Spinal Cord Med* November 2011;**34**(6):547—54. https://doi.org/10.1179/107902611X13186000420242.

21. Kirshblum SC, Burns SP, Biering-Sorensen F, Donovan W, Graves DE, Jha A, Johansen M, Jones L, Krassioukov A, Mulcahey MJ, Schmidt-Read M, Waring W. International standards for neurological classification of spinal cord injury (revised 2011). *J Spinal Cord Med* November 2011;**34**(6):535—46. https://doi.org/10.1179/204577211X13207446293695.

22. Committee Membership, Burns S, Biering-Sørensen F, Donovan W, Graves DE, Jha A, Johansen M, Jones L, Krassioukov A, Kirshblum S, Mulcahey MJ, Read MS, Waring W. International standards for neurological classification of spinal cord injury, revised 2011. *Top Spinal Cord Inj Rehabil* 2012;**18**(1):85—99. https://doi.org/10.1310/sci1801-85. Winter.

23. Schuld C, Franz S, Brüggemann K, Heutehaus L, Weidner N, Kirshblum SC, Rupp R, EMSCI study group. International standards for neurological classification of spinal cord injury: impact of the revised worksheet (revision 02/13) on classification performance. *J Spinal Cord Med* September 2016;**39**(5):504—12. https://doi.org/10.1080/10790268.2016.1180831.

24. Schuld C, Franz S, van Hedel HJ, Moosburger J, Maier D, Abel R, van de Meent H, Curt A, Weidner N, EMSCI study group, Rupp R. International standards for neurological classification of spinal cord injury: classification skills of clinicians versus computational algorithms. *Spinal Cord* April 2015;**53**(4):324—31. https://doi.org/10.1038/sc.2014.221.

25. Walden K, Bélanger LM, Biering-Sørensen F, Burns SP, Echeverria E, Kirshblum S, Marino RJ, Noonan VK, Park SE, Reeves RK, Waring W, Dvorak MF. Development and validation of a computerized algorithm for international standards for neurological classification of spinal cord injury (ISNCSCI). *Spinal Cord* March 2016;**54**(3):197—203. https://doi.org/10.1038/sc.2015.137.

26. ASIA and ISCoS International Standards Committee, Betz R, Biering-Sørensen F, Burns SP, Donovan W, Graves DE, Guest J, Jones L, Kirshblum S, Krassioukov A, Mulcahey MJ, Schmidt Read M, Rodriguez GM, Rupp R, Schuld C, Tansey K, Walden K. The 2019 revision of the international standards for neurological classification of spinal cord injury (ISNCSCI)-what's new? *Spinal Cord* October 2019; **57**(10):815—7. https://doi.org/10.1038/s41393-019-0350-9.

27. Brown PJ, Marino RJ, Herbison GJ, Ditunno Jr JF. The 72-hour examination as a predictor of recovery in motor complete quadriplegia. *Arch Phys Med Rehabil* 1991;**72**: 546—8.

28. Daverat P, Sibrac MC, Dartigues JF, et al. Early prognostic factors for walking in spinal cord injuries. *Spinal Cord* 1988;**26**:255—61.

29. Ko H-Y, Ditunno Jr JF, Graziani V, Little JW. The pattern of reflex recovery during spinal shock. *Spinal Cord* 1999; **37**:402—9.

30. Kirshblum S, Waring 3rd W. Updates for the international standards for neurological classification of spinal cord injury. *Phys Med Rehabil Clin* August 2014;**25**(3): 505—17. https://doi.org/10.1016/j.pmr.2014.04.001. vii, [Review].

31. Scivoletto G, Tamburella F, Laurenza L, Molinari M. Distribution-based estimates of clinically significant changes in the international standards for neurological classification of spinal cord injury motor and sensory scores. *Eur J Phys Rehabil Med* June 2013;**49**(3):373—84.

32. Noonan VK, Soril L, Atkins D, Lewis R, Santos A, Fehlings MG, Burns AS, Singh A, Dvorak MF. The application of operations research methodologies to the delivery of care model for traumatic spinal cord injury: the access to care and timing project. *J Neurotrauma* 2012;**29**:2272—22782.

33. Santos A, Gurling J, Dvorak MF, Noonan VK, Fehlings MG, Burns AS, Lewis R, Soril L, Fallah N, Street JT, Bélanger L, Townson A, Liang L, Atkins D. Modeling the patient journey from injury to community reintegration for persons with acute traumatic spinal cord injury in a Canadian centre. *PLoS One* 2013;**8**: e72552.

34. van Middendorp JJ, Hosman AJ, Donders AR, Pouw MH, Ditunno Jr JF, Curt A, Geurts AC, Van de Meent H, EM-SCI Study Group. A clinical prediction rule for ambulation outcomes after traumatic spinal cord injury: a longitudinal cohort study. *Lancet* March 19, 2011;**377**(9770):1004—10.

35. Scivoletto G, Tamburella F, Laurenza L, Torre M, Molinari M. Who is going to walk? A review of the factors influencing walking recovery after spinal cord injury. *Front Hum Neurosci* 2014;**8**:1—11.

36. Kirshblum SC, O'Connor KC. Levels of spinal cord injury and predictors of neurologic recovery. *Phys Med Rehabil Clin* 2000;**11**:1—27.

37. American Spinal Injury Association. *ASIA e-learning center.* 2020 [Accessed 20 January 18], https://asia-spinalinjury.org/learning/.

38. Srinivasan P. A review of thoracolumbar spine fracture classification systems. *Indian Spine J* 2018;**1**(2):71—8.

39. Vaccaro AR, Koerner JD, Radcliff KE, Oner FC, Reinhold M, Schnake KJ, Kandziora F, Fehlings MG, Dvorak MF, Aarabi B, Rajasekaran S, Schroeder GD, Kepler CK, Vialle LR. AO Spine subaxial cervical spine injury classification system. *Eur Spine J* 2016;**25**(7):2173—84.

40. Vaccaro AR, Oner C, Kepler CK, Dvorak M, Schnake K, Bellabarba C, Reinhold M, Aarabi B, Kandziora F, Chapman J, Shanmuganathan R, Fehlings M, Vialle L, AOSpine Spinal Cord Injury & Trauma Knowledge Forum. AO Spine thoracolumbar spine injury classification system: fracture description, neurological status, and key modifiers. *Spine* November 1, 2013;**38**(23):2028—37.

41. AO Spine. *AO spine injury classification systems.* 2020 [Accessed 20 January 18], https://aospine.aofoundation.org/clinical-library-and-tools/ao-spine-classification-systems.

42. Kepler CK, Vaccaro AR, Koerner JD, Dvorak MF, Kandziora F, Rajasekaran S, Aarabi B, Vialle LR, Fehlings MG, Schroeder GD, Reinhold M, Schnake KJ, Bellabarba C, Cumhur Öner F. Reliability analysis of the AO spine thoracolumbar spine injury classification system by a worldwide group of naïve spinal surgeons. *Eur Spine J* 2016;**25**(4):1082—6.

43. Tan LA, Kasliwal MK, Traynelis VC. Comparison of CT and MRI findings for cervical spine clearance in obtunded patients without high impact trauma. *Clin Neurol Neurosurg* May 2014;**120**:23—6.

44. Kalsi-Ryan S, Karadimas SK, Fehlings MG. Cervical spondylotic myelopathy: the clinical phenomenon and the current pathobiology of an increasingly prevalent and devastating disorder, the neuroscientist: a review journal bringing neurobiology. *Neurol & Psychiatry* 2013;**19**(4):409—21.

45. Fehlings MG, Tetreault L, Hsieh PC, Traynelis V, Wang MY. Introduction: degenerative cervical myelopathy: diagnostic, assessment, and management strategies, surgical complications, and outcome prediction. *Neurosurg Focus* 2016;**40**(6):E1.

46. Nouri A, Tetreault L, Singh A, Karadimas SK, Fehlings MG. Degenerative cervical myelopathy: epidemiology, genetics, and pathogenesis. *Spine* 2015;**40**(12):E675—93.

47. Ray SK, Matzelle DD, Wilford GG, Hogan EL, Banik NL. Increased calpain expression is associated with apoptosis in rat spinal cord injury: calpain inhibitor provides neuroprotection. *Neurochem Res* 2000;**25**(9—10):1191—8.

48. Tetreault L, Goldstein CL, Arnold P, Harrop J, Hilibrand A, Nouri A, et al. Degenerative cervical myelopathy: a spectrum of related disorders affecting the aging spine. *Neurosurgery.* 2015;**77**(Suppl. 4):S51—67.

49. Young WF. Cervical spondylotic myelopathy: a common cause of spinal cord dysfunction in older persons. *Am Fam Physi* 2000;**62**(5):1064—70. 1073.

50. King Jr JT, McGinnis KA, Roberts MS. Quality of life assessment with the medical outcomes study short form-36 among patients with cervical spondylotic myelopathy. *Neurosurgery* 2003;**52**(1):113—20. Discussion 121.

51. Tracy JA, Bartleson JD. Cervical spondylotic myelopathy. *Neurol* 2010;**16**(3):176—87.

52. Fehlings MG, Wilson JR, Karadimas SK, Arnold PM, Kopjar B. Clinical evaluation of a neuroprotective drug in patients with cervical spondylotic myelopathy undergoing surgical treatment: design and rationale for the CSM-Protect trial. *Spine* October 15, 2013;**38**(22 Suppl. 1):S68—75. https://doi.org/10.1097/BRS.0b013e3182a7e9b0 [Review].

53. Nurick S. The pathogenesis of the spinal cord disorder associated with cervical spondylosis. *Brain* 1972;**95**:87—100.

54. Nurick S. The natural history and the results of surgical treatment of the spinal cord disorder associated with cervical spondylosis. *Brain* 1972;**95**(1):101—8.

55. Benzel EC, Lancon J, Kesterson L, Hadden T. Cervical laminectomy and dentate ligament section for cervical spondylotic myelopathy. *J Spinal Disord* 1991;**4**(3):286—95.

56. Bartels RH, Verbeek AL, Benzel EC, et al. Validation of a translated version of the modified Japanese Orthopaedic Association score to assess outcomes in cervical spondylotic myelopathy: an approach to globalize outcomes assessment tools. *Neurosurgery* 2010;**66**:1013—6.

57. Singh A, Crockard HA. Comparison of seven different scales used to quantify severity of cervical spondylotic myelopathy and post- operative improvement. *J Outcome Meas* 2001;**5**:798—818.

58. Kopjar B, Tetreault L, Kalsi-Ryan S, Fehlings M. Psychometric properties of the modified Japanese Orthopaedic Association scale in patients with cervical spondylotic myelopathy. *Spine* January 1, 2015;**40**(1):E23—8. https://doi.org/10.1097/BRS.0000000000000648.

59. Fehlings MG, Wilson JR, Kopjar B, Yoon ST, Arnold PM, Massicotte EM, et al. Efficacy and safety of surgical decompression in patients with cervical spondylotic myelopathy: results of the AOSpine North America prospective multi-center study. American volume *J Bone & Joint Surg* 2013;**95**(18):1651—8.

60. Fehlings MG, Tetreault LA, Riew KD, Middleton JW, Aarabi B, Arnold PM, Brodke DS, Burns AS, Carette S, Chen R, Chiba K, Dettori JR, Furlan JC, Harrop JS, Holly LT, Kalsi-Ryan S, Kotter M, Kwon BK, Martin AR, Milligan J, Nakashima H, Nagoshi N, Rhee J, Singh A, Skelly AC, Sodhi S, Wilson JR, Yee A, Wang JC. A clinical practice guideline for the management of patients with degenerative cervical myelopathy: recommendations for patients with mild, moderate, and severe disease and nonmyelopathic patients with evidence of cord compression. *Global Spine J* September 2017;**7**(3 Suppl. 1):70S—83S. https://doi.org/10.1177/2192568217701914.

61. Kalsi-Ryan S, Singh A, Massicotte EM, Arnold PM, Brodke DS, Norvell DC, Hermsmeyer JT, Fehlings MG. Ancillary outcome measures for assessment of individuals with cervical spondylotic myelopathy. *Spine* October 15, 2013;**38**(22 Suppl. 1):S111—22. https://doi.org/10.1097/BRS.0b013e3182a7f499.

Outcome measures

Jetan H. Badhiwala[1], Christopher D. Witiw[1,2], Hetshree Joshi[1], Omar Khan[1], Sukhvinder Kalsi-Ryan[3,4]

[1]Division of Neurosurgery, Department of Surgery, University of Toronto, Toronto, ON, Canada; [2]Division of Neurosurgery, St. Michael's Hospital, Toronto, ON, Canada; [3]KITE-UHN, Toronto Rehabilitation Institute, Department of Physical Therapy, University of Toronto, Toronto, ON, Canada; [4]KITE Research Institute, University Health Network, Toronto, ON, Canada

Introduction

Spinal cord injury (SCI) results in devastating physical, occupational, and psychosocial consequences for over 600,000 patients worldwide every year.[1,2] Individuals with SCI frequently have traumatic injuries resulting in acute damage to the spinal cord. At a pathophysiological level, this damage triggers a host of molecular cascades that result in ischemia, inflammation, and damage to the spinal cord and its surrounding structures.[3] The damage, combined with the limited recovery potential of nervous tissue, can occasionally cause irreparable neurological harm.[4] As a result, SCI frequently causes lasting functional, neurological, and even cognitive deterioration in patients.[5–7] The long-term—and potentially permanent—decline in the functional abilities of the patient to perform daily activities often warrants extended monitoring and long-term care, resulting in significant economic burden on the health-care system.[8,9]

While SCI can frequently be devastating, there is heterogeneity in the type of functional decline that patients experience.[10] Some patients can have primarily upper extremity motor deficits, while others may have mainly lower extremity motor deficits, and still others experience sensory changes.[11] Moreover, the impact of SCI can extend beyond pure somatic sensory or motor dysfunction and spill over into the autonomic[12] and even neurocognitive domains.[7,13] Given the wide-ranging impact of SCI, as well as the growing degree of clinical research in the area, it is necessary for clinicians and investigators to have a wide and robust armament of outcome measures to track progress in patients with SCI. Such outcome measures can be instrumental for tracking response to treatment, guiding clinical research in SCI, and informing discussions with patients and families.[14]

In this chapter, we will discuss the variety of outcome measures used in SCI. To organize our discussion, we have grouped SCI outcomes into four categories: imaging outcomes, clinically

© 2022 Elsevier Inc. All rights reserved.

determined outcomes, electrophysiologic outcomes, and patient-reported outcomes.

Imaging outcomes

Imaging techniques are becoming increasingly used for the determination of neurological repair and outcome for multiple disease states such as multiple sclerosis.[15,16] Magnetic resonance imaging (MRI) is considered the modality of choice for the assessment of SCI for planning clinical intervention, for determining prognosis, and for determining the success of treatment.[17,18] MRI has been found to be the most capable at defining the extent of injury, the degree of cord compromise, and for detecting additional changes (e.g., hemorrhage, edema) that guide the evaluation of injury severity and aid with prognosis.[19] Further, detection of SCI via MRI can be done in the acute setting,[20] which demonstrates the power of MRI as an imaging modality for SCI patients.

Some SCI patients undergo surgical treatment, which may require fusion and placement of instrumentation to help stabilize the spine.[21] Consequently, postsurgical follow-up often requires careful assessment of the integrity of the fusion and instrumentation applied during surgery as 'outcome measures.' While MRI can be used as a tool to determine the stability of the spine and surrounding instrumentation, the major utility of MRI arises from its ability to detect soft tissue (e.g., ischemia, edema, inflammation) changes at an acceptable resolution.[17,19] Detection of these changes in postsurgical follow-up as 'outcome measures' is best performed with MRI; however, spinal X-rays allow a more dynamic evaluation of technical variables such as degree of fusion, stability of instrumentation, and spinal stability.[22] The added usefulness of spinal X-rays arises from the ability to perform X-rays at multiple positions (e.g., flexion, extension, neutral).[23] It is occasionally at nonneutral positions that subtle changes in these technical variables can be picked up, adding to the clinician's assessment of surgical success.

In addition to conventional MRI, special MRI techniques have been proposed as tools for providing insight into patient status after SCI. For instance, diffusion-weighted imaging may help with ascertainment of axon loss after SCI,[24] while functional MRI may aid in examining sensorimotor activity via imaging of metabolic activity in the brain and spinal cord.[25] Further, microstructural MRI can provide improved granularity and may potentially even serve as a tool to detect SCI before the onset of clinical manifestations.[26] However, these techniques are limited by time constraints and by lower resolution relative to the size of the spinal cord.[27,28] Moreover, there is a paucity of research to support the use of MRI and its companion modalities as potential surrogates for neurological outcome measures. Aside from a few imaging parameters, there exists a lack of evidence to support a strong correlation between imaging features and clinical status.[19] To address this knowledge gap, clinical trials (NCT: 03854214) are currently underway to elucidate the utility of imaging as a potential outcome measure for SCI.

Clinically determined outcome measures

Numerous clinically determined outcomes covering multiple areas of the patient's status have been studied and validated for the assessment of SCI. Here, we will discuss some of those outcomes, including those arising from the International Standards for Neurological Classification of Spinal Cord Injury (ISNCSCI) exam,[29] the Spinal Cord Independence Measure (SCIM),[30] the Functional Independence Measure (FIM),[31] specialized upper and lower extremity outcome measures, and outcome measures of autonomic and sphincter function.

ISNCSCI exam

The ISNCSCI exam[29] is a comprehensive head-to-toe clinical assessment that allows the clinician to extract multiple neurological measures useful for the classification and tracking of a patient's sensory and motor status. It can be divided into a sensory component and motor component. In the sensory portion of the exam, the clinician tests the dermatomes from C_2 (which corresponds to the superior portion of the neck and the posterior region of the head) to S_4–S_5 (which corresponds to the perianal region) for light touch and pinprick sensation (Fig. 5.1). The level of sensation on a particular dermatome is graded on a scale from 0 to 2, where 0 denotes absent sensation, 1 denotes altered or diminished sensation, and 2 denotes normal sensation. For a given sensory modality (i.e., light touch or pinprick), the sensory score is recorded bilaterally and summed, for a total possible single-modality sensory score of 112 (56 for each side). The sensory examination is

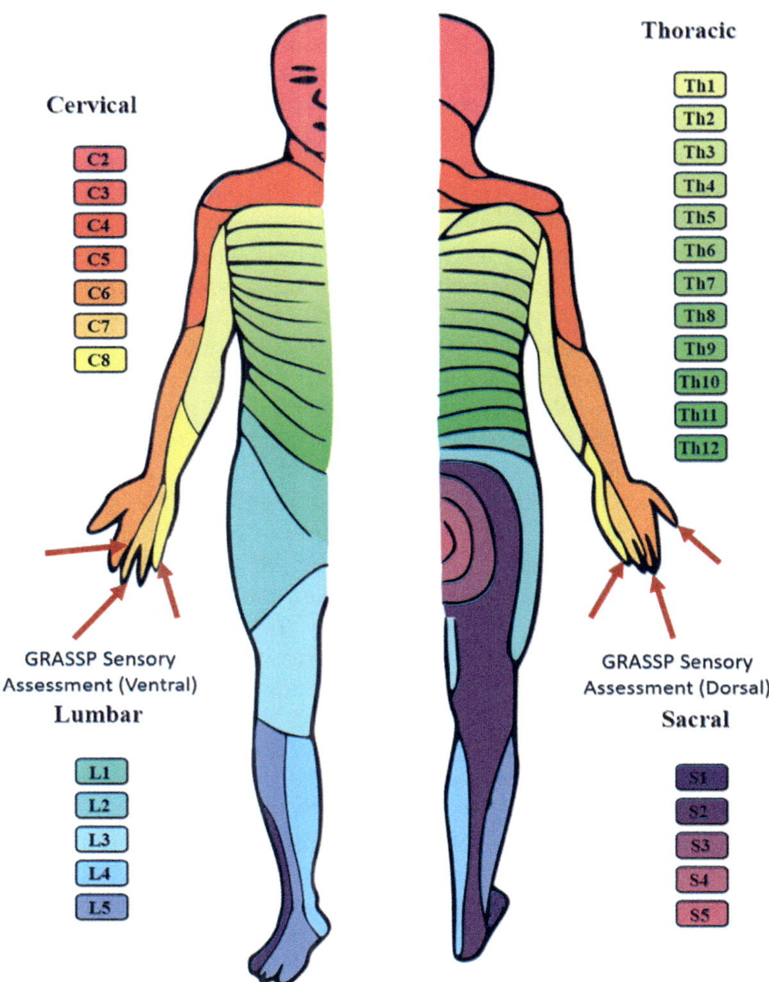

FIGURE 5.1 Dermatomes assessed in the ISNCSCI exam via both light touch and pinprick sensation. For the GRASSP assessment, the sensory domain is evaluated using the dermatomes of the hand as shown in the figure. *GRASSP*, Graded and Redefined Assessment of Strength, Sensibility, and Prehension; *ISNCSCI*, International Standard for the Neurological Classification of Spinal Cord Injury.

then repeated for the other sensory modality, giving an overall picture of the patient's sensory status throughout the body.

Assessment of motor function follows the sensory assessment in the ISNCSCI exam. In this subassessment, 10 key muscle groups are assessed bilaterally on both extremities. In the upper extremity, this includes the five major upper extremity muscle groups innervated by C_5 (elbow flexion), C_6 (wrist extension), C_7 (elbow extension), C_8 (finger flexion), and T_1 (finger abduction). In the lower extremity, the five major lower extremity muscle groups correspond to innervation by L_2 (hip flexion), L_3 (knee extension), L_4 (ankle dorsiflexion), L_5 (long toe extension), and S_1 (ankle plantar flexion). Each muscle group is graded on strength from 0 to 5, where 0 denotes total paralysis, 1 denotes some muscle contraction, 2 denotes active movement without gravity, 3 denotes active movement against gravity, 4 denotes active movement against some resistance, and 5 denotes active movement against full resistance. Muscle strength is assessed using this scale bilaterally for each muscle group and the scores are tallied for each side, such that the total score is 50 for both upper extremity and lower extremity muscles bilaterally. This total score is also described as the American Spinal Injury Association (ASIA) Motor Score, or AMS, which is tallied separately for upper extremities and lower extremities.[29] The AMS is frequently used as an outcome measure for SCI, especially in the research context.

In addition to conventional motor and sensory testing, special tests in the perianal region are also performed to determine sacral neurological integrity. Sensation to deep anal pressure is examined and graded as either absent or present, and voluntary anal contraction is also assessed as either absent or present. These special tests, combined with the aforementioned sensory and motor examination, facilitate the determination of the ASIA Impairment Scale (AIS), which grades the degree of SCI from most severe (A) to normal (E). Crucial components in the determination of the AIS grade include the sensory and motor level of injury, which are the most caudal levels at which sensory and motor functions are intact, respectively. Another important component is the overall neurological level of injury (NLI), defined as the most caudal level at which both sensory and motor functions are intact.

From these levels of injury, the AIS grade may be determined. A complete SCI (AIS grade A) occurs when voluntary anal contraction, sensation to deep anal pressure, S_4–S_5 light touch, and pinprick sensation are all absent. A sensory incomplete SCI (AIS grade B) occurs when sensory function is intact below the NLI and in the S_4–S_5 region; however, motor function is not intact more than three levels below the motor level of injury (i.e., muscle strength is less than 3) or there is an absence of voluntary anal contraction. AIS grade C injuries are motor incomplete injuries and are defined as injuries where motor function is intact in less than half of the key muscles below the NLI or when voluntary anal contraction is preserved. AIS grade D injuries occur when motor function is intact in at least half of the key muscles below the NLI. Finally, AIS grade E describes situations in which sensory and motor function are completely normal.

The AIS injury grade is a validated and oft-applied tool to assess outcome in SCI, both clinically and in the research setting.[32] However, classification of SCI according to SCI grade does carry with it some limitations. First, it is argued that AIS may be an insensitive as an outcome measure for determining the benefits of intervention.[33] Second, the AIS and the ISNCSCI exam in general do not have any components for upper cervical and thoracic motor function, which restricts its scope as a global assessment of neurological status.[34] Alternative measures addressing these limitations are under consideration.[33]

Functional outcomes measures

The lasting impact of SCI on patient function has required the development and application of a series of functional outcome measures to

SCI. The FIM[31] and the SCIM[30] are two such psychometrically validated and frequently used outcome measures. While the SCIM is specific to SCI, the FIM is designed to measure burden of care in a variety of disabling circumstances[35] and is not specific to SCI.

The FIM consists of two general domains: cognitive and motor, within which are multiple subdomains governing different cognitive and motor tasks. Within the motor domain, the subdomains include self-care (eating, grooming, bathing, dressing, and toileting), sphincter control (bowel management, bladder management), transfers (bed, toilet, tub/shower), and locomotion (walking, climbing stairs). Within the cognitive domain, the subdomains include communication (comprehension, expression) and social cognition (social interaction, problem-solving, memory). Independence in each task is scored on a scale from 1 to 7, where 1 denotes total assistance required in performing the task, 7 denotes complete independence, and scores from 2 to 6 describe intermediate levels of dependence.

The SCIM has been through multiple iterations.[36,37] The most recent iteration, SCIM III, was shown by Itzkovich and colleagues[30] to be reliable and valid in an international cohort of patients. Like the FIM, SCIM assesses independence in a variety of tasks across multiple functional domains (Table 5.1). These domains include self-care (feeding, bathing, dressing, grooming), respiration and sphincter management (respiration, bowel management, bladder management, use of toilet), room and toilet mobility (mobility in bed and activity to avoid pressure sores, transfer from bed to wheelchair, transfer from wheelchair to toilet or tub), and indoor and outdoor mobility (indoor mobility, mobility for moderate distances, outdoor mobility, stair management, wheelchair to car transfers, and ground to wheelchair transfer). Each task within these domains is rated on a numerical scale from requiring total assistance in accomplishing the task to requiring no assistance. The extent of this scale varies from one task to another.

TABLE 5.1 Overview of the Spinal Cord Independence Measure (SCIM) and the components used in its evaluation. A maximum score is awarded for functions in which the patient is completely independent, while complete dependence is awarded a score of 0 for that function.

SCIM outcome component	Maximum score (i.e., full independence)
Feeding	3
Bathing (upper body)	3
Bathing (lower body)	3
Dressing (upper body)	4
Dressing (lower body)	4
Grooming	3
Respiration	10
Bladder management	15
Bowel management	10
Toileting	5
Mobility in bed	6
Transfers—bed to wheelchair	2
Transfers—wheelchair to toilet to tub	2
Indoor mobility	8
Moderate distance mobility	8
Outdoor mobility	8
Stair management	3
Transfers—wheelchair to car	2
Transfers—ground to wheelchair	1

Another functional outcome measure developed specifically for SCI patients is the quadriplegia index of function (QIF).[38] Like the SCIM and the FIM, the QIF assesses activities of daily living (ADLs) such as transfers, grooming, bathing, dressing, feeding, mobility, bed activities, bladder management, bowel management, and understanding of personal care. It does this via a 37-item questionnaire[39] in which each of the 10 ADLs is evaluated for independence and scored in order of increasing independence from 0 to 4. While some studies have shown the QIF to be a reliable, valid, and sensitive tool, some groups have posited that compared to FIM and SCIM, there is a lack of evidence to support QIF as a measure of functional independence.[40] For now, QIF remains to be sufficiently investigated and validated as a functional outcome metric for SCI.

Specialized upper and lower extremity outcome measures

While the ISNCSCI exam and functional outcome measures for SCI offer some degree of assessment of upper and lower extremity neurological function, there is a lack of granularity in the degree of upper and lower extremity function they can capture. In addition, generic tests of upper and lower extremity function, such as the Jebsen test of Hand Function, are not sufficiently sensitive to capture the changes that occur with SCI.[41] To overcome these issues, multiple outcome measures have been developed to specifically target upper and lower extremity function in SCI.

For instance, the grasp and release test (GRT) was developed in 1994[42] to test the opening and closing of the hand in patients with SCI, particularly in patients who have neuroprostheses. The GRT consists of grasping, manipulating, and releasing six objects of varying sizes and shapes. These objects include a peg, paperweight, block, can, videotape, and fork. The GRT for the first five objects is performed by having the patient grasp the object, lift the object, move it over a testing board, and release the object into the target area. For the fork, the subject is asked to grasp the fork and depress the fork with a vertical force of at least 4.4 N. The sequence for all objects is completed within 30 s. A complete GRT is performed with all six objects, with five trials for each object undertaken in random order

to minimize systematic error. Patients are scored based on their ability to successfully complete the sequences. On reliability testing and validation, the GRT has been successful in meeting the criteria for an upper extremity SCI measure and has also been demonstrated to have good reliability.[43]

Another tool for assessing upper extremity function in SCI is the Graded and Redefined Assessment of Strength, Sensibility, and Prehension (GRASSP).[44] This tool consists of three domains: sensory, strength, and prehension. To determine function in the sensory domain, monofilaments are used to determine sensation across the dorsal and palmar aspects of the hand at three different points: the tip of the first digit (C_6), the tip of the third digit (C_7), and the tip of the fifth digit (C_8). The six testing locations are each scored on sensation from 0 to 4, for a total sensory score of 24 (Fig. 5.1). The strength domain of GRASSP is assessed by determining the power on a scale from 0 to 5 (much like the ISNCSCI exam) in 10 upper extremity muscles: C_5, Anterior Deltoid and Biceps; C_6, Wrist Extensors; C_7, Triceps and Opponens Pollicis; C_8, Extensor Digitorum, Finger Flexor of digit 3, Flexor Pollicis Longus; T_1, Finger Abductor of digit 5 and the First Dorsal Interossei.

Finally, testing of the prehension domain first involves evaluating the ability to grasp objects in three different ways, with each grasp graded 0–4: 1. Cylindrical Grasp; 2. Lateral Key Pinch; 3. Tip to Tip Pinch. In addition, testing prehension requires assessing patient performance in six prehension tasks: 1. Pouring water. 2. Opening jars. 3. Picking up and turning a key. 4. Transferring nine pegs from board to board. 5. Picking up four coins and placing them in slots. 6. Screwing four nuts onto bolts. Each task is graded from 0 to 5, resulting in a total score of 30 for one side. In a 2012 publication by Kalsi-Ryan and colleagues,[45] the GRASSP tool was demonstrated to show good interrater and test–retest reliability, with a sensitivity exceeding that of the ISNCSCI exam for upper extremity neurological function.

Assessment of lower extremity neurological function is often conducted using ambulation and gait testing. A reliable and psychometrically validated scale that has been developed for SCI is the Walking Index for Spinal Cord Injury II (WISCI II).[46,47] The WISCI II assessment grades a patient's gait over a 10 m flat surface on a scale from 0 to 20 in ascending order of independence based on four parameters: devices used, braces, assistance, and distance walked. Patients who require parallel bars as devices, who require braces, who need assistance from 1 or two people, and who are unable to complete the 10-m walk are assigned a lower level. Meanwhile, patients requiring limited assistance, no braces, or devices that provide less assistance such as walkers, canes, or crutches are assigned a higher level on the WISCI scale.

A simple test that has shown good validity and reliability is the 10-m walk test (10MWT).[48] This test is not specific to SCI, for it is also used in multiple sclerosis, Parkinson's, cerebral palsy, and other neurological impairments.[49–51] However, it has excellent results on psychometric testing,[52] which makes it the recommended test for ambulation assessment along with WISCI II. In the 10MWT, the subject is asked to walk at a comfortable pace for 10 m and timed from the 2-m mark to the 8-m mark. This test can be performed at normal walking speed or at maximum walking speed. Three trials of this test are performed, and the average speed is calculated from the trials to result in a final walking speed.

Outcome measures of autonomic and sphincter function

Autonomic dysfunction is a common occurrence after SCI, which can manifest clinically in many ways.[12] Patients can experience hemodynamic instability secondary to neurogenic shock or cardiac dysrhythmias. In addition, patients can develop sweating disturbances and temperature dysregulation as a result of SCI. Despite these sequelae, there has been little in the way of development of autonomic outcome measures for SCI. However, some groups have proposed an addition to the ISNCSCI exam which includes an assessment of autonomic function.

For instance, Alexander and colleagues[53] developed the autonomic standards assessment form in 2009 to supplement the ISNCSCI exam for SCI. This form includes an evaluation of cardiac arrhythmias, blood pressure, temperature regulation, respiratory control, urinary tract function, bowel function, and sexual function. Each item representing a component of general autonomic function (i.e., arrhythmias, blood pressure, temperature, respiratory control) is clinically evaluated and categorized as either normal, abnormal, or 'unable to assess.' For the urinary, bowel, and sexual categories, functional ability is classified as either normal, reduced, absent, or 'unable to assess.' With this scale, multiple components of autonomic function can be clinically evaluated in SCI patients, adding to the information gleaned from the regular ISNCSCI exam.

Electrophysiological outcomes

In addition to effects on spinal structure and clinically observed functional outcomes, SCI has significant consequences on nerve conduction. These consequences may not necessarily be observed via functional outcomes, which frequently lack the granularity to decipher individual nerve function.[54] As a result, the presence of robust electrophysiological outcomes is essential when putting together a complete picture of the patient's neurological status. Additionally, electrophysiological outcome measures afford the luxury of intraoperative assessment, in addition to greater objectivity, versatility, and diagnostic value.[54,55]

The two major electrophysiological outcome measures include motor evoked potentials (MEPs) and somatosensory evoked potentials (SSEPs), which each test a different arm of the nervous system.[56] MEPs are measured by stimulating the motor cortex and using surface electrodes on muscles to ascertain information about the conduction speed of impulses as they descend the corticospinal tract.[56,57] This information can provide insight into the neurological health of the corticospinal tract, allowing the clinician to stratify the severity of the SCI. In the intraoperative context, MEPs can be useful during critical points of the operation by allowing surgeons to recognize injury and either reverse it or prevent it from progressing. MEPs can also be used as surrogates for the patient's functional outcome, as they have been shown to be correlated with motor recovery and ambulation.[58]

On the other hand, SSEPs are measured in an 'ascending' manner; that is, peripheral nerves are stimulated and electrical impulses are recorded from the spine and scalp.[55] As with MEPs, SSEP recordings yield important information on neuronal conductive properties, allowing the clinician to assess the integrity of sensory pathways. In the operative setting, SSEPs have been shown to help reducing the rates of intraoperative SCI,[59] underscoring their utility in optimizing surgical outcomes. Additionally, like MEPs, SSEPs are useful metrics for predicting hand function and ambulation ability, a finding which has been demonstrated by multiple studies.[60,61]

Patient-reported outcome measures

Clinical literature in recent years has placed a greater emphasis on a patient-centered approach and patient experience, which has resulted in an increased interest in utilizing patient-reported outcome measures (PROMs) to quantify treatment efficacy.[62] One popular outcome measure is the short-form 36 questionnaire (SF-36),[63] a patient-reported health-related quality-of-life outcome which assesses multiple domains related to overall patient well-being. These include physical functioning, role limitations due to physical and emotional problems, pain, general health, vitality, social functioning, and mental health. Multiple research groups in the literature have endeavored to psychometrically validate the SF-36; their studies have demonstrated that the SF-36 score exhibits strong internal consistency and good interrater and intrarater reliability across its subdomains. Variants of the SF-36, such as the SF-12 (determine via a 12-item questionnaire unlike the 36-item questionnaire corresponding to the SF-36), have also been developed,[64] though their use is less prevalent in SCI literature.

Another important PROM for SCI is the SCI secondary condition scale (SCI-SCS). A 2007 study by Kalpakjian and colleagues[65] was the first to introduce and validate this scale against the SF-12 on a preliminary sample of 65 patients. The focus of the SCI-SCS diverges from that of the SF-36; rather than evaluating general patient health and functioning, it assesses the severity of various medical complications of SCI (Table 5.2). These complications can lead to significant morbidity and mortality,[65] and assessing their extent SCI can add an additional dimension to the management and prognosis of SCI. In the SCI-SCS, patients rate the severity of each of the 16 medical complications on a scale from 0 to 3, with a score of 3 denoting a significant impact of that complication on quality-of-life. While the use of this scale is not fully established in the literature given that it is still in its early phases, it has shown promise in psychometric analyses, demonstrating high inter- and intrarater reliability.[65,66]

A recently developed PROM for SCI is the SCI-QOL metric[67]; like the SF-36, this is a health-related quality-of-life metric that examines both the patient's physical and mental health. However, the SCI-QOL is designed to assess aspects of physical and mental health that are specifically affected by SCI. The metric

TABLE 5.2 Overview of the Spinal Cord Injury Secondary Conditions Scale (SCI-SCS) and the components used in its evaluation. For the SCI-SCSs, a maximum score of 3 is awarded for components which significantly affect the patient's functioning, while a minimum score of 0 is awarded for components which the patient is not adversely affected by.

SCI-SCS outcome component	Description
Decubitus ulcers	Measures the effect of decubitus ulcers (i.e., redness/ulceration at pressure-dependent areas)
Injury from loss of sensation	Measures the effect of injuries incurred secondary to sensory loss
Spasticity	Measures the effect of muscle spasms in patient functioning
Contractures	Measures the effect of contractures manifesting as limited range-of-motion in patient functioning
Heterotopic ossification	Measures the effect of limited range-of-motion from heterotopic ossification on patient functioning
Diabetes mellitus	Assesses how diabetes affects patient functioning
Bladder function	Assesses the patient's self-reported bladder function (with dysfunction manifesting as incontinence, retention)
Bowel function	Assesses the patient's self-reported bowel function (with dysfunction manifesting as incontinence, retention)
Urinary tract infections	Measures the effect of cystitis on patient functioning
Sexual dysfunction	Measures the effect of decreased sensation and erectile dysfunction on patient functioning
Autonomic dysreflexia	Measures the effect of temperature dysregulation on patient functioning
Postural hypotension	Measures the effect of orthostatic hypotension (i.e., significant decrease in blood pressure on rising) on patient functioning
Vascular disease	Measures the effect of varicose veins, edema, and deep vein thrombosis on patient functioning
Respiratory disease	Measures the effect of respiratory insufficiency and respiratory infections on patient functioning
Chronic pain	Measures the effect of chronic pain on patient functioning
Joint and muscle pain	Measures the effect of pain in specific joint or muscle groups on patient functioning

examines a total of 22 subdomains spread across four major categories: physical-medical health, emotional health, social participation, and physical functioning (Table 5.3). The 22 subdomains include medical complications of SCI (such as those in the SCI-SCS metric) in addition to specific mental health outcomes commonly impacted by SCI (e.g., anxiety, depression, self-esteem). Initial psychometric analyses of SCI-QOL have revealed favorable results,[68] with the metric demonstrating strong internal consistency. However, given that SCI-QOL was only

recently introduced to the literature, it has not yet been widely applied to clinical SCI research, and its validity is yet to be determined in the broader context.

A proposed addition to the selection of PROMs in SCI includes patient-reported experience measures (PREMs).[69] PREMs are metrics that specifically assess patient experiences in receiving care, such as accessibility and wait times. While the use of PREMs is not fully established in the SCI literature, they can offer important insight into the barriers that patients face,

TABLE 5.3　Overview of the Spinal Cord Injury Quality-of-Life questionnaire (SCI-QOL) and the 22 subdomains used in its evaluation. Abbreviations in parentheses next to each component denote the major domain to which the subdomain belongs. *EH*, emotional health; *PF*, physical functioning; *PMH*, physical—mental health; *SP*, social participation.

SCI-QOL outcome component	Description
Bladder management issues (PMH)	Assess ability to carry out bladder program and impact on daily living
Bowel management issues (PMH)	Assess ability to carry out bowel program and impact on daily living
Bladder complications (PMH)	Assess the impact of bladder complications (e.g., urinary tract infections, retention, incontinence) on sexual functioning and daily living
Decubitus ulcers (PMH)	Measure the effect of decubitus ulcers on patient functioning
Pain interference (PMH)	Assesses the extent to which pain hinders daily activities
Pain behavior (PMH)	Measures how pain manifests itself in the patient
Depression (EH)	Measures sadness and despair on a depression scale
Anxiety (EH)	Measures fearfulness and panic on an anxiety scale
Resilience (EH)	Measures the ability to adapt to changing/adverse circumstances
Positive affect and well-being (EH)	Assesses life satisfaction and sense of purpose in the patient
Grief/Loss (EH)	Assesses emotional reactions to sustaining a spinal cord injury
Self-esteem (EH)	Assesses the patient's perspective on their own self-worth
Stigma (EH)	Assesses the patient's perspective on the discrimination faced due to spinal cord injury
Psychological trauma (EH)	Measures the effect of psychological trauma sustained from spinal cord injury
Ability to participate in social roles and activities (SP)	Evaluates the degree of involvement in social activities (e.g., work, family, friends)
Satisfaction with social roles and activities (SP)	Evaluates the degree of satisfaction with social activities (e.g., work, family, friends)
Independence (SP)	Assesses perceived independence and ability to control one's life
Basic mobility (PF)	Assesses patient perception of ability to change body positions and undertake basic transfers
Self-care (PF)	Assesses patient perception of ability to perform self-care activities such as eating, grooming, and bathing
Fine motor functioning (PF)	Assesses patient perception of manual dexterity and ability to manipulate objects
Wheelchair mobility (PF)	Assesses patient perception of mobility in a wheelchair
Ambulation (PF)	Assesses patient perception of ambulation under different conditions, including climbing stairs

allowing clinicians to identify areas of improvement in health-care delivery. In doing so, there may also be a positive resultant effect on other clinical, functional, and PROM metric, as improved patient experience can motivate increased adherence to treatment.

Conclusions

Outcome measures are key in deciding management, assessing patient response to treatment, and determining prognosis after SCI. To this end, many metrics covering the areas of imaging, clinical and functional outcomes, and patient-reported outcomes have been developed and validated in the literature. Some of these metrics are summarized in Table 5.4. By using a combination of these measures in patient assessment, spine physicians can gain a multidimensional understanding of treatment efficacy. In conjunction with the patient, they can then adjust their recommendations by prioritizing the treatment that best optimizes particular outcomes.

TABLE 5.4 Summary of the descriptions of the significant outcome measures discussed in the chapter. SCI: spinal cord injury, ISNCSCI: International Standard for the Neurological Classification of Spinal Cord Injury.

Imaging outcomes	Description
Spinal X-ray	Can assess degree of fusion, instrumentation, and spinal stability in multiple positions
Conventional magnetic resonance imaging	Provides an assessment of soft tissue integrity in SCI
Functional magnetic resonance imaging	Provides an assessment of spinal cord activity in SCI
Diffusion-weighted imaging	Provides an assessment of axonal integrity in SCI
Clinical and functional outcomes	**Description**
ISNCSCI exam	Comprehensive neurological exam that evaluates motor, sensory, and sphincter function
Spinal cord independence measure	Evaluates multiple functional domains in areas of self-care, respiration and sphincter management, and mobility
Functional independence measure	Evaluates multiple functional domains in areas of motor and cognitive function
Quadriplegia index of function	Assesses multiple activities of daily living
Grasp and release test	Examination of hand opening and closing
Graded and redefined assessment of strength, sensibility, and prehension	Assesses upper extremity sensory function, strength, and prehension (ability to grasp objects in multiple ways)
Walking index for spinal cord injury II	Assesses gait quality over a 10-meter walking distance
10-Meter walk test	Assesses gait speed from 2 to 8 m during a 10-meter walking distance
Autonomic standards assessment form	Assesses multiple areas of autonomic function (e.g., hemodynamics, bowel/bladder function, temperature regulation)
Electrophysiological outcome measures	**Description**
Motor evoked potentials	Electrophysiologic evaluation of nerve conduction along motor pathways
Somatosensory evoked potentials	Electrophysiologic evaluation of nerve conduction along sensory pathways
Patient-reported outcome measures	**Description**
Short-form 36	A health-related quality-of-life questionnaire evaluating patient perception of physical and mental health
SCI secondary condition scale	An SCI-specific questionnaire evaluating patient perception of medical complications of SCI
SCI quality-of-life	An SCI-specific questionnaire evaluating patient perception of physical and mental health

References

1. Kumar R, Lim J, Mekary RA, et al. Traumatic spinal injury: global epidemiology and worldwide volume. *World Neurosurg* 2018;**113**:e345–63.

2. McDonald JW, Sadowsky C. Spinal-cord injury. *Lancet* 2002;**359**(9304):417–25.

3. Ahuja CS, Wilson JR, Nori S, et al. Traumatic spinal cord injury. *Nat Rev Dis Primers* 2017;**3**:17018.

4. Ahuja CS, Nori S, Tetreault L, et al. Traumatic spinal cord injury-repair and regeneration. *Neurosurgery* 2017;**80**(3S):S9–22.

5. Tate DG, Kalpakjian CZ, Forchheimer MB. Quality of life issues in individuals with spinal cord injury. *Arch Phys Med Rehabil* 2002;**83**(12 Suppl. 2):S18–25.

6. Alizadeh A, Dyck SM, Karimi-Abdolrezaee S. Traumatic spinal cord injury: an overview of pathophysiology, models and acute injury mechanisms. *Front Neurol* 2019;**10**:282.

7. Sachdeva R, Gao F, Chan CCH, Krassioukov AV. Cognitive function after spinal cord injury: a systematic review. *Neurology* 2018;**91**(13):611–21.

8. Hachem LD, Ahuja CS, Fehlings MG. Assessment and management of acute spinal cord injury: from point of injury to rehabilitation. *J Spinal Cord Med* 2017;**40**(6): 665–75.

9. Krueger H, Noonan VK, Trenaman LM, Joshi P, Rivers CS. The economic burden of traumatic spinal cord injury in Canada. *Chronic Dis Inj Can* 2013;**33**(3): 113–22.

10. Gupta S, Jaiswal A, Norman K, DePaul V. Heterogeneity and its impact on rehabilitation outcomes and interventions for community reintegration in people with spinal cord injuries: an integrative review. *Top Spinal Cord Inj Rehabil* 2019;**25**(2):164–85.

11. Maynard Jr FM, Bracken MB, Creasey G, et al. International standards for neurological and functional classification of spinal cord injury. American Spinal Injury Association. *Spinal Cord* 1997;**35**(5):266–74.

12. Krassioukov A. Autonomic function following cervical spinal cord injury. *Respir Physiol Neurobiol* 2009;**169**(2): 157–64.

13. Davidoff GN, Roth EJ, Richards JS. Cognitive deficits in spinal cord injury: epidemiology and outcome. *Arch Phys Med Rehabil* 1992;**73**(3):275–84.

14. Tomaschek R, Gemperli A, Rupp R, Geng V, Scheel-Sailer A, German-speaking Medical SCISEGDG. A systematic review of outcome measures in initial rehabilitation of individuals with newly acquired spinal cord injury: providing evidence for clinical practice guidelines. *Eur J Phys Rehabil Med* 2019;**55**(5):605–17.

15. Harel NY, Strittmatter SM. Functional MRI and other non-invasive imaging technologies: providing visual biomarkers for spinal cord structure and function after injury. *Exp Neurol* 2008;**211**(2):324–8.

16. Ontaneda D, Fox RJ. Imaging as an outcome measure in multiple sclerosis. *Neurotherapeutics* 2017;**14**(1):24–34.

17. Bozzo A, Marcoux J, Radhakrishna M, Pelletier J, Goulet B. The role of magnetic resonance imaging in the management of acute spinal cord injury. *J Neurotrauma* 2011;**28**(8):1401–11.

18. Freund P, Seif M, Weiskopf N, et al. MRI in traumatic spinal cord injury: from clinical assessment to neuroimaging biomarkers. *Lancet Neurol* 2019;**18**(12): 1123–35.

19. Kurpad S, Martin AR, Tetreault LA, et al. Impact of baseline magnetic resonance imaging on neurologic, functional, and safety outcomes in patients with acute traumatic spinal cord injury. *Global Spine J* 2017;**7**(3 Suppl. 1):151S–74S.

20. Chandra J, Sheerin F, Lopez de Heredia L, et al. MRI in acute and subacute post-traumatic spinal cord injury: pictorial review. *Spinal Cord* 2012;**50**(1):2–7.

21. Vale FL, Burns J, Jackson AB, Hadley MN. Combined medical and surgical treatment after acute spinal cord injury: results of a prospective pilot study to assess the merits of aggressive medical resuscitation and blood pressure management. *J Neurosurg* 1997;**87**(2):239–46.

22. Nouh MR. Spinal fusion-hardware construct: basic concepts and imaging review. *World J Radiol* 2012;**4**(5): 193–207.

23. Yu Z, Lin K, Chen J, et al. Magnetic resonance imaging and dynamic X-ray's correlations with dynamic electrophysiological findings in cervical spondylotic myelopathy: a retrospective cohort study. *BMC Neurol* 2020; **20**(1):367.

24. Pouw MH, van der Vliet AM, van Kampen A, Thurnher MM, van de Meent H, Hosman AJ. Diffusion-weighted MR imaging within 24 h post-injury after traumatic spinal cord injury: a qualitative meta-analysis between T2-weighted imaging and diffusion-weighted MR imaging in 18 patients. *Spinal Cord* 2012;**50**(6):426–31.

25. Powers JM, Ioachim G, Stroman PW. Ten key insights into the use of spinal cord fMRI. *Brain Sci* 2018;**8**(9).

26. Martin AR, De Leener B, Cohen-Adad J, et al. Can microstructural MRI detect subclinical tissue injury in subjects with asymptomatic cervical spinal cord compression? A prospective cohort study. *BMJ Open* 2018;**8**(4):e019809.

27. Smith SA, Pekar JJ, van Zijl PC. Advanced MRI strategies for assessing spinal cord injury. *Handb Clin Neurol* 2012;**109**:85–101.

28. Kornelsen J, Mackey S. Potential clinical applications for spinal functional MRI. *Curr Pain Headache Rep* 2007; **11**(3):165–70.

29. Kirshblum SC, Burns SP, Biering-Sorensen F, et al. International standards for neurological classification of spinal cord injury (revised 2011). *J Spinal Cord Med* 2011;**34**(6):535—46.

30. Itzkovich M, Gelernter I, Biering-Sorensen F, et al. The Spinal Cord Independence Measure (SCIM) version III: reliability and validity in a multi-center international study. *Disabil Rehabil* 2007;**29**(24):1926—33.

31. Hall KM, Cohen ME, Wright J, Call M, Werner P. Characteristics of the functional independence measure in traumatic spinal cord injury. *Arch Phys Med Rehabil* 1999;**80**(11):1471—6.

32. Roberts TT, Leonard GR, Cepela DJ. Classifications in brief: American Spinal Injury Association (ASIA) impairment scale. *Clin Orthop Relat Res* 2017;**475**(5): 1499—504.

33. van Middendorp JJ, Hosman AJ, Pouw MH, Group E-SS, Van de Meent H. ASIA impairment scale conversion in traumatic SCI: is it related with the ability to walk? A descriptive comparison with functional ambulation outcome measures in 273 patients. *Spinal Cord* 2009; **47**(7):555—60.

34. Franz S, Kirshblum SC, Weidner N, Rupp R, Schuld C, Group ES. Motor levels in high cervical spinal cord injuries: implications for the international standards for neurological classification of spinal cord injury. *J Spinal Cord Med* 2016;**39**(5):513—7.

35. Ring H, Feder M, Schwartz J, Samuels G. Functional measures of first-stroke rehabilitation inpatients: usefulness of the functional independence measure total score with a clinical rationale. *Arch Phys Med Rehabil* 1997; **78**(6):630—5.

36. Catz A, Itzkovich M, Agranov E, Ring H, Tamir A. SCIM–spinal cord independence measure: a new disability scale for patients with spinal cord lesions. *Spinal Cord* 1997;**35**(12):850—6.

37. Catz A, Itzkovich M, Tamir A, et al. [SCIM–spinal cord independence measure (version II): sensitivity to functional changes]. *Harefuah* 2002;**141**(12):1025—31. 1091.

38. Gresham GE, Labi ML, Dittmar SS, Hicks JT, Joyce SZ, Stehlik MA. The quadriplegia index of function (QIF): sensitivity and reliability demonstrated in a study of thirty quadriplegic patients. *Paraplegia* 1986;**24**(1): 38—44.

39. Marino RJ, Goin JE. Development of a short-form quadriplegia index of function scale. *Spinal Cord* 1999;**37**(4): 289—96.

40. Anderson K, Aito S, Atkins M, et al. Functional recovery measures for spinal cord injury: an evidence-based review for clinical practice and research. *J Spinal Cord Med* 2008;**31**(2):133—44.

41. Sears ED, Chung KC. Validity and responsiveness of the Jebsen-Taylor hand function test. *J Hand Surg Am* 2010; **35**(1):30—7.

42. Wuolle KS, Van Doren CL, Thrope GB, Keith MW, Peckham PH. Development of a quantitative hand grasp and release test for patients with tetraplegia using a hand neuroprosthesis. *J Hand Surg Am* 1994;**19**(2): 209—18.

43. Mulcahey MJ, Smith BT, Betz RR. Psychometric rigor of the grasp and release test for measuring functional limitation of persons with tetraplegia: a preliminary analysis. *J Spinal Cord Med* 2004;**27**(1):41—6.

44. Kalsi-Ryan S, Curt A, Verrier MC, Fehlings MG. Development of the graded redefined assessment of strength, sensibility and prehension (GRASSP): reviewing measurement specific to the upper limb in tetraplegia. *J Neurosurg Spine* 2012;**17**(1 Suppl. l):65—76.

45. Kalsi-Ryan S, Beaton D, Curt A, et al. The graded redefined assessment of strength sensibility and prehension: reliability and validity. *J Neurotrauma* 2012;**29**(5): 905—14.

46. Marino RJ, Scivoletto G, Patrick M, et al. Walking index for spinal cord injury version 2 (WISCI-II) with repeatability of the 10-m walk time: inter- and intrarater reliabilities. *Am J Phys Med Rehabil* 2010;**89**(1):7—15.

47. Ditunno Jr JF, Ditunno PL, Scivoletto G, et al. The walking index for spinal cord injury (WISCI/WISCI II): nature, metric properties, use and misuse. *Spinal Cord* 2013;**51**(5):346—55.

48. Amatachaya S, Naewla S, Srisim K, Arrayawichanon P, Siritaratiwat W. Concurrent validity of the 10-meter walk test as compared with the 6-minute walk test in patients with spinal cord injury at various levels of ability. *Spinal Cord* 2014;**52**(4):333—6.

49. Lang JT, Kassan TO, Devaney LL, Colon-Semenza C, Joseph MF. Test-retest reliability and minimal detectable change for the 10-meter walk test in older adults with Parkinson's disease. *J Geriatr Phys Ther* 2016; **39**(4):165—70.

50. Kempen JC, de Groot V, Knol DL, Polman CH, Lankhorst GJ, Beckerman H. Community walking can be assessed using a 10-metre timed walk test. *Mult Scler* 2011;**17**(8):980—90.

51. Bahrami F, Noorizadeh Dehkordi S, Dadgoo M. Inter and intra rater reliability of the 10 meter walk test in the community Dweller Adults with spastic cerebral palsy. *Iran J Child Neurol* 2017;**11**(1):57—64.

52. van Hedel HJ, Wirz M, Dietz V. Assessing walking ability in subjects with spinal cord injury: validity and reliability of 3 walking tests. *Arch Phys Med Rehabil* 2005; **86**(2):190—6.

53. Alexander MS, Biering-Sorensen F, Bodner D, et al. International standards to document remaining autonomic function after spinal cord injury. *Spinal Cord* 2009;**47**(1):36—43.

54. Hubli M, Kramer JLK, Jutzeler CR, et al. Application of electrophysiological measures in spinal cord injury clinical trials: a narrative review. *Spinal Cord* 2019;**57**(11):909—23.

55. Pajewski TN, Arlet V, Phillips LH. Current approach on spinal cord monitoring: the point of view of the neurologist, the anesthesiologist and the spine surgeon. *Eur Spine J* 2007;**16**(Suppl. 2):S115—29.

56. Xie J, Boakye M. Electrophysiological outcomes after spinal cord injury. *Neurosurg Focus* 2008;**25**(5):E11.

57. Curt A, Dietz V. Electrophysiological recordings in patients with spinal cord injury: significance for predicting outcome. *Spinal Cord* 1999;**37**(3):157—65.

58. Petersen JA, Spiess M, Curt A, Dietz V, Schubert M, Group E-SS. Spinal cord injury: one-year evolution of motor-evoked potentials and recovery of leg motor function in 255 patients. *Neurorehabilitation Neural Repair* 2012;**26**(8):939—48.

59. Dawson EG, Sherman JE, Kanim LE, Nuwer MR. Spinal cord monitoring. Results of the scoliosis research society and the European spinal Deformity society survey. *Spine* 1991;**16**(8 Suppl. 1):S361—4.

60. Curt A, Dietz V. Ambulatory capacity in spinal cord injury: significance of somatosensory evoked potentials and ASIA protocol in predicting outcome. *Arch Phys Med Rehabil* 1997;**78**(1):39—43.

61. Curt A, Dietz V. Traumatic cervical spinal cord injury: relation between somatosensory evoked potentials, neurological deficit, and hand function. *Arch Phys Med Rehabil* 1996;**77**(1):48—53.

62. Kingsley C, Patel S. Patient-reported outcome measures and patient-reported experience measures. *BJA Education* 2017;**17**(4):137—44.

63. Ware Jr JE, Sherbourne CD. The MOS 36-item short-form health survey (SF-36). I. Conceptual framework and item selection. *Med Care* 1992;**30**(6):473—83.

64. Whitehurst DG, Engel L, Bryan S. Short form health surveys and related variants in spinal cord injury research: a systematic review. *J Spinal Cord Med* 2014;**37**(2):128—38.

65. Kalpakjian CZ, Scelza WM, Forchheimer MB, Toussaint LL. Preliminary reliability and validity of a spinal cord injury secondary conditions scale. *J Spinal Cord Med* 2007;**30**(2):131—9.

66. Arora M, Harvey LA, Lavrencic L, et al. A telephone-based version of the spinal cord injury-secondary conditions scale: a reliability and validity study. *Spinal Cord* 2016;**54**(5):402—5.

67. Tulsky DS, Kisala PA, Victorson D, et al. Overview of the spinal cord injury–quality of life (SCI-QOL) measurement system. *J Spinal Cord Med* 2015;**38**(3):257—69.

68. Post MW, Adriaansen JJ, Charlifue S, Biering-Sorensen F, van Asbeck FW. Good validity of the international spinal cord injury quality of life basic data set. *Spinal Cord* 2016;**54**(4):314—8.

69. Dvorak MF, Cheng CL, Fallah N, et al. Spinal cord injury clinical registries: improving care across the SCI care continuum by identifying knowledge gaps. *J Neurotrauma* 2017;**34**(20):2924—33.

Imaging: spine trauma

Parthik D. Patel[1], Michael Markowitz[1], Srikanth N. Divi[2], Gregory D. Schroeder[1], Alexander R. Vaccaro[3]

[1]Rothman Orthopaedic Institute, Philadelphia, PA, United States; [2]Northwestern University Feinberg School of Medicine, Chicago, IL, United States; [3]Department of Orthopedic Surgery, Rothman Institute, Thomas Jefferson University, Philadelphia, PA, United States

Introduction

Imaging assessment of acute trauma patients has undergone significant evolution over the years, particularly in the evaluation of the spine. Plain radiography, computed tomography (CT), and magnetic resonance imaging (MRI) are the three most commonly utilized imaging modalities, each of which have unique advantages as well as drawbacks including varying levels of cost, availability, exposure to radiation, and user error. This chapter aims to review the indications for imaging for spinal cord injury and spine trauma in a posttraumatic setting, as well as discuss the available options, function, and benefits/limitations of each modality. In addition, discussion will center on imaging considerations in special patient populations (e.g., geriatrics, pediatrics, etc.) and a variety of different spinal trauma injuries. Finally, we will analyze imaging of the thoracolumbar spine and vascular injury after spinal trauma.

Cervical spine imaging indications

In the setting of acute cervical spine trauma, imaging guides treatment by detailing the extent of soft tissue and bony damage with the use of multiplanar reconstruction. Spinal imaging allows the physician to diagnose cervical spine injury, characterize the type of injury, assess for the potential for instability, and evaluate the integrity of the neural elements.[1] Radiographs have universally been used as a quick and easy evaluation of the spine, but at the same time are frequently overused leading to unnecessary radiation and health-care costs.[2] The following sections will focus on: (1) patients who warrant radiographic imaging following blunt cervical or thoracolumbar trauma and (2) the most appropriate modality for such imaging.

Patients who warrant imaging

Two multicenter, prospective clinical studies have stood at the forefront of clinical decision

© 2022 Elsevier Inc. All rights reserved.

making for the use of cervical spine imaging in the posttraumatic setting: the National Emergency X-radiology Utilization Study (NEXUS) and the Canadian C-Spine Rule (CCR). In 1998, the NEXUS study examined 34,069 patients who underwent selective radiography of the cervical spine for blunt trauma.[3] The study concluded that patients did not need cervical spine radiographic imaging as long as they met the following criteria:

(1) No posterior midline cervical spine tenderness
(2) No focal neurologic deficits
(3) Normal level of alertness
(4) No intoxication
(5) No painful distraction injury

This study reported 99.5% sensitivity and 99.5% negative predictive value for identifying patients with cervical spine fracture when utilizing this criterion. However, the NEXUS study did have its limitations in its low specificity (12.5%), variable interobserver agreement, and possible compromised external validity via patient selection.

Subsequently in 2001, the CCR study examined 8924 cases of alert patients (Glasgow Coma Scale [GCS] = 15) in stable condition at risk of neck injury.[4] This model aimed to identify the necessity of obtaining a cervical radiograph by recognizing high-risk and low-risk factors allowing for safe assessment of active range of motion (AROM), including as follows:

High-risk factors requiring radiography	Low-risk factors for safe AROM
Age ≥65 years	Simple rear-end MVA
Dangerous mechanism of injury[a]	Sitting position in Emergency Department
Sensory deficits in extremities	Ambulatory at any time
	Delayed onset of neck pain
	Absence of midline cervical spinal tenderness

[a] *Includes: Fall from >1 m or 5 flights of stairs, MVA ≥100 km/hour, bicycle collision.*

In contrast to the results of the NEXUS study, the CCR algorithm showed a similar sensitivity (100% vs. 99.5%) but higher specificity (42.5% vs. 12.5%). However, limitations of this study included the use of "clinically important C-spine injury" as their primary outcome and lack of cervical radiographs in all patients.

In 2003, Stiell et al. reported a multicenter, prospective cohort validation study in *The New England Journal of Medicine* comparing the CCR versus NEXUS algorithm in alert patients with acute trauma to the head/neck in a stable condition.[5] Their findings confirmed a higher sensitivity, specificity, and negative predictive value of the CCR algorithm. In addition, physician interrater reliability favored the CCR algorithm over the NEXUS. The study concluded that the CCR clinical decision-making tool had the higher potential to standardize protocol when it came to cervical spine radiographs in the posttraumatic setting.

Analogous recommendations in the evaluation of blunt thoracolumbar spine trauma are not as well defined in the literature. However, studies suggest that similar to the cervical spine criteria, patients who are awake, alert, and without clinical evidence of neurologic compromise do not require thoracolumbar imaging evaluation in the acute setting.[6]

Plain radiography (X-ray)

Ideally three standard radiographs of the cervical spine are obtained in the posttraumatic setting: anterior/posterior (A/P), lateral, and open mouth (odontoid) views, with the lateral view being the most reliable with regard to detection of injury.[2,7]

The A/P (Fig. 6.1A) radiographic view is a coronal assessment of the cervical spine. Coronal imaging allows visualization from C3 to the upper thoracic vertebrae. When evaluating the A/P view, special attention should be given to the alignment of spinous processes, vertebral body height/morphology, vertebral disc height, alignment of the lateral margins via the lateral masses,

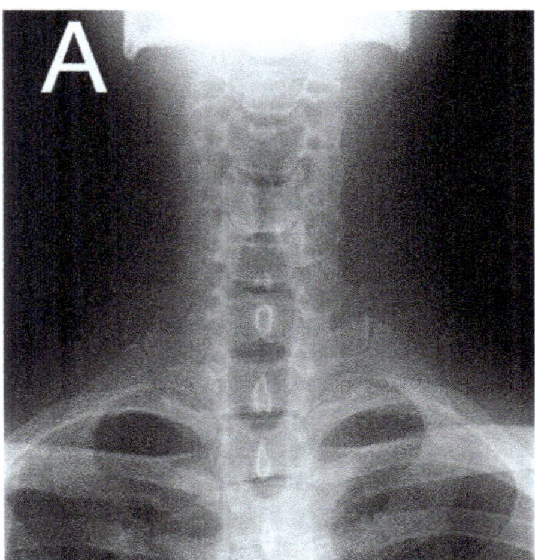

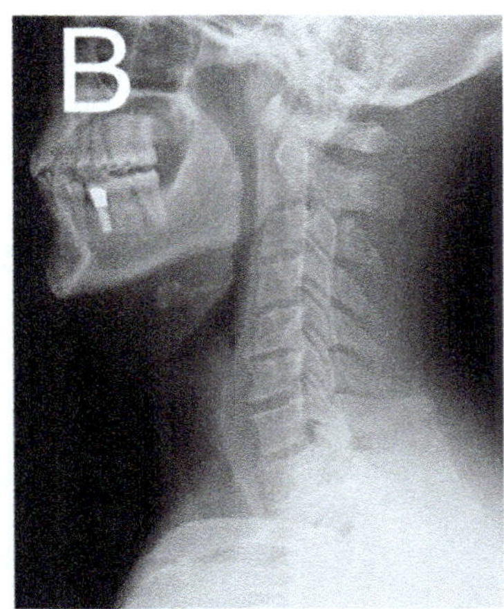

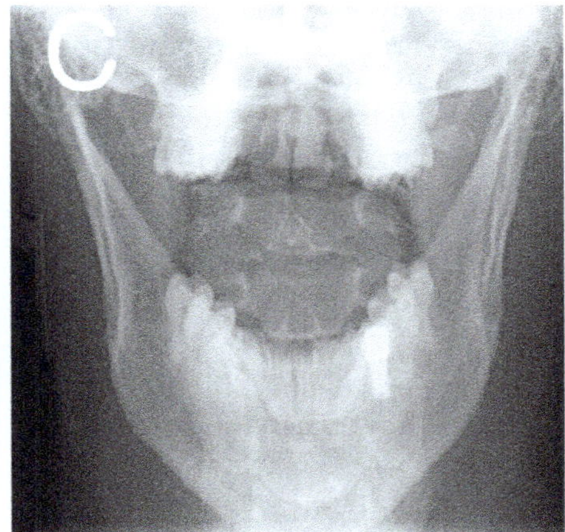

FIGURE 6.1 (A) AP view of the cervical spine, (B) lateral view, and (C) open-mouth (odontoid) view. *Image by Stillwaterrising via WikiCommons, distributed under a CC 1.0 Public Domain Dedication.*

and symmetry of the uncovertebral joints.[8] Misalignment of these structures may suggest a rotation injury to the spine, whereas a loss in height unilaterally or bilaterally may be indicative of a flexion injury. With the patient rotated, A/P imaging can give the false impression of injury in the setting of misalignment. Patient rotation can be assessed by estimating the distance between the spinous processes and pedicles, with any deviation on either side being abnormal.[2]

The lateral view (Fig. 6.1B) is a sagittal assessment of the cervical spine between the occiput and superior endplate of T1. A true lateral view shows all seven cervical vertebrae and the C7–T1 junction, a common location of traumatic injury. Harris et al.[9] described four cardinal lines when studying the lateral radiograph which include:

(1) anterior vertebral body line
(2) posterior vertebral body line
(3) spinolaminar line
(4) posterior spinous process line

stability to the head. In the setting of trauma, this stability can be disrupted and lead to possible bony or ligamentous injury. Initial evaluation should focus on the alignment of the C1–C2 articular masses, followed by analysis of the occipital condyles as well as assessment of possible basilar invagination.[11] In general, the lateral masses of C1 should not hang over the lateral masses of C2; the rule of Spence suggests an overhang >6.9 mm is suspicious of a transverse ligament injury and should be further evaluated with MRI. In addition, the articulation between the lateral masses of C1 and body of C2 as well as the space between the dens and lateral masses of C1 should be symmetric. Any deviation may be indicative of a Jefferson fracture of C1 or dens fracture.

While dynamic cervical spine radiographs including flexion and extension imaging are helpful for identifying instability from degenerative disease, they are not appropriate in the acute trauma setting.[12]

Computed tomography

While plain radiographs are still utilized due to their low cost and ease of access, advances in CT technology such as the advent of the multidetector-row computed tomography (MDCT) has superseded plain radiographic evaluation and even conventional CT scanners in the trauma spine setting.[13] As an added benefit, MDCT examination has become an instrumental part of an acute trauma workup as it is used in the assessment of the chest, abdomen, and pelvis and, as a result, can also provide insight to thoracic and lumbar spine injury.[2]

Although many algorithms of interpreting CT images of the cervical spine exist with slight variations, a thorough investigation of each slice should be performed to rule out injury. This includes analysis of parasagittal, coronal, and axial slices. Each slice should be assessed for alignment, bony abnormalities, joint space

Subtle discrepancies in these lines should prompt scrutiny for further abnormality as these findings in the trauma setting may warrant immobilization or surgical fixation. Attention should also be given to the prevertebral soft tissues as swelling may indicate occult injury about the spine, where the shadow should be less than 7 mm at C2 level and gradually increases to be less than 22 mm by C6.[10] The lateral view also allows visualization of Harris's ring, a ringlike structure that results from the projection of the lateral masses of C2 upon its body. A step-off in this ring may indicate a C2 base or body fracture.

The open mouth or odontoid view (Fig. 6.1C) is an A/P radiographic assessment that can be used to better evaluate for possible C1/C2 injury. The anatomic architecture of the C1/C2 articulation allows for extensive side-to-side rotational motion while providing dynamic

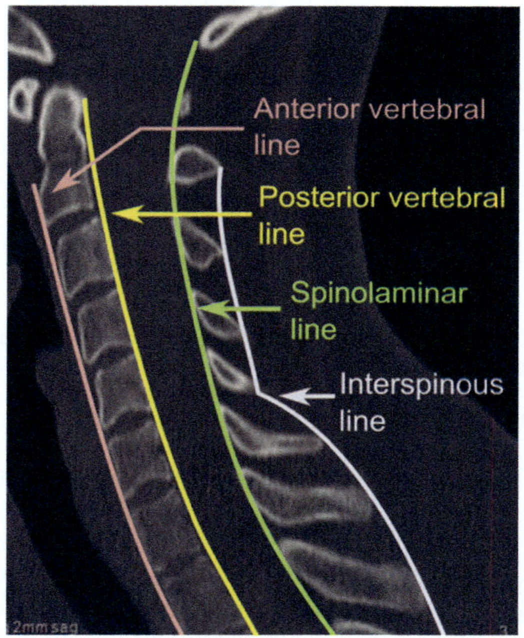

FIGURE 6.2　Sagittal CT of the cervical spine overlaid with the four cardinal lines to help assess alignment of the vertebrae. *Image by Mikael Haggstrom, M.D. via WikiCommons, distributed under a CC 1.0 Universal Public Domain Dedication.*

widening, and soft tissue irregularities. The midsagittal slices allow assessment with Harris's four cardinal lines (Fig. 6.2), as well as examination of the anterior/posterior arches of C1, dens, and vertebral bodies/spinous processes of C2−C7.[9] Furthermore, these slices allow for more fine-tuned evaluation of the BDI, interspinous spaces, and disc spaces. Other parasagittal slices allow assessment of the right/left lateral masses and facet joints. The coronal slices show images of the dens in relation to the C1 lateral masses and C2−C7 lateral masses, as well as evaluation of the ADI and vertebral body disc spaces. Finally, the axial slices show Harris's ring, C2−C7 pedicles/lamina, and facet joints. In the case of a facet dislocation, the inferior articular process of the cranial vertebra is dislocated anterior to the superior articular process of the caudal vertebra, resulting in a readily recognized radiographic abnormality that is commonly referred to as the "hamburger sign".[11]

Additionally, CT has significant superiority in its ability to assess injury in areas of poor visualization. For instance, assessment of the occipitocervical junction, atlantoaxial junction, and subaxial spine can be very difficult on standard radiographs due to the overlap of anatomic landmarks. The most sensitive tool to evaluate the occipitocervical junction is described by Harris's "Rule of Twelves." This rule states that the distance between the basion and dens or the distance between the basion and a vertical line parallel to the posterior cortex of C2 should not exceed 12 mm or occipitocervical dissociation is suspected.[14] The atlantoaxial junction can be more accurately assessed on midsagittal CT slices, with instability defined by an ADI >3 mm in adults. Increased widening of this space suggests injury to the transverse atlantal ligament and indicates instability of C1−2. Lastly, assessment of the subaxial spine (C3−C7 vertebrae) in all three CT orientations allows for detection of numerous injuries that previously were only inferred from radiography. Daffner and Harris simplified the subaxial measurement parameters

to the "Rule of 2s," where interspinous distance, interlaminar distance, interpedicular distance, and facet joint widths are considered abnormal if adjacent segments vary by greater than 2 mm.[15]

Most major trauma centers today rely on MDCT as the primary modality to assess blunt cervical spine trauma due to its superior sensitivity (up to 98%) and increased specificity (52%) to detect fractures.[14,16] However, CT does have limitations such as its relatively high cost and inability to visualize soft tissue injuries such as disc space or posterior ligamentous injury. In addition, interpretation of imaging can be difficult in patients with significant degenerative disease or loss in bone mineral density.

Magnetic resonance imaging for the injured spinal column

MRI is commonly used as an adjunct to MDCT due to its high utility and sensitivity in identifying spinal cord lesions, posterior ligamentous injury, intervertebral disc compromise, and vascular abnormalities.[17,18] In addition, it plays a role in evaluating the acuity of osteoporotic fractures, as it is able to reliably detect edema seen in the bone marrow of compressive injuries.[19] While the best practice guidelines for this modality are still called into question due to its availability, cost, and most importantly ease of use in multitrauma patients, its benefits can help guide treatment management and surgical planning.

The standard MRI in the trauma setting includes sagittal T1-weighted, sagittal T2-weighted, and short tau inversion recovery (STIR) images, in addition to axial sequences in both T1/T2-weighted images.[11] T1-weighted sagittal images are often used to delineate the anatomy of the spine and assess the quality of the bone and marrow space. T2 and STIR sequences are particularly useful in their emphasis of edema in the anatomy surrounding or within

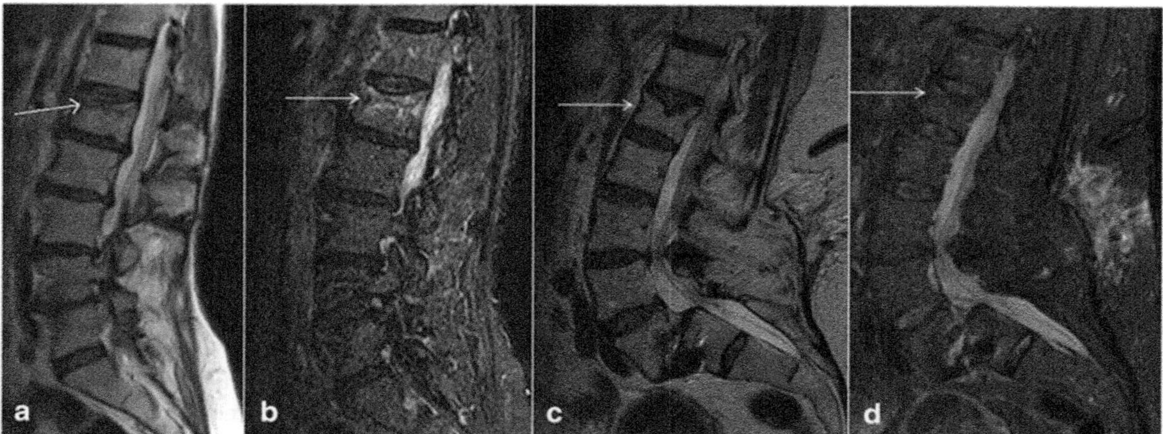

FIGURE 6.3   Sagittal (A) T2-weighted image of a patient with an age-indeterminate fracture (arrow) and (B) short tau inversion recovery (STIR) image of the same patient showing bone marrow edema at the site of the L1 compression fracture, suggestive of an acute injury. Sagittal (C) T2-weighted image of a different patient with an age-indeterminate compression fracture and (D) STIR sequence showing no evidence of bone marrow edema, indicating a chronic injury. *Copyright © Kumar, Hayashi, 2016.*

the spinal cord, discs, and epidural spaces (Fig. 6.3).[20] T2 imaging has the strongest correlation in patient prognosis with acute spinal cord injury because it can show the level of cord injury, as well as amount of hemorrhage and extension throughout the spine.[21]

Axial imaging in either T1/T2 sequences is useful for appreciating acute disc herniation, cord compression, and foraminal/central canal stenosis. Traumatic disc herniation (Fig. 6.4A and B) is usually seen in association with fracture dislocations at the level of injury. On MRI these

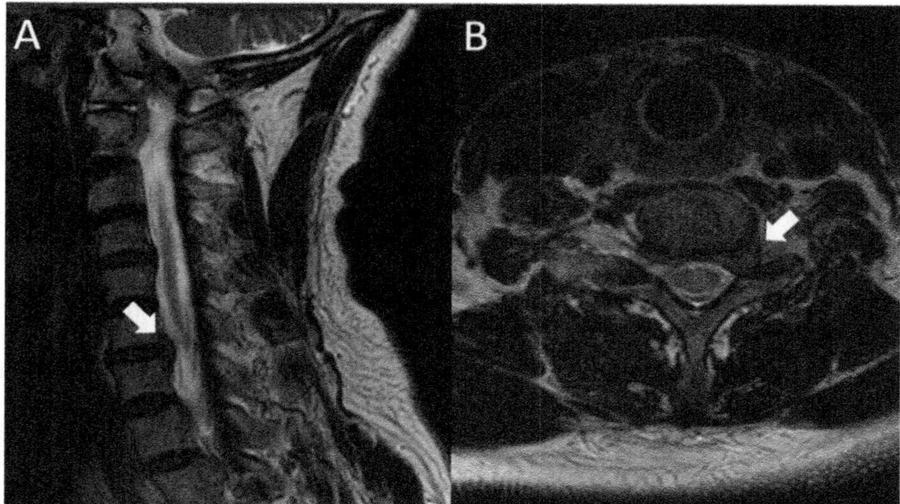

FIGURE 6.4   T2-weighted cervical MRI: (A) sagittal and (B) C6/7 axial cut showing a laterally protruding left disc herniation (arrow). MRI allows detection of disc herniation and other pathologies to a greater extent due to contrast differences of CSF, bone, and soft tissue. *Image by Shenai et al.; license under the CC 2.0 Attribution License.*

are more easily identified due to their variable contrast intensities between the disc, vertebral body, and cerebrospinal fluid.[22] Undetected herniations can cause progressive and permanent neurologic sequelae, which is why prompt evaluation with appropriate imaging is essential.

Extramedullary hemorrhage and fluid collections may also pose a problem in the traumatic setting. While these can be appreciated on CT scans, MRI can further describe the extent of the collection due to a higher level of contrast. Depending on the timing of evaluation, hemorrhage can be difficult to discern as it evolves over time. In the acute stage, hemorrhage may appear hyperintense on T2 imaging due to oxyhemoglobin in the blood. At 1–3 days, as the oxyhemoglobin deoxygenates it will darken on the T2-weighted images when compared to the spinal cord. Finally, by 3–7 days deoxyhemoglobin converts to methemoglobin, where it will continue to remain dark on T2 imaging but appear hyperintense on T1 imaging.[2,23] Extradural bleeds may be seen in 1%–2% of all traumatic injuries to the spine and are typically found in the cervical and thoracic regions and should not be overlooked without monitoring.[24]

In the posttraumatic setting with a negative cervical CT scan, controversy still exists around the utility of further evaluation using MRI. In 2017, Maung et al.[25] published a prospective multicenter study of the Research Consortium of New England Centers for Trauma (ReCONECT) evaluating the utility of cervical spine MRI in patients with negative CT scans. This study found that of 767 adult blunt trauma patients with negative CT scans who underwent MRI, 23.6% had abnormalities consistent with: ligamentous injury (16.6%), soft tissue swelling (4.3%), vertebral disc injury (1.4%), and subdural/epidural hematomas (1.3%). Interestingly, despite these findings, in awake and oriented patients, this information did not alter the treatment course to operative intervention. However,

in the unreliable or obtunded patient, MRI is recommended by many when available as this may change the course of treatment.[26] More recently in 2018, Malhotra et al.[27] published a retrospective cohort study of 1080 patients with CT and MRI in the setting of blunt trauma. Their results found that 712/1080 (65.9%) had a negative CT with 149/712 patients (20.9%) having a subsequent positive MRI; however, only 11 patients had evidence of an unstable injury. Similar to the ReCONECT study, the authors of this study found that the utility of MRI in this setting was low taking into consideration other factors like cost, accessibility, etc.

MRI is not without its limitations. In patients with metallic implants such as pacemakers, aneurysm clips, and other foreign bodies, this study is contraindicated due to dislodgment from the high magnetic fields created.

Cervical spine clearance

In the setting of acute neurologic impairment, or change/decline in neurologic status, and/or continued high clinical suspicion in the absence of radiographic findings, additional imaging in the form of CT or MRI is warranted. CT allows clear identification of bony fractures whereas MRI delineates soft tissue injury to the discoligamentous complex, compression to neurologic elements, and any injury to the posterior ligamentous complex. The appropriate modality for cervical spine for clearance has been debated for years, with some arguing that plain radiography is sufficient while others advocate for routine CT imaging with or without MRI. The Western Trauma Association Trial evaluated 10,276 blunt trauma patients with CT for the detection of cervical spine injuries. Of those evaluated, only three patients with negative CT were found to have a dangerous neurologic injury consistent with cord edema after undergoing MRI evaluation.[28] This study concluded that in an awake and alert patient with a reliable neurologic exam that CT scan alone was enough to clear patients with a suspected cervical spine injury.

In the case of obtunded, geriatric, and pediatric patients, different algorithms are suggested to appropriately assess and diagnose traumatic injury. The optimal imaging strategy for clearing an obtunded patient is still controversial. As a negative CT scan alone may be sufficient to rule out injury, others argue MRI as a necessary adjunct due to its nonzero rate of detecting clinically significant injuries.[29] Tomycz et al. conducted a retrospective review of 180 obtunded/comatose blunt trauma patients with a negative CT of the cervical spine and found that 38/180 (21.1%) had acute findings on MRI in the cervical spine (including fracture, disc herniation, cord contusion or edema, or soft tissue injuries); however, none of these patients had unstable injuries that required surgery.[30] The authors concluded that in obtunded patients, the additional use of MRI was unlikely to change treatment. In 2008, Menaker et al. conducted a retrospective study on 203 patients without obvious neurological deficits with a GCS $\leq$14 with a negative CT.[31] Their results showed that 18/203 (8.9%) patients had an abnormal MRI, 2 of which required operative repair and 14 requiring extended cervical collar use. The authors recommended that patients with unreliable physical exams due to various mental states after trauma should undergo MRI as part of their evaluation. Given the lack of consensus in literature at this time, further investigation is necessary to create a consensus protocol for obtunded patients with blunt trauma spine injuries.

Numerous studies argue that MDCT can be used alone to safely clear the spine; however, controversy still exists as to when one should progress to MRI.[14,30,32] Overall, both imaging modalities present different and essential information and when used in the right setting can prevent long-term neurologic sequelae.

Magnetic resonance imaging for spinal cord injury

While routine sequences of MRI have allowed for visualization of subtle fractures and edema, new MRI technology provides greater insight into the degenerative/regenerative processes of the spinal cord at the microscopic level.

For instance, magnetic resonance diffusion-weighted (MR-DW) and diffusion tensor imaging (MR-DTI) sequences help assess the integrity of nerve fiber tracts based on the principles of osmosis. In intact neurons, such motion is restricted to one direction (anisotropic); however, in the setting of SCI, such motion can be disrupted.[33] Therefore MR-DW or MR-DTI provide greater sensitivity in the early detection of microscopic abnormalities before any structural changes occur. MR spectroscopy functions by comparing concentrations of metabolites in specific anatomic locations in diseased or injured and nondiseased or uninjured states. Studies have shown that its utility can vary from detection of spinal cord lesions to quantifying the chemical composition of the degenerating disc.[34]

Over the past few years, several measurement parameters have been introduced to help assess the degree of spinal cord compression and spinal canal stenosis on MRI. The most commonly measured axial MRI parameters utilized are compression ratio (CR; Fig. 6.5A) and transverse area (TA; Fig. 6.5B), while those on sagittal MRI are the maximum spinal cord compression (MSCC; Fig. 6.6A) and maximum canal compromise (MCC; Fig. 6.6B). All parameters are based on T2-weighted MRI, with the exception of MCC which is measured on T1-weighted MRI. TA is measured as the cross-sectional area of the spinal cord at the site of greatest compression

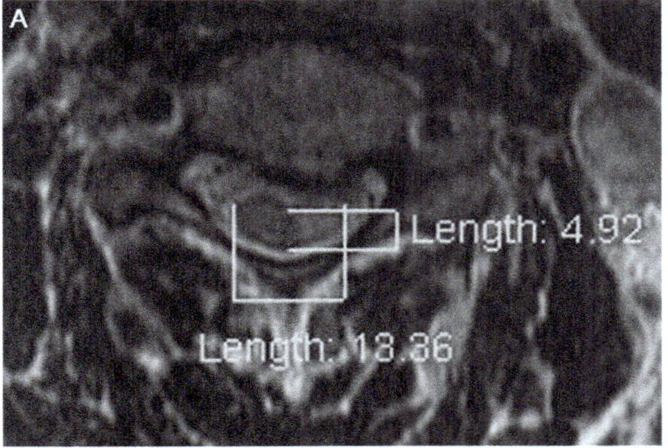

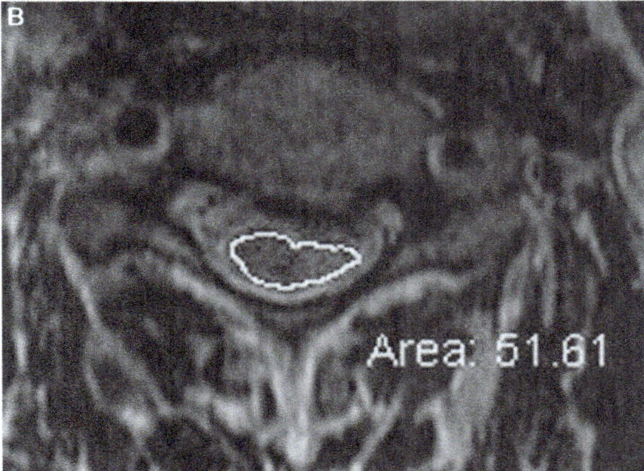

FIGURE 6.5 Axial T2-weighted cervical MRI measurements for (A) compression ratio (CR) and (B) transverse area (TA). *Image by Karpova et al.*

and CR is the ratio of sagittal diameter to transverse diameter of the spinal cord.[35,36] MSCC and MCC are percentages calculated using the following formula[37]:

$$\mathrm{MSCC} = \left(1 - \frac{d_i}{d_a + d_b/2}\right) \times 100\%$$

where d_i is the anterior–posterior spinal canal diameter at the level of MSCC, d_a is the anterior–posterior spinal canal diameter at the first normal vertebral level above, and d_b is the anterior–posterior spinal canal diameter at the first normal vertebral level below. In a prospective study assessing the inter-/intrarater reliability of these measurements, Kaporva et al. noted that TA had the highest inter- and intrarelated reliability. However, CR, MSCC, and MCC correlated to clinical severity more than TA.[38]

Additionally, MRI signal changes can give insight into the pathological processes that occur within the spinal cord after SCI. For instance, spinal cord edema may be represented by high signal intensity on T2-weighted imaging and normal signal intensity on T1-weighted imaging (Fig. 6.7).[39] Spinal cord hemorrhage within the first 24 hours appears as a heterogeneous signal (center hypointensity with peripheral hyperintensity). However, at 72 hours, the area of

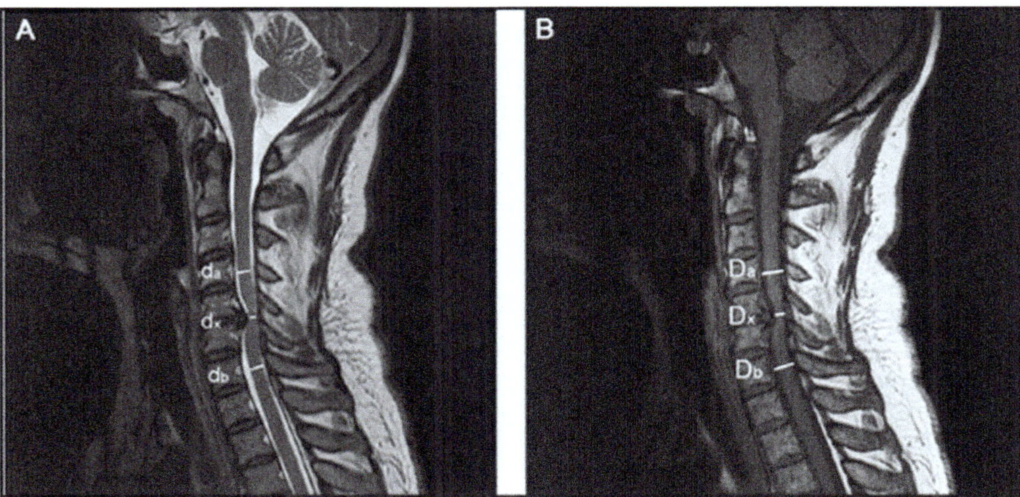

FIGURE 6.6 (A) T2-weighted sagittal cervical MRI measurements for maximum spinal cord compression (MSCC) and (B) T1-weighted sagittal cervical MRI measurements for maximum canal compromise (MCC). *Image by Karpova et al.*

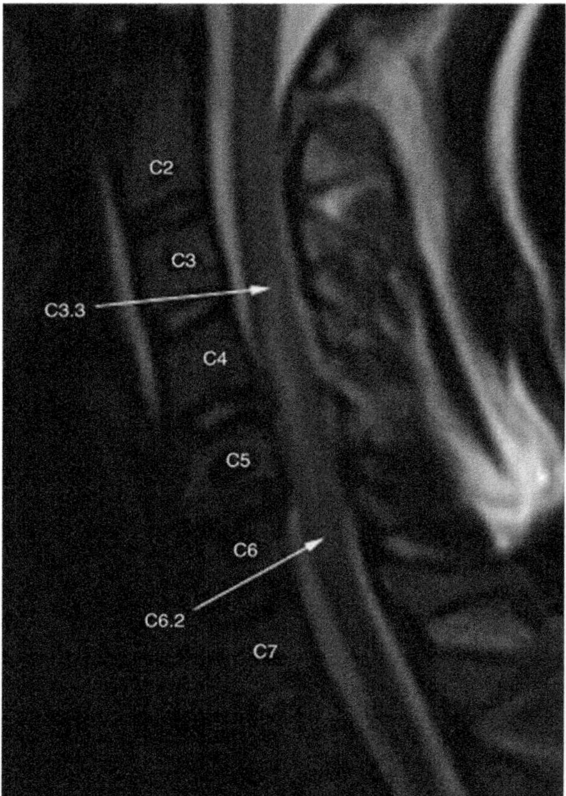

FIGURE 6.7   Sagittal T2-weighted cervical MRI showing uniform hyperintense intramedullary signal from C3 extending down to C6 (as shown by the *two arrows*) indicative of spinal cord edema. *Image by Mahmood et al.*

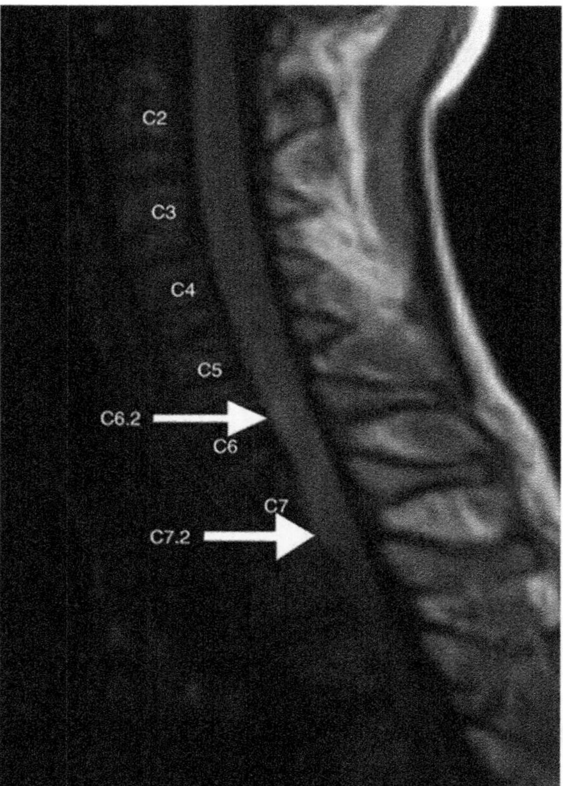

FIGURE 6.8   Sagittal T1-weighted cervical MRI performed 6 days after SCI showing patchy hyperintense spinal cord signal changes from C6 extending down to C7 (as shown by the two arrows) indicative of spinal cord hemorrhage. *Image by Mahmood et al.*

hemorrhage appears hyperintense in both T1- and T2-weighted imaging (Fig. 6.8).[40] Cord contusion appears normal on T1-weighted imaging; however, T2-weighted imaging shows central isointensity with a peripheral rim of hyperintensity (Fig. 6.9).[39] The pathological processes captured on MRI are not static moments, but dynamic changes that evolve rapidly over the acute period after SCI.

Recent literature has established the role of conventional MRI in evaluating the prognosis of spinal cord injury in the posttraumatic setting. A prospective study by Miyanji et al. of 100 patients who experienced cervical SCI (CSCI)

showed that degree of spinal cord compression and existence of spinal cord swelling/hemorrhage were predictors of poor neurologic recovery.[41] Recently, the introduction of the UCSF Brain and Spinal Injury Center (BASIC) score allows utilization of axial T2-weighted imaging to describe five different patterns of intramedullary spinal cord abnormalities. A study by Talbott et al. established a strong correlation between the BASIC score and neurologic symptoms at admission/discharge, as well as improvement in AIS grade.[42] In CSCI, Aarabi et al. established postoperative intramedullary

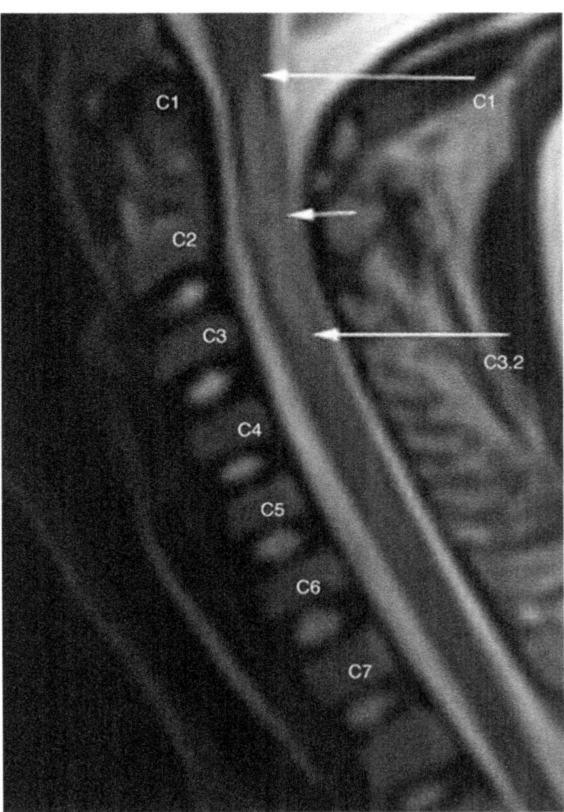

FIGURE 6.9 Sagittal T2-weighted cervical MRI performed 3 days after SCI showing a central hypointense signal change at C2 (*small white arrow*) indicative of spinal cord contusion with associated spinal cord edema from C1 extending down to C3 (as shown by the *two arrows*). *Image by Mahmood et al.*

lesion length on MRI as a strong predictor of improvement in ASIA grade following decompressive surgery.[43,44] In particular, longer intramedullary lesion lengths predicted less improvement in AIS grade postoperatively. Beyond diagnosis, advanced MR techniques continue to provide valuable information on the prognosis of recovery from spinal cord injury. A prospective study by Wang et al. showed that susceptibility-weighted (SW) MRI was more sensitive than routine MRI in detecting petechial hemorrhages in CSCI patients.[45]

Similarly, Shabani et al. conducted a prospective study of 19 CSCI and 4 thoracic SCI (TSCI) patients who underwent MR-DTI. The authors demonstrated a significant correlation in the CSCI group between diffusion anisotropy and injury severity and follow-up function.[46] While more human studies are needed to discern its utility in diagnosis/prognosis of pathology in the posttraumatic thoracolumbar spine, studies have shown MR-DTI metrics correlated with both histologic signs of axonal injury and measures of functional recovery.[47] Advanced imaging techniques can be limited due to cost, convenience, or level of artifact showcased, but they have the potential to assist in the early detection and demonstration of spinal cord regeneration in the setting of spinal cord injury.

Noteworthy cases

Spinal trauma in the geriatric population requires a high index of suspicion and low threshold for advanced imaging. Generally, elderly patients experience different injury patterns secondary to age-related bone loss, degenerative disc and joint disease, and potentially ankylosed spinal segments.[20] Not only do these changes create a more rigid spine susceptible to injury but may also mask more subtle injuries. Injury patterns typically involve more than one level with concomitant instability in the upper cervical levels.[48] They are commonly a result of lower energy trauma such as falls from standing height. Bub et al. conducted a retrospective case–control study of 103 patients above the age of 64 who were evaluated for blunt trauma.[49] While the NEXUS and CCR criteria are helpful in the evaluation of elderly patients, this study demonstrated other clinically important factors such as focal neurologic deficits, severe head injury, and moderate- to high-energy mechanisms that dictate advanced imaging evaluation. In suspected elderly spine trauma, CT imaging remains the gold standard modality. However in this population of patients, subtle degenerative changes may be further elucidated via MRI to help identify acute versus chronic changes and prevent acute exacerbation of injuries.

Imaging the pediatric population in the setting of acute traumatic SCI presents a uniquely challenging scenario as the maturing spinal cord responds differently to external forces compared to an adult spinal cord. This may be attributed to anatomic and biomechanical differences of the immature spine such as incomplete ossification, differences in vertebral configuration, and ligamentous laxity.[50] While CT is the modality of choice in the adult population, children under the age of 14 (age of full spine maturation) are recommended to undergo plain radiography to avoid excessive radiation exposure. Muchow et al. estimated a lifetime risk of thyroid cancer in the pediatric population as high as 25% in those receiving just a single CT scan.[51] Although there are studies arguing the NEXUS criteria utility in determining low-risk patients for acute SCI in pediatric patients, a second set of variables was put forth in a multicenter study by the Pediatric Emergency Care Applied Research Network (PECARN) in identifying patients who are considered high risk for traumatic SCI. The presence of altered mental status, focal neurologic findings, neck pain, torticollis, substantial torso injury, conditions predisposing to SCI, diving, and high-risk crash MVA warranted imaging.[52] Three-view X-rays are considered as an acceptable initial diagnostic evaluation with a demonstrated sensitivity of 90% in pediatric SCI.[53]

Imaging evaluation of vertebral artery injury

Over the past few years, the detection and therefore incidence of vascular injury have increased in the posttraumatic setting with the widespread use of CT angiography. Cothren et al. found that vascular injuries are generally associated with fractures involving C1–C3, particularly those that propagate through the transverse foramen or associated with subluxation injuries.[54] Blunt injury to the vertebral artery or carotid artery may present asymptomatically for extended periods of time or masked in multitrauma patients, potentially leading to significant neurologic compromise and mortality if unrecognized and untreated. Studies have shown improved outcomes with antiplatelet and anticoagulation therapy suggesting the need for evaluation when suspicion exists in the clinical setting.[55]

While dissection of the vertebral artery is more common than the carotid artery with cervical spine subluxation/fracture, studies have shown that 0.5% of all spine traumas will be associated with a vertebral artery injury and 70% of those will have associated cervical spine fracture.[20,56] Historically, angiography was the gold standard test for vascular injury due to its ability to accurately identify arterial flow. However, angiography is not without its limitations such as invasiveness in critically ill patients. This test itself carries a 0.5% risk of stroke and high morbidity due to complications including: bleeding at the catheter entry site, retroperitoneal bleeding, infection, radiation exposure, contrast-induced nephropathy, and pseudoaneurysm formation.[57,58] Magnetic resonance angiography (MRA) is another alternative for evaluating arterial injury. MRA's major drawbacks lie in the timing of test administration again in critically ill patients and contraindication in patients with metallic devices such as pacemakers, aneurysm clips, and other foreign bodies.

Although several studies debate over a wide range of sensitivity and specificity for CTA, advances in technology have demonstrated its accuracy comparable to plain angiography and ability to outperform MRA.[57] CTA use has reduced the time to diagnosis of vascular injury nearly 12-fold and decreased the rate of stroke four-fold in these instances.[59] Despite continued debate over CTA's diagnostic accuracy, many institutions have adopted this imaging modality for the evaluation of blunt spinal trauma with suspected vascular compromise. In the setting of blunt cervical trauma, the reported incidence of vertebral artery injury varies from 0.5% to 2.0%.[57]

Summary

While no true agreed upon algorithm exists to assess spine trauma patients, the evolution of MDCT and MRI has come a long way to aid their diagnosis. MDCT scans have become a quick and accurate way to reconstruct bony injury and have nearly replaced plain radiographic evaluation. MRI plays a key role in conjunction with MDCT to appreciate subtle soft tissue injuries or locate the etiology of neurologic compromise. While controversy continues over the most appropriate imaging modality in the posttraumatic setting, the evolution of imaging and its ability to aid management decisions continues to evolve to this day.

References

1. Guarnieri G, Izzo R, Muto M. The role of emergency radiology in spinal trauma. *Br J Radiol* 2016;**89**:1061. British Institute of Radiology.
2. Browner BD. *Skeletal trauma: basic science, management, and reconstruction*. Saunders/Elsevier; 2008.
3. Hoffman JR, Mower WR, Wolfson AB, Todd KH, Zucker MI. Validity of a set of clinical criteria to rule out injury to the cervical spine in patients with blunt trauma. *N Engl J Med* July 2000;**343**(2):94—9.
4. Stiell IG, et al. The Canadian C-spine rule for radiography in alert and stable trauma patients. *J Am Med Assoc* October 2001;**286**(15):1841—8.
5. Stiell IG, et al. The Canadian C-spine rule versus the NEXUS low-risk criteria in patients with trauma. *N Engl J Med* December 2003;**349**(26):2510—8.
6. Durham RM, Luchtefeld WB, Wibbenmeyer L, Maxwell P, Shapiro MJ, Mazuski JE. Evaluation of the thoracic and lumbar spine after blunt trauma. *Am J Surg* December 1995;**170**(6):681—4. Discussion 684-5.
7. Christenson PC. The radiologic study of the normal spine: cervical, thoracic, lumbar, and sacral. *Radiol Clin* August 1977;**15**(2):133—54.
8. Pope TL, Harris JH, Delgadillo D. *Harris & harris radiología de medicina en emergencias/Editado por Thomas L. Pope, John H. Harris*. tradución David Delgadillo; 2014.
9. Harris JH, Carson GC, Wagner LK. Radiologic diagnosis of traumatic occipitovertebral dissociation: 1. Normal occipitovertebral relationships on lateral radiographs of supine subjects. *Am J Roentgenol* 1994;**162**(4):881—6.
10. Shen FH, Samartzis D, Fessler RG. *Textbook of the cervical spine*.
11. Canale TS, Beaty JH. Cervical discography. In: *Campbell's operative orthopaedics*. Elsevier Health Sciences; 2012. p. 4166.
12. Padayachee L, et al. Cervical spine clearance in unconscious traumatic brain injury patients: dynamic flexion-extension fluoroscopy versus computed tomography with three-dimensional reconstruction. *J Trauma - Inj Infect Crit Care* February 2006;**60**(2):341—5.
13. Tins BJ. Imaging investigations in spine trauma: the value of commonly used imaging modalities and emerging imaging modalities. *J Clin Orthop & Trauma* April 1, 2017;**8**(2):107—15. Elsevier B.V.
14. Harris TJ, Blackmore CC, Mirza SK, Jurkovich GJ. Clearing the cervical spine in obtunded patients. *Spine* June 2008;**33**(14):1547—53.
15. Lee N, Wong B. Clinics in diagnostic imaging (192). *Singap Med J* November 2018:562—6.
16. Holmes JF, Akkinepalli R. Computed tomography versus plain radiography to screen for cervical spine injury: a meta-analysis. *J Trauma* May 2005;**58**(5):902—5.
17. Parizel PM, et al. Trauma of the spine and spinal cord: imaging strategies. *Eur Spine J* 2010;**19**(Suppl. 1). Springer Verlag.
18. Goradia D, Linnau KF, Cohen WA, Mirza S, Hallam DK, Blackmore CC. Correlation of MR imaging findings with intraoperative findings after cervical spine trauma. *Am J Neuroradiol* February 2007;**28**(2):209—15.
19. Brinckman MA, Chau C, Ross JS. Marrow edema variability in acute spine fractures. *Spine J* March 2015;**15**(3):454—60.
20. Shah LM, Ross JS. Imaging of spine trauma. *Neurosurgery* November 2016;**79**(5):626—42.
21. Shimada K, Tokioka T. Sequential MR studies of cervical cord injury: correlation with neurological damage and clinical outcome. *Spinal Cord* 1999;**37**(6):410—5.
22. Kumar Y, Hayashi D. Role of magnetic resonance imaging in acute spinal trauma: a pictorial review. *BMC Muscoskel Disord* July 22, 2016;**17**(1). BioMed Central Ltd.
23. Bradley Jr WG. MR appearance of hemorrhage in the brain. *Radiology* 1993;**189**(1):15—26.

24. Garza-Mercado R. Traumatic extradural hematoma of the cervical spine. *Neurosurgery* 1989;**24**(3):410–4.

25. Maung AA, et al. Cervical spine MRI in patients with negative CT: a prospective, multicenter study of the Research Consortium of New England Centers for Trauma (ReCONECT). *J Trauma & Acute Care Surg* 2017;**82**(2):263–9.

26. Stassen NA, et al. Magnetic resonance imaging in combination with helical computed tomography provides a safe and efficient method of cervical spine clearance in the obtunded trauma patient. *J Trauma Inj Infect Crit Care* January 2006;**60**(1):171–7.

27. Malhotra A, et al. Utility of MRI for cervical spine clearance in blunt trauma patients after a negative CT. *Eur Radiol* July 2018;**28**(7):2823–9.

28. Coimbra R, et al. Cervical spinal clearance. *J. Trauma Acute Care Surg.* 2016;**81**(6):1122–30.

29. Minja FJ, Mehta KY, Mian AY. Current challenges in the use of computed tomography and MR imaging in suspected cervical spine trauma. *Neuroimaging Clin* August 01, 2018;**28**(3):483–93. W.B. Saunders.

30. Tomycz ND, et al. MRI is unnecessary to clear the cervical spine in obtunded/comatose trauma patients: the four-year experience of a level I trauma center. *J Trauma - Inj Infect Crit Care* May 2008;**64**(5):1258–63.

31. *Computed tomography alone for cervical spine clearance in the unreliable patient–are we there yet?* - PubMed - NCBI. [Online]. Available: https://www.ncbi.nlm.nih.gov/pubmed/18404054. [Accessed 18 November 2019].

32. Raniga SB, Menon V, Al Muzahmi KS, Butt S. MDCT of acute subaxial cervical spine trauma: a mechanism-based approach. *Insights Imag* 2014.

33. Nouh MR. Imaging of the spine: where do we stand? *World J Radiol* April 2019;**11**(4):55–61.

34. Vargas MI, et al. Advanced magnetic resonance imaging (MRI) techniques of the spine and spinal cord in children and adults. *Insights Imag* August 01, 2018;**9**(4):549–57. Springer Verlag.

35. Okada Y, Ikata T, Yamada H, Sakamoto R, Katoh S. Magnetic resonance imaging study on the results of surgery for cervical compression myelopathy. *Spine* 1993;**18**(14):2024–9.

36. Chen CJ, Lyu RK, Lee ST, Wong YC, Wang LJ. Intramedullary high signal intensity on T2-weighted MR images in cervical spondylotic myelopathy: prediction of prognosis with type of intensity. *Radiology* 2001;**221**(3):789–94.

37. Furlan JC, et al. A quantitative and reproducible method to assess cord compression and canal stenosis after cervical spine trauma: a study of interrater and intrarater reliability. *Spine* Septemeber 2007;**32**(19):2083–91.

38. Karpova A, et al. Reliability of quantitative magnetic resonance imaging methods in the assessment of spinal canal stenosis and cord compression in cervical myelopathy. *Spine* February 2013;**38**(3):245–52.

39. Ramón S, et al. Clinical and magnetic resonance imaging correlation in acute spinal cord injury. *Spinal Cord* 1997;**35**(10):664–73.

40. Mahmood NS, Kadavigere R, Ramesh AK, Rao VR. Magnetic resonance imaging in acute cervical spinal cord injury: a correlative study on spinal cord changes and 1 month motor recovery. *Spinal Cord* December 2008;**46**(12):791–7.

41. Miyanji F, Furlan JC, Aarabi B, Arnold PM, Fehlings MG. Acute cervical traumatic spinal cord injury: MR imaging findings correlated with neurologic outcome - prospective study with 100 consecutive patients. *Radiology* June 2007;**243**(3):820–7.

42. Talbott JF, et al. The Brain and Spinal Injury Center score: a novel, simple, and reproducible method for assessing the severity of acute cervical spinal cord injury with axial T2-weighted MRI findings. *J Neurosurg Spine* October 2015;**23**(4):495–504.

43. Aarabi B, et al. Intramedullary lesion length on postoperative magnetic resonance imaging is a strong predictor of ASIA impairment scale grade conversion following decompressive surgery in cervical spinal cord injury. *Clin Neurosurg* April 2017;**80**(4):610–20.

44. Aarabi B, et al. Efficacy of ultra-early (< 12 h), early (12-24 h), and late (>24-138.5 h) surgery with magnetic resonance imaging-confirmed decompression in American Spinal Injury Association impairment scale grades A, B, and C cervical spinal cord injury. *J Neurotrauma* August 2019;**37**(3):448–57.

45. Wang M, Dai Y, Han Y, Haacke EM, Dai J, Shi D. Susceptibility weighted imaging in detecting hemorrhage in acute cervical spinal cord injury. *Magn Reson Imaging* April 2011;**29**(3):365–73.

46. Shabani S, Kaushal M, Budde M, Kurpad SN. Correlation of magnetic resonance diffusion tensor imaging parameters with American Spinal Injury Association score for prognostication and long-term outcomes. *Neurosurg Focus* March 2019;**46**(3).

47. Zaninovich OA, Avila MJ, Kay M, Becker JL, Hurlbert RJ, Martirosyan NL. The role of diffusion tensor imaging in the diagnosis, prognosis, and assessment of recovery and treatment of spinal cord injury: a systematic review. *Neurosurg Focus* March 2019;**46**(3).

48. Lomoschitz FM, Blackmore CC, Mirza SK, Mann FA. Cervical spine injuries in patients 65 years old and older: epidemiologic analysis regarding the effects of age and injury mechanism on distribution, type, and stability of injuries. *Am J Roentgenol* 2002;**178**(3):573–7.

49. Bub LD, Blackmore CC, Mann FA, Lomoschitz FM. Cervical spine fractures in patients 65 years and older: a clinical prediction rule for blunt trauma. *Radiology* January 2005;**234**(1):143—9.

50. Platzer P, et al. Cervical spine injuries in pediatric patients. *J Trauma - Inj Infect Crit Care* February 2007; **62**(2):389—94.

51. Muchow RD, Egan KR, Peppler WW, Anderson PA. Theoretical increase of thyroid cancer induction from cervical spine multidetector computed tomography in pediatric trauma patients. *J. Trauma Acute Care Surg.* February 2012;**72**(2):403—9.

52. Leonard JC, et al. Factors associated with cervical spine injury in children after blunt trauma. *Ann Emerg Med* 2011;**58**(2):145—55.

53. Nigrovic LE, et al. Utility of plain radiographs in detecting traumatic injuries of the cervical spine in children. *Pediatr Emerg Care* 2012;**28**(5):426—32.

54. Cothren CC, et al. Cervical spine fracture patterns predictive of blunt vertebral artery injury. *J Trauma* 2003; **55**(5):811—3.

55. *Blunt cerebrovascular injuries:dde does treatment always matter? - PubMed - NCBI.* [Online]. Available: https://www.ncbi.nlm.nih.gov/pubmed/19131816. [Accessed: 18 November 2019].

56. Fassett DR, Dailey AT, Vaccaro AR. Vertebral artery injuries associated with cervical spine injuries: a review of the literature. *J Spinal Disord Tech* June 2008; **21**(4):252—8.

57. De Souza RM, Crocker MJ, Haliasos N, Rennie A, Saxena A. Blunt traumatic vertebral artery injury: a clinical review. *Eur Spine J* September 2011;**20**(9):1405—16.

58. Goodwin RB, et al. Computed tomographic angiography versus conventional angiography for the diagnosis of blunt cerebrovascular injury in trauma patients. *J Trauma - Inj Infect Crit Care* 2009;**67**(5):1046—50.

59. Eastman AL, Muraliraj V, Sperry JL, Minei JP. CTA-based screening reduces time to diagnosis and stroke rate in blunt cervical vascular injury. *J Trauma - Inj Infect Crit Care* September 2009;**67**(3):551—5.

Advanced imaging for spinal cord injury

Muhammad Ali Akbar[1,3], Allan R. Martin[4], Dario Pfyffer[2], David W. Cadotte[5,6], Shekar Kurpad[7], Patrick Freund[2], Michael G. Fehlings[1,3]

[1]Division of Neurosurgery, Department of Surgery, University of Toronto, Toronto, ON, Canada; [2]Spinal Cord Injury Center, Balgrist University Hospital, University of Zurich, Zurich, Switzerland; [3]Spine Program, Krembil Neuroscience Centre, Toronto Western Hospital, University Health Network, Toronto, ON, Canada; [4]Department of Neurological Surgery, University of California, Davis, CA, United States; [5]Division of Neurosurgery, Department of Clinical Neurosciences, University of Calgary, Calgary, AB, Canada; [6]Combined Spine Program, Hotchkiss Brain Institute, University of Calgary, Calgary, AB, Canada; [7]Department of Neurosurgery, Medical College of Wisconsin, Milwaukee, WI, United States

Introduction

Traumatic and nontraumatic spinal cord injury

Spinal cord injury (SCI) can be classified in two ways; a traumatic spinal cord injury (tSCI) occurs due to an acute mechanical (compressive, shear, or distractive) force impacting the spinal cord. This sudden traumatic injury leads to temporary or permanent neurological compromise below the level of injury. Most patients with tSCI will present with permanent motor, sensory, and possibly autonomic dysfunction.[1] Nontraumatic SCI, on the other hand, is of a chronic nature and characterized as a progressive compression of the spinal cord, most commonly due to degenerative changes to intervertebral discs in the spinal column. In the cervical region, degenerative cervical myelopathy (DCM) is an all-encompassing term for these degenerative processes including cervical spondylotic myelopathy (CSM), ossified posterior longitudinal ligament (OPLL), and degenerative disc disease (DDD), among others. DCM as a collective is considered the most common cause of spinal cord dysfunction in adults.[2]

The primary mechanism of injury in DCM and tSCI may differ as mentioned earlier; however, both share several similarities in the mechanisms of secondary injury.[2,3] Animal studies have shown common pathophysiological changes after initial injury in tSCI and DCM, including a cascade of programmed cell death, extracellular as well as macro- and microvascular remodeling. These cascade events lead to demyelination, axonal degeneration, and atrophy.[1–4] There has been growing evidence of both anterograde and

Neural Repair and Regeneration after Spinal Cord Injury and Spine Trauma
https://doi.org/10.1016/B978-0-12-819835-3.00028-9

© 2022 Elsevier Inc. All rights reserved.

retrograde changes in the macro/microstructure of the neural axis corresponding with observations caudal and rostral to the site of injury.[3,5] Imaging of the spinal cord in DCM is considered more pragmatic and easier as compared with tSCI, where patients in the latter group often undergo surgery within hours of injury limiting the opportunity for detailed imaging before metallic instrumentation is inserted. This lends DCM as a more practical population to study new and investigative imaging methods.

Rationale for advanced imaging

Diagnostic imaging has evolved dramatically over the years to now include an array of modalities such as plain X-ray, computed tomography (CT) and magnetic resonance imaging (MRI) at the disposal of clinicians. The role of traditional imaging such as X-ray, CT, and anatomical MRI in spinal cord injury has been highlighted in Chapter 6. The impact of advanced CT techniques, such as multidetector CT, has been limited in SCI research. This is likely because CT techniques cannot visualize neural and soft tissues to the same extent as MRI. MRI is by far one of the greatest achievements in the field of radiology and neuroimaging. Unlike any other imaging modality, MRI allows for parameter selection in a multitude of different combinations to create tissue contrast and visualize not only anatomical structures but also physiological processes such as neuronal activity, perfusion, and metabolism in real time. Typical MRI protocols consist of sequences that allow for structural and anatomical imaging. Conventional MRI plays a very important role in the management of tSCI (see Chapter 6) and DCM. It can help identify compression or impact to the spinal cord and injury to the discoligamentous complex. Macrostructural changes within the spinal cord can also be seen on the basis of T2-weighted (T2W) hyperintensity or T1-weighted (T1W) hypointensity. T2W signal change in the spinal cord can be due to a variety of reversible and nonreversible causes such as edema, gliosis,

demyelination, myelomalacia, or hemorrhage, highlighting its lack of specificity and correlation with neurological impairment or recovery.[6] T2W signal hyperintensity in conjunction with T1W signal hypointensity improves the correlation with neurological impairment and outcomes, albeit the cooccurrence of both together is rare leading to a lack of overall sensitivity.[7] These macrostructural findings can help in initial decision-making and surgical planning in tSCI, but provide little to no information on microstructural changes at the site of injury or remote to the injury, and has little value in monitoring rehabilitation and recovery. In cases of DCM, the presence of signal intensity changes in addition to spinal cord compression can aid in the diagnosis of the condition, but the degree of compression does not always correlate with severity of disease and cannot reliably predict which patients will deteriorate or develop further symptoms in the future (Fig.7.1).[8]

There is a great need for improved imaging and electrophysiological biomarkers in traumatic and nontraumatic SCI that can correlate with neurological assessments. Ideally, these biomarkers would be quantifiable and reflect microstructural properties of the spinal cord including myelination/demyelination, iron content, plaque deposition, and axonal integrity. Having a reliable way of quantifying microstructural changes in the spinal cord could help clinicians understand the neurodegenerative changes in the spinal cord after injury as well as the evolution over time. These markers could identify patients most likely to require early surgery in DCM or most likely to recover after a tSCI. Past attempts at defining such biomarkers using conventional imaging modalities have shown modest results at best. The imaging modality that has shown the most promise in this regard has been quantitative MRI (qMRI). qMRI refers to a collection of sophisticated techniques that provide quantifiable and normative values having direct or indirect correlation with intrinsic tissue properties.

To date, most MRI research and development has been conducted in the brain instead of the

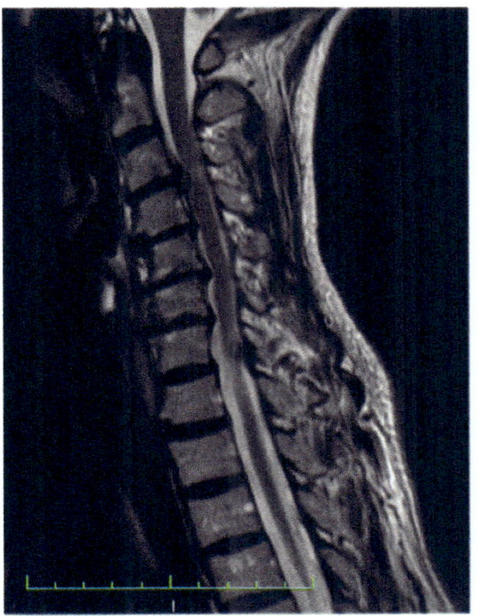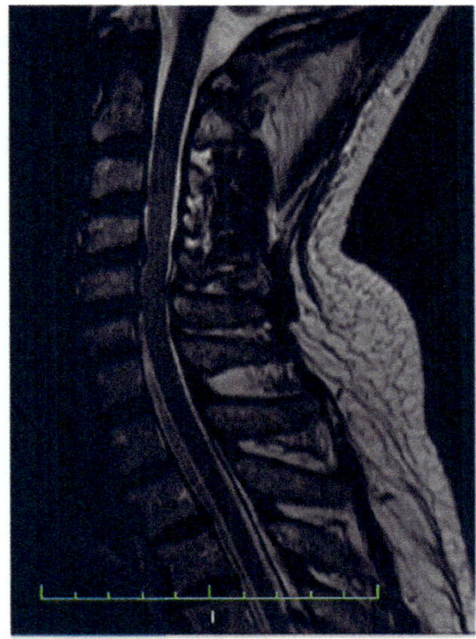

FIGURE 7.1 Representative images of two patients with DCM highlighting the discrepancies between imaging and clinical severity. (A) Patient on the left is seen to have multilevel disease, significant compression of the spinal cord at C3—4 with T2-weighted signal hyperintensity at this level. This patient has a mJOA of 16 and categorized as mildly symptomatic. (B) Patient on the right has single- to two-level disease with cord compression at C4—5 without any signal change. On general observation, the findings would be considered "less" severe compared with patient (A); however, patient (B) is severely symptomatic with a mJOA of 11. *DCM*, degenerative cervical myelopathy; *mJOA*, modified Japanese Orthopedic Association score.

spine. One of the primary reasons is that the human spinal cord is a very difficult structure to image. This difficulty arises due to the small cross-sectional area of the spinal cord as well as its close proximity to cerebrospinal fluid (CSF), bone, and soft tissues. The small area of the cord requires increased resolution and higher strength MRI magnets to reliably image the anatomy. Secondly, the anatomical structures such as bone and soft tissues clustered together in a small space cause disruptions in the magnetic field homogeneity, resulting in susceptibility artifact in many acquisition sequences. Lastly, motion artifacts can also be introduced secondary to physiological movement of the spinal cord during respiratory and cardiac cycles.[9] With increasing availability of high field strength clinical MRIs and improved technology, employing advanced techniques in the human spinal cord has become much easier and more practical. This has led to rapid growth in the field of SCI imaging research over the past decade.

In this chapter, we review advanced postprocessing techniques that can be used in conventional MRI to go beyond just qualitative assessments of pathology and provide robust markers of injury. We then provide an overview of advanced qMRI techniques in the spinal cord, their respective significance, and potential clinical applications in SCI. We conclude with a summary and future directions for the field of spinal cord imaging.

Conventional magnetic resonance imaging

Conventional MRI sequences—including axial[10] and sagittal[11] T2W MR sequences—are capable of detecting macrostructural changes at the lesion epicenter in the very acute setting (<48 h) and provide qualitative information about the spinal cord and injury. Combined with quantitative measures of lesion length and spinal cord compression, these MRI features have been used in the past to distinguish sensorimotor complete from incomplete cervical SCI patients, the latter with lower frequencies of edema and hemorrhage as well as shorter lesions and less cord compression.[12] The true extent of tissue damage, however, can be better quantified 3–4 weeks post-SCI, once the edema and cord swelling has resolved.[13]

Advanced image postprocessing and analysis has allowed for refinement of quantifiable measures to assess the extent of preserved tissue bridges adjacent to the lesion or to perform detailed slice-by-slice morphometric analysis of the spinal cord to provide precise insights on cord injury and atrophy.

Tissue bridges

One way to assess the true extent of damage is through measurement of preserved tissue bridges adjacent to the lesion. Defined as the hypointense intramedullary regions between the relatively hyperintense cystic cavity on one side and the CSF on the other side assessed on midsagittal T2W images, spared ventral and dorsal tissue bridges can be measured separately with regard to the neural tissue being preserved either at the anterior or posterior site of the lesion cavity, respectively (Fig. 7.2).[14,15] An accurate quantitative assessment of preserved midsagittal tissue bridges can be reliably performed with a high inter- and intrarater reliability.[14–16]

The location of spared midsagittal tissue bridges can represent an intrinsic property of the tract-specific motor and sensory recovery potential of SCI patients. Being permissive for electrophysiological information flow, larger spared ventral tissue bridges at 1 month were shown to be associated with shorter MEP latencies and greater MEP amplitude ratios at 12-month post-SCI.[17] Dorsal midsagittal tissue bridges at 1 month, on the other hand, were shown to be larger in patients with shorter SEP latencies and greater SEP amplitudes at 12-month follow-up.[17]

Shape analysis and atrophy

Measurement of the spinal cord cross-sectional area (CSA) with high-resolution axial images has already been used to estimate the degree of atrophy in MS patients.[18] This has been difficult to replicate in tSCI since most patients undergo surgical instrumentation and metal artifacts in imaging prevent accurate and reliable quantification of CSA at the level of injury. However, similar analyses can be done above or below the level of injury, in artifact-free areas of the spinal cord. Multiple prospective longitudinal studies of tSCI patients have shown progressive rostral atrophy corresponding to a 14% decrease in CSA at C2–3 within 2 years of injury[19–21] and a decrease of up to 30% at 5 years after injury.[22,23]

In mild DCM patients who have not undergone surgery, CSA at the level of maximal stenosis showed worsening atrophy at 1 year follow-up.[24] Patients with DCM also exhibit rostral atrophy above the level of compression.[25,26] Gradient echo (GRE) MRI sequences can also be used to segment white and gray matters effectively.[27] In such analysis, DCM patients show atrophy in both white (7.2%) and gray (13.9%) matter when compared with healthy controls providing another sensitive marker of degeneration.[25]

Clinical significance

Conventional MRI has the power to noninvasively and directly detect the location and extent of spinal cord damage and spared neural tissue.[14,15,17] The width of midsagittal tissue bridges at 1 month is predictive of long-term neurological and functional recovery.[14,15,28] Larger baseline ventral and dorsal tissue bridges have been associated with higher American Spinal Injury Association (ASIA) Impairment Scale (AIS) grades and AIS conversion rates at 12 months after the injury.[28] The prognostic value of midsagittal tissue bridges for tract-

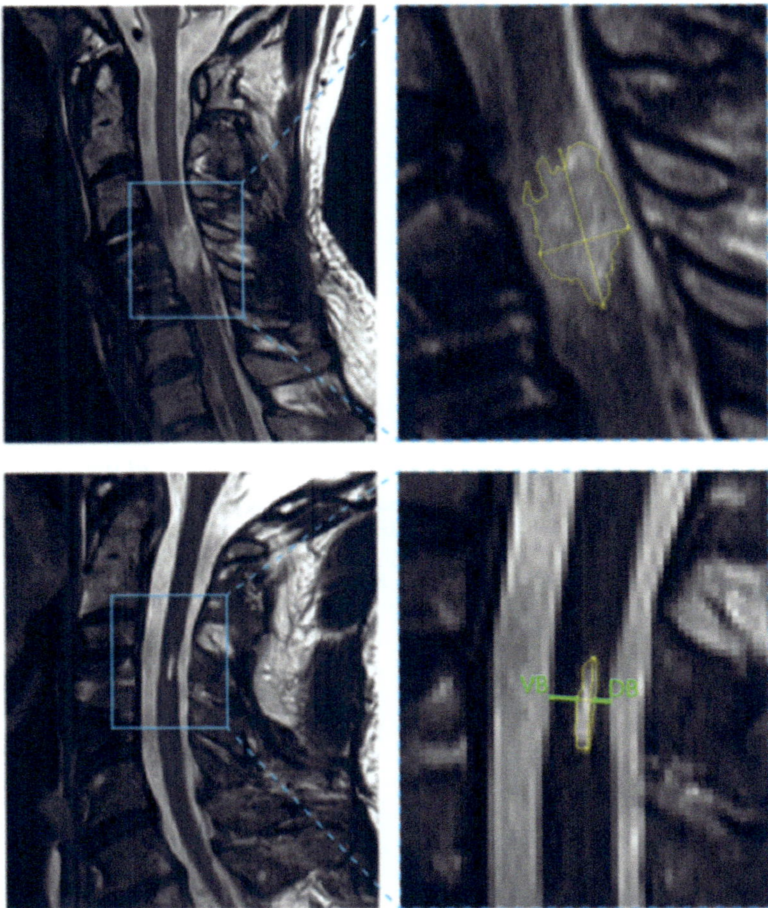

FIGURE 7.2 Representative T2-weighted midsagittal MRI scans of two traumatic SCI patients. The patient of the upper panel presents with no spared midsagittal tissue bridges as seen on the zoomed-in image on the top right with the cystic cavity outlined in yellow and spanning the whole anteroposterior cord. The patient of the lower panel presents with spared ventral (VB) and dorsal (DB) tissue bridges indicated in green and adjacent to the cystic cavity outlined in yellow, as shown on the zoomed-in image on the bottom right image. *MRI*, magnetic resonance imaging; *SCI*, spinal cord injury.

specific recovery has further been explored using conditional inference tree analysis for prediction-based stratification of heterogeneous SCI patient populations into more homogeneous subgroups. In this study by Pfyffer et al.,[28] conditional inference tree analysis identified 1-month ventral tissue bridges as a predictor of 12-month motor scores and pin-prick score, and dorsal tissue bridges at 1 month as a predictor of 12-month

SCIM scores. With regard to functional independence, early midsagittal tissue bridges have also been shown to be predictive of outdoor walking ability 1 year after SCI, yet being outperformed by adding baseline motor scores.[29] In another study investigating the relationship between injury-spared tissue bridges and the presence of neuropathic pain, larger midsagittal T2W-assessed ventral tissue bridges at 1-month

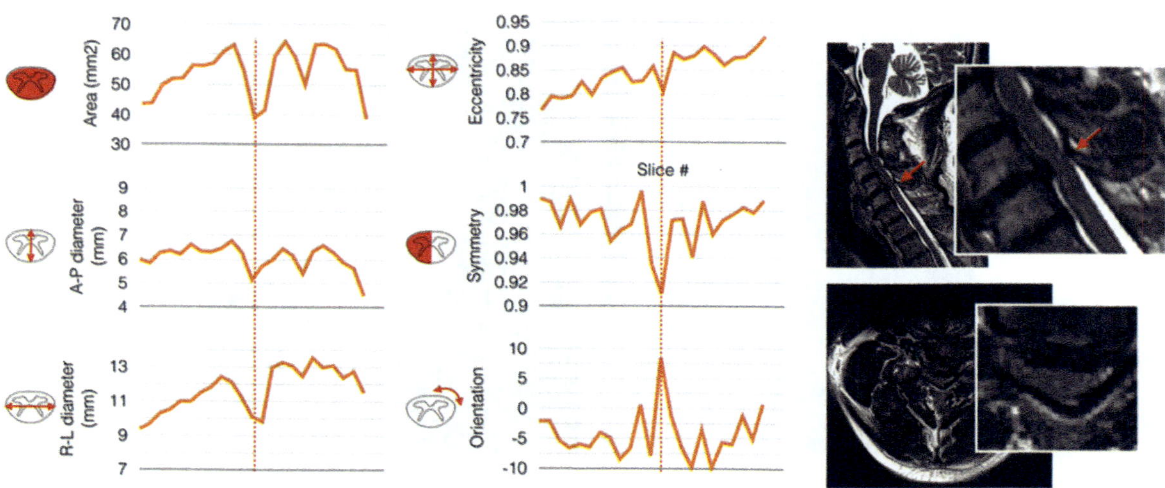

FIGURE 7.3   Morphometric/shape analysis of the spinal cord in a patient with degenerative cervical myelopathy. Morphological metrics across the rostrocaudal extent of the cord are graphed individually. The red dotted line represents the most compressed level corresponding to the red arrow on the T2-weighted sagittal image on the right. This technique can indicate location and extent of compression. A-P, anterior—posterior; R-L, right—left. *Reproduced with permission from Cadotte et.al. What has been learned from magnetic resonance imaging examination of the injured human spinal cord: a Canadian perspective.* Journal of Neurotrauma.[34]

positively correlated with higher mean pain intensities and higher pin-prick scores at 12-month after SCI, independent of dorsal tissue bridges and baseline clinical scores.[30] Moreover, conditional inference tree analysis identified 1-month ventral tissue bridges as a predictor of 12-month mean pain intensity and pin-prick score, splitting up patients into subgroups with a cut-off value of 2.1 mm in ventral tissue bridges' width.[30] Given that ventral tissue bridges cover parts of the anterior spinothalamic tract and this tract has been linked to the experience of neuropathic pain,[31,32] ventral tissue bridges are likely to represent an MRI-derived surrogate marker underlying the emergence of post-SCI NP and hold the potential to be used for pain outcome prediction and patient stratification in interventional clinical trials.[30]

When looking at atrophy in patients with cervical tSCI, dorsal and ventral horn CSA shows positive correlation with sensory and motor clinical outcomes, respectively.[23] In DCM, cross-sectional area at the most compressed level showed a high degree of correlation with neurological impairment; however, in this case, the change in CSA reflects the direct compression on the spinal cord rather than long-term atrophy.[33] CSA in conjunction with other morphometric measurements such as anteroposterior (A-P) diameter, right-left diameter, orientation (PCA approximation of largest and smallest cord diameter), eccentricity, and symmetry can be used to conduct automated shape analysis. This process identifies abnormalities compared with normal spinal cord morphology. It not only detects spinal cord compression with high diagnostic accuracy, but also identifies dynamic compression of the spinal cord.[26] This methodology could potentially allow for presymptomatic diagnosis of myelopathy and serve as a stepping stone for further automated or machine learning analyses, which could use this technique in autonomous identification of the compressed cord levels (Fig. 7.3).

Quantitative magnetic resonance imaging

This section outlines a collection of qMRI techniques that have been used in the spinal cord and summarizes the significant findings from studies in tSCI and DCM patients as well as occasional animal studies, when relevant.

Diffusion tensor imaging

Basic concepts

Conventional MRI primarily exploits the protons in water molecules to generate a nuclear magnetic resonance signal. The different relaxation times of these protons in response to radio frequency pulses help create varied contrast in the image. In biological tissues, especially fibrous structures such as muscles and myelinated axons, water diffusion is anisotropic and nonrandom, a property that can be further exploited. Diffusion-weighted imaging (DWI) and diffusion tensor imaging (DTI) are MRI techniques that are sensitive to the movement of water and utilize this movement to create contrast for imaging.[35] DWI can provide the magnitude of water diffusion averaged over all directions, whereas DTI can detect both magnitude and direction of water diffusion, characterized as a diffusion tensor. A tensor can be visualized as an ellipsoid with the eigenvalues (λ_1, λ_2, λ_3) representing directions and magnitude of diffusion within each voxel of an image (Fig. 7.4). The eigenvalues are used to calculate various indices including axial diffusivity (AD $= \lambda_1$), radial diffusivity (RD $= \frac{\lambda_2 + \lambda_3}{2}$), mean diffusivity (MD $= \frac{\lambda_1 + \lambda_2 + \lambda_3}{3}$), and finally fractional anisotropy (FA), which is a scalar measurement from 0 to 1 representing the overall degree of anisotropy in water diffusion.

DWI and DTI show sensitivity and specificity to the axon and myelin pathology.[36] Animal and clinical research over the past 20 years have validated and correlated DTI metrics with the histopathology seen in SCI (Fig. 7.5), for example,

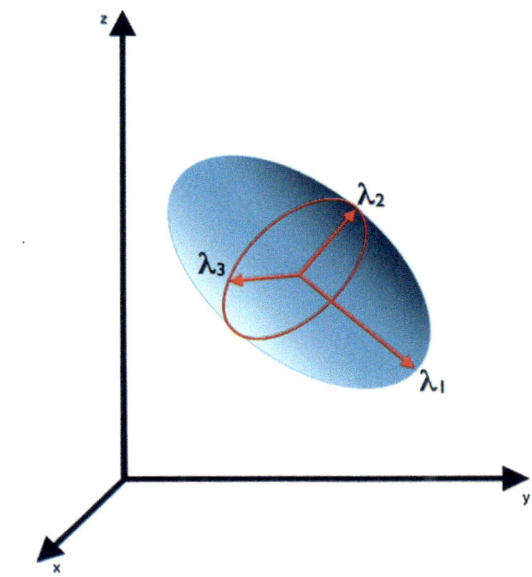

FIGURE 7.4 Visual representation of an anisotropic diffusion tensor. The lengths of the axis within the ellipsoid represent eigenvalues (λ_1, λ_2, λ_3). The primary eigenvalue λ_1 corresponds to the predominant direction of diffusion.

axial diffusivity and its correlation with axonal degeneration and radial diffusivity as a marker for demyelination. Comparison of various DTI metrics in a specific white matter tract can provide important information about the type and degree of degeneration and injury.[36] However, it must be understood that these are surrogate markers and indirect methods of measurement, since other structures including cytoskeletal proteins can play a role in regulating the diffusion of water molecules through an axon[38] and other processes can affect DTI metrics such as seen with inflammation and decreases in radial diffusivity.[39]

Clinical significance

Spinal cord DTI has seen increased uptake in the research and clinical community in the past two decades, especially in the study of multiple sclerosis, amyotrophic lateral sclerosis, DCM/CSM, and tSCI. Within these studies, FA has shown the most robust evidence toward clinical utility.[40] In tSCI patients, multiple studies have

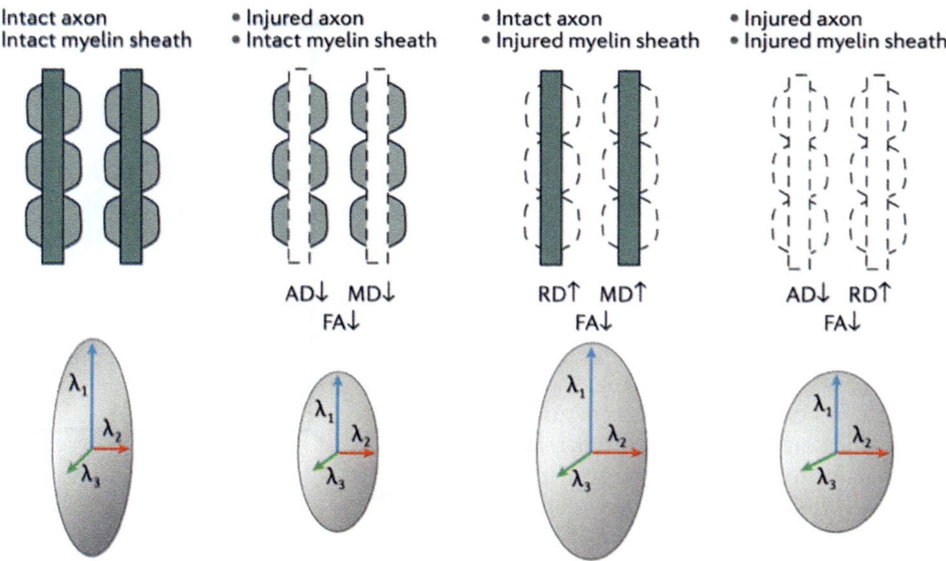

FIGURE 7.5 Illustration of various axonal injury models and the diffusion tensor ellipsoid in the corresponding voxel through which the axon traverses. The shape of the ellipsoid is representative of the degree of anisotropy within the voxel. From left to right: intact axon and myelin show expected diffusion anisotropy with greatest diffusion along the longitudinal axis of the axon λ_1; axonal degeneration with intact myelin leads to decreased AD as well as the overall anisotropy (FA); loss of myelin or demyelination leads to increase in RD but decrease in FA; loss of both structures results in decreased AD and FA but increase in RD. Note that FA decreases in all injury models. *Adapted with permission from David et al. Traumatic and nontraumatic spinal cord injury: Pathological insights from neuroimaging, Nature reviews.*[37]

consistently shown reduced white matter fractional anisotropy at the epicenter of injury compared with healthy controls.[41–43] Imaging the corticospinal tracts and dorsal columns above and below the level of injury also shows a decrease in fractional anisotropy and axial diffusivity with a corresponding increase of radial diffusivity in both acute[42] and chronic tSCI[44,45] patients (Fig. 7.6). These findings suggest both anterograde and retrograde degenerative changes in efferent and afferent pathways early on after injury. Fractional anisotropy values also correlate positively with baseline ISNCSCI motor and sensory scores.[46] The baseline DTI metrics (FA, MD, and AD) at the epicenter of injury also are predictive of ISNCSCI motor scores at 1 year.[47]

In patients with DCM, similar findings have been observed, i.e., fractional anisotropy is reduced (predominantly in the lateral and dorsal columns[48]) at the site of maximal compression[49,50] as well as rostral[25,33,50] and caudal[51] to the stenotic level. An increase in radial diffusivity in conjunction with decreased fractional anisotropy at all levels points to demyelination as the primary degeneration process in this disease.[52] Fractional anisotropy also correlated positively with baseline-modified Japanese Orthopedics Association (mJOA) scores[33] and negatively with baseline Nurick scores.[52,53] Finally, preoperative fractional anisotropy at the level of stenosis strongly correlated with change in functional outcomes (as measured by mJOA) after surgery,[53] and high preoperative mean diffusivity was associated with impaired recovery.[54] These studies highlight the sensitivity with which clinicians can detect microstructural change even before they have affected neurological function, as well as the specificity to the type of pathological process taking place.

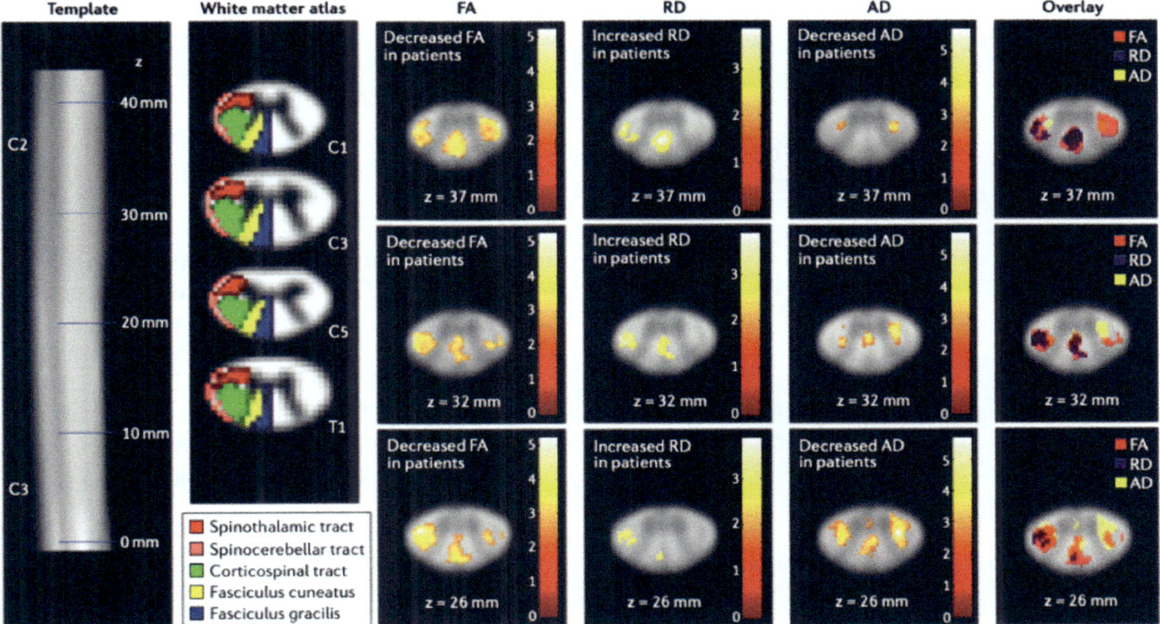

FIGURE 7.6 Voxel-wise analysis of DTI metrics between healthy controls and tSCI patients. (**Column1**) Sagittal spinal cord template. (**Column 2**) Template cross sections of the spinal cord with various white matter tracts overlayed (PAM50 template[3]) for reference. (**Columns 3—5**) Parametric maps of t-statistics for group differences in each DTI metric between the two populations, which are overlayed on the cross-sectional template. The colored bar from 0 to 5 represents t values (yellow or 5 meaning high significance). Patients with tSCI show reduced FA and AD in the lateral and dorsal funiculi, increased RD in the dorsal and left lateral funiculi. (**Column 6**) Overlays all DTI metrics on to the same template (color coded) providing possible clues to the varied pathological processes at play, such as retrograde degeneration in the lateral motor tracts versus. Wallerian degeneration in the dorsal columns. *AD*, axial diffusivity; *DTI*, diffusion tensor imaging; *FA*, fractional anisotropy; *RD*, radial diffusivity; *tSCI*, traumatic spinal cord injury. *Adapted with permission from Huber et al. Dorsal and ventral horn atrophy is associated with clinical outcome after spinal cord injury. Neurology.*[23]

Magnetization transfer and myelin water fraction imaging

Basic concepts

As discussed in the previous section, DTI techniques are an indirect measure of myelin. In contrast, magnetization transfer and myelin water fraction imaging are more specific methods of myelin quantification. Protons in neural tissues exist either bound to macromolecules such as lipids in myelin, as water trapped in myelin sheaths, as water in the intracellular and extracellular environment and as CSF (Fig. 7.7).[55] Protons bound to macromolecules

cannot be seen directly since the T2 relaxation time is too short to be imaged in current MR scanners. However, if saturated using a dedicated radiofrequency (RF) pulse, these protons interact with surrounding free water through direct chemical exchange or dipole—dipole interactions. These processes are generally known as magnetization transfer. The magnetization transfer is proportional to the relative amounts of protons bound to macromolecules.[56] MT imaging can exploit these interactions to indirectly measure the quantity of myelin. One metric used to quantify this interaction is magnetization transfer ratio (MTR), which is essentially the difference between

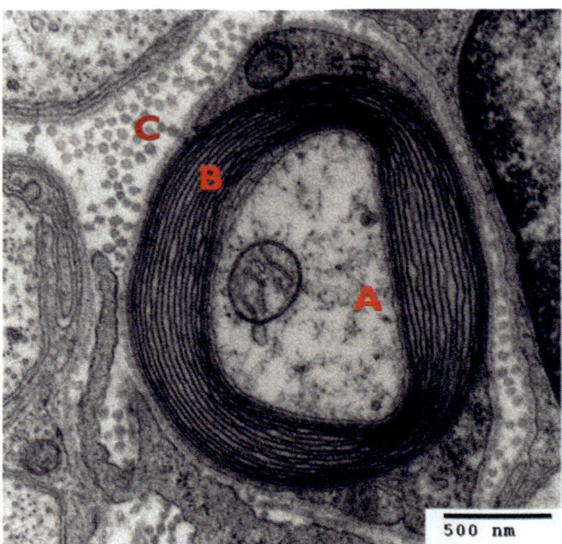

FIGURE 7.7   Cross-sectional view of nerve fiber under an electron microscope. (A) intracellular water, (B) water between myelin sheaths, (C) extracellular water. *Image reproduced with modification by Roadnottaken, licensed under CC BY-SA 3.0 (https://commons.wikimedia.org/wiki/File:Myelinated_ neuron.jpg).*

images acquired with the saturating RF pulse and one without the pulse. MTR is used in the identification of demyelinating diseases such as multiple sclerosis and neuromyelitis optica (NMO).[57] The limitation of MTR is that it is a semiquantitative measure of myelin since it is not quantifying myelin directly and the methodology relies heavily on the acquisition parameters and characteristics of the RF pulse used. Furthermore, it is subject to motion artifact since it is calculated from two separate images.[58] There is evidence that MTR may be correlated with inflammation[59] and axonal integrity[60] as well, raising questions on its specificity to myelin.

Myelin water fraction (MWF) imaging on the other hand uses variable relaxation times to directly measure the ratio of water trapped in myelin sheaths (i.e., myelin water). MWF quantifies the fraction of water bound to myelin compared with the total water signal within a voxel.[55] It is a validated marker for myelin and demyelination.[61]

Clinical significance

Use of MWF in tSCI has been limited, and most clinical research and applications have been related to inflammatory demyelinating diseases such as MS. However, animal studies of demyelination and tSCI murine models have shown that MWF closely follows and reflects histologically confirmed myelin changes in vivo[62] and concludes that MWF is more sensitive to changes in myelin than DTI.[63] In comparison, MT imaging in tSCI has been more commonly utilized, and studies have found lower MTR rostral to the site of injury, once again suggesting demyelination.[64]

In DCM patients, one study showed a reduction in cord MWF in DCM patients with pathologic SSEPs when compared with healthy controls. MWF also showed good correlation with SSEP latencies.[65] One of the earliest utilizations of MT imaging and MTR in DCM was demonstrated by Martin et al., showing significant group differences at the site of stenosis and rostral to the stenosis. MTR also showed correlation with functional outcomes (mJOA) and upper extremity motor scores; however, these were not as strong as seen with FA. In a recent study, Paliwal et al. found significant correlation between preoperative anterior and lateral cord MTR with postoperative mJOA and NDI, respectively. MTR also demonstrated good predictive ability for identifying patients with at least 50% mJOA improvement postoperatively.[66]

Magnetic resonance spectroscopy

Basic concepts

Magnetic resonance spectroscopy (MRS) is quite different from other MRI techniques discussed in this chapter. MRS provides the flexibility to extract information from a varied number of nuclei to obtain chemical information about biological tissues; however, in neural tissue, hydrogen (^{1}H) remains the primary nucleus of choice. MRS can quantify concentrations of metabolites containing the nucleus being targeted and provide insights on the chemical composition of the tissue being imaged. The presence and relative concentrations of these metabolites can serve as surrogate markers of physiological and pathological processes. In MRS of the brain and spine, common metabolites include N-acetylaspartate (NAA) associated with neurodegeneration or metabolic abnormality; myo-inositol (mI) associated with astrocytic activation, gliosis, and inflammation; choline (Cho) associated with cell membrane turnover, found commonly in malignancy and scarcely with demyelination; and creatine (Cr) related to energy metabolism with a relatively stable availability making it a good reference metabolite.[67]

Other common nuclei used for MRS include carbon (^{13}C), sodium (^{23}Na), phosphorus (^{31}P), and fluorine (^{19}F). The latter has been of great interest to scientists because of only trace amounts of ^{19}F present in neural tissues and the human body in general.[68] This, in addition to the favorable resonance properties of ^{19}F, has made it a preferred choice as a tracer to selectively image and assess tissues. ^{19}F-based perfluorocarbons (PFCs) are biologically inert compounds that have been used for various preclinical oncological studies such as tumor oxygenation quantification,[69] targeted drug therapy monitoring,[70] and labeled cell tracking.[71]

Clinical significance

MRS has seen limited use in the brain and spine due to the technical challenges involved.[9] The primary application of MRS has been in multiple sclerosis research and very few studies involving tSCI and DCM exist.[67] Supralesional metabolite abnormalities have been documented in tSCI patients where N-acetylaspartate (NAA) to myo-inositol (mI) and choline (Cho) to myo-inositol (mI) ratios are lower compared with healthy controls. This is relevant because NAA

and choline would be expected to decrease with axonal loss and demyelination, respectively, whereas inflammation and gliosis would increase the myo-inositol levels. NAA/mI and Cho/mI ratios are therefore sensitive markers of neurodegeneration, demyelination, and gliosis. Lower ratios in this study were also associated with lower sensory and motor scores.[72] In DCM patients, Holly et al. demonstrated the feasibility of MRS and found a lower NAA/Cr ratio at C2−3 when compared with healthy controls. This once again shows neuronal loss rostral to chronic compression.[73]

Cellular therapeutics in SCI has been a growing field, and with it has come the need for noninvasive methods to monitor cells after administration. ^{19}F MRI/MRS has been extensively used in cell tracking in vitro and in vivo over the years. One of the earlier demonstrations of ^{19}F MRI for in vivo cell tracking was by Ahrens et al. in 2005, after perfluoropolyether (PFPE)-tagged dendritic cells were injected in mice for application in cancer immunotherapeutics. This evolved to tracking PFPE-tagged stem cells transplanted in stroke-damaged mouse brains.[74] More recently, Richard et al. labeled human glial-restricted progenitor (hGRP) stem cells with a commercial PFC probe in vitro and transplanted them into the spinal cord and brains of a murine ALS model. They were effectively able to evaluate the survival status and migration of the cells in vivo using ^{19}F MRI.[75]

Functional magnetic resonance imaging

Basic concepts

Functional MRI (fMRI) can indirectly measure neuronal activity by detecting changes in blood flow in response to motor tasks or sensory stimuli. Oxygenated and deoxygenated hemoglobin have varied magnetic properties, where the latter is paramagnetic and can cause local variations in the magnetic field strength. With neuronal activation, local blood flow increases, and deoxygenated blood is replaced with oxygenated blood, a phenomenon known as neurovascular coupling. The change in blood flow causes local magnetic perturbations that can in turn be measured with an MRI. This technique is known as blood oxygen level–dependent (BOLD) fMRI.[76] The signal changes with BOLD technique are relatively weak, and therefore, images are acquired quickly and repeatedly in time periods known as "blocks" during which the patient is asked to do a motor task or a sensory stimulus is applied or directed toward a patient (visual, auditory, tactile etc.).

Since the BOLD signal is dependent on several intrinsic and extrinsic factors including blood flow and blood volume, it is susceptible to physiological noise. This can be defined as any signal derived from intrinsic neural metabolism and physiological processes such as cardiac and respiratory cycles producing pulsatility and patient movement.[77] The data, therefore, have to be analyzed through statistical inference to detect true task-related activation in a sea of physiological noise (Fig. 7.8).[79]

BOLD fMRI has been used extensively in imaging and studying brain circuitry. Similar to the brain, the activation of spinal neurons also leads to localized blood flow changes that have been observed in animal studies using implanted microelectrodes in dog, lamb, and sheep models[80] and further confirmed through optical imaging in rat cervical spinal cords.[81] fMRI can also be conducted without a specific motor task or sensory stimulus. This methodology is commonly known as resting state fMRI (rs-fMRI) analysis, and it measures low-frequency changes in the BOLD signal at rest. The neurovascular coupling in the spine, however, may differ from the brain in many ways including a brisker vascular response reducing the BOLD sensitivity in the spinal cord.[82] Some of the other challenges in spinal cord fMRI include knowing where to measure the BOLD signal (i.e., which vertebral level, localization within the cord, ipsilateral vs. contralateral), accounting for signal to noise ratio and task-related motion in the spinal cord.[79,83] Historically, these factors have contributed to issues with reproducibility of data in spinal cord fMRI studies.[84,85] Nonetheless, spinal cord fMRI continues to be a developing field, and ongoing work with repeated studies with robust methodology will help improve its sensitivity and reliability for clinical use.

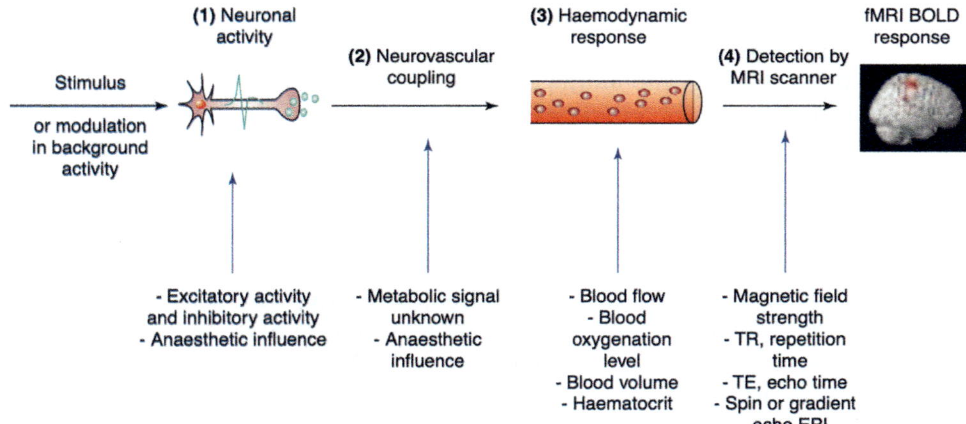

FIGURE 7.8 BOLD contrast mechanism through neurovascular coupling. Factors that can potentially confound the response and the end detected signal are also listed. *BOLD*, blood oxygen level–dependent. *Reproduced with permission from Arthurs et al. How well do we understand the neural origins of the fMRI BOLD signal? Trends Neurosci.*[78]

Clinical significance

Spinal cord fMRI studies in healthy controls have helped establish the current protocols being utilized in patient populations.[84,86,87] One of the earliest studies showed that patients with complete SCI demonstrated BOLD signal in the spinal cord in response to thermal stimuli despite the patients not reporting any active sensation.[88] Since then, multiple studies have shown the presence of neuronal activity rostral and caudal to the site of injury, in response to motor and sensory stimulation.[89–91] Studies have also demonstrated the plasticity of the spinal cord and alterations to afferent sensory pathways after trauma. One such study found increased overall activity in the dorsal columns of patients with incomplete SCI when compared with controls. This study also conducted a connectivity analysis, finding a higher number of intraspinal connections when compared with a healthy spinal cord.[92]

Resting-state fMRI (rs-fMRI) is an intriguing technique for tSCI since it does not require active motor task participation. Most studies involving rs-fMRI in tSCI patients have been on brain and cortical connectivity changes after injury.[93–95] Use of rs-fMRI in the healthy spinal cord has only recently been proven to be feasible, providing evidence of robust networks of interconnected spinal cord regions.[96–99] However, given that rs-fMRI relies on analyzing very small fluctuations in BOLD signal, application in the injured spinal cord has been more challenging due to physiological noise and artifact from metallic instrumentation.

Clinical applications

Imaging biomarkers and implications for clinical trials

The purpose of pursuing advanced spinal cord imaging has been twofold: first to be able to understand and document the structural, physiological, and functional changes after spinal cord injury; and second to identify sensitive and specific imaging biomarkers that can detect subtle changes in spinal cord viability and predict residual function after injury. So far, quantitative MRI (qMRI) and fMRI techniques described before check all the prerequisites and show the most promise in being able to achieve these goals.

In traumatic SCI, early surgical decompression within 24 h of injury is the greatest predictor of sensorimotor recovery.[100] After which, neurorehabilitation provides the highest chance for functional improvement and recovery.[1] MRI methods such as DTI and fMRI have helped elucidate the various changes in the neural axis after injury and provided reliable diagnostic and predictive utility. Evidence of retrograde and anterograde degeneration with the decreases in CSA,[19–21] FA,[23] and MTR[64] above the level of injury have been discussed earlier. We now understand that these changes are part of a neurodegenerative gradient that can be traced from the epicenter of the injury and beyond. For the purposes of their investigation, Azzarito. et al. characterized this gradient by macro- and microstructural qMRI metrics including CSA, spinal cord shape parameters, and spinal cord MT.[101] In parallel to these changes, progressive atrophy and neurodegeneration have also been observed in the brain using a multiparametric mapping (MPM) protocol,[20] combining various qMRI techniques in a single scan.[102] All these changes in the brain and spinal cord are also predictive of outcomes, as patients with less spinal cord and corticospinal tract atrophy, and lower DTI alterations, were more likely to have greater motor recovery.[21,47] Finally, fMRI and rs-fMRI can complement these techniques by elucidating normal spinal cord networks and comparing it with the injured spinal cord. The knowledge of such circuitry in the spinal cord and how they are altered through injury will help differentiate adaptive plasticity from maladaptive plasticity and could help in monitoring progress during rehabilitation and treatment.

Another potential benefit of qMRI is how it can help transform the way clinical trials can be conducted. Historically, the major challenges of introducing novel interventions and therapeutics in human SCI trials have been lack of correlation with preclinical models and lack of sensitive and specific markers in humans to detect minor effects of treatment. With respect to regenerative therapeutics and stem cell therapies, these have been especially difficult.[103] Preclinical techniques of histological staining to confirm cell survival, migration, and differentiation in the spinal cord are not possible in human clinical trials. However, the noninvasive approach of using ^{19}F MRI to track PFC-tagged cells in vivo has been effective in multiple animal studies over the years.[104,105] It is now confirmed that PFCs do not alter the transplanted neural progenitor cells, their capacity to differentiate and their efficacy.[75] Clinical feasibility of ^{19}F MRI has already been established by Ahrens et al., who were able to track and quantify PFC-labeled immunotherapeutic dendritic cells used in colorectal adenocarcinoma patients.[106] It is likely that we will see the use of this technique for stem cell therapy trials in SCI very soon.

Another aspect of treatment monitoring is through endpoints. Currently, most clinical trials in spinal cord injury use clinical outcomes as primary endpoints. It is extremely difficult to assess the true efficacy of any treatment or intervention with only clinical end points since these measures do not carry enough -sensitivity to detect subclinical changes that have not been translated to sensorimotor recovery. qMRI techniques individually and in combination are well suited to detect subtle microstructural changes in response to a medication or treatment. Tissue bridges have now entered clinical trials as biomarkers to track intramedullary structural changes upon therapeutic intervention.[103,107] Similarly, the multiparametric mapping (MPM) protocol is also in use for a phase two clinical trial in traumatic SCI.[107]

Detecting myelopathic progression

The natural history of DCM is poorly understood.[2] In late stages of DCM, clinical examination and conventional imaging are sensitive to pick up the diagnosis, but often the deficits can be quite significant affecting the gait, hand dexterity, sensation, and bladder function. The definitive management in patients with moderate to severe symptomology is surgical decompression.[108] Although most patients do improve neurologically, full recovery is rare. Mild DCM (mJOA 15−17) patients on the other hand are typically managed nonoperatively,[109] but up to 54% of these patients are likely to fail conservative management and require surgery.[2] Mild DCM patients who have progressive neurological impairment are recommended for surgery. However, the caveat here is that in a progressive degenerative disease, microstructural changes in the spinal cord may not immediately translate into clinical impairment despite ongoing injury. To be able to retain as much function and prevent long-term disability and pain, it is important to detect the disease and its progression as early as possible. Reliable biomarkers should have sensitivity and specificity to pick up subclinical changes corresponding to injury and aid in timely intervention. Indeed, reduction in FA at compressed spinal cord levels in asymptomatic patients suggests ongoing subclinical injury before symptoms start to appear.[26]

Kara et al. demonstrated the role of fractional anisotropy in early detection of myelopathy and spinal cord injury, before patients developed T2 signal hyperintensity on conventional imaging.[110] For the detection of myelopathic progression, Martin et al. used a multiparametric qMRI protocol including DTI, MT, and T2* to study DCM patients in a longitudinal study (Fig. 7.9). Composite qMRI scores of FA, MTR, CSA, and T2* white matter to gray matter ratio (T2*WI WM/GM) in the spinal cord were calculated at baseline and follow-up. The study

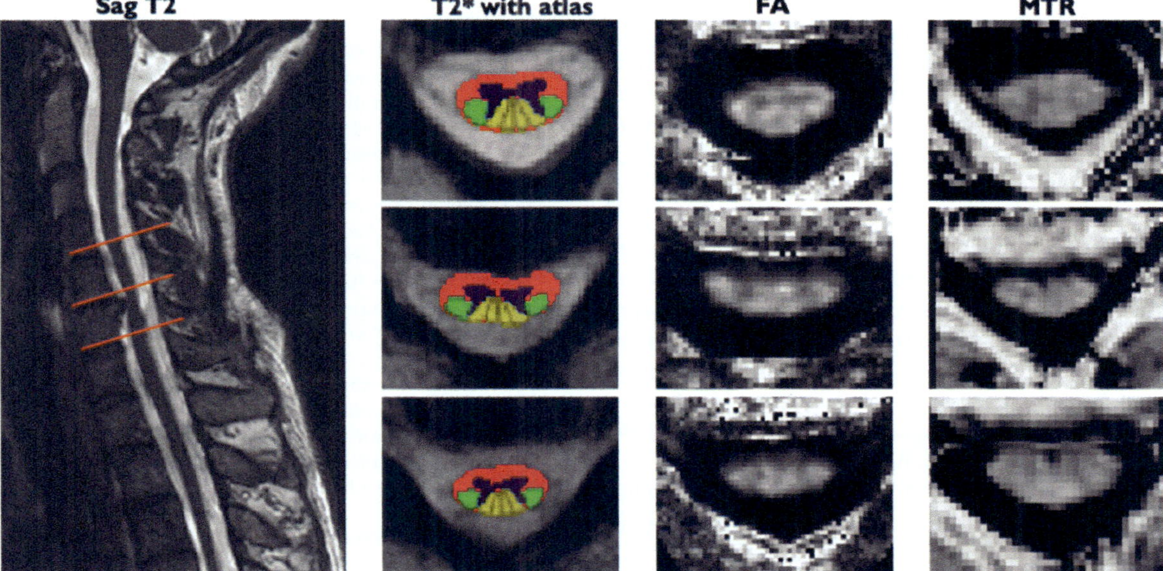

FIGURE 7.9   Voxel-wise analysis of DTI and MT imaging in DCM patient. (**Column 1**) Sagittal T2 weighted MRI of a patient with DCM with three representative cross-sectional cuts. (**Column 2**) Axial T2*-weighted image at corresponding cross-sectional cuts warped to PAM-50 template.[3] Segmentation of the spinal cord performed through Spinal Cord Toolbox (SCT)[111] and overlaid with probabilistic maps of spinal tracts and gray matter. These data can also be used to calculate white and gray matter cross-sectional area. (**Column 3**) Axial fractional anisotropy (FA) map registered to anatomical data in column 2. (**Column 4**) Magnetization transfer ratio (MTR) map registered to anatomical data. This process allows for template-based metric quantification in any spinal tract. Red, spinal cord segmentation; blue, gray matter; green, corticospinal tract; yellow, dorsal column pathways; *DCM*, degenerative cervical myelopathy; *DTI*, diffusion tensor imaging; *MRI*, magnetic resonance imaging; *MT*, magnetization transfer.

found qMRI scores to be more sensitive in detecting myelopathic progression compared with functional outcome scores (mJOA), conventional MRI, or comprehensive clinical exam.[112]

Conclusion and future directions

tSCI and DCM exhibit similar secondary microstructural changes at the site and remote to the injury. qMRI techniques, including DTI, MT, MWF, MRS, and fMRI, can provide valuable structural and functional information in these conditions. Metrics derived from qMRI, especially DTI, have been sensitive and specific to neurodegenerative changes in the spinal cord tissue and show significant correlation with functional, clinical, and electrophysiological outcomes. Among these metrics, FA holds the most promise. Each technique has its own challenges and shortcoming; however, combining multiple methodologies into multiparametric acquisition protocols may help compensate for individual limitations. qMRI could potentially play an important role in the diagnosis, prognostication, and management of these patients. In the future, larger multicenter and multivendor longitudinal studies with a priori hypotheses will be needed to better establish imaging biomarkers. For qMRI metrics to be considered clinically relevant, they must provide specific information and insight on the individual and their disease status.

The clinical translation and adaptation of qMRI methodology has also been challenging partially because of the technical expertise required to acquire and process qMRI data, the extra time needed for scanning acutely ill patients, the presence of metal artifacts secondary to surgical implants, and issues with reproducibility between scan vendors and centers.[107,113] Some of the technical challenges of qMRI are being mitigated with advances in both hardware and software such as high sensitivity coils, shim systems, and modified pulse sequences to improve image quality and reduce artifact and acquisition time.[34] Cross-vendor standardization has been a major focus, and cross-compatible generic qMRI acquisition protocols have been developed for ease of adaptation and improved reproducibility between centers.[114] Advances in image processing include new processing pipelines and automated segmentation of qMRI data with template-based analysis,[111] which have helped reduce inter- and intrarater variability and simplified data analysis.

In conclusion, qMRI has proven to be a collection of effective methods to evaluate neural repair, regeneration, and plasticity in SCI. As we have demonstrated, qMRI and fMRI methods are sensitive to microstructural changes in neural tissues. Thus, they are well suited to not only detect injury but also precisely measure recovery processes such as remyelination using DTI and MTI, or characterize adaptive/maladaptive plasticity in the corticospinal networks through fMRI. There is no doubt that further advances will continue to aid in making qMRI techniques more adaptable and clinically feasible in the near future.

References

1. Ahuja CS, Wilson JR, Nori S, et al. Traumatic spinal cord injury. *Nat Rev Dis Primers* 2017;**3**:17018.
2. Badhiwala JH, Ahuja CS, Akbar MA, et al. Degenerative cervical myelopathy - update and future directions. *Nat Rev Neurol* 2020;**16**(2):108−24.
3. Akter F, Kotter M. Pathobiology of degenerative cervical myelopathy. *Neurosurg Clin* 2018;**29**(1):13−9.
4. Tator CH, Fehlings MG. Review of the secondary injury theory of acute spinal cord trauma with emphasis on vascular mechanisms. *J Neurosurg* 1991; **75**(1):15−26.
5. Buss A, Brook GA, Kakulas B, et al. Gradual loss of myelin and formation of an astrocytic scar during Wallerian degeneration in the human spinal cord. *Brain* 2004;**127**(Pt 1):34−44.
6. Wada E, Ohmura M, Yonenobu K. Intramedullary changes of the spinal cord in cervical spondylotic myelopathy. *Spine* 1995;**20**(20):2226−32.
7. Nouri A, Tetreault L, Dalzell K, Zamorano JJ, Fehlings MG. The relationship between preoperative clinical presentation and quantitative magnetic resonance imaging features in patients with degenerative cervical myelopathy. *Neurosurgery* 2017;**80**(1): 121−8.
8. Nouri A, Martin AR, Kato S, Reihani-Kermani H, Riehm LE, Fehlings MG. The relationship between MRI signal intensity changes, clinical presentation, and surgical outcome in degenerative cervical myelopathy: analysis of a global cohort. *Spine* 2017;**42**(24): 1851−8.
9. Stroman PW, Wheeler-Kingshott C, Bacon M, et al. The current state-of-the-art of spinal cord imaging: methods. *Neuroimage* 2014;**84**:1070−81.
10. Talbott JF, Whetstone WD, Readdy WJ, et al. The Brain and Spinal Injury Center score: a novel, simple, and reproducible method for assessing the severity of acute cervical spinal cord injury with axial T2-weighted MRI findings. *J Neurosurg Spine* 2015;**23**(4):495−504.
11. Schaefer DM, Flanders A, Northrup BE, Doan HT, Osterholm JL. Magnetic resonance imaging of acute cervical spine trauma. Correlation with severity of neurologic injury. *Spine* 1989;**14**(10):1090−5.
12. Miyanji F, Furlan JC, Aarabi B, Arnold PM, Fehlings MG. Acute cervical traumatic spinal cord injury: MR imaging findings correlated with neurologic outcome−prospective study with 100 consecutive patients. *Radiology* 2007;**243**(3):820−7.
13. Farhadi HF, Kukreja S, Minnema A, et al. Impact of admission imaging findings on neurological outcomes in acute cervical traumatic spinal cord injury. *J Neurotrauma* 2018;**35**(12):1398−406.
14. Huber E, Lachappelle P, Sutter R, Curt A, Freund P. Are midsagittal tissue bridges predictive of outcome after cervical spinal cord injury? *Ann Neurol* 2017; **81**(5):740−8.
15. Pfyffer D, Huber E, Sutter R, Curt A, Freund P. Tissue bridges predict recovery after traumatic and ischemic thoracic spinal cord injury. *Neurology* 2019;**93**(16): e1550−60.
16. Cummins DP, Connor JR, Heller KA, et al. Establishing the inter-rater reliability of spinal cord damage manual measurement using magnetic resonance imaging. *Spinal Cord Ser & Cases* 2019;**5**:20.

17. Vallotton K, Huber E, Sutter R, Curt A, Hupp M, Freund P. Width and neurophysiologic properties of tissue bridges predict recovery after cervical injury. *Neurology* 2019;**92**(24):e2793−802.

18. Kearney H, Yiannakas MC, Abdel-Aziz K, et al. Improved MRI quantification of spinal cord atrophy in multiple sclerosis. *J Magn Reson Imag* 2014;**39**(3): 617−23.

19. Freund P, Weiskopf N, Ashburner J, et al. MRI investigation of the sensorimotor cortex and the corticospinal tract after acute spinal cord injury: a prospective longitudinal study. *Lancet Neurol* 2013;**12**(9):873−81.

20. Grabher P, Callaghan MF, Ashburner J, et al. Tracking sensory system atrophy and outcome prediction in spinal cord injury. *Ann Neurol* 2015;**78**(5):751−61.

21. Ziegler G, Grabher P, Thompson A, et al. Progressive neurodegeneration following spinal cord injury: implications for clinical trials. *Neurology* 2018;**90**(14): e1257−66.

22. Freund P, Weiskopf N, Ward NS, et al. Disability, atrophy and cortical reorganization following spinal cord injury. *Brain J Neurol* 2011;**134**(Pt 6):1610−22.

23. Huber E, David G, Thompson AJ, Weiskopf N, Mohammadi S, Freund P. Dorsal and ventral horn atrophy is associated with clinical outcome after spinal cord injury. *Neurology* 2018;**90**(17):e1510−22.

24. Martin AR, De Leener B, Cohen-Adad J, et al. Monitoring for myelopathic progression with multiparametric quantitative MRI. *PLoS One* 2018;**13**(4): e0195733.

25. Grabher P, Mohammadi S, Trachsler A, et al. Voxel-based analysis of grey and white matter degeneration in cervical spondylotic myelopathy. *Sci Rep* 2016;**6**: 24636.

26. Martin AR, De Leener B, Cohen-Adad J, et al. Can microstructural MRI detect subclinical tissue injury in subjects with asymptomatic cervical spinal cord compression? A prospective cohort study. *BMJ Open* 2018;**8**.

27. Prados F, Ashburner J, Blaiotta C, et al. Spinal cord grey matter segmentation challenge. *Neuroimage* 2017; **152**:312−29.

28. Pfyffer D, Vallotton K, Curt A, Freund P. Predictive value of midsagittal tissue bridges on functional recovery after spinal cord injury. *Neurorehabilit Neural Repair* 2021;**35**(1):33−43.

29. Berliner JC, O'Dell DR, Albin SR, et al. The influence of conventional T(2) MRI indices in predicting who will walk outside one year after spinal cord injury. *J Spinal Cord Med* 2021:1−7.

30. Pfyffer D, Vallotton K, Curt A, Freund P. Tissue bridges predict neuropathic pain emergence after spinal cord injury. *J Neurol Neurosurg Psychiatry* 2020; **91**(10):1111−7.

31. Hari AR, Wydenkeller S, Dokladal P, Halder P. Enhanced recovery of human spinothalamic function is associated with central neuropathic pain after SCI. *Exp Neurol* 2009;**216**(2):428−30.

32. Wasner G, Lee BB, Engel S, McLachlan E. Residual spinothalamic tract pathways predict development of central pain after spinal cord injury. *Brain J Neurol* 2008;**131**(Pt 9):2387−400.

33. Martin AR, De Leener B, Cohen-Adad J, et al. A novel MRI biomarker of spinal cord white matter injury: T2*-weighted white matter to gray matter signal intensity ratio. *AJNR Am J Neuroradiol* 2017;**38**(6): 1266−73.

34. Cadotte DW, Akbar MA, Fehlings MG, Stroman PW, Cohen-Adad J. What has been learned from magnetic resonance imaging examination of the injured human spinal cord: a Canadian perspective. *J Neurotrauma* 2018;**35**(16):1942−57.

35. Stejskal EO, Tanner JE. Spin diffusion measurements: spine echoes in the presence of a time-dependent field gradient. *J Phys Chem* 1965;**42**:288−92.

36. Brennan FH, Cowin GJ, Kurniawan ND, Ruitenberg MJ. Longitudinal assessment of white matter pathology in the injured mouse spinal cord through ultra-high field (16.4 T) in vivo diffusion tensor imaging. *Neuroimage* 2013;**82**:574−85.

37. David G, Mohammadi S, Martin AR, et al. Traumatic and nontraumatic spinal cord injury: pathological insights from neuroimaging. *Nat Rev Neurol* 2019; **15**(12):718−31.

38. Pierpaoli C, Jezzard P, Basser PJ, Barnett A, Di Chiro G. Diffusion tensor MR imaging of the human brain. *Radiology* 1996;**201**(3):637−48.

39. Xie M, Wang Q, Wu TH, Song SK, Sun SW. Delayed axonal degeneration in slow Wallerian degeneration mutant mice detected using diffusion tensor imaging. *Neuroscience* 2011;**197**:339−47.

40. Martin AR, Aleksanderek I, Cohen-Adad J, et al. Translating state-of-the-art spinal cord MRI techniques to clinical use: a systematic review of clinical studies utilizing DTI, MT, MWF, MRS, and fMRI. *NeuroImage Clin* 2016;**10**:192−238.

41. Shanmuganathan K, Gullapalli RP, Zhuo J, Mirvis SE. Diffusion tensor MR imaging in cervical spine trauma. *AJNR Am J Neuroradiol* 2008;**29**(4):655−9.

42. Cheran S, Shanmuganathan K, Zhuo J, et al. Correlation of MR diffusion tensor imaging parameters with ASIA motor scores in hemorrhagic and nonhemorrhagic acute spinal cord injury. *J Neurotrauma* 2011;**28**(9): 1881−92.

43. Chang Y, Jung TD, Yoo DS, Hyun JK. Diffusion tensor imaging and fiber tractography of patients with cervical spinal cord injury. *J Neurotrauma* 2010;**27**(11): 2033–40.

44. Ellingson BM, Ulmer JL, Kurpad SN, Schmit BD. Diffusion tensor MR imaging in chronic spinal cord injury. *AJNR Am J Neuroradiol* 2008;**29**(10):1976–82.

45. Koskinen E, Brander A, Hakulinen U, et al. Assessing the state of chronic spinal cord injury using diffusion tensor imaging. *J Neurotrauma* 2013;**30**(18): 1587–95.

46. Freund P, Schneider T, Nagy Z, et al. Degeneration of the injured cervical cord is associated with remote changes in corticospinal tract integrity and upper limb impairment. *PLoS One* 2012;**7**(12):e51729.

47. Shanmuganathan K, Zhuo J, Chen HH, et al. Diffusion tensor imaging parameter obtained during acute blunt cervical spinal cord injury in predicting long-term outcome. *J Neurotrauma* 2017;**34**(21):2964–71.

48. Freund P, Thompson A, Curt A, et al. Author response: progressive neurodegeneration following spinal cord injury: implications for clinical trials. *Neurology* 2018; **91**(21):985.

49. Uda T, Takami T, Tsuyuguchi N, et al. Assessment of cervical spondylotic myelopathy using diffusion tensor magnetic resonance imaging parameter at 3.0 tesla. *Spine* 2013;**38**(5):407–14.

50. Budzik JF, Balbi V, Le Thuc V, Duhamel A, Assaker R, Cotten A. Diffusion tensor imaging and fibre tracking in cervical spondylotic myelopathy. *Eur Radiol* 2011; **21**(2):426–33.

51. Chen X, Kong C, Feng S, et al. Magnetic resonance diffusion tensor imaging of cervical spinal cord and lumbosacral enlargement in patients with cervical spondylotic myelopathy. *J Magn Reson Imag* 2016; **43**(6):1484–91.

52. Rajasekaran S, Yerramshetty JS, Chittode VS, Kanna RM, Balamurali G, Shetty AP. The assessment of neuronal status in normal and cervical spondylotic myelopathy using diffusion tensor imaging. *Spine* 2014;**39**(15):1183–9.

53. Vedantam A, Rao A, Kurpad SN, et al. Diffusion tensor imaging correlates with short-term myelopathy outcome in patients with cervical spondylotic myelopathy. *World Neurosurg* 2017;**97**:489–94.

54. Sato T, Horikoshi T, Watanabe A, et al. Evaluation of cervical myelopathy using apparent diffusion coefficient measured by diffusion-weighted imaging. *AJNR Am J Neuroradiol* 2012;**33**(2):388–92.

55. Whittall KP, MacKay AL, Graeb DA, Nugent RA, Li DK, Paty DW. In vivo measurement of T2 distributions and water contents in normal human brain. *Magn Reson Med* 1997;**37**(1):34–43.

56. Henkelman RM, Stanisz GJ, Graham SJ. Magnetization transfer in MRI: a review. *NMR Biomed* 2001;**14**(2): 57–64.

57. Filippi M, Rocca MA, Martino G, Horsfield MA, Comi G. Magnetization transfer changes in the normal appearing white matter precede the appearance of enhancing lesions in patients with multiple sclerosis. *Ann Neurol* 1998;**43**(6):809–14.

58. Cercignani M, Symms MR, Ron M, Barker GJ. 3D MTR measurement: from 1.5 T to 3.0 T. *Neuroimage* 2006; **31**(1):181–6.

59. Gareau PJ, Rutt BK, Karlik SJ, Mitchell JR. Magnetization transfer and multicomponent T2 relaxation measurements with histopathologic correlation in an experimental model of MS. *J Magn Reson Imag* 2000; **11**(6):586–95.

60. Fisher E, Chang A, Fox RJ, et al. Imaging correlates of axonal swelling in chronic multiple sclerosis brains. *Ann Neurol* 2007;**62**(3):219–28.

61. Laule C, Vavasour IM, Zhao Y, et al. Two-year study of cervical cord volume and myelin water in primary progressive multiple sclerosis. *Mult Scler* 2010;**16**(6):670–7.

62. McCreary CR, Bjarnason TA, Skihar V, Mitchell JR, Yong VW, Dunn JF. Multiexponential T2 and magnetization transfer MRI of demyelination and remyelination in murine spinal cord. *Neuroimage* 2009;**45**(4): 1173–82.

63. Kozlowski P, Raj D, Liu J, Lam C, Yung AC, Tetzlaff W. Characterizing white matter damage in rat spinal cord with quantitative MRI and histology. *J Neurotrauma* 2008;**25**(6):653–76.

64. Cohen-Adad J, El Mendili MM, Lehericy S, et al. Demyelination and degeneration in the injured human spinal cord detected with diffusion and magnetization transfer MRI. *Neuroimage* 2011;**55**(3):1024–33.

65. Liu H, MacMillian EL, Jutzeler CR, et al. Assessing structure and function of myelin in cervical spondylotic myelopathy: evidence of demyelination. *Neurology* 2017;**89**(6):602–10.

66. Paliwal M, Weber 2nd KA, Hopkins BS, et al. Magnetization transfer ratio and morphometrics of the spinal cord associates with surgical recovery in patients with degenerative cervical myelopathy. *World Neurosurg* 2020;**144**:e939–47.

67. Solanky BS, De Vita E. Single voxel MR spectroscopy in the spinal cord: technical challenges and clinical applications. In: Cohen-Adad J, Wheeler-Kingshott C, editors. *Quantitative MRI of the spinal cord*. Elsevier; 2014. p. 267–90.

68. Longmaid 3rd HE, Adams DF, Neirinckx RD, et al. In vivo 19F NMR imaging of liver, tumor, and abscess in rats. Preliminary results. *Invest Radiol* 1985;**20**(2): 141–5.

69. Shi Y, Oeh J, Eastham-Anderson J, et al. Mapping in vivo tumor oxygenation within viable tumor by 19F-MRI and multispectral analysis. *Neoplasia* 2013;**15**(11):1241−50.

70. Luo Z, Jin K, Pang Q, et al. On-demand drug release from dual-targeting small nanoparticles triggered by high-intensity focused ultrasound enhanced glioblastoma-targeting therapy. *ACS Appl Mater Interfaces* 2017;**9**(37):31612−25.

71. Constantinides C, Maguire M, McNeill E, et al. Fast, quantitative, murine cardiac 19F MRI/MRS of PFCE-labeled progenitor stem cells and macrophages at 9.4T. *PLoS One* 2018;**13**(1):e0190558.

72. Wyss PO, Huber E, Curt A, Kollias S, Freund P, Henning A. MR spectroscopy of the cervical spinal cord in chronic spinal cord injury. *Radiology* 2019; **291**(1):131−8.

73. Holly LT, Freitas B, McArthur DL, Salamon N. Proton magnetic resonance spectroscopy to evaluate spinal cord axonal injury in cervical spondylotic myelopathy. *J Neurosurg Spine* 2009;**10**(3):194−200.

74. Boehm-Sturm P, Aswendt M, Minassian A, et al. A multi-modality platform to image stem cell graft survival in the naive and stroke-damaged mouse brain. *Biomaterials* 2014;**35**(7):2218−26.

75. Richard JP, Hussain U, Gross S, et al. Perfluorocarbon labeling of human glial-restricted progenitors for (19) F magnetic resonance imaging. *Stem Cells Transl Med* 2019;**8**(4):355−65.

76. Ogawa S, Lee TM, Kay AR, Tank DW. Brain magnetic resonance imaging with contrast dependent on blood oxygenation. *Proc Natl Acad Sci USA* 1990;**87**(24): 9868−72.

77. Kruger G, Glover GH. Physiological noise in oxygenation-sensitive magnetic resonance imaging. *Magn Reson Med* 2001;**46**(4):631−7.

78. Arthurs OJ, Boniface S. How well do we understand the neural origins of the fMRI BOLD signal? *Trends Neurosci* 2002;**25**(1):27031.

79. Summers P, Brooks J, Cohen-Adad J. Spinal cord fMRI. In: cohen-Adad J, Wheeler-Kingshott C, editors. *Quantitative MRI of the spinal cord*. Elsevier; 2014. p. 221−39.

80. Marcus ML, Heistad DD, Ehrhardt JC, Abboud FM. Regulation of total and regional spinal cord blood flow. *Circ Res* 1977;**41**(1):128−34.

81. Sasaki S, Yazawa I, Miyakawa N, et al. Optical imaging of intrinsic signals induced by peripheral nerve stimulation in the in vivo rat spinal cord. *Neuroimage* 2002; **17**(3):1240−55.

82. Nix W, Capra NF, Erdmann W, Halsey JH. Comparison of vascular reactivity in spinal cord and brain. *Stroke* 1976;**7**(6):560−3.

83. Kornelsen J, Stroman PW. fMRI of the lumbar spinal cord during a lower limb motor task. *Magn Reson Med* 2004;**52**(2):411−4.

84. Bouwman CJ, Wilmink JT, Mess WH, Backes WH. Spinal cord functional MRI at 3 T: gradient echo echo-planar imaging versus turbo spin echo. *Neuroimage* 2008;**43**(2):288−96.

85. Giove F, Garrefa G, Mangia S, Colonnese C, Maraviglia B. Issues about the fMRI of the human spinal cord. *Magn Reson Imag* 2004;**22**(10):1505−16.

86. Yoshizawa T, Nose T, Moore GJ, Sillerud LO. Functional magnetic resonance imaging of motor activation in the human cervical spinal cord. *Neuroimage* 1996;**4**(3 Pt 1):174−82.

87. Madi S, Flanders AE, Vinitski S, Herbison GJ, Nissanov J. Functional MR imaging of the human cervical spinal cord. *AJNR Am J Neuroradiol* 2001;**22**(9): 1768−74.

88. Stroman PW, Tomanek B, Krause V, Frankenstein UN, Malisza KL. Mapping of neuronal function in the healthy and injured human spinal cord with spinal fMRI. *Neuroimage* 2002;**17**(4):1854−60.

89. Stroman PW, Kornelsen J, Bergman A, et al. Noninvasive assessment of the injured human spinal cord by means of functional magnetic resonance imaging. *Spinal Cord* 2004;**42**(2):59−66.

90. Stroman PW, Khan HS, Bosma RL, et al. Changes in pain processing in the spinal cord and brainstem after spinal cord injury characterized by functional magnetic resonance imaging. *J Neurotrauma* 2016;**33**(15): 1450−60.

91. Zhong XP, Chen YX, Li ZY, Shen ZW, Kong KM, Wu RH. Cervical spinal functional magnetic resonance imaging of the spinal cord injured patient during electrical stimulation. *Eur Spine J* 2017;**26**(1): 71−7.

92. Cadotte DW, Bosma R, Mikulis D, et al. Plasticity of the injured human spinal cord: insights revealed by spinal cord functional MRI. *PLoS One* 2012;**7**(9): e45560.

93. Hou JM, Sun TS, Xiang ZM, et al. Alterations of resting-state regional and network-level neural function after acute spinal cord injury. *Neuroscience* 2014; **277**:446−54.

94. Hawasli AH, Rutlin J, Roland JL, et al. Spinal cord injury disrupts resting-state networks in the human brain. *J Neurotrauma* 2018;**35**(6):864−73.

95. Kaushal M, Oni-Orisan A, Chen G, et al. Evaluation of whole-brain resting-state functional connectivity in spinal cord injury: a large-scale network analysis using network-based statistic. *J Neurotrauma* 2017;**34**(6): 1278−82.

96. Sprenger C, Finsterbusch J, Buchel C. Spinal cord-midbrain functional connectivity is related to perceived pain intensity: a combined spino-cortical FMRI study. *J Neurosci* 2015;**35**(10):4248−57.

97. Eippert F, Kong Y, Winkler AM, et al. Investigating resting-state functional connectivity in the cervical spinal cord at 3T. *Neuroimage* 2017;**147**:589−601.

98. Wei P, Li J, Gao F, Ye D, Zhong Q, Liu S. Resting state networks in human cervical spinal cord observed with fMRI. *Eur J Appl Physiol* 2010;**108**(2):265−71.

99. Barry RL, Conrad BN, Smith SA, Gore JC. A practical protocol for measurements of spinal cord functional connectivity. *Sci Rep* 2018;**8**(1):16512.

100. Badhiwala JH, Wilson JR, Witiw CD, et al. The influence of timing of surgical decompression for acute spinal cord injury: a pooled analysis of individual patient data. *Lancet Neurol* 2021;**20**(2):117−26.

101. Azzarito M, Seif M, Kyathanahally S, Curt A, Freund P. Tracking the neurodegenerative gradient after spinal cord injury. *NeuroImage Clin* 2020;**26**:102221.

102. Weiskopf N, Suckling J, Williams G, et al. Quantitative multi-parameter mapping of R1, PD(*), MT, and R2(*) at 3T: a multi-center validation. *Front Neurosci* 2013;**7**:95.

103. Curt A, Hsieh J, Schubert M, et al. The damaged spinal cord is a suitable target for stem cell transplantation. *Neurorehabilit Neural Repair* 2020;**34**(8):758−68.

104. Wu L, Liu F, Liu S, Xiuan X, Zhaoxi L, Xilin S. Perfluorocarbons-based 19F magnetic resonance imaging in biomedicine. *Int J Nanomed* 2021;**15**:7377−95.

105. Modo M. (19)F magnetic resonance imaging and spectroscopy in neuroscience. *Neuroscience* 2021;**474**:37−50.

106. Ahrens ET, Helfer BM, O'Hanlon CF, Schirda C. Clinical cell therapy imaging using a perfluorocarbon tracer and fluorine-19 MRI. *Magn Reson Med* 2014;**72**(6):1696−701.

107. Freund P, Seif M, Weiskopf N, et al. MRI in traumatic spinal cord injury: from clinical assessment to neuroimaging biomarkers. *Lancet Neurol* 2019;**18**(12):1123−35.

108. Fehlings MG, Kwon BK, Tetreault LA. Guidelines for the management of degenerative cervical myelopathy and spinal cord injury: an introduction to a focus issue. *Global Spine J* 2017;**7**(3 Suppl. l):6S−7S.

109. Tetreault LA, Skelly AC, Dettori JR, Wilson JR, Martin AR, Fehlings MG. Guidelines for the management of degenerative cervical myelopathy and acute spinal cord injury: development process and methodology. *Global Spine J* 2017;**7**(3 Suppl. l):8S−20S.

110. Kara B, Celik A, Karadereler S, et al. The role of DTI in early detection of cervical spondylotic myelopathy: a preliminary study with 3-T MRI. *Neuroradiology* 2011;**53**(8):609−16.

111. De Leener B, Levy S, Dupont SM, et al. SCT: spinal cord toolbox, an open-source software for processing spinal cord MRI data. *Neuroimage* 2017;**145**(Pt A):24−43.

112. Martin AR, De Leener B, Cohen-Adad J, et al. Correction: monitoring for myelopathic progression with multiparametric quantitative MRI. *PLoS One* 2018;**13**(9):e0204082.

113. Gulani V, Seiberlich N. Quantitative MRI: rationale and challenges. In: Seiberlich N, Gulani V, Calamante F, et al., editors. *Advances in magnetic resonance technology and applications*, vol. 1. Elsevier; 2020 [xxxvii-li].

114. Cohen-Adad J, Alonso-Ortiz E, Abramovic M, et al. Generic acquisition protocol for quantitative MRI of the spinal cord. *Nat Protoc* 2021;**16**(10):4611−32.

Intraoperative imaging and image guidance

Daipayan Guha[1], Adam A. Dmytriw[2], James D. Guest[3], Victor X.D. Yang[4]

[1]Division of Neurosurgery, McMaster University, Toronto, ON, Canada; [2]Massachusetts General Hospital, Harvard Medical School, Boston, MA, United States; [3]Department of Neurological Surgery, Miller School of Medicine, University of Miami, Miami, FL, United States; [4]Division of Neurosurgery, University of Toronto, Toronto, ON, Canada

Evolution of computer-assisted navigation

In surgery, maintaining the complex 3D relationships between anatomical targets and instruments is paramount to safe and effective interventions. While often done through direct visual feedback, navigation becomes more relevant when anatomic landmarks are not readily visible. The earliest surgical navigation systems were developed in neurosurgery, to correlate external cranial anatomy to underlying internal structures intraoperatively. In 1908 Horsley and Clarke coined the term "stereotactic" in describing a novel device allowing the placement of intracranial electrodes into precise targets in an animal model, using a rigid frame.[1] The first human clinical application of frame-based stereotaxy was by Spiegel and Wycis in the 1940s, using a Cartesian coordinate system.[2] Subsequent development for use with emerging cross-sectional imaging modalities, including computed tomography (CT) and magnetic resonance imaging (MRI), led to the introduction of rigid frame and arc localization systems.

With rapid advances in imaging and computing power over the past three decades, frameless stereotaxy, also termed image-guided surgery (IGS), neuronavigation, or computer-assisted navigation (CAN), was pioneered again first in a neurosurgical context. Frameless stereotaxy, by virtue of its lack of a rigid mechanical linkage between the patient anatomic space and the instrument space, therefore requires the matching of patient and device (or image) spaces, a process termed *registration*.

History of spinal computer-assisted navigation

Prior to the advent of CAN, intraoperative navigation in the spine was typically performed using a combination of anatomic knowledge as

© 2022 Elsevier Inc. All rights reserved.

well as radiographic feedback from serial XR (X-ray) or fluoroscopy. While plain XR remains useful for the initial localization of a skin incision or vertebral levels, it is associated with a significant time lag particularly when digital radiograph processing units are unavailable, and provides only a single temporal snapshot. Poor image quality due to metallic artifact, bony obstruction, or large patient body habitus requires repeated XR and therefore increases this time cost as well as potential for error. C-arm fluoroscopy has therefore traditionally been the imaging modality of choice for many spinal surgeons for intraoperative guidance. C-arms may provide a single XR snapshot for incision and anatomic level localization, and may also acquire continuous images to allow for real-time localization of instruments in the operative field. However, this practice is associated with significant occupational radiation exposure, particularly to the surgical team. Extensive investigation has been performed on the cumulative radiation dose from C-arm fluoroscopy to various parts of the surgeon's body, at various positions around the operative table (i.e., on the side of the detector vs. emitter), with varying patient body habitus, and with the duration of fluoroscopy.[3-6] Moreover, standard C-arm fluoroscopy only provides a single in-plane view, with multiple planes possible only with the introduction of a second orthogonally positioned C-arm or by moving the single C-arm back and forth to the required planes, a cumbersome and ergonomically disruptive exercise.[7,8]

Extension of CAN from cranial to spinal procedures was therefore a natural target. The evolution from early arm-based CAN techniques to armless systems was enabled by the development of dynamic reference frames (DRFs), consisting of an electromagnetic (EM) coil or active or passive IR LED arrays affixed to rigid bony anatomy. This allowed the tracking of instruments in the patient space without rigid anatomic fixation, as is typically accomplished in cranial neurosurgery with the use of rigid head fixation devices, but is not feasible in spinal approaches.[9] Kalfas et al. were the first to adapt

frameless stereotaxy for clinical use in the spine in the mid-1990s, using a wand fitted with ultrasonic emitters allowing tracking using sonic digitizers placed around the operating field, registered to a preoperatively acquired volumetric CT dataset.[10,11] Subsequent development in the late 1990s expanded the scope of spinal CAN to include updating of imaging in real-time intraoperatively using C-arm fluoroscopy, termed "virtual fluoroscopy."[12,13] Unfortunately, virtual fluoroscopy systems remained limited to 2D projection images, without true multiplanar views in the axial, sagittal, and coronal planes.[14] 3D CAN systems, based initially on preoperative CT imaging and subsequently on intraoperative mobile CT as well as isocentric C-arm fluoroscopy, therefore, arose to the forefront starting in the early 2000s.[15-17] These early imaging devices remained limited by poor image quality and cumbersome workflow. A major step forward in spinal intraoperative imaging and navigation was taken in 2006, with the introduction of the O-Arm by Breakaway Imaging (now Medtronic), allowing 360 degrees cone-beam CT (CBCT)-quality imaging with a breakable gantry facilitating movement around the operating table, and automatic registration to the patient spinal anatomy.[14] A full discussion of contemporary spinal CAN imaging and registration techniques is presented in Registration, imaging, and actuation techniques in spinal computer-assisted navigation section of this chapter.

Current applications of spinal computer-assisted navigation

The first spinal CAN systems were used to guide the placement of lumbar pedicle instrumentation.[10,11] In the two decades following, instrumentation placement remains the primary application of contemporary spinal image guidance systems, with a body of literature encompassing over 10,000 pooled pedicle screws.[18] CAN systems have been used to guide pedicle screws from the atlantoaxial (C1—C2) and subaxial cervical spine,[19,20] down to the sacrum and pelvis.[21,22] While there is literature to suggest that the freehand placement of standard posterior thoracolumbar and sacral pedicle screws may be safe in highly trained hands, CAN systems have expanded the accessibility of accurate and safe placement of instrumentation at these levels, particularly in revision and deformity correction cases where typical anatomic landmarks are distorted.[23,24] CAN has also facilitated novel instrumentation approaches that are otherwise feasible only with repeated fluoroscopy, including odontoid screw placement at C2,[25] oblique prepsoas and extreme lateral transpsoas approaches to the lumbar spine,[26,27] as well as percutaneous and minimally invasive instrumentation at all spinal levels.[28-31]

The utility of modern CAN techniques has expanded from instrumentation to guidance and confirmation of the extent of decompression. Navigation guidance for anterior transoral approaches to the craniocervical junction, for inflammatory and neoplastic etiologies, has been reported as early as 2003.[32] More recently, anterior CAN-guided subaxial cervical transcorporal tunnel approaches to treat focal pathology underlying cervical myelopathy have been reported.[33] Ligamentous decompression with MIS epiduroscopic laser ablation in the lumbar spine may also be guided by modern CAN techniques.[34] Osteotomies for correction of spinal alignment, with or without instrumentation, may also be guided by CAN techniques in order to precisely plan preoperatively, and subsequently execute intraoperatively, the specific bony extirpations required to achieve a desired alignment.[35] In an oncologic context, the extent of osseous and soft-tissue tumor decompression may be guided and confirmed by intraoperative CAN, particularly when coupled with CT/MRI fusion techniques and intraoperative imaging.[36] CAN guidance may facilitate less invasive "separation surgery" for metastatic epidural disease, whereby a transpedicular approach is used to resect sufficient tumor lateral and ventral to the spinal cord to allow safe high-dose fractionated radiation therapy with minimal neurotoxicity.[37] At the extremes of minimally invasive surgery, CAN may guide percutaneous catheters for laser interstitial thermal therapy, with the purpose of ablating epidural tumor similar to "separation surgery."[38] By merging preoperative MRI and intraoperative 3D fluoroscopic images, CAN may also facilitate the resection of intradural tumors, by minimizing the extent of soft-tissue and bony exposure and, for intrinsic spinal cord tumors, centering the tumor to more precisely localize the midline myelotomy.[39]

While infusion and neuromodulatory therapies for spinal chronic pain conditions are typically performed safely and easily freehand, such as with the implantation of dorsal root ganglion stimulation electrodes, CAN may be useful in cases of severely distorted or disrupted anatomy, such as in one series of intrathecal baclofen pump catheter implantation in cerebral palsy patients with severe neuromuscular scoliosis.[40]

Rationale for spinal computer-assisted navigation

CAN was applied to spinal procedures initially for the guidance of lumbar pedicle screws. Instrumentation guidance remains the primary application for CAN by most spinal surgeons, with multiple systematic reviews and meta-analyses reporting on the radiographic accuracy of pedicle screws in the cervical, thoracic, and lumbosacral spine, and in multiple clinical contexts including minimally invasive percutaneous instrumentation as well as in adolescent idiopathic scoliosis patients.[41-50] Comparison of radiographic accuracy of pedicle screw placement between navigated and freehand, or conventional fluoroscopy, techniques was made most recently in a meta-analysis by Mason et al.[45] This analysis found an overall radiographic accuracy rate, across all spinal regions, of 68.1% for freehand techniques, versus 84.3% for 2D navigation and 95.5% for 3D navigation, with all 3D techniques pooled (isocentric fluoroscopy, CBCT, fan-beam CT (FBCT)). Further stratification of navigated pedicle screw accuracy was made most recently by Du et al., in a meta-analysis comparing specifically 3D FBCT versus 3D isocentric fluoroscopy (Table 8.1).[41] Interestingly, although the *diagnostic* accuracy

TABLE 8.1 Studies of pedicle screw misplacement.

Study (year)	Study design (Class of evidence)	Participants (Spinal level)	Interventions	Pedicle screw misplacement			Accuracy assessment
				3D FluNAV	2D FluNAV	CT Nav	
Liu et al. (2010)[7]	Prospective cohort study (II)	58 patients: 40 males, 18 females; age: 20–75 years Abnormal cervical curvature, instability trauma, spinal kyphosis, degenerative cervical disease (C2–C7)	G1: 3D FluoroNav $n = 29$ (140 screws) G2: CT Nav $n = 29$ (159 screws)	0/140		4/159	CT
Tan et al. (2013)[5]	Random controlled cadaveric study (I)	Forty adult spine cadavers (C1–T1)	G1: 3D FluoroNav $n = 8$ (80 screws) G2: CT Nav $n = 8$ (80 screws) G3: 2D FluoroNav $n = 8$ (80 screws)	0/80	10/80	0/80	CT
Lee et al. (2007)[4]	Retrospective cohort study (III)	60 patients: sex unstated; age: 21–91 years Cervical spondylotic myelopathy, trauma, deformity, infection, OPLL, RA, osteoporosis (C7, T1, T2)	G1: CT Nav $n = 9$ (45 screws) G2: 2D FluoroNav $n = 19$ (63 screws)		13/98	6/104	CT
Lekovic et al. (2007)[12]	Retrospective cohort study (III)	37 patients: 20 males, 27 females; age: 35–81 years Trauma, degenerative disease, tumor	G1: 3D FluoroNav $n = 12$ (94 screws) G2: 2D Nav $n = 25$ (183 screws)	5/94	16/183		CT
Nottmeier et al. (2009)[8]	Retrospective cohort study (III)	220 patients: sex and age unstated Spinal disease unstated (T1–S1)	Gl: 3D FluoroNav $n = 140$ (637 screws) G2: CT Nav $n = 80$ (314 screws)	42/637		29/314	CT
Gruetzner et al. (2004)[9]	Retrospective cohort study (III)	79 patients: 52 males, 27 females; age: 16–76 years Injuries and degeneration changes to the spine (cervical, thoracic, lumbar)	G1: 3D FluoroNav $n = 24$ (114 screws) G2: CT Nav $n = 27$ (112 screws) G3: 2D FluoroNav $n = 28$ (108 screws)	1/114	3/108	5/112	CT
Kotani et al. (2014)[6]	Retrospective cohort study (III)	61 patients: 11 males, 50 females; age: 12–31 years Scoliosis with a thoracic curve	G1: 3D FluoroNav $n = 32$ (416 screws) G2: CT Nav $n = 29$ (222 screws)	13/416		11/222	CT

Wood et al. (2011)[11]	Prospective cohort study (II)	67 patients: 30 males, 37 females; age: 22–78 years Lumbar radiculopathy, instability with spondylolisthesis, trauma, tumor	G1: 3D FluoroNav $n = 43$ (186 screws) G2: 3D FluoroNav $n = 24$ (110 screws)	3/186	7/110		CT
Fu et al. (2008)[2]	Retrospective cohort study (III)	24 patients: 9 males, 15 females; age: 19–79 years Fracture, spondylolisthesis, tuberculous spondylitis, ankylosing spondylitis, revision (below T8 level)	G1: CT Nav $n = 11$ (76 screws) G2: 2D FluoroNav $n = 13$ (74 screws)		5/74	3/76	CT
Huang et al. (2009)[3]	Retrospective cohort study (III)	42 patients: 29 males, 13 females; age: 24–64 years Fracture, spondylolisthesis, lumbar disk herniation (thoracic, lumbar)	G1: CT Nav $n = 21$ (104 screws) G2: 2D FluoroNav $n = 21$ (98 screws)		13/98	6/104	CT

2D FluoroNav, *2-dimensional fluoroscopy-based navigation*; 3D FluoroNav, *3-dimensional fluoroscopy-based navigation*; CT, *computed tomography*. Articles are stratified by insertion technique, specifically with distinction among 3D CT versus fluoroscopic guidance.
Reprinted from Du et al.,[41] by permission of Elsevier.

of isocentric fluoroscopy for intraoperative pedicle breach identification has compared favorably to postoperative CT in prior studies,[51] the analysis by Du et al. found greater radiographic screw accuracy with isocentric fluoroscopy rather than CT-based navigation. This has potential implications from a hospital/departmental purchasing perspective, whereby less-costly isocentric fluoroscopy units may become more attractive in the context of spinal navigation. Within CT-based systems, there does not appear to be a significant difference in accuracy between systems registering to pre- versus intraoperatively acquired CT; however, in current paradigms the intraoperative CT-based systems register significantly faster due to automatic registration protocols.[52] Nooh et al. report that differences in accuracy may exist even across manufacturers of CAN devices employing similar registration and imaging modalities.[53]

As the benefits of CAN are most evident in MIS and deformity-correcting procedures, where anatomic landmarks are less readily identifiable, the advantage of CAN in potentially reducing intraoperative fluoroscopy and its associated radiation cost in MIS procedures has also come under significant investigation. Intraoperative fluoroscopy, the current gold standard for the evaluation of real-time instrument positioning and spinal alignment, is associated with an average dose to the surgeon of 53.3 mrem/min at the torso in one study, greater in the hand and less in the neck, with variation in dose based on distance from the beam source and patient body habitus.[4] A multitude of subsequent studies has demonstrated reduced occupational radiation dose, i.e., to operating room (OR) personnel, with 3D fluoroscopy—based navigation,[12,54,55] as well as with intraop CBCT,[56–58] relative to standard C-arm fluoroscopy. However, while CAN reduces the radiation exposure to surgical and OR personnel, it does appear that this is more a result of shifting the burden of radiation to the patient rather than a reduction in overall radiation exposure. Early

spinal CAN systems registered to preoperative CT imaging, with the radiation cost to the patient of dedicated spinal imaging exceeding that of any other nonspinal musculoskeletal CT imaging by 10- to 12-fold.[59] With the development of more advanced CAN techniques registered to intraoperative imaging, the argument has been made that intraoperative 3D fluoroscopy or CBCT can both guide instrumentation and provide postimplantation imaging to check hardware accuracy, as a replacement for the otherwise obligate postoperative CT scan. However, particularly with larger patients, with longer instrumentation constructs, or with any inadvertent shifting of the DRF intraoperatively or other source of navigation error, multiple intraoperative imaging sequences may be required. Lange et al. have estimated that three or more intraoperative O-Arm imaging cycles, at standard manufacturer-recommended dosing, result in patient radiation exposure equivalent to that one of standard abdominal CT scan.[60] Therefore, while CAN techniques may reduce occupational radiation exposure for OR personnel, particularly in traditionally fluoroscopy-heavy procedures including MIS and deformity corrections, the burden of radiation exposure remains, and in the current paradigm of CAN techniques is shifted to the patient rather than eliminated entirely.[58]

With an increasing focus on value-based health care and efficiency optimization, CAN techniques have also been purported to improve surgical temporal workflow thereby reducing costly OR time.[61] Prolonged operative times have been associated with increased blood loss and more frequent infectious and ischemic complications,[62] though certainly there remains significant equipoise on this point in the literature, and the majority of surgical morbidity likely remains secondary to patient comorbidities and the treated pathology rather than operative time alone.[63] While perhaps not dramatically reducing operative times compared to traditional fluoroscopy-guided or freehand techniques, the use of CAN appears to be at least time equivalent. In a comparative study of O-Arm (3D CBCT) versus fluoroscopy guidance for MIS lateral interbody lumbar fusions, Zhang et al. demonstrated a statistically insignificant increase in operative time with CAN guidance.[64] However, Sasso et al. demonstrated a statistically significant time savings in posterior L5–S1 fusions with 3D fluoroscopy versus serial XR.[65] In a cadaveric setting, Webb et al. found total operative time equivalence for CAN-guided lateral interbody thoracolumbar fusion versus fluoroscopy.[66] In larger in vivo comparative studies, both Rajasekaran et al. and Tabaraee et al. found time equivalence for 3D CBCT–based navigation versus fluoroscopy for the placement of posterior thoracolumbar pedicle screws.[67,68] While a temporal efficiency benefit to CAN has yet to be demonstrated with current paradigms of navigation, there does appear to be a significant learning curve, with increased operative times early in the curve followed by time equivalence or even modest savings once sufficient familiarity has been achieved. While no specific number of cases to achieve "competence" has been postulated in the literature, a significant improvement in radiographic instrumentation accuracy was observed by Wood et al. after 50 cases of CT CAN–guided MIS lumbar pedicle screw placement.[69] Ryang et al. demonstrated substantial and statistically significant continual improvements in both temporal efficiency and radiographic instrumentation accuracy with 3D fluoroscopy–guided open thoracolumbar pedicle screw placement.[70] In fact, the learning curve in Ryang et al.'s study extended to the radiology technicians operating the 3D fluoroscopy CAN system, with continual improvements in scan time over the duration of the study. It therefore appears from the body of literature that while the current paradigm of spinal CAN does not offer significant workflow improvements relative to traditional fluoroscopy, time equivalence can typically be achieved following a significant learning curve for both surgeons and involved OR personnel.

In part from purported time savings, and in greater part from a potential reduction in complications and subsequent reoperations from misplaced instrumentation, an argument in favor of spinal CAN usage has been made from an economic and cost-effectiveness perspective. The literature on this subject is only recently beginning to expand, partly because comparative data on clinical complications and reoperation rates from CAN versus traditionally guided instrumentation have required long-term follow-up for adequate analysis. In the earliest economic analysis of CAN guidance, Watkins et al. found a nonstatistically significant reduction in revision surgeries for misplaced hardware with 3D fluoroscopy guidance (0.2%, vs. 3% with traditional fluoroscopy), with an associated cost of revision surgery of US $23,762 assuming a hospital stay of two nights.[71] Their navigation system of choice, a 3D fluoroscopy unit, had an upfront cost of US $475,000, not including annualized maintenance costs. In more recent studies, Hodges et al. approximated a 1% rate of revision surgery for thoracolumbar pedicle screws placed with traditional C-arm fluoroscopy, versus 0% with O-Arm CBCT guidance, at an average revision surgery cost of $17,650.[72] Sanborn et al. concluded that intraoperative O-Arm CBCT imaging was a cost-effective alternative to neuromonitoring or postoperative CT scanning for the confirmation of screw placement, albeit with a flawed analysis that accounted only for personnel costs of the imaging or monitoring techniques, and therefore attributed a cost of zero to O-Arm imaging.[73] Costa et al. compared OR costs of instrumented fusion procedures using a preoperative CT-based CAN device versus an O-Arm CBCT-based device, and concluded a nonsignificant cost savings of only 3.8% with CBCT due entirely to an average time savings of 27 min using intraoperative imaging as a result of the automated registration protocol, with no difference in clinical complications.[74] In the most thorough analysis to date, Dea et al. performed a retrospective comparative study of a prospectively maintained cohort of patients undergoing posterior spinal instrumentation with either O-Arm CBCT CAN or standard C-arm fluoroscopy. They concluded a cost of reoperation of CAD $12,618, and a statistically significant reduction in revision surgery rate of 5.2% with CAN guidance, thereby concluding cost-effectiveness of the CAN technique if more than 254 instrumented cases per year are performed at a given institution.[75] In this study, as a result of higher revision surgery costs, cost-effectiveness in the United States was achieved at a fewer number of cases, 168 per year. Recent data supports improved short-term clinical outcomes with CAN usage, with reduced 30-day reoperation rates for hardware malposition-related neurovascular complications as well as wound infections.[44,76,77] When long-term complications of misplaced hardware, including poor osseous fusion and construct loading leading to junctional failure, are taken into account, the economic argument in favor of CAN likely becomes more robust.[78]

Despite the increasing range of applications for spinal CAN described in the literature, adoption of CAN among spinal surgeons remains limited, without establishment of the technology as standard of care.[79] In the only study to date quantifying the current state of navigation usage, Hartl et al. surveyed a worldwide population of 3348 spinal surgeons, predominantly based in Europe, Latin America, and the Asia-Pacific region, and found a worldwide CAN usage rate of only 11%.[80] By contrast, 78% of surgeons in the same survey reported using fluoroscopy as their primary method of intraoperative image guidance. In separate surveys by Hartl et al. and Choo et al., the predominant barriers to spinal CAN adoption were a definitive lack of evidence supporting improved accuracy, workflow disruption primarily from cumbersome registration protocols, high capital costs, increased radiation exposure to either the patient and/or OR personnel, and steep learning curves.[80,81] With a body of literature reporting the safe and accurate placement of thoracic pedicle screws with freehand technique in highly experienced hands,[23] and lack of definitive clinical benefit and complication reduction with the use of CAN, albeit in short-term follow-up,[82] it is unsurprising that significant barriers remain to the widespread adoption of CAN.

Registration, imaging, and actuation techniques in spinal computer-assisted navigation

The basic tenet of intraoperative computer-assisted image guidance is real-time correlation of cross-sectional imaging data to patient anatomy, to provide surgeons with a view of structures that cannot otherwise be visualized directly. A unifying requirement for any frameless stereotactic navigation technique, in the spine or elsewhere, is registration of the imaging and patient spaces. The imaging dataset to be registered to, and the technique for registration itself, varies widely among various CAN techniques. A brief summary of imaging and registration techniques in contemporary CAN systems is presented in this section.

2D navigation

While the first published spinal CAN system provided 3D navigation based on a preoperative CT,[10] the prevalence and relative compactness of C-arm fluoroscopy units rendered 2D navigation, or "virtual fluoroscopy," the next step in evolution. Imaging in 2D navigation is performed using a standard C-arm fluoroscope modified with an attached calibration target. In typical workflow, a DRF is affixed to rigid patient anatomy, and XR images are taken with the C-arm in the planes desired for navigation (typically a cross-table lateral for sagittal views, and an anteroposterior view). As most "virtual fluoroscopy" systems rely on optical instrument tracking, a separate IR camera tracks the relative position of the C-arm (specifically the attached calibration target) and patient-mounted DRF during XR imaging, and automatically computes the transformation matrix required to register patient and image spaces. The position of tracked instruments can then be overlaid on the multiplanar XR images to allow for real-time navigated surgery with only the single fluoroscope.[12] Unfortunately, as their name belies, 2D navigation systems are hampered by their inability to display reconstructed axial views, leaving surgeons to mentally reconstruct the imaged planes into a 3D structure. Moreover, as with all plain radiographs, the quality of navigation images is entirely dependent on the quality of the initial acquisition XR, leaving much to be desired in obese or osteopenic patients.[14,83]

3D navigation

By definition, 3D CAN techniques provide full multiplanar reconstruction (MPR) of cross-sectional imaging, displaying the relevant anatomy in axial, sagittal, and coronal planes and, depending on the software package used, in a 3D reconstruction. All current spinal 3D CAN techniques are reliant on a DRF for maintaining image-to-patient registration and instrument tracking. The imaging modalities and techniques used to register this imaging to real-world anatomy vary among systems, and are discussed further in this section.

Imaging techniques

Imaging for navigation in 3D CAN techniques may be acquired either pre- or intraoperatively. Among systems registering to preoperative imaging, the most common imaging dataset is spiral CT, best if acquired at high-resolution thin slices of thickness <2 mm.[84] The typical workflow for preoperative CT-based techniques involves transferring of Digital Imaging and Communications in Medicine (DICOM) files to the CAN workstation preoperatively. A DRF is affixed to the patient, and manual registration of the preoperative CT dataset to patient anatomy is then performed using one of three techniques, described in greater detail in 'Registration Techniques'. Following successful image-to-patient registration, MPR views of the registered anatomy are displayed on a screen, and IGS may proceed. Preoperative CAN systems remain of value as they obviate the need for bulky intraoperative imaging devices, particularly in the case of intraoperative CT scanners, that are costly to both acquire and maintain, often require specially trained personnel to operate, and image at lower resolution than the fan-beam medical-grade scanners used for preoperative imaging. However, the manual registration protocol of current CAN systems registering to preoperative imaging significantly increases operative times,[85] and necessitates a preoperative CT scan with its associated cost and patient radiation exposure.[83,86] The accuracy of the initial registration is also highly operator dependent, as the quality of selected points for paired-point matching can significantly impact the robustness of the resultant transformation matrix.[87] Moreover, as these systems rely on an imaging dataset acquired with the patient in a supine position, whereas most navigated spinal procedures are performed in the prone position, intervertebral spinal mobility due to positioning may result in significant navigation inaccuracy, particularly when operating at levels distant to which the DRF is affixed.[10,83,86,88] The lack of associated intraoperative imaging also eliminates the ability to assess instrumentation placement or spinal alignment without a separate intraoperative imaging device, or a dedicated postoperative CT scan with its associated cost and patient radiation exposure.

In an effort to overcome some of these drawbacks associated with registration to preoperative imaging, 3D CAN systems with intraoperative imaging devices were developed. 3D intraoperative imaging techniques include isocentric fluoroscopy, or "3D fluoroscopy," CBCT, and FBCT. Isocentric fluoroscopy devices employ a C-arm which is motorized to automatically rotate a fixed amount (typically 190 degrees), centered about a point in the spine chosen by the operator (hence "isocentric"). A DRF is affixed to patient anatomy, as with all modern CAN techniques, and imaging proceeds with subsequent automatic transferring of images to the CAN workstation and image-to-patient registration. The first CBCT device, the O-Arm (Medtronic Sofamor Danek; Memphis, TN, USA), was introduced in 2006 and employs a similar fluoroscopy unit with flat-panel detector as isocentric fluoroscopy devices.[14] The O-Arm is differentiated from isocentric fluoroscopy by its ability to rotate a full 360 degrees, optimizing volume sampling and minimizing reconstruction artifacts due to limited projection views that are prominent with isocentric fluoroscopy devices. This is permitted by its toroid form factor, with a breakable gantry allowing lateral access to the operating table. DRF placement and automatic image transfer and registration otherwise proceed similar to isocentric fluoroscopy devices. The latest evolution in intraoperative imaging devices is mobile multirow FBCT images, more similar in design to the fixed scanners found in radiology departments than to the O-Arm.[14] These devices, examples of which include the BodyTom (Samsung Electronics America; Ridgefield Park, NJ, USA) and Airo (Brainlab AG; Munich, Germany), offer significantly better contrast resolution than flat-panel cone-beam devices (isocentric fluoroscopy and CBCT) with comparable imaging times. With a precalibrated relationship between CT gantry and operating table (and therefore patient) position, image transfer and registration are automatic as with isocentric fluoroscopy and CBCT devices. However, FBCT systems are associated with significantly greater capital cost, due in part to the requirement for a dedicated operating

room with proprietary attached operating table to facilitate ingress/egress from the CT gantry. Without flat-panel detectors as with standard C-arm fluoroscopes, FBCT devices also lack real-time fluoroscopy capabilities that are useful for initial incision and vertebral level localization.[14,86] The adoption of FBCT systems has therefore been limited largely to highly specialized academic institutions, with an unclear role in the future of spinal CAN. Nonetheless, each of the three intraoperative imaging techniques has advanced spinal CAN, by improving workflow through automated registration protocols, and providing real-time feedback on intraoperative spinal alignment and hardware placement through repeat imaging cycles.

There is ample evidence in the literature to suggest that all intraoperative imaging techniques reduce cumulative radiation dose to the surgeon and OR staff, as these personnel are typically able to leave the OR while navigation and verification scans are being performed via automated motorized actuation. Villard et al. demonstrated that the use of isocentric fluoroscopy, to guide lumbar pedicle instrumentation as well as verify hardware position postimplantation intraoperatively, reduces surgeon radiation exposure by almost 10-fold relative to standard fluoroscopy.[89] They also found that the cumulative radiation dose to the patient was halved, largely due to avoidance of a postoperative CT to verify screw position. In more recent studies of CBCT systems, Costa et al. found an intraoperative radiation dose to surgical staff that was essentially negligible, however with increased intraoperative exposure to the patient relative to standard fluoroscopy.[90] Similarly, Mendelsohn et al. found increased patient radiation exposure by 2.77-fold with CBCT navigation relative to literature values for fluoroscopy-guided thoracolumbar instrumentation, however with negligible surgeon radiation exposure.[57] Interestingly, they also found no difference in the need for postoperative XR or CT in patients who had undergone a navigated versus nonnavigated procedure, although their institution did not routinely perform intraoperative postimplantation verification CBCT scans, which might otherwise have reduced the need for postoperative scanning in the navigated cohort. Nonetheless, while all intraoperative imaging devices reduce radiation exposure to OR personnel, the burden of exposure is shifted somewhat to patients, to a significantly greater extent for CBCT/FBCT systems than with isocentric fluoroscopy, with the tradeoff of improved image quality and potentially navigation accuracy with CT systems.

Registration techniques

A fundamental requirement for any frameless stereotaxis is to identify a fixed relationship between patient and image spaces, a process termed registration. A transformation must be computed, which allows the mapping of any point in anatomical (real-world) space to its corresponding point in image space, thereby allowing instrument tracking in both environments.[9] Multiple techniques for image-to-patient registration have been proposed and iteratively refined over the past three decades, and are described in greater detail in this section. Early CAN techniques relied on paired-point matching, surface mapping, or a hybrid of these two techniques for registration. The advent of intraoperative imaging suites, including isocentric fluoroscopy, CBCT, and FBCT, has allowed for automatic registration protocols. Regardless of registration technique, in the current paradigm all techniques of spinal CAN first require a DRF to be rigidly affixed to patient anatomy in the operative position, followed by execution of the relevant registration process or intraoperative imaging with automatic registration. A given transformation linking patient and imaging spaces would remain valid only while the patient remained in the initial position; any movement of the patient or operating table following registration would render the initial

transformation obsolete, and a repeat calibration would have to be performed. The presence of a patient- or operating-table-affixed DRF therefore allows some patient motion to be compensated for, with only movement *relative to the DRF* unaccounted for by the initial registration transformation.

Paired-point matching

Paired-point transformation represents the earliest technique used to register imaging and patient spaces, and was employed in the first cranial and spinal CAN systems.[10] Points in the image space, a minimum of three, are matched to corresponding readily identifiable points in the patient space.[91] Points in patient space may be anatomic bony or skin-surface landmarks, adhesive skin-surface fiducials, or internal bone-affixed fiducials. In spinal surgery, anatomic bony landmarks are most commonly used for paired-point registration, due to the mobility and lack of adhesion of skin fiducials for posterior approaches, in contrast to the relative immobility of the scalp for skin-adhesive fiducials in cranial applications. Where skin-surface fiducials have been employed in spinal navigation, registration errors have been unacceptably large, up to 2 cm at the level of the disc space.[92] Bone-implanted fiducials are demonstrably superior in registration accuracy to anatomic or skin-marker fiducials, however are rarely used in spinal surgery due to their invasiveness.[93] In practice, while paired-point registration works well in principle with readily identifiable corresponding landmarks in both the patient and image spaces, it is time consuming intraoperatively and often tedious if the desired anatomic landmarks cannot be localized precisely, and prone to error whenever fiducials other than radiolucent implanted markers are used, impractical in most common procedures.[93]

Surface contour matching

An alternative method of manual registration, attempting to overcome some of the drawbacks of paired-point methodologies, is surface-based registration, described first by Pelizzari et al. in 1987.[94] Fundamentally, surface registration techniques attempt to align two surfaces, assumed to be rigid bodies, by iteratively applying scaling, translational, and rotational transformations until a particular distance or other error metric is minimized or otherwise optimized.[14,91] Surfaces in the imaging space are typically isolated from cross-sectional imaging by contouring and thresholding algorithms, which are beyond the scope of this discussion. Surfaces in the patient space may be acquired by a number of differing techniques. The earliest methods acquired surfaces point by point using a tracked tool, a highly time-intensive and laborious exercise.[91] Laser contouring devices may also be used to scan surfaces and are often used in commercial applications, however in their current form are typically less accurate than marker-based point-matching techniques.[95] Automated optical surface scanning via deformation of a projected structured light pattern may also be applied toward acquiring depth information in the patient space.[96]

Hybrid matching

The accuracy of surface scanning techniques is known to diminish with distance from the mapped surface.[97] A hybrid registration technique incorporating surface mapping for initial registration, followed by refinement with paired-point matching using deeper anatomic fiducials, has been proposed by Maurer et al. Deep point matching has been described using a standard optically tracked pointer, or with one-dimensional ultrasonic sensors applied percutaneously to reduce invasiveness.[98] As navigation accuracy at significant depth is most relevant in cranial procedures, hybrid techniques have not been widely explored in the spine, and will not be discussed in any greater detail in this chapter.

Automatic registration

The development of sophisticated intraoperative imaging devices, coupled with the tedious and time-consuming nature of manual registration procedures, has driven significant improvements in automatic image transfer and registration techniques. In the context of spinal surgery, the first automatic algorithms were developed to allow registration of patient anatomy to a preoperative CT using multiple intraoperative C-arm XR images (typically one lateral and multiple oblique views).[15] Fundamentally, all automatic registration algorithms employ a known relationship between the intraoperative imaging device and the output images, based on manufacturer/laboratory calibration of the imaging device, to allow subsequent automatic registration of the imaged patient anatomy to the imaging space dataset. With FBCT systems the operating table is fixed to the CT toroid, allowing for factory calibration of the relationship between the CT imaging plane and the operating table, with real-time adjustments made if the gantry angle is altered for the intraoperative scan. This relationship is updated intraoperatively using a patient-affixed DRF and an LED tracking array mounted to the scanner itself.

Automatic registration techniques with modern intraoperative imaging systems have been demonstrated to be more time-efficient than existing paradigms of registration to preoperative imaging,[99] however still requiring on average 8–9 min for a complete registration cycle, from initial clamping of the DRF to being ready to navigate.[100] Moreover, these registrations remain accurate only for the patient and DRF position at the time of intraoperative imaging; any subsequent manipulation or deformation of either patient anatomy or the DRF necessitates a repeat scan, with its associated time and patient radiation cost, as no rapid and radiation-free technique of updating a registration exists to date.

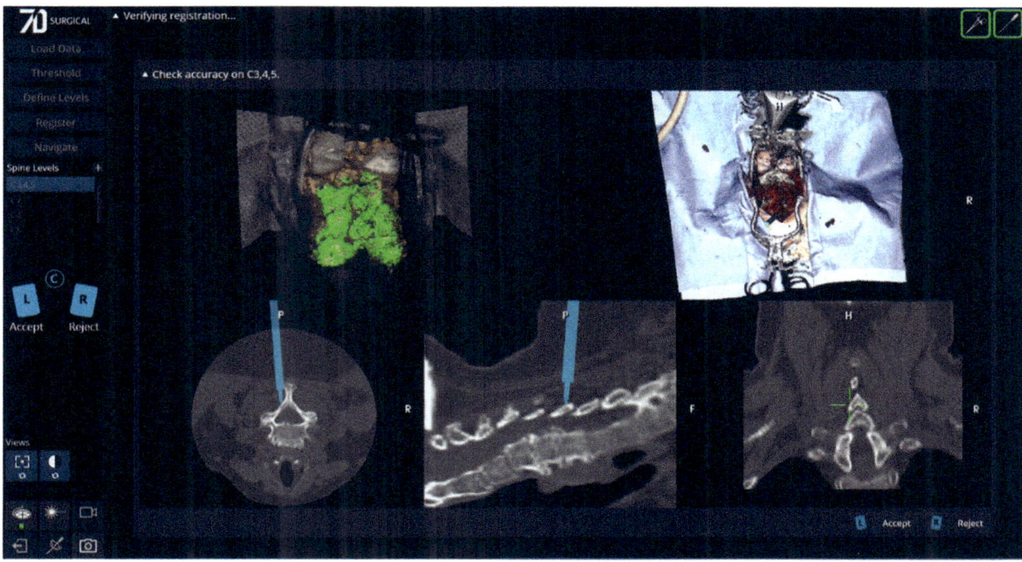

FIGURE 8.1   **Optical topographic imaging.** Screenshot of the user interface from the 7D Surgical Inc. spinal navigation platform, demonstrating optical topographic imaging acquisition of exposed anatomy in a standard open posterior cervical exposure (top left and right), with surgeon manual verification of registration accuracy using a tracked awl (bottom left, middle, and right).

Optical topographic imaging

Despite workflow improvements afforded by automatic registration to intraoperatively acquired imaging, the radiation exposure to patients and OR personnel with intraoperative imaging devices remains of concern. Conversely, while this radiation exposure is minimized by registration to preoperative CT or MRI with conventional techniques, these remain so cumbersome and time consuming as to deter the widespread adoption of navigation for spinal procedures.[101] Intraoperative stereoscopic optical imaging, or computer stereovision, is one technique that has been explored to allow radiation-free registration of a preoperative imaging dataset to an optically acquired surface map of intraoperative anatomy. In neurosurgical navigation applications, stereovision was first explored for the automated resolution of brain shift for updating of registrations for intracranial navigation.[102–104] Applications for open posterior spinal surgery were investigated more recently in the mid-2010s.[105] A commercial device employing active structured light illumination—based surface mapping, coupled with thresholding algorithms to appropriately isolate scanned bony anatomy, is currently marketed by 7D Surgical Inc. (Fig. 8.1).

Instrument tracking and actuation

Regardless of the method by which image-to-patient registration is achieved, all contemporary intraoperative navigation suites aim to subsequently track instruments in the merged coordinate system, to facilitate a given surgical maneuver. Instrument tracking may be accomplished through a number of techniques, which merits some discussion as tool tracking itself is associated with a quantifiable error, termed "jitter," which contributes to the final application accuracy.[106] Furthermore, the process of intraoperative instrument calibration and tracking contributes significantly to the overall usability and workflow of any CAN system; inadvertent bumping of the DRF required for optical instrument tracking, for instance, is a commonly cited pain point in current CAN workflows.[81]

Instruments in the earliest frameless CAN systems were tracked by mechanical articulated arms, in which multidirectional position sensors at each articulation enabled computation of the location and orientation of the instrument at the end effector of the arm.[107–109] Due to relative inaccuracy and sensitivity to ambient noise from other operating room equipment as well as to air temperature, acoustic tracking was largely abandoned in favor of optical or EM tracking, the two primary paradigms in current use.[110]

Contemporary optical tracking systems (OTSs) rely on an infrared camera typically mounted on a mobile platform in the operating room. In a passive OTS, an IR emitter is also located on the camera unit, and the reflection of the IR light from reflective markers on the tracked instruments and DRF are used to triangulate the pose and position of each tool. In an active OTS, IR LEDs are situated on both the DRF and each tracked instrument, with the IR light detected by the mobile camera unit and used to calculate the pose and position of each tool.[111] Both active and passive OTS devices are currently available commercially. There has been some suggestion, particularly among early-generation technologies, that active OTS provides lower tracking error and greater consistency than passive systems. These differences have been mitigated in modern devices largely through more powerful IR emitters and improved filtering techniques for the IR cameras; current optical systems are able to track individual markers with an accuracy of approximately 0.25 mm, and instrument tips at accuracies of 1-2 mm.[112] However, OTS tool tracking relies on a precalibrated relationship between the IR tracker array mounted on an instrument (either

passive reflective markers, or active LEDs) and the instrument tip, hence is unable to track needles and other nonrigid tools. More importantly, and of greater relevance in day-to-day use, accurate and real-time optical tracking necessitates constant line of sight between the IR camera ± emitter platform, and the DRF as well as the tracking array on each monitored instrument. Clutter in the surgical field, and even the physical position of OR personnel and the hand position of the tool operator, can greatly influence optical instrument tracking. Moreover, the relative positioning of the DRF, tracker arrays on each tool, and the camera unit can influence the tracking error of the system. It is also known that optical instrument tracking accuracy in vivo degrades over time as well as with increasing distance from the DRF.[113] Operator-dependent mechanisms to reduce optical tracking error therefore include situating the IR cameras as close to the surgical field as possible, and moreover aligning the camera z-axis with the direction in which clinical accuracy is least important, as well as ensuring that as many markers as possible are visible to the camera when affixing the DRF and tracking a given surgical instrument.

EM tracking systems were developed earlier than OTS technologies, however have come into widespread clinical use only recently due in part to their lack of reliance on line of sight as with optical systems. EM systems are composed of an EM field generator or emitter, typically a bulky device positioned near the operative field in nonsterile fashion, a DRF and a tracked instrument. Both the DRF and instrument contain sensor coils in which a voltage is produced within the EM field generated by the emitter; the magnitude and direction of the voltage generated in these cylindrical coils are used to compute the relative position and two-dimensional orientation of the instrument (pitch, yaw). Computation of the third dimension of orientation (roll, around the sensor coil's longitudinal axis) requires a combination of coils oriented orthogonally.[114] The primary advantage of EM tracking is that the coil sensors are small, on the order of <1 mm in diameter and <10 mm in length. They may therefore be embedded directly into instrument tips, including rigid and flexible needles, catheters, and endoscopes; as the instrument tip is tracked directly given that no line of sight to an external detector is required, this enables the tracking of flexible instruments within tissue cavities where line of sight is unobtainable.[111] However, EM tracking, particularly in older generation units, has been hindered by significant interference from surrounding ferromagnetic materials, including most standard surgical instruments and operating tables.

In all current navigation techniques, with instruments tracked either optoelectronically or electromagnetically, the surgical instruments are typically actuated or manipulated freehand by the surgeon, with continuous visual feedback from a display placing the tracked instrument in the shared coordinate space with the registered imaging dataset. Initial CAN suites tracked only a pointer or similar probe-type instrument; in the context of spinal surgery, the surgeon would then be required to manually follow the trajectory planned with the navigated pointer using untracked instruments for pedicle cannulation, tract preparation, and screw placement. As a result, some error of the final screw placement could result from slight deviations in manually actuated trajectory or entry point from that planned with the tracked pointer noncontemporaneously. Subsequent generations of CAN devices have spawned instrument sets to match evolving capabilities of modern trackers and intraoperative imaging devices, with dedicated tracked instruments for pedicle cannulation, tract tapping, as well as screw placement. This may potentially minimize the final application error of navigated screw placement, however still relies on freehand actuation of the navigated instrument by the operator, with continuous visual feedback from a navigation display. The latest advance in instrument actuation, in an effort to minimize the component of application error resulting from freehand manipulation of a tracked instrument, is robotic guidance. Multiple systems have been developed and are commercially available for spinal applications, all of which rely on optical IR tracking similar to freehand OTS techniques. A mobile platform with a robotically actuated arm is situated adjacent to the patient and, using an affixed optical tracking array, is registered into the same coordinate system as the patient anatomy and imaging dataset. The robotic end effector, which may include a guide for instruments for pedicle drilling/cannulation, tract preparation, and screw placement, is subsequently moved into the appropriate trajectory automatically with navigation guidance.[18] 25 series studying the accuracy of spinal instrumentation placement with various robotic guidance systems have been published to date, demonstrating radiographic accuracy rates of 85.0%–100%.[115] Comparative studies have demonstrated significant improvement in radiographic accuracy, as well as decrease in intraoperative patient and surgeon radiation exposure, with robotic guidance versus freehand or fluoroscopically guided instrumentation placement. However, only one study to date has compared robotically actuated to freehand-actuated navigated pedicle screw placement. In a three-armed prospective trial of freehand versus freehand-navigated versus robotic-navigated thoracolumbar pedicle screw placement, Roser et al. demonstrated no significant difference in screw placement accuracy with standard navigation or robotic guidance, but with significantly greater preoperative preparation time and greater intraoperative radiation dose with robotic versus standard navigation.[116] While robotic guidance may represent the next step in evolution of intraoperative CAN systems, particularly as integration with intraoperative imaging devices improves, in their current form there does not appear to be any discernible benefit relative to freehand-actuated navigation techniques.

Intraoperative ultrasound

Imaging coregistration or static anatomical CT or MRI data has become essential to spine surgery; however, it is traditionally limited by lack or real-time feedback. This type of feedback can inform changes in, or adjustment to, surgical plans during the procedure. In addition, traditional cross-sectional imaging modalities are also at times limited by ionizing radiation, MR-specific equipment limitations, speed, and size. To this end, an advantage of ultrasound has been speed, compact nature, and near-seamless integration without a need for protracted surgical pause.[117,118] In our experience, the ideal benefit is derived from high-frequency microultrasound (μUS) as a large depth of penetration is not required for cord anatomy, and the ensuring 10−20 μm resolution is helpful during spine surgery, as real-time feedback can empower delineations of tumor margins, lesion vascularity, and scope of decompression. With the improved resolution of μUS, white matter tracts delineation, nerve root visualization, and microvascularity can be elaborated with direct epidural or even subdural probe placement.

Coregistration of ultrasound

Ultrasound or μUS may also be fused with anatomic CT or MRI in order to provide real-time elaboration of tissue either not seen with the former or not seen dynamically with the latter. When possible, optical tracking may also be employed to place the target of insonation in the context of the cross-sectional imaging, which can be achieved with submillimetric precision on our current platform, particularly for laminectomy and tumor resection. Optical topographical imaging (OTI) can also be achieved with standard ultrasound platforms, and multimodality fusions (e.g., CT to MRI) can be tracked optically.

This technique is precedented in intracranial surgery brain, where OTI with CT and/or MRI has been implemented successfully. As has been described in this chapter, the OTI system operates via a 3D point cloud of the surface anatomy, and the probe is tracked by a binocular infrared-diode system as with surgical tools. The topographic map identified by infrared cameras is mapped to the anatomy anticipated on presurgical images, and the closest point at which the probe is being placed is deduced by an iterative algorithm.

Technique and use cases

In our practice, following bone removal, the surgical field is irrigated with sterile saline and the probes themselves are protected with sterile covers. At our institutions, sterilization is done with ethylene oxide or activated dialdehyde. The conventional ultrasound system most common employed is a Hitachi Hi Vision (Hitachi Aloka Inc., Twinsburg, OH) with central frequency and bandwidth of 10 and 4 MHz probes. The μUS platform utilized in our practice is the Vevo system (FUJIFILM VisualSonics, Toronto, ON) with central frequencies of 15, 30, and 50 MHz, with higher frequencies producing higher detail, but reducing the depth penetration of the imaging. In our experiences, the probe centered at 30 MHz allowed for visualization of the full spinal cord with impressive detail, potentially opening a new realm of diagnostic capability, allowing for visualization of critical structures including differentiation between gray and white matter, as well as delineation of vascular network details. The highest-frequency probes may allow for additional diagnostic benefit through fine delineation of tumors surrounding critical structures, where iatrogenic damage must be minimized, such as in intramedullary tumors; however, more study is required to determine the potential impact on patient outcomes of such guided resections.

The μUS has been utilized for several indications in craniospinal surgery at our institution, with each frequency providing various benefits with regard to specific pathologies. Demonstrated in the images below is the fine differentiation between gray and white matter, with particular structures clearly visible when using the midrange probe centered at 30 MHz. With regard to the image, we have drawn the visible delineation of various segments of the axial cross section in the C-spine, with the same being visible throughout in various regions. Additionally, a clear and quite impressive view of the arachnoid layer is observed in both longitudinal and axial cross sections. Additionally, visualization of the fibers of the cauda equina, as well as the transition from the conus, is exquisitely detailed. Each of these is detailed in Figs. 8.2 and 8.3.

Beyond visualization of common tissue structures, use in differentiation in spinal is equally observed. In patient with segmental ossification of the posterior longitudinal ligament and osseous cervical neuroforaminal stenosis, both canal decompression and relief after laminectomy can be assessed. In cases of meningioma, the system serves to delineate tumor extent and success of decompression and laminectomy where applicable for neuroforaminal extent. In meningioma cases, psammoma bodies have been grossly apparent on μUS such that diagnostic confidence can be honed intraoperatively. Similarly, the system has been utilized for aggressive hemangioma to assess cord compression and relief thereof. Regarding compression, the fibers of the cauda equina and transition from the conus can also be exquisitely detailed by ultrasound. Our experience comprises epidural lymphoma and suspected metastases at the junction of the conus and cauda where neural tissue and microvascular tumor networks can be seen. The modality also aids in the separation of tumor from the cord proper in preparation for radiotherapy (Fig. 8.4).

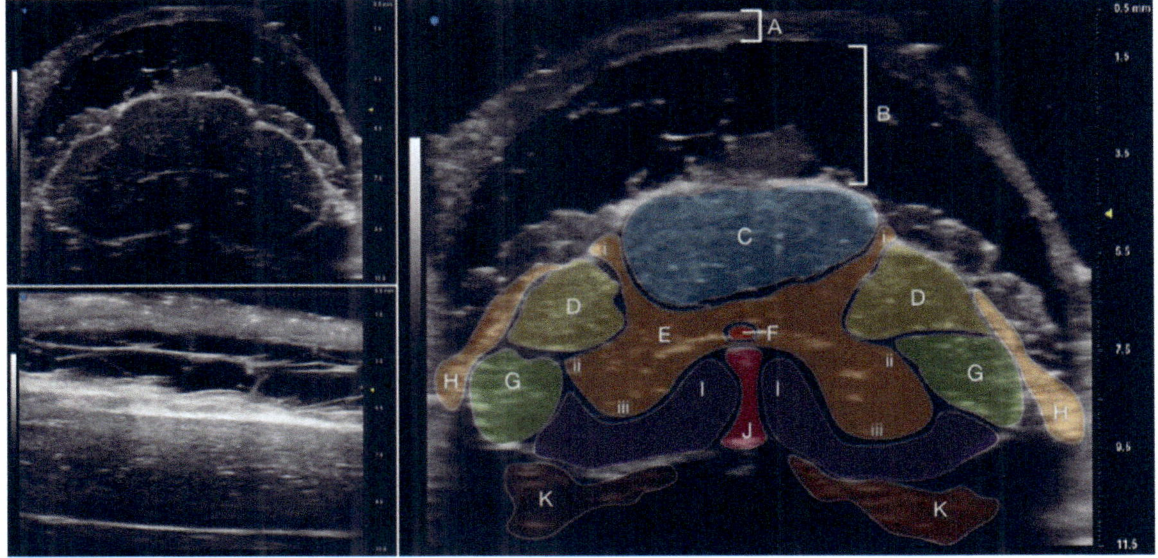

FIGURE 8.2 **Cross sections at C6 taken using μUS probe with central frequency of 30 MHz.** (Top left) An axial cross section of the spinal cord; (right) the same cross section with markup of visible sections delineated; and (bottom left) the longitudinal cross section. Labels for right are as follows: (A) dura mater; (B) arachnoid layer; (C) dorsal/posterior columns (white matter); (D) and (G) lateral columns; (E) gray matter with (i) posterior/dorsal, (ii) lateral, and (iii) ventral horns; (F) central canal; (H) dorsal root ganglion; (I) anterior/ventral columns; (J) anterior median fissure; and (K) ventral root.

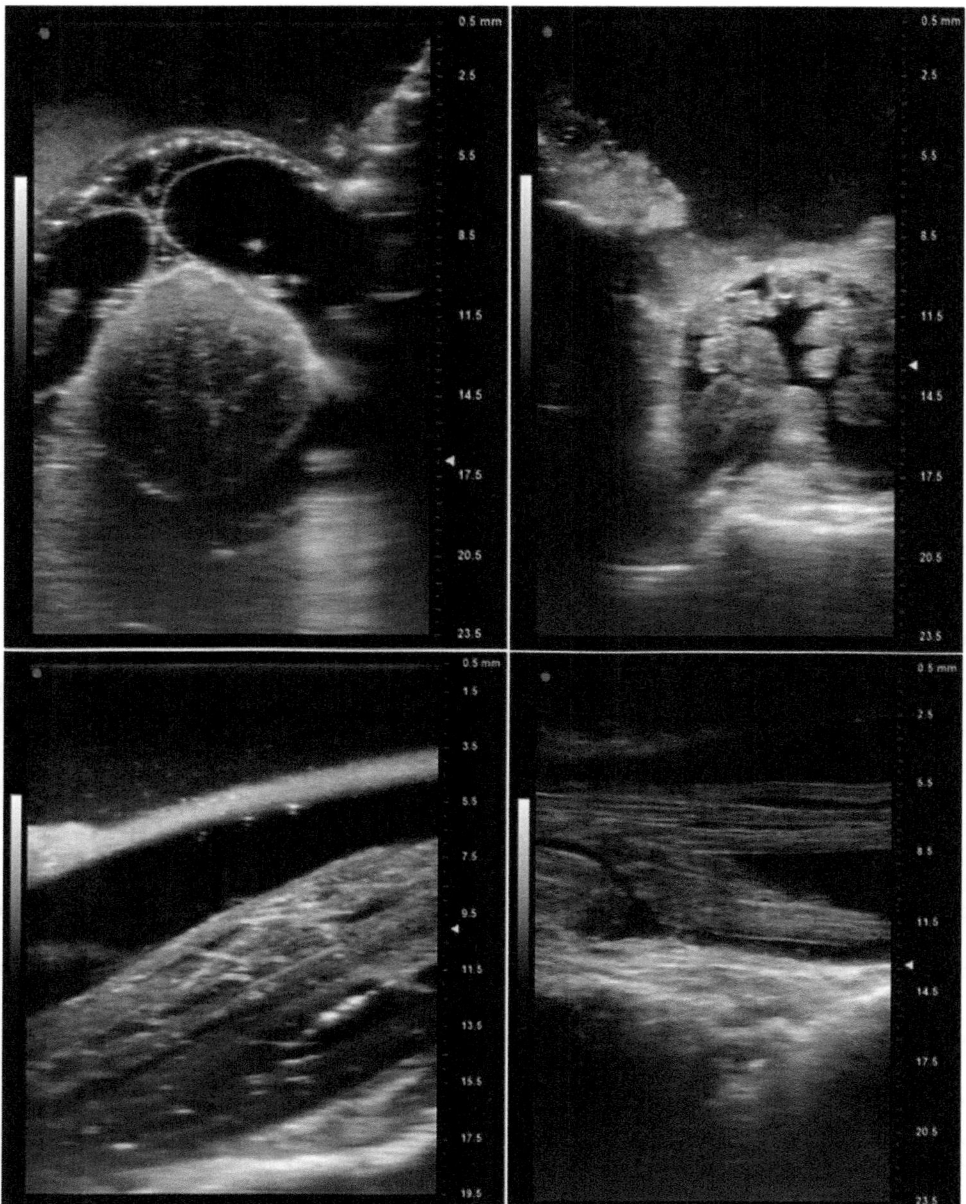

FIGURE 8.3   **Cross sections taken using the μUS probe with a central wavelength of 30 MHz.** (Top left) An axial cross section of the thoracic region; (Top right) an axial cross section of the lumbar region; (Bottom left) a longitudinal cross section at the transition area to the cauda equina; and (Bottom right) the individual fibers of the cauda equina. Note the individual fibers visible in both the axial cross section of the L-spine, as well as in both longitudinal cross sections.

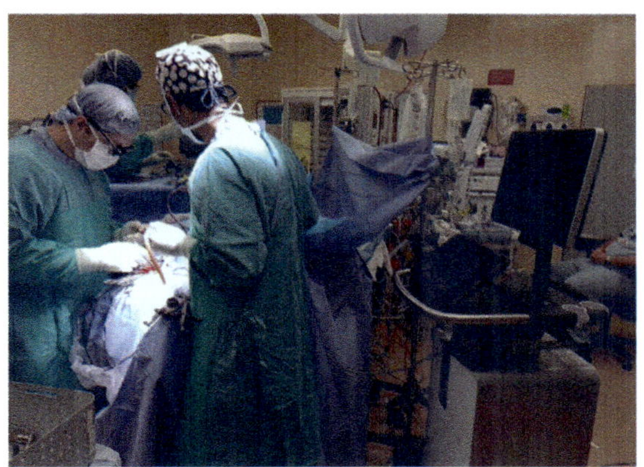

FIGURE 8.4 **Intraoperative utilization of high-frequency µUS.** Scrubbed operators are able to employ sonographic interrogation to a live surgical case with a sterile probe cover and display placed outside of the field. Optical topographical interfaces can also be integrated such that the main unit does not need to be positioned.

Future directions

Use of µUS permits the independent or coregistered tracked use of submillimetric resolution of spinal neural structures. While initial attempts at use have been focused upon benign disease and assessment of decompression, it is anticipated that real-time anatomic information will be able to guide malignant resection margins and provide feedback on safety to structures. Thus, in the case of spinal tumor resection where millimeter precision is required, dynamic imaging feedback can be provided during surgery and essentially guide tumor resection with higher precision. Furthermore, quantification of resection and residua can aid in the prescription of radiotherapy dose fractionation to optimize palliation while observing acceptable cord doses.[119] In addition, real-time information regarding macro- and microvascularity may aid in the assessment of tumor supply, success of embolization, and dangerous anastomoses. These principles may also be applied to vascular lesions and vascular malformations, which is also expected to benefit from multimodality fusion.

References

1. Pereira EAC, Green AL, Nandi D, Aziz TZ. Stereotactic neurosurgery in the United Kingdom: the hundred years from Horsley to hariz. *Neurosurgery* 2008;**63**(3): 594—607. https://doi.org/10.1227/01.NEU.0000316 854.29571.40.
2. Spiegel EA, Wycis HT, Marks M, Lee AJ. Stereotaxic apparatus for operations on the human brain. *Science* 1947;**106**(2754):349—50. https://doi.org/10.1126/ science.106.2754.349.
3. Mulconrey DS. Fluoroscopic radiation exposure in spinal surgery: in vivo evaluation for operating room personnel. *Clin Spine Surg* 2016;**29**(7):E331—5. https://doi.org/10.1097/BSD.0b013e31828673c1.
4. Rampersaud YR, Foley KT, Shen AC, Williams S, Solomito M. Radiation exposure to the spine surgeon during fluoroscopically assisted pedicle screw insertion. *Spine* 2000;**25**(20):2637—45.
5. Mroz TE, Abdullah KG, Steinmetz MP, Klineberg EO, Lieberman IH. Radiation exposure to the surgeon during percutaneous pedicle screw placement. *J Spinal Disord Tech* 2011;**24**(4):264—7. https://doi.org/ 10.1097/BSD.0b013e3181eed618.
6. Smith HE, Welsch MD, Sasso RC, Vaccaro AR. Comparison of radiation exposure in lumbar pedicle screw placement with fluoroscopy vs computer-assisted image guidance with intraoperative three-dimensional imaging. *J Spinal Cord Med* 2008;**31**(5):532—7.
7. Tjardes T, Shafizadeh S, Rixen D, et al. Image-guided spine surgery: state of the art and future directions. *Eur Spine J* 2010;**19**(1):25—45. https://doi.org/ 10.1007/s00586-009-1091-9.
8. Xu R-J, Yan Y-Q, Chen G-X, Zou T-M, Cai X-Q, Wang D-L. A method of percutaneous vertebroplasty under the guidance of two C-arm fluoroscopes. *Pak J Med Sci* 2014;**30**(2):335—8.
9. Grunert P, Darabi K, Espinosa J, Filippi R. Computer-aided navigation in neurosurgery. *Neurosurg Rev* 2003; **26**(2):73—99. https://doi.org/10.1007/s10143-003-0262-0.
10. Kalfas IH, Kormos DW, Murphy MA, et al. Application of frameless stereotaxy to pedicle screw fixation of the spine. *J Neurosurg* 1995;**83**(4):641—7. https:// doi.org/10.3171/jns.1995.83.4.0641.
11. Murphy MA, McKenzie RL, Kormos DW, Kalfas IH. Frameless stereotaxis for the insertion of lumbar pedicle screws. *J Clin Neurosci* 1994;**1**(4):257—60.
12. Foley KT, Simon DA, Rampersaud YR. Virtual fluoroscopy: computer-assisted fluoroscopic navigation. *Spine* 2001;**26**(4):347—51.
13. Fu T-S, Chen L-H, Wong C-B, et al. Computer-assisted fluoroscopic navigation of pedicle screw insertion: an in vivo feasibility study. *Acta Orthop Scand* 2004;**75**(6): 730—5.

14. Helm PA, Teichman R, Hartmann SL, Simon D. Spinal navigation and imaging: history, trends, and future. *IEEE Trans Med Imag* 2015;**34**(8):1738−46. https://doi.org/10.1109/TMI.2015.2391200.

15. Nolte LP, Slomczykowski MA, Berlemann U, et al. A new approach to computer-aided spine surgery: fluoroscopy-based surgical navigation. *Eur Spine J* 2000;**9**(Suppl. 1):S78−88. https://doi.org/10.1007/PL00010026.

16. Waschke A, Walter J, Duenisch P, Reichart R, Kalff R, Ewald C. CT-navigation versus fluoroscopy-guided placement of pedicle screws at the thoracolumbar spine: single center experience of 4,500 screws. *Eur Spine J* 2013;**22**(3):654−60. https://doi.org/10.1007/s00586-012-2509-3.

17. Euler E, Heining S, Fischer T, Pfeifer KJ, Mutschler W. Initial clinical experiences with the siremobil Iso-C^{3D}. *Electromedica-Erlangen* 2002;**70**(1):48−51.

18. Overley SC, Cho SK, Mehta AI, Arnold PM. Navigation and robotics in spinal surgery. *Where Are We Now* 2017;**80**(3). https://doi.org/10.1093/neuros/nyw077.

19. Smith JD, Jack MM, Harn NR, Bertsch JR, Arnold PM. Screw placement accuracy and outcomes following O-Arm-Navigated atlantoaxial fusion: a feasibility study. *Global Spine J* 2016;**6**(4):344−9. https://doi.org/10.1055/s-0035-1563723.

20. Shimokawa N, Takami T. Surgical safety of cervical pedicle screw placement with computer navigation system. *Neurosurg Rev* May 2016. https://doi.org/10.1007/s10143-016-0757-0.

21. Shin JH, Hoh DJ, Kalfas IH. Iliac screw fixation using computer-assisted computer tomographic image guidance: technical note. *Neurosurgery* 2012;**70**(1 Suppl):16−20. https://doi.org/10.1227/NEU.0b013e318230517a. Discussion 20.

22. Ray WZ, Ravindra VM, Schmidt MH, Dailey AT. Stereotactic navigation with the O-arm for placement of S-2 alar iliac screws in pelvic lumbar fixation. *J Neurosurg Spine* 2013;**18**(5):490−5. https://doi.org/10.3171/2013.2.SPINE12813.

23. Kim YJ, Lenke LG, Bridwell KH, Cho YS, Riew KD. Free hand pedicle screw placement in the thoracic spine: is it safe? *Spine* 2004;**29**(3):333−42. Discussion 342.

24. Fridley J, Fahim D, Navarro J, Wolinsky J, Omeis I. Free-hand placement of iliac screws for spinopelvic fixation based on anatomical landmarks: technical note. *Internet J Spine Surg* 2014;**8**(1). https://doi.org/10.14444/1003. 3−3.

25. Pisapia JM, Nayak NR, Salinas RD, et al. Navigated odontoid screw placement using the O-arm: technical note and case series. *J Neurosurg Spine* 2017;**26**(1):10−8. https://doi.org/10.3171/2016.5.SPINE151412.

26. DiGiorgio AM, Edwards CS, Virk MS, Mummaneni PV, Chou D. Stereotactic navigation for the prepsoas oblique lateral lumbar interbody fusion: technical note and case series. *Neurosurg Focus* 2017;**43**(2):E14. https://doi.org/10.3171/2017.5.FOCUS17168.

27. Joseph JR, Smith BW, Patel RD, Park P. Use of 3D CT-based navigation in minimally invasive lateral lumbar interbody fusion. *J Neurosurg Spine* 2016;**25**(September):339−44. https://doi.org/10.3171/2016.2.SPINE151295.

28. Komatsubara T, Tokioka T, Sugimoto Y, Ozaki T. Minimally invasive cervical pedicle screw fixation by a posterolateral approach for acute cervical injury. *Clin spine Surg* July 2016. https://doi.org/10.1097/BSD.0000000000000421.

29. Kim TT, Johnson JP, Pashman R, Drazin D. Minimally invasive spinal surgery with intraoperative image-guided navigation. *BioMed Res Int* 2016;**2016**:1−7. https://doi.org/10.1155/2016/5716235.

30. Kim TT, Drazin D, Shweikeh F, Pashman R, Johnson JP. Clinical and radiographic outcomes of minimally invasive percutaneous pedicle screw placement with intraoperative CT (O-arm) image guidance navigation. *Neurosurg Focus* 2014;**36**(3):E1. https://doi.org/10.3171/2014.1.FOCUS13531.

31. Nakashima H, Sato K, Ando T, Inoh H, Nakamura H. Comparison of the percutaneous screw placement precision of isocentric C-arm 3-dimensional fluoroscopy-navigated pedicle screw implantation and conventional fluoroscopy method with minimally invasive surgery. *J Spinal Disord Tech* 2009;**22**(7):468−72. https://doi.org/10.1097/BSD.0b013e31819877c8.

32. Vougioukas VI, Hubbe U, Schipper J, Spetzger U. Navigated transoral approach to the cranial base and the craniocervical junction: technical note. *Neurosurgery* 2003;**52**(1):247−50. Discussion 251.

33. Quillo-Olvera J, Lin G-X, Suen T-K, Jo H-J, Kim J-S. Anterior transcorporeal tunnel approach for cervical myelopathy guided by CT-based intraoperative spinal navigation: technical note. *J Clin Neurosci* November 2017. https://doi.org/10.1016/j.jocn.2017.11.012.

34. Jeon S, Lee GW, Jeon YD, Park I-H, Hong J, Kim J-D. A preliminary study on surgical navigation for epiduroscopic laser neural decompression. *Proc Inst Mech Eng Part H J Eng Med* 2015;**229**(10):693−702. https://doi.org/10.1177/0954411915599801.

35. Metz LN, Burch S. Computer-assisted surgical planning and image-guided surgical navigation in refractory adult scoliosis surgery: case report and review of the literature. *Spine* 2008;**33**(9):E287−92. https://doi.org/10.1097/BRS.0b013e31816d256e.

36. Bandiera S, Ghermandi R, Gasbarrini A, Barbanti Brodano G, Colangeli S, Boriani S. Navigation-assisted surgery for tumors of the spine. *Eur Spine J* 2013;**22**(Suppl. 6):S919–24. https://doi.org/10.1007/s00586-013-3032-x.

37. Nasser R, Nakhla J, Echt M, et al. Minimally invasive separation surgery with intraoperative stereotactic guidance: a feasibility study. *World Neurosurg* 2018;**109**:68–76. https://doi.org/10.1016/j.wneu.2017.09.067.

38. Tatsui CE, Nascimento CNG, Suki D, et al. Image guidance based on MRI for spinal interstitial laser thermotherapy: technical aspects and accuracy. *J Neurosurg Spine* 2017;**26**(5):605–12. https://doi.org/10.3171/2016.9.SPINE16475.

39. Stefini R, Peron S, Mandelli J, Bianchini E, Roccucci P. Intraoperative spinal navigation for the removal of intradural tumors: technical notes. *Oper Neurosurg* August 2017. https://doi.org/10.1093/ons/opx179.

40. Robinson S, Robertson FC, Dasenbrock HH, O'Brien CP, Berde C, Padua H. Image-guided intrathecal baclofen pump catheter implantation: a technical note and case series. *J Neurosurg Spine* February 2017:1–7. https://doi.org/10.3171/2016.8.SPINE16263.

41. Du JP, Fan Y, Wu QN, Wang DH, Zhang J, Hao DJ. Accuracy of pedicle screw insertion among 3 image-guided navigation systems: systematic review and meta-analysis. *World Neurosurg* 2018;**109**(January):24–30. https://doi.org/10.1016/j.wneu.2017.07.154.

42. Chan A, Parent E, Narvacan K, San C, Lou E. Intraoperative image guidance compared with free-hand methods in adolescent idiopathic scoliosis posterior spinal surgery: a systematic review on screw-related complications and breach rates. *Spine J* April 2017. https://doi.org/10.1016/j.spinee.2017.04.001.

43. Bourgeois AC, Faulkner AR, Bradley YC, et al. Improved accuracy of minimally invasive transpedicular screw placement in the lumbar spine with 3-dimensional stereotactic image guidance: a comparative meta-analysis. *J Spinal Disord Tech* 2015;**28**(9):324–9. https://doi.org/10.1097/BSD.0000000000000152.

44. Luther N, Iorgulescu JB, Geannette C, et al. Comparison of navigated versus non-navigated pedicle screw placement in 260 patients and 1434 screws: screw accuracy, screw size, and the complexity of surgery. *J Spinal Disord Tech* 2015;**28**(5):E298–303. https://doi.org/10.1097/BSD.0b013e31828af33e.

45. Mason A, Paulsen R, Babuska JM, et al. The accuracy of pedicle screw placement using intraoperative image guidance systems. *J Neurosurg Spine* 2014;**20**(2):196–203. https://doi.org/10.3171/2013.11.spine13413.

46. Tian NF, Huang QS, Zhou P, et al. Pedicle screw insertion accuracy with different assisted methods: a systematic review and meta-analysis of comparative studies. *Eur Spine J* 2011;**20**(6):846–59. https://doi.org/10.1007/s00586-010-1577-5.

47. Shin BJ, James AR, Njoku IU, Hartl R, Härtl R. Pedicle screw navigation: a systematic review and meta-analysis of perforation risk for computer-navigated versus freehand insertion. *J Neurosurg Spine* 2012;**17**(2):113–22. https://doi.org/10.3171/2012.5.spine11399.

48. Verma R, Krishan S, Haendlmayer K, Mohsen A. Functional outcome of computer-assisted spinal pedicle screw placement: a systematic review and meta-analysis of 23 studies including 5,992 pedicle screws. *Eur Spine J* 2010;**19**(3):370–5. https://doi.org/10.1007/s00586-009-1258-4.

49. Gelalis ID, Paschos NK, Pakos EE, et al. Accuracy of pedicle screw placement: a systematic review of prospective in vivo studies comparing free hand, fluoroscopy guidance and navigation techniques. *Eur Spine J* 2012;**21**(2):247–55. https://doi.org/10.1007/s00586-011-2011-3.

50. Amiot LP, Lang K, Putzier M, Zippel H, Labelle H. Comparative results between conventional and computer-assisted pedicle screw installation in the thoracic, lumbar, and sacral spine. *Spine* 2000;**25**(5):606–14.

51. Qureshi S, Lu Y, McAnany S, Baird E. Three-dimensional intraoperative imaging modalities in orthopaedic surgery. *J Am Acad Orthop Surg* 2014;**22**(12):800–9. https://doi.org/10.5435/JAAOS-22-12-800.

52. Costa F, Cardia A, Ortolina A, Fabio G, Zerbi A, Fornari M. Spinal navigation: standard preoperative versus intraoperative computed tomography data set acquisition for computer-guidance system: radiological and clinical study in 100 consecutive patients. *Spine* 2011;**36**(24):2094–8. https://doi.org/10.1097/BRS.0b013e318201129d.

53. Nooh A, Lubov AJ, Weber MH, et al. Differences between manufacturers of computed tomography — based computer-assisted surgery systems do exist. *Syst Liter Rev* 2017;**7**(1):83–94. https://doi.org/10.1055/s-0036-1583942.

54. Izadpanah K, Konrad G, Südkamp NP, Oberst M. Computer navigation in balloon kyphoplasty reduces the intraoperative radiation exposure. *Spine* 2009;**34**(12):1325–9. https://doi.org/10.1097/BRS.0b013e3181a18529.

55. Schafer S, Nithiananthan S, Mirota DJ, et al. Mobile C-arm cone-beam CT for guidance of spine surgery: image quality, radiation dose, and integration with interventional guidance. *Med Phys.* 2011;**38**(8):4563–74. https://doi.org/10.1118/1.3597566.

56. Abdullah KG, Bishop FS, Lubelski D, Steinmetz MP, Benzel EC, Mroz TE. Radiation exposure to the spine surgeon in lumbar and thoracolumbar fusions with the use of an intraoperative computed tomographic 3-dimensional imaging system. *Spine* 2012;**37**(17):E1074–8. https://doi.org/10.1097/BRS.0b013e318 25786d8.

57. Mendelsohn D, Strelzow J, Dea N, et al. Patient and surgeon radiation exposure during spinal instrumentation using intraoperative computed tomography-based navigation. *Spine J* 2016;**16**(3):343–54. https://doi.org/10.1016/j.spinee.2015.11.020.

58. Bandela JR, Jacob RP, Arreola M, Griglock TM, Bova F, Yang M. Use of CT-based intraoperative spinal navigation: management of radiation exposure to operator, staff, and patients. *World Neurosurg* 2013;**79**(2):390–4. https://doi.org/10.1016/j.wneu.2011.05.019.

59. Biswas D, Bible JE, Bohan M, Simpson AK, Whang PG, Grauer JN. Radiation exposure from musculoskeletal computerized tomographic scans. *J Bone Jt Surg Am Vol* 2009;**91**(8):1882–9. https://doi.org/10.2106/JBJS.H.01199.

60. Lange J, Karellas A, Street J, et al. Estimating the effective radiation dose imparted to patients by intraoperative cone-beam computed tomography in thoracolumbar spinal surgery. *Spine* 2013;**38**(5):E306–12. https://doi.org/10.1097/BRS.0b013e31 8281d70b.

61. Fan G, Han R, Gu X, et al. Navigation improves the learning curve of transforamimal percutaneous endoscopic lumbar discectomy. *Int Orthop* 2017;**41**(2):323–32. https://doi.org/10.1007/s00264-016-3281-5.

62. Baig MN, Lubow M, Immesoete P, Bergese SD, Hamdy E-A, Mendel E. Vision loss after spine surgery: review of the literature and recommendations. *Neurosurg Focus* 2007;**23**(5):E15. https://doi.org/10.3171/FOC-07/11/15.

63. Fogarty BJ, Khan K, Ashall G, Leonard AG. Complications of long operations: a prospective study of morbidity associated with prolonged operative time (> 6 h). *Br J Plast Surg* 1999;**52**(1):33–6. https://doi.org/10.1054/bjps.1998.3019.

64. Zhang Y-H, White I, Potts E, Mobasser J-P, Chou D. Comparison perioperative factors during minimally invasive pre-psoas lateral interbody fusion of the lumbar spine using either navigation or conventional fluoroscopy. *Global Spine J* 2017;**7**(7):657–63. https://doi.org/10.1177/2192568217716149.

65. Sasso RC, Garrido BJ. Computer-assisted spinal navigation versus serial radiography and operative time for posterior spinal fusion at L5-S1. *J Spinal Disord Tech* 2007;**20**(2):118–22. https://doi.org/10.1097/01.bsd.0000211263.13250.b1.

66. Webb JE, Regev GJ, Garfin SR, Kim CW. Navigation-assisted fluoroscopy in minimally invasive direct lateral interbody fusion: a cadaveric study. *SAS J* 2010;**4**(4):115–21. https://doi.org/10.1016/j.esas.2010.09.002.

67. Rajasekaran S, Vidyadhara S, Ramesh P, Shetty AP. Randomized clinical study to compare the accuracy of navigated and non-navigated thoracic pedicle screws in deformity correction surgeries. *Spine* 2007;**32**(2):E56–64. https://doi.org/10.1097/01.brs.0000 252094.64857.ab.

68. Tabaraee E, Gibson AG, Karahalios DG, Potts EA, Mobasser J-P, Burch S. Intraoperative cone beam–computed tomography with navigation (O-arm) versus conventional fluoroscopy (C-arm). *Spine* 2013;**38**(22):1953–8. https://doi.org/10.1097/BRS.0b013e 3182a51d1e.

69. Wood MJ, McMillen J. The surgical learning curve and accuracy of minimally invasive lumbar pedicle screw placement using CT based computer-assisted navigation plus continuous electromyography monitoring - a retrospective review of 627 screws in 150 patients. *Int J spine Surg* 2014;**8**. https://doi.org/10.14444/1027. 27–27.

70. Ryang Y-M, Villard J, Obermüller T, et al. Learning curve of 3D fluoroscopy image–guided pedicle screw placement in the thoracolumbar spine. *Spine J* 2015;**15**(3):467–76. https://doi.org/10.1016/j.spinee.2014.10.003.

71. Watkins RG, Gupta A, Watkins RG. Cost-effectiveness of image-guided spine surgery. *Open Orthop J* 2010;**4**:228–33. https://doi.org/10.2174/18743250010040 10228.

72. Hodges SD, Eck JC, Newton D. Analysis of CT-based navigation system for pedicle screw placement. *Orthopedics* 2012;**35**(8):e1221–4. https://doi.org/10.3928/01477447-20120725-23.

73. Sanborn MR, Thawani JP, Whitmore RG, et al. Cost-effectiveness of confirmatory techniques for the placement of lumbar pedicle screws. *Neurosurg Focus* 2012;**33**(1):E12. https://doi.org/10.3171/2012.2.focus121.

74. Costa F, Porazzi E, Restelli U, et al. Economic study: a cost-effectiveness analysis of an intraoperative compared with a preoperative image-guided system in lumbar pedicle screw fixation in patients with degenerative spondylolisthesis. *Spine J* 2014;**14**(8):1790–6. https://doi.org/10.1016/j.spinee.2013.10.019.

75. Dea N, Fisher CG, Batke J, et al. Economic evaluation comparing intraoperative cone beam CT-based navigation and conventional fluoroscopy for the placement of spinal pedicle screws: a patient-level data cost-effectiveness analysis. *Spine J* 2016;**16**(1):23–31. https://doi.org/10.1016/j.spinee.2015.09.062.

76. Xiao R, Miller JA, Sabharwal NC, et al. Clinical outcomes following spinal fusion using an intraoperative computed tomographic 3D imaging system. *J Neurosurg Spine* March 2017:1–10. https://doi.org/10.3171/2016.10.SPINE16373.

77. Fichtner J, Hofmann N, Rienmüller A, et al. Revision rate of misplaced pedicle screws of the thoracolumbar spine - comparison of 3D fluoroscopy navigated with freehand placement - a systematic analysis and review of the literature. *World Neurosurg* September 2017. https://doi.org/10.1016/j.wneu.2017.09.091.

78. Acikbas SC, Arslan FY, Tuncer MR, Matge G, Muciejczak A. The effect of transpedicular screw misplacement on late spinal stability. *Acta Neurochir* 2003;**145**(11):949–55. https://doi.org/10.1007/s00701-003-0116-0.

79. Schröder J, Wassmann H. Spinal navigation: an accepted standard of care? *Zentralbl Neurochir* 2006;**67**(3):123–8. https://doi.org/10.1055/s-2006-942146.

80. Hartl R, Lam KS, Wang J, Korge A, Kandziora F, Audige L. Worldwide survey on the use of navigation in spine surgery. *World Neurosurg* 2013;**79**(1):162–72. https://doi.org/10.1016/j.wneu.2012.03.011.

81. Choo AD, Regev G, Garfin SR, Kim CW. Surgeons' perceptions of spinal navigation: analysis of key factors affecting the lack of adoption of spinal navigation technology. *SAS J* 2008;**2**(4):189–94. https://doi.org/10.1016/S1935-9810(08)70038-0.

82. Wagner SC, Morrissey PB, Kaye ID, Sebastian A, Butler JS, Kepler CK. Intraoperative pedicle screw navigation does not significantly affect complication rates after spine surgery. *J Clin Neurosci* 2017;**47**:198–201. https://doi.org/10.1016/j.jocn.2017.09.024.

83. Holly LT, Foley KT. Intraoperative spinal navigation. *Spine* 2003;**28**(Suppl. ment):S54–61. https://doi.org/10.1097/01.BRS.0000076899.78522.D9.

84. Herz T, Franz A, Giacomuzzi SM, Bale R, Krismer M. Accuracy of spinal navigation for magerl screws. *Clin Orthop Relat Res* 2003;**409**(409):124–30. https://doi.org/10.1097/01.blo.0000053345.97749.a6.

85. Parker SL, McGirt MJ, Farber SH, et al. Accuracy of free-hand pedicle screws in the thoracic and lumbar spine: analysis of 6816 consecutive screws. *Neurosurgery* 2011;**68**(1):170–8. https://doi.org/10.1227/NEU.0b013e3181fdfaf4. Discussion 178.

86. Bourgeois AC, Faulkner AR, Pasciak AS, Bradley YC. The evolution of image-guided lumbosacral spine surgery. *Ann Transl Med* 2015;**3**(5):69. https://doi.org/10.3978/j.issn.2305-5839.2015.02.01.

87. Tamura Y, Sugano N, Sasama T, et al. Surface-based registration accuracy of CT-based image-guided spine surgery. *Eur Spine J* 2005;**14**(3):291–7. https://doi.org/10.1007/s00586-004-0797-y.

88. Ringel F, Villard J, Ryang Y-M, Meyer B. Navigation, robotics, and intraoperative imaging in spinal surgery. *Adv Tech Stand Neurosurg* 2014;**41**:3–22. https://doi.org/10.1007/978-3-319-01830-0_1.

89. Villard J, Ryang Y-M, Demetriades AK, et al. Radiation exposure to the surgeon and the patient during posterior lumbar spinal instrumentation: a prospective randomized comparison of navigated versus non-navigated freehand techniques. *Spine* 2014;**39**(13). https://doi.org/10.1097/BRS.0000000000000351.

90. Costa F, Tosi G, Attuati L, et al. Radiation exposure in spine surgery using an image-guided system based on intraoperative cone-beam computed tomography: analysis of 107 consecutive cases. *J Neurosurg Spine* 2016;**25**(5):654–9. https://doi.org/10.3171/2016.3.SPINE151139.

91. Eggers G, Mühling J, Marmulla R. Image-to-patient registration techniques in head surgery. *Int J Oral Maxillofac Surg* 2006;**35**(12):1081–95. https://doi.org/10.1016/j.ijom.2006.09.015.

92. Roessler K, Ungersboeck K, Dietrich W, et al. Frameless stereotactic guided neurosurgery: clinical experience with an infrared based pointer device navigation system. *Acta Neurochir* 1997;**139**(6):551–9.

93. Mascott CR, Sol JC, Bousquet P, Lagarrigue J, Lazorthes Y, Lauwers-Cances V. Quantification of true in vivo (application) accuracy in cranial image-guided surgery: influence of mode of patient registration. *Neurosurgery* 2006;**59**(1 Suppl. 1). https://doi.org/10.1227/01.NEU.0000220089.39533.4E.

94. Pelizzari C, Chen G. The use of computers in radiation therapy. In: Bruinvis IAD, editor. *Proceedings of the ninth international conference on the use of computers in radiation therapy held in Scheveningen, The Netherlands, June 22–25, 1987*; 1987. p. 590. North Holland.

95. Schlaier J, Warnat J, Brawanski A. Registration accuracy and practicability of laser-directed surface matching. *Comput Aided Surg* 2002;**7**(5):284–90. https://doi.org/10.1002/igs.10053.

96. Hoppe H, Däuber S, Kübler C, Raczkowsky J, Wörn H. A new, accurate and easy to implement camera and video projector model. *Stud Health Technol Inf* 2002;**85**:204–6.

97. Maurer Jr CR, Aboutanos GB, Dawant BM, Margolin RA, Maciunas RJ, Fitzpatrick JM. Registration of CT and MR brain images using a combination of points and surfaces. In: Loew MH, editor. *Proc. SPIE. International society for optics and photonics*; 1995. p. 109–23. https://doi.org/10.1117/12.208683.

98. Schmerber S, Chen B, Lavallee S, et al. Markerless hybrid registration method for computer assisted endoscopic ENT surgery. In: Lemke H, Vannier M, Inamura K, editors. *CAR*. Amsterdam: Elsevier; 1997. p. 799–806.

99. Kotani T, Akazawa T, Sakuma T, et al. Accuracy of pedicle screw placement in scoliosis surgery: a comparison between conventional computed tomography-based and O-Arm-Based navigation techniques. *Asian Spine J* 2014;**8**(3):331. https://doi.org/10.4184/asj.2014.8.3.331.

100. Nottmeier EW, Crosby T. Timing of vertebral registration in three-dimensional, fluoroscopy-based, image-guided spinal surgery. *J Spinal Disord Tech* 2009;**22**(5):358–60. https://doi.org/10.1097/BSD.0b013e31817dfcda.

101. Nottmeier EW. A review of image-guided spinal surgery. *J Neurosurg Sci* 2012;**56**(1):35–47.

102. Sun H, Lunn KE, Farid H, et al. Stereopsis-guided brain shift compensation. *IEEE Trans Med Imag* 2005;**24**(8):1039–52. https://doi.org/10.1109/TMI.2005.852075.

103. Paul P, Morandi X, Jannin P. A surface registration method for quantification of intraoperative brain deformations in image-guided neurosurgery. *IEEE Trans Inf Technol Biomed* 2009;**13**(6):976–83. https://doi.org/10.1109/TITB.2009.2025373.

104. DeLorenzo C, Papademetris X, Staib LH, Vives KP, Spencer DD, Duncan JS. Image-guided intraoperative cortical deformation recovery using game theory: application to neocortical epilepsy surgery. *IEEE Trans Med Imag* 2010;**29**(2):322–38. https://doi.org/10.1109/TMI.2009.2027993.

105. Ji S, Fan X, Paulsen KD, Roberts DW, Mirza SK, Lollis SS. Patient registration using intraoperative stereovision in image-guided open spinal surgery. *IEEE Trans Biomed Eng* 2015;**62**(9):2177–86. https://doi.org/10.1109/TBME.2015.2415731.

106. Khadem R, Yeh CC, Sadeghi-Tehrani M, et al. Comparative tracking error analysis of five different optical tracking systems. *Comput Aided Surg* 2000;**5**(2):98–107. https://doi.org/10.3109/10929080009148876.

107. Guthrie BL, Adler JR. Computer-assisted preoperative planning, interactive surgery, and frameless stereotaxy. *Clin Neurosurg* 1992;**38**:112–31.

108. Carney AS, Patel N, Baldwin DL, Coakham HB, Sandeman DR. Intra-operative image guidance in otolaryngology—the use of the ISG viewing wand. *J Laryngol Otol* 1996;**110**(4):322–7.

109. Alshail E, Rutka JT, Drake JM, et al. Utility of frameless stereotaxy in the resection of skull base and Basal cerebral lesions in children. *Skull Base Surg* 1998;**8**(1):29–38.

110. Tatar F, Mollinger J, den Dulk R, van Duyl W, Goosen J, Bossche A. Measurement position and orientation of surgery tools inside the human body using ultrasound. In: Margineanu, editor. *International conference on optimizatoin of electrical and electronic equipments OPTM. Brasov, Romania*. Transilvania University; 2002. p. 721–4.

111. Glossop ND. Advantages of optical compared with electromagnetic tracking. *J Bone Jt Surg Ser A* 2009;**91**(Suppl. 1):23–8. https://doi.org/10.2106/JBJS.H.01362.

112. Wiles AD, Thompson DG, Frantz DD. Accuracy assessment and interpretation for optical tracking systems. *SPIE Med Imag* 2004:421. https://doi.org/10.1117/12.536128.

113. Quiñones-Hinojosa A, Robert Kolen E, Jun P, Rosenberg WS, Weinstein PR. Accuracy over space and time of computer-assisted fluoroscopic navigation in the lumbar spine in vivo. *J Spinal Disord Tech* 2006;**19**(2):109–13. https://doi.org/10.1097/01.bsd.0000168513.68975.8a.

114. Milne AD, Chess DG, Johnson JA, King GJW. Accuracy of an electromagnetic tracking device: a study of the optimal operating range and metal interference. *J Biomech* 1996;**29**(6):791–3.

115. Joseph JR, Smith BW, Liu X, Park P. Current applications of robotics in spine surgery: a systematic review of the literature. *Neurosurg Focus* 2017;**42**(5):E2. https://doi.org/10.3171/2017.2.FOCUS16544.

116. Roser F, Tatagiba M, Maier G. Spinal robotics: current applications and future perspectives. *Neurosurgery* 2013;**72**(Suppl. 1):12–8. https://doi.org/10.1227/NEU.0b013e318270d02c.

117. Prada F, Vetrano IG, Filippini A, et al. Intraoperative ultrasound in spinal tumor surgery. *J Ultrasound* 2014;**17**(3):195–202. https://doi.org/10.1007/s40477-014-0102-9.

118. Jakubovic R, Ramjist J, Gupta S, et al. High-frequency micro-ultrasound imaging and optical topographic imaging for spinal surgery: initial experiences. *Ultrasound Med Biol* 2018;**44**(11):2379–87. https://doi.org/10.1016/j.ultrasmedbio.2018.05.003.

119. Cox BW, Spratt DE, Lovelock M, et al. International spine radiosurgery consortium consensus guidelines for target volume definition in spinal stereotactic radiosurgery. *Int J Radiat Oncol Biol Phys* 2012;**83**(5):e597–605. https://doi.org/10.1016/j.ijrobp.2012.03.009.

Upper cervical spine and spinal cord injuries

Erik Hayman[2], Rod J. Oskouian[1], Jens R. Chapman[1]

[1]Swedish Neuroscience Institute, Seattle, WA, United States; [2]Department of Neurosurgery, University of South Florida, Tampa, FL, United States

Traumatic disruption of the atlas, axis, and their ligamentous attachments to the occiput and the lower cervical spine form a unique class of injuries, collectively termed upper cervical spine injuries. Several features of the upper cervical spine significantly complicate management of upper cervical spinal cord injury, including its complex range of movement, intimate relationship to the posterior cerebral vasculature, and unique osseoligamentous anatomy. Most importantly, injury to the cervicomedullary junction may rapidly compromise respiratory function, leading to high early mortality rates unless promptly diagnosed and treated.

The goal of this chapter is to provide an overview of some of these specific issues. Following a review of diagnostic considerations and a brief epidemiological review, we will address basic preoperative management of patients with upper cervical spine injuries, particularly with regard to differences in the management to subaxial spine injuries. We will then review strategies for definitive management of these injuries, particularly with regard to outcomes. Finally, we will conclude with management of common comorbid conditions in upper cervical spinal cord injury, focusing on blunt cerebrovascular injury (BCVI) and chronic ventilator dependence.

Diagnostic considerations

As with any medical condition, diagnosis of upper cervical spine trauma remains the first and most important step in treatment. Timely diagnosis is especially important in injuries of the craniocervical junction and upper cervical spine, as a failure to properly treat and stabilize injuries in this area may have severely debilitating, if not fatal, consequences.

In the case of upper cervical spine trauma, maintaining sufficient suspicion to evaluate for this condition itself represents a significant clinical challenge. Historically, upper cervical spine injury was a relatively rare clinical entity but a common pathological one. One autopsy study of 146 victims of fatal traffic accidents identified upper cervical spine injuries in nearly one-fifth of these patients. Due to a combination of improvements in resuscitation and changing patterns of injury,

© 2022 Elsevier Inc. All rights reserved.

patients with upper cervical spine injuries are increasingly recognized in a salvageable state. Nevertheless, injuries of the upper cervical cord represent a minority of patients presenting with spinal cord injury.

Aside from its rarity, several other aspects of upper cervical spinal cord injury make its diagnosis challenging. Trauma of sufficiently high magnitude to disrupt the relatively strong ligaments of the craniocervical junction tends to occur via blunt high-energy impact mechanisms, such as motor vehicle collisions. In one series of such patients, patients with evidence of severe craniocervical injury presented with an average of three injured organ systems.[1] These polytraumatized patients can thus have numerous confounding injuries, such as a limb immobilized by a fracture or a visceral injury requiring urgent exploration, that can serve to limit full examination or distract the clinician. Moreover, greater than two-thirds of these patients present with skull fractures, facial fractures, or intracranial injury. The associated loss of global neurologic function increases the challenge of examining these patients. As such, an elevated degree of suspicion for craniocervical injuries should be maintained in the initial evaluation of patients with polytrauma, especially in those with skull or facial fractures. Due to the high rate of neurologic injury associated with delayed diagnosis, maintaining suspicion of injury itself is potentially life-saving. For this reason a formalized secondary trauma assessment has been suggested as a meaningful countermeasure to missed craniocervical injuries in multiply injured patients with incomplete initial assessment.[1]

Although sometimes neglected in polytraumatized patients requiring resuscitation, the initial physical and neurologic examination provides valuable opportunities to identify craniocervical injury. Methodical and observant physical examination can guide a clinician toward early diagnosis of craniocervical injury. Craniocervical deformity, tenderness of the upper cervical spine, or the various stigmata of craniofacial such as lacerations or facial swelling should increase suspicion on the part of the clinician with regard to craniocervical instability. The neurologic exam may also provide valuable clues. Due to its complex anatomy, the injuries at the craniocervical junction manifest in over three-fourths of patients with a variety of incomplete and atypical asymmetric motor and cranial nerve findings, the so-called cervical medullary syndromes summarized in Table 9.1 and elaborated on in a later section of this chapter.[1] These unique neurologic findings are highly suggestive of upper cervical spinal cord injury and should prompt urgent stabilization and focused diagnostic evaluation. Equally important is accurate and concurrent documentation of these neurologic findings to facilitate early recognition of delayed neurologic decline, a common presentation of craniocervical injuries.[2] Finally, due to aforementioned confounding factors associated with initial workup of these injuries, a formal secondary survey remains of paramount importance in this patient population, as craniocervical instability is frequently missed on initial surveys.

Historically, diagnosis of craniocervical injuries on plain films presented a challenge for both clinicians and radiologists. Fortunately, the incorporation of CT scanning into trauma screening algorithms has significantly improved the diagnostic ease and accuracy of pathology. Modern trauma head CT scans routinely incorporate the upper cervical spine, while large trauma centers routinely perform full craniocervical CT scans as part of whole body survey imaging. Unsurprisingly, this improved imaging corresponds to a dramatic increase in the diagnosis of craniocervical pathology. For example, the entirety of the clinical literature on occipital condyle fractures in the late 1980s consisted of approximately 20 cases, whereas modern case series recognize that these fractures are found in roughly 1% of trauma patients undergoing a CT scan of this region.[3] Obviously, CT provides the gold standard for diagnosis of fractures and other osseous lesions. Furthermore, CT findings such as facet widening, significant retropharyngeal hematoma, abnormal angulation, or posterior fossa hematoma provide valuable indirect indications of ligament injury and potential instability; some of these parameters to assess ligamentous instability are depicted in Fig. 9.1. However, CT scanning is not infallible

TABLE 9.1   Cervicomedullary syndromes.

Name	Clinical presentation	Anatomic basis
Cruciate paralysis of bell	Spastic paresis of bilateral upper extremities with relative sparing of lower extremities	The motor tracts of the upper extremities decussate at a more cranial level than the lower extremities. Central compression at this level will affect these more medial structures while sparing the more laterally nondecussated lower extremity tracts.
Wallenberg's/ lateral medullary syndrome	Vertigo, ipsilateral cerebellar signs, loss of pain/ temperature sensation in ipsilateral face and contralateral body, ipsilateral loss of pharyngeal function, ipsilateral Horner's	Compromise of the lateral aspect of the medulla within PICA, most notably due to vascular causes such as vertebral artery injury with sparing of the medial medulla.
Hemiplegia cruciata	Spastic paresis of the ipsilateral upper extremity and contralateral lower extremity	Compression of the motor tracts of the upper and lower extremity after the decussation of the upper extremity fibers but prior to the lower extremity decussation.
Bulbar palsy	Dysphagia, difficulty with handling secretions, dysarthria, loss of gag and jaw jerk reflex	Extension of edema into medulla and lower pons with compromise of the nuclei of cranial nerves IX–XII; direct compression of lower cranial nerves at the skull base.

Summarized are the various neurologic syndromes associated with upper cord injuries, highlighting their often divergent and irregular clinical presentations.

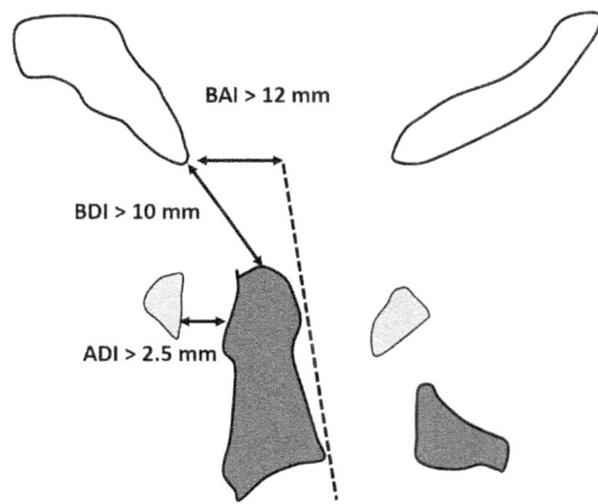

FIGURE 9.1   **Radiographic measures of craniocervical instability.** Depicted are some of the standard radiographic parameters for diagnosing craniocervical instability. These include the distance between the basion and dens (basion dens interval or BDI) of greater than 10 mm, a distance between the basion and the posterior border of the axis (basion axial interval or BAI) of greater than 12 mm, and distance between the atlas and the dens (atlantodental interval) of greater than 2.5 mm. Of note, in highly unstable injuries these parameters may be falsely normal and should not be substituted for MRI or other more direct measures of ligamentous integrity.

in the diagnosis of upper cervical spine pathology; for example, initial CT scan identified only 35% of patients with craniocervical dissociation in one series.[1] As such, indirect indicators of instability such as significant retropharyngeal hematoma should prompt further diagnostic workup even in the absence of abnormal distraction, deformity, or subluxation, as highly unstable craniocervical injuries may assume a transient, deceptively normal alignment on CT imaging. Similarly, a relatively normal CT scan in a patient with neurologic findings concerning for an upper cervical spine injury should prompt further workup.

MRI remains the gold standard for cervical examination in trauma patients, especially those too obtunded to cooperate with clinical evaluation. The advantages of MRI in evaluation of upper cervical spine injuries are numerous. MRI provides a direct assessment of ligamentous injury, with sequences such as STIR readily identifying ligamentous strain and tears. MRI provides direct imaging of relevant neurologic structures, allowing for confirmation and localization of traumatic spinal cord lesions. Most importantly, MRI may identify nonosseous compression of the spinal cord, such as compression from a traumatic disc herniation or epidural hematoma. Despite the risks of MRI scan in a potentially unstable trauma patient, its benefits at very least promote strong consideration in patients with clinical concerns for upper cervical spine injury even with a normal CT scan.

Prior to the widespread clinical availability of MRI, determination of transverse atlantal ligamentous (TAL) integrity relied on indirect markers of stability. In the case of a fractured atlas, the "Rule of Spence" served as the predominant assessment of TAL integrity: if, on an anterior posterior view, the lateral masses of the atlas demonstrate greater than 7 mm of lateral overhang relative to C2, the transverse ligament is disrupted. Despite its convenience, however, recent studies have challenged the validity of this rule, demonstrating clear radiographic disruption of the TAL in the absence of significantly displaced C1 lateral masses.[4] As such, MR evaluation of the transverse ligament is strongly indicated in patients with evidence of significant C1 or C2 injury, even in the absence of spinal cord injury.

Assessment of integrity of the atlantooccipital instability represents another significant clinical challenge, due to the relative subtlety of its associated clinical findings. Although a number of radiographic parameters have been adapted to CT imaging in order to screen for AOD, evaluation of the gap between the occipital condyle and the C1 facet (CCI) represents a simple, reproducible, and highly sensitive measure of craniocervical instability: if either CCI is greater than 2.5 mm or if their combined sum is greater than 5 mm, AOD should be suspected based on CT criteria alone.[5] Alternatively, when AOD is clinically suspected, MRI should be obtained to rule out significant disruption of the ligamentous stabilizers. Finally, in patients with suspected craniocervical dissociation but equivocal radiographic findings, both cadaveric and clinical studies support provocative traction testing to identify unstable craniocervical injuries, with significant enlargement of the condylar gaps after 5—10 lbs of axial traction indicating incompetence of the craniocervical junction and need for fusion.[1,6]

Anatomy and injury classification

Osseous and ligamentous structures

Usually, the upper cervical spine anatomically and functionally consists of the coupled three bone segments of the occiput, the atlas, and the axis (C0, C1, and C2, respectively). Despite representing an anatomically small proportion of the cervical spine, the upper cervical spine plays an outsized functional role in neck motion, accounting for approximately 30 degrees of flexion and extension, two-thirds of which are derived from the C0—C1 segment, and between 20 and 40 degrees of axial rotation, most of which derives from C1—2 segment.[7,8] The unique anatomy of this region reflects this functionality, with a particular importance of ligamentous integrity in maintaining the stability of these uniquely mobile and functionally interconnected segments. The importance of these ligaments is readily demonstrated in the evaluation of traumatic injuries, where degree of ligamentous, rather than osseous, injury provides the basis for determining biomechanical stability (Fig. 9.2A).

The first cervical vertebra, the atlas, consists of a ringlike structure consisting of two large lateral masses connected via anterior and posterior arches. In contrast to other vertebra, the atlas lacks a true vertebral body and associated disc space. As such, the lateral masses of C1 represent its principal structural elements. Motion between the skull and the atlas derives primarily from occipital condyles, which permit a wide degree of flexion and extension while limiting translation, distraction, and rotation at these levels. As such capsular integrity of these joints arguably represents the most important determinant of atlantooccipital stability.[9] Other

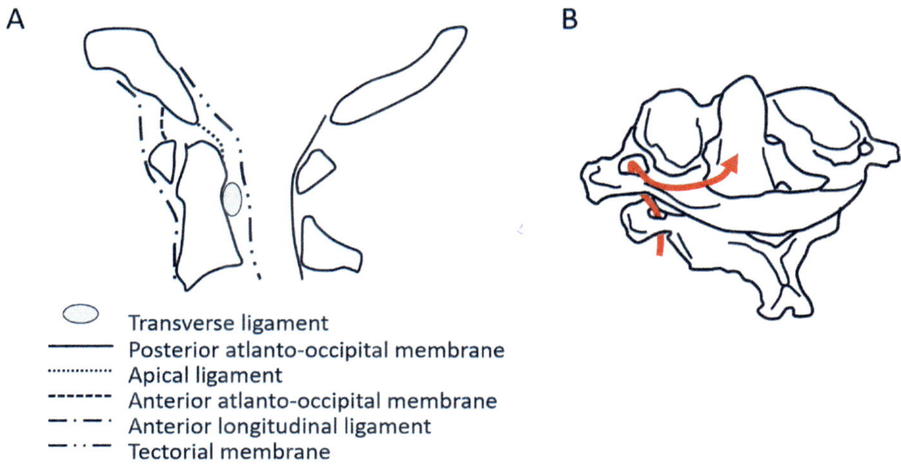

FIGURE 9.2 **Anatomy of the craniocervical junction.** (A) Depiction of the ligamentous anatomy of the craniovertebral junction, sagittal view. (B) Course of the vertebral artery in relation to C1—2. Note that the intimate relationship between the vertebra and the artery places it at risk during various surgical procedures including dissection along C1, drilling the C1—C2 facet, and instrumentation.

important ligamentous stabilizers of C0—C1 include the alar ligaments between the dens and the clivus, which limit rotation of C0 on C1, and the tectorial membrane.[10] These ligamentous structures together provide a strongly redundant stability to the craniocervical junction, with disruption of all three structures required to destabilize the junction.[6] By contrast, the apical ligaments of the dens appear vestigial and most likely do not act as a physiological stabilizer.[11]

The second cervical vertebra, the axis, in many ways resembles the lower cervical vertebrae with the exception of the dens, a large peg-like structure which projects from the body of C2 into the anterior aspect of the C1 ring. The dens serves as a spokelike anchoring structure, providing the vast majority of rotational movement of the head in relation to the body. The transverse ligament of the atlas, a tough tissue band connecting the two lateral masses of C1, serves to confine the dens to the anterior aspect of C1 and prevents translation of C1 on C2 while allowing for relatively free rotation. As such, the transverse ligament represents the principal stabilizer of C1 and C2, although the capsular ligaments of the C1—2 facet also play important roles as stabilizers. Stability of C2 in relation to C3 is largely a product of the facet capsules and the disc space, with facets providing approximately one-third of the stability to torsion and flexion in analogy to the other cervical levels.[12]

The complexity of fracture patterns and ligamentous disruptions in the upper cervical spine necessitates a unique set of injury classification distinct from the subaxial cervical spine. Historically, clinicians have divided upper cervical spine injuries into several functionally separate subcategories, based primarily on fracture morphology and ligamentous stability: (i) condylar fractures and ligamentous disruptions of the craniocervical junction,[3,13] (ii) fractures of the atlas with or without transverse ligament disruption (28816879), (iii) fractures of odontoid process of C2 (dens fracture),[14] (iv) fractures of the pars of C2 (Hangman's fractures),[15] and (v) fractures of the vertebral body of C2 (Benzel fractures).[16] Although each classification system possesses unique advantages and disadvantages, the existence of multiple conflicting schemes impairs communication among physicians and comparisons within the literature; moreover, virtually all classification schemes consider injuries in isolation, whereas a significant proportion of patients exhibit multiple fracture types.

Recent work by the AOSpine group has led to development of a simplified classification system to redress these shortcomings.[17] The AOSpine upper cervical spine classification initially divides these injuries into three anatomic areas: (I) occipital condyle/craniocervical junction (OC), (II) C1 ring/C1−2 joint (C1), and (III) C2/C2−3 joint (C2). Each injury area further divides into an injury morphology. Type A injuries represent pure bone injuries without ligamentous disruption, Type B injuries represent ligamentous injuries with or without bone injury, and Type C injuries represent ligamentous injuries with gross evidence of displacement or translation. Furthermore, each injury is subject to modifiers. M1 signifies a highly unstable injury, such as a displaced C1 fracture with a midsubstance tear of the transverse ligament. M2 signifies an injury at high risk for nonunion with nonoperative treatment such as a widely displaced fracture. M3 signifies patient factors that might affect treatment, such as tobacco use and its associated pseudoarthrosis risks. M4 signifies a vertebral artery injury or aberrant vertebral artery anatomy impacting treatment. Finally, the use of neurology classifiers (N0−4) signifies degree of neurologic impairment. Taken together, the AOSpine classification greatly simplifies communication. As an example, a patient with a complex Hangman's fracture with angulation, a vertebral artery injury, and no neurologic deficit simplifies to C2 Type C N0 M4. This greatly simplified classification system provides a simple and logical means of communicating both fracture morphology while simultaneously providing an implicit treatment algorithm. Although this classification demonstrates well-documented intra- and interobserver reliability, whether general practitioners will adopt this more comprehensive categorization of craniocervical injuries remains to be seen.

Arterial structures

The course of the vertebral artery intimately associates with the upper cervical spine, coursing through the transverse foramina of C2 and C1 in its V3 section, before traveling posteromedially in the sulcus arteriosus on the cranial aspect of C1 and piercing the dura in its V4 segment (Fig. 9.2B). Due to this close association, several forms of upper cervical injury place the vertebral artery at risk for blunt cerebrovascular injury, particularly those involving fractures through the transverse foramen, fractures of either C1 or C2, or vertebral subluxation.[18] Modern trauma survey imaging may serve as a useful screen for BCVI, but dedicated neck CT angiography or catheter angiography should be used in cases with a high index of suspicion.[19] BCVI divides into distinct grades typically characterized via the Biffl scale, characterized by low-grade stenosis of <25% (Grade I), higher grade stenosis (Grade II), pseudoaneurysm formation (Grade III), and complete occlusion (Grade IV).[20] Risk of stroke generally correlates with higher grade. Although arguably less dangerous than carotid BCVI, high-grade vertebral BCVI nevertheless causes strokes in 3%−7% of patients and demonstrates high mortality, mandating clinical vigilance and prompt management where indicated.[21,22]

Spinal cord and nerve roots

As the cranial-most aspect of the spinal cord, the upper cervical spine contains virtually all relevant tracts for motor function and sensation below the head, with the exception of the trapezius and the sternocleidomastoid which derive some innervation from the 11th cranial nerve. As such, sufficiently severe injury at this level will lead to loss of all meaningful motor, sensory, and sympathetic function below then neck, including loss of respiratory function. In contrast, parasympathetic tone via the vagus nerve is largely maintained. Although the loss of motor function and respiration is readily appreciated, the excess parasympathetic tone in these patients can lead to profound persistent hypotension requiring long-term pressors, severe orthostatic hypotension, and cardiac arrest.[23] This global loss of both cardiac and respiratory control accounts for a high mortality associated with upper cervical spinal cord injury.

Incomplete injuries of the upper cervical cord may demonstrate a set of neurological findings atypical for other areas of spinal cord injury. At the cranial cervical junction, the motor fibers of the corticospinal tract undergo a complex decussation at the medullary pyramids, with the upper body fibers decussating at a higher level than the lower body fibers. Because of the complex anatomy of decussation, incomplete injury of the cervical spinal cord at this level can manifest in patterns atypical for spinal cord injury at lower levels, such as loss of upper extremity motor function with preserved lower motor function (cruciate paralysis) or paralysis of one upper extremity and the contralateral lower extremity (hemiplegia cruciata).[24] These cervicomedullary syndromes are summarized in Table 9.1. Finally, extension of edema from a cervicomedullary lesion into the brainstem can lead to compromise of bulbar functions. Given the sometimes radiographically covert nature of upper cervical spine injuries such as atlantooccipital dislocation, clinicians will hopefully scrutinize patients presenting with these atypical spinal cord injury patterns for craniocervical instability.

Upper cervical spine injury

Upper cervical spine injury encompasses a wide variety of subtypes ranging from extremely common injuries, such as type II dens fractures, to rarer entities such as internal decapitation type injuries in the form of craniocervical dissociative injuries. Although an exhaustive overview of upper cervical spine injuries lies beyond the scope of this chapter, we provide a brief overview of common injury modalities to situate our discussion of spinal cord injury.

Fractures involving the first two cervical vertebrae represent the most common form of upper cervical spine injury. Fracture of the axis itself represents the most common level of upper cervical spine injury, with involvement in 9%—18% of patients with spine fractures.[25] Fractures of the axis broadly divide into three subtypes. Fractures of the odontoid process, a large peglike protuberance of the axis, represent a common form of C2 injury, especially in geriatric patients where they represent the single most common form of cervical fracture.[26] Similarly, C2 pars defects (Hangman's fractures) although less common than dens fractures, as a category nevertheless account for approximately 5% of all cervical fractures. Various fractures of the atlas are also common, accounting for one-tenth of cervical spine fractures.[27] Moreover, fractures of C1 and C2 frequently occur in combination with each other as well as other levels of the cervical spine.

Patients less frequently present with pure ligamentous injuries of the upper cervical spine, namely atlantoaxial dislocation (AAD) and atlantooccipital dislocation (AOD). One retrospective review of trauma CT scans at a single center over a 15-year period identified only 31 cases of AOD, with a total incidence of 0.6%.[28] A similar study of patients with AOD undergoing surgical stabilization identified only 17 cases surviving until surgical stabilization over an 8-year period at a major trauma center.[1] AAD represents another relatively rare form of upper cervical spine injury, with one case series identifying only 3 cases out of 300 patients evaluated for cervical spine injury, with only one of those cases representing a pure ligamentous injury.[29] Although rare, these entities demonstrate clinical importance due to their high degree of instability and relatively subtle presentation, which renders misdiagnosis both easy and fatal (Fig. 9.3). In one case series, a quarter of patients with radiologic diagnosis of AOD were never clinically diagnosed; strikingly, all patients with a clinical diagnosis of AOD died.[28] In another series, clinicians only identified AOD in a delayed fashion after neurological deterioration in a third of patients presenting with craniocervical instability.[1] As such, despite their infrequency, AOD and AAD represent important clinical entities in cervical trauma.

Spinal cord injury in upper cervical spine injury

Despite their relatively frequent occurrence, upper cervical spine injuries infrequently present with acute spinal cord injury. Several features of the upper cervical spine injuries may explain this paradoxical observation. First, from an anatomic perspective, the spinal canal widens near the craniocervical junction, allowing the spinal cord to tolerate a larger degree of compression without injury. Second, common unstable fracture patterns either expand the canal, as occurs with traumatic spondylolisthesis, or occur via low-energy mechanisms, as with dens fractures in the elderly, leading to a decreased likelihood of cord injury. Finally, injury at or above the C3 level places patients at high risk of immediate diaphragmatic paralysis, respiratory failure, and death.[30] As such, survivorship bias may obscure the true incidence of upper cervical

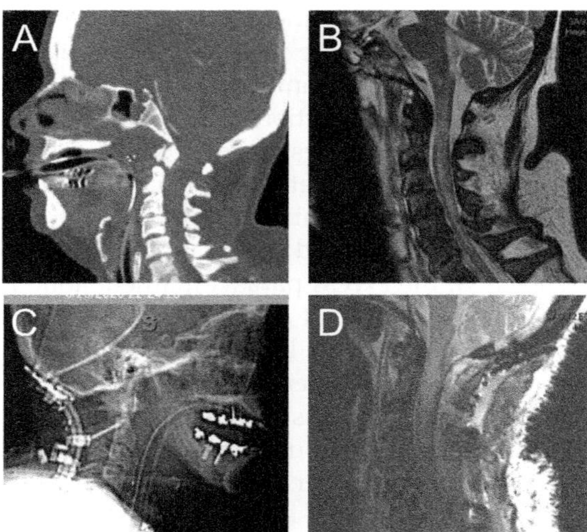

FIGURE 9.3   **A 74-year-old female with unstable dens fracture and complete spinal cord injury.** (A) Initial CT of a 74-year-old female presenting with cardiac arrest, tetraplegia, and respiratory failure after a fall. After resuscitation, patient required pacer placement due to bradycardia. Patient's arrest was initially misattributed to cardiac arrest resulting from a bradyarrhythmia, with the patient's poor neurologic status misattributed to postarrest brain injury, leading to delay in initial CT and diagnosis of the odontoid fracture. Retrospectively, the patient's fall most likely caused the odontoid fracture, leading to a cervicomedullary spinal cord injury, cardiac arrest, tetraplegia, and subsequent bradycardia. (B) MRI obtained approximately 4 days after injury, demonstrating ictal injury at C2 with subsequent spread of edema from the medulla to C5. (C) Postoperative film demonstrating successful stabilization and fusion from occiput to C6. (D) Postoperative film demonstrating progression of edema despite successful decompression and stabilization. Edema demonstrated continued craniocaudal progression despite surgical intervention. The patient continued to demonstrate a poor neurologic recover and family ultimately opted to withdraw care.

cord injury, an observation supported by the predominance of upper cervical spine injury in fatal injuries.[31] Of note, due to the high coincidence of other levels of spine injury with upper cervical spine fractures, spinal cord injury at levels below C2–3 may be considerably more common in this patient population, affecting up to 20% of patients in some series.[32]

Due to their relatively high frequency, fractures of the axis account for a significant proportion of upper cervical spinal cord injury and can serve as a useful paradigm for understanding spinal cord injury in this area. Spinal cord injury following odontoid fracture represents a rare event. Reported incidences of neurologic deficit following odontoid fracture are low, with an incidence of 7.5%—13% in the general population.[33] However, series evaluating these fractures in the elderly, the population most susceptible to odontoid fractures, suggest an increased incidence of neurologic deficit, which may be as high as 30%,[34] despite these fractures generally occurring via low-energy mechanisms such as ground-level falls in the elderly population. Although speculative, survivorship bias may explain these findings. Due to their low-energy mechanism, the spinal cord injuries sustained by the elderly may be less severe when present, allowing a greater proportion of patients to survive until hospitalization. As such, the increased incidence of spinal cord injury in this population may reflect a decreased likelihood of initially fatal spinal cord injury, rather than a greater predilection for injury in this population. Further, the influence of cervical spondylotic disease with some element of preexisting neural compression cannot be excluded.

Traumatic spondylolisthesis of the axis (C2 pars fracture) uncommonly results in spinal cord injury. Aside from the increased diameter of the canal at this level, traumatic spondylolisthesis typically expands the canal in the absence of significant subluxation or angulation. These factors theoretically preclude spinal cord injury in the absence of significant instability. The limited case series evaluating spinal cord injury in this population corroborate this hypothesis, with one study identifying spinal cord injury in only 15% of patients with C2 pars fractures, with injury occurring only in the variants with associated discoligamentous disruption.[35] Moreover, the majority of these patients demonstrated mild, incomplete injuries. As such, the

risk of spinal cord injury after C2 pars fractures may be best characterized as mild. One notable exception to this generality arises when patients present with concurrent dislocations of the C2—3 facet joints in addition to bilateral C2 pars fractures. This much more rare and mechanically unstable constellation, historically termed a 'Type 3 injury' in the Effendi system (AOType C 3), presents a very challenging treatment scenario where rapid surgical intervention is usually needed to avoid pithing of the spinal cord. Of note, these advanced pars injuries are frequently not amenable to closed reduction maneuvers.

Isolated atlas fractures rarely present with spinal cord injury, even in studies including unstable fracture morphologies,[36] although neurologic deficit may result from cranial nerve injury following atlas fracture.[37] Rather, ligamentous injury in the form of atlantooccipital dislocation represents the primary cause of traumatic spinal cord injury above C1. The rarity of craniocervical dissociative injuries like atlantooccipital dislocation (AOD) as a clinical entity complicates precise estimates of associated spinal cord injury in this patient population, with an incidence between 39% and 88% of patients.[1,28] The vast majority of these injuries are usually motor-incomplete, most consistent with a high mortality in patients with complete injury following AOD.

Prognosis after upper cervical spinal cord injury

The prognosis of patients with a motor complete upper cervical spine injury is largely poor due to a combination of complete quadriplegia and near universal ventilator dependency. In one large series of patients with neurologic injury due to odontoid fracture, prognosis in patients with complete injury was poor, with a 70% overall mortality and 100% ventilator dependence in survivors. More recent series find a similarly high degree of mortality in complete injuries of the upper cervical spine, with three-quarters of patients dying from their injury and the remainder demonstrating persistent quadriplegia.[38] Other studies suggest that patients with complete injuries experience a seven- to ninefold risk of mortality.[33,39] The case illustrated in Fig. 9.3 demonstrates this relatively poor prognosis. Of note, however, this seemingly poor prognosis may not apply universally across injury types, as survivors of AOD even with complete spinal cord injury may demonstrate significant improvement in motor strength or even conversion to a lower grade of injury.[1,28]

In contrast to the poor prognosis of their complete counterparts, incomplete injuries of the upper cervical spine tend toward neurologic improvement with timely stabilization. Patients presenting with incomplete injuries following odontoid fractures, for example, demonstrate significant neurologic improvement, with most patients improving at least one ASIA grade after 1 year while demonstrating an average of 46 points on their motor score (16859273). Patients with spinal cord injury surviving until surgical stabilization following AOD demonstrate a similar pattern of improvement, with 85% of patients with incomplete injury improving at least one letter grade on their ASIA impairment score.[1] One large study of neurologic deficit after upper cervical spine injury identified neurologic improvement in over 90% of patients with incomplete upper cervical spinal cord, with complete resolution of deficits in 75% of patients, and with a significant proportion of deficits resolving within 1 month of injury.[35] As such, incomplete injury after upper cervical spine injury enjoys a favorable prognosis (see case vignette in Fig. 9.4).

Several factors may account for the relatively robust neurologic prognosis in upper cervical spine injury. First, survivorship bias will invariably select for patients with less severe forms of injury, a population well poised to neurologic recovery. Second, a significant proportion of patients with upper cervical spine lesions may present with concurrent head trauma, confounding a patient's initial neurologic exam due to their altered mental status and causing patients to present with an apparently more severe degree of injury. Finally, in contrast to the subaxial spine, upper cervical spinal cord injuries more frequently arise from ligamentous instability, resulting in a contusive injury which does not necessarily result in ongoing cord compression; this stands in contrast to injuries in the subaxial spine such as burst fractures or bilateral jumped facets, where a persistent compressive component necessarily leads to ongoing secondary injury prior to surgical decompression. Thus, incomplete upper cervical spine injuries appear to enjoy an enhanced prognosis for neurologic recovery, although the reasons for this are less clear and most likely multifactorial.

Management of upper cervical spinal cord injury

Management of patients with upper cervical spinal cord injury proceeds similarly to patients with traumatic spinal cord injury in general, including prompt diagnosis and stabilization of the spine to prevent secondary injury, close respiratory monitoring and, if needed, rapid intubation to provide further hypoxic injury, and early surgical stabilization and decompression of the spine if indicated. However, several management issues arising in spinal cord injury in the upper cervical spine differ sufficiently from those of the subaxial spine to merit independent discussion. The following sections will focus on these aspects specific to upper cervical spinal cord injury.

Acute resuscitation

As discussed previously, due to their devastating nature, early recognition of cervical cord injuries remains a mainstay in the evaluation of traumatically injured patients, requiring a high

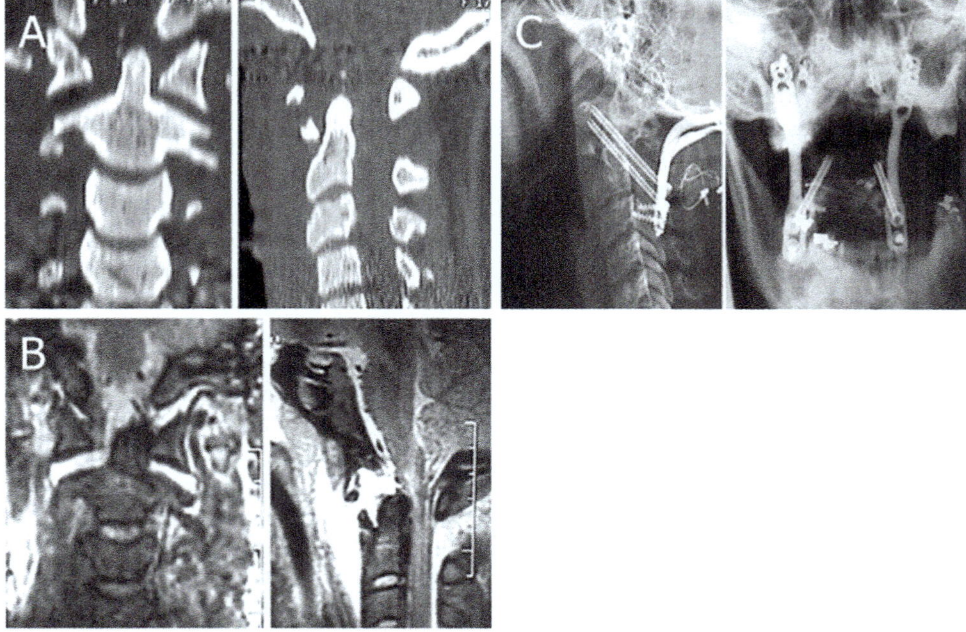

FIGURE 9.4 **A 17-year-old female with atlantooccipital dislocation.** (A) Initial CT of a 17-year-old female presenting with mild upper extremity weakness after a high-speed MVC. Patient was initially diagnosed with a simple atlas fracture despite widening of the atlanto-dens interval and occipital condyle-C1 facets. Patient was taken to MRI, where she experienced respiratory arrest and acute quadriplegia with some preserved sensory function, representing a decline from ASIA D to ASIA B. (B) MRI demonstrating disruption of facets at O—C1 and C2—3, as well as ligamentous attachments of C2 to the clivus, consistent with craniocervical disruption. Contusional injury of the spinal cord by the dens is noted, explaining the patient's sudden deterioration. The patient was urgently taken to the OR, where she underwent reduction and occipital—cervical fusion, as depicted in (C). Patient ultimately did well from a neurologic perspective, eventually making a full, ASIA E recovery.

index of suspicion throughout the initial workup. In the event of a cervical cord injury, either suspected or confirmed, initial steps should focus on the basics of resuscitation, including maintenance of airway, breathing, and circulation.

Cervical spine injury predisposes patients to acute respiratory failure and need for intubation due to loss of respiratory accessory muscles. This predisposition to respiratory failure is especially true above the C5 level, due to development of diaphragm dysfunction, and virtually all patients with severe injuries above C5 will require intubation to support ventilation.[40] As such, intubation must be strongly considered in patients with moderate to severe upper cervical cord injury, even in the absence of clinical respiratory failure, due to the risks of hypoxia, which may severely exacerbate spinal cord injury. However, intubation itself may pose substantial risks in the context of an unstable upper spinal cord injury, as the traditional intubation positions place the atlantoaxial and atlantooccipital joints into full extension. As such, current recommendations include in-line stabilization during urgent intubations and, when possible, use of adjunctive techniques such as video laryngoscopy or fiber-optic intubation to minimize pathologic movement of the spine during intubation.[41]

Maintenance of blood pressure and adequate hemodynamic support also play an essential role in managing spinal cord injury. As discussed previously severe upper cervical cord injury results in a loss of spinal sympathetic tone with preserved or augmented sympathetic tone, resulting in hypotension or even frank circulatory collapse. Given that even brief episodes of hypoperfusion early in the wake of a spinal cord injury may compromise recovery of the injured spinal cord, early initiation of pressors and restoration of mean arterial pressure are important components of resuscitation patients with suspected spinal cord injury.[42] Although a mean arterial blood pressure goal of 80 mmHg or higher maintained for a week after injury remains a current care recommendation, the evidence supporting this practice remains of relatively low quality, especially given potential risks such as cardiac complications or spinal cord contusion expansion arising from prolonged pressor use.[43] Additionally, not all pressors are equivalent, with individual pressors potentially possessing specific beneficial or deleterious effects on the acutely injured cord. Therefore, despite widespread agreement on the importance of avoiding hypotension after spinal cord injury, further study of the optimal role, timing, and type of blood pressure modulation remains an active area of research.

The subject of transfusion and optimal hemoglobin concentration after spinal cord injury remains controversial. Although anemia could exacerbate the spinal cord injury due to decreased oxygen delivery to already poorly perfused tissue, transfusion itself exerts potent pro-inflammatory. Although a hemoglobin goal of greater than 8 g/dL is accepted among practitioners for patients with a neurologic injury, definitive evidence favoring liberal versus restrictive transfusion patterns represents an important gap in the general neurotrauma literature.

Neuroprotection

A large body of research supports the notion that spinal cord injury occurs in two broadly separate phases: an ultra-early injury at time of trauma and a prolonged secondary injury phase due to a variety of proposed mechanisms including inflammation, blood—spinal cord barrier disruption, excitotoxicity, and oxidative injury. Due to the delayed nature of the latter injury phase, a tremendous amount of clinical and basic research has attempted to ameliorate these secondary forms of injury, without any agent or intervention demonstrating clear efficacy. Although a thorough review of these mechanisms and therapeutic agents is well beyond the scope of this chapter, two potential modes of protection, corticosteroids and hypothermia, merit further discussion.

Although their universal clinical application has fallen out of favor, corticosteroids represent the only treatment for traumatic spinal cord injury with some evidence of success in large randomized clinical trials, specifically the NAS-CIS studies. The NASCIS studies consist of three related clinical trials of high-dose intravenous solumedrol administered early after spinal cord injury. The first trial took place in a clinical milieu that presupposed the efficacy of corticosteroids due to widespread clinical use as well as a number of favorable preclinical studies; as such, NASCIS I focused on a low- versus high-dose regimen of IV methylprednisolone after acute SCI. Although the study failed to find a difference between low and high doses of steroids on neurologic outcome, the higher dose experienced significantly higher complications, especially infections. Subsequent trials using placebo or nonsteroid treatments failed to demonstrate a benefit to steroids on planned analysis with regard to motor or sensory function; a small to modest benefit was found for steroids only on *post hoc* stratification of patients.[44]

Given the discord between preclinical results, individual experience, and clinical trials, the subject of steroids is likely to remain controversial in the management of spinal cord injury, with experts both arguing for and against their continued use.

Hypothermia represents another widely considered modality supported less by broad clinical evidence than anecdotal success in the treatment of a prominent athlete. Although initial successful trials of hypothermic neuroprotection after cardiac arrest provided indirect support for the use of hypothermia for spinal cord injury. Subsequent studies of hypothermia in traumatic brain injury, arguably the condition most analogous to traumatic spinal cord injury, have failed to demonstrate a therapeutic benefit to hypothermia.[45] Moreover, subsequent reevaluations of temperature management after cardiac arrest suggest that avoidance of hyperthermic injury via maintenance of normothermia, rather than neuroprotection through hypothermia, represents the principal benefit of cooling patients after neurologic injury.[46] Although meritorious of further study, available evidence does not support the routine use of hypothermia as a neuroprotective therapy after spinal cord injury.

Stabilization

In patients with fractures without evidence of ligamentous instability, such as nondisplaced dens or pars fractures, stabilization in a hard collar or, less commonly, a halo vest typically provides sufficient immobility to stabilize the lesion and prevent neurologic decline; moreover, in the absence of significant displacement or ligamentous disruption, such external immobilization may provide definitive treatment of the lesion.[47] With regard to stabilization of the upper cervical spine, halo vests are markedly more effective from a biomechanical perspective; however, although halo vests may provide better long-term fusion rates, they are also associated with significant complications which may compromise their utility.[48] As such, halo use for management of upper cervical spine fractures appears to have significantly declined based on national registry data.[49,50] Nevertheless, given its efficacy and low rate of major complications, halo vest placement may have significant utility in younger patients.[51]

Although the most common halo complications such as pin loosening, pin infection, and inadvertent intracranial injury may be successfully avoided with careful placement and regular postplacement care, complications of halo use in the elderly deserve particular consideration. Studies evaluating halo placement in geriatric patients appear to independently increase mortality,[48] although these findings are not universal. Although this increase in mortality is most likely multifactorial, halo placement appears compromise swallowing function, leading to increased rates of dysphagia, aspiration, and respiratory dysfunction.[52] Of note, these effects have been noted even in healthy volunteers subjected to halo vest placement.[53] Note that from an applied biomechanical perspective halo vest applications per se are distractive and not compressive devices. Therefore their placement in a distractive type craniocervical injury such as CCD is unlikely to fulfill its intended goal of providing a safe reduction and at least temporary stabilization and in fact may worsen displacement. Given the relatively high complication rate associated with halo placement combined with its reduced success rate in the elderly, several authors have suggested early surgery rather than halo placement in this population, although data supporting this practice are by no means definitive, especially given perioperative morbidity expected in this demographic population.[54]

Within the scope of upper cervical spine stabilization, patients with pure ligamentous disruption deserve special consideration. In patients with suspected ligamentous instability of the upper cervical spine (namely atlantooccipital or atlantoaxial dislocation), clinicians should avoid use of cervical collars. Not only do cervical collars fail to prevent motion in patients with these conditions, application of the collar itself induces a significant degree of pathologic movement and may even lead to compression of the spinal cord.[55] Similarly, traction, a common technique to reduce fractures following cervical trauma, may also cause significant neurologic worsening, contraindicating its use in these injuries.[56] Immobilization strategies should include use of sandbags arranged in a horseshoe pattern around the head and traversing tape to secure the forehead on either side. In the event that definitive surgical stabilization must be delayed, halo vest placement represents a preferable option to a hard collar.

Timing of surgery

Current best evidence argues for surgery within 24 hours of injury for patients with cervical spinal cord injury.[57,58] Although lesions above C2 represent a minority within these cohorts, two aspects of upper cervical cord injury arguably make early surgery even more important for this patient population. First, patients with upper cervical cord injury typically present with incomplete cord injuries. As noted previously, patients with upper cervical cord injury who survive to hospitalization tend to have less severe forms of spinal cord injury.[35] Because incompletely injured patients enjoy increased prospects of neurologic recovery, patients with upper cervical injury are well poised to enjoy the beneficial effects of early decompression following cervical cord injury. Second, the injury patterns that tend to result in spinal cord injury in the upper cervical spine tend to be *highly* unstable modes of injury unsuitable to stabilization with traditional techniques, such as traction or cervical collar. For example, while patients with AOD who undergo rapid diagnosis and stabilization rarely demonstrate neurologic

decline, nearly half of patients with delayed stabilization demonstrate neurologic decline, sometimes profound.[1] Thus achieving early definitive stabilization, usually in the form of surgical intervention, represents a key principle in the management of upper cervical spine injury.

Surgical indications and determination of ligamentous integrity

Most patients with pure osseous injury without a significant ligamentous component may be managed with external orthosis. However, patients managed conservatively with external orthosis benefit from close follow-up, as most patients destined to fail management with external bracing demonstrate evidence of treatment failure within the first weeks of therapy.[51] As discussed previously, patients with total failure of stabilizing ligaments demonstrate high degrees of instability and preferably are treated with early posterior stabilization and fusion with either an occipital–cervical fusion for AOD or a C1–2 fusion for AAD. Management of fractures is more complex. Fractures of the odontoid process and the atlas are common and commonly concurrent injury patterns. Displaced fractures of the dens, once reduced, may undergo repair or fusion either with an anterior odontoid screw fixation or posterior C1–2 fusion in the presence of an intact transverse ligament. Similarly, fractures of C1 may be managed with either an external orthosis or a C1 ring osteosynthesis provided the transverse ligament is intact. Therefore, determination of ligamentous integrity represents the most important decision point in the surgical management of upper cervical spine trauma.

Respiratory care

Cervical spine injury predisposes patients to acute respiratory failure and need for intubation due to loss of respiratory accessory muscles. This predisposition to respiratory failure is especially true above the C5 level, due to development of diaphragm dysfunction, and virtually all patients with severe injuries above C5 will require intubation to support ventilation.[40] As such, intubation must be strongly considered in patients with moderate to severe upper cervical cord injury, even in the absence of clinical respiratory failure, due to the risks of hypoxia, which may severely exacerbate spinal cord injury. However, intubation itself may pose substantial risks in the context of an unstable upper spinal cord injury, as the traditional intubation positions place the atlantoaxial and atlantooccipital joints into full extension. As such, current recommendations include in-line stabilization during urgent intubations and, when possible, use of adjunctive techniques such as video laryngoscopy or fiber-optic intubation to minimize pathologic movement of the spine during intubation.[41]

Due to diaphragmatic compromise, patients with upper cervical cord injuries, especially with ASIA A complete trauma, will invariably require prolonged intubation. In one series, complete injury at the C2 or C3 level led to tracheostomy in all patients, suggesting that early tracheostomy should receive strong consideration in any patient with a high cervical cord injury requiring intubation.[59] Furthermore, the vast majority of patients with high cervical cord injuries demonstrate permanent ventilator dependency in two-thirds of cases, with delayed recovery in a minority of patients.[60] As such, patients and their families should be counseled as to the high probability of prolonged, and potentially permanent, need for mechanical ventilation, which may impact goals of care. Finally, given the significant morbidities and compromised quality of life associated with permanent ventilatory failure, patients should be considered early for adjunctive technologies such as diaphragm pacing when feasible.[61,62]

Management of vertebral artery injury

Given that vertebral artery injury is strongly associated with the same spinal injury patterns as upper cord injury, management of associated BCVI is an important part of managing upper cervical cord injuries. Current best practice recommendations include early anticoagulation or antiplatelet therapy, to be initiated as soon as feasible after diagnosis. However, prompt treatment of vertebral artery injury in patients with an associated traumatic spinal cord injury presents an obvious clinical dilemma. Patients with spinal cord injury from upper cervical spine trauma may present with intramedullary contusions and, as discussed previously, usually require early surgery. Initiation of either anticoagulation or antiplatelet therapy under these circumstances could theoretically result in serious complications, including contusion expansion or postoperative hematoma. Despite representing a common clinical dilemma, no high-quality study to date explicitly evaluates the impact of early versus delayed medical therapy for BCVI in patients with a surgical spinal cord injury. One small retrospective study evaluating treated versus untreated BCVI in patients with either TBI or SCI found a 10-fold higher stroke rate in the untreated cohort with similar rates of hemorrhagic deterioration in both treated and untreated patients with an average time to treatment of 3 days.[63] Another retrospective study evaluating the initiation of medical therapy for BCVI in patients with a relative contraindication to early medical therapy (solid organ trauma, TBI, or SCI) did weakly suggest that therapy may be slightly delayed by 1–2 days without increased incidence of stroke or hemorrhagic complications, although the retrospective nature of the study, heterogeneity of the comparison populations, and relatively early average initiation of therapy render application of these findings in clinical practice difficult.[64] As such, the timing of medical therapy for BCVI in patients with SCI represents an important subject in need of clarification, ideally through a randomized controlled trial mechanism.

Conclusion

Upper cervical spinal cord injury represents an uncommon but important variant of spine trauma. Although traditionally considered rare, improved resuscitation and increased awareness of the condition will undoubtedly continue to increase its clinical incidence. Given that patients with incomplete forms of upper cervical spine enjoy a relatively favorable prognosis if diagnosed and stabilized early and, conversely, the potential for catastrophic deterioration if misdiagnosed, a high degree of both clinical vigilance and familiarity with the unique nuances of managing these patients is mandatory for any clinician involved in the care of trauma patients.

References

1. Bellabarba C, Mirza SK, West GA, Mann FA, Dailey AT, Newell DW, et al. Diagnosis and treatment of craniocervical dislocation in a series of 17 consecutive survivors during an 8-year period. *J Neurosurg Spine* 2006;**4**(6): 429–40.
2. Reis A, Bransford R, Penoyar T, Chapman JR, Bellabarba C. Diagnosis and treatment of craniocervical dissociation in 48 consecutive survivors. *Evid Base Spine Care J* 2010;**1**(2):69–70.
3. Anderson PA, Montesano PX. Morphology and treatment of occipital condyle fractures. *Spine* 1988;**13**(7): 731–6.
4. Radcliff KE, Sonagli MA, Rodrigues LM, Sidhu GS, Albert TJ, Vaccaro AR. Does C(1) fracture displacement correlate with transverse ligament integrity? *Orthop Surg* 2013;**5**(2):94–9.
5. Gire JD, Roberto RF, Bobinski M, Klineberg EO, Durbin-Johnson B. The utility and accuracy of computed tomography in the diagnosis of occipitocervical dissociation. *Spine J* 2013;**13**(5):510–9.
6. Child Z, Rau D, Lee MJ, Ching R, Bransford R, Chapman J, et al. The provocative radiographic traction test for diagnosing craniocervical dissociation: a cadaveric biomechanical study and reappraisal of the pathogenesis of instability. *Spine J* 2016;**16**(9):1116–23.
7. Panjabi MM, Crisco JJ, Vasavada A, Oda T, Cholewicki J, Nibu K, et al. Mechanical properties of the human cervical spine as shown by three-dimensional load-displacement curves. *Spine* 2001; **26**(24):2692–700.
8. Lopez AJ, Scheer JK, Leibl KE, Smith ZA, Dlouhy BJ, Dahdaleh NS. Anatomy and biomechanics of the craniovertebral junction. *Neurosurg Focus* 2015;**38**(4):E2.
9. Pang D, Nemzek WR, Zovickian J. Atlanto-occipital dislocation–part 2: the clinical use of (occipital) condyle-C1 interval, comparison with other diagnostic methods, and the manifestation, management, and outcome of atlanto-occipital dislocation in children. *Neurosurgery* 2007;**61**(5):995–1015 [Discussion].
10. Dvorak J, Schneider E, Saldinger P, Rahn B. Biomechanics of the craniocervical region: the alar and transverse ligaments. *J Orthop Res* 1988;**6**(3):452–61.
11. Tubbs RS, Grabb P, Spooner A, Wilson W, Oakes WJ. The apical ligament: anatomy and functional significance. *J Neurosurg* 2000;**92**(2 Suppl. l):197–200.
12. Zdeblick TA, Abitbol JJ, Kunz DN, McCabe RP, Garfin S. Cervical stability after sequential capsule resection. *Spine* 1993;**18**(14):2005–8.
13. Tuli S, Tator CH, Fehlings MG, Mackay M. Occipital condyle fractures. *Neurosurgery* 1997;**41**(2):368–76. Discussion 76-7.
14. Anderson LD, D'Alonzo RT. Fractures of the odontoid process of the axis. *J Bone Joint Surg Am* 1974;**56**(8): 1663–74.
15. Levine AM, Edwards CC. The management of traumatic spondylolisthesis of the axis. *J Bone Joint Surg Am* 1985;**67**(2):217–26.
16. Benzel EC, Hart BL, Ball PA, Baldwin NG, Orrison WW, Espinosa M. Fractures of the C-2 vertebral body. *J Neurosurg* 1994;**81**(2):206–12.
17. Divi SN, Schroeder GD, Oner FC, Kandziora F, Schnake KJ, Dvorak MF, et al. AOSpine-spine trauma classification system: the value of modifiers: a narrative review with commentary on evolving descriptive principles. *Global Spine J* 2019;**9**(1 Suppl. l):77S–88S.
18. Cothren CC, Moore EE, Biffl WL, Ciesla DJ, Ray Jr CE, Johnson JL, et al. Cervical spine fracture patterns predictive of blunt vertebral artery injury. *J Trauma* 2003;**55**(5): 811–3.
19. Laser A, Kufera JA, Bruns BR, Sliker CW, Tesoriero RB, Scalea TM, et al. Initial screening test for blunt cerebrovascular injury: validity assessment of whole-body computed tomography. *Surgery* 2015;**158**(3):627–35.
20. Foreman PM, Griessenauer CJ, Kicielinski KP, Schmalz PGR, Rocque BG, Fusco MR, et al. Reliability

assessment of the Biffl Scale for blunt traumatic cerebrovascular injury as detected on computer tomography angiography. *J Neurosurg* 2017;**127**(1):32−5.

21. Lauerman MH, Feeney T, Sliker CW, Saksobhavivat N, Bruns BR, Laser A, et al. Lethal now or lethal later: the natural history of Grade 4 blunt cerebrovascular injury. *J Trauma Acute Care Surg* 2015;**78**(6):1071−4. Discussion 4-5.

22. Scott WW, Sharp S, Figueroa SA, Eastman AL, Hatchette CV, Madden CJ, et al. Clinical and radiological outcomes following traumatic Grade 3 and 4 vertebral artery injuries: a 10-year retrospective analysis from a Level I trauma center. The Parkland Carotid and Vertebral Artery Injury Survey. *J Neurosurg* 2015;**122**(5):1202−7.

23. Hagen EM. Acute complications of spinal cord injuries. *World J Orthoped* 2015;**6**(1):17−23.

24. Yayama T, Uchida K, Kobayashi S, Nakajima H, Kubota C, Sato R, et al. Cruciate paralysis and hemiplegia cruciata: report of three cases. *Spinal Cord* 2006;**44**(6):393−8.

25. Robinson AL, Olerud C, Robinson Y. Epidemiology of C2 fractures in the 21st century: a national registry cohort study of 6,370 patients from 1997 to 2014. *Adv Orthop* 2017;**2017**:6516893.

26. Ryan MD, Henderson JJ. The epidemiology of fractures and fracture-dislocations of the cervical spine. *Injury* 1992;**23**(1):38−40.

27. Matthiessen C, Robinson Y. Epidemiology of atlas fractures–a national registry-based cohort study of 1,537 cases. *Spine J* 2015;**15**(11):2332−7.

28. Mendenhall SK, Sivaganesan A, Mistry A, Sivasubramaniam P, McGirt MJ, Devin CJ. Traumatic atlantooccipital dislocation: comprehensive assessment of mortality, neurologic improvement, and patient-reported outcomes at a Level 1 trauma center over 15 years. *Spine J* 2015;**15**(11):2385−95.

29. Pissonnier ML, Lazennec JY, Renoux J, Rousseau MA. Trauma of the upper cervical spine: focus on vertical atlantoaxial dislocation. *Eur Spine J* 2013;**22**(10):2167−75.

30. Miyata K, Mikami T, Koyanagi I, Mikuni N, Narimatsu E. Cervical spinal cord injuries associated with resuscitation from fatal circulatory collapse. *Acute Med Surg* 2016;**3**(2):86−93.

31. Alker GJ, Oh YS, Leslie EV, Lehotay J, Panaro VA, Eschner EG. Postmortem radiology of head neck injuries in fatal traffic accidents. *Radiology* 1975;**114**(3):611−7.

32. Greene KA, Dickman CA, Marciano FF, Drabier JB, Hadley MN, Sonntag VK. Acute axis fractures. Analysis of management and outcome in 340 consecutive cases. *Spine* 1997;**22**(16):1843−52.

33. Patel A, Smith HE, Radcliff K, Yadlapalli N, Vaccaro AR. Odontoid fractures with neurologic deficit have higher mortality and morbidity. *Clin Orthop Relat Res* 2012;**470**(6):1614−20.

34. Ryan MD, Taylor TK. Odontoid fractures in the elderly. *J Spinal Disord* 1993;**6**(5):397−401.

35. Fujimura Y, Nishi Y, Chiba K, Kobayashi K. Prognosis of neurological deficits associated with upper cervical spine injuries. *Paraplegia* 1995;**33**(4):195−202.

36. Kim HS, Cloney MB, Koski TR, Smith ZA, Dahdaleh NS. Management of isolated atlas fractures: a retrospective study of 65 patients. *World Neurosurg* 2018;**111**:e316−22.

37. Dettling SD, Morscher MA, Masin JS, Adamczyk MJ. Cranial nerve IX and X impairment after a sports-related Jefferson (C1) fracture in a 16-year-old male: a case report. *J Pediatr Orthop* 2013;**33**(3):e23−7.

38. Morita T, Takebayashi T, Irifune H, Ohnishi H, Hirayama S, Yamashita T. Factors affecting survival of patients in the acute phase of upper cervical spine injuries. *Arch Orthop Trauma Surg* 2017;**137**(4):543−8.

39. Daneshvar P, Roffey DM, Brikeet YA, Tsai EC, Bailey CS, Wai EK. Spinal cord injuries related to cervical spine fractures in elderly patients: factors affecting mortality. *Spine J* 2013;**13**(8):862−6.

40. Velmahos GC, Toutouzas K, Chan L, Tillou A, Rhee P, Murray J, et al. Intubation after cervical spinal cord injury: to be done selectively or routinely? *Am Surg* 2003;**69**(10):891−4.

41. Austin N, Krishnamoorthy V, Dagal A. Airway management in cervical spine injury. *Int J Crit Illn Inj Sci* 2014;**4**(1):50−6.

42. Haldrup M, Dyrskog S, Thygesen MM, Kirkegaard H, Kasch H, Rasmussen MM. Initial blood pressure is important for long-term outcome after traumatic spinal cord injury. *J Neurosurg Spine* 2020:1−5.

43. Evaniew N, Mazlouman SJ, Belley-Cote EP, Jacobs WB, Kwon BK. Interventions to optimize spinal cord perfusion in patients with acute traumatic spinal cord injuries: a systematic review. *J Neurotrauma* 2020;**37**(9):1127−39.

44. Liu Z, Yang Y, He L, Pang M, Luo C, Liu B, et al. High-dose methylprednisolone for acute traumatic spinal cord injury: a meta-analysis. *Neurology* 2019;**93**(9):e841−50.

45. Lewis SR, Evans DJ, Butler AR, Schofield-Robinson OJ, Alderson P. Hypothermia for traumatic brain injury. *Cochrane Database Syst Rev* 2017;**9**:CD001048.

46. Hayman EG, Patel AP, Kimberly WT, Sheth KN, Simard JM. Cerebral edema after cardiopulmonary resuscitation: a therapeutic target following cardiac arrest? *Neurocrit Care* 2018;**28**(3):276−87.

47. Ryken TC, Hadley MN, Aarabi B, Dhall SS, Gelb DE, Hurlbert RJ, et al. Management of isolated fractures of the axis in adults. *Neurosurgery* 2013;**72**(Suppl. 2):132–50.

48. Sharpe JP, Magnotti LJ, Weinberg JA, Schroeppel TJ, Fabian TC, Croce MA. The old man and the C-spine fracture: impact of halo vest stabilization in patients with blunt cervical spine fractures. *J Trauma Acute Care Surg* 2016;**80**(1):76–80.

49. Grabel ZJ, Armaghani SJ, Vu C, Jain A, Yoon ST. Variations in treatment of C2 fractures by time, age, and geographic region in the United States: an analysis of 4818 patients. *World Neurosurg* 2018;**113**:e535–41.

50. Passias PG, Poorman GW, Segreto FA, Jalai CM, Horn SR, Bortz CA, et al. Traumatic fractures of the cervical spine: analysis of changes in incidence, cause, concurrent injuries, and complications among 488,262 patients from 2005 to 2013. *World Neurosurg* 2018;**110**:e427–37.

51. Bransford RJ, Stevens DW, Uyeji S, Bellabarba C, Chapman JR. Halo vest treatment of cervical spine injuries: a success and survivorship analysis. *Spine* 2009;**34**(15):1561–6.

52. Bradley 3rd JF, Jones MA, Farmer EA, Fann SA, Bynoe R. Swallowing dysfunction in trauma patients with cervical spine fractures treated with halo-vest fixation. *J Trauma* 2011;**70**(1):46–8. Discussion 8-50.

53. Morishima N, Ohota K, Miura Y. The influences of Halo-vest fixation and cervical hyperextension on swallowing in healthy volunteers. *Spine* 2005;**30**(7):E179–82.

54. Smith HE, Kerr SM, Maltenfort M, Chaudhry S, Norton R, Albert TJ, et al. Early complications of surgical versus conservative treatment of isolated type II odontoid fractures in octogenarians: a retrospective cohort study. *J Spinal Disord Tech* 2008;**21**(8):535–9.

55. Liao S, Schneider NRE, Huttlin P, Grutzner PA, Weilbacher F, Matschke S, et al. Motion and dural sac compression in the upper cervical spine during the application of a cervical collar in case of unstable craniocervical junction-A study in two new cadaveric trauma models. *PLoS One* 2018;**13**(4):e0195215.

56. Theodore N, Aarabi B, Dhall SS, Gelb DE, Hurlbert RJ, Rozzelle CJ, et al. The diagnosis and management of traumatic atlanto-occipital dislocation injuries. *Neurosurgery* 2013;**72**(Suppl. 2):114–26.

57. Fehlings MG, Vaccaro A, Wilson JR, Singh A, Cadotte DW, Harrop JS, et al. Early versus delayed decompression for traumatic cervical spinal cord injury: results of the Surgical Timing in Acute Spinal Cord Injury Study (STASCIS). *PLoS One* 2012;**7**(2):e32037.

58. Badhiwala JH, Wilson JR, Witiw CD, Harrop JS, Vaccaro AR, Aarabi B, et al. The influence of timing of surgical decompression for acute spinal cord injury: a pooled analysis of individual patient data. *Lancet Neurol* 2021;**20**(2):117–26.

59. Harrop JS, Sharan AD, Scheid Jr EH, Vaccaro AR, Przybylski GJ. Tracheostomy placement in patients with complete cervical spinal cord injuries: American Spinal Injury Association Grade A. *J Neurosurg* 2004;**100**(1 Suppl. Spine):20–3.

60. Oo T, Watt JW, Soni BM, Sett PK. Delayed diaphragm recovery in 12 patients after high cervical spinal cord injury. A retrospective review of the diaphragm status of 107 patients ventilated after acute spinal cord injury. *Spinal Cord* 1999;**37**(2):117–22.

61. Posluszny Jr JA, Onders R, Kerwin AJ, Weinstein MS, Stein DM, Knight J, et al. Multicenter review of diaphragm pacing in spinal cord injury: successful not only in weaning from ventilators but also in bridging to independent respiration. *J Trauma Acute Care Surg* 2014;**76**(2):303–9. Discussion 9-10.

62. Onders RP, Elmo M, Kaplan C, Schilz R, Katirji B, Tinkoff G. Long-term experience with diaphragm pacing for traumatic spinal cord injury: early implantation should be considered. *Surgery* 2018;**164**(4):705–11.

63. Callcut RA, Hanseman DJ, Solan PD, Kadon KS, Ingalls NK, Fortuna GR, et al. Early treatment of blunt cerebrovascular injury with concomitant hemorrhagic neurologic injury is safe and effective. *J Trauma Acute Care Surg* 2012;**72**(2):338–45. Discussion 45-6.

64. McNutt MK, Kale AC, Kitagawa RS, Turkmani AH, Fields DW, Baraniuk S, et al. Management of blunt cerebrovascular injury (BCVI) in the multisystem injury patient with contraindications to immediate antithrombotic therapy. *Injury* 2018;**49**(1):67–74.

Spine trauma management issues: C-spine

Jared T. Wilcox, Mina Aziz, Rakan Bokhari, Solon Schur, Lior Elkaim, Michael H. Weber, Carlo Santaguida

Montreal University Health Center, McGill University, Montreal, Quebec, Canada

Cervical fracture dislocations are the exemplar of spine injuries. Trainees will frequently rehearse these scenarios and surgeons will often debate the merits of different treatments for this fracture pattern due to the nuances of the management of this patient population. The goal of this chapter is to provide an overview of management issues of cervical spine trauma from the initial assessment to definitive treatment.

Preoperative management

Imaging in cervical trauma

Spinal instability is defined as the spine's inability to maintain normal anatomical relationships between vertebrae under physiological loads, leading to spinal cord or nerve root injury, deformity, and/or pain.[1]

Cervical spinal injuries occur in roughly 2%–6% of patients after blunt injury[2]; if undiagnosed and left untreated, the resulting neurological deficits can lead to significant patient morbidity and mortality.[3] Proper clinical and radiological workup will detect an unstable C-spine, allowing for appropriate treatment, and avoiding unnecessary secondary injury. Imaging modality selection is important for proper decision-making.

X-ray for unstable C-spine

Computed tomography (CT) is widely available and has surpassed plain films for acute evaluation of C-spine injury as scans are significantly more sensitive and specific for identifying fractures. Plain radiographs identify only a third of fractures visible on CT,[4] but when performed, three views should be included: anteroposterior, lateral (including cervicothoracic junction), and open mouth odontoid.[5] Multiple studies have shown that, even when properly administered, flexion–extension radiographs are of limited value when evaluating possible ligamentous injury.[6,7] Plain X-rays are particularly inadequate for evaluating the cervicothoracic region. While it is sometimes possible to obtain a reasonable lateral view of the lower cervical segments by applying gentle traction to the upper extremities, the C7T1 junction cannot be adequately visualized in most cases. However, X-rays are useful in following potentially unstable injuries and are the modality of choice in most follow-up visits.

Neural Repair and Regeneration after Spinal Cord Injury and Spine Trauma
https://doi.org/10.1016/B978-0-12-819835-3.00008-3

© 2022 Elsevier Inc. All rights reserved.

CT for unstable C-spine

The 2013 Guidelines for the Management of Acute Cervical Spine and Spinal Cord Injuries reported level 1 evidence that a symptomatic patient should undergo a CT scan.[8] CT scan is an invaluable tool for rapid and accurate direct assessment of osseous structures and indirect assessment of ligamentous structures after C-spine injury. Clinical situations for when C-spine CTs after trauma are not required have been made by National Emergency X-Radiography Utilization Study (NEXUS)[9] and the Canadian Cervical Rules (CCR).[10] Unstable fractures in the axial spine include burst, flexion teardrop, wedge with posterior ligamentous rupture, extension teardrop, and facet dislocations. To better predict C-spine instability in the context of trauma, Anderson and colleagues described the cervical spine injury severity score.[11] The system assesses the degree of anatomical displacement in four regions of the C-spine (anterior column, posterior column, and two lateral columns). In general, the greater the injuries across all four regions, the greater the chance of instability. The posterior column is the most difficult to assess, with findings of ligamentous injury such as facet diastasis and spinous processes splaying having variable rates of detection.[12] The C-spine should be imaged if there is another fracture identified, as one in five patients will have a concurrent noncontiguous fracture.[13]

Preoperative MRI indications

Magnetic resonance imaging (MRI) is the modality of choice to assess the soft tissues of the cervical spine and ruling out ligamentous injury. MRI has been shown to identify soft-tissue injuries in 5%—24% of blunt trauma patients with negative C-spine CT.[5] Preoperative MRI in the context of cervical spine injury can be used to identify cord compression from an acute disc herniation, epidural or intramedullary hemorrhage, or ligamentous injury.[14,15] While there is a consensus that postoperative MRIs of the cervical spine should be used to confirm adequate decompression and indicate prognosis, preoperative MRI for acute cervical spine injury remains more controversial.[16,17] Fat suppression sequences (i.e., short tau inversion recovery (STIR) or Fat-Sat) have been shown to be highly sensitive for injuries to the intervertebral disk (93%), PLL (93%), and interspinous soft tissues (100%).[18] Specific pathological changes suggestive of underlying unstable C-spine injuries visible on MRI include extramedullary joint diastasis, disc herniation, bone fragments, and hematoma.[5]

Specific structural changes to the spinal cord tissue can be reliably determined in the acute and subacute period, which correlate to spinal cord tissue damage, and long-term functional performance. Primary traumatic injury causes microhemorrhage and tissue laceration. Secondary injury progresses with lengthening of edema and size of hemorrhage beyond the initial segmental level of injury. Chronically, posttraumatic cysts and syringomyelia form, with or without intervening tissue bridges. These lesions can all be properly identified and quantified by T2-weighted, Gradient echo T2 (GRE), and possibly STIR.[19]

MRI as clinical decision-making tool

Preoperative MRI is useful for prognostication and clinical decision-making, but carries risks of worsening neurological injury from positioning, hypotension, and medical stability during long study time, and possible delay to definitive operative management.[14,20] The most consistent imaging sequences corresponding to prognosis (i.e., AIS grade conversion) are T2 and GRE intramedullary signal indicating length of hemorrhage, edema, and cord swelling.[21,22] The high sensitivity of MRI in detecting soft-tissue injuries could lead to treating stable or benign injuries due to overcalling radiographic findings, as the sensitivity and specificity of MRI for soft tissue injuries as read by certified neurosurgeons are 100% and 52%, respectively.[23]

According to the 2017 AO Spine guidelines for the management of acute spinal cord injury (SCI), preoperative MRIs are recommended for patients with nonreducible fracture dislocations, who cannot be examined neurologically (i.e., obtunded), and/or require assessment of the degree of compression after attempted reduction.[14,24] High-energy mechanisms of injury with suspicion of craniocervical injury also require MRI, as the bony intervals and ratios measured by CT often miss significant ligamentous injuries to the craniovertebral junction.[25] An early prospective trial determined that emergent MRI was essential in all acute SCI for decision-making.[26] Preoperative MRI has since been reported to influence surgical decision-making (timing of surgery, planned surgical approach, localization of the injured level) in up to 50% of cases.[19,20,27]

Presence of hemorrhage and length of intramedullary lesion >1 segment (4 cm) are consistently the most valuable for prognostication of neurological outcomes.[19,22,28] Successful surgical decompression is highly dependent on intramedullary lesion length; underestimation of lesion extent could be a worse prognosticator than preop injury severity.[29] With regard to long-term outcome, structural measures provided by MRI have poor predictive power ($\sim$73% successful prediction) for ASIA conversion at 6 months in a prospective trial.[30] Structural MRI parameters were also outperformed by CSF biomarkers in this trial. More accurate evaluation of cord injury in acute-phase MRI has been developed, i.e., the 5-point scale by The Brain and Spinal Injury Centre (BASIC) and subsequent multivariable assessment,[22,31] which has not yet been independently validated. The BASIC scoring system is only reliable <24 hours after injury; however, this aligns to the clinical setting as decompressive surgery should also be performed in this timeframe. In unexplained neurologic deficits, MRI remains the single modality for identification of critical disc herniation and epidural hematomas. For traumatic SCI, MRI has been shown to impact decision-making; however, the benefits need to be weighed against the resources costs, potential delays in surgical management, and the risk of hemodynamic fragility for up to 45 minutes studies in polytrauma patients.[32]

Advanced MRI as a biomarker

Preoperative MRI is suggested for acute cervical spinal trauma, but there are no strong recommendations obviating its use for surgical decision-making.[14,33] Advanced MRI modalities are resource-demanding studies and are not available in most settings. These include diffusion tensor imaging (DTI), magnetization transfer, and MR spectroscopy.[34,35] Identifying disruption of specific microstructural elements of spinal tissue (i.e., myelin, axon, heme deposition) has been shown reliable in acute cervical SCI in a research setting, and may provide more useful detail than the gross T1/T2 signal changes.[34,36] Diffusion tensor imaging measures the freedom of movement of water within tissue. This acts to generate tractography in the cortex, and axonal counts/integrity in the spinal cord.[35] Unfortunately, this modality requires a 3-Tesla MRI, and has not been developed sufficiently for routine use. Magnetization transfer is a modality that can quantify myelin and iron within tissue; however, this has proven to have higher variance and more difficult to standardize than DTI.[36] While these advanced MRI modalities may in future be able to detect microhemorrhages and neurodegeneration, we are still focusing on "classic" structural modalities such as T2, GRE, and STIR to provide information for direct surgical planning.[31,34]

Indications for imaging and treating vascular injuries

Blunt cerebrovascular injury (BCVI; dissection, occlusion, pseudoaneurysm, thrombus, intramural hematoma) involving the internal carotid arteries and vertebral arteries is strongly associated with cervical spine fractures.[37,38] Patients showing signs and symptoms of BCVI require screening with vascular imaging, including: arterial hemorrhage, cervical bruit, expanding cervical hematoma, focal neurological deficit, or stroke.[37] Over two-thirds of BCVIs remain clinically silent, and are associated with increased morbidity and mortality, stroke, seizures, and death.[39–43] Prompt identification and treatment of BCVI with anticoagulation, antiplatelet therapy, or coiling/stenting improves neurological outcomes.[40,44,45]

While it has become increasingly common to screen for vascular injury with CT angiography (CTA) in patients with cervical spine fractures, the clinical yield of detection in all cervical fractures is <1%.[46] More judicious screening protocols have been developed, with a focus on high-risk cervical spine fractures to better identify cases of BCVI.[37,46,47] High-risk features include C1-3 fractures, fractures through the foramen transversarium, and facet subluxation. The AANS/CNS acute cervical spine guidelines recommend CTA screening for patients who meet the modified Denver Screening Criteria for suspected BCVI.[24] Developed in 1994 and revised in 2011, the Denver Screening Criteria were validated by subsequent studies as having a sensitivity of 97%, and a negative predictive value of 99.8%.[43,48] While catheter angiography is the gold standard in identifying vascular injury, the noninvasive nature and high sensitivity and specificity of CTA (up to 98% and 100%, respectively) make it the study of choice.[5]

Classification of subaxial cervical spine fractures

The majority of cervical fractures and 75% of cervical dislocations occur in the subaxial cervical spine; it is important to have familiarity with a subaxial cervical fracture classification[18] (see Chapter 3).[49] Historically, the Allen and Ferguson classification was developed from mechanism of injury inferred in a retrospective analysis of X-rays.[50] This classification system included six types: flexion–compression, vertical compression, flexion–distraction, extension–compression, extension–distraction, and lateral flexion, with additional subtypes described; however, more recently classification systems are in higher utility for injury type and decision-making for clinicians.

Subaxial cervical spine injury classification system

The subaxial cervical spine injury classification system (SLIC) followed the Allen and Ferguson. The system adds discoligamentous complex (DLC) as a component to the scoring system, and attempts to place patients in a nonoperative, indeterminate, and operative category. The scoring system is helpful for trainees as a starting point in decision-making, and a communication tool when conferring with a senior clinician.[18]

The classification has three major categories. Injury morphology (A) includes no abnormality, compression, burst, distraction, and translation/rotation. DLC injury (B) includes intact, indeterminate, and disrupted. Neurological status (C) includes intact, root injury, complete cord injury, incomplete cord injury, and modifier for continued compression. There is also inclusion of confounders like ankylosing spondylitis (AS), diffuse idiopathic skeletal hyperostosis (DISH), osteoporosis, and other entities.

The evaluation of the DLC had the poorest inter- and intrarater agreement. The DLC was considered disrupted if the disk was widened or there was facet dislocation and the indeterminate category was applied if there was abnormal MRI signal change or interspinous widening. The sum of the points assigned resulted in nonoperative (score <4), indeterminate (score = 4), and operative (>4). The total score had an interrater reliability of 0.71 and intrarater reliability of 0.83 when assessed by intraclass correlation.[18] External validation of this system has demonstrated poor agreement of injury morphology, average agreement on treatment verdict, but consistent correlation of higher score (5 or greater) with surgical intervention.[51,52]

AO spine

The AO Spine group has since developed classifications of cervical, thoracolumbar, and sacral fractures. The cervical classification by Vaccaro et al. is more complicated than the TL system, and is composed of four criteria: morphology of the injury, facet injury, neurologic status, and case-specific modifiers (Table 10.1, Fig. 10.1).[53] Morphology of injury group is subdivided into types: A, compression injuries; B, tension band injuries; and C, translational injury. Facet fractures are: F1, nondisplaced facet fractures (<1 cm or $<40\%$ lateral mass); F2, potentially unstable facet fractures (fragment >1 cm or $>40\%$ of the lateral mass); F3, floating lateral mass; and F4, subluxation or perched/dislocated facet.

TABLE 10.1 Imaging modalities and surgical approaches of cervical traumatic pathologies based on AO Spine subaxial injury classification.

Morphology	Class	Extent	Stable	Imaging	Surgical approach[a]
Compression	A1	Wedge	Y		
	A2	Split	Y		
	A3	Partial burst	?	MRI versus X-ray	Post > ant
	A4	Complex	?	MRI versus X-ray	Post > ant
Tension band	B1	Bony posterior	Variable		
	B2	Bony, capsule, ligament	N	MRI, no X-ray	Post
	B3	Anterior	N	MRI	Ant
Translation	C	All columns	N	MRI, no X-ray	[c]Post + ant (360)
Facet	F1	Nondisplaced	Y		
	F2	Displaced (minimal)	Y	X-ray	
	F3	Floating LM	Variable	MRI versus X-ray	[b]Post > ant
	F4	Perched/Dislocated	N		Ant[c] or post
Bilateral			++Reduced	MRI	[b]Likely surgical
Modifiers	M1	Partial B2	Variable		
	M2	Disc herniation (F3-4)		MRI	Ant
	M3	Bony disease	↑Compression	MRI	[d]Ant > post
	M4	Vascular injury		CTA	Coil/stent, ASA

Ant, anterior ACDF ± ligamentotaxis; *LM*, lateral mass; *MRI*, T1/T2/GRE imaging studies; *N*, no; *Post*, posterior decompression and instrumentation/fusion; *X-ray*, dynamic flexion/extension (10 degrees/image); *Y*, yes.
[a] *The authors recommend using SLIC score and criteria.*
[b] *Bilateral obviates/increases need for posterior reduction and fusion.*
[c] *Failure requiring posterior conversion* ∼ $<5\%$.
[d] *Anterior or posterior based on compressive element and sagittal alignment.*
Adapted from Vaccaro et al. Eur Spine J 2016;25:2173–84; Divi et al. Global Spine J 2019;9(15):S77–88; https://aospine.aofoundation.org/clinical-library-and-tools/ao-spine-classification-systems.

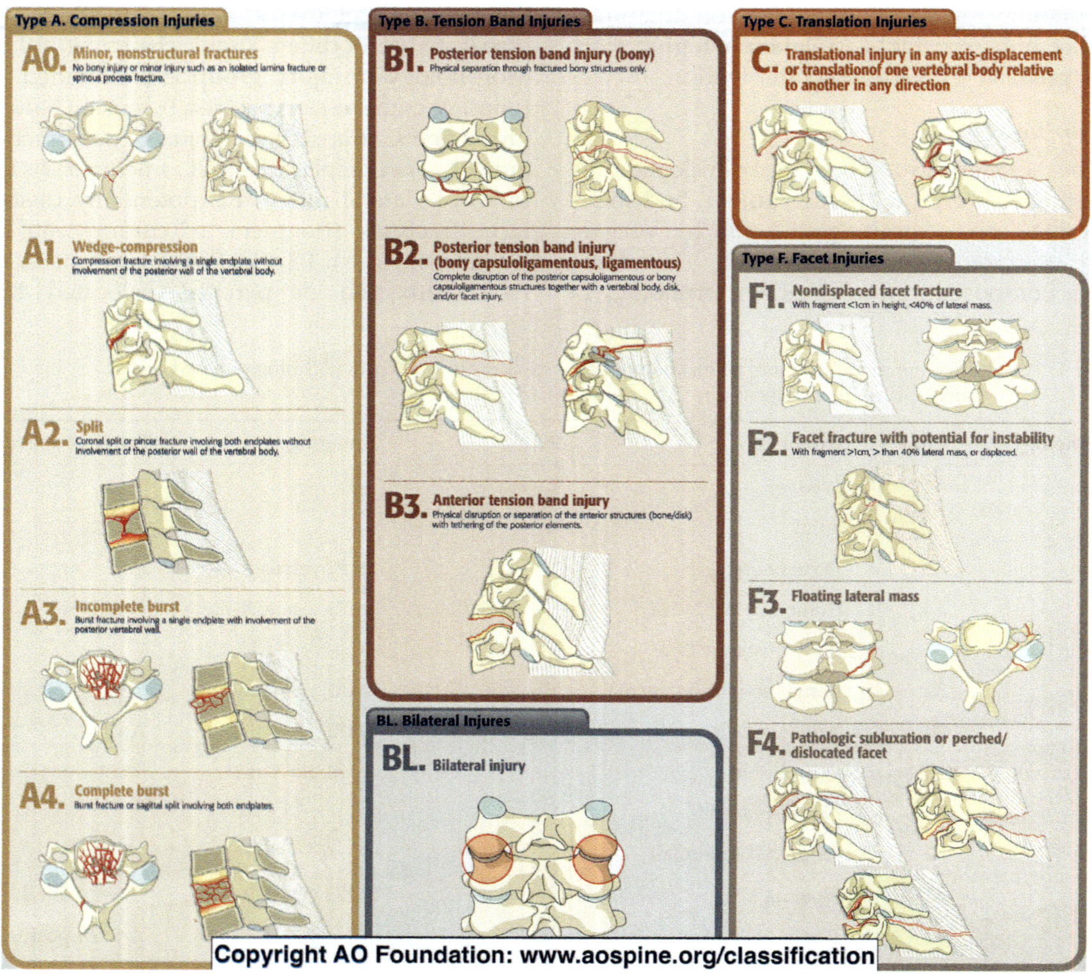

FIGURE 10.1   **AO Spine cervical classification system.** Image provided graciously with permission from AO Spine International and the AO Foundation. © AO Foundation, Switzerland. The AO Spine Injury Classification Systems were developed and funded by AO Spine through the AO Spine Knowledge Forum Trauma, a focused group of international spine trauma experts. AO Spine is a clinical division of the AO Foundation, which is an independent medically guided not-for-profit organization. Study support was provided directly through the AO Spine Research Department. *Retrievable from www.aospine.org/ classification (from 2021). Vaccaro et al. Eur Spine J 2016;25:2173−84; Divi et al. Global Spine J 2019;9(15):S77−88.*

The extent of neurologic injury spans five categories: N0, intact; N1, transient deficit; N2 radiculopathy; N3, incomplete SCI; N4, complete SCI; and NX, cannot be determined properly. The modifiers were designed to highlight features that may influence decision-making that include: M1, posterior capsuloligamentous complex injury without complete disruption; M2, critical disk herniation; M3, stiffening/metabolic bone disease like AS or DISH. The intra- and interobserver reliabilities measured by the kappa statistic were 0.75 and 0.64 for all subtypes,[53] with adequate agreement and validity in external evaluation by different observers.[54]

Cervical immobilization and positioning

Prevention of secondary injury is the primary objective in treatment of spinal cord trauma. Sources of potential secondary injury include mechanical strain on the cord, direct compression by hematoma and edema, and decreased cord perfusion by circulatory or respiratory compromise. The Advanced Trauma Life Support (ATLS) has provided a framework to identify and address these insults in a systematic manner,[55,56] which has greatly reduced mortality and improved outcomes.[57] Neurological compromise due to improper handling of spine trauma patients is evident in both prehospital and hospital settings.[58–60] Spinal immobilization using orthotics and surgical stabilization has become standard of care for any patients with suspected SCI.

In the awake patient, findings that raise the suspicion for spinal injury include presence of midline spinal pain/tenderness, neuropathic pain, sensorimotor deficits, and urinary or fecal incontinence. In the obtunded patient, the threshold of suspicion for fracture is low with presentation of high-energy trauma, falls >5 m (15 feet), severe facial injury, abdominal breathing, severe back pain, or seat belt–associated pain.[61] In the context of these presentations, prehospital spinal immobilization with expeditious in-line external immobilization with neutral neck alignment is essential.

The gold standard for neck immobilization, and current AANS/CNS recommendation, is the combined use of a cervical collar and sandbags that are taped or strapped to a hard spineboard.[27,62] There is high propensity for iatrogenic injury in patient transfers despite adequate immobilization[59,60]; sandbags have a tendency to shift and cervical collars provide nominal protection against vertebral displacement.[63,64] Collars must be specifically

fit to patients, including adjusting the chin rest to limit flexion, avoiding extension that can lead to spinal distraction, and ensuring clearance of internal jugular veins to avoid elevated intracranial pressure. To obtain neutral alignment, patients need anywhere from 1 to 5 cm of padding,[65] which also provides protection against pressure sores.[66]

Cervical collars are not benign and conservative estimates suggest only 1%–2% of patients who are immobilized are found to have significant instability.[27,67,68] Patients with AS may develop catastrophic cord injury if forced into extension from their often-impressive baseline kyphosis (Fig. 10.2), and may require several pillows for proper padding.[59,62,69] Conversely, pediatric patients have unproportionally large skulls, and need support beneath their back to prevent being forced into kyphosis.[27,70] Immobilization can also be extremely uncomfortable and often causes combativeness in confused patients, thus necessitating sedation and intubation. The Congress of Neurological Surgeons taskforce on acute cervical spine and spinal cord injuries concluded that level-2 evidence exists for immobilization of trauma patients to be deferred or cleared if: they are awake, alert, and are not intoxicated; are without neck pain or tenderness; do not have an abnormal motor or sensory examination; and do not have any significant distracting associated injuries.[71]

Intraoperative positioning of cervically injured patients poses specific challenges, and catastrophic injuries have been reported while turning patients prone. There is some suggestion that patient safety has been improved with combined measures including introduction of the radiolucent Jackson table, adjuncts for flipping patients prone, and use of foam pillows compared to the Mayfield pinning device.[72,73] However, the specific processes and procedures needed for safe transfer and positioning are not clear, and a Cochrane review by Kwan et al.[74] was not able to conclude that neck immobilization improved outcomes.

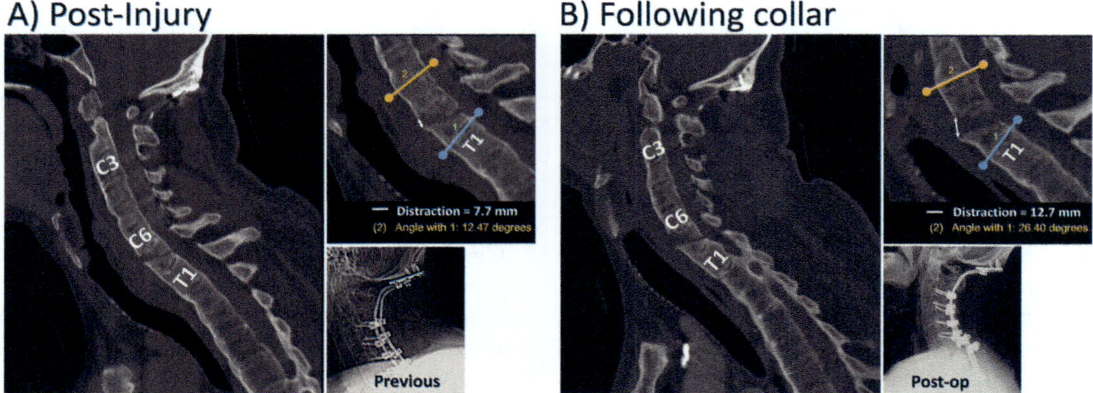

FIGURE 10.2   **Exacerbation of traumatic distraction by cervical collar.** (A) Patient with ankylosing spondylitis on transfer from a peripheral hospital, postinjury angulation at C7 fracture of approximately 12°, and distraction of 7.7 mm is evident by CT. (B) Prehospital collar (deemed undersized) was exchanged for rigid collar at trauma center, increased the angulation to approximately 26°, and distraction of 12.7 mm. Previous occiput-C4 fusion were evident on preoperative X-ray. This was adjoined for definitive construct as seen in postoperative imaging. This obviates the need to avoid overextension during positioning and immobilization and surgical intervention with careful attention to distraction and rotation.

We recommend applying the above procedures and devices in concert with vigilance during positioning and repositioning, and clear closed-loop communication for team cohesion in chaotic trauma and operating rooms. The surgeon should carefully review implications of a patient's deformity on final operative positioning and incorporate standard checks to avoid poor final position and sequelae. Specifically, the surgeon and surgical team must place the neck in a patient-specific neutral position, recognize and eliminate inadvertent head movement, avoid ergonomically impossible positions for the operating team, and limit time spent in eyes-below-heart position (for risk of blindness).[59,62,69]

Airway management of the unstable cervical spine

Ensuring airway patency and adequate ventilation is paramount for any trauma patient. Airway obstruction is not infrequent, and hypoxia is a potent cause of secondary injury and mortality; attempts should be made to avoid this at all costs. Immobilization of the cervical spine after trauma makes airway management significantly more difficult, necessitating modified maneuvers for ventilation and intubation. Multilevel fractures occur in up to 20% of cervically injured patients, and failure to identify unstable fractures may have catastrophic outcomes.[3] Immobilization before or after presentation to a trauma center should include immobilization as detailed above. The nature of the trauma sustained and patient factors such as underlying comorbidities may necessitate additional changes to intubation maneuvers, adjuncts, or pharmacological agents.

Proper evaluation of the airway is often significantly hindered by the applied cervical collar and immobilization measures, which itself worsens Mallampati grade. Cervical injuries are often concomitant with facial and oropharyngeal injuries causing obstructive blood, secretions, or prolapse of tissues (soft palate, tongue, or epiglottis), and inability to protect the airway (i.e., obtunded, or suppressed cough reflex). Even well-appearing C5 cord injury patients sustain a 30%—50% reduction of vital capacity in the first week after their injury.[59] Attempts should be made to identify those patients at immediate need for airway support and likely to develop subsequent ventilatory failure. Additional patient-related factors causing respiratory distress and difficult intubation include tracheobronchial injuries and burns; loose or broken dentition; atlantoaxial instability (i.e., trisomy 21); AS and other arthritides that predispose unstable fractures; and extraskeletal disease.

Orotracheal intubation receives significant consideration and attention by medical practitioners. Typical intubation attempts to align the intubator's line of sight with the oral, pharyngeal, and laryngeal axes. This is typically done by near-full extension of the atlantooccipital joints, flexion of the subaxial spine, with forward translation of the head to adopt the sniffing position. Evidence suggests each aspect of securing the airway is a possible cause for cervical spinal secondary injury. Chin tilt is usually avoided in these patients, but the jaw thrust maneuver has been shown to result in significant motion that can equal or be greater than typical orotracheal intubation.[75,76] Cervical motion also occurs with pressure applied to the face with a bagmask to obtain an air seal during ventilation, as well as cricoid pressure as part of rapid sequence intubation.[77]

Cadaver studies have shown cervical collars restrict motion to a limited degree, and persons performing intubation cannot rely on the collar for any benefit in protecting the cervical spine. In fact, collars interfere with intubation and can increase vertebral manipulation by preventing necessary mouth opening and levering the mouth against a fixed chin causing extension. Medical practitioners that perform intubation, specifically in trauma centers, ought to be familiar with these aspects of cervical trauma and be familiar with airway adjuncts such as oropharyngeal and nasopharyngeal airways, and fiber-optic scopes to reduce cervical motion during intubation.

Traction and closed reduction

The major goals of intervention in cervical trauma are to reduce the fracture segment, decompress the spinal cord, and stabilize the spinal column. External traction is a powerful technique for reduction and decompression in a safe and efficacious manner. Subsequent stabilization is less burdensome since the spinal cord is decompressed and fracture reduced. In clinical and animal studies early reduction of subaxial spinal dislocation increases the likelihood for neurological recovery below or at the area of injury.[78–80] A systematic review of cases series published of closed reduction identifies success in approximately 80% of cases.[81] The efficacy appears to decrease if there is a delay in presentation or management, with small case series suggesting a zero success rate for injuries >5 days old.[82]

The indications for using external traction, such as Gardner–Wells tong skull traction, include unilateral and bilateral facet dislocation, or perched facets. It is imperative to ensure that one can adequately assess the patient both radiologically and clinically prior to the start of the procedure. A patient with a distracting injury or who is intubated or intoxicated may not be appropriate for closed reduction, particularly without a prereduction MRI. In the context of closed reduction, the finding of a herniated disk or disrupted disk may be as high as 55%; however, case series have demonstrated closed reduction in alert patients does not result in neurologic deteriorations.[83] A systematic review by Gelb et al.[81] failed to identify significant permanent complications from closed reduction, and projected the rate to be <1%. Transient neurological deficits were more common, with estimated incidence between 2% and 4%. Some of the reasons cited for neurologic decline were OPLL, epidural hematoma, unrecognized noncontiguous injury, overdistraction, and disk herniation. The requirement for prereduction

MRI to identify a disk herniation is commonly cited as a concern but may not be a relevant precaution prior to attempting closed reduction in an examinable patient.

The appropriate setting for closed reduction is a monitored unit such as an emergency room, ICU, or operating room. This patient population is susceptible for hemodynamic instability, and they require cardiovascular monitoring throughout the procedure. Procedural fluoroscopy is preferred to ensure clinicians can respond rapidly to changes in patient conditions without preventable delays. Briefly, patients are placed in reverse Trendelenburg with the body elevated on blankets to allow for neck extension. Baseline clinical and radiographic examination must be performed prior to applying any weights. There are two standard options when it comes to skull traction: halo traction and Gardner—Wells tongs. The major advantage of Gardner—Wells tongs is the ease of placement and the ability to apply higher weights with less risk of pin site lacerations. Care must be taken when applying the Gardner—Wells tong to not place the pins too anteriorly and cause damage to the temporal muscle and artery.

Overtightening of the pins can result in penetration of the inner table of calvarium which can lead to intracranial hemorrhage. It is recommended to tighten the pins to 1 mm above the surface of the indicator, which is equal to 139 N. The traction vector is initially in flexion, and once the inferior articular process begins to clear the superior articular process of the caudal vertebrae, the vector is then changed to extension. Weights are added in sequential fashion starting with 10 pounds (lbs), and an addition to 5 lbs per level being reduced. After each additional weight is applied, a repeat clinical and radiographic examination should be performed. The procedure should be terminated if there is: successful reduction, radiographic evidence of overdistraction of the facet joints or the occipital-cervical joint, any new neurologic deficits, or the patient cannot tolerate the additional

traction. If traction fails to reduce the fracture an MRI is recommended to identify if there is a disk herniation or fracture fragment that stifled the reduction attempt.[81] Classically, a maximum of 50 lbs traction was recommended to avoid overdistraction. Cotler et al.[84] demonstrated traction of >50 lbs using Gardner—Wells tong skull traction for cervical dislocation in 17 patients, with up to 140 lbs, and none of these patients had neurological deterioration. While there is no maximal cut-off in regard to weight, it is rare to exceed 70 lbs.

Operative management

Neuromonitoring in cervical spine trauma

Cervical surgery for trauma carries inherent risk of neurologic deterioration. Intraoperative neuromonitoring (IONM) has been used in an attempt to prevent surgery-related adverse events; however, there is no consensus or guideline for its use. The goal of IONM is real-time assessment from positioning to closure in order to rapidly identify and correct interruptions to neurologic function. The utility of neuromonitoring is unproven in nontraumatic decompressive surgery[85]; however, traumatic cervical surgery is complex with inherently higher risk, and use of IONM should be considered. The demands of IONM are significant, including surgical time (15—30 min set-up), inclusion of neurology and neurophysiology team members, specialized anesthetics, and associated costs. Anesthetic considerations are nontrivial, requiring reduced inhalant use, partial neuromuscular blockade, and/or total intravenous anesthesia.[86]

Continuous and triggered monitoring of nerve activity allows the operator to identify when conduction is reduced or lost, allowing for real-time response to dangerous positioning or surgical maneuvers. An experienced team will set thresholds that reduce false positives and warning rates, while maintaining sensitivity and specificity. Somatosensory evoked potentials are specific, and >50% reduction in amplitude or >10% increase in latency is a significant alert. Motor evoked potentials are sensitive, and >75% reduction in amplitude is typically significant.[87] Spontaneous or triggered peripheral electromyography is used less frequently in cervical cases. Transcranial measurements are safe and effective in almost all patients without implanted electrical devices.[88] While the predictive value of single modalities in nontraumatic cases has been reportedly low, combining modalities allows event prediction to approach 100%.[89] Electrophysiologists should also alert the surgical team to a loss in one muscle group, consistent change in tracing morphology, or if eliciting responses become increasingly difficult.

While attempts have been made to standardize the workflow following an alert and implement a checklist, there is no consensus on how to manage these events.[90,91] Surgery-related causes of monitoring loss are generally asymmetric and include compressive maneuver, traction or distraction, structural compression by bony fragments or instrumentation, and progressive injury (i.e., expanding hematoma). Nonsurgical causes are more likely symmetric loss and include electrode pull-out, erroneous recording (false positive), hypotension, neuromuscular blockade, patient positioning, and supratherapeutic anesthesia.[86,91] Partial signal loss with less than or equal to 50% reduction in signal can be often corrected with increasing mean arterial pressure and surgical pause of 15 min.[91] Partial and transient losses are less likely to signify a permanent deficit,[92] while complete loss of IONM should be thoroughly investigated postoperatively.

Standards of practice for using IONM have not been determined by guidelines, but should be considered in traumatic cervical surgery, and should generally involve: (i) a surgical team including experienced anesthesiology and neurophysiology, (ii) initiation prior to positioning, (iii) multimodal testing for highest sensitivity/specificity and reducing false alerts, (iv) predetermined procedures and troubleshooting during alerts, and (v) documentation and postoperative follow-up for losses of neural signal events.

Timing of surgery

Intuitively, early decompression and stabilization of a patient with cervical SCI would promote earlier mobilization and reduce the extent of secondary injury. There are myriad preclinical studies that suggest improved outcomes with an earlier decompression, and that effect sizes are smaller in cervical spine injury than thoracolumbar.[93] Clinical trial data are now available. Randomized controlled trials to study how timing of decompression effects outcome have proven difficult to conduct logistically, due to variability in timing it takes to reach the trauma center, having preoperative workup completed, and surgical suite available. Instead, trials have been designed as cohort analyses of participants receiving surgery <8h, 8–24 hours, or >24 hours after injury, instead of the previously studied (i.e., 72-hour) time point.[94,95]

The best evidence to date is three such prospective cohort studies, including the Surgical Trail of Acute Spinal Cord Injury Study (STASCIS) trial specific to cervical injury.[94,96,97] The STASCIS trial found the early decompression group (<24 hours) had an odds ratio (OR) of 2.83 for a 2+ grade improvement in AIS score at 6 months compared with late decompression.[97] Dvorak et al.[94] was the largest of the prospective studies (N = 888); the AIS B, C, and D patients exhibited a 6-point motor score improvement when decompressed early. The remaining prospective cohort study from Wilson et al.[97] (N = 84) found a 13-point AIS motor score improvement and higher likelihood of 2+ grade AIS improvement if decompressed early.[97]

The major counterargument to early decompression is whether there are safety benefits to delaying surgery; however, there were no significant differences in complication rates between the early and late decompression groups in these studies.[95,96] In the context of central cord injury where there is no clear instability or fracture, Lenehan et al.[98] similarly demonstrate early decompression (<24 hours) versus late decompression (>24 hours) conferring a 7.5-point improvement in the total AIS motor score at 6 months. Consequently, the most recent iteration of Clinical Practice Guidelines for the Management of Acute Spinal Cord Injury concluded early surgery supports improved neurologic recovery despite limited and variable quality of evidence.[99]

Surgical management of subaxial C-spine fractures

The specific goals of each surgery should be individualized based on patient factors and injury patterns. However, the goals of surgery can generally be divided into (i) adequate decompression of neural structures to halt secondary injury, and (ii) restoring spinal stability. These aims are not mutually exclusive. They can interact positively, as restoring normal spinal alignment often confers indirect decompression. They can also act negatively, as wide posterior decompression can disrupt the posterior tension band and transfer forces through the facet joints, thereby resulting in progressive instability and postlaminectomy kyphosis.[100]

Surgical approaches for subaxial cervical injuries include using anterior, posterior, or a combination of both approaches (Table 10.1). The choice of surgical approach will depend on injury patterns, patient characteristics, and surgeon preference.[101,102] Preoperative planning using multiplanar CT scans and MRI helps guide the choice of surgical approach based on the specific fracture patterns and cord damage (see above). Important information that should be gleaned from a careful review of the imaging includes — but are not limited to — the presence of

endplate fractures, posterior element involvement, single versus multiple fractures, extent of cord edema/hemorrhage, and neural element compression.[101]

Patient characteristics can greatly impact surgical management of cervical spine injuries, and many cases require some degree of individualization. Surgeons must consider patients' age, frailty, functional status, comorbidities, habits, prior surgeries, and preexisting deformities when planning surgical management.[102] Certain comorbidities warrant special discussion, such as the presence of AS. Einsiedel et al.[103] preformed a retrospective review on 37 patients who suffered a subaxial cervical spine injury in the setting of AS. The authors noted that 50% of the patients treated with just anterior fixation suffered hardware failure by the time of hospital discharge.[102,103] Patients who have undergone prior cervical surgeries present their own set of challenges that need to be considered during preoperative planning. These challenges include the need to revise or retain previous hardware and whether the current instrumentation is compatible with the previous instrumentation. Scarring due to prior anterior cervical procedures causes an increased risk of complications such as recurrent laryngeal nerve injury and vocal cord paralysis.[104,105]

Anterior cervical decompression and fusion

Anterior approaches for cervical facet dislocation are more nuanced than posterior approaches and this option should be selected carefully. The major advantages of the anterior approach is that it is less invasive, and may not require an MRI since the anterior compression can be directly addressed. It utilizes a supine position, which is easier and safer than prone positioning of an unstable neck particularly with polytrauma patients.[102,106] If there is an anterior compression from a retropulsed burst fracture or disk herniation, there is general consensus that the patient should undergo an anterior approach with or without the addition of posterior fixation (Fig. 10.3). Some authors suggest that unilateral dislocations are better candidates for stand-alone anterior discectomy and fusion than bilateral dislocation.[107]

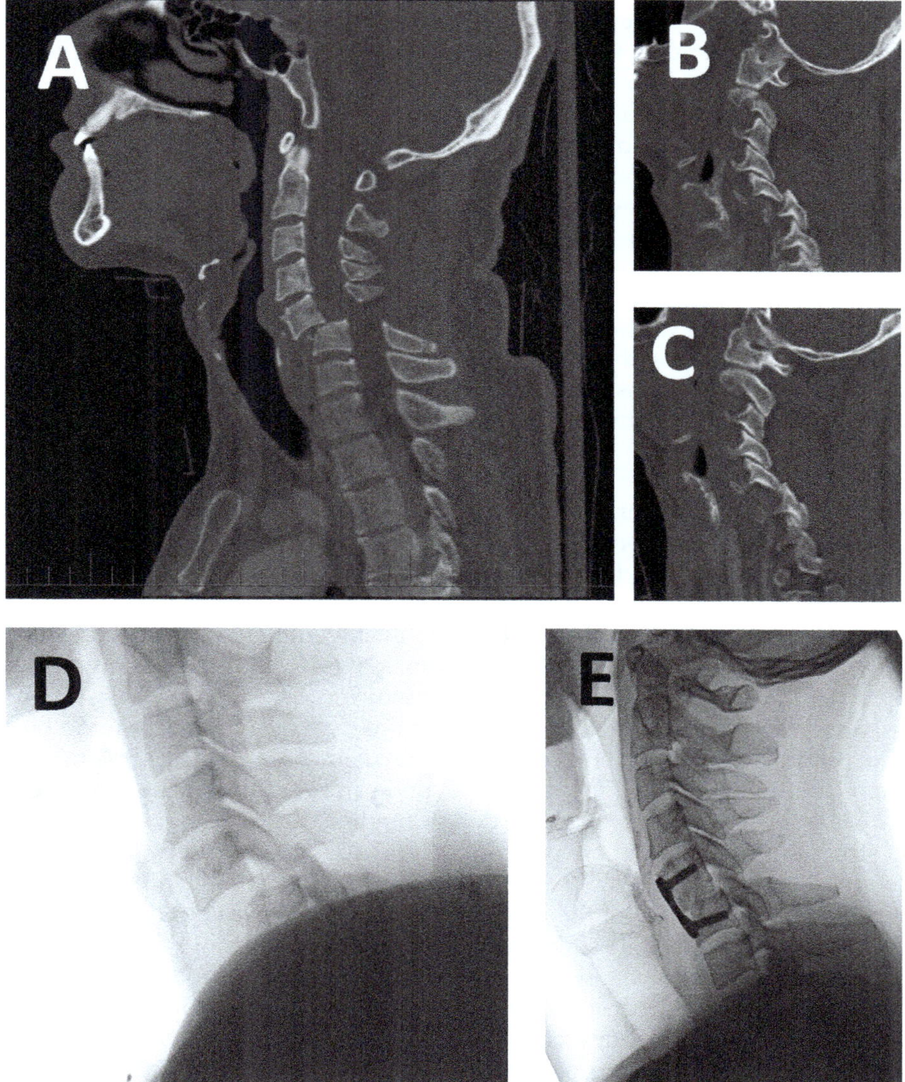

FIGURE 10.3 **Anterior approach for bilateral cervical dislocation.** (A—C) Patient presenting with C5/6 dislocation without fracture following motor vehicle collision. (D—E) Preoperative X-ray was sufficient to identify bilateral jumped facets (D), which was treated definitively with intraoperative traction and a single level anterior cervical decompression and fusion (E), with postoperative X-ray demonstrating maintenance of reduction. *Acknowledgment: Images courtesy of Dr. Peter Jarzem, McGill University.*

Anterior surgical technique

The patient is positioned supine with Gardner—Wells traction. It is important to ensure that the neck is in a relatively neutral position as neck flexion will exacerbate the deformity while extension could exacerbate spinal cord compression. Fluoroscopy should be used to confirm the level of the injury and ensure that all areas of interest are adequately visualized prior to the start of the procedure. A standard anterior approach to the cervical spine is then used to expose the injured level. Weights should be added sequentially to distract the disc space. A discectomy is then performed. Further distraction can be obtained by inserting a laminar spreader or paddle distractor into the disc space and distracting against the endplates. The spreader should be inserted as far back as the posterior wall of the vertebral body but should not enter the spinal canal.

Under fluoroscopic guidance, the facet joint is distracted until the inferior articular process has cleared the superior articular process. At this point the laminar spreader should be angled in a cephalad direction in order to translate the superior vertebra posteriorly and then distraction is released. Alternatively, divergent Casper pins can be inserted in the bodies of the superior and inferior vertebras, and as the pins are brought into a parallel alignment, they distract the disc space.

Once the dislocation is adequately reduced, an interbody structural graft should be inserted into the disc space to provide support and to achieve an indirect decompression of the neural elements. Care should be taken to undersize the graft as facet dislocation causes significant disruption to the ligamentous structures, which can lead to overstuffing of the disc space. If the intention is a stand-alone anterior approach, using an iliac crest bone graft may be preferable; however, this is not supported by evidence. If subsequent posterior fixation for "360 degree" is intended, then any graft material is appropriate. Should anterior maneuvers fail to reduce the dislocation, a posterior approach should be undertaken.

Anterior limitations

Following failure to reduce the dislocation from the anterior approach, the patient requires a posterior surgery for reduction and instrumented fusion, followed by anterior fusion.[108,109] The front-back-front is time-consuming and extends the anesthetic time significantly. In some circumstances the failure rate of the anterior approach may be prohibitively high.

Johnson et al.[110] conducted a retrospective review of patients with single-level unilateral or bilateral facet dislocation who had anterior discectomy, fusion, and plating. They reported a 13% rate of radiographic mechanical failure, with highest rate at C6/7 level. Patients who had superior endplate compression fractures (of the lower vertebra) or facet fractures were more likely to lose their postoperative alignment. Based on these findings, the authors advocated for the use of a posterior approach or combined anterior—posterior approaches to manage unilateral and bilateral facet dislocations with endplate compression or facet fractures.[110]

Careful consideration should be given to the management of multiple fractures with a stand-alone anterior approach and multilevel corpectomy and fusion. One of the main concerns is the loss of reduction and graft migration. Wang et al.[111] conducted a retrospective review of 249 patients who underwent standalone corpectomies and fusions for degenerative cervical stenosis over a period of 25 years. They noted that the risk of graft migration increased as the number of treated levels increased. Similarly, they noted corpectomy to the C7 level increased the risk of graft migration. While there are limitations to this study and the degree to which its results can be extrapolated, it does offer a reason to pause and consider supplementing long constructs with posterior instrumentation.

Posterior decompression and fusion

The posterior approach to the unstable cervical spine offers several advantages when it comes to treating subaxial cervical spine dislocations. It is a ubiquitous approach and most spine surgeons are familiar with its use, principally as it allows surgeons to address the main pathology of facet dislocation and the associated posterior ligamentous injury (Fig. 10.4).[101,102] Conversely, the chief disadvantage to the posterior approach is the inability to address anterior disc herniation, retropulsion, and misalignment. Posterior cervical spine approaches are associated with worse postoperative neck pain,[112] an increased risk of surgical site infections,[113] and obviate the use of prone positioning of an unstable spine. Finally, using the posterior approach typically necessitates fusion of multiple spinal levels to achieve adequate fixation particularly with associated fracture.[109]

Posterior surgical technique

The intubated patient is first placed in the supine position while maintaining vigilant cervical spine precautions. The patient's head is fixed in a Mayfield frame, with care to observe this

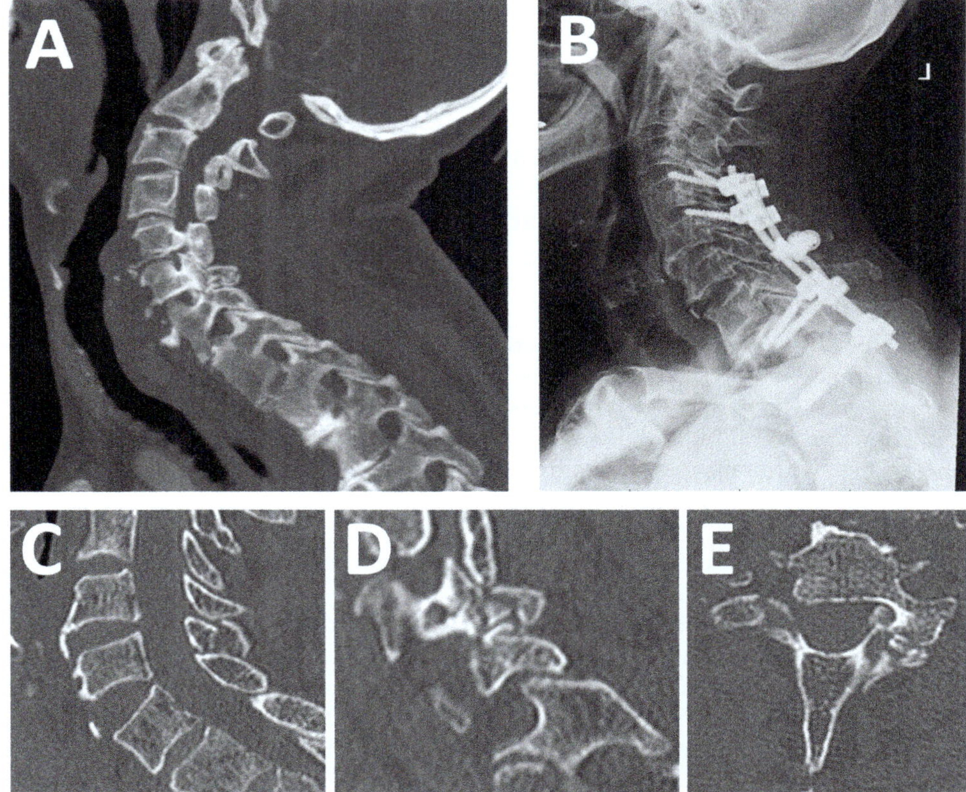

FIGURE 10.4   **Posterior approach for bilateral facet fracture.** (A) A 74-year-old woman with extension and bilateral comminuted facet fracture of C6/7 was evident on CT scan. (B) This was amenable to posterior approach for decompression and stabilization with instrumented fusion with good result. (C) Extension injury of the C6—C7 disk space in a nonankylosed spine. (D—E) Significant comminuted fracture was found on bilateral facets and laminae.

position in all principal axes, and the patient is flipped on a Jackson frame. We will generally perform a wake-up test after intubation and prone positioning. A midline posterior cervical spine approach is utilized, and once the dislocated facet joint is adequately visualized, the surgeon can adjust head positioning as needed to apply appropriate traction. If further distraction is required, laminar spreaders can be applied to the superior and inferior spinous processes and used to distract the joint. If these maneuvers are insufficient to achieve reduction, the tip of the superior articular process of the caudal vertebra can be drilled down judiciously to allow the inferior articular process of the cephalad vertebra to translate posteriorly. Following reduction, the weights and the distractors can be removed, and the adequacy of the reduction should be confirmed with fluoroscopy. Further decompression can be accomplished with a laminectomy if clinically indicated. The affected levels can then be instrumented, and the facet joints decorticated and applied with bone graft to assist with fusion.[109]

Posterior limitations

When using the posterior approach to treat subaxial cervical spine injuries, one must be cognizant of the relative instability of the cervicothoracic junction. Several authors have noted a greater risk of hardware failure and progressive deformity when the fixation does not span the cervicothoracic junction.[114,115] Nagashima et al.[116] conducted a retrospective review of posterior cervical long-segment fusion and found that when C7 was the lowest instrumented level, the risk of hardware failure was as high as 40%, but successfully mitigated with C6 pedicle screws (with buttress threading). A number of studies have compared the mechanical strength of pedicle screws and lateral mass screws. Pedicle screws at C7 tend to have higher pullout strength and more stability; however, lateral mass screws at C6 + C7 were comparable.[116,117] Cervical pedicle screw insertion comes with a

significant risk of vertebral artery injury given the small diameter of the pedicle and the proximity of the vertebral artery.[118,119] Several techniques have been developed to mitigate these risks including the use of anatomical-based landmarks and image-guided navigation.[119–121] If the C7 has a substantial lateral mass, we advocate for C6 + C7 lateral mass with T1 pedicle with a conventional posterior fixation system.

Surgical decision-making guide

Dvorak et al. proposed an evidence-based algorithm for the management of subaxial cervical spine injuries based on the SLIC scoring system, a systematic literature review, and a consensus of experts.[101] *Central cord*: in the setting of an incomplete central cord syndrome, the main decider is the overall alignment of the cervical spine. The authors recommend a laminectomy and fusion or a laminoplasty for patients with an overall lordotic alignment. For those with a kyphotic alignment, an anterior procedure and possible supplementation with a posterior fusion is recommended. *Burst fracture*: for patients who suffer from a vertebral burst fracture, the authors recommend an anterior vertebrectomy and insertion of a structural graft in order to restore anterior column support and ensure adequate spinal cord decompression. *Hyperextension injuries*: the authors recommended an anterior discectomy and fusion except in the setting of a stiff spine such as AS and DISH. In these patient populations, the authors advocated for a combined anterior and posterior fixation. *Facet injury*: with regard to facet subluxation and perched facets, the authors advocated for an anterior discectomy and fusion if there is a disc herniation evident on the MRI. Alternatively, if there is no disc herniation, both an anterior and a posterior approach could be utilized based on surgeon preference and patient factors. For patients who have suffered facet fracture dislocations, the authors take into account the presence and type of vertebral body fractures and presence of disc herniation into the canal. *Disc herniation*: an associated herniation should be treated with an anterior cervical discectomy and fusion with potential to supplement the fixation posteriorly. *In summary*: posterior-alone approach is considered for lordotic central cord without significant anterior compression; 360 (anterior + posterior) approach is considered for translation and rotation injuries, facet fracture with disc herniation or burst fracture, kyphotic central cord, and patients with AS or DISH; and anterior-alone approach is considered for remaining pathologies.

Future considerations

Outcomes following C-spine injury

The natural history following cervical SCI involves significant risk of mortality due to respiratory crises, ventilator dependence and pneumonia, autonomic dysreflexia and cardiac events, or thromboses.[122,123] Mortality rates in-hospital have been reduced to 8%–10% with advancements in critical care and aggressive medical management.[124,125]

Morbidity and overall societal impact have proven more difficult to mitigate. Compounded with frequent fall-related injuries in aging populations, incidence of cervical injuries now approaches 75% of all SCIs.[124,126] Significant advances have been made to reduce morbidity with early surgical decompression, management by intensive care in centers of excellence, and intensive physical rehabilitation, which are now standard of care.[122,127]

Study of outcomes following spinal injuries commonly focuses on mortality, conversion from complete to incomplete injuries on the ASIA Impairment Scale (AIS), return of motor control on Upper Extremity Motor Score (UEMS), and lowering the neurologic level of injury. Recent meta-analysis has reported conversion rates of complete (AIS A) to sensory and motor incomplete (AIS B–D) injuries at approximately 46% of patients with early decompressive surgery, and 33% without surgery.[128] These rates are significantly higher than previously thought, classically stated as 10%–20%, which could be due to improved management, or earlier initial assessment of ASIA grade.[129] Complete (AIS A) and incomplete (AIS B–D) cervical injuries can expect improvement in motor score of roughly 10 and >26 points, respectively.[130,131] Neurologic level of injury is also spontaneously improved/lowered; however, only 20%–30% of patients achieve the target 2-level improvement (i.e., C5 to C7). This corresponds to approximately 18-motor point improvement, which is the range required for meaningfully improved quality of life and independence.[132]

Clinical measures for treatments and clinical trials should relate directly to patient-centered outcomes.[133,134] Patients with incomplete injuries are more likely to regain motor function. Impaired neurologic levels with some motor function within days of injury are more likely to regain antigravity function for up to 1 year.[129,131] Mid-to-low cervical injuries represent the majority of incident spinal trauma cases. Myotome distribution of arm and hand function in these spinal cord segments corresponds to large impacts on quality of life and independence.[130,133,135] Emerging therapies are keenly focused on progressing the historic conversion rates and motor score improvement.

Emergent surgical therapies

Historically, a reduction of cervical dislocation was considered to provide enough decompression, and a laminectomy may have been discouraged to avoid promoting further destabilizing of the patient's spine. However, several studies have shown that, following an SCI, intraparenchymal swelling increases over the first

3 days resulting in increased intrathecal pressure and thus reduced perfusion pressure to the spinal cord.[136,137] The authors support wide posterior decompression in patients with SCI even if indirect decompression is accomplished by reduction of the dislocation.

Recently, there has been an interest in exploring the role that the dura plays in the compression of the spinal cord.[136] Traditionally, removal of epidural structures that cause spinal cord compression has been the treatment of choice for relieving spinal cord compression.[136,137] This is analogous to compartment syndrome of the extremities, where tissue damage leads to increased edema, which is further constrained by tight fascial compartments, thereby leading to progressively increasing intracompartmental pressure and loss of perfusion pressure within the compartment. Smith et al.[138] conducted a study in a cervical SCI rat model to assess the impact of decompressing the intradural compartment on functional recovery. They noted on immunohistochemistry analysis that this group had decreased cavitation and decreased scar formation within the spinal cord. More importantly, rats that underwent duraplasty had better functional recovery after acute SCI.

Phang et al.[139] subsequently conducted an open-label prospective trial for patients with traumatic SCI to study the effect of duraplasty on spinal cord physiology and patient outcomes. They assigned 11 participants to undergo laminectomy alone and 10 to receive both a laminectomy and duraplasty. One out of 10 participants who received duraplasty developed a CSF leak and was treated with a lumbar CSF drain. The authors reported that duraplasty was able to improve intraspinal pressure and spinal cord perfusion pressure (SCPP) compared to the laminectomy-only group. Furthermore, they reported that the duraplasty group had a higher change in ASIA grade, better walking ability, and bladder function. A similar strategy has been employed by Yue et al. using lumbar drainage of CSF to reduce intrathecal flow impedance and raise SCPP in a phase I trial of 15 participants.[140] They reported no drain-associated adverse events and had appreciable monitoring and control of SCPP. The functional outcomes of this laminectomy trial did not reach statistical significance, and the Yue et al. trial was not controlled. Regardless, the authors concluded that these findings showed promise, but need to be confirmed by larger randomized controlled trials before wide adoption.

Concluding remarks

Cervical spinal trauma is a nuanced and difficult entity to treat. Mortality has been greatly reduced with emergency transport to a level 1 trauma center with excellence in SCI. Making advances to reduce morbidity has proven more difficult. Preoperative imaging should be judicious with MRI or vascular studies when indicated. Delaying proper urgent surgical management must be avoided. Clinical studies have used <24 hours as the threshold for early decompression. While this has yet to reach standard of care, we should strive for earliest treatment possible. Perioperative and nonsurgical management should include immobilization, specialized neurointensive care as appropriate, blood pressure control, and managing adverse sequelae. Rapid closed reduction of a cervical fracture dislocation in an appropriate patient is highly recommended prior to surgical stabilization. The choice of surgical treatment for cervical trauma using an anterior, posterior, or combined approach should be based on the nature of spinal column injury, structural instability, neurologic compromise, and patient-specific considerations. Neurologic recovery can be greater in cervical trauma than corresponding thoracic/lumbar injury and return of motor function to the arm and hand can greatly improve patient independence and quality of life.

Acknowledgments

The authors thank Dr. Rudy Reindl and Dr. Oded Rabau for their expert discussions on the content of this chapter.

References

1. Panjabi MM, Thibodeau LL, Crisco JJ, White AA. What constitutes spinal instability? *Clin Neurosurg* 1988;**34**: 313—39.
2. Tan LA, Kasliwal MK, Traynelis VC. Comparison of CT and MRI findings for cervical spine clearance in obtunded patients without high impact trauma. *Clin Neurol Neurosurg* 2014;**120**:23—6.
3. Levi AD, Hurlbert RJ, Anderson P, Fehlings M, Rampersaud R, Massicotte EM, et al. Neurologic deterioration secondary to unrecognized spinal instability following trauma–a multicenter study. *Spine* 2006; **31**(4):451—8.
4. Bailitz J, Starr F, Beecroft M, Bankoff J, Roberts R, Bokhari F, et al. CT should replace three-view radiographs as the initial screening test in patients at high, moderate, and low risk for blunt cervical spine injury: a prospective comparison. *J Trauma* 2009;**66**(6):1605—9.
5. Beckmann NM, West OC, Nunez Jr D, Kirsch CFE, Aulino JM, Broder JS, et al. ACR Appropriateness criteria((R)) suspected spine trauma. *J Am Coll Radiol* 2019;**16**(5s):S264—85.
6. Khan SN, Erickson G, Sena MJ, Gupta MC. Use of flexion and extension radiographs of the cervical spine to rule out acute instability in patients with negative computed tomography scans. *J Orthop Trauma* 2011; **25**(1):51—6.
7. Sim V, Bernstein MP, Frangos SG, Wilson CT, Simon RJ, McStay CM, et al. The (f)utility of flexion-extension C-spine films in the setting of trauma. *Am J Surg* 2013;**206**(6):929—33. Discussion 33-4.
8. Ryken TC, Hadley MN, Walters BC, Aarabi B, Dhall SS, Gelb DE, et al. Radiographic assessment. *Neurosurgery* 2013;**72**(Suppl. 2):54—72.
9. Hoffman JR, Mower WR, Wolfson AB, Todd KH, Zucker MI. Validity of a set of clinical criteria to rule out injury to the cervical spine in patients with blunt trauma. National Emergency X-Radiography Utilization Study Group. *N Engl J Med* 2000;**343**(2):94—9.
10. Stiell IG, Wells GA, Vandemheen KL, Clement CM, Lesiuk H, De Maio VJ, et al. The Canadian C-spine rule for radiography in alert and stable trauma patients. *J Am Med Assoc* 2001;**286**(15):1841—8.
11. Anderson PA, Moore TA, Davis KW, Molinari RW, Resnick DK, Vaccaro AR, et al. Cervical spine injury severity score. Assessment of reliability. *J Bone Jt Surg Am* 2007;**89**(5):1057—65.
12. Goradia D, Linnau KF, Cohen WA, Mirza S, Hallam DK, Blackmore CC. Correlation of MR imaging findings with intraoperative findings after cervical spine trauma. *AJNR Am J Neuroradiol* 2007;**28**(2): 209—15.
13. McGuire Jr RA. Protection of the unstable spine during transport and early hospitalization. *J Miss State Med Assoc* 1991;**32**(8):305—8.
14. Fehlings MG, Martin AR, Tetreault LA, Aarabi B, Anderson P, Arnold PM, et al. A clinical practice guideline for the management of patients with acute spinal cord injury: recommendations on the role of baseline magnetic resonance imaging in clinical decision making and outcome prediction. *Global Spine J* 2017;**7**(3 Suppl. l):221S—30S.
15. Awad BI, Carmody MA, Lubelski D, El Hawi M, Claridge JA, Como JJ, et al. Adjacent level ligamentous injury associated with traumatic cervical spine fractures: indications for imaging and implications for treatment. *World Neurosurg* 2015;**84**(1):69—75.
16. Jug M, Kejzar N, Vesel M, Al Mawed S, Dobravec M, Herman S, et al. Neurological recovery after traumatic cervical spinal cord injury is superior if surgical decompression and instrumented fusion are performed within 8 hours versus 8 to 24 hours after injury: a single center experience. *J Neurotrauma* 2015;**32**(18): 1385—92.
17. Burke JF, Yue JK, Ngwenya LB, Winkler EA, Talbott JF, Pan JZ, et al. Ultra-early (<12 hours) surgery correlates with higher rate of American spinal injury association impairment scale conversion after cervical spinal cord injury. *Neurosurgery* 2019;**85**(2):199—203.
18. Vaccaro AR, Hulbert RJ, Patel AA, Fisher C, Dvorak M, Lehman Jr RA, et al. The subaxial cervical spine injury classification system: a novel approach to recognize the importance of morphology, neurology, and integrity of the disco-ligamentous complex. *Spine* 2007;**32**(21): 2365—74.
19. Grassner L, Wutte C, Zimmermann G, Grillhosl A, Schmid K, Weibeta T, et al. Influence of preoperative magnetic resonance imaging on surgical decision making for patients with acute traumatic cervical spinal cord injury: a survey among experienced spine surgeons. *World Neurosurg* 2019;**131**:e586—92.
20. Kurpad S, Martin AR, Tetreault LA, Fischer DJ, Skelly AC, Mikulis D, et al. Impact of baseline magnetic resonance imaging on neurologic, functional, and safety outcomes in patients with acute traumatic spinal cord injury. *Global Spine J* 2017;**7**(3 Suppl. l): 151S—74S.
21. Seif M, Gandini Wheeler-Kingshott CA, Cohen-Adad J, Flanders AE, Freund P. Guidelines for the conduct of clinical trials in spinal cord injury: neuroimaging biomarkers. *Spinal Cord* 2019;**57**(9):717—28.

22. Talbott JF, Whetstone WD, Readdy WJ, Ferguson AR, Bresnahan JC, Saigal R, et al. The Brain and Spinal Injury Center score: a novel, simple, and reproducible method for assessing the severity of acute cervical spinal cord injury with axial T2-weighted MRI findings. *J Neurosurg Spine* 2015;**23**(4):495–504.

23. Rihn JA, Yang N, Fisher C, Saravanja D, Smith H, Morrison WB, et al. Using magnetic resonance imaging to accurately assess injury to the posterior ligamentous complex of the spine: a prospective comparison of the surgeon and radiologist. *J Neurosurg Spine* 2010;**12**(4):391–6.

24. Hadley MN, Walters BC. Introduction to the guidelines for the management of acute cervical spine and spinal cord injuries. *Neurosurgery* 2013;**72**(Suppl. 2):5–16.

25. Roy AK, Miller BA, Holland CM, Fountain Jr AJ, Pradilla G, Ahmad FU. Magnetic resonance imaging of traumatic injury to the craniovertebral junction: a case-based review. *Neurosurg Focus* 2015;**38**(4):E3.

26. Papadopoulos SM, Selden NR, Quint DJ, Patel N, Gillespie B, Grube S. Immediate spinal cord decompression for cervical spinal cord injury: feasibility and outcome. *J Trauma* 2002;**52**(2):323–32.

27. Walters BC, Hadley MN, Hurlbert RJ, Aarabi B, Dhall SS, Gelb DE, et al. Guidelines for the management of acute cervical spine and spinal cord injuries: 2013 update. *Neurosurgery* 2013;**60**(CN_Suppl. 1_1):82–91.

28. Wilson JR, Grossman RG, Frankowski RF, Kiss A, Davis AM, Kulkarni AV, et al. A clinical prediction model for long-term functional outcome after traumatic spinal cord injury based on acute clinical and imaging factors. *J Neurotrauma* 2012;**29**(13):2263–71.

29. Aarabi B, Olexa J, Chryssikos T, Galvagno SM, Hersh DS, Wessell A, et al. Extent of spinal cord decompression in motor complete (American Spinal Injury Association Impairment Scale Grades A and B) traumatic spinal cord injury patients: post-operative magnetic resonance imaging analysis of standard operative approaches. *J Neurotrauma* 2019;**36**(6):862–76.

30. Dalkilic T, Fallah N, Noonan VK, Salimi Elizei S, Dong K, Belanger L, et al. Predicting injury severity and neurological recovery after acute cervical spinal cord injury: a comparison of cerebrospinal fluid and magnetic resonance imaging biomarkers. *J Neurotrauma* 2018;**35**(3):435–45.

31. Haefeli J, Mabray MC, Whetstone WD, Dhall SS, Pan JZ, Upadhyayula P, et al. Multivariate analysis of MRI biomarkers for predicting neurologic impairment in cervical spinal cord injury. *AJNR Am J Neuroradiol* 2017;**38**(3):648–55.

32. Morais DF, de Melo Neto JS, Meguins LC, Mussi SE, Filho JR, Tognola WA. Clinical applicability of magnetic resonance imaging in acute spinal cord trauma. *Eur Spine J* 2014;**23**(7):1457–63.

33. Bozzo A, Marcoux J, Radhakrishna M, Pelletier J, Goulet B. The role of magnetic resonance imaging in the management of acute spinal cord injury. *J Neurotrauma* 2011;**28**(8):1401–11.

34. Martin AR, De Leener B, Cohen-Adad J, Cadotte DW, Kalsi-Ryan S, Lange SF, et al. Clinically feasible microstructural MRI to quantify cervical spinal cord tissue injury using DTI, MT, and T2*-weighted imaging: assessment of normative data and reliability. *AJNR Am J Neuroradiol* 2017;**38**(6):1257–65.

35. Zaninovich OA, Avila MJ, Kay M, Becker JL, Hurlbert RJ, Martirosyan NL. The role of diffusion tensor imaging in the diagnosis, prognosis, and assessment of recovery and treatment of spinal cord injury: a systematic review. *Neurosurg Focus* 2019;**46**(3):E7.

36. Freund P, Seif M, Weiskopf N, Friston K, Fehlings MG, Thompson AJ, et al. MRI in traumatic spinal cord injury: from clinical assessment to neuroimaging biomarkers. *Lancet Neurol* 2019;**18**(12):1123–35.

37. Cothren CC, Moore EE, Ray Jr CE, Johnson JL, Moore JB, Burch JM. Cervical spine fracture patterns mandating screening to rule out blunt cerebrovascular injury. *Surgery* 2007;**141**(1):76–82.

38. Brommeland T, Helseth E, Aarhus M, Moen KG, Dyrskog S, Bergholt B, et al. Best practice guidelines for blunt cerebrovascular injury (BCVI). *Scand J Trauma Resuscitation Emerg Med* 2018;**26**(1):90.

39. McKevitt EC, Kirkpatrick AW, Vertesi L, Granger R, Simons RK. Blunt vascular neck injuries: diagnosis and outcomes of extracranial vessel injury. *J Trauma* 2002;**53**(3):472–6.

40. Fabian TC, Patton Jr JH, Croce MA, Minard G, Kudsk KA, Pritchard FE. Blunt carotid injury. Importance of early diagnosis and anticoagulant therapy. *Ann Surg* 1996;**223**(5):513–22. Discussion 22-5.

41. Miller PR, Fabian TC, Bee TK, Timmons S, Chamsuddin A, Finkle R, et al. Blunt cerebrovascular injuries: diagnosis and treatment. *J Trauma* 2001;**51**(2):279–85. Discussion 85-6.

42. Biffl WL, Moore EE, Ryu RK, Offner PJ, Novak Z, Coldwell DM, et al. The unrecognized epidemic of blunt carotid arterial injuries: early diagnosis improves neurologic outcome. *Ann Surg* 1998;**228**(4):462–70.

43. Burlew CC, Biffl WL, Moore EE, Barnett CC, Johnson JL, Bensard DD. Blunt cerebrovascular injuries: redefining screening criteria in the era of noninvasive diagnosis. *J Trauma Acute Care Surg* 2012;**72**(2):330–5. Discussion 6-7, quiz 539.

44. Catapano JS, Israr S, Whiting AC, Hussain OM, Snyder LA, Albuquerque FC, et al. Management of extracranial blunt cerebrovascular injuries: experience with an aspirin-based approach. *World Neurosurg* 2019;**133**:e385–90.

45. Brott TG, Halperin JL, Abbara S, Bacharach JM, Barr JD, Bush RL, et al. American College of Cardiology Foundation/American Heart Association Task F, American Stroke A, American Association of Neuroscience N, American Association of Neurological S, American College of R, American Society of N, et al. 2011 ASA/ACCF/AHA/AANN/AANS/ACR/ASNR/CNS/SAIP/SCAI/SIR/SNIS/SVM/SVS guideline on the management of patients with extracranial carotid and vertebral artery disease: executive summary. *J Neurointerventional Surg* 2011;**3**(2):100–30.

46. Lockwood MM, Smith GA, Tanenbaum J, Lubelski D, Seicean A, Pace J, et al. Screening via CT angiogram after traumatic cervical spine fractures: narrowing imaging to improve cost effectiveness. Experience of a Level I trauma center. *J Neurosurg Spine* 2016;**24**(3):490–5.

47. Schneidereit NP, Simons R, Nicolaou S, Graeb D, Brown DR, Kirkpatrick A, et al. Utility of screening for blunt vascular neck injuries with computed tomographic angiography. *J Trauma* 2006;**60**(1):209–15. Discussion 15-6.

48. Beliaev AM, Barber PA, Marshall RJ, Civil I. Denver screening protocol for blunt cerebrovascular injury reduces the use of multi-detector computed tomography angiography. *ANZ J Surg* 2014;**84**(6):429–32.

49. Aarabi B, Walters BC, Dhall SS, Gelb DE, Hurlbert RJ, Rozzelle CJ, et al. Subaxial cervical spine injury classification systems. *Neurosurgery* 2013;**72**(Suppl. 2):170–86.

50. Allen Jr BL, Ferguson RL, Lehmann TR, O'Brien RP. A mechanistic classification of closed, indirect fractures and dislocations of the lower cervical spine. *Spine* 1982;**7**(1):1–27.

51. Samuel S, Lin J-L, Smith MM, Hartin NL, Vasili C, Ruff SJ, et al. Subaxial injury classification scoring system treatment recommendations: external agreement study based on retrospective review of 185 patients. *Spine* 2015;**40**(3):137–42.

52. van Middendorp JJ, Audige L, Bartels RH, Bolger C, Deverall H, Dhoke P, et al. The subaxial cervical spine injury classification system: an external agreement validation study. *Spine J* 2013;**13**(9):1055–63.

53. Vaccaro AR, Koerner JD, Radcliff KE, Oner FC, Reinhold M, Schnake KJ, et al. AOSpine subaxial cervical spine injury classification system. *Eur Spine J* 2016;**25**(7):2173–84.

54. Urrutia J, Zamora T, Yurac R, Campos M, Palma J, Mobarec S, et al. An independent inter- and intraobserver agreement evaluation of the AOSpine subaxial cervical spine injury classification system. *Spine* 2017;**42**(5):298–303.

55. Boström M, Kalm M, Karlsson N, Hellström Erkenstam N, Blomgren K. Irradiation to the young mouse brain caused long-term, progressive depletion of neurogenesis but did not disrupt the neurovascular niche. *J Cerebr Blood Flow Metabol* 2013;**33**(6):935–43.

56. Vioque SM, Kim PK, McMaster J, Gallagher J, Allen SR, Holena DN, et al. Classifying errors in preventable and potentially preventable trauma deaths: a 9-year review using the joint commission's standardized methodology. *Am J Surg* 2014;**208**(2):187–94.

57. Sundheim SM, Cruz M. The evidence for spinal immobilization: an estimate of the magnitude of the treatment benefit. *Ann Emerg Med* 2006;**48**(2):217–8. Author reply 8-9.

58. Prasad VS, Schwartz A, Bhutani R, Sharkey PW, Schwartz ML. Characteristics of injuries to the cervical spine and spinal cord in polytrauma patient population: experience from a regional trauma unit. *Spinal Cord* 1999;**37**(8):560–8.

59. Sciubba DM, Nelson C, Hsieh P, Gokaslan ZL, Ondra S, Bydon A. Perioperative challenges in the surgical management of ankylosing spondylitis. *Neurosurg Focus* 2008;**24**(1):E10.

60. Jeanneret B, Magerl F, Ward JC. Overdistraction: a hazard of skull traction in the management of acute injuries of the cervical spine. *Arch Orthop Trauma Surg* 1991;**110**(5):242–5.

61. Bonner S, Smith C. Initial management of acute spinal cord injury. *BJA Education* 2013;**13**(6):224–31.

62. Podolsky SM, Hoffman JR, Pietrafesa CA. Neurologic complications following immobilization of cervical spine fracture in a patient with ankylosing spondylitis. *Ann Emerg Med* 1983;**12**(9):578–80.

63. Perry SD, McLellan B, McIlroy WE, Maki BE, Schwartz M, Fernie GR. The efficacy of head immobilization techniques during simulated vehicle motion. *Spine* 1999;**24**(17):1839–44.

64. Horodyski M, DiPaola CP, Conrad BP, Rechtine 2nd GR. Cervical collars are insufficient for immobilizing an unstable cervical spine injury. *J Emerg Med* 2011;**41**(5):513–9.

65. Theodore N, Hadley MN, Aarabi B, Dhall SS, Gelb DE, Hurlbert RJ, et al. Prehospital cervical spinal immobilization after trauma. *Neurosurgery* 2013;**72**(Suppl. 2(4)):22–34.

66. Worsley PR, Stanger ND, Horrell AK, Bader DL. Investigating the effects of cervical collar design and fit on

the biomechanical and biomarker reaction at the skin. *Med Devices (Auckl)* 2018;**11**:87—94.

67. Domeier RM, Evans RW, Swor RA, Hancock JB, Fales W, Krohmer J, et al. The reliability of prehospital clinical evaluation for potential spinal injury is not affected by the mechanism of injury. *Prehosp Emerg Care* 1999;**3**(4):332—7.

68. Deasy C, Cameron P. Routine application of cervical collars–what is the evidence? *Injury* 2011;**42**(9):841—2.

69. Thumbikat P, Hariharan RP, Ravichandran G, McClelland MR, Mathew KM. Spinal cord injury in patients with ankylosing spondylitis: a 10-year review. *Spine* 2007;**32**(26):2989—95.

70. Ahmed OZ, Webman RB, Sheth PD, Donnenfield JI, Yang J, Sarcevic A, et al. Errors in cervical spine immobilization during pediatric trauma evaluation. *J Surg Res* 2018;**228**:135—41.

71. Brown LH, Gough JE, Simonds WB. Can EMS providers adequately assess trauma patients for cervical spinal injury? *Prehosp Emerg Care* 1998;**2**(1):33—6.

72. Dipaola CP, Conrad BP, Horodyski M, Dipaola MJ, Sawers A, Rechtine 2nd GR. Cervical spine motion generated with manual versus jackson table turning methods in a cadaveric c1-c2 global instability model. *Spine* 2009;**34**(26):2912—8.

73. DiPaola MJ, DiPaola CP, Conrad BP, Horodyski M, Del Rossi G, Sawers A, et al. Cervical spine motion in manual versus Jackson table turning methods in a cadaveric global instability model. *J Spinal Disord Tech* 2008;**21**(4):273—80.

74. Kwan I, Bunn F, Roberts I. Spinal immobilisation for trauma patients. *Cochrane Database Syst Rev* 2001;**2**: Cd002803.

75. Aprahamian C, Thompson BM, Darin JC. Recommended helmet removal techniques in a cervical spine injured patient. *J Trauma* 1984;**24**(9):841—2.

76. Donaldson WF, Heil BV, Donaldson VP, Silvaggio VJ. The effect of airway maneuvers on the unstable C1-C2 segment. A cadaver study. *Spine* 1997;**22**(11):1215—8.

77. Hauswald M, Sklar DP, Tandberg D, Garcia JF. Cervical spine movement during airway management: cinefluoroscopic appraisal in human cadavers. *Am J Emerg Med* 1991;**9**(6):535—8.

78. Rorabeck CH, Rock MG, Hawkins RJ, Bourne RB. Unilateral facet dislocation of the cervical spine. An analysis of the results of treatment in 26 patients. *Spine* 1987;**12**(1):23—7.

79. Tarlov IM. Spinal cord compression studies. III. Time limits for recovery after gradual compression in dogs. *AMA Arch Neurol Psychiatry* 1954;**71**(5):588—97.

80. Burke DC, Berryman D. The place of closed manipulation in the management of flexion-rotation dislocations of the cervical spine. *J Bone Joint Surg Br* 1971;**53**(2): 165—82.

81. Gelb DE, Hadley MN, Aarabi B, Dhall SS, Hurlbert RJ, Rozzelle CJ, et al. Initial closed reduction of cervical spinal fracture-dislocation injuries. *Neurosurgery* 2013; **72**(Suppl. 2):73—83.

82. O'Connor PA, McCormack O, Noël J, McCormack D, O'Byrne J. Anterior displacement correlates with neurological impairment in cervical facet dislocations. *Int Orthop* 2003;**27**(3):190—3.

83. Rizzolo SJ, Vaccaro AR, Cotler JM. Cervical spine trauma. *Spine* 1994;**19**(20):2288—98.

84. Cotler JM, Herbison GJ, Nasuti JF, Ditunno Jr JF, An H, Wolff BE. Closed reduction of traumatic cervical spine dislocation using traction weights up to 140 pounds. *Spine* 1993;**18**(3):386—90.

85. Ajiboye RM, Zoller SD, Sharma A, Mosich GM, Drysch A, Li J, et al. Intraoperative neuromonitoring for anterior cervical spine surgery. *Spine* 2017;**42**(6): 385—93.

86. Rabai F, Sessions R, Seubert CN. Neurophysiological monitoring and spinal cord integrity. *Best Pract Res Clin Anaesthesiol* 2016;**30**(1):53—68.

87. Schwartz DM, Auerbach JD, Dormans JP, Flynn J, Drummond DS, Bowe JA, et al. Neurophysiological detection of impending spinal cord injury during scoliosis surgery. *J Bone Jt Surg Am* 2007;**89**(11): 2440—9.

88. McDonald JW, Becker D, Sadowsky CL, Jane JA, Conturo TE, Schultz LM. Late recovery following spinal cord injury. Case report and review of the literature. *J Neurosurg* 2002;**97**(2 Suppl. 1):252—65.

89. Kim DH, Zaremski J, Kwon B, Jenis L, Woodard E, Bode R, et al. Risk factors for false positive transcranial motor evoked potential monitoring alerts during surgical treatment of cervical myelopathy. *Spine* 2007;**32**(26): 3041—6.

90. Ziewacz JE, Berven SH, Mummaneni VP, Tu T-H, Akinbo OC, Lyon R, et al. The design, development, and implementation of a checklist for intraoperative neuromonitoring changes. *Neurosurg Focus* 2012;**33**(5): E11.

91. Vitale MG, Skaggs DL, Pace GI, Wright ML, Matsumoto H, Anderson RCE, et al. Best practices in intraoperative neuromonitoring in spine deformity surgery: development of an intraoperative checklist to optimize response. *Spine Deform* 2014;**2**(5):333—9.

92. Appel S, Biron T, Goldstein K, Ashkenazi E. Effect of intra- and extraoperative factors on the efficacy of intraoperative neuromonitoring during cervical spine surgery. *World Neurosurg* 2019;**123**:e646—51.

93. Batchelor PE, Wills TE, Skeers P, Battistuzzo CR, Macleod MR, Howells DW, et al. Meta-analysis of pre-clinical studies of early decompression in acute spinal cord injury: a battle of time and pressure. *PLoS One* 2013;**8**(8):e72659.

94. Dvorak MF, Noonan VK, Fallah N, Fisher CG, Finkelstein J, Kwon BK, et al. The influence of time from injury to surgery on motor recovery and length of hospital stay in acute traumatic spinal cord injury: an observational Canadian cohort study. *J Neurotrauma* 2015;**32**(9):645−54.

95. Bourassa-Moreau É, Mac-Thiong J-M, Ehrmann Feldman D, Thompson C, Parent S. Complications in acute phase hospitalization of traumatic spinal cord injury: does surgical timing matter? *J Trauma Acute Care Surg* 2013;**74**(3):849−54.

96. Fehlings MG, Vaccaro A, Wilson JR, Singh A, Cadotte DW, Harrop JS, et al. Early versus delayed decompression for traumatic cervical spinal cord injury: results of the Surgical Timing in Acute Spinal Cord Injury Study (STASCIS). *PLoS One* 2012;**7**(2):e32037.

97. Wilson JR, Singh A, Craven C, Verrier MC, Drew B, Ahn H, et al. Early versus late surgery for traumatic spinal cord injury: the results of a prospective Canadian cohort study. *Spinal Cord* 2012;**50**(11):840−3.

98. Lenehan B, Fisher CG, Vaccaro A, Fehlings M, Aarabi B, Dvorak MF. The urgency of surgical decompression in acute central cord injuries with spondylosis and without instability. *Spine* 2010;**35**(21 Suppl. 1): S180−6.

99. Wilson JR, Tetreault LA, Kwon BK, Arnold PM, Mroz TE, Shaffrey C, et al. Timing of decompression in patients with acute spinal cord injury: a systematic review. *Global Spine J* 2017;**7**(3 Suppl. 1):95s−115s.

100. Albert TJ, Vacarro A. Postlaminectomy kyphosis. *Spine* 1998;**23**(24):2738−45.

101. Dvorak MF, Fisher CG, Fehlings MG, Rampersaud YR, Oner FC, Aarabi B, et al. The surgical approach to subaxial cervical spine injuries: an evidence-based algorithm based on the SLIC classification system. *Spine* 2007;**32**(23):2620−9.

102. Gelb DE, Aarabi B, Dhall SS, Hurlbert RJ, Rozzelle CJ, Ryken TC, et al. Treatment of subaxial cervical spinal injuries. *Neurosurgery* 2013;**72**(Suppl. 2):187−94.

103. Einsiedel T, Schmelz A, Arand M, Wilke H-J, Gebhard F, Hartwig E, et al. Injuries of the cervical spine in patients with ankylosing spondylitis: experience at two trauma centers. *J Neurosurg Spine* 2006; **5**(1):33−45.

104. Kasimatis GB, Panagiotopoulos E, Gliatis J, Tyllianakis M, Zouboulis P, Lambiris E. Complications of anterior surgery in cervical spine trauma: an overview. *Clin Neurol Neurosurg* 2009;**111**(1):18−27.

105. Tan TP, Govindarajulu AP, Massicotte EM, Venkatraghavan L. Vocal cord palsy after anterior cervical spine surgery: a qualitative systematic review. *Spine J* 2014;**14**(7):1332−42.

106. Woodworth RS, Molinari WJ, Brandenstein D, Gruhn W, Molinari RW. Anterior cervical discectomy and fusion with structural allograft and plates for the treatment of unstable posterior cervical spine injuries. *J Neurosurg Spine* 2009;**10**(2):93−101.

107. Nassr A, Lee JY, Dvorak MF, Harrop JS, Dailey AT, Shaffrey CI, et al. Variations in surgical treatment of cervical facet dislocations. *Spine* 2008;**33**(7):E188−93.

108. Reindl R, Ouellet J, Harvey EJ, Berry G, Arlet V. Anterior reduction for cervical spine dislocation. *Spine* 2006; **31**(6):648−52.

109. Patel VV, Burger E, Brown CW. In: *Spine trauma.* Springer Science & Business Media; August 24, 2010. p. 413.

110. Johnson MG, Fisher CG, Boyd M, Pitzen T, Oxland TR, Dvorak MF. The radiographic failure of single segment anterior cervical plate fixation in traumatic cervical flexion distraction injuries. *Spine* 2004;**29**(24):2815−20.

111. Wang JC, Hart RA, Emery SE, Bohlman HH. Graft migration or displacement after multilevel cervical corpectomy and strut grafting. *Spine* 2003;**28**(10):1016−21. Discussion 21-2.

112. Theodotou CB, Ghobrial GM, Middleton AL, Wang MY, Levi AD. Anterior reduction and fusion of cervical facet dislocations. *Neurosurgery* 2019;**84**(2): 388−95.

113. Cooper K, Glenn CA, Martin M, Stoner J, Li J, Puckett T. Risk factors for surgical site infection after instrumented fixation in spine trauma. *J Clin Neurosci* 2016;**23**:123−7.

114. Lenoir T, Hoffmann E, Thevenin-Lemoine C, Lavelle G, Rillardon L, Guigui P. Neurological and functional outcome after unstable cervicothoracic junction injury treated by posterior reduction and synthesis. *Spine J* 2006;**6**(5):507−13.

115. Truumees E, Singh D, Geck MJ, Stokes JK. Should long-segment cervical fusions be routinely carried into the thoracic spine? A multicenter analysis. *Spine J* 2018; **18**(5):782−7.

116. Nagashima K, Koda M, Abe T, Kumagai H, Miura K, Fujii K, et al. Implant failure of pedicle screws in long-segment posterior cervical fusion is likely to occur at C7 and is avoidable by concomitant C6 or T1 buttress pedicle screws. *J Clin Neurosci* 2019;**63**: 106−9.

117. Rhee JM, Kraiwattanapong C, Hutton WC. A comparison of pedicle and lateral mass screw construct stiffnesses at the cervicothoracic junction: a biomechanical study. *Spine* 2005;**30**(21):E636−40.

118. Kast E, Mohr K, Richter H-P, Börm W. Complications of transpedicular screw fixation in the cervical spine. *Eur Spine J* 2006;**15**(3):327−34.

119. Clifton W, Louie C, Williams DB, Damon A, Dove C, Pichelmann M. Safety and accuracy of the freehand placement of C7 pedicle screws in cervical and cervico-thoracic constructs. *Cureus* 2019;**11**(8):e5304.

120. Reinhold M, Magerl F, Rieger M, Blauth M. Cervical pedicle screw placement: feasibility and accuracy of two new insertion techniques based on morphometric data. *Eur Spine J* 2007;**16**(1):47–56.

121. Desai S, Sethi A, Ninh CC, Bartol S, Vaidya R. Pedicle screw fixation of the C7 vertebra using an anteroposterior fluoroscopic imaging technique. *Eur Spine J* 2010;**19**(11):1953–9.

122. Tator CH. Review of treatment trials in human spinal cord injury: issues, difficulties, and recommendations. *Neurosurgery* 2006;**59**(5):957–82. Discussion 82-7.

123. Frankel HL, Coll JR, Charlifue SW, Whiteneck GG, Gardner BP, Jamous MA, et al. Long-term survival in spinal cord injury: a fifty year investigation. *Spinal Cord* 1998;**36**(4):266–74.

124. Pickett GE, Campos-Benitez M, Keller JL, Duggal N. Epidemiology of traumatic spinal cord injury in Canada. *Spine* 2006;**31**(7):799–805.

125. Sekhon LH, Fehlings MG. Epidemiology, demographics, and pathophysiology of acute spinal cord injury. *Spine* 2001;**26**(24 Suppl. l):S2–12.

126. Ackery A, Tator C, Krassioukov A. A global perspective on spinal cord injury epidemiology. *J Neurotrauma* 2004;**21**(10):1355–70.

127. Dvorak MF, Noonan V, Fallah N, Fisher CG, Finkelstein J, Kwon BK, et al. The influence of time from injury to surgery on motor recovery and length of hospital stay in acute traumatic spinal cord injury: an observational Canadian cohort study. *J Neurotrauma* 2015;**32**(9):645–54.

128. El Tecle NE, Dahdaleh NS, Bydon M, Ray WZ, Torner JC, Hitchon PW. The natural history of complete spinal cord injury: a pooled analysis of 1162 patients and a meta-analysis of modern data. *J Neurosurg Spine* 2018;**28**(4):436–43.

129. Evaniew N, Sharifi B, Waheed Z, Fallah N, Ailon T, Dea N, et al. The influence of neurological examination timing within hours after acute traumatic spinal cord injuries: an observational study. *Spinal Cord* 2020;**58**(2):247–54.

130. Steeves JD, Lammertse D, Curt A, Fawcett JW, Tuszynski MH, Ditunno JF, et al. Guidelines for the conduct of clinical trials for spinal cord injury (SCI) as developed by the ICCP panel: clinical trial outcome measures. *Spinal Cord* 2007;**45**(3):206–21.

131. Khorasanizadeh M, Yousefifard M, Eskian M, Lu Y, Chalangari M, Harrop JS, et al. Neurological recovery following traumatic spinal cord injury: a systematic review and meta-analysis. *J Neurosurg Spine* 2019;**30**(5):683–99.

132. Kramer JLK, Lammertse DP, Schubert M, Curt A, Steeves JD. Relationship between motor recovery and independence after sensorimotor-complete cervical spinal cord injury. *Neurorehabil Neural Repair* 2012;**26**(9):1064–71.

133. Anderson DK, Beattie M, Blesch A, Bresnahan J, Bunge M, Dietrich D, et al. Recommended guidelines for studies of human subjects with spinal cord injury. *Spinal Cord* 2005;**43**(8):453–8.

134. Lo C, Tran Y, Anderson K, Craig A, Middleton J. Functional priorities in persons with spinal cord injury: using discrete choice experiments to determine preferences. *J Neurotrauma* 2016;**33**(21):1958–68.

135. van Hedel HJA, Curt A. Fighting for each segment: estimating the clinical value of cervical and thoracic segments in SCI. *J Neurotrauma* 2006;**23**(11):1621–31.

136. Grassner L, Grillhosl A, Griessenauer CJ, Thome C, Buhren V, Strowitzki M, et al. Spinal meninges and their role in spinal cord injury: a neuroanatomical review. *J Neurotrauma* 2018;**35**(3):403–10.

137. Tykocki T, Poniatowski L, Czyz M, Koziara M, Wynne-Jones G. Intraspinal pressure monitoring and extensive duroplasty in the acute phase of traumatic spinal cord injury: a systematic review. *World Neurosurg* 2017;**105**:145–52.

138. Smith JS, Anderson R, Pham T, Bhatia N, Steward O, Gupta R. Role of early surgical decompression of the intradural space after cervical spinal cord injury in an animal model. *J Bone Joint Surg Am* 2010;**92**(5):1206–14.

139. Phang I, Werndle MC, Saadoun S, Varsos G, Czosnyka M, Zoumprouli A, et al. Expansion duroplasty improves intraspinal pressure, spinal cord perfusion pressure, and vascular pressure reactivity index in patients with traumatic spinal cord injury: injured spinal cord pressure evaluation study. *J Neurotrauma* 2015;**32**(12):865–74.

140. Yue JK, Hemmerle DD, Winkler EA, Thomas LH, Fernandez XD, Kyritsis N, et al. Clinical implementation of novel spinal cord perfusion pressure protocol in acute traumatic spinal cord injury at U.S. Level I Trauma Center: TRACK-SCI study. *World Neurosurg* 2020;**133**:e391–6.

Spine trauma management issues: thoracic and lumbar

David Ben-Israel, W. Bradley Jacobs

Division of Neurosurgery, Department of Clinical Neurosciences, University of Calgary, Calgary, AB, Canada

List of abbreviations

ASIA American Spinal Injury Association
CTO Cervicothoracic Orthosis
CT Computed tomography
DVT Deep venous thrombosis
DISH Diffuse idiopathic skeletal hyperostosis
LSO Lumbosacral orthosis
MRI Magnetic resonance imaging
PLC Posterior Ligamentous Complex
TLSO Thoracolumbosacral orthosis

Introduction

As with injuries to other regions of the spine, the treatment goals for thoracic and lumbar spinal trauma are to preserve or restore neurologic function, maintain or regain normal alignment, and restore spinal stability. While treatment can vary considerably from conservative management to emergent surgical intervention, adherence to basic principles and a focus on these universal treatment goals can help guide clinicians in making appropriate management decisions.

Although this chapter focuses on management considerations, it is important to have an understanding of thoracolumbar spinal epidemiology, anatomy, imaging, and classification systems, and the reader is referred to other chapters within this text (see Chapter 1: Anatomy, Chapter 2: Epidemiology, Chapter 3: Classification Systems: Spine Trauma, Chapter 6: Imaging: Spine Trauma, Chapter 7: Imaging: SCI) for these details.

Clinical evaluation

Clinical history

In addition to the basic elements that are part of any medical history, there are historical elements of specific relevance in patients that have sustained a traumatic thoracolumbar injury. For instance, details concerning the mechanism of injury, whether obtained from the patient or via collateral history, are critical. This mechanistic information can be extremely helpful in understanding the specific spinal injury pattern, as well as to predict the likelihood of

© 2022 Elsevier Inc. All rights reserved.

neurological injury and mechanical instability.[1] Historical information regarding the location, quality, and progression of pain is also important to help guide further investigations and decrease the potential for missed noncontiguous spinal column injuries. Finally, it is imperative to establish the presence of any neurologic deficit, either transient or persistent, to help guide both the nature and urgency of intervention.

Physical exam

A thorough physical examination is critical in all patients that have sustained a traumatic spinal column injury. First, it is important to document a complete set of vital signs. In addition to the potential for hemorrhagic shock in any patient that has sustained a traumatic injury, spinal cord injury to the cervical or upper thoracic cord can result in neurogenic shock. Shock-induced hemodynamic instability should be recognized and treated promptly, as even transient hypotension can exacerbate the severity of spinal cord injury.[2-4]

Examination of the spine should be performed in all patients with suspected spinal trauma. During examination, patients must be log rolled into a lateral decubitus position with careful maintenance of spinal precautions so as to avoid iatrogenic neural injury. The presence of a step deformity or tenderness to midline palpation is suggestive of posterior ligamentous complex (PLC) disruption, and should be further evaluated with appropriate CT and MRI cross-sectional imaging.[5] A full neurologic assessment including motor function, sensory function, and reflexes should be performed on every patient; sacral sensation, deep anal pressure, and voluntary anal contraction are mandatory components of the neurological examination for all patients with potential spinal cord injury. If neurologic deficits are present, a validated neurologic assessment tool should be used, such as the American Spinal Injury Association (ASIA)

impairment scale,[6] for the purposes of prognostication as well as monitoring patient recovery.[7]

Investigations

While screening for spinal trauma can begin with plain radiographs, computed tomography (CT) scans are routinely obtained in the polytrauma patient, as they are more sensitive than plain radiographs at diagnosing bony spinal column injuries.[8,9] CT is mandatory if there is a high-energy mechanism of injury, if the patient has pain or neurologic symptoms suspicious for spinal trauma, or if the patient is found to have pathology on plain radiography that requires further characterization. Spinal magnetic resonance imaging (MRI) is obtained if there is a need to characterize injury to the spinal cord, to evaluate the integrity of the discoligamentous structures, or to assess the intradural or epidural spaces.[10,11] Further, MRI is also useful to assess the integrity of the PLC. Information about PLC integrity is an important component of the comprehensive assessment of the stability of the injured spinal segment and a major factor in the decision-making paradigm when evaluating the need for surgical intervention in a number of thoracolumbar injury patterns.[12] However, while MRI is highly sensitive in the detection of PLC imaging abnormality, it is also highly nonspecific. As such, MRI diagnosis of PLC injury should not be used in isolation to derive treatment recommendations.[13,14]

General management considerations

Initial spinal immobilization

As per practice guidelines, trauma patients at risk for spinal injuries should undergo spinal motion restriction during transport using a combination of a properly sized cervical spinal hard collar and supine positioning on a flat surface, such as a spinal long board.[15] Patients should be moved onto a soft surface such as an ambulance cot or a hospital stretcher and kept in the supine position as soon as feasible.[15] Minimization of the duration of time a patient spends on a spinal board is important to mitigate the risk of pressure ulcer development.[16] Spinal motion restriction should be maintained until the appropriate physician has deemed the spine is stable, at which point the relevant spinal precautions can be relaxed.

Thromboembolic prophylaxis

Patients with thoracolumbar spinal trauma are at high risk of developing deep venous thrombosis (DVT) and pulmonary embolism, with reported rates as high as 100% in patients with spinal cord injuries.[17,18] As such, it is recommended that all spinal trauma patients receive chemical DVT prophylaxis in the form of low-molecular-weight heparin, unless otherwise contraindicated.[19,20] If anticoagulation is contraindicated, for example, if the patient has also sustained a traumatic intracranial hemorrhage, the appropriate consulting service should be involved to advise when chemical DVT prophylaxis is safe to initiate. When chemical DVT prophylaxis is contraindicated, mechanical prophylaxis, such as pneumatic compression devices, should be initiated.

Surgical timing

While the evidence in the published scientific literature regarding the timing of surgical decompression and fixation of thoracolumbar spinal column injuries is equivocal, there are a number of established practice patterns. Typically, intervention should occur as soon as reasonably possible in order to expedite patient mobilization and rehabilitation. Evidence has shown that early surgery, defined as within 24 hours in some studies and within 72 hours in others, has a positive impact on length of

hospital stay, perioperative complications, and length of dependence on mechanical ventilation.[21–28] However, surgery should be reasonably delayed for medical optimization; for example, if a patient is taking anticoagulant medications, surgery may be delayed until the patient has a normal coagulation profile. Patients with neurologic deficits may warrant more expeditious decompression and stabilization in order to preserve neurologic function. While some studies have reported no clear benefit from early surgical intervention,[29–31] many studies have shown improved neurologic recovery.[22,30,32,33] A recent pooled analysis of multiple acute spinal cord injury datasets totaling 1549 patients by Badhiwala et al. found that surgery within 24 hours led to statistically significantly higher motor and sensory scores at 1 year on the standardized neurologic examination, as well as a better overall ASIA Impairment Scale grade.[34] In the setting of acute spinal cord injury, current guidelines suggest that early surgery be offered as a treatment option regardless of the spinal injury level.[35] A more in-depth discussion of surgical timing in the setting of spinal cord injury is available elsewhere (see Chapter 18: SCI management: role and timing of surgical intervention).

Nonspinal injuries

Given the typical high-energy mechanism impacts that result in thoracolumbar spinal column injuries, nonspinal traumatic injuries are quite common in this patient cohort. As such, careful consideration to, and treatment of, these nonspinal injuries is critical to the overall care of the thoracolumbar spinal column trauma patient. In certain cases, other injuries can provide insight into the severity of the mechanism of injury and the subsequent mechanical stability of the spine.

Nonspinal traumatic pathology may also be relevant in treatment planning for thoracolumbar spinal trauma. For instance, certain injuries may interfere with patient positioning in spinal surgery (e.g., pelvic external-fixation).[1] Alternatively, spinal motion restriction may interfere with the treatment of a patient's other injuries. Delay of treatment may also exacerbate other injuries, for example, if a patient with poor tidal volumes from chest injuries must remain supine due to an unstable thoracolumbar spine, respiratory function is highly likely to decompensate. In the setting of polytrauma, it is therefore important that all medical teams involved in a patient's care coordinate with each other to ensure the timing of treatment for each injury is optimized.

Thoracolumbar spinal orthoses

The purpose of external spinal orthoses' use in thoracolumbar spinal trauma is to limit the mobility of the injured spinal level in order to protect the neural elements and facilitate bone healing.[36] In addition to spinal immobilization, spinal orthoses can serve as a reminder to patients of their injuries and encourage caution when mobilizing. Effective use of spinal orthoses may allow patients to avoid the need for surgical intervention for certain spinal fractures; however, studies have reported conflicting results on the efficacy of spinal orthoses compared to surgical intervention.[37] Since these devices are largely unregulated, commercially available spinal orthoses vary considerably in the degree to which they have been clinically tested and caution must be used by the treating surgeon to select an appropriate device. Many different types of spinal orthoses products exist, some prefabricated and others custom-fitted for the individual patient. Common types of thoracolumbar spinal orthoses include the thoracolumbosacral orthosis (TLSO), lumbosacral orthosis (LSO), and thoracolumbar hyperextension orthosis (such as a Jewett brace, a cruciform anterior spinal hyperextension brace, or a Knight-Taylor brace). TLSOs are appropriate for managing injuries between T6 and L4, while LSOs are only used in fractures at L3 or below. Hyperextension orthoses like the Jewett brace are designed to offload the anterior column and are ideal for limiting motion in the sagittal plane. They can be used for anterior column fractures involving T10 to L2. High thoracic fractures, though less commonly seen in thoracolumbar spinal trauma, require the use of cervical orthoses with thoracic extensions, such as a halo or a cervicothoracic (CTO) orthosis. When using spinal orthoses, patients should be monitored regularly to ensure patient compliance, look for signs of skin breakdown, and confirm proper fit (especially as patients recover and become more mobile).

Specific management considerations by fracture subtype

Several thoracolumbar fracture classification systems exist and these are covered in detail elsewhere (see Chapter 3: Classification systems: spine trauma). In the following sections, we review the management considerations for the four major thoracolumbar fracture patterns, as originally described by Denis in 1983.[38] These major fracture patterns should also be considered in the context of the AOSpine morphologic classification system, developed by the AOSpine Spinal Cord Injury & Trauma Knowledge Forum,[39] which divides thoracolumbar fractures into three basic types: Type A (compression/burst injury), Type B (distraction/tension band injury), and Type C (translational/rotational injury).

Compression

These injuries are caused by axial loading through the vertebral column, typically with some associated component of flexion. Compression fractures only involve the anterior column of the spine, with fracturing of the vertebral body superior endplate, inferior endplate, or both (Fig. 11.1). By definition the fracture does

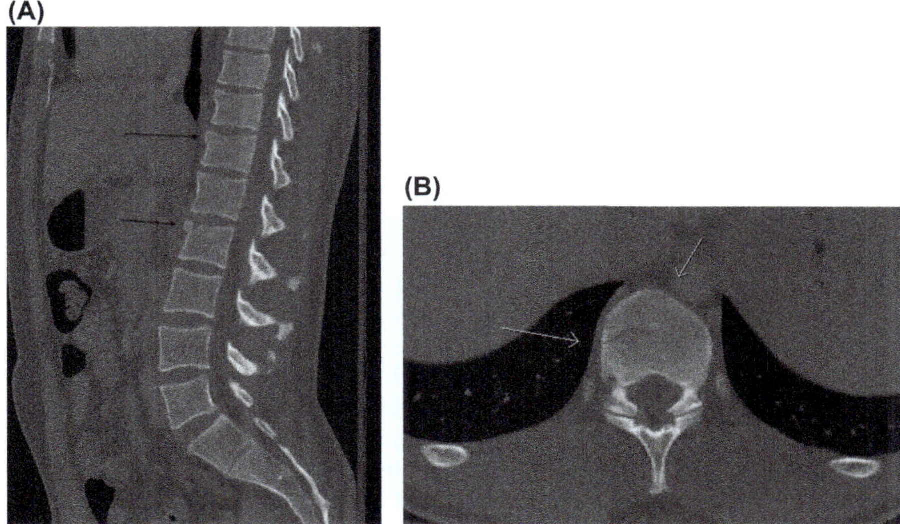

FIGURE 11.1   (A) Sagittal CT reconstruction demonstrating compression fractures (AO A1 type) of T12 and L2 (*black arrows*); (B) axial CT image of T12 vertebral level. Note that fractures (*white arrows*) do not involve the posterior vertebral body wall.

not extend through the posterior wall of the vertebral body. Compression fractures are defined as A1 or A2 fractures in the AOSpine classification system.

While these injuries are typically seen in minor traumas, often in the context of poor bone mineral density, they can also occur with high-energy mechanisms.[40] Care must be taken to carefully evaluate the posterior elements in these fractures as more severe spinal injuries such as flexion–distraction injuries (see below) can be misdiagnosed as compression fractures if the full extent of the injury is not realized.[1]

These fractures are mechanically stable and can be managed nonoperatively. While no specific bracing is required, management with an appropriate spinal orthosis may be considered and has been shown to improve posture, reduce pain, and improve quality of life in selected patients.[41] Pain usually resolves over the course of 8–12 weeks, though some patients will experience persistent back pain.[42,43] Patients should be followed with serial upright spinal radiographs, to monitor for delayed or worsening kyphotic deformity and to confirm fracture healing.[43,44] Depending on the severity, delayed kyphotic deformities may need to be managed surgically. Patients with suspected osteoporotic fractures should have formal bone mineral density testing with subsequent osteoporosis management if indicated.[45] Patients with fractures suspicious for a pathologic process should undergo a bone biopsy to rule out a neoplasm.

Numerous case series have supported the treatment of osteoporotic fractures with kyphoplasty or vertebroplasty to improve pain and alignment.[46–49] However, the efficacy of these treatment modalities is controversial and several randomized controlled trials concluded that vertebroplasty was found to have no significant effect on patients' overall pain when compared with a sham procedure.[50–52]

Mechanical stability becomes compromised in the context of compression fractures with increasing kyphotic deformity and vertebral body height loss. While some older research has suggested thresholds of 30 degrees of kyphosis and 50% height loss as an indication for surgical intervention, due to the low quality of evidence, this has not been adopted formally in practice guidelines.[53] Surgery, though uncommonly required, is usually approached posteriorly, with short-segment pedicle screw fixation.

Burst

Burst fractures, like compression fractures, are caused by axial loading through the vertebral column, but unlike compression fractures, also involve the posterior vertebral body wall. Within the AOSpine classification system, burst fractures are defined as A3 (incomplete burst) or A4 (complete burst) fractures. While the anterior and middle columns are the focus of this fracture pattern, burst fractures may also involve fractures within the posterior column such as pedicles, facets, or laminae. Notably, fracturing of

the middle column can result in retropulsion of dorsal vertebral body fragments, compromising the size of the spinal canal and potentially causing spinal cord or cauda equina compression (Fig. 11.2).

The severity of burst fractures can vary considerably from essentially mild compression fractures with minimal involvement of the middle column to highly unstable fractures with ongoing compression of the neural elements. Mild burst fractures can be managed nonoperatively; external orthoses may be used to help with pain management and to facilitate early mobilization. Despite the potential benefit of external orthoses in the short term, reasonably high-quality clinical studies have demonstrated that long-term outcomes (as measured by functional ability, pain, and quality of life questionnaires) are no different in thoracolumbar burst fractures patients whether treated with or without external bracing.[43,54—56] Patients managed nonoperatively should be followed

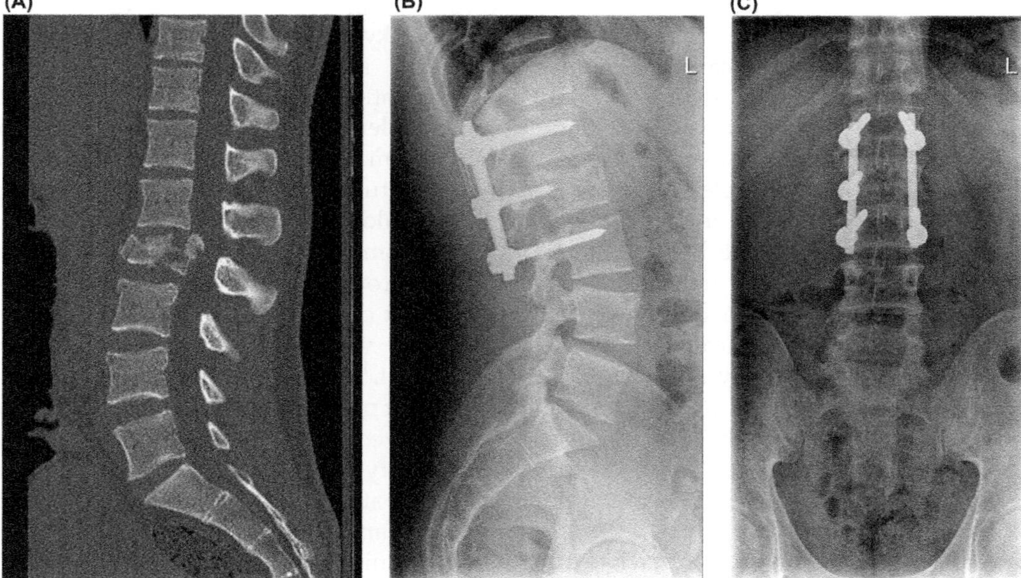

FIGURE 11.2   (A) Sagittal CT reconstruction demonstrating an L2 complete burst fracture (AO A4 type). (B) Lateral and (C) anteroposterior postoperative radiographs demonstrating successful short-segment pedicle screw fracture fixation.

with serial upright plain radiographs to confirm fracture healing, assess for delayed or worsening kyphotic deformity, and clinically assessed to monitor for neurologic deterioration (albeit, an extremely uncommon occurrence).[57,58]

In neurologically intact patients, burst fractures on either end of the severity generally have well-defined management plans, with mild fractures typically treated nonoperative and severe fractures typically managed surgically. It can be difficult, however, to determine the optimal treatment modality for burst fracture that does not lie on the extremes of this spectrum. Several small clinical trials have tested external orthoses versus surgery for neurologically intact patients with burst fractures and produced conflicting results.[59–64] A Cochrane systematic review conducted in 2013 failed to show an advantage of either treatment approach.[37] Ultimately, surgical guidelines currently leave the role of surgical fixation up to the discretion of the consulting surgeon.[65]

In neurologically intact patients, treatment considerations for burst fractures ultimately relate to the spine surgeon's ability to adequately and consistently determine the mechanical stability of the fracture. To help determine this mechanical stability, a number of important factors have been identified: degree of kyphosis, percent of vertebral body height loss, and integrity of the posterior elements.[66] Older research had suggested that kyphosis of greater than 30 degrees, height loss of greater than 50%, and canal compromise of greater than 50% are indications of mechanical instability.[57,67,68] These thresholds, however, were developed in the pre-MRI era and were designed as surrogate measures to predict the integrity of the PLC. At present, such absolute thresholds have been called into question, with some advocating for MRI to be used to assess the PLC prior to the determination of mechanical stability.[12,69,70] Further, the use of canal compromise greater than 50% (in the absence of neurologic deficit) as a surgical indication is also controversial; studies have now

shown patients with canal compromise of greater than 50% have been successfully managed nonoperatively and that retropulsed fragments typically resorb over time.[71–73]

Neurologic status must be considered in the treatment paradigm of burst fractures. In patients with isolated partial nerve root injuries without evidence of ongoing compression on spinal imaging, a number of small studies have concluded that nonoperative management provides similar results to surgery in terms of neurologic recovery, spinal alignment, and long-term pain control.[74–76] More significant neurologic injuries should be managed surgically with the primary goals of decompression of the neural elements and reestablishing spinal stability.[66,77,78]

If operative treatment is warranted, surgery may be approached either from an anterior approach, a posterior approach, or a combination of the two. Anterior approaches are ideal for injuries with considerable canal compromise (particularly in the setting of neurologic deficit), extensive fracture comminution, and significant kyphotic deformity; however, anterior procedures are technically more challenging and generally impart higher perioperative risks.[79–82] Posterior approaches are more familiar to the vast majority of spinal surgeons and thus technically less demanding; however, alleviating neural element compression secondary to ventral osseous fragments can be difficult via a posterior-only approach. Further, prevention of progressive postoperative kyphosis also typically requires a much longer segmental fixation than is necessary with anterior-based procedures.[83–85]

There are currently only limited studies directly comparing outcomes of anterior versus posterior approaches[86–90]; two randomized trials with small sample sizes found no difference for select patients between the two treatments in terms of kyphosis, functional status, pain, quality of life, or incidence of adverse events.[79,91] While studies have shown more extensive canal decompression from an anterior approach, the lack of difference in neurologic recovery calls the significance of this result into question.[66] Studies comparing a combined anterior and posterior approach with a posterior-only approach have also failed to show a difference in correction of spinal deformity, fusion rates, neurologic recovery, pain, and return to work.[86,92–94] One study did show the posterior approach resulted in an additional 4 degrees of kyphosis long-term; however, this was not associated with an increased incidence of pain suggesting there was little clinical relevance to this marginally increased kyphotic angulation.[92] Combined anteroposterior surgery has also been noted to have longer operative times with higher volumes of blood loss.[86] Given the paucity of evidence comparing these three approaches, current guidelines do not recommend a specific approach and leave the decision up to the discretion of the operating surgeon.[95]

The most commonly used posterior approach involves pedicle screw fixation,[66] though hook and rod constructs have also been well described.[96] In terms of pedicle screw construct length, some research has suggested that short-segment fixation (one level above and below the injury) (Fig. 11.2) in patients with minimal fracture comminution or kyphotic deformity yields adequate rates of fusion and spinal alignment.[97,98] Other studies have found high rates of hardware failure and recurrent kyphosis[83,85,99] A randomized trial comparing short- and long-segment (two levels above and below the injury) constructs showed long constructs result in less kyphosis and have lower hardware failure rates.[100] However, when short-segment constructs are combined with vertebral body cement augmentation, kyphotic angle and failure rates become similar to anterior-only approaches.[101–105] Unfortunately, vertebroplasty with cement augmentations can only be offered to patients with isolated anterior

and middle column injuries, with an intact posterior longitudinal ligament, given the risk of cement extravasation into the spinal canal in cases with fracture extension through the posterior vertebral body wall. In neurologically intact patients that do not require decompression, a growing body of literature suggests that results of minimally invasive percutaneous pedicle screw fixation are favorable compared with open posterior approaches.[106–111]

Distraction

Distraction injuries are mechanistically distinctly from compression and burst fractures, as distractive forces are dominant over any axial load force that is applied to the spinal column.[112] This distraction occurs about an axis of rotation which is variable in position; it can either be within the anterior column, leading to an element of ventral vertebral body compression (with or without bony fragment retropulsion into the spinal canal), anterior to the spinal column entirely, resulting in distractive forces through all three spinal columns (the classic "Chance fracture"), or located posteriorly, leading to an extension–distractive injury. These distractive forces result in tension band injury, and accordingly, as per the AOSpine classification system are classified as either: B1 (monosegmental osseous posterior tension band) injuries, B2 (osseous and/or ligamentous posterior tension band) injuries (Fig. 11.3), or B3 (hyperextension) injuries (Fig. 11.4).

B1 and B2 injuries often occur at the thoracolumbar junction and are typically caused by high-energy injury mechanisms. This injury was originally referred to as a "seat-belt type" as the classic etiology is a back-seat passenger involved in a high-speed motor vehicle collision, wearing only a lap belt (without a shoulder strap).[38] Thoracolumbar flexion–distraction

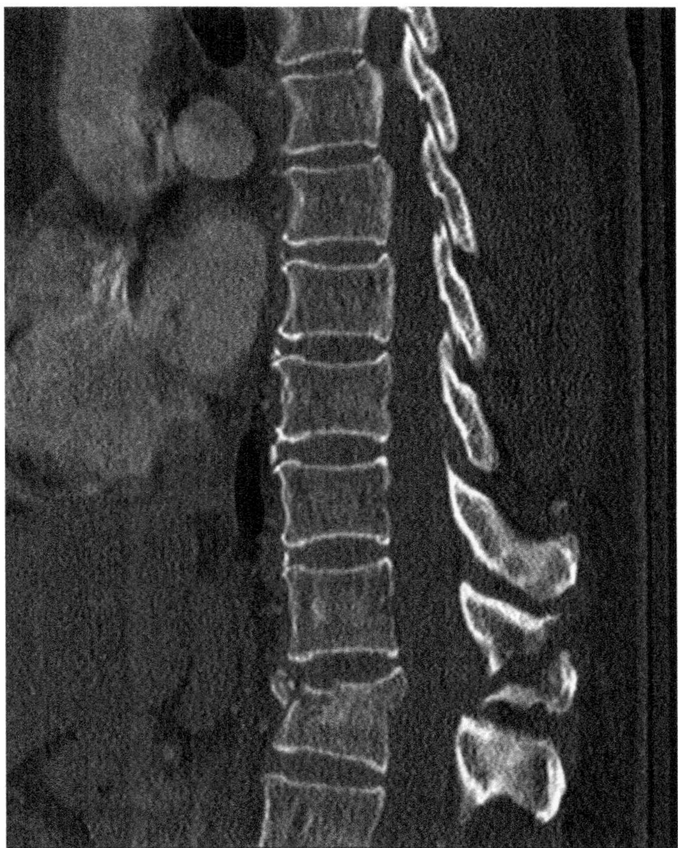

FIGURE 11.3   Sagittal CT reconstruction of a lower thoracic AO B2 type, or flexion—distraction, injury. Note the associated axial load injury (AO A type) to the vertebral body and the fracture through the spinous process of the superior adjacent level.

fractures in which the axis of rotation is anterior to the spinal column (i.e., recognized via the presence of distraction through all three vertebral columns) are associated with a 44%—67% rate of concomitant abdominal injuries[113] and a 25% risk of spinal cord injury, more than half of which are incomplete.[114] B3 injuries (Fig. 11.4) often occur in the setting of ankylosing conditions of the spine (ankylosing spondylitis or diffuse idiopathic skeletal hyperostosis (DISH)), and are typically highly unstable and at high risk for neurological injury (see below).

Because of the tension band disruption that is inherent with distractive injuries, flexion—distraction and extension—distraction injuries should be considered mechanically unstable. Surgery is therefore the main treatment approach for the injury type.[112,113,115,116] If, however, the flexion—distraction injury is restricted to the osseous components of the vertebral column (i.e., a B1 injury), external orthosis is a reasonable treatment option, assuming that close clinical and radiographic follow-up is possible. High fusion rates following treatment with external orthosis were reported for select patients with good apposition of the bony fragments and less than 10 degrees of kyphosis at the injured level.[113,114] Several investigations

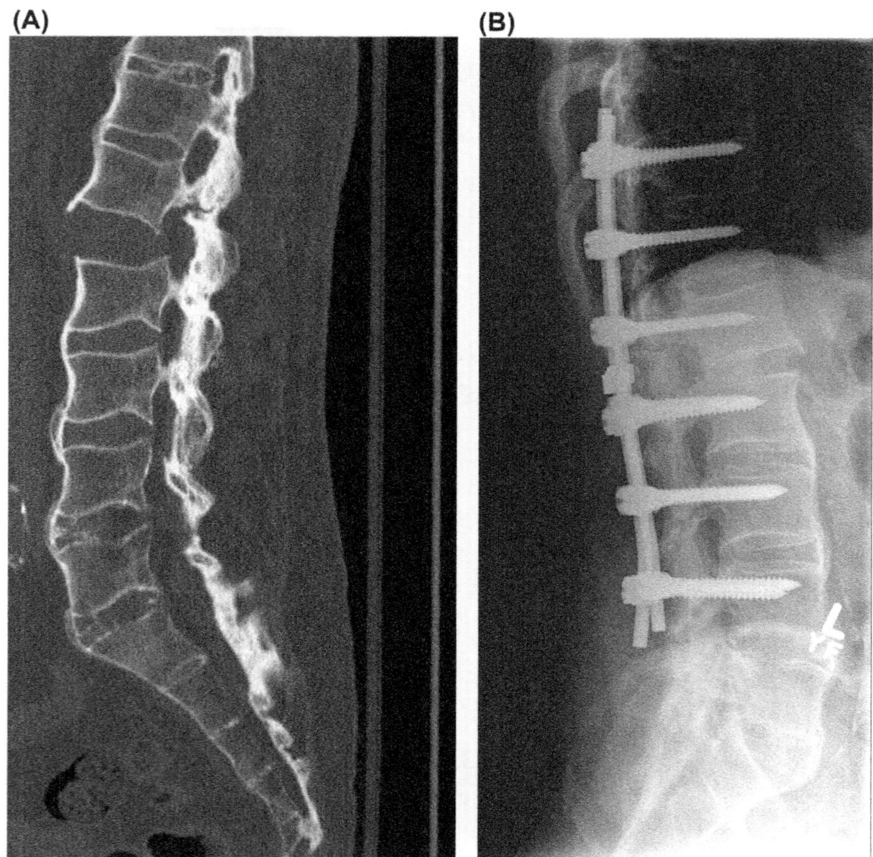

FIGURE 11.4   (A) Sagittal CT reconstruction of an L1−2 extension−distraction (AO B3 type) injury in the setting of ankylosing spondylitis. (B) Postoperative lateral radiograph demonstrating successful long-segment posterior fixation.

have looked at degree of kyphosis as a predictor of overall nonoperative treatment failure due to instability and have suggested thresholds between 12 degrees and 20 degrees.[117−120] Studies comparing operative and nonoperative treatment for thoracolumbar flexion−distraction injuries, however, suggest that nonoperative strategies led to worsened long-term kyphosis, pain, and functional outcomes, though these studies were small nonrandomized case series.[113,121]

Due to the lack of randomized trials comparing surgical treatment approaches, there exists no evidence-based optimal strategy, whether comparing posterior alone versus posterior and anterior or comparing short- versus long-segment constructs.[116] A posterior longsegment approach is the most commonly used and the most established treatment in the literature with the longest follow-up data available.[121−127] There are currently no studies directly comparing long- versus short-segment fixation in thoracolumbar flexion−distraction injuries; however, research has illustrated favorable results using single segment constructs.[122]

Some surgeons have attempted to augment pedicle screw fixation with posterior tension band wiring in an attempt to replicate the physiologic function of the posterior tension band. Studies have not been able to show improved outcomes using posterior tension band wiring augmentation compared with pedicle screw fixation alone.[124,125] In patients where neural element decompression is not indicated, percutaneous pedicle screw fixation has been shown to provide similar results in terms of long-term kyphosis and neurologic recovery and decreased operative time and blood loss, compared to open pedicle screw fixation and interbody fusion.[107]

In flexion—distraction injuries with an associated compression or burst fracture component, an anterior approach can be combined with the posterior fixation in order to reestablish the integrity of the anterior column and if necessary decompress the neural elements.[128,129] Evidence is currently limited on specific radiographic markers signifying the need for the addition of an anterior approach, and this decision is currently up to the discretion of the treating surgeon.[115]

Fracture dislocation

Fracture dislocations (Fig. 11.5) are high-energy injuries and have complex mechanisms

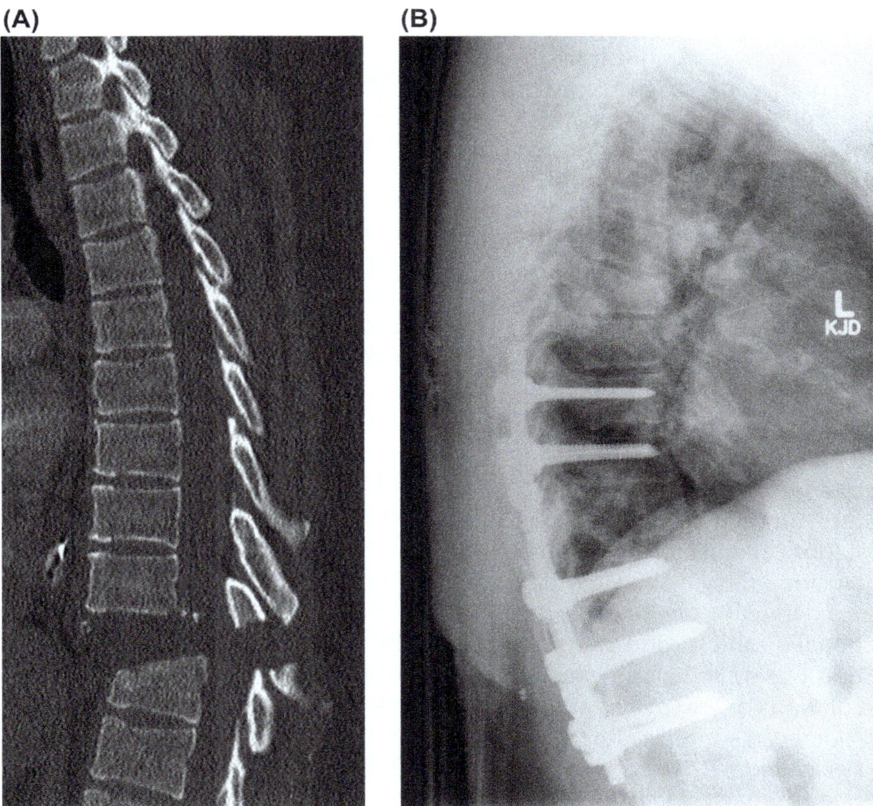

FIGURE 11.5   (A) Sagittal CT reconstruction of a T9/10 fracture dislocation (AO C type), as sustained in a high-speed motor vehicle crash. (B) Postoperative lateral radiograph demonstrating fracture reduction with long-segment posterior fixation.

involving any combination of distraction, flexion, extension, translational, rotation, and shear forces. The AOSpine classification system terms these injuries as C-type fractures. By definition, these injuries will have some element of unilateral or bilateral facet fracture and/or dislocation, involve all three columns, and are highly unstable.[1] A significant proportion of patients with fracture dislocations will have an associated neurologic injury, though neurologically intact patients have been well described in the literature.[130] If a patient is neurologically intact, particular care must be taken in these highly unstable injuries to prevent iatrogenic neural injury.

It is generally agreed that there is no role for external spinal orthoses as the primary treatment modality for in the management of fracture dislocations. The preferred surgical treatment approach is typically a long-segment posterior pedicle screw fixation[1,116,131−138] (Fig. 11.5). Combined posterior and anterior approaches may be indicated when direct ventral decompression is needed or if the integrity of the anterior column is significantly compromised.[89] Research, however, has found that there is no difference in neurologic recovery when comparing anterior direct and posterior indirect decompression.[131,139] Similarly, some studies have found that the integrity of the anterior column can be adequately reconstructed via a posterior approach with interbody fusion.[140,141]

A single center randomized trial comparing combined posterior and anterior approaches with interbody fusion failed to show a difference in fusion rates, extent of decompression, kyphosis, functional disability, and pain; however, the population was very heterogeneous and no power calculations were provided.[131] The study did find, however, a higher rate of surgical complications associated with the combined posterior and anterior approach. As surgical implant quality has improved, small studies have also shown satisfactory results using an isolated anterior approach.[89,142]

Special considerations

Ankylosing conditions

Thoracolumbar spinal trauma in the setting of ankylosing conditions, such as ankylosing spondylitis and DISH, poses unique management challenges. Because these patients have rigid spinal columns with ossified ligaments, they are at risk for extension−distraction type (AOSpine B3) injuries, which typically involve all three columns and are highly unstable (Fig. 11.4). These fractures are at high risk of being not recognized during initial presentation, with between 10% and 37% of fractures missed initially.[143] Furthermore, multiple fractures are seen in 6%−8% of cases thus imaging of the entire spinal column with CT is generally recommended.[143]

Risk in this population for sustaining a spinal cord injury, nerve root injury, or spinal epidural hematoma is reported to be 11.4 times higher than general spine trauma patients.[144] MRI is therefore often obtained in order to assess the neural elements and epidural space. Due to the highly unstable nature of these injuries, as well as the high rate of missed injuries and the often abnormal curvature of the spine at baseline, patients unfortunately suffer secondary neurologic injury in 15% of cases due to improper spinal immobilization.[145]

In general, nonoperative treatment of thoracolumbar spinal trauma for patients with ankylosing conditions is limited and results in inferior clinical outcomes.[146] Reduction is often a critical and challenging component of management and can be accomplished using very judicious and careful closed reduction with traction, preoperative patient positioning, and open reduction intraoperatively. Stabilization, as well as decompression if indicated, is mainly achieved via a posterior approach. Due to the long segments of fused vertebrae on either side of the fracture, significant mechanical stress is placed on the fusion construct, resulting in screw loosening rates as high as 15%.[145] Therefore, long construct lengths (Fig. 11.4) of at least two levels (typically three levels) above and below the level of injury are recommended.[143,147] In the setting of osteoporosis, increased construct length and cement augmentation are also advised. In select patients who do not require decompression and satisfactory closed reduction can be achieved, a growing body of literature has shown good clinical results using percutaneous pedicle screw instrumentation; studies show that minimally invasive approaches have led to shorter operative times, decreased blood loss, lower transfusion rates, and fewer perioperative complications.[148–150]

Penetrating injuries

Penetrating spinal trauma carries very different management considerations compared with blunt trauma. While penetrating trauma can be caused by a variety of etiologies, the most frequent in civilian populations are gunshot injuries. They typically occur in young males and impart a high rate of neurologic injury.[151] These injuries are found to be mechanically unstable in only around 10% of cases.[152] In general, mechanical stability and spinal cord injury are assessed and managed using the same principles described for blunt traumatic injuries.

When managing spinal gunshot wounds, special attention must be paid to the surrounding viscera as soft tissue injuries are extremely common and often fatal.[152] Tetanus vaccination status should be investigated and the appropriate prophylactic agents administered.[153] Patient physical examination should include inspection of entry and exit wounds to help determine the location of ammunition. While some studies have reported the absence of bullet migration during MRI, these scans are typically avoided as the bullet composition is often unknown and exposing patients to this risk is generally unnecessary.[154,155] Intact bullets or residual fragments can be characterized by CT; their location should be qualified by involvement of the spinal canal, synovial joints, and intervertebral disc spaces.

It is recommended that all patients receive a 48- to 72-hour course of prophylactic broad-spectrum antibiotics, with durations extended to 7–14 days in the setting of gastrointestinal tract perforation.[152,153,156,157] Surgical removal of the bullet as a prophylactic measure to prevent infection has been shown in multiple studies to be unnecessary.[158,159] Furthermore, bullet resection has been shown to impart no improvement in long-term neuropathic pain, which is commonly seen in cases involving the cauda equina, as this is typically secondary to the nerve injury induced by the initial ballistic impact and not from ongoing nerve root compression.[160,161]

Patients with gunshot wounds are at risk of developing lead poisoning and should be monitored with serial serum lead levels[162–166]; residual bullets in contact with the facet joints or intervertebral disc spaces are at particular risk as synovial fluid interferes with the body's ability to contain the lead within the encapsulated foreign body.[165,167–169] Some surgeons suggest bullets in these higher-risk locations should be removed to avoid the potential for lead poisoning.[152] In cases where lead poisoning develops, the bullet should be resected for source control.

Apart from treating mechanical instability, surgery is indicated in gunshot wounds in the presence of cutaneous or pleural cerebrospinal fluid fistula, and in the setting of progressive neurologic deficits with suspected ongoing compression of the neural elements.[152] While surgical decompression for patients with complete neurologic injuries is controversial,[170,171] current evidence supports decompression and removal of the projectile in cases of incomplete neurologic injury.[172] Some surgeons advocate for removal of bullets within the spinal canal or intervertebral disc space due to concerns that bullet migration can cause or worsen neurologic injury,[152] although others have demonstrated that bullet migration does not lead to neurologic injury.[173]

Conclusion

Appropriate thoracolumbar spinal trauma management requires careful spinal immobilization during initial transport, early identification of neurologic deficits, and recognition of the underlying fracture pattern to guide appropriate treatment. While certain fracture types may be managed with external orthosis, unstable fractures should undergo early surgical management as soon as permitted within the context of concomitant extraspinal injuries. Care plans that adhere to the universal treatment goals of preservation or restoration of neurologic function, maintenance or restoration of normal alignment, and restoration of spinal stability will allow for the best possible treatment outcome for this patient population.

References

1. Wood KB, et al. Management of thoracolumbar spine fractures. *Spine J* 2014;**14**(1):145–64.
2. Hawryluk G, et al. Mean arterial blood pressure correlates with neurological recovery after human spinal cord injury: analysis of high frequency physiologic data. *J Neurotrauma* 2015;**32**(24):1958–67.
3. Inoue T, et al. Medical and surgical management after spinal cord injury: vasopressor usage, early surgerys, and complications. *J Neurotrauma* 2014;**31**(3):284–91.
4. Vale FL, et al. Combined medical and surgical treatment after acute spinal cord injury: results of a prospective pilot study to assess the merits of aggressive medical resuscitation and blood pressure management. *J Neurosurg* 1997;**87**(2):239–46.
5. Inaba K, et al. Clinical examination is insufficient to rule out thoracolumbar spine injuries. *J Trauma Inj Infect Crit Care* 2011;**70**(1):174–9.
6. Kirshblum S, Waring W. Updates for the international standards for neurological classification of spinal cord injury. *Phys Med Rehabil Clin* 2014;**25**(3):505–17 [vii].
7. Dobran M, et al. Neurological outcome in a series of 58 patients operated for traumatic thoracolumbar spinal cord injuries. *Surg Neurol Int* 2014;**5**(Suppl. 7):S329–32.
8. Hauser CJ, et al. Prospective validation of computed tomographic screening of the thoracolumbar spine in trauma. *J Trauma* 2003;**55**(2):228–35.
9. Inaba K, et al. Visceral torso computed tomography for clearance of the thoracolumbar spine in trauma: a review of the literature. *J Trauma* 2006;**60**(4):915–20.
10. Pizones J, et al. Impact of magnetic resonance imaging on decision making for thoracolumbar traumatic fracture diagnosis and treatment. *Eur Spine J* 2011;**20**(S3):390–6.
11. Winklhofer S, et al. Magnetic resonance imaging frequently changes classification of acute traumatic thoracolumbar spine injuries. *Skeletal Radiol* 2012;**42**(6):779–86.
12. Qureshi S, et al. Congress of neurological surgeons systematic review and evidence-based guidelines on the evaluation and treatment of patients with thoracolumbar spine trauma: radiological evaluation. *Neurosurgery* 2019;**84**(1):E28–31.
13. Mehta G, et al. Evaluation of diagnostic accuracy of magnetic resonance imaging in posterior ligamentum complex injury of thoracolumbar spine. *Asian Spine J* 2021;**15**(3):333–9.
14. Vaccaro AR, et al. Injury of the posterior ligamentous complex of the thoracolumbar spine. *Spine* 2009;**34**(23):E841–7.
15. Fischer PE, et al. Spinal motion restriction in the trauma patient - a joint position statement. *Prehosp Emerg Care* 2018;**22**(6):659–61.
16. Ham W, et al. Pressure ulcers from spinal immobilization in trauma patients: a systematic review. *J Trauma Acute Care Surg* 2014;**76**(4):1131–41.
17. Geerts WH, et al. A prospective study of venous thromboembolism after major trauma. *N Engl J Med* 1994;**331**(24):1601–6.
18. Myllynen P, et al. Deep venous thrombosis and pulmonary embolism in patients with acute spinal cord injury: a comparison with nonparalyzed patients immobilized due to spinal fractures. *J Trauma* 1985;**25**(6):541–3.
19. Geerts WH, et al. Prevention of venous thromboembolism: the seventh ACCP conference on antithrombotic and thrombolytic therapy. *Chest* 2004;**126**(3 Suppl. 1):338S–400S.
20. Rogers FB, et al. Practice management guidelines for the prevention of venous thromboembolism in trauma patients: the EAST practice management guidelines work group. *J Trauma Acute Care Surg* 2002;**53**:142–64.

21. Boakye M, et al. Retrospective, propensity score-matched cohort study examining timing of fracture fixation for traumatic thoracolumbar fractures. *J Neurotrauma* 2012;**29**(12):2220–5.

22. Cengiz ŞL, et al. Timing of thoracolomber spine stabilization in trauma patients; impact on neurological outcome and clinical course. A real prospective (rct) randomized controlled study. *Arch Orthop Trauma Surg* 2008;**128**(9):959–66.

23. Kerwin AJ, et al. Best practice determination of timing of spinal fracture fixation as defined by analysis of the National Trauma Data Bank. *J Trauma* 2008;**65**(4):824–30.

24. Pakzad H, et al. Delay in operative stabilization of spine fractures in multitrauma patients without neurologic injuries: effects on outcomes. *Can J Surg* 2011;**54**(4):270–6.

25. Park K-C, et al. Clinical results of early stabilization of spine fractures in polytrauma patients. *J Crit Care* 2014;**29**(4):694.e7–9.

26. Schinkel C, et al. Timing of thoracic spine stabilization in trauma patients: impact on clinical course and outcome. *J Trauma* 2006;**61**(1):156–60.

27. Schlegel J, et al. Timing of surgical decompression and fixation of acute spinal fractures. *J Orthop Trauma* 1996;**10**(5):323–30.

28. Stahel PF, et al. The impact of a standardized "spine damage-control" protocol for unstable thoracic and lumbar spine fractures in severely injured patients: a prospective cohort study. *J Trauma Acute Care Surg* 2013;**74**(2):590–6.

29. Kerwin AJ, et al. The effect of early surgical treatment of traumatic spine injuries on patient mortality. *J Trauma* 2007;**63**(6):1308–13.

30. Petitjean ME, et al. Thoracic spinal trauma and associated injuries: should early spinal decompression be considered? *J Trauma* 1995;**39**(2):368–72.

31. Rahimi-Movaghar V, et al. Early versus late surgical decompression for traumatic thoracic/thoracolumbar (T1-L1) spinal cord injured patients. Primary results of a randomized controlled trial at one year follow-up. *Neurosciences* 2014;**19**(3):183–91.

32. Dvorak MF, et al. The influence of time from injury to surgery on motor recovery and length of hospital stay in acute traumatic spinal cord injury: an observational Canadian cohort study. *J Neurotrauma* 2015;**32**(9):645–54.

33. Gaebler C, et al. Results of spinal cord decompression and thoracolumbar pedicle stabilisation in relation to the time of operation. *Spinal Cord* 1999;**37**(1):33–9.

34. Badhiwala JH, et al. The influence of timing of surgical decompression for acute spinal cord injury: a pooled analysis of individual patient data. *Lancet Neurol* 2021;**20**(2):117–26.

35. Fehlings MG, et al. A clinical practice guideline for the management of patients with acute spinal cord injury and central cord syndrome: recommendations on the timing (≤24 hours versus >24 hours) of decompressive surgery. *Global Spine J* 2017;**7**(3_Suppl. l):195S–202S.

36. Agabegi SS, Asghar FA, Herkowitz HN. Spinal orthoses. *J Am Acad Orthop Surg* 2010;**18**(11):657–67.

37. Abudou M, et al. Surgical versus non-surgical treatment for thoracolumbar burst fractures without neurological deficit. *Cochrane Database Syst Rev* 2013;**31**(25). 2881 - 33.

38. Denis F. The three column spine and its significance in the classification of acute thoracolumbar spinal injuries. *Spine* 1983;**8**(8):817–31.

39. Vaccaro AR, et al. AOSpine thoracolumbar spine injury classification system. *Spine* 2013;**38**(23):2028–37.

40. Genev IK, et al. Spinal compression fracture management: a review of current treatment strategies and possible future avenues. *Global Spine J* 2017;**7**(1):71–82.

41. Pfeifer M, Begerow B, Minne HW. Effects of a new spinal orthosis on posture, trunk strength, and quality of life in women with postmenopausal osteoporosis. *Am J Phys Med Rehabil* 2004;**83**(3):177–86.

42. Mazanec DJ, et al. Vertebral compression fractures: manage aggressively to prevent sequelae. *Cleve Clin J Med* 2003;**70**(2):147–56.

43. Stadhouder A, et al. Nonoperative treatment of thoracic and lumbar spine fractures: a prospective randomized study of different treatment options. *J Orthop Trauma* 2009;**23**(8):588–94.

44. Francis RM, et al. Back pain in osteoporotic vertebral fractures. *Osteoporos Int* 2008;**19**(7):895–903.

45. Parreira PCS, et al. An overview of clinical guidelines for the management of vertebral compression fracture: a systematic review. *Spine J* 2017;**17**(12):1932–8.

46. Barr JD, et al. Percutaneous vertebroplasty for pain relief and spinal stabilization. *Spine* 2000;**25**(8):923–8.

47. Lee MJ, et al. Percutaneous treatment of vertebral compression fractures: a meta-analysis of complications. *Spine* 2009;**34**(11):1228–32.

48. McGirt MJ, et al. Vertebroplasty and kyphoplasty for the treatment of vertebral compression fractures: an evidenced-based review of the literature. *Spine J* 2009;**9**(6):501–8.

49. Trout AT, et al. Evaluation of vertebroplasty with a validated outcome measure: the Roland-Morris disability questionnaire. *Am J Neuroradiol* 2005;**26**(10):2652–7.

50. Buchbinder R, et al. A randomized trial of vertebroplasty for painful osteoporotic vertebral fractures. *N Engl J Med* 2009;**361**(6):557–68.

51. Firanescu CE, et al. Vertebroplasty versus sham procedure for painful acute osteoporotic vertebral compression fractures (VERTOS IV): randomised sham controlled clinical trial. *Br Med J* 2018;**361**:k1551.

52. Kallmes DF, et al. A randomized trial of vertebroplasty for osteoporotic spinal fractures. *N Engl J Med* 2009;**361**(6):569—79.

53. Day B, Kokan P. Compression fractures of the thoracic and lumbar spine from compensable injuries. *Clin Orthop Relat Res* 1977;**124**:173—6.

54. Bailey CS, et al. Orthosis versus no orthosis for the treatment of thoracolumbar burst fractures without neurologic injury: a multicenter prospective randomized equivalence trial. *Spine J* 2014;**14**(11):2557—64.

55. Post RB, et al. Functional outcome 5 years after nonoperative treatment of type A spinal fractures. *Eur Spine J* 2005;**15**(4):472—8.

56. Shamji MF, et al. A pilot evaluation of the role of bracing in stable thoracolumbar burst fractures without neurological deficit. *J Spinal Disord Tech* 2014;**27**(7):370—5.

57. Krompinger WJ, et al. Conservative treatment of fractures of the thoracic and lumbar spine. *Orthop Clin N Am* 1986;**17**(1):161—70.

58. Willén J, et al. The natural history of burst fractures at the thoracolumbar junction. *J Spinal Disord* 1990;**3**(1):39—46.

59. Landi A, et al. Percutaneous short fixation vs conservative treatment: comparative analysis of clinical and radiological outcome for A. 3 burst fractures of thoraco-lumbar junction and lumbar spine. *Eur Spine J* 2014;**23**(Suppl. 6):S671—6.

60. Medici A, Meccariello L, Falzarano G. Non-operative vs. percutaneous stabilization in Magerl's A1 or A2 thoracolumbar spine fracture in adults: is it really advantageous for a good alignment of the spine? Preliminary data from a prospective study. *Eur Spine J* 2014;**23**(Suppl. 6):S677—683.

61. Shen WJ, Liu TJ, Shen YS. Nonoperative treatment versus posterior fixation for thoracolumbar junction burst fractures without neurologic deficit. *Spine* 2001;**26**(9):1038—45.

62. Siebenga J, et al. Treatment of traumatic thoracolumbar spine fractures: a multicenter prospective randomized study of operative versus nonsurgical treatment. *Spine* 2006;**31**(25):2881—90.

63. Wood K, et al. Operative compared with nonoperative treatment of a thoracolumbar burst fracture without neurological deficit. A prospective, randomized study. *J Bone Joint Surg* 2003;**85-A**(5):773—81.

64. Wood KB, et al. Operative compared with nonoperative treatment of a thoracolumbar burst fracture without neurological deficit: a prospective randomized study with follow-up at sixteen to twenty-two years. *J Bone Joint Surg* 2015;**97**(1):3—9.

65. Rabb CH, et al. Congress of neurological surgeons systematic review and evidence-based guidelines on the evaluation and treatment of patients with thoracolumbar spine trauma: operative versus nonoperative treatment. *Neurosurgery* 2019;**84**(1):E50—2.

66. Dai L-Y, et al. A review of the management of thoracolumbar burst fractures. *Surg Neurol* 2007;**67**(3):221—31. Discussion 231.

67. Cantor JB, et al. Nonoperative management of stable thoracolumbar burst fractures with early ambulation and bracing. *Spine* 1993;**18**(8):971—6.

68. Reid DC, et al. The nonoperative treatment of burst fractures of the thoracolumbar junction. *J Trauma* 1988;**28**(8):1188—94.

69. Öner FC, et al. MRI findings of thoracolumbar spine fractures: a categorisation based on MRI examinations of 100 fractures. *Skeletal Radiol* 1999;**28**(8):433—43.

70. Saifuddin A. MRI of acute spinal trauma. *Skeletal Radiol* 2001;**30**(5):237—46.

71. Dai L-Y. Remodeling of the spinal canal after thoracolumbar burst fractures. *Clin Orthop Relat Res* 2001;**382**:119—23.

72. Mumford J, et al. Thoracolumbar burst fractures. The clinical efficacy and outcome of nonoperative management. *Spine* 1993;**18**(8):955—70.

73. Yazici M, et al. Spinal canal remodeling in burst fractures of the thoracolumbar spine: a computerized tomographic comparison between operative and nonoperative treatment. *J Spinal Disord* 1996;**9**(5):409—13.

74. Andreychik DA, et al. Burst fractures of the second through fifth lumbar vertebrae. Clinical and radiographic results. *J Bone Joint Surg* 1996;**78**(8):1156—66.

75. Mick CA, et al. Burst fractures of the fifth lumbar vertebra. *Spine* 1993;**18**(13):1878—84.

76. Seybold EA, et al. Functional outcome of low lumbar burst fractures. A multicenter review of operative and nonoperative treatment of L3-L5. *Spine* 1999;**24**(20):2154—61.

77. Alpantaki K, et al. Thoracolumbar burst fractures: a systematic review of management. *Orthopedics* 2010;**33**(6):422—9.

78. Bakhsheshian J, et al. Evidence-based management of traumatic thoracolumbar burst fractures: a systematic review of nonoperative management. *Neurosurg Focus* 2014;**37**(1):E1.

79. Esses SI, Botsford DJ, Kostuik JP. Evaluation of surgical treatment for burst fractures. *Spine* 1990;**15**(7):667—73.

80. Haas N, Blauth M, Tscherne H. Anterior plating in thoracolumbar spine injuries indication, technique, and results. *Spine* 1991;**16**(3):S100—11.

81. Kaneda K, et al. Anterior decompression and stabilization with the kaneda device for thoracolumbar burst fractures associated with neurological deficits. *J Bone Jt Surg Am* 1997;**79**(1):69–83.

82. McDonough PW, et al. The management of acute thoracolumbar burst fractures with anterior corpectomy and Z-plate fixation. *Spine* 2004;**29**(17):1901–8.

83. Carl AL, Tromanhauser SG, Roger DJ. Pedicle screw instrumentation for thoracolumbar burst fractures and fracture-dislocations. *Spine* 1992;**17**(8 Suppl. l): 317–24.

84. Ebelke DK, et al. Survivorship analysis of VSP spine instrumentation in the treatment of thoracolumbar and lumbar burst fractures. *Spine* 1991;**16**(8):S433.

85. McLain RF, Sparling E, Benson DR. Early failure of short-segment pedicle instrumentation for thoracolumbar fractures. A preliminary report. *J Bone Joint Surg* 1993;**75**(2):162–7.

86. Danisa OA, et al. Surgical approaches for the correction of unstable thoracolumbar burst fractures: a retrospective analysis of treatment outcomes. *J Neurosurg* 1995; **83**(6):977–83.

87. Hitchon PW, et al. Comparison of anterolateral and posterior approaches in the management of thoracolumbar burst fractures. *J Neurosurg Spine* 2006;**5**(2): 117–25.

88. Lin B, et al. Anterior approach versus posterior approach with subtotal corpectomy, decompression, and reconstruction of spine in the treatment of thoracolumbar burst fractures: a prospective randomized controlled study. *Clin Spine Surg* 2015:1. Publish Ahead of Print.

89. Sasso RC, et al. Unstable thoracolumbar burst fractures. *J Spinal Disord Tech* 2006;**19**(4):242–8.

90. Wu H, et al. Comparison of three different surgical approaches for treatment of thoracolumbar burst fracture. *Chin J Traumatol* 2013;**16**(1):31–5.

91. Wood KB, Bohn D, Mehbod A. Anterior versus posterior treatment of stable thoracolumbar burst fractures without neurologic deficit: a prospective, randomized study. *J Spinal Disord Tech* 2005;**18**(Suppl. l):S15–23.

92. Been HD, Bouma GJ. Comparison of two types of surgery for thoraco-lumbar burst fractures: combined anterior and posterior stabilisation vs. posterior instrumentation only. *Acta Neurochir* 1999;**141**(4):349–57.

93. Korovessis P, et al. Combined anterior plus posterior stabilization versus posterior short-segment instrumentation and fusion for mid-lumbar (L2–L4) burst fractures. *Spine* 2006;**31**(8):859–68.

94. Schmid R, et al. Combined posteroanterior fusion versus transforaminal lumbar interbody fusion (TLIF) in thoracolumbar burst fractures. *Injury* 2012;**43**(4): 475–9.

95. Anderson PA, et al. Congress of neurological surgeons systematic review and evidence-based guidelines on the evaluation and treatment of patients with thoracolumbar spine trauma: surgical approaches. *Neurosurgery* 2019;**84**(1):E56–8.

96. McBride GG. Cotrel-dubousset rods in surgical stabilization of spinal fractures. *Spine* 1993;**18**(4):466–73.

97. Markel DC, Graziano GP. A comparison study of treatment of thoracolumbar fractures using the ACE Posterior Segmental Fixator and Cotrel-Dubousset instrumentation. *Orthopedics* 1995;**18**(7):679–86.

98. Parker JW, et al. Successful short-segment instrumentation and fusion for thoracolumbar spine fractures. *Spine* 2000;**25**(9):1157–70.

99. McNamara MJ, Stephens GC, Spengler DM. Transpedicular short-segment fusions for treatment of lumbar burst fractures. *J Spinal Disord* 1992;**5**(2):183–7.

100. Tezeren G, Kuru I. Posterior fixation of thoracolumbar burst fracture: short-segment pedicle fixation versus long-segment instrumentation. *J Spinal Disord Tech* 2005;**18**(6):485–8.

101. Chen J-F, Lee S-T. Percutaneous vertebroplasty for treatment of thoracolumbar spine bursting fracture. *Surg Neurol* 2004;**62**(6):494–500.

102. Cho D-Y, Lee W-Y, Sheu P-C. Treatment of thoracolumbar burst fractures with polymethyl methacrylate vertebroplasty and short-segment pedicle screw fixation. *Neurosurgery* 2003;**53**(6):1354–61.

103. Christodoulou A, et al. Vertebral body reconstruction with injectable hydroxyapatite cement for the management of unstable thoracolumbar burst fractures: a preliminary report. *Acta Orthop Belg* 2005;**71**(5): 597–603.

104. Li K-C, et al. Transpedicle body augmenter: a further step in treatimg burst fractures. *Clin Orthop Relat Res* 2005;**436**:119–25.

105. Verlaan J-J, et al. Balloon vertebroplasty in combination with pedicle screw instrumentation. *Spine* 2005; **30**(3):E73–9.

106. Dong SH, et al. Effects of minimally invasive percutaneous and trans-spatium intermuscular short-segment pedicle instrumentation on thoracolumbar mono-segmental vertebral fractures without neurological compromise. *Orthop Traumatol Surg Res* 2013;**99**(4): 405–11.

107. Grossbach AJ, et al. Flexion-distraction injuries of the thoracolumbar spine: open fusion versus percutaneous pedicle screw fixation. *Neurosurg Focus* 2013; **35**(2):E2.

108. Jiang XZ, et al. Comparison of a paraspinal approach with a percutaneous approach in the treatment of thoracolumbar burst fractures with posterior ligamentous

complex injury: a prospective randomized controlled trial. *J Int Med Res* 2012;**40**(4):1343–56.

109. Lee J-K, et al. Percutaneous short-segment pedicle screw placement without fusion in the treatment of thoracolumbar burst fractures: is it effective?: comparative study with open short-segment pedicle screw fixation with posterolateral fusion. *Acta Neurochir* 2013;**155**(12):2305–12.

110. Vanek P, et al. Treatment of thoracolumbar trauma by short-segment percutaneous transpedicular screw instrumentation: prospective comparative study with a minimum 2-year follow-up. *J Neurosurg Spine* 2014;**20**(2):150–6.

111. Wang H-W, et al. Percutaneous pedicle screw fixation through the pedicle of fractured vertebra in the treatment of type A thoracolumbar fractures using sextant system: an analysis of 38 cases. *Chin J Traumatol* 2010;**13**(3):137–45.

112. Gertzbein SD, Court-Brown CM. Flexion-distraction injuries of the lumbar spine. Mechanisms of injury and classification. *Clin Orthop Relat Res* 1988;**227**:52–60.

113. Anderson PA, et al. Flexion distraction and chance injuries to the thoracolumbar spine. *J Orthop Trauma* 1991;**5**(2):153–60.

114. Chapman JR, et al. Thoracolumbar flexion-distraction injuries: associated morbidity and neurological outcomes. *Spine* 2008;**33**(6):648–57.

115. Lopez AJ, et al. Management of flexion distraction injuries to the thoracolumbar spine. *J Clin Neurosci* 2015;**22**(12):1853–6.

116. Verlaan JJ, et al. Surgical treatment of traumatic fractures of the thoracic and lumbar spine: a systematic review of the literature on techniques, complications, and outcome. *Spine* 2004;**29**(7):803–14.

117. Nagel DA, et al. Stability of the upper lumbar spine following progressive disruptions and the application of individual internal and external fixation devices. *J Bone Joint Surg* 1981;**63**(1):62–70.

118. Neumann P, Nordwall A, Osvalder AL. Traumatic instability of the lumbar spine. A dynamic in vitro study of flexion-distraction injury. *Spine* 1995;**20**(10):1111–21.

119. Neumann P, et al. The mechanism of initial flexion-distraction injury in the lumbar spine. *Spine* 1992;**17**(9):1083–90.

120. Osvalder AL, et al. A method for studying the biomechanical load response of the (in vitro) lumbar spine under dynamic flexion-shear loads. *J Biomech* 1993;**26**(10):1227–36.

121. Gertzbein SD, Court-Brown CM. Rationale for the management of flexion-distraction injuries of the thoracolumbar spine based on a new classification. *J Spinal Disord* 1989;**2**(3):176–83.

122. Finkelstein JA, et al. Single-level fixation of flexion distraction injuries. *J Spinal Disord Tech* 2003;**16**(3):236–42.

123. Green DA, et al. Flexion-distraction injuries to the lumbar spine associated with abdominal injuries. *J Spinal Disord* 1991;**4**(3):312–8.

124. Hasankhani EG, Omidi-Kashani F. Posterior tension band wiring and instrumentation for thoracolumbar flexion-distraction injuries. *J Orthop Surg* 2014;**22**(1):88–91.

125. LeGay DA, Petrie DP, Alexander DI. Flexion-distraction injuries of the lumbar spine and associated abdominal trauma. *J Trauma Inj Infect Crit Care* 1990;**30**(4):436–44.

126. Liu Y-J, et al. Flexion-distraction injury of the thoracolumbar spine. *Injury* 2003;**34**(12):920–3.

127. Jeanneret B, Ho PK, Magerl F. Burst-shear flexion-distraction injuries of the lumbar spine. *J Spinal Disord* 1993;**6**(6):473–81.

128. Sar C, Bilen FE. Thoracolumbar flexion-distraction injuries combined with vertebral body fractures. *Am J Orthoped* 2002;**31**(3):147–51.

129. Tezer M, et al. Surgical outcome of thoracolumbar burst fractures with flexion-distraction injury of the posterior elements. *Int Orthop* 2005;**29**(6):347–50.

130. Zeng J, et al. Complete fracture-dislocation of the thoracolumbar spine without neurological deficit: a case report and review of the literature. *Medicine* 2018;**97**(9):e0050.

131. Hao D, et al. Two-year follow-up evaluation of surgical treatment for thoracolumbar fracture-dislocation. *Spine* 2014;**39**(21):E1284–90.

132. Phadnis AS, et al. Fracture and complete dislocation of the spine with a normal motor neurology. *Inj Extra* 2006;**37**:479–83.

133. Akay KM, et al. Fracture and lateral dislocation of the T12-L1 vertebrae without neurological deficit. *Neurol Med Chir* 2003;**43**(5):267–70.

134. Enishi T, Katoh S, Sogo T. Surgical treatment for significant fracture-dislocation of the thoracic or lumbar spine without neurologic deficit: a case series. *J Orthop Case Rep* 2014;**4**(3):43–5.

135. Evans LJ. Images in clinical medicine. Thoracolumbar fracture with preservation of neurologic function. *N Engl J Med* 2012;**367**(20):1939.

136. Hsieh C-T, et al. Complete fracture-dislocation of the thoracolumbar spine without paraplegia. *Am J Emerg Med* 2008;**26**(5):633.e5–7.

137. Rahimizadeh A, Asgari N, Rahimizadeh A. Complete thoracolumbar fracture-dislocation with intact

neurologic function: explanation of a novel cord saving mechanism. *J Spinal Cord Med* 2017;**41**(3):1—10.

138. Reinhold M, et al. Operative treatment of 733 patients with acute thoracolumbar spinal injuries: comprehensive results from the second, prospective, internet-based multicenter study of the spine study group of the German Association of Trauma Surgery. *Eur Spine J* 2010;**19**(10):1657—76.

139. Wang F, Zhu Y. Treatment of complete fracture-dislocation of thoracolumbar spine. *J Spinal Disord Tech* 2013;**26**(8):421—6.

140. Machino M, et al. A new thoracic reconstruction technique "transforaminal thoracic interbody fusion": a preliminary report of clinical outcomes. *Spine* 2010;**35**(19):E1000—5.

141. Wu A-M, et al. Transforaminal decompression and interbody fusion in the treatment of thoracolumbar fracture and dislocation with spinal cord injury. *PLoS One* 2014;**9**(8):e105625.

142. Okuyama K, et al. Outcome of anterior decompression and stabilization for thoracolumbar unstable burst fractures in the absence of neurologic deficits. *Spine* 1996;**21**(5):620—5.

143. Caron T, et al. Spine fractures in patients with ankylosing spinal disorders. *Spine* 2010;**35**(11):E458—64.

144. Alaranta H, Luoto S, Konttinen YT. Traumatic spinal cord injury as a complication to ankylosing spondylitis. An extended report. *Clin Exp Rheumatol* 2002;**20**(1):66—8.

145. Reinhold M, et al. Spine fractures in ankylosing diseases: recommendations of the spine section of the German Society for Orthopaedics and Trauma (DGOU). *Global Spine J* 2018;**8**(2 Suppl. l):56S—68S.

146. Westerveld LA, Verlaan JJ, Öner FC. Spinal fractures in patients with ankylosing spinal disorders: a systematic review of the literature on treatment, neurological status and complications. *Eur Spine J* 2009;**18**(2):145—56.

147. Balling H, Weckbach A. Hyperextension injuries of the thoracolumbar spine in diffuse idiopathic skeletal hyperostosis. *Spine* 2015;**40**(2):E61—7.

148. Krüger A, et al. Percutaneous dorsal instrumentation for thoracolumbar extension-distraction fractures in patients with ankylosing spinal disorders: a case series. *Spine J* 2014;**14**(12):2897—904.

149. Moussallem CD, et al. Perioperative complications in open versus percutaneous treatment of spinal fractures in patients with an ankylosed spine. *J Clin Neurosci* 2016;**30**:88—92.

150. Nayak NR, et al. Minimally invasive surgery for traumatic fractures in ankylosing spinal diseases. *Global Spine J* 2015;**5**(4):266—73.

151. Sidhu GS, et al. Civilian gunshot injuries of the spinal cord: a systematic review of the current literature. *Clin Orthop Relat Res* 2013;**471**(12):3945—55.

152. de Barros Filho TEP, et al. Gunshot injuries in the spine. *Spinal Cord* 2014;**52**(7):504—10.

153. Bono CM, Heary RF. Gunshot wounds to the spine. *Spine J* 2004;**4**(2):230—40.

154. Finitsis SN, Falcone S, Green BA. MR of the spine in the presence of metallic bullet fragments: is the benefit worth the risk? *AJNR Am J Neuroradiol* 1999;**20**(2):354—6.

155. Kafadar AM, et al. Intradural migration of a bullet following spinal gunshot injury. *Spinal Cord* 2006;**44**(5):326—9.

156. Kumar A, Wood GW, Whittle AP. Low-velocity gunshot injuries of the spine with abdominal viscus trauma. *J Orthop Trauma* 1998;**12**(7):514—7.

157. Roffi RP, Waters RL, Adkins RH. Gunshot wounds to the spine associated with a perforated viscus. *Spine* 1989;**14**(8):808—11.

158. Lin SS, et al. Low-velocity gunshot wounds to the spine with an associated transperitoneal injury. *J Spinal Disord* 1995;**8**(2):136—44.

159. Velmahos G, Demetriades D. Gunshot wounds of the spine: should retained bullets be removed to prevent infection? *Ann R Coll Surg Engl* 1994;**76**(2):85—7.

160. Spaić M, et al. DREZ surgery on Conus medullaris (after failed implantation of vascular omental graft) for treating chronic pain due to spine (gunshot) injuries. *Acta Neurochir* 1999;**141**(12):1309—12.

161. Waters RL, Adkins RH. The effects of removal of bullet fragments retained in the spinal canal. *Spine* 1991;**16**(8):934—9.

162. Cagin CR, Diloy-Puray M, Westerman MP. Bullets, lead poisoning, and thyrotoxicosis. *Ann Intern Med* 1978;**89**(4):509.

163. Cristante AF, et al. Lead poisoning by intradiscal firearm bullet. *Spine* 2010;**35**(4):E140—3.

164. Gracia RC, Snodgrass WR. Lead toxicity and chelation therapy. *Am J Health Syst Pharm* 2007;**64**(1):45—53.

165. Grogan DP, Bucholz RW. Acute lead intoxication from a bullet in an intervertebral disc space. A case report. *J Bone Joint Surg* 1981;**63**(7):1180—2.

166. Linden MA, et al. Lead poisoning from retained bullets. Pathogenesis, diagnosis, and management. *Ann Surg* 1982;**195**(3):305—13.

167. de Madureira PR, Capitani EMD, Vieira RJ. Lead poisoning after gunshot wound. *Sao Paulo Med J* 2000;**118**(3):78—80.

168. Meyer PA, Brown MJ, Falk H. Global approach to reducing lead exposure and poisoning. *Mutat Res Rev Mutat Res* 2008;**659**(1—2):166—75.

169. Onalaja AO, Claudio L. Genetic susceptibility to lead poisoning. *Environ Health Perspect* 2000;**108**(Suppl. 1):23—8.

170. Stauffer ES, Wood RW, Kelly EG. Gunshot wounds of the spine: the effects of laminectomy. *J Bone Joint Surg* 1979;**61**(3):389—92.

171. Yashon D, Jane JA, White RJ. Prognosis and management of spinal cord and cauda equina bullet injuries in sixty-five civilians. *J Neurosurg* 1970;**32**(2):163—70.

172. Kumar A, et al. Penetrating spinal injuries and their management. *J Craniovertebr Junc Spine* 2011;**2**(2):57—61.

173. Çağavi F, et al. Migration of a bullet in the spinal canal. *J Clin Neurosci* 2007;**14**(1):74—6.

Spine trauma: sacral fractures

Carlo Bellabarba, Haitao Zhou, Richard J. Bransford

Department of Orthopaedics and Sports Medicine, Department of Neurological Surgery, University of
Washington, Harborview Medical Center, Seattle, WA, United States

Introduction

As the keystone at the junction of the pelvis and the spine, the sacrum serves the vital roles of protecting lumbosacral neurologic function while maintaining pelvic and spinal column alignment. Injuries to the sacrum may result in pelvic and spinopelvic deformity, chronic pain, and loss of lower extremity, bowel, bladder, and sexual function. Treatment of sacral fractures therefore needs to optimize biomechanical and neurologic outcomes, a goal that requires a thorough understanding of spinopelvic anatomy, neurologic decompression techniques, and principles of skeletal reconstruction. The type and severity of sacral fractures varies from minimally displaced low-energy fragility fractures to widely displaced, comminuted fractures that result from high-energy injury mechanisms such as motor vehicle collisions or falls from significant heights. Sacral fractures are classified according to the AO Spine Sacral fracture classification according to the presence of posterior pelvic and spinopelvic instability, and treated accordingly. Treatment of patients with sacral fractures must take into account associated injuries in the trauma patient and the underlying medical conditions in patients with metabolic bone disease. Coordination of care between the spine surgeon, orthopedic trauma surgeon, and other surgical, medical, critical care, and rehabilitation specialists is necessary to optimize results.

Anatomy

The sacrum is the lowest functional portion of the spinal column and provides an anchor for the mobile lumbar spine. Lumbosacral motion occurs through the lumbosacral intervertebral disc and the paired zygapophyseal (facet) joints. In addition to the usual intervertebral stabilizing structures, the lumbosacral articulation is further stabilized by the iliolumbar ligaments connecting the L5 transverse processes to the crest of the ilium and the sacrolumbar ligaments that have an origin contiguous with the iliolumbar

© 2022 Elsevier Inc. All rights reserved.

ligament and insert into the anterosuperior aspect of the sacrum and sacroiliac joint. These structures make the L5–S1 motion segment inherently more stable than the more cephalad lumbar motion segments.[1]

The sacrum has a kyphotic inclination that is balanced by a cascade of lumbar lordosis, thoracic kyphosis, and cervical lordosis so that a plumb line from C7 typically passes through the posterior aspect of the L5–S1 intervertebral disc. The sacral inclination angle, defined as the angle between a tangent to the dorsal surface of the sacrum and a vertical reference line, varies between 10 and 90 degrees and is usually between 45 and 60 degrees.[1,2] Pelvic incidence measures the orientation of the lumbosacral junction relative to the pelvis and approximates 50 degrees in the general population.[3] It is normally referred to in the context of sagittal plane deformities, and is valuable in determining sagittal alignment when treating sacral fractures. Sacral fractures that significantly alter spinal sagittal alignment, usually by shifting the C7 plumb line anteriorly, may result in difficulty with maintaining an erect posture, compensatory hyperlordosis, and associated functional deficits[4,5] (Fig. 12.1).

The sacrum forms the posterior segment of the pelvic ring through the sacroiliac joints. Forces transmitted through the lumbosacral articulation into the upper sacrum propagate laterally across the sacroiliac joints to the pelvis. Because of these load-transferring properties, the sacrum has been described as the keystone of the pelvic ring.[6] More specifically, the sacrum functions as a "true keystone" in the pelvic outlet plane, in which the orientation of the sacrum relative to the ilium is such that axial forces lock the sacrum into the pelvic ring and further stabilize the sacroiliac articulation. In the pelvic inlet plane the sacrum is shaped like a "reverse keystone" and its position in this plane is inherently unstable, permitting pelvic ring motion. This pelvic ring instability necessitates substantial intrinsic and extrinsic ligamentous

stabilization of the sacroiliac joints. Intrinsic stability is provided by the substantial interosseous ligaments and posterior sacroiliac ligaments, as well as the relatively weaker anterior sacroiliac ligaments. Extrinsic stability is provided by the sacrospinous and sacrotuberous ligaments of the pelvic floor.

The sacrum is composed of five kyphotically aligned, fused vertebral segments. The upper two sacral segments are similar in size to the lumbar vertebrae, whereas the lower three sacral segments taper rapidly in size. Significant variability in upper sacral anatomy can exist in the form of transitional vertebrae and sacral dysplasia. Forms of transitional vertebrae include a sacralized lumbar segment, an incomplete congenital fusion of L5 to the sacrum that is usually unilateral, and a lumbarized sacrum identified by a persistent articulation between the S1 and S2 segments.[1] Complete or bilateral sacralization may result in dysplastic sacral morphology with the upper sacrum located more cephalad relative to the iliac crests. Because upper sacral variability results in significant alteration in the relationships among the sacrum, pelvis, and spinal column, as well as their adjacent neurovascular structures, anatomic variations must be recognized during the evaluation and treatment of sacral fractures, particularly if surgical treatment is considered as it may have implications on type of surgical approach and instrumentation.[7]

The upper sacral body, which constitutes the base of the mobile spine, consistently has the highest density of cancellous bone, with the S1 segment adjacent to the superior endplate having the greatest density overall. The ventral aspect of the upper S1 body that projects anteriorly and superiorly into the pelvis is termed *the sacral promontory*. The sacral ala, the lateral portion of the sacrum that articulates with the ilium through the sacroiliac joints, is largely cancellous and is formed by the coalescence of the sacral transverse processes. The cancellous alar bone has low density, particularly in older

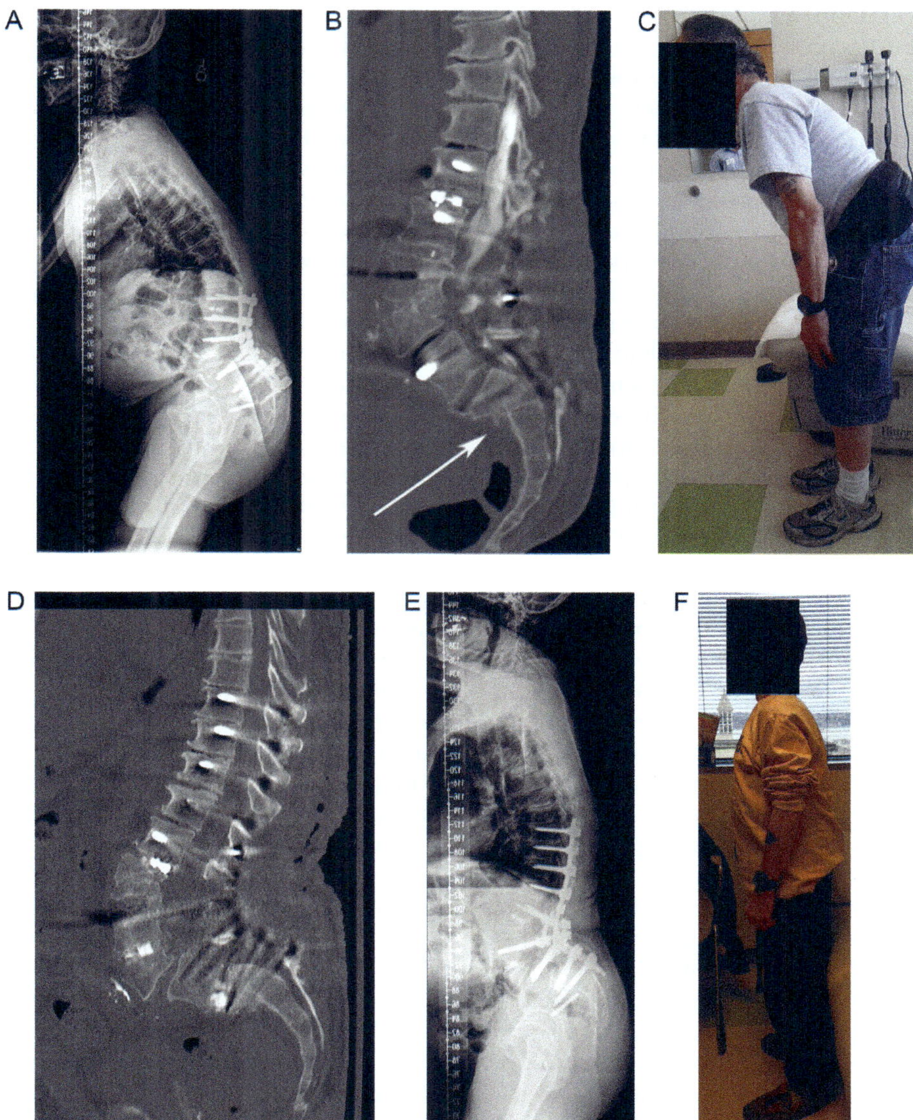

FIGURE 12.1 Sagittal plane malalignment after malunion/nonunion of AO Type C3 sacral fracture. The caudal location of sacral fractures accentuates the effect of kyphotic malalignment on global sagittal spinal alignment. (A) Lateral spine radiograph and (B) sagittal CT image of the thoracolumbosacral spine after L4 pedicle subtraction osteotomy had been done in an unsuccessful attempt at treating severe sagittal imbalance due to kyphotically malunited AO Type C3 sacral fracture (*white arrow*). (C) Severely kyphotic sagittal plane deformity. (D) Sagittal CT image, and (E) lateral radiograph after L2 pedicle subtraction osteotomy had restored physiologic sagittal plane alignment (F). Reduction of the kyphotic deformity and stabilization of AO Type C3 sacral fractures in acceptable alignment helps prevent the need for complex future reconstructive operations.

individuals, and an alar void is a consistent finding in middle-aged and older adults.[8,9] The relative difference in bone density between the upper and lower sacral body predisposes this transitional area to fracture. Similarly, younger individuals injured before complete ossification of the sacrum are predisposed to disruption at the S1–2 level owing to relative weakness at that interval. The ala is predisposed, particularly in the older and osteopenic person, to fracture line propagation. This effect is accentuated by the relative strength of the sacroiliac joint ligaments that stabilize and prevent their displacement or disruption. The relatively sparse alar bone density must also be taken into consideration when planning reconstructive procedures.

The posterior surface of the sacrum is convex and is formed by the coalescence of posterior elements. Accordingly, the middle sacral crest corresponds to the spinous processes, the intermediate sacral crests correspond to the zygapophyseal joints, and the area in between corresponds to the laminae. The lowest one or two sacral segments have incompletely formed bony posterior elements, resulting in an aperture into the sacral spinal canal known as the sacral hiatus. Enlargement of the sacral hiatus may relatively weaken the sacrum and predispose it to fracture.[1,10]

Sacral pedicles and laminae form the lateral and posterior borders of the spinal canal as it tapers in size distally from a triangular cross section at the S1 level to a flat and narrow cross section at the most caudal segments. The dural sac typically ends at the S2 level, and the filum terminale attaches its caudal end to the coccyx. At the junction of the body and the sacral ala are four paired ventral and dorsal neuroforamina through which the ventral and dorsal sacral nerve root rami pass, respectively. The relative space available to the sacral nerve roots in the ventral foramina is lowest at the S1 and S2 levels, where the nerve roots occupy one-third to one-fourth of the foraminal area compared with the S3 and S4 levels, where the nerve roots occupy one-sixth of the available foraminal space. It follows that the lower sacral roots are less likely to be impinged upon in injuries involving displacement and narrowing of the neuroforamina.[11]

Numerous structures positioned adjacent to the sacrum may be injured in sacral fracture-dislocations. The majority are neurovascular structures, with the rectum being the only visceral structure in close proximity. Below the rectosigmoid junction, located at the S3 level, the mesentery disappears and the rectum lies directly adjacent to the sacrum.[12]

The dorsal nerve roots exit through their respective posterior neuroforamina to supply motor branches to the paraspinous muscles and cutaneous sensory branches that form the cluneal nerves. Anteriorly, the L5 nerve root is in close contact with the sacrum as it passes beneath the inferior edge of the sacrolumbar ligament and drapes over the anterosuperior aspect of the sacral ala. It anastomoses with the L4 ventral ramus joined by the exiting ventral sacral nerve roots to form the sacral plexus as it courses caudally adjacent to the ala.[13] Branches of the sacral plexus include the sciatic nerve, pudendal nerve, superior gluteal nerve, and inferior gluteal nerve. In addition, the sympathetic chain lies directly on the ventral sacrum, just medial to the neuroforamina. At each level it distributes a gray ramus communicans to the ventral nerve root. Therefore, the sacral plexus mediates bowel, bladder, and sexual function as well as major lower extremity function. The dual innervation of the perineal structures from both the left and right sacral plexus helps protect bowel, bladder, and sexual function, since these functions are largely preserved in the event of unilateral transection of the sacral nerve roots, whereas bilateral transection causes complete loss of function.[14]

The presacral area has an extensive vascular network that is highly variable. The middle sacral artery typically courses ventrally along the midline of the L5 vertebral body, across the

L5–S1 disc space and the sacral promontory, and down the sacrum after branching from the aorta at the common iliac bifurcation. The common iliac arteries subsequently give rise to the internal iliac arteries that lie anterior to the sacroiliac joints and give off both superior and inferior lateral sacral arteries. The superior lateral sacral artery traverses the upper sacroiliac joint and courses caudally, just lateral to the sacral foramina, giving off spinal arteries that pass through the S1 and S2 ventral foramina to the spinal canal. The inferior lateral sacral artery traverses the inferior aspect of the sacroiliac joint before anastomosing with the middle sacral artery and giving off spinal arteries that pass through the S3 and S4 ventral foramina.[15] The internal iliac veins lie posteromedial to the internal iliac arteries and course caudally. They are located medial to the sacroiliac joint directly adjacent to the sacral ala.[12,15] The internal iliac veins are the repository of an extensive presacral venous plexus. This plexus is formed by extensive anastomoses between the lateral and middle sacral veins and receives transforaminal veins that communicate with epidural veins in the spinal canal.[1,15] This extensive vascular network renders anterior exposures to the sacrum impractical.

Epidemiology

Sacral fractures follow a bimodal distribution age distribution which correlates inversely with the energy of the injury mechanism. High-energy injuries generally occur in younger patients with favorable bone quality, whereas low-energy fragility fractures occur more typically in older patients with metabolic bone disease and may result in severe instability and spinopelvic dissociation in almost 2% of patients over 55 years of age after a ground-level fall.[16]

Classification

Sacral fracture classifications have evolved over the past 75 years to emphasize factors that are believed to have the highest impact on treatment and prognosis, namely the presence of neurologic deficits, posterior pelvic instability, and spinopelvic instability. Despite various other classifications being proposed over the past several decades,[17–19] none was widely adopted until 1988 when Denis and colleagues,[11] based on a series of 236 sacral fractures, formulated a simplified anatomic classification that correlates fracture location with the incidence of neurologic injury. This classification divides the sacrum into three zones (Fig. 12.2). Zone I—alar zone, Zone II—foraminal zone, and Zone III—central zone, because the incidence and the severity of neurologic injury were noted to increase with progressively more central location

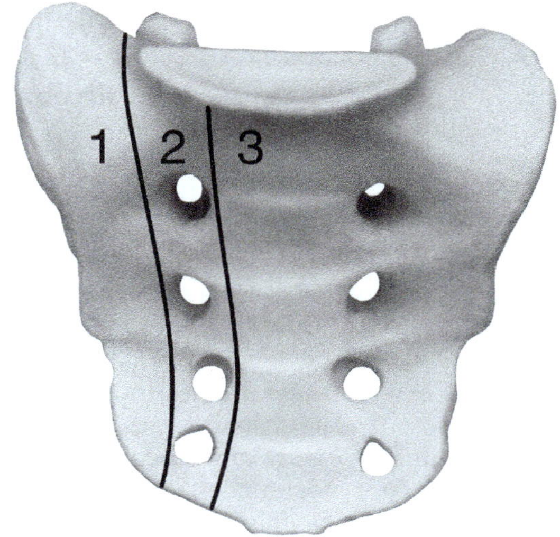

FIGURE 12.2   Denis classification of sacral fractures: Denis and coauthors[11] categorized sacral fractures into three zones according to the location of fractures relative to the sacral foramina. More medially located fractures have a higher risk of neurologic deficits and a worse prognosis.

of the fracture. Accordingly, in the series of Denis and colleagues, zone I fractures had a 5.9% incidence of neurologic injury, primarily to the L5 nerve root as it coursed over the ala. Zone II fractures had a 28.4% incidence of associated neurologic injury occurring as a result of either foraminal displacement and resulting impingement on the exiting nerve root. Zone III fractures had a 56.7% incidence of neurologic injury resulting from injury at the level of the spinal canal, with 76.1% of these individuals having bowel, bladder, and sexual dysfunction, presumably due to bilateral sacral root injury. Denis and colleagues[11] noted a spectrum of zone III sacral fracture-dislocations that included both transverse and longitudinal fracture orientation and described a subtype that they termed *a sacral burst fracture* consisting of retropulsion of the sacral body with intact sacral laminae. It is important to note that more recent studies have demonstrated much lower neurological injury rates than those originally reported by Denis. Khan and coauthors recently reported an overall neurological injury rate of 3.5%, compared to 21.6% in the series by Denis. They categorized their neurological injury according to the Denis zone of injury as: Zone I—1.9% versus 5.9% in the series by Denis; Zone II—5.8% versus 28.4%; and Zone III—8.6% versus 56.7%.[20] Gibbons and colleagues[21] correlated neurologic injury to the severity of sacral fracture according to the Denis classification and found a 34% incidence of neurologic injury overall. Six of 25 patients (24%) with zone I injuries and 2 of 7 patients (28%) with zone II injuries had L5 or S1 nerve root injuries, with no patient reporting bowel or bladder dysfunction. Three of five patients with zone III injuries had neurologic injury, two of which had bladder dysfunction. They used these findings as a basis for classifying neurologic injury from sacral fractures as: 1, no injury; 2, lower extremity paresthesias only; 3, lower extremity motor deficit with intact bowel and bladder function; or 4, impaired bowel and/or bladder control. They also noted

that significant neurologic deficit is rare in transverse sacral fractures below the S4 level. This classification system, however, does not distinguish complete versus incomplete injuries or address sexual function.

Because of the high potential for neurologic injury and spinal column instability, sacral body (Denis zone III) fractures have been studied by several investigators. Early case reports often characterized the injury as a solely transverse fracture pattern, possibly due to imaging limitations.[22–24] Computed tomography (CT) demonstrates that most transverse fractures of the upper sacrum have more complex fracture patterns and are now understood to consist of a transverse fracture of the sacrum with associated longitudinal or "vertical" injury components, usually in the form of bilateral transforaminal fractures that extend as far rostrally as the lumbosacral junction, to form the so-called sacral "U" fracture. As a corollary, bilateral transforaminal fractures have been shown to be associated with a transverse fracture in 87% of cases.[25] Variations in fracture line propagation result in "H," "Y," and "lambda" fracture patterns (Fig. 12.3).

Roy-Camille and colleagues[26] reported on 13 patients with transverse sacral fractures, as a result of attempted suicide by jumping. They classified the fractures as: type 1—flexion deformity of upper sacrum (angulation alone); type 2—flexion deformity with posterior displacement of the upper sacrum (angulation and posterior translation); and type 3—anterior displacement of the upper sacrum without angulation (anterior translation alone). Based on cadaveric studies, they hypothesized that types 1 and 2 were caused by the lumbar spine being in a flexed position at the time of impact, whereas type 3 fractures occurred when the lumbar spine and hips were in extension. Strange-Vognsen and Lebech[27] subsequently reported a case of comminution of the upper sacrum without significant angulation or translation that they named a *type 4 injury relative to*

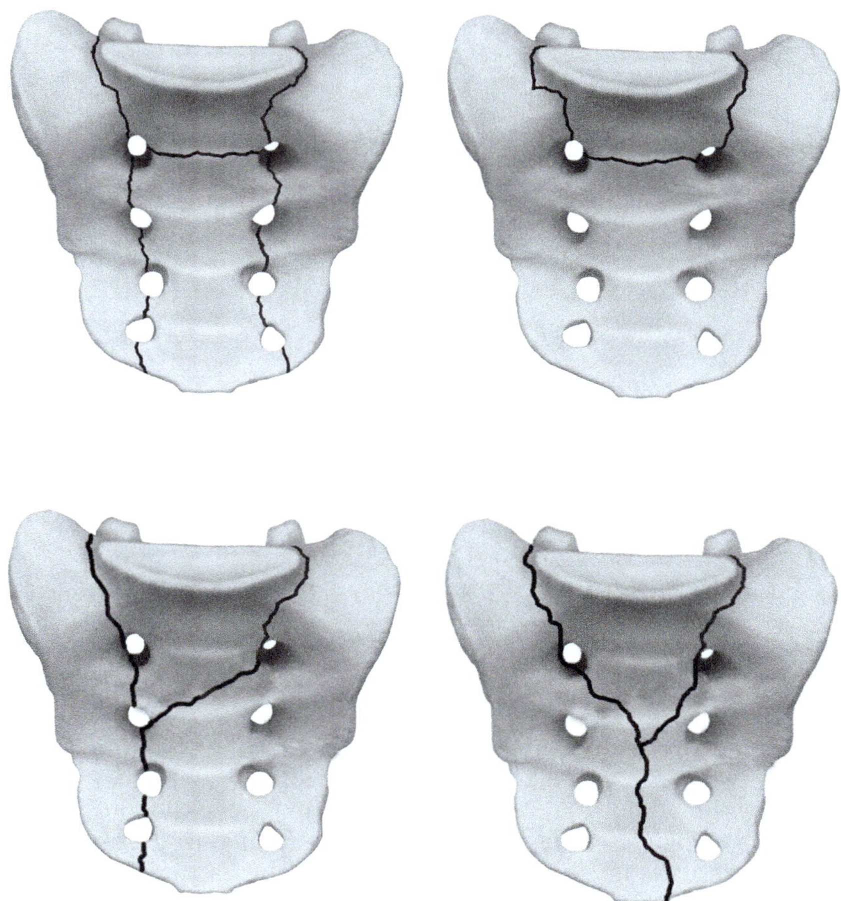

FIGURE 12.3 Descriptive classification of sacral fractures: Sacral "U" variants with spinopelvic instability include (clockwise from top left) "H," "U," "Y," and lambda fractures.

the Roy-Camille classification based on the hypothesis that in type 4 injuries the lumbar spine was in the neutral position at the moment of impact (Fig. 12.4). All these injuries are caused by indirect forces to the lumbosacral junction. The location of the transverse component of the sacral body fracture is not specified by this classification system, which limits its usefulness since defining the transverse sacral fracture as high (involving S2 or above), or low (involving S3 or below) can be helpful in understanding the injury from both the biomechanical and neurologic standpoint.[22] A type 5 injury, which consists of segmental comminution of the sacrum, has been proposed by Schildhauer and coauthors.[30]

Other patterns of sacral fractures have been identified as resulting from specific mechanisms or having predictable patterns of associated injuries. In contrast to transverse Denis zone III sacral fractures, midline longitudinal Denis zone III sacral fractures have a low incidence of neurologic injury and consist of a very different injury from the standpoint of treatment and

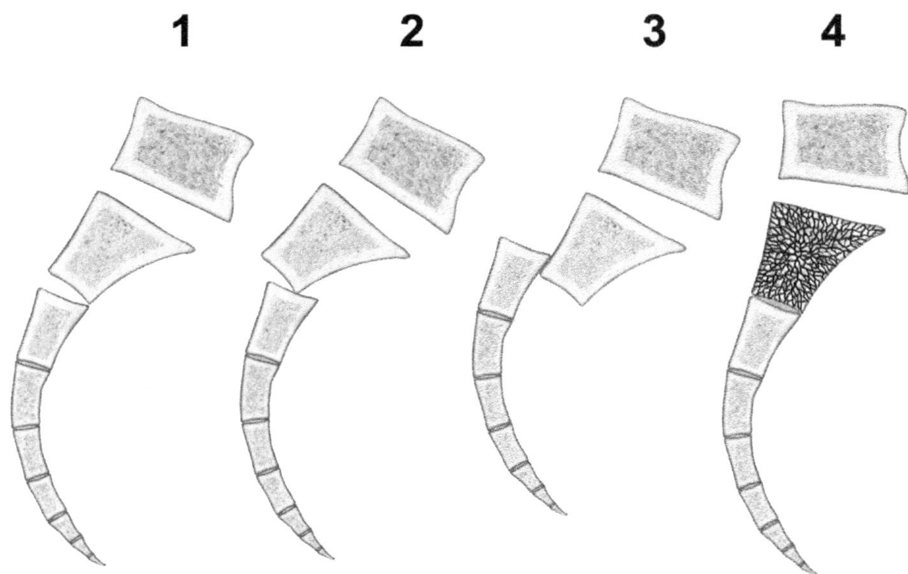

FIGURE 12.4   Roy-Camille classification of fractures involving sacral canal and vertebral body: The Roy-Camille classification[26] as modified by Strange-Vognsen and Lebech[27] categorizes AO Type C3[28,29] fractures according to sagittal plane angulation and displacement. Type 1 − Flexion deformity of upper sacrum without displacement; 2 − Flexion deformity with posterior displacement of the upper sacrum; 3 − Anterior translation of the upper sacrum; 4 − Comminution of the upper sacral body without angulation or displacement.

prognosis (Fig. 12.5).[31−33] This injury was originally reported in 1979 by Wiesel and colleagues who suggested that these fractures may have a lower incidence of neurologic injury than transverse fractures because the nerve roots are subjected to a posterolateral rotation force instead of being subjected to the compression or possibly even transection that can result from shear forces associated with displaced transverse fractures.[33] Bellabarba and colleagues[31] subsequently described this injury as a variant of the anteroposterior compression pelvic ring injury, in which the posterior pelvic ring injury occurs atypically through Denis' sacral Zone III. In their series of 10 patients, none had neurologic deficits, in contrast to the high incidence of neurologic injury reported in patients with predominantly transverse sacral fractures through the spinal canal.[31]

Even in the absence of a transverse fracture, sacral fractures can be associated with spinal column instability. Isler described variations of Denis Zone II longitudinal sacral fractures that result in L5−S1 motion segment instability owing to facet joint disruption (Fig. 12.6).[34] Injuries with the fracture line lateral to the S1 articular process are not associated with instability of the lumbosacral articulation because the L5−S1 articulation remains continuous with the stable fracture component of the sacrum. Fracture that extends into or medial to the S1 articular process, however, may disrupt the associated facet joint and potentially destabilize the lumbosacral junction (Fig. 12.7). Complete displacement of the facet joint can cause a locked facet joint, making it difficult to reduce the sacral fracture with closed methods alone (Fig. 12.8). Facet disruption may also cause posttraumatic arthrosis and late lumbosacral pain.

In contrast to high-energy sacral fractures, the precipitating event in sacral insufficiency fractures is often not identifiable or may be related

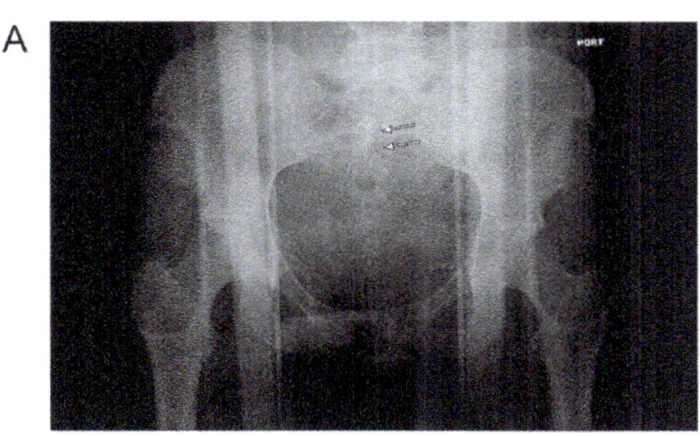 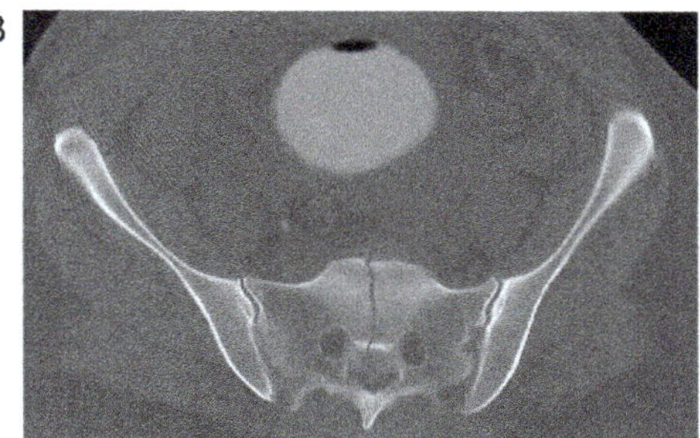

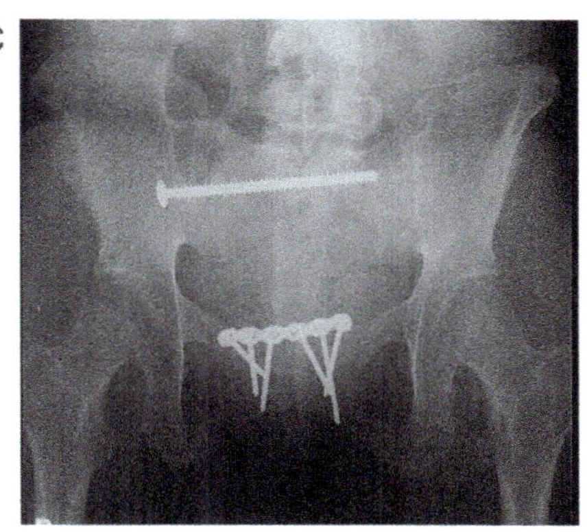

FIGURE 12.5 Anteroposterior radiograph (A) and axial CT image (B) of the pelvis demonstrate an atypical Denis zone III fracture that carries a lower risk of neurologic deficits, consisting of a midline longitudinal fracture of the sacral vertebral bodies (*white arrows*) that does not extend beyond the spinal canal level into the foramina. Because this is typically the most benign of the longitudinal sacral fracture pattern, it has been designated, in the AO Classification, as AO Type B1. (C) AP pelvis 6 months after iliosacral screw fixation and symphyseal plating.

to a ground-level fall onto the buttocks. Insufficiency fractures occur in weakened bone, typically due to decreased bone mineral density in postmenopausal women, but can also be caused by other conditions such as chronic corticosteroid use or a history of radiation therapy to the pelvis.[35] The presenting symptoms are often vague and frequently consist of poorly localized low back pain that is exacerbated by sitting and standing, possibly with lumbosacral radiculopathy. These fractures are typically oriented vertically and occur through the ala adjacent to the sacroiliac joint (Fig. 12.9). There may also be a transverse component resulting in more complex sacral "U" fracture variants. Because of the predisposition of these patients to

osteoporotic fractures, there is high association with insufficiency fractures at other sites, most commonly the pubis and thoracolumbar vertebrae. Although cauda equina dysfunction has been reported, neurologic deficits are uncommon.[36–38]

Mechanical factors related to the transfer of forces from the lumbosacral spine to the pelvis may also cause sacral insufficiency fractures in patients predisposed to osteoporosis.[39] Of particular significance to spine surgeons are sacral insufficiency fractures caudal to previous lumbar fusions that probably represent failure of the sacrum to withstand forces concentrated there as a result of the large cephalad lever arm.[40–44]

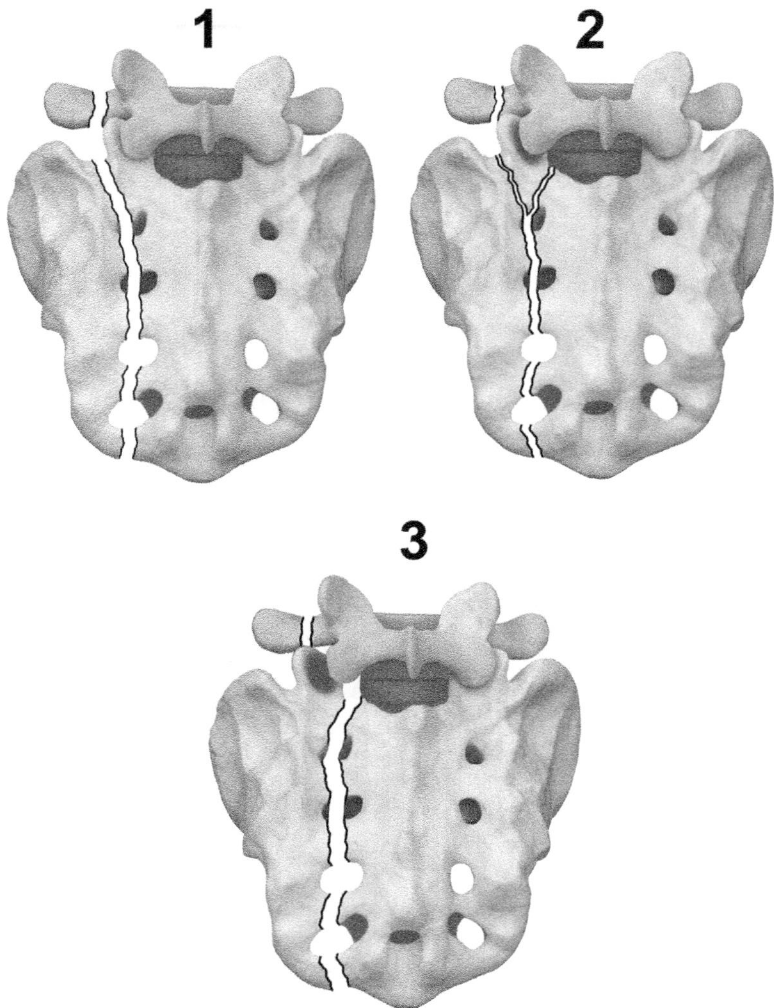

FIGURE 12.6   Isler classification of sacral fractures: Isler[34] classified vertical sacral fractures into three types, according to the location of the fracture relative to the S1 superior facet. Fracture patterns that extend medial to the S1 superior facet (Types 2 and 3) result in spinopelvic instability.

These earlier classifications provide the foundation for the AO Spine sacral fracture classification, which has categorized sacral fractures based primarily on: 1. the extent and type of instability; 2. the severity of neurologic injury; and 3. associated injuries or conditions that may influence treatment or prognosis.[28,45] This was done with the goal of combining the existing classifications and other published observations described above into a single comprehensive classification system that helps guide treatment, correlates with prognosis, and has favorable intra- and interrater reproducibility (Fig. 12.10). This system took into account the perspective of spine and orthopedic trauma surgeons and was designed to complement the separate AO

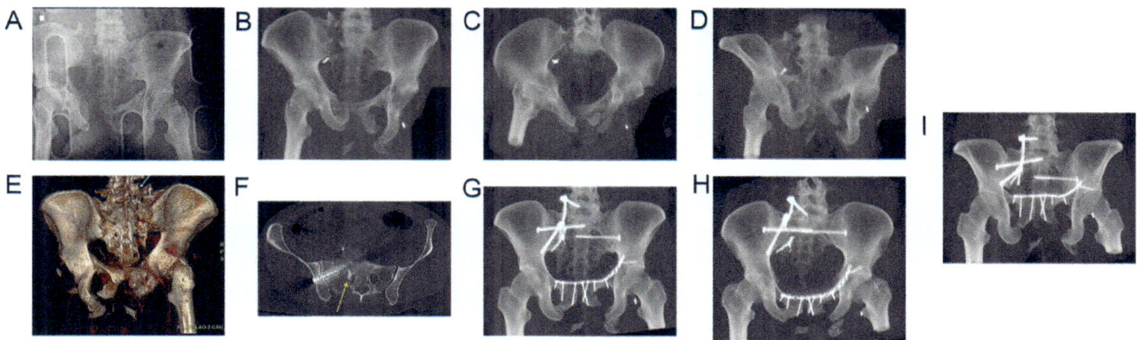

FIGURE 12.7 AO Type C1 M3, M4 sacral fracture. Anteroposterior radiograph (A) demonstrates vertical shear pelvic injury with severe displacement of the right hemipelvis and extensive comminution of the anterior pelvic ring. Resuscitation included arterial embolization and partial closed reduction with application of right femoral traction and circumferential pelvic sheet, with improved provisional alignment on anteroposterior (B), inlet (C), and outlet (D) pelvic reconstructions. Posterior view of a 3D CT reconstruction (E) shows the fracture line extending medial to the S1 superior facet (*white arrow*), dislocating the L5—S1 facet joint and making this an AO C1 injury with spinopelvic dissociation, or Isler Type 3. Because associated craniofacial injuries precluded definitive reconstruction of the spinopelvic injury for over 2 weeks, provisional closed reduction, external fixation of the anterior pelvis, and percutaneous right iliosacral screw fixation of the sacral fracture were performed for damage control purposes, as seen on postoperative (F) axial CT imaging, which also demonstrates S1 root impingement from a foraminal bony fragment, consistent with neurologic deficits in the S1 distribution (*yellow arrow*). Anteroposterior (G), inlet (H), and outlet (I) CT reconstructions show acceptable realignment of the right hemipelvis after open reduction, S1 foraminotomy, and right-sided triangular osteosynthesis. A left-sided iliosacral screw was added due to concern for potential left sacroiliac joint instability.

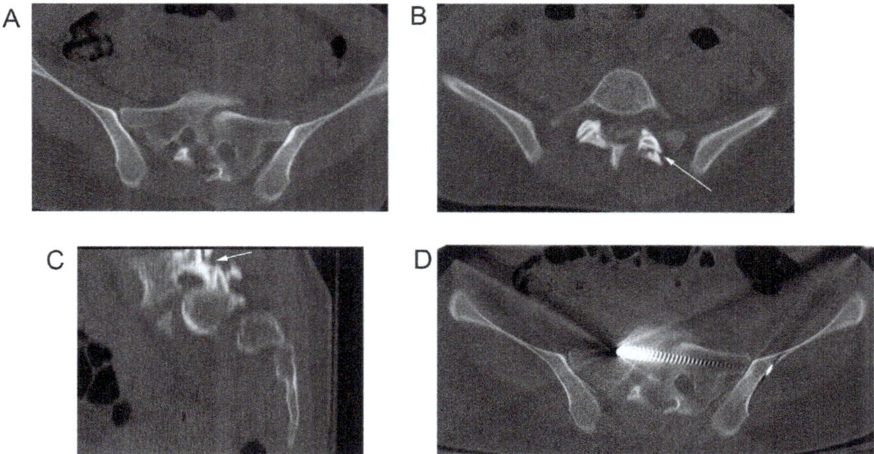

FIGURE 12.8 AO Type C1 M3 sacral fracture with locked facet. Axial (A) and (B) and sagittal (C) CT images of longitudinal sacral fracture with left L5—S1 facet dislocation (*white arrows*). Closed reduction of sacral fractures with locked L5—S1 facet dislocations is unsuccessful, since reduction of the sacral fracture is contingent on successful reduction of the L5—S1 facet joint, which typically cannot be accomplished by closed methods. Closed percutaneous fixation of these injuries therefore results in persistent malalignment of the sacral fracture, as shown on postoperative axial (D) CT image.

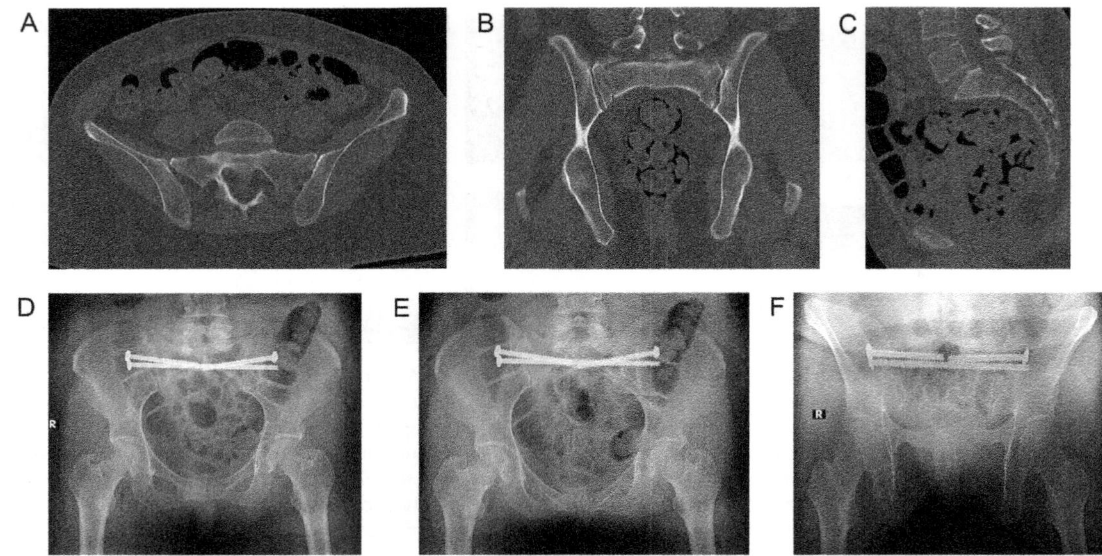

FIGURE 12.9 AO Type C0 fracture. Axial (A), coronal (B), and sagittal (C) demonstrate AO Type C0 sacral U insufficiency fracture diagnosed in a patient with Parkinsonism who had severe pelvic pain with attempts to sit or stand after a ground-level fall. Percutaneous transiliac and bilateral iliosacral screw fixation, illustrated on postoperative anteroposterior (D), inlet (E), and outlet (F) radiographs, allowed the patient to resume their normal level of activity without the need for prolonged bed rest.

Trauma pelvic fracture classification.[46] Type A fractures range from inconsequential injuries to severely displaced transverse fractures that occur below the SI joint, and therefore result in neither posterior pelvic nor spinopelvic instability. Type B fractures are longitudinal or "vertical" fracture patterns that result in posterior pelvic instability, without compromise of the spinopelvic junction. Type C injuries result in spinopelvic instability, with or without posterior pelvic instability. Type A and B fractures are divided into three subtypes, and type C injuries are divided into four subtypes, categorized according to injury severity based on greater risk of neurological deficit or of instability. Neurologic injury is also classified according to the same system used by the AO Spine fracture classification used throughout the entire spine: NX—patient not examinable; N0 — no neurologic deficit; N1—transient neurologic deficit; N2 — nerve root injury; N3 — incomplete spinal cord injury or cauda equina syndrome (either

complete or incomplete); N4 — complete spinal cord injury. Because sacral fractures do not occur at the spinal cord level, they cannot cause neurologic injuries more severe than N3. Modifiers which may affect treatment or prognosis include: M1 — soft tissue injury (either an open or a closed degloving injury); M2 — metabolic bone disease; M3 — anterior pelvic or acetabular fracture; M4 — sacroiliac joint injury.[29,47]

Evaluation

Sacral fractures occur in two different patient groups. Most commonly, sacral fractures are the result of high-energy trauma. They are increasingly common in patients with metabolic bone disorders that predispose to pathologic fractures. Diagnosis is frequently delayed, especially with fragility fractures, which may result in further displacement or neurologic deterioration. In their retrospective analysis of 236

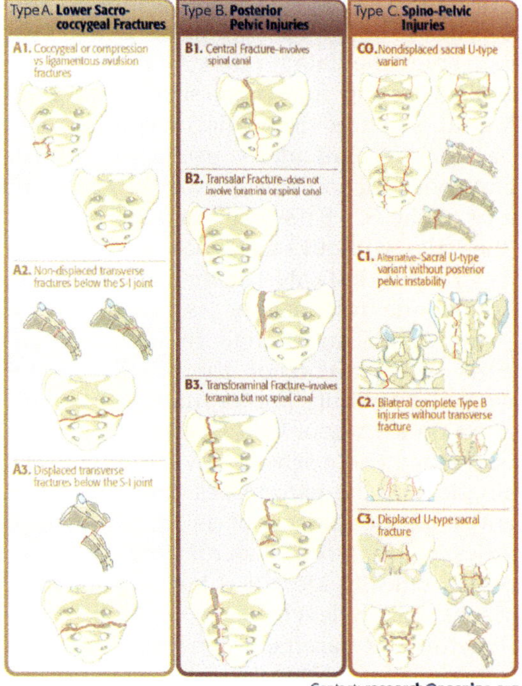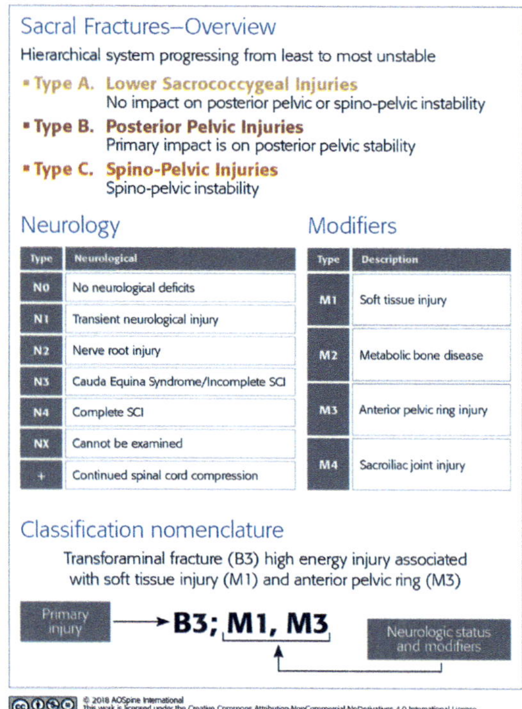

FIGURE 12.10   AO Spine sacral fracture classification. Morphology: Type A: No posterior pelvic or spinopelvic instability. Type B: Posterior pelvic instability without spinopelvic instability. Type C: Spinopelvic instability with or without posterior pelvic instability. Neurology: Nx − N4. Modifiers: M1−M4.

patients with sacral fractures, Denis and colleagues[11] found that in neurologically intact patients the diagnosis of sacral fractures was made during the initial hospitalization only half of the time. The presence of a neurologic deficit increased the diagnostic accuracy to only 70%. As in most cases of diagnostic delay, the etiology of missed sacral fractures is multifactorial and includes the relative difficulty in identifying these fractures on screening anteroposterior pelvis radiographs combined with the presence of associated injuries in the trauma patient and low clinical suspicion in patients with insufficiency fractures.[11,48] With the increased use of full body CT scans to evaluate polytrauma patients, delayed diagnosis of sacral fractures has become much less common in patients with high-energy injuries.

Trauma patients with sacral fractures have often sustained multisystem injuries, including life-threatening head and thoracoabdominal injuries. In these patients, emergent resuscitation is the highest priority. Accordingly, Advanced Trauma Life Support (ATLS) protocol begins with a primary survey during which conditions that are immediately life threatening are addressed. Resuscitation is focused on maintaining cardiopulmonary and hemodynamic stability. Only after this primary goal has been achieved should the secondary survey, composed of examination of the patient to identify additional injuries, be completed.

The secondary survey includes a screening evaluation of both the spinal column and pelvic ring. Precautions are necessary to maintain spinal column integrity, and patients should be initially kept on a flat surface and log-rolled side to side to prevent spinal column displacement. Evaluation includes inspection and palpation of the patient's back from the occiput to the coccyx. Sacral fractures commonly have overlying skin discoloration or lacerations, palpable step-offs or crepitus, localized tenderness, and hematomas, any of which can indicate the presence of a sacral fracture. Significant soft-tissue contusion or internal degloving, analogous to Morel-Lavallee lesions seen with acetabular fractures, can have implications on subsequent treatment.[49] Manual compression over the iliac crests, both anteroposteriorly and mediolaterally, may also help identify a sacral fracture. Perforations of the rectum or vagina can represent open sacral fractures, which can be detected with rectal and vaginal digital examination as well as the use of a speculum and proctoscope.

Because pelvic ring disruption may be associated with significant intrapelvic hemorrhage, temporary methods of pelvic ring stabilization may be necessary to reduce pelvic volume and provide provisional stability. These methods include the application of a circumferential pelvic antishock sheet,[7] pelvic clamp, anterior external fixator, or skeletal traction, depending on the pelvic ring fracture pattern. Associated vascular injury, particularly to the hypogastric arterial system, may require embolization to adequately control arterial hemorrhage.[50]

It is essential to determine neurologic function of patients who have sustained sacral fractures.[51–53] A rectal examination is performed early in the workup of all multiply injured patients, even in the absence of obvious sensorimotor deficits in the extremities, to evaluate perianal sensation, anal sphincter tone, and voluntary perianal contraction and to assess for presence of anal wink and the bulbocavernosus reflex. The bulbocavernosus is a polysynaptic, spine-mediated reflex that is generally useful in testing for spinal shock and gaining information about the state of spinal cord injuries, but is particularly useful in evaluating sacral root function since it will be absent after injury to the S2–S4 spinal nerves. The absence of the reflex without sacral root or conus medullaris trauma indicates spinal shock. Overall severity of neurologic injury is graded according to the American Spinal Injury Association (ASIA) modification of the Frankel grading system.[54] Extremity motor function is further graded on a scale of 0–5 to establish the ASIA motor score, and a sensory level is obtained. Examination of the extremities can only reliably identify injuries as caudal as S1, however, and injuries to the lower sacral roots cannot be more specifically identified beyond obtaining a perianal sensory level.

Fracture displacement can cause neurologic injury from a variety of mechanisms including angulation, translation, and direct compression by displaced bone fragments. Potentially reversible injuries include contusion, compression, and traction. Kyphosis of 20 degrees or more has been correlated with increased risk of neurologic injury.[55] Recovery of transected or avulsed nerve roots, which is not uncommon with more severe AO Type A3 sacral fractures, cannot be expected. Delayed neurologic deficits can occur from epidural hematoma, late fracture displacement, or callus formation[17] and should be promptly reinvestigated to determine potential causes.

Although scrutiny of the anteroposterior pelvic radiograph would allow for a majority of sacral fractures to be identified, sacral fractures can easily be missed with the use of plain radiographs alone, owing to a variety of circumstances. This is particularly true of sacral fractures that present mainly with displacement of the transverse sacral fracture component and spinopelvic instability, in which there is a less pronounced deformity of the pelvic ring. Osteopenic bone and sacral dysmorphism can also obscure landmarks, making the identification of fractures more challenging.

The increasingly routine use of more sophisticated imaging techniques in the initial assessment of the trauma patient's visceral injuries such as CT of the abdomen and pelvis, which allows for reconstructed images of the spine, has improved the detection rate of sacral fractures. The identification of a sacral fracture on these screening studies mandates focused radiographic evaluation, including a dedicated fine cut CT scan of the sacrum and sagittal and coronal reformations to allow for the detail required to determine the fracture configuration, resulting instability pattern, and extent of sacral canal and neuroforaminal compromise.[56] Three-dimensionally reformatted CT scans may add insight into fracture morphology for less experienced clinicians or in the case of highly complex fracture configurations.

Additional plain radiographic projections can yield important information. The pelvic inlet and outlet views, obtained with 45-degree caudal and 60-degree cephalad tilt, respectively, are standard techniques for evaluating pelvic ring injuries. The Ferguson view, a coned-down true anteroposterior view of the sacrum, is obtained with a 30-degree cranially inclined projection. A lateral radiograph is useful in evaluating sacral inclination and the presence of a transverse fracture. Useful radiographic indicators of sacral injuries include abnormalities in the contour of the sacral foramina and sacral arcuate lines and the presence of a "paradoxical inlet" view of the sacrum on the anteroposterior pelvic view. Their presence strongly suggests a sacral fracture, and further investigation should be done. These same radiographic projections can be reconstructed from the 2D CT image, therefore limiting the need for additional imaging studies.

Magnetic resonance imaging is not usually helpful except in cases of unclear neurologic deficits or discrepancies between skeletal and neurologic levels of injury, although it may provide early evidence of lumbosacral nerve root avulsion.[57]

Treatment

General principles

Decision-making in sacral fracture treatment is primarily based on the location and pattern of the fracture and the patient's neurologic status. Surgical indications include the presence of instability, malalignment, and neurologic deficit. Other key factors are the patient's general medical condition and additional injuries. Hemodynamic instability, high intracranial pressures, or compromised pulmonary function may preclude early surgical stabilization in critically injured patients. Conversely, the benefits of early mobilization in trauma patients with pulmonary injuries may make early surgical stabilization advisable.[58,59] Surgical timing in patients with closed-head injuries in particular is controversial, and skilled perioperative management is essential to prevent secondary brain injury.[60,61] Chronic medical conditions also need to be considered and may require an initial period of nonoperative stabilization before surgical intervention to allow for medical comorbidities to be optimized.

Careful examination of the fracture pattern is essential to determine if the sacral fracture is associated with instability of the weight-bearing axis, and whether this involves posterior pelvic instability, spinopelvic instability, or a combination of the two. Unilateral longitudinal fractures through the alar (AO Type B2) or foraminal (AO Type B3) zone maintain continuity of the contralateral weight-bearing axis, allowing weight bearing on that side (Fig. 12.11). Bilateral displaced longitudinal fractures (AO Type C2),

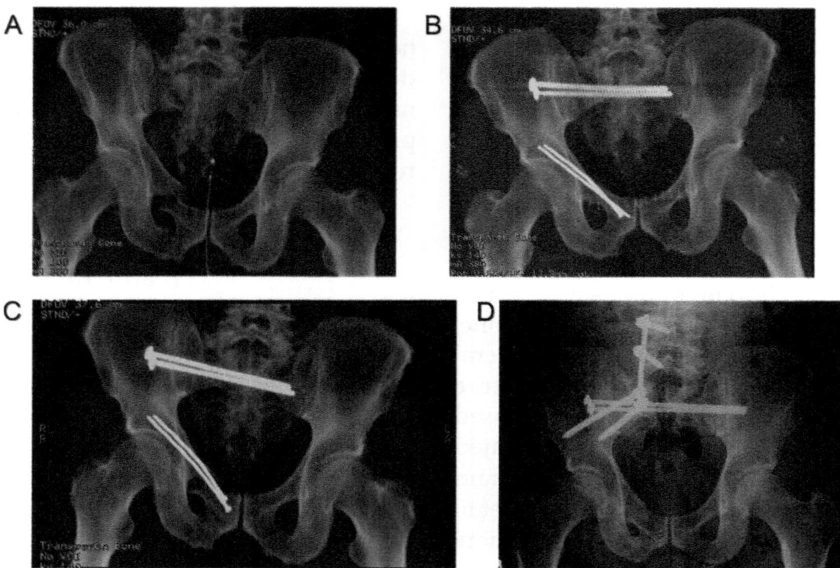

FIGURE 12.11   Failure of transiliac–transsacral screw fixation. Anteroposterior views of the pelvis showing (A) right AO Type B3 (Denis zone II) sacral fracture with displacement of the right hemipelvis, treated with (B) closed reduction and percutaneous iliosacral and superior pubic ramus screw fixation, with restoration of pelvic anatomy. (C) Early fracture displacement required (D) revision with the unilateral, triangular osteosynthesis version of lumbopelvic fixation.

as well as transverse fractures with associated bilateral longitudinal fractures through the upper sacrum (AO Type C3), dissociate the spinal column from the pelvic ring, resulting in disruption of the weight-bearing axis.[26,62–64] Weight bearing on either lower extremity or on the pelvis itself while merely sitting may cause displacement in these instances. Conversely, transverse fractures below the sacroiliac joint (AO Type A) have no implication on the weight-bearing axis and therefore on either posterior pelvic or spinopelvic instability, as long as they are not associated with secondary fracture lines extending rostrally.[65]

Neurologic deficit needs to be correlated with fracture anatomy. Compression of nerve roots at the level of the spinal canal or neuroforamina due to impingement by bony fragments or to malalignment of the spinal canal at the fracture site that may cause the cauda equina to drape over a kyphotic ridge should be identified. Possible neurologic deterioration from persistent fracture instability should also be considered. Although the presence of a neurologic deficit is an indication for operative intervention, the effectiveness of surgery in improving neurologic outcomes after fracture of the sacrum remains unproven, since the literature on this topic consists primarily of small, heterogeneous case series without consistent grading and definitions of neurological dysfunction.[11,18,19,21,22,24,66] Each case therefore needs to be individually considered, and our opinion is that surgical decompression, which usually can also be achieved with fracture reduction and stabilization, should be considered in the presence of a potentially reversible neurologic injury.

Functional outcome studies have demonstrated that most sacral fracture patients were unable to return to their preinjury vocational status over a year after injury. Although the vast majority of patients sustained long-term physical and mental impairment, the severity of these impairments did not correlate directly with fracture characteristics but was more a function of associated injuries and sacral root function.[67,68]

Nonoperative treatment

Minimally displaced, stable sacral fractures can be considered for nonoperative treatment, which typically consists of restricted weight bearing on the injured side for unilateral fractures, or a period of recumbency followed by protected weight bearing for bilateral or sacral U-type fractures. The use of bracing as an attempt to minimize load transfer to the sacrum is no longer common, as the drawbacks typically outweigh its effectiveness. The period of recumbency varies depending on the extent of fracture instability, in order to allow for satisfactory pain control and to allow for sufficient callus formation in order to decrease the possibility of displacement. Recumbency is required if attempting nonoperative treatment of sacral U-type insufficiency fractures.

Although displaced fractures can be treated with skeletal traction to improve alignment, including bifemoral traction for complex sacral fractures with bilateral involvement, these techniques are typically used while awaiting surgical intervention rather than as definitive treatment, due to the disadvantages of prolonged recumbency in both trauma patients and patients with insufficiency fractures. Potential disadvantages of prolonged recumbency include life-threatening pulmonary and thromboembolic events, decubitus ulcers, deconditioning, the inability to reliably relieve sacral canal and neuroforaminal compression, and the potential for late instability causing neurologic deficits and fixed deformities that are difficult to correct (Fig. 12.1).[69]

Nonoperative treatment is most appealing in patients with minimally displaced, unilateral sacral fractures without associated neurologic deficits. Mobilization is usually with a walker or crutches enabling toe-touch weight bearing on the injured side. These patients must be followed carefully for fracture displacement, which may warrant surgical stabilization before the deformity becomes rigid. Although insufficiency fractures are amenable to nonoperative treatment, the refinement of both iliosacral and spinopelvic percutaneous stabilization techniques[70] has made operative intervention a better option for many patients, especially in patients with bilateral sacral fractures who would require prolonged bed rest if treated nonoperatively (Fig. 12.9).[38]

Operative treatment

The goals of surgical treatment are twofold: neurologic decompression in cases of neurologic deficit and restoration of a stable, well-aligned sacrum in cases of significant displacement or instability. Because of the anatomic overlap between the spine and pelvis, a gray area exits between the responsibilities of spine surgeons and orthopedic trauma surgeons in the treatment of these injuries, which are often dictated by institutional resources and expertise. However, because orthopedic traumatologists typically have a more refined appreciation for pelvic reduction and fixation techniques and spine surgeons possess a better understanding of neurologic decompression and spinal stabilization methods, the ideal option for most patients would likely be a collaborative, multidisciplinary approach.[71] The timing of surgical intervention is dictated by many factors. The presence of an open fracture, either through the skin or into the alimentary or genitourinary tracts, requires expeditious operative intervention with irrigation and debridement followed by surgical stabilization in cases where there is potential for recurrent or persistent fracture displacement through the open wound. Neurologic deficit also suggests the need for immediate surgical decompression, though associated injuries and the patient's physiologic state may determine the feasibility of doing so, given that decompression within 2 weeks of injury appears to be an acceptable time frame. Severe angulation of a transverse sacral fracture may also tent the overlying soft tissues and cause skin breakdown, particularly in patients whose body habitus or general physical condition predisposes them to pressure sores. Because of the many factors that influence surgical decision-making in these polytraumatized patients, in many cases the trauma patient's physiologic status is the main determinant in the timing of operative intervention and the standards used for cauda equina decompression in patients with degenerative conditions such as stenosis and disc herniation do not apply.

Neurologic decompression

Neurologic decompression can be achieved by either direct or indirect means. Indirect decompression refers to the nerve root decompression achieved simply by realigning the fracture. An attempt at indirect decompression is best accomplished before consolidation of the fracture hematoma.[72] The problem with relying on indirect decompression alone, particularly in comminuted fractures, is that realignment will not reliably relieve compression caused by comminuted bony fragments that do not realign along with the primary fracture fragments. If neural impingement persists and is associated with a neurologic deficit, direct decompression should be considered.

In patients with sacral root deficits, direct decompression by laminectomy and removal of compressive bone fragments optimizes the environment for neurologic recovery.[22] Direct

posterior decompression without fracture reduction and stabilization has limited utility, however, and is generally not recommended because neural impingement is typically contributed to by fracture displacement. Decompression is therefore difficult to achieve solely with laminectomy and foraminotomy, and it is usually necessary to realign and stabilize the fracture.

Whereas unilateral injuries can be approached with a paramedian exposure to the foramina, which is where the neural compression typically occurs in unilateral injuries, the surgical exposure for bilateral injuries is preferably through a posterior midline approach. The paramedian approach is not recommended for anything more than spinopelvic fixation alone of bilateral injuries because it limits exposure to the spinal canal for decompression and fracture reduction and may result in the need for bilateral parasagittal exposures, which are even less desirable owing to the more limited access to the spinal canal from these approaches and the potential for soft-tissue necrosis. Intraoperative fluoroscopy is useful for orientation and to assess alignment and decompression of the spinal canal. Decompression can be performed focally for selected areas with ventral foraminal impingement or to achieve a more comprehensive decompression of the S1—4 nerve roots. In the instance of L5 root entrapment between the L5 transverse process and an alar bone fragment, decompression is performed by following the root laterally onto the shoulder of the ala and removing the offending fragment.

The possibility of significant epidural bleeding during direct decompression requires experience with techniques of epidural hemostasis. Similarly, experience with neural repair is essential and dural tears that are encountered should be repaired if possible. The dura at the sacral level is often relatively thin and friable, and although direct suture repair is optimal, it is often not possible to do so owing to the severity of the disruption, thus requiring the use of a patch. In many cases the disruption is through individual nerve roots caudal to the termination of the dural sac at the S2 level and is therefore irreparable.

Fracture reduction techniques

Unilateral Injuries: Unilateral, vertical AO Type B sacral fractures are most commonly treated with closed reduction using distal femoral traction, and percutaneous iliosacral or transiliac–transsacral screw fixation in the supine position. In more highly displaced fractures in which an acceptable closed reduction cannot be achieved, or if foraminal nerve root compression requires an open decompression, an open reduction can be performed, typically in the prone position through a posterior paramedian approach. Exposure is typically performed as far medially as the spinous process of the sacrum, on which one end of a reduction clamp can be hooked to provide medial–lateral fracture compression as the other end of the clamp is applied to the lateral aspect of the ilium through a small fascial window. Once the fracture is exposed, the fracture edges are debrided of soft tissue and foraminal debris can often be removed through the fracture gap. Reduction is then achieved by identifying fracture lines that correspond to each other. This can be particularly challenging in more comminuted fractures (Fig. 12.12), but usually appropriate length can be assessed at the sciatic notch by either palpation or direct visualization. Once appropriate length has been established the fracture is realigned and reduction is achieved with clamps placed between the sacral spinous processes and the ilium. Fixation is then undertaken percutaneously, usually with iliosacral of transiliac–

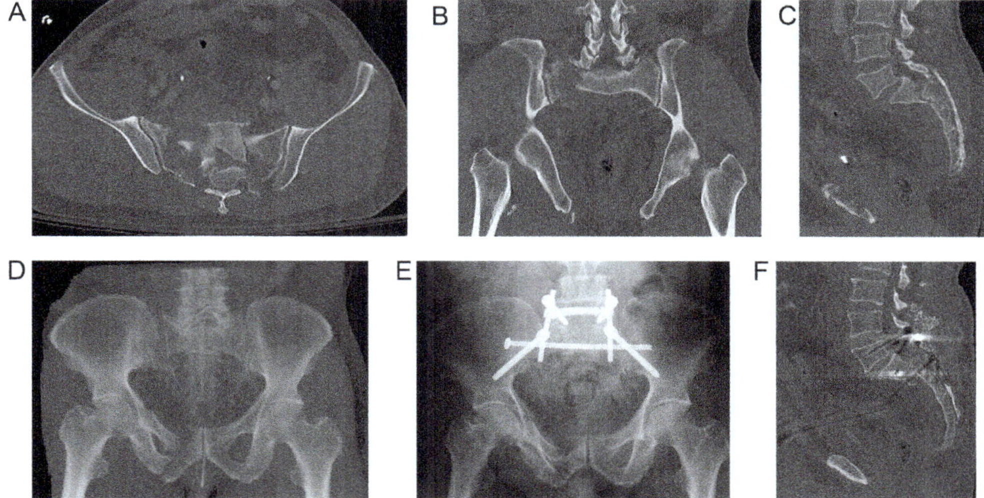

FIGURE 12.12 AO Type C3 M3 sacral fracture with segmental comminution of the sacrum. Axial (A), coronal (B), and sagittal (C) CT imaging and (D) anteroposterior CT reconstruction demonstrate an AO Type C3 sacral fracture with segmental comminution of the sacrum and anterior pelvic ring injury. Segmentally comminuted fractures were added by Schildhauer and coauthors[30] as a separate "Type 5" variant to the Roy-Camille classification,[22] due to the additional challenges involved in achieving and maintaining sacral alignment, as demonstrated by persistent translation of the sacral fracture on postoperative sagittal CT imaging (E), albeit with acceptable decompression and restoration of angular alignment. Extensive comminution required a primarily radiographic assessment of posterior pelvic alignment intraoperatively, rather than relying on directly visualizing the apposition of fracture lines. An anteroposterior radiograph (F) demonstrates progressive fracture healing with acceptable spinopelvic alignment 4 months after open reduction, decompression, anterior pelvic fixation, and posterior transiliac–transsacral and lumbopelvic fixation.

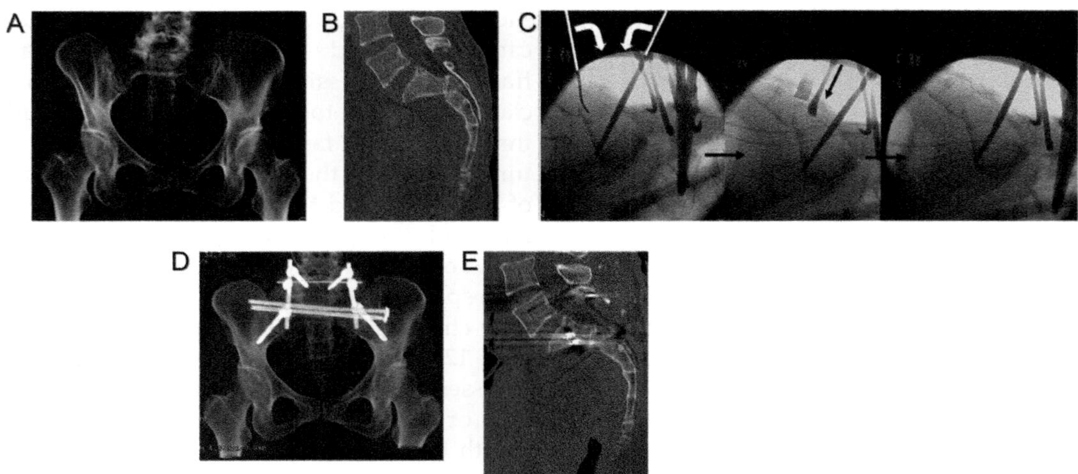

FIGURE 12.13   Open reduction and stabilization of AO Type C3 (Roy-Camille Type 2) sacral fracture. (A) Despite severe sacral fracture displacement with cauda equina syndrome after a fall, anteroposterior radiograph of the pelvis appears deceptively benign in the absence of pelvic ring deformity. (B) Sagittal CT image illustrates kyphosis, retrolisthesis, and spinal canal compromise consistent with Roy-Camille Type 2 variant of AO Type C3 fracture. Note the distended urinary bladder due to sacral root injury. (C) Intraoperative radiographs demonstrate reduction techniques which include mobilization of impacted fracture fragments with an elevator and direct manipulation of the fracture fragments with Schanz pin joystick placed in the upper sacral fracture fragment. (Arrows indicate the direction of reduction forces applied to the instrument). Once the fracture had been reduced, transiliac–transsacral screw fixation was performed for provisional stability, followed by spinopelvic fixation, as illustrated on postoperative (D) anteroposterior CT reconstruction. (E) Sagittal CT images of the sacrum demonstrate restoration of acceptable alignment and decompression of the sacral spinal canal and foramina. This patient regained sacral root function.

transsacral screws. Unilateral spinopelvic fixation can be applied in patients in whom there are concerns for severe instability or insufficient iliosacral or transiliac–transsacral screw fixation. Additional decompression via foraminotomy is performed if still deemed necessary.

Bilateral Injuries: In displaced sacral "U" fracture variants (AO Type C3 injuries), spinal canal and foraminal compromise due to translation and angulation are often the main cause of nerve root compromise. In these cases, decompression, realignment, and stabilization are required. Realignment and stabilization of these fractures can be difficult, and typically involve: 1. restoration of fracture length; 2. mobilization of impacted fracture fragments; 3. fracture reduction by direct manipulation; 4. provisional fracture stabilization with iliosacral or transiliac–transsacral screws; 5. definitive fracture stabilization with spinopelvic

fixation (Figs. 12.13 and 12.14). Fracture length can be obtained through a combination of techniques, including the use of distal bifemoral traction or direct manipulation with various instruments, such as the universal distractor secured to the L5 pedicle and ipsilateral ilium (Fig. 12.15). The use of specialized fracture reduction tables designed for the treatment of pelvic fractures may also assist with reduction. Once length is obtained and access to the anterior column of the transverse fracture has been achieved by laminectomy and sacral root retraction, elevators or a lamina spreader can be used to pry the impacted fracture fragments apart (Figs. 12.13 and 12.14). A combination of instruments can then be used to reduce the fracture, such as a threaded Schanz pin placed into the upper sacral body, which can be used as a joystick to help realign the fracture (Figs. 12.13 and 12.14). The kyphotic segment

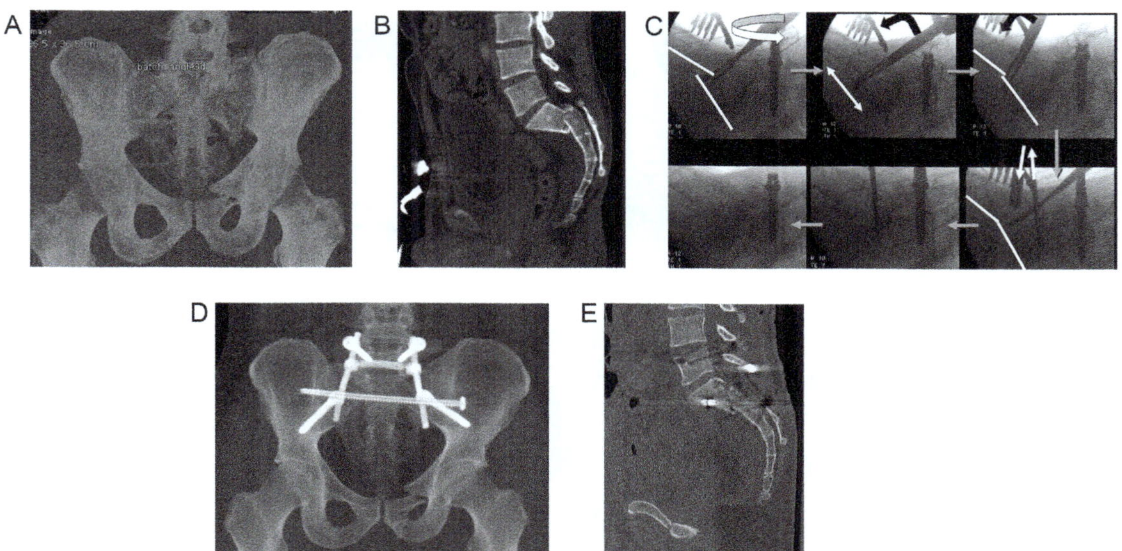

FIGURE 12.14 Open reduction and stabilization of AO Type C3, M3 (Roy-Camille Type 3) sacral fracture. (A) Anteroposterior CT reconstruction of the pelvis and (B) sagittal CT image of the sacrum demonstrate Roy-Camille Type 3 variant of AO Type C3 sacral fracture—characterized by anterior translation at the transverse sacral fracture—with anterior pelvic ring injury and severe urinary bladder distension caused by sacral root injury. (Arrows indicate the direction of reduction forces applied to the instrument). (C) Fracture reduction required prying open the transverse fracture with elevators, followed by fracture manipulation with a Schanz pin inserted as a joystick into the upper sacral body, and direct pressure applied to the posterior lower sacrum by a spike-pusher. Once the fracture had been reduced, a transiliac–transsacral screw was placed for provisional stability, followed by spinopelvic fixation, as illustrated on postoperative (D) anteroposterior CT reconstruction. (E) Postoperative sagittal CT image of the sacrum demonstrates restoration of sagittal plane alignment with both indirect and direct decompression of the spinal canal. Anatomic reduction is typically easier to achieve with Roy-Camille Type 3 compared to Type 2 fractures. This patient regained sacral root function.

may further be reduced by retracting the upper sacral nerve roots and using a bone impactor to directly reduce the apex of angulation by applying an anteriorly directed force to the posterior cortex of the sacral body. Reduction of the bilateral vertical sacral fracture components can be performed in a manner similar to that described above for unilateral sacral fractures. Once reduction is achieved, provisional stabilization can be obtained with either iliosacral or transiliac–transsacral screw fixation (Figs. 12.13 and 12.14). In some situations, particularly when the patient's anatomy allows for placement of multiple screws, this may provide enough stability. However, spinopelvic fixation provides additional stability and

is typically used for definitive fixation of AO C1 through C3 injuries. The typical reduction techniques, which differ slightly due to the differences in fracture displacement and angulation, are illustrated for the Roy-Camille type 2 (Fig. 12.13) and type 3 (Fig. 12.14) variants of AO Type C3 sacral fractures. In situations where anatomic reduction of the sacral kyphosis cannot be achieved or maintained, as is common with Roy-Camille type 2 variants of AO Type C3 fractures in particular, sacral root decompression within the sacral canal can be enhanced by resection of the apex of the kyphosis along the anterior spinal canal, over which the sacral roots would otherwise be draped (Fig. 12.13B).

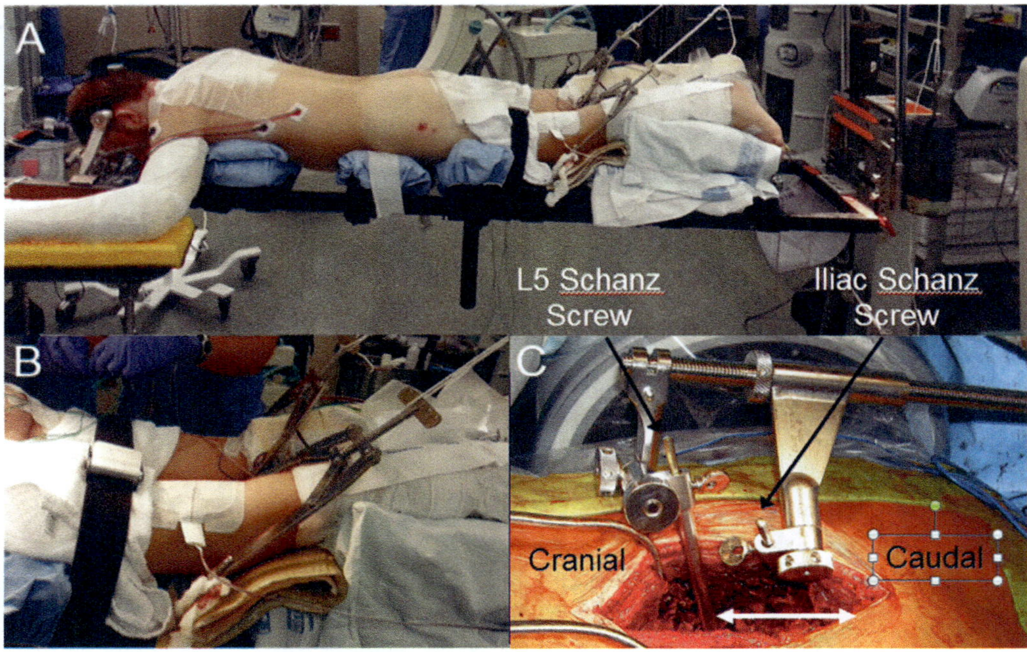

FIGURE 12.15   Restoration of fracture length is often the most challenging part of reducing complex sacral fractures. Techniques that can be used to restore fracture length at the spinopelvic junction include distal bifemoral traction with countertraction applied through Mayfield head clamp (A, B) and the use of a femoral distractor (C) anchored to the lumbar pedicles and the ipsilateral ilium.

Surgical stabilization techniques

The goal of surgical fixation is to avoid prolonged recumbency and to correct fracture malalignment, which may lead to postural difficulties, chronic pain, and nerve compression. Hart et al. reported that restoration of appropriate sagittal alignment of sacral fractures decreases pain by preventing compensatory lumbar hyperlordosis, allowing for a more physiologic alignment of the lumbar spine.[5] Therefore, pelvic incidence can be used as an intraoperative guide to spinopelvic alignment and adequacy of reduction. Lee and coauthors identified higher pelvic incidence in patients with malreduction of their sacral fracture that resulted in sagittal plane malalignment, which correlated with higher Oswestry Disability Index when compared to well-reduced postop spinopelvic dissociation patients with normal sagittal alignment and pelvic incidence (Fig. 12.1).[5]

There are three broad categories of fixation for sacral fractures: 1. direct osteosynthesis of the sacrum; 2. posterior pelvic ring stabilization; and 3. spinopelvic fixation. The appropriate fixation type is based on the fracture pattern and location. Optimal stabilization may require use of multiple sacral fixation techniques. The three situations most commonly encountered in this decision-making process involve: 1. low transverse sacral fracture below the sacroiliac joint, in which there are no implications on either posterior pelvic or spinopelvic instability; 2. vertical sacral fracture with unilateral pelvic instability and stable spinopelvic junction; and 3. complex sacral fracture with spinopelvic and usually also posterior pelvic instability (e.g., sacral U fracture variant or bilateral vertical sacral fracture). These three categories correspond, respectively, to the A, B, and C fracture types described in the AO sacral fracture classification, which therefore helps guide treatment (Fig. 12.10).

AO Sacral Fracture Classification Type A Injuries — Sacral fractures without posterior pelvic or spinopelvic instability: A few fixation options have been described to treat isolated transverse or oblique sacral fractures without pelvic involvement. Roy-Camille[26] described direct osteosynthesis of sacral fractures with sacral alar plates placed lateral to the dorsal foramina and oriented vertically. The orientation of these plates is theoretically optimal to allow for compression loading across a transverse fracture. Although not recommended for stabilization of spinopelvic dissociation injuries, direct plating alone may be useful in maintaining alignment of transverse sacral fractures below the sacroiliac joints because these fractures are not subject to the high loads seen with fractures involving the weight-bearing axis, and the goal of their treatment is primarily to avoid pain due to prominence or nonunion.[65] It can be combined with spinopelvic fixation techniques if there is a need to maintain alignment of the lower sacrum while having to stabilize a contiguous spinopelvic dissociation injury. Sacral alar plating alone therefore has few clinical applications. Its utility is usually in combination with other methods of surgical stabilization in the treatment of sacral fractures involving the weight-bearing axis.[73]

AO Sacral Fracture Classification Type B Injuries — Sacral fractures with posterior pelvic but not spinopelvic instability: Displaced unilateral vertical sacral fractures are generally treated with posterior pelvic stabilization. Sacral bars, tension band plating, iliosacral screws, and, most recently, transiliac–transsacral screws have all been described as a means of stabilizing the posterior pelvic ring.[74–76] Simonian and

Routt,[77] in a biomechanical study of cadaveric specimens, found no difference in the resulting pelvic ring stability between the various constructs. Sacral bars and tension band plating are both predisposed to soft-tissue complications, owing to their location superficial to the nearly subcutaneous dorsal sacrum.[75] The potential for dorsal soft-tissue compromise is minimized by the use of iliosacral or transiliac–transsacral screw fixation, and these methods have become the most widely used for posterior pelvic ring fixation. In all methods of posterior pelvic ring stabilization, however, fixation is perpendicular to the weight-bearing axis and does not provide optimal stabilization for immediate full weight bearing.[76,78,79]

Iliosacral and transiliac–transsacral screw fixation are usually performed percutaneously, with minimal blood loss. The success of an entirely percutaneous approach depends on the surgeon's ability to achieve an acceptable closed reduction under fluoroscopic visualization; otherwise, in addition to the problem of malreduction, a safe trajectory for iliosacral/transiliac–transsacral screw placement cannot be reliably established.[59,80] These techniques can be performed with the patient in either a supine or prone position depending in part on surgeon preference and on the potential need for either concurrent anterior pelvic ring stabilization or open posterior reduction or decompression of foraminal debris. In spite of their advantages, iliosacral and transiliac–transsacral screw fixation have several potential pitfalls. When used for stabilization of comminuted longitudinal fractures through the neuroforamina (AO Type B3), overcompression of the foramina can occur if the screws are placed using interfragmentary compression, which can potentially lead to nerve root injury.[80] However, in experienced centers, the reported rate of neurologic injury is low, even without the use of electrodiagnostic monitoring.[81] Like other methods of closed reduction and posterior pelvic ring stabilization alone, percutaneous iliosacral and transiliac–transsacral screw fixation carry the disadvantage of not allowing for reduction of sacral angulation. Combining these techniques with open reduction can provide the necessary kyphocorrection, but may not provide adequate stability for preventing failure of fixation and recurrent deformity. In the case of sacral comminution or osteoporosis, transiliac–transsacral screws should be used in lieu of iliosacral screws since fixation into the contralateral ilium is likely to provide far better stability than fixation of an iliosacral screw into the compromised bone of the sacrum. As mentioned previously, the orientation of the screw perpendicular to the fracture's deforming forces may also contribute to a higher likelihood of fixation failure (Fig. 12.11). Nevertheless, iliosacral and transiliac–transsacral screw fixation have been shown to be effective in the stabilization of unstable longitudinal sacral fractures.[82] A 3-month period of protected weight bearing is recommended with iliosacral and transiliac–transsacral screw fixation.

AO Sacral Fracture Classification Type C Injuries — Sacral fractures with spinopelvic instability: Unlike sacral bars and posterior tension band plating, which are largely ineffective in stabilizing multiplanar sacral fractures with a transverse component, iliosacral and transiliac–transsacral screws can also be used successfully in patients with minimally displaced sacral "U" fracture variants, whether AO Type C0 or C3, as well as in the less common bilateral vertical sacral fractures (AO Type C2).[25] In a series of 13 patients who had fractures with little enough sacral angulation and displacement to allow for safe in situ screw placement, Nork and colleagues[83] found bilateral percutaneous iliosacral screw fixation to be safe and effective in treating minimally displaced sacral "U-type" fracture patterns. In our experience, this technique has been particularly well suited toward treatment of AO Type C0 sacral "U" variant insufficiency fractures (Fig. 12.9). However, this minimally invasive method does not allow for reduction

of sacral fracture angulation. Moreover, if a near-anatomic reduction cannot be achieved, the safe zone for iliosacral and transiliac–transsacral screw trajectory may be small or absent, thus also precluding its use in more complex and highly displaced injuries.[84] Iliosacral or transiliac–transsacral screw fixation alone are therefore not generally recommended for patients with displaced AO Type C3 (sacral "U") fractures and their variants if they are irreducible by closed manipulation or if neural decompression is required.

Lumbopelvic fixation provides the most rigid fixation of sacral fractures.[85] Sacral fracture fixation is obtained rostrally by pedicle screw fixation in the lumbosacral spine and caudally by screw fixation to the ilium.[86–96] The construct spans, and therefore unloads the sacrum and restores the integrity of load transfer from the lumbar spine to the pelvis (Fig. 12.16). Lumbopelvic fixation can be applied unilaterally (triangular osteosynthesis) to add additional stability to iliosacral/transiliac–transsacral screw fixation for highly displaced or comminuted unilateral AO Type B or AO Type C1 longitudinal sacral fractures (Figs. 12.7 and 12.12), or bilaterally for AO Type C2 and C3 spinopelvic dissociation injuries, and occasionally AO Type C0 injuries. The strength of the construct permits immediate weight bearing without the use of external bracing (Figs. 12.13 and 12.14). Percutaneous lumbopelvic fixation techniques are well suited to situations where the fracture is already well aligned and lumbopelvic fixation is required primarily for added stability, whereas open lumbopelvic fixation techniques are used by default for displaced fractures requiring open reduction and decompression (Fig. 12.17).

Bilateral open lumbopelvic fixation is most commonly indicated for AO Type C2, C3, and some C0 injuries, and is performed through a dorsal midline longitudinal approach with the patient in the prone position. Screw placement may be performed under fluoroscopic guidance or some other form of navigation to ensure

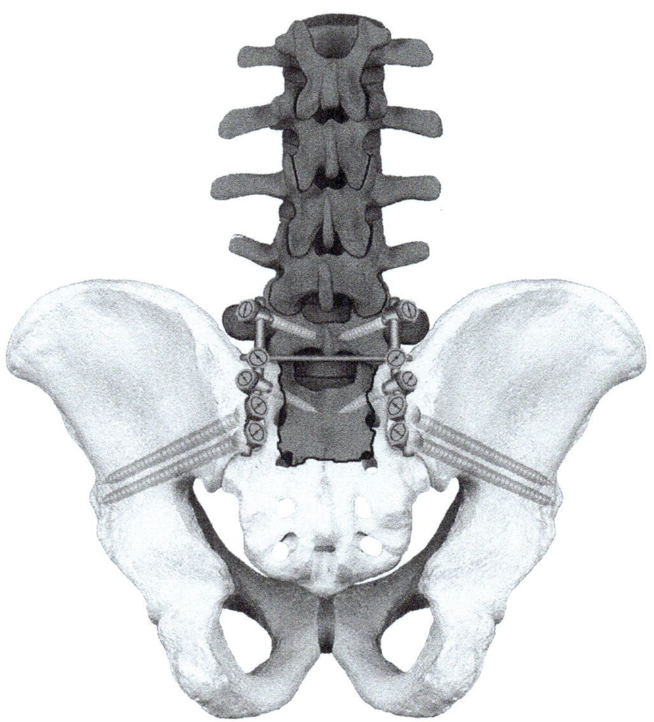

FIGURE 12.16 The two different shades illustrate the two primary fracture "fragments" with spinopelvic dissociation injuries. Lumbopelvic fixation constructs obtain fixation rostrally to the lumbar spine and caudally to the ilium. This allows for the fractured sacrum to be spanned and therefore unloaded, restoring the integrity of load transfer from the lumbar spine to the pelvis.

correct screw orientation. A thorough understanding of pelvic anatomy is necessary for verification of correct iliac screw placement. Screw malposition can be catastrophic, potentially injuring neurovascular structures, the pelvic viscera, or penetrating the acetabulum. Choice of a more anterior iliac screw starting point, such as along the medial (rather than posterior) surface of the posterior ilium, or on the dorsolateral surface of the sacrum, can help prevent excessive prominence of the iliac screw heads. Infection and wound-related problems are common, however, and have been found to approach 20%.[87,92] In situations where formal open reduction is not required but iliosacral/transiliac—transsacral screw fixation is considered to be insufficient, percutaneous lumbopelvic fixation can be used to provide enhanced stability while mitigating the risk of wound-related complications (Fig. 12.17). Formal lumbosacral arthrodesis is not typically performed in the absence of preexisting pathology (e.g., spondylolisthesis) or injury to the lumbosacral facet joints. Hardware removal is generally performed no less than 6 months postoperatively, after CT imaging has confirmed healing of the sacral fracture.[97] Hardware removal is not performed in patients who have had lumbosacral arthrodesis. Because formal sacroiliac joint arthrodesis is not usually performed, in many instances the rod will break above the iliac screw from fatigue failure, an expected consequence of continued sacroiliac joint motion. Late rod fracture is asymptomatic

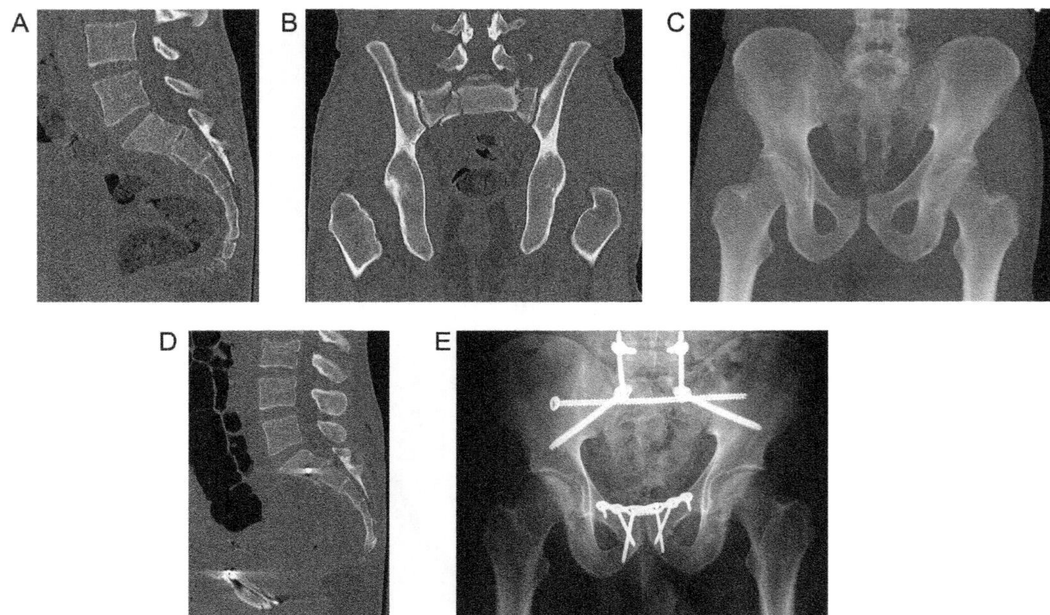

FIGURE 12.17   Percutaneous lumbopelvic fixation of AO Type C3 M1, M3 fracture. This pedestrian was struck by a motor vehicle traveling at moderate speed and sustained blunt trauma to the pelvis severe enough to cause spinopelvic and anterior pelvic injuries. The identification of a transverse fracture at the S1–S2 level on (A) sagittal CT, combined with bilateral transforaminal longitudinal fractures of the sacrum seen on (B) coronal CT images, is consistent with an AO Type C3 fracture. Anteroposterior (C) CT reconstructions demonstrate well-maintained overall pelvic architecture. After fixation of the anterior pelvis injury, percutaneous transiliac–transsacral and lumbopelvic fixation were performed, as demonstrated on (D) postoperative sagittal CT image and (E) anteroposterior radiograph of the pelvis 6 months postoperatively. Patients with spinopelvic instability who do not require open manipulative reduction or direct neurologic decompression, particularly when soft tissues are severely traumatized from blunt trauma, are ideal candidates for percutaneous lumbopelvic fixation.

in the majority of cases, and the need for routine hardware removal in all circumstances is questionable.[87,92]

The optimal stabilization of complex sacral fractures may require the use of a combination of the three categories of fixation described above. For instance, sacral "U" or "H" fractures can be stabilized with the use of iliosacral or transiliac–transsacral screw fixation combined with lumbopelvic fixation to obtain optimal fixation in the "horizontal" direction to stabilize the posterior pelvic ring and in the "vertical" direction along the weight-bearing axis to neutralize spinopelvic instability.[85,91] Lumbopelvic fixation of sacral fractures requiring neural decompression may also benefit from supplementary sacral plating, which in this case is used solely to fine-tune fracture realignment and to prevent recurrent displacement and resulting nerve root compression, allowing the lumbopelvic fixation to neutralize the bulk of the loads being transferred across the sacrum.

Complications

Because of the severity and multisystem nature of high-energy injuries,[98] complications associated with pelvic fractures are common and potentially severe. Gupta and coauthors found that sacral fractures with spinopelvic dissociation were associated with noncontiguous spine fractures at a rate of 86%, lower extremity fractures at a rate of 71%, contiguous pelvic ring fractures at a rate of 67%, and visceral injuries at a rate of 29%.[55] These multiply injured patients are susceptible to life-threatening nonorthopedic conditions such as adult respiratory distress syndrome, thromboembolic disease, pneumonia, and multisystem organ failure. Neurologic injuries and severe subcutaneous degloving lesions can add to the complexities in obtaining favorable outcomes when treating these challenging injuries.

Early complications

Deep and superficial infections have been, respectively, reported in 2%–8% and 3%–26% of operatively treated sacral fractures and are considerably more problematic after complex spinopelvic reconstructions than with percutaneous techniques.[87] Loss of reduction occurs in 0%–10% of patients.[73,87,99–101] Obese patients are 6.9 times more likely to have a complication and 4.7 times more likely to undergo reoperation than patients with BMI less than 30 kg/m^2.[102] Jaeblon and coauthors established a proxy for waist–hip ratio, based on the measurement of soft-tissue diameter at the waist and hips on both the anteroposterior and lateral CT scout views, which they found to be a better predictor than BMI of postoperative infection after pelvic and acetabular surgery.[103]

Infection

Postoperative infection after internal fixation usually occurs secondary to significant soft-tissue integrity or healing problems and correlates with patient comorbidities and iatrogenic factors.[104] Contemporary studies on the posterior approach to the pelvis demonstrate complication rates comparable to those reported with the anterior approach, with an incidence of deep wound infection ranging between 3.4% and 7.1%.[73,105,106] Postoperative infection in the

presence of internal fixation requires early incision and drainage plus debridement. Use of an incisional VAC dressing can accelerate secondary healing or the proliferation of a well-vascularized wound surface. Fixation can be evaluated intraoperatively for stability and, if stable, should generally be left in place at least until healing has occurred. If systemic infection is noted by either laboratory or clinical findings and is not responsive to intravenous antibiotics or if fixation is loose and therefore not contributing to stability, it must be removed and in most cases, also revised.

Loss of fixation

Loss of sacral fracture fixation occurs uncommonly, with an overall incidence of 0%–10%, and an average incidence of 5% with the use of anterior and posterior pelvic stabilization, and an even lower incidence with the use of lumbopelvic fixation.[73,87,99,100,105] Failure of iliosacral or transiliac–transsacral fixation can be treated by more biomechanically rigid fixation techniques such as spinopelvic fixation (Fig. 12.11).

Neurologic injury

Neurologic injuries occur in 8%–25% of pelvic fractures overall, increasing to 60% with sacroiliac joint separation and severe sacral fractures (AO Type C3).[11,21,101,107] The characteristic neurologic injury is a lumbosacral plexus injury in 80% of cases. Fractures through or medial to the sacral foramina are associated with a high incidence of neurologic injury, as are transverse fractures of the sacrum with a kyphotic deformity.[21] Reduction and stabilization of these pelvic injuries may improve recovery.[87,92,99] Decompression of a sacral transverse fracture with a kyphotic deformity or of a burst fracture of the sacrum that appears to compromise the roots posteriorly may be of some value.[92] However, there is no literature providing evidence-based guidelines to help dictate the timing of surgical decompression. Nerve recovery usually begins within 3 months of trauma and plateaus within 2 years. Complete recovery of cauda equina function is more likely in patients with continuity of all sacral roots and incomplete deficits. However, complete neurologic recovery in severe nerve injuries is uncommon.[92,107] Iatrogenic nerve injury may occur secondary to operative treatment. Its incidence has not been reliably decreased by intraoperative electrodiagnostic monitoring.[108]

Thromboembolism

Thromboembolic complications occur commonly in patients with a major pelvic disruption, especially if associated with lower extremity fractures. The incidence ranges from 12% to 61%, depending on the use of prophylaxis and of adequate diagnostic screening.[109,110] Pulmonary embolism is more frequent with pelvic injuries, and surgical fixation of the pelvis is an independent predictor of pulmonary embolism within the first 72 h of admission.[110,111] Despite the use of many different protocols for thromboembolism prophylaxis, however, no single approach has been reliably able to decrease the incidence of fatal pulmonary embolism.[112,113]

Late complications

Pain

Posttraumatic chronic pain is common and may have an identifiable source such as malunion, nonunion, or osteoarthritis of the SI joint. The source of pain, however, can often not reliably be identified. Approximately 25%—40% of patients have neuropathic pain, and 20%—40% have musculoskeletal pain.[114] There is a lower incidence of severe pain in operatively treated patients, averaging 1%—5% versus 27% in nonoperatively treated patients, which supports the hypothesis that anatomic or near-anatomic reduction improves the likelihood of a successful outcome with minimal pain.[105] Nonetheless, some patients continue to complain of discomfort despite having had anatomic reduction

and adequate fracture union. Careful evaluation of the lower lumbar spine must be carried out to ensure the absence of missed lumbosacral injuries, particularly lumbosacral facet injuries that are seen with relative frequency in association with pelvic fractures (Figs. 12.7 and 12.8).

Malunion

Sacral fracture malunion is a common problem that can be a source of chronic pain. It occurs in 30%—42% of pelvic fractures after nonoperative treatment or after solely anterior fixation but in only 7%—10% of cases after combined anterior and posterior fixation, probably because of more anatomic reduction and more stable fixation.[101,105] Both malunion and nonunion are extremely uncommon after spinopelvic fixation despite the complexity of the injuries for which this technique is used.[87,92]

Most clinical consequences of pelvic malunion stem from pelvic obliquity with accompanying sitting imbalance, compensatory scoliosis, relative leg length inequality, and secondary gait abnormalities. Deformities occurring through or adjacent to the SI joint appear to be the most disabling. Kyphotic malunion of AO Type C sacral fractures may result in severe sagittal plane malalignment requiring corrective osteotomy (Fig. 12.1). Complex deformity correction often requires a multistage osteotomy approach with significant surgical risk. Pain relief and deformity correction are usually incomplete, but in carefully selected patients, the likelihood of improvements may warrant the risks. Symptomatic malunion of the SI joint with modest deformity is a more straightforward problem that is usually treated by SI fusion.

Nonunion

Compared with malunion, nonunion is uncommon. Union rates range from 95% to 100% after operative fixation,[105] and 83% after nonoperative treatment.[99] Pelvic pain and instability are the most common symptoms of sacral nonunion.

Complete evaluation of the patient's symptoms and bony pelvic abnormalities is mandatory. The principles of surgical treatment are to achieve stable pelvic ring fixation in proper alignment with debridement and bone grafting of the nonunion. The approach differs depending on the location of the nonunion and the degree of associated pelvic or spinopelvic malalignment, but many cases may require takedown of the nonunion or osteotomy to allow for correction of the deformity followed by stable fixation both anteriorly and posteriorly, similar to that previously discussed for malunions.[115] Sacral fracture nonunions after iliosacral screw fixation techniques can be treated with debridement, bone grafting, and more rigid fixation such as spinopelvic techniques.

Reoperation

Reoperation rates of approximately 40% have been reported after operative treatment of more complex high-energy sacral fractures, many of the reasons for which have already been described. Among the more preventable reasons for reoperation is the need for hardware removal or revision due to prominence, particularly of the iliac screws, which are located in an area with poor soft-tissue coverage. Use of a starting point within the sacrum, between the S1 and S2 foraminal, or of a more anteromedial posterior iliac starting point may decrease the likelihood of symptomatic hardware prominence (Fig. 12.18).

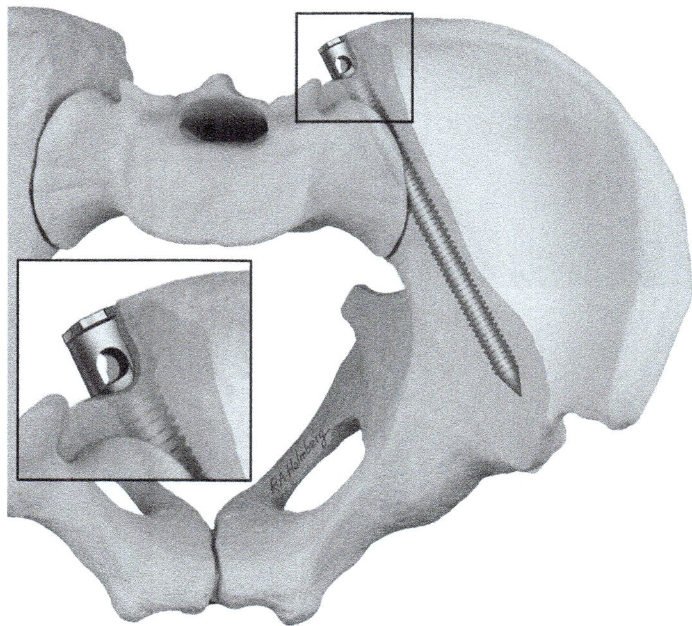

FIGURE 12.18   Use of a more anteromedial posterior iliac starting point that allows the iliac screw to be countersunk helps prevent symptomatic screw prominence, which is a significant cause of reoperation in patients who have had lumbopelvic fixation procedures.

Special considerations

Open sacral fractures

Open sacral fractures, which constitute approximately 2% of all sacral fractures, are severe, high-energy injuries to the bone and soft tissue that have a high association with infection, neurologic injury, and disability (Fig. 12.19). They are defined as any fracture that has the potential for bacterial contamination because of communication with the external environment, which includes the gastrointestinal and genitourinary tracts.[116] The mortality rate has been reported as approximately 20%, an improvement over the 30% mortality observed before 1990.[117,118] Factors that correlate with death include age, injury severity score (ISS), skeletal injury complexity, wound size, and transfusion requirements.[117] Early mortality is almost entirely due to exsanguinating hemorrhage. Late mortality is most commonly due to sepsis, but complications related to associated injuries in these polytraumatized patients, such as traumatic brain injury or respiratory dysfunction, are frequently implicated.[116] Treatment of these challenging injuries must be individualized to the specific situation, with appropriate hemodynamic resuscitation being the foremost priority. Treatment typically involves a coordinated multidisciplinary approach involving a combination of provisional skeletal stabilization, arterial embolization, damage control laparotomy, pelvic packing, wound debridement, fecal and urinary diversion, broad spectrum antibiotics, and definitive skeletal and soft-tissue reconstruction. Severe subcutaneous degloving injuries are common due to shearing and avulsion of the skin and subcutaneous tissue

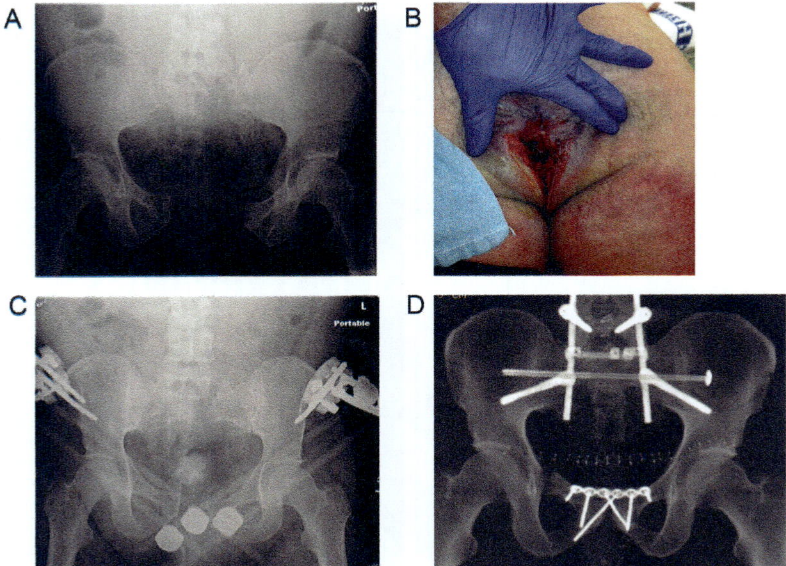

FIGURE 12.19   Open AO Type C3, M1, M3 fracture. (A) Anteroposterior radiograph of severely displaced spinopelvic dissociation with anterior pelvic ring injury, resulting in (B) laceration of the rectum and perianal region that communicated with the fracture. These injuries have high morbidity and mortality, requiring coordinated multidisciplinary care. (C) Anteroposterior radiograph of the pelvis after provisional reduction and external pelvic fixation, which allowed for serial debridement of the open injury and fecal diversion prior to definitive anterior pelvic and posterior transiliac—transsacral and lumbopelvic fixation, as shown with postoperative (D) anteroposterior CT reconstruction of the pelvis.

FIGURE 12.20 Severe subcutaneous degloving injury sustained with spinopelvic trauma, involving the entire torso and gluteal region from shoulders to greater trochanters.

from the underlying muscle fascia, which may require wide debridement of nonviable skin (Fig. 12.20). Neurologic injury is common, which further compromises survivors' long-term outcomes.[11,21,101,107]

Outcomes

Treatment outcomes in patients who sustain factures of the sacrum correlate inversely with the severity of the initial neurologic and musculoskeletal injuries, and directly with restoration of fracture alignment and of neutral sagittal balance.[5] The final functional outcome is usually determined by the associated soft-tissue injury or other nonorthopedic injuries.[67,68] Most authors agree that with anatomic or near-anatomic reduction, 60%—70% of pelvic fracture patients can be expected to have successful outcomes and that complications related to limb length discrepancy, sitting imbalance, and pelvic stability can be reliably avoided. Eighty percent of operatively treated and 68% of nonoperatively treated sacral fracture patients regain normal gait patterns. Return to employment has been reported as 68% without any significant difference in nonoperatively and operatively treated patients. Patients with sacral fractures tended to have worse outcomes despite good reduction because functional outcome was related to the associated nerve injury.[105]

Lindahl and coauthors found that outcomes specifically after spinopelvic dissociation correlated inversely with the severity of displacement of the transverse sacral fracture and of neurologic deficit, and correlated directly with the quality of the radiographic reduction, in particular with correction of the kyphotic deformity. Interestingly, they could not correlate outcomes with Roy-Camille fracture type, ISS, age of the patient, or timing of surgery. Although only 19% of the 36 patients with spinopelvic dissociation in their series recovered complete neurologic function after surgical decompression and stabilization, 94% recovered partially and approximately half of patients with acute cauda equina syndrome (Gibbons 4) recovered bowel and bladder function. They also found treatment outcomes in patients with insufficiency fractures treated nonoperatively versus with spinopelvic or transiliac—transsacral screw fixation has not been reliably assessed.[119]

Summary

Appropriate diagnosis and treatment of sacral fractures and their sequelae can have a profound impact on patients' functional outcomes.

Advancements in diagnostic imaging have improved our understanding and timely diagnosis of these injuries, particularly in cognitively impaired and polytraumatized patients. In situations with neurologic deficit, excessive malalignment or instability, long-term outcomes can be optimized with surgical intervention in the form of neurologic decompression, fracture realignment and surgical fixation. The timing of surgical intervention is contingent on several factors such as the patient's associated injuries, overall physiologic condition, need for further resuscitation, and the presence of either neurologic injury, open fractures, or severe soft-tissue compromise. Surgical options range from percutaneous to open comprehensive stabilization techniques. Many issues pertaining to the evaluation and treatment of patients with sacral fractures remain controversial and, in the absence of comparative treatment trials, conclusions are largely based on anecdotal reports and observations. Treatment decisions must therefore be individualized to a specific patient's situation, based on a combination of sound treatment principles and experience. The challenge to spine surgeons in the future will be to pool resources in order to improve our evidence-based approach to sacral fractures, and to deploy available resources effectively as our understanding of the pathophysiology of these injuries grows.

References

1. Pick TP, Howden R, editors. *Gray's anatomy, classic collector's edition*. New York: Crown Publishers — Bounty Books; 1977. p. 45—50.
2. Wiltse LL, Winter RB. Terminology and measurement of spondylolisthesis. *J Bone Joint Surg Am* 1983;**65**:768—72.
3. Legaye J, Duval-Beaupere G, Hecquet J, et al. Pelvic incidence: a fundamental pelvic parameter for three dimensional regulation of spinal sagittal curves. *Eur Spine J* 1998;**7**:99—103.
4. Hart RA, Badra MI, Madala A, Yoo JU. Use of pelvic incidence as a guide to reduction of H-type spinopelvic dissociation injuries. *J Orthop Trauma* 2007;**21**:369—74.
5. Lee HD, Jeon CH, Won SH, Chung NS. Global sagittal imbalance due to change in pelvic incidence after traumatic spinopelvic dissociation. *J Orthop Trauma* 2017;**31**:e195—9.
6. Schmidek HH, Smith D, Kristiansen TK. Sacral fractures: issues of neural injury, spinal stability, and surgical management. In: Dunsker SB, et al., editors. *The unstable spine*. New York: Harcourt; 1986.
7. Routt ML, Simonian PT, Agnew SG, Mann FA. Radiographic recognition of the sacral alar slope for optimal placement of iliosacral screws: a cadaveric and clinical study. *J Orthop Trauma* 1996;**10**:171—7.
8. Peretz AM, Hipp JA, Heggeness MH. The internal bony architecture of the sacrum. *Spine* 1998;**23**:971—4.
9. Smith SA, Abitbol JJ, Carlson GD, et al. The effects of depth of penetration, screw orientation and bone density on sacral screw fixation. *Spine* 1993;**18**:1006—10.
10. Carter SR. Stress fracture of the sacrum: brief report. *J Bone Joint Surg* 1987;**69**:843—84.
11. Denis F, Davis S, Comfort T. Sacral fractures: an important problem-a retrospective analysis of 236 cases. *Clin Orthop* 1988;**227**:67—81.
12. Mirkovic S, Abitbol JJ, Steinman J, et al. Anatomic consideration for sacral screw placement. *Spine* 1991;**16**:S289—94.
13. Esses SI, Botsford DJ, Huler RJ, Rauschning W. Surgical anatomy of the sacrum: a guide for rational screw fixation. *Spine* 1991;**16**:S283—8.
14. Gunterberg B. Effects of major resection of the sacrum. *Acta Orthop Scand Suppl* 1976;**162**:1—38.
15. Pick TP, Howden R, editors. *Gray's anatomy, classic collector's edition*. New York: Crown Publishers — Bounty Books; 1977. p. 558—71.
16. Wagner D, Ossendorf C, Gruszka D, Hofmann A, Rommens PM. Fragility fractures of the sacrum: how to identify and when to treat surgically. *Eur J Trauma Emerg Surg* 2015;**41**:349—62.
17. Bonnin JG. Sacral fractures. *J Bone Joint Surg Am* 1945;**27**:113—27.
18. Sabiston CP, Wing PC. Sacral fractures. Classification and neurologic implications. *J Trauma* 1986;**26**:1113—5.
19. Schmidek HH, Smith DA, Kristiansen TK. Sacral fractures. *Neurosurgery* 1984;**15**:735—46.
20. Khan JM1, Marquez-Lara A, Miller AN. Relationship of sacral fractures to nerve injury: is the Denis classification still accurate? *J Orthop Trauma* April 2017;**31**(4):181—4.
21. Gibbons KJ, Soloniuk DS, Razack N. Neurological injury and patterns of sacral fractures. *J Neurosurg* 1990;**72**:889—93.
22. Fountain SS, Hamilton RD, Jameson RM. Transverse fractures of the sacrum: a report of six cases. *J Bone Joint Surg Am* 1977;**59**:486—9.

23. Rao SH. Traumatic transverse fracture of sacrum with cauda equina injury: a case report and review of literature. *J Postgrad Med* 1998;**44**:14–5.

24. Phelan ST, Jones DA, Bishay M. Conservative management of transverse fractures of the sacrum with neurological features: a report of four cases. *J Bone Joint Surg Br* 1991;**73**:969–71.

25. Bishop JA, Dangelmajer S, Corcoran-Schwartz I, Gardner MJ, Routt MLC, Castillo TN. Bilateral sacral ala fractures are strongly associated with lumbopelvic instability. *J Orthop Trauma* 2017;**31**:636–9.

26. Roy-Camille R, Saillant G, Gagna G, Mazel C. Transverse fracture of the upper sacrum: suicidal jumper's fracture. *Spine* 1985;**10**:838–45.

27. Strange-Vognsen HH, Lebech A. An unusual type of fracture in the upper sacrum. *J Orthop Trauma* 1991;**5**:200–3.

28. Bellabarba C, Bransford RJ. Spinopelvic fixation. In: Vialle LR, Bellabarba C, Kandziora F, editors. *AOSpine masters series, volume 6: thoracolumbar spine trauma.* New York: Thieme; 2016. p. 152–83.

29. Vaccaro AR, Schroeder GD, Divi SM, Kepler CK, Kleweno CP, Krieg JC, et al. Description and reliability of the AOSpine sacral classification system. *J Bone Joint Surg Am* August 19, 2020;**102**(16):1454–63. https://doi.org/10.2106/JBJS.19.01153.

30. Schildhauer TA, Chapman JR, Mayo KA. Multisegmental open sacral fracture due to impalement: a case report. *J Orthop Trauma* 2005;**19**:134–9.

31. Bellabarba C, Stewart JD, Ricci WM, et al. Midline sagittal sacral fractures in anterior-posterior compression ring injuries. *J Orthop Trauma* 2003;**17**:32–7.

32. Hatem SF, West OC. Vertical fracture of the central sacral canal: plane and simple. *J Trauma* 1996;**40**:138–40.

33. Wiesel SW, Zeide MS, Terry RL. Longitudinal fracture of the sacrum: case report. *J Trauma* 1979;**19**:70–1.

34. Isler B. Lumbosacral lesions associated with pelvic ring injuries. *J Orthop Trauma* 1990;**4**:1–6.

35. Herman MP, Kopetz S, Bhosale PR, et al. Sacral insufficiency fractures after preoperative chemoradiation for rectal cancer: incidence, risk factors, and clinical course. *Int J Radiat Oncol Biol Phys* 2009;**74**:818–23.

36. Gotis-Graham I, McGuigan L, Diamond T, et al. Sacral insufficiency fractures in the elderly. *J Bone Joint Surg Br* 1994;**76**:882–6.

37. Weber M, Haster P, Gerber H, et al. Insufficiency fractures of the sacrum: twenty cases and review of the literature. *Spine* 1993;**18**:2507–12.

38. Jacquot JM, Finiels H, Fardjad S, et al. Neurological complications in insufficiency fractures of the sacrum: three case reports. *Rev Rhum Engl Ed* 1999;**66**:109–13.

39. Wood KB, Geissele AE, Ogilvie JW. Pelvic fractures after long lumbosacral spine fusions. *Spine* 1996;**21**:1357–62.

40. Elias WJ, Shaffrey ME, Whitehill R. Sacral stress fracture following lumbosacral arthrodesis: case illustration. *J Neurosurg* 2002;**96**(Suppl. 1):135.

41. Fourney DR, Prabhu SS, Cohen ZR, et al. Early sacral stress fractures after reduction of spondylolisthesis and lumbosacral fixation: case report. *Neurosurgery* 2002;**51**:1510–1.

42. Klineberg E, McHenry T, Bellabarba C, et al. Sacral insufficiency fractures caudal to instrumented posterior lumbosacral arthrodesis. *Spine* 2008;**33**:1806–11.

43. Mathews V, McCance SE, O'Leary PF. Early fracture of the sacrum or pelvis: an unusual complication after multilevel instrumented lumbosacral fusion. *Spine* 2001;**26**:E571–5.

44. Wood KB, Schendel MJ, Ogilvie JW, et al. Effect of sacral and iliac instrumentation on strains in the pelvis. *Spine* 1996;**21**:1185–91.

45. Bellabarba C, Schroeder GD, Kepler CK, et al. The AOSpine sacral fracture classification. *Global Spine J* 2016;**6** (suppl 1):s-0036-1582696.).

46. Schroeder GD, Kurd MF, Kepler CK, Krieg JC, Wilson JR, Kleweno CP, et al. The development of a universally accepted sacral fracture classification: a survey of AOSpine and AOTrauma members. *Global Spine J* November 2016;**6**(7):686–94.

47. Urrutia J, Meissner-Haecker A, As-tur N, Valencia M, Yurac R, Camino-Willhuber G, Valacco M. An independent inter- and intra-observer agreement assessment of the AOSpine sacral fracture classification system Spine. *J* February 9, 2021:S1529–9430.

48. Laasonen EM. Missed sacral fractures. *Ann Clin Res* 1977;**9**:84–7.

49. Dodwad SNM, Niedermeier SR, Yu E, Ferguson TA, Klineberg EO, Khan SN. The morel-lavallée lesion revisited: management in spinopelvic dissociation. *Spine J* 2015;**15**:e45–51.

50. Ben-Menachem Y, Coldwell DM, Young JW, Burgess AR. Hemorrhage associated with pelvic fractures: causes, diagnosis, and emergent management. *AJR Am J Roentgenol* 1991;**157**:1005–14.

51. Byrnes DP, Russo GL, Duckert TB, Cowley RA. Sacrum fractures and neurological damage: report of two cases. *J Neurosurg* 1977;**47**:459–62.

52. Goodell CL. Neurological deficits associated with pelvic fractures. *J Neurosurg* 1966;**24**:837–42.

53. Lam CR. Nerve injury in fractures of the pelvis. *Ann Surg* 1936;**104**:945–51.

54. American spinal injury association international standards for neurological classification of spinal cord injury.

55. Gupta P, Barnwell JC, Lenchik L, Wuertzer SD, Miller AN. Spinopelvic dissociation: multidetector computed tomographic evaluation of fracture patterns and associated injuries at a single level 1 trauma center. *Emerg Radiol* 2016;**23**:235−40.

56. Kuklo TR, Potter BK, Ludwig SC, et al. Spine trauma study group: radiographic measurement techniques for sacral fractures consensus statement of the spine trauma study group. *Spine* 2006;**31**:1047−55.

57. Sasaka KK, Phisitkul P, Boyd JL, et al. Lumbosacral nerve root avulsions: MR imaging demonstration of acute abnormalities. *AJNR Am J Neuroradiol* 2006;**27**:1944−6.

58. Bone LB, Johnson KD, Weigelt J, Scheinberg R. Early versus delayed stabilization of femoral fractures: a prospective randomized study. *J Bone Joint Surg Am* 1989;**71**:336−40.

59. Johnson KD, Cadambi A, Seibert GB. Incidence of adult respiratory distress syndrome in patients with multiple musculoskeletal injuries: effect of early operative stabilization of fractures. *J Trauma* 1985;**25**:375−84.

60. Jaicks RR, Cohn SM, Moller BA. Early fracture fixation may be deleterious after head injury. *J Trauma* 1977;**42**:1−5.

61. Scalea TM, Scott JD, Brumback RJ, et al. Early fracture fixation may be "just fine" after head injury: no difference in central nervous system outcomes. *J Trauma* 1999;**46**:839−46.

62. Marcus RE, Hansen ST. Bilateral fracture-dislocation of the sacrum: a case report. *J Bone Joint Surg Am* 1984;**66**:1297−9.

63. Pennal GF, Tile M, Waddell JP, Garside H. Pelvic disruption: assessment and classification. *Clin Orthop* 1980;**151**:12−21.

64. Wild J, Hanson GW, Tullas HS. Unstable fractures of the pelvis treated by external fixation. *J Bone Joint Surg Am* 1982;**64**:1010−20.

65. Sommer C. Fixation of transverse fractures of the sternum and sacrum with the locking compression plate system: two case reports. *J Orthop Trauma* 2005;**19**:487−90.

66. Fisher RG. Sacral fracture with compression of cauda equina: surgical treatment. *J Trauma* 1998;**28**:1678−80.

67. Tötterman A, Glott T, Søberg HL, et al. Pelvic trauma with displaced sacral fractures: functional outcome at one year. *Spine* 2007;**32**:1437−43.

68. Tötterman A, Glott T, Madsen JE, Røise O. Unstable sacral fractures: associated injuries and morbidity at 1 year. *Spine* 2006;**31**:E628−35.

69. Latenser BA, Gentilello LM, Tarver AA, et al. Improved outcome with early fixation of skeletally unstable pelvic fractures. *J Trauma* 1991;**31**:28−31.

70. Wang MY, Ludwig SC, Anderson DG, Mummaneni PV. Percutaneous iliac screw placement: description of a new minimally invasive technique. *Neurosurg Focus* 2008;**E17**. https://doi.org/10.3171/FOC/2008/25/8/E17.

71. Lindtner RA, Bellabarba C, Firoozabadi R, et al. Should displaced sacral fractures Be treated by an orthopedic traumatologist or a spine surgeon? *Clin Spine Surg* 2016;**29**:173−6.

72. Pohlemann T, Angst A, Schneider E, et al. Fixation of transforaminal sacrum fractures: a biomechanical study. *J Orthop Trauma* 1993;**2**:107−17.

73. Templeman D, Goulet J, Duwelius PJ, et al. Internal fixation of displaced fractures of the sacrum. *Clin Orthop* 1996;**329**:180−5.

74. Routt Jr ML, Simonian PT. Closed reduction and percutaneous skeletal fixation of sacral fractures. *Clin Orthop* 1996;**329**:121−8.

75. Suzuki T, Hak DJ, Ziran BH, et al. Outcome and complications of posterior transiliac plating for vertically unstable sacral fractures. *Injury* 2009;**40**:405−9.

76. Gardner MJ, Routt Jr ML. Transiliac-transsacral screws for posterior pelvic stabilization. *J Orthop Trauma* 2011;**25**:378−84.

77. Simonian PT, Routt Jr ML. Biomechanics of pelvic fixation. *Orthop Clin N Am* 1997;**28**:351−68.

78. Suzuki K, Mochida J. Operative treatment of a transverse fracture-dislocation at the S1-S2 level. *J Orthop Trauma* 2001;**15**:363−7.

79. Taguchi T, Kawai S, Kaneko K, Yugue D. Operative management of displaced fractures of the sacrum. *J Orthop Sci* 1999;**4**:347−52.

80. Routt MLC, Simonian PT, Swiontkowski MF. Stabilization of pelvic ring disruptions. *Orthop Clin N Am* 1997;**28**:369−88.

81. Gardner MJ, Farrell ED, Nork SE, et al. Percutaneous placement of iliosacral screws without electrodiagnostic monitoring. *J Trauma* 2009;**66**:1411−5.

82. Routt MLC, Nork SE, Mills WJ. Percutaneous fixation of pelvic ring disruptions. *Clin Orthop* 2000;**375**:15−29.

83. Nork SE, Jones CB, Harding SP, et al. Percutaneous stabilization of U-shaped sacral fractures using iliosacral screws: technique and early results. *J Orthop Trauma* 2001;**15**:238−46.

84. Reilly MC, Bono CM, Litkouhi B, et al. The effect of sacral fracture malreduction on the safe placement of iliosacral screws. *J Orthop Trauma* 2003;**17**:88−94.

85. Schildhauer TA, Ledoux WR, Chapman JR, et al. Triangular osteosynthesis and iliosacral screw fixation for unstable sacral fractures: a cadaveric and biomechanical evaluation under cyclic loads. *J Orthop Trauma* 2003;**17**:22−31.

86. Acharya NK, Bijukachhe B, Kumar RJ, Menon VK. Iliolumbar fixation−the Amrita technique. *J Spinal Disord Tech* 2008;**21**:493−9.

87. Bellabarba C, Schildhauer TA, Vaccaro AR, Chapman JR. Complications associated with surgical stabilization of high-grade sacral fracture dislocations with spino-pelvic instability. *Spine* 2006;**31**(11 Suppl. l):S80−8.

88. O'Brien JR, Yu WD, Bhatnagar R, Sponseller P, Kebaish KM. An anatomic study of the S2 iliac technique for lumbopelvic screw placement. *Spine* 2009;**34**:E439−42.

89. Sagi HC. Technical aspects and recommended treatment algorithms in triangular osteosynthesis and spinopelvic fixation for vertical shear transforaminal sacral fractures. *J Orthop Trauma* 2009;**23**:354−60.

90. Schildhauer TA, McCullough P, Chapman JR, Mann FA. Anatomic and radiographic considerations for placement of transiliac screws in lumbopelvic fixations. *J Spinal Disord Tech* 2002;**15**:199−205.

91. Schildhauer TA, Josten C, Muhr G. Triangular osteosynthesis of vertically unstable sacrum fractures: a new concept allowing early weight-bearing. *J Orthop Trauma* 1998;**12**:307−14.

92. Schildhauer TA, Bellabarba C, Nork SE, et al. Decompression and lumbopelvic fixation for sacral fracture-dislocations with spino-pelvic dissociation. *J Orthop Trauma* 2006;**20**:447−57.

93. Schildhauer TA, Bellabarba C, Selznick HS, et al. Unstable pediatric sacral fracture with bone loss caused by a high-energy gunshot injury. *J Trauma* 2007;**63**:E95−9.

94. Strange-Vognsen HH, Kiaer T, Tondevold E. The Cotrel-Dubousset instrumentation for unstable sacral fractures: report of 3 patients. *Acta Orthop Scand* 1994;**65**:219−20.

95. Vilela MD, Gelfenbeyn M, Bellabarba C. U-shaped sacral fracture and lumbosacral dislocation as a result of a shotgun injury: case report. *Neurosurgery* 2009;**64**:E193−4.

96. Lebwohl NH, Cunningham BW, Dmitriev A, et al. Biomechanical comparison of lumbosacral fixation techniques in a calf spine model. *Spine* 2002;**27**:2312−20.

97. König MA, Seidel U, Heini P, et al. Minimal-invasive percutaneous reduction and transsacral screw fixation for U-shaped fractures. *J Spinal Disord Tech* 2013;**26**:48−54.

98. Rodrigues-Pinto R, Kurd MF, Schroeder GD, Kepler CK, Krieg JC, Holstein JH, Bellabarba C, Firoozabadi R, Oner FC, Kandziora F, Dvorak MF, Kleweno CP, Vialle LR, Rajasekaran S, Schnake KJ, Vaccaro AR. Sacral fractures and associated injuries. *Global Spine J* 2017;**7**:609−16.

99. Matta JM, Saucedo T. Internal fixation of pelvic ring fractures. *Clin Orthop Relat Res* 1989;**242**:83−97.

100. Kabak S, Halici M, Tuncel M. Functional outcome of open reduction and internal fixation for completely unstable pelvic ring fractures (type C): a report of 40 cases. *J Orthop Trauma* 2003;**17**:555−62.

101. Lindahl J, Hirvensalo E. Outcome of operatively treated type-C injuries of the pelvic ring. *Acta Orthop* 2005;**76**:667−78.

102. Sems SA, Johnson M, Cole PA, et al. Elevated body mass index increases early complications of surgical treatment of pelvic ring injuries. *J Orthop Trauma* 2010;**24**(5):309−14.

103. Jaeblon T, Perry KJ, Kufera JA. Waist-hip ratio surrogate is more predictive than body mass index of wound complications after pelvic and acetabulum surgery). *J Orthop Trauma* 2018;**32**(4):167−73.

104. Fowler TT, Bishop JA, Bellino MJ. The posterior approach to pelvic ring injuries: a technique for minimizing soft tissue complications. *Injury* 2013;**44**(12):1780−6.

105. Papakostidis C, Kanakaris NK, Kontakis G, et al. Pelvic ring disruptions: treatment modalities and analysis of outcomes. *Int Orthop* 2009;**33**(2):329−38.

106. Stover MD, Sims SH, Matta JM. What is the infection rate of the posterior approach to type C pelvic injuries? *Clin Orthop Relat Res* 2012;**470**:2142−7.

107. Majeed SA. Neurologic deficits in major pelvic injuries. *Clin Orthop Relat Res* 1992;**282**:222−38.

108. Helfet DL, Koval KJ, Hissa EA, et al. Intraoperative somatosensory evoked potential monitoring during acute pelvic fracture surgery. *J Orthop Trauma* 1995;**9**:28−34.

109. Lapidus LJ, Ponzer S, Pettersson H, et al. Symptomatic venous thromboembolism and mortality in orthopaedic surgery—an observational study of 45, 968 consecutive procedures. *BMC Muscoskel Disord* 2013;**14**(1):177.

110. Niikura T, Lee SY, Oe K, et al. Venous thromboembolism in Japanese patients with fractures of the pelvis and/or lower extremities using physical prophylaxis alone. *J Orthop Surg* 2012;**20**(2):196−200.

111. Forsythe RM, Peitzman AB, DeCato T, et al. Early lower extremity fracture fixation and the risk of early pulmonary embolus: filter before fixation? *J Trauma* 2011;**70**(6):1381−8.

112. Steele N, Dodenhoff RM, Ward AJ, et al. Thromboprophylaxis in pelvic and acetabular trauma surgery. The role of early treatment with low-molecular-weight heparin. *J Bone Joint Surg Br* 2005;**87**:209−12.

113. Slobogean GP, Lefaivre KA, Nicolaou S, et al. A systematic review of thromboprophylaxis for pelvic and acetabular fractures. *J Orthop Trauma* 2009;**23**(5): 379–84.
114. Chiodo A. Neurologic injury associated with pelvic trauma: radiology and electrodiagnosis evaluation and their relationships to pain and gait outcome. *Arch Phys Med Rehabil* 2007;**88**(9):1171–6.
115. Ebraheim NA, Biyani A, Wong F. Nonunion of pelvic fractures. *J Trauma* 1998;**4**:202–4.
116. Dente CJ, Feliciano DV, Rozycki GS, et al. The outcome of open pelvic fractures in the modern era. *Am J Surg* 2005;**190**(6):830–5.
117. Grotz MR, Allami MK, Harwood P, et al. Open pelvic fractures: epidemiology, current concepts of management and outcome. *Injury* 2005;**36**(1):1–13.
118. Dong JL, Zhou DS. Management and outcome of open pelvic fractures: a retrospective study of 41 cases. *Injury* 2011;**42**(10):1003–7.
119. Lindahl J, Mäkinen TJ, Koskinen SK, Söderlund T. Factors associated with outcome of spinopelvic dissociation treated with lumbopelvic fixation. *Injury* 2014;**45**: 1914–20.

Spine trauma management issues: polytrauma

Jeremie Larouche[1], Frank Lyons[2]

[1]University of Toronto, Toronto, ON, Canada; [2]Mater Misericordiae University Hospital, Dublin, Ireland

Take home points

- Early spinal clearance, when possible, plays an important role in the management of a polytraumatized patient. Fine cut CT scans read by appropriately trained physicians are essentially capable of ruling out unstable spinal injuries. MRI may provide further information, if required, and help identify discoligamentous injury, presence of a hematoma, or to further assess the neural elements; however, this information must be weighed against the difficulty of obtaining and MRI during active resuscitation efforts.
- In a polytraumatized patient with a known or suspected SCI, permissive hypotension may be required to prioritize patient survival despite the deleterious effects of spinal cord hypoperfusion.
- Once hemostasis has been achieved in the polytraumatized patients, secondary cord injury via hypoperfusion can be avoided by strictly monitoring parameters that effect cord perfusion such as coagulopathy, anemia, and oxygenation.

- Even in the absence of an SCI, early fixation of mechanically unstable spines leads to decrease in complications by allowing early mobilization.
- Early decompression within 24 hours in patients with an SCI leads to greater neurological recovery and improved outcomes as compared to late decompression.

Introduction

The safe and timely care of the polytraumatized patient's spine presents unique considerations and concerns. It may be that the patient has an uninjured stable spine, or that there is injury, either way no other body system causes as much fear, restriction, and delay as the injured or uncleared spine. The incidence of injury to the spinal column necessitating treatment varies according to many factors including the mechanism and energy involved, as well as patient factors such as age or other predisposing conditions. In the overall setting of polytrauma, however, either the provision of continuous total

© 2022 Elsevier Inc. All rights reserved.

protection of the spine or complete clearance are the only two allowable initial options. From this place of safety, evaluation and planning decisions can be made as to the ongoing and definitive care. This may include continuous spinal protection to facilitate continued resuscitation, other nonspine interventions, further spine investigations, or surgical or nonsurgical spine care. Once the patient has been adequately resuscitated it is essential then that the spine be evaluated definitively to enable complete clearance or a plan for definitive care. Keeping a patient in spinal precautions for prolonged periods is rarely acceptable, and will have a detrimental effect on anesthesia, intensive, and nursing care.

Conflicting priorities

Up to 80% of patients with spinal cord injury (SCI) are considered polytrauma patients, defined as an Abbreviated Injury Scale (AIS) score of ≥ 3 in more than one body region, or an Injury Severity Score ≥ 16.[1] In the setting of the polytraumatized patient with established spinal injury, the order and timing of the various surgical interventions is influenced by a number of factors. Patients with SCI suffering polytrauma require special considerations due to the risk of secondary cord injury from hypoperfusion and hypoxemia. In the more severely polytraumatized patients with active bleeding and/or chest or head injury this may raise some very challenging issues; resuscitative efforts may require permissive hypotension strategies, inotropic support, and low-pressure mechanical ventilation, all of which may potentiate secondary injury to the spinal cord.[2] Early, close, and continuous cross-speciality collaboration and communication are most paramount in decision-making such that the correct decisions are made at the correct time. That the patient must survive is uncontestable, but the optimal transition through stages of care from resuscitation and damage control through prevention of secondary injury can only be delivered with all teams working closely

together. While concurrent hemorrhage and neural injury are reliable predictors of bad outcome,[3] the patients that do best are the ones in whom bleeding is rapidly stopped, and are thus able to complete resuscitation and move on to the next phase of care.[4] In other words, it must be considered that deliberate hypotension might be clinically indicated in the overall care of a polytraumatized patient despite a known or suspected SCI. While there exists a recognized body of evidence in support of permissive hypotension in relation to certain isolated traumatic pathologies, it remains an area of trauma and critical care in evolution, as such it behooves the acute care clinician to keep up to date with the current evidence, particularly in the setting of a concomitant SCI.[5–11]

Maintenance of target mean arterial pressure (MAP) has long been considered an important focus for the early management of acute SCI. The practice of MAP maintenance >85 mmHg for 7 days post traumatic SCI (tSCI), while still commonly performed, has key uncertainties lacking higher level evidence in relation to duration, methodology, and by which inotropic agent.[12] It has, however, led to the identification of other modifiable parameters that could be targeted in SCI such as spinal cord blood flow, cerebrospinal fluid drainage, and spinal cord perfusion pressure.[13–15] Expect expansive research in this area in the near future which may direct more targeted strategies for spinal cord perfusion and protection in the acute and SCI polytrauma setting.[13,14,16] While there can be no doubt that the patient must be resuscitated according to evidence-based best practice, it is essential that the spinal surgical team continue to discuss and communicate these issues to critical care and trauma surgical colleagues from the outset.

Spine clearance

Many institutions have an individual policy regarding the clearance of the spine in trauma which may direct treatment at a local level. In recent years, however, there has been significant evolution in evidence-based guidelines and formulae providing algorithms to do so. Many of these center around the use and interpretation of advanced imaging, such as computed tomography (CT) and magnetic resonance imaging (MRI). Further, the selective use of dynamic and *nonsupine* plain radiography remains a valuable adjunct. In a meta-analysis of available evidence, Holmes et al. determined that plain radiographs had a poor sensitivity of only 52% to detect both stable and unstable cervical spine injury.[17] In comparison, CT scan was deemed to be 98% sensitive. Other studies have further corroborated that high-resolution CT scanning is effective at detecting essentially all unstable cervical spine fractures in obtunded patients.[18–20] Although typically unnecessary in the setting of normal CT imaging, the superior soft-tissue resolution of MRI has a valuable role in the setting of an abnormal CT scan. Specifically, this may confirm the disruption of discoligamentous soft tissues; the presence of disc material within the spinal canal prior to a planned reduction maneuver; the presence of epidural hematoma; the appearance of the neural elements in a patient which cannot otherwise be examined, and help delineate acute versus chronic fractures, particularly in older patients or those with preexisting spinal conditions.[21] The benefit of this information must be weighed against the cost of obtaining it, as many trauma teams will find it extremely difficult if not impossible to continue resuscitative efforts in the area of the static magnetic field, and even rapid sequence MRI takes considerably longer than CT.

Primary versus secondary spinal cord injury in the setting of polytrauma

While other chapters in this book will discuss in greater detail the pathophysiological processes occurring after tSCI and where targets for potential protective or regenerative interventions can be directed, it is important to address this issue in the context of SCI and the polytraumatized patient. Neurological injury represents a specific form of trauma with its own considerations. Damage suffered at the time of injury is largely irreversible, but substantial secondary insults frequently occur that are treatable or avoidable. tSCI has been described as a drama that unfolds in two acts of destruction. Part I is the trauma itself, causing destruction of neural tissue through the direct destruction of neuronal and glial cells at the primary lesion site, as well as the transection of axons transiting through the lesioned area. Additionally, damage to the vascular system will provoke hemorrhage and the disruption of the blood—spinal cord barrier. Together, this damage will induce secondary cascades responsible for cell death, enlargement of the lesioned area, and further loss of neurological functions. Edema will develop in the early ischemic period triggering a phase of glutamate excitotoxicity and ionic imbalance. The ensuing mitochondrial failure is thereafter responsible for an energy depletion and oxidative stress state. The rapid inflammatory response to SCI is provided by the resident microglia, but foremost by the infiltrating neutrophils and macrophages. At the end of the acute phase, the lesioned area will become enclosed and stabilized by a fibroglial scar. Such secondary injuries can have serious implications for the long-term outcome of injured patients. Neural element ischemia and inflammatory mediators result in neuronal and astrocytic swelling which, together with vasogenic edema, contribute to central nervous system (CNS) edema.[22,23] These secondary insults are influenced by changes in CNS blood flow (hypo- and hyperperfusion), impairment of CNS vascular autoregulation, metabolic dysfunction, and inadequate oxygenation. Furthermore, excitotoxic cell damage and inflammation may lead to apoptotic and necrotic cell death.[22,23] The contused spinal cord is particularly sensitive to changes in plasma osmotic pressure and the management of neural injury requires an appreciation of the constituents of resuscitation fluids and their possible effects on osmotic swelling of neural tissues. Spinal cord trauma raises additional complications in the form of spinal shock resulting in hypotension disproportionate to the volume of fluid lost.

Cord perfusion and coagulopathy

The early management phase following tSCI aims to optimize cord perfusion and limit secondary insults (e.g., hypotension, hypoxia) prior to definitive management.[24–26] Trauma and critical care have moved away from the administration of crystalloid and/or colloid as part of the resuscitative strategy for polytraumatized patients.[5,11,27,28] A large body of evidence, deriving from both military and civilian research, has emphatically shown that blood should be replaced by blood and blood products.[29–33] Applying this paradigm to the injured spinal cord means greater potential for oxygen delivery while decreased potential for cord edema, coagulopathy, and the broader dysfunction of supporting organs.[4,27] This has a double benefit for the patient, primarily the patient is resuscitated optimally supporting all organs and minimizing metabolic and coagulation derangement. Secondarily, and from the perspective of the spine, it means spinal surgery can be carried out earlier, i.e., less delay waiting for physiologic and metabolic stabilization. While the Advanced Trauma Life Support guidelines still advocate for an initial bolus of intravenous crystalloid before blood product administration,[34] many trauma centers will opt immediately for blood products in the polytraumatized patient.[33,35] The ratio of blood products may vary between centers, and similar to other critical care practices, is under constant review and modification.[33,35] More recently blood product administration guided by patient-specific coagulation analysis has been advocated for; although randomized studies in trauma are currently lacking, future study as well as the evidence base so far may see this become a more

established practice.[36,37] Technologies based on thromboelastography (TEG) and rotational thromboelastometry (ROTEM) have enabled the rapid identification of specific blood product deficiencies enabling a targeted replacement strategy.[38,39] Established practice in cardiac and transplantation surgery, TEG/ROTEM have demonstrated less overall blood product usage and improved patient outcomes by identifying the specific deficiencies that result in hypocoagulability and hyperfibrinolysis.[40,41] For example, a patient in hypovolemic shock who is receiving packed red blood cell replacement can develop a dilutional coagulopathy; using a TEG/ROTEM analysis can identify the blood-specific protein deficiencies enabling immediate targeted replacement.[36,37,42] In the setting of tSCI, research on TEG/ROTEM application has yet to be published; however, there are retrospective studies suggesting clinical value when used in the traumatic brain injury patient, including patients taking regular prescribed oral anticoagulants.[43–45] A clinical trial is underway for major elective spinal surgery aimed at demonstrating similar quality improvements analogous to those observed with major cardiac and transplant surgery.[46,47] Perhaps this will provide the basis for high-level research in spinal trauma surgery and tSCI. For now TEG/ROTEM have proven to improve quality of care in the polytrauma setting and as such its application and continued research will contribute to the management of polytraumatized patients with spinal injury.[48]

Timing of fixation

A point will be reached whereby adequate resuscitation has been achieved and the patients care must move on through damage control and into definitive care. While this too is under continuous study and review, it is irrefutable that the patients' best interest is served by timely intervention allowing discontinuation of spinal precautions, optimal positioning for ventilation, and early mobilization. The decision which must be made is at which point the impact of surgery is warranted in the polytraumatized patients in order to derive the benefits of early fixation without incurring the cost of morbidity and mortality. In the orthopedic literature, the pendulum of surgical timing has had the chance to swing widely over the last 30 years. The *too sick to operate* model of the critical care surgeon underwent an evidence-based shift to a *too sick not to operate* paradigm.[49–51] In 1989 a prospective study of polytraumatized patients by Bone et al. demonstrated decreased incidence of ARDS, sepsis, and emboli following femoral fracture fixation within the first 24 hours[52] This led to the evolution and adoption of the early-total-care (ETC) model of care for the multiply injured patient. While this improved outcomes in overall trauma care, a subset of patients with severe chest and head injuries were observed to do worse.[53,54] Damage control orthopedics (DCO) was born, or moreover, adapted from experiences from critical care surgery; thoraco-abdominal-pelvic organ and vessel injuries are surgically stabilized sufficiently but not definitively to facilitate resuscitative correction of the most critical physiologic, metabolic, and coagulation derangements.[55] Equally, rapid temporary fracture stabilization with external fixators, or similar orthoses, minimizes further insult such that intensive care measures can proceed.[56–59]

In many respects, ETC and DCO represent opposite ends of a surgical philosophical spectrum: "fix everything" or "fix nothing." While much debate exists within the literature in defining and determining objective markers of what patient is deemed appropriate for ETC or DCO, it is evident that a middle ground exists, an approach that captures the protective benefits of ETC without jeopardizing resuscitation or head injury management.[60–62] Early-Appropriate-Care (EAC), Safe-Definitive-Surgery, and Personalized-Individualized-Safe-Management (PRISM) are all concepts put forward as a framework for decision and priority planning

in the trauma setting. When it comes to management of appendicular trauma, the pendulum has now settled on a focus of expedited care while paying close attention to resuscitation markers. Allowing for adequate resuscitation, it has been conclusively demonstrated that early surgical fixation of extremity fractures, even in multiply injured patients, leads to better all-round outcomes as measured by indices such as respiratory infections, ventilator and intensive care unit (ICU) days, overall length of stay, thrombus and embolus incidence, and death.[63–65]

While an optimal timeline of within 24 hours has been reported for extremity fractures, there is, however, a very wide disparity for spine fractures as to what constitutes early fixation, with studies reporting benefits with times of up to 72 hours.[48–51] This broad time range may represent variability in established surgical timeliness practices between trauma centers, or between individual surgeons within centers. It may also be that spinal stabilization is deferred for days while other injuries assume priority; decision-making related to the timing of fixation for those fractures that require it is particularly challenging in the context of other injuries, and the spine has traditionally been pushed to the back of the surgical line. The reasons for this are multifactorial and may have their origins rooted in outdated beliefs anchored in the deep surgical and critical care past. That spinal surgery has grown from two different and at times divisive subspecialties—neurosurgery and orthopedic surgery—has not helped the cause. Acute spinal surgery may have been "covered" by a nonspine fellowship trained neuro- or orthopedic surgeon who may have chosen to defer complex decision-making to a specialist colleague causing delay. Additionally, these spinal surgeries are frequently performed prone on patients with multisystem injuries over prolonged durations, which has provided pause to some anesthesiologists in allowing such cases to go forward. Clearly, a focus on synthesizing the best possible evidence is required when advocating for the health of our patients.

Even in the absence of a neurological injury, the requirement to maintain spinal immobilization in the mechanically unstable spine can lead to many of the same complications as previously identified in long-bone fractures. Almost 20 years ago, Croce et al. reported from a level 1 trauma setting that not only was it safe to perform surgical fixation of nonneurologically impaired spine fractures in severely injured patients, but that it resulted in fewer pulmonary complications, less ICU days, and shorter lengths of stay from a cohort of 291 patients.[66] Subsequently in 2006, in a study of over 1000 patients with injury severity score >18 at Harborview Medical Centre, USA, McHenry et al. reported delay exceeding 48 hours to spinal stabilization was an independent risk factor for the development of respiratory failure ($P < .05$).[67] A more recent similar analysis in 2013 by Stahel et al. of 112 patients proposed establishing a spine damage control operative strategy.[68] This was based on their findings that severely injured patients (ISS>16) with thoracolumbar fractures had statistically significant shorter hospital LOS (14.1 vs. 32.4) and ventilator days (1.5 vs. 2.4) when spine stabilization was carried out in under 72 hours. They went on to report that wound, urinary tract, pulmonary, and pressure sore complications were also significantly less in the earlier surgery group.

A systematic review of 10 studies capturing 2512 spine fracture patients by Xing et al. reported similar results.[69] While high heterogeneity of the included studies was prohibitive of meta-analysis, the authors conclude that the overall strength of evidence to assess whether early surgery in thoracolumbar fractures reduces hospital length of stay, ICU, and ventilator days is "high" and that the estimate of the effect is very certain. However, the overall strength of evidence to assess whether early surgery reduces in-hospital costs and morbidity is "moderate" and they strongly recommend further research to produce an impactful estimate of the effect.

The weakrer recommendation for the second conclusion is likely due to the signigicant heterogeneity in the included studies, which is reflected in the variability in times (<8 hours to <72 hours) and also lower patient numbers with all but one of the studies having less than a total of 200 patients. The included study with the largest cohort (1506 patients), Boakye et al., reported that time to surgery was the strongest independent risk factor for in-hospital complication when spinal stabilization was delayed beyond a 72 hour time point.[70] A 59% increase in the risk of suffering complication was noted, even after controlling for other factors such as age, comorbidity burden, and injury severity score (OR [95%CI] = 1.592 [1.180, 2.148]), and resulted in significantly increased LOS and costs. More recently Bleimel et al. have performed a larger study with 18-year retrospective analysis of the German trauma registry.[71] A total of 2595 thoracic and lumbar fracture patients were included and stratified according to spinal stabilization surgery being carried out also before or after 72 hours of injury. The early surgical group had statistically significantly overall shorter length of stay (thoracic spine $P < .001$; lumbar spine $P < .028$), and nonstatistically significant fewer ICU and total ventilator dependent days ($P < .068$ and $P < .108$). Sepsis rates were decreased in the early surgical group (thoracic $P < .008$; lumbar $P < .001$) but 7-day mortality rates were comparable across all groups. Using the Trauma Quality Improvement Program (TQIP) database of the American College of Surgeons, the authors of this chapter have recently published a review of over 19,000 cases which further supports that, even in the absence of a neurological injury, unstable spine fractures also benefit from early fixation within 24 hours thereby reducing the rate of complications.[72]

In polytraumatized patients with an SCI, there exists an even greater need for urgent spinal surgery, as clinical evidence suggests that greater neurological recovery can be achieved with early decompression. Published in 2012, the STASCIS trial provided compelling evidence for timely surgical decompression in traumatic cervical SCIs.[73] A prospective multicenter cohort study, the STASCIS study put all this together and provided substantive evidence for early surgical intervention in the event of tSCI.[73] This trial determined that patients who underwent surgical decompression within 24 hours of injury were more likely to recover 1 or 2 grades as measured by the American Spinal Injuries Association impairment scale (AIS) compared to those who underwent surgery beyond 24 hours. While some more recent and promising work has explored whether or not "ultra-early" spinal decompression at time points shorter than 24 hours leads to improved outcomes, there is still insufficient evidence to provide this recommendation at this time.[74,75] While this work has provided the spinal surgeon with the evidence to advocate for early surgery in tSCI patients, doubt and dilemma frequently persist with the polytraumatized patient. Chest and head injuries, coagulation disorder, hypothermia, as well as other physiological derangements and injuries common to this patient group can manifest a reluctance to proceed with spinal surgery within the 24-hour time window.

In conclusion, traumatic spinal fractures with and without neurological injury are a common entity after high-energy trauma and often coexist with other life- and limb-threatening injuries. Initial resuscitation efforts must focus on patient survival, and even with a known SCI, permissive hypotension may be clinically warranted. In the setting of trauma, improved functional outcomes have been demonstrated when major in-hospital complications are reduced, and the timing of surgery has been consistently identified as a major predictor of such complications.[60,63,66,69,71,73,76–78] This applies both to patients who are mechanically unstable, or who have an SCI. The challenges of early operative intervention in a critically ill patient with multisystem injuries must be balanced with the disadvantages of prolonged immobilization which may lead to complications such as pneumonia, decubitus ulcers, and venous thromboembolism.

Case examples

Case 1

21-year-old female, highway MVC with rollover. Direct helicopter transfer from scene to level 1 trauma center. On arrival noted to be moving all four extremities on command; however, patient rapidly decompensated physiologically with BP 80/40 mmHg and HR > 150/minute. SatO$_2$ of 84% with 100% supplemental oxygen, speaking coherently but with rapid shallow and distressed breathing. Intubated in trauma bay and following resuscitation CT trauma pan scan was performed.

Open right femur fracture, G and A 3a (Fig. 13.1A).

Highly unstable thoracic spine fracture dislocation, lumbar spine fracture, and massive left-sided hemopneumothorax (Fig. 13.1B–D).

Initial resuscitation efforts included placement of a left-sided chest tube and initiation of a massive transfusion protocol. DCO was initiated, with placement of skeletal traction to the right femur. Twelve hours later, repeat assessment of lactate, coagulation profile, and hemoglobin had normalized. The patient was then initially placed prone for a posterior fixation of the thoracic fracture, followed by minimally invasive plate osteosynthesis of the right femur. Delayed fixation of the tibial plateau was achieved on postadmission day 8 to allow for soft-tissue swelling to subside (Fig. 13.1E–G).

Case 2

28-year-old female, blunt force trauma related to skiing injury.

Presents with grade 4 blunt aortic injury, splenic laceration, left renal laceration, bilateral

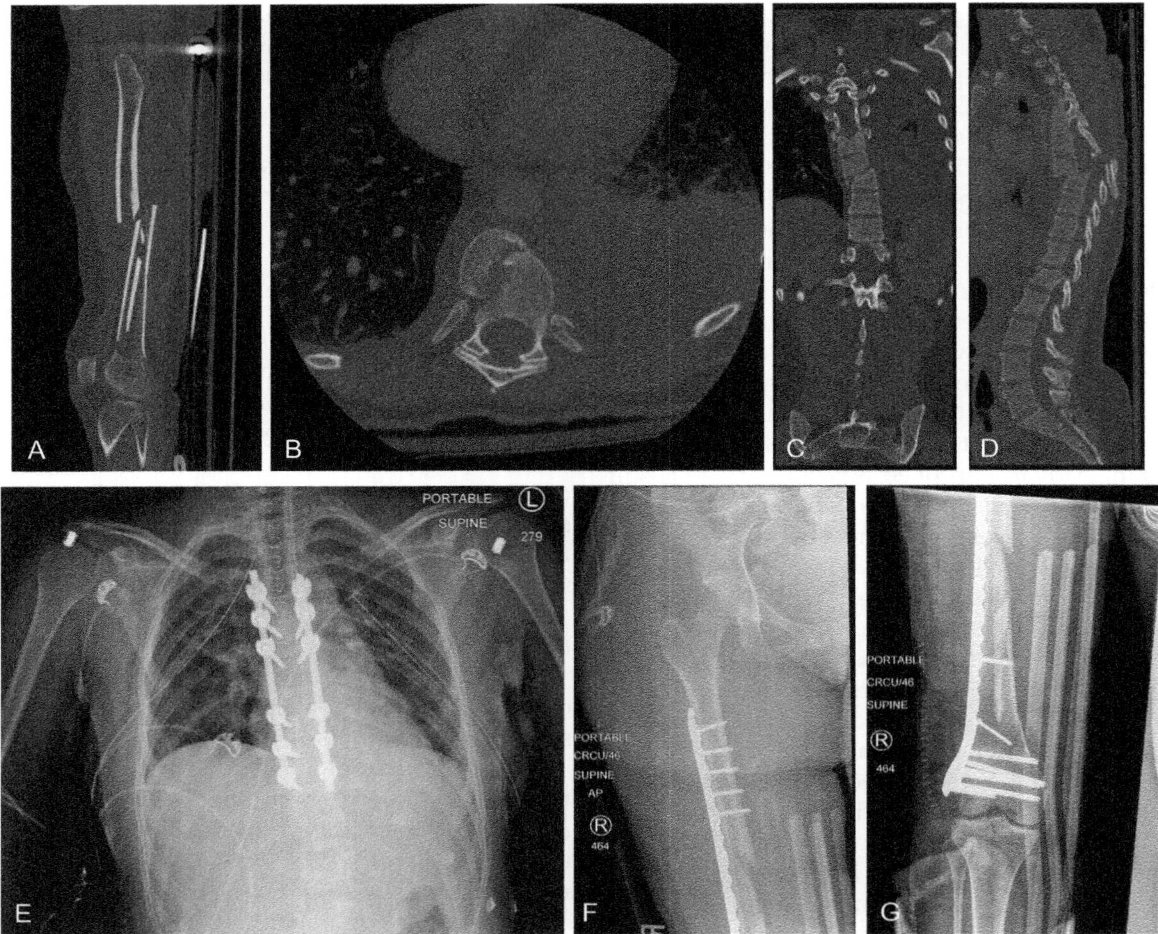

FIGURE 13.1   Initial radiographs of a polytrauma patient with a right open femur fracture (13.1A), fracture dislocation of the thoraric spine (13.1B–D), and left sided hemopneumothorax (13.1B,D). Fig. 13.1E,F, and G reveal post-operative film after fixation of the thoracic spine fracture and right femur fracture.

pneumothoraces, ASIA C SCI at T12, and spino-pelvic dissociation (Fig. 13.2A–C).

Initial resuscitation efforts focused on permissive hypotension for aortic arch injury, two chest tubes, and massive transfusion protocol. Systolic pressure kept under 110 mm/hg initially. Underwent trauma laparotomy, external fixation of left femur, and thoracic endovascular aortic repair 3 hours after arriving to level 1 trauma center. Required 15 units of PRBC. Kept on log-roll precautions. After clearance from vascular and trauma surgery, MAP increased to 85 mm/hg.

On POD 3, cleared by vascular surgery and trauma surgery. Underwent thoracolumbar decompression and fusion, spinopelvic fixation, and retrograde nailing of left femur (Fig. 13.2D–F).

Ultimately, patient regained good strength to lower extremities with exception of bilateral plantar flexion, rated as 2/5. No return of bowel or bladder function.

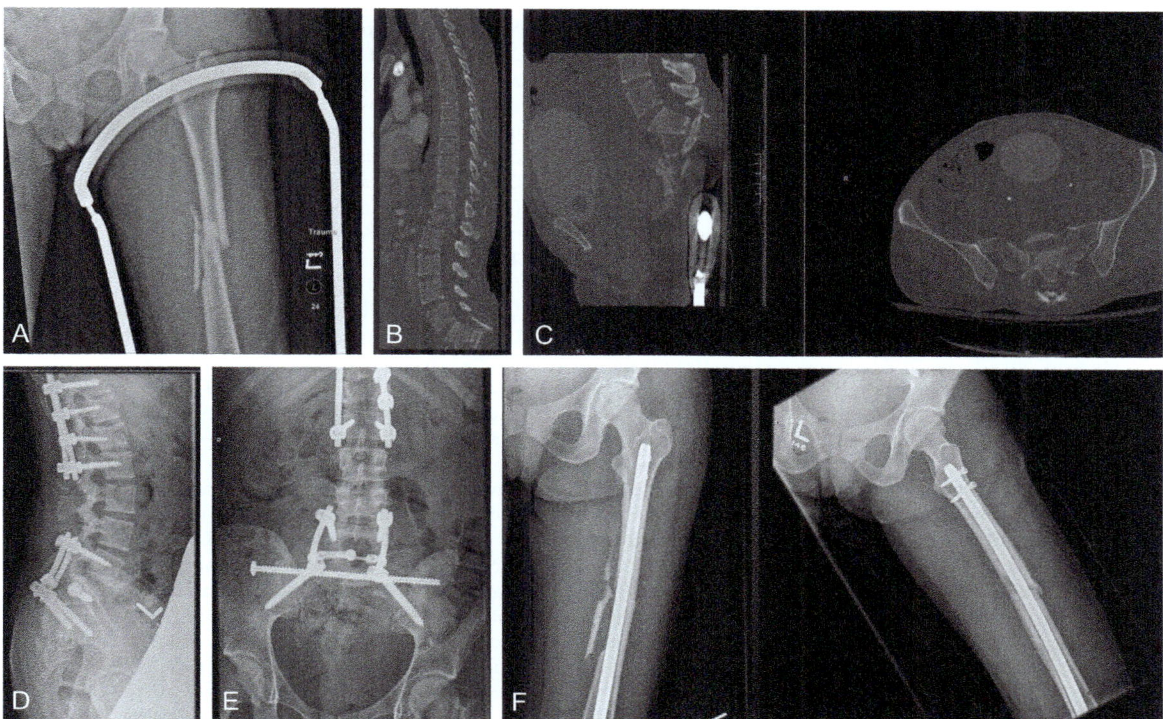

FIGURE 13.2   Radiographs of a polytrauma patient with a left sided femur fracture (Fig. 13.2A), three column thoracolumbar fracture (Fig. 13.2B) and complex spinopelvic dissociations (Fig. 13.2C). Post-operative radiographs (Fig. 13.2 D—F) deonstrated open reduction of the thoracolumbar fracture, triangular osteosynthesis of the spinopelvice fracture, and retrograde nailing of the left femur.

References

1. Burney RE, Maio RF, Maynard F, Karunas R. Incidence, characteristics, and outcome of spinal cord injury at trauma centers in North America. *Arch Surg* 1993; **128**(5):596—9.
2. Tran A, Yates J, Lau A, Lampron J, Matar M. Permissive hypotension versus conventional resuscitation strategies in adult trauma patients with hemorrhagic shock: a systematic review and meta-analysis of randomized controlled trials. *J Trauma & Acute Care Surg* 2018; **84**(5):802—8.
3. Chesnut RM, Marshall LF, Klauber MR, Blunt BA, Baldwin N, Eisenberg HM, et al. The role of secondary brain injury in determining outcome from severe head injury. *J Trauma* 1993;**34**(2):216—22.
4. Tobin JM, Dutton RP, Pittet J-F, Sharma D. Hypotensive resuscitation in a head-injured multi-trauma patient. *J Crit Care* 2014;**29**(2):313. e1—e5.
5. Jansen JO, Thomas R, Loudon MA, Brooks A. Damage control resuscitation for patients with major trauma. *BMJ* 2009;**338**:b1778.
6. Albreiki M, Voegeli D. Permissive hypotensive resuscitation in adult patients with traumatic haemorrhagic shock: a systematic review. *Eur J Trauma Emerg Surg* 2018;**44**(2):191—202.
7. Wiles M. Blood pressure management in trauma: from feast to famine? *Anaesthesia* 2013;**68**(5):445—9.
8. Owattanapanich N, Chittawatanarat K, Benyakorn T, Sirikun J. Risks and benefits of hypotensive resuscitation in patients with traumatic hemorrhagic shock: a meta-analysis. *Scand J Trauma Resuscitation Emerg Med* 2018;**26**(1):107.
9. Bouglé A, Harrois A, Duranteau J. Resuscitative strategies in traumatic hemorrhagic shock. *Ann Intensive Care* 2013;**3**(1):1.
10. Bickell WH, Wall Jr MJ, Pepe PE, Martin RR, Ginger VF, Allen MK, et al. Immediate versus delayed fluid resuscitation for hypotensive patients with penetrating torso injuries. *N Engl J Med* 1994;**331**(17):1105—9.
11. Kudo D, Yoshida Y, Kushimoto S. Permissive hypotension/hypotensive resuscitation and restricted/controlled resuscitation in patients with severe trauma. *Journal of Intensive Care* 2017;**5**(1):11.

12. Menacho ST, Floyd C. Current practices and goals for mean arterial pressure and spinal cord perfusion pressure in acute traumatic spinal cord injury: defining the gaps in knowledge. *J Spinal Cord Med* 2019:1–7.

13. Squair JW, Bélanger LM, Tsang A, Ritchie L, Mac-Thiong J-M, Parent S, et al. Empirical targets for acute hemodynamic management of individuals with spinal cord injury. *Neurology* 2019. https://doi.org/10.1212/WNL.0000000000008125.

14. Squair JW, Bélanger LM, Tsang A, Ritchie L, Mac-Thiong J-M, Parent S, et al. Spinal cord perfusion pressure predicts neurologic recovery in acute spinal cord injury. *Neurology* 2017;**89**(16):1660–7.

15. Saadeh YS, Smith BW, Joseph JR, Jaffer SY, Buckingham MJ, Oppenlander ME, et al. The impact of blood pressure management after spinal cord injury: a systematic review of the literature. *Neurosurg Focus* 2017;**43**(5):E20.

16. Yue JK, Hemmerle DP, Winkler EA, Thomas LH, Fernandez XD, Kyritsis N, et al. Clinical implementation of a novel spinal cord perfusion pressure protocol in acute traumatic spinal cord injury at a US Level I trauma center: a TRACK-SCI study. *World Neurosurg* 2020;**133**: 391–6. https://doi.org/10.1016/j.wneu.2019.09.044.

17. Holmes JF, Akkinepalli R. Computed tomography versus plain radiography to screen for cervical spine injury: a meta-analysis. *J Trauma & Acute Care Surg* 2005;**58**(5):902–5.

18. Hogan GJ, Mirvis SE, Shanmuganathan K, Scalea TM. Exclusion of unstable cervical spine injury in obtunded patients with blunt trauma: is MR imaging needed when multi–detector row CT findings are normal? *Radiology* 2005;**237**(1):106–13.

19. Raza M, Elkhodair S, Zaheer A, Yousaf S. Safe cervical spine clearance in adult obtunded blunt trauma patients on the basis of a normal multidetector CT scan—a meta-analysis and cohort study. *Injury* 2013;**44**(11):1589–95.

20. Panczykowski DM, Tomycz ND, Okonkwo DO. Comparative effectiveness of using computed tomography alone to exclude cervical spine injuries in obtunded or intubated patients: meta-analysis of 14,327 patients with blunt trauma: a review. *J Neurosurg* 2011;**115**(3):541–9.

21. Malhotra A, Durand D, Wu X, Geng B, Abbed K, Nunez DB, et al. Utility of MRI for cervical spine clearance in blunt trauma patients after a negative CT. *Eur Radiol* 2018;**28**(7):2823–9.

22. Couillard-Despres S, Bieler L, Vogl M. Pathophysiology of traumatic spinal cord injury. In: *Neurological aspects of spinal cord injury*. Springer; 2017. p. 503–28.

23. Werner C, Engelhard K. Pathophysiology of traumatic brain injury. *Br J Addiction: Br J Anaesth* 2007;**99**(1):4–9.

24. Witiw CD, Fehlings MG. Acute spinal cord injury. *J Spinal Disord Tech* 2015;**28**(6):202–10.

25. Wilson JR, Tetreault LA, Kwon BK, Arnold PM, Mroz TE, Shaffrey C, et al. Timing of decompression in patients with acute spinal cord injury: a systematic review. *Global Spine J* 2017;**7**(3_Suppl. l):95S–115S.

26. Fehlings MG, Kwon BK, Tetreault LA. *Guidelines for the management of degenerative cervical myelopathy and spinal cord injury: an introduction to a focus issue*. Los Angeles, CA: Sage Publications Sage CA; 2017.

27. James M. Volume therapy in trauma and neurotrauma. *Best Pract Res Clin Anaesthesiol* 2014; **28**(3):285–96.

28. Harada MY, Ko A, Barmparas G, Smith EJ, Patel BK, Dhillon NK, et al. 10-Year trend in crystalloid resuscitation: reduced volume and lower mortality. *Int J Surg* 2017;**38**:78–82.

29. Glassberg E, Nadler R, Erlich T, Klien Y, Kreiss Y, Kluger Y. A decade of advances in military trauma care. *Scand J Surg* 2014;**103**(2):126–31.

30. Haider AH, Piper LC, Zogg CK, Schneider EB, Orman JA, Butler FK, et al. Military-to-civilian translation of battlefield innovations in operative trauma care. *Surgery* 2015;**158**(6):1686–95.

31. McCullough A, Haycock J, Forward D, Moran C. Early management of the severely injured major trauma patient. *Br J Anaesth* 2014;**113**(2):234–41.

32. McCunn M, Dutton RP, Dagal A, Varon AJ, Kaslow O, Kucik CJ, et al. Trauma, critical care, and emergency care anesthesiology: a new paradigm for the "acute care" anesthesiologist? *Anesth Analg* 2015;**121**(6): 1668–73.

33. Chang R, Holcomb JB. Optimal fluid therapy for traumatic hemorrhagic shock. *Crit Care Clin* 2017;**33**(1): 15–36.

34. Henry S. *ATLS 10th edition offers new insights into managing trauma patients*. Bulletin of the American College of Surgeons; June 1, 2018 [Internet].

35. Wong H, Pottle J, Stanworth SJ, Brunskill SJ, Davenport R, Doree C, et al. Strategies for use of blood products for major bleeding in trauma. *Cochrane Database Syst Rev* 2017;**2017**(4).

36. Da Luz LT, Nascimento B, Shankarakutty AK, Rizoli S, Adhikari NK. Effect of thromboelastography (TEG®) and rotational thromboelastometry (ROTEM®) on diagnosis of coagulopathy, transfusion guidance and mortality in trauma: descriptive systematic review. *Crit Care* 2014;**18**(5):518.

37. Hunt H, Stanworth S, Curry N, Woolley T, Cooper C, Ukoumunne O, et al. Thromboelastography (TEG) and rotational thromboelastometry (ROTEM) for trauma-induced coagulopathy in adult trauma patients with bleeding. *Cochrane Database Syst Rev* 2015;**2**.

38. De Nicola P, Mazzetti G. Evaluation of thrombelastography. *Am J Clin Pathol* 1955;**25**(4_ts):447–52.

39. Wikkelsoe A, Afshari A, Wetterslev J, Brok J, Moeller A. Monitoring patients at risk of massive transfusion with thrombelastography or thromboelastometry: a systematic review. *Acta Anaesthesiol Scand* 2011;**55**(10):1174–89.

40. Girdauskas E, Kempfert J, Kuntze T, Borger MA, Enders J, Fassl J, et al. Thromboelastometrically guided transfusion protocol during aortic surgery with circulatory arrest: a prospective, randomized trial. *J Thorac Cardiovasc Surg* 2010;**140**(5):1117–24. e2.

41. Shore-Lesserson L, Manspeizer HE, DePerio M, Francis S, Vela-Cantos F, Ergin MA. Thromboelastography-guided transfusion algorithm reduces transfusions in complex cardiac surgery. *Anesth Analg* 1999;**88**(2):312–9.

42. Holcomb JB, Minei KM, Scerbo ML, Radwan ZA, Wade CE, Kozar RA, et al. Admission rapid thrombelastography can replace conventional coagulation tests in the emergency department: experience with 1974 consecutive trauma patients. *Ann Surg* 2012;**256**(3):476–86.

43. Kay AB, Morris DS, Collingridge DS, Majercik S. Platelet dysfunction on thromboelastogram is associated with severity of blunt traumatic brain injury. *Am J Surg* 2019.

44. Hota S, Ng M, Hilliard D, Burgess J. Thromboelastogram-guided resuscitation for patients with traumatic brain injury on novel anticoagulants. *Am Surg* 2019;**85**(8):861–4.

45. Chen H, Tian H. The utility of thromboelastography for predicting the risk of post-traumatic cerebral infarction in traumatic brain injury. *Neurosurgery* 2019;**66**(Supplement_1):nyz310_688.

46. *Thromboelastography-guided fluid management in spinal surgery*; n.d. https://ClinicalTrials.gov/show/NCT03999086.

47. Li C, Zhao Q, Yang K, Jiang L, Yu J. Thromboelastography or rotational thromboelastometry for bleeding management in adults undergoing cardiac surgery: a systematic review with meta-analysis and trial sequential analysis. *J Thorac Dis* 2019;**11**(4):1170.

48. Wahlen BM, El-Menyar A, Peralta R, Al-Thani H. World Academic Council of Emergency Medicine experience document: implementation of point-of-care thromboelastography at an academic emergency and trauma center. *J Emerg Trauma Shock* 2018;**11**(4):265.

49. Reynolds MA, Richardson JD, Spain DA, Seligson D, Wilson MA, Miller FB. Is the timing of fracture fixation important for the patient with multiple trauma? *Ann Surg* 1995;**222**(4):470.

50. Johnson KD, Cadambi A, Seibert GB. Incidence of adult respiratory distress syndrome in patients with multiple musculoskeletal injuries: effect of early operative stabilization of fractures. *J Trauma* 1985;**25**(5):375–84.

51. Charash WE, Fabian TC, Groce MA. Delayed operative fixation of femur fractures is a risk factor for pulmonary failure independent of thoracic trauma. *J Trauma & Acute Care Surg* 1993;**35**(6):983.

52. Bone L, Johnson K, Weigelt J, Scheinberg R. Early versus delayed stabilization of femoral fractures. A prospective. *J Bone Joint Surg Am* 1989;**71**:336–40.

53. Pape H-C, Auf'm'Kolk M, Paffrath T, Regel G, Sturm JA, Tscherne H. Primary intramedullary femur fixation in multiple trauma patients with associated lung contusion–a cause of posttraumatic ARDS? *J Trauma* 1993;**34**(4):540–7. Discussion 7-8.

54. Townsend RN, Lheureau T, Protetch J, Riemer B, Simon D. Timing fracture repair in patients with severe brain injury (Glasgow Coma Scale score< 9). *J Trauma & Acute Care Surg* 1998;**44**(6):977–83.

55. Stone HH, Strom PR, Mullins RJ. Management of the major coagulopathy with onset during laparotomy. *Ann Surg* 1983;**197**(5):532.

56. Burch JM, Ortiz VB, Richardson RJ, Martin RR, Mattox KL, Jordan Jr GL. Abbreviated laparotomy and planned reoperation for critically injured patients. *Ann Surg* 1992;**215**(5):476.

57. Talbert S, Trooskin SZ, Scalea T, Vieux E, Atweh N, Duncan A, et al. Packing and re-exploration for patients with nonhepatic injuries. *J Trauma* 1992;**33**(1):121–4. Discussion 4-5.

58. Scalea TM, Boswell SA, Scott JD, Mitchell KA, Kramer ME, Pollak AN. External fixation as a bridge to intramedullary nailing for patients with multiple injuries and with femur fractures: damage control orthopedics. *J Trauma & Acute Care Surg* 2000;**48**(4):613–23.

59. Roberts CS, Pape H-C, Jones AL, Malkani AL, Rodriguez JL, Giannoudis PV. Damage control orthopaedics: evolving concepts in the treatment of patients who have sustained orthopaedic trauma. *JBJS* 2005;**87**(2):434–49.

60. Vallier HA, Wang X, Moore TA, Wilber JH, Como JJ. Timing of orthopaedic surgery in multiple trauma patients: development of a protocol for early appropriate care. *J Orthop Trauma* 2013;**27**(10):543–51.

61. Vallier HA, Dolenc AJ, Moore TA. Early appropriate care: a protocol to standardize resuscitation assessment and to expedite fracture care reduces hospital stay and enhances revenue. *J Orthop Trauma* 2016;**30**(6):306–11.

62. Pape H, Andruszkow H, Pfeifer R, Hildebrand F, Barkatali B. Options and hazards of the early appropriate care protocol for trauma patients with major fractures: towards safe definitive surgery. *Injury* 2016;**47**(4):787–91.

63. Byrne JP, Nathens AB, Gomez D, Pincus D, Jenkinson RJ. Timing of femoral shaft fracture fixation following major trauma: a retrospective cohort study

of United States trauma centers. *PLoS Med* 2017;**14**(7): e1002336.

64. Pincus D, Ravi B, Wasserstein D, Huang A, Paterson JM, Nathens AB, et al. Association between wait time and 30-day mortality in adults undergoing hip fracture surgery. *Jama* 2017;**318**(20):1994–2003.

65. Vallier HA, Moore TA, Como JJ, Wilczewski PA, Steinmetz MP, Wagner KG, et al. Complications are reduced with a protocol to standardize timing of fixation based on response to resuscitation. *J Orthop Surg Res* 2015;**10**(1):155.

66. Croce MA, Bee TK, Pritchard E, Miller PR, Fabian TC. Does optimal timing for spine fracture fixation exist? *Ann Surg* 2001;**233**(6):851.

67. McHenry TP, Mirza SK, Wang J, Wade CE, O'keefe GE, Dailey AT, et al. Risk factors for respiratory failure following operative stabilization of thoracic and lumbar spine fractures. *JBJS* 2006;**88**(5):997–1005.

68. Stahel PF, VanderHeiden T, Flierl MA, Matava B, Gerhardt D, Bolles G, et al. The impact of a standardized "spine damage-control" protocol for unstable thoracic and lumbar spine fractures in severely injured patients: a prospective cohort study. *J Trauma & Acute Care Surg* 2013;**74**(2):590–6.

69. Xing D, Chen Y, Ma J-X, Song D-H, Wang J, Yang Y, et al. A methodological systematic review of early versus late stabilization of thoracolumbar spine fractures. *Eur Spine J* 2013;**22**(10):2157–66.

70. Boakye M, Arrigo RT, Gephart MGH, Zygourakis CC, Lad S. Retrospective, propensity score-matched cohort study examining timing of fracture fixation for traumatic thoracolumbar fractures. *J Neurotrauma* 2012; **29**(12):2220–5.

71. Bliemel C, Lefering R, Buecking B, Frink M, Struewer J, Krueger A, et al. Early or delayed stabilization in severely injured patients with spinal fractures? Current surgical objectivity according to the trauma registry of DGU: treatment of spine injuries in polytrauma patients. *J Trauma & Acute Care Surg* 2014;**76**(2):366–73.

72. Guttman M, Larouche J, Lyons F, Nathens A. Early fixation of traumatic spinal fractures reduces complications in the absence of neurologic injury: a retrospective cohort study from ACS-TQIP. *J Neurosurg Spine* 2020 [Published ahead of print].

73. Fehlings MG, Vaccaro A, Wilson JR, Singh A, Cadotte DW, Harrop JS, et al. Early versus delayed decompression for traumatic cervical spinal cord injury: results of the Surgical Timing in Acute Spinal Cord Injury Study (STASCIS). *PLoS One* 2012;**7**(2):e32037.

74. Jug M, Kejžar N, Cimerman M, Bajrović FF. Window of opportunity for surgical decompression in patients with acute traumatic cervical spinal cord injury. *J Neurosurg Spine* 2019;**1**(aop):1–9.

75. Mattiassich G, Gollwitzer M, Gaderer F, Blocher M, Osti M, Lill M, et al. Functional outcomes in individuals undergoing very early (< 5 h) and early (5–24 h) surgical decompression in traumatic cervical spinal cord injury: analysis of neurological improvement from the Austrian Spinal Cord Injury Study. *J Neurotrauma* 2017;**34**(24):3362–71.

76. Bellabarba C, Fisher C, Chapman JR, Dettori JR, Norvell DC. Does early fracture fixation of thoracolumbar spine fractures decrease morbidity or mortality? *Spine* 2010;**35**(9S):S138–45.

77. Frangen TM, Ruppert S, Muhr G, Schinkel C. The beneficial effects of early stabilization of thoracic spine fractures depend on trauma severity. *J Trauma & Acute Care Surg* 2010;**68**(5):1208–12.

78. Holbrook TL, Hoyt DB, Anderson JP. The impact of major in-hospital complications on functional outcome and quality of life after trauma. *J Trauma & Acute Care Surg* 2001;**50**(1):91–5.

Spine trauma in the elderly — management issues and treatment goals

Mark J. Lambrechts[1], Christina L. Goldstein[2], Jamie R.F. Wilson[3]

[1]Rothman Orthopedic Institute at Thomas Jefferson University, Philadelphia, PA, United States; [2]University of Colorado Health - Memorial Hospital, Colorado Springs, CO, United States; [3]Department of Neurosurgery, University of Nebraska Medical Center, Omaha, NE, United States

We acknowledge that the definition of "elderly" or "geriatric" in the domain of spine trauma is open to significant debate. For purposes of this chapter, we have applied the definition of geriatric as an age of 65 years or older.

Pathophysiology

Increased age brings with it an increased incidence of both spondylosis and osteoporosis. Spondylosis is present in almost 100% of patients imaged at age 70 and is implicated in the development of spinal canal and foraminal stenosis. This can increase the risk of spinal cord injury (SCI) even in low-energy impacts. Osteoporosis increases the risk of spine fractures and also has serious implications to the outcomes from surgery. Additionally, the effect of other disorders such as diffuse idiopathic skeletal hyperostosis (DISH) and ankylosing spondylitis (AS) can add further difficulties through intervertebral fusion that predisposes to an increased risk of fracture.[1,2] The etiology of central canal stenosis, and the subsequent myelopathy or radiculopathy that can arise, includes disc spondylosis, hypertrophy of the ligamentum flavum, ossification of the posterior longitudinal ligament, degenerative listhesis, and facet joint hypertrophy (Figs. 14.1A and B). It is well known that degenerative changes in the spine have a high incidence in the elderly population, and this can cause distinct clinical phenotypes depending on the anatomical site.[3–6] In the cervical spine (and to a lesser extent the thoracic spine), progressive canal stenosis leads to cord compression and the clinical syndrome of myelopathy. Canal stenosis in the lumbar spine manifests as spinal claudication, possibly with comorbid mechanical back pain.

Whereas early signs of canal stenosis may not produce clinical symptoms, as these changes progress over time the diameter of the canal decreases.[7] This can lead to the clinical scenario of an asymptomatic, or mildly symptomatic, patient who is at higher risk of SCI due to the existing level of stenosis compared to the general population without stenosis.[8] This may lead to

© 2022 Elsevier Inc. All rights reserved.

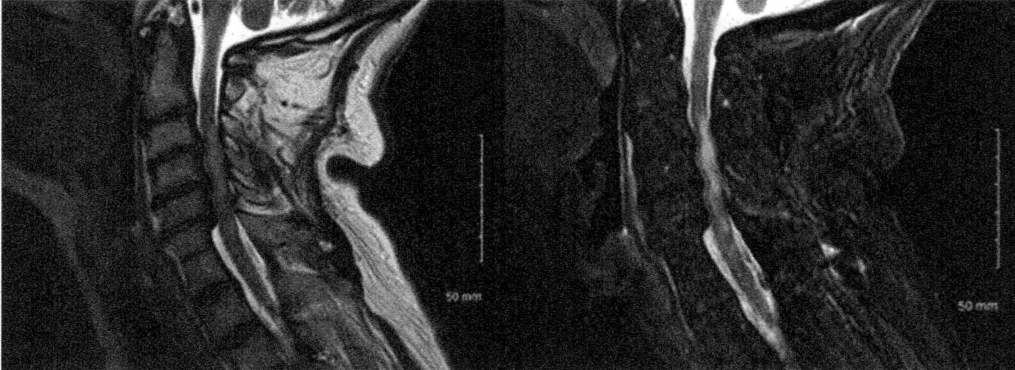

FIGURE 14.1A   T1 (left) and T2 (right) parasagittal cervical spine MRI. This patient has a traumatic central cord syndrome. There is evidence of diffuse cervical spondylosis with posterior longitudinal ligament buckling, ligamentum flavum in-folding, and intervertebral disc bulging causing severe narrowing of the central canal. After a hyperextension injury the patient sustained diffuse cord signal hyperintensity best seen on T2 imaging from C3—C6.

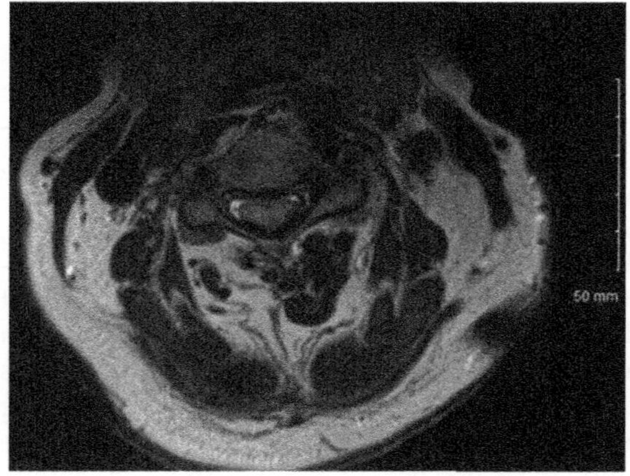

FIGURE 14.1B   T2 axial cervical spine MRI at level of C4. Hyperintensity of the central spinal cord with sparing of the peripheral spinal cord is characteristic of central cord syndrome.

an SCI and severe paralysis with low-energy or minor trauma in the absence of an acute fracture.[9] Historically, these types of injuries have been associated with the label "central cord syndrome"; however, this definition has been under debate in recent years.[10]

Given the high prevalence of spondylosis in the geriatric population, one can therefore assume that the elderly age group is at a higher risk of SCI from low-energy trauma. This raises the question of whether or not patients with moderate—severe canal stenosis warrant prophylactic surgical decompression in the absence of symptoms. Takao[11] tells us that there is a 125 times higher rate of cervical SCI without fracture when patients have severe central canal stenosis. These data are subject to limitations, and the authors suggest that prophylactic decompression would only prevent 0.017% of subjects with canal stenosis from developing a SCI. This is in stark contrast to the complication rate of 1%—5% for surgical decompression of the cervical spinal cord. Until higher quality studies are available clinicians should be aware that geriatric patients, or those with untreated preexisting severe central canal stenosis, are at risk for paralysis if they have a low-energy trauma. This has very particular considerations if these patients are undergoing any form of general anesthesia, and adequate education of the perioperative care team is vital.

Traumatic spinal cord injury

Rates of traumatic and atraumatic SCI have a bimodal distribution in terms of age; one peak occurs between 15 and 29 years of age, and the second occurs at greater than 65 years of age.[12] The mean age of SCI patients has been increasing in recent years as a result of an expansion of the older age group. This phenomenon mirrors the current "epidemiological transition" of the increasing size of the global elderly population.[13] Data extracted from the National Inpatient Sample also show there has been an increase in geriatric SCIs from all causes with an incidence of 84 cases/million in 1993 compared to 131 cases/million (2.7% annual increase) in 2012.[14]

The pattern of injury in geriatric patients is often different than the younger cohort. One in four geriatric patients with a cervical spine fracture has evidence of an SCI with one out of three being a complete injury.[15] Many have an incomplete and less severe SCI, which reflects the lower energy nature of the majority of the injuries. The level of SCI is also usually higher in elderly individuals, with injuries at the C4 level or higher being more likely in geriatric patients (42% compared to 26% in younger adults). Lower cervical or thoracolumbar injuries resulting in paraplegia are more common in the younger cohort (22% vs. 44%) likely secondary to the higher degree of spondylosis, and resulting decrease in motion, in the subaxial spine of geriatric patients.[16,17]

Geriatric patients, although shown to have a similar motor recovery compared to younger patients after SCI, often report worse functional outcomes.[18] Pooled analysis from the North American Clinical Trials Network demonstrates that age differentially affects the degree of recovery after surgery for SCI, with age over 65 having a greater magnitude of effect on the long-term functional independence measure with ASIA B or C patients, when compared with ASIA A or D patients. This has important implications, as geriatric ASIA B or C patients should therefore receive a greater focus on initial rehabilitation measures after surgery.

A retrospective study of patients with existing stenosis who suffer SCI has shown over 60% improvement by at least one ASIA Impairment Scale (AIS) grade after decompressive surgery. However, age did not affect the extent of recovery.[19]

Older age has a significant impact on mortality following SCI with mortality rates in patients over 75 years of age being 19%–29% compared to ~4% in the 16–25 age group. Furthermore, increasing age has a negative effect on AIS recovery after surgery, an effect that appears more pronounced with a higher severity of injury (AIS A or B).[14,18] These findings have largely been ascribed to the presence of age-related comorbidities and the higher rate of secondary complications in the older age group, though the exact factors that account for these discrepancies have not been fully described.

Frailty has been suggested to be an aggregate measure of physiological reserve and is a novel variable that has been correlated to outcomes after SCI. Several frailty indices have been developed secondary to the original Canadian Study of Health and Aging Clinical Frailty Scale, with the most common index in spine surgery being the modified frailty index (mFI), although many disease-specific indices exist in spine surgery.[20] The mFI uses metrics that are primarily derived from data collected in the National Surgical Quality Improvement Program, and previously represented an 11-point score which has been transformed into a shortened 5-point score in recent years (mFI-5).[21] The mFI-5 incorporates a deficit accumulation model using five comorbidities (chronic obstructive pulmonary disease, congestive heart failure, diabetes mellitus, functionally dependent status, and hypertension) and is a validated tool to predict frailty in geriatric spine surgery patients.[20] The severity of frailty has been demonstrated to be predictive of discharge destination, in-hospital adverse events, and mortality in geriatric patients.[22,23] It has also been shown to be independent of chronologic age to determine patient outcomes after spine surgery and SCI; however, geriatric patients are more likely to have worse frailty index scores and are therefore more likely to suffer adverse events and in-hospital mortality.[24] In addition, SCI patients with high frailty indices treated surgically may have worse postoperative outcomes including increased risk of surgical site infections, wound dehiscence, and pseudarthrosis although this has not been specifically studied in the spinal cord injured patient.[25]

For these reasons, older patients are less likely to receive operative management for SCI and have a longer length of hospital stay.[17] Even so, data from the NIS suggest the SCI mortality rate in those over the age of 85 has actually decreased from 24.2% in 1993 to 20.1% in 2012. This improvement is largely attributed to improved medical management and the emergence of geriatric medicine.[14]

Central cord syndrome

Classically, central cord syndrome (CCS) is defined as an incomplete SCI seen in patients with central canal stenosis, a neck hyperextension injury, and the absence of spinal fracture.[26] The upper extremities are affected to a larger extent than the lower limbs, and have a slower recovery. The traditional explanation for this pattern relied on the somatotropic organization of the corticospinal tracts, though this has largely been disproven by recent evidence.[10] The same is true for the long-held belief that operative intervention for CCS is best avoided in the acute injury period, with the majority of evidence now favoring emergency decompression as would be performed for any other form of acute SCI.[10] An understanding of CCS is key for spine surgeons and others providing care to trauma victims as the diagnosis may be easily missed in the acute setting and has a higher prevalence in the elderly given preexisting degenerative changes. It should be strongly considered and excluded in any traumatic patient presenting with motor or sensory disturbance in the limbs in the absence of an acute bony injury on the initial CT. Unfortunately, evidence suggests that motor recovery is worse in the geriatric CCS cohort compared to younger adults, with a concurrent increase in mortality.[27–29]

Medical considerations in the geriatric spinal cord injured patient

SCI patients are at increased risk for postoperative complications, including deep vein thrombosis (DVT), urinary tract infections, dysphagia, pressure ulcers, and pneumonia, all of which may be the cause of mortality.[17] Geriatric patients are more likely to have medical comorbidities, and therefore are more likely to suffer from medical complications after spinal fractures or SCI. The most frequent chronic complication after SCI is pressure ulcer formation. Younger age, complete injuries, pneumonia, atelectasis, mechanical ventilation, and violent injuries all increase the long-term risk of development of pressure ulcers.[30–32] Frequent turning of the patient every 2 hours has not been found to be universally effective in reducing the incidence of pressure ulcers although high-risk patients, including geriatric patients, may benefit most from frequent turning.[33] Options to help heal pressure ulcers include negative pressure wound therapy,[34] but if bone or other vital structures are exposed or if significant necrosis is present, operative debridement with tissue flap coverage is the standard of care.

The second most common complication from SCI is pneumonia and atelectasis resulting from intercostal and diaphragmatic paralysis and impaired clearance of respiratory secretions. Although this typically occurs in the acute period during the initial hospital admission, there is a 3.4% incidence of this complication annually.[31] Pneumonia and atelectasis are seen most frequently in geriatric patients who have decreased pulmonary reserve and complete SCIs, especially cervical spine injuries.[35]

Thromboembolic events are another significant cause of morbidity and mortality among SCI patients, especially in the geriatric inpatient population. The risk of DVT is reduced by appropriate administration of anticoagulation within 48 hours of operative intervention. Use of chemoprophylaxis decreases the incidence of

DVT fivefold.[36] However, most acute and chronic SCI complications are unavoidable, including the rate of pulmonary embolism after SCI, which has not been found to be decreased by timely administration of chemoprophylaxis. Since there has been no significant increase in return to the operative suite for hematoma evacuation or wound complications from early administration of DVT prophylaxis, it is recommended in all geriatric patients with SCI after surgery.[36]

Dysphagia is another common complication after SCI, presenting in 22.5% of patients who progress to rehabilitation. Geriatric patients, tracheostomy and mechanical ventilation, and an anterior cervical spine approach have been found to be independent predictors of dysphagia in SCI patients.[37] Dysphagia also predisposes patients to aspiration. Video fluoroscopy has demonstrated that up to 8% of patients have an aspiration after cervical SCI, of which approximately 20% will be asymptomatic. Therefore, geriatric patients and tracheostomy patients should be closely monitored for evidence of aspiration in the presence of a cervical SCI to minimize the risk of pneumonia.[38]

Finally, geriatric patients are at a higher risk than younger cohorts of having medical comorbidities. Since geriatric patients are known to have twice the delay from SCI to hospital arrival it should be assumed these patients have already exhausted much of their limited hemodynamic and pulmonary reserves early in the triage process. This makes emergent multispecialty comanagement imperative for patient optimization prior to proceeding to the operative suite.[17] Further, a complete evaluation of patient's medical comorbidities needs to be accomplished early in the triage process to determine if urgent reversal of chemoprophylaxis is needed in patients with atrial fibrillation. Although there is evidence supporting early surgery in SCI patients, preoperative optimization is paramount in geriatric patients who are at increased risk of mortality. Shared decision-making on timing of surgery should occur between the medical and surgical teams to optimize patient outcomes.

Multimodality analgesia regimen

Chronic pain is perhaps the most difficult sequelae to manage after SCI, with approximately one in three patients with a SCI experiencing "excruciating" long-term neuropathic pain.[39] Opioid use has been a pillar of pain management over the last 20 years, mainly due to the rapid onset of action, availability of multiple routes of administration, and predictable drug metabolism and side effects. However, opioid administration has come under scrutiny in recent years as dependence and poor prescribing practices have contributed to the well-publicized "Opioid Epidemic" worldwide.[40] Of particular concern, opioid administration in the elderly age group is associated with significant morbidity and adverse neurological and psychological side effects. From 2006 to 2013, elderly patients have seen opioid-related mortality increase in stark contrast to younger adults during the same time period.[41] Given these concerns, multimodality analgesia in the perioperative period has become the most appropriate regimen recommended for geriatric spine patients.[42]

Multimodality analgesia uses combination pharmacotherapy, administered by a number of caregivers in the perioperative process. For instance, the administration of preemptive analgesia at induction can complement instillation of local anesthetic agents to the operative site by the surgeons delivered just prior to waking, followed by sustained release of analgesia given on the ward. Recently, the German-Speaking Medical Society for Spinal Cord Injury (a multispecialty group comprised of predominantly psychiatrists and neurologists) came out with recommendations for pain management exclusively in SCI patients. Pregabalin and gabapentin both were given strong recommendations for the treatment of neuropathic pain and should be considered first-line therapy. Mild to moderate activity also carries a strong recommendation for therapeutic relief of musculoskeletal pain and should occur with directed stretching as first-line therapy in all patients. Baclofen also received a strong recommendation for first-line use when neuropathic pain is associated with spasticity. However, the multispecialty group came to a strong conclusion that opioids should only be considered in the management of neuropathic pain if anticonvulsants have not given adequate pain relief and they should be used as adjunctive therapy as opposed to monotherapy. This recommendation was based on the absence of evidence supporting opioid efficacy in this patient population and the high prevalence of their side effects including nausea, vomiting, constipation, and opioid dependence.[43]

Cost-effectiveness of surgery in SCI

Although complication rates after SCI in the geriatric population are high, surgery and rehabilitation after SCI can produce comparable functional outcomes for these patients, albeit at an increased expense. Baseline utility and costs have been gathered for the young and geriatric cohorts of the prospective, observational Surgical Timing in Acute Spinal Cord Injury Study. No significant baseline utility difference was noted between the two groups, but SCI treatment costs were $193,989.85 per quality-adjusted life year (QALY) gained in patients aged >65, compared to $94,043.42 per QALY gained in the younger cohort. When cost-effectiveness models were applied using surgical management and rehabilitation of the younger cohorts as the baseline, the incremental cost-effectiveness ratio analysis demonstrated that it cost an additional $5,655,557 per QALY gained

in the geriatric patient compared to the younger patient.[44] This is clearly a significant cost that has major implications for public health-care systems and represents a significant barrier to delivering optimal care for elderly spinal cord injured patients. Similarly, it also illustrates the need to manage elderly spinal cord injured patients early and aggressively to ensure the ensuing recovery trajectory is as close to younger adults as possible.

Vertebral compression fractures

Fragility fractures and issues relating to C2 dens fractures are addressed in a separate chapter of this book and will not be extensively reviewed here. Instead, we will focus on osteoporotic compression fractures of the thoracolumbar spine and fractures in patients with DISH as they pertain to SCIs.

Geriatric vertebral compression fractures typically occur in patients with osteoporosis in the setting of minimal trauma. Classically an axial load is imparted to the thoracolumbar spine and the decreased bone mineral density (BMD) is unable to resist the resulting compressive force. What ensues is typically a fracture of the superior vertebral endplate and underlying trabecular bone, which usually does not present with neurological deficits or cord injuries. However, there are reported instances of innocuous appearing thoracic compression fractures resulting in ASIA C SCIs.[45] Further, multilevel compression fractures may lead to sagittal imbalance, progressive kyphosis, and delayed motor deficits (Figs. 14.2A and B).[46] If there is any concern for a motor or sensory deficit in the presence of a compression fracture or suspicious mechanism of injury, a magnetic resonance imaging (MRI) should be urgently obtained for evaluation of the neural elements. Once managed, these patients should then be either referred to a bone fragility clinic or have an appropriate screening examination.

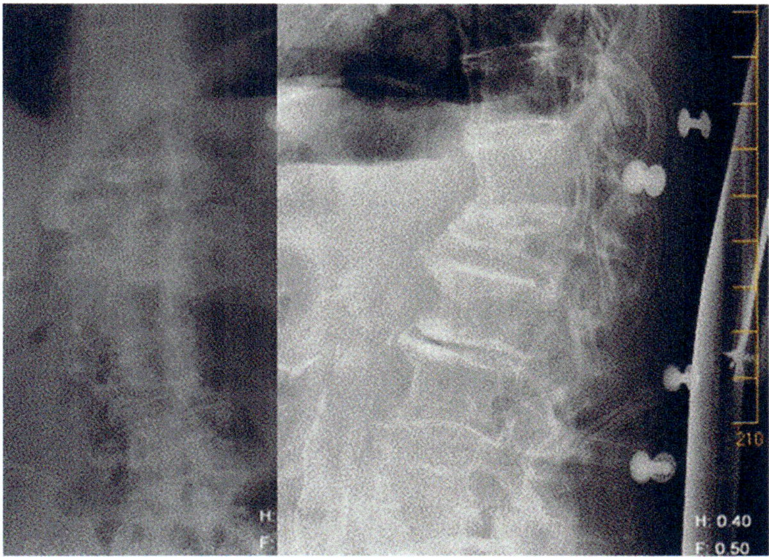

FIGURE 14.2A   AP and lateral upright radiographs of a patient with a chronic T12 compression fracture and an acute nondisplaced T11 chance fracture in the setting of premorbid DISH. Patient was managed nonoperatively into a TLSO brace.

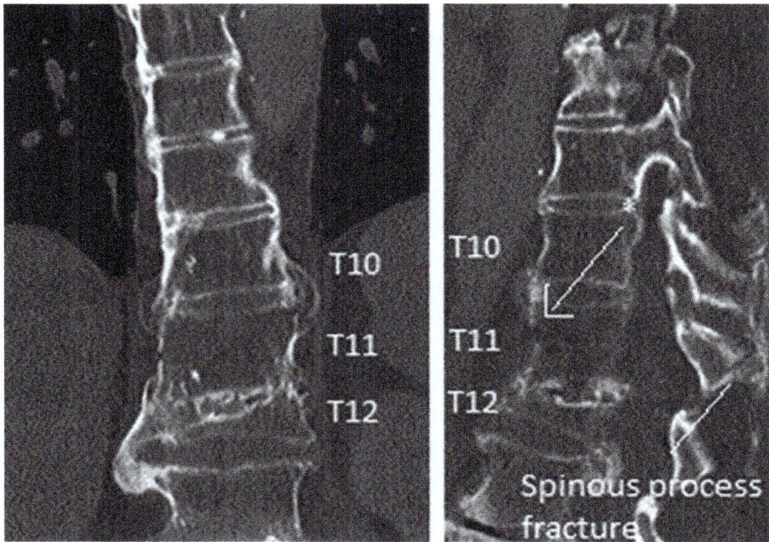

FIGURE 14.2B   Coronal and sagittal CT scan of the thoracic spine in the same patient with underlying DISH as visualized by the multilevel bridging osteophytes. The CT scan better demonstrates the nondisplaced chance fracture in the T11 vertebrae which extends into the spinous process. * represents fracture through the anterior vertebral cortex with fracture extending into the T11 spinous process.

Osteoporosis is common with 15% of Caucasian females and 5% of Caucasian males experiencing a vertebral compression fracture at some point in their lifetime.[47] In elderly patients (aged over 70) the incidence of compression fractures is close to 20%.[48] As the population continues to age, and rates of osteoporosis continue to increase globally, it is expected that the incidence of compression fractures will increase as well. This is alarming, as a 2010 US census estimated that 10.2 million adults have osteoporosis and 43.4 million Americans have low bone mass.[49] Osteoporosis is defined by the World Health Organization based on dual-energy X-ray absorptiometry scans. T-scores of −2.5 or less are consistent with osteoporosis, scores between −1.0 and −2.5 are classified as osteopenia, and scores between 0 and -1.0 are considered normal. The National Osteoporosis Foundation suggests screening investigations for all women 65 years and older and men 70 years and older, any person greater than 50 with a history of a low-energy fracture, and patients aged 50—69 with high-risk factors for osteoporosis including diabetes or other endocrine disorders, autoimmune disorders, sickle cell disease, corticosteroid use, or certain other medications are prone to lower BMD.[48]

Multiple risk factors for vertebral compression fractures have been identified and they closely parallel those for osteoporosis. These factors include increasing age, Caucasian ethnicity, tobacco use, estrogen deficiency, alcoholism, low vitamin D, low BMD, and an inactive lifestyle.[48] Another important risk factor is degenerative disc disease. Intervertebral discs provide load sharing with the bony architecture. However, as disc height decreases with age, the resulting alteration in load sharing between the discs and the vertebral body can lead to bony fatigue and insufficiency fractures. This is caused by changes in local trabecular density as the disc becomes dehydrated with age, thus losing its load-sharing capability.[50]

Patients with osteoporotic vertebral compression fractures should begin immediate medical management to prevent future pathologic fractures. Calcitonin is well known to reduce fracture-related pain through the first 4 weeks after injury, but it has not been shown to increase BMD and it has no role in osteoporotic patients with chronic vertebral compression fractures.[51] The most commonly prescribed medications for osteoporosis patients are calcium and vitamin D. The United States Preventative Services Task Force (USPSTF) has been unable to find sufficient evidence for the efficacy of either of these supplements in preventing subsequent osteoporotic fractures. However, due to the safety of these supplements, it is standard of care to provide them to all osteoporosis patients.[52] The USPSTF recommends dosing greater than 400 IU of vitamin D and greater than 1000 mg calcium. These recommendations are consistent with those of the NIH Osteoporosis National Resource Center, which recommends 1200 mg daily calcium and 800 IU vitamin D in all patients older than 70.

In addition to calcium and vitamin D, bisphosphonates are first-line pharmacologic agents to improve BMD by preventing further bone loss. Multiple bisphosphonates are currently approved for this purpose and they can be administered orally or IV with dosing regimens being daily to yearly depending on the drug. Based on a Cochrane database review of over 12,000 postmenopausal women, if a patient is started on, and compliant with, bisphosphonate treatment, a 45% relative risk reduction for vertebral fractures can be expected.[53] Although there is significant benefit to bisphosphonates, it is important the patient understands they are at increased risk for both osteonecrosis of the jaw and subtrochanteric femur fractures. However, it should be noted, both of these risks are quite rare with osteonecrosis of the jaw having a rate of around 1 in 100,000 patient-years.[54] Further, if new onset pain presents in either location, patients should have X-rays of the affected area and medication should be stopped immediately.

Newer options to improve BMD include denosumab (a human monoclonal antibody which inhibits RANKL) and teriparatide (a recombinant parathyroid hormone). These agents are typically used as second-line therapy, but can be considered first-line in patients with significant kidney disease or those who are unable to tolerate bisphosphonates. Both denosumab and teriparatide are given subcutaneously with dosing regimens of 60 mg every 6 months and 20 μg daily, respectively. Due to their subcutaneous injection route, patients unable to tolerate oral medications may be candidates for denosumab or teriparatide without first failing bisphosphonate treatment. Although a newer drug than bisphosphonates, denosumab has shown increased efficacy in improving lumbar spine BMD (5.3% vs. 4.2%), but it has not been shown to lower the incidence of vertebral fractures compared to alendronate.[55,56] A recent meta-analysis indicated that, unlike denosumab, teriparatide decreased vertebral compression fractures with a risk ratio of 0.57 compared to bisphosphonates.[57] The same authors also found that teriparatide significantly improved BMD in the lumbar spine after 6, 12, and 18 months administration compared to bisphosphonates. It appears teriparatide may also improve union rate after posterolateral spine fusions when compared to oral bisphosphonate treatment (83% vs. 68%).[58]

Diffuse idiopathic skeletal hyperostosis

DISH is often described as bridging osteophytes crossing at least four adjacent spinal levels. Although DISH can affect the lumbar spine it is most commonly present in the thoracic spine and is seen in approximately 20% of patients older than 50.[59] Due to the increased spine rigidity in DISH patients the spine is unable to withstand dynamic energy forces and 70% of fractures in this population are due to low-energy mechanisms.[60,61] For this reason, spine fractures often are not recognizable on plain films due to minimal fracture displacement such that CT is regarded as gold standard for diagnosis (Figs. 14.3A and B). Patients with DISH or AS presenting with new onset mechanical back pain, especially with a hyperextension mechanism of injury, should be evaluated with CT scan upon presentation even if no fracture is identified on plain radiographs. Additionally, 40% of patients with fracture may have a neurologic deficit. Although patients without a neurologic deficit may be managed with a brace, patients with a neurologic deficit should be managed surgically. It should be noted, however, that surgical complications may be as high as one in every three patients with DISH.[62] Further, 75% of patients who sustain a fracture and SCI with DISH will have no neurologic improvement after surgical stabilization. However, with nonsurgical management this number rises to 88%. These results were similar in patients with AS who had a neurologic injury after a spinal column fracture.[61] Thus, despite the high complication rate of surgical stabilization of spine fractures with a neurologic injury in patients with DISH or AS, it is likely preferred over nonsurgical management. Finally, though most patients with DISH and thoracolumbar spine fracture present with acute SCIs, patients without an SCI should be followed closely as delayed neurologic decline has been reported in this population.[63]

Conclusions

Geriatric patients are at a higher risk of both vertebral fractures and SCIs as a result of the increased prevalence of degenerative spine disorders in this patient population. SCI should be considered in all elderly trauma patients who have sensorimotor limb symptoms and should

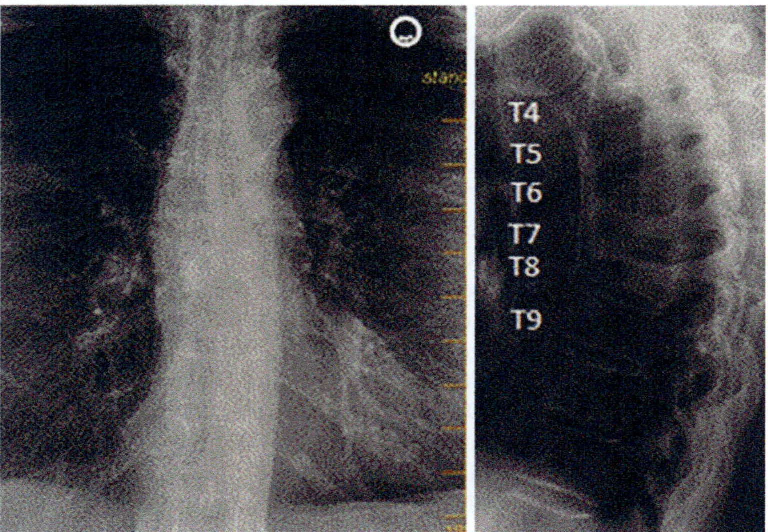

FIGURE 14.3A   AP and lateral thoracic spine radiographs with demonstrating diffuse demineralization of the vertebrae and compression fractures of T4, T5, T6, T7, and T8 in a patient with chronic back pain.

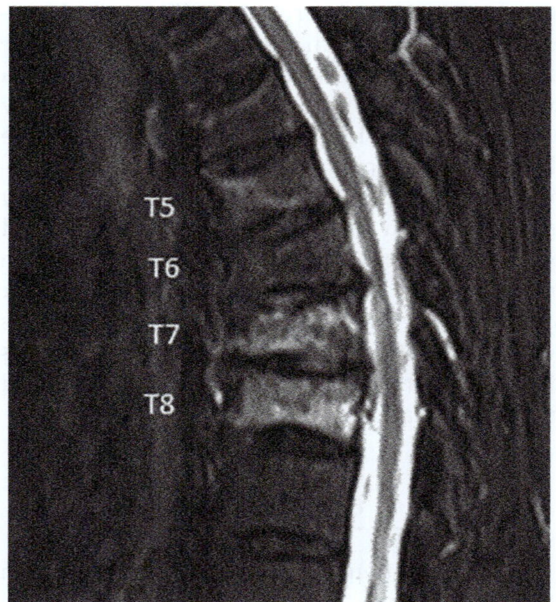

FIGURE 14.3B   Parasagittal MRI of the thoracic spine demonstrating the compression fractures of T4—T8. There is minimal marrow edema of T5 and T6 with profound edema of T7 and T8 representing ongoing inflammation. Evidence of edema suggests a possible role for percutaneous vertebral augmentation.

be expediently ruled out in the absence of spinal fracture with urgent MRI. We recommend timely operative management of all acute SCI, including CCS, in appropriate elderly patients with aggressive postoperative management to reduce secondary complications. Many tools are now available to aid in the preoperative decision-making process and prognostication in SCI, including the use of frailty indices, and their use is encouraged when making treatment decisions in geriatric SCI patients. Finally, spine surgeons must be aware of pathologies that require specific consideration in older patients, such as osteoporosis and DISH, in order to provide optimal outcomes for elderly patients with an SCI.

References

1. Hagen EM, et al. The clinical significance of spinal cord injuries in patients older than 60 years of age. *Acta Neurol Scand* 2005;**112**:42—7.

2. Pray C, et al. Bone mineral density and fracture risk in ankylosing spondylitis: a meta-analysis. *Calcif Tissue Int* August 2017;**101**(2):182–92.

3. Matsumoto M, et al. MRI of cervical intervertebral discs in asymptomatic subjects. *J Bone Surg Br* 1998;**80**:19–24.

4. Lee TH, et al. Prevalence of disc degeneration in asymptomatic Korean subjects. Part 2: cervical spine. *J Korean Neurosurg Soc* 2013;**53**(2):89–95.

5. Teraguchi M, et al. Progression, incidence, and risk factors for intervertebral disc degeneration in a longitudinal population-based cohort: the Wakayama Spine Study. *Osteoarthritis Cartilage* 2017;**25**(7):1122–31.

6. Teresi LM, et al. Asymptomatic degenerative disk disease and spondylosis of the cervical spine: MR imaging. *Radiology* 1982;**164**:83–8.

7. Nakashima H, et al. Narrow cervical canal in 1211 asymptomatic healthy subjects: the relationship with spinal cord compression on MRI. *Eur Spine J* 2016;**25**(7):2149–54.

8. Hayashi K, et al. MRI findings in patients with a cervical spinal cord injury who do not show radiographic evidence of a fracture or dislocation. *Paraplegia* 1995;**33**:212–5.

9. Oichi T, et al. Preexisting severe cervical spinal cord compression is a significant risk factor for severe paralysis development in patients with traumatic cervical spinal cord injury without bone injury: a retrospective cohort study. *Eur Spine J* 2016;**25**(1):96–102.

10. Divi SN, et al. Management of acute traumatic central cord syndrome: a narrative review. *Global Spine J* 2019;**9**(Suppl. l):89S–97S.

11. Takao T, et al. Clinical relationship between cervical spinal canal stenosis and traumatic cervical spinal cord injury without major fracture or dislocation. *Eur Spine J* 2013;**22**:2228–31.

12. van den Berg ME, et al. Incidence of spinal cord injury worldwide: a systematic review. *Neuroepidemiology* 2010;**34**(3):184–92. Discussion 192.

13. Witiw CD, Fehlings MG. Acute spinal cord injury. *J Spinal Disord Tech* 2015;**28**(6):202–10.

14. Jain NB, et al. Traumatic spinal cord injury in the United States, 1993–2012. *J Am Med Assoc* 2015;**313**(22):2236–43.

15. Lomoschitz FM, et al. Cervical spine injuries in patients 65 years old and older. *Am J Roentgenol* 2002;**178**(3):573–7.

16. Fassett DR, et al. Mortality rates in geriatric patients with spinal cord injuries. *J Neurosurg Spine* 2007;**7**(3):277–81.

17. Ahn H, et al. Effect of older age on treatment decisions and outcomes among patients with traumatic spinal cord injury. *CMAJ* 2015;**187**:873–80.

18. Wilson JR, et al. Defining age-related differences in outcome after traumatic spinal cord injury: analysis of a combined, multicenter dataset. *Spine J* 2014;**14**(7):1192–8.

19. Ronzi Y, et al. Spinal cord injury associated with cervical spinal canal stenosis: outcomes and prognostic factors. *Ann Phys Rehabil Med* 2017;**61**:27–32.

20. Chan V, et al. Frailty adversely affects outcomes of patients undergoing spine surgery: a systematic review. *Spine J* 2021;**21**(6):988–1000.

21. Weaver D, et al. The modified 5-item frailty index: a concise and useful tool for assessing the impact of frailty on postoperative morbidity following elective posterior lumbar fusions. *World Neurosurg* 2019;**124**:e626–32.

22. Cheung A, et al. Canadian study of health and aging clinical frailty scale: does it predict adverse outcomes among geriatric trauma patients? *J Am Coll Surg* 2017;**225**:658–65 e3.

23. Ali R, et al. Use of the modified frailty index to predict 30-day morbidity and mortality from spine surgery. *J Neurosurg: Spine SPI* n.d.;**25**(4):537–41.

24. Banaszek D, et al. The effect of frailty on outcome after traumatic spinal cord injury. *J Neurotrauma* 2020;**37**(6):839–45.

25. Miller EK, et al. An assessment of frailty as a tool for risk stratification in adult spinal deformity surgery. *Neurosurg Focus FOC* 2017;**43**(6):E3.

26. Nowak DD, et al. Central cord syndrome. *J Am Acad Orthop Surg* 2009;**17**(12):756–65.

27. Brodell D, et al. National trends in the management of central cord syndrome: an analysis of 16,134 patients. *Spine J* 2015;**15**(3):435–42.

28. Tow AM, Kong KH. Central cord syndrome: functional outcome after rehabilitation. *Spinal Cord* 1998;**36**:156–60.

29. Ishida Y, Tominaga T. Predictors of neurologic recovery in acute central cervical cord injury with only upper extremity impairment. *Spine* 2002;**27**(15):1652–8.

30. Krishnan S, et al. Association between presence of pneumonia and pressure ulcer formation following traumatic spinal cord injury. *J Spinal Cord Med* 2016;**40**:415–22.

31. McKinley WO, et al. Long-term medical complications after traumatic spinal cord injury: a regional model systems analysis. *Arch Phys Med Rehabil* 1999;**80**:1402–10.

32. Brienza D, et al. Predictors of pressure ulcer incidence following traumatic spinal cord injury: a secondary analysis of a prospective longitudinal study. *Spinal Cord* 2018;**56**:28–34.

33. Rich SE, et al. Frequent manual repositioning and incidence of pressure ulcers among bed-bound elderly hip fracture patients. *Wound Repair Regen* 2011;**19**(1):10–8.

34. Argenta LC, et al. Vacuum-assisted closure: state of clinic art. *Plast Reconstr Surg* 2006;**117**(7 Suppl. l). p. 127S–42S.

35. Sezer N. Chronic complications of spinal cord injury. *World J Orthoped* 2015;**6**:24—33.

36. Zeeshan M, et al. Optimal timing of initiation of thromboprophylaxis in spine trauma managed operatively: a nationwide propensity-matched analysis of trauma quality improvement program. *J Trauma Acute Care Surg* 2018;**85**(02):387—92.

37. Kirshblum S, et al. Predictors of dysphagia after spinal cord injury. *Arch Phys Med Rehabil* 1999;**80**:1101—5.

38. Shin CJ, et al. Dysphagia in cervical spinal cord injury. *Spinal Cord* 2011;**49**:1008—13.

39. Siddall PJ, et al. A longitudinal study of the prevalence and characteristics of pain in the first 5 years following spinal cord injury. *Pain* 2003;**103**:249—57.

40. Hagemeier NE. Introduction to the opioid epidemic: the economic burden on the healthcare system and impact on quality of life. *Am J Manag Care* 2018;**24**(10 Suppl. l):S200—6.

41. West NA, et al. Trends in abuse and misuse of prescription opioids among older adults. *Drug Alcohol Depend* 2015;**149**:117—21.

42. a Wilson JRF, et al. *Degenerative cervical myelopathy; a review of the latest advances and future directions in management.* Neurospine; 2019.b Diederichs G, et al. Diffuse idiopathic skeletal hyperostosis (DISH): relation to vertebral fractures and bone density. *Osteoporos Int* 2011;**22**(6):1789—97.

43. Franz S, et al. Management of pain in individuals with spinal cord injury: guideline of the German-speaking medical society for spinal cord injury. *Ger Med Sci* 2019;**17**:Doc05.

44. Furlan JC, et al. Surgical management of the elderly with traumatic cervical spinal cord injury: a cost-utility analysis. *Neurosurgery* 2016;**79**:418—25.

45. Demir SO, et al. Spinal cord injury associated with thoracic osteoporotic fracture. *Am J Phys Med Rehabil* 2007;**86**(3):242—6.

46. Bruno AG, et al. The effect of thoracic kyphosis and sagittal plane alignment on vertebral compressive loading. *J Bone Miner Res* 2012;**27**(10):2144—51.

47. Wehren LE. The epidemiology of osteoporosis and fractures in geriatric medicine. *Clin Geriatr Med* 2003;**19**(2):245—58.

48. Lehman RA, et al. Management of osteoporosis in spine surgery. *J Am Acad Orthop Surg* 2015;**23**(4):253—63.

49. Wright NC, et al. The recent prevalence of osteoporosis and low bone mass in the United States based on bone mineral density at the femoral neck or lumbar spine. *J Bone Miner Res* 2014;**29**(11):2520—6.

50. Adams MA, Dolan P. Biomechanics of vertebral compression fractures and clinical application. *Arch Orthop Trauma Surg* 2011;**131**(12):1703—10.

51. Knopp-Sihota JA, et al. Calcitonin for treating acute and chronic pain of recent and remote osteoporotic vertebral compression fractures: asystematic review and meta-analysis. *Osteoporos Int* 2012;**23**(1):17—38.

52. Moyer VA. Vitamin D and calcium supplementation to prevent fractures in adults: U.S. Preventive Services Task Force recommendation statement. *Ann Intern Med* 2013;**158**(9):691—6.

53. Wells GA, et al. Alendronate for the primary and secondary prevention of osteoporotic fractures in postmenopausal women. *Cochrane Database Syst Rev* 2008;(1): CD001155.

54. Khosla S, et al. Bisphosphonate-associated osteonecrosis of the jaw: report of a task force of the American Society for Bone and Mineral Research. *J Bone Miner Res* 2007; **22**(10):1479—91.

55. Pedersen AB, et al. Comparison of risk of osteoporotic fracture in denosumab vs alendronate treatment within 3 years of initiation. *JAMA Netw Open* 2019; **2**(4):e192416.

56. Brown JP, et al. Comparison of the effect of denosumab and alendronate on BMD and biochemical markers of bone turnover in postmenopausal women with low bone mass: a randomized, blinded, phase 3 trial. *J Bone Miner Res* 2009;**24**(1):153—61.

57. Yuan F, et al. Teriparatide versus bisphosphonates for treatment of postmenopausal osteoporosis: a meta-analysis. *Int J Surg* 2019;**66**:1—11.

58. Ohtori S, et al. Teriparatide accelerates lumbar posterolateral fusion in women with postmenopausal osteoporosis: prospective study. *Spine* 2012;**37**(23):E1464—8.

59. Hiyama A, et al. Prevalence of diffuse idiopathic skeletal hyperostosis (DISH) assessed with whole-spine computed tomography in 1479 subjects. *BMC Muscoskel Disord* 2018;**19**(1):178.

60. Mader R, et al. Diffuse idiopathic skeletal hyperostosis (DISH): where we are now and where to go next. *RMD open* 2017;**3**(1):e000472.

61. Westerveld LA, et al. Spinal fractures in patients with ankylosing spinal disorders: a systematic review of the literature on treatment, neurological status and complications. *Eur Spine J* 2009;**18**(2):145—56.

62. Whang PG, et al. The management of spinal injuries in patients with ankylosing spondylitis or diffuse idiopathic skeletal hyperostosis: a comparison of treatment methods and clinical outcomes. *J Spinal Disord Tech* 2009;**22**(2):77—85.

63. Yamamoto T, et al. Delayed leg paraplegia associated with hyperextension injury in patients with diffuse idiopathic skeletal hyperostosis (DISH): case report and review of the literature. *J Surg Case Rep* 2017;**2017**(3): rjx040.

Spine Trauma: Areas of controversy and Emerging Concepts

Joseph H. McMordie[1], Jamie R.F. Wilson[1], F. Cumhur Oner[2], Alexander R. Vaccaro[3], Michael G. Fehlings[4]

[1]Department of Neurosurgery, University of Nebraska Medical Center, Omaha, NE, United States; [2]Department of Orthopaedics, University Medical Center Utrecht, Utrecht, the Netherlands; [3]Department of Orthopedic Surgery, Rothman Orthopedic Institute, Thomas Jefferson University, Philadelphia, PA, United States; [4]Division of Neurosurgery, Toronto Western Hospital, University Health Network, Toronto, ON, Canada

Introduction

The management of spine trauma has evolved tremendously over the last 20 years. From developing a greater understanding of the pathophysiology of spinal cord injury, advances in imaging modalities, or greater expertise with operative techniques, the vast improvements seen in the care of spine trauma patients have translated to better functional recovery, better quality-of-life outcomes, and improved socioeconomic measures. However, with the improved standards of care in spine trauma, some areas of controversy have emerged that have drawn contrasting opinions. This chapter will serve to highlight and discuss the evidence behind some of the current controversial clinical scenarios: optimal management of A3/A4 thoracolumbar fractures, minimally invasive surgery (MIS) in spine trauma, operative management of C2 peg fractures in the elderly, the timing of surgery in central cord syndrome (CCS), and the role of magnetic resonance imaging (MRI) in the management of spine fractures.

Management of AOSpine A3/A4 fractures: operative versus nonoperative

Thoracolumbar spine fractures are the most common fractures of the axial skeleton with a significant portion occurring at the thoracolumbar junction, positioned between the immobile thoracic spine and flexible lumbar spine.[1,2] Various classifications of thoracolumbar fractures exist and are based largely on bony morphology, mechanism of injury, anatomic stability, and neurological status. The AOSpine Thoracolumbar Spine Injury Classification

Neural Repair and Regeneration after Spinal Cord Injury and Spine Trauma
https://doi.org/10.1016/B978-0-12-819835-3.00022-8

© 2022 Elsevier Inc. All rights reserved.

System is one of the most widely accepted means of classifying thoracolumbar trauma. It was developed by the AOSpine Trauma Knowledge Forum to serve as a comprehensive and clinically applicable classification system. The classification scheme was developed by consensus review of nine fellowship trained spine surgeons grading a selection of cases from the AOSpine database. Evaluation of each injury is based on three parameters: morphologic classification of the fracture, neurological status, and clinical modifiers. Morphologic classification is graded by injury types A, B, or C. Type A includes compression type injuries involving a single vertebral level, Type B relates to fractures with failure of the posterior or anterior tension band without gross translation, and Type C describes failure of all spinal elements leading to dislocation or total disruption of a soft-tissue hinge even if translation is not present.[3]

Type A injuries are subdivided in five subtypes. A0 describes an injury without fracture to the vertebral body or a clinically insignificant fracture to the transverse or spinous process. There is no concern for mechanical instability or neurological deficit. A1 fractures include wedge compression fractures with injury to a single endplate without involvement of the posterior vertebral body wall. A2 fractures describe split-type fractures without involving the posterior vertebral body wall. A3 fractures are considered incomplete burst fractures involving the posterior vertebral body wall and only one endplate. A4 fractures involve the posterior vertebral body wall and both endplates (Figs. 15.1 and 15.2). In all Type A fractures, the posterior ligamentous tension band remains intact although lamina fractures may be present in the Type A3 and A4.

Type B fractures are divided into three subtypes. Type B1 fractures involve only a single vertebral level and describe an osseous failure of the posterior tension band extending into the vertebral body such as a chance fracture. B2 injuries include disruption of the posterior tension band with or without bony fracture and may be found in conjunction with compression type injuries, which are classified according to the appropriate type A injury in addition to the B2 injury. B3 injuries disrupt the anterior tension band and extend through the disc space or vertebral body, leaving the posterior hinge element intact.

Type C injuries do not contain subclassifications as they describe a complete disruption of all spinal elements. Displacement in any plane may be present due to the disassociation of cranial and caudal elements.

Neurological status is also taken into consideration with fracture morphology and is based on five categories. N0 designates neurologically intact patients; N1 represents patients with a

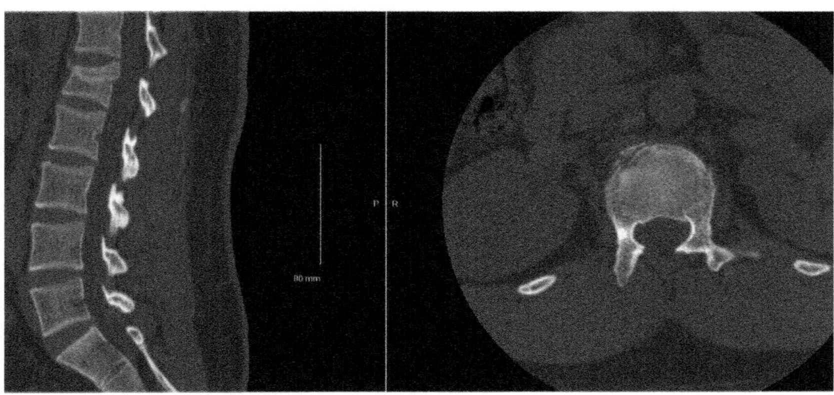

FIGURE 15.1   L1 type A3 fracture.

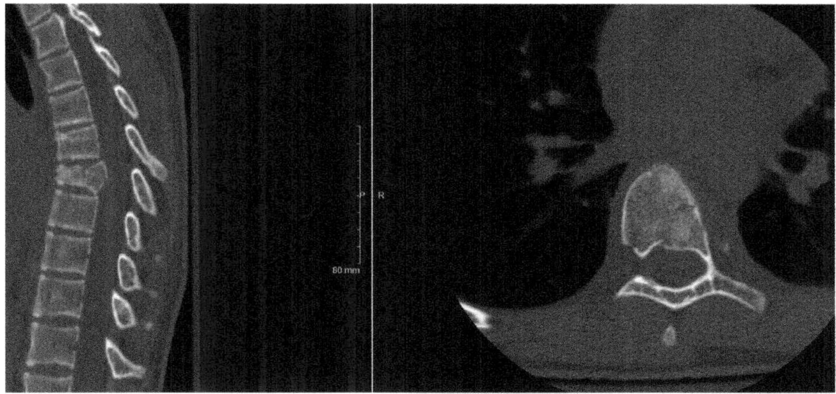

FIGURE 15.2   T7 type A4 fracture.

transient neurological deficit that has resolved; N2 refers to patients with radicular symptoms; N3 denotes incomplete spinal cord injury or cauda equina injury; and N4 indicates a complete spinal cord injury. If the patient is unable to be examined NX is used as the designation.

Two case-specific modifiers were included in the classification although not applicable to all patients. M1 describes fractures with an indeterminate injury to the posterior tension band, which is important for determination of fracture stability and ultimately operative stabilization. M2 denotes patient comorbidities, which may play a role in determining surgical intervention such as osteoporosis, rheumatologic conditions, or diffuse idiopathic skeletal hyperostosis (DISH).

The simplicity of the AO Spine classification system has led to its widespread adoption and clinical application by spine surgeons. Our understanding of the patterns of spine trauma grows with continued improvements in spinal imaging, which further informs surgical decision-making. Despite global adoption of the AO Spine classification system, regional treatment differences exist around the world for management of different fracture types. In A0—A2 fracture types, general agreement exists for conservative management while more severe fracture types B and C or those with neurological deficits benefit from surgical intervention.

However, difficulties arise when recommending definitive treatment algorithms for A3 and A4 fracture types, as prior medical literature and classification schemes have failed to define these fractures as separate entities.[4] Although regional differences and experience have been shown not to affect a spine surgeon's ability to correctly identify A3 and A4 fractures, the management of Type A3 and A4 thoracolumbar fractures remains controversial.[5]

Nonoperative treatment may be considered in patients who are poor surgical candidates due to associated traumatic injuries or existing medical comorbidities. Treatment often consists of a combination of bed rest, thoracolumbar bracing, and rehabilitation with close radiographic follow-up. These treatment options allow for avoidance of the higher costs and morbidity that can be associated with surgical intervention. However, surgery remains an option for patients undergoing conservative management with unsatisfactory clinical outcomes.

Surgical fixation provides immediate stabilization, earlier pain control, faster mobilization, and prevention or correction of deformity. A traumatic kyphotic deformity of >35 degrees is often referenced as an indication of surgical fixation.[6] Surgical approach is variable and largely dependent on fracture morphology and surgeon preference. Potential approaches range from posterior only instrumentation with or without fusion to anterior stand-alone corpectomies or a combination of both. Anterior reconstruction is largely dependent on the degree of vertebral body comminution and need for anterior column reconstruction.[7] Depending on the approach taken, complication risks vary as well as postoperative pain and recovery.

First and foremost, the most important aspect of the initial management of A3/4 fractures is the exclusion of posterior ligamentous injury that would automatically change the classification to a B type injury. The authors strongly recommend the use of MRI for this purpose, which is much more sensitive in identifying posterior element injury compared to CT or standard X-ray. Two randomized control trials have addressed the question of operative versus nonoperative treatment of A3/A4 fractures with conflicting results. Wood et al. studied 47 patients between 1992 and 1998 with thoracolumbar burst fractures randomly assigned to surgical fixation or cast/thoracolumbosacral orthosis. Initially, at an average follow-up of 44 months no significant differences in clinical or radiographic outcomes were found between the two groups. However, when 37 of the original 47 patients were reevaluated at 20-year follow-up, the nonoperative group showed improved pain and functional outcome scores when compared to the operative group. There was no increase in adjacent segment disease or sagittal imbalance noted between the surgically and nonsurgically treated groups.[8] Siebenga et al. randomized 24 patients with thoracolumbar fractures to operative fixation or Jewett hyperextension brace. At 4-year follow-up, the operative group showed lower pain and disability scores as well as less kyphotic deformity.[9] The differing outcomes in these two studies may possibly result from the surgical techniques implemented by each. Woods performed a multicenter study, which employed various surgical techniques at the discretion of the operative surgeon. These techniques ranged from 2- to 5-level posterior stabilizations and fusions as well as anterior stand-alone constructs of 1- or 2-level fibular and rib strut grafts.

Siebenga on the other hand performed a uniformed surgical technique of short-segment posterior stabilizations with pedicle screws.[10] A more recent randomized trial, performed across three Canadian institutions, enrolled 92 patients with acute A0—3 fractures between T11—L3 over the age of 60 years to either thoracolumbar bracing or no orthosis.[11] At 3 months, 1 year, and 2 years the level of disability and quality-of-life outcome scores between both groups was equivalent. Five patients required surgical fixation before discharge, which illustrates the importance of identifying patient factors that may contribute to early treatment failure. Factors such as the presence of other polytrauma, obesity, osteoporosis, hyperostosis (DISH, AS), and significant metabolic deficiencies will significantly increase the risk of failure of nonoperative management. In these cases, early consideration should be given to operative fixation. Similarly, early consideration for surgical management should be considered to patients with significantly comminuted fractures (i.e., Grade 3 on the Gaines load-sharing classification) as this would suggest a much higher risk of treatment failure with nonoperative management.[12]

A recent systematic review and meta-analysis evaluated treatment outcomes of types A3 and A4 fractures in neurologically intact patients through a review of retrospective and prospective data. It included 12 studies each with a minimum of 20 patients. The pooled results did not show any difference for pain and disability in operative and nonoperative groups. The authors note that in the absence of evidence to definitively favor one treatment option, consideration should be given to complication incidence, cost of intervention, and time until full recovery. Of note, the authors reported that surgical complications observed in the studies did clearly favor nonoperative management. However, time required before returning to work was inconclusive due to contradicting results among the studies. Overall, this study was somewhat limited by the heterogeneity in follow-up time ranging from 1 to 10 years and nonoperative treatment modalities.[13]

The clinical decisions for the treatment of A3/A4 fractures remain controversial, and a significant variation exists worldwide regarding the management of these fractures. It has been suggested access to health care, cultural differences toward surgical management, or the reduced cost burden of conservative management may account for this difference.[5] In an effort to clarify treatment for these fractures, AOSpine is currently sponsoring a multicenter cohort analysis investigating surgical versus nonsurgical treatment for neurologically intact patients with A3 and A4 fractures. Enrollment is currently ongoing with an estimated completion date of October 2022.

MIS versus open surgery for fractures without deformity

Thoracolumbar spine fractures carry a broad socioeconomic impact affecting quality of life and prolonged absence from work. As a result, treatment is ever evolving particularly for unstable injuries.[14] The long-standing focus of spine surgery for unstable spinal injuries has been to reduce and stabilize fractures while decompressing neurological structures if necessary. These goals can be achieved through various techniques whether minimally invasive or open. Comparison between open and minimally invasive spinal fracture stabilization is particularly applicable in pedicle screw fixation, which is commonly used as it provides 3-column fixation.

Open treatment of unstable thoracolumbar fractures has long been established as an effective treatment option.[15] Open techniques involve elevation of the paraspinal muscles off of the lamina and facets for identification of the pedicle screw entry point. Extensive bony exposure also allows for arthrodesis of each spinal segment if appropriate. However, open treatment has been associated with high infection rates and significant blood loss. Subperiosteal dissection may compromise the neurovascular supply of the paraspinal muscles resulting in postoperative atrophy and scarring. Muscular injury is also compounded by significant muscular retraction.[16] Recent advances in MIS have sought to reduce these complications by maintaining the soft tissues of the spine.

Originally developed for the treatment of degenerative pathology, minimally invasive procedures have transitioned to the treatment of polytrauma patients with a goal of reducing postoperative complications and recovery time. Minimally invasive procedures aim at reducing blood loss, surgical time, iatrogenic muscle trauma, postoperative infections, and postoperative pain when compared to traditionally open approaches.[17] Minimally invasive fixation is performed in a percutaneous manner with the aid of either direct fluoroscopy or image-guided navigation. It should be noted that significant fluoroscopy time has been associated with percutaneous pedicle screw placement, carrying the potential to exceed a surgeon's occupational exposure limit.[18] Fortunately, with the development of image-guided navigation, surgeon-related radiation exposure has been significantly reduced.[19] Although percutaneous pedicle screw fixation is effective for fracture stabilization, it does not provide arthrodesis between vertebral segments, except in particular pathologies such as the treatment of fractures in ankylosing spondylitis (AS) patients or bony chance fractures where realignment will facilitate spontaneous arthrodesis.

Instrumentation without fusion using minimally invasive techniques has been reported for the effective treatment of traumatic thoracolumbar fractures while maintaining segmental mobility.[20,21] Additionally, hardware may be removed after fracture healing to restore spinal mobility.[22] Fig. 15.3 shows imaging of T12 3-column fracture before and after percutaneous fixation. Unfortunately, MIS has been largely limited in its application due to the associated steep learning curve, the challenges in correction of severe deformities, and the inability to achieve arthrodesis through percutaneous screw fixation

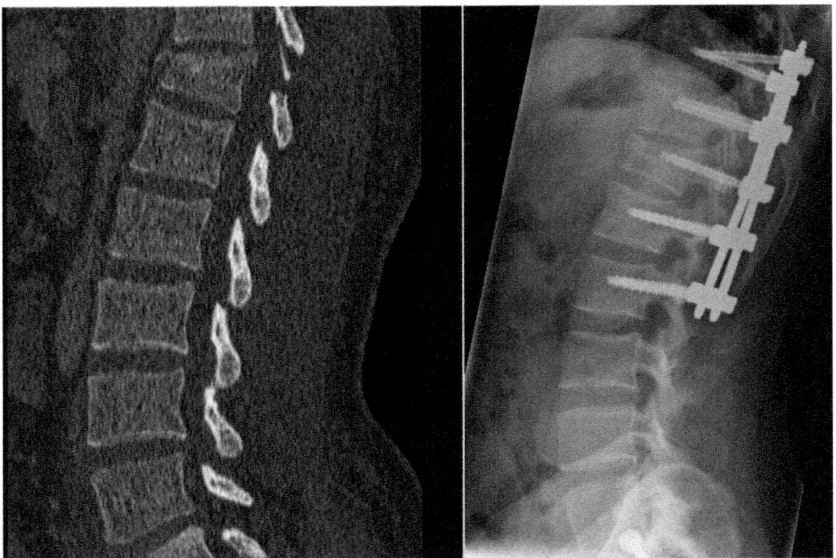

FIGURE 15.3   Percutaneous fixation of T12 3-column fracture.

techniques.[23] However, as technology grows and minimally invasive techniques become more commonplace, learning curves are likely to flatten as residents and fellows develop minimally invasive techniques while still in training.[24]

In 2016, McAnany et al. performed a meta-analysis of six papers comparing open to minimally invasive pedicle screw fixation for the treatment of unstable thoracolumbar fractures. Each study had a minimum of 10 patients with 6 months of follow-up. Significantly less operative time and blood loss were noted for the minimally invasive groups. A difference in postoperative pain reduction was not significant but favored the minimally invasive groups, while changes in vertebral body height and local kyphosis were not significantly different between open and minimally invasive groups. The authors note either minimally invasive or open fixation may effectively manage thoracolumbar fractures with the potential benefit of decreased blood loss and operative time given to minimally invasive techniques.[25]

More recently in 2018, Tian et al. performed a systematic review and meta-analysis to compare the efficacy of percutaneous pedicle screw fixation to standard open procedures. Nine studies comprising a total of 433 patients were reviewed. Of the primary outcomes, blood loss, operative time, and postoperative pain were noted to be lower in the minimally invasive groups. No significant difference was noted between open and minimally invasive for postoperative cobb angle and adverse events. The results showed for secondary outcomes percutaneous procedures reduced the incision size, postoperative drainage, and postoperative hospital stay. However, there were no significant differences in fluoroscopy times, hospitalization costs, or functional outcomes as measured by ODI. The authors do note that fracture reduction and deformity correction is limited with polyaxial pedicle screws and non-adjustable rods used for percutaneous fixation. Overall, the authors report there is positive potential for minimally invasive procedures to replace open surgery without an increase in complications but caution that randomized controlled trials are needed to assess long-term outcomes.[26]

To further address the role of MIS in spine trauma a multicenter randomized control trial is currently ongoing for open versus minimally invasive fixation of thoracolumbar fractures. The primary outcome is postoperative pain followed by several secondary outcomes including intraoperative bleeding, surgery time, length of hospital stay, vertebral segment kyphosis, fractured vertebral body height, accuracy of the pedicle screws, and breakage of implants. The study began in 2017 and is estimated for completion in May 2021 with a total estimated enrollment of 60 participants.[27]

Operative management of type 2 odontoid fractures in the elderly

Odontoid fractures are the most common cervical spine fracture of the elderly and carry a relatively high mortality rate of 18% at 1 year in the geriatric population.[28] Most geriatric patients presenting with odontoid fractures report a history of low impact trauma such as a ground level fall. On evaluation, computed tomography is the preferred imaging techniques due to its high sensitivity for upper cervical spine fractures.[29]

Odontoid fractures typically result from flexion—distraction injuries and are classified into three types depending on fracture location: type I through the odontoid tip, type II through the odontoid base, and type III extends from the odontoid base into the vertebral body (Fig. 15.4).[30] In most cases type I and type III fractures are favorably managed in a rigid collar.[31] Despite type II odontoid fractures being the most common type in the elderly population, management remains controversial.

Type II odontoid fractures are overall more common than types I and III possibly due to a decrease in bone density at the odontoid base that occurs with aging.[32,33] Additionally, nonunion of type II odontoid fractures is high in the elderly population reported well over 50%.[34] Poor healing is thought to result from a watershed area and lack of cancellous bone at

the base of the dens. Additionally, the overall surface area at the base of the dens is relatively small while ligamentous attachments of the dens apply distracting forces to the fracture preventing healing. Due to frailty and high rates of osteoporosis in the elderly, optimization of bone health is extremely important in any treatment plan.

Much debate has occurred over management of type II fractures in the geriatric population given the high rates of comorbidities and increased surgical risk of elderly patients. Recently, studies have shown decreased mortality and improved functional outcomes with operative treatment.[28,35] Ultimately, more consideration is being given to operative stabilization of geriatric type II fractures as a result of multiple factors contributing to nonunions and higher mortality rates observed in the elderly with nonoperative management.[36] Particularly, halo vest application has been associated with high rates of respiratory compromise, failure to thrive, and mortality rates of up to 42% in the geriatric population.[37]

Operative fixation of type II odontoid fractures may be approached anteriorly or posteriorly. When approached anteriorly, the odontoid screw enters at the anterior inferior margin of the C2 body and is directed upward to the apex of the odontoid, penetrating the cortex of the odontoid cap (Fig. 15.5). Advantages of the anterior approach include preserved motion of the C1—2 joint and intraoperative reduction of the fracture. Disadvantages include dysphagia and difficulty with intraoperative imaging as well as instrumentation failure in osteopenic patients. Contraindications to anterior fixation include comminuted fractures, nonreducible fractures, chronic fractures, and fractures extending from the anterior inferior to posterior superior portion of the dens. Patients with barrel chests or short necks are also poor candidates as the appropriate screw trajectory to the odontoid apex is limited.[38] Given the limitations of the anterior approach, many authors recommend its use mainly in younger patients with the appropriate bone quality, fracture morphology, and body habitus.[39]

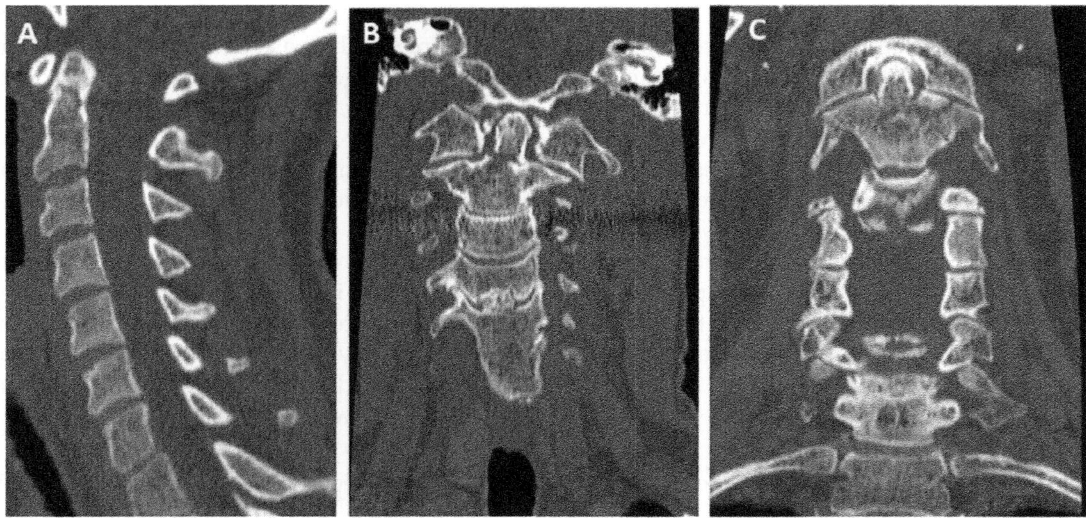

FIGURE 15.4   (A) Type 1 odontoid fracture. (B) Type 2 odontoid fracture. (C) Type 3 odontoid fracture.

Posterior approaches utilize arthrodesis of C1—2 joint, which may be accomplished through wiring, transarticular screws, or C1 lateral mass screws with pars or interlaminar screws at C2 connected by a rod (Fig. 15.6). Posterior fixation has a lower reported nonunion rate when compared to anterior odontoid screws.[40] Disadvantages of the posterior approach include longer reported operative times and 50% loss in axial rotation of the cervical spine.[41] While a 50% loss of axial rotation is significant, it may be better tolerated in the geriatric population due to lower function demands.[31]

Several studies exist when comparing anterior versus posterior approaches in the elderly population. Schroeder et al. performed a systemic review of patients over the age of 60 who underwent treatment of type II odontoid fractures. The review included nine studies on anterior odontoid screw placement and six studies examining posterior cervical fusion. Overall, there were no differences in mortality or complication rates between those who underwent anterior or posterior surgery.[42] Patterson et al. reviewed the National Surgical Quality Improvement Program database for patients older than 65 with odontoid fractures who underwent surgical fixation. Of 141 patients, 93 underwent posterior approaches and 48 underwent anterior approaches. Those having posterior approaches were noted to have longer operative times, while those who underwent anterior fixations had more unplanned hospital readmissions and revision operations.[43]

Management of type II odontoid fractures in the elderly remains difficult with associated high mortality rates regardless of operative or nonoperative management. The current data would support surgery as the preferred option in all but the very elderly, frail, or significantly cognitively impaired.[44,45] Moreover, a posterior C1—2 fusion is presented as a viable option for type II odontoid fractures in the geriatric population when surgical risk is not unacceptably high.[39]

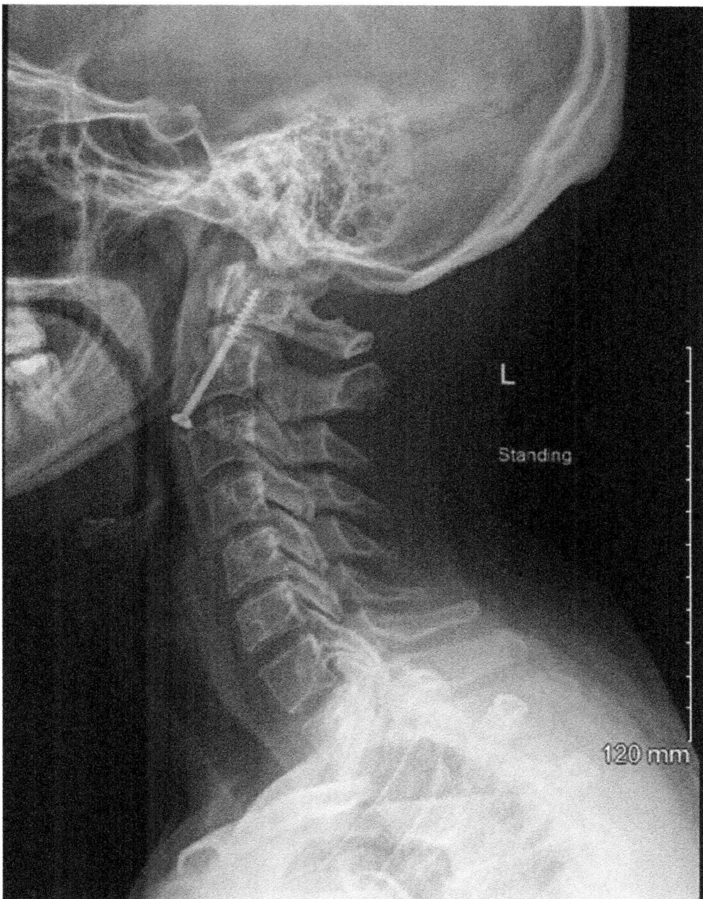

FIGURE 15.5 Odontoid screw.

Timing of surgery for central cord syndrome

CCS results from trauma to the cervical spinal cord and typically presents in patients with underlying cervical stenosis who suffer a hyperextension injury (Fig. 15.7). It is characterized by weakness present in the upper extremities proportionally more than lower extremities with particular weakness noted in hand grip strength, historically "explained" by the central somatotropic organization of the corticospinal tracts disrupted by central hemorrhage.[46] However, recent anatomical and imaging studies have suggested that the lateral corticospinal tracts are crucial for hand dexterity and the phenotype seen in CCS is the result of relative preservation of the extrapyramidal fiber tracts that favor locomotion over upper limb function.[46]

Surgery is recommended for treatment of CCS especially in the setting of spinal instability or

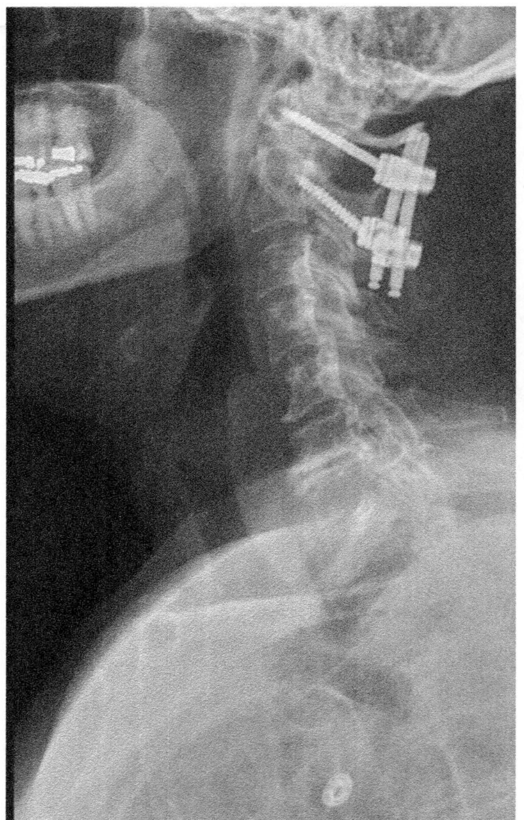

FIGURE 15.6   Posterior C1—2 fixation with C1 lateral mass screws and C2 pars screws.

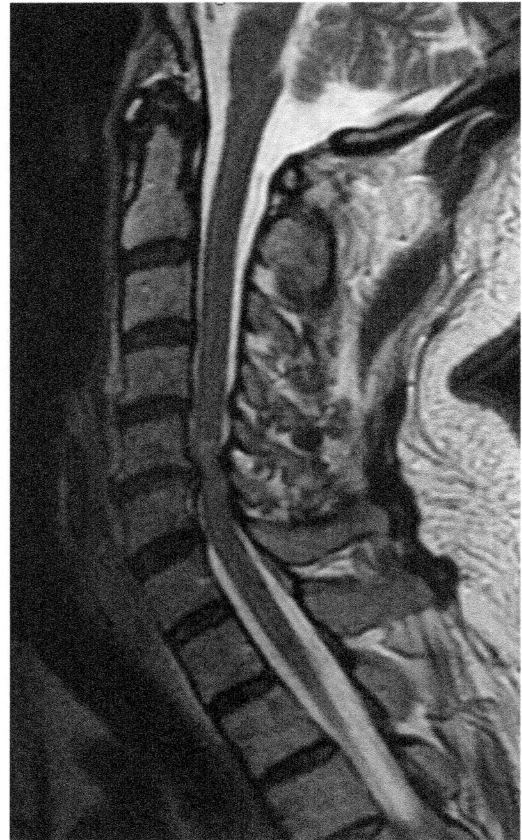

FIGURE 15.7   C5—6 stenosis in a patient presenting with central cord syndrome after hyperextension injury.

spinal stenosis. Patients who undergo surgery for CCS have been shown to have improved outcomes when compared to those undergoing only conservative cares.[47] Timing of surgery for CCS is controversial with some advocating for early surgery before 24 hours while others have cautioned that acute surgical decompression of the fragile spinal cord could cause neurological worsening.[48—50] Anderson et al. reported a systematic study to address the comparison of early versus late surgery for CCS. The paper reported low-level evidence from two prospective studies that showed improved motor scores and functional outcomes when surgery was performed within 24 hours. Anderson also found moderate-level evidence that delayed surgery predicts a lower JOA and recovery rate. There was noted to be insufficient evidence to show any improvement in complication rates, mortality, length of hospital or ICU stays. Overall, the systemic study indicated that patients operated on with 24 hours showed greater improvement in ASIA scores at 6 month and 1 year. Additionally, it is also reported to be more cost-effective to operate during the first hospital admission rather than undergo a second later admission. The authors do acknowledge the evidence supporting early surgery is of low quality; however, a randomized controlled study on the subject would be very unlikely to occur.[51] In 2019 Divi also produced a review of CCS analyzing the current literature and ultimately concluded that early decompression and stabilization should be recommended for patients with spinal instability or neurological deficit. It was noted that patients must be appropriate operative candidates with particular care taken to maintain spinal stability while avoiding hypotension in the operative setting.[52] A recent pooled analysis of 1548 spinal cord injured patients (including CCS) showed operative decompression in less than 24 hours after injury significantly improves the chance of sensorimotor recovery and ASIA Impairment Scale scores at 12 months.[53] This is further evidence that much is to be gained from early intervention, with no demonstrable advantage from delaying surgery. The authors believe the mantra "Time is Spine" should be firmly supported where access to prompt surgical management is available to patients with all forms of incomplete or complete spinal cord injury.

Role of MRI in decision-making for spine fractures

MRI has shown an increasing role in the evaluation of spine fractures through identification of occult fractures and clinically suspected ligamentous injuries.[54] MRI prior to intervention for spinal cord injury is recommended in most cases (when clinically appropriate) in order to identify pathology such as significant disc material in the canal, the extent of signal change or hemorrhage in the spinal cord, the association of spinal cord edema with the point of maximal compression, and possible transection/penetrating injury/vascular injury that all have important surgical or prognostic ramifications.[55,56] In particular, MRI identification of disc material anterior to the cord may have a significant effect in the surgical approach necessitating anterior as well as posterior decompression (see Fig. 15.8).

The role and timing of MRI for evaluation of spine fractures without spinal cord injury are less established. For this reason, diagnosis of occult spinal fractures is often delayed (see Fig. 15.9). This is especially true in patients with rigid spinal pathology such as AS or DISH. As a result of spinal rigidity, a long lever arm is created that acts above and below the fracture acting a fulcrum, making the spine more susceptible to three-column injuries. Early identification of spinal fractures in the AS and DISH population is important as they have higher rates of late-onset neurological deterioration and mortality than the general trauma population.[57] Identification of ligamentous injury is important when assessing spinal stability, particularly when prognostic scoring systems are used. The Subaxial Cervical Spine Injury Classification System (SLICS) and Thoracolumbar Injury Classification and Severity Score (TLICS) are classification systems aimed at guiding surgical treatment of traumatic spinal injuries. Each classification is based on fracture pattern, ligamentous integrity, and neurological function.[58,59] Ligaments provide spinal stability guarding against excessive flexion, rotation, translation, and distraction. For purposes of

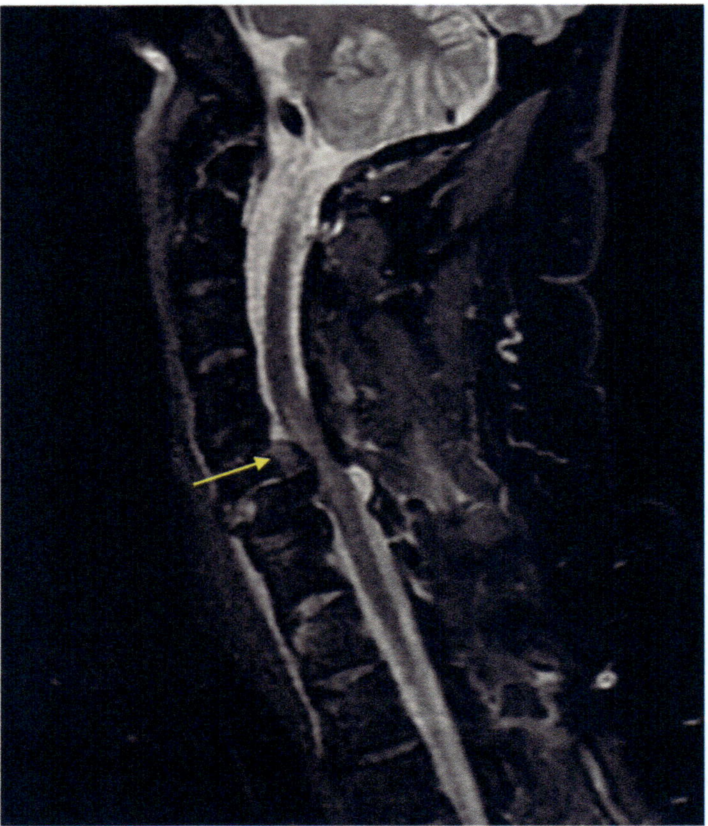

FIGURE 15.8   Sagittal T2 with STIR sequence showing significant disc material in the canal at the site of a fracture dislocation at C5/6. An initial anterior discectomy is required to ensure further cord compression does not occur with a posterior reduction and instrumented fixation.

SLICS and TLICS classification, ligamentous integrity is graded as intact, indeterminate, or disrupted. The potential for MRI to make this determination has been somewhat controversial. The identification of ligament rupture by MRI has been well documented and can be identified as a clear break in the borders of the ligament (Fig. 15.10).[60] However, soft-tissue and ligamentous injury is much less specific if a clear ligamentous rupture is not identified and only STIR signal changes are noted within the posterior soft-tissue elements (Fig. 15.11). The sensitivity of MRI for spinal ligamentous injury has been reported in the 80%—100% range depending on the study and particular region. However, the specificity of MRI for ligamentous injury is much less and reported over a wider range between 37.5% and 95%.[61—63]

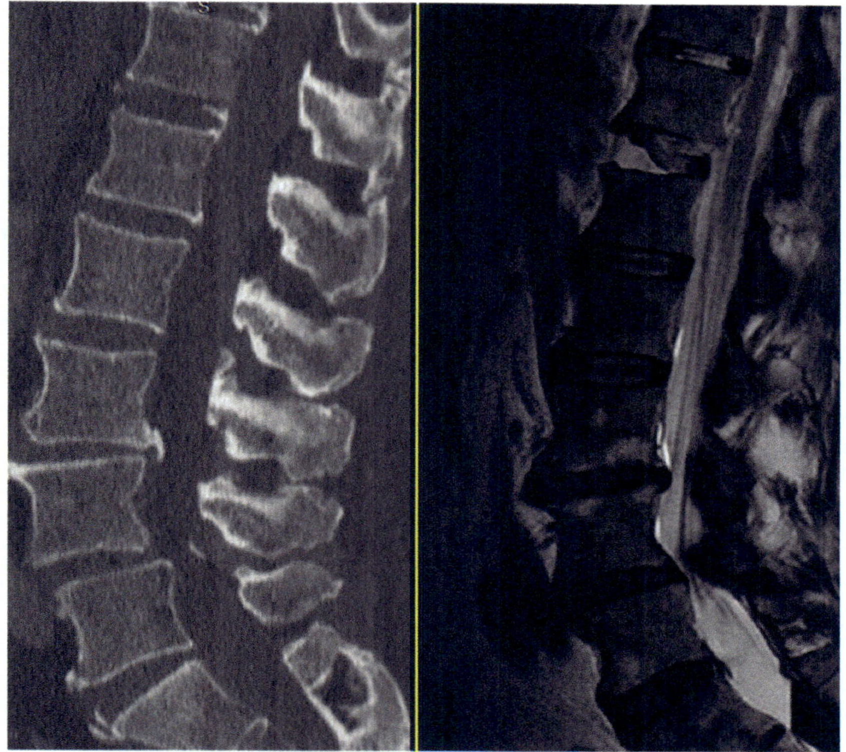

FIGURE 15.9 Occult fracture through T12—L1 disc space not immediately evident on CT imaging, but revealed on MRI.

Determining the ligamentous integrity based solely on MRI is difficult, but if present, should represent a significant factor in the decisions regarding surgical treatment.

One particular role of MRI in the assessment of spine fractures is the determination of the age of injury. MRI with STIR can be useful to determine if the presence of a potential chronic fracture with malunion has an acute component, or whether prior compression fractures still demonstrate evidence of bone marrow edema that allows decisions on further intervention (i.e., vertebroplasty/kyphoplasty).[64] The addition of contrast medium also can aid in establishing the diagnosis of a pathological fracture, if suspected.

Conclusion

The landscape of the management of spine trauma will continue to evolve. Innovation and product development will lead to an increasing prevalence of surgical intervention, but new adopters should use a sense of skepticism to evaluate new technology to see benefit at the patient level. Early surgical decompression should be considered in all patients who present with incomplete spinal cord injury, particularly CCS, as emerging data supports better clinical outcomes in the modern era compared to surgery later than 24—36 hours after injury.

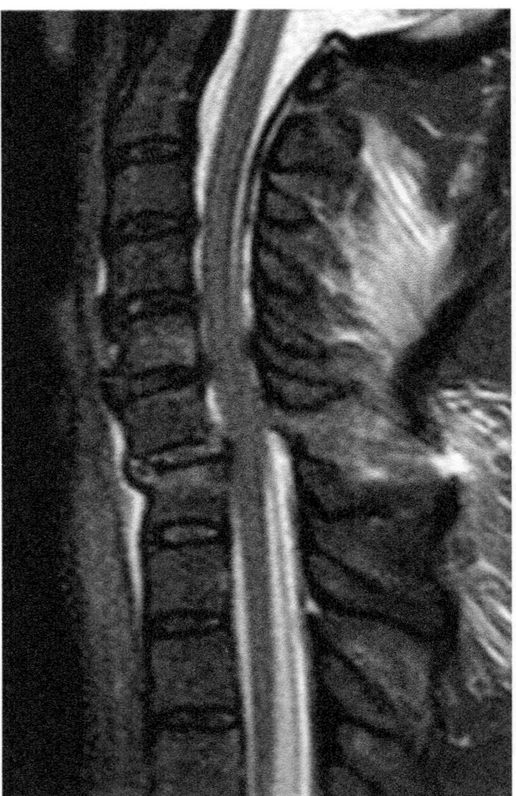

FIGURE 15.10   Rupture of PLL and ligamentum flavum at C6-7.

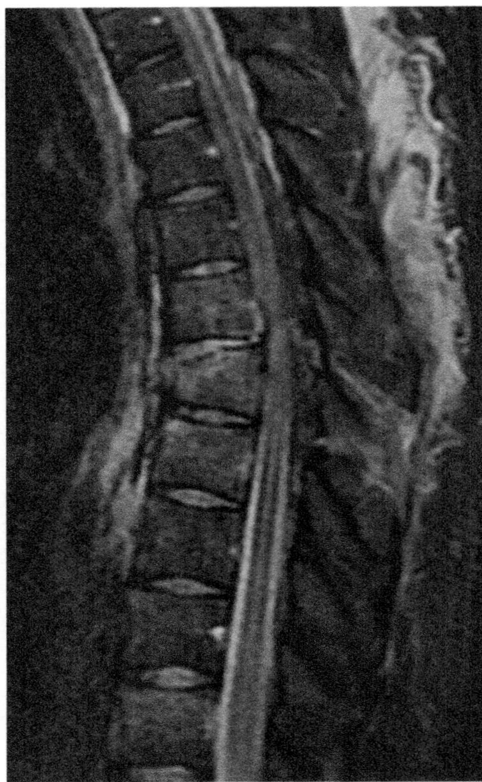

FIGURE 15.11   STIR signal noted in the posterior elements of the mid-thoracic spine.

References

1. Hu R, Mustard CA, Burns C. Epidemiology of incident spinal fracture in a complete population. *Spine* 1996; **21**(4):492–9.   https://doi.org/10.1097/00007632-1996 02150-00016.
2. Denis F. The three column spine and its significance in the classification of acute thoracolumbar spinal injuries. *Spine* 1983;**8**(8):817–31.   https://doi.org/ 10.1097/00007632-198311000-00003.
3. Vaccaro AR, Oner C, Kepler CK, Dvorak M, Schnake K, Bellabarba C, et al. AOSpine thoracolumbar spine injury classification system: fracture description, neurological status, and key modifiers. *Spine* 2013;**38**(23):2028–37. https://doi.org/10.1097/BRS.0b013e3182a8a381.
4. Vaccaro AR, Schroeder GD, Kepler CK, Cumhur Oner F, Vialle LR, Kandziora F, et al. The surgical algorithm for the AOSpine thoracolumbar spine injury classification system. *Eur Spine J* 2016;**25**(4):1087–94.   https:// doi.org/10.1007/s00586-015-3982-2.
5. Schroeder GD, Kepler CK, Koerner JD, Chapman JR, Bellabarba C, Oner FC, et al. Is there a regional difference in morphology interpretation of A3 and A4 fractures among different cultures? *J Neurosurg Spine* 2016;**24**(2):332–9. https://doi.org/10.3171/2015.4.Spine 1584.
6. Cahueque M, Cobar A, Zuñiga C, Caldera G. Management of burst fractures in the thoracolumbar spine. *J Orthop* 2016;**13**(4):278–81. https://doi.org/10.1016/ j.jor.2016.06.007.
7. Vaccaro AR, Lim MR, Hurlbert RJ, Lehman Jr RA, Harrop J, Fisher DC, et al. Surgical decision making for unstable thoracolumbar spine injuries: results of a consensus panel review by the Spine Trauma Study Group. *J Spinal Disord Tech* 2006;**19**(1):1–10.   https:// doi.org/10.1097/01.bsd.0000180080.59559.45.

8. Wood KB, Buttermann GR, Phukan R, Harrod CC, Mehbod A, Shannon B, et al. Operative compared with nonoperative treatment of a thoracolumbar burst fracture without neurological deficit: a prospective randomized study with follow-up at sixteen to twenty-two years. *J Bone Joint Surg Am* 2015;**97**(1):3—9. https://doi.org/10.2106/jbjs.N.00226.

9. Siebenga J, Leferink VJ, Segers MJ, Elzinga MJ, Bakker FC, Haarman HJ, et al. Treatment of traumatic thoracolumbar spine fractures: a multicenter prospective randomized study of operative versus nonsurgical treatment. *Spine* 2006;**31**(25):2881—90. https://doi.org/10.1097/01.brs.0000247804.91869.1e.

10. Oner C, Rajasekaran S, Chapman JR, Fehlings MG, Vaccaro AR, Schroeder GD, et al. Spine trauma-what are the current controversies? *J Orthop Trauma* 2017;**31**(Suppl. 4):S1—6. https://doi.org/10.1097/bot.0000000000000950.

11. Bailey CS, Urquhart JC, Dvorak MF, Nadeau M, Boyd MC, Thomas KC, et al. Orthosis versus no orthosis for the treatment of thoracolumbar burst fractures without neurologic injury: a multicenter prospective randomized equivalence trial. *Spine* 2014;**14**(11):2557—64. https://doi.org/10.1016/j.spinee.2013.10.017.

12. McCormack T, Karaikovic E, Gaines RW. The load sharing classification of spine fractures. *Spine* 1994;**19**(15):1741—4. https://doi.org/10.1097/00007632-199408000-00014.

13. Rometsch E, Spruit M, Härtl R, McGuire RA, Gallo-Kopf BS, Kalampoki V, Kandziora F. Does operative or nonoperative treatment achieve better results in A3 and A4 spinal fractures without neurological deficit?: systematic literature review with meta-analysis. *Global Spine J* 2017;**7**(4):350—72. https://doi.org/10.1177/2192568217699202.

14. McLain RF. Functional outcomes after surgery for spinal fractures: return to work and activity. *Spine* 2004;**29**(4):470—7. https://doi.org/10.1097/01.brs.0000092373.57039.fc. discussion Z476.

15. Denis F, Armstrong GW, Searls K, Matta L. Acute thoracolumbar burst fractures in the absence of neurologic deficit. A comparison between operative and nonoperative treatment. *Clin Orthop Relat Res* 1984;**189**:142—9.

16. Mayer TG, Vanharanta H, Gatchel RJ, Mooney V, Barnes D, Judge L, et al. Comparison of CT scan muscle measurements and isokinetic trunk strength in postoperative patients. *Spine* 1989;**14**(1):33—6. https://doi.org/10.1097/00007632-198901000-00006.

17. Camacho JE, Usmani MF, Strickland AR, Banagan KE, Ludwig SC. The use of minimally invasive surgery in spine trauma: a review of concepts. *J. Spine Surg.* 2019;**5**(Suppl. 1):S91—100. https://doi.org/10.21037/jss.2019.04.13.

18. Mroz TE, Abdullah KG, Steinmetz MP, Klineberg EO, Lieberman IH. Radiation exposure to the surgeon during percutaneous pedicle screw placement. *J Spinal Disord Tech* 2011;**24**(4):264—7. https://doi.org/10.1097/BSD.0b013e3181eed618.

19. Smith HE, Welsch MD, Sasso RC, Vaccaro AR. Comparison of radiation exposure in lumbar pedicle screw placement with fluoroscopy vs computer-assisted image guidance with intraoperative three-dimensional imaging. *J Spinal Cord Med* 2008;**31**(5):532—7. https://doi.org/10.1080/10790268.2008.11753648.

20. Kim YM, Kim DS, Choi ES, Shon HC, Park KJ, Cho BK, et al. Nonfusion method in thoracolumbar and lumbar spinal fractures. *Spine* 2011;**36**(2):170—6. https://doi.org/10.1097/BRS.0b013e3181cd59d1.

21. Diniz JM, Botelho RV. Is fusion necessary for thoracolumbar burst fracture treated with spinal fixation? A systematic review and meta-analysis. *J Neurosurg Spine* 2017;**27**(5):584—92. https://doi.org/10.3171/2017.1.Spine161014.

22. Walker CT, Xu DS, Godzik J, Turner JD, Uribe JS, Smith WD. Minimally invasive surgery for thoracolumbar spinal trauma. *Ann Transl Med* 2018;**6**(6):102. https://doi.org/10.21037/atm.2018.02.10.

23. Mummaneni PV, Shaffrey CI, Lenke LG, Park P, Wang MY, La Marca F, et al. The minimally invasive spinal deformity surgery algorithm: a reproducible rational framework for decision making in minimally invasive spinal deformity surgery. *Neurosurg Focus* 2014;**36**(5):E6. https://doi.org/10.3171/2014.3.Focus1413.

24. Moses ZB, Mayer RR, Strickland BA, Kretzer RM, Wolinsky JP, Gokaslan ZL, Baaj AA. Neuronavigation in minimally invasive spine surgery. *Neurosurg Focus* 2013;**35**(2):E12. https://doi.org/10.3171/2013.5.Focus13150.

25. McAnany SJ, Overley SC, Kim JS, Baird EO, Qureshi SA, Anderson PA. Open versus minimally invasive fixation techniques for thoracolumbar trauma: a meta-analysis. *Global Spine J* 2016;**6**(2):186—94. https://doi.org/10.1055/s-0035-1554777.

26. Tian F, Tu L-Y, Gu W-F, Zhang E-F, Wang Z-B, Chu G, et al. Percutaneous versus open pedicle screw instrumentation in treatment of thoracic and lumbar spine fractures: a systematic review and meta-analysis. *Medicine* 2018;**97**(41):e12535. https://doi.org/10.1097/MD.0000000000012535.

27. Defino HLA, Costa HRT, Nunes AA, Nogueira Barbosa M, Romero V. Open versus minimally invasive percutaneous surgery for surgical treatment of thoracolumbar spine fractures- a multicenter randomized controlled trial: study protocol. *BMC Muscoskel Disord* 2019;**20**(1):397. https://doi.org/10.1186/s12891-019-2763-1.

28. Vaccaro AR, Kepler CK, Kopjar B, Chapman J, Shaffrey C, Arnold P, et al. Functional and quality-of-life outcomes in geriatric patients with type-II dens fracture. *J Bone Joint Surg Am* 2013;**95**(8):729−35. https://doi.org/10.2106/jbjs.K.01636.

29. Antevil JL, Sise MJ, Sack DI, Kidder B, Hopper A, Brown CV. Spiral computed tomography for the initial evaluation of spine trauma: a new standard of care? *J Trauma* 2006;**61**(2):382−7. https://doi.org/10.1097/01.ta.0000226154.38852.e6.

30. Anderson LD, D'Alonzo RT. Fractures of the odontoid process of the axis. *J Bone Joint Surg Am* 1974;**56**(8):1663−74.

31. Iyer S, Hurlbert RJ, Albert TJ. Management of odontoid fractures in the elderly: a review of the literature and an evidence-based treatment algorithm. *Neurosurgery* 2017;**82**(4):419−30. https://doi.org/10.1093/neuros/nyx546.

32. Amling M, Hahn M, Wening VJ, Grote HJ, Delling G. The microarchitecture of the axis as the predisposing factor for fracture of the base of the odontoid process. A histomorphometric analysis of twenty-two autopsy specimens. *J Bone Joint Surg Am* 1994;**76**(12):1840−6. https://doi.org/10.2106/00004623-199412000-00011.

33. Pearson AM, Martin BI, Lindsey M, Mirza SK. C2 vertebral fractures in the medicare population: incidence, outcomes, and costs. *J Bone Joint Surg Am* 2016;**98**(6):449−56. https://doi.org/10.2106/jbjs.O.00468.

34. Raudenbush B, Molinari R. Longer-term outcomes of geriatric odontoid fracture nonunion. *Geriatr Orthopaed Surg Rehabil* 2015;**6**(4):251−7. https://doi.org/10.1177/2151458515593774.

35. Chapman J, Smith JS, Kopjar B, Vaccaro AR, Arnold P, Shaffrey CI, Fehlings MG. The AOSpine North America Geriatric Odontoid Fracture Mortality Study: a retrospective review of mortality outcomes for operative versus nonoperative treatment of 322 patients with long-term follow-up. *Spine* 2013;**38**(13):1098−104. https://doi.org/10.1097/BRS.0b013e318286f0cf.

36. Ryken TC, Hadley MN, Aarabi B, Dhall SS, Gelb DE, Hurlbert RJ, et al. Management of isolated fractures of the axis in adults. *Neurosurgery* 2013;**72**(Suppl. 2):132−50. https://doi.org/10.1227/NEU.0b013e318276ee40.

37. Bednar DA, Parikh J, Hummel J. Management of type II odontoid process fractures in geriatric patients; a prospective study of sequential cohorts with attention to survivorship. *J Spinal Disord* 1995;**8**(2):166−9.

38. Dailey AT, Hart D, Finn MA, Schmidt MH, Apfelbaum RI. Anterior fixation of odontoid fractures in an elderly population. *J Neurosurg Spine* 2010;**12**(1):1−8. https://doi.org/10.3171/2009.7.Spine08589.

39. Wagner SC, Schroeder GD, Kepler CK, Schupper AJ, Kandziora F, Vialle EN, et al. Controversies in the management of geriatric odontoid fractures. *J Orthop Trauma* 2017;**31**(Suppl. 4):S44−s48. https://doi.org/10.1097/bot.0000000000000948.

40. Scheyerer MJ, Zimmermann SM, Simmen HP, Wanner GA, Werner CM. Treatment modality in type II odontoid fractures defines the outcome in elderly patients. *BMC Surg* 2013;**13**:54. https://doi.org/10.1186/1471-2482-13-54.

41. Molinari 3rd WJ, Molinari RW, Khera OA, Gruhn WL. Functional outcomes, morbidity, mortality, and fracture healing in 58 consecutive patients with geriatric odontoid fracture treated with cervical collar or posterior fusion. *Global Spine J* 2013;**3**(1):21−32. https://doi.org/10.1055/s-0033-1337122.

42. Schroeder GD, Kepler CK, Kurd MF, Paul JT, Rubenstein RN, Harrop JS, et al. A systematic review of the treatment of geriatric type II odontoid fractures. *Neurosurgery* 2015;**77**(Suppl. 4):S6−14. https://doi.org/10.1227/neu.0000000000000942.

43. Patterson JT, Theologis AA, Sing D, Tay B. Anterior versus posterior approaches for odontoid fracture stabilization in patients older than 65 years: 30-day morbidity and mortality in a national database. *Clin Spine Surg* 2017;**30**(8):E1033−e1038. https://doi.org/10.1097/bsd.0000000000000494.

44. Fehlings MG, Arun R, Vaccaro AR, Arnold PM, Chapman JR, Kopjar B. Predictors of treatment outcomes in geriatric patients with odontoid fractures: AOSpine North America multi-centre prospective GOF study. *Spine* 2013;**38**(11):881−6. https://doi.org/10.1097/BRS.0b013e31828314ee.

45. Kato S MJ, Fehlings MG. Guidelines for managing the aging cervical spine. *Contemp Neurosurg* 2020;**42**(1).

46. Badhiwala JH, Wilson JR, Fehlings MG. The case for revisiting central cord syndrome. *Spinal Cord* 2020;**58**(1):125−7. https://doi.org/10.1038/s41393-019-0354-5.

47. Bose B, Northrup BE, Osterholm JL, Cotler JM, DiTunno JF. Reanalysis of central cervical cord injury management. *Neurosurgery* 1984;**15**(3):367−72. https://doi.org/10.1227/00006123-198409000-00012.

48. La Rosa G, Conti A, Cardali S, Cacciola F, Tomasello F. Does early decompression improve neurological outcome of spinal cord injured patients? Appraisal of the literature using a meta-analytical approach. *Spinal Cord* 2004;**42**(9):503−12. https://doi.org/10.1038/sj.sc.3101627.

49. Fehlings MG, Vaccaro A, Wilson JR, Singh A, Cadotte DW, Harrop JS, et al. Early versus delayed decompression for traumatic cervical spinal cord injury: results of the Surgical Timing in Acute Spinal Cord Injury Study (STASCIS). *PLoS One* 2012;**7**(2):e32037. https://doi.org/10.1371/journal.pone.0032037.

50. Marshall LF, Knowlton S, Garfin SR, Klauber MR, Eisenberg HM, Kopaniky D, et al. Deterioration following spinal cord injury. A multicenter study. *J Neurosurg* 1987;**66**(3):400–4. https://doi.org/10.3171/jns.1987.66.3.0400.

51. Anderson KK, Tetreault L, Shamji MF, Singh A, Vukas RR, Harrop JS, et al. Optimal timing of surgical decompression for acute traumatic central cord syndrome: a systematic review of the literature. *Neurosurgery* 2015;**77**(Suppl. 4):S15–32. https://doi.org/10.1227/neu.0000000000000946.

52. Divi SN, Schroeder GD, Mangan JJ, Tadley M, Ramey WL, Badhiwala JH, et al. Management of acute traumatic central cord syndrome: a narrative review. *Global Spine J* 2019;**9**(Suppl. 1):89S–97S. https://doi.org/10.1177/2192568219830943.

53. Badhiwala JH, Wilson JR, Witiw CD, Harrop JS, Vaccaro AR, Aarabi B, et al. The influence of timing of surgical decompression for acute spinal cord injury: a pooled analysis of individual patient data. *Lancet Neurol* 2021;**20**(2):117–26. https://doi.org/10.1016/S1474-4422(20)30406-3.

54. Kumar Y, Hayashi D. Role of magnetic resonance imaging in acute spinal trauma: a pictorial review. *BMC Muscoskel Disord* 2016;**17**:310. https://doi.org/10.1186/s12891-016-1169-6.

55. Kurpad S, Martin AR, Tetreault LA, Fischer DJ, Skelly AC, Mikulis D, et al. Impact of baseline magnetic resonance imaging on neurologic, functional, and safety outcomes in patients with acute traumatic spinal cord injury. *Global Spine J* 2017;**7**(Suppl. 3):151S–74S. https://doi.org/10.1177/2192568217703666.

56. Miyanji F, Furlan JC, Aarabi B, Arnold PM, Fehlings MG. Acute cervical traumatic spinal cord injury: MR imaging findings correlated with neurologic outcome–prospective study with 100 consecutive patients. *Radiology* 2007;**243**(3):820–7. https://doi.org/10.1148/radiol.2433060583.

57. Westerveld LA, Verlaan JJ, Oner FC. Spinal fractures in patients with ankylosing spinal disorders: a systematic review of the literature on treatment, neurological status and complications. *Eur Spine J* 2009;**18**(2):145–56. https://doi.org/10.1007/s00586-008-0764-0.

58. Vaccaro AR, Hulbert RJ, Patel AA, Fisher C, Dvorak M, Lehman Jr RA, et al. The subaxial cervical spine injury classification system: a novel approach to recognize the importance of morphology, neurology, and integrity of the disco-ligamentous complex. *Spine* 2007;**32**(21): 2365–74. https://doi.org/10.1097/BRS.0b013e3181557b92.

59. Vaccaro AR, Lehman Jr RA, Hurlbert RJ, Anderson PA, Harris M, Hedlund R, et al. A new classification of thoracolumbar injuries: the importance of injury morphology, the integrity of the posterior ligamentous complex, and neurologic status. *Spine* 2005;**30**(20): 2325–33. https://doi.org/10.1097/01.brs.0000182986.43345.cb.

60. Benedetti PF, Fahr LM, Kuhns LR, Hayman LA. MR imaging findings in spinal ligamentous injury. *Am J Roentgenol* 2000;**175**(3):661–5. https://doi.org/10.2214/ajr.175.3.1750661.

61. Vaccaro AR, Rihn JA, Saravanja D, Anderson DG, Hilibrand AS, Albert TJ, et al. Injury of the posterior ligamentous complex of the thoracolumbar spine: a prospective evaluation of the diagnostic accuracy of magnetic resonance imaging. *Spine* 2009;**34**(23):E841–7. https://doi.org/10.1097/BRS.0b013e3181bd11be.

62. Haba H, Taneichi H, Kotani Y, Terae S, Abe S, Yoshikawa H, et al. Diagnostic accuracy of magnetic resonance imaging for detecting posterior ligamentous complex injury associated with thoracic and lumbar fractures. *J Neurosurg* 2003;**99**(Suppl. 1):20–6. https://doi.org/10.3171/spi.2003.99.1.0020.

63. Lee HM, Kim HS, Kim DJ, Suk KS, Park JO, Kim NH. Reliability of magnetic resonance imaging in detecting posterior ligament complex injury in thoracolumbar spinal fractures. *Spine* 2000;**25**(16):2079–84. https://doi.org/10.1097/00007632-200008150-00012.

64. Minja FJ, Mehta KY, Mian AY. Current challenges in the use of computed tomography and MR imaging in suspected cervical spine trauma. *Neuroimaging Clin* 2018; **28**(3):483–93. https://doi.org/10.1016/j.nic.2018.03.009.

Traumatic central cord injury

Jetan H. Badhiwala[1], Laureen D. Hachem[1], Bizhan Aarabi[2], Brian K. Kwon[3], Michael G. Fehlings[1,4]

[1]Division of Neurosurgery, Department of Surgery, University of Toronto, Toronto, ON, Canada; [2]Department of Neurosurgery, University of Maryland Medical Center and R Adams Cowley Shock Trauma Center, Baltimore, MD, United States; [3]Vancouver Spine Surgery Institute, Vancouver General Hospital, Department of Orthopaedics, UBC, Vancouver, BC, Canada; [4]Division of Neurosurgery, Toronto Western Hospital, University Health Network, Toronto, ON, Canada

Introduction

Acute traumatic cervical spinal cord injury (SCI) is a heterogeneous clinical entity. In addition to the distinction between complete and incomplete SCI, several incomplete injury subtypes, or "spinal cord syndromes," have historically been recognized, each with a purported unique etiopathogenesis, clinical presentation, natural history, and response to treatment.[1] Of these, central cord syndrome (CCS) is the most common, classically characterized by disproportionately greater upper than lower limb weakness, often occurring in the elderly patient with spinal stenosis after a cervical hyperextension injury due to low-velocity trauma, such as a fall.[2–4]

The topic of CCS is especially relevant today. First, with the aging population, the incidence of this disorder has steadily risen; in fact, CCS is expected to soon become the most common form of acute traumatic SCI.[5,6] Second, more recent evidence has evolved our understanding of the pathophysiology of CCS and secondary injury

after SCI,[7–17] and the natural history and surgical outcomes of CCS.[4,9,18–33] Combined with advances in surgical techniques (e.g., microsurgery),[34] operative adjuncts (e.g., intraoperative neurophysiological monitoring, operative microscope),[35] imaging technology (e.g., conventional and quantitative magnetic resonance imaging [MRI]),[36,37] perioperative and critical care,[38] and systems infrastructure (e.g., specialized centers of care for SCI, transport services),[39] this has led to a shift in clinical practice patterns for CCS, moving toward proactive treatment paradigms that integrate surgical decompression with rehabilitative strategies.[9,18,20,23,24,28,31,33,40,41]

Given that CCS has become an important and growing public health problem, and understanding and treatment of CCS have evolved, this chapter aims to provide a detailed contemporary review of the epidemiology, pathophysiology, diagnosis, clinical course, and operative and nonoperative management of CCS based on the latest evidence.

© 2022 Elsevier Inc. All rights reserved.

Diagnosis and definition

Clinical

Central cord syndrome is a *clinical diagnosis*, and its identification relies on a thorough neurological examination. As for any patient with SCI, this examination should be performed in accordance with International Standards for Neurological Classification of Spinal Cord Injury (ISNCSCI)/American Spinal Injury Association (ASIA) standards.[42] CCS was first described in detail by Schneider et al.[43] in a 1954 case series as involving "disproportionately more motor impairment of the upper than of the lower extremities, bladder dysfunction, usually urinary retention, and varying degrees of sensory loss below the level of the lesion." This description has by and large remained unchanged, with the cardinal feature being disproportionate weakness of the upper limb compared with the lower limb, and other secondary criteria of lesser importance including variable sensory loss, bladder dysfunction, sacral sparing, and symmetric motor impairment of the upper limbs.[2] Nonetheless, this is a rather subjective definition that relies on nonspecific criteria and qualitative interpretation of physical examination findings. Hence, more recent attempts have been made at operationalizing diagnostic criteria for CCS, particularly the definition of "greater arm than leg weakness." In a systematic review of the literature, Pouw et al.[2] found the mean ASIA upper extremity motor score (UEMS) to be 10.5 points lower than the mean ASIA lower extremity motor score (LEMS) in studies reporting these parameters for patients with CCS. A global survey of surgeon members of AOSpine International then confirmed that a majority (61%) of surgeons considered a 10-point difference in favor of the lower extremities as an acceptable cutoff criterion for diagnosis of CCS.[3] Patients demonstrating greater upper than lower limb weakness not meeting this threshold (difference of 1–9 points) have been described as having "mild" or "intermediate" CCS.[4] However, these definitions remain controversial, and although useful for research purposes, they are perhaps less relevant in clinical contexts, particularly since they ignore other relevant details, including injury mechanism and anatomical injury characteristics gleaned from imaging investigations, such as computed tomography (CT) and magnetic resonance imaging (MRI). "Burning hands syndrome," characterized by burning dysesthesias and paresthesias affecting the hands, is thought to be a mild variant of CCS that results from disruption of the spinothalamic tract.

Radiological

There are no radiological criteria for the diagnosis of CCS; however, imaging forms an important cornerstone in the evaluation of patients with CCS. High-quality (multidetector)CT and/or three-view plain radiographs of the cervical spine should be obtained to rule out fracture or dislocation. In the current era, we would recommend the former, owing to the greater sensitivity and specificity of CT for injury compared with X-rays.[44] Importantly, fractures and subluxations of the vertebral column are possible but occur with lesser frequency in CCS than other forms of traumatic SCI. An MRI of the cervical spine should be obtained expeditiously thereafter to assess the integrity of the spinal cord, degree of spinal cord compression, as well as potential disruption of the disc and ligaments. Together, these imaging studies provide valuable information relating to the degree of instability, biomechanical failure, the urgency of spinal cord decompression, and the need for internal or external fixation, hence dictating the management of patients with CCS. Furthermore, MRI can provide data of prognostic relevance, as certain radiological parameters, such as length of intramedullary lesion and canal diameter, have been shown to be predictive of outcomes.[9,45]

Epidemiology

The epidemiology of acute traumatic SCI is evolving, and with it, so too are injury patterns. Specifically, the aging population has led to a shift toward an increased proportion of injuries in older patients from low-impact falls. In the United States, from 1993 to 2012, the mean age of patients suffering an acute spinal cord injury rose from 40.5 years to 50.5 years.[46] The incidence of SCI among the younger male population (16–44 years old) declined. This is attributed to several public health interventions, including public education, improved motor vehicle safety features, and stricter safety belt and drunk driving laws. Meanwhile, there was a high rate of increase in SCI incidence among males aged 65–74 years, from 84 cases per million in 1993 to 131 cases per million in 2012. The proportion of SCI cases that were due to unintentional falls increased from 19.3% to 40.4%. Similar trends have been reported in many developed countries,[47,48] including Canada,[49] Japan,[50] Iceland,[51] Spain,[52] and the United Kingdom.[53] Indeed, while the prototypical SCI was once a severe fracture dislocation in a young person after a motor vehicle collision, high-impact sports injury, or other high-energy trauma, today it is a cervical incomplete SCI (e.g., CCS), often without disruption of the spinal column, in an older person from a low-energy or fall-related mechanism.[54,55]

Large epidemiological studies have found cervical SCI to account for approximately 60% of acute traumatic SCI.[48] Of cervical SCI, 70% are incomplete injuries. Traumatic central cord syndrome is the most common form of incomplete SCI, reported to account for 40% of cervical SCI and upward of 25% of all traumatic SCI.[5] By contrast, other SCI syndromes represent a minority, with Brown-Séquard syndrome accounting for as few as 1% of incomplete cervical SCI cases.[56] Age distribution in CCS follows a bimodal distribution: while the largest incidence peak corresponds to SCI in older patients secondary to low-energy hyperextension mechanisms (e.g., minor falls) in the setting of a spondylotic spinal canal, a smaller but significant peak corresponds to younger patients (<50 years) who present with cervical incomplete SCI in the setting of severe spinal column injuries (e.g., fracture, dislocation, disc herniation) from high-velocity trauma.[57]

Pathophysiology

Central cord syndrome commonly occurs after a hyperextension neck injury in the context of central canal stenosis—either congenital, or acquired secondary to degenerative pathology (e.g., spondylosis, ossification of the posterior longitudinal ligament [OPLL]).[43,58] With hyperextension, the ligamentum flavum is thought to buckle, resulting in sudden impingement of the spinal cord in the anteroposterior plane between a disc osteophyte complex anteriorly and the ligamentum flavum posteriorly (Fig. 16.1).[43,58,59] Nonetheless, flexion and vertical compression mechanisms are also compatible with CCS.[60]

In their original description, Schneider et al.[43] posited that a hematomyelic cavity formed within the central spinal cord gray matter, disrupting the medially positioned corticospinal tract (CST) fibers that control hand and upper limb function, but relatively sparing the more lateral fibers that supply the lower limbs (Fig. 16.2, upper panel). This hypothesis was predicated upon the work of Foerster,[61] who supposed that the lateral CST, like the fasciculus cuneatus and gracilis and the lateral spinothalamic tract, had a somatotopic organization. Several landmark experiments have since evolved our understanding of spinal cord functional anatomy and the etiopathogenesis of CCS. First, neuroanatomic tracer[7] and Marchi degeneration[8] studies in monkeys have indicated lack of laminar (i.e., somatotopic) organization to the CST caudal to the pons, at the level of the medullary pyramids and the posterolateral funiculus of the spinal cord. Second,

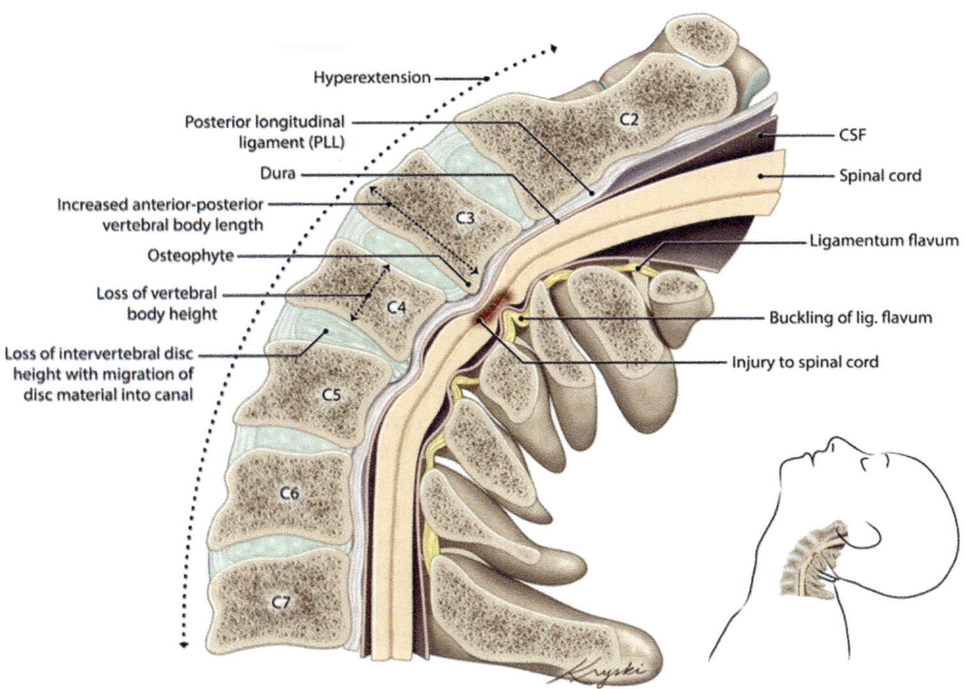

FIGURE 16.1   Etiology of central cord syndrome. *Medical illustration drawn by Diana Kryski, Kryski Biomedia.*

autopsy and MRI investigations have found central hemorrhage to occur rarely in CCS; instead, histologically, there is diffuse axonal injury in the white matter of the lateral columns.[9,10,17] Third, monkey transection experiments have demonstrated nonrecovering hand dysfunction—more specifically, loss of independent finger movements—but a striking paucity of lasting gross locomotor deficits, following complete bilateral pyramidotomy.[62] Indeed, it is currently thought that the lateral CST is critical for hand function and dexterity, providing the capacity for fractionation of movements (e.g., individual finger control), but less so for maintenance of erect posture and locomotion, which may be subserved by the extrapyramidal descending tracts.[11,13,63] Fourth, autopsy studies have demonstrated that a primary injury to the lateral CST, with Wallerian degeneration distal to the epicenter of injury, is consistently observed in CCS.[17] On the other hand, loss of C7, C8, and

T1 α-motoneurons is only seen in cases where the epicenter of injury is adjacent to this area. These data suggest that CCS primarily results from injury to the large fibers of the lateral CST; although injury to α-motoneurons may be an important contributor to hand dysfunction in patients with a low cervical injury, CCS may develop even without any apparent loss in the number of lower motor neurons supplying the hand musculature.

Current conceptualization of the pathophysiology of central cord syndrome

Central cord syndrome typically results from a low-energy hyperextension neck injury in a patient with central canal stenosis, often an older person with cervical spondylosis (Fig. 16.1). Younger patients are more likely to have associated spinal column injuries from major trauma. There is resultant diffuse white matter injury to

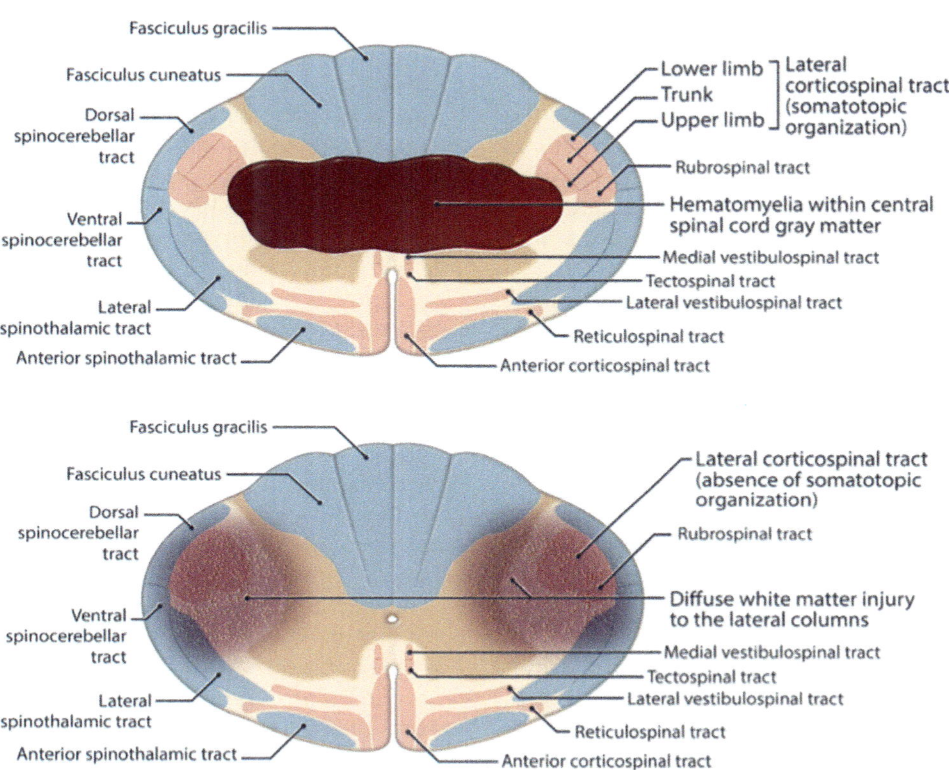

FIGURE 16.2 Pathophysiology of central cord syndrome. (Upper panel) "Classical conceptualization." (Lower panel) "Current conceptualization." *Medical illustration drawn by Diana Kryski, Kryski Biomedia.*

the lateral columns housing the CST (Fig. 16.2, lower panel). Because the dominant function of the lateral CST is fine motor action of the distal upper extremity, there is disproportionate weakness of the upper limb, and in particular, the hand. Damage to the α-motoneurons supplying the hand musculature may be a further contributor in injuries centered at C7, C8, and T1. Intramedullary hematoma is possible, but very uncommon.

Clinical course

Recovery in CCS classically follows an ascending, proximal-to-distal pattern, with return of function occurring first in the lower limbs, followed by bladder, proximal upper limbs, and finally the hand. The clinical course of CCS has traditionally been viewed favorably, and operative intervention has been discouraged out of fear of causing further insult to the spinal cord and derailing potential for natural neurological recovery. Nonetheless, these views are largely based on outdated historical reports. In their original description of CCS, Schneider et al.[43] suggested, "Surgical decompression of the spinal cord is contraindicated because spontaneous improvement or complete recovery may occur. Furthermore, operation has actually been known to harm these patients rather than improve them." In addition to standard decompression through a laminectomy, the authors performed a dural opening, dentate ligament resection, and transdural discectomy. In the past 60 years since the report of Schneider et al.[43], improvements in surgical techniques and technology, coupled with advances in our understanding of SCI pathophysiology and spinal biomechanics, have rendered many of the original concepts surrounding the natural course and ideal treatment of CCS antiquated and obsolete.

Despite the significant heterogeneity of acute traumatic SCI, the prediction of clinical outcomes has proven feasible in both complete and incomplete SCI.[64,65] The baseline neurological status of the patient, as assessed by the ASIA Impairment Scale (AIS) and/or ASIA total motor score (AMS), has been shown to be a key predictor of neurological and functional outcomes in patients with CCS, highlighting that recovery is heterogeneous and dependent by and large on the severity of injury.[4,9,30] In a cohort of CCS patients, Thompson et al.[30] found an AMS of ≥60 on admission or discharge correlated with an 80% probability of being able to walk at discharge from hospital. By contrast, an AMS ≤50 on admission was associated with an 80% chance of not walking. Dvorak et al.[66] found initial AMS, formal education, absence of comorbidities, absence of spasticity, and younger age to be positively correlated with functional status, as evaluated by the Functional Independence Measure (FIM).

Age is another important determinant of the degree of neurological recovery, with several studies reporting poorer neurological outcomes in older patients, particularly those over 60–70 years of age.[20] Furthermore, in studies of large national data sets, increasing age and comorbidities have been associated with higher rates of inpatient mortality.[31] Older age is also associated with greater degree of bowel dysfunction, with lower rates of continence and independence for bowel care.[21]

There has been growing recognition that patients with CCS with and without fracture or subluxation may represent different populations.[27,67] Schroeder et al.[27] found CCS patients with a fracture had a more severe injury at initial presentation, but tended to have neurological improvement in the first week following injury. On the other hand, patients without fracture had a less severe initial neurological injury, but generally had a slight decline in neurological function over the first week. This suggests that secondary injury mechanisms may play a more substantive role in the pathophysiology of CCS without fracture. Another imaging parameter of prognostic relevance is the presence of increased T2 signal intensity within the spinal cord, which has been associated with more severe initial neurological deficit, but relatively minimal early neurological deterioration.[26] Furthermore, length of parenchymal damage on T2 MRI, midsagittal diameter, and percentage maximum canal compromise have been associated with motor recovery and functional outcomes in CCS.[9]

The clinical course of CCS in patients treated conservatively may involve delayed neurological deterioration.[28,68,69] This results from persistent spinal cord compression, analogous to the natural history of progressive degenerative cervical myelopathy.[70] If conservative treatment is opted, patients should therefore be monitored closely for signs and symptoms of neurological worsening. Prompt intervention in these cases may result in satisfactory clinical outcomes.[68,69]

Management

Nonoperative management

The initial evaluation and management of a patient with suspected CCS, as with any SCI, follows standard Advanced Trauma Life Support (ATLS) protocol.[71] Cervical immobilization is maintained, beginning in the field, using a rigid collar. Once the patient has been resuscitated and the primary survey is complete, the secondary survey should include a detailed neurological examination according to ISNCSCI/ASIA standards.[42] Appropriate imaging is obtained, beginning with CT, followed by MRI. Current guidelines recommend patients with CCS be managed in an intensive care unit (ICU) setting, with cardiac, hemodynamic, and respiratory monitoring, particularly if neurological deficits are severe.[41] A mean arterial blood pressure (MAP) of 85–90 mm Hg should be targeted for the first 7 days after injury to prevent propagation of secondary injury from ischemia. Despite clinical guidelines recommending maintenance of these hemodynamic parameters, there is no high-quality evidence clearly supporting that such goal-directed therapy improves clinical outcomes in a CCS-specific SCI population. As a result, clinical judgment must be exercised with regard to vasopressor use, which is associated with a high risk of cardiogenic complications, particularly in older patients.[72] For each unique clinical scenario, a tailored approach should weigh the patient's cardiac risk factors against the need for maintenance of spinal cord perfusion.

Operative management

Operative intervention in the setting of CCS consists of (1) open (i.e., intraoperative) or closed (e.g., with halo traction) reduction of cervical dislocation, if present; (2) decompression of the spinal cord; and (3) internal fixation (i.e., instrumented fusion), if necessary, based on the degree of biomechanical instability. The most important element common to any operation for acute traumatic SCI is relief of any residual compression on the injured spinal cord. This may be achieved through anterior (e.g., anterior cervical discectomy, corpectomy), posterior (e.g., laminectomy), or combined 360 degrees approaches; the optimal approach depends on the specific clinical scenario and injury pattern.

In recent years, there has been a growing body of evidence favoring early surgical decompression for spinal cord injury, and this has been reflected in guideline recommendations.[73–78] The pathophysiology of acute SCI involves both primary and secondary mechanisms that

lead to neurological injury.[79] At the time of trauma, kinetic energy is transmitted to the spinal column and spinal cord; there is usually rapid spinal cord compression and contusion, leading to immediate, irreversible injury and disruption of neural tissue—this is the primary injury. Because the primary injury is permanent and irreversible, it is not amenable to any intervention beyond primary prevention measures. The primary injury, in turn, instigates a signaling cascade of downstream events that leads to progressive cell death and spinal cord damage in the ensuing days to weeks—this is secondary injury. Prevention and mitigation of secondary injury is the target of neuroprotective strategies for SCI. The rationale for early surgery after SCI is to expeditiously relieve compression on the spinal cord so as to attenuate secondary injury mechanisms and ultimately improve long-term neurological outcome.

Recently, a landmark individual patient data metaanalysis including over 1500 patients demonstrated superior sensorimotor recovery at 1 year postinjury in patients with acute spinal cord injury who underwent surgical decompression within 24 hours, compared with those undergoing decompressive surgery later than 24 hours.[80] Furthermore, this study demonstrated a steep decline in neurological recovery with increasing time to decompression during the first 24–36 hours after injury, after which recovery appeared to plateau. This suggests the existence of a critical time window immediately following acute SCI during which earlier surgical decompression improves neurological recovery. The findings of this study bolstered the concept that "time is spine" and highlighted the critical importance of expeditious intervention for acute spinal cord injury.

The evolution of evidence and understanding of the role of early surgery for CCS has closely followed suit with that of acute traumatic SCI, almost as a microcosm of this larger umbrella. That is, there has been a paradigm shift away from conservative treatment and toward surgical intervention, in particular, early surgical decompression, within 24 hours of injury, for CCS.[19,81,82] This is reflected in studies of national trends in the management of CCS.[31,33] Using the NIS, Brodell et al.[31] found surgical treatment in cases of CCS increased by an average of 40% each year from 2003 to 2010 in the United States. Approximately half of surgical procedures for CCS from 2000 to 2009 were performed within 48 hours.[33] This shift in thinking has been fueled by growing recognition that (1) the clinical course of CCS is not substantially different from other cervical incomplete SCI syndromes[4]; (2) even if the recovery profile of CCS is generally favorable, surgical decompression may be expected to work in synergy with, rather than in opposition to, this natural recovery potential, resulting in superior neurological outcomes; and (3) CCS may in fact be the quintessential target for neuroprotective strategies aimed at mitigating secondary injury, namely early

surgical decompression, given the low kinetic energy mechanisms and less severe primary insult typically seen in these injuries, and the presumed ongoing secondary injury to an edematous spinal cord from compression from a narrow spinal canal. Certainly, more recent evidence would suggest that early surgical management of patients with CCS is safe and effective, with surgical decompression prior to 24 hours having shown to result in superior neurological and functional recovery at long-term follow-up.[19] However, this remains controversial.

Published studies evaluating the outcomes of early surgery for CCS are presented in Table 16.1. Recommendations regarding the operative management of CCS are derived primarily from clas III evidence from retrospective case series.[19] Indeed, CCS has been difficult to study clinically, mostly because of sample size issues. Using the minimum clinically important difference for

TABLE 16.1   Evidentiary table for early operative management of CCS.

Study	Description	Level of evidence	Key findings
Schroeder et al. (2016)[26,a]	* Retrospective review of 75 patients with CCS stratified by the presence ($N = 32$) or absence ($N = 43$) of increased T2 signal intensity within the spinal cord	III	* Patients with increased signal intensity had a more severe initial neurological injury (AMS 57.6 vs. 75.3, $P = .01$), but less severe change in AMS over the first week after injury (-0.85 vs. -4.3, $P = .07$). * Age, sex, ISS, congenital stenosis, surgery within 24 hours, or surgery during initial hospitalization did not significantly affect change in AMS on ANOVA.
Kepler et al. (2015)[23,a]	* Retrospective review of 68 patients who underwent surgical decompression for CCS; patients were separated into early (<24 hours) ($N = 19$) and delayed ($\geq$24 hours) ($N = 49$) surgery groups	III	* There was no difference in change in AMS at 7 days between early and delayed surgery groups (-2.9 vs. -4.2, $P = .34$). * The number of patients who had early improvement was similar between early and delayed surgery groups (50% vs. 48%, $P = .94$). * On multiple linear regression, early surgery was not associated with AMS at day 7 or change in AMS; older age was negatively associated with change in AMS.

TABLE 16.1 Evidentiary table for early operative management of CCS.—cont'd

Study	Description	Level of evidence	Key findings
Samuel et al. (2015)[25]	* Retrospective review of 1060 patients with CCS from the National Trauma Data Bank Research Data Set	III	* After controlling for preexisting comorbidity (CCI) and injury severity (ISS), delayed surgery was associated with decreased odds of inpatient mortality (OR 0.81, $P = .04$).
Thompson et al. (2015)[30]	* Retrospective review of 51 patients with CCS after suffering a hyperextension injury	III	* Surgery within 2 weeks of injury was associated with higher rates of clinical improvement than operation after 2 weeks (47% vs. 30%). * Of 20 patients treated nonoperatively, 10 (50%) showed neurological improvement; however, this reflected selection bias for surgery among more severely injury patients.
Anderson et al. (2012)[18,a]	* Retrospective review of 69 patients with CCS treated surgically within 24 hours ($N = 14$), at 24—48 hours ($N = 30$), or after 48 hours ($N = 25$)	III	* Mean AMS improved from 63.2 at presentation to 89.9 at final follow-up; 74% of patients improved at least one AIS grade. * A history of loss of consciousness, decreased rectal tone at presentation, presence of a fracture, timing of surgery, and surgical approach did not significantly impact improvement in AMS.
Aarabi et al. (2011)[9]	* Retrospective review of 42 patients with CCS and spinal stenosis treated surgically within 24 hours ($N = 9$), at 24—48 hours ($N = 10$), or after 48 hours ($N = 23$)	III	* Mean AMS improved from 63.8 at admission to 94.1 at 1-year follow-up. * AMS at final follow-up was associated with admission AMS ($P = .003$), percentage maximum canal compromise ($P = .02$), and midsagittal diameter at the point of maximum compression ($P = .02$). * FIM at final follow-up was associated with admission AMS ($P = .03$), percentage maximum canal compromise ($P = .02$), and age ($P = .02$). * Manual dexterity at final follow-up was associated with admission AMS ($P = .0002$) and length of parenchymal damage on T2 MRI ($P = .002$). * Pain level at final follow-up was associated with age ($P = .02$) and length of parenchymal damage on T2 MRI ($P = .04$). * Surgical approach, mechanism of injury, number of stenotic segments, percentage maximum spinal cord compression, and timing of surgery were not significant predictors of outcomes.
	* Retrospective review of 126 patients with CCS treated nonoperatively ($N = 59$) or surgically	III	* Patients who underwent surgery had an improvement in Frankel grade compared to

Continued

TABLE 16.1   Evidentiary table for early operative management of CCS.—cont'd

Study	Description	Level of evidence	Key findings
Stevens et al. (2010)[28]	($N = 67$) within 24 hours ($N = 16$), after 24 hours during initial hospital admission ($N = 34$), or on a second hospital admission (mean 137 days) ($N = 17$)		those who received medical management alone ($P = .001$). * There was no statistically significant difference in outcome (Frankel grade) between surgical groups based on timing of surgery. * There was a trend toward decreased mortality and complications in patients who received surgery within 24 hours or during the initial hospital admission.
Lenehan et al. (2010)[24]	* Retrospective review of prospectively collected data on 73 patients with CCS (AIS C or D, sacral sensory sparing, LEMS > UEMS) treated surgically within 24 hours ($N = 17$) or after 24 hours ($N = 56$)	II	* Patients who underwent early surgery had a 6.31-point greater improvement in AMS than those who underwent late surgery at 12-month follow-up ($P = .0358$). * Patients who were operated on within 24 hours had 7.79 U more improvement in FIM than those treated with late surgery at 6-month follow-up ($P = .0474$). * Early surgery provided higher probability of improvement in AIS grade at 6-month (OR 3.39) and 12-month (OR 2.81) follow-up compared with late surgery.

AMS, ASIA motor score; *ANOVA*, analysis of variance; *CCI*, Charlson Comorbidity Index; *CCS*, central cord syndrome; *ICU*, intensive care unit; *ISS*, injury severity score; *JOA*, Japanese Orthopaedic Association; *LEMS*, lower extremity motor score; *MRI*, magnetic resonance imaging; *OR*, odds ratio; *UEMS*, upper extremity motor score.
[a] Overlapping cohorts.

AMS (5 points)[83] and a standard deviation for change in AMS of 19.3 for cervical motor incomplete SCI,[84] a simple calculation reveals a sample size of 470 patients (235 in each group) would have 80% power to detect a 5-point difference in change in AMS with a two-sided $\alpha = 0.05$. Yet, the largest contemporary series of CCS have had under 20 patients in the early surgery group.[23,24] Insufficient statistical power may hence explain why many past studies have failed to identify a significant difference in outcomes with early versus late surgical decompression.

Much of the available evidence supports superior outcomes with surgery compared with conservative management for CCS.[20,29,33] Nonetheless, the evidence is more divided when it comes to early surgery. In a systematic review of the literature, Anderson et al.[19] found there

was low level evidence that patients operated on within 24 hours of injury had significantly greater improvements in AMS and functional independence at 1 year than those who underwent surgery more than 24 hours after injury; and there was moderate evidence that patients operated on within 2 weeks of injury had higher JOA score and recovery rate than those who underwent surgery more than 2 weeks after injury. Arguably the highest-quality evidence (class II) comes from an ambispective review of 73 patients with CCS conducted by Lenehan and colleagues.[24] This is one of the largest surgical case series of CCS; 17 patients (23%) were operated on within 24 hours of injury, while 56 patients (77%) underwent delayed surgery ($\geq$24 hours). The authors used multicenter, prospectively collected observational data from the

Spine Trauma Study Group. Enrollment was consecutive, limiting selection bias. Patients who underwent early surgery (<24 hours) had significantly greater improvements in AMS (effect: 6.31 points, $P = .036$) and FIM (effect: 7.79 U, $P = .047$) at 1 year compared with delayed ($\geq$24 hours) surgery. Nonetheless, other class III studies have failed to find a significant difference in outcomes with early versus delayed surgery.[9,18,23] The latest clinical guidelines therefore recommend that early surgery be considered as a treatment option in patients with CCS.[78] Notably, greater uncertainty exists with regard to the management of patients with present with "milder" symptoms of central cord syndrome (e.g., exacerbation of underlying cervical myelopathy); this will require prospective comparative effectiveness studies using more sensitive outcome instruments.

Conclusion

Central cord syndrome refers to a clinical diagnosis following acute traumatic SCI whose hallmark is disproportionately greater upper than lower limb weakness (LEMS > UEMS); a threshold difference of 10 points in ASIA motor score favoring the lower limbs has been proposed (LEMS − UEMS $\geq$10), although this has yet to gain global acceptance. CCS often occurs in an older patient secondary to low-energy (e.g., fall-related) hyperextension mechanisms that do not usually produce unstable spinal column injury. There is diffuse white matter injury, particularly involving the lateral columns harboring the corticospinal tract, which plays a more important role in hand than leg or locomotor function. There has been a shift in treatment paradigms toward operative management and early surgical decompression within 24 hours of injury, although the latter remains a subject of debate. Future studies using high-quality prospective data are needed to more definitively establish the safety and efficacy of early surgical decompression.

References

1. Bosch A, Stauffer ES, Nickel VL. Incomplete traumatic quadriplegia. A ten-year review. *J Am Med Assoc* 1971;**216**(3):473−8.
2. Pouw MH, van Middendorp JJ, van Kampen A, et al. Diagnostic criteria of traumatic central cord syndrome. Part 1: a systematic review of clinical descriptors and scores. *Spinal Cord* 2010;**48**(9):652−6.
3. van Middendorp JJ, Pouw MH, Hayes KC, et al. Diagnostic criteria of traumatic central cord syndrome. Part 2: a questionnaire survey among spine specialists. *Spinal Cord* 2010;**48**(9):657−63.
4. Pouw MH, Van Middendorp JJ, Van Kampen A, Curt A, Van De Meent H, Hosman AJF. Diagnostic criteria of traumatic central cord syndrome. Part 3: Descriptive analyses of neurological and functional outcomes in a prospective cohort of traumatic motor incomplete tetraplegics. *Spinal Cord* 2011;**49**(5):614−22.
5. Thompson C, Mutch J, Parent S, Mac-Thiong JM. The changing demographics of traumatic spinal cord injury: an 11-year study of 831 patients. *J Spinal Cord Med* 2015;**38**(2):214−23.
6. Fehlings MG, Tetreault L, Nater A, et al. The aging of the global population: the changing epidemiology of disease and spinal disorders. *Neurosurgery* 2015;**77**(Suppl. 4):S1−5.
7. Pappas CT, Gibson AR, Sonntag VK. Decussation of hind-limb and fore-limb fibers in the monkey corticospinal tract: relevance to cruciate paralysis. *J Neurosurg* 1991;**75**(6):935−40.
8. Coxe WS, Landau WM. Patterns of Marchi degeneration in the monkey pyramidal tract following small discrete cortical lesions. *Neurology* 1970;**20**(1):89−100.
9. Aarabi B, Alexander M, Mirvis SE, et al. Predictors of outcome in acute traumatic central cord syndrome due to spinal stenosis. *J Neurosurg Spine* 2011;**14**(1):122−30.
10. Quencer RM, Bunge RP, Egnor M, et al. Acute traumatic central cord syndrome: MRI-pathological correlations. *Neuroradiology* 1992;**34**(2):85−94.
11. Baker SN. The primate reticulospinal tract, hand function and functional recovery. *J Physiol* 2011;**589**(Pt 23):5603−12.
12. Zaaimi B, Edgley SA, Soteropoulos DS, Baker SN. Changes in descending motor pathway connectivity after corticospinal tract lesion in macaque monkey. *Brain* 2012;**135**(Pt 7):2277−89.
13. Benglis D, Levi AD. Neurologic findings of craniovertebral junction disease. *Neurosurgery* 2010;**66**(3 Suppl. 1):13−21.
14. Aarabi B, Simard JM, Kufera JA, et al. Intramedullary lesion expansion on magnetic resonance imaging in patients with motor complete cervical spinal cord injury. *J Neurosurg Spine* 2012;**17**(3):243−50.

15. Le E, Aarabi B, Hersh DS, et al. Predictors of intramedullary lesion expansion rate on MR images of patients with subaxial spinal cord injury. *J Neurosurg Spine* 2015;**22**(6):611–21.

16. Petersen JA, Spiess M, Curt A, et al. Upper limb recovery in spinal cord injury: involvement of central and peripheral motor pathways. *Neurorehabil Neural Repair* 2017;**31**(5):432–41.

17. Jimenez O, Marcillo A, Levi AD. A histopathological analysis of the human cervical spinal cord in patients with acute traumatic central cord syndrome. *Spinal Cord* 2000;**38**(9):532–7.

18. Anderson DG, Sayadipour A, Limthongkul W, Martin ND, Vaccaro A, Harrop JS. Traumatic central cord syndrome: neurologic recovery after surgical management. *Am J Orthoped* 2012;**41**(8):E104–8.

19. Anderson KK, Tetreault L, Shamji MF, et al. Optimal timing of surgical decompression for acute traumatic central cord syndrome: a systematic review of the literature. *Neurosurgery* 2015;**77**(Suppl. 4):S15–32.

20. Dahdaleh NS, Lawton CD, El Ahmadieh TY, et al. Evidence-based management of central cord syndrome. *Neurosurg Focus* 2013;**35**(1):E6.

21. Furusawa K, Tokuhiro A, Ikeda A, et al. Effect of age on bowel management in traumatic central cord syndrome. *Spinal Cord* 2012;**50**(1):51–6.

22. Hohl JB, Lee JY, Horton JA, Rihn JA. A novel classification system for traumatic central cord syndrome: the central cord injury scale (CCIS). *Spine* 2010;**35**(7):E238–43.

23. Kepler CK, Kong C, Schroeder GD, et al. Early outcome and predictors of early outcome in patients treated surgically for central cord syndrome. *J Neurosurg Spine* 2015;**23**(4):490–4.

24. Lenehan B, Fisher CG, Vaccaro A, Fehlings M, Aarabi B, Dvorak MF. The urgency of surgical decompression in acute central cord injuries with spondylosis and without instability. *Spine* 2010;**35**(21 Suppl. l):S180–6.

25. Samuel AM, Grant RA, Bohl DD, et al. Delayed surgery after acute traumatic central cord syndrome is associated with reduced mortality. *Spine* 2015;**40**(5):349–56.

26. Schroeder GD, Hjelm N, Vaccaro AR, Weinstein MS, Kepler CK. The effect of increased T2 signal intensity in the spinal cord on the injury severity and early neurological recovery in patients with central cord syndrome. *J Neurosurg Spine* 2016;**24**(5):792–6.

27. Schroeder GD, Kepler CK, Hjelm N, Vaccaro AR, Weinstein MS. The effect of vertebral fracture on the early neurologic recovery in patients with central cord syndrome. *Eur Spine J* 2015;**24**(5):985–9.

28. Stevens EA, Marsh R, Wilson JA, Sweasey TA, Branch Jr CL, Powers AK. A review of surgical intervention in the setting of traumatic central cord syndrome. *Spine J* 2010;**10**(10):874–80.

29. Stevenson CM, Dargan DP, Warnock J, et al. Traumatic central cord syndrome: neurological and functional outcome at 3 years. *Spinal Cord* 2016;**54**(11):1010–5.

30. Thompson C, Gonsalves JF, Welsh D. Hyperextension injury of the cervical spine with central cord syndrome. *Eur Spine J* 2015;**24**(1):195–202.

31. Brodell DW, Jain A, Elfar JC, Mesfin A. National trends in the management of central cord syndrome: an analysis of 16,134 patients. *Spine J* 2015;**15**(3):435–42.

32. Chen LF, Chang HK, Chen YC, et al. Five-year medical expenses of central cord syndrome: analysis using a national cohort. *J Neurosurg Sci* 2020;**64**(2):147–53.

33. Yoshihara H, Yoneoka D. Trends in the treatment for traumatic central cord syndrome without bone injury in the United States from 2000 to 2009. *J Trauma Acute Care Surg* 2013;**75**(3):453–8.

34. Hilibrand A, Smith JS. Cervical spine surgery: anterior microsurgery. *Instr Course Lect* 2012;**61**:451–9.

35. Nuwer MR, Emerson RG, Galloway G, et al. Evidence-based guideline update: intraoperative spinal monitoring with somatosensory and transcranial electrical motor evoked potentials: report of the Therapeutics and Technology Assessment Subcommittee of the American Academy of Neurology and the American Clinical Neurophysiology Society. *Neurology* 2012;**78**(8):585–9.

36. Sun LQ, Shen Y, Li YM. Quantitative magnetic resonance imaging analysis correlates with surgical outcome of cervical spinal cord injury without radiologic evidence of trauma. *Spinal Cord* 2014;**52**(7):541–6.

37. Martin AR, Aleksanderek I, Cohen-Adad J, et al. Translating state-of-the-art spinal cord MRI techniques to clinical use: a systematic review of clinical studies utilizing DTI, MT, MWF, MRS, and fMRI. *Neuroimage Clin* 2016;**10**:192–238.

38. Jia X, Kowalski RG, Sciubba DM, Geocadin RG. Critical care of traumatic spinal cord injury. *J Intensive Care Med* 2013;**28**(1):12–23.

39. Parent S, Barchi S, LeBreton M, Casha S, Fehlings MG. The impact of specialized centers of care for spinal cord injury on length of stay, complications, and mortality: a systematic review of the literature. *J Neurotrauma* 2011;**28**(8):1363–70.

40. Lenehan B, Street J, Kwon BK, et al. The epidemiology of traumatic spinal cord injury in British Columbia, Canada. *Spine* 2012;**37**(4):321–9.

41. Aarabi B, Hadley MN, Dhall SS, et al. Management of acute traumatic central cord syndrome (ATCCS). *Neurosurgery* 2013;**72**(Suppl. 2):195–204.

42. Kirshblum SC, Burns SP, Biering-Sorensen F, et al. International standards for neurological classification of spinal cord injury (revised 2011). *J Spinal Cord Med* 2011;**34**(6):535–46.

43. Schneider RC, Cherry G, Pantek H. The syndrome of acute central cervical spinal cord injury; with special reference to the mechanisms involved in hyperextension injuries of cervical spine. *J Neurosurg* 1954;**11**(6):546–77.

44. Ryken TC, Hadley MN, Walters BC, et al. Radiographic assessment. *Neurosurgery* 2013;**72**(Suppl. 2):54–72.

45. Kurpad S, Martin AR, Tetreault LA, et al. Impact of baseline magnetic resonance imaging on neurologic, functional, and safety outcomes in patients with acute traumatic spinal cord injury. *Global Spine J* 2017;**7**(3 Suppl. l):151S–74S.

46. Jain NB, Ayers GD, Peterson EN, et al. Traumatic spinal cord injury in the United States, 1993-2012. *J Am Med Assoc* 2015;**313**(22):2236–43.

47. Lee BB, Cripps RA, Fitzharris M, Wing PC. The global map for traumatic spinal cord injury epidemiology: update 2011, global incidence rate. *Spinal Cord* 2014;**52**(2):110–6.

48. Singh A, Tetreault L, Kalsi-Ryan S, Nouri A, Fehlings MG. Global prevalence and incidence of traumatic spinal cord injury. *Clin Epidemiol* 2014;**6**:309–31.

49. Couris CM, Guilcher SJ, Munce SE, et al. Characteristics of adults with incident traumatic spinal cord injury in Ontario, Canada. *Spinal Cord* 2010;**48**(1):39–44.

50. Tafida MA, Wagatsuma Y, Ma E, Mizutani T, Abe T. Descriptive epidemiology of traumatic spinal injury in Japan. *J Orthop Sci* 2018;**23**(2):273–6.

51. Knutsdottir S, Thorisdottir H, Sigvaldason K, Jonsson Jr H, Bjornsson A, Ingvarsson P. Epidemiology of traumatic spinal cord injuries in Iceland from 1975 to 2009. *Spinal Cord* 2012;**50**(2):123–6.

52. Van Den Berg M, Castellote JM, Mahillo-Fernandez I, de Pedro-Cuesta J. Incidence of traumatic spinal cord injury in Aragon, Spain (1972-2008). *J Neurotrauma* 2011;**28**(3):469–77.

53. McCaughey EJ, Purcell M, McLean AN, et al. Changing demographics of spinal cord injury over a 20-year period: a longitudinal population-based study in Scotland. *Spinal Cord* 2016;**54**(4):270–6.

54. van den Berg ME, Castellote JM, Mahillo-Fernandez I, de Pedro-Cuesta J. Incidence of spinal cord injury worldwide: a systematic review. *Neuroepidemiology* 2010;**34**(3):184–92.

55. Devivo MJ. Epidemiology of traumatic spinal cord injury: trends and future implications. *Spinal Cord* 2012;**50**(5):365–72.

56. Friedli L, Rosenzweig ES, Barraud Q, et al. Pronounced species divergence in corticospinal tract reorganization and functional recovery after lateralized spinal cord injury favors primates. *Sci Transl Med* 2015;**7**(302):302ra134.

57. Molliqaj G, Payer M, Schaller K, Tessitore E. Acute traumatic central cord syndrome: a comprehensive review. *Neurochirurgie* 2014;**60**(1–2):5–11.

58. Schneider RC, Thompson JM, Bebin J. The syndrome of acute central cervical spinal cord injury. *J Neurol Neurosurg Psychiatry* 1958;**21**(3):216–27.

59. Taylor AR. The mechanism of injury to the spinal cord in the neck without damage to vertebral column. *J Bone Joint Surg Br* 1951;**33-B**(4):543–7.

60. Li XF, Dai LY. Acute central cord syndrome: injury mechanisms and stress features. *Spine* 2010;**35**(19):E955–64.

61. Foerster O. Symptomatologie der erkrankungen des rückenmarks und seiner wurzeln. In: Bumke O, Foerster O, editors. *Handbook of neurology*, vol. 5. Berlin: Springer; 1936. p. 83.

62. Lawrence DG, Kuypers HG. The functional organization of the motor system in the monkey. I. The effects of bilateral pyramidal lesions. *Brain* 1968;**91**(1):1–14.

63. Lawrence DG, Kuypers HG. The functional organization of the motor system in the monkey. II. The effects of lesions of the descending brain-stem pathways. *Brain* 1968;**91**(1):15–36.

64. Tanadini LG, Steeves JD, Hothorn T, et al. Identifying homogeneous subgroups in neurological disorders: unbiased recursive partitioning in cervical complete spinal cord injury. *Neurorehabil Neural Repair* 2014;**28**(6):507–15.

65. Tanadini LG, Hothorn T, Jones LA, et al. Toward inclusive trial protocols in heterogeneous neurological disorders: prediction-based stratification of participants with incomplete cervical spinal cord injury. *Neurorehabil Neural Repair* 2015;**29**(9):867–77.

66. Dvorak MF, Fisher CG, Hoekema J, et al. Factors predicting motor recovery and functional outcome after traumatic central cord syndrome: a long-term follow-up. *Spine* 2005;**30**(20):2303–11.

67. Paquet J, Rivers CS, Kurban D, et al. The impact of spine stability on cervical spinal cord injury with respect to demographics, management, and outcome: a prospective cohort from a national spinal cord injury registry. *Spine J* 2018;**18**(1):88–98.

68. Jin W, Sun X, Shen K, et al. Recurrent neurological deterioration after conservative treatment for acute traumatic central cord syndrome without bony injury: seventeen operative case reports. *J Neurotrauma* 2017; **34**(21):3051−7.

69. Liu Y, Wang Z, Yang S, Yang H, Zou J. The effect of surgical intervention for delayed cervical central cord syndrome. *BioMed Res Int* 2017;**2017**:7979850.

70. Badhiwala JH, Wilson JR. The natural history of degenerative cervical myelopathy. *Neurosurg Clin N Am* 2018; **29**(1):21−32.

71. *Advanced trauma life support (ATLS) student course manual.* 9th ed. American College of Surgeons; 2012.

72. Readdy WJ, Whetstone WD, Ferguson AR, et al. Complications and outcomes of vasopressor usage in acute traumatic central cord syndrome. *J Neurosurg Spine* 2015:1−7.

73. Fehlings MG, Vaccaro A, Wilson JR, et al. Early versus delayed decompression for traumatic cervical spinal cord injury: results of the Surgical Timing in Acute Spinal Cord Injury Study (STASCIS). *PLoS One* 2012;**7**(2):e32037.

74. Furlan JC, Noonan V, Cadotte DW, Fehlings MG. Timing of decompressive surgery of spinal cord after traumatic spinal cord injury: an evidence-based examination of pre-clinical and clinical studies. *J Neurotrauma* 2011;**28**(8):1371−99.

75. Wilson JR, Singh A, Craven C, et al. Early versus late surgery for traumatic spinal cord injury: the results of a prospective Canadian cohort study. *Spinal Cord* 2012; **50**(11):840−3.

76. Fehlings MG, Rabin D, Sears W, Cadotte DW, Aarabi B. Current practice in the timing of surgical intervention in spinal cord injury. *Spine* 2010;**35**(21 Suppl. l):S166−73.

77. van Middendorp JJ, Hosman AJ, Doi SA. The effects of the timing of spinal surgery after traumatic spinal cord injury: a systematic review and meta-analysis. *J Neurotrauma* 2013;**30**(21):1781−94.

78. Fehlings MG, Tetreault LA, Wilson JR, et al. A clinical practice guideline for the management of patients with acute spinal cord injury and central cord syndrome: recommendations on the timing (</=24 hours versus >24 hours) of decompressive surgery. *Global Spine J* 2017; **7**(3 Suppl. l):195S−202S.

79. Ahuja CS, Wilson JR, Nori S, et al. Traumatic spinal cord injury. *Nat Rev Dis Primers* 2017;**3**:17018.

80. Badhiwala JH, Wilson JR, Witiw CD, et al. The influence of timing of surgical decompression for acute spinal cord injury: a pooled analysis of individual patient data. *Lancet Neurol* 2021;**20**(2):117−26.

81. Riew KD, Kang DG. Central cord syndrome: is operative treatment the standard of care? *Spine J* 2015;**15**(3): 443−5.

82. Ho A, Chi JH. Early surgery recommended for acute central cord injury. *Neurosurgery* 2010;**67**(6):N21−2.

83. Scivoletto G, Tamburella F, Laurenza L, Molinari M. Distribution-based estimates of clinically significant changes in the International standards for neurological classification of spinal cord injury motor and sensory scores. *Eur J Phys Rehabil Med* 2013;**49**(3):373−84.

84. Casha S, Zygun D, McGowan MD, Bains I, Yong VW, Hurlbert RJ. Results of a phase II placebo-controlled randomized trial of minocycline in acute spinal cord injury. *Brain* 2012;**135**(Pt 4):1224−36.

Management and pathophysiology

James Hong[1], Noah Poulin[2], Brian K. Kwon[3], Michael G. Fehlings[1]

[1]Krembil Research Institute, Toronto, ON, Canada; [2]University of Cambridge, Cambridge, United Kingdom; [3]International Collaboration on Repair Discoveries, University of British Columbia, Vancouver, BC, Canada

Epidemiology

Incidence

Global incidence of traumatic spinal cord injury (SCI) in 2007 was estimated to be between 133,000 and 226,000, or 23 cases per million.[1,2] The incidence by country was found to range from 8.0 per million in Spain to 49.1 per million in New Zealand.[3] The Global Burden of Disease Study reported a global estimate of 0.93 million new cases of SCI in 2016.[4] There is a consistent global trend of higher rates of SCI in males than females, although the ratio varies widely across countries.[3] From 1993 to 2012, the overall incidence of acute SCI in the United States has slightly increased from 53 cases per 1 million (95%CI: 52–54) to 54 cases per 1 million (95% CI: 53–55). The incidence in the young male population (16–24 years) has decreased from 144 cases per million to 87 cases per million. A similar trend is seen for males aged 24–44, decreasing from 96 cases per million to 71 cases per million. Conversely, there has been an increase in incidence for men aged 65–74 from 84 cases per million in 1993 to 131 cases per million in 2012.[5] The National Spinal Cord Injury Statistical Center reports a similar overall incidence rate of 54 cases per million in 2018, increasing from an estimated 40 per million in 2011. The average age at injury increased from 29 during 1973–79 to 37.6 in 2000 and 43 in 2018.[6–8] In a recent analysis of traumatic pediatric cervical spinal fracture patients in the United States, the incidence of SCI was found to have slightly increased between 2003 and 2012, from 2.39% to 3.12%.[9] In Canada, there is a cost of approximately \$2.7 billion associated with the approximately 1389 new traumatic SCI patients every year.[10] The initial incidence of traumatic SCI was estimated to be approximately 53 per million in 2010, which includes cases which did not survive to hospital, while the discharge incidence was estimated to be 41 per million.[11]

Prevalence

The global prevalence of SCI was estimated to be 27.04 million (24.98–30.15 million) in 2016.[4]

The prevalence in the United States is estimated to be approximately 853 per million (range: 721–4187)[1] or 906 per million.[3] The Canadian prevalence of traumatic SCI has been estimated at 1184 per million.[1] Overall, the

Neural Repair and Regeneration after Spinal Cord Injury and Spine Trauma
https://doi.org/10.1016/B978-0-12-819835-3.00024-1

© 2022 Elsevier Inc. All rights reserved.

prevalence of SCI in Canada is approximately 85,556 persons, and this figure is split almost evenly across traumatic SCI (51%) and nontraumatic SCI (49%).[11]

Complications and level-specific data

The mortality rate for lesions in the cervical region is significantly higher than any other region.[12] In the United States, the in-hospital mortality decreased slightly among those aged 85 or older: (24.2% in 1993—96 to 20.1% in 2010—12). However, overall in-hospital mortality increased slightly from 6.6% in 1993—96 to 7.5% in 2010—12.[5] Poorman et al. report that infants and children most frequently fractured at C2, while young adults frequently fractured at C7, although overall, upper cervical SCI was less common at 5.8% compared with 10.9% lower cervical SCI. The most prevalent type of injury was unspecified SCI, and the next most common was complete lesions, which was more likely in lower-level fractures.[9]

Etiology

The etiology of traumatic SCI varies globally according to local cultural factors, in addition to demographic and economic factors. Generally, younger patients sustain spinal injury through high-energy mechanisms (i.e., motor vehicle collision, fall from height) while elderly patients endure lower-energy mechanisms (i.e., falls from a standing height). Overall, traffic accidents are the primary cause of SCI.[3] In Canada, land transport-related SCI has been reported as the primary etiology at 47%, ranging from 34% to 56% across regions. Similarly, the most common etiology in the United States is land transport at 48%.[12] In developed countries, the incidence of traumatic SCI due to motor vehicle accidents has mostly decreased or stabilized; however, this figure is increasing in developing countries. This is likely due to an ongoing transition to motorized vehicles in the latter. In comparison, higher developed regions generally have safer vehicles, roads, and more advanced alternative transport infrastructure, although there are significant variations within countries.[1] A pattern displaying elevated rates of violence-related traumatic SCI can be seen through North and South America, Southern Africa, and the Middle East. The largest proportion of gunshot-related SCI is present in South Africa, but this etiology is also elevated in Brazil and the United States.[1] Violence was responsible for 15% of SCI in the United States, with 14% gunshot-related and 1% stabbing-related.[12] Greenland has an extremely elevated proportion of suicide-related traumatic SCI relative to other countries, at 23%.[1,3] In the United States, the proportion of SCI related to falls in those older than 65 years increased from 28% during 1997—2000 to 66% during 2010—12.[5] Overall, approximately 23% of traumatic SCI are due to falls in the United States,[1] while in Canada, 17% of SCI are due to falls.[12] The widely reported trend toward low fall—related SCI in elderly populations is especially important in developed regions with an increasingly aging population.[5,13] Notably, Japan has the highest proportion of tetraplegia in the world, likely due to an aging population and possible genetic factors.[1,14,15]

Clinical management of spinal cord injury—standard of care in North America

Spine immobilization and stabilization

All patients with a suspected spinal injury should have their spine immobilized during emergency transportation and in early in-hospital management, until ruled out according to neurological assessment and/or diagnostic imaging, when appropriate. Prehospital immobilization (more accurately termed spinal motion restriction) has the aim of preventing further mechanical damage to the cord and remains standard procedure in most centers. The 2012 AANS/CNS guidelines provide a level II recommendation for spinal immobilization of all trauma patients with an SCI or a mechanism with potential for SCI. However, trauma patients with a normal neurological evaluation who are awake alert and not intoxicated with no neck pain should not be immobilized. Notably, spinal immobilization is not recommended in patients with penetrating trauma due to increased mortality from delayed resuscitation. Prolonged immobilization also increases morbidity, including pressure sores and airway complications. In awake and symptomatic patients, the 2012 AANS/CNS guidelines for the management of cervical spinal injuries provide a level I recommendation for radiographic evaluation using high-quality CT. Alternatively, in the absence of high-quality CT imaging, three-way cervical spine radiographs (AP/lateral/odontoid) are recommended. These also provide a level I recommendation against radiographic evaluation of the cervical spine in awake and asymptomatic patients with a normal neurological examination, without neck pain or tenderness and who can complete a range of motion evaluation. Alternatively, the Canadian C-spine Rule determines the need for radiography based first on the following high-risk factors: age >65 years, a dangerous mechanism (fall from elevation, axial load to head, high speed motor vehicle collision, motorized recreational vehicle, bicycle struck, or collision) or paresthesia in extremities. In the absence of a high risk factor, the inability to safely assess range of motion or inability to actively rotate the neck 45° both left and right indicates a requirement for radiography.[16,17] The 2018 AANS/CNS guidelines on thoracolumbar spine injuries provide a grade B recommendation for the use of MRI to assess posterior ligamentous complex integrity when determining the need for surgery, but note that there is insufficient evidence that radiographic findings can predict clinical outcomes in thoracolumbar fractures.[18]

Hemodynamic management

Neurogenic shock may worsen secondary SCI through hypoperfusion and ischemia. Spinal cord perfusion pressure (SCPP) is defined as the difference between intrathecal pressure (ITP) and mean arterial pressure (MAP). After SCI, a decrease in MAP may occur through either loss of sympathetic vasomotor tone, leading to pooling of blood in the extremities, or interruption of sympathetic input to cardiac fibers, which results in arrhythmias due to exclusive parasympathetic contribution. Furthermore, hypovolemia from concurrent injuries can result in hypotension, although this is usually associated with tachycardia, rather than the bradycardic hypotension commonly developing from neurogenic shock. These complications are most effectively recognized and managed in an intensive care setting.[19] Initially, restoration of normal hemodynamic parameters should be attempted using crystalloid-based volume resuscitation, although this may not be effective if cardiac output is compromised. Vasopressors should be used to obtain normotension, and those with both alpha- and beta-adrenergic effects are preferable as the lack of beta-adrenergic stimulation can result in reflex bradycardia. In 2008, the Consortium for Spinal Cord Medicine included a recommendation for either dopamine or norepinephrine in cervical and high thoracic SCI due to dual alpha- and beta-adrenergic effects. In lower regions, phenylephrine was suggested.[20] With aims to increase SCPP, several other clinical practice guidelines have recommended maintenance of an MAP of 85—90 mmHg for up to 5—7 days after SCI.[19,21] Regarding thoracolumbar injuries specifically, the AANS/CNS guidelines found insufficient evidence to recommend for or against maintenance of arterial blood pressure. They instead note that results from pooled (cervical + thoracolumbar) populations indicate that clinicians may choose to maintain MAP >85 mmHg. However, in the 2012 AANS/CNS guidelines on management of cervical injuries, a level III recommendation is provided to maintain MAP between 85 and 90 mmHg in the first week after an acute spinal cord injury to improve cord perfusion.[22] A recent metaanalysis[23] reported that in two prospective studies, neurological outcomes were stable to be improved with management of MAP above 85 and 90 mmHg, although the clinical evidence to support these MAP targets is weak.[23,24] These authors note that norepinephrine is preferred for cervical and upper thoracic injuries and phenylephrine or norepinephrine for mid- to lower thoracic injuries.[23] In 2016, a crossover evaluation was undertaken in a series of 11 patients with cervical or thoracic SCI, using norepinephrine and dopamine. While maintaining a similar MAP, a significantly higher spinal cord perfusion pressure (SCPP) was obtained using norepinephrine when compared with dopamine. This was achieved through a significantly lower intrathecal pressure (ITC) with the norepinephrine administration. A recent systematic review suggests that dopamine is associated with higher rates of complications than phenylephrine and supports the view that norepinephrine slightly increases perfusion pressure relative to dopamine and phenylephrine.[25] Although no significant differences in neurological outcomes were observed, other reports suggest that cardiogenic complications were more strongly associated with the use of dopamine (OR 8.97, $P < .001$) than phenylephrine (OR 5.92, $P = .004$). Both types of vasopressors were associated with increased complications in SCI.[26] This is supported by Readdy et al., who reported that cardiac complications were associated with both dopamine and phenylephrine, but that the differences were not statistically significant.[27]

Neurological assessment

The 2018 AANS/CNS guidelines on the evaluation and treatment of patients with thoracolumbar spine trauma contain a level C recommendation for the use of the functional independence measure, Sunnybrook cord injury scale, Frankel scale for SCI while noting that the American Spinal Injury Association International Standards for Spinal Cord Injury Classification scale (ASIA) has not specifically been validated in thoracolumbar injuries. However, a level B recommendation is given on the use of Entry ASIA scale grade, sacral sensation, ankle spasticity, urethral and rectal sphincter function, and AbH motor function in predicting neurological function and outcome in patients with thoracic and lumbar fractures.[28] The 2012 AANS/CNS guidelines on the management of acute cervical spine and spinal cord injury contain a level II recommendation for the use of ASIA as the preferred tool for assessment of neurological status and a level I recommendation for the use of the Spinal Cord Independence Measure (SCIM III) for assessment of functional status.[29] Despite insufficient evidence supporting the use of AIS scale in thoracolumbar injury specifically, it remains the most widely used assessment tool for determining the level of injury, extent of injury, and functional status. The ASIA system involves sensory and motor examinations to determine the neurological level of injury (NLI) and a score on the ASIA impairment scale (AIS). The NLI is defined as the most caudal cord segment with intact sensory and antigravity muscle function. The level of sensation on each side is defined as the most caudal intact dermatome for both pinprick (sharp/dull discrimination) and light touch sensation. The motor level is determined by the lowest key muscle function rated at least 3 (full ROM against gravity without resistance). AIS grade A is used to describe complete injuries, in which no sensory or motor function is preserved in the sacral segments S4—S5A. This requires the absence of voluntary rectal sphincter function, no deep rectal pressure sensation, and a score of 0 in all the S4—5 sensory tests. AIS grade B describes sensory incomplete injuries, in which sensory but not motor function is preserved below the NLI, including the sacral segments S4—S5 and no motor function is preserved more than three levels below the motor level (lowest level with full ROM against gravity) on either side of the body. AIS grade C indicates a motor incomplete injury, in which motor function is preserved below the NLI and more than half of key muscle functions below the single NLI have a muscle grade <3. AIS grade D indicates a motor incomplete injury in which at least half of key muscle functions below the NLI have a muscle grade >3. A grade of E refers to normal sensation and motor function in all segments.[30]

Pharmacological management of spinal cord injury

The use of methylprednisolone (MPSS) in SCI has remained contentious over the past several decades. Its originally reported clinical benefits have been questioned, especially given the increased risk of serious side effects. In the National Acute Spinal Cord Injury series of trials,[31–33] the increased risk of infection-related complications using a high dose 48 hour protocol outweighed potential neurological benefits. The 24 h protocol (30 mg/kg bolus + 5.4 mg/kg/hours × 23 hours) had a lower rate of complications. In a widely disputed[34,35] post hoc subgroup analysis of patients (that the authors contend was planned a priori) given MPSS within 8 hours of injury, MPSS was found to improve neurological outcomes. This conclusion was recently supported by a Cochrane review, reporting a significantly increased ASIA motor score in patients who received MPSS within 8 hours of injury.[36]

Recently, the 2017 AOSpine Guidelines provided a weak recommendation to offer as a treatment option the NASCIS 2 protocol of 24-hour infusion of high-dose MPSS within 8 hours of acute SCI and suggest not offering high-dose MPSS in a 24-hour infusion after 8-hours postinjury. These guidelines also recommend that the 48-hour infusion of high-dose MPSS not be administered to adult patients with SCI. The 2018 AANS/CNS Guidelines on thoracolumbar SCI found insufficient evidence to make a recommendation on whether the administration of a

pharmacologic agent improves outcomes in patients with thoracolumbar fractures and spinal cord injury; however, they noted that the complication profile of methylprednisolone should be carefully considered when deciding on administration. Previous guidelines on cervical SCI management from the AANS/CNS have varied on the recommendation for use of MPSS. The 2002 guidelines suggested either 24-hour or 48-hour infusion while noting that the evidence of serious infection-related side effects is more consistent than the evidence of neurological improvement. In contrast, the 2013 guidelines provided a level I recommendation against the use of MPSS based on a lack of approval for SCI by the FDA, the lack of class I and II evidence supporting clinical benefits, and class I, II, and III evidence of harmful side effects in high-dose MPSS.

Operative versus nonoperative management and timing of decompression

Surgical fusion and/or decompression are required when injuries to the spine result in instability. Spinal instability may be quantified using several different scales, however, in the cervical region the Subaxial Spine Injury Classification (SLIC) scale delineates a need for surgical intervention at a score of 5 or more, while a score of 4 is equivocal and a score of 3 or less should be treated nonsurgically. This system demonstrates an interrater intraclass reliability coefficient of 0.71, and its use is supported by a level I recommendation from the AANS/CNS 2012 guidelines. These guidelines also recommend use of the Cervical Spine Injury Severity Score (CSISS), which has higher interrater reliability (0.883), but is more complicated and may be more challenging to implement clinically.[37] Analogously, in the thoracolumbar region, the Thoracolumbar Spine Injury Classification (TLIC) or AO Spine Classification systems can be used to characterize the injury[38] and predict the need for surgical management. Similar to the SLIC, the TLIC also defines the need for surgical intervention at a score of 5 or above, with 4 being equivocal and 3 or less treated nonoperatively.[39]

Thus far, the Surgical Timing in Acute Spinal Cord Injury Study (STASCIS) provides the highest-quality evidence supporting early surgical decompression (<24 hours). In a prospective, multicenter cohort study, patients who had surgery within 24 hours showed a higher rate of 2-or-more point improvement in AIS grades at 6 months follow-up. However, these results should be evaluated in the context of the study's limitations stemming from the cohort study design and the reported lack of statistical difference between treatment groups in a 1-point grade improvements on the AIS scale.[40] A 2017 systematic review noted that the results and quality of evidence for early surgical decompression were variable depending on level, timing of follow-up, and outcome but suggested that sufficient evidence existed to support the improved neurological recovery for cervical SCI patients with early surgical intervention.[41] The 2017 AO Spine Guidelines suggest that surgical decompression within 24 hours be considered as a treatment option in adults with traumatic central cord syndrome. They also suggest that early surgery be offered to adults with acute SCI regardless of level. The quality of evidence was low, and the strength of these recommendations was weak. Recently, a study applying the AO Spine Subaxial cervical spine trauma classification system concluded that early decompression (<72 hours) is not required for type A and F1-3 fractures. In contrast, early surgical treatment was especially beneficial in type B and type C/F4 fractures.[42]

The 2018 AANS/CNS guidelines state that in patients with thoracolumbar burst fractures who are neurologically intact, the conflicting evidence is not sufficient to make a recommendation and the discretion of the treating provider be used. These guidelines also state that there is insufficient evidence to recommend for or against surgical intervention in nonburst thoracic or lumbar injuries, and similarly recommend that the decision be at the discretion of the treating physician. The 2012 AANS/CNS guidelines give a level III recommendation to reduce subaxial cervical fractures or dislocations with the goal of spinal cord decompression/restoration of the spinal canal. Stable immobilization by internal fixation or external immobilization is recommended. If surgical treatment is considered, either anterior or posterior fixation and fusion is acceptable if a particular approach is not required. If these treatment options are not available, prolonged bed rest in traction is the recommended treatment. In patients with ankylosing spondylitis, posterior long segment instrumentation and fusion or a combined dorsal/anterior procedure is recommended as anterior stand-alone instrumentation and fusion is associated with a failure rate of up to 50%.

Venous thromboembolism prophylaxis and treatment

Pulmonary embolism (PE) and deep vein thrombosis (DVT) are commonly associated with spinal cord injuries. The majority of these occur within the first 3 months after injury. Patients with severe motor deficits following SCI should be treated prophylactically for venous thromboembolism (VTE) using a combination of anticoagulation and pneumatic compression devices. The use of oral coagulation and low-dose heparin alone is not recommended. The prophylactic treatment should be initiated within 72 hours and continued for 3 months (level II).

Pathophysiology of spinal cord injury

The timeline of pathological effects after SCI can be organized into acute, subacute, intermediate, and chronic phases. The acute phase spans approximately the first 48 hours after injury. The subacute phase lasts from 48 hours postinjury to approximately 14 days postinjury, while the intermediate phase lasts from 14 days to 3 months, and chronic lasts for >3 months. The effects of SCI can also be divided into primary and secondary mechanisms.[43]

The pathophysiology of SCI is highly diverse and dependent on the biological and physical conditions of the primary injury. Primary injury refers to the initial mechanical impact that may entail varying degrees of compression, distraction (stretching), laceration, or contusion that results in structural damage.[44] Commonly, however, it is due to contusion with or without accompanying persistent compression from bone fragments due to vertebral burst fractures.[45] Particularly in traumatic SCI, the precise mechanism of compression can be difficult to ascertain, as patients may present with a combination of lateral, posterior, and anterior compression. Disc herniation, vertebral body fracture, and spinal column dislocation can result in anterior compression, while the intact lamina/posterior elements can result in posterior compression. The complete transection of the SC is rarely clinically relevant,[46] and for this reason, animal models that focus on contusion have been developed and shown to successfully mimic human pathological changes after SCI. Energy transfer through the subdural cerebrospinal fluid (CSF) and the resultant impact on the cord varies according to morphology and thus partially depends on the level of injury as described earlier.[45–47] Across the range of underlying etiologies, mechanical insults cause instantaneous physical changes by destroying neural cells and vascular structures at the site of injury.[48] Both the duration and severity of compression influence the functional outcome and at a critical threshold, significant, irreversible, and immediate neurological damage is produced, which compromises the physiological transmission through the spinal cord. The extent of the primary injury is predictive of functional recovery.[43] The secondary injury in SCI includes a multitude of physical, biochemical, and cellular effects that progressively damage tissue. These begin immediately after injury and last for several weeks to several months.[49,50] These mechanisms of injury have traditionally been the focus of SCI therapeutics research, as the narrow timeline of primary injury does not allow for a sufficient therapeutic window. The cells that survive the initial injury but are damaged may undergo necrosis or apoptosis.[48] At the injury epicenter, cystic cavities known as syringomyelia form and spread over time. Astrocytes and pericyte-derived cells surround the cavity to attenuate the spread of the lesion while a fibrous scar (collagen and ECM molecules) is also formed.[51] The secondary injuries are a significant contributor to functional impairment and physical changes in the spinal cord and expand the area of primary injury.[52]

Vascular disruption

Hemorrhage in the spinal cord can occur due to nontraumatic or traumatic etiology and can be categorized based on compartmental location. Overall, the presence of hemorrhage is predictive of reduced motor function after SCI,[53] with toxic effects from the blood itself leading to necrosis, apoptosis, and impaired regeneration.[54,55] Losey and colleagues[55] note that arterial injuries are rare in the arteries supplying the spinal cord and those adjacent to the spinal column. However, the vertebral arteries are frequently injured in cervical injuries, but these do not result in neurological deficits. Rather, intramedullary hemorrhage is believed to be a significant contributor to secondary damage after SCI. Microvascular intramedullary hemorrhage (hematomyelia) can be detected almost immediately after contusion SCI. Notably, as the gray matter (GM), which contains cell bodies, is more heavily vascularized, it displays more hemorrhagic necrosis after SCI.[45,56] Figley[57] notes that hemorrhage is narrowly limited to the immediate site of physical deformation. However, secondary microcirculatory hemorrhage has been reported to occur in nearby tissue, particularly in the cranial and caudal directions, and is linked with secondary lesion expansion.[58–61] Ischemia can be defined as insufficient oxygen availability due to reduced blood flow to tissue. Neurons and glial cells in the CNS display high metabolic activity and therefore are especially sensitive to disruptions in oxygen and glucose supply. With primary SCI, the instantaneous mechanical damage to vasculature is often limited to smaller intramedullary vessels and capillaries rather than the larger arteries.[43,44,62] The immediate destruction of microcirculation results in local anoxia and thus directly leads to CNS cell death through lack of oxygen and nutrients.[52] In the hours after primary injury, local ischemia increases.[62] Beyond the primary injury rupturing blood vessels in the spinal cord, perfusion is further restricted due to vasospasm, which can reduce blood flow by up to 80%.[43,63] Hypoperfusion in the injured spinal cord can also occur due to systemic hypotension as a result of damage to sympathetic circuits.[64] Additionally, thrombosis may play a role in secondary ischemia.[59,65]

Ischemia increases local vascular permeability, compromising the integrity of the BSCB.[66] This, in addition to the high colloid pressure of the blood itself, contributes to local edema, which is especially prevalent in contusion injuries relative to transection or laceration injuries. Edema damages the cord by mechanically enlarging internal cavitations, leading to syringomyelia, which is a major barrier to axon regeneration and synaptic formation. Additionally, it has been reported that the increased intrasyringal pressure compresses surrounding tissue, resulting in regional ischemia and a local blood flow reduction of over 80%.[46] This suggests the existence of a detrimental feedback mechanism between ischemia and edema after initial vascular damage. There is a delayed rise in cytoplasmic levels of arachidonic acid associated with tissue edema and Na^+/K^+

ATPase inhibition.[67] The activation of intracellular calcium-dependent phospholipase A2 and its subsequent action on membrane phospholipids provides a source of arachidonic acid. Cyclooxygenases 1 and 2 then convert arachidonic acid to prostaglandin H2, which is further processed by platelets to produce thromboxane A2, which has prothrombic properties, further contributing to hypoxic ischemia.[68] Thromboxane B2 levels have been shown to increase disproportionally to levels of 6-keto-PGF10 in response to SCI, which may contribute to local hypoxia.[69] Furthermore, Sharma et al.[70] showed that inhibition of prostaglandins results in a decreased microvascular permeability after SCI. This suggests that prostaglandins may contribute to edema and associated ischemia. Ischemia has been linked to mitochondrial disfunction through irreversibly increased membrane permeability. The intensity of ischemia corresponds with the mechanism of cell death, with subacute ischemia allowing for progression of the apoptotic pathway by maintaining a low level of ATP production. In contrast, fulminant ischemia rapidly disrupts mitochondrial function and leads to necrosis.[71] With the progression of the ischemic process, a late-hypoxic state is reached in which the mitochondria permeability increases drastically, leading to swelling and eventual plasma membrane rupture, which irreversibly damages the cell and causes necrotic cell death.

Ionic imbalance and excitotoxicity

The intracellular accumulation of Na^+ and depletion of K^+ resulting from failure of the Na^+/K^+ ATPase due to prolonged ischemia and anoxia results in membrane depolarization and Na^+/Ca^{2+} exchange, leading to increased intracellular calcium ion concentration. This activates a range of apoptotic enzymes, including phospholipases, calpain, and protein kinase C (PKC).[72] The glutamate concentration in the spinal cord increases excessively after injury, including at the injury site, where it damages the CNS directly through excitotoxic cell death.[73,74] In the spinal cord, uptake of glutamate has been shown to be largely mediated by astroglial rather than neuronal transporters.[75] Excitotoxicity is a result of excessive activation of glutamate receptors, to which neurons and oligodendrocytes are particularly vulnerable, given their expression of all glutamate receptor types.[76] The axonal white matter was believed to be spared from excitotoxicity, due to its lack of synapses within the CNS.[77] However, Xu et al. showed that after SCI, glutamate could reach toxic concentrations in white matter.[74] This supports the finding by Agrawal and Fehlings that axonal injury can occur through activation of non-NMDA ionotropic glutamate receptors.[78] Ionotropic glutamate receptor activation and intracellular Na^+ accumulation lead to intracellular acidosis and cytotoxic edema.[79–82] Additionally, the increase in cytosolic Na^+ increases intracellular Ca^{2+} concentrations to damaging levels by modulating activity of the Ca^{2+}/Na^+ exchanger.[83,84] Intracellular calcium is transported into the mitochondria via the potential driven Ca^{2+} uniporter. The levels of intramitochondrial Ca^{2+} remain elevated and continue to accumulate with successive NMDA receptor stimulation.[85,86] Mitochondrial uptake of cytosolic calcium is linked to mitochondrial dysfunction and resultant cell death, as changes in mitochondrial polarization inhibit ATP production. This metabolic deficiency further exacerbates the hypoxia induced by primary and secondary vascular disfunction. Epstein also suggested a role for calcium dysregulation in reperfusion injury, where Ca^{2+} may contribute to the formation of membrane-damaging free radicals.[87]

Oxidative stress

Reactive oxygen species (ROS) cause apoptosis and necrosis of neural cells through damage of macromolecular cellular structures, including DNA, phospholipids, and proteins.[88,89] ROS production peaks 12 hours after SCI and remains elevated before return to baseline levels within 4–5 weeks after SCI.[90] Structural and metabolic dysfunction in mitochondria due to ischemia and elevated intracellular Ca^{2+} cause ROS production.[91] Additionally, phagocytic leukocytes and microglia contribute to ROS due to increased oxygen consumption and production of superoxide through NADPH oxidase (NOX2).[88,92] The elevated ROS results in extensive and damaging lipid peroxidation of fatty acids in cell membranes and myelin.[93,94] Notably, oxidative stress may contribute to BSCB dysfunction by promoting downregulation of tight junction proteins and activating matrix metalloproteinases.[95,96]

Neuroinflammation

Inflammation arising from cellular and molecular immunity occurs immediately after SCI and continues for several weeks.[97] The BSCB is disrupted mechanically and chemically during primary injury. The immediate damage to the BSCB leads to extravasation of large molecules and cells within minutes of injury[98,99] Damage to the BSCB increases the inflammatory response by allowing inflammatory cells into the site of injury.[100,101] Locally, microglia are activated through necrotic by-products from the primary injury and release proinflammatory cytokines, recruiting neutrophils and monocytes from the systemic circulation. These, in turn, may further increase the permeability of the BSCB upon extravasation.[102,103] Initially, the immunological response to lesions in the brain and spinal cord is dominated by microglia, which secrete proinflammatory cytokines, recruiting leukocytes, which also upregulate proinflammatory mediators. Approximately 24 hours after SCI, neutrophils infiltrate the lesion site. Blood monocytes differentiate into macrophages and remain at the site of injury.[104–106] These chronically release proinflammatory cytokines, chemokines, nitric oxide, ions, and proteases, resulting in fibrosis and cell death through apoptosis and necrosis.[100,107] Neutrophils are not found natively in the spinal cord. After SCI, they enter the site of injury passively through both microvascular damage and active recruitment to the site of injury, where they phagocytose damaged tissue and produce inflammatory cytokines and proteases, which also serve as chemoattractants for macrophages.[105] Importantly, edema may be related to leukocyte recruitment and infiltration as demonstrated in a collagenase-induced mouse model of intracerebral hemorrhage (ICH).[108] However, anatomical differences between the brain and spinal cord vasculature may affect the extrapolation of brain-based ICH studies to SCI. Importantly, leukocytes are more often recruited through the venous system than the arterial system, and in rats, there are significantly more sulcal (central) veins than sulcal arteries.[105] Meanwhile, there are similar proportions of veins and arteries in the rat brain. This implies greater effects of neutrophil extravasation in SCI relative to the cerebral parenchyma. Losey et al. also noted the potential influence of elevated neutrophil recruitment in the spinal cord relative to the brain.[55]

Additionally, Wallerian degeneration, the retrograde degradation of axons away from the primary site of injury, is associated with the infiltration of monocytes. When in direct contact with damaged axons and myelin debris, macrophages undergo a process of polarization from antiinflammatory to a proinflammatory state. In the CNS, macrophages may remain near the area of axonal degeneration for several months.[109] Thus, the increased infiltration of these cells due to hemorrhage, extravasation and a disrupted BSCB contributes to neural degradation and the associated functional deficits.[110,111] Microglia have classically been described as existing in either a resting or activated state, although recently, it has been demonstrated that this relationship is more a spectrum than a dichotomy and that many intermediate types exist.[112] After SCI, microglia

migrate to the site of injury and proliferate and phagocytose injured and dying cells. Additionally, they secrete cytokines and proteases that contribute to axonal damage and edema.[113] In the acute phase, microglia are distributed across a spectrum between the M1 and M2 phenotypes. However, with prolonged inflammation, the M1 phenotype becomes more prevalent, which produces more proinflammatory cytokines, resulting in a feedback cycle with an increase in M1-oriented microglial phenotypes. Meanwhile, the M2 phenotype is broadly antiinflammatory and shows neuroprotective effects through increased phagocytic activity, matrix deposition, and wound healing.[112] The neuroprotective effects of immune cells in SCI have been widely recognized. Notably, the astrocyte-derived glial scar prevents the spread of the lesion, and the macrophages and microglia clear neuronal debris.[110,112,114–116]

The early management of SCI intends to minimize secondary injury, and there is evidence that approaches such as early decompression and hemodynamic management can influence outcome. The pathophysiology of secondary injury is complex and includes vascular disruption, ionic imbalance, oxidative stress, and inflammation. Obtaining a more complete understanding of these processes in human SCI will be necessary to developing therapeutics, which have to date not shown convincing efficacy.

References

1. Lee BB, Cripps RA, Fitzharris M, Wing PC. The global map for traumatic spinal cord injury epidemiology: update 2011, global incidence rate. *Spinal Cord* 2014; **52**(2):110–6. https://doi.org/10.1038/sc.2012.158.
2. Fitzharris M, Cripps RA, Lee BB. Estimating the global incidence of traumatic spinal cord injury. *Spinal Cord* 2014;**52**(2):117–22. https://doi.org/10.1038/sc.2013.135.
3. Singh A, Tetreault L, Kalsi-Ryan S, Nouri A, Fehlings MG. Global prevalence and incidence of traumatic spinal cord injury. *Clin Epidemiol* 2014;**6**:309–31. https://doi.org/10.2147/CLEP.S68889.

4. James SL, Theadom A, Ellenbogen RG, et al. Global, regional, and national burden of traumatic brain injury and spinal cord injury, 1990–2016: a systematic analysis for the Global Burden of Disease Study 2016. *Lancet Neurol* 2019;**18**(1):56–87. https://doi.org/10.1016/S1474-4422(18)30415-0.

5. Jain NB, Ayers GD, Peterson EN, et al. Traumatic spinal cord injury in the United States, 1993–2012. *JAMA J Am Med Assoc* 2015;**313**(22):2236–43. https://doi.org/10.1001/jama.2015.6250.

6. Stover SL, Fine PR. The epidemiology and economics of spinal cord injury. *Paraplegia* 1987;**25**(3):225–8. https://doi.org/10.1038/sc.1987.40.

7. National Spinal Cord Injury Statistical Center. Spinal cord injury facts and figures at a glance. *J Spinal Cord Med* 2013. https://doi.org/10.1179/1079026813Z.000000000136.

8. National Spinal Cord Injury Statistical Center. Spinal cord injury facts and figures at a glance. *J Spinal Cord Med* 2018;**36**. https://doi.org/10.1179/1079026813Z.000000000136.

9. Poorman GW, Segreto FA, Beaubrun BM, et al. Traumatic fracture of the pediatric cervical spine: etiology, epidemiology, concurrent injuries, and an analysis of perioperative outcomes using the kids' inpatient database. *Internet J Spine Surg* 2019;**13**(1):68–78. https://doi.org/10.14444/6009.

10. Krueger H, Noonan VK, Trenaman LM, Joshi P, Rivers CS. The economic burden of traumatic spinal cord injury in Canada. *Chronic Dis Inj Can* 2013;**33**(3):113–22. http://www.ncbi.nlm.nih.gov/pubmed/23735450. [Accessed 8 September 2019].

11. Noonan VK, Fingas M, Farry A, et al. Incidence and prevalence of spinal cord injury in Canada: a national perspective. *Neuroepidemiology* 2012;**38**(4):219–26. https://doi.org/10.1159/000336014.

12. Cripps RA, Lee BB, Wing P, Weerts E, MacKay J, Brown D. A global map for traumatic spinal cord injury epidemiology: towards a living data repository for injury prevention. *Spinal Cord* 2011;**49**(4):493–501. https://doi.org/10.1038/sc.2010.146.

13. Montoto-Marqués A, Ferreiro-Velasco ME, Salvador-De La Barrera S, Balboa-Barreiro V, Rodriguez-Sotillo A, Meijide-Failde R. Epidemiology of traumatic spinal cord injury in Galicia, Spain: trends over a 20-year period. *Spinal Cord* 2017;**55**(6):588–94. https://doi.org/10.1038/sc.2017.13.

14. Katoh S, Enishi T, Sato N, Sairyo K. High incidence of acute traumatic spinal cord injury in a rural population in Japan in 2011 and 2012: an epidemiological study. *Spinal Cord* 2014;**52**(4):264–7. https://doi.org/10.1038/sc.2014.13.

15. Kudo D, Miyakoshi N, Hongo M, et al. An epidemiological study of traumatic spinal cord injuries in the fastest aging area in Japan. *Spinal Cord* 2019;**57**(6):509–15. https://doi.org/10.1038/s41393-019-0255-7.

16. Stiell IG, Clement CM, Grimshaw J, et al. Implementation of the Canadian C-spine rule: prospective 12 centre cluster randomised trial. *BMJ* 2009;**339**(7729):1071. https://doi.org/10.1136/bmj.b4146.

17. Stiell IG, Clement CM, McKnight RD, et al. The Canadian C-spine rule versus the NEXUS low-risk criteria in patients with trauma. *N Engl J Med* 2003;**349**(26):2510–8. https://doi.org/10.1056/NEJMoa031375.

18. Qureshi S, Dhall SS, Anderson PA, et al. Congress of neurological surgeons systematic review and evidence-based guidelines on the evaluation and treatment of patients with thoracolumbar spine trauma: radiological evaluation. *Neurosurgery* 2019;**84**(1):E28–31. https://doi.org/10.1093/neuros/nyy373.

19. Ryken TC, Hurlbert RJ, Hadley MN, et al. The acute cardiopulmonary management of patients with cervical spinal cord injuries. *Neurosurgery* 2013;**72**(Suppl. 2):84–92. https://doi.org/10.1227/NEU.0b013e318276ee16.

20. Consortium for Spinal Cord Medicine. Early acute management in adults with spinal cord injury. *J Spinal Cord Med* 2008;**31**. https://doi.org/10.1043/1079-0268-31.4.408.

21. Wing PC. Early acute management in adults with spinal cord injury: a clinical practice guideline for healthcare providers. *J Spinal Cord Med* 2008;vol. 31. https://doi.org/10.1080/10790268.2008.11760737.

22. Aarabi B, Hadley MN, Dhall SS, et al. Management of acute traumatic central cord syndrome (ATCCS). *Neurosurgery* 2013;**72**(Suppl. 2):195–204. https://doi.org/10.1227/NEU.0b013e318276f64b.

23. Saadeh YS, Smith BW, Joseph JR, et al. The impact of blood pressure management after spinal cord injury: a systematic review of the literature. *Neurosurg Focus* 2017;**43**(5):1–7. https://doi.org/10.3171/2017.8.FOCUS17428.

24. Evaniew N, Mazlouman SJ, Belley-Côté EP, Jacobs WB, Kwon BK. Interventions to optimize spinal cord perfusion in patients with acute traumatic spinal cord injuries: a systematic review. *J Neurotrauma* 2020;**37**(9):1127–39. https://doi.org/10.1089/neu.2019.6844.

25. Yue JK, Tsolinas RE, Burke JF, et al. Vasopressor support in managing acute spinal cord injury: current knowledge. *J Neurosurg Sci* 2019;**63**(3):308–17. https://doi.org/10.23736/S0390-5616.17.04003-6.

26. Inoue T, Manley GT, Patel N, Whetstone WD. Medical and surgical management after spinal cord injury: vasopressor usage, early surgerys, and complications.

J Neurotrauma 2014;**31**(3):284—91. https://doi.org/10.1089/neu.2013.3061.

27. Readdy WJ, Whetstone WD, Ferguson AR, et al. Complications and outcomes of vasopressor usage in acute traumatic central cord syndrome. *J Neurosurg Spine* 2015;**23**(5):574—80. https://doi.org/10.3171/2015.2.SPINE14746.

28. O'Toole JE, Kaiser MG, Anderson PA, et al. Congress of neurological surgeons systematic review and evidence-based guidelines on the evaluation and treatment of patients with thoracolumbar spine trauma: executive summary. *Clin Neurosurg* 2019;**84**(1):2—6. https://doi.org/10.1093/neuros/nyy394.

29. Hadley MN, Walters BC, Aarabi B, et al. Clinical assessment following acute cervical spinal cord injury. *Neurosurgery* 2013;**72**(Suppl. 2):40—53. https://doi.org/10.1227/NEU.0b013e318276edda.

30. Kirshblum SC, Waring W, Biering-Sorensen F, et al. International standards for neurological classification of spinal cord injury (revised 2011). *J Spinal Cord Med* 2011;**34**(6):547—54. https://doi.org/10.1179/107902611X13186000420242.

31. Bracken MB, Shepard MJ, Holford TR, et al. Administration of methylprednisolone for 24 or 48 hours or tirilazad mesylate for 48 hours in the treatment of acute spinal cord injury: results of the third national acute spinal cord injury randomized controlled trial. *J Am Med Assoc* 1997;**277**(20):1597—604. https://doi.org/10.1001/jama.1997.03540440031029.

32. Bracken MB, Collins WF, Freeman DF, et al. Efficacy of methylprednisolone in acute spinal cord injury. *JAMA J Am Med Assoc* 1984;**251**(1):45—52. https://doi.org/10.1001/jama.1984.03340250025015.

33. Bracken MB, Shepard MJ, Collins WF, et al. A randomized, controlled trial of methylprednisolone or naloxone in the treatment of acute spinal-cord injury: results of the second national acute spinal cord injury study. *N Engl J Med* 1990;**322**(20):1405—11. https://doi.org/10.1056/NEJM199005173222001.

34. Hextrum S, Bennett S. A critical examination of subgroup analyses: the national acute spinal cord injury studies and beyond. *Front Neurol* February 2018;**9**. https://doi.org/10.3389/fneur.2018.00011.

35. Sayer FT, Kronvall E, Nilsson OG. Methylprednisolone treatment in acute spinal cord injury: the myth challenged through a structured analysis of published literature. *Spine J* 2006;**6**(3):335—43. https://doi.org/10.1016/j.spinee.2005.11.001.

36. Bracken MB. Cochrane review NASCIS (steroids for acute spinal cord injury) by same author as Nascis. n.d. https://doi.org/10.1002/14651858.CD001046.pub2.

37. Aarabi B, Walters BC, Dhall SS, et al. Subaxial cervical spine injury classification systems. *Neurosurgery* 2013;**72**(Suppl. 2):170—86. https://doi.org/10.1227/NEU.0b013e31828341c5.

38. Dailey AT, Arnold PM, Anderson PA, et al. Congress of neurological surgeons systematic review and evidence-based guidelines on the evaluation and treatment of patients with thoracolumbar spine trauma: classification of injury. *Neurosurgery* 2019;**84**(1):E24—7. https://doi.org/10.1093/neuros/nyy372.

39. Lee JY, Vaccaro AR, Lim MR, et al. Thoracolumbar injury classification and severity score: a new paradigm for the treatment of thoracolumbar spine trauma. *J Orthop Sci* 2005;**10**(6):671—5. https://doi.org/10.1007/s00776-005-0956-y.

40. Fehlings MG, Vaccaro A, Wilson JR, et al. Early versus delayed decompression for traumatic cervical spinal cord injury: results of the surgical timing in acute spinal cord injury study (STASCIS). *PLoS One* 2012;**7**(2). https://doi.org/10.1371/journal.pone.0032037.

41. Wilson JR, Tetreault LA, Kwon BK, et al. Timing of decompression in patients with acute spinal cord injury: a systematic review. *Global Spine J* 2017;**7**(3_Suppl.):95S—115S. https://doi.org/10.1177/2192568217701716.

42. Du JP, Fan Y, Zhang JN, Liu JJ, Meng YB, Hao DJ. Early versus delayed decompression for traumatic cervical spinal cord injury: application of the AOSpine subaxial cervical spinal injury classification system to guide surgical timing. *Eur Spine J* 2019. https://doi.org/10.1007/s00586-019-05959-6.

43. Tator CH, Fehlings MG. Review of the secondary injur theory of acute spinal cord trauma with emphasis on vascular mechanisms. *J Neurosurg* 1991;**75**(1):15—26. https://doi.org/10.3171/jns.1991.75.1.0015.

44. Alizadeh A, Dyck SM, Karimi-Abdolrezaee S. Traumatic spinal cord injury: an overview of pathophysiology, models and acute injury mechanisms. *Front Neurol* 2019;**10**. https://doi.org/10.3389/fneur.2019.00282.

45. Sekhon LHS, Fehlings MG. Epidemiology, demographics, and pathophysiology of acute spinal cord injury. *Spine* 2001;**26**(24 Suppl. L):S2—12. https://doi.org/10.1097/00007632-200112151-00002.

46. Stokes BT, Jakeman LB. Experimental modelling of human spinal cord injury: a model that crosses the species barrier and mimics the spectrum of human cytopathology. *Spinal Cord* 2002;**40**(3):101—9. https://doi.org/10.1038/sj.sc.3101254.

47. DeVivo MJ, Rutt RD, Black KJ, Go BK, Stover SL. Trends in spinal cord injury demographics and treatment outcomes between 1973 and 1986. *Arch Phys Med Rehabil* 1992;**73**(5):424—30. http://www.ncbi.nlm.nih.gov/pubmed/1580768. [Accessed 8 September 2019].

48. Oyinbo CA. Secondary injury mechanisms in traumatic spinal cord injury: a nugget of this multiply cascade. *Acta Neurobiol Exp* 2011;**71**(2):281–99. http://www.ncbi.nlm.nih.gov/pubmed/21731081. [Accessed 9 September 2019].

49. Ahuja CS, Nori S, Tetreault L, et al. Traumatic spinal cord injury - repair and regeneration. *Clin Neurosurg* 2017;**80**(3):S22–90. https://doi.org/10.1093/neuros/nyw080.

50. Tanhoffer RA, Yamazaki RK, Nunes EA, et al. Glutamine concentration and immune response of spinal cord-injured rats. *J Spinal Cord Med* 2007;**30**(2):140–6. https://doi.org/10.1080/10790268.2007.11753925.

51. Göritz C, Dias DO, Tomilin N, Barbacid M, Shupliakov O, Frisén J. A pericyte origin of spinal cord scar tissue. *Science* 2011;**333**(6039):238–42. https://doi.org/10.1126/science.1203165.

52. Tator CH, Koyanagi I. Vascular mechanisms in the pathophysiology of human spinal cord injury. *Neurosurg Focus* 1997;**2**(1):E2. https://doi.org/10.3171/foc.1997.2.1.2.

53. Losey P, Anthony DC. Impact of vasculature damage on the outcome of spinal cord injury: a novel collagenase-induced model may give new insights into the mechanisms involved. *Neural Regen Res* 2014;**9**(20):1783–6. https://doi.org/10.4103/1673-5374.143422.

54. Whalley K, O'Neill P, Ferretti P. Changes in response to spinal cord injury with development: vascularization, hemorrhage and apoptosis. *Neuroscience* 2006;**137**(3):821–32. https://doi.org/10.1016/j.neuroscience.2005.07.064.

55. Losey P, Young C, Krimholtz E, Bordet R, Anthony DC. The role of hemorrhage following spinal-cord injury. *Brain Res* 2014;**1569**:9–18. https://doi.org/10.1016/j.brainres.2014.04.033.

56. Wolman L. The disturbance of circulation in traumatic paraplegia in acute and late stages: a pathological study. *Paraplegia* 1965;**2**(4):213–26. https://doi.org/10.1038/sc.1964.39.

57. Figley SA, Khosravi R, Legasto JM, Tseng Y-FF, Fehlings MG. Characterization of vascular disruption and blood-spinal cord barrier permeability following traumatic spinal cord injury. *J Neurotrauma* 2014;**31**(6):541–52. https://doi.org/10.1089/neu.2013.3034.

58. Tomko P, Farkaš D, Čížková D, Vanický I. Longitudinal enlargement of the lesion after spinal cord injury in the rat: a consequence of malignant edema? *Spinal Cord* 2017;**55**(3):255–63. https://doi.org/10.1038/sc.2016.133.

59. Nelson E, Gertz SD, Rennels ML, Ducker TB, Blaumanis OR. Spinal cord injury: the role of vascular damage in the pathogenesis of central hemorrhagic necrosis. *Arch Neurol* 1977;**34**(6):332–3. https://doi.org/10.1001/archneur.1977.00500180026005.

60. Gerzanich V, Woo SK, Vennekens R, et al. De novo expression of Trpm4 initiates secondary hemorrhage in spinal cord injury. *Nat Med* 2009;**15**(2):185–91. https://doi.org/10.1038/nm.1899.

61. Henke D, Gorgas D, Doherr MG, Howard J, Forterre F, Vandevelde M. Longitudinal extension of myelomalacia by intramedullary and subdural hemorrhage in a canine model of spinal cord injury. *Spine J* 2016;**16**(1):82–90. https://doi.org/10.1016/j.spinee.2015.09.018.

62. Fehlings MG, Tator CH, Linden RD. The relationships among the severity of spinal cord injury, motor and somatosensory evoked potentials and spinal cord blood flow. *Electroencephalogr Clin Neurophysiol Evoked Potentials* 1989;**74**(4):241–59. https://doi.org/10.1016/0168-5597(89)90055-5.

63. Anthes DL, Theriault E, Tator CH. Ultrastructural evidence for arteriolar vasospasm after spinal cord trauma. *Neurosurgery* 1996;**39**(4):804–14. https://doi.org/10.1097/00006123-199610000-00032.

64. Furlan JC, Fehlings MG. Cardiovascular complications after acute spinal cord injury: pathophysiology, diagnosis, and management. *Neurosurg Focus* 2008;**25**(5):E13. https://doi.org/10.3171/FOC.2008.25.11.E13.

65. Alshareef M, Krishna V, Ferdous J, et al. Effect of spinal cord compression on local vascular blood flow and perfusion capacity. In: Fehlings M, editor. *PLoS one*, vol. 9; 2014. e108820. https://doi.org/10.1371/journal.pone.0108820 (9).

66. Jacobs TP, Kempski O, McKinley D, Dutka AJ, Hallenbeck JM, Feuerstein G. Blood flow and vascular permeability during motor dysfunction in a rabbit model of spinal cord ischemia. *Stroke* 1992;**23**(3):367–73. https://doi.org/10.1161/01.STR.23.3.367.

67. Faden AI, Chan PH, Longar S. Alterations in lipid metabolism, Na+,K+-ATPase activity, and tissue water content of spinal cord following experimental traumatic injury. *J Neurochem* 1987;**48**(6):1809–16. https://doi.org/10.1111/j.1471-4159.1987.tb05740.x.

68. Ley K. The microcirculation in inflammation. In: *Microcirculation.* Elsevier Inc.; 2008. p. 387–448. https://doi.org/10.1016/B978-0-12-374530-9.00011-5.

69. Hsu CY, Halushka PV, Hogan EL, Banik NL, Lee WA, Perot PL. Alteration of thromboxane and prostacyclin levels in experimental spinal cord injury. *Neurology* 1985;**35**(7). https://doi.org/10.1212/WNL.35.7.1003. 1003-1003.

70. Sharma HSS, Olsson Y, Nyberg F, Dey PKK. Prostaglandins modulate alterations of microvascular permeability, blood flow, edema and serotonin levels following spinal cord injury: an experimental study in the rat. *Neuroscience* 1993;**57**(2):443–9. https://doi.org/10.1016/0306-4522(93)90076-R.

71. Hu Z, Tu J. The roads to mitochondrial dysfunction in a rat model of posttraumatic syringomyelia. *BioMed Res Int* 2015;**2015**. https://doi.org/10.1155/2015/831490.

72. Stys PK, Lopachin RM. Mechanisms of calcium and sodium fluxes in anoxic myelinated central nervous system axons. *Neuroscience* 1998;**82**(1):21–32. https://doi.org/10.1016/s0306-4522(97)00230-3.

73. Liu D, Xu GY, Pan E, McAdoo DJ. Neurotoxicity of glutamate at the concentration released upon spinal cord injury. *Neuroscience* 1999;**93**(4):1383–9. https://doi.org/10.1016/S0306-4522(99)00278-X.

74. XU GY, Hughes MG, Ye Z, Hulsebosch CE, McAdoo DJ. Concentrations of glutamate released following spinal cord injury kill oligodendrocytes in the spinal cord. *Exp Neurol* 2004;**187**(2):329–36. https://doi.org/10.1016/j.expneurol.2004.01.029.

75. Rothstein JD, Dykes-Hoberg M, Pardo CA, et al. Knockout of glutamate transporters reveals a major role for astroglial transport in excitotoxicity and clearance of glutamate. *Neuron* 1996;**16**(3):675–86. https://doi.org/10.1016/S0896-6273(00)80086-0.

76. Kolodziejczyk K, Saab AS, Nave KA, Attwell D. Why do oligodendrocyte lineage cells express glutamate receptors? *F1000 Biol Rep (Wash D C)* 2010;**2**(1). https://doi.org/10.3410/B2-57.

77. Imaizumi T, Kocsis JD, Waxman SG. Anoxic injury in the rat spinal cord: pharmacological evidence for multiple steps in Ca^{2+}-dependent injury of the dorsal columns. *J Neurotrauma* 1997;**14**:299–311. https://doi.org/10.1089/neu.1997.14.299. Mary Ann Lieber, Inc.

78. Agrawal SK, Fehlings MG. Role of NMDA and non-NMDA ionotropic glutamate receptors in traumatic spinal cord axonal injury. *J Neurosci* 1997;**17**(3):1055–63.

79. Faden AI, Simon RP. A potential role for excitotoxins in the pathophysiology of spinal cord injury. *Ann Neurol* 1988;**23**(6):623–6. https://doi.org/10.1002/ana.410230618.

80. Choi DW. Ionic dependence of glutamate neurotoxicity. *J Neurosci* 1987;**7**(2):369–79.

81. Fehlings MG, Agrawal S. Role of sodium in the pathophysiology of secondary spinal cord injury. *Spine* 1995;**20**(20):2187–91. https://doi.org/10.1097/00007632-199510001-00002.

82. Ates O, Cayli SR, Gurses I, et al. Comparative neuroprotective effect of sodium channel blockers after experimental spinal cord injury. *J Clin Neurosci* 2007;**14**(7):658–65. https://doi.org/10.1016/j.jocn.2006.03.023.

83. Agrawal SK, Fehlings MG. The effect of the sodium channel blocker QX-314 on recovery after acute spinal cord injury. *J Neurotrauma* 1997;**14**(2):81–8. https://doi.org/10.1089/neu.1997.14.81.

84. Li S, Jiang Q, Stys PK. Important role of reverse Na^+-Ca^{2+} exchange in spinal cord white matter injury at physiological temperature. *J Neurophysiol* 2000;**84**(2):1116–9. www.jn.physiology.org.

85. Stout AK, Raphael HM, Kanterewicz BI, Klann E, Reynolds IJ. Glutamate-induced neuron death requires mitochondrial calcium uptake. *Nat Neurosci* 1998;**1**(5):366–73. https://doi.org/10.1038/1577.

86. Peng TI, Jou MJ, Sheu SS, Greenamyre JT. Visualization of NMDA receptor-induced mitochondrial calcium accumulation in striatal neurons Tsung-I. *Exp Neurol* 1998;**149**(1):1–12. https://doi.org/10.1006/exnr.1997.6599.

87. Epstein FH, Cheung JY, Bonventre JV, Malis CD, Leaf A. Calcium and ischemic injury. *N Engl J Med* 1986;**314**(26):1670–6. https://doi.org/10.1056/NEJM198606263142604.

88. Khayrullina G, Bermudez S, Byrnes KR. Inhibition of NOX2 reduces locomotor impairment, inflammation, and oxidative stress after spinal cord injury. *J Neuroinflammation* 2015;**12**(1). https://doi.org/10.1186/s12974-015-0391-8.

89. Poh Loh K, Hong Huang S, De Silva R, Tan B H, Zhun Zhu Y. Oxidative stress: apoptosis in neuronal injury. *Curr Alzheimer Res* 2006;**3**(4):327–37. https://doi.org/10.2174/156720506778249515.

90. Donnelly DJ, Popovich PG. Inflammation and its role in neuroprotection, axonal regeneration and functional recovery after spinal cord injury. *Exp Neurol* 2008;**209**(2):378–88. https://doi.org/10.1016/j.expneurol.2007.06.009.

91. Jia Z, Zhu H, Li J, Wang X, Misra H, Li Y. Oxidative stress in spinal cord injury and antioxidant-based intervention. *Spinal Cord* 2012;**50**(4):264–74. https://doi.org/10.1038/sc.2011.111.

92. Robinson JM. Reactive oxygen species in phagocytic leukocytes. *Histochem Cell Biol* 2008;**130**(2):281–97. https://doi.org/10.1007/s00418-008-0461-4.

93. Kamencic H, Griebel RW, Lyon AW, Paterson PG, Juurlink BHJ. Promoting glutathione synthesis after spinal cord trauma decreases secondary damage and promotes retention of function. *FASEB J* 2001;**15**(1):243–50. https://doi.org/10.1096/fj.00-0228com.

94. Christie SD, Comeau B, Myers T, Sadi D, Purdy M, Mendez I. Duration of lipid peroxidation after acute spinal cord injury in rats and the effect of methylprednisolone: laboratory investigation. *Neurosurg Focus* 2008;**25**(5). https://doi.org/10.3171/FOC.2008.25.11.E5.

95. Okamoto T, Akaike T, Sawa T, Miyamoto Y, Van der Vliet A, Maeda H. Activation of matrix metalloproteinases by peroxynitrite-induced protein S-glutathiolation via disulfide S-oxide formation. *J Biol Chem* 2001;**276**(31):29596–602. https://doi.org/10.1074/jbc.M102417200.

96. Lochhead JJ, McCaffrey G, Quigley CE, et al. Oxidative stress increases blood-brain barrier permeability and induces alterations in occludin during hypoxia-reoxygenation. *J Cerebr Blood Flow Metabol* 2010;**30**(9): 1625—36. https://doi.org/10.1038/jcbfm.2010.29.

97. Fehlings MG, Nguyen DH. Immunoglobulin G: a potential treatment to attenuate neuroinflammation following spinal cord injury. *J Clin Immunol* 2010; **30**(Suppl. 1):S109—12. https://doi.org/10.1007/s10875-010-9404-7.

98. Maikos JT, Shreiber DI. Immediate damage to the blood-spinal cord barrier due to mechanical trauma. *J Neurotrauma* 2007;**24**(3):492—507. https://doi.org/10.1089/neu.2006.0149.

99. Reyes-Alva HJ, Franco-Bourland RE, Martinez-Cruz A, Grijalva I, Madrazo I, Guizar-Sahagun G. Characterization of spinal subarachnoid bleeding associated to graded traumatic spinal cord injury in the rat. *Spinal Cord* 2014;**52**(Suppl. 2). https://doi.org/10.1038/sc.2014.93.

100. Muldoon LL, Alvarez JI, Begley DJ, et al. Immunologic privilege in the central nervous system and the blood-brain barrier. *J Cerebr Blood Flow Metabol* 2013;**33**(1): 13—21. https://doi.org/10.1038/jcbfm.2012.153.

101. Engelhardt B, Sorokin L. The blood-brain and the blood-cerebrospinal fluid barriers: function and dysfunction. *Semin Immunopathol* 2009;**31**(4):497—511. https://doi.org/10.1007/s00281-009-0177-0.

102. Larochelle C, Alvarez JI, Prat A. How do immune cells overcome the blood-brain barrier in multiple sclerosis? *FEBS Lett* 2011;**585**(23):3770—80. https://doi.org/10.1016/j.febslet.2011.04.066.

103. Bolton SJ, Anthony DC, Perry VH. Loss of the tight junction proteins occludin and zonula occludens-1 from cerebral vascular endothelium during neutrophil-induced blood-brain barrier breakdown in vivo. *Neuroscience* 1998;**86**(4):1245—57. https://doi.org/10.1016/S0306-4522(98)00058-X.

104. Gensel JC, Zhang B. Macrophage activation and its role in repair and pathology after spinal cord injury. *Brain Res* 2015;**1619**:1—11. https://doi.org/10.1016/j.brainres.2014.12.045.

105. Schnell L. Acute inflammatory responses to mechanical lesions in the CNS: differences between brain and spinal cord. *Eur J Neurosci* 1999;**11**(10):3648—58. https://doi.org/10.1046/j.1460-9568.1999.00792.x.

106. Tzekou A, Fehlings MG. Treatment of spinal cord injury with intravenous immunoglobulin G: preliminary evidence and future perspectives. *J Clin Immunol* 2014;**34**(Suppl. 1):132—8. https://doi.org/10.1007/s10875-014-0021-8.

107. Gaudet AD, Popovich PG. Extracellular matrix regulation of inflammation in the healthy and injured spinal cord. *Exp Neurol* 2014;**258**:24—34. https://doi.org/10.1016/j.expneurol.2013.11.020.

108. Titova E, Ostrowski RP, Kevil CG, et al. Reduced brain injury in CD18-deficient mice after experimental intracerebral hemorrhage. *J Neurosci Res* 2008;**86**(14): 3240—5. https://doi.org/10.1002/jnr.21762.

109. Dusart I, Schwab ME. Secondary cell death and the inflammatory reaction after dorsal hemisection of the rat spinal cord. *Eur J Neurosci* 1994;**6**(5):712—24. https://doi.org/10.1111/j.1460-9568.1994.tb00983.x.

110. Kong X, Gao J. Macrophage polarization: a key event in the secondary phase of acute spinal cord injury. *J Cell Mol Med* 2017;**21**(5):941—54. https://doi.org/10.1111/jcmm.13034.

111. Katoh H, Yokota K, Fehlings MG. Regeneration of spinal cord connectivity through stem cell transplantation and biomaterial scaffolds. *Front Cell Neurosci* 2019;**13**. https://doi.org/10.3389/fncel.2019.00248.

112. Cherry JD, Olschowka JA, O'Banion MK. Neuroinflammation and M2 microglia: the good, the bad, and the inflamed. *J Neuroinflammation* 2014;**11**. https://doi.org/10.1186/1742-2094-11-98.

113. Gao Z, Wang J, Thiex R, Rogove AD, Heppner FL, Tsirka SE. Microglial activation and intracerebral hemorrhage. *Acta Neurochir Suppl* 2008;(105):51—3. https://doi.org/10.1007/978-3-211-09469-3_11.

114. Mautes AE, Weinzierl MR, Donovan F, Noble LJ. Vascular events after spinal cord injury: contribution to secondary pathogenesis. *Phys Ther* 2000;**80**(7): 673—87. https://doi.org/10.1093/ptj/80.7.673.

115. Anderson MA, Burda JE, Ren Y, et al. Astrocyte scar formation aids central nervous system axon regeneration. *Nature* 2016;**532**(7598):195—200. https://doi.org/10.1038/nature17623.

116. Bradbury EJ, Burnside ER. Moving beyond the glial scar for spinal cord repair. n.d. https://doi.org/10.1038/s41467-019-11707-7.

SCI management: role and timing of surgical intervention

Paula Valerie ter Wengel[1,2,3], *Fan Jiang*[4], *Jefferson R. Wilson*[5], *Michael G. Fehlings*[4]

[1]Leiden University Medical Center, Leiden, the Netherlands; [2]HMC, The Hague, the Netherlands; [3]Amsterdam UMC, Amsterdam, the Netherlands; [4]Division of Neurosurgery, Toronto Western Hospital, University Health Network, Toronto, ON, Canada; [5]St. Michael's Hospital, Toronto, ON Canada

Introduction

Surgical intervention aims to prevent further damage from ongoing compression of the spinal cord, prevention of secondary damage to the spinal cord, and restoration of the spinal integrity. The effects of spinal cord decompression and its timing have been widely investigated in animal studies. Numerous preclinical studies have identified a relationship linking improved clinical outcomes with shorter duration of compression.[1–4] However, these results have not repeatedly been proven in a clinical setting.

Natural history

To increase our understanding of the impact of surgery on neurological recovery, it is important to have an understanding of the spontaneous neurological improvement in traumatic spinal cord injury (tSCI) patients. Multiple large prospective data sets exist from which neurological recovery patterns in tSCI patients were extracted, to assess the natural history in tSCI patients. It is important to note that some of these patients also received surgery at a certain point in time. Since surgical timing appears to play a role in neurological recovery, the neurological recovery rates from these data sets can only roughly estimate the actual natural history in tSCI.

Cervical spine

Approximately 17.3%–34% of patients with cervical tSCI will present with complete spinal cord injury.[5,6] In polytrauma patients who have sustained a cervical fracture, the incidence of complete tSCI is even higher, respectively 65.4%.[7] A longitudinal cohort study in 1393 patients evaluated the neurological recovery pattern in patients with cervical tSCI from the National Spinal Cord Injury Database (NSCISC).[8] Most of the patients from the NSCISC underwent surgery at some point.[9]

© 2022 Elsevier Inc. All rights reserved.

TABLE 18.1 Natural history cervical spinal cord injury.

Cervical tSCI		Neurologic examination at 1-year follow-up (NSCISC)				
		AIS A	AIS B	AIS C	AIS D	AIS E
Baseline	AIS A	236 (70.2%)	49 (14.6%)	27 (8%)	24 (7.1%)	0
	AIS B	11 (8.8%)	31 (24.8%)	37 (29.6%)	46 (36.8%)	0
	AIS C	1 (0.9%)	3 (2.8%)	15 (13.8%)	88 (80.7%)	2 (1.8%)
	AIS D	0	1 (0.7%)	1 (0.7%)	114 (84.4%)	19 (14.1%)

Data from Marino et al.[8]

One year after injury, neurological recovery was calculated in 705 patients (Table 18.1). The majority of patients with initial AIS A or AIS D injury did not recover, 70.2% and 84.8%, respectively. It appears that patients with an initial AIS C injury have the greatest ability to improve at least one grade, since 80.7% will improve to AIS D and 1.8% even to AIS E. Contrarily, only 14.1% of AIS D-injured patients will improve one AIS grade, i.e., to AIS E.

Thoracic spine

Thoracic spinal cord injury is less frequent than cervical tSCI. Most of the studies have grouped together patients with thoracic and thoracolumbar injuries. One study in 661 patients with thoracic tSCI (T2-T12) from the Spinal Cord Injury Model Systems (SCIMS) evaluated the neurological recovery rate.[10] Roughly half of all patients in the SCIMS data set received surgery at some point.[11] One year after injury, neurologic examination was performed in 265 patients. The majority of AIS A patients, as well as the majority of AIS D patients, did not recover (Table 18.2).

Another study in AIS A thoracic (T2-T12) tSCI patients evaluated the neurological recovery

TABLE 18.2 Natural history in thoracic and thoracolumbar spinal cord injury.

Thoracic tSCI		Neurologic examination at 1 year follow up (SCIMS)				
		AIS A	AIS B	AIS C	AIS D	AIS E
Baseline	AIS A	84.5%	7.7%	3.1%	4.6%	0
	AIS B	20.6%	20.6%	26.5%	29.4%	0
	AIS C	0	9.5%	4.8%	81%	4.8%
	AIS D	0	0	0	87.5%	12.5%

Data from Lee et al.[10]

patterns from three large data sets.[12] Almost all patients underwent surgery at some point in time. In this study, 76.6%—83.3% of all AIS A patients did not recover, 8.5%—10.9% recovered to AIS B, 7.8%—16.7% recovered to AIS C, and 5.3% recovered to AIS D.

In line with the cervical tSCI studies, patients with an initial AIS B or AIS C injury appear to have the greatest potential to recover neurologically in contrast to patients with AIS A or AIS D injuries.

The nature of surgical decompression for spinal cord injury

Studies have shown that spinal cord decompression can prevent further direct and ischemic damage to the spinal cord from continuous compression.[13] Decompression can be achieved through various methods, such as surgical decompression with or without internal fixation or in case of facet dislocation, a simple closed reduction. Additionally, surgical decompression can be achieved though anterior, posterior, or combined approaches. Currently, there are no studies comparing the type of intervention in relation to the neurological outcome; therefore the ideal approach will need to be assessed case by case. Additionally, only a few studies describe postoperative imaging to evaluate the extent to which the surgical intervention has "successfully" decompressed the injured spinal cord.[14-22] This is of particular importance, as 78% of cervical tSCI patients with less than 25% spinal canal narrowing will show spinal cord compression on MRI.[23] Traumatic herniated discs after closed reduction and expansion of spinal cord edema to levels adjacent to the level of injury can further impede adequate decompression of the spinal canal.[22,24-26]

One study in 184 motor complete cervical tSCI patients investigated the rate of successful decompression after various techniques of spinal cord decompression.[21] They performed preoperative (<8 hours) and postoperative MRI images to verify the rate of adequate spinal cord decompression. In patients who underwent anterior cervical discectomy (ACDF) and fusion, the rate of successful decompression was 46.8% compared with 58.6% when a corpectomy (ACCF) was performed. When a laminectomy was added to an ACDF or ACCF, the rate of successful decompression increased to 72% and 73.1%, respectively. The number of laminectomies appeared to be significantly associated with a successful decompression of the injured spinal cord. A successful decompression of the spinal cord in single, two, three, four, or five level laminectomies could be performed in, respectively, 58.3%, 68%, 78%, 80% and 100% of the cases.

The importance of adequate decompression of the spinal cord is portrayed by findings from a recent study in 100 cervical tSCI patients.[22] This study analyzed the relationship between AIS grade conversion and adequacy of surgical decompression. Adequate decompression on postoperative imaging was defined as reestablishment of a cerebrospinal fluid signal in the surrounding subarachnoid space around the spinal cord on a postoperative MRI. When MRI showed evidence of adequate spinal cord decompression, 58.9% of the patients conversed in AIS grade in contrast to only 18.5% when there was inadequate decompression.

Timing of surgical decompression in cervical spinal cord injury

The cervical spine is the most frequently affected level in tSCI. Over the past decades, there has been an increase in severe (AIS A–C) high (C1–C4) cervical spinal cord injuries and decrease in severe lower (C5–C8) cervical spinal cord injuries.[27] Cervical spinal cord injury can have a devastating impact on a patient's life. Studies have investigated whether surgical timing can reverse neurological injury and prevent further secondary damage to the spinal cord. The definition of what early surgery exactly is differs in various studies.[14,15,28-30] While some define <48 or 72 hours as early, most studies and recent cervical tSCI guidelines define early surgery as surgery performed within 24 hours after injury.[14,31] There are multiple factors involved, which make it not only difficult to study but also to interpret the effect of surgical timing in tSCI patients. One of them is the relative low incidence of spinal cord injury, which makes it difficult to investigate the role of surgical timing in large homogeneous patient groups. Another factor is heterogeneity of the type of injury, intervention, and patient-related characteristics. The majority of studies are based on relatively small heterogeneous patient groups with variable levels and severities of tSCI. This is of particular importance, since the neurological outcome is affected not only by the initial severity but also by the level of neurological injury.[32-35] Because the level of injury plays an important role in neurological outcome, we focused on clinical outcome in cervical tSCI patients specifically in the following part, as this level is the most often affected.

Early surgical timeframe (<24 hours)

The Surgical Timing in Acute Spinal Cord Injury Study (STASCIS) was the first large, multicenter, prospective cohort study in 313 cervical tSCI patients to investigate the effect of surgical timing on neurological outcome.[14] Patients were divided into early (<24 hours) and late (>24 hours) surgical groups according to the time elapsed from injury to spinal cord decompression. Surgical timing depended on transfer time from injury to the treating hospital, obtaining all diagnostic investigations, and the treating surgeon's decision on surgical timing. From the 313 patients, 182 underwent surgery within 24 hours. The main significant differences between both cohorts were age, initial severity of neurological injury, and administration of steroids, where patients in the early surgical cohort tended to be younger and had more severe neurological injuries at presentation (i.e., AIS A, B) and a higher proportion received steroids at admission compared with the late surgical cohort. After 6 months of follow-up, the patients in the early surgical cohort developed a significantly higher AIS grade improvement, where 19.8% experienced at least two AIS grades improvement compared with only 8.8% in the late surgical cohort (odds ratio [OR]: 2.57; 95% credibility interval [CI]: 1.1–7.28). In addition, early surgery did not lead to a difference in postoperative complications nor mortality. Other smaller studies have also examined the impact of surgical decompression within 24 hours for patients with cervical tSCI, with variable results.[36–39] One prospective cohort study, including 92 patients with similar demographics at baseline, found a significant difference in improvement of ≥1 AIS grade favoring early surgery (OR: 3.12; 95% CI: 1.21–8.02). This effect was, however, nonsignificant when a success was defined as an improvement of ≥2 AIS grades.[39] In contrast, other smaller studies did not find a significant beneficial effect of surgical decompression within 24 hours, compared with after 24 hours.[36,37]

Ultraearly surgical timeframes

Recently, various studies have investigated even earlier timeframes within 24 hours. One prospective cohort study in 48 cervical tSCI patients investigated whether surgical decompression within 8 hours after injury increased the likelihood of neurological recovery compared with surgery within 8–24 hours after injury.[40] At baseline, there were no significant differences between both groups; however, at 6-month follow-up, patients who were surgically decompressed <8 hours showed significantly more neurological recovery of ≥2 AIS grades compared with patients treated within 8–24 hours. This beneficial effect was still present after adjustment for the initial severity of neurological injury and preoperative degree of spinal canal compromise. Similarly, Grassner et al. observed a beneficial effect of surgical decompression within 8 hours for cervical tSCI.[15] In this retrospective study, 70 patients were divided into early (<8 hours; mean 4:22 hours) and late (>8 hours; mean 89:27 hours) surgical groups. After 1-year follow-up, significant differences were found between motor levels and neurological recovery, which were in favor of early surgery (<8 hours).

Another study including 57 male rugby players with facet dislocations suggested 4 hours as the optimal time window to restore neurological function after tSCI.[41] This study observed a negative association between neurological recovery and longer time to reduction of the spinal cord. From the patients with complete cervical tSCI which were reduced within 4 hours, 5/8 AIS A patients regained normal function (i.e., AIS E). In contrast, 0/24 patients with complete cervical tSCI recovered to AIS E when they underwent reduction >4 hours after injury.

By contrast, another study in 49 patients with cervical tSCI did not demonstrate superiority of surgical decompression within 5 hours after injury compared with 5–24 hours.[30] In this study, neurological improvement of ≥1 AIS grade was not significantly different between both groups, where 42% in the <5 hour group improved at least one AIS grade compared with 31% in the 5–24 hour group. In addition, an improvement of ≥ 2 AIS grades was significantly higher in patients undergoing surgery within 5–24 hours. In evaluating the neurological outcomes after ultraearly and even early surgery, it is important to bear in mind that the timing of the baseline neurological examination might also negatively influence the neurological outcome.[42,43] One study in 85 cervical tSCI patients evaluated the effect of the timing of baseline neurological investigation and corrected this for surgical timing among others.[43] This study observed an increased likelihood of AIS conversion when the baseline neurological investigation was performed within 4 hours after injury compared with thereafter (4–48 hours). This effect remained after controlling for surgical timing. Consequently, the optimal surgical timeframe within 24 hours remains unclear.

Severity of neurological injury and surgical timing

Patients who present with complete cervical tSCI will have worse neurological outcomes compared with patient who present with incomplete injuries.[44] Several studies have evaluated the effect of early decompression in patients with complete tSCI.[38,45–47] Bourassa-Moreau et al. addressed the effect of surgical timing in patients with complete tSCI and observed a significant ($P = .008$) benefit of surgical decompression within 24 hours on ≥ 1 AIS grade improvement.[38] 9 (64.3%) out of 14 patients in the early group improved at least one grade, compared with none out of 6 in the late group. Likewise, a study in 20 complete tSCI patients

evaluated the effect of early surgical decompression in combination with perioperative regional epidural cooling within 8 hours of injury.[47] 14 of these patients had cervical injuries, from which 9 (64.3%) improved at least one AIS grade.

A recent metaanalysis in 1.126 patients from 15 publications also confirmed a beneficial effect of early surgery in patients with complete cervical tSCI (Table 18.3).[45] In patients with complete cervical tSCI (n = 422), improvement was significantly more frequent after early surgery than after late surgery (respectively, 22.6%, 95% CI: 16.6%–28.7% and 10.4%, 95% CI: 5.6%–15.8%; OR: 2.6 [95% CI: 1.4–5.1] Fig. 18.1), whereas in patients with incomplete cervical tSCI (n = 636), improvement was similar between early and late surgery (respectively, 30.4%, 95% CI: 19.8%–41.6% and 32.5%, 95% CI: 21.4% –45.8%; OR 0.9 [95% CI: 0.4–1.9] Fig. 18.2). The latter should be interpreted with caution, since small differences in neurological improvement might not always lead to an improvement in AIS grade.[48] This is particularly true for ASIA D patients, where a "ceiling" effect has been suggested, preventing patients to improve further to an ASIA E. Due to the overrepresentation of ASIA D injuries in the incomplete tSCI group, it remains unclear why patients with more severe incomplete tSCI would not benefit from early surgery. Since a large number of incomplete tSCI patients will somewhat recover neurologically, regardless of surgical timing, small differences in recovery based on timing might be underpowered in the current studies and therefore not lead to the same effect as in complete tSCI.[34,42]

TABLE 18.3 Neurological recovery in complete and incomplete cervical tSCI.

	Odds ≥2 AIS grade improvement	
Cervical tSCI	<24 hours surgery	>24 hours surgery
AIS A (n = 422)	22.6% (95% CI: 16.6%–28.7%)	10.4% (95% CI: 5.6%–15.8%)
AIS BCD (n = 636)	30.4% (95% CI: 19.8%–41.6%)	32.5% (95% CI: 21.4%–45.8%)

The CI means Credibility Interval due to the Bayesian analysis
Data from ter Wengel et al.[45]

first author of study	total nr of patients	nr of patients with improvement	percentage improvement	lower 95%CI	higher 95%CI
Fehlings [30]	44	8	21.7	14.2	29.3
Levi [34]	22	5	22.7	14.9	31.8
Umerani [41]	12	3	23.0	14.3	34.3
Newton [37]	24	7	23.7	16.0	34.8
Randle [39]	20	3	21.7	12.7	30.0
Bourassa–Moreau [28]	14	4	23.4	15.1	35.5
Papadopoulos [38]	38	8	22.4	15.2	30.5
Jug [33]	26	4	21.5	12.8	29.4
Mattiassich [36]	20	5	23.0	15.0	33.0
Hansebout [32]	14	4	23.3	15.1	34.2
Grassner [31]	14	5	24.2	16.0	37.4
Early surgery	**248**	**56**	**22.6**	**16.6**	**28.7**
Fehlings [30]	27	3	10.6	5.3	17.1
Levi [34]	14	4	11.9	5.9	21.8
Umerani [41]	20	2	10.5	5.2	17.1
Randle [39]	12	1	10.5	5.1	17.2
Liu [35]	66	8	10.8	5.7	16.3
Benzel [23]	35	0	9.2	3.5	15.1
Late surgery	**174**	**18**	**10.4**	**5.6**	**15.8**

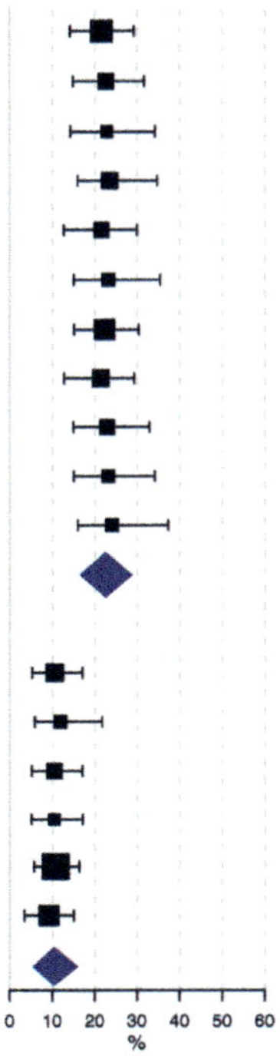

FIGURE 18.1 Surgical timing and >2 American Spinal Injury Association (ASIA) grade improvement in complete cervical traumatic spinal cord injury (tSCI). *Data from ter Wengel et al.*[45] *by the courtesy of* Journal of Neurotrauma, *Mary Ann Liebert inc.*

To date, it seems that in cervical tSCI, early decompression within 24 hours has a positive effect on neurological recovery. An additional beneficial effect of ultraearly intervention is not clear. It has yet to be elucidated which cervical tSCI patients will benefit from early surgical decompression, and who will not recover despite early intervention.

Timing of decompression in the "central cord" pattern of cervical incomplete tetraplegia

Traumatic central cord injury (TCCI) is the most common form of an incomplete (AIS C or D) tSCI and often presents after low energetic falls in patients with preexisting cervical spinal canal narrowing. TCCI is typically characterized by more severe motor impairment in upper extremities than in lower extremities. The description of TCCI and incomplete tSCI has been variable in the literature; therefore, recently a definition for TCCI has been proposed to describe TCCI more uniformly.[49] The proposed definition for TCCI was at least a 10-point difference between ASIA upper extremity motor score (UEMS) and lower extremity motor score (LEMS). It is important to note that this definition does not encompass all variable degrees of TCCI, making a clear distinction between TCCI and incomplete cervical tSCI still difficult.

first author of study	total nr of patients	nr of patients with improvement		percentage improvement	lower 95%CI	higher 95%CI
Fehlings [30]	87	30		33.4	25.1	43.2
Umerani [41]	19	8		35.6	22.1	53.9
Newton [37]	17	6		32.5	18.8	49.6
Papadopoulos [38]	28	6		26.4	13.6	38.3
Jug [33]	16	8		38.3	23.6	59.5
Mattiassich [36]	29	9		30.7	18.9	44.2
Grassner [31]	21	1		20.3	5.7	35.2
Early surgery	**217**	**68**		**30.4**	**19.8**	**41.6**
Fehlings [30]	64	17		28.8	19.0	38.2
Umerani [41]	41	8		25.3	13.8	36.4
Randle [39]	12	6		38.6	22.9	59.7
Liu [35]	251	86		34.0	28.6	39.6
Benzel [23]	51	21		37.9	27.4	50.8
Late surgery	**419**	**138**		**32.5**	**21.4**	**45.8**

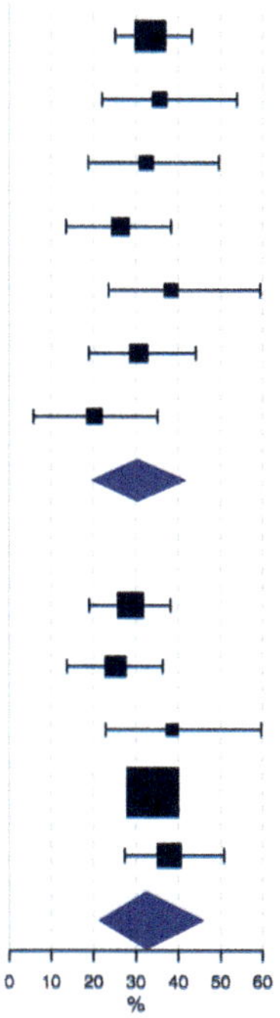

0 10 20 30 40 50 60 %

FIGURE 18.2 Surgical timing and >2 American Spinal Injury Association (ASIA) grade improvement in incomplete cervical traumatic spinal cord injury (tSCI). *Data from ter Wengel et al.*[45] *by the courtesy of* Journal of Neurotrauma, *Mary Ann Liebert inc.*

Possibly one could argue whether TCCI is a complete different entity compared with incomplete cervical tSCI.[49,50] One study has argued that the severity of initial neurological injury predominantly affects neurological outcome, rather than the presence of a central cord type injury pattern versus incomplete tSCI pattern.[49] Additionally, in contrast to our previous understanding of the pathophysiology, TCCI does not appear to originate from centrally located injury to the spinal cord. Rather there is evidence that the lateral corticospinal tracts do not have a somatotopic organization and TCCI is marked by diffuse white matter injury to the lateral columns.[51,52]

Since many of the TCCIs occur without spinal fractures or instability, surgery may only be indicated to decompress the spinal cord with the intent to potentiate neurological recovery. The impact of surgery and in particular the optimal surgical timeframe to potentiate neurological recovery in TCCI is still controversial. Stevens et al. observed more profound neurological recovery when patients with TCCI had undergone surgery, regardless of its urgency, compared with patients managed nonoperatively.[53] In this retrospective cohort of 126 patients with TCCI, 16 patients underwent surgery within 24 hours, 34 patient after 24 hours within the same admission (range: 5–52 days), and 17 patients underwent surgery within a subsequent hospital admission (range: 2–209 days). The authors did not find a difference in neurological recovery rate within the cohorts. However, in contrast to the 59 patients managed nonoperatively, they observed a significant difference in neurological recovery. In contrast, another study in 50 TCCI patients did observe a beneficial effect of surgical decompression within 24 hours.[54] Remarkably, this effect was only observed in TCCI patients with spinal cord compression due to spinal dislocation, fractures, or traumatic herniated discs and not in TCCI patients with preexisting spinal canal narrowing or spondylosis. Surgical urgency in TCCI patients without a fracture or herniated disc did not appear to influence neurological recovery rates. Contrary to these results, Lenehan et al. evaluated the neurological outcome in TCCI patients with preexisting spondylosis and observed favorable neurological results of early surgical decompression in these patients.[55]

In conclusion, due to the variable interpretation of TCCI and the underlying mechanism of injury, studies have been difficult to compare.[53–57] While some show beneficial effects of early surgery, others do not. Nonetheless recent guidelines have recommended performing surgical decompression within 24 hours for patients sustaining TCCI due to preexisting canal narrowing, even when instability or fractures are lacking.[31]

Timing of decompression in thoracic and thoracolumbar spinal cord injury

The impact of surgical timing in neurological recovery in thoracic and thoracolumbar traumatic tSCI is still a subject of discussion. Whereas in cervical tSCI a beneficial effect of early intervention within 24 hours has been suggested, this has not yet been demonstrated clearly for thoracic tSCI. Because of the low prevalence of thoracic tSCI, studies often contain even smaller patient cohorts. Moreover, most studies on neurological outcome include patients with both complete and incomplete thoracic and thoracolumbar tSCI. This is in particular important since the majority of conus medullaris injuries from thoracolumbar fractures have concomitant dysfunction of the cauda equina and have a better prognosis compared with thoracic tSCI.[12,58] Due to these small heterogeneous groups, it is difficult to interpret the effect of early surgical decompression.

Early surgical timeframe (<24 hours)

In contrast to cervical tSCI cohorts, there are fewer studies evaluating the effect of surgical timing on neurological outcome. Furthermore, there are only two studies evaluating the effect of surgical timing in thoracic tSCI specifically.[18,59] The majority of the studies contain patients with either thoracic or thoracolumbar tSCI. One prospective study in 27 patients with thoracolumbar tSCI (T8-L2) randomly assigned patients to an early or late surgical intervention group, depending on the weekday of admission. Both groups had similar baseline demographics. At follow-up (12–20 months), patients who underwent surgical decompression within 8 hours showed significantly more neurological improvement compared with patients treated between 3 and 15 days after injury. This favorable effect of early surgery was also present in a large single center cohort study in 721 patients with incomplete thoracolumbar tSCI (T1–L1).[20] Patients who underwent surgical decompression within 24 hours significantly showed an improvement of ≥ 1 and ≥ 2 AIS grades. Similarly, other studies found a similar effect in favor of early surgery within 24 hours.[16,19,60,61] On the other hand, two other studies failed to show a beneficial effect of early surgery. One of them investigated the effect of early surgical decompression in AIS A thoracolumbar tSCI patients (T2–L2)[38]; the other study evaluated neurological recovery in conus medullaris tSCI patients (T12–L1).[17]

A recent review and metaanalysis, combining studies on thoracic and thoracolumbar tSCI only, addressed neurological improvement after early and late surgery.[62] In the qualitative analysis, six of seven studies, which investigated the effect of surgical timing, observed a significant effect of early surgery on at least one AIS grade improvement. The quantitative analysis in 948 patients with thoracic and thoracolumbar tSCI injuries, however, did not reveal a significant increase in odds of ≥ 1 AIS grade recovery in early surgery (66.8% [95% CI: 45.0%–87.8%] compared with late surgery (48.9% [95% CI: 25.1% –70.7%; OR: 2.2 (95% CI: 0.6–14.0]). The rate of ≥ 2 AIS grades improvement in the early surgery group was 42.0% (95% CI: 22.7%–64.6%) compared with the late surgery group of 27.3% (95% CI: 12.5%–47.3%; OR: 1.9 (95% CI: 0.6–7.3)). This study did not observe a significant beneficial effect of surgical decompression within 24 hours in patients with thoracic and thoracolumbar tSCI.

To date, there is uncertainty about the role of early surgical decompression in thoracic and thoracolumbar tSCI. From a biological standpoint, there is no reason to believe that early decompression in thoracic SCI would not afford the same potential benefits as it does in cervical SCI; however, it is quite conceivable that the degree of primary damage due to the significant mechanical trauma in thoracic SCI is so severe that measurable gains in neurologic recovery are harder to achieve.

Timing in cauda equina injury

Cauda equina syndrome (CES) occurs in approximately 1 per 100,000 per year and accounts for 1%–6% of surgically treated lumbar disk herniations.[63–66] This rare condition is the result of a compressive lesion to the nerve roots below the conus medullaris.[65,67,68] The etiology of CES is quiet variable and comprises trauma, infection, disk herniation, tumor, and epidural hematoma.[65,67–71] Due to the variable severity of the compression, the presenting symptoms can include back pain, motor weakness and sensory loss of the lower extremities, saddle anesthesia, as well as bowel and bladder dysfunction.[65,67,69] The current definition of CES remains inconsistent in the literature; however, most spine surgeons agree that a component of bladder dysfunction is critical for the diagnosis.[65,72,73] Although CES is typically not considered as spinal cord injury, it is briefly discussed in this chapter given it shares a similar principle in its relative urgency in requiring decompression of the nerve roots.

Early surgical timeframe

There is no argument that in CES, surgical intervention is advised with the goal to decompress the nerve roots to maximize the potential for neurological and functional recoveries. Shephard in 1959 proposed early surgical intervention for patients with CES to maximize clinical recovery.[74] Since then, considerable research efforts have been put forth from the scientific community attempting to establish the appropriate timing for surgical decompression with contradictory results.[68,69,75–80] While, it is generally accepted that diagnosis of acute CES requires urgent surgical decompression; however, the exact definition of "early surgery" remains unclear. In an attempt to rectify the timing of surgical intervention in CES, Ahn et al. performed a systematic review of the literature.[81] The result of their metaanalysis, based on a cohort of 332 cases from 42 studies, revealed an advantage of urgent decompression within 48 hours of the onset of symptoms. On the contrary, surgical timing ≤ 24 hours did not appear to have a significant impact. The lack of effect of ≤ 24 hours surgical intervention baffled the clinicians and hinted the existence of subgroup within the CES patient population with variable severity and response to treatment.

Severity of injury and timing

Given historically patients with CES were viewed as one single group with similar prognosis, the previous inconsistency in the literature in regard to the impact of early surgical intervention led to speculation of the existence of heterogeneity in this patient population.[64] Gleave and Macfarlan (2002) proposed a classification system to categorize the presentation of patients into complete and incomplete CES based on the severity of urinary symptoms.[64] The patients presenting with urinary incontinence due to overflow, painless retention, and extensive saddle anesthesia were defined as complete CES (CES-R), whereas ones presenting with signs and symptoms of neurogenic bladder dysfunction, including decreased sensation during micturition, increased effort, reduced desire, and low stream, are considered as incomplete CES (CES-I).

With the differentiation of incomplete and complete CES, previous studies were able to demonstrate the variable effect of timing on surgical outcomes. It has been shown that patients presenting with CES-I, if decompression performed promptly, lead to a more favorable outcome as compared with those with CES-R.[67,80,82] Hence, it was suggested that early surgical intervention before the conversion of CES-I to CES-R would maximize the prognosis. However, in delayed cases where incomplete injury is converted to a complete deficit, the benefit of operative decompression is significantly reduced.[64] Therefore, patients with CES should not be viewed as a homogeneous group. Subgroups within this population can have a variable trajectory of recovery. Hence, it is imperative clinically to make the distinction when assessing the outcomes of surgical intervention in patients with CES.[83]

Chau et al. (2014) systematically reviewed the updated literature for surgical timing in patients with the diagnosis of CES.[84] Due to the heterogeneity of the patient population in most identified studies, the authors performed a metaanalysis limiting only to CES-R patients reported. Unfortunately, the result was unable to provide significant evidence to support operation within 48 hours. However, based on their systematic review of the literature, the authors suggested that the extent of neurological deficit (CES-I vs. CES-R) is likely the most predictive of postoperative outcomes.

Thakur et al. (2017) performed an analysis of the Nationwide Inpatient Sample (NIS) database identifying 4066 CES patients who underwent surgical decompression in the United States between 2005 and 2011.[73] To improve the understanding of the role of early surgical intervention, the authors categorized the patients into surgical intervention within 24 hours, 24—48 hours, and 48 hours and beyond. Interestingly, the authors reported a higher risk of unfavorable disposition, extended stay, and more significant hospital charges in both incomplete and complete CES patients with delayed (>48 hours) surgical intervention. Additionally, the timing of surgery did not appear to play a role in the traumatic CES in this cohort in regard to discharge destination or length of stay. However, delayed surgical intervention seems to associate with increased hospital charges.[73]

The notion for early ($\leq$48 hours) surgical intervention was further supported by Hogan et al. 2019 in a reanalysis of the updated NIS database.[85] In their study of 20,924 patients with the clinical diagnosis of CES treated operatively from 2000 to 2014, it was found that delayed surgical intervention beyond 48 hours after admission significantly increased the risk of mortality, adverse events, and unfavorable discharge. Unfortunately, given the nature of the database, the timing of surgical intervention in relation to the onset of symptoms was not available and timing was based on time of admission to the hospital. Furthermore, the authors acknowledged the limitation in that the incomplete and complete CES were not differentiated for the analysis. Despite this, the need for early decompression is nevertheless clearly presented in this study. Additionally, the authors reported that despite evidence in the literature for early decompression, over the years, the proportion of patients operated within 48 hours remain unchanged.

Surgical timing <24 hours is not adequately studied, and literature evidence is conflicting in its effect on clinical outcome in CES.[67,77,80,86] Todd in 2005 reviewed the literature and performed a metanalysis based on three studies on surgical timing in CES.[87] It was reported a significant benefit in decompression within 24 hours in improving the recovery of bladder function. However, Todd's analysis was later criticized for drawing its conclusion based on three small heterogeneous studies and also methodological issues with the metaanalysis model used.[84]

More recently, Heyes et al. (2018) reported in a single institution retrospective analysis of 136 individuals with a diagnosis of CES that improvement of bowel and bladder functions is observed in both complete and incomplete CES patients with surgical intervention.[86] However, the timing of surgical intervention (within 24 hours or beyond 24 hours) did not have a significant impact on the outcome.

In conclusion, given the growing evidence that recovery in CES is associated with the extent or severity of the damage, careful clinical assessment is imperative to identify incomplete from complete injuries accurately. Although the literature for timing for surgery in CES remains inconclusive, and likely related to lack of differentiation of CES-I and CES-R patients, in the authors' opinion, early surgical intervention should be implemented for both subgroups of patients whenever possible to mitigate damage and maximize clinical recovery. Future research should be aimed at a more detailed subgroup analysis with a larger patient population.

Current guidelines

Two recent international guidelines on surgical timing have been published.[31,88] The first guideline was developed by a multidisciplinary team under the auspices of the AOSpine in 2017.[31] This guideline addresses the surgical timing in TCCI and tSCI patients. Based on the available literature, the working group made the following weak recommendations:

1. "We suggest that early surgery be considered as a treatment option in adult patients with traumatic central cord syndrome."
2. "We suggest that early surgery be offered as an option for adult acute SCI patients regardless of level."

Overall the quality of evidence was considered low for the both recommendations.

The second guideline on surgical timing was published in 2018.[88] This guideline was established by the Congress of Neurological Surgeons and focuses exclusively on the surgical timing in thoracic tSCI. Based on the available literature, there is insufficient evidence to propose a specific surgical timeframe; however, the working group suggest that early surgery (<8 to <72 hours) should be offered as an option for patients with thoracic tSCI.

Knowledge gaps surgical timing and recommendation for future research

To date, there is still insufficient statistically robust evidence for the optimal surgical timeframe to reverse neurological injury after tSCI. Surgical decompression within 24 hours after injury appears to improve neurological outcome in cervical tSCI; however, the effect is less clear for thoracic and thoracolumbar injuries. In addition, while some patients appear to have improved neurologically from early intervention, still a large proportion will not recover despite early surgical intervention. Future studies should therefore focus to identify subgroups which benefit from early surgical decompression. Moreover, future studies should evaluate the impact of ultraearly surgical time-frames to elucidate the optimal surgical time-frame in which neurological injury potentially can be reversed. Recently the SCI-POEM study has finished patient recruitment.[89] This European multicenter study in 309 tSCI patients will investigate the effect of surgical decompression within 12 hours after injury versus thereafter.

As tSCI is relatively uncommon and many factors such as severity and level of injury, patient, and treatment heterogeneity affect neurological outcomes, large prospective international collaborations are needed to investigate patient-tailored care properly.[90,91] In addition, future studies should be encouraged to further explore other functional recoveries and incorporate more detailed neurological recoveries to improve study designs and interpretation of meaningful neurological recovery.[92]

References

1. Rabinowitz RS, Eck JC, Harper CMJ, et al. Urgent surgical decompression compared to methylprednisolone for the treatment of acute spinal cord injury: a randomized prospective study in beagle dogs. *Spine* 2008;**33**(21):2260–8. https://doi.org/10.1097/BRS.0b013e31818786db.

2. Dimar JR, Glassman SD, Raque GH, Zhang YP, Shields CB. The influence of spinal canal narrowing and timing of decompression on neurologic recovery after spinal cord contusion in a rat model. *Spine* 1999;**24**(16):1623–33. https://doi.org/10.1097/00007632-199908150-00002.

3. Carlson GD, Gorden CD, Oliff HS, Pillai JJ, LaManna JC. Sustained spinal cord compression. Part I: time-dependent effect on long-term pathophysiology. *J Bone Jt Surg Ser A* 2003;**85**(1):86–94. http://www.embase.com/search/results?subaction=viewrecord&from=export&id=L36537978.

4. Batchelor PE, Wills TE, Skeers P, et al. Meta-analysis of pre-clinical studies of early decompression in acute spinal cord injury: a battle of time and pressure. *PLoS One* 2013;**8**(8). https://doi.org/10.1371/journal.pone.0072659.

5. Nijendijk JHB, Post MWM, van Asbeck FWA. Epidemiology of traumatic spinal cord injuries in the Netherlands in 2010. *Spinal Cord* 2014;**52**(4):258–63. https://doi.org/10.1038/sc.2013.180.

6. Pickett GE, Campos-Benitez M, Keller JL, Duggal N. Epidemiology of traumatic spinal cord injury in Canada. *Spine* 2006;**31**(7):799–805. https://doi.org/10.1097/01.brs.0000207258.80129.03.

7. Stephan K, Huber S, Häberle S, et al. Spinal cord injury - incidence, prognosis, and outcome: an analysis of the TraumaRegister DGU. *Spine J* 2015;**15**(9):1994–2001. https://doi.org/10.1016/j.spinee.2015.04.041.

8. Marino RJ, Burns S, Graves DE, Leiby BE, Kirshblum S, Lammertse DP. Upper- and lower-extremity motor recovery after traumatic cervical spinal cord injury: an update from the national spinal cord injury database. *Arch Phys Med Rehabil* 2011;**92**(3):369–75. https://doi.org/10.1016/j.apmr.2010.09.027.

9. National Institute on Disability Independent Living and Rehabilitation Research. *Spinal cord injury model systems.* HHS-2016-ACL-NIDILRR-SI-0158. NSCISC Natl SCI Stat Cent. 2017. p. 24.

10. Lee BA, Leiby BE, Marino RJ. Neurological and functional recovery after thoracic spinal cord injury. *J Spinal Cord Med* 2016;**39**(1):67–76. https://doi.org/10.1179/2045772314Y.0000000280.

11. Waters RL, Meyer PR, Adkins RH, Felton D. Emergency, acute, and surgical management of spine trauma. *Arch Phys Med Rehabil* 1999;**80**(11):1383–90. https://doi.org/10.1016/S0003-9993(99)90248-4.

12. Aimetti AA, Kirshblum S, Curt A, Mobley J, Grossman RG, Guest JD. Natural history of neurological improvement following complete (AIS A) thoracic spinal cord injury across three registries to guide acute clinical trial design and interpretation. *Spinal Cord* 2019. https://doi.org/10.1038/s41393-019-0299-8.

13. Dimar 2nd JR, Glassman SD, Raque GH, Zhang YP, Shields CB. The influence of spinal canal narrowing and timing of decompression on neurologic recovery after spinal cord contusion in a rat model. *Spine* 1999;**24**(16):1623–33.

14. Fehlings MG, Vaccaro A, Wilson JR, et al. Early versus delayed decompression for traumatic cervical spinal cord injury: results of the surgical timing in acute spinal cord injury study (STASCIS). *PLoS One* 2012;**7**(2). https://doi.org/10.1371/journal.pone.0032037.

15. Grassner L, Wutte C, Klein B, et al. Early decompression (< 8 h) after traumatic cervical spinal cord injury improves functional outcome as assessed by spinal cord independence measure after one year. *J Neurotrauma* 2016;**33**(18):1658–66. https://doi.org/10.1089/neu.2015.4325.

16. Clohisy JC, Akbarnia BA, Bucholz RD, Burkus JK, Backer RJ. Neurologic recovery associated with anterior decompression of spine fractures at the thoracolumbar junction (T12−L1). *Spine* 1992;**17**(8 Suppl. L.): 325−30. https://doi.org/10.1097/00007632-199208001-00019.

17. Rahimi-Movaghar V, Vaccaro AR, Mohammadi M. Efficacy of surgical decompression in regard to motor recovery in the setting of conus medullaris injury. *J Spinal Cord Med* 2006;**29**(1):32−8. https://doi.org/10.1080/10790268.2006.11753854.

18. Rahimi-Movaghar V. Efficacy of surgical decompression in the setting of complete thoracic spinal cord injury. *J Spinal Cord Med* 2005;**28**(5):415−20. https://doi.org/10.1080/10790268.2005.11753841.

19. Rath SA, Kahamba JF, Kretschmer T, Neff U, Richter H-P, Antoniadis G. Neurological recovery and its influencing factors in thoracic and lumbar spine fractures after surgical decompression and stabilization. *Neurosurg Rev* 2005;**28**(1):44−52. https://doi.org/10.1007/s10143-004-0356-3.

20. Du JP, Fan Y, Liu JJ, et al. Decompression for traumatic thoracic/thoracolumbar incomplete spinal cord injury: application of AO spine injury classification system to identify the timing of operation. *World Neurosurg* 2018; **116**:e867−73. https://doi.org/10.1016/j.wneu.2018.05.118.

21. Aarabi B, Olexa J, Chryssikos T, et al. Extent of spinal cord decompression in motor complete (American Spinal Injury Association Impairment Scale Grades A and B) traumatic spinal cord injury patients: post-operative magnetic resonance imaging analysis. *J Neurotrauma* 2019;**876**:862−76. https://doi.org/10.1089/neu.2018.5834.

22. Aarabi B, Ibrahimi DM, Simard JM, et al. Intramedullary lesion length on postoperative magnetic resonance imaging is a strong predictor of ASIA impairment scale grade conversion following decompressive surgery in cervical spinal cord. *Neurosurgery* 2017;**80**(4):610−20. https://doi.org/10.1093/neuros/nyw053.

23. Fehlings MG, Rao SC, Tator CH, et al. The optimal radiologic method for assessing spinal canal compromise and cord compression in patients with cervical spinal cord injury. *Spine* 1999;**24**(6):605−13. https://doi.org/10.1097/00007632-199903150-00023.

24. Grant GA. Risk of early closed reduction in cervical spine subluxation injuries. *J Neurosurg* 1999;**90**(Spine 1):13−8. doi:spi.1999.90.1.0013.

25. Gelb DE, Hadley MN, Aarabi B, et al. Initial closed reduction of cervical spinal fracture-dislocation injuries. *Neurosurgery* 2013;**72**(Suppl. 2):73−83. https://doi.org/10.1227/NEU.0b013e318276ee02.

26. Aarabi B, Simard JM, Kufera JA, et al. Intramedullary lesion expansion on magnetic resonance imaging in patients with motor complete cervical spinal cord injury. *J Neurosurg Spine* 2012;**17**(3):243−50. https://doi.org/10.3171/2012.6.SPINE12122.

27. Chen Y, He Y, DeVivo MJ. Changing demographics and injury profile of new traumatic spinal cord injuries in the United States, 1972−2014. *Arch Phys Med Rehabil* 2016;**97**(10):1610−9. https://doi.org/10.1016/j.apmr.2016.03.017.

28. Liu Y, Shi CG, Wang XW, et al. Timing of surgical decompression for traumatic cervical spinal cord injury. *Int Orthop* 2015;**39**(12):2457−63. https://doi.org/10.1007/s00264-014-2652-z.

29. Aarabi B, Danesh NA, Chryssikos T, et al. Efficacy of ultra - early (< 12 hours), early (12−24 hours), and late (> 24−138. 5 hours) surgery with MRI - confirmed decompression in AIS grades A, B, and C cervical spinal cord injury. *J Neurotrauma* 2019:1−44. https://doi.org/10.1089/neu.2019.6606.

30. Mattiassich G, Gollwitzer M, Gaderer F, et al. Functional outcomes in individuals undergoing very early (< 5 h) and early (5−24 h) surgical decompression in traumatic cervical spinal cord injury: analysis of neurological improvement from the Austrian spinal cord injury study. *J Neurotrauma* 2017:3371. https://doi.org/10.1089/neu.2017.5132.

31. Fehlings MG, Tetreault LA, Wilson JR, et al. A clinical practice guideline for the management of patients with acute spinal cord injury and central cord syndrome: recommendations on the timing (≤ 24 hours versus >24 hours) of decompressive surgery. *Global Spine J* 2017; **7**(3_Suppl. l):195S−202S. https://doi.org/10.1177/2192568217706367.

32. Wilson JR, Cadotte DW, Fehlings MG. Clinical predictors of neurological outcome, functional status, and survival after traumatic spinal cord injury: a systematic review. *J Neurosurg Spine* 2012;**17**(Suppl. 1):11−26. https://doi.org/10.3171/2012.4.AOSPINE1245.

33. El Tecle NE, Dahdaleh NS, Bydon M, Ray WZ, Torner JC, Hitchon PW. The natural history of complete spinal cord injury: a pooled analysis of 1162 patients and a meta-analysis of modern data. *J Neurosurg Spine* 2018;**28**(4):436−43. https://doi.org/10.3171/2017.7.SPINE17107.

34. Fawcett JW, Curt A, Steeves JD, et al. Guidelines for the conduct of clinical trials for spinal cord injury as developed by the ICCP panel: spontaneous recovery after spinal cord injury and statistical power needed for therapeutic clinical trials. *Spinal Cord* 2007;**45**(3): 190−205. https://doi.org/10.1038/sj.sc.3102007.

35. Geisler FH, Coleman WP, Grieco G, Poonian D, Sygen Study Group. Measurements and recovery patterns in a multicenter study of acute spinal cord injury. *Spine* 2001;**26**(24 Suppl. l):S68−86. http://www.ncbi.nlm.nih.gov/pubmed/11805613.

36. Sewell MD, Vachhani K, Alrawi A, Williams R. Results of early and late surgical decompression and stabilization for acute traumatic cervical spinal cord injury in patients with concomitant chest injuries. *World Neurosurg* 2018;**118**:e161—5. https://doi.org/10.1016/j.wneu.2018.06.146.

37. Levi L, Wolf A, Rigamonti D, Ragheb J, Mirvis S, Robinson W. Anterior decompression in cervical spine trauma: does the timing of surgery affect the outcome? *Neuro* 1991;**29**(2):216—22.

38. Bourassa-Moreau É, Mac-Thiong J-M, Li A, et al. Do patients with complete spinal cord injury benefit from early surgical decompression? Analysis of neurological improvement in a prospective cohort study. *J Neurotrauma* 2016;**33**(3):301—6. https://doi.org/10.1089/neu.2015.3957.

39. Umerani MS, Abbas A, Sharif S. Clinical outcome in patients with early versus delayed decompression in cervical spine trauma. *Asian Spine J* 2014;**8**(4):427—34. https://doi.org/10.4184/asj.2014.8.4.427.

40. Jug M, Kejžar N, Vesel M, et al. Neurological recovery after traumatic cervical spinal cord injury is superior if surgical decompression and instrumented fusion are performed within 8 hours versus 8 to 24 hours after injury: a single center experience. *J Neurotrauma* 2015;**32**(18):1385—92. https://doi.org/10.1089/neu.2014.3767.

41. Newton D, England M, Doll H, Gardner BP. The case for early treatment of dislocations of the cervical spine with cord involvement sustained playing rugby *Bone Joint Lett J* 2011;**93-B**(12):1646—52. https://doi.org/10.1302/0301-620X.93B12.27048.

42. Kalsi-Ryan S, Wilson J, Yang JM, Fehlings MG. Neurological grading in traumatic spinal cord injury. *World Neurosurg* 2014;**82**(3):509—18. https://doi.org/10.1016/j.wneu.2013.01.007.

43. Evaniew N, Sharifi B, Waheed Z, et al. The influence of neurological examination timing within hours after acute traumatic spinal cord injuries: an observational study. *Spinal Cord* 2019. https://doi.org/10.1038/s41393-019-0359-0.

44. Khorasanizadeh M, Yousefifard M, Eskian M, et al. *Neurological recovery following traumatic spinal cord injury: a systematic review and meta-analysis.* 2019. p. 1—17. https://doi.org/10.3171/2018.10.SPINE18802.

45. ter Wengel PV, de Witt Hamer PC, Pauptit JC, Van der Gaag NA, Oner FC, Vandertop WP. Early surgical decompression improves neurological outcome after complete traumatic cervical spinal cord injury. *A Meta-Anal J Neurotrauma* September 2018. https://doi.org/10.1089/neu.2018.5974.

46. Nagata K, Inokuchi K, Chikuda H, et al. Early versus delayed reduction of cervical spine dislocation with complete motor paralysis: a multicenter study. *Eur Spine J* 2017;**26**(4):1272—6. https://doi.org/10.1007/s00586-017-5004-z.

47. Hansebout RR, Hansebout CR. Local cooling for traumatic spinal cord injury: outcomes in 20 patients and review of the literature. *J Neurosurg Spine* 2014;**20**(5):550—61. https://doi.org/10.3171/2014.2.SPINE13318.

48. Gündoļdu Akyüz M, Öztürk EA, Cąkc FA. Can spinal cord injury patients show a worsening in ASIA impairment scale classification despite actually having neurological improvement the limitation of ASIA Impairment Scale Classification. *Spinal Cord* 2014;**52**(9):667—70. https://doi.org/10.1038/sc.2014.89.

49. Pouw MH, Van Middendorp JJ, Van Kampen A, Curt A, Van De Meent H, Hosman AJF. Diagnostic criteria of traumatic central cord syndrome. Part 3: descriptive analyses of neurological and functional outcomes in a prospective cohort of traumatic motor incomplete tetraplegics. *Spinal Cord* 2011;**49**(5):614—22. https://doi.org/10.1038/sc.2010.171.

50. Badhiwala JH, Wilson JR, Fehlings MG. The case for revisiting central cord syndrome. *Spinal Cord* 2019:1—3. https://doi.org/10.1038/s41393-019-0354-5.

51. Collignon F, Martin D, Lénelle J, Stevenaert A. Acute traumatic central cord syndrome: magnetic resonance imaging and clinical observations. *J Neurosurg* 2002;**96**(1 Suppl. L.):29—33.

52. Levi ADO, Tator CH, Bunge RP. Clinical syndromes associated with disproportionate weakness of the upper versus the lower extremities after cervical spinal cord injury. *Neurosurgery* 1996;**38**(1):179—85. https://doi.org/10.1097/00006123-199601000-00039.

53. Stevens EA, Marsh R, Wilson JA, Sweasey TA, Branch CLJ, Powers AK. A review of surgical intervention in the setting of traumatic central cord syndrome. *Spine J* 2010;**10**(10):874—80. https://doi.org/10.1016/j.spinee.2010.07.388.

54. Guest J, Eleraky MA, Apostolides PJ, Dickman CA, Sonntag VKH. Traumatic central cord syndrome: results of surgical management. *J Neurosurg Spine* 2002;**97**(1):25—32. https://doi.org/10.3171/spi.2002.97.1.0025.

55. Lenehan B, Fisher CG, Vaccaro A, Fehlings M, Aarabi B, Dvorak MF. The urgency of surgical decompression in acute central cord injuries with spondylosis and without instability. *Spine* 2010;**35**(21 Suppl. l):S180—6. https://doi.org/10.1097/BRS.0b013e3181f32a44.

56. Anderson KK, Tetreault L, Shamji MF, et al. Optimal timing of surgical decompression for acute traumatic central cord syndrome: a systematic review of the literature. *Neurosurgery* 2015;**77**(Suppl. 4):S15—32. https://doi.org/10.1227/NEU.0000000000000946.

57. Kepler CK, Kong C, Schroeder GD, et al. Early outcome and predictors of early outcome in patients treated surgically for central cord syndrome. *J Neurosurg Spine* 2015;**23**(4):490—4. https://doi.org/10.3171/2015.1.SPINE 141013.

58. Doherty JG, Burns AS, O'Ferrall DM, Ditunno JF. Prevalence of upper motor neuron vs lower motor neuron lesions in complete lower thoracic and lumbar spinal cord injuries. *J Spinal Cord Med* 2002;**25**(4): 289—92. https://doi.org/10.1080/10790268.2002. 11753630.

59. Dobran M, Iacoangeli M, Di Somma LG, et al. Neurological outcome in a series of 58 patients operated for traumatic thoracolumbar spinal cord injuries. *Surg Neurol Int* 2014;**5**(8):329. https://doi.org/10.4103/2152-7806. 139645.

60. Ramírez-Villaescusa J, Hidalgo JL-T, Ruiz-Picazo D, Martin-Benlloch A, Torres-Lozano P, Portero-Martinez E. The impact of urgent intervention on the neurologic recovery in patients with thoracolumbar fractures. *J Spine Surg* 2018;**4**(2):388—96. https:// doi.org/10.21037/jss.2018.06.07.

61. Wilson JR, Jaja BNR, Kwon BK, et al. Natural history, predictors of outcome, and effects of treatment in thoracic spinal cord injury: a multi-center cohort study from the north American clinical trials network. *J Neurotrauma* 2018;**35**(21):2554—60. https://doi.org/ 10.1089/neu.2017.5535.

62. ter Wengel PV, Martin E, De Witt Hamer PC, et al. Impact of early (<24 h) surgical decompression on neurological recovery in thoracic spinal cord injury: a meta-analysis. *J Neurotrauma* 2019:1—37. https:// doi.org/10.1089/neu.2018.6277.

63. Kostuik JP. Medicolegal consequences of cauda equina syndrome: an overview. *Neurosurg Focus* 2004;**16**(6): 39—41. https://doi.org/10.3171/foc.2004.16.6.7.

64. Gleave JRW, Macfarlane R. Cauda equina syndrome: what is the relationship between timing of surgery and outcome? *Br J Neurosurg* 2002;**16**(4):325—8. https://doi.org/10.1080/0268869021000032887.

65. Spector LR, Madigan L, Rhyne A, et al. Cauda equina syndrome. *J Am Acad Orthop Surg* 2008;**16**(8):471—9. http://www.embase.com/search/results?subaction= viewrecord&from=export&id=L352214520.

66. Gardner A, Gardner E, Morley T. Cauda equina syndrome: a review of the current clinical and medicolegal position. *Eur Spine J* 2011;**20**(5):690—7. https:// doi.org/10.1007/s00586-010-1668-3.

67. Dinning TAR, Schaeffer HR. Discogenic compression of the cauda equina: a surgical emergency. *Aust N Z J Surg* 1993;**63**(12):927—34. https://doi.org/10.1111/j.1445-2197.1993.tb01721.x.

68. Jennett WB. A study of 25 cases of compression of the cauda equina by prolapsed intervertebral discs. *J Neurol Neurosurg Psychiatry* 1956;**19**(2):109—16. https://doi.org/10.1136/jnnp.19.2.109.

69. Shapiro S. Cauda equina syndrome secondary to lumbar disc herniation. *Neurosurgery* 1993;**32**(5):743—7. https://doi.org/10.1227/00006123-199305000-00007.

70. Kebaish KM, Awad JN. Spinal epidural hematoma causing acute cauda equina syndrome. *Neurosurg Focus* 2004;**16**(6):e1.

71. Bagley CA, Gokaslan ZL. Cauda equina syndrome caused by primary and metastatic neoplasms. *Neurosurg Focus* 2004;**16**(6):11—8. https://doi.org/10.3171/foc. 2004.16.6.3.

72. Brouwers E, van de Meent H, Curt A, Starremans B, Hosman A, Bartels R. Definitions of traumatic conus medullaris and cauda equina syndrome: a systematic literature review. *Spinal Cord* 2017;**55**(10):886—90. https://doi.org/10.1038/sc.2017.54.

73. Thakur JD, Storey C, Kalakoti P, et al. Early intervention in cauda equina syndrome associated with better outcomes: a myth or reality? Insights from the Nationwide Inpatient Sample database (2005—2011). *Spine J* 2017; **17**(10):1435—48. https://doi.org/10.1016/j.spinee.2017. 04.023.

74. Shephard RH. Diagnosis and prognosis of cauda equina syndrome produced by protrusion of lumbar disk. *Br Med J* 1959;**2**(5164):1434—9. https://doi.org/10.1136/ bmj.2.5164.1434.

75. Kostuik JP, Harrington I, Alexander D, Rand W, Evans D. Cauda equina syndrome and lumbar disc herniation. *J Bone Joint Surg Am* 1986;**68**(3):386—91.

76. Shapiro S. Medical realities of cauda equina syndrome secondary to lumbar disc herniation. *Spine* 2000;**25**(3): 348—52. https://doi.org/10.1097/00007632-200002010-00015.

77. Kennedy JG, Soffe KE, McGrath A, Stephens MM, Walsh MG, McManus F. Predictors of outcome in cauda equina syndrome. *Eur Spine J* 1999;**8**(4):317—22. https:// doi.org/10.1007/s005860050180.

78. Hussain SA, Gullan RW, Chitnavis BP. Cauda equina syndrome: outcome and implications for management. *Br J Neurosurg* 2003;**17**(2):164—7. https://doi.org/ 10.1080/0268869031000109098.

79. McCarthy MJH, Aylott CEW, Grevitt MP, Hegarty J. Cauda equina syndrome: factors affecting long-term functional and sphincteric outcome. *Spine* 2007;**32**(2):207—16. https://doi.org/10.1097/01.brs.0000251750.20508.84.

80. Qureshi A, Sell P. Cauda equina syndrome treated by surgical decompression: the influence of timing on surgical outcome. *Eur Spine J* 2007;**16**(12):2143—51. https:// doi.org/10.1007/s00586-007-0491-y.

81. Ahn UM, Ahn NU, Buchowski JM, Garrett ES, Sieber AN, Kostuik JP. Cauda equina syndrome secondary to lumbar disc herniation: a meta-analysis of surgical outcomes. *Spine* 2000;**25**(12):1515—22. https://doi.org/10.1097/00007632-200006150-00010.

82. Gleave JRW, Macfarlane R. Prognosis for recovery of bladder function following lumbar central disc prolapse. *Br J Neurosurg* 1990;**4**(3):205—9. https://doi.org/10.3109/02688699008992725.

83. DeLong WB, Polissar N, Neradilek B. Timing of surgery in cauda equina syndrome with urinary retention: meta-analysis of observational studies. *J Neurosurg Spine* 2008;**8**(4):305—20. https://doi.org/10.3171/SPI/2008/8/4/305.

84. Chau AM, Xu LL, Pelzer NR, Gragnaniello C. Timing of surgical intervention in cauda equina syndrome: a systematic critical review. *World Neurosurg* 2014;**81**(3—4): 640—50. https://doi.org/10.1016/j.wneu.2013.11.007.

85. Hogan WB, Kuris EO, Durand WM, Eltorai AEM, Daniels AH. Timing of surgical decompression for cauda equina syndrome. *World Neurosurg* 2019;**132**: e732—8. https://doi.org/10.1016/j.wneu.2019.08.030.

86. Heyes G, Jones M, Verzin E, McLorinan G, Darwish N, Eames N. Influence of timing of surgery on Cauda equina syndrome: outcomes at a national spinal centre. *J Orthop* 2018;**15**(1):210—5. https://doi.org/10.1016/j.jor.2018.01.020.

87. Todd NV. Cauda equina syndrome: the timing of surgery probably does influence outcome. *Br J Neurosurg* 2005;**19**(4):301—6. https://doi.org/10.1080/02688690500305324.

88. Eichholz KM, Rabb CH, Anderson PA, et al. Congress of neurological surgeons systematic review and evidence-based guidelines on the evaluation and treatment of patients with thoracolumbar spine trauma: timing of surgical intervention. *Neurosurgery* September 2018. https://doi.org/10.1093/neuros/nyy362.

89. Van Middendorp JJ, Barbagallo G, Schuetz M, Hosman AJF. Design and rationale of a prospective, observational european multicenter study on the efficacy of acute surgical decompression after traumatic spinal cord injury: the SCI-POEM study. *Spinal Cord* 2012;**50**(9):686—94. https://doi.org/10.1038/sc.2012.34.

90. Wilson JR, Witiw CD, Badhiwala J, Kwon BK, Fehlings MG, Harrop JS. Early surgery for traumatic spinal cord injury: where are we now? *Global Spine J* 2020;**10**(1_Suppl. l):84S—91S. https://doi.org/10.1177/2192568219877860.

91. Kelley K, Clark B, Brown V, et al. The case for the future role of evidence-based medicine in the management of cervical spine injuries, with or without fractures. *Int J Qual Health Care* 2019;**31**(4):457—63. https://doi.org/10.3171/2019.6.SPINE19652.

92. ter Wengel PV, Post MWM, Martin E, et al. Neurological recovery after traumatic spinal cord injury: what is meaningful? A patients' and physicians' perspective. *Spinal Cord* 2020. https://doi.org/10.1038/s41393-020-0436-4.

Intensive care and drugs after spinal cord injury

Anton Fomenko[1,a], Alwyn Gomez[1,a], Gregory W.J. Hawryluk[1]

[1]Section of Neurosurgery, Health Sciences Centre, Winnipeg, MB, Canada

Introduction

Traumatic spinal cord injury (SCI) has profoundly adverse and lifelong impacts on the physical, psychological, and social well-being of affected individuals. With a global annual incidence of 12−60 cases per million per year, SCI contributes to significant patient morbidity and mortality, a chronic burden on caregivers, a tremendous cost to the health-care system, and lost productivity in the workforce.[1] Young men aged 20−34 are particularly affected, and approximately 8% succumb to their injuries during hospitalization.[1,2] Without a definitive cure, historical treatment of SCI was limited to supportive care, which led to rapid death due to secondary complications such as pressure ulcers and urinary tract infections.[3] In the modern era of evidence-based medicine, the clinician's armamentarium has expanded to neuroprotective and neuroregenerative treatments that, along with surgical stabilization and decompression and realignment, have paved the way for modern management guidelines. Despite the lack of a cure or effective treatment, research has revealed pharmacologic therapeutic windows of intervention seeking to minimize the devastating effects of secondary injury.[4]

Initial management

The initial management of acute SCI falls into the broader category of trauma and should therefore follow the principles of Advanced Trauma Life Support (ATLS). This framework places a focus on early immobilization of the spine, to avoid potential secondary injury, and correction of any airway, respiratory, and cardiovascular compromise.[5] The recently injured spinal cord is particularly susceptible to secondary insults from hypoxia and hypotension; therefore prioritization of cardiopulmonary stabilization in the early management of SCI is paramount to maximizing recovery.

[a] These authors contributed equally to the work.

© 2022 Elsevier Inc. All rights reserved.

As per ATLS protocols initial airway assessment and management is prioritized in all traumas. In the setting of acute SCI, additional challenges become present. The inline stabilization required to maintain spinal stability can increase the difficulty of endotracheal intubation by reducing the available range of motion of the neck. Additionally, SCI patients often present with and are sensitive to hemodynamic instability. Avoidance of hypotension during induction is therefore crucial and warrants careful selection of pharmaceutical agents. Propofol is a commonly used induction agent but can often worsen hypotension. Ketamine does not have this hemodynamic effect but is limited by its propensity to elevate intracranial pressure, a characteristic that makes it less than idea in patients with concomitant traumatic brain injury. Finally, patients with cervical or high thoracic SCIs may be at risk of bradycardia, hypotension, and even cardiac arrest during intubation and airway management. These actions cause vagal stimulation that is unopposed in patients with higher SCI.[6]

Respiratory complications from dysfunctional ventilation are common in the setting of acute SCI. These are largely related to the level of the injury, particularly secondary to SCI occurring between C3 and C5, which are associated with disruption of the phrenic innervation and resultant diaphragmatic dysfunction. Injuries at lower levels can still have significant effects on ventilation as intercostal (T1—11) and abdominal (T12—L1) innervation is disrupted and may present with a more insidious onset of respiratory compromise.

Early evidence from the 1980's indicated that vigorous pulmonary therapy initiated in the acute period following traumatic SCI was associated with increased survival and a reduced incidence of pulmonary complications.[7] This was followed by work by Leadsom and Sharp in 1981 who analyzed pulmonary function in patients with complete cervical SCI. They found that in the acute period there was a substantial reduction in forced vital capacity (FVC) and expiratory flow rate. An FVC <25% of expected was identified as being associated with a significantly higher rate of respiratory failure requiring ventilator support.[8]

In 2005, work done by Como et al. characterized the need for mechanical ventilation for patients with acute SCI. In their study, all patients with a complete injury at the level of C5 or above eventually required a definitive airway and tracheostomy. Indeed, of those patient with a complete cervical SCI below C5, 79% eventually required intubation and 50% required tracheostomy.[9] Berlly et al. published a report in 2007 that found that respiratory complications accounted for 36% of morbidity in acute SCI patients and respiratory failure accounted for 86% of the deaths. Notably, ventilatory failure occurred on average 4.5 days after injury.[10] These results point to strongly considering early intubation in patients with complete cervical SCI and enforce the need for prolonged monitoring in the acute period following SCI. It is no surprise that the 2013 Guidelines for the Management of Acute Cervical Spine and Spinal Cord Injuries recommend the use of respiratory monitoring devices to detect respiratory insufficiency in patients following acute SCI.

Maintaining hemodynamic stability during the initial management of SCI is crucial in minimizing secondary injury to the spinal cord. This can be difficult to achieve given the perturbance in the complex interaction between the central and autonomic nervous system and the cardiovascular system found in SCI. This is particularly true in cervical and high thoracic SCI and can be observed in the early stages of SCI. Sinus bradycardia is the most common of these cardiac changes with one study showing 71% of patients with ASIA A or B cervical SCI experienced a bradycardia with heart rate less than 45 beats/min. in the first 14 days of hospitalization and 29% requiring treatment. Of the 31 patients in the severe cervical injury group, 5 experienced primary cardiac arrest with 3 being fatal.

Notably, this same study found no significant cardiac rate disturbances or episodes of spontaneous hypotension after the 14-day postinjury mark in their cohort of 71 acute SCI patients.[6]

Management of hemodynamically unstable sinus bradycardia, whether provoked by airway management or spontaneous, should be managed with atropine as a first-line agent. Its mechanism of action is through an anticholinergic effect on parasympathetic fibers that improves sinoatrial node pacing and atrioventricular conduction. Second-line therapy includes transcutaneous or temporary endocardial pacing in the setting of failed medical management.

The polytrauma patient is subject to a multitude of possible causes of hypotension and shock. In the setting of acute SCI neurogenic shock is a unique cause that warrants consideration. Neurogenic shock occurs in cervical or high thoracic (T6 or above) severe SCI as a result of the patient's loss of supraspinal sympathetic control of cardiovascular functions including cardiac contractility and heart rate. Additionally, decreased systemic vascular resistance results in pooling of blood in the arterioles and venous system. In this population of patients, the parasympathetic cardiac responses via the vagus nerve are the only supraspinal control of the heart leading to bradycardia and other cardiac arrythmias. Hypotension and bradycardia are now recognized at the key features of neurogenic shock.[11]

The prevalence of neurogenic shock is highest with complete cervical SCI. One study showed that 25% of cervical SCI patients experience a component of neurogenic shock. Patients with complete SCI are 5.5 times more likely to develop refractory hypotension as compared to those with incomplete SCI, and this patient population warrants careful monitoring.[12] The diagnosis of neurogenic shock, however, should only be made once other possible causes of hypotension, such as hypovolemic or cardiogenic shock, are excluded.

Initial management of neurogenic shock should center around replenishing intravascular volume until euvolemia. Ongoing hypoperfusion may manifest in the form of elevated serum lactate, base deficit, or decreased central venous oxygenation, and therefore these parameters should be monitored. In the setting of persistent hypoperfusion following adequate fluid resuscitation vasopressor should be considered early in order to avoid the complications of fluid overload. These include pulmonary edema, congestive heart failure, and spinal cord edema.[11] Patients with complete cervical SCI are at highest risk of developing the aforementioned complications. One study showed that 90% of these patients required vasopressor support while incomplete cervical and thoracic SCI patients only required vasopressor support in 52% and 31% of cases, respectively.[13] When selecting a vasopressor, it is important to consider the combination of bradycardia and vasodilation found in neurogenic shock. Thus, vasopressors with both alpha- and beta-adrenergic receptor activity are preferred over those with pure alpha-adrenergic receptor activity.

While in the acute period of SCI, much of the focus is on avoiding and reversing hypotension, patients with acute SCI at T6 or above can develop episodic extreme elevations in blood pressure as well as other signs and/or symptoms of autonomic overactivity in response to noxious or nonnoxious stimuli below the level of SCI. This has been termed "autonomic dysreflexia" (AD) and can be more specifically defined as an increase in systolic blood pressure of at least 20% with a change in heart rate and accompanied sweating, piloerection, facial flushing, headache, blurred vision, and/or stuffy nose.[11] AD is a common cardiovascular complication of SCI in the chronic stage, but recent evidence has shown that episodes of AD can also occur during the acute phase. In a case series in 2003 Krassioukov et al. found that the frequency of AD in the acute stage following SCI was over 5%.[14]

Management of AD in the acute phase following SCI is treated just as it is in the chronic phase and should be taken out in a systematic and stepwise fashion. If initially supine, the patient should be placed in a sitting position and constrictive clothing and devices must to be loosened or removed. Next, potential triggers such as bladder distension and bowel impaction should be investigated and corrected. Systolic blood pressures of 150 mmHg or higher should be treated with rapid-onset and short-duration antihypertensive agents. Blood pressure management needs to be expedited so as to avoid potential devastating complications of untreated AD including intracranial hemorrhage, retinal detachment, and seizures. Monitoring following an episode of AD is necessary for at least 2 hours so as to identify recurrent episodes.[11]

Once the patient is stabilized from a cardiopulmonary perspective, characterization of extent of disability and level of injury can commence. This is done through a thorough clinical examination of the patient and radiographic assessment.

Clinical assessment of the patient can be done using the American Spinal Injury Association (ASIA) International Standards for Neurological Classification of Spinal Cord Injury (ISNCSCI). The classification system focuses on three main components: one motor and two sensory. The motor aspect is tested in 10 myotomes bilaterally and graded from 0 (no movement) to 5 (full strength). The two sensory components are pinprick and light touch. Each of these is tested on 28 dermatomes bilaterally and graded from 0 (no sensation) to 2 (normal sensation). The Neurologic Level of Injury (NLI) is defined as the most caudal segment with intact sensation (2/2) and antigravity muscle strength (3/5). The levels rostral to the NLI have normal sensation and motor function.[15]

The ASIA Impairment Scale (AIS) grades SCI into fives grades from A to E. Grade A represents a complete SCI where no sensation or motor function is preserved in the sacral segments S4–S5. Grade B is sensory incomplete where sensory function is preserved below the NLI and includes S4–S5. Grades C and D represent motor incomplete injuries where motor function is preserved below the NLI. If greater than half of the key myotomes below the NLI have at least antigravity strength then a grade of D is given, otherwise it is a grade C. Grade E is given to those with normal sensory and motor function throughout.[15]

Radiographic assessment is a vital component of the initial management of acute SCI. While plain film X-rays had been utilized to evaluate the spine in trauma, CT imaging has now largely replaced them in most large centers. This is in large part due to its extremely high sensitivity when it comes to detecting fracture, in some studies approaching 100%.[16] Additional imaging modalities such as CT angiography (CTA) also have a role in SCI as they are useful in determining vascular injuries such as vertebral artery dissection such as those found in 24% of patients with a fracture that spans the foramen transversarium.[17] The 2013 Guidelines for Management of Acute Cervical Spine and Spinal Cord Injuries states that CTA is the gold standard reference test for vertebral artery injury in patients who have sustained blunt trauma.[18] Indeed, Eastman et al. in 2006 concluded that CTA had an accuracy of 99.3% for a cervical vascular injury following blunt trauma.[19] MRI can also play an important role in the evaluation of acute SCI. It can provide a useful adjunct to the clinical exam when trying to determine the level of injury which is often indicated by T2 signal change. Additionally, MRI can provide information about traumatic disc herniation, and ligamentous and soft tissue injury, both of which can inform subsequent operative management of the patient.

Timing of surgical decompression

Surgery in the setting of SCI has two distinct but equally important goals. The first is to decompress the spinal cord in those with ongoing compression and the second is to realign the spine and restore spinal stability. The role of stabilization of the spine is clear as it enables early mobilization and rehabilitation of the patient. The role of decompression, especially early decompression, has been an area on significantly more study and controversy, and is addressed in other chapters in this text. While there is still much more research that needs to be done, at present it seems that early surgery is safe and likely provides some clinical benefit in SCI.

Hemodynamic targets after SCI

Following acute SCI, hypotension can become a major challenge in management. The cause of hypotension can vary, especially since acute SCI often happens in the setting of polytrauma. Neurogenic shock, marked by hypotension and bradycardia, is a cause of hypotension unique to SCI and was discussed previously. Hemorrhagic shock, however, is the most common cause of hypotension in trauma patients and so clinicians should be prepared to encounter either, or both, types of shock in the acute SCI patient. The role of hypotension in poor clinical outcomes following SCI has been elucidated slowly over time but is still not fully understood.

As with most clinical research, the physiologic basis for hemodynamic manipulation following SCI can be found in animal research. Work done by Streijger et al. utilized a T10 porcine contusion model and analyzed regional biochemical markers to conclude that there is an expanding area of ischemia and hypoxia from the injury site up to 7 days later.[20] This area is analogous to a penumbra in stroke, and region at risk for damage that can be salvaged

if systemic and regional parameters are optimized. Further work done by Martirosyan et al. examined vasopressor response and complications. The group evaluated 15 pigs who received mean arterial pressure (MAP) augmentation. After traumatic SCI, blood flow decreased by an average of 56% but with MAP augmentation, a decrease of 34% was seen.[21] This suggests that manipulation of systemic hemodynamic parameters such as MAP significantly impacts regional blood flow to the spinal cord following SCI.

The main clinical evidence base for MAP-directed therapy comes from prospective reports published in the 1990s by Levi et al. and Vale et al. suggesting that prevention of hypotension after SCI and even augmentation of MAPs may benefit clinically relevant neurologic recovery.[12,13] Much research performed since has added to this growing body of evidence resulting in hemodynamic management following SCI becoming a key aspect in the medical management of SCI. Avoidance of hypotension (defined as a systolic blood pressure <90 mmHg) and augmentation of MAPs to $> 85-90$ mmHg have been suggested in both the CNS 2013 Acute Cervical Spine and Spinal Cord Injuries and 2019 Thoracolumbar Spine Trauma Guidelines.[22,23] While these guidelines do make clinical recommendations, they both acknowledge that the quality of evidence for which they are based is low. A recent systematic review of the literature published by Saadeh et al. found a number of conflicting retrospective studies and still identified the work done by Levi et al. and Vale et al. as the highest quality evidence in the literature to date.[24]

The senior author of this chapter recently utilized a big data approach to analyze targeted blood pressure management of MAP values in a series of 100 patients with SCI. Within the first 7 days of hospitalization a total of 28.8% of measurements were below target values of 85 mmHg.[25] This may explain why there has been a failure to identify a clear clinical improvement with MAP-directed therapy in SCI. Capatano et al. further analyzed a subpopulation of the previous study studied and looked at 62 patients with minute-by-minute MAP measurements after injury. Their results showed significantly higher mean MAPs of AIS A patients who improved versus those that did not (96.6 vs. 94.4 mmHg). Similarly, MAP values <85 mmHg were significantly lower in patients that did not improve (13.5% vs. 25.6%).[26] These results support the current guidelines and indicate that even small perturbances in MAP may impact neurologic recovery.

Recent published data have helped guide optimal vasopressor selection. Further work done by Streijger et al. in a porcine model of SCI directly compared norepinephrine and phenylephrine effect on improving spinal cord perfusion. Following decompression, norepinephrine was found to be more effective at increasing spinal cord blood flow and oxygenation that phenylephrine. In addition, phenylephrine was associated with greater hemorrhage through the injury site.[27] Inoue et al. evaluated 131 patients with SCI who were treated with vasopressors found that dopamine and phenylephrine use was independently associated with higher rate of major complications including ST elevation, troponin elevation, atrial fibrillation, and ventricular tachycardia. Norepinephrine and epinephrine use had not such association.[28] In a subgroup analysis of a study done by Readdy et al. found that in patients >55 years of age, dopamine increased cardiogenic complication rates compared with phenylephrine (83.3% vs. 50%); however, no difference in neurologic outcomes was found between the two vasopressors.[29] A recent review of vasopressors in SCI examined seven studies and found higher complication rates with dopamine than phenylephrine along with a slight improvement in cord perfusion pressure with norepinephrine. Subgroup analysis of these studies showed that elderly patients had more complications and that no specific vasopressor correlated with

improved neurologic outcome.[30] While the strength of the evidence around vasopressors is weak there seems to be no one vasopressor has been shown to improve neurologic outcomes over another and selection of agents should be focused on mitigating side effects.

Significantly more research is needed to be done before a strong recommendation can be made about hemodynamic endpoints in SCI management. It is likely that these recommendations will be guided not only by MAPs but by other physiologic parameters such as intraspinal perfusion pressure.

Spinal cord perfusion pressure monitoring

Considerations of hemodynamic management of patients with SCI are centered around providing the spinal cord with adequate perfusion so as to allow for the greatest chance to recover and minimize secondary injury. However, as in traumatic brain injury, perfusion of the injured spinal cord is not solely dependent on MAPs. In recent years significant research has been done around the area of spinal cord perfusion pressure (SCPP). This analogous concept to cerebral perfusion pressure (CPP) is defined as the difference between intraspinal pressure (ISP) and mean arterial pressure (SCPP = MAP − ISP).

The concept of SCPP is not a new one and has been studied for years in the setting of thoracoabdominal aortic aneurysm surgery where a reduction in SCPP has been shown to cause ischemic damage to the spinal cord resulting in paraplegia.[31] In this setting, the practice of lowering intraspinal pressure by draining of CSF has been shown to prevent spinal cord ischemia but also reverse late-onset paraplegia when performed after ischemic paralysis has occurred.[32,33] Only recently has this concept been studied in the setting of SCI and has provided both a new prognostic marker in SCPP and a new therapeutic target in the form of ISP.

Some of the earliest clinical work done in studying ISP in patients with SCI was reported by Kwon et al. in 2009. They performed a prospective randomized trial in which 22 patients with SCI had their ISP monitored with lumbar intrathecal catheter. In the treatment arm patients had CSF drained from the lumbar catheter in an attempt to reduce ISP. The results demonstrated a reduction of intrathecal pressure after spinal decompression (13.8–7.9 mmHg, $P < .0001$) but no difference in pressure for patients with postoperative drainage compared with no drainage (28.1 vs. 30.6 mmHg, $P = .15$).[31] While this limited study failed to show a difference in its primary outcome it did suggest that ISP was significantly affected by spinal decompression.

A follow-up multicenter prospective observational clinical trial by Squair et al. examined the effect of SCPP in the first 5 days postinjury on the neurologic recovery of 92 patients with SCI. This study also utilized a lumbar catheter to derive SCPP and found that it was independently associated with positive neurologic recovery. In particular it identified an SCPP of 50 mmHg as an important threshold as patients who were exposed to an SCPP below this were significantly less likely to demonstrate neurologic recovery.[34] This foundational work indicated that SCPP may find itself being a future target for goal-directed therapy in the management of SCI.

Concurrent work performed by the Papadopoulos group has shown that another safe method of monitoring ISP is by placing subdural pressure probes at the time of decompressive surgery.[35] In 2017, the group reported further research from an observational cohort study that monitoring of ISP and SCPP predicts neurological improvement at 9–12 months. Specifically mean ISPs <10 mmHg and mean SCPP of >90 mmHg were associated with the best neurologic recovery.[36] The same group recently reported on data that indicated that performing an expansion duroplasty at the time spinal decompression may help significantly lower ISP.

There is still much more research that needs to be done in the area of ISP and SCPP but recent data seem to point at these as more clinically relevant than previously recommended MAP goals found in current guidelines. Some intuitions are already reporting on their experience of using SCPP-guided therapy as a standard of care.[37] Future guidelines will likely incorporate SCPP and ISP management as more clinical experience is gained with these promising metrics.

Pharmacologic management of SCI

The initial traumatic event causing neuronal damage in SCI is termed *primary injury*, and involves impact to spinal cord via blunt contusion, compression, distraction, shear, or laceration of the neural elements. Most commonly, the spinal cord suffers a contusion from displaced ligaments or bony elements in the vertebral column, or in the case of penetrating SCI, direct impact by a foreign body. *Secondary injury* occurs days to weeks later and involves a prolonged cascade of vascular and biochemical processes which contribute to cell loss and dysfunction. These destructive processes include inflammation, lipid peroxidation, mitochondrial dysfunction, vasospasm of the microvasculature, scar formation, Wallerian degeneration, and alterations in ionic milieu leading to cytotoxic cell death. The driving principle of modern research into SCI management is to prevent or inhibit these mechanisms of secondary SCI via neuroprotective agents, prevent secondary insults and complications, and possibly enhance neural tissue regrowth via neuroregenerative treatments.[4,38] Although no agent is currently strongly supported for neuroprotection or neuroregeneration after SCI,[39–41] we overview the current landscape of clinical trials and present the up-to-date guidelines for according to their levels of evidence.[42]

Neuroprotective agents

Corticosteroids

After their isolation, corticosteroids were liberally prescribed in the treatment of acute SCI, after a high-dose trial in dogs in 1969 suggested a functional restorative benefit.[43] Potential molecular mechanisms include its immunosuppressive and antiinflammatory properties that may inhibit secondary injury insults such as oxidative stress.[44–46] Thereafter, the National Spinal Cord Injury Study (NASCIS) research consortium conducted a series of three high-quality trials to evaluate the efficacy of corticosteroids in a large multicenter patient population. The first double-blind multicenter randomized trial (NASCIS I, 1984) investigated low- versus high-dose methylprednisolone sodium succinate (MPSS) (100 mg vs. 1000 mg bolus, then daily for 10 days) in 330 patients with acute SCI[47] presenting within 48 hours of injury. This trial failed to detect a difference in neurological outcome between groups, with the high-dose group incurring a statistically significant increase in wound infection, and a trend toward increased sepsis, pulmonary embolism (PE), and death. Notably, the absence of a placebo ground and a nonstandard neurologic grading scale limited the interpretation of these results.

NASCIS II, 1990, randomized 487 patients presenting within 12 hours after SCI to three arms: placebo, high-dose MPSS, and naloxone infusions for 23 hours. No difference in the standardized American SCI Assessment (ASIA) was found after neurological assessment at 1 year. A post hoc assessment showed that patients receiving MPSS within 8 hours fared better in the motor score than those receiving steroids in a delayed fashion. However, differences in baseline characteristics between these subgroups have questioned the validity of this analysis.

A later trial (NASCIS III, 1997) investigated whether MPSS infusion for 24, 48 hours, or tirilazad mesylate for 48 hours conferred therapeutic benefit to 499 SCI patients randomized within 12 hours of injury.[48] The study did not meet significance for any of the primary endpoints at 1 year, but those receiving methylprednisone 3–8 hours after injury improved by five motor points on the ASIA scale compared to the other groups. As before, prolonged MPSS use was associated with increased morbidity, namely pneumonia.

Since NASCIS II-III suggested a possible modest benefit in motor recovery if administered promptly after injury, later studies have reexamined the safety profile of a single 24-hour MPSS infusion. One such review in elective and acute surgical patients demonstrated no increased risk of wound complications, pneumonia, or death in 2500 patients from 51 clinical trials using high-dose MPSS.[49] Furthermore, a recent systematic review of three RCTs[50–52] and one prospective cohort trial[53] showed no difference between sham and 24-hour MPSS in the pooled risk of death, wound infection, gastrointestinal hemorrhage, or PE, urinary tract infection, or pneumonia.[54]

The latest 2017 guidelines from several consortia (AOSpine North America, AOSpine International, and the AANS/CNS) reflect a waning enthusiasm for corticosteroids in the setting of acute SCI. Current guidelines suggest not offering MPSS to adult patients who present in a delayed fashion (>8 hours) with acute SCI.[40,54] For those presenting within 8 hours of injury, a 24-hour infusion of high-dose MPSS can be offered to adult patients as a treatment option. However, the risk of steroid-associated adverse events must be weighted carefully against the overall clinical gestalt of the patients and the severity of the injury. The use of 48-hour infusions of MPSS is not recommended because of the increased risk of severe pneumonia and sepsis.[40,54] All recommendations are Level I OCEBM and are weak recommendations with moderate evidence.

Riluzole

Currently approved by the FDA and Canadian/Australian regulatory bodies for the treatment of amyotrophic lateral sclerosis (ALS), riluzole is a sodium channel–blocking anticonvulsant. As an important pathologic mechanism in the toxic secondary injury cascade, trauma-induced accumulation of sodium ions via voltage-gated channels can promote intracellular acidosis, leading to excitotoxic cell death. Riluzole was thus investigated as a means to inhibit excitotoxicity by reducing voltage-gated sodium entry after SCI. A small phase I clinical trial with 36 patients investigating the safety and pharmacokinetics of riluzole compared to standard of care in acute SCI was completed in 2011.[55] While motor scores improved at 3 months for riluzole-treated cervical injury patients, there was no difference at 6 months in terms of ISNCSCI motor scores.[55] A phase IIB/III double-blind RCT evaluating the efficacy and safety of riluzole administered for 2 weeks compared to placebo in cervical SCI (Riluzole in Acute Spinal Cord Injury Study) was initiated in 2014. We eagerly await the results of this study, which will be the largest of its kind, enrolling 351 participants and assessing ISNCSCI motor scores at 6 months after injury.[56]

Minocycline

Minocycline is a tetracycline antibiotic routinely used in the clinical treatment of acne and pneumonia, which also crossed the blood–brain barrier.[4,38] In animal models of chronic degenerative disease, a neuroprotective benefit was seen with minocycline in Parkinson's disease, ALS, multiple sclerosis, and Huntington's disease.[57–59] Minocycline's wide spectrum of activity includes inhibition of several toxic cascades in secondary SCI, including reduction in mitochondrial-mediated neural and glial apoptosis, reduced calcium ion chelation, inhibition of microglial proliferation, and antiinflammatory properties. A double-blind phase II

study of 52 patients with heterogenous SCI randomized to placebo or minocycline administered for 1 week, within 12 hours of injury, was published in 2012.[60] This study was notable for employing a rapid IV loading dose, and pharmacokinetic monitoring to ensure sufficient drug concentrations in both serum and CSF. Although no statistically significant difference between groups was shown, the subgroup of cervical injury patients receiving minocycline approached significance in motor recovery. A phase III clinical trial with enrollment of 248 participants is nearing completion (NCT01828203) which will confirm or deny the promise of this neuroprotective agent.

Neuroregenerative agents

In addition to neuroprotection, there is ongoing research aimed at discovering agents with regenerative properties. One of the significant barriers to functional recovery after SCI is the absence of axon regeneration and neural regrowth.[61] Neuroregenerative agents such as gangliosides and myelin-associated inhibitor blockers may promote promoting neurite outgrowth, neural repair, and improve functional recovery after cord injury. The preclinical and clinical evidence for neuroregenerative agents after SCI is covered in other chapters of this work. Nevertheless, current Level I OCEBM evidence recommends against the administration of these experimental agents in acute cervical SCI.[41]

Furthermore, promoting reconstitution of neural and glial milieu at the site of SCI via transplantation of stem cells is currently an area of burgeoning investigation. Several mechanisms involved in secondary cascade such as scarring, inhibitory myelin-associated proteins, and an inhospitable extracellular matrix devoid of neurotrophins prevent the regeneration of axons below the level of injury which might otherwise contribute to functional recovery. Grafting a critical mass of progenitor stem cells with robust sprouting potential and insensitivity to inhibitors at the site of injury represents a promising therapeutic avenue. Cell types that have been trialled include neural progenitor stem cells, oligodendrocyte precursor cells, pluripotent stem cells, mesenchymal stem cells, Schwann cells, and olfactory ensheathing cells, and are covered in other chapters of this work.

Extraspinal complications of SCI

Anticoagulation prophylaxis

Patients with SCI are at increased risk of deep venous thrombosis (DVT) and PE due to venous stasis with prolonged immobilization, transient hypercoagulability, and traumatic endothelial damage. The insidious presentation, high mortality, and long-term clinical consequences of undetected venous thromboembolism make it a prime target for intervention after SCI. The timing and choice of anticoagulation must be carefully weighed against the risks of bleeding from concomitant injuries, exacerbating spinal cord contusion, or surrounding the perioperative period of spinal instrumentation.

Guidelines by the Paralyzed Veterans of America emphasized the early application of mechanical compression devices for all patients. In cases of intracranial, perispinal, or intrathoracic bleeding, anticoagulation should be withheld until primary hemostasis is evident, followed by low-molecular-weight heparin (LMWH) and intermittent pneumatic compression. Routine use of vena cava filters is not recommended, except for those with bleeding expected to persist over 72 hours[62] Additional guidelines by the AANS/CNS suggest supplementing LMWH with rotating beds or electrical calf stimulation, as well as maintaining treatment for 3 months.[63] Expanding on these, new 2017 guidelines by the AOSpine consortium suggest that routine anticoagulant thromboprophylaxis should be offered within 72 hours of SCI

consisting of either subcutaneous LMWH or fixed, low-dose unfractionated heparin, to reduce the risk of thromboembolic events.[54] Adjusted-dose unfractionated heparin is not recommended, given the increased risk of bleeding events. These recommendations were either Level I or II, OCEBM, Weak Recommendation; Low Evidence.

Dermatologic and nutritional

Daily antibiotic prophylaxis should not be used in patients with acute SCIs, despite their elevated risk of acquiring urinary tract infections, pneumonia, and pressure ulcers. With respect to nutrition, SCI patients are often at caloric deficit and nutritional consultation should be initiated as soon as feasible to ensure the recommended protein (1.25–1.5 g/kg) and caloric (30–35 kcal/kg) daily intake. Indirect calorimetry is recommended to assess energy expenditure in both the acute and chronic phases of SCI (Class II evidence).[64] For those patients with injuries that are prohibitive of oral feeding, early enteral nutrition initiated within 72 hours appears to be safe (Level III evidence).

References

1. Van Den Berg MEL, Castellote JM, Mahillo-Fernandez I, De Pedro-Cuesta J. Incidence of spinal cord injury worldwide: a systematic review. *Neuroepidemiology* 2010;**34**(3):184–92.
2. Pickett GE, Campos-Benitez M, Keller JL, Duggal N. Epidemiology of traumatic spinal cord injury in Canada. *Spine* 2006;**31**(7):799–805.
3. Nas K, Yazmalar L, Şah V, Aydin A, Öneş K. Rehabilitation of spinal cord injuries. *World J Orthoped* 2015; **133**:e391–6.
4. Karsy M, Hawryluk G. Pharmacologic management of acute spinal cord injury. *Neurosurg Clin* 2017.
5. Collicott PE, Hughes I. Training in advanced trauma life support. *J Am Med Assoc* 1980;**243**(11):1156–9. https://doi.org/10.1001/jama.1980.03300370030022. Available from:.

6. Lehmann KG, Lane JG, Piepmeier JM, Batsford WP. Cardiovascular abnormalities accompanying acute spinal cord injury in humans: incidence, time course and severity. *J Am Coll Cardiol* 1987;**10**(1):46—52.

7. McMichan JC, Michel L, Westbrook PR. Pulmonary dysfunction following traumatic quadriplegia. Recognition, prevention, and treatment. *J Am Med Assoc* 1980;**243**(6):528—31.

8. Ledsome JR, Sharp JM. Pulmonary function in acute cervical cord injury. *Am Rev Respir Dis* 1981;**124**(1):41—4.

9. Como JJ, Sutton ERH, McCunn M, Dutton RP, Johnson SB, Aarabi B, et al. Characterizing the need for mechanical ventilation following cervical spinal cord injury with neurologic deficit. *J Trauma* 2005;**59**(4):912—6.

10. Berlly M, Shem K. Respiratory management during the first five days after spinal cord injury. *J Spinal Cord Med* 2007;**30**(4):309—18.

11. Furlan JC, Fehlings MG. Cardiovascular complications after acute spinal cord injury: pathophysiology, diagnosis, and management. *Neurosurg Focus FOC* 2008;**25**(5):E13. Available from: https://thejns.org/focus/view/journals/neurosurg-focus/25/5/article-pE13.xml.

12. Levi L, Wolf A, Belzberg H. Hemodynamic parameters in patients with acute cervical cord trauma: description, intervention, and prediction of outcome. *Neurosurgery* 1993;**33**(6):1007.

13. Vale FL, Burns J, Jackson AB, Hadley MN. Combined medical and surgical treatment after acute spinal cord injury: results of a prospective pilot study to assess the merits of aggressive medical resuscitation and blood pressure management. *J Neurosurg* 1997;**87**(2):239—46.

14. Krassioukov AV, Furlan JC, Fehlings MG. Autonomic dysreflexia in acute spinal cord injury: an under-recognized clinical entity. *J Neurotrauma* 2003;**20**(8):707—16.

15. Kirshblum SC, Burns SP, Biering-Sorensen F, Donovan W, Graves DE, Jha A, et al. International standards for neurological classification of spinal cord injury (revised 2011). *J Spinal Cord Med* 2011;**34**(6):535—46. Available from: https://www.ncbi.nlm.nih.gov/pubmed/22330108.

16. Ryken TC, Hadley MN, Walters BC, Aarabi B, Dhall SS, Gelb DE, et al. Radiographic assessment. *Neurosurgery* 2013;**72**(Suppl. 3):54—72. https://doi.org/10.1227/NEU.0b013e318276edee. Available from:.

17. Oetgen ME, Lawrence BD, Yue JJ. Does the morphology of foramen transversarium fractures predict vertebral artery injuries? *Spine* 2008;**33**(25):E957—61.

18. Harrigan MR, Hadley MN, Dhall SS, Walters BC, Aarabi B, Gelb DE, et al. Management of vertebral artery injuries following non-penetrating cervical trauma. *Neurosurgery* 2013;**72**(Suppl. 2):234—43.

19. Eastman AL, Chason DP, Perez CL, McAnulty AL, Minei JP. Computed tomographic angiography for the diagnosis of blunt cervical vascular injury: is it ready for primetime? *J Trauma* 2006;**60**(5):925—9.

20. Streijger F, So K, Manouchehri N, Tigchelaar S, Lee JHT, Okon EB, et al. Changes in pressure, hemodynamics, and metabolism within the spinal cord during the first 7 days after injury using a porcine model. *J Neurotrauma* 2017;**34**(24):3336—50.

21. Martirosyan NL, Kalani MYS, Bichard WD, Baaj AA, Gonzalez LF, Preul MC, et al. Cerebrospinal fluid drainage and induced hypertension improve spinal cord perfusion after acute spinal cord injury in pigs. *Neurosurgery* 2015;**76**(4):461—9.

22. Ryken TC, Hurlbert RJ, Hadley MN, Aarabi B, Dhall SS, Gelb DE, et al. The acute cardiopulmonary management of patients with cervical spinal cord injuries. *Neurosurgery* 2013;**72**(Suppl. 3):84—92. https://doi.org/10.1227/NEU.0b013e318276ee16. Available from:.

23. Dhall SS, Dailey AT, Anderson PA, Arnold PM, Chi JH, Eichholz KM, et al. Congress of neurological surgeons systematic review and evidence-based guidelines on the evaluation and treatment of patients with thoracolumbar spine trauma: hemodynamic management. *Neurosurgery* 2019;**84**(1):E43—5.

24. Saadeh YS, Smith BW, Joseph JR, Jaffer SY, Buckingham MJ, Oppenlander ME, et al. The impact of blood pressure management after spinal cord injury: a systematic review of the literature. *Neurosurg Focus FOC* 2017;**43**(5):E20. Available from: https://thejns.org/focus/view/journals/neurosurg-focus/43/5/article-pE20.xml.

25. Hawryluk G, Whetstone W, Saigal R, Ferguson A, Talbott J, Bresnahan J, et al. Mean arterial blood pressure correlates with neurological recovery after human spinal cord injury: analysis of high frequency physiologic data. *J Neurotrauma* 2015;**32**(24):1958—67.

26. Catapano JS, John Hawryluk GW, Whetstone W, Saigal R, Ferguson A, Talbott J, et al. Higher mean arterial pressure values correlate with neurologic improvement in patients with initially complete spinal cord injuries. *World Neurosurg* 2016;**96**:72—9.

27. Streijger F, So K, Manouchehri N, Gheorghe A, Okon EB, Chan RM, et al. A direct comparison between norepinephrine and phenylephrine for augmenting spinal cord perfusion in a porcine model of spinal cord injury. *J Neurotrauma* 2018;**35**(12):1345—57.

28. Inoue T, Manley GT, Patel N, Whetstone WD. Medical and surgical management after spinal cord injury: vasopressor usage, early surgeries, and complications. *J Neurotrauma* 2014;**31**(3):284—91.

29. Readdy WJ, Whetstone WD, Ferguson AR, Talbott JF, Inoue T, Saigal R, et al. Complications and outcomes

of vasopressor usage in acute traumatic central cord syndrome. *J Neurosurg Spine* 2015;**23**(5):574—80.

30. Yue JK, Tsolinas RE, Burke JF, Deng H, Upadhyayula PS, Robinson CK, et al. Vasopressor support in managing acute spinal cord injury: current knowledge. *J Neurosurg Sci* 2019;**63**(3):308—17.

31. Kwon BK, Curt A, Belanger LM, Bernardo A, Chan D, Markez JA, et al. Intrathecal pressure monitoring and cerebrospinal fluid drainage in acute spinal cord injury: a prospective randomized trial. *J Neurosurg Spine* 2009;**10**(3):181—93. Available from: https://thejns.org/spine/view/journals/j-neurosurg-spine/10/3/article-p181.xml.

32. Coselli JS, LeMaire SA, Koksoy C, Schmittling ZC, Curling PE. Cerebrospinal fluid drainage reduces paraplegia after thoracoabdominal aortic aneurysm repair: results of a randomized clinical trial. *J Vasc Surg* 2002;**35**(4):631—9.

33. Ackerman LL, Traynelis VC. Treatment of delayed-onset neurological deficit after aortic surgery with lumbar cerebrospinal fluid drainage. *Neurosurgery* 2002;**51**(6):1412—4.

34. Squair JW, Bélanger LM, Tsang A, Ritchie L, Mac-Thiong J-M, Parent S, et al. Spinal cord perfusion pressure predicts neurologic recovery in acute spinal cord injury. *Neurology* 2017;**89**(16):1660 LP—1667. Available from: http://n.neurology.org/content/89/16/1660.abstract.

35. Werndle MC, Saadoun S, Phang I, Czosnyka M, Varsos GV, Czosnyka ZH, et al. Monitoring of spinal cord perfusion pressure in acute spinal cord injury: initial findings of the injured spinal cord pressure evaluation study. *Crit Care Med* 2014;**42**(3):646—55.

36. Saadoun S, Chen S, Papadopoulos MC. Intraspinal pressure and spinal cord perfusion pressure predict neurological outcome after traumatic spinal cord injury. *J Neurol Neurosurg Psychiatry* 2017;**88**(5). 452 LP—453 LP. Available from: http://jnnp.bmj.com/content/88/5/452.abstract.

37. Yue JK, Hemmerle DD, Winkler EA, Thomas LH, Fernandez XD, Kyritsis N, et al. Clinical implementation of novel spinal cord perfusion pressure protocol in acute traumatic spinal cord injury at U.S. Level I Trauma Center: TRACK-SCI study. *World Neurosurg* 2019. Available from: http://www.sciencedirect.com/science/article/pii/S187887501932474X.

38. Wang S, Hawryluk GWJ, Spano S, Fehlings MG. Pharmacologic protocols for spinal cord-injured athletes. *Semin Spine Surg* 2010;**22**(4):181—92.

39. Rabinstein AA. Traumatic spinal cord injury. *Continuum* 2020:271—80.

40. O'Toole JE, Kaiser MG, Anderson PA, Arnold PM, Chi JH, Dailey AT, Dhall SS, Eichholz KM, Harrop JS, Hoh DJ, Qureshi S. Congress of neurological surgeons systematic review and evidence-based guidelines on the evaluation and treatment of patients with thoracolumbar spine trauma: executive summary. *Neurosurgery* 2019 Jan 1;**84**(1):2—6.

41. Walters BC. Methodology of the guidelines for the management of acute cervical spine and spinal cord injuries. *Neurosurgery* 2013;**72**(3):17—21.

42. Phillips B, et al. *Oxford Centre for evidence-based Medicine — levels of evidence (March 2009) for evidence-based Oxford Centre Medicine — levels of evidence.* Centre for Evidence Based Medicine; 2014.

43. Ducker TB, Hamit HF. Experimental treatments of acute spinal cord injury. *J Neurosurg* 1969;**30**(6):693—7.

44. Campbell JB, DeCrescito V, Tomasula JJ, Demopoulos HB, Flamm ES, Ortega BD. Effects of antifibrinolytic and steroid therapy on the contused spinal cord of cats. *J Neurosurg* 1974;**40**(6):726—33.

45. Black P, Markowitz RS. Experimental spinal cord injury in monkeys: comparison of steroids and local hypothermia. *Surg Forum* 1971;**22**:409—11.

46. Burns CM. The history of cortisone discovery and development. *Rheum Dis Clin N Am* 2016;**42**(1):1—4.

47. Bracken MB, Collins WF, Freeman DF, Shepard MJ, Wagner FW, Silten RM, et al. Efficacy of methylprednisolone in acute spinal cord injury. *J Am Med Assoc* 1984;**251**(1):45—52.

48. Bracken MB, Shepard MJ, Holford TR, Leo-Summers L, Aldrich EF, Fazl M, et al. Administration of methylprednisolone for 24 or 48 hours or tirilazad mesylate for 48 hours in the treatment of acute spinal cord injury: results of the Third National Acute Spinal Cord Injury randomized controlled trial. *J Am Med Assoc* 1997;**277**(20):1597—604.

49. Sauerland S, Nagelschmidt M, Mallmann P, Neugebauer EAM. Risks and benefits of preoperative high dose methylprednisolone in surgical patients: a systematic review. *Drug Saf* 2000;**23**(5):449—61.

50. Otani K, Abe H, Kadoya S, Nakagawa H, Ikata TTS. Beneficial effect of methylprednisolone sodium succinate in the treatment of acute spinal cord injury. *Sekitsui Sekizui J* 1994;**7**:633—47.

51. Pointillart V, Petitjean M, Wiart L, Vital J, Lassié P, Thicoipé M, et al. Pharmacological therapy of spinal cord injury during the acute phase. *Spinal Cord* 2000;**38**(2):71—6.

52. Bracken MB, Jo SM, Collins WF, Holford TR, Young W, Baskin DS, et al. A randomized, controlled trial of methylprednisolone or naloxone in the treatment of acute spinal-cord injury: results of the second national acute spinal cord injury study. *N Engl J Med* 1990;**322**(20):1405—11.

53. Evaniew N, Noonan VK, Fallah N, Kwon BK, Rivers CS, Ahn H, et al. Methylprednisolone for the treatment of patients with acute spinal cord injuries: a propensity score-matched cohort study from a Canadian multi-center spinal cord injury registry. *J Neurotrauma* 2015; **32**(21):1674—83.

54. Fehlings MG, Tetreault LA, Wilson JR, Kwon BK, Burns AS, Martin AR, et al. A clinical practice guideline for the management of acute spinal cord injury: introduction, rationale, and scope. *Global Spine J* 2014;**31**(3): 239—55.

55. Grossman RG, Fehlings MG, Frankowski RF, Burau KD, Chow DSL, Tator C, et al. A prospective, multicenter, Phase i matched-comparison group trial of safety, pharmacokinetics, and preliminary efficacy of riluzole in patients with traumatic spinal cord injury. *J Neurotrauma* 2014.

56. Fehlings MG, Nakashima H, Nagoshi N, Chow DSL, Grossman RG, Kopjar B. Rationale, design and critical end points for the Riluzole in Acute Spinal Cord Injury Study (RISCIS): a randomized, double-blinded, placebo-controlled parallel multi-center trial. *Spinal Cord* 2016;**54**(1):8—15.

57. Yong VW. Prospects for neuroprotection in multiple sclerosis. *Front Biosci* 2004;**9**(864):1.

58. Matsukawa N, Yasuhara T, Hara K, Xu L, Maki M, Yu G, et al. Therapeutic targets and limits of minocycline neuroprotection in experimental ischemic stroke. *BMC Neurosci* 2009;**10**(1):1—6.

59. Plane JM, Shen Y, Pleasure DE, Deng W. Prospects for minocycline neuroprotection. *Arch Neurol* 2010;**67**(12): 1442—8.

60. Casha S, Zygun D, McGowan MD, Bains I, Yong VW, John Hurlbert R. Results of a phase II placebo-controlled randomized trial of minocycline in acute spinal cord injury. *Brain* 2012;**135**(4):1224—36.

61. Prinjha R, Moore SE, Vinson M, Blake S, Morrow R, Christie G, et al. Inhibitor of neurite outgrowth in humans. *Nature* 2000;**403**(6768):383—4.

62. Statements PC. Prevention of venous thromboembolism in individuals with spinal cord injury: clinical practice guidelines for health care providers. *Topics in Spinal Cord Injury Rehabilitation* 2016;**22**(3):209—40.

63. Dhall SS, Hadley MN, Aarabi B, Gelb DE, Hurlbert RJ, Rozzelle CJ, et al. Deep venous thrombosis and thromboembolism in patients with cervical spinal cord injuries. *Neurosurgery* 2013;**72**(1):244—54.

64. Dhall SS, Hadley MN, Aarabi B, Gelb DE, John Hurlbert R, Rozzelle CJ, et al. Nutritional support after spinal cord injury. *Neurosurgery* 2013;**72**(3):255—9.

Further reading

1. McKinley W, Meade MA, Kirshblum S, Barnard B. Outcomes of early surgical management versus late or no surgical intervention after acute spinal cord injury. *Arch Phys Med Rehabil* 2004;**85**(11):1818—25. https://doi.org/10.1016/j.apmr.2004.04.032. Available from:.

2. Fehlings MG, Vaccaro A, Wilson JR, Singh A, Cadotte DW, Harrop JS, et al. Early versus delayed decompression for traumatic cervical spinal cord injury: results of the surgical timing in acute spinal cord injury study (STASCIS). *PLoS One* 2012;**7**(2):e32037. Available from: https://www.ncbi.nlm.nih.gov/pubmed/22384132.

3. Wilson JR, Singh A, Craven C, Verrier MC, Drew B, Ahn H, et al. Early versus late surgery for traumatic spinal cord injury: the results of a prospective Canadian cohort study. *Spinal Cord* 2012;**50**(11):840—3.

4. Jug M, Kejzar N, Vesel M, Al Mawed S, Dobravec M, Herman S, et al. Neurological recovery after traumatic cervical spinal cord injury is superior if surgical decompression and instrumented fusion are performed within 8 hours versus 8 to 24 hours after injury: a single center experience. *J Neurotrauma* 2015;**32**(18): 1385—92.

5. Geisler FH, Coleman WP, Grieco G, Poonian D. The Sygen® multicenter acute spinal cord injury study. *Spine* 2001.

6. Lozano AM, Lipsman N. Probing and regulating dysfunctional circuits using deep brain stimulation. *Neuron* 2013;**77**(3):406—24.

7. Cox A, Varma A, Banik N. Recent advances in the pharmacologic treatment of spinal cord injury. *Metab Brain Dis* 2015.

8. Geisler FH, Coleman WP, Dorsey FC, Coleman WP. Recovery of motor function after spinal-cord injury — a randomized, placebo-controlled trial with GM-1 ganglioside. *N Engl J Med* 1991.

9. Hunt D, Coffin RS, Anderson PN. The Nogo receptor, its ligands and axonal regeneration in the spinal cord. *Rev J Neurocytol* 2002.

10. Schwab ME, Strittmatter SM. Nogo limits neural plasticity and recovery from injury. *Curr Opin Neurobiol* 2014.

11. Dergham P, Ellezam B, Essagian C, Avedissian H, Lubell WD, McKerracher L. Rho signaling pathway targeted to promote spinal cord repair. *J Neurosci* 2002.

12. Lord-Fontaine S, Yang F, Diep Q, Dergham P, Munzer S, Tremblay P, et al. Local inhibition of Rho signaling by cell-permeable recombinant protein BA-210 prevents secondary damage and promotes functional recovery following acute spinal cord injury. *J Neurotrauma* 2008.

13. Fehlings MG, Theodore N, Harrop J, Maurais G, Kuntz C, Shaffrey CI, et al. A phase I/IIa clinical trial of a recombinant Rho protein antagonist in acute spinal cord injury. *J Neurotrauma* 2011.

14. Donovan J, Kirshblum S. Clinical trials in traumatic spinal cord injury. *Neurotherapeutics* 2018.

15. Jain KK. Neuroprotection in spinal cord injury. In: *The handbook of neuroprotection*. New York, NY: Springer; 2019. p. 337—67.

16. Rosenzweig ES, Brock JH, Lu P, Kumamaru H, Salegio EA, Kadoya K, et al. Restorative effects of human neural stem cell grafts on the primate spinal cord. *Nat Med* 2018.

17. Knoller N, Auerbach G, Fulga V, Zelig G, Attias J, Bakimer R, et al. Clinical experience using incubated autologous macrophages as a treatment for complete spinal cord injury: phase I study results. *J Neurosurg Spine* 2005.

18. Lammertse DP, Jones LAT, Charlifue SB, Kirshblum SC, Apple DF, Ragnarsson KT, et al. Autologous incubated macrophage therapy in acute, complete spinal cord injury: results of the phase 2 randomized controlled multicenter trial. *Spinal Cord* 2012.

SCI management: rehabilitation

Julio C. Furlan[1,2,3,4,5], *B. Catharine Craven*[1,2,3]

[1]Lyndhurst Centre, Toronto Rehabilitation Institute, University Health Network, Toronto, ON, Canada; [2]KITE Research Institute, University Health Network, Toronto, ON, Canada; [3]Department of Medicine, Division of Physical Medicine and Rehabilitation, Institute of Medical Science, University of Toronto, Toronto, ON, Canada; [4]Rehabilitation Sciences Institute, University of Toronto, Toronto, ON, Canada; [5]Institute of Health Policy, Management and Evaluation, University of Toronto, Toronto, ON, Canada

Introduction

Traumatic spinal cord injury (SCI) is a potentially catastrophic event for individuals and their family members, with major medical, social, and financial implications for society. Historically, cases of traumatic SCI were documented in the Edwin Smith papyrus that was written between 2500 and 3000 BC, with further documentation and elaboration of the medical experiences with traumatic SCI during the American Civil War.[1,2] While the surgical and medical advancements in the World War I facilitated the maturation of the Physical Medicine and Rehabilitation as specialty in the 20th century, the turning point in the management of SCI truly occurred during World War II.[2–4] The impact of the innovative scientific and clinical contributions of Sir Ludwig Guttmann led to the founding of the National Spinal Injuries Centers in the United Kingdom.[5] Sir Ludwig Guttmann was a neurologist with particular interest in the rehabilitation of individuals with traumatic SCI who became a cofounder of the field of interprofessional neurorehabilitation.[5]

The development of comprehensive multidisciplinary neurorehabilitation and SCI units created the landscape for the modern Spinal Cord Medicine.[3,4] The promotion of preventive measures and early recognition and proper management of secondary health conditions specific to SCI were key determinants in the improvement of survival and quality of life among individuals with traumatic SCI.[6] Most notable among these were prevention of pressure injury, recognition and management of urinary tract infections (UTIs), and provision of wheelchairs for community mobility. In addition, the prehospital recognition of SCI and acute surgical management of individuals with acute traumatic SCI has improved with timely spinal cord decompression and maintenance of spinal cord perfusion.[7]

These advances are reflected in the survival rates of individuals with SCI. For instance, survival rates of individuals with SCI in Australia from 1995 to 2006 were reported to be 91.2%

© 2022 Elsevier Inc. All rights reserved.

for those with tetraplegia and 95.9% for those with paraplegia.[8] The global 40-year survival rate of individuals with SCI was 47% for those with paraplegia, and 62% for persons with tetraplegia. Once the spine is stable, and the individual is afebrile, breathing independently and hemodynamically stable they can be transferred to a rehabilitation setting.

Overview of the rehabilitation principles

The clinical outcomes of SCI rehabilitation are a composite outcome of the individual's neurological level of injury, severity of injury (or ASIA Impairment Scale [AIS]), access to appropriate and timely rehabilitation therapy and adequate devices, concurrent secondary health conditions and their associated multimorbidity, and the individual's drive, financial resources, social supports, and self-management skills over time. Many authors have tried to explain the complex interplay between the individual, their social and environmental context, their impairment, activity, and participation. In 2012, Craven et al.[9] presented a hybrid model of the International Classification of Functioning, Disability, and Health framework.[10] The model displays the 37 domains of SCI rehabilitation and explains how application of the domains is customized to the individual and their personal and environmental context and desired rehabilitation goals, which are typically derived from the individual's impairments, health beliefs, and life situation. The delivery of rehabilitation services is best provided in an environment which includes an interprofessional team (including neurology, physiatry, nursing, physical therapy, occupational therapy, psychology, psychiatry, urology, social work, speech language pathology, recreation therapy, kinesiology, peer and family education, and support services) invested in delivering transdisciplinary care in a coordinated fashion to achieve common mutually negotiated goals within a realistic time frame.

Realistic patient-driven goals should be discussed and established by the patients in concert with their health-care team upon admission to an SCI rehabilitation program. The goals should be specific, measurable, achievable, relevant, and time limited as in the mnemonic: SMART.[11] Integration of SMART goal setting which reflects the individual's priorities and potential abilities is essential to the rehabilitation process.

The current rehabilitation therapy protocols for management of individuals with traumatic SCI aim to maximize restoration of the remaining function and minimize secondary health conditions through provision of rehabilitation and related education and training.[7] Prior surveys indicated that restoration of motor function (which includes walking for individuals living with paraplegia, and arm/hand function for individuals living with tetraplegia), bowel function, bladder function, and sexual function are the health priorities for individuals living with SCI.[12] The anticipated functional outcome based on neurological level of injury and degree of impairment is well described in the Paralyzed Veterans of America guidelines for outcomes after SCI.[13] The best practice for management of individuals with SCI begins with learning how to maintain continence of their bowel and bladder, maintain skin integrity, avoid orthostatic hypotension, and recognize episodes of autonomic dysreflexia that gradually progresses to address their mobility and independence in activities of daily living through routine participation in physical therapy, occupational therapy, while learning to use assistive devices as appropriate, in a multidisciplinary approach. The rehabilitative therapies should be tailored to the patient's goals and available resources with the purpose of maximizing recovery, minimizing current or future secondary health conditions, and optimizing the individual's quality of life. Application of early rehabilitation principles and procedures should be considered in the acute care spine trauma when patient's clinical condition allows and necessary resources are available prior to transfer to a rehabilitation SCI unit.[14]

Restoration of motor and sensory function

There is a growing body of evidence to support the notion that timely rehabilitation using the optimal type, volume, and intensity of therapy maximizes motor recovery after SCI.[14,15]

Typically, inpatient rehabilitation regimen can last up to 6 months for individuals with acute tetraplegia and up to 3 months for individuals with acute paraplegia, followed by outpatient rehabilitation that could extend beyond 1 year after SCI.[16] Despite the common sense is that rehabilitative therapies should be applied for a prolonged period of time early after SCI; there have been recent unparalleled health system pressures in many jurisdictions including Canada to reduce: (a) the time period from SCI onset to admission to an SCI rehabilitation center; and (b) length of inpatient rehabilitation stay.[15] A major reason is that the costs to care for individuals living with SCI are substantial in the context of financial constraints of the health-care system. As a consequence, the health-care systems are required to develop the structure, processes, and services that facilitate access to the effective modalities and intensity of therapies applicable in tertiary outpatient rehabilitation settings.[15] Health system providers have been challenged to facilitate the earlier transition from inpatient rehabilitation to community living in a continuum, without compromising the delivery of a sufficient volume, intensity, and type of therapy services.[15] While the evolving evidence suggests that the volume and intensity of motor training have a significant impact on the outcomes the optimal volume and intensity of therapy necessary to maximize restoration of each person's motor function remain poorly understood in part due to the heterogeneity of impairment and multiple potential confounders including age at SCI onset, sex/gender, levels and severity of SCI, cause of

injury, and preexisting medical comorbidities. The following rehabilitation modalities or therapies are herein categorized into activity-based therapies, neuromodulation, and reconstructive strategies for upper-extremity function.

Early phase of rehabilitation

The initial processes of rehabilitation care should be initiated in the early stages after SCI when the patient may be still admitted in an acute care spine trauma center provided the patient is clinically stable and required resources are available.[17] Specific activities include routine use of lung volume augmentation following decannulation to prevent pneumonia due to atelectasis, routine turning in bed at 2 hour intervals to mitigate the risk of tissue injury, removal of an indwelling urinary catheter and initiation of intermittent catheterization for patients with upper motor neuron injury and lower urinary tract dysfunction, early mobilization and bed mobility, initiation of thromboprophylaxis, and careful monitoring of temperature and hemodynamic status to maintain adequate cerebral and cord perfusion.[18] In general transfer to a SCI rehabilitation center is appropriate once the person is able to sit for at least 2 hours to allow for participation in rehabilitation therapy and accrual of increased functional independence based on their degree of impairment and anticipated neurorecovery.

Following admission to the rehabilitation center, the individual with SCI should be provided with a tour of the facility, introductions to their interprofessional team members and their roles and skills, education regarding their impairment, anticipated recovery, and rehabilitation length of stay prior to engaging in the goal setting process. Initial education and training for patients and families should include appropriate positioning and pressure relief techniques in bed and wheelchair independently or with assistance of caregivers to prevent tissue injury.[16] Furthermore, patients with tetraplegia or high paraplegia are recommended to be gradually trained for the upright position ("verticalization") with assistance of the health-care team (e.g., physical and occupational therapists).[16] Verticalization training should be started with a minimum of 30 minutes of upright positioning every day that can be increased in a stepwise manner.[16] Following mobilization, inspection of the skin is recommended for early identification of areas overlying bony prominences susceptible to tissue injury, as well as for adjustment or replacement of any objects that may cause a tissue injury to progress to a pressure sore (e.g., bed mattress, seat cushion, wheelchair back seat).[16]

Conventional rehabilitation therapies

The literature on natural history of recovery after traumatic SCI indicates the potential for motor recovery within the first year following injury when most of the compensation, neural plasticity, and repair occur.[19] In addition, prior studies emphasized sensory recovery is also enhanced within the first year after injury and significantly influences the outcomes after traumatic SCI.[19,20] The continuity of therapy is also a key determinant of rehabilitation outcomes. Many rehabilitation service interruptions are due to changes in health status or occurrence of secondary health conditions. Service interruptions resulted in a mean of 9 days loss of inpatient rehabilitation services reportedly resulting in a 50% reduction in Functional Independence Measure (FIM) efficiency.[21] Although there are many financial and human resource constraints in the current health system, changes in the patient's health status behoove rehabilitation professionals to advocate for adequate therapy intensity despite changes in health status during rehabilitation service delivery.

The current understanding of the values of clinical rehabilitation is that performing repetitive physical tasks promotes strengthening of residual neural projections, stabilization and strengthening of newly formed sprouting fibers, connections between cortex and brainstem, connections between brainstem and spinal cord, and connections within the spinal cord.[22,23] Prior preclinical and clinical studies have revealed that sensory inputs substantially influence the sophisticated spinal networks, which can also be modulated by task-specific training.[24] It is postulated that a sufficient degree of supraspinal influence associated with physical activities promotes adaptive neuroplasticity by modulating central excitability and facilitating task-specific retraining.[24] The latter is achieved by weight bearing of the extremities, optimizing sensory input, improving kinematics, minimizing compensation, and maximizing recovery.[24] These training principles are useful to guide the selection of activities for the physical and occupational therapy, as well as for home-based exercises driven by the patient and/or their family members.[24] There is emerging evidence for the role of rhythmic cardiorespiratory endurance activity and intermittent hypoxia for augmenting gait training outcomes and longitudinal cardiorespiratory health. There is significant interinstitutional variation in terms of the volume and intensity of the conventional rehabilitation protocols due to staff training, variations in clinic, and device accessibility and affordability.

The most comprehensive, multiinstitutional analysis of the processes and outcomes of rehabilitation services for patients with SCI was the SCIRehab investigation study.[25,26] This unique initiative prospectively collected data on the multidisciplinary services provides a good overview of the actual conventional rehabilitation therapies in six inpatient SCI rehabilitation programs in the United States.[25,26] The SCIRehab investigation study included data from 1376 patients with acute traumatic SCI who received 151,172 treatment sessions including physical therapy (45.7%), occupational therapy (39.7%), speech therapy (5.6%), and recreational therapy (8.9%). The average time from SCI to transfer to inpatient rehabilitation SCI center was 31 days, and the average length of stay in the inpatient SCI rehabilitation center was

55 days.[27] During this period, the individuals with SCI received a total of 110 sessions of conventional therapy including a mean of 50 sessions of physical therapy (approximately 57 ± 36 hours [average $\pm$ standard deviation]), 44 sessions of occupational therapy (54 ± 37 hours), 6 sessions of speech therapy (5 ± 11 hours), and 10 sessions of recreational therapy (18 ± 16 hours).[26,27] Overall, the SCIRehab investigation study showed that at least one factor affected the objective and/or content of the conventional rehabilitation therapy occurred in 30.8% of the sessions.[27] The most common factors affecting therapy was of a medical nature (mainly, pain, fatigue, spasticity, and "other medical complications") followed by behavioral factors (e.g., fatigue [also counted as medical factor], need for psychological support, behavioral issues, refusal to participate in therapy, or a component of therapy) and "other factors" (e.g., cognitive issue, communication or language issue, equipment malfunction, visual or hearing impairment, and cultural issues).[27] A negative impact of these factors was seen in the therapist's rating of the patient's participation in the treatment sessions (as assessed using the Pittsburg Rehabilitation Participation Scale) but there was a weaker relationship between these factors and the length of stay in inpatient SCI rehabilitation center.[27] Of note, the potential effects of those factors on other outcome measures were not reported, even though one may speculate that poorer participation in rehabilitation therapy would adversely affect the individual's neurological and functional recovery after SCI.[21]

In a separate retrospective analysis of SCIRehab investigation study database, 1032 participants were randomly selected for a regression analysis of the factors associated with missed time of physical therapy, occupational therapy, speech therapy, recreational therapy, and psychological therapy.[28] Overall, 153 ± 100 minutes (average $\pm$ standard deviation) of conventional rehabilitation therapy were missed on a weekly basis; this represented an average total of 20 ± 19 hours for the overall inpatient rehabilitation admission after acute traumatic SCI.[28] While patients most commonly missed physical, occupational, and speech therapy due to the lack of patient readiness and medical reasons, patient's refusal was the most frequent reason for missing recreational therapy sessions.[28] The factors significantly associated with a greater total hours of therapy missed for any reason included: low tetraplegia (at C5 to C8 levels) or paraplegia, higher Comprehensive Severity Index score, higher admission FIM motor and cognitive scores, violence as the cause of SCI, longer rehabilitation LOS, treatment site, and a greater proportion of sessions limited by fatigue.[28] Given the potential impact on rehabilitation outcomes and costs related to lack of therapy delivery, healthcare teams and rehabilitation hospital administrators should develop procedures and policies to reduce missed therapy.[21,28]

Activity-based restorative therapies

By definition, *activity-based restorative therapies* are "a group of multimodal interventions that seek to minimize compensatory mechanisms of functional recovery and provide activation of the neuromuscular system below the level of injury with the goal of retraining the nervous system in order to facilitate the recovery of a specific motor task."[29] The techniques used in the activity-based restorative therapies can be grouped into those which promote patterned motor activation (e.g., locomotor training, functional electrical stimulation (FES), and/or leg ergometry), nonpatterned motor activation (e.g., recruitment and strengthening, task-specific training), or sensory stimulation (e.g., sensorimotor therapy).[29]

In a recent systematic review and meta-analysis, Quel de Oliveira et al.[30] examined prior studies on interventions using activity-based restorative therapies for rehabilitation of individuals with SCI with regard to their impact on their mobility, degree of independence, and quality of life. The literature search strategy captured 19 randomized or nonrandomized clinical trials that compared activity-based therapy with either no intervention (n = 3) or conventional rehabilitation therapies (n = 16), which were applied for rehabilitation of the upper extremities (n = 6), lower extremities (n = 11), or multimodal interventions (n = 2).[30] The authors of the systematic review reported that activity-based restorative therapy improved patients' independence and lower-extremity mobility to a similar degree to the control group without intervention; however, a large positive effect on upper-extremity function was found among those individuals who underwent activity-based restorative therapy.[30] The authors also documented that activity-based therapy had no significant effect on lower-extremity mobility, degree of independence, or quality of life when compared to those individuals who underwent conventional rehabilitation therapy; however, activity-based therapy improved upper-extremity function beyond convention rehabilitation therapy.[30] Overall, the results of this recent systematic review including 19 clinical trials with moderate scientific quality (mean PEDro score of 5.3) suggest that activity-based restorative therapy has benefits for upper-extremity rehabilitation, but it was not superior to conventional rehabilitation therapy when applied to the lower extremity in individuals with SCI.[30] Nonetheless, the authors emphasized the major methodological limitations including the potential risk for type II error in the data analyses due to relatively small sample size in most of the clinical trials, and differences in the type of training modality and methodology used across the studies.[30] Furthermore, Behrman et al.[24] underlined some of the limitations of the classification and longitudinal assessments of the lower-extremity function after SCI when using International Standards for Neurological Classification of SCI (ISNCSCI), which reinforces the need to evaluate the results of the locomotor training at regular intervals with a reliable, responsive, and validated outcome measure.

According to Behrman et al.[24] intense rehabilitation regimens typically include a minimum of 60 sessions of 1.5 hours each of daily activity—based therapies where there is repetitive practice of the task. Given the SCIRehab reports, one would anticipate that somewhere between 52 and 60 sessions of therapy is appropriate for individuals with motor incomplete SCI. Regardless of the duration and intensity of therapy, routine implementation of standardized assessments at regular intervals following 15 to 20 therapy sessions is recommended. Nonetheless, further investigations are required in order to determine the optimal and cost-effective, volume, and intensity of the activity-based therapy that maximizes outcomes.

Neuromodulation

The use of electrical stimulation to modify neuronal activity is a neuromodulatory alternative for management of neurogenic bladder, respiratory function, neuropathic pain, and motor restoration after SCI.[31] While functional electrical stimulation therapy (FES-t) protocols are currently available for optimization of the functional outcomes of physical and occupational therapy in many rehabilitation facilities worldwide, other promising neuromodulation techniques that may promote motor restoration after SCI have been investigated: transmagnetic stimulation, transcranial direct current stimulation, brain—machine interface technology, epidural cord stimulation, and deep brain stimulation.[31] The clinical use of most neuromodulation techniques has been restricted to research settings due to concerns with regard to durability of the devices, accessibility, and affordability.[31] Our poor understanding of the mechanisms of action of these neuromodulation techniques also represents an important barrier to their widespread suitability for longitudinal clinical practice implementation.[31]

FES uses short low-energy pulses to produce muscle contractions that can be sequenced and synchronized in order to create a functional movement such as grasping or walking.[32] FES is a neuroprosthesis that delivers electrical pulses using transcutaneous, percutaneous, or implanted electrodes as a lifetime substitute for activities of daily living in those individuals who cannot perform those activities without assistance.[32] FES-t requires the temporary use of a neuroprosthesis for a short-term therapeutic intervention to retrain voluntary motor function of the upper or lower extremities.[32] FES-t can also be used in activity-based restorative therapy, when the patient is able to voluntarily initiate the movement as previously mentioned.

There is a growing body of evidence that endorses the efficacy of FES-t in combination with task-specific training in the rehabilitation of individuals living with SCI.[33–36] In a prior randomized controlled trial, 40 sessions (1 hour duration each) of multichannel FES-t in addition to conventional occupational therapy significantly reduced disability and improved voluntary grasping to a greater extent than conventional therapy among individuals with tetraplegia.[35] Moreover, this technology-based therapy in combination with task-specific training provided superior results compared to conventional therapy alone; these functional gains persisted for at least 6 months after completion of the FES-t.[34] The rationale for the 40 sessions of FES-t was based on previous data on the period for maximum recovery after onset of SCI (i.e., initial 6 months) in combination with feasibility data on the mean time period SCI onset to SCI to rehabilitation admission (approximately 30 days) and data on the mean length of stay for inpatient rehabilitation of individuals with tetraplegia (approximately 3 months).[37] With this, FES-t protocols were designed to accommodate the therapy during the period where in the individuals with SCI would be admitted for inpatient rehabilitation. FES-t in combination with task-specific training

putatively promotes adaptive neuroplasticity because the spinal cord has the ability to generate motor responses using sensory input; however, further investigations are needed to understand the pathophysiological mechanisms regarding how this technology influences the neuromuscular systems.[38]

Spinal stimulation is another evolving technique that uses electrical stimulation to promote neurorehabilitation.[31,39] Spinal electrical stimulation can be delivered transcutaneously, epidurally, and intraspinally.[39] The location of the electrodes determines the stimulation parameters to properly activate the neural pathways; however, transcutaneous spinal stimulation is less likely to accurately stimulate the targeted neural pathways when compared to epidural and intradural spinal stimulation.[31] Initial reports suggest the promise of epidural stimulation; however, long-term evaluation of the downstream implications of epidural stimulation is needed prior to routine deployment. Future investigation is required to determine the role of spinal stimulation in the rehabilitation of individuals after traumatic SCI.

Robotic-assisted rehabilitation

Several rehabilitation technologies of variable width, speed, weight support, and exoskeletons have been developed to restore standing and locomotion after SCI including but not limited to: long-leg braces and forearm crutches, Parastep, and Body Weight Support Treadmill Training (BWSTT).

Exoskeletons typically provide a patient with an absence of voluntary movement of the lower extremities, confined to a manual wheelchair for household and community mobility, with the ability to change position from sit to stand, and to walk with the supervision of a physiotherapist (contact or guard) after donning the device. The exoskeleton does so by providing motorized assistance with hip and knee flexion through dynamic ankle–foot orthoses supporting to the lower leg, ankle, and foot. The EKSO GT exoskeleton offers several training modes: (1) *First step mode* allows the therapist to manually trigger and control steps; (2) *Pro-step mode* offers complete assistance, typically for patients with motor complete SCI; and (3) *Pro-step + mode* provides adaptive and variable assist features for persons with motor incomplete SCI experiencing motor recovery. In addition, the ankle–foot orthosis can be adjusted to accommodate the users' available ankle range of motion.

There is a plethora of literature describing the benefits of exoskeletons in terms of improvements in standing time, step number, step length, and walking endurance.[24] In addition, exoskeletons may reduce the risk of tissue injury, spasticity, extremity joint contractures, osteoporosis, and edema among other benefits.[40] However, initial assistive devices such as braces, orthotics, and hybrid systems incorporating FES technology showed limited efficiency due to high-energy demand, prolonged periods of time for donning and doffing, and a dysfunctional gait pattern.[40] With this, exoskeletons, which were initially developed in the military setting, became an attractive emerging rehabilitation modality because, according to preliminary studies, they may potentially facilitate functional standing and/or ambulation, decrease secondary health conditions after SCI, and enhance neuroplasticity following SCI.[40] Nonetheless, there are multiple obstacles for the widespread use of the exoskeletons including the elevated costs for acquisition and maintenance, limited battery endurance (at most 8 hours of life to date), community barriers (e.g., stairs and ramps), need for upper-extremity support, and need for caregiver assistance/supervision, among others.[40] Furthermore, the guidelines for use of the devices for walking are variable depending upon the manufacturer

that commonly precludes their use if the individual is pregnant or has spasticity, pressure ulcer, heterotopic ossification, cognitive or psychiatric condition, uncontrolled autonomic dysreflexia, spinal instability, unresolved deep venous thrombosis (DVT), or concurrent medical diagnosis.[40]

In a recent systematic review and meta-analysis, Cheung et al.[41] reported that robotic-assisted rehabilitation is "an adjunct therapy in functional recovery of patients with incomplete SCI." Those authors claimed that there is sufficient evidence to "support the use of robotic training for promoting walking ability in patients with SCI in both the acute and chronic stage."[41] Nevertheless, those authors also recognize the need for further, high-quality studies including randomized clinical trials in order to clarify the effectiveness of exoskeletons on the rehabilitation of the upper-extremity function, gait, and cardiopulmonary function in individuals living with SCI.[41] Overall, the current state suggests that exoskeletons remain a promising technology for robotic-assisted rehabilitation in a supervised setting, especially of the lower extremity.[40–42] Current devices require further refinement and safety evaluations prior to dissemination and widespread independent use in the community.

Surgical strategies for restoration of the upper-extremity function

Novel surgical options for restoration of the upper-extremity function in individuals with tetraplegia include tendon transfer and peripheral nerve transfer, which have distinct advantages and disadvantages.

Tendon transfer surgery

Tendon transfer procedures are usually considered only when the patient reached their plateau of neurologic recovery (approximately 9 months after SCI). However, patients in the chronic stage (even after 10 years) after cervical

SCI are also potential candidates for tendon transfer surgery.[43,44] According to guidelines from the International Classification for Surgery of the Hand in Tetraplegia, the effectiveness of the reconstructive procedure relies on wrist extension torque and the availability of active muscles for tendon transfers.[45] Given that a tenodesis action is initiated by the wrist, individuals with tetraplegia require voluntary active wrist extension in order to generate good grasp. The tenodesis occurs when an active extension of the wrist promotes passive tension of the flexor muscles at the wrist causing a gross grip pattern of the fingers and a lateral pinch of the thumb.[43] On the contrary, wrist flexion promotes increased passive force in the tendons of the extensor muscles causing the opening of the fingers and thumb.[43] The tendon transfers that are commonly performed include: (a) transfer of the posterior deltoid or biceps to triceps in order to facilitate elbow extension; (b) creation of pinch and/or finger—palm grasp; and (c) provision of wrist extension.[43] Generally speaking, individuals with SCI at C5 to C8 neurological levels are considered candidates for tendon transfer procedures if they are able to reach their mouth with the hand. Otherwise, the proximal attachment of any potential donor muscles in the upper extremity is considered too weak to generate action if their distal tendon is transferred.[43]

Although the period of postoperative immobilization of the upper extremity depends on the surgical technique, it is generally accepted that the period of tendon healing is between 3 and 4 weeks.[43] Rehabilitation after tendon transfer surgery is usually recommended to activate the muscle in its usual role, electrically stimulate the muscle, or trigger the muscle using biofeedback. Furthermore, rehabilitation is focused on regaining range of motion of the involved joints, reeducating of new muscle action, and subsequently, learning and practicing functional tasks and activities of daily living.[43] Despite the functional gains and good patient satisfaction in most cases of the tendon transfer surgery, 60%—70% of the individuals with tetraplegia are eligible for a surgical restoration of the upper-extremity function; however, less than 10% of them actually undergo tendon transfer surgery.[46] Barriers related to the population, provider, and healthcare system are main reasons for the underutilization of surgical upper-extremity restoration interventions.[46]

Nerve transfer surgery

Restoration of upper-extremity function after SCI using surgical techniques for nerve transfer has gained popularity because the tendon transfer may be complicated by loss of muscle strengthening, tendon rupture, or failure of tenodesis.[47] The general principle of the nerve transfer technique is that a peripheral nerve with regenerating axons under voluntary cortical control is surgically placed near to target motor endplates for restoration of function in individuals with tetraplegia.[48]

In a recent systematic review, Khalifeh et al.[47] found 21 publications on nerve transfer surgery using a diversity of strategies for restoration of upper-extremity function (i.e., thumb and finger flexion, elbow extension, and wrist and finger extension) in individuals in the subacute or chronic stage after cervical SCI (mean time from injury to surgery of 18.7 months, varying from 4 months to 13 years). The authors reported that the recipient muscle reached a power of at least three out of five in the Medical Research Council scale. While the larger case series showed a great variation in the results, patients who underwent nerve transfer surgery for restoration of wrist/finger extension and elbow extension had the most consistent outcomes.[47] Khalifeh et al. also emphasized that further studies are needed to determine the optimal timing and long-term outcomes of the nerve transfer surgery to restore upper-extremity function after tetraplegia.[47] The nerve transfer surgery should be carried out allowing enough time for the muscle reinnervation to occur before the irreversible motor endplate degeneration that usually takes place 18 months after lower motor neuron injury due to SCI.[49] Ideally, assessment of potential candidates for the nerve transfer surgery should be referral to a skilled and experienced surgeon at earliest time within 12 weeks after injury, which would allow the surgical intervention earlier than 1 year following SCI.[49] Friden and Gohritz[50] also highlighted that surgical transfer of axillary, musculocutaneous, or radial nerve fascicles cranial to the injury can be considered valuable options to enhance motor outcome and sensory preservation in individuals with tetraplegia.

Referral for and access to experienced and skilled surgeons and therapists able to appropriately select candidates with tetraplegia for nerve transplant and/or tendon transfer at the appropriate time period during natural recovery and provide perioperative care is a hallmark of high-quality tertiary rehabilitation centers. Essential elements for the success of those surgical treatment strategies include careful selection of the potential candidates, and full disclosure of the risks and benefits including understanding of what represents a satisfactory outcome.

Key determinants of outcomes of SCI rehabilitation

Prior studies indicated that several factors may influence the outcomes and prognosis of individuals with traumatic SCI who undergo rehabilitation as follows: level and severity of SCI, pattern of SCI or spinal cord syndrome, age at the onset of SCI, sex/gender, preexisting medical comorbidities, and secondary health conditions after SCI.

More cranial (e.g., cervical) levels of injury and more severe SCI (i.e., AIS A or B) are associated with poorer neurological recovery and lower degree of independence due to the patient's physical limitations that may have an adverse effect on their participation in rehabilitation programs. Specific examples include: (1) Individuals with motor severe SCI at C1 to C3 are most likely ventilator dependent, and hence, they are at a greater risk of pneumonia; (2) individuals with motor severe SCI at C1 to C5 have reduced or absent upper-extremity motor function in addition to regional hypo- or hyperesthesia; (3) individuals with severe SCI at T6 or more cranial levels are more prone to develop significant cardiovascular dysfunction including orthostatic hypotension, bradycardia, and episodes of autonomic dysreflexia; and (4) individuals with severe motor SCI at T9 or more cranial levels usually experience trunk instability[13,51] which impedes the ability to perform independent. Furthermore, neurological recovery after traumatic SCI is dependent upon the level and severity of the injury as extensively documented in the literature.[13,19] Functional recovery may also be affected by the pattern of SCI or spinal cord syndrome. For instance, McKinley et al.[52] reported that individuals with Brown-Sequard syndrome achieved a greater degree of functional improvement after rehabilitation than individuals with central cord syndrome. Overall, the authors documented that 20.9% of 839 individuals admitted for inpatient rehabilitation were diagnosed with a spinal cord syndrome such as central cord syndrome (9.2% of all cases of SCI), cauda equina syndrome (5.2%), Brown-Sequard syndrome (3.6%), conus medullaris syndrome (1.7%), anterior cord syndrome (<1%), and posterior cord syndrome (<1%).[52]

Older age at the time of injury was previously suggested as a potential predictor of poorer outcomes after traumatic SCI. In fact, survival in the elderly population within the first year (mainly acute stage) after injury is lower than younger adults following traumatic SCI, even though the higher frequency of preexisting medical comorbidities in the former group appear to be the primary reason for their elevated mortality.[53–56] Contrary to the results of prior studies based on unadjusted data analyses, there is a growing body of evidence suggesting that chronological age is not a reliable predictor of neurological and functional outcome after SCI, when data analyses are adjusted for major potential confounders.[53,54,57,58] However, elderly patients sometimes require a longer rehabilitation length of stay to achieve the same functional outcomes. Additionally, health-care professionals may have ageistic attitudes when caring for individuals with SCI in the acute care and/or rehabilitation settings that may adversely affect the older patients' outcomes during rehabilitation.[59]

Outstanding surgical and appropriate rehabilitation care is required for management of all individuals with traumatic SCI including the geriatric group that has the potential to achieve favorable outcomes.[54,60] The understanding of the complexity and nuances of the care for older individuals with SCI needs to be tailored into the transdisciplinary rehabilitation models, which support the notion that a more prolonged inpatient rehabilitation program is necessary for most of the elderly individuals in order to reach their optimal neurorecovery after SCI.[60] Despite the paucity of publications of protocols on rehabilitation for older individuals with traumatic SCI, the general principles of geriatric rehabilitation for other neurological disorders should be considered in the SCI population. Weber et al.[61] underlined the key principles for geriatric rehabilitation including: (a) function-oriented approach; (b) primary focus on the disease diagnosis and treatment, but "assessment of the functional effect of the pathologic change is emphasized"; and (c) interdisciplinary team approach aiming to address the individual and their family goals.

The literature has consistently documented that traumatic SCI is more common among males than females.[62,63] Sex may also influence the epidemiological profile of traumatic SCI and its outcomes.[64–66] In a recent retrospective study including 504 consecutive cases of acute spine trauma, females present with a greater

number of preexisting comorbidities, a higher frequency of thoracolumbar trauma, less severe neurological impairment, and a greater proportion of injuries related to motor vehicle accidents than males.[64] In a retrospective cohort study using data of 14,433 individuals with traumatic SCI, Sipski et al.[65] documented that females may have more neurologic recovery than males, but functional recovery in some male subgroups may be better than their female counterparts. Furthermore, sex-related or other unconscious biases (heritage, religion, culture, etc.) may exist and have a consequent impact on outcomes after SCI rehabilitation, even though these issues are largely understudied.[67]

Minimizing secondary health conditions

Secondary health conditions are biopsychosocial consequences of injury which are increased in frequency or severity due to SCI which result in impaired function. Secondary health conditions may include autonomic dysreflexia, respiratory infection, UTIs, venous thromboembolism, heart disease, osteoporosis, overuse upper-extremity injuries, sleep disorders, sexual disorders, depression and suicidal ideation, tissue injury or pressure sores, chronic pain, and fatigue.[66,68–70] These conditions are key contributors to inappropriate Emergency Department visits, rehospitalizations, and/or death in the postacute phase.[71–74] In the acute and subacute settings after injury secondary health conditions in acute care can predict an extended rehab length of stay or result in rehabilitation service interruptions.[21,75] Based on the knowledge of an individual's age at injury, degree of spinal cord impairment (e.g., level and severity), prior ventilation or requirement for an endotracheal tube, presence of an indwelling bladder catheter, and lower-extremity motor scores at rehabilitation admission, hospital administrators can identify patients with a likely extended rehabilitation length of stay due to service interruptions or concurrent illnesses. Moreover, secondary health conditions have been implicated in delay of rehabilitation program with potential, adverse effects of the patients' recovery.[21] Dijkers and Zanca[27] reported that neuropathic pain, fatigue, and spasticity were frequent causes of delay in rehabilitation sessions of physical, occupational, speech, and recreational therapies, but many other health conditions also affected the patients' participation on those therapies (e.g., pressure sores, cardiovascular dysfunction, respiratory dysfunction, neurogenic bladder, neurogenic bowel, heterotopic ossification). Individuals with indwelling catheters and detrusor sphincter dyssynergia and high intravesicular pressures are at an increased risk of UTIs.

Preventable secondary health conditions after SCI

Although several secondary health conditions following motor complete SCI are considered nonpreventable, the risk for pressure ulcers, venous thromboembolism, and infections can be reduced if preventive measures as described in the current clinical practice guidelines are implemented by the patients and health-care professionals.[18,76,77]

Pressure ulcers

In a recent systematic review, the frequency of pressure ulcers in individuals with SCI among developing countries varied from 43.1% to 94.1% at admission in rehabilitation centers, and 12%–91% after discharge from rehabilitation centers.[78] In a recent study in the United States, Brienza et al.[79] documented that 37.5% of the patients with SCI developed at least one pressure sore during acute care hospitalization or inpatient rehabilitation. The authors identified higher neurological level and more severe SCI, as well as pneumonia as predicting factors for the development of pressure ulcers during inpatient management of acute traumatic SCI.[79] All

patients with SCI should be deemed high risk for tissue injury and fit to receive greater monitoring of those with low Spinal Cord Independence Measure (SCIM) mobility subscores and incontinence[80,81]; interventions to modify the intrinsic wound characteristics and extrinsic risk factors with new surfaces and dressing are some of the new advances in the prevention and treatment of pressure ulcers in individuals living with SCI.[82]

Venous thromboembolism

DVT and pulmonary embolism (PE) in untreated individuals with SCI were diagnosed in 12%–64% only on the basis of clinical criteria.[83] Venous thromboembolism includes a continuum of presentation from small, asymptomatic thrombi to massive, fatal PE.[84] Venous thromboembolism is serious life-threatening health condition that occurs in the presence or absence of thromboprophylaxis. Pharmacologic thromboprophylaxis should be started within 72 hours following acute SCI if there is no contraindication.[84,85] Paciaroni et al.[86] reported results of a systematic review of thromboprophylaxis in individuals with SCI patients demonstrating that low-molecular-weight heparin was associated with a significant decrease in PE and a trend for fewer DVT and major bleeding compared to low-dose unfractionated heparin. In a prior evidence-based analysis, Furlan and Fehlings[83] concluded that there is insufficient evidence to support (or refute) a recommendation for routine screening for DVT in adults with acute traumatic SCI prescribed thromboprophylaxis.

Infections

Pneumonia, UTI, wound infection, and sepsis are the most frequent infectious complications that may occur in the acute, subacute, and chronic stages following SCI. Jaja et al.[87] documented that 11.3% of the patients were diagnosed with pneumonia, 1.8% had a wound infection, and 0.8% developed sepsis during initial hospitalization after acute SCI. The authors also reported that the strongest predictors for those post-SCI infectious complications were level of injury, severity of injury (i.e., AIS), and "premorbid medical status" (including high blood pressure, diabetes, heart attack, malignancy, pulmonary disease, cerebrovascular disease, smoking history, and substance abuse).[87] Infectious complications also occur during inpatient rehabilitation but in lower frequencies than those documented during acute care hospitalization.[88] Risk factors for symptomatic UTI include incomplete voiding, increased intravesical pressure, using intermittent and/or indwelling catheters, and an upper motor neuron bladder.[89] Among the signs and symptoms, cloudy urine had the highest accuracy (83.1%) and leukocytes in the urine had the highest sensitivity (82.8%) for the presence of UTI, and fever had very high specificity (99%) but very low sensitivity (6.9%) for diagnosis of UTI.[90] Fever during a UTI suggests the presence of the inflammation, which may lead to severe sepsis or risk of developing an acquired immunodeficiency syndrome.[91] Nonetheless, the risk for infectious complications (e.g., UTI and pneumonia) remains within the first year following acute SCI.[92] Prevention, early diagnosis, and effective management of respiratory dysfunction, neurogenic bladder, and pressure ulcers represent critical measures for prevention of the main infectious complications after SCI.[18,77,93–95]

Sepsis of any etiology (fever and hypotension with an infection) increases the risk of developing an acquired immune deficiency syndrome and adversely effects rehabilitation outcomes out to 5 years postinjury.[96,97] For this reason, most rehabilitation programs implement standard procedures to reduce hospital acquired infections including pneumonia, line sepsis, and UTI. Furthermore, neurological recovery after traumatic SCI is dependent upon the level and severity of the injury as extensively documented in the literature.[13,19]

Other secondary health conditions following SCI

Individuals living with SCI often develop many other secondary health conditions (Table 20.1). Early diagnosis and proper management of each secondary health condition may reduce its short-term and long-term consequences, improve patient's quality of life and social participation, and reduce its economic burden. Although many secondary health conditions after SCI are promptly identified and managed during rehabilitation, there are some secondary health conditions that are at times initially underrecognized. For instance, episodes of autonomic dysreflexia were reported in 5.7% of the individuals in the acute stage after traumatic SCI at T6 or more cranial level.[98] Episodes of autonomic dysreflexia were documented in 26%–90% of the individuals in the subacute and chronic stages following traumatic SCI (Table 20.1). Autonomic dysreflexia was also reported in nontraumatic SCI including spinal cord neoplasm,[99] metastatic spine cancer,[100] neuromyelitis optica,[101] and multiple sclerosis.[102] Autonomic dysreflexia is commonly associated with other more commonly recognized cardiovascular dysfunctions such as low baseline arterial blood pressure, orthostatic hypotension, and cardiac arrhythmias (Table 20.1). Yet, a prior cross-sectional study revealed that 41% of the Canadian individuals living with SCI at a high risk for development of autonomic dysreflexia never heard about this secondary medical condition.[103] Moreover, 22% of those individuals living with SCI reported symptoms consistent with unrecognized autonomic dysreflexia.[103] Given the knowledge gaps on secondary health conditions (particularly related to autonomic dysfunction) after SCI, the "International Standards Assessment Form" was developed in order to promote early recognition of effects of traumatic SCI on cardiovascular, bronchopulmonary, sudomotor, bladder, bowel, and sexual function.[104]

TABLE 20.1　Other common secondary health conditions after spinal cord injury.

Secondary health conditions	Frequency	Treatment	References
Neurogenic bowel	39%—74%	Clinical and surgical management, and neuromodulation	117—120
Neurogenic bladder	68%—84%	Clinical and surgical management, and neuromodulation	18,31,94,121—123
Neuropathic pain	18%—92%	Clinical and surgical management, and neuromodulation	31,69,88,124,125
Spasticity	9%—60%	Clinical and surgical management	69,88,124,126
Low baseline blood pressure	44%—51%	Clinical management	51,69,98,124,127,128
Orthostatic hypotension	59%—74%	Clinical management	51,69,98,124,127,128
Autonomic dysreflexia	26%—90%	Clinical management	51,69,98,105,106,124,127,128
Bradycardia and other cardiac arrhythmias	17%—77%	Clinical management	51,69,98,124,127—129
Mood disorders	29%—46%	Clinical management	130—134
Anxiety disorders	6%—30%	Clinical management	66,130—133
Sleep-related breathing disorders	13%—91%	Clinical management	135,136
Heterotopic ossification	2%—7%	Clinical and surgical management, and radiotherapy	92,137
Sublesional osteoporosis	2%—33%	Clinical management and technology-based therapy	138—142

Moreover, effective management of one secondary health condition after SCI may also have a positive impact on other secondary health conditions. For instance, episodes of autonomic dysreflexia can be triggered by inappropriate management of neurogenic bladder (causing urinary retention or UTI), poor control of neurogenic bowel (causing fecal impaction), or lack of preventive measures for skin integrity preservation (causing pressure ulcers).[105,106] Similarly, untreated moderate-to-severe sleep-related breathing disorders are associated with cardiovascular dysfunction (causing episodes of autonomic dysreflexia, cardiac arrhythmias), mood and anxiety disorders, and hypovitaminosis D that is a risk factor for osteoporosis.[107—109]

Secondary health conditions and lifetime care

The importance of long-term follow-up for individuals with chronic SCI and secondary health conditions over the course of their lifetime cannot be overstated. Long-term follow-up is essential to address new rehabilitation goals as they emerge, and maintain functional abilities over time, through access to new or developing therapies and new devices over time. Furthermore, it is valuable to monitor the individual for changes in neurologic impairment, prevent worsening of secondary health conditions, and assist with accommodation to aging with a disability and living with multimorbidity after

SCI. The burden of multimorbidity often results in a substantial functional decline and an increased need for attendant care services in the last 10 years of an individual's life. Although survival and life expectancy after SCI have increased, most individuals with 10 years postinjury report a mean of seven concurrent secondary health conditions with one in four of these patients being hospitalized each year. Secondary health conditions can be life-threatening (e.g., sepsis, cardiac arrhythmias, autonomic dysreflexia), often reduce the individuals' social participation and quality of life, and impose significant economic burden for the individuals living with SCI and their family members, as well as for the society.

The impact of secondary health conditions and degree of impairment after SCI was evaluated in a Canadian cross-sectional, telephone-based survey study that estimated the utilities for different secondary health conditions in a group of adults with traumatic and nontraumatic SCI, using Health Utilities Index-Mark III (HUI-III) questionnaire.[110] Of note, utility is a measure of individual's preference for a health state based on quantity and quality of life; by convention, utility values are anchored at 0 (death) and 1 (perfect health). While "Standard Gamble" and "Time Trade-off" are the most accurate methods to estimate utility, questionnaires such as EQ-5D-5L, SF-6D, and HUI-III can be used to derivate utility values. In that Canadian study,[110] the mean utility value was 0.27, which is less than, or equal to, the mean utility values reported for other populations with different neurological conditions: 0.58 to 0.68 for stroke[111]; 0.57 for multiple sclerosis[112]; and 0.42 for Parkinson's disease.[113] In another study using data from that Canadian cross-sectional, telephone-based survey study, older age, employment, and high SCIM scores were associated with greater utility values, whereas a higher secondary health condition impact score was associated with lower utility values.[114]

In summary, traumatic SCI results in diverse, often devastating motor, sensory, and autonomic impairments, including absence or limitations in one's involuntary ability to breathe, regulate blood pressure and temperature, and voluntary ability to evacuate one's bladder or bowel, dress, bathe, feed oneself, or move about one's home or community requiring an interprofessional team to provide transdisciplinary care and mitigate the impact of these impairments on the individual's long-term health and well-being.[115]

Knowledge gaps and research opportunities

In a prior review of the scientific literature, the top-cited articles among clinicians and scientists mostly matched with the perspective from the individuals living with SCI; however, most of the research has been focused on motor function assessment and recovery following SCI.[116] This reinforces the general sense that further research is needed in the field of rehabilitation of individuals with SCI, more specifically on prevention and effective management of secondary health conditions following SCI. Innovative technology—based therapies and novel peripheral nerve transfer strategies are promising but future research is needed to elucidate their real role in the restoration of motor and sensory function after SCI.

References

1. Eltorai IB. History of spinal cord medicine. In: Lin VW, Cardenas DD, Cutter NC, Frost FS, Hammond MC, Lindblom LB, et al., editors. *Spinal cord medicine: principles and practice*. 1st ed. New York: Demos Medical Publishing; 2003. p. 3—14.
2. Dowdy J, Pait TG. The influence of war on the development of neurosurgery. *J Neurosurg* 2014;**120**(1):237—43.
3. Eldar R, Jelic M. The association of rehabilitation and war. *Disabil Rehabil* 2003;**25**(18):1019—23.

4. Lanska DJ. Historical perspective: neurological advances from studies of war injuries and illnesses. *Ann Neurol* 2009;**66**(4):444—59.

5. Stahnisch FW, Tynedal JD. Sir Ludwig Guttmann (1899-1980). *J Neurol* 2012;**259**:1512—4.

6. van den Berg ME, Castellote JM, de Pedro-Cuesta J, Mahillo-Fernandez I. Survival after spinal cord injury: a systematic review. *J Neurotrauma* 2010;**27**(8):1517—28.

7. Ramer LM, Ramer MS, Bradbury EJ. Restoring function after spinal cord injury: towards clinical translation of experimental strategies. *Lancet Neurol* 2014;**13**(12):1241—56.

8. Middleton JW, Dayton A, Walsh J, Rutkowski SB, Leong G, Duong S. Life expectancy after spinal cord injury: a 50-year study. *Spinal Cord* 2012;**50**(11):803—11.

9. Craven BC, Verrier M, Balioussis C, Wolfe D, Hsieh J, Noonan V, et al. *Rehabilitation environmental e-scan Atlas: capturing capacity in Canadian SCI rehabilitation.* Vancouver, BC: Rick Hansen Institute and Toronto Rehabilitation Insititute; 2012.

10. WHO. *How to use the ICF: a practical manual for using the International Classification of Functioning, disability and health (ICF).* Geneva: World Health Organization; 2013 2013.

11. Barnes MP. Principles of neurological rehabilitation. *J Neurol Neurosurg Psychiatry* 2003;**74**(Suppl. 4):iv3—7.

12. Simpson LA, Eng JJ, Hsieh JT, Wolfe DL, Spinal Cord Injury Rehabilitation Evidence Scire Research Team. The health and life priorities of individuals with spinal cord injury: a systematic review. *J Neurotrauma* 2012;**29**(8):1548—55.

13. PVA. *Outcomes following traumatic spinal cord injury: clinical practice guidelines for health-care professionals.* 1st ed. Washington, DC: Paralyzed Veterans of America; 1999.

14. Harkema SJ, Hillyer J, Schmidt-Read M, Ardolino E, Sisto SA, Behrman AL. Locomotor training: as a treatment of spinal cord injury and in the progression of neurologic rehabilitation. *Arch Phys Med Rehabil* 2012;**93**(9):1588—97.

15. Craven BC, Balioussis C, Verrier MC, Team E-SI. The tipping point: perspectives on SCI rehabilitation service gaps in Canada. *Int J Phys Med Rehabil* 2013;**1**(8):1—8.

16. Kessler TM, Traini LR, Welk B, Schneider MP, Thavaseelan J, Curt A. Early neurological care of patients with spinal cord injury. *World J Urol* 2018;**36**(10):1529—36.

17. Fehlings MG, Tetreault LA, Aarabi B, Anderson P, Arnold PM, Brodke DS, et al. A clinical practice guideline for the management of patients with acute spinal cord injury: recommendations on the type and timing of rehabilitation. *Global Spine J* 2017;**7**(Suppl. 3):231S—8S.

18. Kavanagh A, Baverstock R, Campeau L, Carlson K, Cox A, Hickling D, et al. Canadian Urological Association guideline: diagnosis, management, and surveillance of neurogenic lower urinary tract dysfunction - full text. *Can Urol Assoc J* 2019;**13**(6):E157—76.

19. Curt A, Van Hedel HJ, Klaus D, Dietz V, Group E-SS. Recovery from a spinal cord injury: significance of compensation, neural plasticity, and repair. *J Neurotrauma* 2008;**25**(6):677—85.

20. Grabher P, Callaghan MF, Ashburner J, Weiskopf N, Thompson AJ, Curt A, et al. Tracking sensory system atrophy and outcome prediction in spinal cord injury. *Ann Neurol* 2015;**78**(5):751—61.

21. Bhide R, Rivers C, Kurban D, Chen J, Noonan V, Farahani F, et al. Service interruptions and their impact on rehabilitation length of stay among Ontarians with traumatic, subacute spinal cord injury. *Crit Rev Phys Rehabil Med* 2018;**30**(1):45—66.

22. Hofer AS, Schwab ME. Enhancing rehabilitation and functional recovery after brain and spinal cord trauma with electrical neuromodulation. *Curr Opin Neurol* 2019;**32**(6):828—35.

23. Hilton BJ, Tetzlaff W. A brainstem bypass for spinal cord injury. *Nat Neurosci* 2018;**21**(4):457—8.

24. Behrman AL, Ardolino EM, Harkema SJ. Activity-based therapy: from basic science to clinical application for recovery after spinal cord injury. *J Neurol Phys Ther* 2017;**41**(Suppl. 3):S39—45.

25. Whiteneck G, Gassaway J, Dijkers M, Jha A. New approach to study the contents and outcomes of spinal cord injury rehabilitation: the SCIRehab Project. *J Spinal Cord Med* 2009;**32**(3):251—9.

26. Whiteneck GG, Gassaway J. SCIRehab uses practice-based evidence methodology to associate patient and treatment characteristics with outcomes. *Arch Phys Med Rehabil* 2013;**94**(Suppl. 4):S67—74.

27. Dijkers MP, Zanca JM. Factors complicating treatment sessions in spinal cord injury rehabilitation: nature, frequency, and consequences. *Arch Phys Med Rehabil* 2013;**94**(Suppl. 4):S115—24.

28. Hammond FM, Lieberman J, Smout RJ, Horn SD, Dijkers MP, Backus D. Missed therapy time during inpatient rehabilitation for spinal cord injury. *Arch Phys Med Rehabil* 2013;**94**(Suppl. 4):S106—14.

29. Sadowsky CL, McDonald JW. Activity-based restorative therapies: concepts and applications in spinal cord injury-related neurorehabilitation. *Dev Disabil Res Rev* 2009;**15**(2):112—6.

30. Quel de Oliveira C, Refshauge K, Middleton J, de Jong L, Davis GM. Effects of activity-based therapy interventions on mobility, independence, and quality of life for people with spinal cord injuries: a systematic review and meta-analysis. *J Neurotrauma* 2017;**34**(9):1726—43.

31. James ND, McMahon SB, Field-Fote EC, Bradbury EJ. Neuromodulation in the restoration of function after spinal cord injury. *Lancet Neurol* 2018;**17**(10): 905—17.

32. Craven BC, Hadi SC, Popovic MR. Functional electrical stimulation therapy: enabling function through reaching and grasping. In: Söderback I, editor. *International handbook of occupational therapy interventions.* Switzerland Springer; 2015. p. 587—605.

33. Kapadia N, Masani K, Catharine Craven B, Giangregorio LM, Hitzig SL, Richards K, et al. A randomized trial of functional electrical stimulation for walking in incomplete spinal cord injury: effects on walking competency. *J Spinal Cord Med* 2014;**37**(5): 511—24.

34. Kapadia NM, Zivanovic V, Furlan JC, Craven BC, McGillivray C, Popovic MR. Functional electrical stimulation therapy for grasping in traumatic incomplete spinal cord injury: randomized control trial. *Artif Organs* 2011;**35**(3):212—6.

35. Popovic MR, Kapadia N, Zivanovic V, Furlan JC, Craven BC, McGillivray C. Functional electrical stimulation therapy of voluntary grasping versus only conventional rehabilitation for patients with subacute incomplete tetraplegia: a randomized clinical trial. *Neurorehabil Neural Repair* 2011;**25**(5):433—42.

36. Popovic MR, Keller T, Pappas IP, Dietz V, Morari M. Surface-stimulation technology for grasping and walking neuroprosthesis. *IEEE Eng Med Biol Mag* 2001;**20**(1):82—93.

37. Burns AS, Ditunno JF. Establishing prognosis and maximizing functional outcomes after spinal cord injury: a review of current and future directions in rehabilitation management. *Spine* 2001;**26**(Suppl. 24): S137—45.

38. Roy RR, Harkema SJ, Edgerton VR. Basic concepts of activity-based interventions for improved recovery of motor function after spinal cord injury. *Arch Phys Med Rehabil* 2012;**93**(9):1487—97.

39. Ievins A, Moritz CT. Therapeutic stimulation for restoration of function after spinal cord injury. *Physiology* 2017;**32**(5):391—8.

40. Palermo AE, Maher JL, Baunsgaard CB, Nash MS. Clinician-focused overview of bionic exoskeleton use after spinal cord injury. *Top Spinal Cord Inj Rehabil* 2017;**23**(3):234—44.

41. Cheung EYY, Ng TKW, Yu KKK, Kwan RLC, Cheing GLY. Robot-assisted training for people with spinal cord injury: a meta-analysis. *Arch Phys Med Rehabil* 2017;**98**(11):2320—2331 e12.

42. Mekki M, Delgado AD, Fry A, Putrino D, Huang V. Robotic rehabilitation and spinal cord injury: a narrative review. *Neurotherapeutics* 2018;**15**(3):604—17.

43. Dunn JA, Sinnott KA, Rothwell AG, Mohammed KD, Simcock JW. Tendon transfer surgery for people with tetraplegia: an overview. *Arch Phys Med Rehabil* 2016; **97**(Suppl. 6):S75—80.

44. Dunn JA, Hay-Smith EJ, Keeling S, Sinnott KA. Decision-making about upper limb tendon transfer surgery by people with tetraplegia for more than 10 years. *Arch Phys Med Rehabil* 2016;**97**(Suppl. 6):S88—96.

45. McDowell CL, Moberg EA, House JH. The second international conference on surgical rehabilitation of the upper limb in tetraplegia (quadriplegia). *J Hand Surg* 1986;**11**(4):604—8.

46. Punj V, Curtin C. Understanding and overcoming barriers to upper limb surgical reconstruction after tetraplegia: the need for interdisciplinary collaboration. *Arch Phys Med Rehabil* 2016;**97**(Suppl. 6):S81—7.

47. Khalifeh JM, Dibble CF, Van Voorhis A, Doering M, Boyer MI, Mahan MA, et al. Nerve transfers in the upper extremity following cervical spinal cord injury. Part 1: systematic review of the literature. *J Neurosurg Spine* 2019:1—12.

48. Ray WZ, Chang J, Hawasli A, Wilson TJ, Yang L. Motor nerve transfers: a comprehensive review. *Neurosurgery* 2016;**78**(1):1—26.

49. Simcock JW, Dunn JA, Buckley NT, Mohammed KD, Beadel GP, Rothwell AG. Identification of patients with cervical SCI suitable for early nerve transfer to achieve hand opening. *Spinal Cord* 2017;**55**(2):131—4.

50. Friden J, Gohritz A. Tetraplegia management update. *J Hand Surg Am* 2015;**40**(12):2489—500.

51. Furlan JC, Fehlings MG. Cardiovascular complications after acute spinal cord injury: pathophysiology, diagnosis, and management. *Neurosurg Focus* 2008;**25**(5):E13.

52. McKinley W, Santos K, Meade M, Brooke K. Incidence and outcomes of spinal cord injury clinical syndromes. *J Spinal Cord Med* 2007;**30**(3):215—24.

53. Furlan JC, Bracken MB, Fehlings MG. Is age a key determinant of mortality and neurological outcome after acute traumatic spinal cord injury? *Neurobiol Aging* 2010;**31**(3):434—46.

54. Furlan JC, Fehlings MG. The impact of age on mortality, impairment, and disability among adults with acute traumatic spinal cord injury. *J Neurotrauma* 2009;**26**(10):1707—17.

55. Kattail D, Furlan JC, Fehlings MG. Epidemiology and clinical outcomes of acute spine trauma and spinal cord injury: experience from a specialized spine trauma center in Canada in comparison with a large national registry. *J Trauma* 2009;**67**(5):936—43.

56. Furlan JC, Kattail D, Fehlings MG. The impact of comorbidities on age-related differences in mortality after acute traumatic spinal cord injury. *J Neurotrauma* 2009;**26**(8):1361—7.

57. Furlan JC, Craven BC, Fehlings MG. Surgical management of the elderly with traumatic cervical spinal cord injury: a cost-utility analysis. *Neurosurgery* 2016;**79**(3): 418—25.

58. Furlan JC, Hitzig SL, Craven BC. The influence of age on functional recovery of adults with spinal cord injury or disease after inpatient rehabilitative care: a pilot study. *Aging Clin Exp Res* 2013;**25**(4):463—71.

59. Furlan JC, Craven BC, Ritchie R, Coukos L, Fehlings MG. Attitudes towards the older patients with spinal cord injury among registered nurses: a cross-sectional observational study. *Spinal Cord* 2009; **47**(9):674—80.

60. Strasser DC, Solomon DH, Burton JR. Geriatrics and physical medicine and rehabilitation: common principles, complementary approaches, and 21st century demographics. *Arch Phys Med Rehabil* 2002;**83**(9): 1323—4.

61. Weber DC, Fleming KC, Evans JM. Rehabilitation of geriatric patients. *Mayo Clin Proc* 1995;**70**(12): 1198—204.

62. Furlan JC, Tator CH. Global epidemiology of traumatic spinal cord injury. In: Morganti-Kossman C, Raghupathi R, Maas A, editors. *Traumatic brain and spinal cord injury: challenges and developments*. 1st ed. Cambridge: Cambridge University Press; 2012. p. 360.

63. Jain NB, Ayers GD, Peterson EN, Harris MB, Morse L, O'Connor KC, et al. Traumatic spinal cord injury in the United States, 1993-2012. *JAMA* 2015;**313**(22):2236—43.

64. Furlan JC, Craven BC, Fehlings MG. Sex-related discrepancies in the epidemiology, injury characteristics and outcomes after acute spine trauma: a retrospective cohort study. *J Spinal Cord Med* 2019;**42**(Suppl. 1): 10—20.

65. Sipski ML, Jackson AB, Gomez-Marin O, Estores I, Stein A. Effects of gender on neurologic and functional recovery after spinal cord injury. *Arch Phys Med Rehabil* 2004;**85**(11):1826—36.

66. Furlan JC, Krassioukov AV, Fehlings MG. The effects of gender on clinical and neurological outcomes after acute cervical spinal cord injury. *J Neurotrauma* 2005; **22**(3):368—81.

67. Furlan JC, Craven BC, Fehlings MG. Is there any gender or age-related discrepancy in the waiting time for each step in the surgical management of acute traumatic cervical spinal cord injury? *J Spinal Cord Med* 2019;**42**(Suppl. 1):233—41.

68. Hitzig SL, Tonack M, Campbell KA, McGillivray CF, Boschen KA, Richards K, et al. Secondary health complications in an aging Canadian spinal cord injury sample. *Am J Phys Med Rehabil* 2008;**87**(7):545—55.

69. Noreau L, Proulx P, Gagnon L, Drolet M, Laramee MT. Secondary impairments after spinal cord injury: a population-based study. *Am J Phys Med Rehabil* 2000; **79**(6):526—35.

70. Krassioukov AV, Furlan JC, Fehlings MG. Medical co-morbidities, secondary complications, and mortality in elderly with acute spinal cord injury. *J Neurotrauma* 2003;**20**(4):391—9.

71. Guilcher SJ, Craven BC, Calzavara A, McColl MA, Jaglal SB. Is the emergency department an appropriate substitute for primary care for persons with traumatic spinal cord injury? *Spinal Cord* 2013;**51**(3):202—8.

72. Jaglal SB, Munce SE, Guilcher SJ, Couris CM, Fung K, Craven BC, et al. Health system factors associated with rehospitalizations after traumatic spinal cord injury: a population-based study. *Spinal Cord* 2009;**47**(8):604—9.

73. Anson CA, Shepherd C. Incidence of secondary complications in spinal cord injury. *Int J Rehabil Res* 1996; **19**(1):55—66.

74. Dryden DM, Saunders LD, Rowe BH, May LA, Yiannakoulias N, Svenson LW, et al. Utilization of health services following spinal cord injury: a 6-year follow-up study. *Spinal Cord* 2004;**42**(9):513—25.

75. Catharine Craven B, Kurban D, Farahani F, Rivers CS, Ho C, Linassi AG, et al. Predicting rehabilitation length of stay in Canada: it's not just about impairment. *J Spinal Cord Med* 2017;**40**(6):676—86.

76. Schurch B, Iacovelli V, Averbeck MA, Stefano C, Altaweel W, Finazzi Agro E. Urodynamics in patients with spinal cord injury: a clinical review and best practice paper by a working group of the International Continence Society Urodynamics Committee. *Neurourol Urodyn* 2018;**37**(2):581—91.

77. Houghton PE, Campbell KE, Panel C. *Canadian best practice guidelines for the prevention and management of pressure ulcers in people with spinal cord injury. A resource handbook for clinicians*. Mississauga, Canada: Ontario Neurotrauma Foundation; 2013.

78. Zakrasek EC, Creasey G, Crew JD. Pressure ulcers in people with spinal cord injury in developing nations. *Spinal Cord* 2015;**53**(1):7—13.

79. Brienza D, Krishnan S, Karg P, Sowa G, Allegretti AL. Predictors of pressure ulcer incidence following traumatic spinal cord injury: a secondary analysis of a prospective longitudinal study. *Spinal Cord* 2018;**56**(1): 28—34.

80. Scovil CY, Delparte JJ, Walia S, Flett HM, Guy SD, Wallace M, et al. Implementation of pressure injury prevention best practices across 6 Canadian rehabilitation sites: results from the spinal cord injury knowledge mobilization network. *Arch Phys Med Rehabil* 2019;**100**(2):327—35.

81. Salzberg CA, Byrne DW, Cayten CG, Kabir R, van Niewerburgh P, Viehbeck M, et al. Predicting and preventing pressure ulcers in adults with paralysis. *Adv Wound Care* 1998;**11**(5):237—46.

82. Bogie K, Powell HL, Ho CH. New concepts in the prevention of pressure sores. *Handb Clin Neurol* 2012;**109**: 235—46.

83. Furlan JC, Fehlings MG. Role of screening tests for deep venous thrombosis in asymptomatic adults with acute spinal cord injury: an evidence-based analysis. *Spine* 2007;**32**(17):1908−16.

84. PVA. *Prevention of venous thromboembolism in individuals with spinal cord injury: clinical practice guidelines for health-care professionals.* 3rd ed. Washington, DC: Paralyzed Veterans of America; 2016.

85. Fehlings MG, Tetreault LA, Aarabi B, Anderson P, Arnold PM, Brodke DS, et al. A clinical practice guideline for the management of patients with acute spinal cord injury: recommendations on the type and timing of anticoagulant thromboprophylaxis. *Global Spine J* 2017;**7**(Suppl. 3):212S−20S.

86. Paciaroni M, Ageno W, Agnelli G. Prevention of venous thromboembolism after acute spinal cord injury with low-dose heparin or low-molecular-weight heparin. *Thromb Haemostasis* 2008;**99**(5):978−80.

87. Jaja BNR, Jiang F, Badhiwala JH, Schar R, Kurpad S, Grossman RG, et al. Association of pneumonia, wound infection, and sepsis with clinical outcomes after acute traumatic spinal cord injury. *J Neurotrauma* 2019;**36**(21): 3044−50.

88. Wahman K, Nilsson Wikmar L, Chlaidze G, Joseph C. Secondary medical complications after traumatic spinal cord injury in Stockholm, Sweden: towards developing prevention strategies. *J Rehabil Med* 2019;**51**(7): 513−7.

89. Cardenas DD, Moore KN, Dannels-McClure A, Scelza WM, Graves DE, Brooks M, et al. Intermittent catheterization with a hydrophilic-coated catheter delays urinary tract infections in acute spinal cord injury: a prospective, randomized, multicenter trial. *Pharm Manag PM R* 2011;**3**(5):408−17.

90. Massa LM, Hoffman JM, Cardenas DD. Validity, accuracy, and predictive value of urinary tract infection signs and symptoms in individuals with spinal cord injury on intermittent catheterization. *J Spinal Cord Med* 2009;**32**(5):568−73.

91. Brommer B, Engel O, Kopp MA, Watzlawick R, Muller S, Pruss H, et al. Spinal cord injury-induced immune deficiency syndrome enhances infection susceptibility dependent on lesion level. *Brain* 2016;**139**(Pt 3): 692−707.

92. Stillman MD, Barber J, Burns S, Williams S, Hoffman JM. Complications of spinal cord injury over the first year after discharge from inpatient rehabilitation. *Arch Phys Med Rehabil* 2017;**98**(9): 1800−5.

93. PVA. *Pressure ulcer prevention and treatment following spinal cord injury: clinical practice guidelines for health-care professionals.* 2nd ed. Washington, DC: Paralyzed Veterans of America; 2014.

94. PVA. *Bladder management for adults with spinal cord injury: : a clinical practice guideline for health-care professionals.* 2nd ed. Washington, DC: Paralyzed Veterans of America; 2006.

95. PVA. *Respiratory management following spinal cord injury: a clinical practice guideline for health-care professionals.* 2nd ed. Washington, DC: Paralyzed Veterans of America; 2005.

96. Pruss H, Tedeschi A, Thiriot A, Lynch L, Loughhead SM, Stutte S, et al. Spinal cord injury-induced immunodeficiency is mediated by a sympathetic-neuroendocrine adrenal reflex. *Nat Neurosci* 2017;**20**(11):1549−59.

97. Kopp MA, Watzlawick R, Martus P, Failli V, Finkenstaedt FW, Chen Y, et al. Long-term functional outcome in patients with acquired infections after acute spinal cord injury. *Neurology* 2017;**88**(9):892−900.

98. Krassioukov AV, Furlan JC, Fehlings MG. Autonomic dysreflexia in acute spinal cord injury: an under-recognized clinical entity. *J Neurotrauma* 2003;**20**(8): 707−16.

99. Furlan JC, Fehlings MG, Halliday W, Krassioukov AV. Autonomic dysreflexia associated with intramedullary astrocytoma of the spinal cord. *Lancet Oncol* 2003;**4**(9): 574−5.

100. Furlan JC, Robinson LR, Murray BJ. Clinical reasoning: stepwise paralysis in a patient with adenocarcinoma of lung. *Neurology* 2016;**86**(12):e122−7.

101. Furlan JC. Autonomic dysreflexia following acute myelitis due to neuromyelitis optica. *Mult Scler Relat Disord* 2018;**23**:1−3.

102. Bateman AM, Goldish GD. Autonomic dysreflexia in multiple sclerosis. *J Spinal Cord Med* 2002;**25**(1):40−2.

103. McGillivray CF, Hitzig SL, Craven BC, Tonack MI, Krassioukov AV. Evaluating knowledge of autonomic dysreflexia among individuals with spinal cord injury and their families. *J Spinal Cord Med* 2009;**32**(1):54−62.

104. Alexander MS, Biering-Sorensen F, Bodner D, Brackett NL, Cardenas D, Charlifue S, et al. International standards to document remaining autonomic function after spinal cord injury. *Spinal Cord* 2009;**47**(1):36−43.

105. Furlan JC. Autonomic dysreflexia: a clinical emergency. *J Trauma Acute Care Surg* 2013;**75**(3):496−500.

106. PVA. *Acute management of autonomic dysreflexia: individuals with spinal cord injury presenting to health-care facilities.* 2nd ed. Washington, DC: Paralyzed Veterans of America; 2011.

107. Brown JP, Bauman KA, Kurili A, Rodriguez GM, Chiodo AE, Sitrin RG, et al. Positive airway pressure therapy for sleep-disordered breathing confers short-term benefits to patients with spinal cord injury despite widely ranging patterns of use. *Spinal Cord* 2018;**56**(8): 777−89.

108. Neighbors CLP, Noller MW, Song SA, Zaghi S, Neighbors J, Feldman D, et al. Vitamin D and obstructive sleep apnea: a systematic review and meta-analysis. *Sleep Med* 2018;**43**:100—8.

109. Garbarino S, Bardwell WA, Guglielmi O, Chiorri C, Bonanni E, Magnavita N. Association of anxiety and depression in obstructive sleep apnea patients: a systematic review and meta-analysis. *Behav Sleep Med* 2018:1—23.

110. Craven C, Hitzig SL, Mittmann N. Impact of impairment and secondary health conditions on health preference among Canadians with chronic spinal cord injury. *J Spinal Cord Med* 2012;**35**(5):361—70.

111. Schultz SE, Kopec JA. Impact of chronic conditions. *Health Rep* 2003;**14**(4):41—53.

112. Warren SA, Turpin KVL, Pohar SL, Jones CA, Warren KG. Comorbidity and health-related quality of life in people with multiple sclerosis. *Int J MS Care* 2009;**11**:6—16.

113. Kleiner-Fisman G, Stern MB, Fisman DN. Health-related quality of life in Parkinson disease: correlation between health utilities Index III and unified Parkinson's disease rating Scale (UPDRS) in U.S. Male veterans. *Health Qual Life Outcome* 2010;**8**:91.

114. Mittmann N, Hitzig SL, Catharine Craven B. Predicting health preference in chronic spinal cord injury. *J Spinal Cord Med* 2014;**37**(5):548—55.

115. Hitzig SL, Romero Escobar EM, Noreau L, Craven BC. Validation of the reintegration to normal living index for community-dwelling persons with chronic spinal cord injury. *Arch Phys Med Rehabil* 2012;**93**(1):108—14.

116. Furlan JC, Fehlings MG. A Web-based systematic review on traumatic spinal cord injury comparing the "citation classics" with the consumers' perspectives. *J Neurotrauma* 2006;**23**(2):156—69.

117. Qi Z, Middleton JW, Malcolm A. Bowel dysfunction in spinal cord injury. *Curr Gastroenterol Rep* 2018;**20**(10):47.

118. PVA. *Neurogenic bowel management in adults with spinal cord injury*. 2nd ed. Whashington, DC: Paralyzed Veterans of America; 1998.

119. Furlan JC, Urbach DR, Fehlings MG. Optimal treatment for severe neurogenic bowel dysfunction after chronic spinal cord injury: a decision analysis. *Br J Surg* 2007;**94**(9):1139—50.

120. Stoffel JT, Van der Aa F, Wittmann D, Yande S, Elliott S. Neurogenic bowel management for the adult spinal cord injury patient. *World J Urol* 2018;**36**(10):1587—92.

121. Pavese C, Schneider MP, Schubert M, Curt A, Scivoletto G, Finazzi-Agro E, et al. Prediction of bladder outcomes after traumatic spinal cord injury: a longitudinal cohort study. *PLoS Med* 2016;**13**(6). e1002041.

122. Kreydin E, Welk B, Chung D, Clemens Q, Yang C, Danforth T, et al. Surveillance and management of urologic complications after spinal cord injury. *World J Urol* 2018;**36**(10):1545—53.

123. Ginsberg D. The epidemiology and pathophysiology of neurogenic bladder. *Am J Manag Care* 2013;**19**(Suppl. 10):s191—6.

124. Noreau L, Noonan VK, Cobb J, Leblond J, Dumont FS. Spinal cord injury community survey: a national, comprehensive study to portray the lives of canadians with spinal cord injury. *Top Spinal Cord Inj Rehabil* 2014;**20**(4):249—64.

125. Loh E, Guy SD, Mehta S, Moulin DE, Bryce TN, Middleton JW, et al. The CanPain SCI clinical practice guidelines for rehabilitation management of neuropathic pain after spinal cord: introduction, methodology and recommendation overview. *Spinal Cord* 2016;**54**(Suppl. 1):S1—6.

126. Chang E, Ghosh N, Yanni D, Lee S, Alexandru D, Mozaffar T. A review of spasticity treatments: pharmacological and interventional approaches. *Crit Rev Phys Rehabil Med* 2013;**25**(1—2):11—22.

127. Furlan JC, Fehlings MG, Shannon P, Norenberg MD, Krassioukov AV. Descending vasomotor pathways in humans: correlation between axonal preservation and cardiovascular dysfunction after spinal cord injury. *J Neurotrauma* 2003;**20**(12):1351—63.

128. Hector SM, Biering-Sorensen T, Krassioukov A, Biering-Sorensen F. Cardiac arrhythmias associated with spinal cord injury. *J Spinal Cord Med* 2013;**36**(6):591—9.

129. Furlan JC, Verocai F, Palmares X, Fehlings MG. Electrocardiographic abnormalities in the early stage following traumatic spinal cord injury. *Spinal Cord* 2016;**54**(10):872—7.

130. Migliorini CE, New PW, Tonge BJ. Comparison of depression, anxiety and stress in persons with traumatic and non-traumatic post-acute spinal cord injury. *Spinal Cord* 2009;**47**(11):783—8.

131. Migliorini C, Sinclair A, Brown D, Tonge B, New P. Prevalence of mood disturbance in Australian adults with chronic spinal cord injury. *Intern Med J* 2015;**45**(10):1014—9.

132. Ullrich PM, Smith BM, Blow FC, Valenstein M, Weaver FM. Depression, healthcare utilization, and comorbid psychiatric disorders after spinal cord injury. *J Spinal Cord Med* 2014;**37**(1):40—5.

133. Lim SW, Shiue YL, Ho CH, Yu SC, Kao PH, Wang JJ, et al. Anxiety and depression in patients with traumatic spinal cord injury: a nationwide population-based cohort study. *PLoS One* 2017;**12**(1):e0169623.

134. PVA. *Depression following spinal cord injury: a clinical practice guideline for primary care physicians*. 2nd ed. Washington, DC: Paralyzed Veterans of America; 1998.

135. Chiodo AE, Sitrin RG, Bauman KA. Sleep disordered breathing in spinal cord injury: a systematic review. *J Spinal Cord Med* 2016;**39**(4):374–82.

136. Fuller DD, Lee KZ, Tester NJ. The impact of spinal cord injury on breathing during sleep. *Respir Physiol Neurobiol* 2013;**188**(3):344–54.

137. Brady RD, Shultz SR, McDonald SJ, O'Brien TJ. Neurological heterotopic ossification: current understanding and future directions. *Bone* 2018;**109**:35–42.

138. Morse LR, Biering-Soerensen F, Carbone LD, Cervinka T, Cirnigliaro CM, Johnston TM, et al. Bone mineral density testing in spinal cord injury: the 2019 ISCD official positions. *J Clin Densitom* 2019;**22**(4):554–66.

139. Bauman WA, Cardozo CP. Osteoporosis in individuals with spinal cord injury. *Pharm Manag PM R* 2015;**7**(2):188–201.

140. Craven BC, Lynch CL, Eng JJ. Bone health following spinal cord injury. In: *Spinal cord injury rehabilitation evidence*. Vancouver: SCIRE. Version 5.0; 2014. p. 1–37.

141. Otom A, Al-Ahmar MR. Osteoporosis following spinal cord injury. *J Royal Me Ser* 2012;**19**(1):68–71.

142. Vestergaard P, Krogh K, Rejnmark L, Mosekilde L. Fracture rates and risk factors for fractures in patients with spinal cord injury. *Spinal Cord* 1998;**36**(11):790–6.

Economic impact of traumatic spinal cord injury

Julio C. Furlan[1,2,3,4,5,6], *Brian C.F. Chan*[2,6], *Vivien K.Y. Chan*[8], *Michael G. Fehlings*[4,7,8]

[1]Lyndhurst Centre, Toronto Rehabilitation Institute, University Health Network, Toronto, ON, Canada; [2]KITE Research Institute, University Health Network, Toronto, ON, Canada; [3]Department of Medicine, Division of Physical Medicine and Rehabilitation, Institute of Medical Science, University of Toronto, Toronto, ON, Canada; [4]Institute of Medical Science, University of Toronto, Toronto, ON, Canada; [5]Rehabilitation Sciences Institute, University of Toronto, Toronto, ON, Canada; [6]Institute of Health Policy, Management and Evaluation, University of Toronto, Toronto, ON, Canada; [7]Division of Neurosurgery, Toronto Western Hospital, University Health Network, Toronto, ON, Canada; [8]Division of Neurosurgery, Department of Surgery, University of Toronto, Toronto, ON, Canada

Introduction

Healthcare spending and the need for decision-making

Health care represents a sizable portion of government spending. Among the members of the Organisation for Economic Co-operation and Development (OECD), healthcare expenditures represent 8.8% of gross domestic product (GDP) of which the total healthcare expenditures are funded by the public payer varying from 5.7% (Slovenia) to 85.5% (Norway) with an average of 46%.[1] With limited government funding and greater scrutiny on public spending, healthcare systems worldwide are tasked with improving the health of the population on a limited budget. At the same time, there is a constant stream of new healthcare interventions entering the market promising better outcomes, improved efficiencies, or greater patient satisfaction. In the past 5 years, 219 new active substances were introduced by the pharmaceutical industry.[2] For medical devices, 51 first-time device approvals were granted, while another 3248 received clearance to enter the market (510(k)) from the Food and Drug Administration in the United States in 2017 alone.[3] Constraint to fixed budgets and a growing list of treatment options, healthcare decision-makers are pressure to determine whether new interventions represent good value for money and should be funded.

Economic evaluation of healthcare interventions utilizes a set of analytical methods that

© 2022 Elsevier Inc. All rights reserved.

set out to compare the costs and benefits of alternative uses of healthcare resources. In most cases, this is a comparison of a healthcare intervention to other treatment options. In the comparison of costs and benefits, the benefits include all improvements that are expected to occur related to the intervention both monetary and nonmonetary. On the other hand, costs relate to the benefits that are lost as a result of the decision to receive a treatment. Foundational to economic evaluations is the understanding that the decision to use resources to administer a certain healthcare intervention is a decision to forgo the benefits of the next best alternative use of those resources. This is also known as opportunity cost. Thus the economic evaluation assists the decision-maker in understanding the value of choosing an intervention as it relates to alternative approaches to treatment (most often another intervention). The intervention of interest is not studied in isolation but within the myriad of options that a patient is faced with as they are seeking a solution to their health condition. This explanation of the definition of benefit and cost is key in our understanding of the importance and usefulness of economic evaluations. However, as we move on toward the description of the methods of economic evaluation, cost is used, in these instances, to describe the monetary amount or equivalent that is provided to receive something in return.

The main types of economic evaluations include cost-minimization analysis, cost-effectiveness analysis, cost—utility analysis, and cost—benefit analysis. There are several key considerations that are important when conducting these analyses.

The perspective of the analysis is important in determining what healthcare resources are included. The researcher may be interested in the hospital perspective, for instance. In this analysis, only the costs and outcomes that impact the hospital would be considered. For instance, the cost for the operating room bed, laboratory test, inpatient pharmacy, and nursing/pharmacy/administrative staff may

be included, but the cost of patient parking and travel would be excluded. If the perspective was expanded to the public healthcare payer perspective, then the cost of healthcare services provided/reimbursed by the public payer may be included in the analysis. The most expansive is the societal perspective that includes the medical care—related costs borne by the patient and informal caregivers.

The timeframe in which the economic evaluation is based is an important factor in determining the costs and outcomes that will be included in an analysis. For example, is the economic evaluation examining the 30-days postdischarge, the first year after an individual initiates treatment, the first 5 years, or lifetime? The options for the study timeframe are endless. However, the recommended duration of the analysis is one that ensures that all the benefits associated with the intervention are accounted for. If the intervention only has an impact on shorter outcomes, a shorter timeframe may be acceptable.

The intervention of interest should be compared with alternative treatment options for the same condition in the same population. The alternative should represent the next best treatment option and reflect current medical practice. In many cases, multiple treatment options are available, and all options should be considered in the economic evaluation. This may include a "do-nothing" option.

Cost-minimization analysis

In the situation where an intervention of interest has the same clinical outcomes as the alternatives a cost-minimization analysis may be conducted to compare the costs of intervention to the cost of the alternatives. In this analysis, only the difference in costs between the intervention of interest and the alternative interventions is of importance and the results are presented as an incremental cost. However, in most situations, a new intervention will have some benefit that differentiates it from alternative treatments. In this case, a more comprehensive economic evaluation that includes a comparison of clinical outcomes is required.

Cost-effectiveness analysis

The cost-effectiveness analysis incorporates both the comparison of costs and clinical outcome in a single analysis. Incremental cost is calculated by comparing the cost of the intervention with the cost of the alternative; similar to the cost-minimization analysis. The clinical outcome is likewise compared between the intervention and the alternative. The choice of outcome to include in the cost-effectiveness analysis is up to the researcher's discretion. It is recommended that a well-recognized, clinically important outcome be selected to improve the relevance of the analysis. The results are presented as an incremental cost-effectiveness ratio with the incremental cost divided by the incremental effect and reported as an incremental cost per unit of outcome improvement. Unfortunately, a single disease area may have numerous important outcomes, and in this circumstance, it may be difficult to choose the most important outcome. Also, many clinical outcomes are important to a narrow range of disease areas. As such, many cost-effectiveness analyses will have limited comparability to similar analyses for other health conditions. This represents a significant limitation especially for healthcare administrators who need to make decisions on interventions in various disease areas. To address this limitation, a standard outcome that has broad application between different disease areas and addressed important outcomes in health was developed and accepted by practitioners of economic evaluations for health interventions. This outcome, the quality-adjusted life year (QALY), is incorporated into a category of economic evaluations called cost-utility analysis.

Cost-utility analysis

QALY is an outcome that measures the quantity of life experienced by an individual adjusted for the quality of that life. This outcome values interventions that improve the longevity of life or improves the quality of life of the individual. The benefit of an intervention is reduced if it improves life expectancy but results in a poor quality of life and vice versa. Quality of life is measured through health utilities values that evaluate preferences for health states in a scale between 0 and 1 with 1 representing "perfect health" and 0 representing "death". These values are often calculated through generic quality of life measures that capture a health state and have translated the state to a health utility value. The sum of the product of each health utility value by the duration of time spent in the corresponding health state represents the QALY. The primary outcome of interest in a cost-utility analysis is the incremental cost per QALY. In this outcome, the incremental cost of the intervention of interest less the cost of the comparator is divided by the incremental QALY of the intervention less the comparator. In many circumstances, the intervention of interest is a new treatment that exhibits improved outcomes but at a higher cost compared with other treatment options. For example, a new intervention may have an incremental cost per QALY of \$20,000/QALY. Whether this is considered cost-effective depends on the willingness-to-pay of the decision-maker for an additional improvement in QALY.

Cost–benefit analysis

In cost–benefit analysis, the clinical outcomes are converted to a monetary value so that both cost and effects are measured in monetary units. The conversion of healthcare benefits to monetary units is achieved through willingness-to-pay studies. By subtracting the benefits by the costs, the resulting value may be positive, meaning the benefits are greater than the cost in monetary value or the result may be negative meaning the costs are greater than the benefits.

In summary, economic evaluations are analyses that assist the healthcare decision-maker in determining the value for money of healthcare interventions. There are various types of economic evaluations that can help in this regard.

Economic burden of diseases

Global healthcare costs

Health services are costly to the society regardless of the model of the healthcare system.[4] The estimated global health spending increased from \$3.5 trillion USD in 1995 (or \$4.3 trillion in purchasing-power parity-adjusted USD that corresponds to 6.9% of GDP) to \$8.0 trillion USD in 2016 (or \$10.3 trillion in purchasing-power parity-adjusted USD that corresponds to 8.6% of global GDP).[5] This represents a health spending per capita of \$1077 USD and a health spending per GDP of 8.6% in 2016.[5] The increase in health spending over time is annualized at a rate of 4%, which corresponds to an annualized rate of change in health spending per GDP of 1.02% from 1995 to 2016.[5] The global healthcare costs are expected to continue to rise for different reasons including: (1) the burden of communicable diseases remain elevated in some parts of the world; (2) the prevalence of noncommunicable diseases (e.g., heart disease, cancer, chronic diseases such as obesity) continue increasing along with the aging of the population; and (3) the development of more sophisticated diagnostic tools and therapeutic options.[4]

The distribution of the health spending varies considerably among different jurisdictions in the world. The health spending per capita in high-income countries was 130.2 times greater than low-income countries in 2016.[5] The high-income countries were responsible for 81.0% of the global health spending in 2016, whereas upper middle—income countries were accountable for 15.7% followed by lower middle—income countries (3.0%) and low-income countries (0.4%), even though the latter included approximately 10% of the worldwide population.[5] The United States alone was accounted for 41.7% of the total health spending in the world while representing 4.4% of the global population.[5]

The economic burden of spinal cord injury

Personal healthcare spending in the United States summed \$2.1 trillion USD in 2013, with the top three most costly conditions being diabetes mellitus (\$101.4 billion USD), ischemic heart disease (\$88.1 billion USD), and low back and neck pain (\$87.6 billion USD).[6] Spinal cord injury (SCI) was presumably included in the group of neurological disorders other than cerebrovascular disorders, which comprised an overall personal healthcare spending of \$43.7 billion USD.[6]

The collaborators of the Global Burden of Disease Study 2016[7] recently analyzed the worldwide burden diseases that included 15 neurological disorder categories using disability-adjusted life years (DALYs) and mortality. Overall, neurological disorders were the leading cause of burden accounting for a total of 276 million DALYs (11.6% of global DALYs for all diseases) and second leading cause of mortality with 9.0 million deaths (16.5% of total global deaths) in 2016.[7] Globally, the four neurological disorders that contributed to the most DALYs were stroke (42.2%), migraine (16.3%), Alzheimer's and other dementias (10.4%), and meningitis (7.9%), whereas SCIs comprised 9.5 million DALYs that correspond to 3.4% of the overall DALYs related to neurological disorders.[7] Nonetheless, SCIs occupied the sixth place in the global ranking of age-standardized DALY rates for all neurological disorders after stroke, migraine, Alzheimer's disease and other dementias, meningitis, and epilepsy.[7]

Traumatic SCI is associated with substantial lifetime costs and annual costs in high-income countries. In the United States, the lifetime cost for a person injured at 25 years of age is estimated to be \$4.6 million USD for tetraplegia and \$2.3 million USD for paraplegia.[8] In Australia, the lifetime cost is estimated to be \$9.5 million AUD for tetraplegia and \$5.0 million AUD for paraplegia.[8] According to the Victorian Neurotrauma Initiative, the total annual cost for traumatic SCI in Australia is estimated to be $2.0 billion AUD.[8] In Canada, the lifetime cost is estimated to be $3.0 million CAD for tetraplegia and $1.5 million CAD for paraplegia with an overall estimated annual economic burden estimated to be $2.7 billion CAD in 2011.[9] Given the significant economic burden associated with traumatic SCI, much effort is necessary to better understand the factors that impact the costs of caring for individuals living with SCI.

Components of the costs of caring for individuals with spinal cord injury

Indirect costs may exceed direct costs associated with traumatic SCI.[8] Berkowitz et al. estimated that indirect costs account for 65% of total aggregate costs of traumatic SCI to society.[9] For instance, the estimated annual economic burden for individuals with SCI due to indirect costs in Canada was estimated to be $1.1 billion CAD in 2003.[10] However, it is difficult to accurately estimate the total indirect costs (e.g., lost income, fringe benefits, productivity, and leisure time) associated with traumatic SCI due to the numerous factors that influence this estimate such as age, injury level, severity of injury, education level, and environment.[9,11,12] As a consequence, most of the focus in prior research has been on direct costs.

Direct costs are highest in the first year due to hospitalization costs, rehabilitation costs, and cost of equipment, but the annual cost decreases over time after the first year.[8] Costs of hospitalization after the index event constitute a large component of the first-year expenses. The largest components of the total first-year cost are acute inpatient care following the index event and inpatient rehabilitation after the index event in the United States.[13,14] In Canada, where the total first-year cost for the province of Ontario in 2005 was estimated to be $23 million CAD, acute inpatient care and rehabilitation accounted for 30% ($6.6 million CAD) and 58% ($13.4 million CAD) of the cost, respectively.[15] Other costs

include emergency department visits ($36 720 CAD), home care ($473,302 CAD), physician visits ($590,291 CAD), and inpatient readmission ($844,676 CAD).[15] Therefore, hospitalization cost has been the focus of many studies, with the aim of identifying modifiable determinants of cost and length of stay.

The costs of caring for individuals with traumatic SCI are mainly influenced by many factors, including nature of initial injury, level and severity of injury, premorbid condition, timeliness of treatment, length of stay in hospital, rehospitalization, and secondary complications after SCI.[8,12,14,16–18] Prior studies reported higher cranial injury level and complete SCI to be associated with greater treatment costs due to various reasons including more severe cardiovascular dysfunction, respiratory complications, and need for mechanical ventilation support.[8,14,19,20] There is a high frequency of secondary complications in patients admitted for management of acute traumatic SCI (frequency range from 39% to 77%), which commonly cause prolonged length of stay and higher costs.[21–25] A large prospective registry study reported that mean time to discharge in patients with complications was 51.5 days compared with 23.5 days in those without complications.[26]

Current knowledge on cost-effective opportunities

Acute spine trauma: surgical treatment

Given the substantial economic impact of acute traumatic SCI, there has been an increasing demand for better understanding the opportunities and challenges to reach excellence in clinical practice tailored to optimization of relatively scarce resources. Healthcare professionals including clinicians and surgeons are also accountable to provide the best care for individuals with SCI by utilizing sufficient resources without waste and by choosing the most cost-effective alternatives for their management. With this background of demand for knowledge on health economics, a recent systematic review with scoping synthesis examined 11 original articles published between 1946 and 2017 that were focused on cost-effectiveness, cost—utility, cost—benefit, cost minimization, cost comparison, and economic analysis related to surgical management of acute spine trauma, including traumatic SCI.[27] The authors identified four cost—utility analyses, five cost analyses that compared the cost of intervention with a comparator, and two studies examining direct costs without a comparator.[27] None of those economic studies raised methodological issues when they were scrutinized using reporting checklists to evaluate their quality.[27] More recently, at least two more economic analyses on the surgical management of individuals with acute spine trauma were published in the literature.[28,29] Overall, those prior economic studies underline a few potentially cost-effective strategies in the surgical management of patients with acute spine trauma; however,

further economic analyses are desirable in several areas of the surgical care of individuals with acute traumatic SCI (Table 21.1).

Cost-effectiveness studies

Aras et al.[36] compared surgery to conservative management for thoracolumbar (T11—L2) incomplete burst fractures using a cost-effectiveness study under the perspective of a tax-funded state-run universal healthcare system in Denmark. Both study groups showed no differences in effectiveness, but the healthcare costs for the surgery group were $10,734 (in 2010 EUR) higher than those treated conservatively matched for age and sex.[36] The authors concluded that surgical treatment for thoracolumbar burst fractures was not cost-effective compared with conservative strategy even after the probabilistic sensitivity analysis.[36]

Barlow et al.[37] carried out a cost—utility analysis comparing surgical versus nonsurgical management of geriatric individuals with type II odontoid fractures that were stratified into the age groups of 65—74 years, 75—84 years, and

TABLE 21.1 Summary of the economic studies on surgical management of acute spine trauma.

References	Year	Jurisdiction	Spine disorder	Study type
Watts et al.[30]	1993	United States	Unstable SCI	Cost analysis
van der Roer et al.[31]	2005	Netherlands	Thoracolumbar spine fracture	Cost analysis
Siebenga et al.[32]	2007	Netherlands	Thoracolumbar spine fracture	Cost comparison
Boakye et al.[33]	2012	United States	Thoracolumbar spine fracture	Cost comparison
Nandyala et al.[34]	2013	United States	Cervical spine trauma	Cost comparison
Medress et al.[35]	2015	United States	Cervical fracture	Cost comparison
Aras et al.[36]	2016	Netherlands	Thoracolumbar burst fracture	Cost comparison
Barlow et al.[37]	2016	United States	Geriatric odontoid fracture	Cost-utility analysis
Furlan et al.[38]	2016	Canada	Cervical SCI in elderly	Cost-utility analysis
Furlan et al.[39]	2016	Canada	Cervical SCI	Cost-utility analysis
Lee et al.[40]	2017	South Korea	Thoracolumbar burst fracture	Cost-utility analysis
Richard-Denis et al.[28]	2017	Canada	Traumatic motor complete SCI	Cost comparison
Furlan et al.[29]	2019	Canada	Acute spine trauma in elderly	Cost comparison

SCI, spinal cord injury.

85 years or older at the SCI onset. This cost-effectiveness analysis was presumably undertaken under the perspective of multiple private insurers in the United States.[37] The authors concluded that surgical treatment was likely cost-effective for individuals between 65 and 84 years of age; nonetheless, nonsurgical management was considered the dominant approach when compared with surgical treatment for individuals older than 84 years, for a willingness-to-pay of $100,000 USD in 2013.[37]

Furlan et al.[38] performed a cost–utility comparing younger adults versus elderly individuals (65 years of age or older) with acute traumatic cervical SCI with respect to their initial surgical treatment and inpatient rehabilitation within the first 6 months after SCI onset. This cost-effectiveness analysis was carried out under the perspective of a publicly funded insurer in Canada.[38] Clinical and utility data were derived from a prior multicenter observational. The average cost for treatment of cervical SCI in the elderly group was $56,600 USD (in 2014) higher than that for the younger group and a mean utility difference of 0.01 lower in the elderly group, which resulted in an enormous incremental cost-effectiveness ratio of $5,655,557 USD per QALY gained when managing elderly patients compared with younger individuals with SCI.[38] The authors concluded that surgical management and rehabilitation of acute traumatic cervical SCI in the elderly are costlier but similarly effective when compared with younger adults.[38] The lack of age-related effectiveness of the management of individuals with SCI in their economic study is consistent with many prior clinical and histopathological studies that showed elderly individuals have similar potential to recover neurologically and functionally to younger adults after acute traumatic SCI.[41–44]

In another cost–utility study, Furlan et al.[39] examined the cost-effectiveness of timing of surgical decompression after acute traumatic SCI. The authors compared individuals who underwent early (within 24 hours from SCI onset)

with individuals who underwent delayed surgical decompression of the spinal cord after SCI.[39] This cost-effectiveness analysis included data of surgical management and rehabilitation within the first 6 months following SCI that were analyzed under the perspective of a publicly funded insurer in Canada.[39] For individuals with motor complete SCI, the incremental cost-effectiveness ratio analysis indicated a saving of $58,368,024 USD per QALY gained for early decompression group.[39] For motor incomplete SCI, the incremental cost-effectiveness ratio analysis showed a saving of $536,217 USD per QALY for the early spinal decompression group.[39] The authors concluded that early surgical decompression of the spinal cord decreases healthcare costs and may slightly increase the patient's quality of life.[39] Prior preclinical and clinical studies have mostly shown the benefits of early surgical decompression of the spinal cord for management of individuals with acute SCI, which led to this recommendation in recent guidelines.[45–47] Also, Furlan et al. performed a benchmarking analysis to define what factors could preclude the option for early decompression of the spinal cord.[48] The authors found that time in the general hospital and time of waiting for a surgical decision were the most important causes of delay of surgical decompression of the spinal cord.[48] The authors concluded that early surgical decompression is possible in the vast majority of the cases because health-related factors, not patient-related factors, are key determinants of the timing from SCI to spinal cord decompression.[48]

Lee et al. carried out a cost-utility analysis comparing individuals of whom the pedicle screw was removed after successful posterior instrumented fusion with individuals of whom the screw was kept after surgery for thoracolumbar burst fractures.[40] This cost-effectiveness analysis was undertaken under the perspective of a publicly funded insurer in South Korea.[40] The authors reported that removal of the pedicle screw resulted in an incremental cost-effectiveness ratio of $26,072 USD per QALY at

1 year and $13,125 USD per QALY at 2 years when compared with the patients who retained the pedicle screws.[40] The authors concluded that removal of the pedicle screws between 1 and 2 years after the initial surgical treatment of thoracolumbar burst fractures was more cost-effective than not removing the screws.[40]

Costing studies comparing treatment options

Using a propensity score method, Medress et al.[35] documented that initial hospital costs for individuals who underwent early surgery (within 72 hours of hospital admission) were significantly lower than for individuals who had delayed surgery ($63,065 and $77,049 USD in 2009, respectively) for management of traumatic cervical fractures in the United States. Nandyala et al.[34] reported that hospital costs for patients who were admitted on the weekend were significantly higher than for those who were admitted on a weekday (on average, $10,045 USD higher for those who underwent anterior cervical fusion, $10,227 USD higher for those who underwent posterior cervical fusion, and $11,301 USD higher for those who underwent anterior and posterior cervical fusion) in the United States. However, the study groups significantly differ from each other with regard to their mean age, sex, and number of preexisting medical comorbidities, which raises concerns about the validity of their results.[34]

Using a propensity score method, Boakye et al.[33] showed that hospital charges for individuals who underwent early fracture fixation surgery (within 72 hours of hospital admission) were $38,120 USD lower than hospital charges for individuals who had delayed surgery ($213,031 USD vs. $251,151 USD, respectively) for management of traumatic thoracolumbar fractures in the United States. Siebenga et al.[32] reported that the mean direct costs for the surgical treatment of individuals with traumatic thoracolumbar spine fractures ($21,960 USD) were significantly higher than those ones who were conservatively managed ($11,880 USD) in the Netherlands. Nonetheless, the costs associated with general practitioner visits, private expenditures, and absenteeism were significantly lower in the surgical group ($13 USD, $550 USD, and $6630 USD, respectively) than in the nonsurgical group ($34 USD, $816 USD, and $10,329 USD, respectively).[32]

Other costing studies

Watts et al.[30] analyzed the costs for supplies used in posterior fusion with sublaminar wiring for surgical treatment of individuals with acute traumatic, unstable cervical spine injury in the United States, which were estimated to be $2700 USD in 1993 for the Haid plate, $2640 USD for the Halifax clamp, and $5.94 USD for the wiring system.

Van der Roer et al.[31] reported that the mean total inpatient and outpatient costs for management of individuals with traumatic thoracolumbar spine fractures in the Netherlands were $19,700 EUR in 2001 for patients who underwent surgical treatment, and $12,500 EUR for those individuals who underwent conservative management.

Richard-Denis et al.[28] analyzed early transfer to a spine trauma care center prior to surgical treatment for acute traumatic motor complete SCI as a potential cost saving strategy under the perspective of a publicly funded, single insurer in Canada. The median costs for those patients who were transferred to a spine trauma center only after surgery significantly higher ($17,920 CAD) than those ones who were managed in a spine trauma care center promptly after the trauma SCI center ($10,522 CAD).[28]

In a recent cost analysis, Furlan et al.[29] compared younger adults with elderly individuals (65 years of age or older) with regard to their mean total hospital costs and utilization of inpatient services during the initial admission after acute spine trauma in Canada. The mean total hospital costs for initial admission after acute spine trauma in the elderly ($19,338 USD in 2017) were significantly higher than among younger adults ($13,775 USD). However, elderly individuals had significantly lower *per diem* total, fixed, direct, and indirect costs for acute spine trauma than younger individuals.[29] Both age groups utilized similar proportions of services during their initial admission in the acute care hospital.[29]

Acute spine trauma: rehabilitation

The main goals of rehabilitation protocols for management of individuals with traumatic SCI include minimizing secondary complications and maximizing restoration of the remaining function.[49] While minimization of secondary complications after SCI presumably reduces healthcare costs, maximization of restoration of the remaining function is more likely linked to increased initial healthcare costs due to lengthier regimens of motor-assisted therapies that can be associated with technology-based strategies. In the context of the increasing healthcare costs and limited resources, economic studies examining cost-effective rehabilitation strategies are a high priority. A systematic review from the Spinal Cord Injury Rehabilitation Evidence (SCIRE) project included six comparative economic studies on rehabilitation for SCI that were published between 2004 and 2013.[50] However, the study by Bensmail et al.[51] compared options for treatment of spasticity in

a heterogeneous group of individuals with different diseases of the central nervous system (e.g., SCI, highly dependent multiple sclerosis, traumatic brain injury, cerebral palsy, and stroke). Subsequently, at least four other economic studies on rehabilitation for SCI have been published[52–55] (Table 21.2).

Economic analysis on secondary complications after spinal cord injury

Kadyan et al.[56] performed a cost-effectiveness analysis of duplex ultrasound screening for deep vein thrombosis in individuals with SCI who were admitted to rehabilitation facilities in the United States. The incremental cost of admission duplex ultrasound was calculated to be $313 USD (in 2001–2002) per person when compared with no ultrasound-based screening at admission in a rehabilitation facility, which was associated with a decrease in mortality of 0.51%.[56] The cost for one life saved was estimated to be $61,542 USD, and the cost per life year gained varied from $1193 to $9050 USD.[56] The authors suggested that duplex ultrasound screening may be a cost-effective strategy for individuals with SCI who are admitted to a rehabilitation facility.[56] However, current standard of care includes pharmacological thromboprophylaxis for individuals with paralysis after

TABLE 21.2 Summary of the economic studies on rehabilitation for acute spine trauma.

References	Year	Jurisdiction	Focus	Study type
Kadyan et al.[56]	2004	United States	Deep venous thrombosis	Cost-effectiveness analysis
Mittmann et al.[57]	2005	Canada	Erectile dysfunction	Cost-utility analysis
Christensen et al.[58]	2009	United States	Neurogenic bowel	Cost comparison
Mittman et al.[59]	2011	Canada	Pressure sore	Cost-effectiveness analysis
Bermingham et al.[60]	2013	United States	Neurogenic bladder	Cost-effectiveness analysis
Miller et al.[53]	2016	United States	Physical activity	Cost comparison
Arora et al.[52]	2017	India and Bangladesh	Pressure sore	Cost-effectiveness analysis
Welk et al.[54]	2018	Canada	Neurogenic bladder	Cost-effectiveness analysis
Skelton et al.[55]	2019	United States	Neurogenic bladder	Cost analysis

SCI, spinal cord injury.

SCI during their admission in a rehabilitation facility and, according to the results of a subsequent systematic review and critical appraisal of the literature, "there is insufficient evidence to support (or refute) a recommendation for routine screening for deep venous thrombosis in adults with acute traumatic SCI under thromboprophylaxis."[61]

Mittmann et al.[57] carried out a cost-effectiveness analysis comparing oral treatment (i.e., sildenafil citrate) with nonoral therapies for erectile dysfunction after SCI that included: intracavernous injections of papaverine prostadil, alprostadil with papaverine, and phentolamine (triple mix); transurethral suppository; vacuum erection devices; and surgically implanted rigid, semirigid, or inflatable prosthetic devices. The cost-utility analysis was undertaken under the perspective of a publicly funded insurer in Canada over a period of time of 1 year hypothesizing that all options would be covered by the government healthcare plan.[57] The authors reported that sildenafil citrate was either a dominant strategy or a cost-effective treatment for erectile dysfunction in patients with SCI using willingness-to-pay threshold of $20,000 CAD.[57]

Christensen et al.[58] compared conservative bowel management with a transanal irrigation using a self-administered irrigation system for management of neurogenic bowel after SCI. This European, industry-funded, nonconventional cost-effectiveness analysis was undertaken under the societal perspective over a period of time of 10 weeks, but no incremental cost-effectiveness analysis was reported.[58] The authors concluded that transanal irrigation significantly reduced symptoms of neurogenic bowel dysfunction (as assessed using St Mark's fecal incontinence score, Cleveland Clinic constipation score, and neurogenic bowel dysfunction score) when compared with conservative bowel management, and the total costs of transanal irrigation ($38 EUR in 2007, for every 2 days of treatment) were lower than the costs of conservative bowel management ($39 EUR, for every 2 days).[58]

Mittman et al.[59] performed a cost-effectiveness analysis comparing electrical stimulation therapy for 3 months in addition to standard wound care with standard wound care alone for management of pressure ulcers of grade III or IV among individuals with SCI. This cost-effectiveness analysis was undertaken under the perspective of a publicly funded, single insurer in Canada over a period of time of 1 year, using a willingness-to-pay threshold of $50,000 CAD in 2009.[59] The authors reported that electrical stimulation therapy plus standard wound care was less costly ($29,549 vs. $ 29,773 CAD per year, respectively) and more effective (average overall pressure ulcers healed per year of 0.208 vs. 0.045) than standard wound care alone strategy with a decrement cost-effectiveness ratio of $1365.85 CAD per pressure ulcers healed per year, which means the combined approach was dominant.[59]

Bermingham et al.[60] carried out a cost-effectiveness analysis comparing gel reservoir catheters with clean noncoated catheters for management of neurogenic bladder after SCI. This cost-utility analysis was undertaken under the perspectives of the United Kingdom National Health Service and personal social services, for the lifetime horizon.[60] The authors reported that gel reservoir catheters were £28,369 more costly and resulted in an average gain of 0.522 QALYs per patient compared with clean noncoated catheters, which yielded an incremental cost-effectiveness ration of £54,350 per QALY gained exceeding the proposed willingness-to-pay threshold of £30,000.[60]

Welk et al.[54] compared the use of hydrophilic-coated intermittent catheters with the use of uncoated catheters for management of neurogenic bladder after SCI using a cost-effectiveness analysis for a case base of a 50-year-old individual with SCI. This cost-utility analysis was undertaken under the perspective of a publicly funded, single insurer in Canada, using a willingness-to-pay threshold range from $20,000 to $100,000 CAD per QALY for a lifetime horizon in the base case.[54] The authors reported that an incremental cost-effectiveness ratio of $66,634 CAD per QALY if individuals with SCI used hydrophilic-coated intermittent catheters instead of uncoated catheters for management of their neurogenic bladder, which means use of hydrophilic-coated intermittent catheters could be more effective than the use of uncoated catheters depending upon the willingness-to-pay threshold assumed.[54]

Those last two cost-effectiveness analyses on alternative methods of management of neurogenic bladder after SCI are particularly relevant in the context of the costs of this secondary complication. Skelton et al.[55] estimated that the overall annual costs associated with hospitalizations and visits to emergency departments in the United States related to genitourinary complication after SCI were approximately $4.3 billion USD from 2006 to 2015.

Arora et al.[52] performed a cost-effectiveness analysis that compared a telephone-based support with the usual care for the management of pressure ulcers in individuals with SCI in India and Bangladesh. Of note, individuals in the intervention group received weekly telephone calls (duration up to 25 min) over 12 weeks from a healthcare professional who was trained to emphasize self-help strategies relevant for handling pressure ulcers, mitigating psychological stress and improving engagement with life.[52] This cost-utility analysis was undertaken under the societal perspective over a period of time of 12 weeks using a willingness-to-pay threshold of $506 USD for each additional $1\,cm^2$ or threshold of less than three times per capita national gross domestic product (that was $1808 USD in India in 2015) per QALY gained.[52] The authors documented that the telephone-based support strategy was more cost-effective than the usual care of pressure ulcers in 87% of the times with an incremental cost-effectiveness

ratio of US$ 130 (in 2015) per additional cm^2 reduction in the size of the pressure ulcer, and $2523 USD per QALY gained.[52]

Costing analysis on rehabilitation modalities for restoration after spinal cord injury

Miller et al.[53] estimated the potential cost savings associated with routine physical activity within the first year after acute traumatic SCI based on previously reported data stating that approximately 363,000 individuals with SCI (or 65% of the entire SCI population in the United States) engage in insufficient physical activity. The authors documented that individuals who initiate routine physical activity within the first year after SCI and experience typical motor function improvements would save from $290,000 to $435,000 USD in lifetime costs, mainly due to fewer hospitalizations and less dependence on assistive care.[53]

Knowledge gaps and research opportunities

The incidence and prevalence of traumatic SCI remain relatively low; however, its economic burden is substantial to individuals living with SCI and to society.[62,63] To date, many opportunities for more cost-effective management of individuals with SCI in the acute care and rehabilitation settings have been identified. However, there is still a great need for research on knowledge generation and translation on potential cost-effective practices in the continuum for the management of individuals with traumatic SCI from prehospital care to outpatient prevention of secondary complications in the community. There is also lack of studies on cost-effective preventative interventions in the area of spine trauma that represents a great opportunity to reduce disability and healthcare costs. Overall, health economics remains an overt field for research that can also be impactful in the health care and society.

References

1. *OECD health statistics 2019 [internet]*. OECD; 2019 [cited September 23, 2019]. Available from: http://www.oecd.org/els/health-systems/health-data.htm.

2. Aitken M, Kleinrock M. *Medicine use and spending in the U.S.: a review of 2018 and outlook to 2023. Parsippany, NJ.* May 2019.

3. EvaluateMedTech. *World preview 2018, outlook to 2024.* Boston, MA: Evaluate Ltd.; September 2018.

4. WHO. *World Health Report 2010—health systems financing: the path to universal coverage.* WHO; 2010.

5. Global Burden of Disease Health Financing Collaborator Network. Past, present, and future of global health financing: a review of development assistance, government, out-of-pocket, and other private spending on health for 195 countries, 1995-2050. *Lancet* 2019;**393**(10187):2233–60.

6. Dieleman JL, Baral R, Birger M, Bui AL, Bulchis A, Chapin A, et al. US spending on personal health care and public health, 1996-2013. *JAMA* 2016;**316**(24):2627–46.

7. Collaborators GBDN. Global, regional, and national burden of neurological disorders, 1990-2016: a systematic analysis for the Global Burden of Disease Study 2016. *Lancet Neurol* 2019;**18**(5):459–80.

8. WHO. *International perspectives on spinal cord injury.* Geneva, Swtizerland: World Health Organization and International Spinal Cord Society; 2013.

9. Berkowitz M, O'Leary P, Kruse DL, Harvey C. *Spinal cord injury: an analysis of medical and social costs.* 1st ed. New York: Demos Medical Publishing; 1998. 188 pp.

10. Krueger H, Noonan VK, Trenaman LM, Joshi P, Rivers CS. The economic burden of traumatic spinal cord injury in Canada. *Chronic Dis Inj Can* 2013;**33**(3):113–22.

11. Berkowitz M, Harvey C, Greene CG, Wilson SE. *The economic consequences of traumatic spinal cord injury.* New York: Demos; 1992.

12. Devivo MJ, Chen Y, Mennemeyer ST, Deutsch Y. Costs of care following spinal cord injury. *Top Spinal Cord Inj Rehabil* 2011;**16**(4):1–9.

13. Radhakrishna M, Makriyianni I, Marcoux J, Zhang X. Effects of injury level and severity on direct costs of care for acute spinal cord injury. *Int J Rehabil Res* 2014;**37**(4):349–53.

14. Furlan JC, Gulasingam S, Craven BC. The Health Economics of the spinal cord injury or disease among veterans of war: a systematic review. *J Spinal Cord Med* 2017:1–16.

15. Munce SE, Wodchis WP, Guilcher SJ, Couris CM, Verrier M, Fung K, et al. Direct costs of adult traumatic spinal cord injury in Ontario. *Spinal Cord* 2013;**51**(1):64–9.

16. Dryden DM, Saunders LD, Jacobs P, Schopflocher DP, Rowe BH, May LA, et al. Direct health care costs after traumatic spinal cord injury. *J Trauma* 2005;**59**(2):464–7.

17. Harvey C, Wilson SE, Greene CG, Berkowitz M, Stripling TE. New estimates of the direct costs of traumatic spinal cord injuries: results of a nationwide survey. *Paraplegia* 1992;**30**(12):834–50.

18. Johnson RL, Brooks CA, Whiteneck GG. Cost of traumatic spinal cord injury in a population-based registry. *Spinal Cord* 1996;**34**(8):470–80.

19. Cooke CR. Economics of mechanical ventilation and respiratory failure. *Crit Care Clin* 2012;**28**(1). 39–55, [vi].

20. Tator CH, Duncan EG, Edmonds VE, Lapczak LI, Andrews DF. Complications and costs of management of acute spinal cord injury. *Paraplegia* 1993;**31**(11):700–14.

21. Aito S, Gruppo Italiano Studio Epidemiologico Mielolesioni GG. Complications during the acute phase of traumatic spinal cord lesions. *Spinal Cord* 2003;**41**(11):629–35.

22. Krassioukov AV, Furlan JC, Fehlings MG. Medical comorbidities, secondary complications, and mortality in elderly with acute spinal cord injury. *J Neurotrauma* 2003;**20**(4):391–9.

23. Dryden DM, Saunders LD, Rowe BH, May LA, Yiannakoulias N, Svenson LW, et al. Utilization of health services following spinal cord injury: a 6-year follow-up study. *Spinal Cord* 2004;**42**(9):513–25.

24. Fletcher DJ, Taddonio RF, Byrne DW, Wexler LM, Cayten CG, Nealon SM, et al. Incidence of acute care complications in vertebral column fracture patients with and without spinal cord injury. *Spine* 1995;**20**(10):1136–46.

25. New PW, Jackson T. The costs and adverse events associated with hospitalization of patients with spinal cord injury in Victoria, Australia. *Spine* 2010;**35**(7):796–802.

26. Wilson JR, Arnold PM, Singh A, Kalsi-Ryan S, Fehlings MG. Clinical prediction model for acute inpatient complications after traumatic cervical spinal cord injury: a subanalysis from the Surgical Timing in Acute Spinal Cord Injury Study. *J Neurosurg Spine* 2012;**17**(Suppl. 1):46–51.

27. Chan BCF, Craven BC, Furlan JC. A scoping review on health economics in neurosurgery for acute spine trauma. *Neurosurg Focus* 2018;**44**(5):E15.

28. Richard-Denis A, Ehrmann Feldman D, Thompson C, Bourassa-Moreau E, Mac-Thiong JM. Costs and length of stay for the acute care of patients with motor-complete spinal cord injury following cervical trauma: the impact of early transfer to specialized acute SCI center. *Am J Phys Med Rehabil* 2017;**96**(7):449–56.

29. Furlan JC, Fehlings MG, Craven BC. Economic impact of aging on the initial spine care of patients with acute spine trauma: from bedside to teller. *Neurosurgery* 2019;**84**(6):1251–60.

30. Watts C, Smith H, Knoller N. Risks and cost-effectiveness of sublaminar wiring in posterior fusion of cervical spine trauma. *Surg Neurol* 1993;**40**(6):457−60.

31. van der Roer N, de Bruyne MC, Bakker FC, van Tulder MW, Boers M. Direct medical costs of traumatic thoracolumbar spine fractures. *Acta Orthop* 2005;**76**(5): 662−6.

32. Siebenga J, Segers MJ, Leferink VJ, Elzinga MJ, Bakker FC, Duis HJ, et al. Cost-effectiveness of the treatment of traumatic thoracolumbar spine fractures: nonsurgical or surgical therapy? *Indian J Orthop* 2007; **41**(4):332−6.

33. Boakye M, Arrigo RT, Hayden Gephart MG, Zygourakis CC, Lad S. Retrospective, propensity score-matched cohort study examining timing of fracture fixation for traumatic thoracolumbar fractures. *J Neurotrauma* 2012;**29**(12):2220−5.

34. Nandyala SV, Marquez-Lara A, Fineberg SJ, Schmitt DR, Singh K. Comparison of perioperative outcomes and cost of spinal fusion for cervical trauma: weekday versus weekend admissions. *Spine* 2013; **38**(25):2178−83.

35. Medress Z, Arrigo RT, Hayden Gephart M, Zygourakis CC, Boakye M. Cervical fracture stabilization within 72 hours of injury is associated with decreased hospitalization costs with comparable perioperative outcomes in a propensity score-matched cohort. *Cureus* 2015;**7**(1):e244.

36. Aras EL, Bunger C, Hansen ES, Sogaard R. Cost-effectiveness of surgical versus conservative treatment for thoracolumbar burst fractures. *Spine* 2016;**41**(4): 337−43.

37. Barlow DR, Higgins BT, Ozanne EM, Tosteson AN, Pearson AM. Cost effectiveness of operative versus non-operative treatment of geriatric type-II odontoid fracture. *Spine* 2016;**41**(7):610−7.

38. Furlan JC, Craven BC, Fehlings MG. Surgical management of the elderly with traumatic cervical spinal cord injury: a cost-utility analysis. *Neurosurgery* 2016;**79**(3): 418−25.

39. Furlan JC, Craven BC, Massicotte EM, Fehlings MG. Early versus delayed surgical decompression of spinal cord after traumatic cervical spinal cord injury: a cost-utility analysis. *World Neurosurg* 2016;**88**:166−74.

40. Lee HD, Jeon CH, Chung NS, Seo YW. Cost-utility analysis of pedicle screw removal after successful posterior instrumented fusion in thoracolumbar burst fractures. *Spine* 2017;**42**(15):E926−32.

41. Furlan JC, Hitzig SL, Craven BC. The influence of age on functional recovery of adults with spinal cord injury or disease after inpatient rehabilitative care: a pilot study. *Aging Clin Exp Res* 2013;**25**(4):463−71.

42. Furlan JC, Bracken MB, Fehlings MG. Is age a key determinant of mortality and neurological outcome after acute traumatic spinal cord injury? *Neurobiol Aging* 2010;**31**(3):434−46.

43. Furlan JC, Fehlings MG. The impact of age on mortality, impairment, and disability among adults with acute traumatic spinal cord injury. *J Neurotrauma* 2009; **26**(10):1707−17.

44. Furlan JC, Kattail D, Fehlings MG. The impact of co-morbidities on age-related differences in mortality after acute traumatic spinal cord injury. *J Neurotrauma* 2009; **26**(8):1361−7.

45. Furlan JC, Noonan V, Cadotte DW, Fehlings MG. Timing of decompressive surgery of spinal cord after traumatic spinal cord injury: an evidence-based examination of pre-clinical and clinical studies. *J Neurotrauma* 2011;**28**(8):1371−99.

46. Fehlings MG, Vaccaro A, Wilson JR, Singh A, Cadotte D, Harrop JS, et al. Early versus delayed decompression for traumatic cervical spinal cord injury: results of the surgical timing in acute spinal cord injury study (STASCIS). *PLoS One* 2012;**7**(2):1−8.

47. Fehlings MG, Tetreault LA, Wilson JR, Aarabi B, Anderson P, Arnold PM, et al. A clinical practice guideline for the management of patients with acute spinal cord injury and central cord syndrome: recommendations on the timing (≤24 hours versus >24 hours) of decompressive surgery. *Global Spine J* 2017;**7**(Suppl. 3): 195S−202S.

48. Furlan JC, Tung K, Fehlings MG. Process benchmarking appraisal of surgical decompression of spinal cord following traumatic cervical spinal cord injury: opportunities to reduce delays in surgical management. *J Neurotrauma* 2013;**30**(6):487−91.

49. Ramer LM, Ramer MS, Bradbury EJ. Restoring function after spinal cord injury: towards clinical translation of experimental strategies. *Lancet Neurol* 2014;**13**(12): 1241−56.

50. Chan B, McIntyre A, Mittmann N, Teasell RW, Wolfe DL. Economic evaluation of spinal cord injury. In: *Spinal cord injury rehabilitation evidence [internet]. Vancouver, BC: SCIRE. Version 5.0*; 2014. p. 1−21. Available from: www.scireproject.com.

51. Bensmail D, Ward AB, Wissel J, Motta F, Saltuari L, Lissens J, et al. Cost-effectiveness modeling of intrathecal baclofen therapy versus other interventions for disabling spasticity. *Neurorehabil Neural Repair* 2009; **23**(6):546−52.

52. Arora M, Harvey LA, Glinsky JV, Chhabra HS, Hossain MS, Arumugam N, et al. Cost-effectiveness analysis of telephone-based support for the management of pressure ulcers in people with spinal cord injury in India and Bangladesh. *Spinal Cord* 2017;**55**(12):1071−8.

53. Miller LE, Herbert WG. Health and economic benefits of physical activity for patients with spinal cord injury. *Clinicoecon Outcomes Res* 2016;**8**:551—8.

54. Welk B, Isaranuwatchai W, Krassioukov A, Husted Torp L, Elterman D. Cost-effectiveness of hydrophilic-coated intermittent catheters compared with uncoated catheters in Canada: a public payer perspective. *J Med Econ* 2018;**21**(7):639—48.

55. Skelton F, Salemi JL, Akpati L, Silva S, Dongarwar D, Trautner BW, et al. Genitourinary complications are a leading and expensive cause of emergency department and inpatient encounters for persons with spinal cord injury. *Arch Phys Med Rehabil* 2019;**100**(9):1614—21.

56. Kadyan V, Clinchot DM, Colachis SC. Cost-effectiveness of duplex ultrasound surveillance in spinal cord injury. *Am J Phys Med Rehabil* 2004;**83**(3):191—7.

57. Mittmann N, Craven BC, Gordon M, MacMillan DH, Hassouna M, Raynard W, et al. Erectile dysfunction in spinal cord injury: a cost-utility analysis. *J Rehabil Med* 2005;**37**(6):358—64.

58. Christensen P, Andreasen J, Ehlers L. Cost-effectiveness of transanal irrigation versus conservative bowel management for spinal cord injury patients. *Spinal Cord* 2009;**47**(2):138—43.

59. Mittmann N, Chan BC, Craven BC, Isogai PK, Houghton P. Evaluation of the cost-effectiveness of electrical stimulation therapy for pressure ulcers in spinal cord injury. *Arch Phys Med Rehabil* 2011;**92**(6):866—72.

60. Bermingham SL, Hodgkinson S, Wright S, Hayter E, Spinks J, Pellowe C. Intermittent self catheterisation with hydrophilic, gel reservoir, and non-coated catheters: a systematic review and cost effectiveness analysis. *BMJ* 2013;**346**:e8639.

61. Furlan JC, Fehlings MG. Role of screening tests for deep venous thrombosis in asymptomatic adults with acute spinal cord injury: an evidence-based analysis. *Spine* 2007;**32**(17):1908—16.

62. Furlan JC, Sakakibara BM, Miller WC, Krassioukov AV. Global incidence and prevalence of traumatic spinal cord injury. *Can J Neurol Sci (Le journal canadien des sciences neurologiques)* 2013;**40**(4):456—64.

63. Furlan JC, Tator CH. Global epidemiology of traumatic spinal cord injury. In: Morganti-Kossman C, Raghupathi R, Maas A, editors. *Traumatic brain and spinal cord injury: challenges and Developments.* 1st ed. Cambridge: Cambridge University Press; 2012. p. 360.

Complications and adverse events following traumatic spinal cord injury

Zaid Salaheen[1], Nader Hejrati[2], Ian H.Y. Wong[3], Fan Jiang[4], Michael G. Fehlings[1,2,3,4]

[1]Temerty Faculty of Medicine, University of Toronto, Toronto, ON, Canada; [2]Division of Genetics and Development, Krembil Brain Institute, University Health Network, Toronto, ON, Canada; [3]Division of Neurosurgery, Department of Surgery, University of Toronto, Toronto, ON, Canada; [4]Division of Neurosurgery, Toronto Western Hospital, University Health Network, Toronto, ON, Canada

Introduction

Traumatic spinal cord injury (tSCI) is a devastating event with the potential to profoundly impair patients' sensory, motor, and autonomic functions. In addition to the primary injury, tSCI patients also suffer from a range of complications which impact patient outcome and quality of life.[1,2] It is important for clinicians to be aware of the complications likely to arise following tSCI, in order to implement strategies that reduce the burden these adverse events (AEs) pose to patients' overall health. With the prevalence of tSCI predicted to increase as global populations age and more elderly patients are susceptible to fall-related tSCI, it has become even more crucial for physicians to be cognizant of the AEs post-tSCI.[3]

This chapter will describe the different types of complications in tSCI, how they impact patient outcomes, and conclude with a discussion on the role of prediction models in evaluating patients' risk for developing complications following tSCI.

Incidence and prevalence of complications in tSCI

It is well known among clinicians that tSCI is a devastating condition that may cause severe medical complications.[2,4–7] Such complications often result in significant morbidity and mortality that leads to impairment and reduced quality of life.[8,9] Complications occur from the primary injury mechanism, characterized by the direct destruction of neuronal cells and axonal tracts, and the secondary injury mechanism, which occurs as a result of cord bleeding and edema leading to hypoperfusion, ischemia, inflammatory

Neural Repair and Regeneration after Spinal Cord Injury and Spine Trauma
https://doi.org/10.1016/B978-0-12-819835-3.00002-2

385

© 2022 Elsevier Inc. All rights reserved.

changes, and demyelination.[10] Both injury mechanisms have an accumulated effect on the severity of the tSCI, and ultimately the patient outcome.

Aito et al.[4] performed a prospective 2-year survey that assessed 588 patients admitted to 37 spinal units and rehabilitation centers within 60 days from the traumatic injury. Out of 588 patients with American Spinal Injury Association Impairment Scale (AIS) grades A–E, 82% were male and 18% were female. Overall high rates of complications were recorded following tSCI, reportedly 25.5% of the patients developed at least one event on admission to the spine unit, 39.9% on admission to the rehabilitation center, and 25% on admission to rehabilitation service with no beds. The most common complication found in the authors' study was trophic skin changes, which occurred in 15.5% and 29.6% of patients admitted to spine units and rehabilitation centers, respectively. This was followed by heterotopic ossification (6.9%, 1.3%), urinary complications (2.1%, 8.7%), respiratory complications (9.4%, 12.5%), pulmonary embolism (PE) (0.4%, 1.6%), and deep vein thrombosis (DVT) (1.3%, 5.1%).

Grossman et al.[5] performed a prospective review of the spinal cord injury registry of the North American Clinical Trials Network (NACTN), which included 315 consecutive patients with AIS grades A–D presenting at nine university-affiliated hospitals, to determine the spectrum and incidence of complications in tSCIs. Out of the 315 patients, 182 (57.8%) developed at least one complication and 145 (46%) developed more than one complication. In their study, the most reported complications post-tSCI are pulmonary (38.4%), infectious (35.6%), hematological (23.2%), cardiac (23.8%), gastrointestinal/genitourinary (16.8%), skin (15.6%), and neuropsychiatric (18.1%). Distribution of complications by severity included severe (11.2%), moderate (70.9%), and mild (17.9%). Pulmonary complications included respiratory failure, pleural effusion, acute respiratory distress/acute lung injury, and pulmonary embolus. The

cardiac system included bradycardia, cardiac arrest, shock, and myocardial infarction. In the infectious category, pneumonia and wound infections were listed. Gastrointestinal and genitourinary complications included severe ileus and acute renal failure. Skin issues included sacral and operative wound problems. Neuropsychiatric complications included cognitive deterioration and seizure. Finally, hematological issues included anemia and thrombocytopenia.

Jiang et al.[11] performed an analysis of the updated NACTN database, which was expanded to include 11 university-affiliated North American neurosurgery departments and a total of 801 patients with AIS grades A–E. In their study, 502 (63%) patients experienced at least one complication during the acute admission period following tSCI and 352 (44.4%) experienced more than one. Similar to the findings of Grossman et al., the most commonly detected complication was of pulmonary nature (25.1%), followed by infections (19.2%), hematological (17.5%), cardiac (14.5%), gastrourinary (9.6%), skin (7.7%), and neuropsychiatric complications (6.5%). Differences in the rate of AEs among studies can be partially attributed to the methods used to collect data. For example, Street et al. used the Spine Adverse Events Severity System (SAVES) and reported a higher rate of AEs (77%) following tSCI compared to previous studies.[12] Another study by Street et al. compared the SAVES system and ICD-10 coding to identify AEs post-tSCI and found the SAVES to capture twice as many AEs per patient as compared to ICD-10 coding.[13] Hence, it is important to be cognizant of the methods used when comparing rates of AEs among studies. Therefore, patients who sustain a tSCI are at high risk for developing complications during the postinjury period. The goal of the health-care team is to reduce complications associated with the initial primary injury with early surgical stabilization methods[14,15] and to anticipate or minimize secondary complications that arise from the trauma that occurred. Leading reported complications of tSCI include pneumonia, respiratory failure, anemia, and trophic skin changes.[4,5,11] However, the spectrum of complications post-tSCI varies by its level, severity, and duration postinjury, and is likely to affect multiple organ systems, such as pulmonary, hematological, and cardiac systems.[5,11]

Spectrum of complications in traumatic spinal cord injury

As mentioned previously, profound patho-physiologic changes occur in tSCI, often manifesting as complications affecting various organ systems (Table 22.1). These pathological events can be classified into different phases: an early acute phase (<48 hours postinjury), a secondary subacute phase (<14 days), an intermediate phase (<6 months), and a chronic phase (>6 months).[16] As a result of the dynamic pathophysiology of tSCI, the characteristics of complications following injury change over time. The present section will discuss these complications and how their features present at different time points postinjury.

Cardiovascular complications

Patients with tSCIs in the upper thoracic (above the sixth thoracic vertebrae) level can have disruption of descending pathways to the sympathetic trunk, creating an imbalance in the autonomic nervous system (ANS) from unopposed parasympathetic innervation.[17] This can lead to neurogenic shock, which is characterized by signs and symptoms including hypotension, bradyarrhythmia, and impaired temperature regulation.[17] 19%, 7%, and 3% of patients sustaining a cervical, thoracic, and lumbar SCI, respectively, have been reported to experience a neurogenic shock.[18] This complication can present within minutes to hours postinjury and last up to 5 weeks.[17,19]

Also, the disruption of the ANS following tSCI can lead to cardiac arrhythmias.[20] This includes bradyarrhythmias (atrioventricular blocks, asystole, sinus bradycardia) and

tachyarrhythmias (sinus tachycardia, atrial flutter, atrial fibrillation).[19,20] These complications are most pronounced 2–6 weeks postinjury and the risk of arrythmias decreases with time postinjury.[19,20] However, patients suffering from complete cervical tSCI appear to have a long-term risk of recurrent asystole.[21] Damage to the autonomic control in tSCI can also lead to orthostatic hypotension. This complication often improves within the first month post-tSCI because of compensatory mechanisms, such as increased muscle tone in the lower limbs.[22,23] However, orthostatic hypotension may persist beyond the first month in patients with motor complete cervical and upper thoracic SCI.[22] Additionally, patients suffering from tSCI are at increased risk of DVT and PE because of hypercoagulability, stasis, and endothelial cell injury, also known as Virchow's triad.[24] Following tSCI, 15% of patients will eventually develop a DVT and 5% a PE.[25] The risk of a patient developing DVT is greatest 7–10 days postinjury and decreases to levels seen in the general population within 3 months.[23,26] Additionally, damage to the ANS following tSCI can lead to autonomic dysreflexia (AD), a medical emergency that can be caused by the inappropriate response of the sympathetic nervous system to noxious stimuli.[1] These stimuli include bladder distention, bone fracture, fecal impaction, and pressure sores.[27] In AD, there is vasoconstriction and hypertension below the level of injury, followed by a compensatory response by the parasympathetic nervous system. This causes vasodilation above the level of injury, leading to headache, flushing, diaphoresis, nasal congestion, and nausea.[1] The prevalence of AD ranges from 20% to 70%, with rates of over 90% when the lesion is above T6.[27,28] AD generally presents 3–6 months post-SCI; however, it can present at any time postinjury.[23,29]

Cardiometabolic syndrome (CMS) is a significant long-term complication of tSCI, characterized by abdominal obesity, insulin-resistant glucose metabolism, dyslipidemia, and hypertension.[30,31] Based on the definitions used, CMS prevalence in SCI varies between 13% and 38%.[30,32] The conditions comprising this syndrome contribute to SCI patients having a 2- and 3-fold increase in the odds of heart disease and stroke compared to those without SCI, respectively.[33]

Respiratory complications

Respiratory complications are a significant consideration in the care of tSCI patients as they represent the most common acute systemic AE in this population.[34] These patients can have impaired innervation of the diaphragm and intercostal muscles, contributing to impaired cough effectiveness, reduced vital capacity, and decreased lung and chest wall compliance.[34] These changes in the physiologic parameters of the respiratory system make tSCI patients susceptible to atelectasis, hypersecretion of bronchial mucus, bronchospasm, pulmonary edema, pneumonia, pulmonary thromboembolism, and respiratory failure.[35] These complications contribute significantly to the patient's length of stay and hospital costs.[36] Approximately 67% of acute SCI patients suffer from respiratory complications in the days following their injury (atelectasis 36.4%, pneumonia 31.4%, ventilatory failure 22.6%).[37] Although these complications are commonly seen in the days postinjury, there is a long-term risk of pneumonia and atelectasis in tSCI patients.[8] In particular, pneumonia remains one of the leading causes of death in chronic SCI.[38,39] Long-term ventilation following tSCI may be required in some individuals, where prevalence is reported to range from 6% to 8%.[40–42]

Pressure ulcers

Paralysis and sensory impairment following tSCI can predispose patients to pressure ulcers. They are defined as a soft-tissue injury due to unrelieved pressure over a bony prominence, resulting in ischemia, cell death, and tissue necrosis.[43] Patients with tSCI can develop pressure ulcers within hours postinjury as they are immobilized by the health-care team and face a long-term risk of this complication developing due to the lasting effects of SCI on mobility.[8,44] Over one in five patients with a SCI will suffer from a pressure ulcer, which vary in severity from grade 1 (nonblanchable erythema of intact skin) to grade 4 (full-thickness skin and tissue loss).[45,46] Infection of grade 4 pressure ulcers is particularly dangerous as it may lead to overwhelming sepsis.[43] The most common locations of pressure ulcers in order of decreasing prevalence are the ischium (31%), trochanter (26%), sacrum (18%), heel (5%), malleolus (4%), and feet (2%).[8] Patients with SCI and pressure ulcers not only experience a prolonged length of stay for the initial hospitalization, but also experience a disproportionate number of rehospitalization days compared to SCI patients without pressure ulcers.[13,47,48] There are also significant financial costs associated with this complication to the health-care systems. For example, in Ontario, Canada, the mean lifetime cost per person with SCI is $336,100 versus $479,600 for SCI patients with pressure ulcers at initial hospitalization.[49]

Gastrointestinal complications

Patients suffering from tSCI can suffer from a range of gastrointestinal complications including constipation, distention, rectal bleeding, abdominal pain, hemorrhoids, and incontinence.[50] In the 4 weeks following injury, 4.7% of patients have been reported to suffer from acute abdominal pathology, with pancreatitis and upper GI hemorrhage occurring as early as 3 days post-SCI.[51] Additionally, 3.5% of patients with tSCI have been reported to suffer from acute gastroduodenal ulceration and hemorrhage within the first 4 weeks post-tSCI.[52] Dysfunction of the GI tract in tSCI due to damage of nerves innervating the colon is known as neurogenic bowel, which can be classified into two categories: upper motor neuron (UMN) and lower motor neuron (LMN) bowel syndrome. UMN bowel syndrome occurs when there is an injury above the conus medullaris. This leads to hyperreflexia of the pelvic muscles and impairs voluntary control of the external anal sphincter, ultimately resulting in constipation and fecal retention.[1] In LMN bowel syndrome, there is injury to parasympathetic ganglia at the conus medullaris, cauda equina, or pelvic nerve.[1] This contributes to reduced motility of the gastrointestinal tract and decreased external anal sphincter tone, overall leading to constipation and incontinence.[50] Although the risk of serious abdominal complications (pancreatitis, appendicitis, GI hemorrhage) is greatest in the first few months postinjury, the effect of neurogenic bowel on a patient's quality of life will persist throughout their lifetime.[53]

Genitourinary complications

In the acute phase post-tSCI, the bladder and sphincter are hypotonic.[54] Following this, bladder dysfunction can be classified as either a UMN neurogenic bladder or an LMN neurogenic bladder. In UMN neurogenic bladder, there is detrusor-sphincter dyssynergia, defined by detrusor and urinary sphincter hyperactivity. This contributes to increased bladder pressure and vesicoureteral reflux that can result in kidney damage.[55] In LMN neurogenic bladder, there is detrusor areflexia, which can lead to urinary retention contributing to overflow incontinence and urinary tract infections (UTIs).[55] Patients are also at long-term risk for renal calculi, pyelonephritis, UTIs, and renal insufficiency as a result of chronic indwelling catheters and vesicoureteral reflux.[8,56,57] In particular, UTIs are common following SCI, with a rate of 2.5 episodes per patient per year.[58] Additionally, they represent the most frequent cause of septicemia and hospital readmission in SCI.[48,59,60] There are also significant financial costs associated with this complication. For example, in British Columbia, Canada, the presence of a UTI increased the average cost of acute SCI admission from $15,276 to $23,066.[61]

Depending on the degree and the level of injury, tSCI can substantially affect sexual function. The sexual response of arousal in men (that is, penile erection) and women (that is, vaginal lubrication and clitoral swelling) can occur in two different ways: first, reflexive arousal through sacral stimulation and a parasympathetic neurologic pathway, and second, psychogenic under control of the hypogastric plexus originating at T11–L2. Hence patients with UMN lesions above the level of T11 are still able to experience reflexive, but not psychogenic arousals, whereas patients with complete LMN injuries are capable of experiencing psychogenic but not reflexive arousals through sacral stimulation.[62,63] Approximately 95% of males with SCI face persistent ejaculatory problems, with about 80% regaining some erectile function 2 years postinjury.[64] Women also experience sexual dysfunction, with approximately 59%–87% of those with SCI experiencing this complication.[65]

Bone-related complications

Heterotopic ossification is a known complication of tSCI and pertains to the formation of extraosseous bone in soft tissue surrounding joints.[66] Depending on the study, the incidence varies between 10% and 53%.[66] This complication occurs below the level of neurologic injury and most commonly affects the hip and knee.[67] This can lead to a reduction in a patient's range of motion and possibly lead to ankylosis.[67] The symptoms of heterotopic ossification may present 3–12 weeks postinjury.[68]

Osteoporosis is another complication of tSCI. Bone loss is most rapid in the 12- to 18-month period postinjury; however, a new steady state between bone resorption and formation has been reported about 2 years post-SCI.[1,69] This bone loss puts patients at greater risk of fractures and calcium nephrolithiasis.[69] Long bone fractures are a prominent secondary complication in tSCI patients.[70] Loss of bone mineral density from paralysis of the lower limbs increases the likelihood of fracture due to fragility. Most common fracture locations include the diaphyseal or distal femur and proximal lower leg.[71,72] Zehnder et al. reported in a cohort of 100 male patients that the average time from SCI to first bone fracture was 8.9 years, while incidence per year is 2.2%.[73] Women are 1.6 times more likely to fracture a bone compared to men after sustaining tSCI, and lesions at the lumbar level have higher fracture rates compared to those at the cervical level.[74] To prevent fracture events, rehabilitation focused on independent mobility should be of utmost importance.[75]

Spasticity

Immediately postinjury, tSCI patients suffer from spinal shock, which is characterized by the loss of tendon reflexes below the level of injury, muscle paralysis, and reduced muscle tone.[76] Spasticity then presents months after injury as a velocity-dependent increase in the stretch reflex, manifesting as hyperreflexia, muscle spasms, and clonus.[1,76] Functional impairment due to spasticity is reported by 27%, 25%, and 20% of patients with a tSCI, at 1, 2, and 5 years postinjury, respectively.[77] In particular, those with AIS grade C injures have the greatest prevalence of long-term spasticity treatment and generally have poorer functional outcomes compared to other tSCI patients.[77]

Syringomyelia

The term syringomyelia describes an abnormal fluid-filled cavity within the spinal cord that can span over multiple segments and occurs months to years following tSCI. The fluid-filled cavity, also known as the syrinx, is distinct from the more common posttraumatic injury cavity, which is localized to the injury site and arises as a consequence of posttraumatic necrotic and apoptotic tissue loss.[78] While the exact mechanism of a syrinx formation remains poorly understood, disturbed cerebrospinal fluid flow seems to be key to the development of syringomyelia.[78,79] Progressive fluid accumulation increases the pressure on the spinal cord putting it at risk for progressive myelopathy. While 21%−28% of patients suffering from tSCI

have been found to show a syrinx, only 1%−9% of the tSCI population will eventually suffer from symptomatic syringomyelia.[78] Common initial symptoms include pain and sensory loss, while further progression can potentially lead to motor weakness and spasticity.[78] Current treatment options depend on the clinical presentation. While asymptomatic patients are usually monitored, the common goal of treatment in symptomatic patients is to relieve the pressure on the spinal cord by surgical decompression or by connecting the syrinx cavity with the intrathecal space, also known as shunting (Fig. 22.1).[78,80]

Neuropathic arthropathy (Charcot arthropathy)

SCI-associated loss of proprioception and pain sensitivity impairs the protective mechanisms of the joints. As a consequence, unnoticed repetitive microtrauma can potentially lead to progressive degenerative destruction of the affected joints resulting in deformity, dislocation, instability, or pain.[81−83] Charcot arthropathy can affect both the peripheral joints and the spine, and occurs within years to decades following tSCI.[81−84] The spinal involvement, also known as Charcot spinal arthropathy (CSA), most commonly occurs at the thoracolumbar junction or the lumbar spine and frequently presents with spinal deformity, sitting imbalance, paradoxical localized back pain below the level of injury, pressure sores, changes in spasticity, or cracking noises.[81−83] Based on case reports and small case series, the estimated prevalence of CSA following SCI is 1 in 220.[85] In cases of exhausted conservative (by means of clinical and radiologic follow-ups) or symptomatic therapeutic measures (such as bracing or analgesic therapy), surgical treatment may be warranted to halt the progressive and destructive nature of the disease. Surgical treatment of the CSA aims at correcting the spinal deformity and to fuse the affected segments (Fig. 22.2).[82−84]

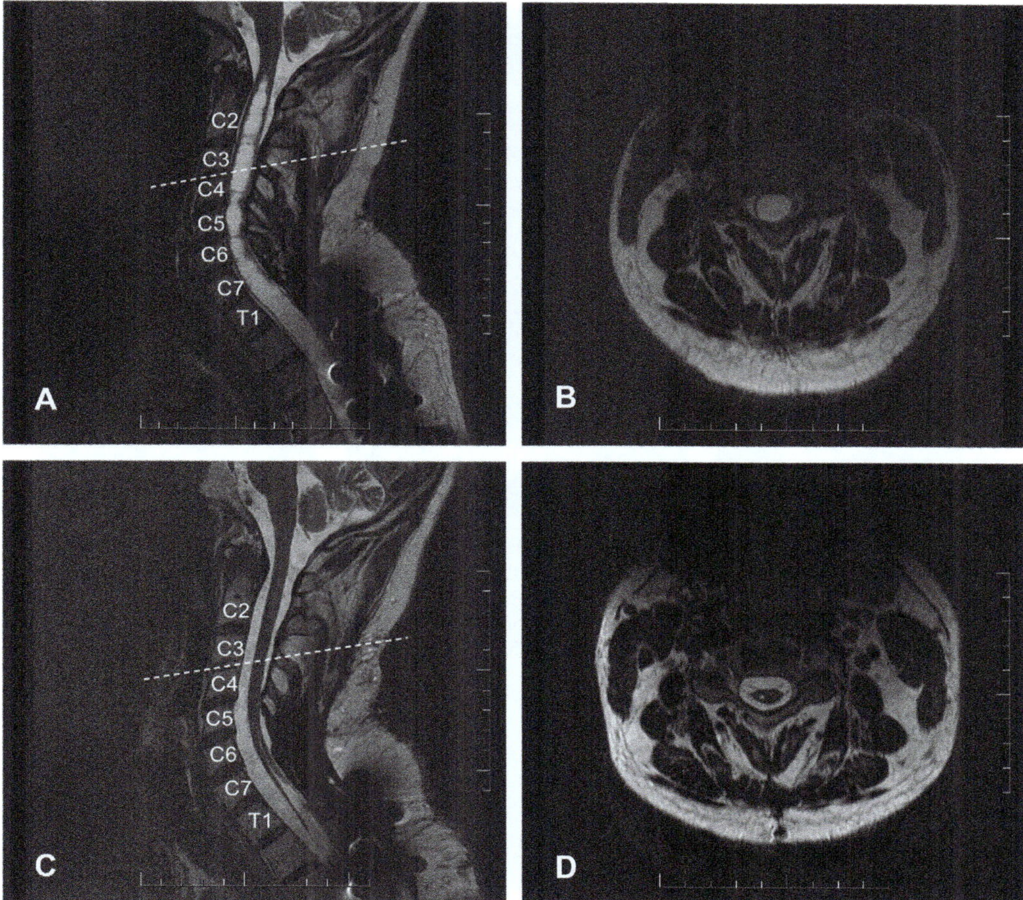

FIGURE 22.1   **Syringomyelia.** Illustrative case of a 38-year-old male with a history of traumatic midthoracic spinal cord injury. The patient developed progressive loss of manual dexterity and loss of sensation in the upper extremities as well as neuropathic pain 12 months after the initial trauma. Due to the progression of neurological deficits, the patient underwent surgical treatment with an implantation of a syringosubarachnoid shunt. (A) The preoperative sagittal T2 magnetic resonance imaging (MRI) study demonstrates an extensive syringomyelia extending proximally until the lower brainstem. (B) The axial T2 image highlights the extension of the syrinx at an exemplary level of C3/4. (C) The postoperative sagittal MRI shows a marked reduction of the syrinx. (D) The axial image shows reduced distention of the cervical spinal cord with a subtotal regressive syrinx at the level of C3/4.

Cognitive impairment

A growing body of evidence supports the potential role of tSCI in causing long-term cognitive dysfunctions with deficits in memory, attention, concentration, abstract reasoning, verbal learning, and psychomotor speed.[86–90] Estimates of frequencies of cognitive deficits among SCI patients vary and have been reported to range between 10% and 60%.[86,87,90] The wide range of reported

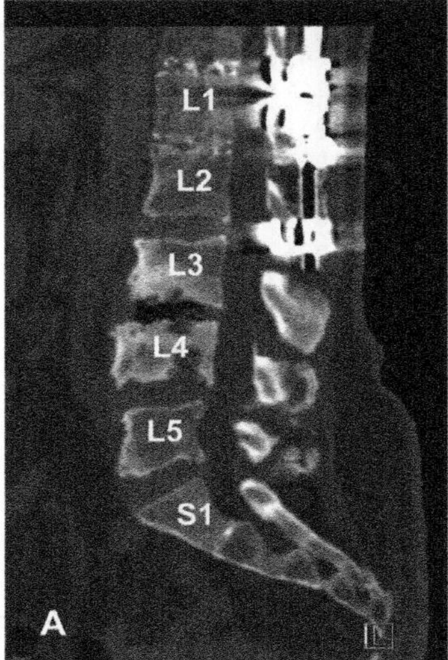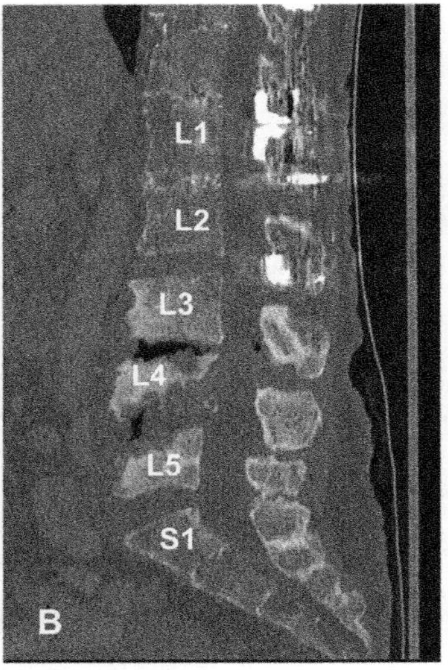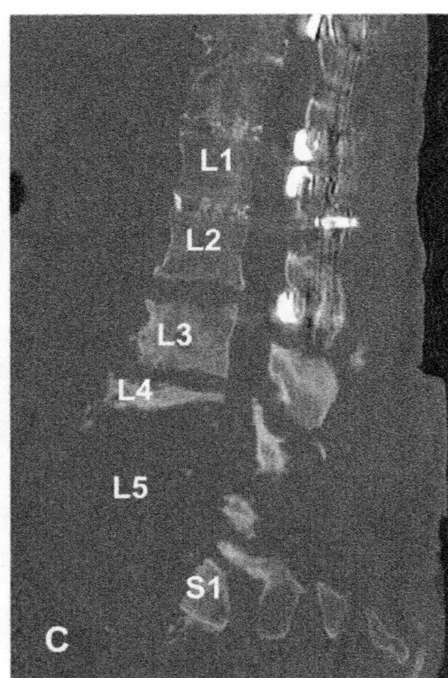

FIGURE 22.2 **Charcot Spinal Arthropathy.** Illustrative case of a 64-year-old woman with a history of traumatic lower thoracic spinal cord injury who developed a Charcot spine arthropathy 20 years after the initial traumatic incident. (A) The sagittal computed tomography (CT) imaging study demonstrates remote initial bony destruction of the L4 vertebra. (B) A 2-year follow-up CT study shows progressive bone resorption of the L4 vertebral body and initial resorption of the L5 vertebra. (C) One year thereafter, the CT scan demonstrates complete L5 and subtotal L4 and sacrum bony resorption causing lumbosacral instability and deformity.

incidences is possibly attributed to the heterogeneity of the patient population and sample sizes, study designs, different time points since SCI, the used screening tools, and varied presence of comorbidities.[86] While the exact underlying pathomechanism in the development of cognitive dysfunction remains unclear, potential factors that have been discussed are (1) concomitant traumatic brain injuries, (2) psychological or somatic comorbidities, and (3) decentralized cardiovascular control causing chronic hypotension and reduced cerebrovascular blood flow.[86,87,91] These data suggest that patients with a history of tSCI should be routinely screened for cognitive impairments in an effort to optimize rehabilitation and functional outcomes.

Impact of complications on neurological and functional recoveries

The complications of tSCI can have a significant impact on patients' neurological and functional recoveries. In a multicenter prospective study, Jiang et al. compared long-term outcomes of tSCI patients who developed AEs with those that did not develop AEs.[11] The patients who developed AEs were found to have less neurological recovery and were more likely to need assisted breathing, dependent ambulation, and had impaired bladder or bowel function compared to those without AEs. Similar results were presented by Jaja et al., where tSCI patients with pneumonia, wound infection, or sepsis (PWS) had greater odds of requiring respiratory and ambulatory support at 6-month follow-up.[92] Additionally, complications can pose further long-term effects on patients' health, as the presence of pneumonia or postoperative wound infections in tSCI patients is associated with reduced AIS upward conversion and American Spinal Injury Association (ASIA) motor score improvement 1-year postinjury.[93] Additionally, these two complications are associated with reduced improvement in Functional Independence Measure (FIM$_{motor}$) up to 5 years postinjury.[94]

Ultimately, complications can have significant long-term effects on patients' recovery following tSCI. Therefore, high clinical vigilance, aggressive preventative measures, and prompt treatment may affect long-term outcomes and improve the quality of life in these patients.

Mortality following traumatic spinal cord injury

tSCI and its associated complications contribute to mortality during acute hospital care and in the years following injury. Overall, patients with tSCI are two to five times more likely to die prematurely than those without tSCI.[95] There are several risk factors for in-hospital mortality post-tSCI, including older age, level of injury, comorbid injuries, complications, traumatic brain injury, and polytrauma.[96] Rates of in-hospital mortality in thoracolumbar tSCI range from 0% to 10.4%.[97] However, patients with high cervical tSCI (C1—C3) have a greater risk of mortality. Kamp et al. document 43.5% of patients with high cervical tSCI dying within 24 hours of hospital admission and 63.9% during hospital stay.[98] Other reports of mortality for cervical tSCI report 4.2%—26.2% of patients dying in hospital.[99] Furthermore, Shibahashi et al. investigated in-hospital mortality for 8069 tSCI patients and reported 5.6% mortality.[100] Using these data, a prediction model for in-hospital mortality post-tSCI was made with a receiver operating characteristic curve of 0.88 (95% confidence internal, 0.86—0.90). The eight predictors of in-hospital mortality in the model were: age, sex, Glasgow Coma Scale on arrival, hypotension on arrival, bradycardia on arrival, severe head injury, Injury Severity Score, and neurological severity.[100] Therefore, patient demographics and injury characteristics on admission are critical in prognosticating mortality in tSCI patients. Inglis et al. further reinforced the role of variables such as age on mortality by reporting the in-hospital mortality of 1018 tSCI patients over 65, using the Canadian Rick Hansen SCI Registry.[101] In the registry, 75% of patients in the surgery and nonsurgery group died within 50 and 20 days post-tSCI, respectively. Additionally, the odds of dying in the 50 days following surgery was six times greater for patients >77 than those 65—76, five times greater in patients with AIS A compared to AIS, B, C, or D, and seven times greater for patients on a ventilator.[101] In addition to patient characteristics, in-hospital mortality also varies based on region, with mortality being reported to be 10.8%—34.6% in Africa, 3.1%—12.6% in the Americas, and 1.1%—15.9% in Europe.[99] Ultimately, in-hospital mortality post-tSCI varies widely based on patient demographics, level of injury, and geographic location. It is critical for clinicians to recognize patients at greatest risk of mortality during the critical period of acute hospital care.

tSCI also contributes to long-term mortality, with rates varying based on the study population. A systematic review by Chamberlain et al. looked at 68 papers discussing mortality following tSCI.[99] The median survival following tSCI ranged from 2.8 to 43 years. Among the studies, 1-year survival ranged from 79% to 100%, 5-year survival from 85% to 96%, and 10-year survival between 81% and 93%.[99] Survival is particularly poor for cases of cervical tSCI. Higashi et al. reported an overall mortality of 19% in patients with cervical tSCI with an average follow-up period of $19.8 + 20$ months.[102] Additionally, Bank et al. reported a 10-year mortality of 33% for cervical tSCI.[103] In contrast, a systematic review on thoracolumbar tSCI reported mortality at 14-year follow-up ranging from 13% to 21%.[97] Furthermore, there are many causes of mortality following tSCI, with the most common being pneumonia and heart disease.[99] However, tSCI patients in developing countries are more likely to die from preventable secondary conditions, such as urologic complications and infections from untreated pressure ulcers.[95] Risk factors associated with mortality include age, tracheostomy, ASIA A, pneumonia, kidney calculus, and complete tSCI.[7,102,104]

Life expectancy is lower among patients with tSCI compared to the rest of the population.[99] Middleton et al. calculated estimated percentage life expectancies from 25 to 65 years for tSCI patients to range between 69%−64%, 74%−65%, 88%−91%, and 97%−96% for C1−C4 AIS A−C, C5−C8 A−C, T1−S5 A−C, and all AIS D lesions, respectively.[105] As this estimate shows, tSCI carries a high rate of mortality; however, life expectancy has increased in the past few decades.[106] For example, a US-based study by Strauss et al. reported a 40% mortality reduction between 1973 and 2004 in the 2 years post-tSCI.[107] Another US study by Saunders et al. reported a significant 3% decrease in tSCI mortality annually from 1981 to 1998.[108] Moving forward, continued progress in patient care and rehabilitation will be critical in further reducing mortality and improving life expectancy in tSCI, particularly in cases of C-spine injuries.

The role of prediction models in evaluating patients' risk for complications

Prediction models have become increasingly used in health care to facilitate clinical decision-making and guide patient management.[109] These models have additionally provided valuable information to patients and their families by addressing the prognostic uncertainty that often accompanies traumatic injury. Furthermore, they have utility in research, as prediction models can stratify patients into different risk categories, improving study design and analysis.[110] In the context of tSCI, there is significant value in using prediction models to identify patients at risk for complications as they are more likely to experience less neurologic recovery and poorer long-term functional outcomes.[11] Prediction models identifying these at-risk patients can

TABLE 22.1   Overview of complications after traumatic spinal cord injury.

Cardiovascular:	Musculoskeletal and dermatologic:
• Neurogenic shock • Autonomic dysreflexia • Cardiac arrhythmia • Orthostatic hypotension • Deep vein thrombosis • Coronary artery disease	• Spasticity • Heterotopic ossification • Osteoporosis • Charcot arthropathy • Pressure sores
Gastrointestinal:	**Respiratory:**
• Gastrointestinal bleeding • Pancreatitis • Neurogenic bowel syndrome	• Respiratory compromise • Pneumonia
Genitourinary:	**Central nervous system:**
• Neurogenic bladder syndrome • Sexual dysfunction	• Syringomyelia • Cognitive impairment

therefore lead to more aggressive complication prevention plans, possibly improving outcomes for these patients. This section will highlight the current predictors and prediction models for tSCI complications, their limitations, and future directions in the development and validation of prediction models.

Several studies have documented associations between tSCI complications and patient factors.[34,111,112] Jaja et al. investigated tSCI patients from the NACTN for SCI and the Surgical Timing in Acute Spinal Cord Injury Study (STASCIS), and identified predictors for PWS.[92] The three most significant predictors of PWS were injury level, baseline AIS grade, and prior medical status (high BP, diabetes, heart attack, malignancy, pulmonary disease, cerebrovascular disease, smoking history, or drug abuse). The results of this study provide valuable insight into which patients may develop PWS. This is useful for clinicians as tSCI patients with PWS are more likely to require respiratory and ambulatory support 6 months postinjury.[92] Although many studies have documented associations between tSCI complications and patient characteristics, there are a limited number of prediction models available.

Prediction models involve using robust statistical methods to develop an algorithm that computes the probability of an outcome occurring. Unlike the studies documenting predictors of tSCI complications separately, these models combine individual parameters to develop more accurate predictions. Wilson et al. developed a prediction model using data on cervical tSCI patients from STASCIS.[113] In their multivariate logistic regression model, an increased likelihood of acute inpatient complications was predicted by more severe initial AIS grade, a high-energy injury mechanism, older age, absence of steroid administration, and comorbid illness. The area under the curve for this prediction model was 0.75, demonstrating reasonable predictive ability. Although this model has been internally validated, external validation is still necessary prior to its clinical use for prognosis of tSCI patients. Furthermore, it is limited to cervical tSCI and does not predict specific complications, but rather predicts complication development in general.

Other prediction models for tSCI complications exist in the literature.[114–117] However, a problem facing much of the scientific literature is that there is greater emphasis on the development of new prediction models rather than the validation or implementation of current models into practice.[109] More external validation is necessary to provide clinicians with models that have sufficient evidence to be used in practice. This can be facilitated by increased collaboration between research groups.[110] By sharing data, prediction models can be tested and refined through a number of different patient registries. Collaboration across different institutions also promotes the development of a single set of accurate and validated prediction models, rather than several different nonvalidated models for tSCI complications. Moving forward, large multicentered efforts will improve data collection and validation of models, which can eventually be implemented into clinical practice. The insight afforded by these models is promising to the future of tSCI patient care.

Conclusion

Complications following tSCI are common, with over 50% of patients experiencing at least one complication during the acute admission period.[5,11] These complications involve several organ systems and can have a devastating impact on patient outcomes and quality of life.[8,93,94] Prediction models are promising tools to identify patients at the greatest risk for complications, but more external validation is needed for their widespread use.[109,113] Current and future spine surgeons must be knowledgeable in the diagnosis and management of tSCI complications as they undoubtedly will encounter them in practice.

References

1. Sezer N, Akkuş S, Uğurlu FG. Chronic complications of spinal cord injury. *World J Orthoped* 2015;6(1):24–33.
2. Hagen EM. Acute complications of spinal cord injuries. *World J Orthoped* 2015;6(1):17–23.
3. Chamberlain JD, et al. Epidemiology and contemporary risk profile of traumatic spinal cord injury in Switzerland. *Inj Epidemiol* 2015;2(1):28.
4. Aito S, Group GISEMG. Complications during the acute phase of traumatic spinal cord lesions. *Spinal Cord* 2003;41(11):629–35.
5. Grossman RG, et al. Incidence and severity of acute complications after spinal cord injury. *J Neurosurg Spine* 2012;17(Suppl. 1):119–28.
6. Barker RN, et al. The relationship between quality of life and disability across the lifespan for people with spinal cord injury. *Spinal Cord* 2009;47(2):149–55.
7. Hagen EM, et al. Mortality after traumatic spinal cord injury: 50 years of follow-up. *J Neurol Neurosurg Psychiatry* 2010;81(4):368–73.
8. McKinley WO, et al. Long-term medical complications after traumatic spinal cord injury: a regional model systems analysis. *Arch Phys Med Rehabil* 1999;80(11): 1402–10.
9. Dijkers MP. Quality of life of individuals with spinal cord injury: a review of conceptualization, measurement, and research findings. *J Rehabil Res Dev* 2005; 42(Suppl. 3):87–110.

10. Tator CH, Fehlings MG. Review of the secondary injury theory of acute spinal cord trauma with emphasis on vascular mechanisms. *J Neurosurg* 1991;**75**(1):15−26.

11. Jiang F, et al. Acute adverse events after spinal cord injury and their relationship to long-term neurologic and functional outcomes: analysis from the North American clinical Trials Network for spinal cord injury. *Crit Care Med* 2019;**47**(11):e854−62.

12. Street JT, et al. Use of the Spine Adverse Events Severity System (SAVES) in patients with traumatic spinal cord injury. A comparison with institutional ICD-10 coding for the identification of acute care adverse events. *Spinal Cord* 2013;**51**(6):472−6.

13. Street JT, et al. Incidence of acute care adverse events and long-term health-related quality of life in patients with TSCI. *Spine J* 2015;**15**(5):923−32.

14. Fehlings MG, et al. Current practice in the timing of surgical intervention in spinal cord injury. *Spine* 2010;**35**(Suppl. 21):S166−73.

15. Fehlings MG, et al. Early versus delayed decompression for traumatic cervical spinal cord injury: results of the surgical timing in Acute Spinal Cord Injury Study (STASCIS). *PLoS One* 2012;**7**(2):e32037.

16. Rowland JW, et al. Current status of acute spinal cord injury pathophysiology and emerging therapies: promise on the horizon. *Neurosurg Focus* 2008;**25**(5):E2.

17. Taylor MP, Wrenn P, O'Donnell AD. Presentation of neurogenic shock within the emergency department. *Emerg Med J* 2017;**34**(3):157−62.

18. Guly HR, et al. The incidence of neurogenic shock in patients with isolated spinal cord injury in the emergency department. *Resuscitation* 2008;**76**(1):57−62.

19. Furlan JC, Fehlings MG. Cardiovascular complications after acute spinal cord injury: pathophysiology, diagnosis, and management. *Neurosurg Focus* 2008;**25**(5): E13.

20. Grigorean VT, et al. Cardiac dysfunctions following spinal cord injury. *J Med Life* 2009;**2**(2):133−45.

21. Franga DL, et al. Recurrent asystole resulting from high cervical spinal cord injuries. *Am Surg* 2006;**72**(6): 525−9.

22. Sidorov EV, et al. Orthostatic hypotension in the first month following acute spinal cord injury. *Spinal Cord* 2008;**46**(1):65−9.

23. Popa C, et al. Vascular dysfunctions following spinal cord injury. *J Med Life* 2010;**3**(3):275−85.

24. Aito S, et al. Primary prevention of deep venous thrombosis and pulmonary embolism in acute spinal cord injured patients. *Spinal Cord* 2002;**40**(6):300−3.

25. Waring WP, Karunas RS. Acute spinal cord injuries and the incidence of clinically occurring thromboembolic disease. *Paraplegia* 1991;**29**(1):8−16.

26. Ploumis A, et al. Thromboprophylaxis in patients with acute spinal injuries: an evidence-based analysis. *J Bone Joint Surg Am* 2009;**91**(11):2568−76.

27. Bycroft J, et al. Autonomic dysreflexia: a medical emergency. *Postgrad Med J* 2005;**81**(954):232−5.

28. Lee ES, Joo MC. Prevalence of autonomic dysreflexia in patients with spinal cord injury above T6. *BioMed Res Int* 2017;**2017**:2027594.

29. Eldahan KC, Rabchevsky AG. Autonomic dysreflexia after spinal cord injury: systemic pathophysiology and methods of management. *Auton Neurosci* 2018; **209**:59−70.

30. Nash MS, et al. Cardiometabolic syndrome in people with spinal cord injury/disease: guideline-derived and nonguideline risk components in a pooled sample. *Arch Phys Med Rehabil* 2016;**97**(10):1696−705.

31. Myers J, Lee M, Kiratli J. Cardiovascular disease in spinal cord injury: an overview of prevalence, risk, evaluation, and management. *Am J Phys Med Rehabil* 2007; **86**(2):142−52.

32. Szlachcic Y, et al. Cardiometabolic changes and disparities among persons with spinal cord injury: a 17-year cohort study. *Top Spinal Cord Inj Rehabil* 2014;**20**(2): 96−104.

33. Cragg JJ, et al. Cardiovascular disease and spinal cord injury: results from a national population health survey. *Neurology* 2013;**81**(8):723−8.

34. Aarabi B, et al. Predictors of pulmonary complications in blunt traumatic spinal cord injury. *J Neurosurg Spine* 2012;**17**(Suppl. 1):38−45.

35. Berlly M, Shem K. Respiratory management during the first five days after spinal cord injury. *J Spinal Cord Med* 2007;**30**(4):309−18.

36. Winslow C, et al. Impact of respiratory complications on length of stay and hospital costs in acute cervical spine injury. *Chest* 2002;**121**(5):1548−54.

37. Jackson AB, Groomes TE. Incidence of respiratory complications following spinal cord injury. *Arch Phys Med Rehabil* 1994;**75**(3):270−5.

38. van den Berg ME, et al. Survival after spinal cord injury: a systematic review. *J Neurotrauma* 2010;**27**(8):1517−28.

39. Waddimba AC, et al. Predictors of cardiopulmonary hospitalization in chronic spinal cord injury. *Arch Phys Med Rehabil* 2009;**90**(2):193−200.

40. Carter RE. Experience with ventilator dependent patients. *Paraplegia* 1993;**31**(3):150−3.

41. Leelapattana P, et al. Predicting the need for tracheostomy in patients with cervical spinal cord injury. *J Trauma Acute Care Surg* 2012;**73**(4):880−4.

42. Quesnel A, et al. What are the perspectives for ventilated tetraplegics? A French retrospective study of 108 patients with cervical spinal cord injury. *Ann Phys Rehabil Med* 2015;**58**(2):74−7.

43. Kruger EA, et al. Comprehensive management of pressure ulcers in spinal cord injury: current concepts and future trends. *J Spinal Cord Med* 2013;**36**(6):572−85.

44. Jia X, et al. Critical care of traumatic spinal cord injury. *J Intensive Care Med* 2013;**28**(1):12−23.

45. Chen HL, et al. Incidence of pressure injury in individuals with spinal cord injury: a systematic review and meta-analysis. *J Wound Ostomy Continence Nurs* 2020;**47**(3):215−23.

46. Edsberg LE, et al. Revised national pressure ulcer advisory panel pressure injury staging system: revised pressure injury staging system. *J Wound Ostomy Continence Nurs* 2016;**43**(6):585−97.

47. DeJong G, et al. Rehospitalization in the first year of traumatic spinal cord injury after discharge from medical rehabilitation. *Arch Phys Med Rehabil* 2013;**94**(Suppl. 4):S87−97.

48. Middleton JW, et al. Patterns of morbidity and rehospitalisation following spinal cord injury. *Spinal Cord* 2004;**42**(6):359−67.

49. Chan BC, et al. The lifetime cost of spinal cord injury in Ontario, Canada: a population-based study from the perspective of the public health care payer. *J Spinal Cord Med* 2019;**42**(2):184−93.

50. Ebert E. Gastrointestinal involvement in spinal cord injury: a clinical perspective. *J Gastrointestin Liver Dis* 2012;**21**(1):75−82.

51. Berlly MH, Wilmot CB. Acute abdominal emergencies during the first four weeks after spinal cord injury. *Arch Phys Med Rehabil* 1984;**65**(11):687−90.

52. Kewalramani LS. Neurogenic gastroduodenal ulceration and bleeding associated with spinal cord injuries. *J Trauma* 1979;**19**(4):259−65.

53. McKinley WO, et al. Spinal cord injury medicine. 2. Medical complications after spinal cord injury: identification and management. *Arch Phys Med Rehabil* 2002;**83**(Suppl. 3). S58−S64, S90−S98.

54. Hagen EM, et al. Cardiovascular and urological dysfunction in spinal cord injury. *Acta Neurol Scand* 2011;(191):71−8.

55. Dorsher PT, McIntosh PM. Neurogenic bladder. *Adv Urol* 2012;**2012**:816274.

56. Ku JH, et al. Complications of the upper urinary tract in patients with spinal cord injury: a long-term follow-up study. *Urol Res* 2005;**33**(6):435−9.

57. Weld KJ, et al. Influences on renal function in chronic spinal cord injured patients. *J Urol* 2000;**164**(5):1490−3.

58. Siroky MB. Pathogenesis of bacteriuria and infection in the spinal cord injured patient. *Am J Med* 2002;**113**(Suppl. 1A):67S−79S.

59. Waites KB, et al. Bacteremia after spinal cord injury in initial versus subsequent hospitalizations. *J Spinal Cord Med* 2001;**24**(2):96−100.

60. Savic G, et al. Hospital readmissions in people with chronic spinal cord injury. *Spinal Cord* 2000;**38**(6):371−7.

61. White BAB, et al. The economic burden of urinary tract infection and pressure ulceration in acute traumatic spinal cord injury admissions: evidence for comparative economics and decision analytics from a matched case-control study. *J Neurotrauma* 2017;**34**(20):2892−900.

62. Benevento BT, Sipski ML. Neurogenic bladder, neurogenic bowel, and sexual dysfunction in people with spinal cord injury. *Phys Ther* 2002;**82**(6):601−12.

63. Reitz A, et al. Impact of spinal cord injury on sexual health and quality of life. *Int J Impot Res* 2004;**16**(2):167−74.

64. Hess MJ, Hough S. Impact of spinal cord injury on sexuality: broad-based clinical practice intervention and practical application. *J Spinal Cord Med* 2012;**35**(4):211−8.

65. Hajiaghababaei M, et al. Female sexual dysfunction in patients with spinal cord injury: a study from Iran. *Spinal Cord* 2014;**52**(8):646−9.

66. van Kuijk AA, Geurts AC, van Kuppevelt HJ. Neurogenic heterotopic ossification in spinal cord injury. *Spinal Cord* 2002;**40**(7):313−26.

67. Sullivan MP, et al. Heterotopic ossification after central nervous system trauma: a current review. *Bone Joint Res* 2013;**2**(3):51−7.

68. Shehab D, Elgazzar AH, Collier BD. Heterotopic ossification. *J Nucl Med* 2002;**43**(3):346−53.

69. Jiang SD, Dai LY, Jiang LS. Osteoporosis after spinal cord injury. *Osteoporos Int* 2006;**17**(2):180−92.

70. Frotzler A, et al. Long-bone fractures in persons with spinal cord injury. *Spinal Cord* 2015;**53**(9):701−4.

71. Cochran TP, Bayley JC, Smith M. Lower extremity fractures in paraplegics: pattern, treatment, and functional results. *J Spinal Disord* 1988;**1**(3):219−23.

72. Comarr AE, Hutchinson RH, Bors E. Extremity fractures of patients with spinal cord injuries. *Am J Surg* 1962;**103**:732−9.

73. Zehnder Y, et al. Long-term changes in bone metabolism, bone mineral density, quantitative ultrasound parameters, and fracture incidence after spinal cord injury: a cross-sectional observational study in 100 paraplegic men. *Osteoporos Int* 2004;**15**(3):180−9.

74. Vestergaard P, et al. Fracture rates and risk factors for fractures in patients with spinal cord injury. *Spinal Cord* 1998;**36**(11):790−6.

75. Nas K, et al. Rehabilitation of spinal cord injuries. *World J Orthoped* 2015;**6**(1):8−16.

76. Elbasiouny SM, et al. Management of spasticity after spinal cord injury: current techniques and future directions. *Neurorehabil Neural Repair* 2010;**24**(1):23−33.

77. Holtz KA, et al. Prevalence and effect of problematic spasticity after traumatic spinal cord injury. *Arch Phys Med Rehabil* 2017;**98**(6):1132–8.

78. Brodbelt AR, Stoodley MA. Post-traumatic syringomyelia: a review. *J Clin Neurosci* 2003;**10**(4):401–8.

79. Klekamp J. The pathophysiology of syringomyelia - historical overview and current concept. *Acta Neurochir* 2002;**144**(7):649–64.

80. Schurch B, Wichmann W, Rossier AB. Post-traumatic syringomyelia (cystic myelopathy): a prospective study of 449 patients with spinal cord injury. *J Neurol Neurosurg Psychiatry* 1996;**60**(1):61–7.

81. Alpert SW, Koval KJ, Zuckerman JD. Neuropathic arthropathy: review of current knowledge. *J Am Acad Orthop Surg* 1996;**4**(2):100–8.

82. Aebli N, Pötzel T, Krebs J. Characteristics and surgical management of neuropathic (Charcot) spinal arthropathy after spinal cord injury. *Spine J* 2014;**14**(6):884–91.

83. Standaert C, Cardenas DD, Anderson P. Charcot spine as a late complication of traumatic spinal cord injury. *Arch Phys Med Rehabil* 1997;**78**(2):221–5.

84. Brown CW, et al. Neuropathic (Charcot) arthropathy of the spine after traumatic spinal paraplegia. *Spine* 1992;**17**(Suppl. 6):S103–8.

85. Urits I, et al. A comprehensive review of the treatment and management of Charcot spine. *Ther Adv Musculoskelet Dis* 2020;**12**. 1759720X20979497.

86. Sachdeva R, et al. Cognitive function after spinal cord injury: a systematic review. *Neurology* 2018;**91**(13):611–21.

87. Nightingale TE, et al. Diverse cognitive impairment after spinal cord injury is associated with orthostatic hypotension symptom burden. *Physiol Behav* 2020;**213**:112742.

88. Distel DF, et al. Cognitive dysfunction in persons with chronic spinal cord injuries. *Phys Med Rehabil Clin* 2020;**31**(3):345–68.

89. Li Y, et al. Dementia, depression, and associated brain inflammatory mechanisms after spinal cord injury. *Cells* 2020;**9**(6).

90. Davidoff GN, Roth EJ, Richards JS. Cognitive deficits in spinal cord injury: epidemiology and outcome. *Arch Phys Med Rehabil* 1992;**73**(3):275–84.

91. Jegede AB, et al. Cognitive performance in hypotensive persons with spinal cord injury. *Clin Auton Res* 2010;**20**(1):3–9.

92. Jaja BNR, et al. Association of pneumonia, wound infection, and sepsis with clinical outcomes after acute traumatic spinal cord injury. *J Neurotrauma* 2019;**36**(21):3044–50.

93. Failli V, et al. Functional neurological recovery after spinal cord injury is impaired in patients with infections. *Brain* 2012;**135**(Pt 11):3238–50.

94. Kopp MA, et al. Long-term functional outcome in patients with acquired infections after acute spinal cord injury. *Neurology* 2017;**88**(9):892–900.

95. Bickenbach J, et al. *International perspectives on spinal cord injury*. World Health Organization; 2013.

96. Selassie AW, et al. Determinants of in-hospital death after acute spinal cord injury: a population-based study. *Spinal Cord* 2013;**51**(1):48–54.

97. Azarhomayoun A, et al. Mortality rate and predicting factors of traumatic thoracolumbar spinal cord injury; A systematic review and meta-analysis. *Bull Emerg Trauma* 2018;**6**(3):181–94.

98. Kamp O, et al. Survival among patients with severe high cervical spine injuries - a TraumaRegister DGU® database study. *Scand J Trauma Resuscitation Emerg Med* 2021;**29**(1):1.

99. Chamberlain JD, et al. Mortality and longevity after a spinal cord injury: systematic review and meta-analysis. *Neuroepidemiology* 2015;**44**(3):182–98.

100. Shibahashi K, et al. Epidemiological state, predictors of early mortality, and predictive models for traumatic spinal cord injury: a multicenter nationwide cohort study. *Spine* 2019;**44**(7):479–87.

101. Inglis T, et al. In-hospital mortality for the elderly with acute traumatic spinal cord injury. *J Neurotrauma* 2020;**37**(21):2332–42.

102. Higashi T, et al. Analysis of the risk factors for tracheostomy and decannulation after traumatic cervical spinal cord injury in an aging population. *Spinal Cord* 2019;**57**(10):843–9.

103. Bank M, et al. Age and other risk factors influencing long-term mortality in patients with traumatic cervical spine fracture. *Geriatr Orthop Surg Rehabil* 2018;**9**. 2151459318770882.

104. Cao Y, Krause JS, DiPiro N. Risk factors for mortality after spinal cord injury in the USA. *Spinal Cord* 2013;**51**(5):413–8.

105. Middleton JW, et al. Life expectancy after spinal cord injury: a 50-year study. *Spinal Cord* 2012;**50**(11):803–11.

106. Frankel HL, et al. Long-term survival in spinal cord injury: a fifty year investigation. *Spinal Cord* 1998;**36**(4):266–74.

107. Strauss DJ, et al. Trends in life expectancy after spinal cord injury. *Arch Phys Med Rehabil* 2006;**87**(8):1079–85.

108. Saunders LL, et al. Traumatic spinal cord injury mortality, 1981-1998. *J Trauma* 2009;**66**(1):184–90.

109. Hendriksen JM, et al. Diagnostic and prognostic prediction models. *J Thromb Haemostasis* 2013;**11**(Suppl. 1):129–41.

110. Steyerberg EW, et al. Prognosis Research Strategy (PROGRESS) 3: prognostic model research. *PLoS Med* 2013;**10**(2):e1001381.

111. Agostinello J, Battistuzzo CR, Batchelor PE. Early clinical predictors of pneumonia in critically ill spinal cord injured individuals: a retrospective cohort study. *Spinal Cord* 2019;**57**(1):41–8.

112. Ihalainen T, et al. Risk factors for laryngeal penetration-aspiration in patients with acute traumatic cervical spinal cord injury. *Spine J* 2018;**18**(1):81–7.

113. Wilson JR, et al. Clinical prediction model for acute inpatient complications after traumatic cervical spinal cord injury: a subanalysis from the Surgical Timing in Acute Spinal Cord Injury Study. *J Neurosurg Spine* 2012;**17**(Suppl. 1):46–51.

114. Hou YF, et al. Development and validation of a risk prediction model for tracheostomy in acute traumatic cervical spinal cord injury patients. *Eur Spine J* 2015;**24**(5):975–84.

115. Green D, et al. Spinal cord injury risk assessment for thromboembolism (SPIRATE study). *Am J Phys Med Rehabil* 2003;**82**(12):950–6.

116. Durga P, et al. Development and validation of predictors of respiratory insufficiency and mortality scores: simple bedside additive scores for prediction of ventilation and in-hospital mortality in acute cervical spine injury. *Anesth Analg* 2010;**110**(1):134–40.

117. Croce MA, Tolley EA, Fabian TC. A formula for prediction of posttraumatic pneumonia based on early anatomic and physiologic parameters. *J Trauma* 2003;**54**(4):724–9.

Neurochemical biomarkers of acute spinal cord injury

Andrea J. Mothe[1], Steven Casha[2], Brian K. Kwon[3,4]

[1]Krembil Research Institute, Toronto Western Hospital, University Health Network, Toronto, ON, Canada; [2]Hotchkiss Brain Institute, University of Calgary Spine Program, Department of Clinical Neurosciences, University of Calgary, Calgary, AB, Canada; [3]International Collaboration on Repair Discoveries, University of British Columbia, Vancouver, BC, Canada; [4]Vancouver Spine Surgery Institute, Vancouver General Hospital, Department of Orthopaedics, UBC, Vancouver, BC, Canada

Introduction

The basic definition of a biomarker is *"a defined characteristic that is measured as an indicator of normal biological processes, pathogenic processes, or responses to an exposure or intervention."* In the context of SCI, such defined characteristics could include molecular/biochemical/cellular features in the blood and CSF, imaging features on MRI or CT scanning, or electrophysiologic features on neurophysiologic testing. Unlike the standard neurologic assessment of acute SCI, all of these have the ability to provide objective measures of the injury, although the great challenge is in relating these quantifiable features with discernible clinical outcomes such as "motor score recovery" or "AIS grade conversion." Biomarkers have many potential applications in biomedical research and clinical care, including diagnosis, monitoring of disease progression or of response to treatment or environmental exposure, prediction of response to treatment or susceptibility/ risk, and prognosis of outcome. This chapter will focus on neurochemical biomarkers in blood and CSF and how these may have application in diagnosing injury severity, prognosticating outcome, and monitoring treatment response after acute SCI.

Why are biomarkers needed in acute SCI?

Because the paralysis of acute SCI represents such a sudden and dramatic event, it is not intuitively obvious why a biomarker would even be necessary for the diagnosis of acute SCI. The history, imaging studies, and even a cursory/ limited clinical examination are usually sufficient to make the diagnosis. However, ambiguous cases do occur with some regularity. Many patients experience polytrauma and confounding neurological injuries such as brachial plexus injury or traumatic brain injury can make the diagnosis less clear. Patients are often obtunded, intoxicated, or ventilated, making

© 2022 Elsevier Inc. All rights reserved.

the neurological exam difficult and imprecise. Imaging in central cord SCI is often not specific for the acute injury and may be reflective of chronic myelopathy. Furthermore, a diagnostic biomarker may be useful in atraumatic cases such as an evolving ischemic injury secondary to vascular compromise during aortic aneurysm surgery. Most often the more practical challenge that clinicians face is the discussions with the acute SCI patient and his/her family regarding the prognosis of the injury: *How much neurologic recovery is to be expected? Will independent walking be possible? What functional limitations will there be?* A prognostic biomarker to inform that discussion would be extremely useful and may provide more objectivity to the discussion. Irrespective of possible management changes, there is no question that the uncertainty about neurologic prognosis is very challenging for patients and clinicians.

Aside from this potential utility in clinical practice, biomarkers are sorely needed for the translation of preclinical therapeutic discoveries through human clinical trials in acute SCI. Clinical trials that seek to determine if "treatment A" improves neurologic function over "control" (whether that be randomized controls, matched case controls, or historical controls) encounter a number of obstacles. At the heart of these obstacles is the reliance of clinical trials upon the neurologic examination to determine the degree of neurologic impairment. The International Standards for Neurological Classification of SCI (ISNCSCI) examination is the "gold standard" for characterizing neurologic impairment in a spinal cord injured individual — and having such a standardized exam is indeed highly valuable for the SCI field. Performing a full ISNCSCI exam, however, is very time-consuming and requires considerable training on the part of the examiner to obtain consistent, reliable assessments. While the ISNCSCI examination was never meant to be an "outcome measure" per se, in the context of clinical trials that seek to determine therapeutic efficacy of a particular treatment,

the improvement of AIS grade and change in motor score in subjects treated has traditionally been used as an important indicator of therapeutic effect (and in most cases represents the primary outcome). This reliance upon the ISNCSCI creates many challenges for clinical trials of acute SCI.

Firstly, in many acute SCI patients the ISNCSCI exam is impossible to actually perform. Such patients often arrive in the emergency room with multiple injuries and an altered level of consciousness due to intoxication, and/or sedation from analgesics. A detailed, time-consuming neurologic assessment that requires conscious participation from the patient becomes impossible, and without the ability to establish a baseline injury severity, such patients cannot be enrolled into acute clinical trials. It is estimated that approximately a third of acute SCI patients fall into this category,[1] so acute SCI clinical trials that are trying to recruit subjects with a condition that is already relatively rare are further burdened with this additional limitation. An objective biomarker that could be used to stratify injury severity in the acute setting would help immensely in one of the major difficulties that all acute clinical trials face: the recruitment of a sufficient number of subjects.

Another obstacle is the variability of spontaneous neurologic recovery that occurs after injury necessitating a relatively large number of subjects to detect a potentially small (albeit meaningful) therapeutic effect. This is further compounded by inaccuracies in performing the ISNCSCI examination, when the assignment of an AIS impairment grade may depend upon very subtle differences during the exam such as the presence of deep anal pain sensation, or a tepid voluntary anal contraction or toe flexion. While these differences are subtle (and understandably difficult to discern in the acute trauma setting), the general prognosis for neurologic recovery is vastly different between these grades, and misclassification can have dramatic effects on a clinical trial, adding variability and decreasing power.

Nonetheless, even with a "reliable" ISNCSCI examination there is still considerable variability in spontaneous recovery. The resultant challenge to recruitment is reflected in the paucity of phase 3 clinical trials that have been completed and in the number of acute clinical trials that have been terminated prematurely. Biomarkers reflective of secondary injury progression or regeneration could provide a more reliable and less variable outcome measure.

A further challenge is related to the spontaneous neurologic recovery variability, which makes it extremely difficult for small, early-stage clinical trials in particular, to detect a signal of therapeutic efficacy. Such early-stage clinical trials are vital for the decision to proceed with a large-scale definitive prospective randomized study. It seems very likely that neurological and functional outcomes will remain the key outcomes for efficacy studies. Biomarkers would be most useful for early-phase studies (phase 1 and phase 2 studies), increasing efficiency and enabling studies that may otherwise not be feasible. For example, motor recovery as measured by the ISNCSCI exam in a study sample plateaus around 3 months, consequently most studies include 6-month or 1-year measurements of the outcome. Early structural biomarkers combined with therapy-specific markers of biological activity of the treatment may allow for faster throughput of these early-phase clinical trials. In the absence of the ability to see a "therapeutic signal" in such a trial, a biomarker that served as a surrogate biological outcome measure would be highly informative. For example, if a potential treatment for acute SCI were thought to have an antiinflammatory mechanism of action, it would be highly useful in an early-stage clinical trial to have a biomarker of spinal cord inflammation and to determine whether the treatment was actually having the expected biological effect (even if a functional neurologic improvement was impossible to detect in such a small cohort). Phase 0 trials in particular, and markers of biological activity,

have been largely absent in human SCI investigations. The ability to reproduce a biological effect that had been seen in preclinical studies would better inform the decision to pursue larger trials and aid in determining human dosing, ultimately improving the likelihood of success. In summary, biomarkers would be extremely valuable for clinical trials for a number of reasons: 1. objectively stratifying injury severity in acute SCI subjects in whom it is impossible to do a reliable ISNCSCI examination; 2. providing a more precise prediction of neurologic recovery, so that variability in spontaneous recovery can be interpreted and controlled for in determining therapeutic effect; and 3. serving as biological surrogate outcome measures to indicate if the treatment is having the intended biological effect (Table 23.1). Aside from this, there is a strong rationale to study the biological responses to SCI in search of biomarkers that could serve as potential targets of therapeutic intervention or serve to identify select patients who would benefit from a treatment (and those who would not). Of course, different clinical trials may require different biomarkers for distinct purposes. Acute clinical trials may primarily need biomarkers for early inclusion and stratification of subjects when the initial neurologic impairment is uncertain, while chronic clinical trials enrolling subjects with an established neurologic impairment may require biomarkers to better determine who is actually a candidate for the treatment in question. Either way, the search for biomarkers has a compelling rational for translational research in spinal cord injury.

Promising neurochemical biomarkers of acute SCI

Following acute SCI, the CNS/blood and CNS/CSF barriers transiently open, allowing proteins to enter the bloodstream and CSF where they are not normally found.[2] The primary

TABLE 23.1   Potential roles for biomarkers for acute SCI.

Diagnostic	A biological measure of SCI and its severity would be very useful given that an ISNSCI examination cannot be performed acutely in many patients and that the current AIS grades poorly capture the heterogeneity of injury.
Prognostic	A biological measure of SCI that predicts recovery more precisely would assist clinicians and patients and also facilitate clinical trials by reducing the variability in outcome.
Surrogate outcome	A biological measure of treatment response would be very helpful for treatments undergoing human clinical trials to determine if they were having the intended effect on the cord.

mechanical injury to the spinal cord damages neural cells, releasing their intracellular contents into the extracellular space which subsequently enter into the CSF and systemic circulation. Intracellular proteins released upon damage to glia and neurons include GFAP (glial fibrillary acidic protein), S100-β (S100 calcium-binding protein-β), neurofilaments (NFs), neuron-specific enolase (NSE), and tau. The detection of these proteins above basal levels within the CSF and blood reflects biomarkers of acute SCI, and their concentration may directly relate to the extent of neural tissue damage that has occurred.[3,4] Moreover, secondary injury mechanisms induce increased circulating inflammatory mediators such as cytokines; these include interleukin-1β (IL-1β), interleukin-6 (IL-6), interleukin-8 (IL-8), and monocyte chemoattractant protein 1 (MCP-1), which are also detectable in biofluids after acute SCI.[2,5–7] Collectively, clinical and experimental SCI studies have identified several candidate biomarkers including structural proteins, inflammatory cytokines, and microRNAs.[2,3,5–9] Herein we will highlight the most promising biomarker candidates for acute SCI, describing their function and identification in experimental and clinical studies.

Glial-enriched structural protein biomarkers

GFAP and S100-β are structural protein biomarkers released from glial cells after injury. After SCI, the cytoskeleton of damaged astrocytes disintegrates, releasing GFAP intermediate filaments which have been detected acutely in both CSF and serum following SCI.[2,5] S100-β is a calcium-binding protein found primarily in astrocytes and Schwann cells, although it is also found in adipocytes, chondrocytes, and melanocytes.[10,11] Thus, the release of S100-β from adipose tissue may confound the interpretation of increased serum concentrations as a measure of injury severity.[12] However, CSF levels of S100-β from SCI patients have shown good correlation with injury severity and neurological recovery.[6]

Neuronal-enriched enzyme and structural protein biomarkers

Damaged neurons and axons release structural proteins including NFs and tau, and enzymes such as UCH-L1 (ubiquitin C-terminal hydrolase L1) and NSE. UCH-L1 is a highly abundant neuron-specific enzyme essential for axonal function, and NSE is a cytoplasmic enzyme involved in glycolytic energy metabolism. However, the clinical utility of NSE as a blood biomarker for SCI is more limited due to the lack of specificity, given that it is also expressed by erythrocytes, platelets, and cells of neuroendocrine origin.[13]

NFs are intermediate filaments and the most abundant proteins in the neuronal cytoskeleton, regulating axonal transport and neuronal signaling.[14] The presence of extracellular NF protein is considered an indicator of axonal damage in several other neurological conditions,[2,4] and the most widely studied NF subunits as biomarkers of SCI are NF light-chain (NFL) and NF heavy-chain (NFH). NF proteins have long half-lives, and the phosphorylated form of NFH is resistant to protease degradation, thus extending the time window of detection and potentially enhancing its utility as a biomarker.[9] Tau is a microtubule-associated protein that promotes the formation and stabilization of microtubules and is involved in anterograde axoplasmic transport. After SCI, tau undergoes various posttranslational modifications, primarily hyperphosphorylation of serine and threonine amino acid residues to form p-tau (phosphorylated tau), and also cleavage into smaller fragments known as cleaved tau.[15] Following axonal injury, tau and p-tau leak into the CSF or blood and are sensitive biomarkers of SCI.[3]

Experimental studies of acute SCI biomarkers

Experimental models of SCI provide the opportunity to measure differences in biomarker concentration in the blood or CSF from sham control versus animals injured under controlled experimental settings. Although biomarker levels and time course of protein expression will vary depending on the SCI model used and methods of analysis, experimental studies have identified several promising biomarker candidates for acute SCI with the distinct advantage of being able to correlate biomarker levels to precise injury severities induced in the laboratory setting. Key experimental studies of acute SCI biomarkers are summarized in Table 23.2.

In rats with graded thoracic contusion SCI produced with the Infinite Horizon Impactor (Precision Systems and Instrumentation, Fairfax Station, VA), Loy et al.[16] found a significant increase in acute serum concentrations of NSE and S100-β protein at 6 hours, followed by a decrease by 24 hours postinjury compared to sham animals. Although functional recovery data were

not included, no association was found between serum concentration of either NSE or S100-β and the severity of SCI.[16] In terms of NSE, this may reflect the aforementioned lack of specificity of NSE as a serum biomarker since it is also expressed by cells other than neurons. Alternatively, the enzyme-linked immunosorbent assay (ELISA) method used may have lacked sufficient sensitivity to detect small differences of these proteins in serum. In rats with graded thoracic contusions via weight drop, serum and CSF levels of both NSE and S100-β significantly increased by 2 hours and peaked at 6 hours post-SCI with differences between injury severities.[17] However, ELISA methods were not included and correlations were not assessed.[17] Similarly, serum levels of S100-β were shown to increase in SCI rats compared to uninjured controls within 3 hours after a spinal cord compression lesion at level T9 produced with a bulldog type vessel clamp.[18]

Preliminary data demonstrated the concentration of GFAP and UCH-L1 peaked at 4 hours post-SCI in CSF from rats with weight-drop contusion injury.[2] In comparison, maximal serum concentration of hyperphosphorylated NFH (p-NFH) was not seen until 3 days postinjury in individual rats with either hemisection or contusion SCI.[19] Likewise, Ueno et al.[20] showed that p-NFH levels in plasma peaked at 3 days in rats with moderate contusion SCI produced with the Infinite Horizon Impactor. Moreover, plasma p-NFH levels at 3 days postinjury negatively correlated with hind limb function at 28 days, indicating that animals with lower acute levels of p-NFH achieved a higher motor score.[20] More recently, a profile of protein biomarkers including S100-β, GFAP, UCH-L1, and p-NFH was examined in rat CSF and serum using sandwich ELISA at 4h, 24h, and 7d following moderate or severe thoracic SCI produced with the NYU-MASCIS impactor.[21] Significant differences in CSF UCH-L1 levels were detected at 4 hours after moderate versus severe SCI. Differences in serum GFAP concentration were also found at 4 and 24 hours postinjury,[21] although

TABLE 23.2 Key experimental biomarker studies for acute SCI.

Reference	Biomarker; detection method	Species	SCI model; level; severity	Biofluid	Biofluid sampling time Post-SCI	Significant findings
Loy et al.[16]	NSE, S100-β; ELISA	Rat	Contusion (Infinite Horizons impactor); T9; moderate (150 kdyn), severe (200 kdyn)	Serum	6, 24 hours	Serum NSE higher at 6 and 24 hours; serum S100-β higher at 6 hours after severe injury; found no association between serum protein levels and injury severity
Cao et al.[17]	NSE, S100-β; ELISA	Rat	Contusion (weight drop 10 g); T8; mild (2.5 cm), moderate (5 cm), severe (15 cm)	CSF serum	30 minutes, 2, 6, 12, 24 hours	CSF and serum levels of NSE and S100-β peaked at 6 hours; higher levels of proteins in biofluids with more severe injury
Ma et al.[18]	S100-β; immunoluminometric assay	Rat	Compression (bulldog vessel clamp); T9; severe (50 g)	Serum	0, 3, 6, 12, 24 hours, 3, 6, 10 days	Serum S100-β levels increased from 3 hours to 3 days
Yokobori et al.[2]	GFAP, UCH-L1; ELISA	Rat	Contusion (weight drop); T8; not indicated	CSF	4, 24, 48 hours	CSF levels of GFAP and UCH-L1 peaked at 4 hours
Shaw et al.[19]	p-NFH; ELISA	Rat	Contusion (NYU-impactor, 10 g weight); T11; severe (25 mm); spinal hemisection at T11	Serum	5 minutes, 2, 8, 16, 24 hours, 2, 3–21 days	Serum p-NFH levels peaked at 3 days in both injury models
Ueno et al.[20]	p-NFH; ELISA	Rat	Contusion (Infinite Horizons impactor); T10; moderate (150 kdyn)	Plasma	1, 2, 3, 4 days	Plasma p-NFH levels peaked at 3 days and negatively correlated with hind limb function
Yang et al.[21]	S100-β, GFAP, UCH-L1, p-NFH; ELISA	Rat	Contusion (NYU-impactor, 10 g weight); T9; moderate (12.5 mm), severe (25 mm)	CSF serum	4, 24 hours, 7 days	CSF levels of all proteins peaked at 4 hours; CSF UCH-L1 levels higher at 4 hours in severe versus moderate injury; serum S100-β and UCH-L1 peaked at 4 hours; serum GFAP higher at 4 and 24 hours and after severe injury; serum p-NFH levels increased from 4 hours to 7 days, and higher after severe injury at 24 hours and 7 days
Roerig et al.[22]	Total tau; ELISA	Dog	Intervertebral disk herniation; thoracolumbar and cervical	CSF	CSF sampled on admission to veterinary hospital	CSF total tau levels higher in dogs with more severe injury compared to normal or less severely injured; dogs that improved showed lower tau levels

TABLE 23.2 Key experimental biomarker studies for acute SCI.—cont'd

Reference	Biomarker; detection method	Species	SCI model; level; severity	Biofluid	Biofluid sampling time Post-SCI	Significant findings
Caprelli et al.[23]	Total tau, p-tau; Simoa digital immunoassay	Rat	Impact-compression (clip); T8; moderate/severe (35 g clip closing force)	CSF serum	1, 7 days	CSF total tau and CSF and serum p-tau levels increased at 1 day and decreased to baseline levels by 7 days
Tang et al.[24]	p-tau; ELISA	Rat	Contusion (NYU-impactor, 10 g weight); T8; mild (12.5 mm), moderate (25 mm), severe (50 mm)	CSF serum	1, 6, 12, 24 hours, 3, 7, 14, 28 days	CSF and serum p-tau levels peaked at 12 hours; higher p-tau levels with more severe injury

Abbreviations: *CSF*, cerebrospinal fluid; *ELISA*, enzyme-linked immunosorbent assay; *GFAP*, glial fibrillary acidic protein; *IHI*, Infinite Horizons Impactor; *NSE*, neuron-specific enolase; *p-NFH*, phosphorylated-neurofilament heavy-chain; *p-tau* = phosphorylated-tau; *S100-β*, S100 calcium-binding protein-β; *UCH-L1*, ubiquitin C-terminal hydrolase L1.

correlations between the concentration of biomarkers and injury severity were not assessed.

Few studies have examined tau and its derivatives as biomarkers of acute SCI. A study in dogs with intervertebral disk herniation collected CSF immediately on admission to the veterinary hospital, and showed that CSF tau levels were higher in dogs with more severe injury, and dogs that improved had lower levels of CSF tau.[22] This study suggests that acute increases in CSF total tau may correlate to decreased likelihood of functional recovery. Using a clinically relevant model of thoracic impact-compression SCI in rats, we recently found a 550-fold increase in total tau levels in the CSF at 1 day postinjury compared to naïve animals where total tau was not detected in the CSF with Simoa digital immunoassay analysis.[23] We also showed for the first time the potential of p-tau as a specific biomarker for acute SCI. At 1 day post-SCI, significantly elevated levels of p-tau were detected in CSF and serum which diminished to baseline levels by 7 days postinjury.[23] Complementary to the tau analysis in biofluids, p-tau immunostaining of lesioned spinal cord tissue showed a similar time course and distribution pattern to β-APP, a marker of axonal injury. Both p-tau and β-APP positive swollen axons and terminal end-bulbs were evident at the lesion site with maximal immunoreactivity at 1 day post-SCI which was significantly reduced by 7 days postinjury.[23] In a recent study, rats with a thoracic weight drop SCI of varying severities showed higher tau levels in the CSF and serum from more severely injured rats.[24]

Collectively, there are few experimental studies of biomarkers that use clinically relevant and consistent models of SCI which reflect human pathology to maximize translational application. Thus, further studies investigating specific and reliable biomarkers for acute SCI are warranted using clinically relevant animal models and technological advancements have allowed more sensitive methods of analysis. Ultimately, biofluid biomarkers may also serve as surrogate indicators of the efficacy of therapeutic agents in preclinical studies of acute SCI which should also inform clinical trials.

Clinical studies of CSF and serum biomarkers

Several biomarkers quantified from the CSF of SCI patients with indwelling subarachnoid catheters inserted for monitoring spinal cord perfusion pressure,[5,6,25] including cytokines and the structural proteins GFAP, S100-β, NFL, and NFH, correlated to clinical parameters including motor function and recovery.[6] These biomarkers predicted injury severity and neurological recovery on a group basis, in addition to gross recovery changes among groups of patients such as from Grade A to Grade B on the ASIA scale. In individual patients, discriminant analysis based on levels of several biomarkers (IL-6, IL-8, MCP-1, tau, S100-β, and GFAP) demonstrated an 83.3% accuracy in predicting AIS grade improvement at 6 months. Moreover, all six of these protein biomarkers in CSF significantly correlated with ASIA motor score improvement, particularly in cervical SCI patients.

Casha et al.[26] also measured a number of interleukins and other proteins in the CSF of 29 patients undergoing treatment with minocycline, and found significant elevations of several biomarkers including IL-1β, GFAP, MMP-9, and C-X-C motif chemokine10 (CXCL10) whose levels correlated with injury severity and with worse neurological recovery at 1 year. Higher concentration of biomarkers has been found in CSF relative to blood.[5,6] Higher levels of biomarkers in the CSF reflect its close proximity to the spinal cord, while other proteins abundant in serum may interfere with the detection of relatively smaller amounts of the biomarker proteins. While CSF therefore may be most representative of the injured spinal cord parenchyma, blood-based biomarkers would obviously be advantageous because of relative safety and ease of collection. Increased serum levels of GFAP, NSE, and p-NFH were detected in 35 patients with acute traumatic SCI relative to the control group, and GFAP levels correlated with injury severity.[27] Kuhle et al.[28] found that serum NFL closely correlated with ASIA motor score at baseline and after 24 hours, and at 3–12 month motor outcome in 27 patients with acute SCI. Treatment with minocycline was also associated with a decreased level of NFL in patients with motor complete SCI, illustrating

the potential for biomarkers to be utilized as a surrogate outcome measure for evaluating the biological effect of a treatment for acute SCI.[29]

CSF is obviously of interest to others for exploring biomarkers, as exemplified by Yokobori et al. in a literature review that included seven acute SCI patients from whom CSF samples could be obtained through lumbar puncture for measurement of GFAP, UCH-L1, alphaII-spectrin breakdown products (SBDPs), and myelin basic protein (MBP).[2] Like others, they too found correlations between GFAP levels and injury severity, and were the first to show that CSF UCH-L1 is elevated in human SCI. This also illustrates how challenging it is to obtain sufficient numbers of acute SCI patients to do such exploratory biomarker investigations in human SCI. This group notably has employed broader proteomics approaches to biomarker discovery on these valuable human biospecimens and has endeavored to link this to experimental rodent data. This has identified other potential biomarker candidates such as PEA-15, transferring, and triosephosphate isomerase-1.[29] We also have been interested in expanding the scope of what can be gleaned from these very valuable human CSF and serum biospecimens with proteomic,[30] metabolomic,[31] and genomic studies.[8] These approaches have the potential not only to identify novel biomarkers of injury but to also provide empirical evidence about the basic biological responses to human SCI — not a trivial matter given the paucity of studies that describe the pathophysiology of injury in human subjects.

Aside from these CNS-specific biomarkers that reflect damage to the spinal cord parenchyma, there are also systemic changes that occur after SCI that may not be directly attributable to spinal cord but may nonetheless serve as biomarkers that reflect aspects of the condition such as injury severity, outcome, or response to treatment. For example, Tong et al. found that higher serum albumin following SCI was associated with neurological outcomes, irrespective of initial injury severity.[32]

Future applications

Unfortunately, the study of novel therapeutics in acute SCI will not get any easier or faster in the future if we continue to perform clinical trials in the same manner as we have for the past 30 years. The field of neurochemical biomarkers for SCI is still in development and by virtue of the relatively low incidence of acute SCI is similarly challenged by small research study cohorts. However, given what is known to date, there are biomarker candidates that could be utilized in clinical trials to at least explore their utility for stratifying injury severity and predicting outcome. For example, our previous demonstration that the combination of inflammatory (IL-6, IL-8, MCP-1) and "structural" (tau, S100β, GFAP) biomarkers in CSF could be used to correctly predict AIS grade conversion would suggest that this could be harnessed in a clinical trial.[6] We have recently embarked on a prospective trial of spinal cord perfusion pressure management in which CSF samples will continue to be collected through lumbar intrathecal catheters (the Canadian-American Spinal Cord Perfusion and Biomarker (CASPER) study — ClinicalTrials.gov NCT03911492). This trial design therefore affords the opportunity to evaluate how such CSF biomarkers might help to predict outcome and thus interpret whether observed neurologic recovery is related to the investigational treatment or simply expected based on the initial biological assessment.

Clearly, work in the future will need to refine what the most useful and practical biomarkers are (e.g., CSF vs. blood-based biomarkers) and determine what combinations of these are the most promising. In an effort to get the most complete objective assessment of the cord injury, it is almost intuitive that neurochemical biomarkers could be powerfully utilized in conjunction with MRI biomarkers in the future. Already, there are measurable aspects on MRI scans such as "intramedullary lesion length" and presence of hematoma that have been described as being predictive of neurologic recovery[33,34] and conceivably the combination of these features with neurochemical biomarkers will be useful.

Of note, advances in both MRI technology (e.g., diffusion tensor imaging) and in the neurochemical biomarker field will likely establish more precise objective measures for documenting injury severity and predicting outcome. Ultimately, such efforts will help to facilitate therapeutic advances for this devastating injury.

Take home messages

- Neurochemical biomarkers may play a critical role in the evaluation of novel therapies for SCI by reducing our reliance upon the ISNCSCI examination for classifying injury severity and determining efficacy.
- Many candidate biomarkers have been identified in animal studies of traumatic SCI, where analysis of CSF and serum samples has shown that proteins such as S100-β, GFAP, NSE, UCH-L1, and p-NFH are associated with injury severity and outcome.
- Human studies of biomarkers are much more limited by the availability of CSF and serum samples, but have also revealed promise for such proteins as GFAP, S100-b, tau, and NSE.
- At this stage, more research is needed to validate neurochemical biomarkers and establish them as useful tools in clinical trials or in clinical decision-making.

References

1. Lee RS, et al. Feasibility of patient recruitment into clinical trials of experimental treatments for acute spinal cord injury. *J Clin Neurosci* 2012;**19**(10):1338–43.
2. Yokobori S, et al. Acute diagnostic biomarkers for spinal cord injury: review of the literature and preliminary research report. *World Neurosurg* 2015;**83**(5):867–78.
3. Caprelli MT, Mothe AJ, Tator CH. CNS injury: post-translational modification of the tau protein as a biomarker. *Neuroscientist* 2019;**25**(1):8–21.
4. Zetterberg H, Smith DH, Blennow K. Biomarkers of mild traumatic brain injury in cerebrospinal fluid and blood. *Nat Rev Neurol* 2013;**9**(4):201–10.
5. Kwon BK, et al. Cerebrospinal fluid inflammatory cytokines and biomarkers of injury severity in acute human spinal cord injury. *J Neurotrauma* 2010;**27**(4):669–82.
6. Kwon BK, et al. Cerebrospinal fluid biomarkers to stratify injury severity and predict outcome in human traumatic spinal cord injury. *J Neurotrauma* 2017;**34**(3):567–80.

7. Pouw MH, et al. Biomarkers in spinal cord injury. *Spinal Cord* 2009;**47**(7):519–25.

8. Tigchelaar S, et al. MicroRNA biomarkers in cerebrospinal fluid and serum reflect injury severity in human acute traumatic spinal cord injury. *J Neurotrauma* 2019;**36**(15):2358–71.

9. Hulme CH, et al. The developing landscape of diagnostic and prognostic biomarkers for spinal cord injury in cerebrospinal fluid and blood. *Spinal Cord* 2017;**55**(2):114–25.

10. Nylen K, et al. Serum levels of S100B, S100A1B and S100BB are all related to outcome after severe traumatic brain injury. *Acta Neurochir* 2008;**150**(3):221–7. discussion 227.

11. Rainey T, et al. Predicting outcome after severe traumatic brain injury using the serum S100B biomarker: results using a single (24 h) time-point. *Resuscitation* 2009;**80**(3):341–5.

12. Chatfield DA, et al. Discordant temporal patterns of S100beta and cleaved tau protein elevation after head injury: a pilot study. *Br J Neurosurg* 2002;**16**(5):471–6.

13. Johnsson P, et al. Neuron-specific enolase increases in plasma during and immediately after extracorporeal circulation. *Ann Thorac Surg* 2000;**69**(3):750–4.

14. Miller CC, et al. Axonal transport of neurofilaments in normal and disease states. *Cell Mol Life Sci* 2002;**59**(2):323–30.

15. Hung KS, et al. Calpain inhibitor inhibits p35-p25-Cdk5 activation, decreases tau hyperphosphorylation, and improves neurological function after spinal cord hemisection in rats. *J Neuropathol Exp Neurol* 2005;**64**(1):15–26.

16. Loy DN, et al. Serum biomarkers for experimental acute spinal cord injury: rapid elevation of neuron-specific enolase and S-100beta. *Neurosurgery* 2005;**56**(2):391–7. discussion 391-7.

17. Cao F, et al. Elevation of neuron-specific enolase and S-100beta protein level in experimental acute spinal cord injury. *J Clin Neurosci* 2008;**15**(5):541–4.

18. Ma J, et al. Plexus avulsion and spinal cord injury increase the serum concentration of S-100 protein: an experimental study in rats. *Scand J Plast Reconstr Surg Hand Surg* 2001;**35**(4):355–9.

19. Shaw G, et al. Hyperphosphorylated neurofilament NF-H is a serum biomarker of axonal injury. *Biochem Biophys Res Commun* 2005;**336**(4):1268–77.

20. Ueno T, et al. Hyperphosphorylated neurofilament NF-H as a biomarker of the efficacy of minocycline therapy for spinal cord injury. *Spinal Cord* 2011;**49**(3):333–6.

21. Yang Z, et al. Temporal profile and severity correlation of a panel of rat spinal cord injury protein biomarkers. *Mol Neurobiol* 2018;**55**(3):2174–84.

22. Roerig A, et al. Cerebrospinal fluid tau protein as a biomarker for severity of spinal cord injury in dogs with intervertebral disc herniation. *Vet J* 2013;**197**(2):253–8.

23. Caprelli MT, Mothe AJ, Tator CH. Hyperphosphorylated tau as a novel biomarker for traumatic axonal injury in the spinal cord. *J Neurotrauma* 2018;**35**(16):1929–41.

24. Tang Y, et al. Serum and cerebrospinal fluid tau protein level as biomarkers for evaluating acute spinal cord injury severity and motor function outcome. *Neural Regen Res* 2019;**14**(5):896–902.

25. Kwon BK, et al. Intrathecal pressure monitoring and cerebrospinal fluid drainage in acute spinal cord injury: a prospective randomized trial. *J Neurosurg Spine* 2009;**10**(3):181–93.

26. Casha S, et al. Cerebrospinal fluid biomarkers in human spinal cord injury from a phase II minocycline trial. *J Neurotrauma* 2018;**35**(16):1918–28.

27. Ahadi R, et al. Diagnostic value of serum levels of GFAP, pNF-H, and NSE compared with clinical findings in severity assessment of human traumatic spinal cord injury. *Spine* 2015;**40**(14):E823–30.

28. Kuhle J, et al. Serum neurofilament light chain is a biomarker of human spinal cord injury severity and outcome. *J Neurol Neurosurg Psychiatry* 2015;**86**(3):273–9.

29. Moghieb A, et al. Differential neuroproteomic and Systems biology analysis of spinal cord injury. *Mol Cell Proteomics* 2016;**15**(7):2379–95.

30. Streijger F, et al. A targeted proteomics analysis of cerebrospinal fluid after acute human spinal cord injury. *J Neurotrauma* 2017;**34**:2054–68.

31. Wu Y, et al. Parallel metabolomic profiling of cerebrospinal fluid and serum for identifying biomarkers of injury severity after acute human spinal cord injury. *Sci Rep* 2016;**6**:38718.

32. Tong B, et al. Serum albumin predicts long-term neurological outcomes after acute spinal cord injury. *Neurorehabil Neural Repair* 2018;**32**(1):7–17.

33. Aarabi B, et al. Intramedullary lesion length on postoperative magnetic resonance imaging is a strong predictor of ASIA impairment scale grade conversion following decompressive surgery in cervical spinal cord injury. *Neurosurgery* 2017;**80**(4):610–20.

34. Kurpad S, et al. Impact of baseline magnetic resonance imaging on neurologic, functional, and safety outcomes in patients with acute traumatic spinal cord injury. *Global Spine J* 2017;**7**(Suppl. 3):151s–74s.

Clinical trials: pharmacological approaches to enhance neural repair and regeneration after spinal cord injury

Armaan K. Malhotra[1], Laureen D. Hachem[1], Jetan H. Badhiwala[1], Mark R.N. Kotter[2], Michael G. Fehlings[1]

[1]Division of Neurosurgery, Department of Surgery, University of Toronto, Toronto, ON, Canada;
[2]Division of Neurosurgery, Department of Clinical Neurosciences, University of Cambridge, Cambridge, United Kingdom

Introduction

Spinal cord injury (SCI) is associated with significant physical and psychological morbidity among patients and carries major economic ramifications to both individuals and health-care systems. The ability to mitigate the damage caused by acute SCI and impact the natural history of this condition with pharmacotherapy has been a naturally enticing, yet immensely complex, prospect for many years. Advancements on this front would serve to reduce disability and have major clinical, social, and economic impact.

Over the past 40 years there have been major efforts to find an effective drug to improve functional recovery after SCI. This goal has stemmed directly from advancements in our understanding of the pathobiological mechanisms of SCI secondary injury. A number of agents have undergone preclinical examination in animal models, with a small subset of these reaching clinical trials. Even among those examined in patients, few have shown a meaningful clinical improvement in motor function. Generally, candidate drugs have been classified into either neuroprotective agent based on their mechanistic role in enhancing cell survival or neuroregenerative agents for their role in enhancing repair mechanisms following SCI.

In this chapter, we outline candidate drugs that have previously or are currently undergoing clinical trials for use in SCI patients. We first discuss agents that have completed clinical trial testing in SCI, followed by those under active clinical testing, and lastly highlight agents that have just entered clinical trials. For each agent, we highlight the mechanism of action, preclinical data, and clinical trial design and outcomes (Table 24.1). Where relevant we outline pitfalls in trial design and limitations in order to inform future clinical trials.

© 2022 Elsevier Inc. All rights reserved.

TABLE 24.1   Drugs tested for use in SCI.

Agent	Mechanism of action	Clinical trial results
Methylprednisolone sodium succinate (MPSS)	MPSS is a synthetic glucocorticoid with numerous mechanisms of action. It has been described in several physiological capacities such as modulating postinjury peroxidation of myelin, vascular endothelium, glial and neuronal membrane components.	Five studies have examined MPSS in human SCI. Refer to Table 24.2 for details regarding these clinical trial results.
Nimodipine	Nimodipine is a dihydropyridine calcium channel blocker that binds L-type voltage-gated calcium channels. In acute SCI, calcium signaling has been implicated in driving secondary neuronal and glutamate-induced excitotoxicity.	Nimodipine was evaluated through prospective placebo-controlled randomized trial examining the safety and efficacy of nimodipine compared to methylprednisolone and control in acute SCI patients. There was no significant improvement between placebo, suggesting a lack of therapeutic benefit of nimodipine in SCI.
Minocycline	Minocycline (7-dimethylamino-6-dimethyl-6-deoxytetracycline) is a semisynthetic tetracycline antibiotic with antioxidant properties. This antioxidant quality is believed to be the reason behind its ability to modulate components of the secondary injury cascade in SCI.	Minocycline was assessed through Phase II blinded placebo-controlled randomized trial in acute SCI. There were no statistically significant differences in functional outcome among 52 patients between placebo and minocycline groups despite a trend toward improved function at 6 months in cervical motor-incomplete subgroup. Minocycline in acute Spinal Cord Injury (MASC) is a Phase III trial that was recently terminated because of enrollment issues.
GM-1 ganglioside	GM-1 ganglioside refers to a specific ganglioside molecule (monosialotetrahexosylganglioside). The mechanism of action of GM-1 is related to promotion of endogenous neurotrophin action.	Multicenter prospective randomized blinded placebo-controlled trial included 760 patients from 28 centers. Significant difference in functional improvement at 8 weeks across injury groups with GM-1 ganglioside, but no difference compared to placebo by 6 months.
VX-210/Cethrin	VX-210 is an enzyme derivative of C3 transferase with Rho-inhibiting properties. Numerous endogenous mechanisms act to inhibit axonal regrowth through activation of intracellular Rho pathways. Inhibition of Rho has therefore been shown to enhance axon regeneration.	Spinal Cord Injury Rho Inhibition Investigation (SPRING) examined the efficacy of VX-210 in acute cervical traumatic SCI. At study termination only 37 patients had progressed to 6-month follow-up with no statistically significant difference in primary outcome between VX-210 and placebo therapy arms.
Gacyclidine		
Riluzole	Synthetic benzothiazole with neuroprotective action through sodium channel blockade and modulation of glutamatergic excitotoxicity.	Riluzole in Acute Spinal Cord Injury Study (RISCIS) has concluded enrollment. Awaiting clinical trial results.
Anti-Nogo-A antibody	Anti-Nogo-A is an IgG recombinant human antibody that binds the Nogo-A protein-binding domain and inhibits its endogenous activity. Nogo-A is a myelin-associated membrane protein with neurite growth inhibiting properties.	Nogo Inhibition in Spinal Cord Injury (NISCI) is currently ongoing. Awaiting clinical trial results.

TABLE 24.1 Drugs tested for use in SCI.—cont'd

Agent	Mechanism of action	Clinical trial results
Anti-RGM-A antibody (Elezanumab)	Elezanumab is an IgG human monoclonal antibody with indirect neuroregenerative properties through binding and inhibition of repulsive guidance molecule A (RGMa). The endogenous RGMa activity includes axonal growth inhibition.	Ongoing Phase II placebo-controlled randomized control trial enrollment.
Thyrotropin-releasing hormone	Thyrotropin-releasing hormone is endogenously secreted peptide. Numerous described in vivo mechanisms of neuroprotection.	Prospective single-center placebo-controlled randomized-controlled trial with 20 patients showed no difference in primary outcome among TRH-treated patients compared to placebo. In a subgroup of incomplete injury patients, TRH was associated with significantly improved motor and sensory scores, though cohort size greatly limits the applicability of this evidence.
AC105	AC105 is a formulation made up of magnesium chloride in polyethylene glycol 3350 as a stabilizing agent. Magnesium itself is an NMDA receptor antagonist and therefore has been shown to exert neuroprotective benefits through reduction in excitotoxicity.	Phase II double-blinded placebo-controlled trial terminated after 15 patients were enrolled. Insufficient rate of enrollment as well as low patient retention.

Completed clinical trials

Methylprednisolone sodium succinate

Mechanism of action

Methylprednisolone succinate (MPSS, Solu-medrol) is a synthetic glucocorticoid with several physiological mechanisms of action. Glucocorticoid function is dictated by binding with the glucocorticoid receptor, an intracellular nuclear receptor that plays a role in transcriptional regulation of metabolism, inflammation, and development.[1] Specific mechanisms of glucocorticoids on CNS tissue involve reducing postinjury peroxidation of myelin, vascular endothelium along with glial and neuronal membrane components. Hall et al. studied the mechanisms of MPSS action in a cat model of SCI, demonstrating attenuation of lipid peroxy products and improved Na^+/K^+ ATPase functionality in the injured spinal cord following administration of MPSS at 15 mg/kg and 30 mg/kg 1 hour after injury.[2] The peroxidation process is triggered by free radicals generated after SCI, hemorrhage and ischemia, the combined effect in SCI resulting in a loss of enzymatic activity and cell destruction. MPSS has also been shown to reduce inflammatory signaling by attenuation of the pro-inflammatory mediator TNF-alpha, attenuate BBB permeability, reduce cord edema, and enhance blood flow to the injured cord following SCI.[2,3]

Preclinical data

The role of MPSS in animal models of SCI has been extensively studied. In a T8 cord contusion model in rats, intravenous MPSS administered 5 minutes, 2, and 6 hours following SCI resulted in improved open-field walking scores between days 8—43 postinjury.[4] Subsequent dose—response assessment showed improved motor recovery at day 29 postinjury in rats treated with a 60 mg/kg intravenous bolus at 5 minutes, 2, 4, and 6 hours following injury compared to 30 mg/kg or control. There was also greater tissue sparing in the 60 mg/kg subgroup. Rabchevsky et al. examined MPSS at a similar dose of 60 mg/kg followed by 5.4 mg/kg/hr for 24 hours in a T10 rat impaction contusion model of SCI.[5] There was no significant difference in lesion size at 6 weeks between MPSS and control groups. Furthermore, there was no time period at which a significant difference in locomotor score or functional outcome was present between saline and MPSS-treated rats. In a systematic review and meta-analysis specifically examining animal model SCI outcome data of MPSS therapy,[6] among 62 studies, it was found that 34% showed beneficial MPSS effect on functional outcome, 58% showed no effect, and 8% described mixed effects.

Clinical data

Five major human SCI trials have been conducted on the use of MPSS in acute traumatic SCI. These trials are detailed below and also summarized in Table 24.2.

NASCIS I

The National Acute Spinal Cord Injury Study I (NASCIS I) was the first trial to investigate the potential neuroprotective benefits of MPSS in SCI. NASCIS I was a multicenter, double-blinded randomized trial which compared (1) high-dose MPSS 1,000 mg bolus on admission and 1000 mg daily for 10 days (total 11,000 mg)

TABLE 24.2 Summary of clinical trials examining methylprednisolone sodium succinate (MPSS) in human spinal cord injury.

Trial/lead author name	Summary of clinical trial result
NASCIS I	The National Acute Spinal Cord Injury Study I (NASCIS I) was a multicenter double-blind randomized trial comparing high- and low-dose MPSS for acute traumatic spinal cord injury. There were 330 patients recruited from 9 centers. The major criticisms in this trial included lack of a placebo treatment arm. No significant differences between high- and low-dose treatment groups were found.
NASCIS II	NASCIS II was a multicenter randomized double-blinded placebo-controlled trial investigating naloxone and MPSS in acute SCI. There were 487 patients included in the trial. The primary outcome examined was neurological recovery at 6 months. There was sensory improvement but no motor benefit among patients treated with MPSS compared to placebo. Despite the primary study outcome was negative, subgroup analysis showed patients receiving infusions of MPSS within 8 hours exhibited significantly improved motor and sensory scores at 6 months in complete and incomplete patients.
NASCIS III	The NASCIS III trial aimed to identify the optimal duration of MPSS therapy. NASCIS III was a multicenter double-blinded randomized-controlled trial involving 499 patients with acute SCI within 8 hours of injury. The primary outcome measure was change in motor score and functional independence measure (FIM) at 6 weeks and 6 months postinjury. At the 6-month trial endpoint postinjury, there was a nonsignificant trend toward improved motor outcome in MPSS-treated patients treated for 48 hours compared to 24 hours ($P = .07$). There was an increased observed severe sepsis rate with MPSS administration ($P = .07$) and severe pneumonia rate ($P = .02$) in the 48-hour treatment group compared to the 24-hour MPSS treatment group.
Otani et al.	Using the NASCIS II protocol, 158 patients were randomly assigned to NASCIS II MPSS protocol following SCI or standard medical care. At 6 months posttherapy there was increased motor recovery in the group of patients treated with MPSS. Lack of blinded study design and length of follow-up were criticisms of this trial.
Matusmoto et al.	NASCIS II MPSS administration was utilized in a randomized prospective double-blinded placebo-controlled study. There were 46 adult cervical SCI patients recruited within 8 hours of injury. The study showed an increase in pulmonary complications ($P = .009$), particularly in patients over the age of 60 ($P = .029$), and increased gastrointestinal complications ($P = .036$) with MPSS therapy. Methodological critique of this study cited lack of internal matching between placebo and treatment arms regarding neurological injury severity.

with (2) standard dose MPSS 100 mg bolus on admission and 100 mg daily for 10 days (1100 mg).[7] Three hundred and thirty patients were recruited across nine centers. Patients with acute traumatic spinal cord injury presenting within 48 hours were included. Those with cauda equina or nerve root injury, severe comorbidities, prior receipt of MPSS, age less than 13 years, or concomitant major traumatic injuries were excluded. Over 60% of patients presented within 6 hours of SCI. Major methodological concerns arose from the absence of a placebo arm, which was precluded by a lack of clinical equipoise. There was no significant difference in motor function, sensory function, or mortality in low- versus high-dose treatment groups at 6 weeks and 6 months. There was an increase in wound infections in the higher treatment group (RR 3.55). While negative for its primary outcome, NASCIS I raised the question of MPSS efficacy in SCI. This finding was contradictory to the near-dogmatic utilization of MPSS in clinical management of SCI at the time, which was based solely on preceding animal studies.

NASCIS II

NASCIS II was designed to overcome pitfalls of NASCIS I and to specifically include a placebo-control group.[8] NASCIS II was a multi-center randomized double-blinded placebo-controlled trial examining naloxone, an emerging alternate drug candidate for SCI, and MPSS in acute SCI. Patients were randomly allocated to three treatment arms: (1) intravenous MPSS 30 mg/kg in 1 hour followed by 23 hours of 5.4 mg/kg MPSS infusion, (2) intravenous naloxone 5.4 mg/kg in 1 hour followed by 23 hours of 4.0 mg/kg naloxone, and (3) placebo. Exclusion criteria were similar to NASCIS I with further exclusion of patients with opiate addiction. Neurological assessments were performed at initial presentation, in addition to 6 weeks and 6 months postinjury. There were 487 included patients, of which approximately 60% had complete injuries at presentation and average

time from injury to bolus dose was 8.7 ± 3.0 hours. There was a high degree of adherence to study design with 80% of patients receiving medication within the designated time period and 92.1% completing dosing schedules as planned. The primary outcome of neurological recovery at 6 months showed improvement in sensory testing ($P = .012$ pinprick score difference, $P = .042$ touch score difference) but no motor benefit among patients treated with MPSS compared to placebo. Naloxone did not provide any significant improvements in motor or sensory scores at 6 months compared to placebo. Following subgroup analysis of patients receiving infusion within 8 hours, there were significant improvements in motor and sensory scores at 6 months in both complete and incomplete SCI patients. While there was a similar major morbidity and mortality profile across the three treatment arms, there was a nonsignificant trend toward a higher infectious complication rate in the MPSS group (7.6%) compared to placebo (3.6%). These findings generated substantial debate in the clinical community and while there was early adoption of the NASCIS II protocol for acute SCI, significant criticisms arose regarding misinterpretation of findings and concerns that the primary study outcome was negative.[9,10]

NASCIS III

Following NASCIS II, the NASCIS III trial aimed to delineate the optimal duration of MPSS therapy in acute SCI.[11] The trial also included trilazad mesylate, a direct lipid peroxidation inhibitor, with the hypothesis that trilazad mesylate would exhibit fewer complications compared to MPSS.[12] This double-blinded randomized-controlled trial involved 499 patients with acute SCI within 8 hours of injury recruited from 16 centers across North America. Primary outcome measures were change in motor scores and Functional Independence Measure (FIM) at 6 weeks and 6 months postinjury. All patients received MPSS bolus infusion at 30 mg/kg and underwent subsequent randomization between the following treatments: (1) 5.4 mg/kg MPSS hourly for 24 hours, (2)

5.4 mg/kg MPSS hourly for 48 hours, (3) trilazad mesylate 2.5 mg/kg bolus infusion every 6 hours for 48 hours. Similar inclusion and exclusion criteria were used to the prior NASCIS trials, with an additional exclusion criterion for patients weighing more than 109 kg. At 6 months postinjury there was a nonsignificant trend toward improvement in motor scores among patients treated with MPSS for 48 hours compared to 24 hours ($P = .07$). In subgroup analysis, patients presenting 3–8 hours postinjury receiving the 48-hour MPSS protocol showed significantly improved motor scores at 6 months compared to those in the 24 hour MPSS protocol ($P = .01$). Furthermore, greater improvement in FIM was noted at 6 months ($P = .08$). This effect came with an increase in severe sepsis rates ($P = .07$) and severe pneumonia ($P = .02$) compared to the 24-hour MPSS treatment group. Trilazad treatment for 48 hours showed similar outcomes to patients treated with MPSS for 24 hours. It was concluded from this study that patients presenting within 3 hours of SCI should undergo treatment with MPSS for 24 hours and those presenting 3–8 hours postinjury should continue therapy for 48 hours, though the benefit of motor recovery at 6 months must be measured against potential risks of sepsis and pneumonia in the shorter term.

Japanese MPSS trials

Otani et al. utilized the NASCIS II treatment protocol to validate its efficacy in a subsequent trial in SCI patients.[13] In this study, 158 patients were randomly assigned to NASCIS II protocol MPSS within 8 hours of injury or standard medical care for SCI. There was a nonsignificant increase in motor recovery noted in the MPSS arm (3.9 points) at the 6-month trial endpoint. This study was the subject of major criticism given lack of rigorous blinding between treatment groups and significant loss to follow-up. Matsumoto et al. further assessed the complication profile of NASCIS II MPSS administration in a randomized prospective double-blinded placebo-controlled study. Forty-six adult patients with cervical SCI were recruited and

MPSS treatment was administered within 8 hours of injury. There was reported increase in pulmonary complications ($P = .009$), especially in patients over 60 years of age ($P = .029$), as well as increased gastrointestinal complications ($P = .036$) with MPSS therapy. This study suffered from a lack of internal matching between placebo and treatment arms regarding neurological injury severity. Specifically, among the eight patients with Frankel grade D injuries, 7 were included in the placebo arm and thus differences in neurological injury severity could confound results.

The use of MPSS in acute traumatic SCI has remained an area of active debate, with systematic reviews and meta-analyses yielding differing recommendations following inclusion of similar primary studies.[14] Based on the current literature, application of MPSS to human SCI must be done in a measured fashion with acknowledgment and understanding of biases in the body of evidence available to clinicians. Furthermore, the utility of MPSS in elderly patients requires further assessment given the unique clinical considerations within this growing patient population. Fehlings et al. conducted a systematic review and meta-analysis to specifically investigate the role of MPSS therapy in acute SCI compared to no pharmacological treatment. There were 13 articles that met inclusion criteria, 7 of which were other systematic reviews. They concluded that there was moderate evidence supporting the 24-hour MPSS protocol had no impact on long-term neurological recovery when all postinjury outcomes were considered. Another conclusion drawn with moderate evidence was that MPSS administration within 8 hour received 3.2 motor recovery points ($P = .04$) compared to placebo.

The most current evaluation of the literature conducted by a Multidisciplinary Guideline Development Group found that MPSS administered within 8 hours of injury may lead to modest improvements in motor scores in adult patients.[15,16] Further recommendations made by this guideline document included suggesting not to offer MPSS therapy to patients presenting outside the 8-hour time window based on moderate quality evidence. Finally, a weak recommendation was made not to offer 48-hour MPSS administration protocol following acute traumatic SCI. Current practice guidelines thus recommend a 24-hour infusion of MPSS as an option in patients presenting with acute SCI within 8 hours postinjury.

Minocycline

Minocycline (minocycline hydrochloride) is a semisynthetic tetracycline antibiotic that has been utilized for over 30 years in the context of infectious diseases. In addition to its traditional use as an antibiotic, minocycline has numerous nonantibiotic applications related to its antiinflammatory, antiapoptotic, and tumor metastasis inhibiting properties.[17] In light of these properties, minocycline has been extensively studied as a potential neuroprotective agent in the setting of SCI.

Mechanism of action

Minocycline (7-dimethylamino-6-dimethyl-6-deoxytetracycline) is a highly lipophilic molecule and therefore has a propensity to pass through the blood—brain barrier.[18] Its antioxidant properties have been linked to its ability to scavenge free radicals through its phenolic hydroxyl group in addition to actions on nitric oxide synthase and inflammatory pathways.[19] Minocycline has been shown to modulate many components of the secondary cascade of SCI in vitro. In spinal cord tissue culture experiments, treatment with minocycline mitigated the effects of glutamate excitotoxicity, microglial activation, and subsequent interleukin and nitric oxide metabolite production on neuronal cell death.[20] Minocycline has also been shown to reduce NMDA-mediated excitotoxicity, calcium influx, and mitochondrial uptake of calcium ions.[21] In a murine macrophage tissue culture model, minocycline reduced the effect of stimulated inducible nitric oxide synthase resulting in reduced nitrite accumulation.[22] Furthermore,

minocycline has demonstrated direct inhibitory effects on MMP-9 by way of interacting with zinc moieties that are critical for enzymatic activity.[23] MMP-9 is an inflammatory mediator in the pathogenesis of SCI leading to abnormal vascular permeability, where its inhibition improves blood—spinal barrier disruption and reduces apoptotic cell death.[25] While the exact downstream molecular mediators of minocycline's effect in the setting of SCI is subject to ongoing research, it is posited that minocycline action is linked to modulation of inflammation through regulating mitogen-activated protein kinase (MAPK) and phosphoinositide 3-kinase (PI3K) pathways, resulting in downstream effects on arachidonic acid metabolism, cytokine production, and reactive oxygen species propagation.[24]

Preclinical data

Minocycline has been extensively studied in preclinical animal models of SCI. Using a C7/8 dorsal column transection model in rats, Stirling et al.[26] found a reduction in cavitation size, lesional microglia, and macrophage density as well as reduced oligodendrocyte apoptosis in proximal and distal ascending sensory tracts in minocycline-treated animals compared to controls (intraperitoneal injection 30 minutes prior to SCI, 8 hours following injury, and twice daily for 48 hours). Furthermore, minocycline was associated with improved functional outcomes. This finding was further corroborated by numerous subsequent studies. Intraperitoneal minocycline administered 1 hour postinjury for a duration of 5 days in a T3/4 clip compression rodent model of SCI[27] was associated with reduced mortality (20%, N = 5 of 25 mice vs. 61.5%, N = 16 of 26 mice), increased BBB locomotor scores, reduced lesion size, and increased axonal sparing. Minocycline (administered every 12 hours for 3 days after SCI) led to a reduction in oligodendrocyte apoptosis which was mediated in part by reduction in p38 MAPK phosphorylation and subsequent decrease in proNGF synthesis by microglia. Furthermore, in an intraaortic balloon catheter rat model of spinal cord ischemia, minocycline treatment significantly increased BBB locomotor scores and reduced activated microglia and white matter vacuolations.[28]

Importantly, not all preclinical studies have demonstrated minocycline efficacy following SCI. Two notable studies failed to show significant functional improvement or histopathological evidence of reduced secondary injury parameters following minocycline treatment in contusive models of SCI.[29,30] The conflicting reports may be due to numerous factors including differences in animal models, injury severities, dosing regimens, and routes of administration across studies. Nevertheless, a large body of favorable preclinical evidence ultimately formed the impetus for subsequent testing of minocycline in human clinical trials of SCI.

Clinical data

Minocycline had been extensively tested in other clinical settings prior to its application to SCI. Therefore, by the time the wealth of preclinical data had accumulated to fuel subsequent SCI trials, there were well-described safety data regarding minocycline use in humans. Over an 8-week placebo-controlled safety and tolerability study of 100 and 200 mg daily minocycline in 60 patients with Huntington's disease, minocycline was concluded to be safe and well tolerated with minimal adverse events.[31] Specifically, there were no significant differences in adverse events between placebo and minocycline groups with the only serious adverse event reported as a fall with head trauma. In trials conducted in Parkinson's disease patients, minocycline was safe and well tolerated over a period of 18 months at 200 mg daily.[32] In a placebo-controlled Phase I/II trial in amyotrophic lateral sclerosis patients, a cohort of 23 patients received minocycline doses up to 400 mg daily for 8 months demonstrating an upper limit of oral tolerated dosing at 387 mg daily.[33] At this dose, there was a trend toward gastrointestinal side effects with elevation in liver enzymes and blood urea

nitrogen. Furthermore, intravenous high-dose (10 mg/kg) minocycline administration was assessed in the stroke population for 72 hours without major side effects.[34]

Phase II trial: MASC

A Phase II placebo-controlled randomized trial of minocycline administered within 12h of acute SCI was conducted by Casha et al. between 2004 and 08.[35] Inclusion criteria consisted of patients with age greater than 16 years, cervical or thoracic SCI, and English language speaking. Exclusion criteria included preexisting hypersensitivity to minocycline, systemic lupus erythematosus, renal or hepatic disease, leukopenia, pregnancy, or presence of concomitant injuries precluding accurate assessment or neurological examinations. Patients were stratified into motor-complete (American Spinal Injury Association (ASIA) A or B) injuries, motor-incomplete (ASIA C or D) injuries, and central cord injuries (ASIA C or D with mean lower extremity score - > upper extremity score). As part of the study protocol, all patients received lumbar catheters for CSF sampling and monitoring of spinal perfusion pressure. Surgical decompression and/or stabilization occurred within 24 hours of injury if indicated. All study participants and personnel were blinded to either minocycline or equal volume saline intravenous administration through subclavian lines. The first five patients enrolled in the minocycline arm received a dose of 200 mg twice daily and subsequently this was increased to target a steady state of $7-10$ µg/mL serum concentrations. The regimen consisted of an 800 mg loading dose, tapered by 100 mg each 12 hours to a dose of 400 mg, then completed at this dose for the remainder of 14 total doses or 7 days following the first dose. Serial neurological examinations were conducted throughout the study period with a final time point of 12 months post-SCI.

There were 52 subjects included in the trial, 27 of whom were randomized to receive minocycline; 36 patients had motor-complete SCI (ASIA

A or B), 10 patients had motor-incomplete SCI (ASIA C or D), and 6 patients had central cord syndrome. There were two mortalities in this study: one case of high cervical quadriplegia in the placebo arm secondary to respiratory distress syndrome and another case of narcotic overdose at 6 months in the high-dose minocycline group. There were no differences between adverse events of low-dose minocycline (200 mg twice daily), high-dose minocycline (400 mg twice daily), and placebo. The central cord cohort was not included in the final outcome analysis due to small sample size. Using the ASIA motor score assessment, there was a composite 6-point greater motor score in patients receiving minocycline than those receiving placebo among 44 patients at 1 year follow-up ($P = .20$). There was no benefit shown in thoracic SCI ($N = 17$). In cervical SCI ($N = 25$), there was a 14-point motor score difference in the minocycline treatment arm ($P = .05$). Importantly there were more motor-complete injury cases in the placebo cohort compared to minocycline cohort (68% vs. 44%). The majority of improvements were noted in lower extremities, with minimal differences in upper-extremity motor scores (UEMSs) between placebo and treatment arms. There were no significant differences in sensory neurological outcomes between treatment groups, though there was a trend toward improved sensory testing in cervical motor-incomplete patients in the minocycline group. There were no statistically significant differences in functional recovery between treatment groups, although a trend was observed toward improved function at 6 months in cervical motor-incomplete cases.

While the phase II randomized-controlled trial was well designed and achieved good adherence to study protocols, it was limited by a small sample size. As a result, the groundwork was laid for larger multicenter randomized trials to ascertain whether a true treatment effect exists. Minocycline in Acute Spinal Cord Injury (MASC) was a Phase III trial with an aim of 248 patients for recruitment among four treatment centers in Canada and Australia (NCT01828203).

Unfortunately, the MASC trial was recently terminated due to enrollment challenges.

Gacyclidine

Mechanism of action

Early work demonstrated augmented neuroprotective effect of PCP derivatives through replacement of a phencyclidine phenyl group by 2-thiophene (TCP).[36] Gacyclidine is a synthetic derivative of TCP with NMDA receptor noncompetitive antagonist pharmacodynamics.[37,38] Based on postulated neuroprotective roles related to glutamate inhibition, gacyclidine became the subject of further research in the field of SCI. Gacyclidine is a racemic mixture and both enantiomers exhibit rapid and extensive penetration into spinal cord tissue, with no significant pharmacokinetic differences detectable via microdialysis experiments after drug administration.[39] The (−) enantiomer, however, has been shown to exhibit greater neuroprotective effects in vitro.[40]

Preclinical data

Gaviria et al. examined the in vivo neuroprotective effect of gacyclidine in a photochemical lesioning rodent model of SCI.[41,42] Using an open-field walking test, there was significantly improved function following gacyclidine administration, most pronounced at a dose of 1 mg/kg administered 10 minutes after SCI. There was no significant difference in inclined-plane stability testing between treated and untreated groups. Gacyclidine administration was associated with a significantly smaller region of cord damage at the lesion core and proximal to the lesion. The degree of cord damage was significantly reduced in the 1 mg/kg cohort compared to higher doses of 2.5 mg/kg and 5 mg/kg. In the time-dependent model there were significant improvements in open field walking test in early neuroprotection groups (10- and 30-minute gacyclidine administration) compared to late groups (60- and 120-minute gacyclidine administration) or control. These findings demonstrate gacyclidine exhibits dose-dependent and time-dependent neuroprotective effect in SCI in vivo with improvement in both histological and functional outcomes.

A subsequent study replicating previous findings was conducted in a clinically relevant contusive model of SCI.[41,42] In this model intravenous gacyclidine was administered at 1 mg/kg in time intervals from 10 to 120 minutes after SCI compared to placebo. There was significant improvement in walking recovery ($P < .0125$) in the 10-minute posttreatment group compared to untreated animals at 18 days following injury. Once again, there was histopathological correlation with reduced epicenter size in gacyclidine-treated animals. Optimal results were observed in the 10-minute post-SCI injury cohort thus demonstrating the acuity of excitotoxic insults after SCI. These preclinical studies were the foundation for subsequent clinical trials with gacyclidine in human SCI.

Clinical data

In 2003, Tadie et al. conducted a clinical trial consisting of 272 patients examining the efficacy of gacyclidine in SCI.[43] Inclusion criteria comprised patients with SCI weighing less than 110 kg, aged 18–65, and with maximum ASIA motor score of 15 for most significantly injured lower extremity. There was requirement for decompressive surgery within 8 hours and no hypotension less than systolic blood pressure of 90 mmHg for more than 15 minutes. Furthermore, exclusion criteria included pregnancy, altered level of consciousness, and patients unable to receive gacyclidine within 2 hours of injury. Randomization took place between placebo and three gacyclidine doses (0.005, 0.010, and 0.020 mg/kg). The drug was administered within 2 hours of injury and again 4 hours after the first dose. The trial results showed no significant difference between ASIA scores or FIM at 1 year. Despite the preclinical rationale and efficacy in animal models as a neuroprotective

agent, a meaningful therapeutic benefit did not materialize when applied in the clinical setting.

Nimodipine

Mechanism of action

Nimodipine is a Food and Drug Administration (FDA)-approved dihydropyridine calcium channel blocker that binds L-type voltage-gated calcium channels and acts as a negative allosteric modulator of calcium channel function.[44] L-type calcium channels are ubiquitously expressed in CNS neurons and have been implicated in gene expression, excitability, and signaling homeostasis.[45,46] Dysregulated calcium signaling is a driver of secondary neuronal injury in acute SCI. In glutamate-mediated excitotoxicity, initial cell membrane depolarization can lead to downstream L-type calcium channel activation and subsequent amplified calcium influx. Through its role in calcium channel blockade, nimodipine may protect neuronal cells from glutamate-mediated excitotoxicity.[47] Beyond its role as a calcium channel blocker, nimodipine has also been studied for its modulatory properties in neuroinflammation. L-type calcium channels are found on microglia and nimodipine has been shown to inhibit interleukin release from amyloid-stimulated microglial cells.[48] Furthermore, nimodipine attenuates lipopolysaccharide-induced degeneration of dopaminergic neurons in vitro.[49,50]

Nimodipine has also been studied for its potential role in attenuating ischemic insults in the setting of SCI. Following injury, blood flow to the spinal cord is impaired[51,52] with one purported mechanism involving calcium channel–induced relaxation of blood vessels. Guha et al. demonstrated a dose–response relationship between healthy rat spinal cord blood flow and nimodipine administration[51] and subsequent experiments revealed that nimodipine improves posttraumatic spinal cord blood flow in a clip compression model of SCI.[52] Nimodipine has been the subject of extensive interest in SCI stemming from its antineuroinflammatory action, ability to supplement spinal cord blood flow and prevent ischemia as well as mitigate sequelae of excitotoxic calcium release.

Preclinical data

Early experiments examining nimodipine combined with dextran in a rat model of SCI demonstrated improved spinal cord blood flow measurements and recovery of evoked potential activity.[53] This was argued to be either an effect of dextran causing volume expansion, or that correction of posttraumatic hypotension was necessary for nimodipine activity. A subsequent study found no effect on spinal cord blood flow or electrophysiological monitoring when varying doses of nimodipine alone were used in a rat SCI model.[54]

Ross et al. performed a blinded randomized trial of nimodipine and MPSS in a T1 clip compression model of SCI.[55] Rats (n = 44) were randomized to four treatment groups: (1) intravenous placebo bolus followed by 0.02 mg/kg/h nimodipine for 8 hours and 20 mg/kg enteral three times daily for 7 days, (2) intravenous placebo bolus followed by 0.02 mg/kg/h nimodipine for 8 hours and placebo three times daily for 7 days, (3) 30 mg/kg MPSS bolus then 5.4 mg/kg/hr MPSS for 8 hours and placebo three times daily for 7 days, (4) placebo IV, infusion and three times daily for 7 days. There was no significant survival benefit across treatment groups nor was there a significant difference in retrogradely labeled red nucleus neurons between groups. Functional outcomes assessed by inclined-plane scores were significantly improved in nimodipine group 1 compared to group 3 ($P < .05$), but not difference when compared to group 4. Group 1 appeared to have the highest magnitude of functional outcome scores. The investigators tallied the various histological, functional, electrophysiological measures and survival rates into a composite score for each treatment group. The magnitude of the composite score was highest in group 1, though not statistically significant. Overall, while there was some suggestion in favor of nimodipine therapy, the composite score did not significantly favor nimodipine over placebo.

In contrast, Jia et al. showed strongly favorable results of nimodipine treatment in a rat SCI model.[56] Following injury, 12 rats were treated with nimodipine 1.0 mg/kg or equivalent amount of saline intraperitoneally. Treatments occurred three times daily for 1 week. There was significant improvement in BBB scores ($P < .01$) along with significantly reduced evidence of oxidative injury and inflammation in histopathological analysis in nimodipine-treated rats. The body of preclinical evidence provided mixed evidence for nimodipine efficacy in rodent SCI and may be in part secondary to methodological and dosing variations across studies.

Clinical data

Pointillart et al. conducted a prospective randomized trial in France evaluating the safety and efficacy of nimodipine, methylprednisolone, or a combination compared to control in acute SCI patients with a primary outcome of ASIA score at 1-year follow-up.[57] Patients were prospectively randomized over a 5-year period into four groups: MPSS at 30 mg/kg over 1 hour followed by 5.4 mg/kg for 23 hours, nimodipine at 0.15 mg/kg for 2 hours followed by 0.03 mg/kg for 7 days, combined methylprednisolone and nimodipine at the same dosing regiments, and placebo group. Inclusion criteria consisted of patients with acute traumatic SCI aged 15–65 years and presentation within 8 hours of trauma. Patients with polytrauma, concomitant head injury or GCS <13, pulmonary contusions, open spinal injury, refractory hypotension (MAP <60 mm Hg), or major medical comorbidities were excluded. The protocol was discontinued if patients were unable to maintain an MAP above 60 mm Hg for 1 hour. Over the recruitment period, there were 106 patients that met study criteria including 48 paraplegic patients and 58 tetraplegic patients with 48% of patients sustaining a complete injury.

76% of patients underwent surgical decompression and/or stabilization within 24 hours and 61% of operated patients underwent surgery within 8 hours. There were no major differences in trauma severity, age, ASIA scores, or treatment delays between cohorts in the trial. There were 5 mortalities in the group and 100 patients were examined at 1-year follow-up. There was no significant improvement between any treatment arm in neurological outcome, and specifically with nimodipine treatment either alone or in combination with MPSS. As observed previously, MPSS treatment portended a slightly higher infectious complication rate, though this did not reach statistical significance. Hyperglycemia and insulin infusion requirement were more commonly encountered in the MPSS treatment arm. The overall findings of this study suggested a lack of evidence to support a therapeutic benefit of nimodipine for SCI.

VX-210/Cethrin

VX-210 otherwise known as Cethrin or BA-210 is a bacterial enzyme derivative of C3 transferase with Rho-inhibiting properties. Specific interest in this agent was generated by evidence suggesting its role in facilitating axonal repair and regeneration. Herein we discuss available mechanistic, preclinical, and clinical data regarding VX-210 in SCI.

Mechanism of action

Following SCI, numerous endogenous mechanisms act to inhibit axonal regrowth and functional regeneration including growth inhibitory signals released from the glial scar (e.g., chondroitin sulfate proteoglycans, semaphorins), myelin debris (neurite outgrowth inhibitory protein A, oligodendrocyte-myelin glycoprotein), and inflammatory cytokines (tumor necrosis factor, interleukins). Together these pathways converge in the activation of intracellular Rho (GTPase) through specific interactions with cell membrane proteins[58,59,60,61] resulting in

axonal growth cone collapse and neuronal apoptosis.[62,63] Rho activation results in axonal growth arrest, while inhibition of Rho enhances axon regeneration in vivo.[64] Rho and Rho-associated protein kinase (ROCK), a downstream effector kinase whose activity is modulated by Rho, play a role in modulating the actin cytoskeleton in neurite outgrowth, axon guidance, and neuronal development. Activated Rho inhibits axonal elongation through blocking turnover of actin subunits.[65] Furthermore, Rho-ROCK activation triggers a molecular cascade that results in myosin light-chain phosphorylation and collapse of growth cones. In the context of SCI, Rho upregulation in primary and secondary injury phases is well documented[66] and thus blockade of Rho-ROCK signaling has emerged as an attractive pharmacotherapeutic target in acute SCI.

Early work by Lehmann et al. showed that direct inhibition of Rho GTPase resulted in axonal regeneration in the presence of inhibitory substrates such as myelin breakdown products in cell culture experiments.[67] In this work, *Clostridium botulinum* C3 transferase was isolated and utilized for its ability to selectively target ADP-ribosylate, the Rho effector domain. This effectively binds RhoA in an inactive state and prevents further participation in downstream signaling. BA-210 and VX-210 are recombinant engineered variants of C3 transferase as the bacterial-derived version does not readily cross cell membranes. Lord-Fontaine et al. reported on molecular alterations to BA-210 including linking a transport sequence to facilitate receptor-independent passage across cell membranes and modifications to reduce breakdown secondary to proteolysis. In their models, this agent was shown to readily pass through the dura and deposit in injured cord regions with higher Rho activity.[62] The body of evidence and rationale for BA-210's role in downregulating mechanisms that block axonal regeneration in SCI serve to underscore subsequent preclinical experiments.

Preclinical data

Early experiments targeting Rho signaling in SCI initially used wild type C3, but with the improved pharmacodynamics of BA-210, the latter agent has been the focus of modern preclinical studies. Lord-Fontaine et al. demonstrated the impact of BA-210 in a rodent SCI model.[62] Both direct hemisection (T7) and cord contusion (T9) models were used to assess histological Rho expression/inhibition and BBB locomotion scores, respectively. Direct BA-210 administration with a fibrin sealant was performed on exposed dura. Levels of activated Rho were reduced to basal amounts after immediate and 24 hour delayed application of BA-210 in the hemisection model of SCI. Improved BBB locomotor scores were obtained in the BA-210 treatment group (immediate and 24-hour delay) compared to placebo at 16-days postinjury. There appeared to be diminished magnitude of effect with delayed treatment with nonsignificant results in animals treated 72-hour postinjury. In the cord contusion model, which better reflects human SCI, there was a significant improvement in locomotor function in rats receiving BA-210 compared to vehicle. Histopathological analyses corroborated functional data and demonstrated reduced gray and white matter volume loss at the lesion epicenter and increased tissue sparring at the level of injury.

Clinical data

Phase I/IIa trial

Fehlings et al. conducted a Phase I/IIa trial examining the safety and tolerability of BA-210 (Cethrin) in human SCI. Forty-eight patients aged 16–70 years were enrolled from 9 centers (32 thoracic SCI and 16 cervical SCI). Complete AIS A cord injuries were included, and all patients underwent surgical decompression within 7 days of injury. Cethrin (with fibrin sealant) was directly applied on the dural surface at a dose of either 0.3, 1, 3, 6, or 9 mg. Exclusion criteria comprised of significant organ system comorbidity or malignancy, medical conditions or neurological conditions precluding an accurate ASIA assessment, insulin-dependent diabetes, coagulopathy, traumatic brain injury, ankylosing spondylitis, or cognitive impairment.

Thirty-five patients completed the trial (12 complete cervical SCI and 23 complete thoracic SCI) with mean administration of BA-210 occurring 53 hours after injury. Cethrin was found to be safe and well tolerated at all doses administered via direct topical dural administration with no associated serious adverse events. There were two mortalities, both of which were not associated with Cethrin administration. There was minimal improvement in motor scores in thoracic SCI patients receiving Cethrin (1.8 ± 5.1) with a comparatively larger effect seen in cervical SCI patients with an average change of 18.6 ± 19.3 at 12 months. In the 3 mg Cethrin group with cervical SCI, the magnitude of motor gains at 12 months was 27.3 ± 13.3, arguably more than expected in true ASIA A injuries.[68] Similar effects were seen in sensory scores among thoracic and cervical groups. The ASIA conversion effect was greatest in the 3 mg cervical group, where approximately 66% of patients improved to ASIA C. Overall this safety and feasibility study demonstrated Cethrin was well tolerated with no major adverse events and suggested improved neurological outcomes in cervical SCI compared to thoracic SCI patients. Nevertheless, any efficacy statements are not possible as the trial lacked a placebo arm. Furthermore, initial clinical assessments were conducted 24 hours postinjury raising concerns of examination accuracy at such an early time point which may inflate the magnitude of motor score recovery seen.

Phase IIb/III trial

In light of these promising results, a multicenter Phase IIb/III randomized double-blind placebo-controlled trial was conducted to investigate VX-210 (by Vertek).[69] Spinal Cord Injury Rho Inhibition Investigation (SPRING) was a multicenter effort to provide evidence for the efficacy of VX-210 in acute cervical traumatic SCI (NCT02669849). The trial targeted an enrollment of 100 patients from 45 North American sites with a primary outcome of UEMS change at 6 months. Patients age 14–75 with ASIA A or B cervical cord injuries at C4–7 and undergoing decompression/stabilization surgery within 72 hours were included in the study. Patients with gunshot or penetrating SCI, head injuries/neurological comorbidities precluding accurate assessment, immunodeficiency, significant medical or psychiatric comorbidity, one or more untestable upper-extremity muscle group, or BMI >40 were excluded. The intended randomization was 1:1 stratified by age <30 and >30 years as well as ASIA grade A or B. Treatment arms included a single dose of 9 mg VX-210 in fibrin sealant at time of surgery applied topically on the dura or placebo in fibrin sealant. An interim analysis after enrollment of 33% of patients demonstrated that the predefined study measures were not met to justify study completion.[70] At study termination, 67 patients were randomized and received either VX-210 or placebo, 37 patients had progressed to 6-month follow-up with no statistically significant difference in UEMS between VX-210 and placebo therapy arms. Results from this Phase IIb/III trial suggested that Cethrin/VX-210/BA-210 does not contribute to meaningful improvement in motor function among patients with acute traumatic cervical SCI. Despite a compelling body of in vitro and in vivo preclinical data suggesting its efficacy, this highlights the challenges in clinical translation of pharmacotherapies in a condition as heterogeneous as SCI.

GM-1 ganglioside

Mechanism of action

Gangliosides are amphipathic components of predominantly the outer leaflet of cell membrane phospholipid bilayers in the central nervous system. Structurally they are comprised of glycosphingolipid and sialic acid. The ganglioside family has been proposed to play both neuroprotective and neuroregenerative roles in SCI. GM-1 ganglioside refers to a specific prototypic ganglioside known as monosialotetrahexosylganglioside (trade name Sygen).

The mechanism of action of GM-1 is posited to rely in part on its role in promoting endogenous neurotrophin action.[71] GM-1 has been shown to potentiate BDNF activity[72,73] which in turn facilitates dendritic growth. Specifically, the group of ganglioside molecules has shown both neuritogenic and neuronotrophic properties resulting in increased branching, length, and neuronal number, as well as enhanced neuronal survival.[74] In a fetal mouse spinal cord tissue culture injury model, treatment with GM-1 ganglioside resulted in reduced release of lactate dehydrogenase, reduced tissue injury and cell loss.[75] Injurious effects of glutamate exposure in vitro were attenuated by GM-1 ganglioside, suggesting an antiexcitotoxic mechanism of action.[76] Regenerative properties were demonstrated in vitro by increased neuronal axonal sprouting in the presence of ganglioside medium.[77]

Preclinical data

Early experiments by Bose et al. demonstrated that GM-1 ganglioside treatment contributes to the reestablishment of axonal continuity in a transection model of SCI. GM-1 (3.5 mg/kg administered daily for 42 days) was associated with increased axonal sprouting and transport channels in surviving axons at the site of SCI.[78] Despite these promising molecular and histological findings, the effects of GM-1 ganglioside on functional improvement have been inconsistent. In 1994 Constantini and Young examined the therapeutic potential of GM-1 ganglioside either alone or in combination with MPSS in a cord contusion model of SCI in rats. There was no significant difference in functional outcomes or histopathological evidence of injury in rats treated with GM-1 ganglioside compared to control and in fact, GM-1 ganglioside appeared to inhibit the central and peripheral therapeutic effects of MPSS in vivo.[79]

Marcon et al. further showed data that did not support efficacy of GM-1 ganglioside in improving functional outcomes in a rat model of SCI.[80] In this study, 50 rats underwent SCI with groups randomized to receive sham laminectomy, GM-1 ganglioside 30 mg/kg intraperitoneally, erythropoietin, and saline control. BBB scores were higher in the erythropoietin cohort, but combined erythropoietin and GM-1 ganglioside showed the greatest improvement at 6 weeks ($P < .05$). There was no significant difference between GM-1 ganglioside versus saline groups at 6 weeks. More recently, Yuan et al. assessed GM-1 ganglioside (30 mg/kg infusion for 7 days) treatment in a rat model of SCI.[81] It was found that SCI led to an increase in caspase-3 expression and decreased nerve growth factor. GM-1 ganglioside treatment reduced caspase-3 activity and increased nerve growth factor expression after SCI, which were postulated as additional mechanisms of GM-1 therapeutic effect. Despite these purported mechanisms of attenuating the secondary injury of SCI, the body of preclinical evidence for GM-1 ganglioside in the setting of SCI was not as robust as other candidate drugs, and its translation to clinical trials was in part backed by literature supporting GM-1 ganglioside in other central nervous system pathologies.

Clinical data

Phase II trial

Geisler et al. conducted a prospective double-blinded randomized control pilot trial examining the effect of GM-1 ganglioside on human SCI.[82] This single-center trial included 37 patients with a minimum age of 18 years, spinal cord injury with a major motor deficit in upper or lower extremity that were followed for 1 year. Patients with major medical or neurological comorbidity and those unable to receive their first dose of medication within 72 hours of injury were excluded. At the time of the trial, the maximum FDA-approved daily dose was 100 mg GM-1 ganglioside (Sygen). Patients were randomized to receive placebo or 100 mg GM-1 ganglioside intravenous daily for 18–32 days and first doses were administered within 72 hours of injury. Three patients were excluded from subsequent analysis, one of whom passed away from acute respiratory distress syndrome. Within the remaining cohort, 27 patients underwent decompressive surgery within 72 hours and the other 7 had delayed decompression. There were no serious adverse events or side effects attributable to GM-1 ganglioside and the overall profile of complications did not differ between placebo or treatment groups. At 1-year follow-up there was significant improvement in Frankel scores ($P = .034$) and ASIA motor scores ($P = .047$) in GM-1 ganglioside–treated patients compared to the placebo group. This trial provided data supporting a favorable safety profile in human SCI patients and a sound rationale for subsequent larger scale studies to examine the therapeutic role of GM-1 ganglioside in SCI.

The above trial served as the impetus for subsequent multicenter prospective randomized-controlled trials examining GM-1 ganglioside in acute traumatic SCI.[83] Between 1992 and 97, 760 patients were included from a total of 28 centers in North America. All patients were administered NASCIS II dosing of MPSS within 8 hours of SCI. Once MPSS administration was complete (within 72 hours), patients received either a low-dose protocol (loading dose of 300 mg followed by 100 mg/day of Sygen for 56 days), high-dose protocol (loading dose of 600 mg Sygen followed by 200 mg/day for 56 days), or placebo (loaded with placebo followed by 56 days of placebo infusions). ASIA grade at presentation as well as both ASIA grade and Modified Benzel Classification at 4, 6, 16, 26, and 52 weeks were measured. Patients were stratified into six groups: either cervical or thoracic neurological level or injury and ASIA

A, B or C and D neurological levels of impairment.

Primary efficacy outcome was defined as an improvement of at least two grades on the Benzel Classification by 26 weeks. The majority of patients in both cervical and thoracic injury groups were ASIA grade A, with fewer incomplete injuries. Median time of GM-1 ganglioside administration was 54.9 hours. Patients in both treatment groups were well balanced regarding comorbidities, other injuries, demographics, level of injury, and study drug administration. There was no significant difference in mortality between placebo and Sygen treatment groups. Despite an early significance in the primary effect outcome 8 weeks in favor of Sygen treatment, the analysis at 26 weeks (and 52 weeks) showed no significant difference in recovery between treatment and placebo groups. This suggested a plateau effect with delayed recovery in the placebo arm. In subgroup analysis, ASIA B patient cohorts in the Sygen treatment arms showed greater marked recovery rates (47%) than placebo (25%). In the ASIA C and D group there was also a more rapid rate of recovery attained in the Sygen treatment group, but the rates normalized in placebo and treatment groups by 52 weeks. There were more functional bladder outcomes in the Sygen treatment arm, but this did not reach significance at 26 weeks ($P = .0584$). By week 52, approximately twice the ASIA B patients had normal bladder function in Sygen treatment arm compared to placebo. Despite some elevation in cholesterol and triglycerides in Sygen treatment cohorts, there were no adverse events attributable to Sygen therapy.

Overall, while this trial demonstrated a significant difference in functional improvement at 8 weeks across injury groups with Sygen treatment, this difference normalized by week 26, resulting in a negative primary trial effect outcome. The trial did demonstrate that Sygen treatment is well tolerated and safe in acute adult SCI and subgroup analysis showed a trend toward improved outcome in ASIA grade B patients in the Sygen treatment arm. Nevertheless, a Cochrane review of available literature found no evidence to support improved survival or quality of life among patients with SCI treated with gangliosides.[84]

Thyrotropin-releasing hormone

Mechanism of action

Thyrotropin-releasing hormone (TRH) is an endogenous tripeptide released by the hypothalamus that plays a role in pituitary thyrotropin secretion and associated downstream physiological effects. Early animal experiments demonstrated that TRH can increase the excitability of motor neurons by inducing depolarization resulting in enhanced hindlimb electromyographic activity and increased spinal cord monosynaptic reflexes. Furthermore, TRH has shown antiinflammatory, antioxidant, and membrane-stabilizing properties.[85] There is also evidence suggesting TRH acts through antagonizing endogenous leukotrienes, endorphins, and platelet-activating factor, all of which are compounds and signaling molecules associated with the SCI secondary injury cascade.[86,87,88] Together, these multiple findings suggested a potential neuroprotective role for TRH in the setting of neuronal injury and in enhancing neuronal excitability and signaling.

Preclinical data

Faden et al. studied the effect of TRH on neurological recovery in a C7 cat model of SCI.[89] In this study, three treatment groups were examined: (1) TRH 2 mg/kg bolus and 2 mg/kg per hour infusion for 4 hours, (2) dexamethasone 0.5 mg/kg bolus and 0.5 mg/kg per hour infusion for 4 hours, or (3) saline control. TRH was associated with improved neurological outcomes at 1 week following treatment compared to dexamethasone and saline groups. Specifically, the difference in forelimb and hindlimb motor function persisted throughout the entire study duration and the magnitude of improvement was greatest at 6 weeks (<0.01). A follow-up dose-dependent study was later conducted to corroborate these initial experiments demonstrating functional improvements in animals when treated after 24 hours of SCI.[90] Additional animal studies demonstrated that TRH administered either acutely (at 24h) or subacutely (at 7 days) after injury improves neurological recovery in a dose-dependent fashion.[91]

Clinical data

TRH was initially examined in a cohort of 13 patients with SCI during the spinal shock phase.[92] The study protocol involved 1 mg IV of TRH administration as a bolus followed by 1 mg infusion over 5 minutes. Baseline preinfusion anal sphincter electromyography (EMG) and basal cystometry were compared to post-treatment measurements. Subsequently TRH was administered 1 mg intravenously every 12 hours for 3 days and repeat EMG and cystometry were conducted along with detrusor pressure and bladder compliance measurements. TRH treatment was associated with increased detrusor pressure and reduced bladder compliance following injury-induced spinal shock.

Phase I trial

The safety and efficacy of TRH for acute SCI was examined in a prospective randomized double-blinded clinical trial of 20 patients assigned to TRH or placebo.[93] TRH was administered as an initial bolus of 0.2 mg/kg IV and continued by a 0.2 mg/kg/h infusion over 6 hours with placebo group receiving saline. Patients presenting within 12 hours of injury were included and stratified on the basis of injury severity (complete vs. incomplete). Inclusion criteria comprised individuals over the age of 13 years, no thyroid-related comorbidities, concomitant injury, or comorbidity affecting neurological examinations. Patients were examined for motor and sensory scores as well as

Sunnybrook score for functional outcome. There were no major adverse events and TRH was well tolerated in the cohort. No significant differences in outcome scores in the complete injury cohort was found between TRH and placebo treatment groups. However, among incomplete injury patients, TRH was associated with significantly improved motor scores ($P = .043$), sensory scores ($P = .031$), and Sunnybrook scores ($P = .044$) compared to placebo controls. Caution must be taken in interpreting results of this trial due to the small numbers of included patients (6 treated and 3 placebo cases of incomplete injury) and lack of analysis comparing demographic, neurological, and comorbid differences between patients in placebo and treatment arms.

AC105

Mechanism of action

The cellular regulatory functions of magnesium are ubiquitous. Magnesium is an NMDA receptor antagonist, blocking the ion channel and reducing its activation.[94] Therefore, it can play a role in mitigating secondary injury pathways and detrimental effects of excitotoxicity. Other mechanisms of magnesium action include reduction of apoptosis through caspase-3 inhibition, modulation of intracellular calcium release through the IP3 pathway, and reduction in free radical release.[95,96,97,98] Magnesium infusion has been studied in traumatic brain injury and stroke models with mixed results. High doses of magnesium sulfate have been shown to exert neuroprotective effects in animal SCI models; however, the doses required to obtain this effect were far beyond those that are tolerable for translation to clinical practice (i.e., 600 mg/kg).[99,100,101]

Polyethylene glycol (PEG) was utilized to enhance delivery of magnesium to central nervous system tissue by acting as a pharmacological excipient.[102] AC105 is a formulation comprised of magnesium chloride in PEG 3350. The enhanced delivery of Mg^{2+} to spinal cord tissue determined by microdialysis measurements following administration of magnesium sulfate PEG formulation was demonstrated in an in vivo SCI model.[103] Following SCI there was a reduction in basal Mg^{2+} at the core and rostral perilesional area. While infusion of saline and magnesium sulfate did not change extracellular Mg^{2+} concentrations, delivery of AC105 significantly increased the core and perilesional Mg^{2+} concentration and multiple infusions corrected the magnesium content to pre-SCI levels. Furthermore, this was associated with a reduction in extracellular glutamate levels at the lesion epicenter.

Preclinical data

Ditor and colleagues first described the use of PEG with magnesium sulfate to potentiate its penetration into the spinal cord.[102] In a T4-clip compression model of SCI in rats, PEG and magnesium sulfate delivered 6 hours after injury resulted in lower mechanical allodynia and improved locomotor recovery at 6 weeks compared to saline-treated animals. BBB scores were higher with PEG treatment and magnesium sulfate treatment groups compared to saline, though the combination of magnesium and PEG did not lead to improved outcome than each agent alone. Histopathological analysis showed a reduction in lesional volume and greater dorsal myelin sparing in PEG—magnesium sulfate and magnesium sulfate—treated animals.

Following this study, Kwon et al.[104] sought to revalidate and refine the administration of PEG—magnesium sulfate in a preclinical model of SCI. Magnesium, PEG, and magnesium—PEG combination were tested in a T10 cord contusion model utilizing a calibrated impactor. Rats were randomized to sham control, saline, magnesium sulfate (500 umol/kg), PEG (1 g/kg), or combined PEG—magnesium sulfate. There was a 51% reduction in lesional volume following administration of PEG—magnesium combination compared to saline control, while PEG alone did not significantly reduce lesion size. PEG—magnesium resulted in significantly improved BBB locomotor function at 6 weeks compared to saline or PEG alone. Subsequent experiments demonstrated an increased magnitude of treatment effect with greater infusions of a magnesium chloride formulation (6 infusions compared to 2). Furthermore, dosing magnesium chloride infusions at 254 umol/kg compared to 127 resulted in greater reduction of lesion volume and improved locomotor scores. Finally, it was demonstrated that outcomes at 6 weeks were improved with PEG—magnesium treatment compared to methylprednisolone. Overall this study demonstrated a robust in vivo effect of PEG—magnesium formulation reducing lesion size and improving locomotor outcomes in rat SCI. It also served to establish a dose—response relationship with shortened time between infusions and increased dose resulting in greater therapeutic effect. From this study, a 4-hour interval was proposed for dosing PEG—magnesium post-SCI. The body of evidence provided was promising for magnesium as a neuroprotective drug in SCI and formed the evidence to proceed to clinical trials.

Clinical data

Phase II trial

A phase II double-blind, randomized, placebo-controlled study was initiated in 2012 by Acorda Therapeutics Inc. (Chelsea, MA, USA) comparing treatment with AC105 (magnesium/PEG formulation) versus placebo following acute SCI (NCT01750684). The trial aimed to recruit 40 patients and to determine the safety, tolerability, and potential efficacy of AC105 in SCI. The protocol included patients aged 18—65 with acute traumatic SCI with neurological level between C4 and T11, nonpenetrating mechanisms, and ASIA impairment A, B, or C. Psychiatric or major medical comorbidities precluding accurate neurological assessments, digoxin use, mean arterial pressure less than 60 despite pressor therapy, significant renal disease, and pregnancy were exclusion criteria. Drug administration was stratified into groups based on first dose postinjury at 6, 9, and 12 hours followed by five additional doses administered every 6 hours. The investigators aimed to follow serum biomarkers, pharmacokinetics, side effects and compare neurological outcomes at 6 months. Fifteen patients were enrolled; however, the study was terminated by the study sponsor due to insufficient rate of enrollment as well as low patient retention. As such, the neuroprotective role of AC105 in patients for SCI remains inconclusive. Furthermore, tolerability studies are necessary to determine the safety profile of AC105 prior to proceeding with any future larger scale Phase III trials.

Drugs under active clinical investigation

Riluzole

Riluzole (2-amino-6-trifluoromethoxy benzothiazole, Rilutek) is a synthetic benzothiazole that was designed for use as a centrally acting muscle relaxant in the 1950s[105] and subsequently used for its anticonvulsant and neuroprotective properties. In the mid-1990s, the FDA approved riluzole as a therapeutic agent for amyotrophic lateral sclerosis[106] after demonstrating modest improvements in survival in a series of small randomized-controlled trials. Application of riluzole to SCI was initially spurred by its ability to block Na^+ channels and modulate sodium ion homeostasis[107,108] — critical mediators of the acute secondary injury cascade of SCI.

Mechanism of action

The neuroprotective effects of riluzole principally derive from sodium channel blockade and modulation of glutamatergic excitotoxicity.[109] SCI induces aberrant upregulation of neuronal voltage-gated Na^+ channels[107] leading to dysregulated osmotic gradients, acidosis (through disrupted hydrogen sodium antiporter mechanisms in axon membranes), and calcium entry through Na^+/Ca^{2+} membrane exchangers,

which contribute to excitotoxicity through extracellular glutamate release.[110,111] By reducing the magnitude of sodium ion membrane homeostasis dysregulation, the downstream effects of intracellular acidosis, influx of calcium ions, and propagation of glutamate-mediated secondary neuronal toxicity may be mitigated by the therapeutic actions of riluzole.[112] Furthermore, beyond the indirect presynaptic sodium-mediated mechanism of modulating downstream glutamate release, riluzole is believed to directly contribute antiglutamatergic effects. Reduction in glutamate release and enhanced clearance of glutamate from the synaptic cleft through increased glutamate uptake are both proposed mechanisms of action.[113,114] Models of riluzole biochemistry have also demonstrated an effect in antagonizing acetylcholine release through N-methyl-D-aspartate (NMDA) pathways in vitro and reducing downstream effects of glutamate activity via cyclic guanosine monophosphate blockade.[115,116]

Preclinical studies

The elucidation of sodium ion dysregulation and pathological axonal permeability after SCI laid the framework for subsequent application of Na^+ channel blocking agents in preclinical models. An examination of different sodium channel blocking agents in a clip compression model of SCI including riluzole, phenytoin, and CNS5546A demonstrated significant functional improvements with all agents, with the most pronounced effects seen in the riluzole treatment group. Specifically, riluzole-treated animals demonstrated improved motor recovery as well as sparing of gray and white matter regions with reduced tissue necrosis. Further studies demonstrated a reduction in cord edema, improvement in neurobehavioral scores, and histopathological recovery with riluzole after SCI.[117]

The pharmacokinetics and therapeutic time window for riluzole was examined in a cervical SCI rodent model.[118] Rats were treated with riluzole at either 1 hour (P1), 3 hours (P3), or with control vehicle following cervical SCI. The P1 and P3 groups went on to have riluzole every 12 hours for 7 days. Both P1 and P3 groups exhibited significantly improved locomotor scores, somatosensory evoked potentials, and improved axonal integrity and function. Furthermore, it was found that the spinal cord was penetrated within 15 minutes of administration and acute SCI slowed the elimination and redistribution of riluzole from the spinal cord, resulting in prolonged drug half-life. While extrapolation of therapeutic windows from rodent models to humans is challenging, overall this study suggested a 12-hour window for riluzole administration.

Clinical data

Riluzole was a promising agent for translation to clinical trials of SCI due to its demonstrated efficacy in several preclinical models. Furthermore, its use in patients with amyotrophic lateral sclerosis provided a body of clinical data suggestive of a fairly well-tolerated safety profile with few adverse effects.

Phase I trial

A Phase I prospective single-arm, open-label trial was conducted between 2010 and 2012 to assess the safety profile and preliminary efficacy of riluzole in acute traumatic SCI. The study was performed at six centers through the North American Clinical Trial Network (NACTN) for Treatment of Spinal Cord Injury and had a target enrollment of 36 patients.[119] Inclusion criteria were age between 18 and 70, lack of concomitant life-threatening injuries, ASIA impairment grade A, B, or C, and nonpenetrating SCI with neurological level C4–T11. Patients with hepatic or renal disease and comorbidities precluding accurate neurological assessment were excluded. Given the nature of a Phase I trial and lack of a control arm in the prospective collection, matched controls were used from the NACTN registry for 36 patients who did not receive riluzole but had similar profiles of injury to meet inclusion and exclusion criteria. Riluzole at a dose

of 50 mg was administered within 12 hours of injury and followed with a twice daily dosing for 14 days. This dosing regimen was taken from scaling preclinical animal model data as well as utilizing clinical and pharmacodynamic results from preceding amyotrophic lateral sclerosis clinical trials.[120]

Both the riluzole and control registry cohorts consisted of 28 cervical and 8 thoracic cord injuries. Within the riluzole and registry cohorts, 33 and 36 patients, respectively, underwent surgical decompression and stabilization, most of which were performed within 24 hours. Preexisting medical comorbidities were also similar across groups. In the riluzole cohort, 71% of patients received the full indicated course of therapy (28 doses) and an additional 26% received 27 doses. Pharmacokinetic profiles of riluzole were assessed on day 3 and day 14 of therapy. There were differences in pharmacokinetics in the acute and subacute phases of spinal cord injury with significantly higher drug serum concentrations from 0 to 12 hours in the acute phase, likely due to lower clearance and volumes of distribution in the acute SCI phase.[121]

The 30-day complication rate was not significantly different between riluzole and registry cohorts and there were no serious adverse events attributable to riluzole. However, riluzole therapy was associated with an elevation in liver enzymes and bilirubin, with over half of patients having mild (upper limit of normal) to moderate (5 times upper limit of normal) elevations in either AST (63%), ALT (70%), or GGT (53%). Overall, the study demonstrated pharmacokinetic data in a cohort of acute SCI patients and established the safety profile of riluzole with an emphasis on future monitoring of liver enzymes and bilirubin during therapy.

While the study also aimed to assess the preliminary efficacy of riluzole after SCI, it was not powered to detect significant changes in neurological outcome. In subgroup analysis, ASIA B patients showed the greatest magnitude of motor recovery (4.13-fold increase at 180 days compared to time of admission) followed by groups C and A. The mean motor score difference between riluzole and registry groups at 90 days (n = 24 in riluzole group, n = 26 in registry group) was 2.4 points (A; $P = .787$), 27.9 points (B; $P = .037$), 13.7 points (C; $P = .194$). At 180 days there was no difference across grades A, B, and C. Gains in pinprick and light touch scores were not significant between registry and riluzole groups at 180 days. The conversion proportion of ASIA impairment grades was higher in the riluzole compared to registry cohorts. Grade B impairment subgroup once again exhibited the most robust conversion in the riluzole cohort. At 180 days all ASIA B riluzole-treated patients (n = 8) converted to a better functional grade compared to three out of five in the registry group.

Overall, while this Phase I trial was nonblinded and not powered to identify significant neurological differences between treated and control groups, it did produce important pharmacokinetic data and drug safety profile of riluzole in SCI patients and provided early data to suggest a promising effect on functional improvement.

Phase IIB/III trial: RISCIS

Results from the Phase I trial served as the foundation for a subsequent Phase IIB/III trial known as Riluzole in Acute Spinal Cord Injury Study (RISCIS). This randomized double-blind placebo-controlled parallel-center study (NCT01597518) aimed to more rigorously evaluate the therapeutic impact of riluzole administration in acute SCI.[122] The study was initiated in January 2014 and has now recently concluded enrollment following a promising interim analysis after an enrollment of 193 patients was achieved. Pharmacokinetic lessons from the initial Phase I trial served as the groundwork for the RISCIS trial design. There were 26 participating centers including 23 in North America and 3 in Australia. The protocol includes a regimen of 100 mg PO/NG twice daily riluzole in the first 24 hours following SCI followed by 50 mg twice daily for 13 days in patients with C4—8 level of neurological injury and grade A, B, or C impairment on initial assessment with a final

endpoint of 180 days.[123] Patients aged 18—75 are randomized 1:1 to riluzole or placebo groups. Exclusion criteria include factors precluding accurate neurological examination such as concomitant head injury, preexisting neurological comorbidity or mental disorder, prior spinal cord injury, as well as evidence of hepatic or renal dysfunction. Overall, RISCIS aims to provide level 1 evidence for the role of riluzole in acute traumatic SCI.

Anti-NOGO-A

Mechanism of action

Nogo-A is a myelin-associated oligodendrocyte membrane protein that interacts with neuronal receptors and inhibits neurite growth.[124,125] This endogenous cascade has been described in neuronal maturation and implicated in development whereby limited neuronal turnover and stabilizing networks are physiologically prioritized.[126] Nogo-A activation triggers a Rho-mediated cascade that results in growth cone collapse along with downregulation of transcriptional growth pathways.[127] The anti-neuronal growth activity of Nogo-A was effectively inhibited in vitro by treatment with anti-Nogo-A antibodies as demonstrated in cultured neuronal experiments and through demonstration of injured axon regeneration in vivo.[128,129] Anti-Nogo-A (NG-101 or ATI355) is an IgG recombinant human antibody with affinity to the Nogo-A protein-binding domain that results in reduced endogenous Nogo-A protein activity.[130] Based on the potential neuroregenerative implications of anti-Nogo-A activity, there was significant interest in applying this agent in the setting of SCI.

Preclinical data

Schnell et al. provided the first in vivo evidence supporting the neuroregenerative properties of antimyelin-associated neurite growth inhibitor antibodies.[131] Antibody-producing tumors were implanted into young rats; there was significantly more sprouting observed in the rats implanted with antibody-producing tumor following corticospinal transection at the site of lesion compared to control. Following this experiment, a similar methodology was applied in rats to assess functional outcome following cord transection.[132] Treatment with antineurite growth inhibitor antibody-secreting tumors resulted in corticospinal and brain stem axonal regeneration and plasticity. Furthermore, locomotive and reflex arc functional gains were seen in these rats compared to control groups.

Anti-Nogo-A antibodies were then directly administered via intrathecal catheter delivery over a two-week period in a thoracic cord SCI model.[134] Improved motor function two weeks following antibody administration was demonstrated compared to control cohorts. This was shown to correlate to functional MRI cortical responses and histopathological assessment of corticospinal axons and length of sprouting in anti-Nogo-A treatment rats. Subsequently, this in vivo finding was applied to SCI functional outcomes in nonhuman primates.[133] Following unilateral cervical lesioning, functional recovery was significantly greater in the anti-Nogo-A antibody treatment group. There was a trend toward improved rate of extremity limb recovery in the treatment group compared to control as well as increased axonal sprouting in both partially and completely transected spinal cord cohorts. Similar results have been demonstrated in nonhuman primates following cervical unilateral spinal cord lesioning and subsequent recovery of function performing standardized upper extremity dexterity tests.[135] Specifically, the rate of recovery and magnitude of recovery were significantly improved in the anti-Nogo-A treatment groups. The above series of selected preclinical experiments serve as evidence supporting the therapeutic application of anti-Nogo-A antibodies to SCI.

Clinical data

Phase I trial

A Phase I multicenter open-label cohort study was conducted to assess the feasibility, pharmacokinetics, safety profile, and efficacy of intrathecal anti-Nogo-A (ATI355) delivery to patients with acute traumatic cord injuries through either continuous versus intermittent administration methods.[130] Inclusion criteria were ages 18—65 years and acute traumatic ASIA grade A injuries. Patients with imaging evidence of cord transection or penetration spinal trauma were also excluded. Primary outcomes included assessment of tolerability, safety profile, and pharmacokinetics of ATI355 delivery to paraplegic and tetraplegic patients. Cohorts were divided to receive either continuous (24 hours—28 days) or intermittent (6 injections over 4 weeks) intrathecal drug boluses and dose escalation was applied using varying doses and durations. There were 52 patients that completed the study. The most common side effects included urinary tract infection, decubitus ulcers, and headache. The number of adverse events was similar across intrathecal intermittent and continuous infusion groups. Across all cohorts, 16 patients reported 15 serious adverse events with none attributable to ATI355 at 1-year follow-up (mostly related to underlying SCI, other medications, or in one case, bacterial meningitis secondary to the intrathecal access device). There was dose dependency with serum drug levels and variability in cerebrospinal fluid drug concentrations following administration. This study was not controlled nor powered to detect clinical recovery over the duration of follow-up. Within the cohort, 7/19 patients converted from complete to incomplete tetraplegia. Overall, this Phase I trial demonstrated the safety and pharmacokinetics of intrathecal anti-Nogo-A administration in acute traumatic SCI.

Phase II/III trial

Nogo Inhibition in Spinal Cord Injury (NISCI) is an ongoing Phase II/III trial (NCT03935321). It is designed as a placebo-controlled randomized double-blinded multicenter trial aimed at assessing the efficacy of early (days 4—28 postinjury) anti-Nogo-A administration in acute traumatic cervical SCI. Primary outcome measure is change in

UEMS with numerous secondary outcome measures including quality of life, electrophysiological, motor, and sensory endpoints. Investigators estimate 114 participants and current recruitment criterion is listed as age 18—70 years. Other inclusion criteria include C1—8 neurological level of injury, ASIA A—D severity, lack of mechanical ventilatory support, and hemodynamic stability. Exclusion criteria include history of meningitis within 6 months, comorbidities precluding accurate neurological assessment such as severe brain injury, psychiatric comorbidity, plexus injury, and preexisting severe comorbid systemic conditions. The exposure arms include six intrathecal bolus injections of 45 mg NG-101 compared to a similar number of placebo injections. Informed by preceding Phase I and preclinical data, this study aims to assess the impact of anti-Nogo-A antibody therapy in acute traumatic cervical SCI.

Drugs entering clinical trials

Elezanumab

Mechanism of action

Elezanumab (ABT-555) is an IgG human monoclonal antibody with inhibitory activity against Repulsive Guidance Molecule A (RGMa), an endogenous axonal growth inhibitor implicated in secondary SCI pathophysiology.[136,137] RGMa is a glycoprotein that binds to the neuronal receptor Neogenin. This receptor—ligand complex subsequently exerts its activity by localizing to cell membrane lipid raft domains and triggering downstream guidance receptor activity resulting in neuronal inflammatory cascades and neurite growth inhibition.[138] RGMa inhibitory effects have been implicated in many forms of CNS injury including stroke, TBI, MS, and SCI.[139] RGMa is upregulated in both animal and human SCI with as high as 15-fold increase in its expression in oligodendrocytes, astrocytes, microglia, and macrophages at the site of animal SCI.[140] Human monoclonal antibodies targeting the N-terminal domain of RGMa along with Neogenin function have been developed and used in multiple preclinical models of CNS injury including SCI.[141] These antibodies have been shown to block RGMa-induced neurite growth inhibition and chemorepulsion in multiple models of CNS injury.

Preclinical data

Several studies have demonstrated the role of RGMa-mediated inhibition in neurite development and regrowth pathways in SCI in vitro and in vivo. In 2006, Hata et al. showed neurite growth inhibition through RGMa activity in vitro through a RhoA-Rho Kinase-dependent pathway and that RGMa was expressed in CNS myelin of rat spinal cord specimens.[136] Following dorsal cord hemisection in adult rats, intrathecal delivery of rabbit-derived RGMa monoclonal antibodies at the site of injury resulted in improved BBB locomotion scores at 6—9 weeks postinjury compared to control. Histological analysis demonstrated reduction in lesion size and increased corticospinal tract regrowth in the antibody treatment cohort.

Human-derived RGMa antibody was subsequently tested in a clinically relevant clip compression model of rodent SCI.[140] RGMa antibody was delivered weekly through direct injection or intravenous administration starting immediately after SCI. Intravenous administration of RGMa monoclonal antibodies resulted in antibody uptake into spinal cord tissue, cerebrospinal fluid, and serum, suggesting that this route is a feasible method of administration. The study also corroborated previous findings on the efficacy of RGMa antibodies, demonstrating improved BBB locomotion scores at 6-week postinjury compared to placebo with increased serotonergic pathway and corticospinal tract axonal regeneration. A subsequent study[139] demonstrated that even delayed initiation of elezanumab therapy with first dose at 3 or 24 hours after SCI was still effective in significantly improving locomotor scores and fine motor function at 9 weeks following injury. Furthermore, all treatment groups showed increased perilesional neuronal sparing and increased corticospinal axonal plasticity. Importantly, for the first time described following RGMa antibody treatment, there was earlier spontaneous voiding in all elezanumab treatment groups. Overall, these preclinical data provided strong in vivo support of elezanumab efficacy in acute SCI, demonstrating that delayed administration is effective and intravenous administration is a feasible mode of drug delivery.

Clinical data

Phase I trial

A Phase I double-blind, placebo-controlled randomized trial was conducted to evaluate the safety of elezanumab in patients with multiple sclerosis.[142] There were 20 randomized to 3 elezanumab dose groups (150 mg, 600 mg, and 1800 mg) and 1 placebo group. Elezanumab was administered intravenously every 4 weeks for four total doses with a loading dose on day 1 of the protocol which was double the dose of subsequent maintenance doses. While headache was the most common adverse effect (25%), the study found no major adverse events from elezanumab administration and it was well tolerated in the studied cohort.

Phase II trial

Currently there is ongoing recruitment for a Phase II trial randomized placebo-controlled double-blinded proof-of-concept trial aimed to determine the safety and efficacy of elezanumab in acute cervical traumatic SCI (NCT04295538). This study will utilize intravenous administration of elezanumab versus placebo with a primary outcome measure of 52-week UEMS subsection of the International Standards for Neurological Classification of Spinal Cord Injury scale. Secondary outcome measures include Spinal Cord Independence Measure change from baseline and change in UEMS from baseline from 0 to 52 weeks. Pertinent inclusion criteria include drug administration within 24 hours, ASIA grade A or B cervical (C4-7) injury without thoracic or lumbar spinal injury, and maximum UEMS score 32. There are similar exclusion criteria as other previous trials in SCI including major distracting head or extremity injuries precluding accurate neurological assessment, penetrating spinal injury, major psychiatric comorbidity, or evidence of complete spinal cord transection. The study aims to recruit 54 participants across 36 participating centers and is currently ongoing.

Conclusion

Several decades of research in search for a drug to cure paralysis has resulted in a number of candidate agents that hold promise but lack significant efficacy in improving motor recovery after SCI. The identification of secondary injury cascades and pathobiological mechanisms following SCI have been essential groundwork for subsequent drug discovery. One of the greatest challenges in drug design is failure of translation from animal models to patients. Throughout this chapter we have highlighted the available human trial data for numerous candidate drugs and identified methodological flaws that limit interpretation of results. Sensitive and meaningful outcome measures must also be used to detect differences in treatment effect. The development of imaging, serological, and genetic biomarkers would represent a significant advance. Microstructural MRI holds promise as a potential surrogate biomarker to enhance translational efforts.[143] ASIA score alone may not detect subtle but meaningful changes. Incorporation of parameters such as GRASSP, functional impairment, or quality-of-life metrics is required to ensure robust outcome measures. Furthermore, there is significant heterogeneity that is challenging to capture in lab-based settings when translating animal SCI models to human trials. In some cases, we have seen negative primary outcome measures, but subsequent stratification of patients to account for heterogeneity results in identification of subgroups that could benefit from a drug. Inferences from this mode of analysis must be made with caution as there may be a tendency to overgeneralizability. Collaboration across sites and institutions will be an essential element of moving forward with studies examining novel potential therapeutic candidates in SCI.

References

1. Timmermans S, Souffriau J, Libert C. A General introduction to glucocorticoid biology. *Front Immunol* 2019;**10**:1545.
2. Hall ED, Braughler JM. Effects of intravenous methylprednisolone on spinal cord lipid peroxidation and $Na^+ + K^+$-ATPase activity. Dose-response analysis during 1st hour after contusion injury in the cat. *J Neurosurg* 1982;**57**(2):247−53.
3. Xu J, Qu ZX, Hogan EL, Perot Jr PL. Protective effect of methylprednisolone on vascular injury in rat spinal cord injury. *J Neurotrauma* 1992;**9**(3):245−53.
4. Behrmann DL, Bresnahan JC, Beattie MS. Modeling of acute spinal cord injury in the rat: neuroprotection and enhanced recovery with methylprednisolone, U-74006F and YM-14673. *Exp Neurol* 1994;**126**(1):61−75.
5. Rabchevsky AG, Fugaccia I, Sullivan PG, Blades DA, Scheff SW. Efficacy of methylprednisolone therapy for the injured rat spinal cord. *J Neurosci Res* 2002;**68**(1):7−18.
6. Akhtar AZ, Pippin JJ, Sandusky CB. Animal studies in spinal cord injury: a systematic review of methylprednisolone. *Altern Lab Anim* 2009;**37**(1):43−62.
7. Bracken MB, Collins WF, Freeman DF, et al. Efficacy of methylprednisolone in acute spinal cord injury. *J Am Med Assoc* 1984;**251**(1):45−52.
8. Bracken MB, Shepard MJ, Collins WF, et al. A randomized, controlled trial of methylprednisolone or naloxone in the treatment of acute spinal-cord injury. Results of the Second National Acute Spinal Cord Injury Study. *N Engl J Med* 1990;**322**(20):1405−11.
9. Coleman WP, Benzel D, Cahill DW, et al. A critical appraisal of the reporting of the National Acute Spinal Cord Injury Studies (II and III) of methylprednisolone in acute spinal cord injury. *J Spinal Disord* 2000;**13**:185−99.
10. Hugenholtz H. Methylprednisolone for acute spinal cord injury: not a standard of care. *CMAJ* 2003;**168**:1145−6.
11. Bracken MB, Shepard MJ, Holford TR, et al. Administration of methylprednisolone for 24 or 48 hours or tirilazad mesylate for 48 hours in the treatment of acute spinal cord injury. Results of the third national acute spinal cord injury randomized controlled trial. National acute spinal cord injury study. *J Am Med Assoc* 1997;**277**(20):1597−604.
12. Anderson DK, Hall ED, Braughler JM, McCall JM, Means ED. Effect of delayed administration of U74006F (tirilazad mesylate) on recovery of locomotor function after experimental spinal cord injury. *J Neurotrauma* 1991;**8**:187−92.
13. Otani K, Abe H, Kadoya S. Beneficial effect of methylprednisolone sodium succinate in the treatment of acute spinal cord injury. *Sekitsui Sekizui* 1994;**7**:633−47.
14. Evaniew N, Belley-Côté EP, Fallah N, et al. Methylprednisolone for the treatment of patients with acute spinal cord injuries: a systematic review and meta-analysis. *J Neurotrauma* 2016;**33**(5):468−81. https://doi.org/10.1089/neu.2015.4192.
15. Fehlings MG, Wilson JR, Harrop JS, et al. Efficacy and safety of methylprednisolone sodium succinate in acute spinal cord injury: a systematic review. *Global Spine J* 2017;**7**(Suppl. 3):116S−37S.
16. Fehlings MG, Wilson JR, Tetreault LA, et al. A clinical practice guideline for the management of patients with acute spinal cord injury: recommendations on the use of methylprednisolone sodium succinate. *Global Spine J* 2017;**7**:203s−11s.
17. Garrido-Mesa N, Zarzuelo A, Gálvez J. Minocycline: far beyond an antibiotic. *Br J Pharmacol* 2013;**169**(2):337−52.
18. Brogden RN, Speight TM, Avery GS. Minocycline: a review of its antibacterial and pharmacokinetic properties and therapeutic use. *Drugs* 1975;**9**(4):251−91.
19. Kraus RL, Pasieczny R, Lariosa-Willingham K, et al. Antioxidant properties of minocycline: neuroprotection in an oxidative stress assay and direct radical-scavenging activity. *J Neurochem* 2005;**94**:819−27.
20. Tikka T, Fiebich BL, Goldsteins G, Keinanen R, Koistinaho J. Minocycline, a tetracycline derivative, is neuroprotective against excitotoxicity by inhibiting activation and proliferation of microglia. *J Neurosci* 2001;**21**(8):2580−8.
21. Garcia-Martinez EM, Sanz-Blasco S, Karachitos A, et al. Mitochondria and calcium flux as targets of neuroprotection caused by minocycline in cerebellar granule cells. *Biochem Pharmacol* 2010;**79**:239−50.

22. Amin AR, Attur MG, Thakker GD, et al. A novel mechanism of action of tetracyclines: effects on nitric oxide synthases. *Proc Natl Acad Sci USA* 1996;**93**(24):14014—9.

23. Paemen L, Martens E, Norga K, et al. The gelatinase inhibitory activity of tetracyclines and chemically modified tetracycline analogues as measured by a novel microtiter assay for inhibitors. *Biochem Pharmacol* 1996;**52**:105—11.

24. Shultz RB, Zhong Y. Minocycline targets multiple secondary injury mechanisms in traumatic spinal cord injury. *Neural Regen Res* 2017;**12**(5):702—13.

25. Yu F, Kamada H, Niizuma K, Endo H, Chan PH. Induction of MMP-9 expression and endothelial injury by oxidative stress after spinal cord injury. *J Neurotrauma* 2008;**25**:184—95.

26. Stirling DP, Khodarahmi K, Liu J, et al. Minocycline treatment reduces delayed oligodendrocyte death, attenuates axonal dieback, and improves functional outcome after spinal cord injury. *J Neurosci* 2004;**24**(9):2182—90.

27. Wells JE, Hurlbert RJ, Fehlings MG, Yong VW. Neuroprotection by minocycline facilitates significant recovery from spinal cord injury in mice. *Brain* 2003;**126**(Pt 7):1628—37.

28. Takeda M, Kawaguchi M, Kumatoriya T, et al. Effects of minocycline on hind-limb motor function and gray and white matter injury after spinal cord ischemia in rats. *Spine* 2011;**36**:1919—24.

29. Lee JH, Tigchelaar S, Liu J, et al. Lack of neuroprotective effects of simvastatin and minocycline in a model of cervical spinal cord injury. *Exp Neurol* 2010;**225**(1):219—30.

30. Pinzon A, Marcillo A, Quintana A, et al. A re-assessment of minocycline as a neuroprotective agent in a rat spinal cord contusion model. *Brain Res* 2008;**1243**:146—51.

31. Huntington Study Group. Minocycline safety and tolerability in Huntington disease. *Neurology* 2004;**63**(3):547—9.

32. NINDS NET-PD Investigators. A pilot clinical trial of creatine and minocycline in early Parkinson disease: 18-month results. *Clin Neuropharmacol* 2008;**31**(3):141—50.

33. Gordon PH, Moore DH, Gelinas DF, et al. Placebo-controlled phase I/II studies of minocycline in amyotrophic lateral sclerosis. *Neurology* 2004;**62**(10):1845—7.

34. Fagan SC, Waller JL, Nichols FT, et al. Minocycline to improve neurologic outcome in stroke (MINOS): a dose-finding study. *Stroke* 2010;**41**(10):2283—7.

35. Casha S, Zygun D, McGowan MD, et al. Results of a phase II placebo-controlled randomized trial of minocycline in acute spinal cord injury. *Brain* 2012;**135**(Pt 4):1224—36.

36. Levallois C, Calvet MC, Kamenka JM, Petite D, Privat A. TCP enhances the survival of human fetal spinal cord cells in culture. *Brain Res* 1992;**573**(2):327—30.

37. Hirbec H, Gaviria M, Vignon J. Gacyclidine: a new neuroprotective agent acting at the N-methyl-D-aspartate receptor. *CNS Drug Rev* 2001;**7**(2):172—98.

38. Kamenka JM, Chiche B, Goudal R, et al. Chemical synthesis and molecular pharmacology of hydroxylated 1-(1-phenylcyclohexyl-piperidine) derivatives. *J Med Chem* 1982;**25**(4):431—5.

39. Hoizey G, Kaltenbach ML, Dukic S, et al. Pharmacokinetics of gacyclidine enantiomers in plasma and spinal cord after single enantiomer administration in rats. *Int J Pharm* 2001;**229**(1—2):147—53.

40. Drian MJ, Kamenka JM, Privat A. In vitro neuroprotection against glutamate toxicity provided by novel noncompetitive N-methyl-D-aspartate antagonists. *J Neurosci Res* 1999;**57**(6):927—34.

41. Gaviria M, Privat A, d'Arbigny P, et al. Neuroprotective effects of gacyclidine after experimental photochemical spinal cord lesion in adult rats: dose-window and time-window effects. *J Neurotrauma* 2000;**17**(1):19—30.

42. Gaviria M, Privat A, d'Arbigny P, et al. Neuroprotective effects of a novel NMDA antagonist, Gacyclidine, after experimental contusive spinal cord injury in adult rats. *Brain Res* 2000;**874**(2):200—9.

43. Tadie M, Gaviria M, Mathe JF. Early care and treatment with a neuroprotective drug, gacyclidine, in patients with acute spinal cord injury. *Rachis* 2003;**15**:363—76.

44. Tang L, Gamal El-Din TM, Swanson TM, et al. Structural basis for inhibition of a voltage-gated Ca^{2+} channel by Ca^{2+} antagonist drugs. *Nature* 2016;**537**(7618):117—21.

45. Carlson AP, Hänggi D, Macdonald RL, Shuttleworth CW. Nimodipine reappraised: an old drug with a future. *Curr Neuropharmacol* 2020;**18**(1):65—82.

46. Ludwig A, Flockerzi V, Hofmann F. Regional expression and cellular localization of the $\alpha 1$ and β subunit of high voltage-activated calcium channels in rat brain. *J Neurosci* 1997;**17**(4):1339—49.

47. Abele A, Scholz KP, Scholz WK, Miller RJ. Excitotoxicity induced by enhanced excitatory neurotransmission in cultured hippocampal pyramidal neurons. *Neuron* 1990;**4**(3):413—9.

48. Sanz JM, Chiozzi P, Colaianna M, et al. Nimodipine inhibits IL-1β release stimulated by amyloid β from microglia. *Br J Pharmacol* 2012;**167**(8):1702—11.

49. Espinosa-Parrilla JF, Martínez-Moreno M, Gasull X, Mahy N, Rodríguez MJ. The L-type voltage-gated calcium channel modulates microglial pro-inflammatory activity. *Mol Cell Neurosci* 2015;**64**:104—15.

50. Li Y, Hu X, Liu Y, Bao Y, An L. Nimodipine protects dopaminergic neurons against inflammation-mediated degeneration through inhibition of microglial activation. *Neuropharmacology* 2009;**56**(3):580—9.

51. Guha A, Tator CH, Piper I. Increase in rat spinal cord blood flow with the calcium channel blocker, nimodipine. *J Neurosurg* 1985;**63**(2):250—9.

52. Guha A, Tator CH, Piper I. Effect of a calcium channel blocker on posttraumatic spinal cord blood flow. *J Neurosurg* 1987;**66**:423—30.

53. Fehlings MG, Tator CH, Linden RD. The effect of nimodipine and dextran on axonal function and blood flow following experimental spinal cord injury. *J Neurosurg* 1989;**71**(3):403—16.

54. Ross IB, Tator CH. Further studies of nimodipine in experimental spinal cord injury in the rat. *J Neurotrauma* 1991;**8**(4):229—38.

55. Ross IB, Tator CH, Theriault E. Effect of nimodipine or methylprednisolone on recovery from acute experimental spinal cord injury in rats. *Surg Neurol* 1993;**40**(6):461—70.

56. Jia YF, Gao HL, Ma LJ, Li J. Effect of nimodipine on rat spinal cord injury. *Genet Mol Res* 2015;**14**(1):1269—76.

57. Pointillart V, Petitjean ME, Wiart L, et al. Pharmacological therapy of spinal cord injury during the acute phase. *Spinal Cord* 2000;**38**(2):71—6.

58. Dubreuil CI, Winton MJ, McKerracher L. Rho activation patterns after spinal cord injury and the role of activated Rho in apoptosis in the central nervous system. *J Cell Biol* 2003;**162**:233—43.

59. Filbin MT. Myelin-associated inhibitors of axonal regeneration in the adult mammalian CNS. *Nat Rev Neurosci* 2003;**4**(9):703—13.

60. Madura T, Yamashita T, Kubo T, et al. Activation of Rho in the injured axons following spinal cord injury. *EMBO Rep* 2004;**5**:412—7.

61. McKerracher L, David S. Easing the brakes on spinal cord repair. *Nat Med* 2004;**10**(10):1052—3.

62. Lord-Fontaine S, Yang F, Diep Q, et al. Local inhibition of Rho signaling by cell-permeable recombinant protein BA-210 prevents secondary damage and promotes functional recovery following acute spinal cord injury. *J Neurotrauma* 2008;**25**(11):1309—22.

63. Shibata A, Wright MV, David S, McKerracher L, Braun PE, Kater SB. Unique responses of differentiating neuronal growth cones to inhibitory cues presented by oligodendrocytes. *J Cell Biol* 1998;**142**:191—202.

64. Dergham P, Ellezam B, Essagian C, et al. Rho signaling pathway targeted to promote spinal cord repair. *J Neurosci* 2002;**22**(15):6570—7.

65. Schmidt A, Hall A. Guanine nucleotide exchange factors for Rho GTPases: turning on the switch. *Genes Dev* 2002;**16**(13):1587—609.

66. Forgione N, Fehlings MG. Rho-ROCK inhibition in the treatment of spinal cord injury. *World Neurosurg* 2014;**82**(3—4):e535—9.

67. Lehmann M, Fournier A, Selles-Navarro I, et al. Inactivation of Rho signaling pathway promotes CNS axon regeneration. *J Neurosci* 1999;**19**:7537—47.

68. Steeves JD, Lammertse D, Curt A, et al. Guidelines for the conduct of clinical trials for spinal cord injury (SCI) as developed by the ICCP panel: clinical trial outcome measures. *Spinal Cord* 2007;**45**:206—21.

69. Fehlings MG, Kim KD, Aarabi B, et al. Rho inhibitor VX-210 in acute traumatic subaxial cervical spinal cord injury: design of the SPinal cord injury Rho INhibition InvestiGation (SPRING) clinical trial. *J Neurotrauma* 2018;**35**(9):1049—56.

70. Fehlings MG, Tator CH, Linden RD, et al. Relationship between spinal cord blood flow and axonal function in the motor and sensory tracts of the cord after experimental spinal cord injury. *Surg Forum* 1987;**38**:508—9.

71. Magistretti PJ, Geisler FH, Schneider JS, Li PA, Fiumelli H, Sipione S. Gangliosides: treatment avenues in neurodegenerative disease. *Front Neurol* 2019;**10**:859.

72. Fadda E, Negro A, Facci L, Skaper SD. Ganglioside GM1 cooperates with brain-derived neurotrophic factor to protect dopaminergic neurons from 6-hydroxy-dopamine-induced degeneration. *Neurosci Lett* 1993;**159**:147—50.

73. Lim ST, Esfahani K, Avdoshina V, Mocchetti I. Exogenous gangliosides increase the release of brain-derived neurotrophic factor. *Neuropharmacology* 2011;**60**:1160—7.

74. Ledeen RW. Biology of gangliosides: neurotogenic and neuronotrophic properties. *J Neurosci Res* 1984;**12**:147—59.

75. Bonheur JL, Laev H, Vorwerk C, Karpiak SE. Traumatic injury of spinal cord cells in vitro reduced by GM1 ganglioside. *Restor Neurol Neurosci* 1994;**6**(2):127—33.

76. Vorwerk CK, Bonheur J, Kreutz MR, Dreyer EB, Laev H. GM1 ganglioside administration protects spinal neurons after glutamate excitotoxicity. *Restor Neurol Neurosci* 1999;**14**(1):47—51.

77. Roisen EJ, Bartfelt J, Nagele R, Yorke G. Ganglioside stimulation of axonal sprouting in vitro. *Science* 1981;**214**:577.

78. Bose B, Osterholm JL, Kalia M. Ganglioside-induced regeneration and reestablishment of axonal continuity in spinal cord-transected rats. *Neurosci Lett* 1986;**63**(2):165—9.

79. Constantini S, Young W. The effects of methylprednisolone and the ganglioside GM1 on acute spinal cord injury in rats. *J Neurosurg* 1994;**80**(1):97−111.

80. Marcon RM, Cristante AF, de Barros TE, Filho Ferreira R, Dos Santos GB. Effects of ganglioside G(M1) and erythropoietin on spinal cord lesions in rats: functional and histological evaluations. *Clinics* 2016;**71**(6):351−60.

81. Yuan B, Pan S, Zhang WW. Effects of gangliosides on expressions of caspase-3 and NGF in rats with acute spinal cord injury. *Eur Rev Med Pharmacol Sci* 2017;**21**(24):5843−9.

82. Geisler FH, Dorsey FC, Coleman WP. Recovery of motor function after spinal-cord injury–a randomized, placebo-controlled trial with GM-1 ganglioside. *N Engl J Med* 1991;**324**(26):1829−38.

83. Geisler FH, Coleman WP, Grieco G, Poonian D, Sygen Study Group. The Sygen multicenter acute spinal cord injury study. *Spine* 2001;**26**(Suppl. 24):S87−98.

84. Chinnock P, Roberts I. Gangliosides for acute spinal cord injury. *Cochrane Database Syst Rev* 2005;**2005**(2): CD004444.

85. Faden AI, Vink R, McIntosh TK. Thyrotropin-releasing hormone and central nervous system trauma. *Ann N Y Acad Sci* 1989;**553**:380−4.

86. Feuerstein G, Lux Jr WE, Ezra D, et al. Reversal of leukotriene D4 hypotension by thyrotropin-releasing hormone. *Neurosci Res* 1984;**2**(1−2):121−4.

87. Feuerstein G, Lux Jr WE, Snyder F, et al. Hypotension produced by platelet-activating factor is reversed by thyro- tropin-releasing hormone. *Circ Shock* 1984; **13**(3):255−60.

88. Naftchi NE. Prevention of damage in acute spinal cord injury by peptides and pharmacologic agents. *Peptides* 1982;**3**(3):235−47.

89. Faden AI, Jacobs TP, Holaday JW. Thyrotropin-releasing hormone improves neurologic recovery after spinal trauma in cats. *N Engl J Med* 1981;**305**(18): 1063−7.

90. Faden AI, Jacobs TP, Smith MT. Thyrotropin-releasing hormone in experimental spinal injury: dose response and late treatment. *Neurology* 1984; **34**(10):1280−4.

91. Hashimoto T, Fukuda N. Effect of thyrotropin-releasing hormone on the neurologic impairment in rats with spinal cord injury: treatment starting 24 h and 7 days after injury. *Eur J Pharmacol* 1991;**203**(1): 25−32.

92. Vaidyanathan S, Rao MS, Sharma PL, Sachdeva NK. Modulation of urinary bladder function with thyrotropin-releasing hormone in patients with spinal cord injuries during the spinal shock phase. *Ann Clin Res* 1983;**15**(2):66−70.

93. Pitts LH, Ross A, Chase GA, Faden AI. Treatment with thyrotropin-releasing hormone (TRH) in patients with traumatic spinal cord injuries. *J Neurotrauma* 1995; **12**(3):235−43.

94. Nikolaev MV, Magazanik LG, Tikhonov DB. Influence of external magnesium ions on the NMDA receptor channel block by different types of organic cations. *Neuropharmacology* 2012;**62**(5−6):2078−85.

95. Lee JS, Han YM, Yoo DS, et al. A molecular basis for the efficacy of magnesium treatment following traumatic brain injury in rats. *J Neurotrauma* 2004;**21**:549−61.

96. Mami AG, Ballesteros J, Mishra OP, Ivoria-Papadopoulos M. Effects of magnesium sulfate administration during hypoxia on Ca(2þ) influx and IP(3) receptor modification in cerebral cortical neuronal nuclei of newborn piglets. *Neurochem Res* 2006;**31**:63−70.

97. Solaroglu I, Kaptanoglu E, Okutan O, et al. Magnesium sulfate treatment decreases caspase-3 activity after experimental spinal cord injury in rats. *Surg Neurol* 2005; **64**(Suppl. 2):17−21.

98. Vink R, Cernak I. Regulation of intracellular free magnesium in central nervous system injury. *Front Biosci* 2000;**5**:D656−65.

99. Gok B, Okutan O, Beskonakli E, Kilinc K. Effects of magnesium sulphate following spinal cord injury in rats. *Chin J Physiol* 2007;**50**:93−7.

100. Kaptanoglu E, Beskonakli E, Solaroglu I, Kilinc A, Taskin Y. Magnesium sulfate treatment in experimental spinal cord injury: emphasis on vascular changes and early clinical results. *Neurosurg Rev* 2003;**26**:283−7.

101. Suzer T, Coskun E, Islekel H, Tahta K. Neuroprotective effect of magnesium on lipid peroxidation and axonal function after experimental spinal cord injury. *Spinal Cord* 1999;**37**:480−4.

102. Ditor DS, John SM, Roy J, et al. Effects of polyethylene glycol and magnesium sulfate administration on clinically relevant neurological outcomes after spinal cord injury in the rat. *J Neurosci Res* 2007;**85**:1458−67.

103. Huang Z, Filipovic Z, Mp N, et al. AC105 Increases extracellular magnesium delivery and reduces excitotoxic glutamate exposure within injured spinal cords in rats. *J Neurotrauma* 2017;**34**(3):685−94.

104. Kwon BK, Roy J, Lee JH, et al. Magnesium chloride in a polyethylene glycol formulation as a neuroprotective therapy for acute spinal cord injury: preclinical refinement and optimization. *J Neurotrauma* 2009;**26**(8): 1379−93.

105. Domino EF, Unna KR, Kerwin J. Pharmacological properties of benzazoles: relationship between structure and paralyzing action. *J Pharmacol Exp Therapeut* 1952;**105**:486−97.

106. Miller RG, Mitchell JD, Lyon M, Moore DH. Riluzole for amyotrophic lateral sclerosis (ALS)/motor neuron disease (MND). *Cochrane Database Syst Rev* 2002;**2**: CD001447.

107. Agrawal SK, Fehlings MG. Mechanisms of secondary injury to spinal cord axons in vitro: role of Na^+, Na(+)-K(+)-ATPase, the Na(+)-H^+ exchanger, and the Na(+)-Ca^{2+} exchanger. *J Neurosci* 1996;**16**:545–52.

108. Schwartz G, Fehlings MG. Evaluation of the neuroprotective effects of sodium channel blockers after spinal cord injury: improved behavioral and neuroanatomical recovery with riluzole. *J Neurosurg* 2001;**94**(Suppl. 2):245–56.

109. Wilson JR, Fehlings MG. Riluzole for acute traumatic spinal cord injury: a promising neuroprotective treatment strategy. *World Neurosurg* 2014;**81**(5–6):825–9.

110. Li S, Mealing GA, Morley P, Stys PK. Novel injury mechanism in anoxia and trauma of spinal cord white matter: glutamate release via reverse Na^+-dependent glutamate transport. *J Neurosci* 1999;**19**:RC16.

111. Stys PK, Waxman S, Ransom BR. Na^+-Ca^{2+} Exchanger mediates Ca^{2+} influx during anoxia in mammalian central nervous system white matter. *Ann Neurol* 1991;**30**:375–80.

112. Koek W, Woods JH. 2-Amino-6-trifluoromethoxy benzothiazole (PK 26124), a proposed antagonist of excitatory amino acid neurotransmission, does not produce phencyclidine-like behavioral effects in pigeons, rats and rhesus monkeys. *Neuropharmacology* 1988;**27**:771–5.

113. Fumagalli E, Funicello M, Rauen T, et al. Riluzole enhances the activity of glutamate transporters GLAST, GLT1 and EAAC1. *Eur J Pharmacol* 2008;**578**:171–6.

114. Wang SJ, Wang KY, Wang WC. Mechanisms underlying the riluzole inhibition of glutamate release from rat cerebral cortex nerve terminals (synaptosomes). *Neuroscience* 2004;**125**:191–201.

115. Benavides J, Camelin JC, Mitrani N, et al. 2-Amino-6-trifluoromethoxy benzothiazole, a possible antagonist of excitatory amino acid neurotransmission-II. *Neuropharmacology* 1985;**24**:1085–92.

116. Bissaro M, Moro S. Rethinking to riluzole mechanism of action: the molecular link among protein kinase CK1δ activity, TDP-43 phosphorylation, and amyotrophic lateral sclerosis pharmacological treatment. *Neural Regen Res* 2019;**14**(12):2083–5.

117. Ates O, Cayli SR, Gurses I, et al. Comparative neuroprotective effect of sodium channel blockers after experimental spinal cord injury. *J Clin Neurosci* 2007;**14**(7):658–65.

118. Wu Y, Satkunendrarajah K, Teng Y, Chow DS, Buttigieg J, Fehlings MG. Delayed post-injury administration of riluzole is neuroprotective in a preclinical rodent model of cervical spinal cord injury. *J Neurotrauma* 2013;**30**(6):441–52.

119. Fehlings MG, Wilson JR, Frankowski RF, et al. Riluzole for the treatment of acute traumatic spinal cord injury: rationale for and design of the NACTN Phase I clinical trial. *J Neurosurg Spine* 2012;**17**(Suppl. 1): 151–6.

120. Lacomblez L, Bensimon G, Leigh PN, Guillet P, Meininger V. Dose-ranging study of riluzole in amyotrophic lateral sclerosis. Amyotrophic Lateral Sclerosis/Riluzole Study Group II. *Lancet* 1996;**347**(9013): 1425–31.

121. Chow DS, Teng Y, Toups EG, et al. Pharmacology of riluzole in acute spinal cord injury. *J Neurosurg Spine* 2012;**17**(Suppl. 1):129–40.

122. Fehlings MG, Nakashima H, Nagoshi N, et al. Rationale, design and critical end points for the Riluzole in Acute Spinal Cord Injury Study (RISCIS): a randomized, double-blinded, placebo-controlled parallel multi-center trial. *Spinal Cord* 2016;**54**(1):8–15.

123. Badhiwala JH, Wilson JR, Kwon BK, Casha S, Fehlings MG. A review of clinical trials in spinal cord injury including biomarkers. *J Neurotrauma* 2018; **35**(16):1906–17.

124. Chen MS, Huber AB, van der Haar ME. Nogo-A is a myelin-associated neurite outgrowth inhibitor and an antigen for monoclonal antibody IN-1. *Nature* 2000; **403**:434–9.

125. Chen M, Ona VO, Li MW, et al. Minocycline inhibits caspase-1 and caspase-3 expression and delays mortality in a transgenic mouse model of Huntington disease. *Nat Med* 2000;**6**:797–801.

126. Akbik FV, Bhagat SM, Patel PR, Cafferty WB, Strittmatter SM. Anatomical plasticity of adult brain is titrated by Nogo receptor 1. *Neuron* 2013;**77**(5): 859–66.

127. Chivatakarn O, Kaneko S, He Z, Tessier-Lavigne M, Giger RJ. The Nogo-66 receptor NgR1 is required only for the acute growth cone-collapsing but not the chronic growth-inhibitory actions of myelin inhibitors. *J Neurosci* 2007;**27**(27):7117–24.

128. Gonzenbach RR, Schwab ME. Disinhibition of neurite growth to repair the injured adult CNS: focusing on Nogo. *Cell Mol Life Sci* 2008;**65**:161–76.

129. Zorner B, Schwab ME. Anti-Nogo on the go: from animal models to a clinical trial. *Ann N Y Acad Sci* 2010; **1198**(Suppl. 1):E22–34.

130. Kucher K, Johns D, Maier D, et al. First-in-man intrathecal application of neurite growth-promoting anti-Nogo-A antibodies in acute spinal cord injury. *Neurorehabil Neural Repair* 2018;**32**(6—7):578—89.

131. Schnell L, Schwab ME. Axonal regeneration in the rat spinal cord produced by an antibody against myelin-associated neurite growth inhibitors. *Nature* 1990;**343**(6255):269.

132. Bregman BS, Kunkel-Bagden E, Schnell L, Dai HN, Gao D, Schwab ME. Recovery from spinal cord injury mediated by antibodies to neurite growth inhibitors. *Nature* 1995;**378**(6556):498—501.

133. Freund P, Schmidlin E, Wannier T, et al. Nogo-A-specific antibody treatment enhances sprouting and functional recovery after cervical lesion in adult primates [published correction appears in Nat Med. 2006 Oct; 12(10):1220]. *Nat Med.* 2006;**12**(7):790—2. https://doi.org/10.1038/nm1436.

134. Liebscher T, Schnell L, Schnell D, et al. Nogo-A antibody improves regeneration and locomotion of spinal cord-injured rats. *Ann Neurol* 2005;**58**(5):706—19.

135. Freund P, Schmidlin E, Wannier T, et al. Anti-Nogo-A antibody treatment promotes recovery of manual dexterity after unilateral cervical lesion in adult primates—re-examination and extension of behavioral data. *Eur J Neurosci* 2009;**29**(5):983—96.

136. Hata K, Fujitani M, Yasuda Y, et al. RGMa inhibition promotes axonal growth and recovery after spinal cord injury. *J Cell Biol* 2006;**173**(1):47—58.

137. Schwab JM, Conrad S, Monnier PP, et al. Spinal cord injury-induced lesional expression of the repulsive guidance molecule (RGM). *Eur J Neurosci* 2005;**21**(6):1569—76.

138. Tassew NG, Mothe AJ, Shabanzadeh AP, et al. Modifying lipid rafts promotes regeneration and functional recovery. *Cell Rep* 2014;**8**(4):1146—59.

139. Mothe AJ, Coelho M, Huang L, et al. Delayed administration of the human anti-RGMa monoclonal antibody elezanumab promotes functional recovery including spontaneous voiding after spinal cord injury in rats. *Neurobiol Dis* 2020;**143**:104995.

140. Mothe AJ, Tassew NG, Shabanzadeh AP, et al. RGMa inhibition with human monoclonal antibodies promotes regeneration, plasticity and repair, and attenuates neuropathic pain after spinal cord injury. *Sci Rep* 2017;**7**(1):10529.

141. Demicheva E, Cui YF, Bardwell P, et al. Targeting repulsive guidance molecule A to promote regeneration and neuroprotection in multiple sclerosis. *Cell Rep* 2015;**10**(11):1887—98.

142. Ziemann A, Rosebraugh M, Barger B, Cree B. A Phase 1, multiple-dose study of Elezanumab (ABT-555) in patients with relapsing forms of multiple sclerosis. *Neurology* 2019;**92**(Suppl. 15):S56.001.

143. Martin AR, De Leener B, Cohen-Adad J, et al. Clinically feasible microstructural MRI to quantify cervical spinal cord tissue injury using DTI, MT, and T2-Weighted imaging: assessment of normative data and reliability. *AJNR Am J Neuroradiol* 2017;**38**(6):1257—65.

Clinical trials: cellular regenerative approaches

Nayaab Punjani[1,2,a], Vjura Senthilnathan[2,3,a], Christopher S. Ahuja[1,2,4], Michael G. Fehlings[2,4,5]

[1]Institute of Medical Science, University of Toronto, Toronto, ON, Canada; [2]Division of Genetics and Development, Krembil Research Institute, University Health Network, Toronto, ON, Canada; [3]University of Toronto Scarborough, Toronto, ON, Canada; [4]Division of Neurosurgery, Department of Surgery, University of Toronto, Toronto, ON, Canada; [5]Division of Neurosurgery, Toronto Western Hospital, University Health Network, Toronto, ON, Canada

Introduction

Primary injury in traumatic SCI often results in a lesion or compression of the spinal cord, thus resulting in damage to the cells[1] within the CNS. This is followed by secondary cascades, which involve the loss of neurons, glial cells (i.e., oligodendrocytes, astrocytes, etc.),[1] and extracellular matrix at the site of injury. This is accompanied by neuroinflammatory processes and the formation of cysts and an inhibitory glial scar.[1] Cells in the CNS, however, have a low regenerative and proliferation capacity; thus, lost neural networks are not efficiently reformed, often resulting in a permanent loss of functionality.[2]

Regenerative medicine is a research field that aims to replace cellular loss through various methods. This chapter will focus on the use of cell-based therapies, often achieved through cell transplantation, to provide the required cells at the injury site. The end goal is to improve or restore neural networks and help patients regain sensorimotor control.

Stem cells are often utilized in regenerative medicine due to their ability to differentiate into various different cell types, as well as proliferate and self-renew, thus yielding additional stem cells.[3,4] These cells may be classified in four ways depending on their differentiation capabilities. Totipotent stem cells are the least differentiated and are present at initial zygote formation.[3,5,6] These cells are capable of forming embryonic and extraembryonic cells.[4,6] This is followed by pluripotent stem cells, which are

[a] Nayaab Punjani and Vjura Senthilnathan designating co-first authorship.

© 2022 Elsevier Inc. All rights reserved.

capable of differentiating into cells of the three germ layers—endoderm, mesoderm, and ectoderm.[4,5,7] Then, there are multipotent stem cells, which are able to form cells of a particular type.[3,4,6] This includes neural stem cells (NSCs), which differentiate into various CNS cell subtypes. Lastly, the most differentiated is unipotent cells, which are able to produce a single type of cell.[3,4,6]

Stem cells may be obtained through three means. The first of which is primary cell lines or direct extraction of cells from living tissue. This includes fetal stem cells, obtained from the developing fetus,[4] and adult stem cells. Adult stem cells may be derived from two sources. Autologous stem cells are obtained directly from the patient's own tissue.[4] In contrast, allogenic stem cells are obtained from another individual whose human leukocyte antigen (HLA) is matched to the patient.[4,8]

In regenerative medicine, pluripotent stem cells are often used as they can be guided to a particular cell lineage through the use of molecular engineering and transcription factors. Pluripotent stem cells may be procured from human embryonic stem cells (hESCs), and they are derived from the blastocyst[5,7,9,10] in unutilized in vitro fertilization (IVF) embryos.[10] This has been a controversial issue, with some contending that the use of the ESCs for this purpose resulted in a loss of the embryo.[10] New evidence, however, has found a method for obtaining hESCs in xeno-free conditions, whereby the embryo is not destroyed.[10]

Another method involves taking adult somatic fully differentiated cells and converting them back to the pluripotent state, thus forming induced pluripotent stem cells (iPSCs).[11] Evidence has demonstrated that human iPSCs are similar to ESCs in various means such as proliferative capacity and morphology.[11,12] Furthermore, iPSCs have been successfully produced from adult fibroblast cells.[11,13] The process of iPSC development, however, is often lengthy and costly and has the potential for tumorigenesis if the iPSCs do not differentiate into the correct cell product.[14]

A newer method, termed direct cell reprogramming or transdifferentiation, aims to convert a somatic cell directly into another fully differentiated cell, without an intermediate pluripotent event, thereby lessening the potential for tumor formation.[14] Methods for transdifferentiation involve CRISPR/Cas9 genetic modification and the use of transcription factors,[14] which is further highlighted in Fig. 25.1.

Clinical trial design for cell-based therapies involves a three-stage process as referenced in Fig. 25.2. Stage 1 involves the development of an in vitro cell line, which consists of characterizing and modifying the cell line followed by small animal testing. Stage 2 involves good manufacturing practices (GMP) grade cell line development as well as further testing in large animal models to optimize regulatory requirements. The last stage of the regenerative trial design involves conducting clinical trials to assess the safety and efficacy of the GMP grade cell line in the target human population. This follows a four-phase clinical trial process with manufacturing optimization occurring between phases 1 and 2.[15,16]

Clinical trials assess the severity of SCI in patients using the International Standards for Neurological Classification of Spinal Cord Injury (ISNCSCI) to assess the motor and sensory capabilities of SCI patients.[17] Included under these standards is the often-utilized ASIA Impairment Score (AIS).[18] This classification scale varies from an impairment grade of A to E, with decreasing severity. Level A indicates complete SCI with a lack of all sensorimotor functioning including lower-sacral segments of the spinal cord. Level E indicates normal spinal cord functioning with motor and sensory abilities remaining intact.[18]

This chapter will be discussing various clinical trials examining different cellular regenerative therapies currently being conducted for the treatment of SCI. The trials being discussed were obtained using *ClinicalTrials.gov*, with a primary focus on trials conducted in North America, Europe, Japan, and Australia.

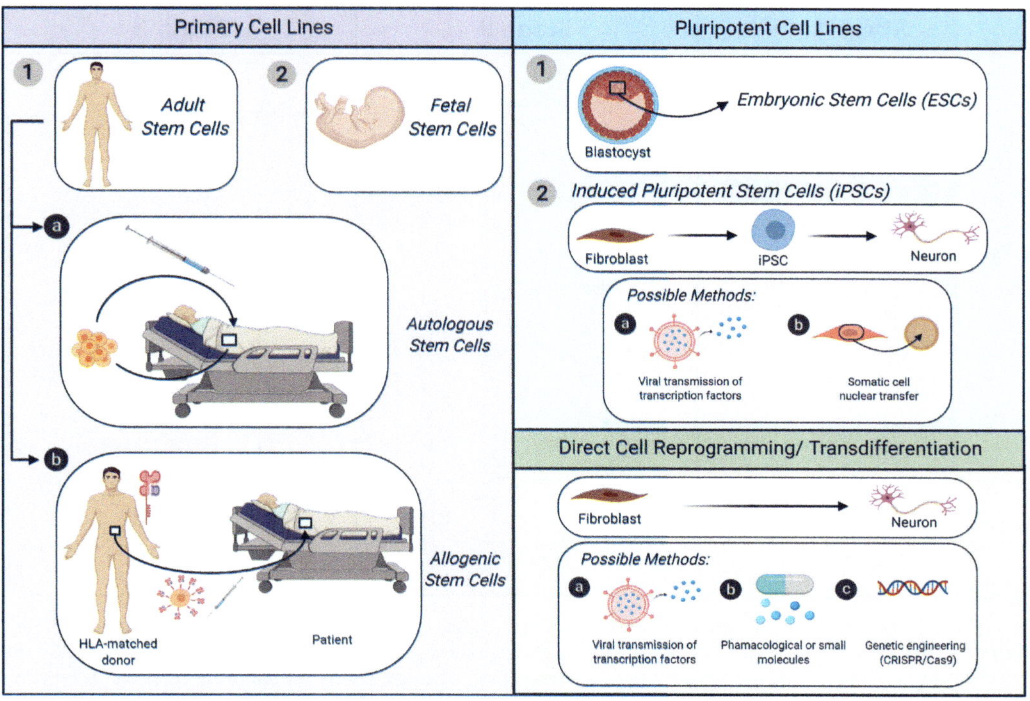

FIGURE 25.1 **Sources of stem cells for cell-based regenerative therapies.** Stem cells may be derived from primary cell lines, pluripotent cell lines, or direct cell reprogramming. Primary cell lines involve direct extraction of the required cells from live tissue. This may be obtained from the developing fetus[4] or an adult. Adult stem cells may be extracted from an autologous source, from the patient,[4] or an allogenic source, through an HLA-matched donor.[4,8] The cell of interest can also be obtained from pluripotent cells that are later differentiated, and these cells are derived from two possible sources. Embryonic stem cells (ESCs) are extracted from the blastocyst.[5,7,9,10] Induced pluripotent stem cells (iPSCs) involve converting a fully differentiated somatic cell into an induced pluripotent state, which can later be differentiated into another cell type.[11] This may be achieved through viral transmission of transcription factors[11,12] or somatic cell nuclear transfer to an oocyte.[12] Lastly, direct cell reprogramming or transdifferentiation consists of directly converting a fully differentiated somatic cell into another somatic cell without the need for an intermediate pluripotent state.[14] Some possible methods for this process include the viral transmission of transcription factors, pharmacological or small molecular interventions, or genetic engineering techniques such as CRISPR/Cas9.[14] *Created with BioRender.com.*

Neural stem cells

Neural stem cells (NSCs) are multipotent with self-renewing capabilities and the potential for differentiation into the three major cell types of the central nervous system—neurons, astrocytes, and oligodendrocytes,[2,19] and this is illustrated in Fig. 25.3. These cells are found in vivo in certain regions of the nervous system and may be cultured in vitro using ependymal growth factor.[19]

Short-term and long-term Canadian and Switzerland multicenter clinical trials examined the use of human CNS stem cells (HuCNS-SC) for transplantation in the spinal cord (NCT01321333; NCT03069404), sponsored by StemCell Inc. Firstly, an interventional short-term 1-year phase I/II clinical trial completed in 2015 was conducted on 12 participants with subacute T2-T11 SCI, as defined by magnetic

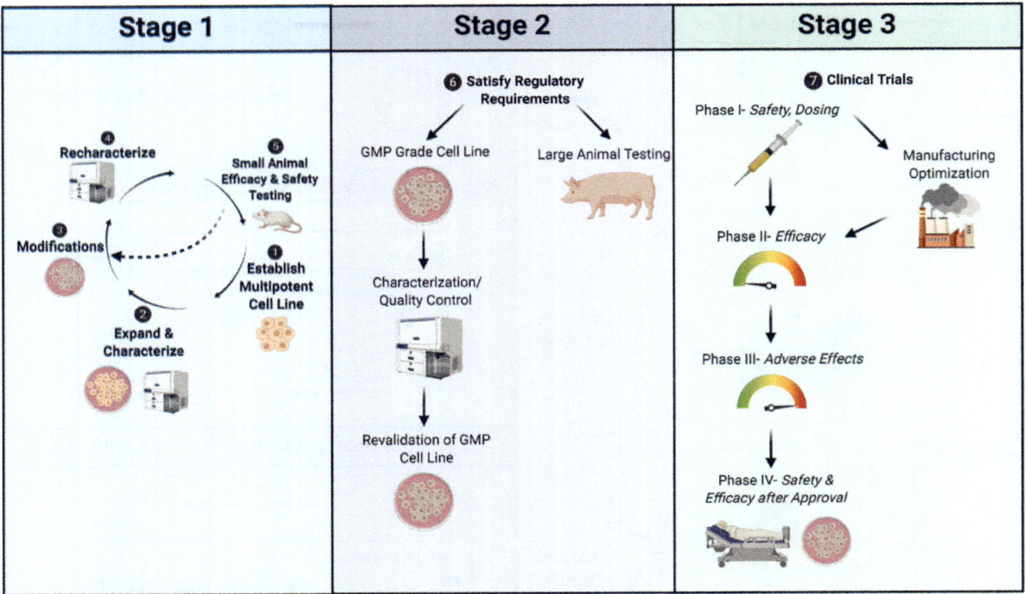

FIGURE 25.2   **Steps involved in regenerative trial design.** Stage 1 represents the establishment of a cell line through in vitro techniques. When tested in small animals, modifications to this process can be made as demonstrated by the dotted line and cyclic nature of stage 1. Stage 2 represents the steps involved in satisfying regulatory requirements, which consists of creating a good manufacturing practices (GMP) cell line and large animal testing. Stage 3 represents the clinical testing phases in humans[15,16] with spinal cord injury along with manufacturing optimization. If stage 3 is successful, the regenerative treatment in question may be implemented as an approved treatment protocol. *Created with BioRender.com.*

resonance imaging (MRI) or computed tomography (CT). Inclusion criteria further involved patients having an AIS grade of A, B, or C, while being at least 6-week postinjury at the onset of the study. After an open-label intramedullary allogenic HuCNS-SC transplantation or bone marrow transplantation[96] into the thoracic spine, patients were provided with 9 months of posttransplant immunosuppression with up to 1-year posttransplantation monitoring for safety and detection of adverse side effects (NCT01321333 or CL-N02-SC). All 12 patients from the CL-N02-SC trial were then enrolled in a current 5-year long-term observational phase I/II clinical trial, which was completed in April 2020. The primary outcome measure of this trial is whether participants will improve in their AIS-grade posttransplantation. Results have not been posted to date (NCT03069404 or INSTrUCT-SCI).

A phase I open-label interventional clinical trial is currently being conducted in the United States at the UCSD Medical Center to assess the short-term and long-term safety of the surgical implantation of human spinal cord—derived neural stem cells (HSSCs) in chronic SCI patients, sponsored by Neuralstem Inc. Group A contains four patients with T2-T12 level SCI, and Group B is recruiting four patients with C5—C7 level SCI. Inclusion criteria consist of patients having an AIS grade A, with SCI occurring within 1—2 years the study, and the presence of bone fusion. Study duration will be 6-month posttransplant to examine any adverse side effects. This will be followed by 54 months of additional follow-up, examining the secondary measures of transplant graft survival via MRI and autopsy, as well as monitoring the maintenance of immunosuppression. This study is scheduled for completion in 2022 (NCT01772810).

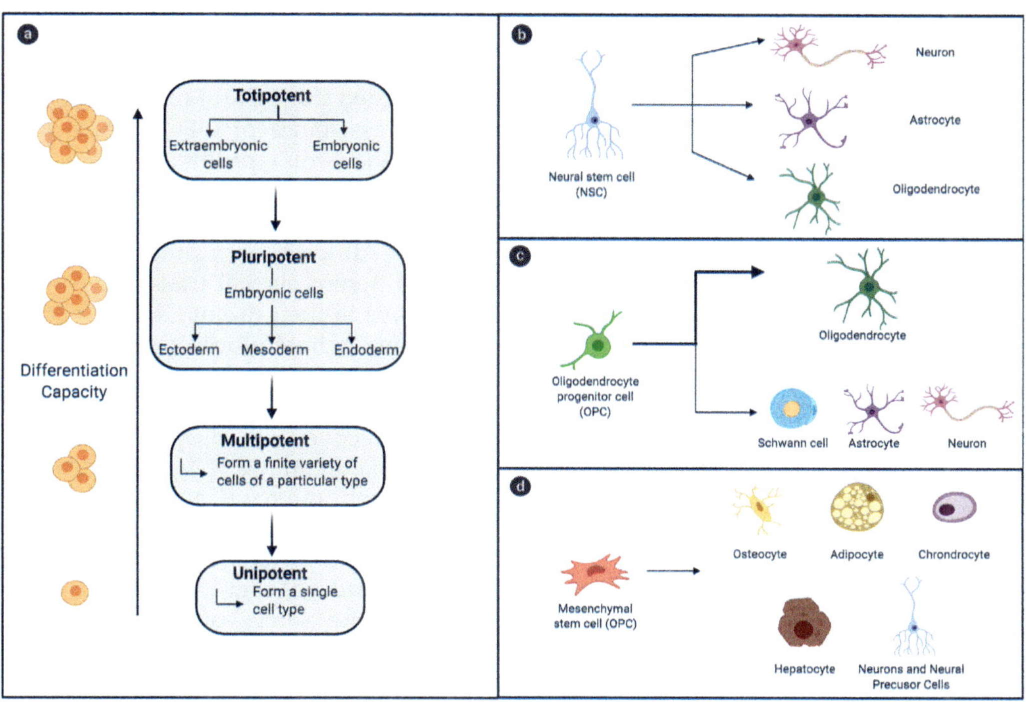

FIGURE 25.3   **Stem cell differentiation and relevant multipotent stem cells for spinal cord injury regenerative therapies.**
(A) Stem cell differentiation from totipotent stem cells to unipotent cells, with decreasing differentiation capacity as demonstrated on the figure.[3–7] (B) Differentiation of multipotent neural stem cells (NSCs) into neurons, astrocytes, and oligodendrocytes.[2,19] (C) Differentiation of multipotent oligodendrocyte progenitor cells (OPCs) into primarily oligodendrocytes,[20–22] but possible differentiation into Schwann cells,[21,23,92] astrocytes,[23] and neurons.[93,94] (D) Differentiation of multipotent mesenchymal stem cells (MSCs) into various cell types such as osteocytes, adipocytes, chondrocytes, hepatocytes,[57] as well as neurons and neural precursor cells.[57,95] *Created with BioRender.com.*

Oligodendrocyte progenitor cells

Oligodendrocyte progenitor cells (OPCs) have the ability to preferentially differentiate into oligodendrocytes,[20–22] as well as possible production of Schwann cells,[21,23] thus helping to restore the myelination of neuronal axons damaged during SCI,[23] which is demonstrated in Fig. 25.3. A phase I/IIa interventional trial commonly known as the SCiStar study was completed in December 2018 to assess the safety of AST-OPC1 in the treatment of subacute cervical SCI (NCT02302157). AST-OPC1 is an OPC cell line derived from human embryonic stem cells.[24,25] Preclinical animal testing of these cells

in a cervical SCI rat model demonstrated improved locomotion and no safety complications.[25]

This trial sponsored by Asterias Biotherapeutics Inc. took place in various medical centers across the United States and involved 25 participants split into five cohorts. They assessed the safety of increasing doses of AST-OPC1 cells administered at a single timepoint 21–42 days postinjury. Participants either received one injection of 2 or 10 million cells or two injections of 10 million cells each. Three of the cohorts had complete sensorimotor SCI (AIS grade A) and two cohorts had incomplete

SCI (AIS grade B). Furthermore, the participants had to undergo an MRI to determine the epicenter and bounds of the lesion for future monitoring, and C4–C7 must be the last single neurological level (SNL) with retained sensorimotor functioning. Participants were monitored over the course of 1-year posttransplant to assess any adverse events and track neurological function, such as their motor level on the ISNCSCI (NCT02302157).

Results were provided in an update by Lineage Cell Therapeutics, the parent organization of Asterias Biotherapeutics, Inc., in 2019. Motor ability was improved by at least one neurological level in 96% of patients, whereby levels were determined by upper extremity motor scores indicating the level of the spine above which motor function was restored.[26] Some patients were followed for up to 2 years, and upper extremity motor function was improved in those patients.[27]

GRNOPC1 cells are OPCs derived from hESCs by Geron Corporation (United States) that received approval from the Food and Drug Administration (FDA) in 2009.[28,29] However, it was put on clinical hold due to the increased incidence of cysts in an animal experiment.[29] In 2010, the clinical hold was lifted after further animal studies were conducted to rectify these concerns.[29] This is the world's first human clinical trial using ESC therapy.[29] The primary outcome measure for this phase I clinical trial was to evaluate the safety of GRNOPC1 cells when administered 7 or 14 days after injury (NCT01217008). This would be measured by the occurrence and the severity of adverse events during the first-year posttransplant resulting from the injection of GRNOPC1s, transplant procedure, and immunosuppressant administration. The inclusion criteria for this trial were patients with complete traumatic SCI (AIS grade A). This trial was completed in July 2013; however, no results have been released (NCT01217008). According to California's Stem Cell Agency, this particular trial was cut short due to lack of funding; however, this was picked up by Asterias Biotherapeutics.[30]

Olfactory ensheathing cells

The olfactory mucosa that is found in the nose is regenerated throughout life through stem cells with the assistance of OECs. These OECs are found in the lamina propria which is below the sensory epithelium. OECs can be extracted from the mucosa or the olfactory bulb (OB) and may have different morphology and function. Hence, different clinical trials may not be fully comparable in their use of OECs for SCI.[31]

Previous animal studies have demonstrated the ability of OECs obtained from the OB to improve regeneration of spinal axons[32–35] and potential recovery of proprioception,[36] movement, and breathing.[34,37] In contrast, OECs obtained from the nasal mucosa lack the ability to regenerate severed neural fibers.[32,38]

A phase I single blind clinical trial conducted by Féron and colleagues aimed to assess the feasibility and safety of autologous OECs 1-year posttransplantation. Three male patients between the ages 18 and 55 years, with complete thoracic SCI that occurred between 6 and 32 months prior to enrolling into this trial were included. The inclusion criteria consisted of having complete SCI occurring between T4 and T10 with no evidence of ongoing recovery (AIS grade A), and injury occurring between 6 months and 3 years. When these participants were evaluated 1-year posttransplant, there were no adverse medical or surgical complications to render this procedure unsafe. This included no formation of cysts, tumors, or syrinx, as well as no neuropathic pain. Féron and colleagues concluded that transplanting autologous OECs into SCI patients is feasible and safe up to 1 year after transplantation.[39]

Following this clinical trial, a phase I/IIa single blind design clinical trial was conducted by Mackay-Sim and colleagues published in 2008 using autologous transplantation of OECs into the spinal cord of six male patients with complete thoracic paraplegia to assess the feasibility and safety of autologous OEC transplantation during a period of 3 years. The primary inclusion criteria for this clinical trial included having SCI between T4 and T10 occurring between 6 months and 3 years prior to enrolling in the trial, with sustained and complete loss of sensorimotor function below the level of injury (AIS grade A). Mackay-Sim and colleagues did not find any evidence of tumor formation, posttraumatic syringomyelia, nor neuropathic pain for 3 years posttransplantation of autologous OECs. However, they did not find significant functional recovery in these patients. Hence, they concluded that transplanting autologous OECs in SCI patients was feasible and safe up to 3 years posttransplant. These results should be considered preliminary due to the small sample size of the trial.[40]

A previous phase I clinical trial examined the use of autologous mucosal ensheathing cells, along with olfactory nerve fibroblasts (ONFs), in three complete thoracic SCI patients. No side effects were perceived, and two of the three patients showed an improved AIS grade.[41] A subsequent clinical trial involving a single patient with a T9 complete SCI used OECs and ONFs extracted from the OB, along with autologous sural nerve implants. This first successful attempt enhanced recovery over mucosal derived OECs,[32,41] with AIS grade improvement to level C, and partial improvement of sensation and voluntary movement.[35] This experimental design provides the foundation for the subsequent clinical trial.

A current phase I/II trial is being conducted at the Wroclaw Medical University in Poland and is sponsored by the Nicholls Spinal Injury Foundation. This interventional open-label, nonrandomized, single group assignment trial is currently recruiting two participants with a single complete SCI between C5 and T10, as assessed through AIS grade A criteria and physiological measurements (e.g., electromyography (EMG)) (NCT03933072).

The goal of this trial is to examine the safety of autologous human OECs transplanted as part of a glial neuropatch with ONFs. This two-part surgical procedure will first involve the use of a minimally invasive transnasal endoscopy-assisted approach[42,43] to harvest one OB from each patient and prepare each patient's sural nerve for grafting. OECs and OFCs will be extracted from the OB and embedded into a collagen scaffold or OEC/ONF glial neuropatch. The second surgery will involve exposure of the transected spinal cord through a laminectomy and midline durotomy, followed by the transplantation of the tissue-engineered graft at the injury site and the use of autologous sural nerve grafts to bridge the gap. This surgery will be followed by a minimum of 2-year postoperative neurorehabilitation and monitoring (NCT03933072).

Primary outcome measures involve improvement in AIS grade, assessments of trunk muscle strength and risk of falling using the Berg Balance scale, improved deep pressure sensation, as well as the evaluation of vibratory and joint position sensation. This trial is projected for completion in 2022 (NCT03933072).

There were a small number of trials that are focused on the olfactory mucosa to access OECs. The olfactory mucosa consists of a variety of cell types including OECs and other stem-like progenitor cells.

A pilot clinical trial to assess the feasibility and safety of transplanting olfactory mucosa autografts was conducted in Portugal in patients with chronic SCI. The inclusion criteria consisted of patients with SCI occurring between C4 and T10 for more than 6 months with an AIS grade A or B, under the age of 35 years with no significant nasal and paranasal sinus pathology. The outcome measures consisted of ISNCSCI examinations, EMG, spinal MRI scans, and otolaryngological evaluations. Patients showed improvements in AIS motor scores with two patients going from AIS grade A to C. All patients except one showed improvements in AIS sensory scores. Two patients had returned sensation of the bladder, and one of these patients was also able to voluntarily contract the anal sphincter. Adverse events consisted of transient pain, which was treated with medication. One patient showed a decrease in AIS sensory scores, which the researchers attributed to difficulty locating the injury site during surgery. These results suggest that olfactory mucosa autograft transplantation is feasible, relatively safe, and has the potential to show efficacy in SCI patients. However, long-term monitoring is required to fully assess safety.[31,44]

Following this study, another pilot study was conducted in India to assess autologous mucosal transplantation in chronic SCI 18 months posttransplant using the methods described by Lima and colleagues. Patients with chronic, motor complete SCI between C5-T12, with AIS grade A or B between the ages 18 and 40 years, with SCI occurring more than 18 months prior to enrollment were included. No significant improvements in the assessment criteria were observed. Hence, they concluded that although this procedure is safe and feasible, they were not able to demonstrate efficacy.[45]

Finally, a third report about the safety and feasibility of autologous transplantation of olfactory mucosa 12—45 months posttransplant were observed in a phase I/II nonrandomized, noncontrolled prospective open-label study. Patients with SCI between C4 and T10, 18—189 months prior to enrolling with an AIS grade A or B, between the ages 18 and 40 years, and no significant nasal and paranasal sinus pathology were included in this trial. Outcome measures include AIS grade changes, MRI of the spinal cord, Functional Independence Measure (FIM), Walking Index for Spinal Cord Injury (WISCI), otolaryngological, and psychological assessments. This trial was different compared with the aforementioned trials as participants underwent preoperative rehabilitation exercise programs 4 months prior to transplant, but no improvement was observed during this phase. After transplantation, these rehabilitation regiments were continued. 11 out of the 20 participants showed improvement in AIS grade; however, one patient deteriorated from AIS grade B to A. All patients showed improvements in FIM and WISCI scales. MRI scans showed that the transplanted cells filled the lesion cavity with no presence of overgrowth and syringomyelia. The researchers concluded that autologous transplantation of olfactory mucosa is feasible and relatively safe with having the potential to alleviate symptoms in chronic SCI when combined with rehabilitation exercise programs.[46]

Schwann cells

Schwann cells (SCs) produce myelin sheaths in the PNS but are also known to assist in the regeneration of CNS axons.[48] This is mediated through providing growth factors, supplying extracellular matrix components (ECM),[49] and delaying the production of the inhibitory glial scar.[50] Furthermore, SCs may reduce neuroinflammation through the release of antiinflammatory cytokines,[51,52] and recent rodent studies demonstrate the ability of SCs to reduce demyelination and thus improve motor functioning.[47]

Two completed clinical trials have been conducted to assess the safety of autologous human Schwann cell (ahSC) transplantation for the treatment of SCI. Both trials were conducted at the University of Miami, as part of the Miami Project to Cure Paralysis.

A phase I clinical trial completed in 2016 examined the safety and efficacy of these ahSCs across various timepoints up to 12 months posttransplantation in nine participants, in which six participants completed the trial and received cell transplants (NCT01739023).[53] Participants had acute traumatic SCI within the past 30 days at T3—T11 of AIS grade A. ahSCs were obtained via the participant's sural nerve prior to 30 days postinjury and transplanted into the injury site by 72 days postinjury through intramedullary injection.[53] Outcome measures included AIS grade changes, variations in MRI lesion measurements, and degree of neuropathic pain. Published results demonstrate that no severe adverse effects or safety concerns were perceived 1-year posttransplant,[53] although neuropathic pain was experienced by 50% of participants,[54] but this is expected.[54–56]

A second phase I trial completed in 2019 examined the safety of ahSCs paired with fitness conditioning and rehabilitation prior to and posttransplantation in eight chronic SCI participants (NCT02354625). Participants with an AIS grade of A, B, or C were split into two cohorts: thoracic SCI between T2 and T12 or lower cervical SCI between C5 and T1. Outcome measures were monitored over the course of 6 months posttransplant including AIS grade improvements, sensory testing, and timed walking tests. The results have not yet been published at the time of this writing (NCT02354625).

Mesenchymal stem cells

Mesenchymal stem cells (MSCs) are a type of multipotent stem cell that has the ability to differentiate into multiple types of specialized cells such as osteocytes, adipocytes, and neurocytes,[57] which can be seen in Fig. 25.3. These cells can be harvested from a variety of tissues such as bone marrow, umbilical cord, and adipose tissue.[57] MSCs also have the ability to release cytokines that have been known to have neuroprotective effects, which allow for the regeneration of neurons as well as strengthened axonal growth and renewal of damaged neurons.[58,59] Clinical trials investigating MSCs are occurring in a number of countries, described here.

The United States

A one-arm phase I clinical trial conducted by the Mayo Clinic in the United States is evaluating the safety of administering MSCs harvested from adipose tissue through an intrathecal injection in patients with severe traumatic SCI (NCT03308565). Adipose tissue was harvested from these patients via a small incision in their abdomen or thigh. Then, a single dose of the adipose-derived MSCs (approximately 100 million cells) was injected into the L4–5 level of the spine through an intrathecal injection. Patients are then evaluated at varying time points posttransplant. The primary outcome measure for this trial is to determine the occurrence of acute adverse events up to 4 weeks posttransplant. Some of the inclusion criteria include having blunt nondegenerative traumatic SCI, which occurred between 2 weeks and 1 year of the trial with an AIS grade A or B (NCT03308565). The first report of this clinical trial was in February 2020, which involved a 53-year-old patient with an initial AIS grade A who improved to an AIS grade C and did not experience severe adverse effects after this intervention.[60] They also reported that this patient showed signs of improvement in sensory and motor function using the ISNSCSI.[60] In closing, the authors of this trial suggested that intrathecal injection of adipose-derived MSCs could be safe and practical but requires further evaluation.[60] This study is currently active, but not recruiting, with full study completion estimated for November 2023 (NCT03308565).

Spain

The first clinical trial that will be discussed in this section was a phase I/IIa crossover clinical trial sponsored by Banc de Sang i Teixits, which assessed the safety and benefits of an intrathecal injection of MSCs derived from Wharton's jelly (XCEL-UMC-BETA) in patients with chronic traumatic SCI (NCT03003364). After the administration of MSCs, patients were followed up to 6 months postintervention. At the 6-month point, patients who obtained the placebo received the MSCs injection and patients who received the MSCs injection obtained the placebo, thus making this clinical trial a crossover study. At the 12-month time point after the first injection, these patients were randomized again to receive a second dose to assess long-term safety and benefits. The primary outcome measure for this trial was to determine the occurrence of adverse effects due to the intervention at 12 months. Some of the inclusion criteria for this clinical trial consisted of having a single lesion between T2 and T11 that occurred between 1 and 5 years prior to the trial, with an AIS grade A. Although this study was completed in February 2020, no results have been published to date (NCT03003364).

The second clinical trial is a phase I/II randomized study sponsored by Ferrer Internacional S.A (NCT02917291). They are interested in determining the safety and feasibility of human allogenic adipose-derived MSCs that were expanded and pulsed with hydrogen peroxide (FAB117-HC) in patients with acute traumatic SCI. In phase I, patients with a T1—T12 injury of AIS grade A will be administered one dose of either 20 million or 40 million cells between 72 and 120 hours postinjury. In phase II, patients with a T1-T12 SCI of AIS grades A or B will be used to determine the maximum tolerated dose within 96 hours postinjury. This intramedullary administration of MSCs will occur during decompression/stabilization surgery. The primary outcome measure for this clinical trial is to determine the frequency of adverse events as a result of the course of treatment within 1-year postinjection. This clinical trial is actively recruiting patients at this time; thus, no results have been published (NCT02917291).

The next three clinical trials sponsored by the Puerta de Hierro University Hospital examined the safety and feasibility of autologous bone marrow MSCs, as well as their administration methods.

The first clinical trial was a phase I open-label study where they examined the clinical benefits of autologous adult bone marrow MSCs that were expanded in vitro in chronic, incomplete SCI (NCT02165904). These patients received intrathecal administration of MSCs into the subarachnoid space with two additional doses administered in 3-month intervals. The primary outcome measures included improvements in sensitivity observed using the AIS grade, changes in the FIM Scale, Barthel score, International Association of Neurorestoratology Spinal Cord Injury Functional Rating Scale (IANR-SCIFRS), and others. Patients with incomplete SCI were confirmed through neurophysiological examination, in addition to having normal creatinine, serum glutamic oxaloacetic transaminase, among many other measurements (NCT02165904). The results indicated improvements in neuropathic pain, bladder, and bowel control.[61] Some patients witnessed decreased spasticity, thus suggesting that repeated dosage of MSCs has the potential to improve the quality of life in patients with incomplete SCI.[61]

The next clinical trial was a phase II open-label study that further examined the clinical benefits of intrathecal administration of in vitro expanded autologous bone marrow—derived MSCs (NCT02570932) using the same approach as the previous phase I trial (NCT02165904). The primary outcome measure was a change in IANR-SCIFRS score from baseline to the end of the follow-up period at 24 months. Inclusion criteria consisted of patients with SCI who were medically stable for a minimum of 6 months, who had normal creatinine and serum glutamic oxaloacetic transaminase, as well as many other criteria (NCT02570932). The results of this study showed that this method of treatment appears to be safe as no adverse effects were observed.[62] Improvement was seen in neuropathic pain, motor power, spasticity, sphincter dysfunction, and others.[62] The authors of this study concluded that this intervention is safe and potentially beneficial due to the aforementioned signs of improvement.[62]

The final clinical trial was a phase I open-label pilot study that examined the effectiveness and safety of different methods of administrating autologous bone marrow—derived MSCs (NCT01909154). The methods of administration that were used were subarachnoid intrathecal injection, intramedullary intrathecal injection, and subarachnoid injection via lumbar puncture. Patients received two injections that were

3 months apart. The primary outcome measure for this clinical trial was the occurrence of adverse events up to 1-year postsurgery. The inclusion criteria involved having an AIS grade A (NCT01909154). The results indicate that improvements were observed in sensitivity and sphincter control, although a decrease in spasms and spasticity was common.[63] The authors concluded that the aforementioned methods of administrating MSCs are safe and improve the quality of life in patients with SCI.[63]

Brazil

The three clinical trials in Brazil sponsored by the Hospital Sao Rafael are examining MSCs transplanted in patients with chronic SCI.

The first clinical trial was an open-label study in which the researchers examined the safety and clinical effectiveness of transplanting MSCs in patients with chronic SCI using percutaneous injections (NCT02152657). The primary outcome measure for this trial was conducting MRI scans along with clinical and laboratory examinations to determine complications as a result of the intervention. The inclusion criteria consisted of patients with either open or closed SCI below T8 with an AIS grade A (NCT02152657). The results of this pilot study suggest that percutaneous injection of MSCs is a safe procedure as no adverse events occurred.[64] Participants demonstrated improvements in the Spinal Cord Independence Measure (SCIM) and FIM as a result of amelioration in bowel movements.[64] The researchers concluded that MSCs transplanted through percutaneous injection are safe and feasible; however, further research is required to elucidate the clinical benefits of this intervention.[64]

The second clinical trial is a phase II study that focuses on examining the safety and clinical benefits of MSCs derived from autologous bone marrow in patients with complete thoracolumbar SCI (NCT02574585). Patients will receive two percutaneous injections of these cells in a span of 3 months and will be compared with patients who do not receive this treatment. The primary outcome measure for this study is to determine the occurrence of adverse effects using MRI scans. Patients who have blunt or penetrating SCI between T1 and L2 with an AIS grade A and injury occurring at least 12 months prior to the study will be included in this trial. This trial is currently not yet recruiting participants and is expected to be completed by January 2022 (NCT02574585).

The final clinical trial is a pilot phase I open-label study that focuses on examining the safety and benefits of MSCs derived from bone marrow with complete cervical SCI (NCT02574572). The primary and secondary outcome measures, as well as the inclusion and exclusion criteria, were the same as outlined in the clinical trial NCT02574585. This recruitment status is currently unknown (NCT02574572).

Jordan

A phase I/II open-label clinical trial sponsored by the University of Jordan compared the safety and clinical benefits of using autologous bone marrow–derived MSCs (BM-MSCs) or autologous adipose tissue–derived MSCs (AT-MSCs) in patients with SCI (NCT02981576). These cells were transplanted into patients via an intrathecal injection three times. The primary outcome measure for this study was to observe the number of side effects and improvements in AIS grades between the two experimental groups 12 months postintervention. Patients who had a complete SCI at least 2 weeks prior to the start of the study with an AIS grade A, B or C were included in this trial. Although this trial was completed in January 2019, results have yet to be released at the time of writing (NCT02981576).

Pakistan

The Armed Forces Bone Marrow Transplant Centre in Pakistan sponsored a phase I open-label clinical trial completed in 2016 that focused on assessing the safety and effectiveness of autologous BM-MSCs delivered through an intrathecal injection as a potential treatment for SCI patients (NCT02482194). The primary outcome measure for this study was to determine the occurrence of adverse events. The inclusion criteria for this trial consisted of patients with subacute or chronic traumatic SCI at the thoracic level with an AIS grade A, and injury occurring at least 2 weeks prior to enrollment (NCT02482194). The researchers of this clinical trial demonstrated that no adverse effects were observed, suggesting that autologous BM-MSCs administered via intrathecal injection are safe.[65] Since this clinical trial was not randomized, nor placebo controlled, further experimentation is required to elucidate the safety and clinical benefits of BM-MSCs.[65]

China

There are four clinical trials based in China that are attempting to elucidate the safety and clinical benefits of MSCs.

The first clinical trial is a phase III study sponsored by the General Hospital of Chinese Armed Police Forces, which attempted to examine the use and effectiveness of umbilical cord—derived MSCs (UCMSCs) (NCT01873547). This trial was a three-armed study, whereby patients receiving cell therapy were compared with patients who received rehabilitation, and those who received neither. The primary outcome measure for this study was to observe changes in ISNCSCI scores. The inclusion criteria for this trial consisted of SCI patients with injury occurring more than 12 months prior to enrollment, along with confirmation of SCI using MRI, EMG, and electrophysiology (NCT01873547). Confano et al. suggested that this trial did not show satisfactory outcomes.[66]

There are currently three clinical trials sponsored by the Third Affiliated Hospital at Sun Yat-Sen University that are investigating the safety and effectiveness of UCMSC intrathecal injections in patients with subacute and chronic SCI. These three clinical trials are currently recruiting participants, with an expected completion in December 2021 (NCT03521336, NCT03521323, NCT03505034).

The first clinical trial is a phase II randomized study of intrathecal transplantation of UCMSCs in patients with subacute SCI, which will be compared with a placebo group (NCT03521336). The primary outcome measure for this trial is to observe changes in AIS grade up to 1-year posttransplant. Patients who have traumatic SCI with an AIS grade between A to D and with an injury occurring between 2 weeks and 2 months prior to enrollment were included in the trial (NCT03521336).

The second clinical trial is a phase II randomized study similar to NCT03521336, but in patients with early-stage chronic SCI (NCT03521323). This study has the same primary and secondary outcome measures, as well as exclusion criteria outlined in NCT03521336. The only difference is that patients who had an SCI between 2 and 12 months prior to enrollment were included in this study (NCT03521323).

The final clinical trial is a phase II randomized study similar to NCT03521336, but in patients with late-stage chronic SCI (NCT03505034). This study is a one-armed study unlike the other trials, which were two-armed. This trial shares the same primary and secondary outcome measures, as well as exclusion criteria that were mentioned in NCT03521336. However, the inclusion criteria for this specific trial require that SCI patients have an injury of greater than 12 months prior to enrollment (NCT03505034).

South Korea

A phase II/III open-label clinical trial sponsored by Pharmicell Co. Ltd. in South Korea aims to examine the safety and clinical benefits of directly injecting autologous BM-MSCs into the injured spinal cord through a laminectomy (NCT01676441). The primary outcome measure for this trial was to observe changes in motor scores in the ISNCSCI assessments at varying timepoints. To be eligible for this clinical trial, patients had to have a traumatic cervical SCI with an AIS grade B that occurred 12 months or more prior to enrollment, along with stable neurological examination after at least 1 month of rehabilitation. The current trial was recently terminated (NCT01676441), although the primary investigator has published a case report suggesting patients with traumatic cervical SCI who received autologous MSCs showed improvement in motor power and electrophysiology, as well as decreases in cavity size.[67]

There are two completed clinical trials that were sponsored by R-Bio in South Korea, where they attempted to elucidate the safety and clinical benefits of autologous adipose tissue—derived MSCs (AT-MSCs).

The first clinical trial was a phase I/II study that examined intrathecal injections of varying doses of AT-MSCs in patients with SCI (NCT01769872). The primary outcome measure was to observe changes in the AIS grade up to 6 months posttransplant. The inclusion criteria for this trial consisted of patients with SCI occurring greater than 3 months prior to enrollment with an AIS grade A, B, or C. This study was completed in 2016; however, the results have yet to be released at the time of writing (NCT01769872).

The second clinical trial, completed in 2010, was a phase I, open-label study that investigated the safety of intravenous administration of AT-MSCs known as Astrostem in SCI patients (NCT01274975). The primary outcome measure for this trial was to assess the safety of the treatment via physical examination, vital signs, and laboratory tests. The inclusion criteria for this trial consisted of male patients with SCI occurring at least 2 months prior to enrollment with an AIS grade A, B, or C (NCT01274975). The results for this trial demonstrated that none of the participants observed adverse events as a result of the intervention, suggesting that AT-MSCs seem to be safe.[68]

The final clinical trial sponsored by Bukwang Pharmaceutical was a phase I trial that studied the effect of AT-MSCs through intrathecal transplant in patients with SCI (NCT01624779). The primary outcome measure of this trial was to observe changes in MRI before and after transplant. Patients who did not show signs of improvement in neurological function for 4 weeks after treatment for SCI were included in this study. Although this clinical trial was completed in May 2014, results have yet to be published (NCT01624779).

Umbilical cord blood cells

Umbilical cord blood cells (UCBCs) derived from the umbilical cord postnatally contain multipotent stem cells that have the capacity to differentiate into many types of specialized cells, such as neurons and glial cells.[69–71] These cells also release cytokines that appear to be beneficial in the recovery of SCI.[72,73] UCBCs have been shown to be a viable therapeutic for neurological disorders, as they have demonstrated reduced sensorimotor and cognitive deficits such as neonatal hypoxic ischemia encephalopathy (HIE).[74] Thus, this has been implemented in a variety of different clinical trials under various conditions to determine UCBCs regenerative therapeutic potential for SCI.

One phase II, three-armed, double-blind study based in the United States is using umbilical cord blood mononuclear stem cells (UCBMNCs) with intensive locomotor training and administration of oral lithium carbonate in patients with chronic complete SCI (NCT03979742). The UCBMNCs will be injected into the dorsal root entry zones that are above and below the site of injury via a laminectomy. Patients will receive either oral lithium carbonate or a placebo for 6 weeks as well as locomotor training for 3—6 months. The primary outcome measure of this study is to determine a mean change from baseline in the WISCI II at 48 weeks. Some of the inclusion criteria for this trial include patients with chronic traumatic

SCI between C5 and T11, who have an AIS grade A and are able to stand for at least 1-hour a day by using a standing frame, tilt table, or an equivalent. The study is expected to be completed in February 2023 (NCT03979742).

There are three clinical trials based in China that investigated the feasibility, efficacy, and safety of UCBCs and the usage of lithium as part of the treatment protocol in acute, subacute, and chronic SCI (NCT01471613, NCT01354483, NCT01046786). Their primary outcome measure was to assess the degree of improvement in AIS motor and sensory scores at varying time points. The inclusion criteria for these studies included a neurological status of AIS grade A as well as the injury site being within three vertebral levels (NCT01471613, NCT01354483, NCT01046786).

NCT01354483 and NCT01046786 use methylprednisolone, a corticosteroid that is used to decrease inflammation[75] in addition to the transplantation of UCBCs. Based on current research, methylprednisolone confers modest functional benefits and is a practice option, though one that is not universally accepted.[75–77] Many of these clinical trials use lithium or lithium carbonate, as previous studies have suggested that lithium has therapeutic properties to treat neuropathic pain, which is a symptom seen in SCI patients.[78,79] No results have been released for any of the three aforementioned clinical trials; thus the efficacy of UCBCs in humans in SCI is still in question.

Bone marrow stem cells

Bone marrow stem cells (BMSCs) were previously thought to be limited in their differentiation potential.[80] However, the breakthrough discovery of multipotent adult stem cells demonstrated that these cells have the capacity to differentiate into neurons and glial cells,[81] which is potentially a viable cell source for a regenerative therapeutic in SCI. Clinical trials of BMSCs are being undertaken in a number of countries.

The United States

A clinical trial sponsored by MD Stem Cells is currently recruiting participants to elucidate the efficacy and benefits of autologous BMSCs in SCI and spinal cord diseases (NCT03225625). The treatment protocol for this clinical trial involves bilateral paraspinal injections of the cells at three different sites: the injury site, and at two spinal segments above and below the injury. Paraspinal injections are done by injecting the stem cells to adjacent spinal nerves that will enter the spinal canal. These patients may also be subjected to using exoskeletal stimulation, virtual reality, or an equivalent to improve upper motor neuronal firing, as well as the reception of sensory neurons. The BMSCs are harvested from the posterior iliac crest by separating them from the aspirate using an FDA cleared class II device. The primary outcome measure for this trial is to observe improvements in motor or sensory functioning as assessed by the ISNCSCI up to 12 months posttransplant. This study is estimated to be completed in July 2023 (NCT03225625).

Denmark and Spain

A phase II/III, randomized, double-blind clinical trial conducted by Neuroplast is examining the safety and benefits of Neuro-Cells, which are autologous BMSCs in patients with subacute SCI (NCT03935724). All patients participating in this trial will have undergone BMSC harvesting at the beginning of the trial. Both phases II and III consist of two arms: with patients receiving either the Neuro-Cell transplant or placebo 6–8 weeks postinjury, followed by a Neuro-Cell transplant at 32–34 weeks postinjury for the placebo group. The primary outcome measures for this clinical trial include physical changes observed at 3 months posttransplant through conducting a physical exam, as well as an increase in ISNCSCI motor scores 6 months posttransplant. To be eligible for this clinical study, patients must have a complete or incomplete traumatic SCI between C6 and T12. This study aims to be completed by December 2022 (NCT03935724).

India

There are currently two completed clinical trials that evaluated the safety and benefits of autologous BMSCs in varying severities of SCI (NCT02260713, NCT01186679).

A phase I/II clinical trial conducted by the International Stemcell Services Limited at the Sita Bhateja Speciality Hospital assessed the safety and capacity of autologous BMSCs through surgical transplantation of these cells into the injury site in eight chronic SCI patients (NCT01186679). Four acute and subacute SCI patients also received intrathecal stem cell injections or direct injection of the cells into the CSF through a lumbar puncture. Inclusion criteria for patients involved SCI occurring between C4 and T12, with acute patients having an injury of less than 2 weeks, subacute between 2 and 8 weeks, and chronic of longer than 6 months postinjury. The primary outcome measure was to see improvement in the AIS grade and to observe any adverse effects at 18 months. Although this study was completed in August 2010, no results have been released (NCT01186679).

The final phase I/II clinical trial was a pilot study sponsored by the Indian Spinal Injuries Centre that aimed to assess the safety and capacity of autologous BMSCs in acute complete SCI either through a single direct injection into the injury site or through an intrathecal injection (NCT02260713). The primary outcome measure for this trial was to observe any improvements in the AIS grade up to 5 years after the study. To be included in the trial, patients must have had a complete traumatic SCI (AIS grade A) between T1 and T12 that occurred 10–14 days prior to the onset of the trial (NCT02260713). The results from this clinical trial suggest that there were no significant adverse effects from the procedure.[82] However, no improvement was observed that was contributed by the course of treatment.[82]

Brazil

A clinical trial completed at the Hospital Sao Rafael also studied the safety and benefits of using autologous BMSCs transplanted into the injury site of patients with SCI (NCT01325103). The primary outcome measure for this study was to determine the practicality and safety of the transplanted cells 6 months posttransplant. The inclusion criteria involved having patients with traumatic SCI between the thoracic or lumbar regions with an AIS grade of A (NCT01325103). The results from this clinical trial indicated that BMSCs are safe, and subjects demonstrated improvement in tactile sensitivity.[83] Some subjects showed improvements on the AIS grade as well as urologic function.[83] The authors of this trial concluded that autologous BMSCs in patients with chronic SCI are safe and practical and could be a potential therapeutic for SCI as it improved neurological function.[83]

Egypt

The final BMSC trial is a completed phase I/II clinical trial sponsored by Cairo University, where they wanted to determine the safety of autologous BMSCs in chronic SCI patients, with the addition of physical therapy (NCT00816803). They hypothesized that the BMSCs that reach the injury site will promote regeneration locally. The primary outcome measures for this clinical trial were to test the safety of the autologous BMSCs through the absence of infections, intracranial tension, abnormal growth, and tumors using MRI 18 months posttransplant. Some of the inclusion criteria for this trial included having patients with traumatic SCI occurring between 10 months and 3 years prior to the trial, as well as no improvement with physiotherapy for a minimum of 6 months (NCT00816803). The results of this trial showed that patients who received both the BMSC transplant and physical therapy showed functional recovery compared with controls who only underwent physical therapy.[84] They also demonstrated that patients with thoracic SCI had a higher rate of improvement compared with patients with cervical SCI.[84] In closing, the authors of this study suggest that BMSCs are safe and can be used as a potential therapeutic in traumatic SCI.[84]

Comparison of different cell sources

Comparing different stem cell sources for regenerative therapies for SCI is important as optimization of these therapies can allow for full recovery in SCI patients. One clinical trial that attempts to examine this is being conducted in Jordan, where they are comparing purified autologous CD34[+] and CD133[+] stem cells derived from either leukapheresis or from bone marrow in chronic complete SCI (NCT02687672). This phase II double-armed trial's primary outcome measure is to observe sensory and motor improvement using the ISNCSCI assessment 5 years after transplant. This study is currently recruiting with estimated completion in December 2021 (NCT02687672). All clinical trials discussed are summarized in Tables 25.1–25.5.

Conclusion

Various cell-based therapies were discussed in the clinical trials examined in this book chapter. From those trials with results, it appears that most cell types evaluated demonstrated appropriate safety and modest efficacy, with some improvements in AIS score. However, to date, cell-based treatments remain at an investigational stage due to the small size and preliminary nature of the trials.

It is important to discuss the limitations of clinical trials in general as these factors have the potential to influence our interpretation of the results presented. One limitation of clinical trials, especially in randomized trials, is the difficultly of interpreting and generalizing the results to the larger SCI population as the study sample does not always reflect the characteristics of a typical SCI patient.[85] This can be rectified to a certain extent by using a multicentered approach, which allows researchers to gain a more heterogeneous sample, thus increasing external validity.[86,87] Many of the clinical trials that were discussed in this chapter have used this approach. Other limitations that clinical trials face include the lack of participants due to the complexity of the study, stringent inclusion criteria, long-term patient retention during clinical trials, as well as the lack of awareness about clinical trials.[88] There have been several strategies that have been used to increase recruitment such as patient registries, direct mail, and the media.[89]

Stem cell therapies have demonstrated potential to provide various methods to improve cellular damage caused by SCI.[87] However, these trials are often conducted using various cell dosages, types, and transplantation methods at different injury severities.[87] Furthermore, standardization and regulatory procedures may also vary between regulatory agencies.[87] Overall, this results in no general consensus for establishing optimal treatment procedure.[87]

In terms of future directions in the field of regenerative medicine in SCI, the optimization of neuroregeneration using the combinatory effects of biomaterials, stem cells, and small molecules appears to be a practical solution to such a complex disorder.[90] Furthermore, scaffolds may reduce surgical invasiveness and improve effectiveness over direct cell transplantation.[87] Additionally, elucidating the role of the immune system from a regenerative medicine perspective would allow us to create the ideal environment for these cells to grow and thrive.[91] Although we have made progress thus far, there is still a long way for the fruition of such therapies as we are currently still in the infancy of regenerative medicine with respect to SCI.[87]

TABLE 25.1 Clinical trials with unspecified phase (N/A).

Phase	NCT number	Other identifiers	Name of the trial	Sponsor	Treatment or cell source	Severity of SCI	Countries of testing	Completion status	Completion or estimated completion date
N/A	NCT02152657	SCI-002	Evaluation of Autologous Mesenchymal Stem Cell Transplantation in Chronic Spinal Cord Injury: a Pilot Study	Hospital Sao Rafael	Mesenchymal stem cells	Open or closed chronic (below T8) AIS grade A	Brazil	Completed	December 2016
N/A	NCT03225625	SciExVR	Stem Cell Spinal Cord Injury Exoskeleton and Virtual Reality Treatment Study (SciExVR)	MD Stem Cells	Autologous bone marrow−derived stem cells	N/A	United States	Recruiting	July 2023
N/A	NCT01325103	SCI-001	Autologous Bone Marrow Stem Cell Transplantation in Patients with Spinal Cord Injury	Hospital Sao Rafael	Autologous bone marrow stem cells	Thoracic or lumbar SCI Frankel A	Brazil	Completed	December 2012
Pilot study	N/A	N/A	Olfactory Mucosa Autografts in Human Spinal Cord Injury: A Pilot Clinical Study	N/A	Olfactory mucosa autografts	Chronic SCI (C4-T10) with ASIA grade A	Portugal	Completed	October 2005
Pilot study	N/A	N/A	Autologous Mucosal Transplant in Chronic Spinal Cord Injury	N/A	Autologous olfactory mucosa	Chronic SCI (C5-T12) ASIA grade A	India	Completed	June 2009

TABLE 25.2 Phase I clinical trials.

Phase	NCT Number	Other Identifiers	Name of the Trial	Sponsor	Treatment or cell source	Severity of SCI	Countries of testing	Completion status	Completion or estimated completion date
I	NCT01772810	NS2010-1	Safety Study of Human Spinal Cord-derived Neural Stem Cell Transplantation for the Treatment of Chronic SCI (SCI)	Neuralstem Inc.	Human spinal cord—derived neural stem cells	Chronic SCI of AIS grade A Group A: T2 —T12 Group B: C5 —C7	United States	Recruiting	December 2022
I	NCT01739023	20081061	Safety of Autologous Human Schwann Cells (ahSC) in Subjects with Subacute SCI	W. Dalton Dietrich	Autologous human Schwann cells	Subacute thoracic (T3-T11) SCI of ISNCSCI grade A	United States	Completed	August 2016
I	NCT02354625	20140846	The Safety of ahSC in Chronic SCI With Rehabilitation	W. Dalton Dietrich	Autologous human Schwann cells with Fitness conditioning and rehabilitation	Chronic SCI of AIS grades A, B, or C Cohort 1: Thoracic (T2 —T12) Cohort 2: Cervical (C5 —T1)	United States	Completed	August 2019
I	NCT03308565	17-004621; CELLTOP Study	Adipose Stem Cells for Traumatic Spinal Cord Injury	Allan Dietz	Adipose-derived mesenchymal stem cells	AIS grade A or B	United States	Active, not recruiting	November 2023
I	NCT02165904	CME-LEM2; 2011-005684-24 (Registry Identifier: 2011-005684-24)	Subarachnoid Administration of Adults Autologous Mesenchymal Stromal Cells in SCI	Puerta de Hierro University Hospital	Adult autologous mesenchymal bone marrow cells	Chronic incomplete SCI	Spain	Completed	May 2016
I	NCT01909154	CME-LEM1; 2010-023285-46 (EudraCT Number)	Safety Study of Local Administration of Autologous Bone Marrow Stromal Cells in Chronic Paraplegia (CME-LEM1)	Puerta de Hierro University Hospital	Autologous bone marrow stromal cells	Chronic (C6-L1) SCI AIS grade A	Spain	Completed	March 2015

I	NCT02574572	SCI-003	Autologous Mesenchymal Stem Cells Transplantation in Cervical Chronic and Complete Spinal Cord Injury	Hospital Sao Rafael	Autologous mesenchymal stem cells	Chronic cervical (C5–C7) SCI AIS grade A	Brazil	Recruiting	June 2020
I	NCT02482194	AFBMTC-SCI-2013	Autologous Mesenchymal Stem Cells Transplantation for Spinal Cord Injury—A Phase I Clinical Study	Armed Forces Bone Marrow Transplant Center, Rawalpindi, Pakistan	Mesenchymal stem cells (MSCs)	Subacute, chronic thoracic SCI AIS grade A	Pakistan	Completed	March 2016
I	NCT01274975	Astrostem	Autologous Adipose Derived MSCs Transplantation in Patient with Spinal Cord Injury	R-Bio	Autologous adipose-derived mesenchymal stem cells	Injury occurring >2 months AIS A, B, or C	South Korea	Completed	February 2010
I	NCT01624779	KOR-01	Intrathecal Transplantation of Autologous Adipose Tissue Derived MSC in the Patients with Spinal Cord Injury	Bukwang Pharmaceutical	Autologous adipose-derived mesenchymal stem cells	Not specified	South Korea	Completed	May 2014
I	NCT01217008	CP35A007	Safety Study of GRNOPC1 in Spinal Cord Injury	Asterias Biotherapeutics Inc.	OPCs derived from human ESCs (GRNOPC1 cells)	Subacute complete SCI; preserved neurological level between T3-T11 AIS grade A	United States	Completed	July 2013
I	N/A	N/A	Autologous Olfactory Ensheathing Cell Transplantation in Human Spinal Cord Injury	Not specified	Autologous olfactory ensheathing cells	Chronic complete thoracic (T4-T10) SCI with ASIA grade A	Australia	Completed	October 2005

TABLE 25.3   Phase I/II and I/IIa clinical trials.

Phase	NCT number	Other identifiers	Name of the trial	Sponsor	Treatment or cell source
I/II	NCT01321333	CL-N02-SC	Study of Human Central Nervous System Stem Cells (HuCNS-SC) in Patients with Thoracic Spinal Cord Injury	StemCells Inc.	Allogenic human central nervous system stem cells
I/II	NCT03069404	INSTrUCT-SCI	INSTrUCT-SCI: INdependent Observational STUdy of Cell Transplantation in SCI	University of Zurich	Allogenic human central nervous system stem cells
I/IIa	NCT02302157	AST-OPC1-01	Dose Escalation Study of AST-OPC1 in Spinal Cord Injury	Asterias Biotherapeutics Inc.	Human embryonic stem cell-derived oligodendrocyte progenitor cells (AST-OPC1)
I/II	NCT03933072	Wroclaw Walk Again Project	Autologous Bulbar Olfactory Ensheathing Cells and Nerve Grafts for Treatment of Patients With Spinal Cord Transection	Nicholls Spinal Injury Foundation	Autologous bulbar olfactory ensheathing cells paired with olfactory nerve Fibroblasts in a glial neuropatch with a cell suspension in a collagen scaffold Autologous sural nerve graft
I/IIa	NCT03003364	XCEL-SCI-01	Intrathecal Administration of Expanded Wharton's Jelly Mesenchymal Stem Cells in	Banc de Sang i Teixits	Expanded Wharton's jelly mesenchymal

I/II	NCT02917291	FAB117-CT-01	Chronic Traumatic Spinal Cord Injury Safety and Preliminary Efficacy of FAB117-HC in Patients with Acute Traumatic Spinal Cord Injury (SPINE)	Ferrer Internacional S.A	stem cells Human allogenic adipose-derived adult mesenchymal stem cells (HC016 cells)
I/II	NCT02981576	SCIUJCTC	Safety and Effectiveness of BM-MSC vs AT-MSC in the Treatment of SCI patients	University of Jordan	Autologous BM-MSC and AT-MSC
I/II	NCT01769872	KSC-MSCs-SPI	Safety and Effect of Adipose Tissue Derived Mesenchymal Stem Cell Implantation in Patients with Spinal Cord Injury	R-Bio	Adipose-derived mesenchymal stem cells
I/II	NCT01471613	CN102c	Lithium, Cord Blood Cells and the Combination in the Treatment of Acute and Subacute Spinal Cord Injury	China Spinal Cord Injury Network	Umbilical cord blood mononuclear cells
I/II	NCT01354483	CN102B_KM	Umbilical Cord Blood Mononuclear Cell Transplant to Treat Spinal Cord Injury	China Spinal Cord Injury Network	Umbilical cord blood mononuclear cells
I/II	NCT01046786	CN102B	Safety and Feasibility of Umbilical Cord Blood Cell Transplant into Injured Spinal Cord	China Spinal Cord Injury Network	Umbilical cord blood mononuclear cells
I/II	NCT02260713	ISIC-BMC	Autologous Bone Marrow Cell Transplantation in Persons with Acute Spinal Cord Injury - An Indian Pilot Study	Indian Spinal Injuries Centre	Autologous bone marrow stem cells
I/II	NCT01186679	ISSL-AuBM-SCI	Safety and Efficacy of Autologous Bone Marrow Stem Cells in Treating Spinal Cord Injury (ABMST-SCI)	International Stemcell Services Limited	Autologous bone marrow stem cells

I/II	NCT00816803	EGY-SCI-1	Cell Transplant in Spinal Cord Injury Patients	Cairo University	Autologous bone marrow–derived stem cells
I/IIa	N/A	N/A	Autologous Olfactory Ensheathing Cell Transplantation in Human Paraplegia: a 3-year Clinical Trial	Not specified	Autologous OECs
I/II	N/A	N/A	Olfactory Mucosal Autografts and Rehabilitation for Chronic Traumatic Spinal Cord Injury	Not specified	Olfactory mucosal Autografts

Severity of SCI	Countries of testing	Completion status	Completion or estimated completion date	Primary outcome measures	Notes
Thoracic (T2-T11) subacute SCI of AIS grade A, B, or C	Canada and Switzerland	Completed	April 2015	Number of occurrences of adverse side effects	N/A
Thoracic (T2-T11) subacute SCI of AIS grade A, B, or C	Switzerland	Active, not recruiting	April 2021	Amelioration in AIS grade	Participants completed transplant in CL-N02-SC study
Cervical subacute SCI Cohorts 1–3: AIS grade A Cohorts 4–5: AIS grade B	United States	Completed	December 2018	Number of occurrences of adverse side effects as a result of cell transplantation	N/A
Single C5-T10 complete spinal cord transection of AIS grade A	Poland	Recruiting	March 2022	Amelioration in AIS grade, trunk muscle strength, deep sensation, as well as vibration and joint position detection	N/A
Single lesion chronic thoracic (T2-T11) SCI AIS grade A	Spain	Completed	February 2020	Number of occurrences of adverse side effects as a result of the treatment	N/A

Single acute thoracic (D1-D12) SCI AIS grade A	Spain	Recruiting	March 2022	Number of occurrences of adverse side effects as a result of intramedullary cell transplantation	N/A
Complete SCI AIS grade A, B or incomplete C	Jordan	Completed	January 2019	Number of occurrences of adverse events and amelioration in AIS grade	N/A
Injury occurring >3 months AIS grade A, B, or C	South Korea	Completed	January 2016	Variation in AIS grade	N/A
Acute or subacute SCI with neurological level between C5 and T11 AIS grade A	China	Completed	January 2014	Variation in AIS grade	N/A
Chronic SCI with neurological level between C5 and T11 AIS grade A	China	Completed	December 2013	Variation in AIS grade	N/A
Chronic SCI with neurological level between C5 and T11 AIS grade A	China	Completed	December 2013	Amelioration and variation in AIS grade	N/A
Complete acute (T1-T12) SCI AIS grade A	India	Completed	November 2017	Amelioration of ASIA grade	N/A
Subacute or chronic (C4–T12) complete SCI	India	Completed	August 2010	Amelioration of ASIA grade as well as frequency of adverse events	N/A
SCI occurring between 10 months and 3 years	Egypt	Completed	December 2008	Safety of cell transplant through absence of infections, intracranial tension, and abnormal growths/tumors	N/A
Chronic complete thoracic paraplegia (T4–T10), with ASIA grade A	Australia	Completed	September 2008	Clinical Outcome Variable Scale, Functional Independence Measure, ASIA grade, pain, neurological, and neuropsychological assessments	N/A
Chronic complete SCI (C4–T10), ASIA grade A or B	Portugal	Completed	January 2010	ASIA grades, EMG, SSEP, urodynamic studies, MRI, otolaryngology and psychological assessments	N/A

TABLE 25.4 Phase II clinical trials.

Phase	NCT number	Other identifiers	Name of the trial	Sponsor	Treatment or cell source	Severity of SCI	Countries of testing	Completion status	Completion or estimated completion date	Primary outcome measures
II	NCT02570932	CME-LEM3 2014-005613-24 (EudraCT Number)	Administration of Expanded Autologous Adult Bone Marrow Mesenchymal Cells in Established Chronic Spinal Cord Injuries	Puerta de Hierro University Hospital	Expanded autologous bone marrow mesenchymal stem cells	Chronic SCI AIS grades A −D	Spain	Completed	December 2017	Variation in IANR-SCIFRS score
II	NCT02574585	SCI-005	Autologous Mesenchymal Stem Cells Transplantation in Thoracolumbar Chronic and Complete Spinal Cord Injury	Hospital Sao Rafael	Autologous mesenchymal stem cells	Chronic thora-columbar (T1−L2) SCI AIS grade A	Brazil	Not yet recruiting	January 2022	Number of individuals with adverse side effects
II	NCT03521336	IT-UCMSC-SASCI	Intrathecal Transplantation of UC-MSC in Patients with Sub-Acute Spinal Cord Injury	Third Affiliated Hospital at Sun Yat-Sen University	Umbilical cord −derived mesenchymal stem cells	Subacute SCI AIS grades A −D	China	Recruiting	December 2021	Variations in AIS grade
II	NCT03521323	IT-UCMSC-ECSCI	Intrathecal Transplantation of UC-MSC in Patients with Early Stages of Chronic Spinal Cord Injury	Third Affiliated Hospital at Sun Yat-Sen University	Umbilical cord −derived mesenchymal stem cells	Early chronic SCI AIS grades A −D	China	Recruiting	December 2021	Variations in AIS grade

II	NCT03505034	IT-UCMSC-LCSCI	Intrathecal Transplantation of UC-MSC in Patients with Late Stage of Chronic Spinal Cord Injury	Third Affiliated Hospital at Sun Yat-Sen University	Umbilical cord–derived mesenchymal stem cells	Late chronic SCI AIS grades A–D	China	Recruiting	December 2021	Variations in AIS grade
II	NCT03979742	US105d	Umbilical Cord Blood Cell Transplant Into Injured Spinal Cord with Lithium Carbonate or Placebo Followed by Locomotor Training	Stem Cyte Inc.	Umbilical cord blood mono-nuclear cells	Chronic SCI with neuro-logical level between C5–T11 AIS grade A	United States	Not yet recruiting	September 2021	Variations in WISCI II scores
II	NCT02687672	SCA-SCI1	Transplantation of Autologous Bone Marrow or Leukapheresis-Derived Stem Cells for Treatment of Spinal Cord Injury	Stem Cells Arabia	$CD34^+$ and $CD133^+$ stem cells	Chronic SCI	Jordan	Recruiting	December 2021	Amelio-ration in ASIA grade

TABLE 25.5 Phase II/III and III clinical trials.

Phase	NCT number	Other identifiers	Name of the trial	Sponsor	Treatment or cell source	Severity of SCI	Countries of testing	Completion status	Completion or estimated completion date	Primary outcome measures	Notes
III	NCT01873547	2013-03-04SCI-III	Different Efficacy Between Rehabilitation Therapy and Stem Cells Transplantation in Patients with SCI in China (SCI-III)	General Hospital of Chinese Armed Police Forces	Umbilical cord –derived mesenchymal stem cells	Not specified	China	Completed	December 2015	Variation in ISNCSCI score	Unsatisfactory outcomes
II/III	NCT01676441	Cerecellgram-spine	Safety and Efficacy of Autologous Mesenchymal Stem Cells in Chronic Spinal Cord Injury	Parmicell Co., Ltd	Autologous mesenchymal stem cells	Chronic cervical AIS grade B	South Korea	Recruiting	March 2020	Variation in ASIA motor score	
II/III	NCT03935724	A2017SCI03	Clinical Study of an Autologous Stem Cell Product in Patients with a (Sub)Acute Spinal Cord Injury (SCI2)	Neuroplast	Neuro-Cells (autologous bone marrow-derived stem cells)	Complete (AIS grade A)/ incomplete SCI (AIS grade B or C) between C6 and T12	Spain and Demark	Not yet recruiting	June 2022	Physical variations and amelioration in ISNCSCI motor scores	

References

1. Ahuja CS, Wilson JR, Nori S, Kotter MRN, Druschel C, Curt A, Fehlings MG. Traumatic spinal cord injury. *Nat Rev Dis Primers* 2017;**3**(1):17018. https://doi.org/10.1038/nrdp.2017.18.

2. Sabelström H, Stenudd M, Frisén J. Neural stem cells in the adult spinal cord. *Exp Neurol* 2014;**260**:44–9. https://doi.org/10.1016/j.expneurol.2013.01.026.

3. Jaenisch R, Young R. Stem cells, the molecular circuitry of pluripotency and nuclear reprogramming. *Cell* 2008;**132**(4):567–82. https://doi.org/10.1016/j.cell.2008.01.015.

4. Bindu A H, B S. Potency of various types of stem cells and their transplantation. *J Stem Cell Res Ther* 2011;**01**(03). https://doi.org/10.4172/2157-7633.1000115.

5. Hanna JH, Saha K, Jaenisch R. Pluripotency and cellular reprogramming: facts, hypotheses, unresolved issues. *Cell* 2010;**143**(4):508–25. https://doi.org/10.1016/j.cell.2010.10.008.

6. Singh VK, Saini A, Kalsan M, Kumar N, Chandra R. Describing the stem cell potency: the various methods of functional assessment and in silico diagnostics. *Front Cell Dev Biol* 2016;**4**. https://doi.org/10.3389/fcell.2016.00134.

7. Thomson JA. Embryonic stem cell lines derived from human blastocysts. *Science* 1998;**282**(5391):1145–7. https://doi.org/10.1126/science.282.5391.1145.

8. Svenberg P, Remberger M, Svennilson J, Mattsson J, Leblanc K, Gustafsson B, Aschan J, Barkholt L, Winiarski J, Ljungman P, Ringdén O. Allogenic stem cell transplantation for nonmalignant disorders using matched unrelated donors. *Biol Blood Marrow Transplant* 2004;**10**(12):877–82. https://doi.org/10.1016/j.bbmt.2004.08.002.

9. Kang L, Kou Z, Zhang Y, Gao S. Induced pluripotent stem cells (iPSCs)—a new era of reprogramming. *J Genet Genom* 2010;**37**(7):415–21. https://doi.org/10.1016/S1673-8527(09)60060-6.

10. Damdimopoulou P, Rodin S, Stenfelt S, Antonsson L, Tryggvason K, Hovatta O. Human embryonic stem cells. *Best Pract Res Clin Obstet Gynaecol* 2016;**31**:2–12. https://doi.org/10.1016/j.bpobgyn.2015.08.010.

11. Takahashi K, Tanabe K, Ohnuki M, Narita M, Ichisaka T, Tomoda K, Yamanaka S. Induction of pluripotent stem cells from adult human fibroblasts by defined factors. *Cell* 2007;**131**(5):861–72. https://doi.org/10.1016/j.cell.2007.11.019.

12. Yu J, Vodyanik MA, Smuga-Otto K, Antosiewicz-Bourget J, Frane JL, Tian S, Nie J, Jonsdottir GA, Ruotti V, Stewart R, Slukvin II, Thomson JA. Induced pluripotent stem cell lines derived from human somatic cells. *Science* 2007;**5**.

13. Takahashi K, Yamanaka S. Induction of pluripotent stem cells from mouse embryonic and adult fibroblast cultures by defined factors. *Cell* 2006;**126**(4):663–76. https://doi.org/10.1016/j.cell.2006.07.024.

14. Grath A, Dai G. Direct cell reprogramming for tissue engineering and regenerative medicine. *J Biol Eng* 2019;**13**(1):14. https://doi.org/10.1186/s13036-019-0144-9.

15. U.S. Food and Drug Administration (FDA). *The drug development process — step 3: clinical research.* 2018. Retrieved May 7, 2020, from, https://www.fda.gov/patients/drug-development-process/step-3-clinical-research.

16. GSK Canada. Clinical trial phases; n.d. Retrieved May 7, 2020, from https://ca.gsk.com/en-ca/research/trials-in-people/clinical-trial-phases/.

17. Kirshblum SC, Burns SP, Biering-Sorensen F, Donovan W, Graves DE, Jha A, Johansen M, Jones L, Krassioukov A, Mulcahey MJ, Schmidt-Read M, Waring W. International standards for neurological classification of spinal cord injury (revised 2011). *J Spinal Cord Med* 2011;**34**(6):535–46. https://doi.org/10.1179/204577211X13207446293695.

18. Roberts TT, Leonard GR, Cepela DJ. Classifications in brief: American spinal injury Association (ASIA) impairment scale. *Clin Orthopaed Rel Res* 2017;**475**(5):1499–504. https://doi.org/10.1007/s11999-016-5133-4.

19. Weiss S, Dunne C, Hewson J, Wohl C, Wheatley M, Peterson AC, Reynolds BA. Multipotent CNS stem cells are present in the adult mammalian spinal cord and ventricular neuroaxis. *J Neurosci* 1996;**16**(23):7599–609. https://doi.org/10.1523/JNEUROSCI.16-23-07599.1996.

20. Kang SH, Fukaya M, Yang JK, Rothstein JD, Bergles DE. NG^{2+} CNS glial progenitors remain committed to the oligodendrocyte lineage in postnatal life and following neurodegeneration. *Neuron* 2010;**68**(4):668–81. https://doi.org/10.1016/j.neuron.2010.09.009.

21. Zawadzka M, Rivers LE, Fancy SPJ, Zhao C, Tripathi R, Jamen F, Young K, Goncharevich A, Pohl H, Rizzi M, Rowitch DH, Kessaris N, Suter U, Richardson WD, Franklin RJM. CNS-resident glial progenitor/stem cells produce Schwann cells as well as oligodendrocytes during repair of CNS demyelination. *Cell Stem Cell* 2010;**6**(6):578–90. https://doi.org/10.1016/j.stem.2010.04.002.

22. Young KM, Psachoulia K, Tripathi RB, Dunn S-J, Cossell L, Attwell D, Tohyama K, Richardson WD. Oligodendrocyte dynamics in the healthy adult CNS: evidence for myelin remodeling. *Neuron* 2013;**77**(5):873–85. https://doi.org/10.1016/j.neuron.2013.01.006.

23. Duncan GJ, Manesh SB, Hilton BJ, Assinck P, Plemel JR, Tetzlaff W. The fate and function of oligodendrocyte progenitor cells after traumatic spinal cord injury. *Glia* 2019;**68**(2):227–45. https://doi.org/10.1002/glia.23706.

24. Nistor GI, Totoiu MO, Haque N, Carpenter MK, Keirstead HS. Human embryonic stem cells differentiate into oligodendrocytes in high purity and myelinate after spinal cord transplantation. *Glia* 2005;**49**(3):385–96. https://doi.org/10.1002/glia.20127.

25. Manley NC, Priest CA, Denham J, Wirth ED, Lebkowski JS. Human embryonic stem cell-derived oligodendrocyte progenitor cells: preclinical efficacy and safety in cervical spinal cord injury: hESC-derived OPCs for cervical spinal cord injury. *Stem Cell Trans Med* 2017;**6**(10):1917—29. https://doi.org/10.1002/sctm.17-0065.

26. Lineage Cell Therapeutics. *OPC1.* 2019. Retrieved April 27, 2020, from, https://lineagecell.com/products-pipeline/opc1/.

27. Lineage Cell Therapeutics. *Lineage cell therapeutics provides update on SCiStar clinical study and OPC1 spinal cord injury program.* 2019. Retrieved April 27, 2020, from, https://investor.lineagecell.com/news-releases/news-release-details/lineage-cell-therapeutics-provides-update-scistar-clinical-study.

28. Wirth E, Lebkowski JS, Lebacqz K. Response to Frederic Bretzner et al. "Target populations for first-in-human embryonic stem cell research in spinal cord injury". *Cell Stem Cell* 2011;**8**(5):476—8. https://doi.org/10.1016/j.stem.2011.04.008.

29. Lebkowski J. GRNOPC1: the world's first embryonic stem cell-derived therapy. *Regen Med* 2011;**6**(6s):11—3. https://doi.org/10.2217/rme.11.77.

30. Evaluation of safety and preliminary efficacy of escalating doses of GRNOPC1 in subacute spinal cord injury. California's Stem Cell Agency; n.d. Retrieved May 10, 2020, from https://www.cirm.ca.gov/search/site/clinical%2520trial%2520evaluation%2520safety%2520preliminary%2520efficacy%2520escalating%2520doses%2520grnopc1%2520subacute%2520spinal.

31. Mackay-Sim A, St John JA. Olfactory ensheathing cells from the nose: clinical application in human spinal cord injuries. *Exp Neurol* 2011;**229**(1):174—80. https://doi.org/10.1016/j.expneurol.2010.08.025.

32. Ibrahim A, Li D, Collins A, Tabakow P, Raisman G, Li Y. Comparison of olfactory bulbar and mucosal cultures in a rat rhizotomy model. *Cell Transplant* 2014;**23**(11):1465—70. https://doi.org/10.3727/096368913X676213.

33. Li Y, Field PM, Raisman G. Regeneration of adult rat corticospinal axons induced by transplanted olfactory ensheathing cells. *J Neurosci* 1998;**18**(24):10514—24. https://doi.org/10.1523/JNEUROSCI.18-24-10514.1998.

34. Keyvan-Fouladi N, Raisman G, Li Y. Functional repair of the corticospinal tract by delayed transplantation of olfactory ensheathing cells in adult rats. *J Neurosci* 2003;**23**(28):9428—34. https://doi.org/10.1523/JNEUROSCI.23-28-09428.2003.

35. Tabakow P, Raisman G, Fortuna W, Czyz M, Huber J, Li D, Szewczyk P, Okurowski S, Miedzybrodzki R, Czapiga B, Salomon B, Halon A, Li Y, Lipiec J, Kulczyk A, Jarmundowicz W. Functional regeneration of supraspinal connections in a patient with transected spinal cord following transplantation of bulbar olfactory ensheathing cells with peripheral nerve bridging. *Cell Transplant* 2014;**23**(12):1631—55. https://doi.org/10.3727/096368914X685131.

36. Collins A, Li D, Liadi M, Tabakow P, Fortuna W, Raisman G, Li Y. Partial recovery of proprioception in rats with dorsal root injury after human olfactory bulb cell transplantation. *J Neurotrauma* 2018;**35**(12):1367—78. https://doi.org/10.1089/neu.2017.5273.

37. Li Y, Decherchi P, Raisman G. Transplantation of olfactory ensheathing cells into spinal cord lesions restores breathing and climbing. *J Neurosci* 2003;**23**(3):727—31. https://doi.org/10.1523/JNEUROSCI.23-03-00727.2003.

38. Yamamoto M, Raisman G, Li D, Li Y. Transplanted olfactory mucosal cells restore paw reaching function without regeneration of severed corticospinal tract fibres across the lesion. *Brain Res* 2009;**1303**:26—31. https://doi.org/10.1016/j.brainres.2009.09.073.

39. Féron F, Perry C, Cochrane J, Licina P, Nowitzke A, Urquhart S, Geraghty T, Mackay-Sim A. Autologous olfactory ensheathing cell transplantation in human spinal cord injury. *Brain* 2005;**128**(12):2951—60. https://doi.org/10.1093/brain/awh657.

40. Mackay-Sim A, Féron F, Cochrane J, Bassingthwaighte L, Bayliss C, Davies W, Fronek P, Gray C, Kerr G, Licina P, Nowitzke A, Perry C, Silburn PAS, Urquhart S, Geraghty T. Autologous olfactory ensheathing cell transplantation in human paraplegia: a 3-year clinical trial. *Brain* 2008;**131**(9):2376—86. https://doi.org/10.1093/brain/awn173.

41. Tabakow P, Jarmundowicz W, Czapiga B, Fortuna W, Miedzybrodzki R, Czyz M, Huber J, Szarek D, Okurowski S, Szewczyk P, Gorski A, Raisman G. Transplantation of autologous olfactory ensheathing cells in complete human spinal cord injury. *Cell Transplant* 2013;**22**(9):1591—612. https://doi.org/10.3727/096368912X663532.

42. Czyż M, Tabakow P, Gheek D, Miś M, Jarmundowicz W, Raisman G. The supraorbital keyhole approach via an eyebrow incision applied to obtain the olfactory bulb as a source of olfactory ensheathing cells — radiological feasibility study. *Br J Neurosurg* 2014;**28**(2):234—40. https://doi.org/10.3109/02688697.2013.817534.

43. Czyz M, Tabakow P, Hernandez-Sanchez I, Jarmundowicz W, Raisman G. Obtaining the olfactory bulb as a source of olfactory ensheathing cells with the use of minimally invasive neuroendoscopy-assisted supraorbital keyhole approach—cadaveric feasibility study. *Br J Neurosurg* 2015;**29**(3):362—70. https://doi.org/10.3109/02688697.2015.1006170.

44. Lima C, Pratas-Vital J, Escada P, Hasse-Ferreira A, Capucho C, Peduzzi JD. Olfactory mucosa autografts in human spinal cord injury: a pilot clinical study. *J Spinal Cord Med* 2006;**29**(3):191−203. https://doi.org/10.1080/10790268.2006.11753874.

45. Chhabra HS, Lima C, Sachdeva S, Mittal A, Nigam V, Chaturvedi D, Arora M, Aggarwal A, Kapur R, Khan TaH. Autologous mucosal transplant in chronic spinal cord injury: an Indian Pilot study. *Spinal Cord* 2009;**47**(12):887−95. https://doi.org/10.1038/sc.2009.54.

46. Lima C, Escada P, Pratas-Vital J, Branco C, Arcangeli CA, Lazzeri G, Maia CA, Capucho C, Hasse-Ferreira A, Peduzzi JD. Olfactory mucosal autografts and rehabilitation for chronic traumatic spinal cord injury. *Neurorehabil Neural Repair* 2010;**24**(1):10−22. https://doi.org/10.1177/1545968309347685.

47. Mousavi M, Hedayatpour A, Mortezaee K, Mohamadi Y, Abolhassani F, Hassanzadeh G. Schwann cell transplantation exerts neuroprotective roles in rat model of spinal cord injury by combating inflammasome activation and improving motor recovery and remyelination. *Metab Brain Dis* 2019;**34**(4):1117−30. https://doi.org/10.1007/s11011-019-00433-0.

48. Deng L-X, Walker C, Xu X-M. Schwann cell transplantation and descending propriospinal regeneration after spinal cord injury. *Brain Res* 2015;**1619**:104−14. https://doi.org/10.1016/j.brainres.2014.09.038.

49. Mirsky R, Jessen KR, Brennan A, Parkinson D, Dong Z, Meier C, Parmantier E, Lawson D. Schwann cells as regulators of nerve development. *J Physiol Paris* 2002;**96**(1−2):17−24. https://doi.org/10.1016/S0928-4257(01)00076-6.

50. Kuo H-S, Tsai M-J, Huang M-C, Chiu C-W, Tsai C-Y, Lee M-J, Huang W-C, Lin Y-L, Kuo W-C, Cheng H. Acid fibroblast growth factor and peripheral nerve grafts regulate Th2 cytokine expression, macrophage activation, polyamine synthesis, and neurotrophin expression in transected rat spinal cords. *J Neurosci* 2011;**31**(11):4137−47. https://doi.org/10.1523/JNEUROSCI.2592-10.2011.

51. Taskinen HS, Olsson T, Bucht A, Khademi M, Svelander L, Roytta M. Peripheral nerve injury induces endoneurial expression of IFN-g, IL-10 and TNF-a mRNA. *J Neuroimmunol* 2000;**9**.

52. Zhang K-J, Zhang H-L, Zhang X-M, Zheng X-Y, Quezada HC, Zhang D, Zhu J. Apolipoprotein E isoform-specific effects on cytokine and nitric oxide production from mouse Schwann cells after inflammatory stimulation. *Neurosci Lett* 2011;**499**(3):175−80. https://doi.org/10.1016/j.neulet.2011.05.050.

53. Anderson KD, Guest JD, Dietrich WD, Bartlett Bunge M, Curiel R, Dididze M, Green BA, Khan A, Pearse DD, Saraf-Lavi E, Widerström-Noga E, Wood P, Levi AD. Safety of autologous human Schwann cell transplantation in subacute thoracic spinal cord injury. *J Neurotrauma* 2017;**34**(21):2950−63. https://doi.org/10.1089/neu.2016.4895.

54. Siddall PJ, McClelland JM, Rutkowski SB, Cousins MJ. A longitudinal study of the prevalence and characteristics of pain in the first 5 years following spinal cord injury. *Pain* 2003;**103**(3):249−57. https://doi.org/10.1016/S0304-3959(02)00452-9.

55. Cruz-Almeida Y, Martinez-Arizala A, Widerström-Noga EG. Chronicity of pain associated with spinal cord injury: a longitudinal analysis. *J Rehabil Res Dev* 2005;**42**(5):585−94.

56. Finnerup NB, Norrbrink C, Trok K, Piehl F, Johannesen IL, Sørensen JC, Jensen TS, Werhagen L. Phenotypes and predictors of pain following traumatic spinal cord injury: a prospective study. *J Pain* 2014;**15**(1):40−8. https://doi.org/10.1016/j.jpain.2013.09.008.

57. Ullah I, Subbarao RB, Rho GJ. Human mesenchymal stem cells—current trends and future prospective. *Biosci Rep* 2015;**35**(2). https://doi.org/10.1042/BSR20150025.

58. Lee H-L, Lee HY, Yun Y, Oh J, Che L, Lee M, Ha Y. Hypoxia-specific, VEGF-expressing neural stem cell therapy for safe and effective treatment of neuropathic pain. *J Contr Release* 2016;**226**:21−34. https://doi.org/10.1016/j.jconrel.2016.01.047.

59. Itakura G, Kobayashi Y, Nishimura S, Iwai H, Takano M, Iwanami A, Toyama Y, Okano H, Nakamura M. Control of the survival and growth of human glioblastoma grafted into the spinal cord of mice by taking advantage of immunorejection. *Cell Transplant* 2015;**24**(7):1299−311. https://doi.org/10.3727/096368914X681711.

60. Bydon M, Dietz AB, Goncalves S, Moinuddin FM, Alvi MA, Goyal A, Yolcu Y, Hunt CL, Garlanger KL, Del Fabro AS, Reeves RK, Terzic A, Windebank AJ, Qu W. CELLTOP clinical trial: first report from a phase 1 trial of autologous adipose tissue-derived mesenchymal stem cells in the treatment of Paralysis due to traumatic spinal cord injury. *Mayo Clin Proc* 2020;**95**(2):406−14. https://doi.org/10.1016/j.mayocp.2019.10.008.

61. Vaquero J, Zurita M, Rico MA, Bonilla C, Aguayo C, Fernández C, Tapiador N, Sevilla M, Morejón C, Montilla J, Martínez F, Marín E, Bustamante S, Vázquez D, Carballido J, Rodríguez A, Martínez P, García C, Ovejero M, Fernández MV. Repeated subarachnoid administrations of autologous mesenchymal stromal cells supported in autologous plasma improve quality of life in patients suffering incomplete spinal cord injury. *Cytotherapy* 2017;**19**(3):349−59. https://doi.org/10.1016/j.jcyt.2016.12.002.

62. Vaquero J, Zurita M, Rico MA, Aguayo C, Bonilla C, Marin E, Tapiador N, Sevilla M, Vazquez D, Carballido J, Fernandez C, Rodriguez-Boto G, Ovejero M, Vaquero J, Zurita M, Bonilla C, Rico MA, Aguayo C, Rodríguez A, et al. Intrathecal administration of autologous mesenchymal stromal cells for spinal cord injury: safety and efficacy of the 100/3 guideline. *Cytotherapy* 2018;**20**(6):806−19. https://doi.org/10.1016/j.jcyt.2018.03.032.

63. Vaquero J, Zurita M, Rico MA, Bonilla C, Aguayo C, Montilla J, Bustamante S, Carballido J, Marin E,

Martinez F, Parajon A, Fernandez C, Reina LD. An approach to personalized cell therapy in chronic complete paraplegia: the Puerta de Hierro phase I/II clinical trial. *Cytotherapy* 2016;**18**(8):1025—36. https://doi.org/10.1016/j.jcyt.2016.05.003.

64. Larocca TF, Macêdo CT, Souza BS de F, Andrade-Souza YM, Villarreal CF, Matos AC, Silva DN, da Silva KN, de Souza CLEM, Paixão D da S, Bezerra M da R, Alves RL, Soares MBP, dos Santos RR. Image-guided percutaneous intralesional administration of mesenchymal stromal cells in subjects with chronic complete spinal cord injury: a pilot study. *Cytotherapy* 2017;**19**(10):1189—96. https://doi.org/10.1016/j.jcyt.2017.06.006.

65. Satti HS, Waheed A, Ahmed P, Ahmed K, Akram Z, Aziz T, Satti TM, Shahbaz N, Khan MA, Malik SA. Autologous mesenchymal stromal cell transplantation for spinal cord injury: a Phase I pilot study. *Cytotherapy* 2016;**18**(4):518—22. https://doi.org/10.1016/j.jcyt.2016.01.004.

66. Cofano F, Boido M, Monticelli M, Zenga F, Ducati A, Vercelli A, Garbossa D. Mesenchymal stem cells for spinal cord injury: current options, limitations, and future of cell therapy. *Int J Mol Sci* 2019;**20**(11):2698. https://doi.org/10.3390/ijms20112698.

67. Park JH, Kim DY, Sung IY, Choi GH, Jeon MH, Kim KK, Jeon SR. Long-term results of spinal cord injury therapy using mesenchymal stem cells derived from bone marrow in humans. *Neurosurgery* 2012;**70**(5):1238—47. https://doi.org/10.1227/NEU.0b013e31824387f9.

68. Ra JC, Shin IS, Kim SH, Kang SK, Kang BC, Lee HY, Kim YJ, Jo JY, Yoon EJ, Choi HJ, Kwon E. Safety of intravenous infusion of human adipose tissue-derived mesenchymal stem cells in animals and humans. *Stem Cell Dev* 2011;**20**(8):1297—308. https://doi.org/10.1089/scd.2010.0466.

69. Weiss ML, Troyer DL. Stem cells in the umbilical cord. *Stem Cell Rev* 2006;**2**(2):155—62. https://doi.org/10.1007/s12015-006-0022-y.

70. Chua SJ, Bielecki R, Wong CJ, Yamanaka N, Rogers IM, Casper RF. Neural progenitors, neurons and oligodendrocytes from human umbilical cord blood cells in a serum-free, feeder-free cell culture. *Biochem Biophys Res Commun* 2009;**379**(2):217—21. https://doi.org/10.1016/j.bbrc.2008.12.045.

71. Chen N, Hudson JE, Walczak P, Misiuta I, Garbuzova-Davis S, Jiang L, Sanchez-Ramos J, Sanberg PR, Zigova T, Willing AE. Human umbilical cord blood progenitors: the potential of these hematopoietic cells to become neural. *Stem Cells (Dayton, Ohio)* 2005;**23**(10):1560—70. https://doi.org/10.1634/stemcells.2004-0284.

72. Yao L, He C, Zhao Y, Wang J, Tang M, Li J, Wu Y, Ao L, Hu X. Human umbilical cord blood stem cell transplantation for the treatment of chronic spinal cord injury: electrophysiological changes and long-term efficacy. *Neural Regen Res* 2013;**8**(5):397—403. https://doi.org/10.3969/j.issn.1673-5374.2013.05.002.

73. Newman MB, Willing AE, Manresa JJ, Sanberg CD, Sanberg PR. Cytokines produced by cultured human umbilical cord blood (HUCB) cells: implications for brain repair. *Exp Neurol* 2006;**199**(1):201—8. https://doi.org/10.1016/j.expneurol.2006.04.001.

74. Pimentel-Coelho PM, Rosado-de-Castro PH, Barbosa da Fonseca LM, Mendez-Otero R. Umbilical cord blood mononuclear cell transplantation for neonatal hypoxic—ischemic encephalopathy. *Pediatr Res* 2012;**71**(2):464—73. https://doi.org/10.1038/pr.2011.59.

75. Cabrera-Aldana EE, Ruelas F, Aranda C, Rincon-Heredia R, Martínez-Cruz A, Reyes-Sánchez A, Guizar-Sahagún G, Tovar-y-Romo LB. Methylprednisolone administration following spinal cord injury reduces Aquaporin 4 Expression and Exacerbates Edema. *Mediators inflamm* 2017. https://doi.org/10.1155/2017/4792932.

76. Fehlings MG, Wilson JR, Tetreault LA, Aarabi B, Anderson P, Arnold PM, Brodke DS, Burns AS, Chiba K, Dettori JR, Furlan JC, Hawryluk G, Holly LT, Howley S, Jeji T, Kalsi-Ryan S, Kotter M, Kurpad S, Kwon BK, Marino RJ, et al. A clinical practice guideline for the management of patients with acute spinal cord injury: recommendations on the use of methylprednisolone sodium succinate. *Global Spine J* 2017;**7**(Suppl. 3):203S—11S. https://doi.org/10.1177/2192568217703085.

77. Fehlings MG, Wilson JR, Harrop JS, Kwon BK, Tetreault LA, Arnold PM, Singh JM, Hawryluk G, Dettori JR. Efficacy and safety of methylprednisolone sodium succinate in acute spinal cord injury: a systematic review. *Global Spine J* 2017;**7**(Suppl. 3):116S—37S. https://doi.org/10.1177/2192568217706366.

78. Banafshe HR, Mesdaghinia A, Arani MN, Ramezani MH, Heydari A, Hamidi GA. Lithium attenuates pain-related behavior in a rat model of neuropathic pain: possible involvement of opioid system. *Pharmacol Biochem Behav* 2012;**100**(3):425—30. https://doi.org/10.1016/j.pbb.2011.10.004.

79. Hagen EM, Rekand T. Management of neuropathic pain associated with spinal cord injury. *Pain Ther* 2015;**4**(1):51—65. https://doi.org/10.1007/s40122-015-0033-y.

80. Hassan HT, El-Sheemy M. Adult bone-marrow stem cells and their potential in medicine. *J R Soc Med* 2004;**97**(10):465—71.

81. J Yang Y, Li X-L, Xue Y, Zhang C-X, Wang Y, Hu X, Dai Q. Bone marrow cells differentiation into organ cells using stem cell therapy. *Eur Rev Med Pharmacol Sci* 2016. Retrieved May 10, 2020, from, https://www.europeanreview.org/article/11118.

82. Chhabra HS, Sarda K, Arora M, Sharawat R, Singh V, Nanda A, Sangodimath GM, Tandon V. Autologous bone marrow cell transplantation in acute spinal cord injury—an Indian pilot study. *Spinal Cord* 2016;**54**(1):57—64. https://doi.org/10.1038/sc.2015.134.

83. Mendonça MVP, Larocca TF, de Freitas Souza BS, Villarreal CF, Silva LFM, Matos AC, Novaes MA, Bahia CMP, de Oliveira Melo Martinez AC, Kaneto CM, Furtado SBC, Sampaio GP, Soares MBP, dos Santos RR. Safety and neurological assessments after autologous transplantation of bone marrow mesenchymal stem cells in subjects with chronic spinal cord injury. *Stem Cell Res Ther* 2014;**5**(6):126. https://doi.org/10.1186/scrt516.

84. El-Kheir WA, Gabr H, Awad MR, Ghannam O, Barakat Y, Farghali HAMA, El Maadawi ZM, Ewes I, Sabaawy HE. Autologous bone marrow-derived cell therapy combined with physical therapy induces functional improvement in chronic spinal cord injury patients. *Cell Transplant* 2014;**23**(6):729—45. https://doi.org/10.3727/096368913X664540.

85. Collet JP. Limitations of clinical trials. *La Revue Du Praticien* 2000;**50**(8):833—7.

86. Kleiderman E, Boily A, Hasilo C, Knoppers BM. Overcoming barriers to facilitate the regulation of multicentre regenerative medicine clinical trials. *Stem Cell Res Ther* 2018;**9**. https://doi.org/10.1186/s13287-018-1055-2.

87. Yamazaki K, Kawabori M, Seki T, Houkin K. Clinical trials of stem cell treatment for spinal cord injury. *Int J Mol Sci* 2020;**21**(11):3994. https://doi.org/10.3390/ijms21113994.

88. Kadam RA, Borde SU, Madas SA, Salvi SS, Limaye SS. Challenges in recruitment and retention of clinical trial subjects. *Perspect Clin Res* 2016;**7**(3):137—43. https://doi.org/10.4103/2229-3485.184820.

89. Lovato LC, Hill K, Hertert S, Hunninghake DB, Probstfield JL. Recruitment for controlled clinical trials: literature summary and annotated bibliography. *Contr Clin Trials* 1997;**18**(4):328—52. https://doi.org/10.1016/S0197-2456(96)00236-X.

90. Tsintou M, Dalamagkas K, Seifalian AM. Advances in regenerative therapies for spinal cord injury: a biomaterials approach. *Neural Regene Res* 2015;**10**(5):726—42. https://doi.org/10.4103/1673-5374.156966.

91. Mao AS, Mooney DJ. Regenerative medicine: current therapies and future directions. *Proc Natl Acad Sci USA* 2015;**112**(47):14452—9. https://doi.org/10.1073/pnas.1508520112.

92. Assinck P, Duncan GJ, Plemel JR, Lee MJ, Stratton JA, Manesh SB, Liu J, Ramer LM, Kang SH, Bergles DE, Biernaskie J, Tetzlaff W. Myelinogenic plasticity of oligodendrocyte precursor cells following spinal cord contusion injury. *J Neurosci* 2017;**37**(36):8635—54. https://doi.org/10.1523/JNEUROSCI.2409-16.2017.

93. Guo F, Maeda Y, Ma J, Xu J, Horiuchi M, Miers L, Vaccarino F, Pleasure D. Pyramidal neurons are generated from oligodendroglial progenitor cells in adult piriform cortex. *J Neurosci* 2010;**30**(36):12036—49. https://doi.org/10.1523/JNEUROSCI.1360-10.2010.

94. Rivers LE, Young KM, Rizzi M, Jamen F, Psachoulia K, Wade A, Kessaris N, Richardson WD. PDGFRA/NG2 glia generate myelinating oligodendrocytes and piriform projection neurons in adult mice. *Nat Neurosci* 2008;**11**(12):1392—401. https://doi.org/10.1038/nn.2220.

95. Kruminis-Kaszkiel E, Osowski A, Bejer-Oleńska E, Dziekoński M, Wojtkiewicz J. Differentiation of human mesenchymal stem cells from Wharton's jelly towards neural stem cells using a feasible and repeatable protocol. *Cells* 2020;**9**(3):739. https://doi.org/10.3390/cells9030739.

96. Levi AD, Okonkwo P, Jenkins III AL, Kurpad SN, Parr AM, Ganju A, et al. Emerging safety of intramedullary transplantation of human neural stem cells in chronic cervical and thoracic spinal cord injury. *Neurosurgery* 2017;**82**(4):562—75. https://doi.org/10.1093/neuros/nyx250.

Clinical trials: noncellular regenerative approaches

Vjura Senthilnathan[1,2,a], *Nayaab Punjani*[1,3,a], *Narihito Nagoshi*[4], *Christopher S. Ahuja*[1,3,5], *Michael G. Fehlings*[1,5,6]

[1]Division of Genetics and Development, Krembil Research Institute, University Health Network, Toronto, ON, Canada; [2]University of Toronto Scarborough, Toronto, ON, Canada; [3]Institute of Medical Science, University of Toronto, Toronto, ON, Canada; [4]Department of Orthopaedic Surgery, Keio University School of Medicine; [5]Division of Neurosurgery, Department of Surgery, University of Toronto, Toronto, ON, Canada; [6]Division of Neurosurgery, Toronto Western Hospital, University Health Network, Toronto, ON, Canada

Introduction

Spinal cord injury (SCI) involves an overall loss of central nervous system (CNS) cells (i.e., neurons and glia),[1] as well as the surrounding extracellular matrix (ECM), which is involved in the structural integrity of the cord, thus reducing sensorimotor functioning. SCI involves two general steps, from the primary source of injury to further long-term secondary injury involving inflammation, as well as cyst and glial scar formation,[1] which together inhibit the ability of CNS cells to regenerate and replace the loss of cells at the injury site and reform neural networks.[2]

This chapter will examine the use of noncellular therapies utilized in regenerative medicine, as highlighted in Fig. 26.1. These methods include antibodies, growth factors, minerals, drugs, biomaterials, and electrical stimulation to enhance cellular regeneration. These approaches aim to improve the quality of life and autonomy of patients with SCI through improved sensorimotor ability.

After efficient development of laboratory treatments, clinical trials are conducted to assess the safety and efficacy of a drug or treatment for a particular population of patients by proceeding through four phases. Participants in the majority of these trials are between the ages of 18 and 80

---

[a] Vjura Senthilnathan and Nayaab Punjani designating co-first authorship.

© 2022 Elsevier Inc. All rights reserved.

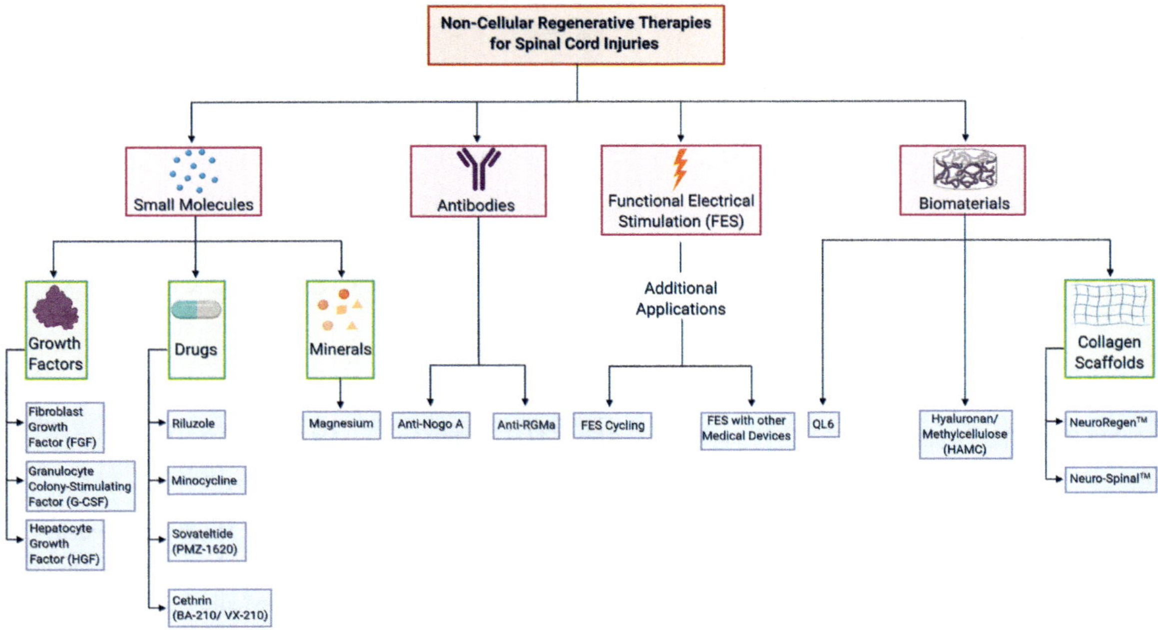

FIGURE 26.1   **Overview of noncellular regenerative therapies being tested for spinal cord injury (SCI).** The therapies are broken down into four major categories—small molecules, antibodies, functional electrical stimulation (FES), and biomaterials. Small molecules are further categorized into growth factors, drugs, and minerals. The specific individual treatments to be addressed in this chapter are provided at the bottom of the figure in blue. *Created with BioRender.com.*

years. Phase I clinical trials are often tested in a small subgroup of patients or healthy volunteers to determine the overall safety and appropriate dosage of a treatment, as well as gather preliminary data on the therapeutic potential of this treatment.[3,4] Phase II clinical trials are generally longer term with more patients, and their goal is to assess the efficacy of the proposed treatment and determine any adverse effects.[3,4] Phase III clinical trials further test the efficacy of the treatment and monitor side effects for a longer period of time, generally up to 4 years,[3] and compare the proposed treatment with existing treatments.[4] Lastly, phase IV clinical trials attempt to test thousands of patients for safety and efficacy of the treatment in question.[3] Treatments in the United States (US) may also be classified as a humanitarian use device (HUD) and be examined through humanitarian device exemption (HDE) studies.[5] This designation is in place for treatments that only aim to treat less than 8000 US patients yearly and thus have more lenient standards on effectiveness due to the inability to test a large group of patients.[5]

The International Standards for Neurological Classification of Spinal Cord Injury (ISNCSCI) developed by the American Spinal Injury Association (ASIA) has been used in the development of the ASIA Impairment Score (AIS) to determine the degree of sensorimotor functioning in SCI patients.[6,7] This is continually modified yearly,[8] although there is a general classification scale that is denoted by letters A—E with increasing functionality.[7]

This chapter discusses clinical trials using noncellular regenerative therapies to assess their therapeutic potential in SCI. ClinicalTrials.gov was used as the source of clinical trials to be examined, with clinical trials performed in North America, Europe, Japan, and Australia primarily emphasized.

Antibodies

Anti-Nogo antibody

Nogo-A is a crucial myelin inhibitory molecule expressed by oligodendrocytes.[9] This molecule binds to the Nogo receptor, which is expressed by neurons, leading to growth cone collapse and inhibition of neurite outgrowth by activating the downstream Rho-ROCK pathway,[10] as demonstrated in Fig. 26.2. Treatment using the anti-Nogo antibody demonstrated axonal sprouting and functional recovery in an animal SCI model.[12] A preliminary study demonstrated that the Nogo-A antibody (AT1355) was safe and potentially effective in patients with acute complete SCI.[13] Now, clinical trials using this antibody are ongoing for both acute (NCT03935321) and chronic SCI (NCT03989440).

A phase II trial termed the Nogo Inhibition in Spinal Cord Injury (NISCI) is a multicenter European trial sponsored by the University of Zurich (NCT03935321). The purpose of this trial is to determine if NG-101 antibody administration will help ameliorate the quality of life and locomotion of tetraplegic patients

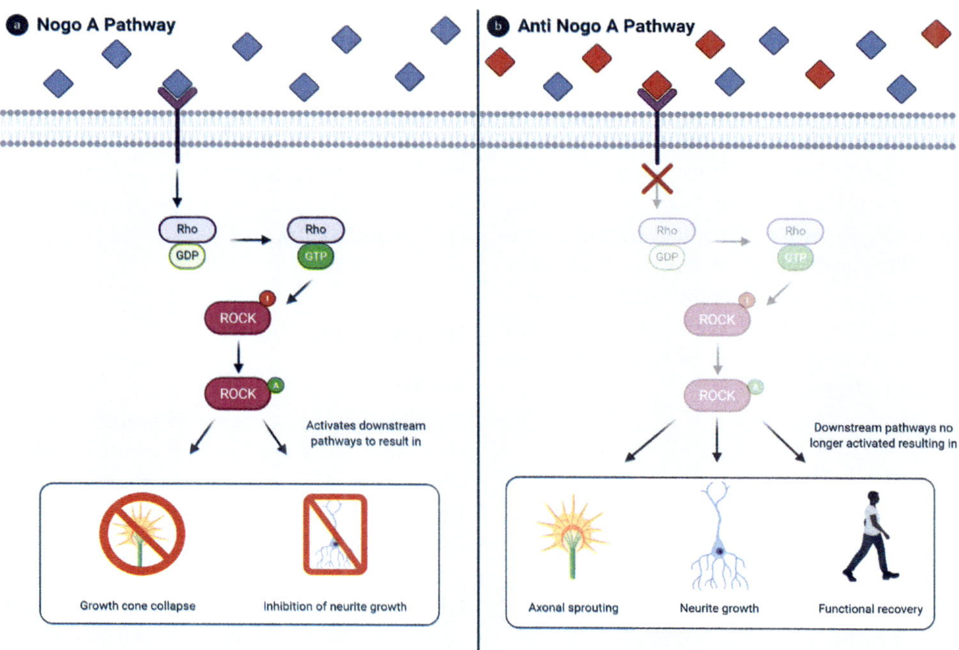

FIGURE 26.2 **Mechanism of action for Nogo-A and anti-Nogo A with regard to SCI.** (A) The Nogo-A molecule activates the Rho-Rock pathway resulting in growth cone collapse and the inhibition of neurite growth.[10,11] (B) In contrast, the administration of anti-Nogo-A inhibits the activation of the Rho-Rock pathway, leading to regeneration of lost neural connections by promoting axonal sprouting and neurite growth, with the optimal end result of functional recovery.[12] *Created with BioRender.com.*

compared with a placebo. This study is recruiting participants with acute cervical SCI at level C1—C8, 4—28 days postinjury with an AIS grades A—D and predicted upper extremity motor score (UEMS) of less than 41/50. Patients will receive six intrathecal repeated bolus administrations of either NG-101 or placebo. The primary outcome measure is a variation in UEMS scores assessed by the ISNCSCI. All outcome measures will be assessed over the course of 168 days. This full trial is estimated to be completed in 2023 (NCT03935321).

A phase I/II American trial sponsored by ReNetX Bio Inc. is currently recruiting chronic cervical SCI patients for a two-part trial assessing the safety, tolerability, and pharmacokinetics of AXER-204; a human fusion protein that aims to trap Nogo-A (NCT03989440). Part 1 of the trial will involve four cohorts of six participants each receiving single increasing doses of AXER-204 through lumbar puncture and slow bolus infusion. Part 2 of the trial will involve repeated doses of AXER-204 or a phosphate buffered saline (PBS) placebo, the dose of which will be determined experimentally from part 1. Inclusion criteria involve having an AIS A traumatic SCI, greater than 1-year postinjury, in the cervical region. The primary outcome measures involve assessing adverse events and the cerebrospinal fluid (CSF) drug concentration over time. The trial is expected to be completed in June 2022 (NCT03989440).

Anti-RGMa antibody

Repulsive guidance molecule A (RGMa) binds to neuronal Neogenin receptors resulting in the inhibition of neurite axon growth.[14,15] Following traumatic SCI through the use of a clip impact compression model in rats, increased RGMa production was observed in various cell types including neurons and oligodendrocytes near the injury site. RGMa was also upregulated in human spinal cord tissue following SCI. The administration of anti-RGMa antibodies has been demonstrated to improve neuronal growth in vitro and functional locomotor recovery in rats.[14]

Elezanumab (ABT-555) is a human immunoglobulin monoclonal antibody that binds to RGMa and is a proposed therapeutic to improve regeneration and motor function following SCI.[15] This antibody is currently being tested in an international multicenter randomized, double-blind, placebo-controlled phase II trial sponsored by Abbvie. This trial aims to assess the safety and efficacy of intravenous ABT-555 in cervical traumatic SCI patients. The first dose of ABT-555 or placebo will be provided within 24 h postinjury and continued monthly for 48 weeks, amounting to 13 doses. The inclusion criteria involve having acute traumatic SCI at C4—C7 within 24 h postinjury, as well as an AIS grade of A or B. The primary outcome measure is upper extremity motor score at 52 weeks. This trial is currently recruiting participants and is scheduled for completion in March 2023 (NCT04295538). This treatment is also being provided through expanded access should it be deemed appropriate by a doctor, and the patient is excluded from the aforementioned trial (NCT04278235).

Small molecules

Riluzole

Riluzole is a benzothiazole anticonvulsant that acts as a sodium channel blocker, and its effect in SCI is further demonstrated in Fig. 26.3. It can be administrated orally, and the Food and Drug Administration (FDA) has approved this drug as a safe and well-tolerated treatment for patients with amyotrophic lateral sclerosis (ALS).[18] The senior author's laboratory demonstrated significant enhancement of functional neurological recovery when riluzole was administrated in a rodent SCI model.[16] The functional recovery has been corroborated by histological analysis, indicating significant long-term

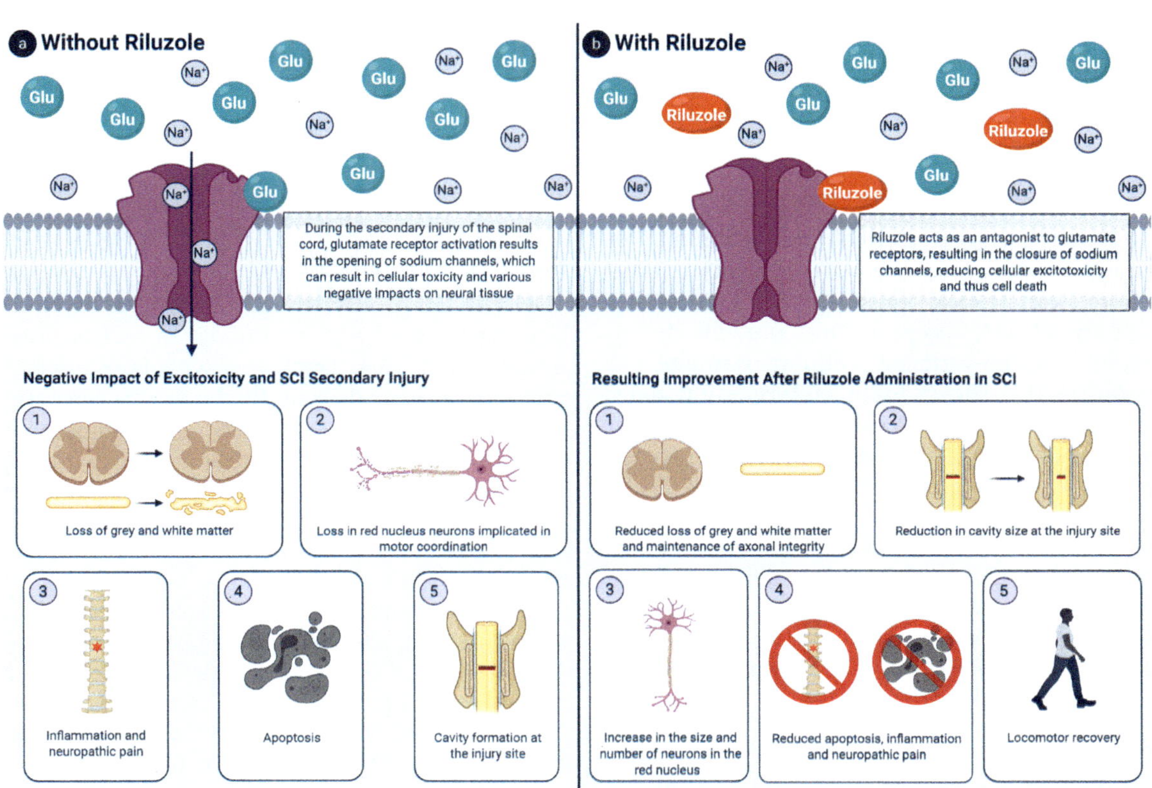

FIGURE 26.3 **Potential effects of riluzole on SCI pathobiology.** (A) Without riluzole, the secondary injury cascade of SCI causes glutamate excitotoxicity resulting in the opening of sodium channels[16] with various negative impacts on the regenerative potential of neural tissue. Resulting negative impacts include the loss of gray and white matter, loss of neurons in the red nucleus,[16] as well as inflammation, neuropathic pain, apoptosis, and cavity formation at the injury site. (B) Riluzole acts as an antagonist to the glutamate receptors,[17] reducing the impact of cellular excitotoxicity. This leads to sparring of gray and white matter, reduced cavity formation, increased neurons in the red nucleus, as well as reduced apoptosis, inflammation, and neuropathic pain.[17] Taken together, these effects have important implications for improving locomotor recovery.[16] *Created with BioRender.com.*

tissue sparing at and around the site of the injured spinal cord, an increase in the number and size of neurons in the red nucleus, and a reduction of cavity area. Subsequently, delayed administration of riluzole 3 h after injury also exerted improvements in sensorimotor function as well as a reduction in apoptosis and inflammation without increased neuropathic pain.[17] Pharmacokinetic data clarified the prolonged half-life and delayed elimination in the injured rats compared with uninjured rats.

An international clinical trial in phase II/III aimed to assess riluzole in cervical SCI patients (NCT01597518). The RISCIS (riluzole in acute spinal cord injury) trial was supported by AO Spine North America in partnership with AO Spine International, NACTN, the Rick Hansen Institute, and the Ontario Neurotrauma

Foundation. The primary outcome measure was a change in AIS Motor Score between baseline and 180 days. This study started in October 2013 has recently been terminated due to slow enrollment.

Magnesium

Magnesium is a physiologic antagonist of N-methyl-D-aspartate (NMDA) receptors. Magnesium can block massive calcium influx through NMDA receptors and prevent downstream calcium-induced cellular damage.[19,20] Regarding the efficacy for SCI intervention, magnesium protects the blood—spinal cord barrier and attenuates lipid peroxidation and ultrastructural damage, leading to enhancement of SSEPs and functional recovery.[21–23]

A phase I clinical study for healthy human volunteers was conducted in 2009, and no adverse effects were observed.[20] A phase II clinical trial for patients with acute SCI began in July 2013 (NCT01750684) but was terminated due to insufficient enrollment.

Minocycline

Minocycline is a second-generation semisynthetic tetracycline that is clinically used as an antibacterial agent. This medication can permeate the CNS through the blood—brain barrier with a longer half-life.[24] Minocycline also has a neuroprotective role in reducing posttraumatic neural inflammation and preventing cell death.[24] In 2003, minocycline was first applied in a rodent SCI model as a therapeutic agent. The results demonstrated the presence of axonal sparing and improved motor functional recovery, along with a significant reduction in lesion size.[25] Using a rat cervical model of SCI, another group showed more detailed results indicating that minocycline treatment reduced apoptotic oligodendrocytes and microglia, activated microglia/macrophage density, and corticospinal tract dieback, leading to functional recovery.[26]

A phase I/II clinical trial was completed for patients with acute SCI in 2010 (NCT00559494).

Some beneficial efficacy was identified in this trial; therefore, a clinical trial in phase III was conducted for patients with acute cervical SCI (NCT01828203). This study started from June 2013, and the estimated completion year was 2018. The current status of the trial is unclear as enrollment has ceased.

Fibroblast growth factor

Fibroblast growth factor (FGF) is a heparin-binding protein, and its impact in SCI is illustrated in Fig. 26.4. Treatment with FGF rescued motor neurons adjacent to the injury site and reduced acute respiratory deficits resulting from loss of ventral horn neurons, by reducing glutamate-mediated excitotoxicity.[29,31] Asubio Pharmaceuticals, Inc. developed recombinant FGF2, which does not possess proliferative actions (SUN13837).

A phase II placebo-controlled clinical trial aiming to assess the efficacy, pharmacokinetics, and overall safety of SUN13837 for patients with acute cervical SCI began in 2012 and was completed in 2014 (NCT01502631). This international trial sponsored by Daiichi Sankyo, Inc. took place across 66 locations in the United States, Canada, France, Poland, Spain, the United Kingdom, and the Czech Republic. The inclusion criteria for the selected participants consisted of having acute traumatic cervical SCI within 12 h of the first SUN13837 injection, with either an AIS grade A injury at the level of C4—C7 or AIS grade B, or C at the level of C3—C8, with additional criteria depending on the AIS grade. Participants were provided with either SUN13837 or a volume-matched placebo once a day for 28 total administrations. This trial was quadruple-masked whereby the participant, healthcare professional, study investigator, and outcome assessor were blinded to the provided treatment. Outcome measures involved comparing the number of participants classified as responders between the experimental and placebo groups, as well as the responder means in the Spinal Cord

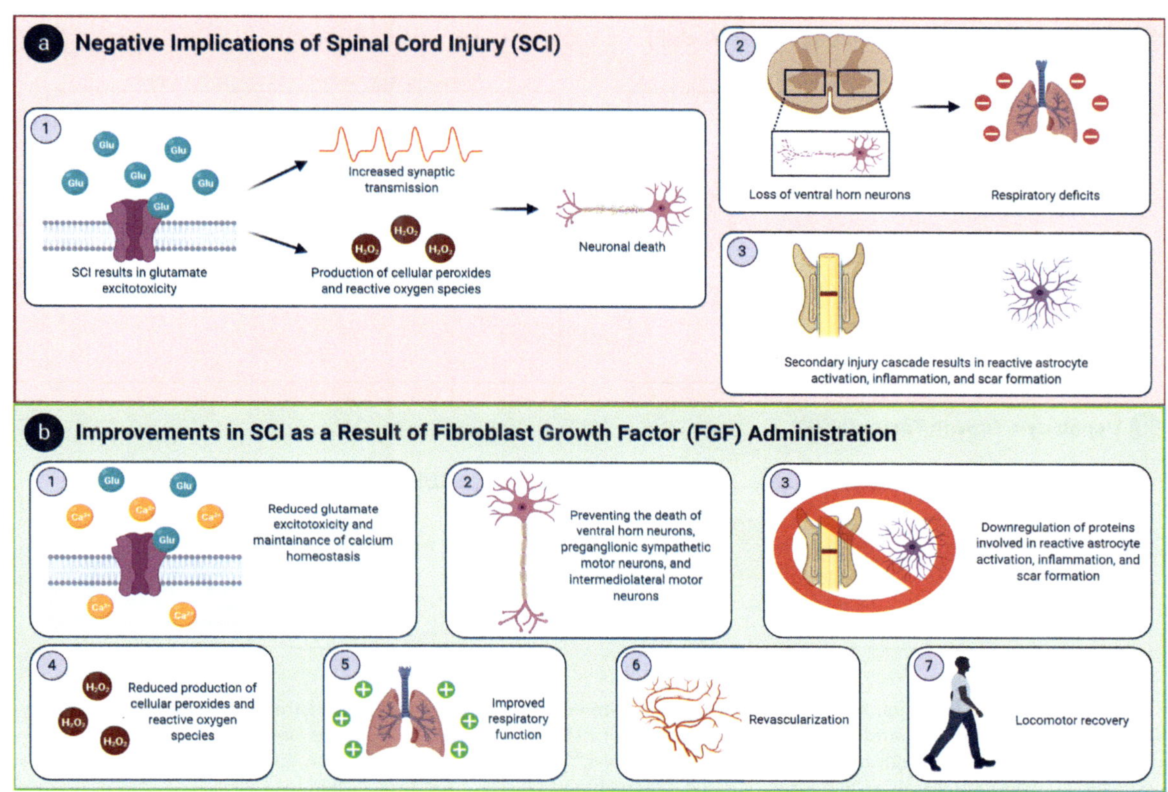

FIGURE 26.4 **Contrasting the impact of fibroblast growth factor (FGF) administration in SCI.** (A) SCI glutamate excitotoxicity results in increased excitatory synaptic transmission[27] and the production of reactive oxygen species,[27,28] resulting in neuronal death. Furthermore, the loss of ventral horn neurons can lead to respiratory deficits,[29] and the secondary injury cascade of SCI may result in scar formation, the production of reactive astrocytes, and inflammation.[30] (B) FGF helps maintain calcium homeostasis[28] by reducing glutamate excitotoxicity, prevents the death of various neuronal subtypes,[29,31] downregulates the production of proteins involved in the secondary injury cascade,[30] reduces the formation of reactive oxygen species,[27,28] and improves respiratory function[29] and revascularization.[32] The end goal is the improvement of locomotor ability.[30,33] *Created with BioRender.com.*

Independence Measure (SCIM) III self-case subscale score for up to 112 ± 7 days. The results of this trial have not been published thus far (NCT01502631).

Granulocyte Colony-stimulating factor

Granulocyte colony-stimulating factor (G-CSF) is a growth factor that affects the hematopoietic system. G-CSF exerts neuroprotective roles for SCI by mobilizing bone marrow–derived stem cells to the lesion site, inhibiting cell death and inflammatory cytokine expression, as well as promoting neovascularization,[34,35] as shown in Fig. 26.5. A phase I/IIa clinical study has already been completed and confirmed the feasibility of G-CSF for treating acute SCI patients.[41] Following these results, a phase III clinical trial was conducted, and patient registration was completed in 2019, and the results have not been published to date.[42]

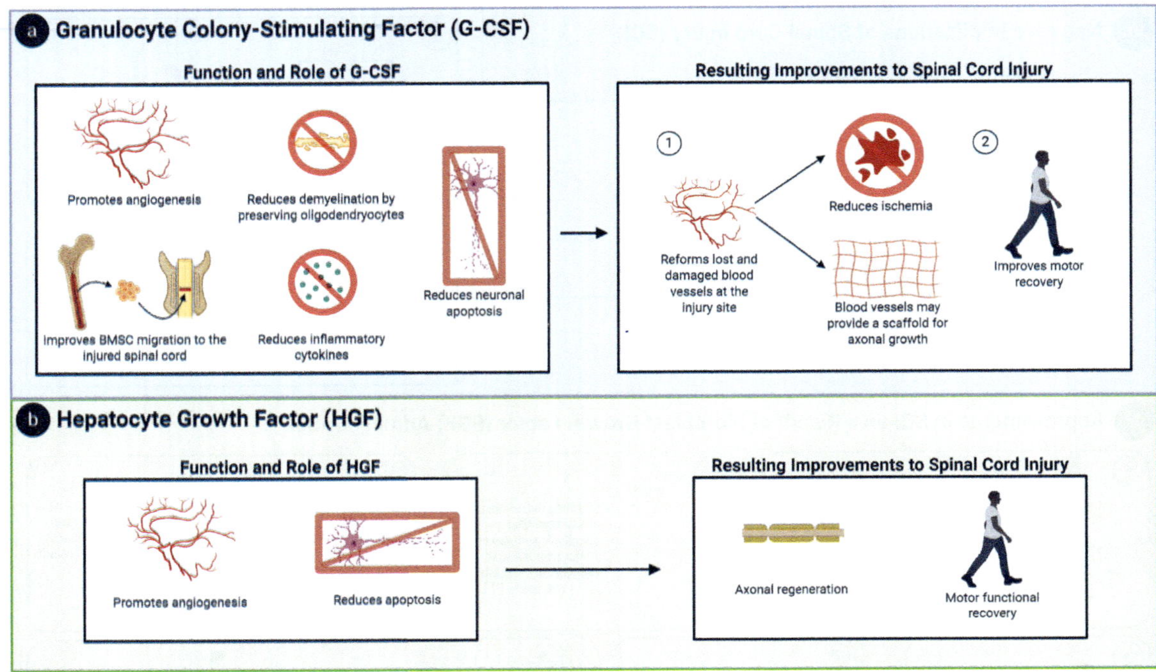

FIGURE 26.5   **The function, role, and resulting improvements in SCI from the administration of granulocyte colony-stimulating factor (G-CSF) and hepatocyte growth factor (HGF).** (A) G-CSF promotes angiogenesis, improves bone-marrow stem cell (BMSC) migration to the injured spinal cord,[35] and reduces demyelination, inflammation,[36] and neuronal apoptosis.[34,37] This may result in the reformation of damaged blood vessels,[38] reducing ischemia,[39,40] and providing a scaffold for axonal growth, with the aim of improving motor recovery.[34,41,42] (B) HGF promotes angiogenesis and reduces apoptosis, thus allowing for axonal regeneration and motor functional recovery.[43–45] *Created with BioRender.com.*

Hepatocyte growth factor

Hepatocyte growth factor (HGF) was identified as a potent mitogen for mature hepatocytes.[46] Administration of HGF for a rodent SCI model reduced apoptosis and promoted angiogenesis, which resulted in axonal regeneration and motor functional recovery,[43,44] as demonstrated in Fig. 26.5. A more severe contusion model of cervical SCI was studied in nonhuman primates, which demonstrated significant motor restoration after intrathecal recombinant human HGF administration.[45]

A phase I/II quadruple-masked, placebo-controlled clinical trial using KP-100IT or HGF was completed in 2018 (NCT02193334).

This Japanese trial sponsored by Kringle Pharma Inc. involved participants with a cervical SCI below C3 of AIS grade A or B, assessed within 72 h of injury. Intrathecal administration of 0.6 mg HGF was provided to the experimental group at 72 h postinjury and continued each week for a total of five administrations. The primary outcome measures were the presence of adverse events and ASIA motor score variations until 24 weeks postbaseline. The results have not been released to date (NCT02193334).

Cethrin (VX-210)

During SCI, activation of the Rho-ROCK pathway results in growth cone collapse, preventing the regeneration of axons.[47,48] Previously, it has been shown that providing the bacterial-derived C3 transferase, an inhibitor to the Rho-GTPase that initiates the Rho-ROCK pathway and helps to improve axon growth.[49] VX-210, previously BA-210, is commonly known by the tradename Cethrin. This is a cell-permeable C3 transferase-derived fusion protein that provides dose-dependent inactivation of Rho. When administered through a fibrin sealant topically at the site of injury, it was observed to help increase tissue sparing and improve functional recovery in rats with both immediate and delayed administration, with no adverse effects observed.[50]

A phase I/IIa Canadian and American multicenter clinical trial aimed to assess the safety of a single administration of Cethrin following acute SCI. Doses of 0.3–9 mg were tested in patients with complete SCI, AIS grade A. This trial involved participants from the ages of 16–70 years with either thoracic level T2–T12 or cervical level C4–T1 SCI with spinal surgery (e.g., decompression) up to 7 days postinjury. The drug was administered with a noninvasive fibrin sealant to the dura at the site of injury during the surgery. Thoracic patients were assessed for dose safety prior to cervical SCI patients and were matched on various factors including gender. The primary outcome measure was to assess any decreased neurological function 8 weeks posttreatment. Observed adverse events included general pain and gastrointestinal issues, which are common for acute SCI; thus, Cethrin was determined to be safe at all doses. Improvements in motor scores were observed in cervical SCI patients, with minimal improvements for thoracic SCI patients. These results suggest that Cethrin may help improve neurological recovery[48,51] (NCT00500812).

Following this trial, a phase IIb/III multicenter placebo-controlled trial, also termed the SPRING trial,[48] was initiated to further assess the safety of Cethrin, as well as its efficacy. Patients with acute traumatic cervical SCI at the level of C4–C7 on each side at a severity of AIS grade A or B, with spine surgery scheduled within 3 days after SCI, were included. A single dose of 9 mg of the drug was administered through the same fibrin sealant. The higher end of the dose spectrum was used to allow for maximum effect, as dose–response was not possible to assess in cervical SCI patients due to the smaller sample size in the phase I/IIa trial.[51] Follow-up assessments were scheduled to be conducted up to 1-year postadministration of Cethrin. The primary outcome measure was upper extremity motor score. A futility analysis was put in place when a third of the expected sample size achieved a 6-month follow-up, whereby the study would be stopped should the conditional power and treatment effect size be small. Following the interim analysis, the study was terminated. Upper extremity motor scores did not significantly change from baseline to 6 months posttreatment in the drug or placebo conditions. Thus, the primary efficacy endpoint was not achieved, which was accompanied with no statistical differences in the secondary endpoints and final assessments. No drug-associated adverse effects were observed. Despite not replicating the motor improvements from the previous phase I/IIa study, this study opens the discussion for possible adjustments to study parameters such as the administration method[52] (NCT02669849).

Sovateltide (PMZ-1620)

Sovateltide is an endothelin-B (ETB) receptor agonist.[53,54] Previous research in cerebral ischemia (stroke) rodent models demonstrated that sovateltide may improve neurogenesis and promote the differentiation of neural progenitor cells (NPCs).[53] Improved recovery has also been perceived in trials with stroke patients.[54]

A current phase II clinical trial is recruiting patients with acute incomplete sensorimotor SCI at the level C5 or below with an AIS grade B, C, or D. This multicenter Indian trial is examining the use of sovateltide (PMZ-1620) as a treatment for acute SCI when compared with a saline control group. The experimental group will be administered three doses of 0.3 μg/kg bodyweight of sovateltide daily, 3 h apart for 3 days (day 1, 3, and 6). The patients will then be monitored for 90 days postdrug administration to track any adverse side effects, tolerability of the drug, AIS grade improvement, and MRI/CT amelioration, among other measures (NCT04054414). This trial aims to be completed in June 2021.

Biomaterials

Biomaterials are attractive tools to fill the cavity at the lesion area and are frequently used as scaffolds to deliver drugs, secrete trophic factors, and diffuse transplanted cells, as illustrated in Fig. 26.6. There are some clinically relevant scaffolds for SCI that have been developed so far.

NeuroRegen collagen scaffold

Five clinical trials sponsored by the Chinese Academy of Sciences are currently being conducted to examine the use of functional neural regeneration collagen scaffolds or NeuroRegen scaffolds as a treatment for SCI. An in vitro study demonstrated that preparing a scaffold out of linear ordered collagen (LOC) fibers may act a directional guide for new neuronal formation and it is not very immunogenic.[55] Furthermore, rodent in vivo studies with collagen scaffolds have demonstrated their ability to improve neuronal regeneration in the injury microenvironment resulting in recovered function,[56,57] possibly through bridging nerve lesions.[58] Pairing of the collagen scaffold with brain-derived neurotrophic factor (BDNF) and an antibody that prevents the activity of the epidermal growth factor receptor (EGFR) may reduce scar formation during SCI.[59]

Five trials are scheduled for completion in December 2021, three of which are enrolling by invitation (NCT02688049; NCT02352077; NCT02688062) and two are recruiting participants (NCT02510365; NCT03966794). Some of these trials have results for a subset of their participants, and these will be discussed accordingly.

A phase I trial is examining the use of the functional collagen scaffold in patients with acute AIS grade A SCI at level C4–T12 with injury within the previous 3 weeks (NCT02510365). The primary outcome measure involves the assessment of adverse events. Results have been published for two patients (T11 and C4 complete SCI) who had the NeuroRegen scaffold[55] functionalized with umbilical cord MSCs from a human donor.[60] Some sensory and motor recovery was perceived including an improvement from an AIS grade A to C, recovered bowel and bladder function, and undisrupted neural conduction.[60]

A phase I/II trial is examining the safety and efficacy of epidural electrical stimulation paired with NeuroRegen transplantation (NCT03966794). Epidural electrical stimulation in animal studies has demonstrated its potential for promoting locomotion[61] and forelimb activity.[62] Some of the participants selected will have previously received scaffold transplantation with observed motor recovery and will just receive electrical stimulation. Other participants will receive both interventions and will have complete AIS grade A SCI at C4–T12/L1 of either acute (less than 14 days postinjury) or chronic (greater than 6 months postinjury) nature. Primary outcome measures include assessing adverse events, as well as changes in AIS grade. Results have yet to be published following the completion of the trial.

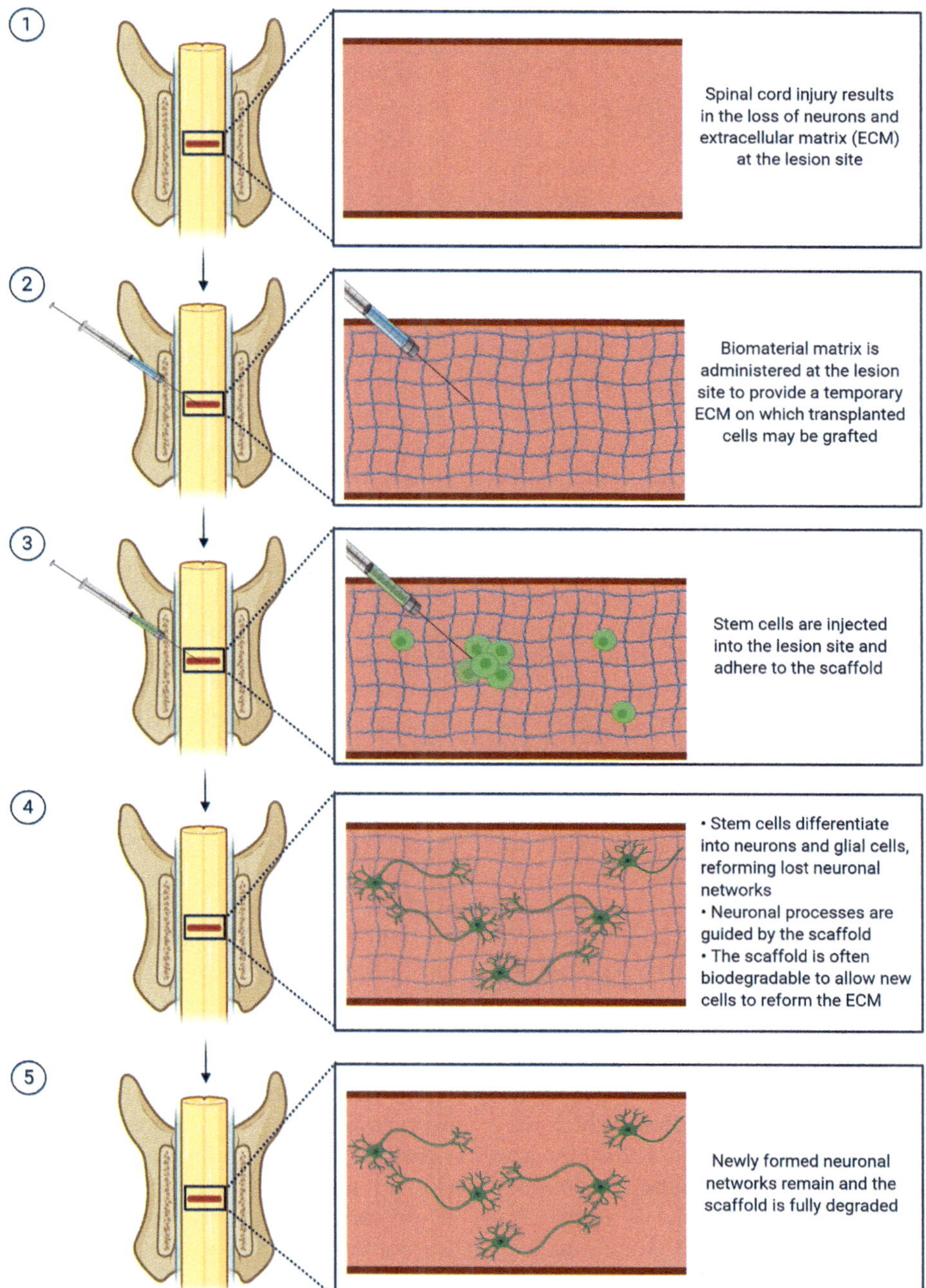

FIGURE 26.6 **The role of biomaterials in providing a scaffold for neuronal growth in SCI.** Spinal cord injury results in the loss of neurons and extracellular matrix (ECM) at the lesion site. A biomaterial matrix may be administered at the lesion site to mimic the ECM. Stem cells may then be injected into the lesion site and will adhere to the scaffold. Stem cells will differentiate into neurons and glial cells, with the scaffold acting as a guide for neuronal processes. The scaffold is often biodegradable and, once fully degraded, will be replaced by newly formed ECM from neurons. *Created with BioRender.com.*

Two trials examined the safety and efficacy of the NeuroRegen scaffold in patients with chronic AIS grade A SCI between C5-T12 (NCT02688049; NCT02352077). The phase I/II trial involved the scaffold functionalized with MSCs or NSCs (NCT02688049), whereas the phase I open-label trial functionalized the scaffold with bone marrow mononuclear cells (BMMCs) or MSCs (NCT02352077). For both trials, the functionalized scaffold administration will be followed by comprehensive rehabilitation, as well as psychological and nutritional support. The phase I/II trial primary outcome measures include improvement in AIS grade, somatosensory evoked potentials (SSEPs), and motor evoked potentials (MEPs), with no results published to date (NCT02688049).

The phase I trial primary outcome measure aims to examine the presence of adverse events up to 6 months posttransplant. Some results have been published for a subset of participants from this trial (NCT02352077). For the patients assessed, no safety concerns were perceived at 1-year posttransplant, and improved sensation, finger activity, and trunk stability were experienced by some patients.[63] Some patients also experienced partial recovery of autonomic nervous system functioning and SSEPs of the lower extremities.[64] These results demonstrate the safety and efficacy of both BMMCs and MSCs paired with the NeuroRegen scaffold and more results are pending as the study is still active (NCT02352077).

Following spinal cord injury, earlier surgical intervention often translates to better recovery.[65] Postsurgical implications may involve the formation of epidural fibrosis or scar tissue, which may bind to dura and nerve roots via adhesions and thus adhesiolytic procedures can be employed.[66] One phase I/II double-blind trial is comparing the safety and efficacy of BMMC-functionalized NeuroRegen scaffold transplantation compared with the SCI treatment of surgical decompression and adhesiolysis (NCT02688062). This will be followed by comprehensive rehabilitation as well as psychological and nutritional procedures. All participants must have chronic and complete thoracic SCI of AIS grade A. All primary and secondary outcome measures are identical to the NCT02688049 trial. No preliminary results have been disclosed at the time of writing.

Neuro-Spinal scaffold

Two HDE studies sponsored by InVivo Therapeutics are examining the use of a poly (lactic-co-glycolic acid)-b-poly (L-lysine) (PLGA-PLL) or Neuro-Spinal scaffold, for the treatment of acute traumatic thoracic SCI, and they are collectively known as the INSPIRE trials (NCT02138110; NCT03762655). Animal SCI model studies assessing the use of the Neuro-Spinal scaffold demonstrated its ability to help ameliorate functional recovery and decrease the formation of a cyst postinjury,[67,68] thus providing evidence for it to be tested through clinical trials. This scaffold is also capable of self-degrading within 4–8 weeks of implantation (NCT03762655).

Selected participants in the INSPIRE studies must have complete nonpenetrating SCI defined as an AIS grade of A at T2–T12 and be willing to undergo open spine surgery/laminectomy within 7 days of injury (NCT02138110; NCT03762655). These trials are currently taking place across a variety of medical centers in the United States.

The first INSPIRE trial started in 2014 is active but not recruiting and is estimated for completion in 2024 (NCT02138110). This study aims to administer the Neuro-Spinal scaffold to 20 participants with the main outcome measures conducted until 6 months postimplantation, with in-person follow-up until 24 months postimplantation and long-term phone follow-ups for 3–10 years postimplantation. The primary outcome measure is the percentage of patients with an improvement in AIS grade at 6 months postimplantation (NCT02138110). Clinical pilot study results for the NCT02138110 trial involved one 25-year-old male patient with a T11—T12 dislocation fracture resulting in a complete AIS grade A injury at T11.[69] At 3 months postimplantation, an AIS grade improvement from A to C was observed. Furthermore, no severe procedural or safety complications were perceived at 6 months postinjury.[69] This illustrates the safety and efficacy of the scaffold in the short term.

Some results for the NCT02138110 study have been published for up to 24 months postimplantation. Of the 20 participants selected, 19 were eligible for scaffold implantation and 16 were assessed at the 6-month follow-up.[70] Three participants died within 2 weeks of initial implantation for reasons were proven to be unrelated to scaffold implantation.[71] 44% of the 16 participants demonstrated an AIS grade improvement of one to two levels.[70,71] Some of the participants who achieved an AIS grade of B at 6 months progressed to improve to level C at either 12 or 24 months postimplantation. No participants observed any AIS grade deterioration or adverse side effects.[70] Hence, this scaffold shows promise to improve the sensorimotor recovery of thoracic SCI patients, but further annual follow-ups have yet to be conducted.

The results of this study have prompted the initiation of another HDE study, the INSPIRE 2 (NCT03762655). This study is currently recruiting acute thoracic SCI patients with inclusion criteria similar to the INSPIRE 1. This randomized, participant-blinded study will have participants receiving standard-of-care open spine surgery alone or followed by Neuro-Spinal scaffold implantation. The same primary outcome measure, a change in AIS grade, is being assessed. This study was initiated in May 2019 and is estimated for completion in July 2028.

Functional electrical stimulation

Functional electrical stimulation (FES) is a technique that uses electrical current to excite existing tissue to potentially restore lost neurological function due to traumatic or nontraumatic SCI.[72] FES and electrical stimulation (ES) are both used for therapeutic applications; however, there are significant differences that need to be noted.[72] ES induces physiological changes in tissue that improves tissue health or function, as these physiological changes do not fade away after ES.[72] On the contrary, FES regenerates lost voluntary function in patients through replacement or assistance.[72] FES presents various beneficial effects to maintain or increase the range of motion, reduce spasticity, improve circulation, and promote functional restoration.[73] This can be further demonstrated in Fig. 26.7.

The FES stimulus system has been demonstrated to improve walking ability for incomplete SCI patients by activating the central pattern generator mechanism.[74,75] A phase III multicenter randomized trial (NCT01292811) is evaluating the efficacy of FES for upper limb function in patients with subacute traumatic incomplete SCI between C5 and C7 of less than 6 months postinjury. The primary outcome measure is to assess the change in burden of care from baseline at 8 weeks and 6-month follow-up, as evaluated by the Functional Independence Measure (FIM). Recruitment status is currently stated as unknown, with no update or results at the time of writing (NCT01292811).

A clinical trial in Estonia conducted by Tartu University Hospital assessed FES in combination with therapeutic exercise in patients

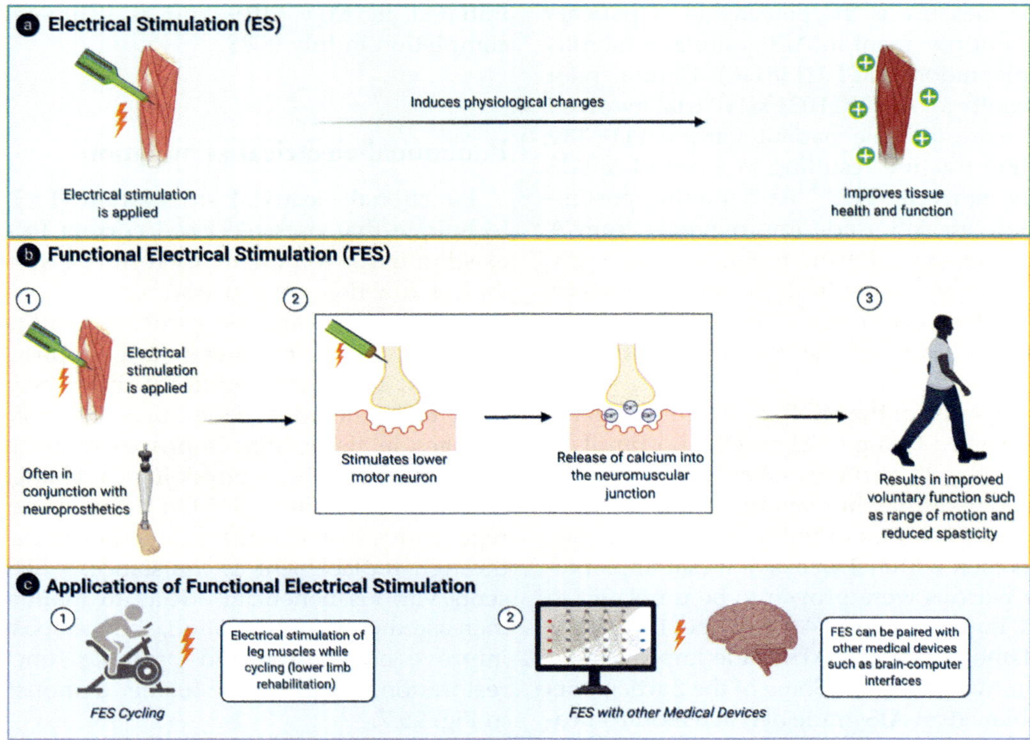

FIGURE 26.7　**The differences between electrical stimulation (ES) and functional electrical stimulation (FES) paired with a discussion of the potential applications of FES.** (A) ES induces physiological changes in muscle tissues, resulting in improvements in tissue health and function.[72] (B) FES is often provided in conjunction with neuroprosthetics and aims to promote the release of calcium into the neuromuscular junction, to improve the range of motion, and to reduce spasticity.[72,73] (C) Applications of FES include FES cycling that is used in lower limb rehabilitation, as well as the pairing of FES with other medical devices such as brain—computer interfaces (BCI). *Created with BioRender.com.*

with traumatic SCI to see if it aids in the improvement of skeletal muscles and lung function as well as quality of life (NCT03517787). This four-armed trial consisted of the following groups: patients who received FES with therapeutic exercise, patients who received therapeutic exercise, healthy adults who received one session of FES and therapeutic exercise, and healthy adults who received one session of therapeutic exercise. The primary outcome measure was to observe changes in the median power frequency and amplitude from baseline to 6 weeks postintervention. Patients with chronic cervical SCI of greater than 1 year were included in the trial (NCT03517787). The researchers concluded that combining FES with therapeutic exercise increased trunk muscle tone and sitting balance, in comparison with patients who only had therapeutic exercise.[76] After 7 months, however, there was a slight decrease in trunk muscle tone and a substantial decrease in sitting balance.[76]

There are currently two clinical trials in progress at the Swiss Paraplegic Centre Nottwil in Switzerland that are assessing the usage of FES in SCI. One of these trials utilizes FES cycling, which will be detailed in the subsequent FES Cycling section.

The first trial aimed to determine the impact of FES on the upper extremities as many studies only focus on lower extremities (NCT03048331). Additionally, there has not been any evaluation of FES in combination with reconstructive hand and arm surgery in patients, thus outlining the primary goals of this clinical trial. The aim of this trial is to determine if conducting FES before and after reconstructive hand and arm surgery improves the strength and function of the muscle through multiple assessments at four different time points. This two-armed study consists of an experimental group that will receive FES before and after surgery in addition to standard care, which will then be compared with the control group that only receives standard therapy. The primary outcome measures for this trial include detecting changes in evoked force and power output. The inclusion criteria for this trial consists of patients with either traumatic or non-traumatic SCI between C4 and T1 levels that occurred greater than 6 months prior to the onset of the study with as AIS grade between A to D. This trial is currently recruiting and is estimated to be completed in December 2023 (NCT03048331).

Functional electrical stimulation cycling

There are currently four trials that focus on the usage of FES with cycling in patients with varying degrees of SCI.

One randomized open-label trial that was completed in the United States in 2008 aimed to determine the effectiveness of FES cycling in children between the ages of 5 and 15 with SCI by looking for improvements in their neurologic status, as well as cardiovascular and musculoskeletal systems (NCT00245726). Participants were subjected to the following conditions: passive cycling, where subjects only cycled; ES therapy, where subjects' leg muscles were electrically stimulated, and FES cycling, where subjects had electrical stimulation of leg muscle while cycling. Their primary outcome measures included evaluating muscle strength, bone mineral density, and AIS grade. The inclusion criteria for this study consisted of having a cervical or thoracic level injury that occurred at least 12 months prior to the study with an AIS grade A or B (NCT00245726). Two articles regarding this clinical trial were published in 2009 and 2011. Johnston and colleagues suggest that children's risk for cardiovascular diseases and type 2 diabetes may decrease if children receive electrically stimulated exercises.[77] They also concluded that FES cycling increased oxygen uptake and ES alone decreased cholesterol levels.[78]

A phase II/III clinical trial that was also completed in the United States in 2013 examined the effects of FES cycling on the blood and spinal cord in patients with SCI (NCT01217047). This trial also aimed to optimize the usage of a device called the RT300-SL Cycle Ergometer as a potential method for functional recovery. Participants were subjected to varying durations of FES cycling. Participants also had to undergo lumbar punctures at the beginning and end of the trial, as well as daily mood assessments. The primary outcomes for this trial were to determine BDNF levels in the CSF before and after FES cycling. The inclusion criteria for this trial consisted of having an AIS grade A traumatic SCI at any level with injury occurring at least 6 months prior to the onset of the study (NCT01217047). This study has been terminated, and no results have been published to date.

Another trial based in the United Kingdom is assessing the efficacy of FES cycling in acute SCI patients with paraplegia or tetraplegia (NCT04064385). This trial consists of two groups: the experimental group where participants undergo FES cycling and a control group. The primary outcome measure for this clinical trial is to observe changes for up to 12 weeks in the Spinal Cord Injury Functional Ambulation Inventory (SCI-FAI), which is a scale used to measure walking ability. The inclusion criteria for this clinical trial comprises of having an incomplete SCI due to trauma, stroke, or surgery that is nonprogressive with an AIS grade B, C, or D occurring within 6 weeks prior to study enrollment (NCT04064385). The trial is recruiting, although the expected end date is unknown.

The final FES cycling trial began in August 2018 at the Swiss Paraplegic Centre Nottwil in Switzerland and wanted to elucidate the effectiveness of two different methods of FES leg exercises in patients with subacute SCI, while also optimizing the treatments (NCT03621254). This two-armed study consists of participants being randomized into one of the two methods. The first method called HI-SHORT consists of high intensity exercise for a short period of time, where participants complete 10×2 min exercises with $1-2$ min rest in between each session. The second method called LO-LONG consists of low- to moderate-intensity exercise that is continuous. Patients who are a part of the LO-LONG group will undergo at least 20 min of continuous exercise. Participants will use the Hasomed RehaStim exercise ergometer $3-4$ days a week for $6-8$ weeks. The primary outcome measure for this particular trial is to determine if there is a change in aerobic fitness before and after the trial by measuring peak oxygen uptake. The inclusion criteria for this trial consist of patients with a C7-T10 SCI at $6-12$ weeks postinjury with an AIS grade A or B. This trial is estimated to be completed in August 2022 (NCT03621254).

Functional electrical stimulation in combination with other medical devices

A collaboration between the Miami Project to Cure Paralysis, Sheba Medical Centre, and the Kessler Institute for Rehabilitation is currently conducting a clinical trial to determine the safety and effectiveness of a medical device called BQ 1.2 in patients with incomplete chronic SCI at the cervical level (NCT04050696). BQ 1.2 is a brain—computer interface that is used for motor rehabilitation as it creates low-frequency, low-intensity electromagnetic fields that stimulate the patient's CNS. The primary objective for this trial is to determine how effective the BQ 1.2 device is in improving the function of the higher extremities in chronic SCI patients. This one-armed study involves patients who will receive this BQ treatment with physical therapy. The primary outcome measure for this trial is to observe changes in the Graded and Redefined Assessment of Strength, Sensibility and Prehension (GRASSP) at 22 weeks to assess upper limb functionality. The inclusion criteria involve patients with chronic cervical SCI between $C1-C8$ with an AIS grade of $B-D$, occurring $12-30$ months prior to the onset of the trial (NCT04050696). This trial is currently recruiting and is expected to be completed by July 2021. All clinical trials are summarized in Tables 26.1—26.5.

TABLE 26.1 Clinical trails with unspecified phase (N/A).

Phase	NCT number	Other identifiers	Name of the trial	Sponsor	Treatment or cell source	Severity of SCI	Countries of testing	Completion status	Completion or estimated completion date	Notes
N/A	NCT02138110	INSPIRE InVivo-100-101	The INSPIRE study: probable benefit of the neuro-spinal scaffold for treatment of AIS a thoracic acute spinal cord injury	InVivo Therapeutics	Poly (lactic-co-glycolic acid)-b-poly(ʟ-lysine) (PLGA-PLL) scaffold (Neuro-Spinal scaffold) implantation	Acute thoracic (T2–T12) SCI of AIS grade A	The United States	Active, not recruiting	August 2024	Humanitarian device exemption (HDE) study
N/A	NCT03762655	INSPIRE 2 InVivo-100-101	Study of probable benefit of the neuro-Spinal scaffold in subjects with complete thoracic AIS a spinal cord injury as compared to standard of care (INSPIRE 2)	InVivo Therapeutics	Standards of care open spine surgery with or without poly (lactic-co-glycolic acid)-b-poly(ʟ-lysine) (PLGA-PLL) scaffold (Neuro-Spinal scaffold) implantation	Acute thoracic (T2–T12) SCI of AIS grade A	The United States	Recruiting	July 2028	Humanitarian device exemption (HDE) study
N/A	NCT03517787	Spinal	The effect of FES and therapeutic exercise on SCI patients skeletal muscles, sitting balance, posture and quality of life	Tartu University Hospital	Therapeutic exercise with or without functional electrical stimulation (FES)	Chronic cervical SCI resulting in tetraplegia or Healthy adult volunteers	Estonia	Unknown	May 2020	N/A
N/A	NCT03048331	2015–06	Functional electrical stimulation (FES) and reconstructive tetraplegia hand and arm surgery (FES)	Swiss Paraplegic Centre, Nottwil	Reconstructive arm and/or hand surgery followed by standard physiotherapy and occupational therapy Some patients will also receive functional electrical stimulation (FES) prior to and following surgery	C4–T1 SCI (>6 months) of AIS grades A–D	Switzerland	Active, not recruiting	December 2020	N/A

Continued

TABLE 26.1 Clinical trails with unspecified phase (N/A).—cont'd

Phase	NCT number	Other identifiers	Name of the trial	Sponsor	Treatment or cell source	Severity of SCI	Countries of testing	Completion status	Completion or estimated completion date	Notes
N/A	NCT00245726	SHC-8540	Functional electrical stimulation (FES) cycling for children with spinal cord injuries (SCI)	Shriners Hospitals for Children	Lower extremity cycling with functional electrical stimulation (FES) versus passive (motor assist) leg cycling versus FES of the lower extremities	Cervical or thoracic SCI at 12 months postinjury with an AIS grade A or B	The United States	Completed	June 2008	N/A
N/A	NCT04064385	GN19NE123	Functional electrical stimulation cycling in SCI	Glasgow Caledonian University	Cycling with functional electrical stimulation (FES) and usual physiotherapy care Versus Usual physiotherapy care alone	Acute SCI of AIS grades B–D	The United Kingdom	Recruiting	July 2020	N/A
N/A	NCT03621254	2018–11	Investigation into optima l FES training characteristics after sub-acute spinal cord injury	Swiss Paraplegic Centre, Nottwil	Cycling with functional electrical stimulation (FES) with varied intensity of exercise	Subacute SCI between C7 and T10 of AIS grade A or B	Switzerland	Active, not recruiting	August 2022	N/A
N/A	NCT04050696	BQ4	The use of electromagnetic field (EMF) treatment in chronic spinal cord injury (SCI) patients	BrainQ Technologies Ltd.	BQ 1.2 electromagnetic field (EMF) with physical therapy (PT)	Chronic cervical SCI between C1 and C8 with AIS grades B–D	The United States and Israel	Not yet recruiting	September 2020	N/A
N/A	NCT04278235	C20-203 C20-284 C20-285	Expanded access to Elezanumab	AbbVie	Elezanumab (ABT-555)	N/A	N/A	Available for expanded access	N/A	Not a clinical trial, expanded access to treatment for patients who are not eligible for phase 2 trial

TABLE 26.2 Phase I clinical trials.

Phase	NCT number	Other identifiers	Name of the trial	Sponsor	Treatment or cell source	Severity of SCI	Countries of testing	Completion status	Completion or estimated completion date
I	NCT02352077	CAS-XDA-SCI/IGDB	NeuroRegen scaffold with stem cells for chronic spinal cord injury repair	Chinese Academy of Sciences	NeuroRegen scaffold transplant with either: mesenchymal cells or bone marrow mononuclear cells As well as comprehensive rehabilitation, psychological, and nutritional measures	Chronic SCI between C5 and T12 of AIS grade A	China	Enrolling by Invitation	December 2020
I	NCT02510365	CAS-XDA-ACSCI/IGDB	Functional neural regeneration collagen scaffold transplantation in acute spinal cord injury patients	Chinese Academy of Sciences	Functional neural regeneration collagen scaffold transplant As well as comprehensive rehabilitation and psychological measures	Acute SCI between C5 and T12 of AIS grade A	China	Recruiting	December 2020

TABLE 26.3 Phase I/II and I/IIa clinical trials.

Phase	NCT number	Other identifiers	Name of the trial	Sponsor	Treatment or cell source	Severity of SCI	Countries of testing	Completion status	Completion or estimated completion date	Primary outcome measures	Notes
I/II	NCT03989440	RNX-AX204-101	AXER-204 in participants with chronic spinal cord injury (RESET)	ReNetX Bio Inc.	Nogo-A trap human fusion protein (AXER-204) versus phosphate buffered saline (PBS) placebo	Chronic cervical SCI of AIS grade A	The United States	Recruiting	December 2021	Frequency of adverse events as well as variation in ISNCSCI-UEMS score and concentration of AXER-204	N/A
I/II	NCT00559494	17,007 PVA2414	Minocycline and perfusion pressure augmentation in acute spinal cord injury	University of Calgary	Minocycline versus placebo	Motor complete or incomplete acute SCI between C0 and T11	Canada	Completed	August 2010	Effectiveness, compliance as well as frequency of adverse events of protocol	N/A
I/II	NCT02193334	KP-100-ND002	Phase I/II study of KP-100IT in acute spinal cord injury	Kringle Pharma Inc.	Hepatocyte growth factor (KP-100IT) versus placebo	Acute cervical SCI below C3 of Frankel scale grade A, B1, or B2	Japan	Completed	July 2018	Variation in ASIA motor score as well as frequency and severity of adverse effects	N/A
I/II	NCT02688049	CAS-XDA-MNSCI/IGDB	NeuroRegen scaffold combined with stem cells for chronic spinal cord injury repair	Chinese Academy of Sciences	NeuroRegen scaffold transplant with 10 million cells of either: Mesenchymal cells or neural stem cells As well as comprehensive rehabilitation, psychological, and nutritional measures	Chronic SCI between C5 and T12 of AIS grade A	China	Enrolling by Invitation	December 2020	Amelioration in ASIA grade, SSEP, and MEP	N/A
I/II	NCT02688062	CAS-XDA-SDSCI/IGDB	NeuroRegen Scaffold with bone marrow mononuclear cells transplantation versus intradural decompression and adhesiolysis in SCI	Chinese Academy of Sciences	NeuroRegen scaffold transplant with bone marrow mononuclear cells versus surgical intradural decompression and adhesiolysis As well as comprehensive rehabilitation, psychological, and nutritional measures	Chronic thoracic SCI of AIS grade A	China	Enrolling by Invitation	December 2020	Amelioration in ASIA grade, SSEP, and MEP	N/A

| I/II | NCT03966794 | CAS-XDA-EESCI/IGDB | Functional scaffold transplantation combined with epidural electrical stimulation for spinal cord injury repair | Chinese Academy of Sciences | Epidural electrical stimulation versus functional neural regeneration collagen scaffold transplant with epidural electrical stimulation | Acute or chronic SCI between C4 and T12/L1 of AIS grade A or previous complete functional scaffold transplant in a prior trial with motor functional recovery perceived | China | Recruiting | December 2021 | Variation in ASIA grade and occurrence of adverse effects to determine safety | Epidural electrical stimulation only group will have obtained the functional neural regeneration collagen scaffold in a previous clinical trial |
| I/IIa | NCT00500812 | BA-210-101 | A safety study for cethrin (BA-210) in the treatment of acute thoracic and cervical spinal cord injuries | Vertex Pharmaceuticals Incorporated | BA-210 (Cethrin) | Complete thoracic (T2–T12) or cervical (C4–T1) SCI with ASIA grade A | Canada and the United States | Completed | February 2009 | Safety and tolerability of cethrin | N/A |

TABLE 26.4 Phase II clinical trials.

Phase	NCT number	Other identifiers	Name of the trial	Sponsor	Treatment or cell source	Severity of SCI	Countries of testing	Completion status	Completion or estimated completion date	Primary outcome Measures	Notes
II	NCT03935321	NISCI trial EudraCT No. 2016,-001227,-31	NISCI - Nogo inhibition in spinal cord injury (NISCI)	University of Zurich	Anti-Nogo-A antibody (NG-101) versus placebo	Acute cervical (C1–C8) SCI of AIS grades A–D	Czechia, Germany, Italy, Spain, and Switzerland	Recruiting	September 2023	Variation in ISNSCSI-UEMS scores	N/A
II	NCT01750684	ACPM-SI-1009	AC105 in patients with acute traumatic spinal cord injury (AC105)	Acorda Therapeutics	AC105 versus saline placebo	Acute SCI between C4 and T11 of AIS grade A, B, or C	The United States	Terminated	N/A	Frequency of adverse events	Terminated due to low enrollment
II	NCT01502631	ASBI 603	Study to evaluate the efficacy, safety, and pharmacokinetics of SUN13837 injection in adult subjects with acute spinal cord injury (ASCI)	Daiichi Sankyo Inc.	SUN13837 versus placebo	Acute cervical SCI: Between C4 and C7 with AIS grade A Between C3 and C8 with AIS grade B or C	The United States, Canada, Czech Republic, France, Poland, Spain, and the United Kingdom	Completed	March 2015	Assessing SUN13837 treated patients in comparison with placebo patients	N/A
II	NCT04054414	PMZ-1620/ CLINICAL-2.3/2017 CTRI/2018/ 12/016,667 (Registry Identifier: Clinical Trials Registry - India)	PMZ-1620 (Sovateltide) in patients of acute spinal cord injury	Pharmazz Inc.	Standard support care along with: Sovateltide (PMZ-1620) or saline control	Acute SCI between C5 and C8, T1 and T12, L1 and L5, or S1 and S5 with an ISNCSCI impairment grade B, C, or D	India	Recruiting	October 2020	Frequency of adverse events and tolerability of treatment (PMZ-1620)	N/A
II	NCT04295538	ELASCI M16-077 2019,-003,752,-36	Safety and efficacy study of intravenous (IV) administration of elezanumab to assess change in upper extremity motor score (UEMS) in adult participants with acute traumatic cervical spinal cord injury (SCI) (ELASCI)	AbbVie	Elezanumab (ABT-555)	Acute traumatic cervical SCI between C4 and C7 within 24 h after injury	The United States, Australia, Canada, Japan, Spain	Recruiting	March 2023	Upper extremity motor score	Also available for expanded access for those not eligible for this trial

TABLE 26.5 Phase I/II and III clinical trials.

Phase	NCT number	Other identifiers	Name of the trial	Sponsor	Treatment or cell source	Severity of SCI	Countries of testing	Completion status	Completion or estimated completion date	Primary outcome measures
II/III	NCT01597518	RISCIS SPN-12-001	Riluzole in spinal cord injury study (RISCIS)	AOSpine North America Research Network	Riluzole versus placebo	Acute cervical (C4–C8) SCI of ISNCSCI grade A, B, or C	The United States, Australia, Canada, and New Zealand	Active, not recruiting	May 2025	Variation in ISCSCI total motor score
III	NCT01828203	MASC RHI-1005	Minocycline in acute spinal cord injury (MASC)	Rick Hansen Institute	Surgical spinal cord decompression followed by: Minocycline versus Saline placebo	Acute cervical (C0–C8) SCI	Australia and Canada	Unknown	June 2018	Variation in ASIA motor scores
III	NCT01292811	10–048	Restoration of upper limb function in individuals with sub-acute spinal cord injury (MCRCT)	Toronto Rehabilitation Institute	Functional electrical stimulation (FES) using a complex motion or HEWHS stimulator versus conventional occupational therapy	Subacute incomplete cervical SCI between C4 and C7	Canada	Unknown	January 2018	Assessment of burden of care using the functional independence measure
II/III	NCT01217047	CURE-SCI NA_00036348	Study on functional electrical stimulation (FES) cycling following spinal cord injury (CURE-SCI)	Hugo W. Moser Research Institute at Kennedy Krieger, Inc.	Cycling with or without functional electrical stimulation (FES) of the lower extremities This will also involve lumbar punctures and mood assessments	SCI of at least 6 months prior to the study with an AIS grade A	The United States	Completed	August 2013	BDNF levels in CSF during various timepoints
IIb/III	NCT02669849	SPRING Trial VX15-210-101	Study to assess the efficacy and safety of VX-210 in subjects with acute traumatic cervical spinal cord injury	Vertex Pharmaceuticals Inc.	VX-210 (Cethrin)	Acute cervical SCI within 3 days after injury with ASIA grade A or B	Canada and the United States	Terminated	January 2020	Upper extremity motor score

Conclusion

Most of the noncellular-based therapies addressed in this chapter had preliminary evidence for the safety, efficacy, and feasibility of these treatments for SCI patients. However, the majority of the trials discussed are still ongoing; thus, results are not yet published.

As SCI is a complex disorder and its impact is felt from multiple levels, it is important to acknowledge these effects and to treat them holistically.[79] Small molecules or other targeted treatments may not be adept in alleviating the full range of deficits.[80] Thus, using a combinatory approach, specifically combining noncellular approaches with cellular transplantation approaches may be an optimal treatment method for SCI.[80] This can include using stem cell therapies in combination with noncellular therapies such as biomaterial to act as a graft to replenish the extracellular matrix. This has been observed in a few clinical trials such as the NeuroRegen Collagen Scaffold, which has demonstrated the safety and feasibility of this approach, thus showing promise for a clinical therapeutic for SCI (NCT02688049, NCT02352077, NCT02688062, NCT02510365, NCT03966794). Using such an approach enables us to address the factors in the pathobiology of SCI such as neuronal cell loss and glial cell activation.[1] Furthermore, considering the pathobiological phases of SCI and their associated activation of secondary cascade mechanisms from the acute, subacute to chronic timepoints,[1] developing treatments would need to consider the safety and efficacy at those different severities of SCI.

The heterogenous nature of SCI in humans may contribute to some of the challenges faced in acute clinical trial administration. This poses significant challenges in trials with broad inclusion criteria, resulting in the inability to detect positive changes for the treatment in question.[81] However, using stringent inclusion criteria may prevent the enrollment of enough patients for a statistically powered trial within a reasonable time frame.[81] Furthermore, preliminary animal models of SCI are often assessed in a very controlled environment, preventing its generalizability to human patient populations.[81]

Another limitation of SCI clinical trials is the often subjective measures to assess SCI severity and prognosis by using tools such as the AIS and arbitrary time frames that may not reflect individual variability. Hence, we need to develop more objective methods to evaluate the severity of SCI, which can include biomarkers and quantitative imaging.[81]

Accounting for the heterogenous nature of SCI and developing more objective measures of SCI severity will allow for better treatment development in regenerative medicine to appropriately target individual variability in SCI. Lastly, combining noncellular regenerative therapies with cell transplantation may allow for optimal targeting of the full spectrum of cellular damage in SCI.[81]

References

1. Ahuja CS, Wilson JR, Nori S, Kotter MRN, Druschel C, Curt A, Fehlings MG. Traumatic spinal cord injury. *Nat Rev Dis Prim* 2017;3(1):17018. https://doi.org/10.1038/nrdp.2017.18.
2. Sabelström H, Stenudd M, Frisén J. Neural stem cells in the adult spinal cord. *Exp Neurol* 2014;260:44—9. https://doi.org/10.1016/j.expneurol.2013.01.026.
3. U.S. Food and Drug Administration (FDA). *The drug development process — step 3: clinical research.* January 4, 2018. Retrieved May 7, 2020, from https://www.fda.gov/patients/drug-development-process/step-3-clinical-research.
4. GSK Canada (n.d.). Clinical trial phases. Retrieved May 7, 2020, from https://ca.gsk.com/en-ca/research/trials-in-people/clinical-trial-phases/.
5. U.S. Food and Drug Administration (FDA). *Humanitarian device exemption.* FDA; September 5, 2019. https://www.fda.gov/medical-devices/premarket-submissions/humanitarian-device-exemption.
6. Kirshblum SC, Burns SP, Biering-Sorensen F, Donovan W, Graves DE, Jha A, Johansen M, Jones L,

Krassioukov A, Mulcahey MJ, Schmidt-Read M, Waring W. International standards for neurological classification of spinal cord injury (Revised 2011). *J Spinal Cord Med* 2011;**34**(6):535−46. https://doi.org/10.1179/204577211X13207446293695.

7. Roberts TT, Leonard GR, Cepela DJ. Classifications in brief: American Spinal Injury Association (ASIA) impairment scale. *Clin Orthop Relat Res* 2017;**475**(5):1499−504. https://doi.org/10.1007/s11999-016-5133-4.

8. American Spinal Injury Association (ASIA). *New) ISNCSCI 2019 revision released*. American Spinal Injury Association; 2019. https://asia-spinalinjury.org/isncsci-2019-revision-released/.

9. Chen MS, Huber AB, van der Haar ME, Frank M, Schnell L, Spillmann AA, Christ F, Schwab ME. Nogo-A is a myelin-associated neurite outgrowth inhibitor and an antigen for monoclonal antibody IN-1. *Nature* 2000;**403**:434−9.

10. Fournier AE, GrandPre T, Strittmatter SM. Identification of a receptor mediating Nogo-66 inhibition of axonal regeneration. *Nature* 2001;**409**:341−6.

11. Liu J, Gao HY, Wang XF. The role of the Rho/ROCK signaling pathway in inhibiting axonal regeneration in the central nervous system. *Neur Regen Res* 2015;**10**(11):1892−6. https://doi.org/10.4103/1673-5374.170325.

12. Freund P, Wannier T, Schmidlin E, Bloch J, Mir A, Schwab ME, Rouiller EM. Anti-Nogo-A antibody treatment enhances sprouting of corticospinal axons rostral to a unilateral cervical spinal cord lesion in adult macaque monkey. *J Comp Neurol* 2007;**502**:644−59.

13. Kucher K, Johns D, Maier D, Abel R, Badke A, Baron H, Thietje R, Casha S, Meindl R, Gomez-Mancilla B, Pfister C, Rupp R, Weidner N, Mir A, Schwab ME, Curt A. First-in-man intrathecal application of neurite growth-promoting anti-Nogo-A antibodies in acute spinal cord injury. *Neurorehab Neural Rep* 2018;**32**(6−7):578−89. https://doi.org/10.1177/1545968318776371.

14. Mothe AJ, Tassew NG, Shabanzadeh AP, Penheiro R, Vigouroux RJ, Huang L, Grinnell C, Cui Y-F, Fung E, Monnier PP, Mueller BK, Tator CH. RGMa inhibition with human monoclonal antibodies promotes regeneration, plasticity and repair, and attenuates neuropathic pain after spinal cord injury. *Sci Rep* 2017;**7**(1):10529. https://doi.org/10.1038/s41598-017-10987-7.

15. AbbVie. *AbbVie receives orphan drug and fast track designations from the U.S. Food and drug administration for elezanumab, an investigational monoclonal antibody RGMa inhibitor, for the treatment of spinal cord injury.* September 28, 2020. https://news.abbvie.com/news/press-releases/abbvie-receives-orphan-drug-and-fast-track-designations-from-us-food-and-drug-administration-for-elezanumab-an-investigational-monoclonal-antibody-rgma-inhibitor-for-treatment-spinal-cord-injury.htm.

16. Schwartz G, Fehlings MG. Evaluation of the neuroprotective effects of sodium channel blockers after spinal cord injury: improved behavioral and neuroanatomical recovery with riluzole. *J Neurosurg* 2001;**94**:245−56.

17. Wu Y, Satkunendrarajah K, Teng Y, Chow DS, Buttigieg J, Fehlings MG. Delayed post-injury administration of riluzole is neuroprotective in a preclinical rodent model of cervical spinal cord injury. *J Neurotrauma* 2013;**30**:441−52.

18. Bensimon G, Lacomblez L, Delumeau JC, Bejuit R, Truffinet P, Meininger V, Riluzole ALSSGII. A study of riluzole in the treatment of advanced stage or elderly patients with amyotrophic lateral sclerosis. *J Neurol* 2002;**249**:609−15.

19. Palmer GC. Neuroprotection by NMDA receptor antagonists in a variety of neuropathologies. *Curr Drug Targets* 2001;**2**:241−71.

20. Kwon BK, Sekhon LH, Fehlings MG. Emerging repair, regeneration, and translational research advances for spinal cord injury. *Spine* 2010;**35**:S263−70.

21. Kaptanoglu E, Beskonakli E, Okutan O, Selcuk Surucu H, Taskin Y. Effect of magnesium sulphate in experimental spinal cord injury: evaluation with ultrastructural findings and early clinical results. *J Clin Neurosci* 2003;**10**:329−34.

22. Kaptanoglu E, Beskonakli E, Solaroglu I, Kilinc A, Taskin Y. Magnesium sulfate treatment in experimental spinal cord injury: emphasis on vascular changes and early clinical results. *Neurosurg Rev* 2003;**26**:283−7.

23. Suzer T, Coskun E, Islekel H, Tahta K. Neuroprotective effect of magnesium on lipid peroxidation and axonal function after experimental spinal cord injury. *Spinal Cord* 1999;**37**:480−4.

24. Stirling DP, Koochesfahani KM, Steeves JD, Tetzlaff W. Minocycline as a neuroprotective agent. *Neuroscientist: Rev J Bring Neurobiol Neurol Psychiat* 2005;**11**:308−22.

25. Wells JE, Hurlbert RJ, Fehlings MG, Yong VW. Neuroprotection by minocycline facilitates significant recovery from spinal cord injury in mice. *Brain J Neurol* 2003;**126**:1628−37.

26. Stirling DP, Khodarahmi K, Liu J, McPhail LT, McBride CB, Steeves JD, Ramer MS, Tetzlaff W. Minocycline treatment reduces delayed oligodendrocyte death, attenuates axonal dieback, and improves functional outcome after spinal cord injury. *J Neurosci* 2004;**24**:2182−90.

27. Hausmann ON. Post-traumatic inflammation following spinal cord injury. *Spinal Cord* 2003;**41**(7):369−78. https://doi.org/10.1038/sj.sc.3101483.

28. Mattson MP, Lovell MA, Furukawa K, Markesbery WR. Neurotrophic factors attenuate glutamate-induced accumulation of peroxides, elevation of intracellular Ca^{2+} concentration, and neurotoxicity and increase antioxidant enzyme activities in hippocampal neurons. *J Neurochem* 1995;**65**(4):1740–51. https://doi.org/10.1046/j.1471-4159.1995.65041740.x.

29. Teng YD, Mocchetti I, Taveira-DaSilva AM, Gillis RA, Wrathall JR. Basic fibroblast growth factor increases long-term survival of spinal motor neurons and improves respiratory function after experimental spinal cord injury. *J Neurosci* 1999;**19**:7037–47.

30. Tsai, M.-C., Shen, L.-F., Kuo, H.-S., Cheng, H., & Chak, K.-F. (n.d.). Involvement of acidic Fibroblast growth factor in spinal cord injury repair processes revealed by a proteomics approach. S. 20.

31. Teng YD, Mocchetti I, Wrathall JR. Basic and acidic fibroblast growth factors protect spinal motor neurones in vivo after experimental spinal cord injury. *Eur J Neurosci* 1998;**10**:798–802.

32. Mocchetti I, Wrathall JR. Neurotrophic factors in central nervous system trauma. *J Neurotrauma* 1995;**12**(5): 853–70. https://doi.org/10.1089/neu.1995.12.853.

33. Rabchevsky AG, Fugaccia I, Fletcher-Turner A, Blades DA, Mattson MP, Scheff SW. Basic fibroblast growth factor (bFGF) enhances tissue sparing and functional recovery following moderate spinal cord injury. *J Neurotrauma* 1999;**16**(9):817–30. https://doi.org/10.1089/neu.1999.16.817.

34. Kawabe J, Koda M, Hashimoto M, Fujiyoshi T, Furuya T, Endo T, Okawa A, Yamazaki M. Neuroprotective effects of granulocyte colony-stimulating factor and relationship to promotion of angiogenesis after spinal cord injury in rats: laboratory investigation. *J Neurosurg Spine* 2011;**15**:414–21.

35. Koda M, Nishio Y, Kamada T, Someya Y, Okawa A, Mori C, Yoshinaga K, Okada S, Moriya H, Yamazaki M. Granulocyte colony-stimulating factor (G-CSF) mobilizes bone marrow-derived cells into injured spinal cord and promotes functional recovery after compression-induced spinal cord injury in mice. *Brain Res* 2007;**1149**:223–31.

36. Kadota R, Koda M, Kawabe J, Hashimoto M, Nishio Y, Mannoji C, Miyashita T, Furuya T, Okawa A, Takahashi K, Yamazaki M. Granulocyte colony-stimulating factor (G-CSF) protects oligodendrocyte and promotes hindlimb functional recovery after spinal cord injury in rats. *PLoS One* 2012;**7**(11):e50391. https://doi.org/10.1371/journal.pone.0050391.

37. Nishio Y, Koda M, Kamada T, Someya Y, Kadota R, Mannoji C, Miyashita T, Okada S, Okawa A, Moriya H, Yamazaki M. Granulocyte colony-stimulating factor attenuates neuronal death and promotes functional recovery after spinal cord injury in mice. *J Neuropathol Exp Neurol* 2007;**66**(8):724–31. https://doi.org/10.1097/nen.0b013e3181257176.

38. Casella GT, Marcillo A, Bunge MB, Wood PM. New vascular tissue rapidly replaces neural parenchyma and vessels destroyed by a contusion injury to the rat spinal cord. *Exp Neurol* 2002;**173**(1):63–76. https://doi.org/10.1006/exnr.2001.7827.

39. Loy DN, Crawford CH, Darnall JB, Burke DA, Onifer SM, Whittemore SR. Temporal progression of angiogenesis and basal lamina deposition after contusive spinal cord injury in the adult rat. *J Comp Neurol* 2002;**445**(4): 308–24. https://doi.org/10.1002/cne.10168.

40. Tator CH, Fehlings MG. Review of the secondary injury theory of acute spinal cord trauma with emphasis on vascular mechanisms. *J Neurosurg* 1991;**75**(1):15–26. https://doi.org/10.3171/jns.1991.75.1.0015.

41. Inada T, Takahashi H, Yamazaki M, Okawa A, Sakuma T, Kato K, Hashimoto M, Hayashi K, Furuya T, Fujiyoshi T, Kawabe J, Mannoji C, Miyashita T, Kadota R, Someya Y, Ikeda O, Hashimoto M, Suda K, Kajino T, Ueda H, Ito Y, Ueta T, Hanaoka H, Takahashi K, Koda M. Multicenter prospective nonrandomized controlled clinical trial to prove neurotherapeutic effects of granulocyte colony-stimulating factor for acute spinal cord injury: analyses of follow-up cases after at least 1 year. *Spine* 2014;**39**:213–9.

42. Koda M, Hanaoka H, Sato T, Fujii Y, Hanawa M, Takahashi S, Furuya T, Ijima Y, Saito J, Kitamura M, Ohtori S, Matsumoto Y, Abe T, Watanabe K, Hirano T, Ohashi M, Shoji H, Mizouchi T, Takahashi I, Kawahara N, Kawaguchi M, Orita Y, Sasamoto T, Yoshioka M, Fujii M, Yonezawa K, Soma D, Taneichi H, Takeuchi D, Inami S, Moridaira H, Ueda H, Asano F, Shibao Y, Aita I, Takeuchi Y, Mimura M, Shimbo J, Someya Y, Ikenoue S, Sameda H, Takase K, Ikeda Y, Nakajima F, Hashimoto M, Ozawa T, Hasue F, Fujiyoshi T, Kamiya K, Watanabe M, Katoh H, Matsuyama Y, Yamamoto Y, Togawa D, Hasegawa T, Kobayashi S, Yoshida G, Oe S, Banno T, Arima H, Akeda K, Kawamoto E, Imai H, Sakakibara T, Sudo A, Ito Y, Kikuchi T, Osaki S, Tanaka N, Nakanishi K, Kamei N, Kotaka S, Baba H, Okudaira T, Konishi H, Yamaguchi T, Ito K, Katayama Y, Matsumoto T, Matsumoto T, Idota M, Kanno H, Aizawa T, Hashimoto K, Eto T, Sugaya T, Matsuda M, Fushimi K, Nozawa S, Iwai C, Taguchi T, Kanchiku T, Suzuki H, Nishida N, Funaba M, Yamazaki M. Study protocol for the G-SPIRIT trial: a randomised, placebo-controlled, double-blinded phase III trial of granulocyte colony-stimulating factor-mediated neuroprotection for acute spinal cord injury. *BMJ Open* 2018;**8**:e019083.

43. Kitamura K, Fujiyoshi K, Yamane J, Toyota F, Hikishima K, Nomura T, Funakoshi H, Nakamura T, Aoki M, Toyama Y, Okano H, Nakamura M. Human hepatocyte growth factor promotes functional recovery in primates after spinal cord injury. *PLoS One* 2011;**6**: e27706.

44. Kitamura K, Iwanami A, Nakamura M, Yamane J, Watanabe K, Suzuki Y, Miyazawa D, Shibata S, Funakoshi H, Miyatake S, Coffin RS, Nakamura T, Toyama Y, Okano H. Hepatocyte growth factor promotes endogenous repair and functional recovery after spinal cord injury. *J Neurosci Res* 2007;**85**:2332—42.

45. Kitamura K, Iwanami A, Iwai H, Toyama Y, Matsumoto M, Okano H, Nakamura M. Therapeutic time window and preclinical efficacy of intrathecal administration of recombinant human hepatocyte growth factor for acute spinal cord injury. *J Spine Res* 2016;**7**:934—9.

46. Nakamura T, Nishizawa T, Hagiya M, Seki T, Shimonishi M, Sugimura A, Tashiro K, Shimizu S. Molecular cloning and expression of human hepatocyte growth factor. *Nature* 1989;**342**:440—3.

47. Forgione N, Fehlings MG. Rho-ROCK inhibition in the treatment of spinal cord injury. *World Neurosurg* 2014; **82**(3—4):e535—9. https://doi.org/10.1016/j.wneu. 2013.01.009.

48. Fehlings MG, Kim KD, Aarabi B, Rizzo M, Bond LM, McKerracher L, Vaccaro AR, Okonkwo DO. Rho inhibitor VX-210 in acute traumatic subaxial cervical spinal cord injury: design of the SPinal cord injury Rho INhibition InvestiGation (SPRING) clinical trial. *J Neurotrauma* 2018;**35**(9):1049—56. https://doi.org/10.1089/ neu.2017.5434.

49. Monnier PP, Sierra A, Schwab JM, Henke-Fahle S, Mueller BK. The Rho/ROCK pathway mediates neurite growth-inhibitory activity associated with the chondroitin sulfate proteoglycans of the CNS glial scar. *Mol Cell Neurosci* 2003;**22**(3):319—30. https://doi.org/ 10.1016/S1044-7431(02)00035-0.

50. Lord-Fontaine S, Yang F, Diep Q, Dergham P, Munzer S, Tremblay P, McKerracher L. Local inhibition of Rho signaling by cell-permeable recombinant protein BA-210 prevents secondary damage and promotes functional recovery following acute spinal cord injury. *J Neurotrauma* 2008;**25**(11):1309—22. https://doi.org/ 10.1089/neu.2008.0613.

51. Fehlings MG, Theodore N, Harrop J, Maurais G, Kuntz C, Shaffrey CI, Kwon BK, Chapman J, Yee A, Tighe A, McKerracher L. A phase I/IIa clinical trial of a recombinant Rho protein antagonist in acute spinal cord injury. *J Neurotrauma* 2011;**28**(5):787—96. https:// doi.org/10.1089/neu.2011.1765.

52. Fehlings MG, Chen Y, Aarabi B, Ahmad F, Anderson KD, Dumont T, Fourney DR, Harrop JS, Kim KD, Kwon BK, Lingam HK, Rizzo M, Shih LC, Tsai EC, Vaccaro A, McKerracher L. A randomized controlled trial of local delivery of a Rho inhibitor (VX-210) in patients with acute traumatic cervical spinal cord injury. *J Neurotrauma* 2021:7096. https://doi.org/ 10.1089/neu.2020.7096.

53. Ranjan A, Briyal S, Hussain M, Giometti A, Gulati A. 746: augmentation of mitochondrial biogenesis and neural regeneration in an ischemic brain by sovateltide. *Crit Care Med* 2020;**48**(1).

54. Gulati A, Agrawal N, Vibha D, Misra UK, Paul B, Kumar R, Pandian J, Varshney S, Lavhale M. 35: a phase II study to determine efficacy of sovateltide in patients with cerebral ischemic stroke. *Crit Care Med* 2020;**48**(1).

55. Lin H, Chen B, Wang B, Zhao Y, Sun W, Dai J. Novel nerve guidance material prepared from bovine aponeurosis. *J Biomed Mater Res A* 2006;**79A**(3):591—8. https://doi.org/10.1002/jbm.a.30862.

56. Fan J, Xiao Z, Zhang H, Chen B, Tang G, Hou X, Ding W, Wang B, Zhang P, Dai J, Xu R. Linear ordered collagen scaffolds loaded with collagen-binding Neurotrophin-3 promote axonal regeneration and partial functional recovery after complete spinal cord transection. *J Neurotrauma* 2010;**27**(9):1671—83. https://doi.org/10.1089/neu.2010.1281.

57. Li X, Han J, Zhao Y, Ding W, Wei J, Han S, Shang X, Wang B, Chen B, Xiao Z, Dai J. Functionalized collagen scaffold neutralizing the myelin-inhibitory molecules promoted neurites outgrowth in vitro and facilitated spinal cord regeneration in vivo. *ACS Appl Mater Interfaces* 2015;**7**(25):13960—71. https://doi.org/10.1021/ acsami.5b03879.

58. Bozkurt A, Boecker A, Tank J, Altinova H, Deumens R, Dabhi C, Tolba R, Weis J, Brook GA, Pallua N, van Neerven SGA. Efficient bridging of 20 mm rat sciatic nerve lesions with a longitudinally micro-structured collagen scaffold. *Biomaterials* 2015;**75**:112—22. https:// doi.org/10.1016/j.biomaterials.2015.10.009.

59. Han Q, Jin W, Xiao Z, Ni H, Wang J, Kong J, Wu J, Liang W, Chen L, Zhao Y, Chen B, Dai J. The promotion of neural regeneration in an extreme rat spinal cord injury model using a collagen scaffold containing a collagen binding neuroprotective protein and an EGFR neutralizing antibody. *Biomaterials* 2010;**31**(35): 9212—20. https://doi.org/10.1016/j.biomaterials. 2010.08.040.

60. Xiao Z, Tang F, Zhao Y, Han G, Yin N, Li X, Chen B, Han S, Jiang X, Yun C, Zhao C, Cheng S, Zhang S, Dai J. Significant improvement of acute complete spinal cord injury patients diagnosed by a combined criteria implanted with NeuroRegen scaffolds and mesenchymal stem cells. *Cell Transplant* 2018;**27**(6):907—15. https://doi.org/10.1177/0963689718766279.

61. Musienko PE, Pavlova NV, Selionov VA, Gerasimenko YP. Locomotion induced by epidural stimulation in a decerebrated cat after spinal cord injury. *Biophysics* 2009;**54**(2):208—13. https://doi.org/ 10.1134/S000635090902016X.

62. Alam M, Garcia-Alias G, Jin B, Keyes J, Zhong H, Roy RR, Gerasimenko Y, Lu DC, Edgerton VR. Electrical neuromodulation of the cervical spinal cord facilitates forelimb skilled function recovery in spinal cord injured rats. *Exp Neurol* 2017;**291**:141—50. https://doi.org/10.1016/j.expneurol.2017.02.006.

63. Zhao Y, Tang F, Xiao Z, Han G, Wang N, Yin N, Chen B, Jiang X, Yun C, Han W, Zhao C, Cheng S, Zhang S, Dai J. Clinical study of NeuroRegen scaffold combined with human mesenchymal stem cells for the repair of chronic complete spinal cord injury. *Cell Transplant* 2017;**26**(5):891—900. https://doi.org/10.3727/09636891 7X695038.

64. Xiao Z, Tang F, Tang J, Yang H, Zhao Y, Chen B, Han S, Wang N, Li X, Cheng S, Han G, Zhao C, Yang X, Chen Y, Shi Q, Hou S, Zhang S, Dai J. One-year clinical study of NeuroRegen scaffold implantation following scar resection in complete chronic spinal cord injury patients. *Sci China Life Sci* 2016;**59**(7):647—55. https://doi.org/10.1007/s11427-016-5080-z.

65. Lee D-Y, Park Y-J, Song S-Y, Hwang S-C, Kim K-T, Kim D-H. The importance of early surgical decompression for acute traumatic spinal cord injury. *Clin Orthop Surg* 2018;**10**(4):448. https://doi.org/10.4055/cios.2018.10.4.448.

66. Manchikanti L, Boswell MV, Rivera JJ, Pampati VS, Damron KS, McManus CD, Brandon DE, Wilson SR. A randomized, controlled trial of spinal endoscopic adhesiolysis in chronic refractory low back and lower extremity pain [ISRCTN 16558617]. *BMC Anesthesiol* 2005;**5**(1):10. https://doi.org/10.1186/1471-2253-5-10.

67. Teng YD, Lavik EB, Qu X, Park KI, Ourednik J, Zurakowski D, Langer R, Snyder EY. Functional recovery following traumatic spinal cord injury mediated by a unique polymer scaffold seeded with neural stem cells. *Proc Natl Acad Sci USA* 2002;**99**(5):3024—9. https://doi.org/10.1073/pnas.052678899.

68. Pritchard CD, Slotkin JR, Yu D, Dai H, Lawrence MS, Bronson RT, Reynolds FM, Teng YD, Woodard EJ, Langer RS. Establishing a model spinal cord injury in the African green monkey for the preclinical evaluation of biodegradable polymer scaffolds seeded with human neural stem cells. *J Neurosci Methods* 2010;**188**(2):258—69. https://doi.org/10.1016/j.jneumeth.2010.02.019.

69. Theodore N, Hlubek R, Danielson J, Neff K, Vaickus L, Ulich TR, Ropper AE. First human implantation of a bioresorbable polymer scaffold for acute traumatic spinal cord injury. *Neurosurgery* 2016;**79**(2):E305—12. https://doi.org/10.1227/NEU.0000000000001283.

70. Kim KD, Lee KS, Chang JJ, Toselli RM. "Neuro-spinal scaffold"! Implantation in patients with acute traumatic thoracic complete spinal cord injury: 24 Month results from the INSPIRE study. *Neurosurgery* 2019;**66**(nyz310_334). https://doi.org/10.1093/neuros/nyz310_334.

71. InVivo Therapeutics. *InVivo therapeutics announces enrollment of first two patients into the INSPIRE 2.0 study for the treatment of acute spinal cord injury — InVivo therapeutics.* May 29, 2019. https://www.invivotherapeutics.com/press-releases/invivo-therapeutics-announces-enrollment-of-first-two-patients-into-the-inspire-2-0-study-for-the-treatment-of-acute-spinal-cord-injury/.

72. Peckham PH, Knutson JS. Functional electrical stimulation for neuromuscular applications. *Annu Rev Biomed Eng* 2005;**7**(1):327—60. https://doi.org/10.1146/annurev.bioeng.6.040803.140103.

73. Martin R, Sadowsky C, Obst K, Meyer B, McDonald J. Functional electrical stimulation in spinal cord injury: from theory to practice. *Top Spinal Cord Inj Rehabil* 2012;**18**:28—33.

74. Kapadia N, Masani K, Catharine Craven B, Giangregorio LM, Hitzig SL, Richards K, Popovic MR. A randomized trial of functional electrical stimulation for walking in incomplete spinal cord injury: effects on walking competency. *J Spinal Cord Med* 2014;**37**:511—24.

75. Tefertiller C, Gerber D. Step ergometer training augmented with functional electrical stimulation in individuals with chronic spinal cord injury: a feasibility study. *Artif Organs* 2017;**41**:E196—202.

76. Bergmann M, Zahharova A, Reinvee M, Asser T, Gapeyeva H, Vahtrik D. The effect of functional electrical stimulation and therapeutic exercises on trunk muscle tone and dynamic sitting balance in persons with chronic spinal cord injury: a crossover trial. *Medicina* 2019;**55**(10). https://doi.org/10.3390/medicina55100619.

77. Johnston TE, Modlesky CM, Betz RR, Lauer RT. Muscle changes following cycling and/or electrical stimulation in pediatric spinal cord injury. *Arch Phys Med Rehabil* 2011;**92**(12):1937—43. https://doi.org/10.1016/j.apmr.2011.06.031.

78. Johnston TE, Smith BT, Mulcahey MJ, Betz RR, Lauer RT. A randomized controlled trial on the effects of cycling with and without electrical stimulation on cardiorespiratory and vascular health in children with spinal cord injury. *Arch Phys Med Rehabil* 2009;**90**(8):1379—88. https://doi.org/10.1016/j.apmr.2009.02.018.

79. Tsintou M, Dalamagkas K, Seifalian AM. Advances in regenerative therapies for spinal cord injury: a biomaterials approach. *Neural Regen Res* 2015;**10**(5):726—42. https://doi.org/10.4103/1673-5374.156966.

80. Yamazaki K, Kawabori M, Seki T, Houkin K. Clinical trials of stem cell treatment for spinal cord injury. *Int J Mol Sci* 2020;**21**(11):3994. https://doi.org/10.3390/ijms21113994.

81. Ahuja CS, Nori S, Tetreault L, Wilson J, Kwon B, Harrop J, Choi D, Fehlings MG. Traumatic spinal cord injury—repair and regeneration. *Neurosurgery* 2017;**80**(3S):S9—22. https://doi.org/10.1093/neuros/nyw080.

Clinical trials: rehabilitation approaches

Newton Cho[1], Paul A. Koljonen[2], Anthony S. Burns[3]

[1]University of Toronto, Toronto, ON, Canada; [2]University of Hong Kong, Hong Kong Special Administrative Region, China; [3]Division of Physical Medicine & Rehabilitation, Department of Medicine, University of Toronto, Toronto, ON, Canada

Background

Spinal cord injuries (SCIs) and disorders have a devastating and debilitating impact on affected individuals. Trauma is the leading cause of loss of human potential globally, and among traumatic injuries, brain injury and SCI are the leading causes of death and disability.[1] SCI has devastating physical, emotional, and psychological consequences. Injured persons suffer from a number of different issues including deficits in locomotion, respiratory complications, cardiovascular disease, neurogenic bowel and bladder dysfunction, sexual dysfunction, and pressure ulcers.[2–5] Moreover, SCI places a significant economic burden on society. One study reported that the total annual cost attributed to SCI was $14.5 billion in the United States.[6] Another Canadian study found that the net lifetime cost of SCI was $336,000 per person.[7] As a result, there is a clear need for interventions to improve neurologic function following SCI.

Despite decades of research, there remains a paucity of treatments that can reliably improve neurologic function after injury, and attempts to ameliorate SCI and accompanying secondary damage have proven to be limited in regard to generating functional improvement and translatable therapies to humans. Early decompressive surgery has been advocated as a means to reduce ongoing compression and prevent worsening of secondary injury. The STASCIS (Surgical Timing in Acute Spinal Cord Injury Study) trial demonstrated modest improvements in neurologic function if surgery is performed within 24 hours of injury, although some question whether the study was designed and powered specifically to detect the reported outcomes.[8,9] Another study found that decompressive surgery performed within 8 hours of thoracolumbar SCI resulted in significantly improved American Spinal Injury Association Impairment Scale (AIS) grades and improved bladder outcomes although these effects were also modest.[10] Several large human clinical trials have also assessed the impact of the administration of methylprednisolone, a glucocorticoid, on recovery of function. Glucocorticoids especially in high doses have been thought to reduce free radical–catalyzed lipid peroxidation in the injured cord.[11] However, three large randomized clinical trials under the

© 2022 Elsevier Inc. All rights reserved.

umbrella NASCIS (National Acute Spinal Cord Injury Study) with different doses and lengths of intravenous methylprednisolone administration post-SCI failed to demonstrate consistent robust improvements in neurologic function.[12–15] Studies intended to regenerate injured neurons and axons across the spinal cord lesion have also demonstrated mixed results. Previous work has demonstrated that degrading the glial scar that forms post-SCI improves regeneration and functional recovery.[16] However, more recently, new evidence from genetic models in mice demonstrated that preventing astrocyte scar formation, attenuating scar-forming astrocytes, or removing chronic scars all resulted in the failure of axons to regrow across a spinal cord lesion.[17] To add to this complexity, the function of regenerating axons remains unclear. A relatively recent study demonstrated robust regeneration of propriospinal axons across an anatomically complete SCI in rats but failed to demonstrate significant improvements in locomotor function.[18]

In summary, methods to halt or reverse spinal injury have largely proven to be ineffective, and current treatments and interventions still focus on early rehabilitation and prevention of secondary complications. If one conceptualizes the modulation of injury mechanisms as a *"physical"* approach to intervention, the modulation of the spinal cord itself after injury can be seen as a more *"functional"* approach to intervention. These interventions take what is remaining of the spinal cord anatomically and functionally after injury and attempt to optimize its function through the use of intensive rehabilitation and associated technologies. This chapter will review recent clinical trials that have addressed rehabilitation approaches and associated technologies to improve function after SCI, in order to provide a sense of the human data available to guide therapy.

Rehabilitative approaches to improve function after SCI

Individuals with SCI suffer from a wide range of functional deficits including ambulation, of functional deficits including ambulation, autonomic dysfunction, and bowel and bladder dysfunction. Various clinical trials have attempted to assess the efficacy of rehabilitative approaches for improving functional outcomes after SCI. However, due to clinical heterogeneity, relatively small cohort sizes, and variability in rehabilitation intervention strategies, there remains significant difficulty generating high-quality evidence to support changes in practice.[19] Furthermore, the direct comparison of different rehabilitative approaches has also proven to be very difficult. Nonetheless, in this chapter, we summarize different themes associated with rehabilitative approaches for functional recovery after SCI and the recent trials that have attempted to address these themes.

Regaining locomotor and upper-extremity function

Neurological prognosis and rehabilitation potential

In general, the capacity for spontaneous locomotor recovery in humans after SCI depends on the severity of injury. The AIS provides a standardized classification system for SCI severity depending on the degree of motor and sensory function[20] (Table 27.1). In a query of three major SCI databases (NACTN, EMSCI, SCIMS) to investigate the natural history of SCI, individuals with complete thoracic SCI (AIS A) demonstrated a pooled 21.1% AIS conversion rate from complete injury to incomplete injury (AIS grades B, C, D), although motor improvement was uncommon in AIS grade A patients.[21] As expected, this study demonstrated that the more severe the injury the less likely the recovery of locomotor function. van Middendorp et al. formalized this concept further by developing a prediction model that accurately predicts whether an individual will be able to walk independently at 1 year after traumatic SCI.[22] The model was based on age (cut-off at 65 years old) and, as expected, preserved muscle strength and sensation: more specifically, the quadriceps femoris muscle grade (L3), gastrocsoleus muscles grade (S1), light touch

TABLE 27.1 American Spinal Injury Association (ASIA) impairment scale.

Completeness of injury	Grade	Description
Complete	A	Absence of motor or sensory function in S4–5
Sensory incomplete	B	Below the neurologic level of injury, have preservation of sensory function (including S4–5) Absence of motor function more than three levels below the motor level on either side of body
Motor incomplete	C	Voluntary anal contraction preserved, or below the neurologic level of injury, have preservation of sensory function (including S4–5) and have motor function more than three levels below ipsilateral motor level on either side of body Less than half of key muscles below neurologic level of injury have grade ≥3
Motor incomplete	D	Motor incomplete as defined with "AIS C" but at least half of key muscles below neurologic level of injury have grade ≥3
Normal	E	Normal sensory and motor function

Adapted from American Spinal Injury Association, International Standards for Neurological Classification of SCI (ISNCSCI) Worksheet, https://asia-spinalinjury.org/international-standards-neurological-classification-sci-isncsci-worksheet/, 2019 (accessed May 27, 2021).

sensation at L3, and light touch sensation at S1. These prognostic factors were also validated using a Canadian multicenter SCI database (Rick Hansen Spinal Cord Injury Registry), and the model remained highly accurate even when S1 muscle grade and L3 light touch sensation assessment were removed.[23] Within the limitations of these studies, it appears that baseline motor and sensory functional status play a significant role in determining future locomotor ability after SCI. However, other evidence suggests that significant improvement is possible even for the most severe SCIs. In a meta-analysis of 11 randomized control trials and 9 observational studies, which included studies examining surgical, biological, and medical treatments for AIS grade A SCIs including cervical SCI (n = 1162), El Tecle et al. observed an overall conversion rate of 28.1%.[24] The authors state that the data bring into question the previously pessimistic view of the recovery potential of severe SCIs, and that rehabilitative therapies could result in functional improvement regardless of the injury grade. Overall, while spontaneous recovery after SCI is often limited, rehabilitative strategies aim to alter the natural history of an SCI and optimize the recovery trajectory of affected individuals.

Rehabilitation of locomotion and upper-extremity function

Repetitive training is a staple of the rehabilitative process after SCI. Activity-based therapy (ABT) has been shown to improve locomotor outcomes in SCI patients. In a randomized-controlled trial, adults with chronic motor incomplete SCI (AIS grade C or D) received 24 weeks of ABT for 9 hours per week,[25] which consisted of various muscle strengthening exercises and task-specific locomotor training. Compared to the control group, the ABT group improved to a greater extent in lower extremity motor score, International Standards for Neurological Classification of SCI (ISNCSCI) total motor score, and the Spinal Cord Injury Functional Ambulation Index, supporting the value of locomotor training for improving functional outcomes after SCI.[25] In regard to upper-extremity function, a randomized-controlled trial of 34 individuals with traumatic SCI (C4–L1), AIS grade A–D, examined the effects of a 9-month exercise training paradigm (twice weekly), which involved progressively intense bouts of arm ergometry and resistance training.[26] The intervention group demonstrated significant improvements in arm ergometry power output and upper-body muscle strength while no significant changes occurred in the control group.

However, when one examines the entire body of available literature, there is considerable variability in the methods used to provide rehabilitative training. For example, Piira et al. performed a single-blinded randomized controlled trial of body weight–supported locomotor training (BWSLT) for individuals with incomplete SCI (AIS grade C–D) of duration greater than 2 years.[27] BWSLT consisted of body weight–supported treadmill training (BWSTT) with manual assistance for 60 days combined with overground training and home exercises compared to control treatment which consisted of passive joint movements, independent gym training, and some overground training. Overall, there were no significant differences in walking speed, distance walked during 6 minutes, and lower extremity motor score improvement between groups.[27] Another group performed a single group, partially blinded crossover trial comparing treadmill training to multimodal training to engage subcortical and corticospinal circuits.[28] Multimodal training consisted of training patients on a balance platform in addition to performing skilled arm or hand manipulations such as picking up cards, tightening/loosening screws, or typing on a keypad, among others. In this trial, only 9 of 24 participants completed both phases of the study, and none demonstrated significant differences in function between interventions.[28] In another trial, individuals with chronic cervical motor incomplete SCI (AIS grade C) within 2 years of injury were randomized to receive either body weight–supported overground training or BWSTT. Both groups demonstrated increases in lower extremity muscle strength and Walking Index for Spinal Cord Injury II scores after 8 weeks of training[29]; however, there were no statistical differences in the change scores between the two groups. Finally, another group assessed the efficacy of a goal-oriented and task-specific approach to rehabilitation in persons with cervical SCI using the task-oriented client-centered upper extremity skilled performance training (ToCUEST) module.[30] The trial failed to demonstrate a statistically significant difference between the ToCUEST and control groups in terms of general arm hand skilled performance improvement. However, comparing 3 months after the start of rehabilitation to the time of discharge, the ToCUEST group demonstrated a statistically significant improvement in the Van Lieshout test, which assesses arm hand skills for basic activities, compared to the control group. In summary, while the available clinical trial data suggest training improves function in individuals with SCI, there is still uncertainty regarding which training methodologies provide the best results.

Rehabilitation dosage and intensity

Theoretically changes in the intensity of training could potentially affect the results of rehabilitation, although the ideal standards in this regard are unclear. Existing research into the intensity of locomotor training has demonstrated that increasing the intensity of stepping practice improves peak walking speeds. For example, following randomization into a 20-week training regimen of high- versus low-intensity motor training (guided by the training heart rate), Brazg et al. showed that ambulatory individuals with chronic incomplete SCI (AIS grade C or D, neurological level T10 or above) walked faster on a treadmill after a period of high-intensity training. This suggests that the monitoring of stepping intensity may play an important role in guiding therapy.[31]

Similarly, available evidence suggests that the dose of training affects outcomes of rehabilitation. Sandler et al. randomized persons with chronic SCI AIS grade C or D (injury level T10 or higher) into the following groups of: (1) treadmill training, (2) treadmill training with electrical stimulation of the peroneal nerve, (3) overground training with functional electrical stimulation (FES) to assist with dorsiflexion during swing, and (4) treadmill training with a robotic gait orthosis. After the training period, a greater distance traveled during overground training with FES resulted in a significantly greater distance walked in 2 minutes and increased walking speed. This was not observed in any of the treadmill-based approaches.[32] In another trial, Wu et al. randomized 14 individuals with chronic motor incomplete SCI (AIS grade C or D) to receive either swing resistance or swing assistance during treadmill training, three times a week for 6 weeks.[33] Within group, there was a statistically significant increase in step length after resistance training; however, there was no significant difference in step length between the two groups. The addition of strength training to rehabilitation has also been proposed for the upper and lower extremities. Bye et al. randomized participants to receive unilateral strength training combined with conventional rehabilitation for the following muscle groups: elbow flexors, elbow extensors, knee flexors, or knee extensors. The side that received combinatorial training was randomized. The other side did not receive strength training. Experimental limbs were trained three times a week for 12 weeks while receiving rehabilitative gait training and functional training for activities of daily living. While additional strength training did improve the strength of participants, it was unclear whether the treatment effect was clinically meaningful.[34]

The potential efficacy of Lokomat-applied resistance during BWSTT has also been investigated. The Lokomat is a robotic device that facilitates assisted gait training. Lam et al. showed that by adding resistance during robotic training, greater functional improvements can be gained.[35] In the study, participants with chronic motor incomplete SCI (AIS grade C or D) were randomized to receive either conventional Lokomat-assisted training or Lokomat-assisted training with additional resistance (Loko-R) against the hip and knee. Walking performance was assessed using the SCI-FAP (Spinal Cord Injury Functional Ambulation Profile).

The Loko-R group demonstrated greater post-training improvement in walking compared to the control group.

Existing studies suggest that increasing the intensity and dose of training can improve rehabilitative outcomes after SCI. These findings are consistent with general training principles and are expected. Individual tolerance ultimately plays a significant role in how much of an increase is possible. As a result, intensity and dosing has to be tailored to the individual, which in turn affects overall outcomes. Advanced rehabilitation techniques such as use of electrical stimulation devices, robotics, or exoskeletons may facilitate the delivery of increased intensity and dosage. These interventions are described in subsequent sections of this chapter.

Adjunctive rehabilitation interventions

With advancing technology, many researchers and clinicians have sought to assess the effects of conventional rehabilitation in combination with other adjunctive strategies. These strategies may not only be useful for increasing training quality and quantity but also potentially be beneficial for enhancing neuroplasticity and overall neurological recovery.

Epidural electrical stimulation

One particularly intensive area of research has been the combination of spinal epidural electrical stimulation (EES) with rehabilitative training. Preclinical data in rodents suggest that this combination can significantly improve locomotor outcomes after SCI and improve supraspinal control of locomotion by promoting plasticity of the motor cortex as well as locomotor centers in the brain stem and the spinal cord.[36,37] Parallel studies in humans also suggest that the application of EES can improve locomotor outcomes. Tonic EES applied at L1–S1 to one individual with motor-complete but sensory-incomplete SCI restored full weight–bearing standing.[38] The same strategy was able to restore full weight–bearing standing in two additional individuals with chronic motor- and sensory-complete SCI.[39] In another study, tonic EES restored volitional movements in the paralyzed lower limbs of individuals with chronic complete SCI.[40] More recently, EES (delivered via electrode implantation at L1–S2) was combined with intense locomotor training and administered to a series of four individuals with chronic motor-complete SCI. Two participants with motor-complete and sensory-incomplete SCI recovered the ability to walk overground with assistive devices, while the two other participants with motor- and sensory-complete SCI were able to achieve some independent stepping on a treadmill with body weight support.[41] Another study of one person with a chronic T6 complete SCI (AIS grade A) demonstrated that the combination of EES and intensive rehabilitation over 43 weeks resulted in EES-enabled stepping on a treadmill, stepping overground with a front-wheeled walker with intermittent trainer assistance, and independent standing.[42]

Despite these encouraging results, the mechanisms by which EES improves locomotor function are poorly understood. Previous research has attempted to shed light on the mechanisms of EES. Computer simulations in combination with experiments in rats have demonstrated that EES primarily activates proprioceptive feedback circuits through the recruitment of myelinated afferent fibers rather than directly influencing motoneurons.[43,44] Studies in mice also confirm the importance of muscle spindle–mediated proprioceptive feedback in the recovery of hindlimb function after injury.[45] Furthermore, previous research has also demonstrated that continuous tonic stimulation protocols may not necessarily provide optimal spinal cord modulation for hindlimb function post-SCI. A study in rats demonstrated that the application of EES with spatial and temporal dynamics matching natural motoneuron activation improved gait parameters after both complete thoracic transection and dorsal contusion SCI to a greater degree than continuous EES stimulation.[46] Furthermore,

computer simulations and additional experiments conducted in humans have shown that EES blocks a significant amount of proprioceptive input in humans but not in rats.[47] This is due to antidromic activity along proprioceptive afferents. Spatiotemporal stimulation paradigms can mitigate the cancellation of proprioceptive input unlike continuous stimulation, which can only act effectively within a narrow range of parameters.[47] This paradigm was applied to humans with chronic SCI in a feasibility trial in which three individuals with chronic traumatic SCI were implanted with technology enabling spatiotemporal lumbar spinal cord segment EES modulation.[48] For this purpose, a gravity-assist algorithm that personalizes multidirectional forces to the trunk was also designed to support natural walking in people with SCI.[49] The combination of these technologies with active rehabilitation resulted in the restoration of locomotor movements during EES in three participants who were completely paralyzed or unable to ambulate prior to therapy and rehabilitation.[48] Their step elevations increased three- to fivefold while also being able to cover 1 km of continuous locomotion without deterioration. Study participants also regained voluntary control of paralyzed muscles without stimulation. There is therefore mounting evidence that the combination of EES with intensive rehabilitation can significantly improve locomotor function in experimental settings after SCI in humans. Given the intensive nature of the intervention, future studies will need to assess its implications for function in household and community settings.

Compared to the lower limbs, the use of EES for the improvement of upper limb function in humans is not as well established and is currently a growing field of research. Gad et al. examined the effect of transcutaneous electrical stimulation, a noninvasive modality for stimulating the spinal cord, combined with rehabilitation on hand grip in individuals with traumatic SCI at C7 or higher with an AIS grade of B and C for 1 year or longer.[50] Although not the same as EES, which is applied epidurally, the study did demonstrate an average 225% increase in grip strength compared to baseline measurements (without stimulation) after 4 weeks of treatment, suggesting the potential value of spinal cord stimulation for cervical SCI. Recent research in primates has also attempted to adapt the spatiotemporal stimulation paradigm in the lower extremities to reaching and grasping.[51] In their study, Barra and colleagues (1) mapped the activation of various segments of the cervical spinal cord during a reaching and grasping task and (2) recorded muscle activity in response to epidural stimulation of different segments of the cord. In this way, the authors were able to selectively recruit different muscles of the upper limb in order to reproduce motoneuron activation patterns for reaching and grasping. It is anticipated that future clinical trials will continue to assess the efficacy of cervical cord stimulation, potentially combined with spatiotemporal EES paradigms, for the rehabilitation of upper-extremity function.

Motor cortex stimulation

Motor cortex stimulation has been proposed as another adjunct to improve the recovery process with rehabilitation. Cortes et al. assessed the effect of transcranial direct current stimulation (tDCS) applied to the motor cortex of individuals with chronic incomplete cervical SCI.[52] Eleven participants were randomized to receive 1 mA, 2 mA, or sham anodal tDCS for 20 minutes over three sessions, after which a small but statistically significant effect on the grasp mean to peak speed ratio was observed in the 2 mA group. In another trial of 9 individuals with chronic cervical motor incomplete SCI (AIS grade C or D), participants were randomized to receive 10 sessions of either tDCS of the primary motor cortex or sham stimulation combined with robot-assisted arm training.[53] There were no significant differences observed between groups for the change in upper-extremity motor scores, but

there was a larger absolute improvement (not statistically significant), as measured using the Jebsen—Taylor Hand Function Test (JTHFT), in the active tDCS group compared to the sham group (40.7 ± 29.1% vs. 6.7 ± 4.5%). The JTHFT measures the time to perform seven everyday activities such as writing or feeding.

Kumru and colleagues assessed the effects of tDCS for lower extremity function, when combined with Lokomat-assisted gait training.[54] Twenty-four study participants who suffered motor incomplete cervical or thoracic SCI (AIS grade C or D) of less than 1 year duration were randomized to receive either tDCS at 2 mA for 20 minutes to the leg motor cortex or sham tDCS for 20 days while undergoing Lokomat gait training, after which Lokomat gait training continued for an additional 4 weeks for a total of 8 weeks of gait training. There were no significant differences between groups in the changes in lower extremity muscle strength or gait assessments including gait velocity, step length, and cadence. The authors concluded that while the application of tDCS is safe, its benefits for lower extremity function improvement are not clear. Raithatha and colleagues examined the effects of tDCS on 15 individuals with incomplete traumatic SCI (AIS grade B, C, or D) of at least 1 year duration prior to enrollment.[55] Study participants were randomized to receive either active tDCS or sham treatment for 20 minutes prior to receiving Lokomat gait training for a total of 36 sessions. Trial results were mixed. Greater improvement in manual muscle testing (muscle strength) was observed in the right lower extremity in the active tDCS group compared to the sham group; however, the sham group demonstrated greater improvements in the 6-minute walk test and the Time Up and Go Test, which measures the ability to stand up from a chair, walk 3 m then walk back to the chair, and sit down. These small trials of tDCS for lower extremity function are therefore inconclusive and its potential role in rehabilitation remains unclear.

Motor cortex stimulation through repetitive transcranial magnetic stimulation (rTMS) has also been investigated in various clinical trials. Kumru et al. assessed the combination of rTMS and Lokomat training.[56] Study participants with motor incomplete (AIS grade C or D) cervical or thoracic SCI of ≤ 6-month duration were randomized to receive either real or sham high-frequency rTMS at 90% of the resting motor threshold of the muscle with the lowest threshold. This was administered for 20 daily sessions over a 4-week period. The change in total motor score was significantly higher in the participants that received real rTMS compared to the sham group when assessed after the completion of 4 weeks of therapy and again at 8 weeks from the beginning of study participation. In addition, the proportion of participants who could perform the 10-min walk test was greater in the real rTMS group compared to the sham group, although this did not reach statistical significance.[56] Gomes-Osman and Field-Fote compared 11 persons with cervical SCI at least 1 year postinjury to 10 neurologically intact individuals.[57] Using a double-blind crossover design, subjects were randomized to receive 10 Hz rTMS to the corticomotor hand area with repetitive task practice (RTP) or sham rTMS with RTP for 1 week then received the alternate treatment in the second week. At the conclusion of the trial the change in time to complete the JTHFT in the trained hand was larger for the rTMS group compared to the sham group, for the participants with SCI, although this difference failed to reach statistical significance. There was also no significant difference in pinch or grasp force in the trained hand. Overall, there is some evidence to suggest that rTMS may augment rehabilitation for both the upper and lower extremities, although more trials are needed to confirm its efficacy.

Intermittent hypoxia

Currently, intermittent hypoxia is not used routinely as an adjunct to rehabilitation after SCI, and the number of trials that have addressed intermittent hypoxia is limited. There is, however, preliminary evidence suggesting that intermittent hypoxia can promote neuroplasticity.[58] In one study, individuals with AIS grade C or D SCI and neurological levels between C5 and T12 were randomized to either receive intermittent hypoxia with BWSTT or continuous normoxia and BWSTT for a total of 4 weeks. The group that received intermittent hypoxia demonstrated greater walking speed and walking endurance.[58] Similarly, another trial of individuals with AIS grade C and D SCI greater than 1 year postinjury found that the addition of intermittent hypoxia to BWSTT over 4 weeks significantly improved dynamic balance (i.e., turning duration, turning cadence, and turn-to-sit duration).[59] Trumbower et al. recently assessed the use of intermittent hypoxia for the improvement of hand function.[60] In this double-blind crossover study, 6 participants with C5 AIS grade C or D SCI received 5 consecutive days each consisting of intermittent hypoxia followed by 20 repetitions of hand opening practice and normoxia. Hand function was assessed using the Box and Block Test for dexterity and the JTHFT for daily living activities. Intermittent hypoxia resulted in statistically significant improvements in Box and Block Test scores compared to sham treatment and reduced the JTHFT time for four of the six participants. Despite its promise, more research will be required before intermittent hypoxia is routinely incorporated into rehabilitative strategies.

Autonomic dysfunction

In addition to motor and sensory deficits, SCI is associated with significant impairments in autonomic function. SCI can cause hypotension and cardiac dysrhythmias, especially bradycardia, and less frequently cardiac arrest or tachyarrhythmias. Individuals with SCI and neurological levels at or above T6 are at risk for a potentially life-threatening syndrome called autonomic dysreflexia (AD). AD is due to an imbalance of parasympathetic and sympathetic activity leading to uncontrolled constriction of splanchnic vessels and malignant hypertension. Additional signs and symptoms of AD include bradycardia, piloerection, headache, visual changes, and general malaise. Common triggers for AD include noxious visceral or somatic stimulation, particularly overdistension of the bowel or bladder. Other noxious stimuli include skin lacerations, ingrown toenails, pressure sores, tight clothing, and certain medical procedures such as bladder catheterization and cystometry.[61–63] Hubli et al. showed that this type of fluctuation in blood pressure can occur up to 40 times or more in 24 hours in some individuals, and the repetitive surges in blood pressure from recurrent AD may cause structural damage to the peripheral vascular system, leading to injury-induced peripheral vascular dysfunctions.[64–66]

In addition to cardiovascular anomalies, abhorrent autonomic function is also linked to abnormalities in the functioning of the immune system, termed SCI-induced immune depression syndrome.[67] When compared to the general population, individuals with chronic SCI tend to have weaker immune systems.[68] In a mouse SCI model, Brommer et al. demonstrated that when induced with pneumonia, high thoracic lesions that interrupt sympathetic innervation to major immune organs, but not low thoracic lesions, significantly increased bacterial load in the lungs.[69] Furthermore, using a human SCI database and regression analysis to determine independent risk factors for infection, it has been shown that a high thoracic (T1–T4) relative to midthoracic (T4–T8) and midthoracic relative to low thoracic (T9–T12) injury is independently associated with increased risk for pneumonia following motor-complete SCI (odds ratio = 1.35, $P < .001$).[69]

Alterations in autonomic function following SCI are also proposed to have a significant effect on the regulation of metabolic diseases, including insulin resistance and dyslipidemia. Bluvshtein et al. showed that postprandial insulin resistance is present in individuals with cervical SCI but not thoracic SCI.[70] It has been proposed that the preservation of some supraspinal control over sympathetic outflow can have a significant influence on energy metabolism in various tissues and organs such as the liver, spleen, and pancreas and is reflected by the amount of circulating catecholamines.[71] Frequent daily bouts of AD accompanied by abnormal surges in glucocorticoids and catecholamines may also contribute to the high reported incidence of metabolic abnormalities in persons with SCI.[72]

Several human studies have attempted to address autonomic dysfunction in persons with SCI, mainly in the context of hypotension. More specifically, the use of EES to modulate blood pressure after SCI has been investigated. West et al. described an individual with a chronic C5 motor-complete SCI (AIS grade B) who was implanted with an epidural spinal cord stimulator at T11–L1.[73] Epidural stimulation resolved the patient's orthostatic hypotension.[73] Aslan et al. investigated seven persons with chronic C5–T4 SCI (AIS grade A or B) implanted with an epidural spinal stimulators at T11–L1.[74] In this trial, persons who normally experienced significant drops in arterial blood pressure in response to standing demonstrated maintenance of blood pressure in the presence of spinal stimulation. In another group of four individuals with chronic cervical motor-complete SCI, epidural spinal cord stimulators were implanted over spinal cord segments L1–S1.[75] Study participants then completed an average of 89 2-hour sessions of daily spinal stimulation calibrated to increase systolic blood pressure, which resulted in the resolution of orthostatic hypotension both during stimulation and after training even in the absence of stimulation.[75,76] In another study of two female participants with chronic thoracic SCI (AIS grade A) who were implanted with spinal epidural stimulators at vertebral level T12, the administration of stimulation was able to resolve orthostatic hypotension in the one individual who demonstrated significant hypotension during tilt-table testing.[77] Stimulation did not affect blood pressure in the other individual in whom baseline hypotension was absent. Despite these encouraging preliminary results, larger clinical trials are needed to fully elucidate the effects of rehabilitation in combination with EES on cardiovascular function post-SCI, especially in terms of hypotension and the potential modulation of blood pressure in AD.

Bowel and bladder dysfunction

Neurogenic bladder and bowel dysfunction is common following SCI. The loss of autonomic and somatic control of detrusor and urinary sphincter function contributes to serious sequelae such as recurrent urinary tract infections, urosepsis, bladder stones, and autonomic dysregulation in addition to urinary incontinence.[78,79] Furthermore, detrusor—sphincter dyssynergia, vesicoureteral reflux, upper renal tract dilation, and renal stone formation may lead to renal failure.[78] Individuals with SCI frequently require lifelong medications such as oral anticholinergics, beta-3 adrenergic agonist (mirabegron), and intradetrusor botulinum toxin, which have undesirable side effects and can be costly. AD triggered by infections or blocked catheters may lead to dangerous and life-threatening hypertensive episodes. In regard to bowel function, changes in gut transit times and evacuation can cause intractable constipation and fecal impaction. Adjuncts to achieve bowel evacuation, such as manual disimpaction and/or digital rectal stimulation, can be laborious and time-consuming, and often require the assistance of another person.[80] Bowel distention can also lead to AD.

The standard of care for neurogenic bladder dysfunction is intermittent catheterization, preferably performed by the affected individual.[78] Alternatively, indwelling catheters are often used in individuals who are unable to self-catheterize or lack the supports for the regular performance of intermittent catheterization. In a subset of individuals, surgical procedures (i.e., sphincterotomy, bladder augmentation, urinary diversion) can improve independence and/or reduce the management burden; however, these are major procedures and carry significant risks. Regardless of treatment approach, there is often a high rate of recurrent lower and upper urinary tract infection.

Rehabilitative approaches for improving the management of neurogenic bowel and bladder have been explored; however, to date none are routinely employed in clinical practice. Elmelund et al. randomized 27 women with incomplete SCI (AIS grade A and B patients were excluded) to receive either pelvic floor muscle training (PFMT) or a combination of PFMT and intravaginal electrical stimulation, for 12 weeks.[79] It is postulated that intravaginal electrical stimulation may stimulate the pudendal nerve innervating the bladder. No differences in episodes of urinary incontinence were found between the group receiving PFMT alone and electrical stimulation combined with PFMT. Brindley et al. utilized electrical stimulation of the bladder and bowel to improve emptying and provide some measure of control to the affected individual with SCI.[81] This is achieved by attaching electrodes to the sacral nerve roots, which in turn are connected to a subcutaneous stimulator. Bladder and bowel evacuation are then initiated and controlled via a remote controlled activator. The surgical implantation is often combined with posterior sacral rhizotomies which allow urine to be stored in the bladder at a low pressure.[82–85] This has the added benefit of ameliorating AD originating from the bladder and bowel. Stimulation of the sacral parasympathetic nerves also increases motility in the lower bowel, reduces constipation, and with appropriate adjustment of stimulus parameters, produces defecation. Stimulation of the S2 nerves can also produce penile erection in some men.[84]

Addressing detrimental secondary biophysical changes with rehabilitation

SCI can give rise to a series of secondary biophysical complications in the long term. These include recurrent pressure ulcers and infections, joint contractures, myoclonus and spasticity, osteoporosis and fractures, cardiopulmonary deconditioning, weight gain, deranged lipid profiles, and vascular diseases. Used consistently, rehabilitative strategies, such as ABT, not only can enhance postural control, balance, and strength for some; they can also be therapeutic and mitigate many secondary biophysical problems. Training that focuses on repetitive and progressive practice of a desired task is believed to cause activity-dependent plasticity through the reorganization of neurological pathways, hence improving pain, reducing spasticity, and regulating autonomic functions. While historically rehabilitation has focused on rehabilitating uninjured components of the neuromuscular system to facilitate and compensate for lost function, in the future it is anticipated that an increased priority will be placed on the retraining of afferent input and stimulating new synaptic formation to improve motor tasks such as standing and stepping.[86–88]

Bone density

Secondary osteoporosis is a well-recognized complication of SCI,[89,90] and persons with SCI are at increased risk of fragility fractures, especially at the proximal tibia and distal femur. The primary drivers of secondary sublesional osteoporosis are prolonged disuse and lack of mechanical stimuli from muscle contractions which lead to a marked increase in osteoclastic bone resorption and a decrease in osteoblastic activity, although hormonal changes following SCI can also contribute.[91,92] The levels of biochemical markers of resorption start increasing weeks after injury and reach a maximum between week 0 to week 16 postinjury. After 1 year, resorption markers remain high, and bone formation markers are low. Some studies suggest that a steady state is achieved around the second year following injury, but others note a continuous loss of bone.[93]

Many of the interventions that are usually employed for SCI attempt to counteract the detrimental consequences of immobilization, the reduction in muscle contraction, and the redistribution of the gravitational forces applied to the bone segments that normally support weight.[93] Electrical stimulation can be applied to nerves or muscles to generate muscle contractions, and is termed neuromuscular electrical stimulation (NMES). The delivery of NMES can also be paired with the performance of a functional activity (i.e., cycling), and is termed FES. In two small observational studies of individuals with recent onset complete (AIS grade A) SCI, Shields and colleagues observed that NMES of the ankle plantar flexors attenuated bone mineral density loss in the tibia.[94,95] There is additional evidence from two small prospective nonrandomized studies of motor-complete (AIS grade A or B) SCI that FES (cycling, rowing) can ameliorate bone mineral density at the distal femur and proximal tibia following an SCI.[96,97] This effect occurred preferentially at sites where muscle contractions exert forces on bone, as opposed to remote sites. Two nonrandomized observational studies also found that NMES can partially reverse established low bone mineral density at the distal femur and proximal tibia.[98,99] Nonrandomized observational studies have also confirmed that FES-assisted cycling or standing can partially reverse bone mineral density loss at the distal femur and proximal tibia.[100–103] A recent study by Draghici and colleagues provides some evidence that the total volume of training may be related to improvements in bone mineral density.[104] It is important to note that following the cessation of NMES and FES, bone mineral density loss resumes. Additional research is needed to define optimal dosing including training volumes, loads, and stimulation parameters.

Loading through the application of forces to bones might also be beneficial for bone mineral density preservation. Alekna et al. studied passive standing utilizing a standing frame in individuals with motor-complete (AIS grade A or B) SCI.[105] Participants (n = 54) were followed for up to 2 years and divided into standing (n = 27) and nonstanding groups (n = 27). While bone mineral density loss was observed in both groups, the magnitude was less in individuals who continued standing following rehabilitation discharge.

Cardiometabolic changes after SCI

There are a growing number of studies suggesting that exercise can improve cardiovascular fitness in persons with SCI. While the quality of evidence is limited, the trend in reporting suggests that exercise is effective for improving aspects of fitness. For example, there is strong evidence that exercise, performed 2–3 times per week at moderate-to-vigorous intensity, increases physical capacity and muscular strength in individuals with chronic SCI. A meta-analysis performed by Hicks et al. concluded that FES exercise (cycling and ambulation), arm ergometry, and wheelchair exercise produced significant improvements in aerobic capacity following chronic SCI.[106]

From a metabolic point of view, there are also considerable risks that need to be addressed when rehabilitating persons with SCI. Epidemiological studies suggest that mortality and morbidity rates associated with cardiovascular disease and type 2 diabetes are elevated in individuals with SCI relative to adults without physical impairments. Cardiovascular risk factors include increased central adiposity, reduced HDL cholesterol, and impaired glucose tolerance.[107] Additional factors that might contribute to an increased risk of cardiovascular disease following SCI include altered autonomic regulation, a reduced cardiovascular response to exercise, and loss of sympathetic innervation of the adrenal medulla which may diminish the metabolic responses to exercise in patients with an SCI at a level higher than T6.[108] In able-bodied individuals, it is well documented that exercise improves whole-body insulin sensitivity and glucose homeostasis. Adaptations of skeletal muscle are essential to this, as this tissue is responsible for the majority of glucose utilization.[109] In a study investigating the effects of a moderate-intensity upper-body exercise training intervention on biomarkers of cardiometabolic component risks, adipose tissue metabolism, and cardiorespiratory fitness, Nightingale et al. randomly allocated 21 inactive men and women with chronic SCI to either a 6-week prescribed home-based exercise intervention or control group.[107] Using blood (i.e., insulin and insulin resistance markers and biomarkers of cardiovascular risk), adipose tissue gene expression, body composition measures, and functional performance measures such as VO2 max, the investigators showed that 6 weeks of home-based moderate-intensity arm-crank exercise significantly improved functional capacity (VO2 peak and power output), fasting insulin concentrations, and insulin resistance patterns.[107]

Spasticity

Spasticity is a common and debilitating secondary complication following SCI. Adams et al. showed that spasticity and related problems afflict up to 66%–98% of individuals with SCI, and many of them report a lower quality of life because of this.[110] Spasticity is associated with upper motor neuron lesions and is characterized by intermittent or sustained involuntary activation of muscles.[111] Common signs and symptoms of spasticity include clonus, paroxysmal involuntary activation of muscles (spasms), cocontractions of agonists and antagonists, and tonic muscle contraction and accompanying rigidity. It has been shown that intensive and repetitive tasks can drive beneficial neuroplasticity and improve spasticity and enhance functional outcomes.[112] New rehabilitative techniques, such as electromechanical orthoses on treadmills or exoskeletons, have focused on devices that support gait training. These rehabilitation strategies originate from the concept that gait is a physiological and automatic movement, and exercise by repetition — albeit passive — can cause activation of neuronal circuits which can reduce nonphysiological and unregulated hyperactivation of tonic responses.[113] Stampacchia et al. investigated subjective and objective changes in pain and spasticity in 21 persons with SCI following participation in an overground walking session assisted by a powered robotic exoskeleton.[114] Immediately following completion of the session, the authors demonstrated a reduction of spasticity, assessed in terms of participant perception and objective evaluations of tonic spasticity and spasm frequency. This interesting observation suggests that rehabilitative strategies, such as assisted gait training, could play a role in reducing spasticity in persons with SCI. Future studies will need to further define the magnitude and persistence of these effects, as well as the carry over to practical, day-to-day activities.

Technological breakthroughs in rehabilitation

Developing technologies are playing an increasingly important role in supporting and promoting the health and quality of life of persons with SCI. Used appropriately and judiciously, they can often be cost-effective for health-care systems. Furthermore, devices developed for SCI often have applications for other conditions and impairments.

Mobility with robotic exoskeletons

The concept of rehabilitating locomotion has evolved from animal and human studies focused on the neural plasticity of the spinal cord.[115–117] It is now widely believed that repetition and habituation, locomotor training, and neuromuscular activation can promote meaningful functional reorganization of spinal neural networks below the level of the lesion,[118–120] leading to substantial functional improvement. A rapidly evolving and exciting technology is the development of powered exoskeletons for gait. Powered exoskeletons can serve as orthoses to achieve walking in otherwise nonambulatory individuals. In theory, they might also facilitate neuroplasticity, although the degree of active participation varies by the specific device and settings. Meta-analysis studies have shown that powered exoskeletons are generally safe, and allow patients with SCI both to rehabilitate in hospital settings and safely ambulate in community/home settings at a physical activity intensity suitable for prolonged use.[121] Despite the interest in powered exoskeletons, future studies still need to establish their efficacy, long-term usage patterns, and relative advantages and disadvantages compared to conventional approaches to gait training and other efficient modes of mobility (i.e., ultralightweight manual wheelchair). This includes an objective assessment of cost-effectiveness. Locomotor training with exoskeletons has been shown to be able to improve upper-body muscular fitness, improve bowel movement regularity, and reduce pain and spasticity,[121] but more trials are required in the future.

Implantable neurostimulators

Phrenic nerve pacing and direct diaphragmatic pacing

The loss of neurological function that accompanies a high-level SCI is particularly damaging to the respiratory system. The standard treatment for persons with high-level SCI who have lost volitional breathing is long-term mechanical ventilation with a tracheostomy. However, this approach is associated with significant adverse effects, such as ventilator-associated respiratory infections, reduced mobility, prolonged hospital stays, and psychosocial stigma.[122]

Phrenic nerve (PNP) and direct diaphragmatic pacing (DP) are technologies that have been approved for use as viable alternatives to long-term mechanical ventilation in suitable individuals. These devices have been shown to be safe and effective for reducing respiratory secretions, reducing hospital length of stay, improving quality of speech, regaining sense of taste and olfaction, ease of eating and drinking, and overall improved quality of life. Candidates for PNP and DP need to be carefully screened and meet specific criteria. Individuals need to be free of chest wall deformity, medically stable, and have viable phrenic nerves (determined with phrenic nerve conduction velocity studies or diaphragm electromyograms). Cervical spine injuries with damage to phrenic nerve nuclei at C3–5 segmental levels may preclude the use of this technology.

PNP involves placing electrodes on the phrenic nerves either through a neck dissection or thoracotomy (open or videoscopic-assisted thoracoscopy). The electrodes are connected to an implanted receiver that is controlled and powered by an external transmitter via radiofrequency signals. When the nerves are stimulated, the diaphragm contracts and moves downward, creating negative pressure within the lung which causes inspiration. It is indicated for high-level SCI and other diagnoses leading to diaphragmatic paralysis.[123]

DP uses a device that applies stimulation to the phrenic nerve motor points on the abdominal surface of the diaphragm using percutaneous wires connected to an external stimulator. It was developed as a less invasive laparoscopic procedure. In the 2014 multicenter study by Posluszny et al., 29 candidates were identified with an average age of 31 years (range 17–65 years).[124] Seven (24%) of the 29 individuals had nonstimulatable diaphragms from either phrenic nerve damage or infarction of the involved phrenic motor neurons and were not implanted. Of the 22 individuals who underwent DP, 72% (16 of 22) were completely free of ventilator support in an average of 10.2 days. For the remaining six study participants, two had delayed weans of 180 days, three had partial weans using DP at times during the day, and one patient went to a long-term facility and subsequently had life-prolonging measures withdrawn. Eight participants (36%) had complete recovery of respiration, and the DP wires were eventually removed. No complications were reported from the laparoscopic procedure or electrode placement. In a review of worldwide experiences, Onders et al. reported that over 18 years and 486 patients implanted, 88% of patients achieved at least 4 hours of pacing, and 60% of patients used DP 24 hours per day.[125]

Brain–machine interface

Brain–machine interfaces (BMIs) are devices that detect the implanted user's intentions by recording electrical signals from the brain and translating them into meaningful actions performed by external computerized devices. Signal detections are usually focused on the user's motor intentions, and even in the absence of actual movement, BMIs are able to translate this brain activity into executable motor commands to control robots, computers, or other external devices.[126]

The voltage potentials being recorded in a BMI are summations of electrical currents picked up from the extracellular medium of active brain cells. When recorded from the scalp, these are referred to as encephalograms (EEGs), whereas signals from subdural electrodes are electrocorticograms (ECoGs), and signals from electrodes inserted into the brain (including the premotor or motor cortices and thalamic motor areas) are referred to as local field potentials (LFPs) or intracranial EEG.[127] Naturally, the accuracy of recordings improves with increasing invasiveness of electrodes.[126]

Translating neural signals from the brain to neuroprostheses allows users to execute real-time control of robot devices. In 2013, Collinger et al. showed that an individual with tetraplegia was able to use her thoughts to accurately control a high-performance anthropomorphic prosthetic limb to perform reaching and grasping functions.[128] Similarly, utilizing a BMI it is also possible to reanimate limbs by transducing commands into direct electrical stimulation of peripheral nerves or muscle groups, thus restoring volitional limb function and preventing muscle atrophy.[129] Currently BMIs are a research tool mainly of interest to persons with the highest levels of SCI.

As the field of BMIs develops, it is anticipated that a wider range of functions will become possible by incorporating the design of hand shape into the repertoire of volitional upper limb control. Addition of tactile feedback from fingertip sensors transmitted to the sensory cortex may create bidirectional links between the brain and the body, and telemetry can eliminate the need for physical leads and connectors.[128] Challenges for BMIs include improving accurate capture of neural signals with a greater signal-to-noise ratio from the scalp and cranium, as well as improving biocompatibility and longevity for implantable cortical electrodes. In order for a neural interface to effectively bypass the injured spinal cord, accuracy of electrical stimulations to maintain the integrity, stability, and equilibrium of movement that aligns with the users' intentions will be of utmost importance.[130] In the future, it is hoped that these technologies will be used in combination with existing rehabilitative paradigms to further improve outcomes.

Conclusions

SCI is a complex process that results in dysfunction across a wide range of body systems including locomotion and hand function, the autonomic system, the cardiovascular system, the respiratory system, and bowel and bladder function. Rehabilitative interventions aim to augment and maximize the function of preserved circuitry after SCI. Intensive longitudinal training programs increasingly employ a variety of technologies including body weight support systems, epidural spinal stimulators and other nerve stimulators, exoskeletons, and BMIs. Aside from the potential direct effects of the rehabilitation, secondary benefits associated with rehabilitation are also possible. To date, clinical trials have generally been small and heterogenous in terms of method of rehabilitation and type of intervention. However, the clinical evidence for these various interventions will likely continue to grow as technological advancements and our understanding of SCI increase.

References

1. Rubiano AM, Carney N, Chesnut R, Puyana JC. Global neurotrauma research challenges and opportunities. *Nature* 2015;**527**:S193−7. https://doi.org/10.1038/nature16035.
2. Claydon VE, Steeves JD, Krassioukov A. Orthostatic hypotension following spinal cord injury: understanding clinical pathophysiology. *Spinal Cord* 2006;**44**:341−51. https://doi.org/10.1038/sj.sc.3101855.
3. Courtois F, Charvier K. Sexual dysfunction in patients with spinal cord lesions. In: *Handbook of clinical neurology*. Elsevier; 2015. p. 225−45. https://doi.org/10.1016/B978-0-444-63247-0.00013-4.
4. Qi Z, Middleton JW, Malcolm A. Bowel dysfunction in spinal cord injury. *Curr Gastroenterol Rep* 2018;**20**:47. https://doi.org/10.1007/s11894-018-0655-4.
5. Sezer N. Chronic complications of spinal cord injury. *World J Orthoped* 2015;**6**:24. https://doi.org/10.5312/wjo.v6.i1.24.
6. Ma VY, Chan L, Carruthers KJ. Incidence, prevalence, costs, and impact on disability of common conditions requiring rehabilitation in the United States: stroke, spinal cord injury, traumatic brain injury, multiple sclerosis, osteoarthritis, rheumatoid arthritis, limb loss, and back pain. *Arch Phys Med Rehabil* 2014;**95**. https://doi.org/10.1016/j.apmr.2013.10.032. 986.e1−995.e1.
7. Chan BC-F, Cadarette SM, Wodchis WP, Krahn MD, Mittmann N. The lifetime cost of spinal cord injury in Ontario, Canada: a population-based study from the perspective of the public health care payer. *J Spinal Cord Med* 2019;**42**:184−93. https://doi.org/10.1080/10790268.2018.1486622.

8. Fehlings MG, Vaccaro A, Wilson JR, Singh A, Cadotte DW, Harrop JS, Aarabi B, Shaffrey C, Dvorak M, Fisher C, Arnold P, Massicotte EM, Lewis S, Rampersaud R. Early versus delayed decompression for traumatic cervical spinal cord injury: results of the surgical timing in acute spinal cord injury study (STASCIS). *PLoS One* 2012;**7**:e32037. https://doi.org/10.1371/journal.pone.0032037.

9. van Middendorp JJ. Letter to the editor regarding: "Early versus delayed decompression for traumatic cervical spinal cord injury: results of the Surgical Timing in Acute Spinal Cord Injury Study (STASCIS)". *Spine J* 2012;**12**:540. https://doi.org/10.1016/j.spinee.2012.06.007.

10. Wutte C, Klein B, Becker J, Mach O, Panzer S, Strowitzki M, Maier D, Grassner L. Earlier decompression (< 8 hours) results in better neurological and functional outcome after traumatic thoracolumbar spinal cord injury. *J Neurotrauma* 2018. https://doi.org/10.1089/neu.2018.6146.

11. Hall ED, Braughler JM. Glucocorticoid mechanisms in acute spinal cord injury: a review and therapeutic rationale. *Surg Neurol* 1982;**18**:320−7. https://doi.org/10.1016/0090-3019(82)90140-9.

12. Bracken MB. Efficacy of methylprednisolone in acute spinal cord injury. *J Am Med Assoc* 1984;**251**:45. https://doi.org/10.1001/jama.1984.03340250025015.

13. Bracken MB, Shepard MJ, Collins WF, Holford TR, Young W, Baskin DS, Eisenberg HM, Flamm E, Leo-Summers L, Maroon J, Marshall LF, Perot PL, Piepmeier J, Sonntag VKH, Wagner FC, Wilberger JE, Winn HR. A randomized, controlled trial of methylprednisolone or naloxone in the treatment of acute spinal-cord injury: results of the second national acute spinal cord injury study. *N Engl J Med* 1990;**322**:1405−11. https://doi.org/10.1056/NEJM199005173222001.

14. Bracken MB. Administration of methylprednisolone for 24 or 48 hours or tirilazad mesylate for 48 hours in the treatment of acute spinal cord injury: results of the third national acute spinal cord injury randomized controlled trial. *J Am Med Assoc* 1997;**277**:1597. https://doi.org/10.1001/jama.1997.03540440031029.

15. Fehlings MG, Wilson JR, Harrop JS, Kwon BK, Tetreault LA, Arnold PM, Singh JM, Hawryluk G, Dettori JR. Efficacy and safety of methylprednisolone sodium succinate in acute spinal cord injury: a systematic review. *Global Spine J* 2017;**7**:116S−37S. https://doi.org/10.1177/2192568217706366.

16. Bradbury EJ, Moon LDF, Popat RJ, King VR, Bennett GS, Patel PN, Fawcett JW, McMahon SB. Chondroitinase ABC promotes functional recovery after spinal cord injury. *Nature* 2002;**416**:636.

17. Anderson MA, Burda JE, Ren Y, Ao Y, O'Shea TM, Kawaguchi R, Coppola G, Khakh BS, Deming TJ, Sofroniew MV. Astrocyte scar formation aids central nervous system axon regeneration. *Nature* 2016;**532**:195−200. https://doi.org/10.1038/nature17623.

18. Anderson MA, O'Shea TM, Burda JE, Ao Y, Barlatey SL, Bernstein AM, Kim JH, James ND, Rogers A, Kato B, Wollenberg AL, Kawaguchi R, Coppola G, Wang C, Deming TJ, He Z, Courtine G, Sofroniew MV. Required growth facilitators propel axon regeneration across complete spinal cord injury. *Nature* 2018;**561**:396−400. https://doi.org/10.1038/s41586-018-0467-6.

19. Burns AS, Marino RJ, Kalsi-Ryan S, Middleton JW, Tetreault LA, Dettori JR, Mihalovich KE, Fehlings MG. Type and timing of rehabilitation following acute and subacute spinal cord injury: a systematic review. *Global Spine J* 2017;**7**:175S−94S. https://doi.org/10.1177/2192568217703084.

20. Roberts TT, Leonard GR, Cepela DJ. Classifications in brief: American spinal injury association (ASIA) impairment Scale. *Clin Orthop Relat Res* 2017;**475**:1499−504. https://doi.org/10.1007/s11999-016-5133-4.

21. Aimetti AA, Kirshblum S, Curt A, Mobley J, Grossman RG, Guest JD. Natural history of neurological improvement following complete (AIS A) thoracic spinal cord injury across three registries to guide acute clinical trial design and interpretation. *Spinal Cord* 2019;**57**:753−62. https://doi.org/10.1038/s41393-019-0299-8.

22. van Middendorp JJ, Hosman AJ, Donders ART, Pouw MH, Ditunno JF, Curt A, Geurts AC, Van de Meent H. A clinical prediction rule for ambulation outcomes after traumatic spinal cord injury: a longitudinal cohort study. *Lancet* 2011;**377**:1004−10. https://doi.org/10.1016/S0140-6736(10)62276-3.

23. Hicks KE, Zhao Y, Fallah N, Rivers CS, Noonan VK, Plashkes T, Wai EK, Roffey DM, Tsai EC, Paquet J, Attabib N, Marion T, Ahn H, Phan P. A simplified clinical prediction rule for prognosticating independent walking after spinal cord injury: a prospective study from a Canadian multicenter spinal cord injury registry. *Spine J* 2017;**17**:1383−92. https://doi.org/10.1016/j.spinee.2017.05.031.

24. El Tecle NE, Dahdaleh NS, Bydon M, Ray WZ, Torner JC, Hitchon PW. The natural history of complete spinal cord injury: a pooled analysis of 1162 patients and a meta-analysis of modern data. *J Neurosurg Spine* 2018;**28**:436−43. https://doi.org/10.3171/2017.7.SPINE17107.

25. Jones ML, Evans N, Tefertiller C, Backus D, Sweatman M, Tansey K, Morrison S. Activity-based therapy for recovery of walking in individuals with chronic spinal cord injury: results from a randomized clinical trial. *Arch Phys Med Rehabil* 2014;**95**. https://doi.org/10.1016/j.apmr.2014.07.400. 2239.e2−2246.e2.

26. Hicks AL, Martin KA, Ditor DS, Latimer AE, Craven C, Bugaresti J, McCartney N. Long-term exercise training in persons with spinal cord injury: effects on strength, arm ergometry performance and psychological well-being. *Spinal Cord* 2003;**41**:34−43. https://doi.org/10.1038/sj.sc.3101389.

27. Piira A, Lannem A, Sørensen M, Glott T, Knutsen R, Jørgensen L, Gjesdal K, Hjeltnes N, Knutsen S. Manually assisted body-weight supported locomotor training does not re-establish walking in non-walking subjects with chronic incomplete spinal cord injury: a randomized clinical trial. *J Rehabil Med* 2019;**51**: 113−9. https://doi.org/10.2340/16501977-2508.

28. Martinez SA, Nguyen ND, Bailey E, Doyle-Green D, Hauser HA, Handrakis JP, Knezevic S, Marett C, Weinman J, Romero AF, Santiago TM, Yang AH, Yung L, Asselin PK, Weir JP, Kornfeld SD, Bauman WA, Spungen AM, Harel NY. Multimodal cortical and subcortical exercise compared with treadmill training for spinal cord injury. *PLoS One* 2018;**13**:e0202130. https://doi.org/10.1371/journal.pone.0202130.

29. Senthilvelkumar T, Magimairaj H, Fletcher J, Tharion G, George J. Comparison of body weight-supported treadmill training versus body weight-supported overground training in people with incomplete tetraplegia: a pilot randomized trial. *Clin Rehabil* 2015;**29**:42−9. https://doi.org/10.1177/0269215514538068.

30. Spooren AIF, Janssen-Potten YJM, Kerckhofs E, Bongers HMH, Seelen HAM. Evaluation of a task-oriented client-centered upper extremity skilled performance training module in persons with tetraplegia. *Spinal Cord* 2011;**49**:1049−54. https://doi.org/10.1038/sc.2011.54.

31. Brazg G, Fahey M, Holleran CL, Connolly M, Woodward J, Hennessy PW, Schmit BD, Hornby TG. Effects of training intensity on locomotor performance in individuals with chronic spinal cord injury: a randomized crossover study. *Neurorehabil Neural Repair* 2017;**31**:944−54. https://doi.org/10.1177/1545968317731538.

32. Sandler EB, Roach KE, Field-Fote EC. Dose-response outcomes associated with different forms of locomotor training in persons with chronic motor-incomplete spinal cord injury. *J Neurotrauma* 2017;**34**:1903−8. https://doi.org/10.1089/neu.2016.4555.

33. Wu M, Landry JM, Kim J, Schmit BD, Yen S-C, McDonald J, Zhang Y. Repeat exposure to leg swing perturbations during treadmill training induces long-term retention of increased step length in human SCI: a pilot randomized controlled study. *Am J Phys Med Rehabil* 2016;**95**:911−20. https://doi.org/10.1097/PHM.0000000000000517.

34. Bye EA, Harvey LA, Gambhir A, Kataria C, Glinsky JV, Bowden JL, Malik N, Tranter KE, Lam CP, White JS, Gollan EJ, Arora M, Gandevia SC. Strength training for partially paralysed muscles in people with recent spinal cord injury: a within-participant randomised controlled trial. *Spinal Cord* 2017;**55**:460−5. https://doi.org/10.1038/sc.2016.162.

35. Lam T, Pauhl K, Ferguson A, Malik RN, Kin B, Krassioukov A, Eng JJ. Training with robot-applied resistance in people with motor-incomplete spinal cord injury: pilot study. *J Rehabil Res Dev* 2015;**52**: 113−30. https://doi.org/10.1682/JRRD.2014.03.0090.

36. Asboth L, Friedli L, Beauparlant J, Martinez-Gonzalez C, Anil S, Rey E, Baud L, Pidpruzhnykova G, Anderson MA, Shkorbatova P, Batti L, Pagès S, Kreider J, Schneider BL, Barraud Q, Courtine G. Cortico−reticulo−spinal circuit reorganization enables functional recovery after severe spinal cord contusion. *Nat Neurosci* 2018;**21**:576−88. https://doi.org/10.1038/s41593-018-0093-5.

37. van den Brand R, Heutschi J, Barraud Q, DiGiovanna J, Bartholdi K, Huerlimann M, Friedli L, Vollenweider I, Moraud EM, Duis S, Dominici N, Micera S, Musienko P, Courtine G. Restoring voluntary control of locomotion after paralyzing spinal cord injury. *Science* 2012;**336**:1182−5. https://doi.org/10.1126/science.1217416.

38. Harkema S, Gerasimenko Y, Hodes J, Burdick J, Angeli C, Chen Y, Ferreira C, Willhite A, Rejc E, Grossman RG, Edgerton VR. Effect of epidural stimulation of the lumbosacral spinal cord on voluntary movement, standing, and assisted stepping after motor complete paraplegia: a case study. *Lancet* 2011;**377**: 1938−47. https://doi.org/10.1016/S0140-6736(11)60547-3.

39. Rejc E, Angeli C, Harkema S. Effects of lumbosacral spinal cord epidural stimulation for standing after chronic complete paralysis in humans. *PLoS One* 2015;**10**:e0133998. https://doi.org/10.1371/journal.pone.0133998.

40. Angeli CA, Edgerton VR, Gerasimenko YP, Harkema SJ. Altering spinal cord excitability enables voluntary movements after chronic complete paralysis in humans. *Brain* 2014;**137**:1394−409. https://doi.org/10.1093/brain/awu038.

41. Angeli CA, Boakye M, Morton RA, Vogt J, Benton K, Chen Y, Ferreira CK, Harkema SJ. Recovery of overground walking after chronic motor complete spinal cord injury. *N Engl J Med* 2018;**379**:1244—50. https://doi.org/10.1056/NEJMoa1803588.

42. Gill ML, Grahn PJ, Calvert JS, Linde MB, Lavrov IA, Strommen JA, Beck LA, Sayenko DG, Van Straaten MG, Drubach DI, Veith DD, Thoreson AR, Lopez C, Gerasimenko YP, Edgerton VR, Lee KH, Zhao KD. Neuromodulation of lumbosacral spinal networks enables independent stepping after complete paraplegia. *Nat Med* 2018;**24**:1677—82. https://doi.org/10.1038/s41591-018-0175-7.

43. Lavrov I, Dy CJ, Fong AJ, Gerasimenko Y, Courtine G, Zhong H, Roy RR, Edgerton VR. Epidural stimulation induced modulation of spinal locomotor networks in adult spinal rats. *J Neurosci* 2008;**28**:6022—9. https://doi.org/10.1523/JNEUROSCI.0080-08.2008.

44. Capogrosso M, Wenger N, Raspopovic S, Musienko P, Beauparlant J, Bassi Luciani L, Courtine G, Micera S. A computational model for epidural electrical stimulation of spinal sensorimotor circuits. *J Neurosci* 2013;**33**: 19326—40. https://doi.org/10.1523/JNEUROSCI.1688-13.2013.

45. Takeoka A, Vollenweider I, Courtine G, Arber S. Muscle spindle feedback directs locomotor recovery and circuit reorganization after spinal cord injury. *Cell* 2014;**159**:1626—39. https://doi.org/10.1016/j.cell.2014.11.019.

46. Wenger N, Moraud EM, Gandar J, Musienko P, Capogrosso M, Baud L, Le Goff CG, Barraud Q, Pavlova N, Dominici N, Minev IR, Asboth L, Hirsch A, Duis S, Kreider J, Mortera A, Haverbeck O, Kraus S, Schmitz F, DiGiovanna J, van den Brand R, Bloch J, Detemple P, Lacour SP, Bézard E, Micera S, Courtine G. Spatiotemporal neuromodulation therapies engaging muscle synergies improve motor control after spinal cord injury. *Nat Med* 2016;**22**:138—45. https://doi.org/10.1038/nm.4025.

47. Formento E, Minassian K, Wagner F, Mignardot JB, Le Goff-Mignardot CG, Rowald A, Bloch J, Micera S, Capogrosso M, Courtine G. Electrical spinal cord stimulation must preserve proprioception to enable locomotion in humans with spinal cord injury. *Nat Neurosci* 2018;**21**:1728—41. https://doi.org/10.1038/s41593-018-0262-6.

48. Wagner FB, Mignardot J-B, Le Goff-Mignardot CG, Demesmaeker R, Komi S, Capogrosso M, Rowald A, Seáñez I, Caban M, Pirondini E, Vat M, McCracken LA, Heimgartner R, Fodor I, Watrin A, Seguin P, Paoles E, Van Den Keybus K, Eberle G, Schurch B, Pralong E, Becce F, Prior J, Buse N, Buschman R, Neufeld E, Kuster N, Carda S, von Zitzewitz J, Delattre V, Denison T, Lambert H, Minassian K, Bloch J, Courtine G. Targeted neurotechnology restores walking in humans with spinal cord injury. *Nature* 2018;**563**:65—71. https://doi.org/10.1038/s41586-018-0649-2.

49. Mignardot J-B, Le Goff CG, van den Brand R, Capogrosso M, Fumeaux N, Vallery H, Anil S, Lanini J, Fodor I, Eberle G, Ijspeert A, Schurch B, Curt A, Carda S, Bloch J, von Zitzewitz J, Courtine G. A multidirectional gravity-assist algorithm that enhances locomotor control in patients with stroke or spinal cord injury. *Sci Transl Med* 2017;**9**:eaah3621. https://doi.org/10.1126/scitranslmed.aah3621.

50. Gad P, Lee S, Terrafranca N, Zhong H, Turner A, Gerasimenko Y, Edgerton VR. Non-invasive activation of cervical spinal networks after severe paralysis. *J Neurotrauma* 2018;**35**:2145—58. https://doi.org/10.1089/neu.2017.5461.

51. Barra B, Roux C, Kaeser M, Schiavone G, Lacour SP, Bloch J, Courtine G, Rouiller EM, Schmidlin E, Capogrosso M. Selective recruitment of arm motoneurons in nonhuman primates using epidural electrical stimulation of the cervical spinal cord. In: *2018 40th Annual international conference of the IEEE Engineering in Medicine and Biology Society (EMBC)*. Honolulu, HI: IEEE; 2018. p. 1424—7. https://doi.org/10.1109/EMBC.2018.8512554.

52. Cortes M, Medeiros AH, Gandhi A, Lee P, Krebs HI, Thickbroom G, Edwards D. Improved grasp function with transcranial direct current stimulation in chronic spinal cord injury. *NeuroRehabilitation* 2017;**41**:51—9. https://doi.org/10.3233/NRE-171456.

53. Yozbatiran N, Keser Z, Davis M, Stampas A, O'Malley MK, Cooper-Hay C, Frontera J, Fregni F, Francisco GE. Transcranial direct current stimulation (tDCS) of the primary motor cortex and robot-assisted arm training in chronic incomplete cervical spinal cord injury: a proof of concept sham-randomized clinical study. *NeuroRehabilitation* 2016;**39**:401—11. https://doi.org/10.3233/NRE-161371.

54. Kumru H, Murillo N, Benito-Penalva J, Tormos JM, Vidal J. Transcranial direct current stimulation is not effective in the motor strength and gait recovery following motor incomplete spinal cord injury during Lokomat® gait training. *Neurosci Lett* 2016b;**620**:143—7. https://doi.org/10.1016/j.neulet.2016.03.056.

55. Raithatha R, Carrico C, Powell ES, Westgate PM, Chelette II KC, Lee K, Dunsmore L, Salles S, Sawaki L. Non-invasive brain stimulation and robot-assisted gait training after incomplete spinal cord injury: a randomized pilot study. *NeuroRehabilitation* 2016;**38**:15—25. https://doi.org/10.3233/NRE-151291.

56. Kumru H, Benito-Penalva J, Valls-Sole J, Murillo N, Tormos JM, Flores C, Vidal J. Placebo-controlled study of rTMS combined with Lokomat® gait training for treatment in subjects with motor incomplete spinal cord injury. *Exp Brain Res* 2016a; **234**:3447–55. https://doi.org/10.1007/s00221-016-4739-9.

57. Gomes-Osman J, Field-Fote EC. Improvements in hand function in adults with chronic tetraplegia following a multiday 10-Hz repetitive transcranial magnetic stimulation intervention combined with repetitive task practice. *J Neurol Phys Ther* 2015;**39**: 23–30. https://doi.org/10.1097/NPT.00000000 00000062.

58. Navarrete-Opazo A, Alcayaga J, Sepúlveda O, Rojas E, Astudillo C. Repetitive intermittent hypoxia and locomotor training enhances walking function in incomplete spinal cord injury subjects: a randomized, triple-blind, placebo-controlled clinical trial. *J Neurotrauma* 2017a;**34**:1803–12. https://doi.org/ 10.1089/neu.2016.4478.

59. Navarrete-Opazo A, Alcayaga JJ, Sepúlveda O, Varas G. Intermittent hypoxia and locomotor training enhances dynamic but not standing balance in patients with incomplete spinal cord injury. *Arch Phys Med Rehabil* 2017b;**98**:415–24. https://doi.org/10.1016/ j.apmr.2016.09.114.

60. Trumbower RD, Hayes HB, Mitchell GS, Wolf SL, Stahl VA. Effects of acute intermittent hypoxia on hand use after spinal cord trauma: a preliminary study. *Neurology* 2017;**89**:1904–7. https://doi.org/ 10.1212/WNL.0000000000004596.

61. Canon S, Shera A, Phan NMH, Lapicz L, Scheidweiler T, Batchelor L. Autonomic dysreflexia during urodynamics in children and adolescents with spinal cord injury or severe neurologic disease. *J Pediatr Urol* 2015;**11**:1–4.

62. Lindan R, Joiner E, Freehafer A, Hazel C. Incidence and clinical features of autonomic dysreflexia in patients with spinal cord injury. *Spinal Cord* 1980;**18**.

63. Snow J, Sideropoulos H, Kripke B, Freed M, Shah N, Schlesinger R. Autonomic hyperreflexia during cystoscopy in patients with high spinal cord injuries. *Spinal Cord* 1978;**15**.

64. Alan N, Ramer LM, Inskip JA, Golbidi S, Ramer MS, Laher I. Recurrent autonomic dysreflexia exacerbates vascular dysfunction after spinal cord injury. *Spine J* 2010;**10**.

65. Hubli M, Gee CM, Krassioukov AV. Refined assessment of blood pressure instability after spinal cord injury. *Am J Hypertens* 2014;**28**.

66. West CR, Squair JW, McCracken L, Currie KD, Somvanshi R, Yuen V. Cardiac consequences of autonomic dysreflexia in spinal cord injury. *Hypertension* 2016;**68**.

67. Riegger T, Conrad S, Liu K, Schluesener HJ, Adibzahdeh M, Schwab JM. Spinal cord injury-induced immune depression syndrome (SCI-IDS. *Eur J Neurosci* 2007;**25**.

68. Campagnolo DI, Keller SE, DeLisa JA, Glick TJ, Sipski ML, Schleifer SJ. Alteration of immune system function in tetraplegics. A pilot study. *Am J Phys Med Rehabil* 1994;**73**.

69. Brommer B, Engel O, Kopp MA, Watzlawick R, Müller S, Prüss H. Spinal cord injury-induced immune deficiency syndrome enhances infection susceptibility dependent on lesion level. *Brain* 2016;**139**.

70. Bluvshtein V, Korczyn AD, Pinhas I, Vered Y, Gelernter I, Catz A. Insulin resistance in tetraplegia but not in mid-thoracic paraplegia: is the mid-thoracic spinal cord involved in glucose regulation? *Spinal Cord* 2011;**49**:648–52. https://doi.org/ 10.1038/sc.2010.152.

71. Barth E, Albuszies G, Baumgart K, Matejovic M, Wachter U, Vogt J. Glucose metabolism and catecholamines. *Crit Care Med* 2007;**35**.

72. Karlsson A. Autonomic dysreflexia. *Spinal Cord* 1999; **37**:383–91. https://doi.org/10.1038/sj.sc.3100867.

73. West CR, Phillips AA, Squair JW, Williams AM, Walter M, Lam T, Krassioukov AV. Association of epidural stimulation with cardiovascular function in an individual with spinal cord injury. *JAMA Neurol* 2018; **75**:630. https://doi.org/10.1001/jamaneurol.2017.5055.

74. Aslan SC, Legg Ditterline BE, Park MC, Angeli CA, Rejc E, Chen Y, Ovechkin AV, Krassioukov A, Harkema SJ. Epidural spinal cord stimulation of lumbosacral networks modulates arterial blood pressure in individuals with spinal cord injury-induced cardiovascular deficits. *Front Physiol* 2018;**9**. https:// doi.org/10.3389/fphys.2018.00565.

75. Harkema SJ, Legg Ditterline B, Wang S, Aslan S, Angeli CA, Ovechkin A, Hirsch GA. Epidural spinal cord stimulation training and sustained recovery of cardiovascular function in individuals with chronic cervical spinal cord injury. *JAMA Neurol* 2018a;**75**:1569. https://doi.org/10.1001/jamaneurol.2018.2617.

76. Harkema SJ, Wang S, Angeli CA, Chen Y, Boakye M, Ugiliweneza B, Hirsch GA. Normalization of blood pressure with spinal cord epidural stimulation after severe spinal cord injury. *Front Hum Neurosci* 2018b;**12**. https://doi.org/10.3389/fnhum. 2018.00083.

77. Darrow D, Balser D, Netoff TI, Krassioukov A, Phillips A, Parr A, Samadani U. Epidural spinal cord stimulation facilitates immediate restoration of dormant motor and autonomic supraspinal pathways after chronic neurologically complete spinal cord injury. *J Neurotrauma* 2019. https://doi.org/10.1089/neu.2018.6006.

78. Burns AS, Rivas DA, Ditunno JF. The management of neurogenic bladder and sexual dysfunction after spinal cord injury. *Spine* 2001;**26**:S129–36. https://doi.org/10.1097/00007632-200112151-00022.

79. Elmelund M, Biering-Sørensen F, Due U, Klarskov N. The effect of pelvic floor muscle training and intravaginal electrical stimulation on urinary incontinence in women with incomplete spinal cord injury: an investigator-blinded parallel randomized clinical trial. *Int Urogynecol J* 2018;**29**:1597–606. https://doi.org/10.1007/s00192-018-3630-6.

80. Burns AS, St-Germain D, Connolly M, Delparte JJ, Guindon A, Hitzig SL, Craven BC. Phenomenological study of neurogenic bowel from the perspective of individuals living with spinal cord injury. *Arch Phys Med Rehabil* 2015;**96**. https://doi.org/10.1016/j.apmr.2014.07.417. 49.e1–55.e1.

81. Brindley G. The first 500 patients with sacral anterior root stimulator implants: general description. *Spinal Cord* 1994;**32**.

82. Koldewijn EL, Van Kerrebroeck PE, Rosier PF, Wijkstra H, Debruyne FM. Bladder compliance after posterior sacral root rhizotomies and anterior sacral root stimulation. *J Urol* 1994;**151**.

83. Robinson L, Grant A, Weston P, Stephenson T, Lucas M, Thomas D. Experience with the Brindley anterior sacral root stimulator. *BJU Int* 1988;**62**.

84. Van Kerrebroeck P, Koldewijn E, Debruyne F. Worldwide experience with the Finetech-Brindley sacral anterior root stimulator. *Neurourol Urodyn* 1993;**12**.

85. Van Kerrebroeck P, E K, H W, F D. Urodynamic evaluation before and after intradural posterior sacral rhizotomies and implantation of the Finetech-Brindley anterior sacral root stimulator. *Urodinamica* 1992;**1**:7–12.

86. Barbeau H, Rossignol S. Recovery of locomotion after chronic spinalization in the adult cat. *Brain Res* 1987;**412**.

87. de Leon R, Hodgson J, Roy R, Edgerton V. Locomotor capacity attributable to step training versus spontaneous recovery after spinalization in adult cats. *J Neurophysiol* 1998;**79**.

88. Jankowska E. Interneuronal relay in spinal pathways from proprioceptors. *Prog Neurobiol* 1992;**38**.

89. Craven BC, Giangregorio L, Robertson L, Delparte JJ, Ashe MC, Eng JJ. Sublesional osteoporosis prevention, detection, and treatment: a decision guide for rehabilitation clinicians treating patients with spinal cord injury. *Crit Rev Phys Rehabil Med* 2008;**20**:277–321. https://doi.org/10.1615/CritRevPhysRehabilMed.v20.i4.10.

90. Garland DE, Adkins RH, Stewart CA, Ashford R, Vigil D. Regional osteoporosis in women who have a complete spinal cord injury. *J Bone Jt Surg Am* 2001;**83**.

91. Jiang SD, Jiang LS, Dai LY. Mechanisms of osteoporosis in spinal cord injury. *Clin Endocrinol (Oxf)* 2006;**65**.

92. Shapiro J, Smith B, Beck T, Ballard P, Dapthary M, BrintzenhofeSzoc K. Treatment with zoledronic acid ameliorates negative geometric changes in the proximal femur following acute spinal cord injury. *Calcif Tissue Int* 2007;**80**.

93. Maïmoun L, Fattal C, Micallef JP, Peruchon E, Rabischong P. Bone loss in spinal cord-injured patients: from physiopathology to therapy. *Spinal Cord* 2005;**44**.

94. Shields RK, Dudley-Javoroski S, Law LAF. Electrically induced muscle contractions influence bone density decline after spinal cord injury. *Spine* 2006;**31**:548–53. https://doi.org/10.1097/01.brs.0000201303.49308.a8.

95. Shields RK, Dudley-Javoroski S. Musculoskeletal plasticity after acute spinal cord injury: effects of long-term neuromuscular electrical stimulation training. *J Neurophysiol* 2006;**95**:2380–90. https://doi.org/10.1152/jn.01181.2005.

96. Lai C, Chang W, Chan W, Peng C, Shen L, Chen J, Chen S. Effects of functional electrical stimulation cycling exercise on bone mineral density loss in the early stages of spinal cord injury. *J Rehabil Med* 2010;**42**:150–4. https://doi.org/10.2340/16501977-0499.

97. Lambach RL, Stafford NE, Kolesar JA, Kiratli BJ, Creasey GH, Gibbons RS, Andrews BJ, Beaupre GS. Bone changes in the lower limbs from participation in an FES rowing exercise program implemented within two years after traumatic spinal cord injury. *J Spinal Cord Med* 2020;**43**:306–14. https://doi.org/10.1080/10790268.2018.1544879.

98. Bélanger M, Stein RB, Wheeler GD, Gordon T, Leduc B. Electrical stimulation: can it increase muscle strength and reverse osteopenia in spinal cord injured individuals? *Arch Phys Med Rehabil* 2000;**81**:1090–8. https://doi.org/10.1053/apmr.2000.7170.

99. Shields RK, Dudley-Javoroski S. Musculoskeletal adaptations in chronic spinal cord injury: effects of long-term soleus electrical stimulation training. *Neurorehabil Neural Repair* 2007;**21**:169–79. https://doi.org/10.1177/1545968306293447.

100. Chen S-C, Lai C-H, Chan WP, Huang M-H, Tsai H-W, Chen J-JJ. Increases in bone mineral density after functional electrical stimulation cycling exercises in spinal cord injured patients. *Disabil Rehabil* 2005;**27**:1337–41. https://doi.org/10.1080/09638280500164032.

101. Dudley-Javoroski S, Saha PK, Liang G, Li C, Gao Z, Shields RK. High dose compressive loads attenuate bone mineral loss in humans with spinal cord injury. *Osteoporos Int* 2012;**23**:2335—46. https://doi.org/10.1007/s00198-011-1879-4.

102. Frotzler A, Coupaud S, Perret C, Kakebeeke TH, Hunt KJ, Donaldson N de N, Eser P. High-volume FES-cycling partially reverses bone loss in people with chronic spinal cord injury. *Bone* 2008;**43**:169—76. https://doi.org/10.1016/j.bone.2008.03.004.

103. Mohr T, Pødenphant J, Biering—Sørensen F, Galbo H, Thamsborg G, Kjær M. Increased bone mineral density after prolonged electrically induced cycle training of paralyzed limbs in spinal cord injured man. *Calcif Tissue Int* 1997;**61**:22—5. https://doi.org/10.1007/s002239900286.

104. Draghici AE, Taylor JA, Bouxsein ML, Shefelbine SJ. Effects of FES-rowing exercise on the time-dependent changes in bone microarchitecture after spinal cord injury: a cross-sectional investigation: FES-rowing alters bone loss after. *SCI JBMR Plus* 2019;**3**:e10200. https://doi.org/10.1002/jbm4.10200.

105. Alekna V, Tamulaitiene M, Sinevicius T, Juocevicius A. Effect of weight-bearing activities on bone mineral density in spinal cord injured patients during the period of the first two years. *Spinal Cord* 2008;**46**:727—32. https://doi.org/10.1038/sc.2008.36.

106. Hicks A, Ginis KM, Pelletier C, Ditor D, Foulon B, Wolfe D. The effects of exercise training on physical capacity, strength, body composition and functional performance among adults with spinal cord injury: a systematic review. *Spinal Cord* 2011;**49**.

107. Nightingale TE, Walhin J-P, Thompson D, Bilzon JLJ. Impact of exercise on cardiometabolic component risks in spinal cord—injured humans: med. *Sci Sports Exerc* 2017;**49**:2469—77. https://doi.org/10.1249/MSS.0000000000001390.

108. Paulson TA, Goosey-Tolfrey VL, Lenton JP, Leicht CA, Bishop NC. Spinal cord injury level and the circulating cytokine response to strenuous exercise. *Med Sci Sports Exerc* 2013;**45**.

109. DeFronzo RA. Lilly lecture. *Diabetes 1988* 1987;**37**(6):667—87.

110. Adams MM, Hicks AL. Spasticity after spinal cord injury. *Spinal Cord* 2005;**43**.

111. Burns AS, Lanig I, Grabljevec K, New PW, Bensmail D, Ertzgaard P, Nene AV. Optimizing the management of disabling spasticity following spinal cord damage: the ability network—an international initiative. *Arch Phys Med Rehabil* 2016;**97**:2222—8. https://doi.org/10.1016/j.apmr.2016.04.025.

112. Morawietz C, Moffat F. Effects of locomotor training after incomplete spinal cord injury: a systematic review. *Arch Phys Med Rehabil* 2013;**94**.

113. Effects of robot-assisted locomotor training in patients with gait disorders following neurological injury: an integrated EMG and kinematic approach. In: Mazzoleni S, Battini E, Stampacchia G, Tombini T, editors. *2015 IEEE International Conference on Rehabilitation Robotics (ICORR)*. IEEE; 2015.

114. Stampacchia G, Rustici A, Bigazzi S, Gerini A, Tombini T, Mazzoleni S. Walking with a powered robotic exoskeleton: subjective experience, spasticity and pain in spinal cord injured persons. *NeuroRehabilitation* 2016;**39**:277—83. https://doi.org/10.3233/NRE-161358.

115. Barbeau H. Locomotor training in neurorehabilitation: emerging rehabilitation concepts. *Neurorehabil Neural Repair* 2003;**17**.

116. Barbeau H, Ladouceur M, Mirbagheri MM, Kearney RE. The effect of locomotor training combined with functional electrical stimulation in chronic spinal cord injured subjects: walking and reflex studies. *Brain Res Brain Res Rev* 2002;**40**.

117. Yang JF, Fung J, Edamura M, Blunt R, Stein RB, Barbeau H. H-reflex modulation during walking in spastic paretic subjects. *Can J Neurol Sci* 1991;**18**.

118. Edgerton VR, Tillakaratne NJ, Bigbee AJ, Leon RD, Roy RR. Plasticity of the spinal neural circuitry after injury. *Annu Rev Neurosci* 2004;**27**:145—67.

119. Harkema SJ. Plasticity of interneuronal networks of the functionally isolated human spinal cord. *Brain Res Rev* 2008;**57**.

120. Wolpaw J. *Spinal cord plasticity in acquisition and maintenance of motor skills*. Oxford; 2007. p. 155—69.

121. Miller LE, Zimmermann AK, Herbert WG. Clinical effectiveness and safety of powered exoskeleton-assisted walking in patients with spinal cord injury: systematic review with meta-analysis. *Med Devices* 2016;**9**.

122. Craven DE, Kunches LM, Kilinsky V, Lichtenberg DA, Make BJ, McCabe WR. Risk factors for pneumonia and fatality in patients receiving continuous mechanical ventilation. *Am Rev Respir Dis* 1986;**133**:792—6.

123. Hirschfeld S, Exner G, Luukkaala T, Baer GA. Mechanical ventilation or phrenic nerve stimulation for treatment of spinal cord injury-induced respiratory insufficiency. *Spinal Cord* 2008;**46**:738—42. https://doi.org/10.1038/sc.2008.43.

124. Posluszny Jr JA, Onders R, Kerwin AJ, Weinstein MS, Stein DM, Knight J. Multicenter review of diaphragm pacing in spinal cord injury: successful not only in weaning from ventilators but also in bridging to independent respiration. *J Trauma Acute Care Surg* 2014;**76**.

125. Onders RP, Elmo M, Kaplan C, Schilz R, Katirji B, Tinkoff G. Long-term experience with diaphragm pacing for traumatic spinal cord injury: early implantation should be considered. *Surgery* 2018;**164**:705—11. https://doi.org/10.1016/j.surg.2018.06.050.

126. Alam M, Rodrigues W, Pham BN, Thakor NV. Brain-machine interface facilitated neurorehabilitation via spinal stimulation after spinal cord injury: recent progress and future perspectives. *Brain Res* 2016;**1646**:25—33.

127. György B, Costas AA, Christof K. The origin of extracellular fields and currents — EEG, ECoG, LFP and spikes. *Nat Rev Neurosci* 2012;**13**.

128. Collinger JL, Wodlinger B, Downey JE, Wang W, Tyler-Kabara EC, Weber DJ. High-performance neuroprosthetic control by an individual with tetraplegia. *Lancet* 2013;**381**.

129. Bensmaia SJ, Miller LE. Restoring sensorimotor function through intracortical interfaces: progress and looming challenges. *Nat Rev Neurosci* 2014;**15**:313—25. https://doi.org/10.1038/nrn3724.

130. Alam M, He J. Lower-limb neuroprostheses: restoring walking after spinal cord injury. *Emerg Theory Pract Neuroprosthetics IGI Glob* 2014:153—80.

Neuroprotective strategies: translational perspectives

James Hong[1], Noah Poulin[2], Michael G. Fehlings[3]

[1]Krembil Research Institute, Toronto, ON, Canada; [2]University of Cambridge, Cambridge, United Kingdom; [3]Division of Genetics and Development, Krembil Research Institute, University Health Network, Toronto, ON, Canada

Introduction

Despite a number of clinical trials in therapeutics for spinal cord injury, effective neuroprotective strategies remain elusive, highlighting the importance of continued research in basic and preclinical areas. Over the past 50 years, only methylprednisolone (MPSS) has been a pharmacological standard of care after SCI, despite concerns about limited efficacy and widespread debate about its use. Neuroprotective strategies in SCI may be classified into different mechanistic categories: apoptosis prevention, excitotoxicity mediation, and modulation of inflammatory processes.[1] It is important to note, however, that physical, immunological, and biochemical pathophysiology after SCI are interrelated. Many potential therapies (Table 28.1), therefore, exert multiple neuroprotective effects and combinatorial strategies may maximize functional outcomes.

Excitotoxicity

Extracellular glutamate concentration increases excessively after spinal cord injury, damaging tissue through excitotoxic cell death.[35,36] The repeated ionotropic glutamate receptor stimulation results in intracellular Na^+ accumulation, intracellular acidosis, and cytotoxic edema,[37–40] in addition to mitochondrial dysfunction.[41–44] Riluzole (2-amino-6-(tri-fluoromethoxy)benzothiazole) is a sodium channel blocker approved by the FDA for the treatment of amyotrophic lateral sclerosis (ALS). Its neuroprotective effects have been widely reported in preclinical models (reviewed by Nagoshi et al.[2]) and are fundamentally achieved by inhibition of Na^+ influx, reducing subsequent cytosolic Ca^{2+} accumulation. This reduces the activation of pro-inflammatory cytosolic enzymes, namely calpains and phospholipases. Inhibition of Na^+ influx also reduces the excessive

Neural Repair and Regeneration after Spinal Cord Injury and Spine Trauma
https://doi.org/10.1016/B978-0-12-819835-3.00020-4

© 2022 Elsevier Inc. All rights reserved.

TABLE 28.1   Summary of neuroprotective strategies for spinal cord injury.

Drug	Mechanism	References
Antiexcitotoxicity		
Riluzole	Na^{2+} channel antagonist	2−4,5,6
Magnesium	NMDA antagonist, vasoconstriction	7
Immune modulation		
DMF	Microglial modulation	8,9,10,11
IFNB	Decreased immune cell infiltration, induces antiinflammatory phenotype	12−14
Azithromycin	Antiinflammatory phenotype induction in phagocytic cells	15
VX-210	Rho inhibition	16,17
MSCs	Macrophage phenotype modulation, decreased immune cell infiltration BSCB maintenance	18,19,20,21
Flufenamic acid	Unknown	22
Glibenclamide	SUR1 inhibition	23
Enzastaurin	PKC-β inhibition	24,25
Midostaurin	PKC-β inhibition	26
IVIG	Immune modulation and BSCB maintenance (undefined mechanism)	27−29
DHA	Multiple/unknown	30,31
Progesterone	Unknown	32,33, 34

glutamate release responsible for neurotoxicity, both through impedance of glutamate release and activation of glutamate transporters.[3−5] A 2010 Phase I/IIA trial enrolled 36 patients with SCI who received riluzole in 28 doses of 50 mg.[5] The ensuing 2014 Phase IIB/III trial (NCT01597518) will enroll approximately 351 patients and results are expected in 2021. Riluzole's preclinical efficacy supports the targeting of ion channels in reducing excitotoxicity after SCI.

BK channels

In axons, large conductance, calcium- and voltage-dependent K^+ channels (BK) are covered by myelin in the juxtaparanodal regions and only exposed to the extracellular environment in pathological states of demyelination, including chronic SCI.[46] These authors suggest that pharmacological inhibition of BK channels could provide a novel neuroprotective strategy after SCI, as their activation decreased high-frequency axonal conductance.[46] In contrast, Jacobsen et al. showed that increased activation of BK channels with isopimaric acid was neuroprotective.[47] Although the mechanism is not clearly understood, these authors proposed that isopimaric acid counters ROS-induced excitotoxic inhibition of BK channels through a decrease in activation potential. The corresponding increase in BK activity decreases Ca^{2+} influx through hyperpolarization, inactivating Ca^{2+} channels.[47,48]

Magnesium

Magnesium is a physiological NMDA receptor antagonist and has been shown to decrease excitotoxicity and tissue loss while improving functional recovery in animal models of SCI.[49,50] Importantly, however, the high concentrations (300–600 mg/kg) sometimes administered in animal models are above the safety threshold in humans.[2] Coadministration of magnesium salts in polyethylene glycol reduces the concentration of Mg^{2+} needed for effective neuroprotection.[51] Kwon et al.[7] used a contusion rat SCI model and established maximum efficacy within a 4-hour window of SCI with an $MgCl_2$ dose of 254 μmol/kg in 5 infusions, 8 hours apart. Lee et al.[52] report both functional improvements with $MgCl_2$ in PEG and higher white matter sparing when compared to controls. Interestingly, $MgCl_2$ alone was also shown to improve epicenter blood flow in a contusive SCI rat model,[53] increasing the number of perfused vessels without increasing hemorrhage. The authors suggest that the protective effects on blood vessels were related to reduction of vasoconstriction surrounding the lesion. In this study, $MgCl_2$ (760 μmol/kg) was infused every 6 hours for 7 days and did not improve neuron, oligodendrocyte, or white matter loss nor improve functional recovery.[53] In a porcine model of SCI, Streijger et al.[54] reported no improvements in functional outcomes or tissue sparing using either AC105 ($MgCl_2$ in PEG) or $MgSO_4$ alone. A Phase I/II trial (NCT01750684) evaluating $MgCl_2$ was terminated due to insufficient enrollment in 2015.

Immunological modulation

Immunological modulation is an important neuroprotective strategy, as inflammation arising from cellular and molecular immunity occurs immediately after SCI and continues for several weeks.[55] Initially, this response is dominated by microglia, which migrate to the site of injury, proliferate, and phagocytose injured and dying cells, secreting cytokines and proteases that contribute to cellular damage and edema.[56] Approximately 24 hours after SCI, neutrophils infiltrate the lesion site and blood monocytes differentiate into macrophages, which remain at the site of injury,[57–59] chronically releasing pro-inflammatory cytokines, chemokines, nitric oxide, ions, and proteases.[60,61]

Microglial phenotype modulation

With extended inflammation, there is an increase in detrimental M1 microglial phenotypes. Meanwhile, the M2 phenotype displays neuroprotective effects through increased phagocytic activity, matrix deposition, and wound healing.[62] Microglial phenotype modulation has implications for many neurodegenerative disorders, and translation of therapeutic strategies in multiple sclerosis may hold promise for immunomodulation in SCI.

Dimethyl fumarate (DMF) is used for treatment of psoriasis and multiple sclerosis. It activates the Nrf2 antioxidant transcription factor, decreasing peripheral immune cell infiltration and promoting M2 microglial phenotypes. It decreases iNOS and TNFα levels while increasing antiinflammatory IL-10. DMF also inhibits glyceraldehyde 3-phosphate dehydrogenase (GADPH) in macrophages, further promoting the M2 phenotype.[8,9] Additionally, DMF inhibits inflammatory cytokine productions using Nrf2-independent pathways, including inhibition of polyubiquitin chain formation, which reduces NFkB and ERK1/2 activation.[63] In experimental autoimmune encephalitis, DMF was equally protective in inducing M2 monocytes in the Nrf2−/− as the wild-type mice.[10] Lim[11] showed that both DMF and monomethyl fumarate (MF) induced Nrf2 pathways, in addition to reducing monocyte infiltration across the BBB. These authors suggested that the latter function was likely mediated by a reduction in vascular cell adhesion molecule expression.

Interferon Beta (IFNβ) is an antiinflammatory cytokine which reduces immune cell infiltration across the BBB[12,13] and increases polarization to antiinflammatory phenotypes through upregulation of IL-10 receptors in monocytes and macrophages.[14]

Macrolide antibiotics bind bacterial ribosomes and have a 14-, 15-, or 16-membered lactone ring. Azithromycin is a second-generation macrolide erythromycin derivative, which has been shown to accumulate in phagocytic cells, where it increases IL-10, reduces IL-12, and reduces M1 phenotype macrophages.[15]

Rho inhibition

Glycoproteins resulting from destroyed myelin signal through the Nogo receptor to activate Rho,[64] which leads to the destruction of growth cones through cytoskeletal remodeling.[65,16] Inactivation of this pathway is associated with improved axonal regeneration.[66,67] VX-210, a Rho inhibitor, has been applied to acute cervical spinal cord injuries in a Phase IIb/III trial[17] where it was delivered using fibrin sealant (FS) on the surface of the cord, as this method was shown to significantly improve motor recovery in a mouse model.[16] The trial was terminated by the sponsor in 2018. In addition to the benefits for regeneration, inhibition of Rho-associated kinase (ROK) by Fasudil may also be neuroprotective, as it has been shown to decrease the inflammatory response and secondary injury cascade.[68]

Mesenchymal stromal cells

Mesenchymal stromal cells (MSCs) are an intensely studied population of multipotent mesodermal progenitors, which are relatively easy to isolate, and display homing properties to the site of SCI. They differentiate quickly into chondrocytes, adipocytes, myocytes, and osteoblasts.[18,69] MSCs can be derived from various sources including bone marrow (BM-MSC), umbilical cord (UC-MSC), and adipose-derived MSC (AD-MSC). These display antiinflammatory properties, notably through macrophage phenotype modulation, and provide trophic support for neuroprotection and regeneration.[19,20]

Spejo et al.[19] report that after axotomy, FS in conjunction with MSCs increased macrophage recruitment and brain-derived neurotrophic factor (BDNF) while increasing neuronal survival and gait recovery. Han and colleagues[70] used a contusion SCI rat model and showed that BMSC infusion decreased TLR4 expression as well as IL-1β and TNF-α levels. Both IL-1β and TNF-α stimulate the production of more proinflammatory cytokines, leading to apoptosis. Interestingly, in both humans and animal models, MSCs also decrease peripheral inflammatory cell infiltration through protection of the brain–spinal cord barrier (BSCB).[69] Various factors have been reported to contribute to this function, notably IL-10, TSG-6, and TIMP-6.[21–23] Recently, MSCs have been reported to reduce intraparenchymal hemorrhage and increase systemic levels of IL-10 in a spleen-dependent manner, which was further supported by altered splenic cytokine levels after MSC infusion.[73] Vawda et al. reported that infusion of BM-MSC reduced glial scarring and increased vascular density without decreasing cavity volume significantly.[74] In a Phase I/II controlled single-blind trial (ID: NCT00816803) where chronic cervical and thoracic SCI patients were administered autologous adherent BM-MSCs, smaller lesions and greater functional improvements were seen in thoracic SCI in comparison with cervical injuries.[75] Currently, there are several clinical trials assessing MSCs: a Phase II/II trial for BM-MSCs in chronic SCI (NCT01676441) has results expected in 2020, and a Phase I trial for AD-MSCs (NCT03308565). UC-MSCs are also being tested for use in subacute and chronic injury in Phase I/II trials (NCT03521323, NCT03505034).

Inflammasome modulation

IL-1β processing in the injured spinal cord requires inflammasome complexes, which are influenced by the purinergic P2X receptor subtype $P2X_4$.[76] In the first step of inflammasome activation in the CNS, the ATP-gated cation channel $P2X_4$ is activated by the ATP from damaged cells.[77] Despite development of $P2X_1$, $P2X_3$, and $P2X_7$ modulators, few effective inhibitors of $P2X_4$ have been described.[78,79] Notably, $P2XR_4$ knockout mice displayed reduced microglial BDNF signaling, reducing allodynia.[80] Importantly, BDNF also modulates synaptic plasticity.[81,82] Adverse effects have been reported in a $P2XR_4$-knockout cerebral ischemia mouse model.[83] The NLRP1 and NLRP3 inflammasomes eventually associate with the adaptor protein ASC and activate caspase-1, which activates IL-1β.[84] Activation of peroxisome proliferator–activated receptor-gamma (PPARγ) was shown to improve functional outcomes in rats after SCI through suppression of the NLRP3 inflammasome.[85] Neutralizing antibodies against the adaptor protein-ASC (ASC)[86] and against inflammasome complexes[87] decreases inflammation after CNS injury. Inhibition of IL-1β, the IL-1 Receptor A (IL-1RA), IL-18, and caspase-1 may also present effective antiinflammatory strategies in CNS injury.[77] Targeting the activity of inflammasomes after SCI remains a promising neuroprotective strategy in SCI.

Blood–spinal cord barrier maintenance

PKC-beta inhibition

Protein kinase C-β (PKC-β) is a member of the conventional (cPKC) group of the PKC serine/threonine kinase family, which is regulated by diacylglycerol, phospholipids, and Ca^{2+}. PKC regulates multiple physiological processes in the CNS, including signaling cascades for actin and microtubule dynamics and NMDA pathways.[88] Activation of PKCs increase the microvascular permeability[88–90] and therefore PKC inhibitors may counteract BSCB disruption induced by neutrophils in SCI, among other neurovascular conditions.[89–94] Additionally, PKC inhibition may attenuate the inhibitory effects of chondroitin sulfate proteoglycans and CNS myelin on neurite outgrowth.[26] This is one of several previously reported potential effects on neural regeneration.[95–97] Midostaurin was originally developed as a PKC inhibitor and was also found to inhibit VEGF receptors. Its VEGF-related inhibition of angiogenesis led to trials in diabetic retinopathy, although limited efficacy and gastrointestinal toxicity suppressed further development. Based on its inhibition of the fms-like tyrosine kinase 3 (FTL3) and the KIT proto-oncogene receptor tyrosine kinase, it is approved by the FDA and EMA for FTL3-mutated acute myeloid leukemia and systemic mastocytosis (SM), respectively.[98] Interestingly, no clinical trials using midostaurin in SCI are currently listed in the *clinicaltrials.gov* database. The preclinical evidence of neuroprotection discussed here may warrant further exploration of midostaurin's potential BSCB protection in SCI. Enzastaurin is a PKC-β inhibitor that has showed antiangiogenesis activity, as well as preferred cytotoxicity toward subsets of tumor cells.[99] Notably, at specific concentrations, administration of enzastaurin increased platelet aggregation, which may be relevant to thrombosis-derived ischemia and microvascular protection after SCI.[24] Stranahan et al. report that enzastaurin restored BBB integrity in a leptin receptor–deficient mouse model, significantly reducing macrophage infiltration and activation.[100] Carducci et al. reported toxicity at doses above 700 mg daily and observed only mild neurologic effects (sensory neuropathy).[101] A subsequent Phase II trial reported a relatively favorable safety profile, with no grade 4 toxic effects, at doses of 500 mg daily.[25] In this study, fatigue (grade 2) and syncope (grade 3) were observed in two patients and one patient, respectively. Interestingly, in their work with rats, Willeman et al. report significant memory impairment after enzastaurin administration.[102]

IgG infusion

Intravenous immunoglobulin G (IVIG) consists of pooled serum immunoglobulin G (IgG) from healthy donors. Despite a relatively undefined mechanism of action, it is currently approved by the FDA for treatment of autoimmune and immunodeficiency disorders, namely primary humoral deficiency idiopathic, thrombocytopenic purpura, and chronic inflammatory demyelinating polyneuropathy.[4,27] Brennan et al. report that IVIG decreased lesion enlargement and axonal degeneration after contusion SCI in a mouse model and suggest that IVIG may achieve these effects through attenuation of the complement response and a decrease in the phagocytic activity of macrophages.[28] In a rat model of cervical compression injury, IVIG administration reduced pro-inflammatory signaling (IL-1β and IL-6), neutrophil infiltration, and the level of BSCB-damaging metalloproteinase-9 (MMP-9).[29] In another study, IVIG improved BSCB integrity but was associated with increased serum levels of inflammatory cytokines (IL-8, MIP-1α, CCL-2/MCP-1, and IL-5). IgG also colocalized with vascular cell adhesion molecule 1 (VCAM-1) without reducing its expression. VCAM-1 is used by neutrophils in extravasation.[27] Notably, IgG colocalized with astrocytes and pericytes, which suggests a role in modulating immune cell infiltration through the BSCB. Administration of 2 g/kg resulted in increased tissue preservation, blood flow, and behavioral recovery, with the effect sizes comparable with MPSS, the former standard of care for acute SCI.[27]

TRMP4/SUR1

Upregulation of transient receptor potential melastatin 4 (Trmp4) and sulfonylurea receptor (SurI) after SCI is associated with secondary hemorrhage. Flufenamic acid (FFA) administration in a contusion SCI model decreased Trmp4 expression and capillary fragmentation. These authors reported smaller lesions and decreased secondary hemorrhage, in addition to decreased expression of MMP-9 and MMP-2. Overall, these results suggest that FFA may act as an inhibitor of BSCB disruption after SCI.[22] Glibenclamide is FDA approved for the treatment of diabetes and blocks the sulfonylurea-receptor-1-(SUR1)-regulated-Ca^{2+}-activated-[ATP]-sensitive nonspecific cation channel (NC_{Ca-ATP}). Simard et al. demonstrated its importance for capillary fragmentation following SCI and the use of glibenclamide in animal models showed decreased lesion volumes and improved behavioral outcomes.[23] A Phase I/II clinical trial administering glibenclamide within 8 hours in patients with cervical SCI (C2–C8) is currently underway and is expected to be completed in 2020 (ID: NCT02524379).

Remote ischemic preconditioning

In remote ischemic preconditioning (RIPC), cycles of ischemia and reperfusion are administered to a remote organ, which protects the injured organ from ischemic/reperfusion injury. This has been applied in the protection of several tissue types in animal models, including the brain, kidney, and heart, among others.[103] In an aortic occlusion spinal ischemia model in rabbits, Dong et al. reported improved neurological outcomes after RIPC. This study also suggests that the ROS produced in the remote organ during RIPC might increase the effects of endogenous antioxidant enzymes, which impart a neuroprotective effect in spinal cord ischemia.[104] Using a porcine model of cord ischemia, Haapanen et al. show that RIPC increased motor-evoked potentials in comparison to the control group, suggesting neuroprotection from nontraumatic ischemia of the spinal cord. Interestingly, however, the histopathological scores of edema, hemorrhage, infarction, and neuronal degeneration did not differ significantly between groups.[105] The same group reports that RIPC increased protection against oxidative stress as demonstrated by increased nuclear factor erythroid 2–related factor (Nrf2), which regulates the cellular antioxidant response. Interestingly, histopathological findings of edema, neuron degeneration, hemorrhage, and infarction were again not significantly different between groups.[106] Fukui et al. report that direct ischemic preconditioning (DIPC) of the rabbit spinal cord before occlusion-based ischemia improved neurological and histopathological outcomes, while kidney and limb RIPC resulted in no similarly protective effects. Notably, the number of morphologically normal neurons in the kidney and limb RIPC did not differ significantly from the control, and pretreatment with the ROS scavenger dimethylthiourea (DMTU) did not attenuate the neuroprotective effects of DIPC.[107] This appears to be in contrast with previously reported effects of oxidative stress.[104] Importantly, however, Dong et al.[104] showed that ROS was implicated in a systemic response to RIPC and did not explore oxidative stress in DIPC. RIPC may also exert its neuroprotective effects through maintenance of BSCB integrity after SCI. Jing et al. showed that it preserved the BSCB and improved motor function in rats after spinal cord occlusion ischemia. The effects on the BSCB were related to increased occludin expression, dependent on increased expression of cannabinoid 1 and 2 receptors (CB1/CB2).[108] RIPC's neuroprotective effects were also associated with an increased concentration of plasma VEGF through downregulation of miRNA-762 and miR-3072-5 in a mouse model of occlusive ischemia.[109] The effects of VEGF in neuroprotection are widely reported.[110] In a systematic review, Dezfulian and colleagues suggest that pre- and postconditioning may provide similar neuroprotection in cerebral ischemia.[111] These authors also highlight the differentiation between early and late effects of preconditioning, associated with posttranslational regulation and gene expression modulation, respectively.[111]

Angiopoietin-1 (Ang-1) induces expression of occludin and zona occludens-1 (ZO1) tight junction proteins, which are essential components of the BSCB.[112–114] Ang-1 protein levels are decreased after CNS injury, and this is associated with increased BSCB permeability after SCI.[115] Han et al.[116] showed functional improvement using coadministration of angiopoietin 1 with C16, which target the Tie2 and αvβ3 receptors, respectively. Both receptors regulate endothelial cell survival through the PI3K/AKT pathway and were shown to reduce leukocyte infiltration in vitro. Simultaneously increasing VEGF and Ang-1 was shown to stabilize the BSCB and improve functional recovery after contusion SCI in rats.[116]

Inhibition of platelet-derived growth factor receptor alpha (PDFR-α) by the receptor tyrosine kinase inhibitor imatinib helps decrease BBB permeability in stroke models.[117] Imatinib has been shown to increase BSCB integrity, tissue preservation, and improve functional outcomes after contusion SCI.[118] Imatinib was approved in 2001 for use in chronic myelogenous leukemia, SM, gastrointestinal stromal tumors, hypereosinophilia, and dermatofibrosarcoma protuberans. Imatinib's established safety profile and efficacy in preclinical models is encouraging and provides a considerable foundation for clinical trials, in addition to further research on PDFR inhibition for neuroprotection after SCI.[119]

Other

Docosahexaenoic acid

Recently, using a laminectomy-free contusion injury model of SCI, Marinelli et al.[30] examined the neuroprotective effects of docosahexaenoic acid (DHA). In this study, 0.082 mg/kg of intraperitoneal DHA was administered within 1 hour of SCI and repeated daily for 5 days. The

DHA-treated group displayed reduced astrocyte activation in the perilesional area, glial scar formation, and neuronal and oligodendrocyte apoptosis. Previously, in a compression SCI mouse model, DHA in a bolus dose of 250 nmol/kg reduced inflammation and apoptosis while improving limb function.[120] Lim et al.,[121] using a bolus of 500 ng/kg 30 minutes after compression SCI in mice, reported improved locomotor recovery, reduced cell loss, and reduced microglial activation. Satkunendrarajah and Fehlings[31] note that the multiple targets of DHA should be explored in relation to its effects on functional recovery after SCI. The varying injury types and severities used in animal models, in conjunction with a range of administration routes, have made the precise mechanism and beneficial effects of DHA difficult to ascertain.[31]

Progesterone

Progesterone administration after SCI has been reported to have beneficial effects on locomotor recovery and white matter preservation,[32,122] potentially acting through local regulation of glial myelin production.[123,124] A 2011 systematic review of 10 studies in traumatic SCI found that 8 used a complete thoracic transection model while the remaining 2 used contusion models.[125] In these 10 studies, progesterone was administered at 4, 8, or 16 mg/kg in rats. The two contusion studies produced different histological results, with Fee et al.[33] reporting no improvement while Thomas et al.[32] reported increased white matter sparing.

Combinatorial strategies

Nanoparticle delivery of ferulic acid conjugated glycol chitosan has been shown to reduce excitotoxicity both in vitro and in contusion SCI in rats, where it reduced cavity volume and the inflammatory response while preserving neurons.[126] QL6 is a self-assembling peptide biomaterial and shows neuroprotective effects both alone and in conjunction with NSCs, reducing inflammation, scarring, and motor function. Other polymers like Hyaluronan/methylcellulose (HAMC) can support the engraftment of cellular therapies and deliver neuroprotective growth factors. An engineered collagen-binding neurotrophic hepatocyte growth factor (CBD-HGF) improved functional and histological outcomes in mouse models of compressive SCI.[127] When combined with gelatin furfurylamine hydrogel (FA) scaffold, CBD-HGF improved outcomes in a complete transection model as compared to CBD-HGF alone or FA with naive HGF.[127] Other neurotrophic factors may present opportunities for neuroprotection after SCI. For example, BDNF has established neuroprotective effects through antioxidative, antiapoptosis, and antiautophagic mechanisms through ERK1, NO, and PI3K/AKT.[128]

Using a combination of UC-MSCs and NSCs increased the number of surviving cells after 2 weeks and maximized functional outcomes in comparison with MSC and NSC treatments alone or with control.[129] In another study, administration of both NSCs and lithium increased locomotor scores and increased the number of proliferated ED1 cells after transplantation.[130]

In a contusive mouse model of thoracic SCI, combination therapy using riluzole and magnesium chloride did not improve behavioral outcomes or the extent of lesion and histological measures when compared to riluzole alone. Riluzole alone displayed significant improvements when compared to the control, while there was no significant difference in tissue sparing between the saline control and the Mg group or the Mg + RIL group.[6]

Martins et al.[131] used a combination of riluzole and dantrolene in rat SCI and showed that these have a synergistic effect, with a larger number of neurons remaining in the perilesional area. In the TUNEL measurements of apoptosis, uninjured controls did not differ significantly from the SCI animals treated with riluzole and dantrolene.

Grosso et al.[132] examined a combination of liposomal Clodronate, Rolipram, and Chondroitinase ABC (ChABC) after hemisection SCI in rats. Liposomal clodronate is a nitrogenous bisphosphonate that depletes peripheral macrophages. Rolipram inhibits cAMP hydrolysis and decreases pro-inflammatory cytokines. ChABC is a bacterial enzyme that reduces the inhibitory chondroitin sulfate glycosaminoglycans, which is beneficial for axon regeneration. The histological analysis in this study showed that the combined treatments reduced lesion size and reduced macrophage accumulation at the site of injury. The combination therapy also induced greater functional recovery compared to individual and control treatment groups.

References

1. Siddiqui AM, Khazaei M, Fehlings MG. Translating mechanisms of neuroprotection, regeneration, and repair to treatment of spinal cord injury. *Prog Brain Res* 2015;**218**.
2. Nagoshi N, Fehlings MG. Investigational drugs for the treatment of spinal cord injury: review of preclinical studies and evaluation of clinical trials from Phase I to II. *Expet Opin Invest Drugs* 2015;**24**:645—58.
3. Coderre TJ, Kumar N, Lefebvre CD, Yu JSC. A comparison of the glutamate release inhibition and anti-allodynic effects of gabapentin, lamotrigine, and riluzole in a model of neuropathic pain. *J Neurochem* 2007;**100**:1289—99.
4. Fehlings MG, Ulndreaj A, Badner A. Promising neuroprotective strategies for traumatic spinal cord injury with a focus on the differential effects among anatomical levels of injury. *F1000Research* 2017;**6**.
5. Grossman RG, et al. A prospective, multicenter, phase I matched-comparison group trial of safety, pharmacokinetics, and preliminary efficacy of riluzole in patients with traumatic spinal cord injury. *J Neurotrauma* 2014;**31**:239—55.
6. Vasconcelos NL, et al. Combining neuroprotective agents: effect of riluzole and magnesium in a rat model of thoracic spinal cord injury. *Spine J* 2016;**16**:1015—24.
7. Kwon BK, et al. Magnesium chloride in a polyethylene glycol formulation as a neuroprotective therapy for acute spinal cord injury: preclinical refinement and optimization. *J Neurotrauma* 2009;**26**:1379—93.
8. Angiari S, O'Neill LA. Dimethyl fumarate: targeting glycolysis to treat MS. *Cell Res* 2018;**28**:613—5.
9. Nally FK, De Santi C, McCoy CE. Nanomodulation of macrophages in multiple sclerosis. *Cells* 2019;**8**:543.
10. Schulze-Topphoff U, et al. Dimethyl fumarate treatment induces adaptive and innate immune modulation independent of Nrf2. *Proc Natl Acad Sci USA* 2016;**113**:4777—82.
11. Lim JL, et al. Protective effects of monomethyl fumarate at the inflamed blood-brain barrier. *Microvasc Res* 2016;**105**:61—9.
12. Kraus J, Oschmann P. The impact of interferon-β treatment on the blood-brain barrier. *Drug Discov Today* 2006;**11**:755—62.
13. Veldhuis WB, et al. Interferon-beta prevents cytokine-induced neutrophil infiltration and attenuates blood-brain barrier disruption. *J Cerebr Blood Flow Metabol* 2003;**23**:1060—9.
14. Liu BS, Janssen HLA, Boonstra A. Type I and III interferons enhance IL-10R expression on human monocytes and macrophages, resulting in IL-10-mediated suppression of TLR-induced IL-12. *Eur J Immunol* 2012;**42**:2431—40.
15. Zhang B, et al. Macrolide derivatives reduce proinflammatory macrophage activation and macrophage-mediated neurotoxicity. *CNS Neurosci Ther* 2019;**25**:591—600.
16. Lord-Fontaine S, et al. Local inhibition of rho signaling by cell-permeable recombinant protein BA-210 prevents secondary damage and promotes functional recovery following acute spinal cord injury. *J Neurotrauma* 2008;**25**:1309—22.
17. Fehlings MG, et al. Rho inhibitor VX-210 in acute traumatic subaxial cervical spinal cord injury: design of the SPinal cord injury rho INhibition InvestiGation (SPRING) clinical trial. *J Neurotrauma* 2018;**35**:1049—56.
18. Cofano F, et al. Mesenchymal stem cells for spinal cord injury: current options, limitations, and future of cell therapy. *Int J Mol Sci* 2019;**20**.
19. Spejo AB, et al. Neuroprotection and immunomodulation following intraspinal axotomy of motoneurons by treatment with adult mesenchymal stem cells. *J Neuroinflammation* 2018;**15**:230.

20. Nakajima H, et al. Transplantation of mesenchymal stem cells promotes an alternative pathway of macrophage activation and functional recovery after spinal cord injury. *J Neurotrauma* 2012;**29**:1614—25.

21. Badner A, et al. Early intravenous delivery of human brain stromal cells modulates systemic inflammation and leads to vasoprotection in traumatic spinal cord injury. *Stem Cells Transl Med* 2016;**5**:991—1003.

22. Yao Y, et al. Flufenamic acid inhibits secondary hemorrhage and BSCB disruption after spinal cord injury. *Theranostics* 2018;**8**:4181—98.

23. Simard JM, et al. Endothelial sulfonylurea receptor 1-regulated NC Ca-ATP channels mediate progressive hemorrhagic necrosis following spinal cord injury. *J Clin Invest* 2007;**117**:2105—13.

24. Lesyk G, Fong T, Ruvolo PP, Jurasz P. The potential of enzastaurin to enhance platelet aggregation and growth factor secretion: implications for cancer cell survival. *J Thromb Haemostasis* 2015;**13**:1514—20.

25. Morschhauser F, et al. A phase II study of enzastaurin, a protein kinase C beta inhibitor, in patients with relapsed or refractory mantle cell lymphoma. *Ann Oncol* 2008;**19**:247—53.

26. Sivasankaran R, et al. PKC mediates inhibitory effects of myelin and chondroitin sulfate proteoglycans on axonal regeneration. *Nat Neurosci* 2004;**7**:261—8.

27. Chio JCT, et al. The effects of human immunoglobulin G on enhancing tissue protection and neurobehavioral recovery after traumatic cervical spinal cord injury are mediated through the neurovascular unit. *J Neuroinflammation* 2019;**16**.

28. Brennan FH, et al. IVIg attenuates complement and improves spinal cord injury outcomes in mice. *Ann Clin Transl Neurol* 2016;**3**:495—511.

29. Nguyen DH, et al. Immunoglobulin G (IgG) attenuates neuroinflammation and improves neurobehavioral recovery after cervical spinal cord injury. *J Neuroinflammation* 2012;**9**.

30. Marinelli S, et al. Innovative mouse model mimicking human-like features of spinal cord injury: efficacy of Docosahexaenoic acid on acute and chronic phases. *Sci Rep* 2019;**9**:1—16.

31. Satkunendrarajah K, Fehlings MG. Do omega-3 polyunsaturated fatty acids ameliorate spinal cord injury?: Commentary on: Lim et al., Improved outcome after spinal cord compression injury in mice treated with docosahexaeonic acid. *Exp Neurol* 2013;**249**:104—10. 239:13-27.

32. Thomas AJ, Nockels RP, Pan HQ, Shaffrey CI, Chopp M. Progesterone is neuroprotective after acute experimental spinal cord trauma in rats. *Spine* 1999;**24**:2134—8.

33. Fee DB, et al. Effects of progesterone on experimental spinal cord injury. *Brain Res* 2007;**1137**:146—52.

34. Gonzalez SL, Labombarda F, Gonzalez Deniselle MC, Mougel A, Guennoun R, Schumacher M, et al. Progesterone neuroprotection in spinal cord trauma involves up-regulation of brain-derived neurotrophic factor in motoneurons. *J Steroid Biochem Mol Biol* 2005;**94**(1—3): 143—9. https://doi.org/10.1016/j.jsbmb.2005.01.016. PMID: 15862959.

35. Liu D, Xu GY, Pan E, McAdoo DJ. Neurotoxicity of glutamate at the concentration released upon spinal cord injury. *Neuroscience* 1999;**93**:1383—9.

36. XU GY, Hughes MG, Ye Z, Hulsebosch CE, McAdoo DJ. Concentrations of glutamate released following spinal cord injury kill oligodendrocytes in the spinal cord. *Exp Neurol* 2004;**187**:329—36.

37. Faden AI, Simon RP. A potential role for excitotoxins in the pathophysiology of spinal cord injury. *Ann Neurol* 1988;**23**:623—6.

38. Choi DW. Ionic dependence of glutamate neurotoxicity. *J Neurosci* 1987;**7**:369—79.

39. Fehlings MG, Agrawal S. Role of sodium in the pathophysiology of secondary spinal cord injury. *Spine* 1995;**20**:2187—91.

40. Ates O, et al. Comparative neuroprotective effect of sodium channel blockers after experimental spinal cord injury. *J Clin Neurosci* 2007;**14**:658—65.

41. Agrawal SK, Fehlings MG. The effect of the sodium channel blocker QX-314 on recovery after acute spinal cord injury. *J Neurotrauma* 1997;**14**:81—8.

42. Li S, Jiang Q, Stys PK. Important role of reverse Na^+-Ca^{2+} exchange in spinal cord white matter injury at physiological temperature. *J Neurophysiol* 2000;**84**:1116—9.

43. Stout AK, Raphael HM, Kanterewicz BI, Klann E, Reynolds IJ. Glutamate-induced neuron death requires mitochondrial calcium uptake. *Nat Neurosci* 1998;**1**:366—73.

44. Peng TI, Jou MJ, Sheu SS, Greenamyre JT. Visualization of NMDA receptor-induced mitochondrial calcium accumulation in striatal neurons Tsung-I. *Exp Neurol* 1998;**149**:1—12.

45. Badhiwala JH, Ahuja CS, Fehlings MG. Time is spine: a review of translational advances in spinal cord injury. *J Neurosurg Spine* 2019;**30**:1—18.

46. Ye H, et al. Expression and functional role of BK channels in chronically injured spinal cord white matter. *Neurobiol Dis* 2012;**47**:225—36.

47. Jacobsen M, Lett K, Barden JM, Simpson GL, Buttigieg J. Activation of the large-conductance, voltage, and Ca^{2+}- activated K^+ (BK) channel in acute spinal cord injury in the Wistar rat is neuroprotective. *Front Neurol* 2018;**9**.

48. Griguoli M, Sgritta M, Cherubini E. Presynaptic BK channels control transmitter release: physiological relevance and potential therapeutic implications. *J Physiol* 2016;**594**:3489–500.

49. Süzer T, Coskun E, Islekel H, Tahta K. Neuroprotective effect of magnesium on lipid peroxidation and axonal function after experimental spinal cord injury. *Spinal Cord* 1999;**37**:480–4.

50. Kaptanoglu E, Beskonakli E, Solaroglu I, Kilinc A, Taskin Y. Magnesium sulfate treatment in experimental spinal cord injury: emphasis on vascular changes and early clinical results. *Neurosurg Rev* 2003;**26**:283–7.

51. Huang Z, et al. AC105 increases extracellular magnesium delivery and reduces excitotoxic glutamate exposure within injured spinal cords in rats. *J Neurotrauma* 2017;**34**:685–94.

52. Lee JHT, et al. Magnesium in a polyethylene glycol formulation provides neuroprotection after unilateral cervical spinal cord injury. *Spine* 2010;**35**:2041–8.

53. Muradov JM, Hagg T. Intravenous infusion of magnesium chloride improves epicenter blood flow during the acute stage of contusive spinal cord injury in rats. *J Neurotrauma* 2013;**30**:840–52.

54. Streijger F, et al. The evaluation of magnesium chloride within a polyethylene glycol formulation in a porcine model of acute spinal cord injury. *J Neurotrauma* 2016;**33**:2202–16.

55. Fehlings MG, Nguyen DH, Immunoglobulin G. A potential treatment to attenuate neuroinflammation following spinal cord injury. *J Clin Immunol* 2010;**30**(Suppl. 1):S109–12.

56. Gao Z, et al. Microglial activation and intracerebral hemorrhage. *Acta Neurochir* 2008;(Suppl.):51–3. https://doi.org/10.1007/978-3-211-09469-3_11.

57. Gensel JC, Zhang B. Macrophage activation and its role in repair and pathology after spinal cord injury. *Brain Res* 2015;**1619**:1–11.

58. Schnell L. Acute inflammatory responses to mechanical lesions in the CNS: differences between brain and spinal cord. *Eur J Neurosci* 1999;**11**:3648–58.

59. Tzekou A, Fehlings MG. Treatment of spinal cord injury with intravenous immunoglobulin G: preliminary evidence and future perspectives. *J Clin Immunol* 2014;**34**:132–8.

60. Gaudet AD, Popovich PG. Extracellular matrix regulation of inflammation in the healthy and injured spinal cord. *Exp Neurol* 2014;**258**:24–34.

61. Muldoon LL, et al. Immunologic privilege in the central nervous system and the blood-brain barrier. *J Cerebr Blood Flow Metabol* 2013;**33**:13–21.

62. Cherry JD, Olschowka JA, O'Banion MK. Neuroinflammation and M2 microglia: the good, the bad, and the inflamed. *J Neuroinflammation* 2014;**11**.

63. McGuire VA, et al. Dimethyl fumarate blocks pro-inflammatory cytokine production via inhibition of TLR induced M1 and K63 ubiquitin chain formation. *Sci Rep* 2016;**6**.

64. Yiu G, He Z. Glial inhibition of CNS axon regeneration. *Nat Rev Neurosci* 2006;**7**:617–27.

65. Niederöst B, Oertle T, Fritsche J, McKinney RA, Bandtlow CE. Nogo-A and myelin-associated glycoprotein mediate neurite growth inhibition by antagonistic regulation of RhoA and Rac1. *J Neurosci* 2002;**22**:10368–76.

66. Dergham P, et al. Rho signaling pathway targeted to promote spinal cord repair. *J Neurosci* 2002;**22**:6570–7.

67. Fournier AE, Takizawa BT, Strittmatter SM. Rho kinase inhibition enhances axonal regeneration in the injured CNS. *J Neurosci* 2003;**23**:1416–23.

68. Impellizzeri D, Mazzon E, Paterniti I, Esposito E, Cuzzocrea S. Effect of fasudil, a selective inhibitor of rho kinase activity, in the secondary injury associated with the experimental model of spinal cord trauma. *J Pharmacol Exp Therapeut* 2012;**343**:21–33.

69. Ahuja CS, et al. Traumatic spinal cord injury - repair and regeneration. *Clin Neurosurg* 2017;**80**:S22–90.

70. Han D, Wu C, Xiong Q, Zhou L, Tian Y. Anti-inflammatory mechanism of bone marrow mesenchymal stem cell transplantation in rat model of spinal cord injury. *Cell Biochem Biophys* 2015;**71**:1341–7.

71. Watanabe J, et al. Administration of TSG-6 improves memory after traumatic brain injury in mice. *Neurobiol Dis* 2013;**59**:86–99.

72. Li R, et al. TSG-6 attenuates inflammation-induced brain injury via modulation of microglial polarization in SAH rats through the SOCS3/STAT3 pathway. *J Neuroinflammation* 2018;**15**.

73. Badner A, et al. Splenic involvement in umbilical cord matrix-derived mesenchymal stromal cell-mediated effects following traumatic spinal cord injury. *J Neuroinflammation* 2018;**15**.

74. Vawda R, et al. Early intravenous infusion of mesenchymal stromal cells exerts a tissue source age-dependent beneficial effect on neurovascular integrity and neurobehavioral recovery after traumatic cervical spinal cord injury. *Stem Cells Transl Med* 2019;**8**:639–49.

75. El-Kheir WA, et al. Autologous bone marrow-derived cell therapy combined with physical therapy induces functional improvement in chronic spinal cord injury patients. *Cell Transplant* 2014;**23**:729–45.

76. de Rivero Vaccari JP, et al. P2X 4 receptors influence inflammasome activation after spinal cord injury. *J Neurosci* 2012;**32**:3058—66.

77. de Rivero Vaccari JP, Dietrich WD, Keane RW. Therapeutics targeting the inflammasome after central nervous system injury. *Transl Res* 2016;**167**:35—45.

78. Suurväli J, Boudinot P, Kanellopoulos J, Rüütel Boudinot S. P2X4: a fast and sensitive purinergic receptor. *Biomed J* 2017;**40**:245—56.

79. North RA, Jarvis MF. P2X receptors as drug targets. *Mol Pharmacol* 2013;**83**:759—69.

80. Ulmann L, et al. Up-regulation of P2X4 receptors in spinal microglia after peripheral nerve injury mediates BDNF release and neuropathic pain. *J Neurosci* 2008;**28**:11263—8.

81. Lu B. Acute and long-term synaptic modulation by neurotrophins. *Prog Brain Res* 2004:135—50. https://doi.org/10.1016/S0079-6123(03)46010-X.

82. Crozier RA, Bi C, Han YR, Plummer MR. BDNF modulation of NMDA receptors is activity dependent. *J Neurophysiol* 2008;**100**:3264—74.

83. Verma R, et al. Deletion of the P2X4 receptor is neuroprotective acutely, but induces a depressive phenotype during recovery from ischemic stroke. *Brain Behav Immun* 2017;**66**:302—12.

84. Freeman LC, Ting JP-Y. The pathogenic role of the inflammasome in neurodegenerative diseases. *J Neurochem* 2016;**136**:29—38.

85. Meng QQ, et al. PPAR-γ Activation exerts an anti-inflammatory effect by suppressing the NLRP3 inflammasome in spinal cord-derived neurons. *Mediat Inflamm* 2019;**2019**.

86. Vaccari JPDR, Lotocki G, Marcillo AE, Dietrich WD, Keane RW. A molecular platform in neurons regulates inflammation after spinal cord injury. *J Neurosci* 2008. https://doi.org/10.1523/JNEUROSCI.0157-08.2008.

87. De Rivero Vaccari JP, et al. Therapeutic neutralization of the NLRP1 inflammasome reduces the innate immune response and improves histopathology after traumatic brain injury. *J Cerebr Blood Flow Metabol* 2009;**29**:1251—61.

88. Pastore D, et al. Age-dependent levels of protein kinase Cs in brain: reduction of endogenous mechanisms of neuroprotection. *Int J Mol Sci* 2019;**20**:3544.

89. Tang Y, et al. Protein kinase C-delta inhibition protects blood-brain barrier from sepsis-induced vascular damage. *J Neuroinflammation* 2018;**15**.

90. Qi X, Inagaki K, Sobel RA, Mochly-Rosen D. Sustained pharmacological inhibition of δPKC protects against hypertensive encephalopathy through prevention of blood-brain barrier breakdown in rats. *J Clin Invest* 2008;**118**:173—82.

91. Lanz TV, et al. Protein kinase Cβ as a therapeutic target stabilizing blood-brain barrier disruption in experimental autoimmune encephalomyelitis. *Proc Natl Acad Sci USA* 2013;**110**:14735—40.

92. Aiello LP, et al. Vascular endothelial growth factor-induced retinal permeability is mediated by protein kinase C in vivo and suppressed by an orally effective β-isoform-selective inhibitor. *Diabetes* 1997;**46**:1473—80.

93. Habib A, et al. Sirolimus-FKBP12.6 impairs endothelial barrier function through protein kinase C-α activation and disruption of the p120-vascular endothelial cadherin interaction. *Arterioscler Thromb Vasc Biol* 2013;**33**:2425—31.

94. Lucke-Wold BP, et al. Common mechanisms of Alzheimer's disease and ischemic stroke: the role of protein kinase C in the progression of age-related neurodegeneration. *J Alzheim Dis* 2015;**43**:711—24.

95. Yang Q, et al. Protein kinase C activation decreases peripheral actin network density and increases central nonmuscle myosin II contractility in neuronal growth cones. *Mol Biol Cell* 2013;**24**:3097—114.

96. Yang P, Li ZQ, Song L, Yin YQ. Protein kinase C regulates neurite outgrowth in spinal cord neurons. *Neurosci Bull* 2010;**26**:117—25.

97. Wang X, Hu J, She Y, Smith GM, Xu XM. Cortical PKC inhibition promotes axonal regeneration of the corticospinal tract and forelimb functional recovery after cervical dorsal spinal hemisection in adult rats. *Cerebr Cortex* 2014;**24**:3069—79.

98. Stone RM, Manley PW, Larson RA, Capdeville R. Midostaurin: its odyssey from discovery to approval for treating acute myeloid leukemia and advanced systemic mastocytosis. *Blood Adv* 2018;**2**:444—53.

99. Ouaret D, Larsen AK. Protein kinase C β inhibition by enzastaurin leads to mitotic missegregation and preferential cytotoxicity toward colorectal cancer cells with chromosomal instability (CIN). *Cell Cycle* 2014;**13**:2697—706.

100. Stranahan AM, Hao S, Dey A, Yu X, Baban B. Blood-brain barrier breakdown promotes macrophage infiltration and cognitive impairment in leptin receptor-deficient mice. *J Cerebr Blood Flow Metabol* 2016;**36**:2108—21.

101. Carducci MA, et al. Phase I dose escalation and pharmacokinetic study of enzastaurin, an oral protein kinase C beta inhibitor, in patients with advanced cancer. *J Clin Oncol* 2006;**24**:4092—9.

102. Willeman MN, et al. The PKC-β selective inhibitor, Enzastaurin, impairs memory in middle-aged rats. *PLoS One* 2018;**13**.

103. Yu Q, Huang J, Hu J, Zhu H. Advance in spinal cord ischemia reperfusion injury: blood—spinal cord barrier and remote ischemic preconditioning. *Life Sci* 2016;**154**:34—8.

104. Dong HL, et al. Limb remote ischemic preconditioning protects the spinal cord from ischemia-reperfusion injury: a newly identified nonneuronal but reactive oxygen species-dependent pathway. *Anesthesiology* 2010;**112**:881—91.

105. Haapanen H, et al. Remote ischemic preconditioning protects the spinal cord against ischemic insult: an experimental study in a porcine model. *J Thorac Cardiovasc Surg* 2016;**151**:777−85.

106. Herajärvi J, et al. Exploring spinal cord protection by remote ischemic preconditioning: an experimental study. *Ann Thorac Surg* 2017;**103**:804−11.

107. Fukui T, et al. Comparison of the protective effects of direct ischemic preconditioning and remote ischemic preconditioning in a rabbit model of transient spinal cord ischemia. *J Anesth* 2018;**32**:3−14.

108. Jing N, Fang B, Wang ZL, Ma H. Remote ischemia preconditioning attenuates blood-spinal cord barrier breakdown in rats undergoing spinal cord ischemia reperfusion injury: associated with activation and upregulation of CB1 and CB2 receptors. *Cell Physiol Biochem* 2017;**43**:2516−24.

109. Ueno K, et al. Increased plasma VEGF levels following ischemic preconditioning are associated with downregulation of miRNA-762 and miR-3072-5p. *Sci Rep* 2016;**6**.

110. Jin KL, Mao XO, Greenberg DA. Vascular endothelial growth factor: direct neuroprotective effect in in vitro ischemia. *Proc Natl Acad Sci USA* 2000;**97**:10242−7.

111. Dezfulian C, Garrett M, Gonzalez NR. Clinical application of preconditioning and postconditioning to achieve neuroprotection. *Transll Stroke Res* 2013;**4**:19−24.

112. Sun J, et al. Vascular expression of angiopoietin1, α5β1 integrin and tight junction proteins is tightly regulated during vascular remodeling in the post-ischemic brain. *Neuroscience* 2017;**362**:248−56.

113. Michinaga S, Koyama Y. Dual roles of astrocyte-derived factors in regulation of blood-brain barrier function after brain damage. *Int J Mol Sci* 2019;**20**.

114. Wang YL, Hui YN, Guo B, Ma JX. Strengthening tight junctions of retinal microvascular endothelial cells by pericytes under normoxia and hypoxia involving angiopoietin-1 signal way. *Eye* 2007;**21**:1501−10.

115. Yin J, Gong G, Liu X. Angiopoietin: a novel neuroprotective/neurotrophic agent. *Neuroscience* 2019;**411**:177−84.

116. Han S, et al. Rescuing vasculature with intravenous angiopoietin-1 and alpha v beta 3 integrin peptide is protective after spinal cord injury. *Brain* 2010;**133**:1026−42.

117. Su EJ, et al. Activation of PDGF-CC by tissue plasminogen activator impairs blood-brain barrier integrity during ischemic stroke. *Nat Med* 2008;**14**:731−7.

118. Abrams MB, et al. Imatinib enhances functional outcome after spinal cord injury. *PLoS One* 2012;**7**.

119. Kjell J, Olson L. Repositioning imatinib for spinal cord injury. *Neural Regen Res* 2015;**10**:1591.

120. Paterniti I, et al. Docosahexaenoic acid attenuates the early inflammatory response following spinal cord injury in mice: in-vivo and in-vitro studies. *J Neuroinflammation* 2014;**11**:6.

121. Lim SN, Huang W, Hall JCE, Michael-Titus AT, Priestley JV. Improved outcome after spinal cord compression injury in mice treated with docosahexaenoic acid. *Exp Neurol* 2013;**239**:13−27.

122. Garcia-Ovejero D, et al. Progesterone reduces secondary damage, preserves white matter, and improves locomotor outcome after spinal cord contusion. *J Neurotrauma* 2014;**31**:857−71.

123. Labombarda F, et al. Progesterone increases the expression of myelin basic protein and the number of cells showing NG2 immunostaining in the lesioned spinal cord. *J Neurotrauma* 2006;**23**:181−92.

124. González SL, et al. Progesterone up-regulates neuronal brain-derived neurotrophic factor expression in the injured spinal cord. *Neuroscience* 2004;**125**:605−14.

125. Kwon BK, et al. A systematic review of directly applied biologic therapies for acute spinal cord injury. *J Neurotrauma* 2011;**28**:1589−610.

126. Wu W, et al. Neuroprotective ferulic acid (FA)-glycol chitosan (GC) nanoparticles for functional restoration of traumatically injured spinal cord. *Biomaterials* 2014;**35**:2355−64.

127. Yamane K, et al. Collagen-binding hepatocyte growth factor (HGF) alone or with a gelatin-furfurylamine hydrogel enhances functional recovery in mice after spinal cord injury. *Sci Rep* 2018;**8**:917.

128. Chen S Der, Wu CL, Hwang WC, Yang DI. More insight into BDNF against neurodegeneration: anti-apoptosis, anti-oxidation, and suppression of autophagy. *Int J Mol Sci* 2017;**18**:1−12.

129. Sun L, et al. Co-transplantation of human umbilical cord mesenchymal stem cells and human neural stem cells improves the outcome in rats with spinal cord injury. *Cell Transplant* 2019;**28**:893−906.

130. Mohammadshirazi A, et al. Combinational therapy of lithium and human neural stem cells in rat spinal cord contusion model. *J Cell Physiol* 2019;**234**:20742−54.

131. Martins BDC, et al. Association of riluzole and dantrolene improves significant recovery after acute spinal cord injury in rats. *Spine J* 2018;**18**:532−9.

132. Grosso MJ, et al. Effects of an immunomodulatory therapy and chondroitinase after spinal cord hemisection injury. *Neurosurgery* 2014;**75**:461−71.

Translational perspective: neuroregenerative strategies and therapeutics for traumatic spinal cord injury

Andrea J. Santamaria[1], *Pedro M. Saraiva*[1], *Juan P. Solano*[2], *James D. Guest*[3]

[1]The Miami Project to Cure Paralysis, University of Miami Miller School of Medicine, Miami, FL, United States; [2]Pediatric Critical Care, University of Miami Miller School of Medicine, Miami, FL, United States; [3]The Miami Project to Cure Paralysis and Neurosurgery, University of Miami Miller School of Medicine, Miami, FL, United States

Introduction

To be able to feel the lightest touch really is a gift.
Christopher Reeve.

The last three decades have witnessed transformative advances in biology and neuroscience. Still, regeneration of axons after traumatic spinal cord injury (SCI) remains a significant medical challenge and need. The central nervous system (CNS) has a limited capacity for endogenous repair as compared to other tissues such as nerve, bone, skin, and liver. Key knowledge areas contributing to the multidisciplinary field of neuroregeneration therapeutics include molecular and developmental neurobiology, stem cell and peripheral nerve biology, experimental neuropathology, neurophysiology, and biomedical and materials engineering. Key experimental techniques include cell culture, microfluidics, high-throughput screening, animal modeling, and human clinical trials. This chapter reviews key concepts with a particular view of the field's evolution, what has been tested in people with SCI, and what therapeutic applications seem feasible in the near future to accomplish more effective regeneration. We briefly describe some unifying key molecular pathways as currently understood, such as PTEN inhibition of the PI3K−Akt−mTOR pathway, and while we emphasize specific interactions such as reducing axonal growth inhibition, few of the molecules discussed have only one activity. The reader is

© 2022 Elsevier Inc. All rights reserved.

referred to online resources such as the Kegg Pathway database (https://www.genome.jp/kegg/pathway.html) for greater detail and omics integrated systems analyses focused on SCI.[1]

After SCI, epicenter neural parenchymal loss disrupts motor, sensory, and autonomic axons in long tracts and segmental spinal cord circuit neurons. This results in changes throughout the body and nervous system with permanent functional deficits of bladder control, the limbs, sensory function, breathing, and blood pressure regulation. Disturbed homeostatic functions impact the immune, metabolic, and neural—gut axis,[2,3] indicating SCI's distributed effects[1] at a multi-organ system level.

In mammals, the known capability of peripheral nerves to regenerate lead to the seminal experiment of testing if transected CNS axons could grow into peripheral nerves applied to the spinal cord. The successful growth of the spinal cord axons into the grafts established that it was the spinal cord environment that was nonpermissive for regeneration.[4,5] This critical evidence focused a subsequent generation of researchers on the causes of regeneration inhibition and also led to attempts to bridge the completely transected spinal cord with peripheral nerve grafts both in animal models[6–8] and people.[9,10] The experiments have consistently shown that some CNS axons can readily grow into nerve grafts, but reentry of regenerated axons to the spinal cord is limited,[11] indicating potent growth inhibition. The regenerative *bridging strategies* also derived their rationale from animal and human studies showing that a small residual fraction of axonal connectivity through the injury epicenter can allow for relatively normal-appearing functions.[12,13] This observation has been extrapolated to infer that functional recovery could occur if only a few hundred or thousand new useful connections could be created. Although the concept that limited spinal axonal regeneration could support functional recovery has some evidence in animals, human evidence is lacking. This is because we have few methods to study CNS regeneration in people compared to animal models.

For this review, *regeneration* refers to a new process extension from a growth cone at a site of axotomy that leads to synapse formation on another neuron, especially to span the injury site. Axonal *neuroplasticity* involves the growth of a new neurite process from an axon whose distal terminal remains intact. This axonal plasticity may compensate for the loss of other axonal connections as observed for the corticospinal tract (CST) after hemisection injuries in primates.[14,15] Neurite growth alone may not have a functional consequence, but both plasticity and regeneration can create new spinal cord circuits. Regeneration of lengthy axons requires the production of substantial amounts of structural components as some human CNS neurons span from the cortex to the distal spinal cord. Regeneration rate is also a limit with a maximum of 2—3 mm/day observed in peripheral nerves. The term "sprouting" is often found in regeneration studies as an index of regenerative activity and refers to multiple new neurites deriving from injured axons but does not establish that functional connections are formed.

To understand effective regeneration, we learn from the peripheral nervous system (PNS)[16] and studies in fish and amphibians where spinal cord regeneration can occur robustly.[17,18] The determinants and mechanisms of axonal regeneration require an interplay between neuron intrinsic genetic, epigenetic, and other molecular signals together with extrinsic cues in the spinal cord tissue environment. Early animal models in rodents mainly used different spinal cord sectioning lesions to study regeneration experimentally: transection, hemisection, and frequently dorsal cord sections. Contusion injuries are now emphasized to emulate human SCI. Earlier studies also generally delivered the test therapeutics before or immediately after injury, whereas the necessity to establish the therapeutic window for effectiveness is increasingly incorporated into experiments to assess clinical feasibility. In therapeutics development, the sequence is often in vitro testing, in vivo preclinical testing, and

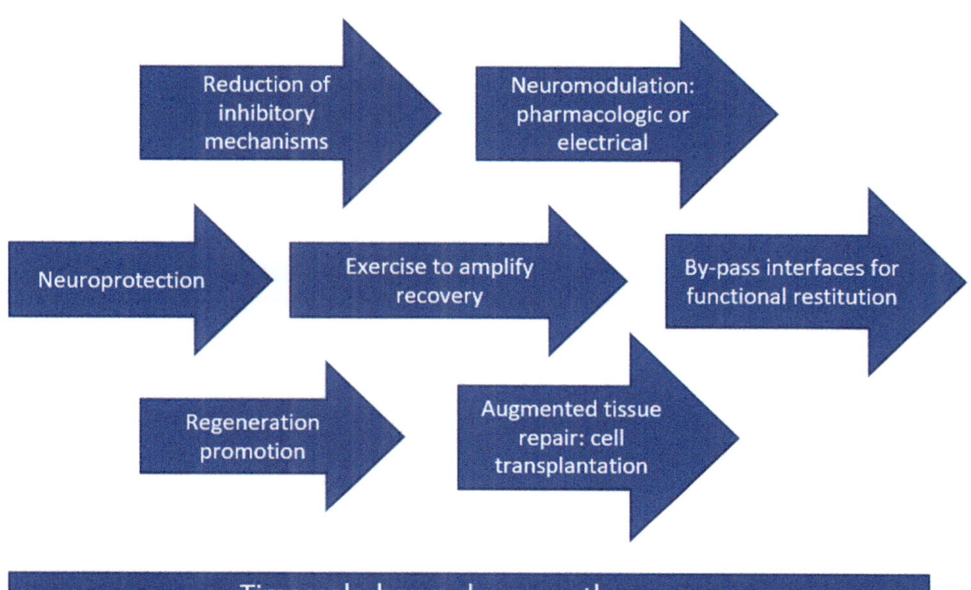

FIGURE 29.1 **Potential combination therapeutics illustrated by Timeframe after injury.** It is broadly accepted that multiple targets will be needed to achieve effective spinal cord repair linked to important functional recovery. There are two conceptual frameworks: the longitudinal temporal (combination in sequence) and the multiple mechanistic (simultaneous multiple therapies). In this diagram the general mechanistic approaches are shown over a nonlinear timeline that extends from hours to years after injury.

then initial clinical studies. The overall themes in functional restitution after SCI are illustrated in Fig. 29.1.

Biologic therapeutics

Biologics include proteins like insulin that can be produced artificially and delivered to the human body. Several biologic therapies utilize monoclonal antibodies due to their very high specificity and ability to interfere with molecular signaling. The first approved monoclonal therapeutic was muromonab, an immune suppressant.[19] Biologics, by definition, are manufactured in living systems, whereas drugs are manufactured through chemical synthesis. Biologics are regulated in the United States by the FDA branch known as CBER, the Center for Biologics Evaluation and Research, which also regulates cell therapies. Currently, several early cell

transplantation trials have been concluded, and antibody biologics are being tested clinically in SCI.[15]

Extrinsic determinants of neuronal regeneration

The function of the axonal cytoskeleton, critical for regeneration, is strongly regulated by extrinsic signals. Peripheral nerve regeneration studies have indicated the importance of the cellular membrane and extracellular matrix (ECM) milieu presented to axonal growth cone receptors such as integrins,[20] Schwann cells (SCs), and their laminin, collagen, and fibronectin ECM. In nerves, SCs can dramatically change their phenotype to support axonal regeneration.[21] This regenerative phenotypic switch appears not to be activated in adult CNS cells, although glia such as oligodendrocyte precursors and astrocytes undergo other phenotypic changes and divide after SCI generating new oligodendroglia and astrocytes.[22]

In recent decades several causes of regeneration failure have been identified. Historically, the *glial scar* was considered to be the primary impediment.[23] After SCI, there is a strong tissue inflammatory response[24] and responding cells remove axonal and myelin fragments. Possibly, to delimit this inflammation, a network of proliferating and reactive astrocytes, NG2+ve cells, fibroblasts, and ECM contain the local inflammatory milieu.[25,26] Historically, histologic observation of hypertrophic astrocytes in apposition to axonal dystrophic end bulbs led to the association between reactive gliosis and impaired axonal regeneration.[27,28]

Further, the scar has been interpreted to restore the glial limiting membrane,[29] a component of the blood—brain barrier that separates the CNS from peripheral structures such as blood vessels. Today, we recognize that this apparent scar barrier is more complex than simply a mechanical obstacle and may not impede regeneration in all contexts.[30–34] There is also in-migration of connective tissue cells during SCI wound healing, including meningeal and vascular fibroblasts, pericytes, ependymal cells, and macrophages that further promote astrocytic reactivity.[35,36] Known axonal growth inhibitors include proteoglycans, myelin-associated growth inhibitors (MAGS), and dysregulated molecules that normally guide axons during development, such as repulsive guidance molecule (RGM), semaphorins, ephrins, Wnts, and netrins.[37–41]

Proteoglycans and neuroplasticity regulation

Perineuronal nets are ECM structures that stabilize synapses[42] and contain chondroitin sulfate proteoglycans (CSPGs). After SCI, injury zone astrocytes release CSPGs that form dense inhibitory nets[43] consisting of a protein core and sulfated glycosaminoglycan (GAG) side chains.[44] The high concentrations of CSPGs can inhibit neuronal survival, synaptogenesis, axonal sprouting, regeneration, and remyelination.[45–47] Versican and neurocan are found in the epicenter core; NG2 and phosphacan are more concentrated in the evolving glial scar margins.[48] Large numbers of activated macrophages further induce astrocyte migration potentiating gliosis, cavitation, and proteoglycan upregulation.[49] However, in keeping with the theme that molecular interactions usually have multiple effects, CSPGs also have beneficial roles in the acute postinjury phase by reducing resident microglia and monocyte toxicity and increasing IGF-1 activity via the cell surface glycoprotein CD44 receptor.[50]

Chondroitinase ABC (ChABC or Ch'ase) is a bacterial enzyme that degrades the GAG side chains of CSPGs,[51] permitting increased plasticity and regeneration. The plasticity promoting effects have been validated in several paradigms and tested in rodent, porcine, canine, and primate experiments. Early interest in chondroitinase arose as a therapeutic to degrade herniated intervertebral disc material. As an enzyme, the delivery methodology is critical in terms of safe tissue distribution and duration of activity. The 2B6 antibody is a reliable histologic biomarker of GAG side-chain cleavage to index CSPG activity. Studies in the brain nigrostriatal tract of adult rats evaluated repair after tract sectioning. Bolus injections of ChABC delivered via a cannula at 3-day intervals enhanced axonal regeneration, with the growth of dopaminergic neurons back to their targets.[52] After an excitotoxic chemical injury to the same pathway, ChABC was neuroprotective.[53] After a C4 dorsal hemisection injury in the spinal cord, ChABC was delivered every 2 days for 10d via intrathecal bolus injections. Local CSPG degradation was correlated with upregulation of growth-associated protein 43 (GAP-43) in DRGs, and regeneration of some axons in both the dorsal sensory and motor (CST) tracts. Neurophysiology testing indicated a recovery of cord dorsum evoked potentials above the lesion with evidence of sensorimotor recovery in treated animals.[54] Other notable examples of the efficacy of ChABC include recovery of skilled locomotion after spinal cord hemisection in cats,[55] improved locomotion and bladder function after compressive injury in rats,[56] improvement of hand function after hemisection injury in primates,[57] and improved gait symmetry in dogs sustaining "natural" SCI.[58] After cervical hemisection in primates, a single-time point injection resulted in extensive GAG absence at 2 weeks postdelivery and recovery of hand function for up to 4 months. However, the same chondroitinase molecule delivered intrathecally in pigs did not penetrate the spinal cord gray matter. Since this is a bacterially derived enzyme, important considerations for clinical translation include potential immunogenicity, enzyme stability, and half-life of activity. Methods to stabilize ChABC delivery include molecular editing to increase temperature stability,[59] combining the enzyme into linked hydrogels for sustained release,[60,61] modifying transplantable cell lines to secrete ChABC,[62] and delivering ChABC continuously via viral vectors.[63] For delivery from mammalian cells, the enzyme requires modification of glycosylation sites.[64] The abundance of preclinical evidence supporting axonal regeneration,[65–68] sprouting,[54,69–71] and neuroprotection[72,73] supports the clinical testing of chondroitinase.

Proteoglycan receptors

Reducing the inhibitory activity of CSPGs can also be achieved by blocking neuronal growth cone receptor activation. The leukocyte antigen–related (LAR) family protein tyrosine phosphatases (PTPs) include PTP sigma (σ), a transmembrane protein receptor with an immunoglobin-like ectodomain, and two cytoplasmic catalytic domains,[74–76] one of which contains a critical signaling "wedge" domain. PTPσ binds with high affinity to CSPGs leading to inhibition of regeneration. When the PTP receptor was knocked out, neurite outgrowth in vitro was increased. In vivo[77] studies showed increased

CST regeneration after both hemisection and contusion injuries in mice.[78] A synthetic compound mimetic of the wedge domain called intracellular sigma peptide (ISP) was designed and combined with an HIV TAT domain to facilitate membrane permeability. This peptide allosterically interferes with the receptor catalytic domain, blocking CSPG signaling. Peptide delivery after acute contusion injury in rodents improved sensory, locomotor, and urinary function[79] correlated to axonal regeneration and plasticity evidence. A subsequent study found that ISP binding to the wedge domain also increased Cathepsin B protease activity to digest CSPGs.[80] Another CSPG mechanism that reduces recovery through the LAR and PTPσ receptor systems is reduced oligodendrocyte survival and precursor generation. The LAR receptor also contains a wedge domain that can be inhibited via a different intracellular TET-linked cell-permeable peptide (ILP). The combination of both ILP and ISP increased the population of M2 macrophages and regulated T-lymphocytes toward a proregenerative immune response,[81] and also promoted neural precursor survival and endogenous oligodendrogenesis. Survival of mature oligodendrocytes was promoted by reducing caspase-3-mediated apoptosis.[82] The use of PTPσ modulating peptides in clinical trials is currently in the planning stage.

CSPGs also act as neurite growth inhibitors by activating the Rho GTPase/ROCK pathway.[46,83] Another approach to reducing inhibition is the inactivation of Rho via the bacterial enzyme-C3 transferase. BA-210 (Cethrin) is a cell-permeable fusion protein derived from C3 transferase. Cethrin was formulated for application with a fibrin sealant extradurally, a practical delivery method compatible with acute decompression surgery. In preclinical studies, administration was associated with improved locomotor outcomes in rats and mice with contusive SCI.[84] The biologic Cethrin was tested in a phase I/IIa multicenter open-label study in cervical and thoracic SCI ASIA Impairment

Scale (AIS) A subjects. In the study, the cervical SCI group showed a 27.3 ± 13.3-point motor score improvement at 12 months post-SCI.[85] This recovery is greater than expected from natural history studies.[86] Subsequently, a pivotal study was stopped at an interim analysis due to the absence of a therapeutic effect.[87] This result underscores the importance of well-controlled studies in order to evaluate efficacy, as open-label studies may result in biased results.

Myelin-associated proteins inhibit regeneration: Nogo, MAG, OMgp, PirB

Myelination is essential to rapid saltatory conduction in the nervous system. Its structure is linked dynamically to axonal activity.[88,89] Whereas previously considered to be relatively static in structure and function, myelination is highly responsive to experience.[90] The onset of myelination in some animals marks the neurodevelopmental end of regeneration capability.[91,92] Important discoveries have revealed mechanisms by which myelin limits aberrant neuroplasticity and inhibits axonal regeneration,[93,94] especially after myelin damage that increases the exposure of myelin debris. Studies of CNS Wallerian degeneration revealed that macrophages engorged with myelin debris were cleared at a much slower rate than in peripheral nerves.[95] When biochemical methods were used to separate CNS myelin protein fractions, some caused CNS axonal growth collapse in neuron culture assays.[96] Some monoclonal antibodies to these CNS myelin fractions increased axonal regeneration in early experiments, presumably by blocking then-unknown inhibitors.[94] Ingeniously, the monoclonal antibody IN-1 was delivered via brain-implanted hybridoma cell tumors in rats, and following thoracic dorsal hemisections CST sprouting extending over $7-11$ mm beyond the caudal injury margin

was observed.[97] IN-1 antibody "treatment" also enabled the plasticity of brainstem axons,[88] with enhanced axonal sprouting observed when delivery was combined with embryonic spinal cord tissue transplants.[98]

In 2000, after more than a decade of focused research, the target of IN-1, Nogo-A, was identified as a critical neurite outgrowth inhibitor expressed by oligodendrocytes and neurons.[99] The Nogo gene encodes three major proteins (Nogo-A, Nogo-B, and Nogo-C),[99] of which Nogo-A has a 66 amino acid (Nogo-66) extracellular loop on the oligodendrocyte surface[100,101] that is responsible for growth cone inhibition. Increased axonal sprouting and improved locomotion have been reported in transgenic mice lacking this Nogo-66 segment. Subsequently, several laboratories have confirmed the inhibitory effects of Nogo-A, and the complexity of a multimolecular signaling complex has been elucidated. The Nogo receptor (NgR)[102–104] binds to three known myelin inhibitors: NogoA, myelin-associated glycoprotein (MAG), and OMgp. NgR is a GPI-linked protein that requires a p75 coreceptor for signaling and activates Rho kinase resulting in Cofilin-mediated actin depolymerization and growth cone collapse. Receptor blockade has been shown to promote axonal regeneration after spinal cord hemisection in rats[102] and rostral axonal sprouting in cervically hemisected primates.[105]

Given the great expense and effort of therapeutics development, elucidation of specific mechanisms and independent replications is essential. Not surprisingly, some controversies arose regarding the effects of Nogo knock-out in mouse models. One research group showed extensive CST sprouting in Nogo-A-/B- mice,[106] but other laboratories failed to confirm these results,[106] detecting evidence of axonal growth only in the absence of Nogo-A, Nogo-B, and Nogo-C[107] and only modest increases in regenerative capacity after disruption of the presumed critical molecule Nogo-66.[108] Regarding the animal models, the effects of Nogo knockouts may differ between mouse strains,[109] an important consideration when planning human translation. It is also notable that the deletion of the inhibitors Nogo, MAG, and OMgp may enhance CST or raphespinal sprouting, but this may not translate into functional recovery.[110] Thus, histologically evident sprouting and regeneration effects may not lead to detectable effective new circuit connections. It is important to bear in mind that effects in animal models incompletely predict both safety and efficacy in people.[111]

MAG, a member of the immunoglobulin superfamily, present in CNS myelin, was also discovered to cause growth cone collapse in neuron culture assays[112] by inhibiting the cAMP protein kinase A pathway.[113] Although MAG was shown to be a potential axonal growth inhibitor in vitro,[114,115] neurite elongation was limited in experiments examining optic nerve and CST growth postinjury in MAG-deficient mice.[116] This again indicates the molecular complexity of axonal growth inhibition. MAG is also found in nerves and may also act as an inhibitory molecule in SCs regulating Wallerian degeneration.[117] The third myelin-derived axonal growth inhibitor, oligodendrocyte myelin glycoprotein (OMgp), is a glycosylphosphatidylinositol (GPI)-anchored CNS myelin protein that also acts through the NgR.[118] Autoimmunity against OMgp and myelin oligodendrocyte glycoprotein have been implicated in transverse myelitis and multiple sclerosis.[119]

The PirB receptor

So far, three NgRs have been identified: NgR1, NgR2, and NgR3.[120] MAG and OMgp bind to these receptors.[112,115,121,122–128] The paired immunoglobulin-like receptor PirB, also called p91, is also a coreceptor of Nogo, MAG, and OMgp. Expression of PirB increases in the spinal cord after injury.[129] PirB exerts its inhibitory effects on neural and axonal regeneration by binding to the three inhibitory molecules MAG, Nogo, and OMGP, and the receptor MHC 1.[130,131] PirB is also considered a potential therapeutic target to enhance axonal regeneration after SCI.[132] Together these "inhibitory" proteins have been found to have important roles in experience-dependent neuroplasticity.[133]

Repulsive guidance molecule

Neogenin is a GPI-linked netrin receptor, important for axonal steering guidance and apoptotic signaling during development. The repulsive guidance molecule A (RGMa)[41,134] is upregulated after SCI[135] and conveys a repulsive signal to the neogenin receptor increasing axonal growth cone GTP-RhoA. This causes Cofilin dephosphorylation leading to actin depolymerization and growth cone collapse. RGMa may be either membrane bound as a component of myelin or soluble and is another molecule that converges on the important Rho kinase signaling pathway.[136] Intrathecal administration of an anti-RGMa antibody increased the distance at which CST sprouting was observed and improved locomotor behavior.[137] Intravenous administration was also observed to increase recovery, an important issue for clinical translation.[138] In one primate study, improved hand function was reported after immediate application to a C6/7 hemisection continued for 30d. In another, injury region CST and serotonergic axon sprouting were increased with a therapeutic window of 24 h.[15] An anti-RGMa antibody, ABT-055 has been tested in multiple sclerosis,[139] and there are currently two separate anti-RGMa clinical trials for acute SCI.

Clinical translation of antibodies to block myelin inhibitors

As therapeutic reagents, antibodies can have a long serum half-life and high specificity. Novartis conducted a Phase I trial of antihuman Nogo-A antibody (ATI 355) in Europe

(NCTO00406016). The study enrolled 52 ASIA A—C cervical and thoracic injured individuals between 4 and 60 days after SCI into sequential dose-escalation cohorts for lumbar intrathecal delivery of the biologic.[140] The Nogo trial results indicated safety with a range of motor score improvements, some of which exceeded the expected natural history.[141] Recently a second (Phase II) study was launched in Europe (Nogo Inhibition in Spinal Cord Injury (NISCI) NCT03935321). The trial aims to enroll 132 cervical ASIA A—D SCI participants randomized to six intrathecal bolus injections of 45 mg of NG-101 (human monoclonal antibody against the human Nogo-A protein) or placebo injections.

In a separate program in development, AXER-204 (NgR1-Fc) is a "Nogo trap" that binds to all three myelin inhibitors: Nogo-A, MAG, and OMgp. After extensive preclinical rodent studies,[142] a preclinical primate study assessed initiating delivery 1 month after C5/6 hemisection injury and reported improvements in hand function and stepping with increased CST sprouting distal to the injury.[143] ReNetX Bio has initiated a clinical trial of the Nogo trap antibody known as the RESET trial.

Neurotrophic factors

Neurotrophic factors (NTFs) regulate neuronal survival and axonal growth during development. They were among the first classes of molecules to be tested as therapeutic biologics in the CNS. The ability of NTFs to be purified biochemically and later cloned supported product development for testing in several neurological diseases. The initial therapeutic applications of NTFs established concepts that currently underlie the use of biologics in CNS diseases and SCI. Typically, NTFs are target derived and can be retrogradely transported from axonal terminals to the cell soma. In the adult CNS NTF expression remains high in regions of ongoing plasticity such as the hippocampus but lower in other regions and declines with age.[144] Exogenously delivered NTFs can increase axonal regenerative responses[145] and neuronal survival.[146–148]

The classical NTFs are NGF, BDNF, NT-3, and NT-4 and signal through the p75 receptor, TrkA, TrkB, and TRkC. Glial-derived neurotrophic factor (GDNF) is a different NTF of the TGFβ family whose receptor is GPI-linked to the RET tyrosine kinase (Table 29.1). Nerve growth factor (NGF), discovered in the 1950s, is the prototype NTF and is both soluble and diffusible. Brain-derived neurotrophic factor (BDNF) was discovered in 1982 using classic biochemical methods[149] and neurotrophin 3 (NT-3) in 1990[150] at Regeneron Pharmaceuticals. NGF is essential in developing basal forebrain cholinergic neurons, and since these neurons are prominently lost in Alzheimer's disease (AD), they became targets for the first NTF clinical trials.[151] Initially, NGF was delivered to the ventricles, but systemic side effects were significant[152] due to extensive nociceptive axonal sprouting in DRGs within the CSF distribution.[153] Other more locally constrained iterations included delivery from implanted viral vector transduced cells,[154] AAV2-mediated gene therapy from direct intraparenchymal injection,[155] and release from biomaterial encapsulated cells.[156]

After SCI, NTF levels may transiently increase near the injury but then decline,[157] leading to chronic atrophy[158] of some neurons. In addition, nonfunctional receptors such as truncated TrkB may have increased expression and "tie-up" NTFs from binding to active receptors.[159]

The temporospatial distribution of NTFs is critical[160] for therapeutic applications as they vary in their diffusion and stability properties, and regenerating axons exhibit growth responses into a concentration gradient, generally steering to the maximal concentration. Further, the delivery method must ensure that the delivered NTFs achieve target engagement, more challenging if the NTF's diffusion distance is small. The transplantation of NTF-producing cell grafts can achieve spatially designed

TABLE 29.1 Regenerative therapeutic mechanisms tested in clinical trials.

Molecular therapeutics		
Rho kinase inhibition	BA-210 (Cethrin)/VX-210	Phase I/II, NCT02669849
NogoA	Receptor blocking antibody	Phase I/II (NISCI) current
Nogo A, MAG, OMgp	Nogo receptor decoy (AXER-204)	RESET trial, Phase I, Phase II
Repulsive guidance molecule	Elezanumab antibody	Phase I/II (ELASCI), NCT03737851
Repulsive guidance molecule	MT-3921 antibody	Phase I/II, NCT04683848
Neurotrophin	FGF-1 mimic, SUN 13837	Phase I/II, NCT01502631
Cell therapeutics		
Remyelination	OPC-1 ES-derived cells	Phase I/II, NCT02302157
Remyelination, regeneration	Autologous Schwann cells	Phase I (2 trials), NCT01739023
Inflammation modulation	Proneuron autologous—activated macrophages	Phase I/II, NCT00073853
Neural stem cell repair	Allogeneic NSC, stem cells, etc.	Phase I/II, NCT02163876
Neural stem cell repair	NSI-566	Phase I, NCT01772810
Olfactory ensheathing glia	Various autologous studies	Phase I studies
Mesenchymal stem cells	Various autologous/allogeneic	Phase I studies
Biomaterial scaffold		
PLGA/PLL cylinder	INSPIRE study	Phase I/II, NCT02138110
Calcium sulfate matrix + FGF-1	Bioartic SC0806 study	Phase I, NCT02490501
Proposed clinical trials		
CSPG-inhibition	Chondroitinase enzyme	Extensive preclinical testing
CSPG-inhibition	PTP receptor, NVG-291	Extensive preclinical testing

paracrine delivery. For example, the longitudinal injection of a continuous pathway of SCs expressing GDNF provided both an NTF gradient and an axonal growth substrate for the regeneration of descending propriospinal neurons through and beyond a spinal cord hemisection.[161] NTFs can also elicit structural plasticity, such as collateral sprouting, which can contribute to neurological recovery. For example, NT-3 delivery increased CST sprouting across the midline.[162] Exemplary experiments include that the injection of lentiviral vectors expressing NT-3 into brainstem targets led to reinnervation and formation of axodendritic synapses in a dose-related manner.[163] Other examples are that NGF-transfected fibroblasts supported sympathetic and sensory neuronal growth after transplantation.[164] BDNF and NT-4/5 prevent atrophy of rat rubrospinal neurons after cervical axotomy,[165] NT-3 promoted the growth of ascending dorsal column sensory axons,[166] and BDNF promotes sprouting of CST fibers.[167] NTFs remain critical neural regeneration tools, and some synthetic NTFs can target multiple receptors.[168] Advances in technology should incorporate methods to regulate precise dosing to optimize the pharmacological effects and avoid excess receptor saturation and desensitization.[169,170]

Intrinsic determinants of neuronal regeneration: the conditioning lesion paradigm

By intrinsic, we refer to transcriptional programs within neurons that regulate axonal growth during development and could theoretically be reactivated after adult CNS injury to initiate and amplify regeneration. As development reaches completion in mammals, CNS connections mature, growth-associated genes are downregulated, and the system's regenerative capacity is reduced[171] in favor of stability. Some neurons, such as motor neurons, serotonergic and propriospinal axons, have shown greater regenerative responses than others, such as CST neurons in experiments.[145] Neurons have complex spatial distributions, for example, having the cell body in the brain cortex with extensive local dendrites and a long axon within the spinal cord. This means that neuron compartments may be quite distant, creating a need for rapid signaling and transport mechanisms along axons (Fig. 29.2). The classic experiment to demonstrate that intrinsic axonal growth potential can be manipulated compartmentally is the *conditioning lesion* paradigm.[172–174] Conditioning experiments use the fact that dorsal root ganglion (DRG) sensory neurons have both a peripheral nerve and a spinal cord axon.[175] The central DRG process enters the spinal cord through the dorsal root entry zone (DREZ), transitioning from SC ensheathment and PNS myelination to astrocyte ensheathment and oligodendroglial CNS myelination, a more inhibitory environment. To create conditioning, an initial injury to PNS axons potentiates the axonal growth responses of a later injury to the centrally projecting axons overcoming the intrinsic inhibitory environment. A CNS sensory axonal injury alone evokes little regenerative response. So, the regenerative potential of the peripheral axons is "transferred" to the central axons. Conditioning effects are transcription dependent and upregulate genes such as GAP-43 that are considered regeneration-associated genes (RAGs).[176–178] GAP-43 transduces signals in the growth cone region to regulate cytoskeletal dynamics.

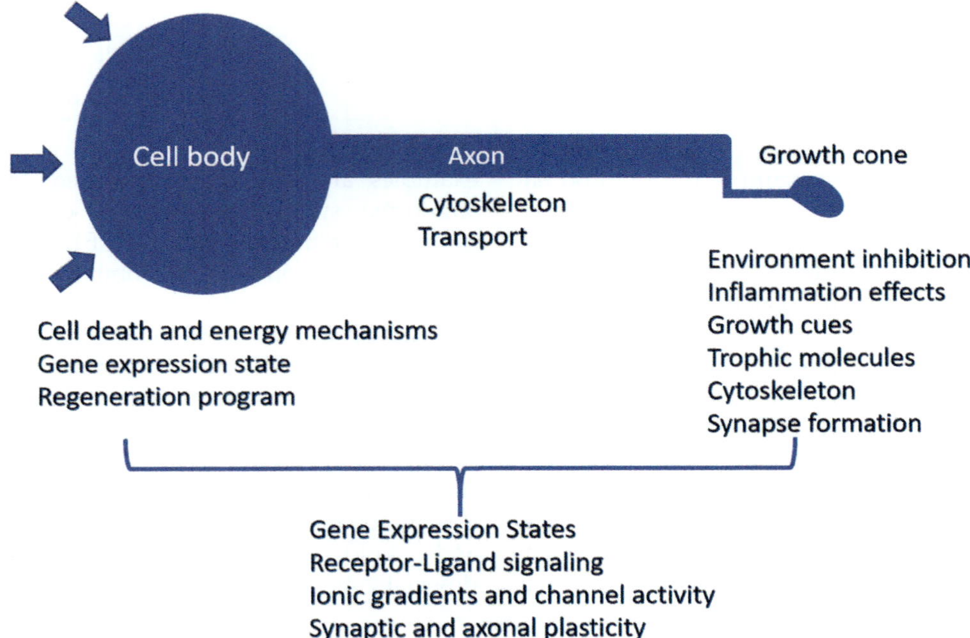

FIGURE 29.2   The major regeneration signaling mechanisms described by anatomic compartment in a schematic neuron.

Calcium and cyclic AMP

One of the earliest events after axonal injury is an influx of calcium to the axoplasm. This transient calcium current, a stress signal, rapidly propagates to the cell body and is critical for responsive membrane stabilization, protein synthesis, and intracellular cascade activation.[179] Signaling through phosphate ester linkages is a fundamental motif in biology,[180] and in neurons, the increased calcium leads to cAMP production by activation of Ca^{2+}-dependent adenyl cyclase.[181] Elevated cAMP levels transition the neuronal state from action potential transmission to de novo process growth,[182] increasing growth cone formation and overcoming extrinsic inhibitory signals such as myelin inhibitors.[183—185] Injection of the cAMP analog dibutyril-cAMP (db-cAMP) into DRG neurons mimics the regenerative effects of a conditioning lesion.[185—187]

Cyclic AMP levels may also be increased by blocking its degradation by phosphodiesterase 4 (PDE4). Rolipram is a clinically approved CNS-specific PDE4 inhibitor that crosses the blood—brain barrier and has been tested in several experiments. In cervical hemisected rats transplanted with embryonic spinal tissue, Rolipram delivery resulted in increased axonal growth into transplants, attenuation of the glial scar, and increased locomotor recovery.[188] Similar favorable effects were found after contusion injury with the combination of oral Rolipram, injected db-cAMP, and transplanted SCs.[189] Rubrospinal neuronal regeneration was observed after combined olfactory ensheathing cell (OEC) transplantation and Rolipram administration.[190] Although cAMP has shown promising neuroregenerative effects, it is also involved in mechanical hyperalgesia and allodynia induced by intradermal capsaicin,[191] and systemic administration of some PDE4 inhibitors is associated with severe nausea and emesis.[192] The need to carefully modulate effects and lack of selectivity of cAMP-increasing drugs has made translation difficult.

Histone—DNA modifications and gene expression

An important gene expression motif is that "relaxed" DNA has higher gene expression. DNA is negatively charged and histones are postively charged. Acetylation "cancels" positive charges and opens DNA wheras methylation "tightens" DNA. Thus, patterns of protein expression critical for axonal regeneration are regulated epigenetically by histone modifications, including histone acetylation and DNA methylation as well as microRNAs, and messenger RNA modifications.[193] Four known histones organize and pack DNA into nucleosomes whose chromatin structure regulates accessibility to promoter regions. Increased access to DNA promoter sites is facilitated by histone lysine acetylation[194,195] via histone acetyltransferases (HATs). Deacetylases (HDACs)[196] reduce DNA access silencing genes. Due to abnormalities of acetylation in cancers, a number of HDAC inhibitor (HDACi) drugs such as vorinostat have been tested and approved, establishing a robust framework for drug development.[197] Epigenetic studies of axonal regeneration using the conditioned DRG model identified that histone 4 hypoacetylation was linked to poor axonal growth responses. Peripheral axonal injury increased H4 acetylation, and the promoter region of several RAGs was immunoprecipitated by acetylated H4, suggesting H4 as a key response control. Some conditioning effects on sensory neurites were replicated using a HDACi (MS-275) leading to increased RAG expression[198] and axonal sprouting of the transected sensory fasciculus gracilis. In an in vitro study comparing sensory and retinal ganglion cell (RGC) neurons, it was observed that after axotomy, a calcium wave injury signal propagated from the injury site to the cell soma and nucleus. In response in sensory but not in RGCs, protein kinase C was phosphorylated and translocated from the cytoplasm to the nucleus. This resulted in a loss of acetylase

HDAC5 binding to histone H3, exit from the nucleus, and transport to the injured axonal tip. With reduced HDAC5 binding to H3, it undergoes acetylation and binding of regenerative transcription factors. In the RGCs, H3 acetylation was not increased.[199] However, epigenetic changes can also contribute to less favorable plasticity, such as neuropathic pain, that can be attenuated by HAT inhibitors.[200]

cAMP-response-binding protein (CREBBP) is an HAT that both remodels chromatin and forms gene transcription complexes. An exciting recent discovery was that environmental enrichment alone could increase CREBBP-mediated histone acetylation, increasing behavioral recovery in rodents with SCI.[201] This recovery was linked to increased proprioceptive activity in DRG neurons, and regenerating fibers appeared to have vGlut1 excitatory contacts on motor neurons.[201] This study shows that epigenetic regulation of gene expression is influenced by environmental conditions relevant to SCI. A blood—brain barrier permeable therapeutic that activates CREB is being tested experimentally.[202]

There is a variation of the efficacy of histone acetylation modifying drugs among different neuronal types. Trichostatine A, another HDACi, produces hyperacetylation and transcription-dependent axonal outgrowth in cerebellar granule neuronal cells growing on permissive and nonpermissive substrates[203]; however, it does not have the same effect in RGCs,[204] or postnatal cortical neurons[205] in similar cultures. The regeneration program of RGCs requires p300 expression, which in turn increases histone and nonhistone gene acetylation and RAG induction.[204] However, the expression of p300 is developmentally regulated and remains repressed even after optic nerve axonal injury.[204]

The pharmacologic use of HDACis is also complicated because HDACs also have nonhistone substrates. HDAC5 and HDAC6 have cytoplasmic transcription—independent activities that control the function, activity, and stability of various proteins other than histones.[206] Other nontranscriptional pathways will be altered when inhibiting them, and thus the balance of activities may be complex. In development and after injury, growth cones explore the tissue environment through a dynamic actin cytoskeleton stabilized by microtubules regulated by several transcriptional programs.[207] A reduction in acetylation of microtubule lysines is associated with the more dynamic activity of a growth cone. The postinjury calcium influx initiates tubulin-deacetylation by HDAC activation.[208]

DNA methylation and hydroxymethylation

DNA methylation is another epigenetic mechanism that influences chromatin structure to regulate gene expression and is relevant to differences in peripheral and central axon injury responses. A CPG island is a region of DNA enriched in cytosine—guanine sequences generally found in promoters. When cytosine is methylated to 5-methylcytosine (5mC), it reduces the gene's expression.[209] For demethylation, an intermediate base, 5-hydroxymethylcytosine (5hmC), is catalyzed by the Ten-eleven translocation (Tet) methylcytosine dioxygenase family (Tet1—3).[209] After peripheral axotomy, Tet3 is upregulated in the "conditioned" DRG along with an increase in 5hmC.[210] Since DRGs can switch to a regenerative stage, genomic analysis has identified that peripheral axotomy triggers 5hmC changes in multiple DRG RAGs, including Atf3, BDNF, and Smad1, known to regulate the axon growth potential.[210] Notably, although central lesions also trigger 5hmC modifications, these had little overlap with those triggered by peripheral lesions, indicating that these differing methylation patterns in RAGs have implications in the failure of central regeneration.[210]

There is also evidence that complete DNA demethylation, not only 5hmC increase, mediates RAG induction of axon growth potential in adult DRG neurons and that there are differences among neuronal groups. Tet3 may be required for peripheral regeneration in DRG neurons, but Tet1 is necessary for PTEN-deletion-induced RGC regeneration, indicating cell-specific DNA demethylation pathways.[211] Additionally, there is recent evidence indicating that the DNA methylation pathways blocking axon regeneration are dependent on the target genes.[212] The expression of RAGs is critical for the activation of the axon regeneration program, but the subtleties of DNA methylation and hydroxymethylation and their specific roles are still poorly understood. Defining the specific loci changes and the differences within particular types of neurons remains to be discovered. In the future, identification of the critical genomic loci may enable the use of CRISPR/Cas9-mediated epigenetic editing systems to facilitate central axonal regeneration.

MicroRNAs

MicroRNAs (miRNAs) are noncoding 22 nucleotide RNAs that act posttranscriptionally as gene regulatory switches[213] to control protein expression. miRNAs interact with protein complexes and mRNA to silence expression and accelerate mRNA degradation. miRNA and DNA methylation and histone changes have reciprocal interactions. miRNA is commonly communicated between cells in exosomes. Given the increasing complexity of our knowledge of molecular biology, databases that specify known interactions and binding sites of miRNA are useful, such as miRbase.org.

As an example, miR-210 is a key signal for cellular responses to hypoxia and regulates BDNF levels.[214] Intermittent hypoxia is emerging as a strategy to increase regeneration by suppressing PTEN with resulting increased

mTOR activity.[215] Returning to the DRG model, miR-21 is upregulated in DRGs after axotomy. In vitro proof-of-concept experiments showed that DRG neurons with lentiviral vector–mediated overexpression of miR-21 promoted neurite outgrowth.[216] miR-21 has several targets, one of which is the important PTEN-PI3K/Akt axis.[217] miR-138 expression increases as cortical neurons mature, and in adult uninjured DRGs, high levels of miR-138 inhibit axon growth through suppression of the SIRT1 deacetylase. After peripheral axotomy, reduced miR-138 transcription is necessary for axonal regeneration[218]; upregulated SIRT1 represses miR-138 transcription. A large number of miRNAs have been identified to have regulatory roles in axonal growth including miR-9, miR-17 to miR-92, miR-21, miR-26a, miR-30b, miR-133b, miR-135a, miR-135b, miR-138, miR-210, miR-222, and miR-431.[219] Highlighted examples include those that target PTEN, a negative regulator of the mTOR pathway involved in the regulation of protein synthesis.[220] miR-26 and miR-22 repress PTEN and increase neurite outgrowth in adult sensory neurons.[221,222]

In SCI experiments of the acute injury stage, the expression of miRNAs was assessed in a contusive model with samples collected at 4 h and 1- and 7-day postinjury. The study showed 30 miRNAs consistently increased across these time points and 16 miRNAs decreased. In the acute time window, most identified miRNAs were associated with inflammation, oxidative stress, and apoptosis.[223] At later time points, miR-21 regulates the proliferation of neural progenitor cells in the spinal cord gray matter.[224] For spinal cord regenerative functions, there is also evidence indicating the involvement of miR-133b. In a zebrafish spinal cord transection model, elevated levels of miR-133b in the nucleus of the medial longitudinal fascicle are essential for locomotor recovery and axonal regeneration.[225] miR-133b targets the small GTPase RhoA, an well-known inhibitor of axonal growth, strongly upregulated following SCI.[226,227] miRNAS are transported between cells and tissues by exosomes. The first miRNA-targeted drug to enter clinical trials is Miravirsen, which blocks the production of miR-122, an anti-hepatitis C therapy.

Signaling pathways + master regulators

Protein phosphorylation is a key posttranslational regulatory motif. As discussed, PTEN and mTOR are "master" regulators of axonal regeneration and other critical processes. mTOR is a serine/threonine-protein kinase involved in cell physiological processes, including transcription, mRNA turnover and translation, ribosomal biogenesis, vesicular trafficking, autophagy, and cytoskeletal organization.[228] mTORC2 signaling is dysregulated in numerous neurodegenerative conditions, including Parkinson's, stroke, Huntington's, and Alzheimer's.[229–232] Disruption of negative regulators of mTOR and PTEN occurs in many cancers but also puts neurons into a more regenerative phenotype. Inhibition of mTOR using rapamycin reduces neural tissue damage and locomotor impairment.[233] PTEN negatively regulates mTOR,[234] suppressing adult mouse CST regeneration, and PTEN deletion increases mTOR and increases CST regenerative capability.[235] In vivo conditional knockout of PTEN also promotes robust axon regeneration after optic nerve injury in adult RGCs.[234] Studies using conditional deletion of PTEN in the sensorimotor cortex of developing mice (day 1 postnatal) resulted in enhanced compensatory sprouting of uninjured CST axons and enabled injured CST axon regeneration after T8 dorsal hemisection.[235] Drugs and biologics to transiently suppress PTEN expression are in development but face concerns of potentially amplifying development of cancers. Another powerful gene, suppressor of cytokine signaling 3 (SOCS3), is a negative regulator of Janus kinase/signal transducers and activators of transcription (JAK/STAT) pathway. SOCS3 deletion in RGCs

promotes dramatic axon regeneration following optic nerve injury.[236] Moreover, the deletion of PTEN and SOCS3 synergistically affects regeneration enabling robust and sustained axon regeneration in adult RGCs.[237] The Krüppel-like factor (KLF) family of zinc-finger transcription factors regulates intrinsic axon growth in RGCs.[238] KLF9 expression is increased with the developmental loss of axonal regeneration[238] and affects neurite growth, branching, and elongation.[239–241] Interfering with a JNK3-binding domain or mutating serine phosphorylation sites abolishes KLF9's neurite growth suppression in vitro and promoted axon regeneration in vivo.[242]

Cellular and acellular (biomaterial) transplant interventions

Tissue and cellular transplantation have long been important themes in regeneration research. Tissue loss and cavitation follow SCI,[243] and neurons, oligodendroglia, and astrocytes are lost together with disruption of vasculature and the pia mater. Limited endogenous cell replacement occurs, but neurons specifically are not replaced.[244] Due to the breached glial limiting boundary, peripheral cells such as fibroblasts and SCs enter the injury sites.[245] SCs may contribute to limited repair[246] but also form nonregenerative neuroma-like structures. One approach tested to replace lost tissue, including neurons, was fetal CNS tissue transplantation initiated in the brain for Parkinson's disease.[247] Fetal tissue contains all CNS cell types and exhibits reduced immunogenicity as compared to adult tissue. Ethical considerations hampered these studies, as well as small available quantities, inconsistent efficacy, and side effects.[248,249]

Modern cellular transplants mainly use the injection of fluid suspensions of cells. Suspensions can conform to tissue boundaries and may be placed in multiple locations. The suspensions can be standardized by setting the cell culture conditions for cGMP manufacturing, cell numbers can be expanded exponentially, and some lineage differentiation may be promoted in culture. Further, unlike tissue, cell suspensions can be readily cryopreserved. Precise surgical systems for cell delivery have been developed[250,251] with cell suspensions delivered to the spinal cord parenchyma, the intrathecal space, and intravenously.

The popularity of cell transplantation rests on the ability to use cell culture techniques to expand and modify the cell population. A generally useful distinction is between neural stem cells (NSCs) and other cell types such as peripheral glial cells, totipotent stem cells, and mesenchymal stem cells. Glial cells are used for tissue bridging and myelin repair and can reconstitute tissue structures to support axon growth in tissue voids, at critical interfaces,[252] provide growth substrates, and ensheath or myelinate axons. Actual neuron circuit replacement requires sufficient residual structure to provide cues for the integration of transplanted stem cells.[253] Important issues in the clinical translation of cell transplantation are the merits of using autologous or allograft cells, the need for immune suppression and its duration; cell dose and delivery methods; assessment of cell survival after transplantation, differentiation, proliferation, migration, and integration into the host tissues. Further, its importnat to determine if cell integration into the host neural circuits results in functional synapses and not in aberrant connections that could exacerbate neuropathic pain and spasticity. Whether the goal is to remyelinate axons, replace neuronal and glial populations, or serve as neurotrophin or cytokine release sources, several cell types have been extensively explored in preclinical studies, with some clinical studies completed. As this is a new field, a great deal is learned from each clinical trial. The efficacy of cell therapy in SCI remains unclear, although a number of significant challenges have been addressed, and evidence of safety is accumulating.

Activated macrophages

An early cell therapy trial in SCI was the delivery of "activated" macrophages to the subacutely injured spinal cord. Macrophages have important roles in the wound healing response after SCI, are polarized to phenotypes known as M1 and M2 (pro-inflammatory and reparative, respectively),[254,255] and dynamically adapt to injury microenvironments.[256,257] This trial was based on observations that myelin debris-laden macrophages were found at lengthy times postinjury in spinal tracts[258–260] undergoing Wallerian degeneration after SCI, whereas in peripheral nerves such debris is rapidly cleared. When blood-derived macrophages were comparatively exposed in culture to peripheral nerve or CNS optic nerve fragments, those exposed to nerve tissue displayed greater debris-processing phagocytic activity,[261] and peripheral nerve–activated macrophages applied to the optic nerve increased axonal sprouting and growth.[262] In a thoracic transection rodent model, functional, electrophysiological, and histological improvement was reported after intraspinal transplantation of macrophages activated in peripheral nerve cocultures.[263] A subsequent contusion study of nerve coculture-activated transplanted macrophages reported motor recovery, reduction of cystic cavities, and secretion of BDNF and interleukin-1ß.[264] During further product development, it was also reported that "activation" via exposure to an incubated skin biopsy exposure was as effective as nerve segments eliminating the need for an invasive nerve biopsy. Once "activated," preclinical evidence suggested that macrophages would induce neuroprotection, neuroregeneration, and plasticity.[265]

Based on these studies, Proneuron biotechnologies received the first FDA IND for cell therapy after SCI and conducted the 2 "ProCord" clinical trials in which autologous macrophages were isolated from blood and activated in cocultures with autologous skin biopsies. The Phase I study recruited eight AIS A subjects with SCI levels C5 to T11 within a 14-day post-SCI time frame. Subjects received four microinjections of activated peripheral macrophages, two 20 µL injections into white matter corresponding to the CST location, and two 10 µL injections into the dorsal columns totaling 4 million cells. After 1 year follow-up, the trial reported conversion from AIS A to C in 3/8 subjects and stable MRI imaging. Electrophysiological findings included recovery of conductivity in 5/8 subjects with initial negative assessments. Overall, the study indicated the safety of the surgical procedures and no serious adverse events directly related to macrophage transplantation.[266] The observed changes in this open-label study could have been due to spontaneous recovery. Thus, a subsequent multicenter randomized Phase II study was initiated in 2003 (NCT00073853) with 2:1 treatment: control (standard of care) intended to recruit 61 participants at six centers in the United States and Israel. The studies were suspended in 2006. Efficacy analyses performed with 43 randomized subjects reported AIS conversion from A to B in 7 treatment and 10 controls, and from A to C in 2 treatment and 2 control subjects. Although no statistical significance was found in treatment versus controls, the trend favored the control group, results that did not support the use of autologous macrophage therapy for SCI.[267] Subsequent experimental studies have shown an important difference in macrophage polarization in which the M2 phenotype supports repair.[268] Thus inflammation modulation remains a promising direction.

Olfactory ensheathing cells

For a period of time OECs were considered the ideal glial cell for reparative spinal cord transplantation. This is because they normally have a unique integrative organization between their peripheral neuron cell bodies in the nasal cavity and CNS terminals in the brain olfactory bulb. The transition within the olfactory bulb from entering olfactory nerve roots to the CNS glomerulus has a seamless interface of OEC and astrocyte processes without intervening basal lamina.[269] This integration differs from the spinal cord root entry zone that contains considerable basal lamina. The term "ensheathing" refers to the organization of numerous small unmyelinated axons wrapped by thin glial slips. Olfactory neurons are replaced throughout life in the basal layer of the olfactory epithelium[270,271] and migrate ahead of the new axons as they project toward the olfactory bulb,[272] where they form effective connections.[273] Some microtransplant experiments suggested the favorable olfactory system anatomical relationships could be reproduced by transplanted cells interacting with CST axons.[274–276] Thus, theoretically, OECs have multiple properties favorable for CNS repair, although they do not normally form myelin. Transplantation studies of OECs derived from the olfactory bulb reported impressive recoveries in rats.[277] Other preclinical studies in OEC-transplanted rodents with experimental SCI reported reduced cavity formation,[278] axonal sprouting and regeneration,[277,279,280] modulation of neuroinflammation with increased phagocytosis,[281,282] improved sensory conduction,[283] and sensory locomotor improvements.[284] However, derivation and purification of OEC cultures from the olfactory bulb are challenging due to the small quantities of source tissue[285] and the limited possible expansion. In people, a biopsy of the olfactory bulb is more risky than e.g. a peripheral nerve.

Researchers in Australia explored the use of purified OECs derived from olfactory mucosal biopsies in a Phase I/IIa blinded study that enrolled six chronic thoracic AIS A subjects, 3 implanted and 3 as controls. The procedures involved the development and use of a stereotactic circular surgical frame to deliver multiple 1.1 µL injections (80,000/uL) distributed in a gridline fashion (covering X, Y, and Z planes) rostral, within, and caudal to the injury epicenter. The three transplanted subjects received an escalating cell dose (12, 24, and 28 million) in a surgical procedure performing 270, 545, and 630 penetrating injections per subject, respectively. The large number of injections was based on an expectation of limited migration. At 1-year follow-up, there were no reported adverse events related to the procedures,[285] and at 3 years, no unexpected changes on MRI and stable neurological outcomes.[286] Although the trial outcomes were considered indicative of safety, no therapeutic effect related to the transplant procedures was discerned.[287]

The enthusiasm for the potential use of OECs in SCI arose at the same time as emerging knowledge of NSCs and early in the peak of unregulated medical stem cell tourism. There was a plethora of reports of various forms of clinical studies, and hundreds if not thousands of people with SCI were implanted with cells claimed to be OECs. Several clinical studies were conducted in China. One report described allograft transplantation of bulb-derived fetal OECs in 300 subjects with chronic (6 months to 31 years, average 3.1 years), complete (n = 222), and incomplete (n = 78) injuries. The transplant procedure included injections of 50 μL OEC suspensions (1×10^6) caudal and rostral to the injury epicenter, with follow-up until 8 weeks postprocedure.[288] Authors reported no adverse events related to the transplants or evidence of rejection, and a "rapid improvement" phase beginning 2−3 days posttransplant with overall AIS score conversion of 17.1% from A to B, 20% from A to C, 50% from AIS B to C, 50% from AIS C to D, and 22.7% from AIS D to E. An independent assessment of the trial did report a transient improvement in one subject,[289] although seven subjects evaluated before and 1 year after transplantation were found to have little change in their neurological function. Extensive commitment to rehabilitation may have accounted for some improvements. The fetal olfactory bulb is tiny, and there is the potential to incorporate other cell types in such cultures.[290] A systematic

review of the preclinical literature found that reports of recovery after OEC transplantation indicated biological plausibility.[291] Studies in the naturally injured dog model indicated a benefit on gait control.[292] Controversy has been focused on whether the olfactory bulb or the easier-to-reach mucosa is the most suitable cell source. Small craniotomy approaches to the olfactory bulb have been described[293] and used to derive the OB cell source for transplantation. Few studies have compared the use of nasal versus bulb-derived cells,[294] and the superiority of one source remains unclear.[10] The great enthusiasm generated around OECs leads to some misinformed human transplantation of olfactory mucosa[287] itself, a complex tissue containing glands with various reports of some efficacy, but also resulting in serious reports of expansile masses and loss of neurological function in persons with SCI.[295] Implantation of nasal mucosa to the spinal cord appears neither safe nor well-grounded scientifically forming an important cautionary lesson in the field of SCI.

Schwann cells

Multiple lines of evidence have supported the testing of SC transplants for repair in the spinal cord. These include that peripheral nerves have a greater regenerative capacity than the spinal cord, and transplants of peripheral nerve segments to gaps after nerve injuries can lead to functional repair. After SCI, SCs migrate into the injury zone and can form myelin sheaths around axons[245,296] in regions devoid of astrocytes. For autologous transplantation, donor nerve segments can be dissociated to component cells, expanded, purified, and transplanted as cellular suspensions. Tissue reconstitution with axonal growth and myelin restoration has been shown in rodent models[297,298] of SCI. SCs produce trophic factors[299] and laminin[300] and directly support axonal growth by ensheathment and myelination,[189,297,301−304] expression of cell adhesion molecules with some evidence of functional recovery in animal models.[298,305,306] SC are capable of enabling vascularized tissue formation. The authors and collaborators conducted two clinical trials of intraparenchymal autologous SC transplantation at the subacute and chronic post-SCI periods in the last decade. The subacute study was a Phase I open-label, dose-escalation trial study that recruited six thoracic AIS A subjects to receive autologous doses of 50, 100, and 150 µL of SCs at 100,000 cells/µL at 4–7 weeks after injury (NCT01739023). Subject assessments included ISNCSCI, MRI changes, MAS, ISS, FIM, SCIM II, EMG, and detailed motor, sensory, and autonomic evoked potentials. There were no notable adverse events related to the cells or medical procedures.[307] One subject converted from AIS A to B at 6 months posttransplant, and some voluntary EMG activity and motor evoked potentials below the injury level were detected in all participants. These findings are important as no changes in clinical measures were detected.[308] Subsequently, a second trial was launched, including eight patients enrolled in two cohorts (NCT02354625). A first cohort of thoracic level 2–12 AIS grade A, B, or C (n = up to 4), and a second cohort of cervical patients' level C5 through T1 AIS A, B, or C (n = up to 6), with two AIS grade A participants enrolled before enrolling a participant with AIS B or C, applicable to both cohorts. Follow-up assessments of this second study were similar to the subacute group. In addition, all patients were exposed to fitness conditioning and rehabilitation boot camp periods pre- and posttransplantation procedures to promote potential neuroplasticity. The endpoint was at 6 months posttransplantation procedures.[309] The potential for allogenic SC transplantation remains unclear but could make trials easier and allow more acute cell delivery as autolgous preparations are laborious. Future research should address increasing SC migration and integration with astrocytic parenchyma.

Neural stem cells

The ideal cell type for repair after SCI would integrate and form replacement cells of both glial and neuronal lineages. They would be well characterized, able to be cryopreserved, and require minimal immune suppression. The discovery of stem cells in the adult brain revolutionized neural repair research.[310] Subsequently, methods to passage primate embryonic stem cells (ESCs) apparently indefinitely were described for primate cells in 1998.[311] Stem cells exhibit the capability for cell division throughout the host lifespan and can be induced to phenotypic differentiation. NSCs may differentiate into neural lineage cell types such as astrocytes and oligodendroglia and under certain conditions neurons.

Whereas NSCs are multipotential cells with both glial and neuronal differentiation potential, ESCs are totipotential with the capacity to form all cells of the body if suitably manipulated. It was reported that cultured ESCs could be differentiated to become oligodendrocytes and that transplanted ESC-derived oligodendroglia remyelinated axons and improved function when transplanted after SCI.[312] In 2010, the Geron Corporation initiated a Phase I open-label trial of ESC-derived human oligodendrocyte precursor cells (OPC1). This was the first medical application of ESC-derived cells in humans and the first NSC study in SCI. The H9 cell line used for derivation was initially developed at the University of Wisconsin.[311] The multicenter study also included developing and using a stereotactic device for the injection procedures, a platform known as the syringe positioning device. After the approval, the lead investigators self-reported the formation of some cystlike structures in additional preclinical studies, and the trial was placed on hold until methods to eliminate the cyst-forming cells were developed.[313]

After the recruitment of 6 of 10 planned subjects, the trial was halted for financial reasons, but the subjects continued in long-term follow-up. In 2013, Geron's stem cell assets, including the Phase I trial, were acquired by BioTime Inc., and with new funding, a subsidiary company named Asterias was created.[314] The Phase I subjects related followed by Asterias have reported no serious adverse effects related to the cells or the immunosuppression protocols used. A new Phase I/IIa dose-escalation study enrolled motor complete AIS A and B, C4 to C7 injuries for transplantation at the subacute post-SCI time point (21–42 days) (NCT02302157). This SCiStar trial enrolled five sequential cohorts (1: AIS A—2 million dose n = 3; 2: AIS A—10 million dose n = 6; 3: AIS A—20 million dose n = 6; 4: AIS B—10 million dose n = 6; 5: AIS B—20 million dose n = 4). Outcome measures included upper-extremity motor scores (UEMSs) and ISNCSCI examinations up to the 1-year posttransplant. Results have been reported in three broad categories: (1) Safety, where all 25 patients from the SCiStar study and the previous 5 patients from the Geron trial have had no evidence of adverse changes reviewed by MRI; (2) cell engraftment such that most subjects have MRI findings consistent with the formation of a tissue matrix at the injury site "preventing" the development of cystic cavities; (3) improved motor function, with no subject having decreased motor function after the transplantation procedures.[315] In 2019, all Asterias proceeds were acquired by Lineage Cell Therapeutics Inc., and OPC1 has received a regenerative medicine advanced therapy (RMAT) and Orphan Drug designations from the FDA. Hopefully, there will be further development of the OPC1 program.[316] However, the foregoing results have not yet been published.

NSI-566 is a neural human stem cell line authorized by the FDA for clinical testing, derived from a single postmortem spinal cord of an 8-week gestational fetus. It was previously tested in ALS,[317] and preclinical SCI studies included rodents and minipigs.[318–320] A human SCI study enrolled four chronic thoracic AIS A individuals with injury levels T2 to T12 who received six stereotactic-guided bilateral injections of NSI-566. Observations reported from this trial including changes in one or two levels in the ISNCSCI in two subjects, detection of voluntary EMG from the rectus abdominus and paraspinal muscles in a subject with initial complete T7 neurological level, and increased caudal sensation at 18 months posttransplant. No serious complications were reported, the small changes did not alter the subject's quality-of-life scores, neither were there visible changes in MRI nor improvement in DTI sequences.[321]

A third clinical trial explored intramedullary perilesional injections of human fetal CNS-derived stem cells (HuCNS-SCs). The Phase I/II open-label study explored manual injection of HuCNS-SCs into the spared parenchyma rostral and caudal to the injury epicenter. Twelve thoracic and 31 cervical AIS A and B patients were enrolled. Overall, all subjects tolerated the procedures, the tacrolimus immunosuppression protocols, and no safety concerns related to the cell transplants were identified after the first-year posttransplant.[322] Interim analysis of the cohorts identified trends toward improvement in UEMS and the GRASSP test; however, the magnitude of the effects did not reach the prospectively defined efficacy threshold, and the study was terminated.[323] Given initial concerns about stem cells potentially forming abberant tissue, the overall extant results with NSCs are favorable.

Induced pluripotency

The idea that any cell in the body has the potential to be dedifferentiated to a totipotential state initially seemed impossible because of the entrenched concept of "terminally differentiated" somatic cells. Work with ESCs lead some researchers to question if key transcription factors involved in their maintenance in an undifferentiated state might reverse phenotype in differentiated fibroblasts.[324] The promise of induced pluripotency is great, but some limitations have been encountered, such as retained epigenetic changes and insertional mutagenesis with the originally used transcription factors.[325] Approaches such as using synthetic mRNA for reprogramming have been shown to be safer and more efficient.[326] Clinical application of induced pluripotency—derived dopaminergic neurons for Parkinson's disease is reaching the clinical trial phase.[327] Such cells are also useful individualized disease models.

Mesenchymal stromal cells

Mesenchymal stromal cells (MSCs) are adult stem cells that normally contribute to tissue repair and have limited differentiation potential. First discovered in the bone marrow, they have subsequently been identified in most organs and tissues[328,329] and applied in numerous diseases and injuries, notably stroke and myocardial infarction. There is considerable therapeutic interest in the biologic effects of MSCs due to their reproducible and relatively simple cell culture paradigms, autotransplant capability, and apparent reduced host response to allografts. The stringency of regulation and oversight of MSC products varies by country, and there have been some insufficiently substantiated efficacy claims.[330–335] Standard criteria to define cultured MSCs are: (1) plastic adherence, (2) $\geq$95% cell population expressing CD105, CD73, and CD90, (3) $\leq$2% population expressing CD45, CD34, CD14 or CD11b, CD79α or CD19, and HLA-DR, and (4) demonstrated ability for in vitro differentiation into osteoblasts, adipocytes, and chondroblasts.[336] Common sources for MSC isolation include the bone marrow,[337–340] adipose tissue,[341,342] and the umbilical cord.[343] MSCs are the most extensively tested cell type in preclinical and clinical SCI studies. Initially, it was thought that MSCs could generate neurons because Y chromosomes were found in cerebellar neurons in female bone marrow recipients.[344] This was later attributed to cell fusion.[345] Current evidence supports that MSCs can home to some tissue injury environments[346,347] and release immunomodulatory cytokines and trophic factors[348] in

perivascular spaces[349] with a limited survival period. Other interesting possible therapeutic mechanisms include secretion of exosomes coating regulatory miRNA to promote the M2 macrophage phenotype,[350] transfer of mitochondria to injured cells observed in cell culture.[351]

A large number of studies have reported delivery of mesenchymal stem cells after SCI, and an autologous MSC product was approved for conditional use in Japan.[352,353] These cells may not survive long-term but can have trophic effects and potentially deliver contents such as mitochondria. Many reported SCI studies are of single cases[354] or open-label with limited methods to reduce bias. The result is a confusing literature from which one can glean that the absence of major adverse events suggests safety, and while there might be a benefit, this remains inconclusive. Delivery of MSC for SCI has been performed intravenously, intrathecally,[354] and directly into the spinal cord. The numbers of injections have been single or multiple over time, and the delivered doses have varied. In general, the most rigorous studies have reported either no apparent effect[355] or minor effects, reported recoveries are quite variable[356] and are more significant with incomplete SCI.[357] Complications have included neuropathic pain and encephalomyelitis, indicating that caution in patient selection is essential.[355] In controlled studies, improvements in sensation are a commonly observed effect.[358] When studies report zero adverse events, it is difficult to trust them, as adverse events are common in all SCI clinical studies.[359] A single patient study of multiple cervical intrathecal injections conducted by the authors employed extensive neurophysiological and neurological outcome measures and found notable plasticity of respiratory neurons to upper-extremity muscles.[360] The use of MSCs is affected by lack of harmonization in cell type, cell dose, delivery routes, and level and time of injury, and control cohorts making it difficult to consolidate available data. MSCs offer advantages over other cell types as they have alternative delivery routes (IV or intrathecal) and fewer ethical constraints versus fetal or embryonic stem cells.[361] However, cautionary lessons, including rare neoplastic growth, vision loss, and acute encephalitis related to these transplants, are reported.[362–364] Taken in aggregate, the literature suggests a therapeutic benefit to some MSC paradigms, but much greater rigor in cell characterization and biomarker assessments are needed to advance this therapeutic.

Exosomes and cellular nanoparticles

The discoveries of exosomes and nanoparticles with signaling functions have opened up exciting new possibilities to utilize cells to produce subcellular products with biological activity that could obviate the need to implant cells themselves. For this purpose, MSCs are promising sources of exosomes that can "educate" other cells through cytokines and RNA molecules to change their phenotype. As an example, miR-133b can promote neurite outgrowth, possibly through RhoA inhibition.[365] Further, this exosomal communication may occur through cellular gap junctions.[366] Hypoxic preconditioned MSCs secrete exosomes containing miR-216a that can be neuroprotective after SCI[367] by shifting microglial polarization to the M2 phenotype. Thus, cultured cells may be used to derive secreted particle therapeutics.

Limitations of cell therapy

A lack of biomarkers of survival, migration, and target engagement in clinical applications are important gaps in current clinical cell therapy for SCI. Most allogenic NSCs are used with immune suppression, but the efficacy regarding transplanted cell survival is not well understood.

Gene therapy

Gene therapy is a highly promising and broad therapeutic approach. It encompasses a number of potential treatment approaches from direct gene transfer to ex vivo transduction of cells to be subsequently transplanted. Several studies have shown enhanced axonal regeneration into grafts of transfected neurotrophin releasing SCs[368,369] or fibroblasts.[370] The recent approval of systemic AAV-9 delivery in children with spinal muscular atrophy is an important demonstration of the balance of efficacy and toxicity in an otherwise lethal neurological disease.[371,372] Several studies have tested gene transfer of neurotrophic proteins and also gene silencing using expressed antisense mRNA in ALS to reduce toxic SOD-1 expression.[373] Gene therapy can lead to prolonged protein production that may be regulated at the promoter level by on-off constructs responsive to innocuous systemic drugs. One promising therapeutic is the delivery of a modified chondroitinase enzyme by gene therapy.[63] Innovations to reduce the immune response and to regulate delivery with a tetracycline switch increase the safety of such an approach by providing temporal control of enzyme activity.[374] Safety is a paramount consideration in gene therapy for SCI due to the potential impact of losses of neurological function if gene activity impairs preserved functions.

Biomaterials and scaffolds

Modern successful biomaterials include artificial skin and bone substrates.[375] Early CNS biomaterials were formed from ECM such as polymerized collagen and prepared in SC culture to mimic nerves.[376] These were followed by the use of polymeric non-degradable guidance channels with limited diffusion to concentrate trophic and other factors.[301] Favorable characteristics for spinal cord scaffolds are: (1) pliability and mechanical properties to provide needed structure without damaging surrounding residual tissues, (2) adequate permeability, (3) axonal or cellular growth supportive matrix for implanted cells and biomaterials, or to promote endogenous repair, (4) biocompatibility with minimal inflammatory response, and (5) biodegrading characteristics that would ideally favorably interact with the inflammatory wound healing host responses. Biomaterials for spinal cord repair can be constructed from natural substances (agarose, alginate, collagen, chitosan, fibrin, hyaluronic acid) or artificial components (poly-β-hydroxybutyrate, poly-ε-caprolactone, polyethylene glycol, poly(lactic acid), poly(lactic-co-glycolic acid), poly(2-hydroxyethyl methacrylate), poly[N-(2-hydroxypropyl) methacrylamide], and polyvinyl alcohol), or a combination of both. Additionally, biomaterials can be shaped in purpose-suited configurations. Numerous preclinical experiments have tested biomaterials that form conduits or channels,[377–383] sheets,[384] three-dimensional scaffolds,[385,386] fibers,[386,387] or hydrogels.[388–392] Clinical translation of biomaterials in the CNS is still in early phases. A linearly ordered collagen scaffold, NeuroRegen, prepared from bovine aponeurosis[393] and seeded with autologous BMSC was tested in five chronic AIS A individuals.[394] Complete transections with scar tissue resection at the epicenter were performed, and authors report scar dissection lengths ranging from 0.5 to 4.5 cm with no changes in sensory or motor levels reported postoperatively. The removed scar tissue was positive for vimentin and CSPG indicating scarring. Subjects were assessed at 1, 3, 6, and 12 MPI with reported safety. Recovery of low amplitude SSEPs was reported, although the normal latencies reported are unusual compared to other studies.[394] In a subsequent study in the subacute time period, myelotomy was performed and necrotic material removed and multiple scaffolds placed together with acutely harvested bone marrow mononuclear cells in several AIA A thoracic injury subjects. No motor improvements were detected.[395] A different scaffold approach that employed slow-release of NT-3 from a Chitosan channel was validated collaboratively by an international expert group.[396] Complications reported include seroma formation and meningitis.[290] The most extensive North American clinical study has tested a biomaterial called the Neuro-Spinal scaffold in INSPIRE studies 1.0 and 2.0 in acute SCI. These studies are conducted under an FDA humanitarian device exemption construct[397–399] is made from poly(lactic-co-glycolic acid) and poly(L-lysine), and being tested in ongoing clinical trials (NCT02138110), with a multicenter study planned to recruit 20 participants. A case report of a T11 AIS A individual described the procedure of the microscaffold insertion via DREZ myelotomy and reported neurological improvement to AIS C with an L1 level 3 months postinjury.[400] In the first clinical report from 19 AIS A thoracic SCI implanted subjects, the rate of AIS conversion exceeded that of historical controls.[401] The second ongoing study employs a randomized controlled design.

The BioArctic scaffold study (SC0806) was based on a biodegradable calcium sulfate matrix engineered to contain autologous nerve graft fascicles and to release FGF-1[402–404] slowly. The dimensional properties were based on detailed spinal cord studies and constructed for a precise fit into a surgically created tissue gap of the spinal cord. The trial enrolled nine subjects with complete thoracic injury and was terminated after interim analysis at 18 months postimplant when no evidence of neurophysiological conduction through the repair site was found in the subjects.

In summary, clinical studies of neuroscaffolds have been initiated and appear relatively safe, but efficacy data to this point are not striking. Further innovation with cells, biologics and electrical currents are being tested preclincally.

Electrical stimulation

The potential for neuromodulation to improve function after SCI is increasingly recognized.[402] Due to the fundamental electrical nature of axonal impulse conduction, there has long been interest in applying electromagnetic fields to increase regeneration.[403] Due to greater geometrical and structural simplicity, many studies have been conducted on peripheral nerves.[404] Molecular pathways responsive to electrical stimulation include calcium-dependent Erk activation and BDNF upregulation in cultured neurons at 20Hz, 100μs pulse duration.[405] Immature neurons implanted to nerves have shown increased muscle innervation after 1 h of 20 Hz stimulation.[406] A consistent mechanistic theme is that electrical stimulation promotes calcium flux and activates cyclic AMP.[407] Electrical stimulation of the medullary pyramid was also shown to increase OPC proliferation and differentiation,[408] and motor cortex stimulation increased CST sprouting.[409] Clinical studies of epidural and transcutaneous stimulation have shown promising effects,[410,411] although whether axonal regeneration has occurred is unknown.

Circuit formation

The formation of new effective neural circuits is essential for effective spinal cord repair. Currently, observed regenerative responses in rodent models are on the order of millimeters or less, and it is thought that actual long-distance regeneration to original targets in humans is unlikely without additional advances. Thus, the formation of relay circuits that utilize new terminations on intraspinal neurons is essential.[412] Such new connections may utilize propriospinal relays,[413] and motor axons from redirected peripheral nerves can create a lesion bypass[414] in rodent studies. In nonhuman primates with a cervical spinal cord hemisection, newly elaborated CST fibers show extensive growth from the uninjured into denervated gray matter that is correlated with recovery.[415] Such new axonal projections can innervate transplanted stem cells, increasing relay potential.[416]

Combination strategies

In this chapter, we have described multiple molecular pathways associated with axonal regeneration. The probability that any single target can support clinically measurable neurologic recovery appears small, and thus methods

to target several pathways and processes are needed. Such combinations will need to be in the correct sequence and likely tailored to the individual injury. Combined effects should complement and not interfere with each other. A regeneration "interactome" derived from bioinformatics and validated experimentally could aid creation of combinations substantially. Currently, datasets such as Kegg (genome.jp), Consensuspathdb.org, and reactome.org contain pathway and interaction maps but lack interfacing with SCI pathomechanisms.

Further advances might allow individualized and temporal interactions during acute and chronic injury to be predicted and validated with biomarkers.[16] As discussed therapeutics that can target several different regeneration cascades are being elucidated. Without such mechanistic insight, empirical discovery and testing is required, such as that afforded by high-throughput screening[417] with advanced devices.[418]

Summary

Axonal regeneration is a complex process in which multiple mechanisms interact. Effective regeneration in people will likely require manipulation of several processes in concert. Current clinical testing primarily focuses on assessing single agents to reduce interpretation complexity. In the last three decades, substantial progress has been made. High-throughput screens have led to the discovery of regeneration-promoting drugs. Biologics such as blocking antibodies have reached clinical testing. Extensive studies of cell transplantation have been initiated. Biomaterials are beign tested. The knowledge base is reaching a point where it is possible to model multiple interactions and rationally combine therapeutics. Despite this progress, the development cycle of therapies is long, usually at least a decade from discovery to clinical testing. Greater data sharing is needed including negative experimental data that are infrequently published,[419] in order to reduce redundancy and provide sufficient confirmation of experimental results.

Conclusions

Axonal regeneration remains a central goal to restore function after SCI. The complexity of interactions indicates that successful repair continues to require rigorous preclinical and clinical science combined with bioinformatics and in silico modelling. Multiple lessons indicate that rigorous study designs that can provide clear answers is essential to reduce confusion and translational inefficiency. Knowledge accumulation is accelerating and therapeutic breakthroughs are increasingly probable.

In memoriam. Several scientists who contributed critical observations reviewed above have passed away recently. These include Drs. Geoffrey Raisman, Marie Filbin, Ron Doucette, and Peter Richardson. We are indebted to their pioneering efforts.

References

1. Squair JW, et al. Integrated systems analysis reveals conserved gene networks underlying response to spinal cord injury. *Elife* 2018;7.
2. Sun X, et al. Multiple organ dysfunction and systemic inflammation after spinal cord injury: a complex relationship. *J Neuroinflammation* 2016;13(1):260.
3. Kigerl KA, et al. The spinal cord-gut-immune axis as a master regulator of health and neurological function after spinal cord injury. *Exp Neurol* 2020;323:113085.
4. Aguayo AJ, David S, Bray GM. Influences of the glial environment on the elongation of axons after injury: transplantation studies in adult rodents. *J Exp Biol* 1981;95:231−40.
5. Richardson PM, McGuinness UM, Aguayo AJ. Axons from CNS neurons regenerate into PNS grafts. *Nature* 1980;284(5753):264−5.
6. Cheng H, Cao YH, Olson L. Spinal cord repair in adult paraplegic rats: partial restoration of hind limb function. *Science* 1996;273(5274):510−3.
7. Houle JD, et al. Combining an autologous peripheral nervous system "bridge" and matrix modification by chondroitinase allows robust, functional regeneration beyond a hemisection lesion of the adult rat spinal cord. *J Neurosci* 2006;26(28):7405−15.
8. Wrathall JR, et al. Reconstruction of the contused cat spinal cord by the delayed nerve graft technique and cultured peripheral non-neuronal cells. *Acta Neuropathol* 1982;57(1):59−69.
9. Carlstedt T, et al. Return of function after spinal-cord implantation of avulsed spinal nerve roots. *Lancet* 1995;346(8986):1323−5.
10. Tabakow P, et al. Functional regeneration of supraspinal connections in a patient with transected spinal cord following transplantation of bulbar olfactory ensheathing cells with peripheral nerve bridging. *Cell Transplant* 2014;23(12):1631−55.
11. Guest JD, et al. The ability of human Schwann cell grafts to promote regeneration in the transected nude rat spinal cord. *Exp Neurol* 1997;148(2):502−22.
12. Windle WF, Smart JO, Beers JJ. Residual function after subtotal spinal cord transection in adult cats. *Neurology* 1958;8(7):518−21.
13. Pfyffer D, et al. Predictive value of midsagittal tissue bridges on functional recovery after spinal cord injury. *Neurorehabil Neural Rep* 2021;35(1):33−43.
14. Friedli L, et al. Pronounced species divergence in corticospinal tract reorganization and functional recovery after lateralized spinal cord injury favors primates. *Sci Transl Med* 2015;7(302):302ra134.
15. Jacobson PB, et al. Elezanumab, a human anti-RGMa monoclonal antibody, promotes neuroprotection, neuroplasticity, and neurorecovery following a thoracic hemicompression spinal cord injury in non-human primates. *Neurobiol Dis* 2021;155:105385.
16. Chandran V, et al. A systems-level analysis of the peripheral nerve intrinsic axonal growth program. *Neuron* 2016;89(5):956−70.
17. Ferretti P, Zhang F, O'Neill P. Changes in spinal cord regenerative ability through phylogenesis and development: lessons to be learnt. *Dev Dynam* 2003;226(2):245−56.
18. Hanslik KL, et al. Regenerative capacity in the lamprey spinal cord is not altered after a repeated transection. *PLoS One* 2019;14(1):27.
19. Cosimi AB, et al. A randomized clinical trial comparing OKT3 and steroids for treatment of hepatic allograft rejection. *Transplantation* 1987;43(1):91−5.
20. Eva R, Fawcett J. Integrin signalling and traffic during axon growth and regeneration. *Curr Opin Neurobiol* 2014;27:179−85.
21. Jessen KR, Mirsky R. The success and failure of the Schwann cell response to nerve injury. *Front Cell Neurosci* 2019;13:33.
22. Levine J. The reactions and role of NG2 glia in spinal cord injury. *Brain Res* 2016;1638(Pt B):199−208.

23. Reier PJ, Houle JD. The glial scar: its bearing on axonal elongation and transplantation approaches to CNS repair. *Adv Neurol* 1988;**47**:87—138.

24. Hausmann ON. Post-traumatic inflammation following spinal cord injury. *Spinal Cord* 2003;**41**(7): 369—78.

25. Sofroniew MV. Astrocyte barriers to neurotoxic inflammation. *Nat Rev Neurosci* 2015;**16**(5):249—63.

26. Hackett AR, Lee JK. Understanding the NG2 glial scar after spinal cord injury. *Front Neurol* 2016;**7**:199.

27. Rudge JS, Silver J. Inhibition of neurite outgrowth on astroglial scars in vitro. *J Neurosci* 1990;**10**(11): 3594—603.

28. Silver J, Miller JH. Regeneration beyond the glial scar. *Nat Rev Neurosci* 2004;**5**(2):146—56.

29. Sasaki M, Ide C. Demyelination and remyelination in the dorsal funiculus of the rat spinal cord after heat injury. *J Neurocytol* 1989;**18**(2):225—39.

30. Anderson MA, et al. Astrocyte scar formation aids central nervous system axon regeneration. *Nature* 2016; **532**(7598):195—200.

31. Faulkner JR, et al. Reactive astrocytes protect tissue and preserve function after spinal cord injury. *J Neurosci* 2004;**24**(9):2143—55.

32. Herrmann JE, et al. STAT3 is a critical regulator of astrogliosis and scar formation after spinal cord injury. *J Neurosci* 2008;**28**(28):7231—43.

33. Okada S, et al. Conditional ablation of Stat3 or Socs3 discloses a dual role for reactive astrocytes after spinal cord injury. *Nat Med* 2006;**12**(7):829—34.

34. Wanner IB, et al. Glial scar borders are formed by newly proliferated, elongated astrocytes that interact to corral inflammatory and fibrotic cells via STAT3-dependent mechanisms after spinal cord injury. *J Neurosci* 2013;**33**(31):12870—86.

35. Cregg JM, et al. Functional regeneration beyond the glial scar. *Exp Neurol* 2014;**253**:197—207.

36. Wanner IB, et al. A new in vitro model of the glial scar inhibits axon growth. *Glia* 2008;**56**(15):1691—709.

37. Giger RJ, Hollis 2nd ER, Tuszynski MH. Guidance molecules in axon regeneration. *Cold Spring Harb Perspect Biol* 2010;**2**(7):a001867.

38. Goldshmit Y, McLenachan S, Turnley A. Roles of Eph receptors and ephrins in the normal and damaged adult CNS. *Brain Res Rev* 2006;**52**(2):327—45.

39. Niclou SP, Ehlert EM, Verhaagen J. Chemorepellent axon guidance molecules in spinal cord injury. *J Neurotrauma* 2006;**23**(3—4):409—21.

40. Onishi K, Hollis E, Zou Y. Axon guidance and injury-lessons from Wnts and Wnt signaling. *Curr Opin Neurobiol* 2014;**27**:232—40.

41. Monnier PP, et al. RGM is a repulsive guidance molecule for retinal axons. *Nature* 2002;**419**(6905):392—5.

42. Fawcett JW, Oohashi T, Pizzorusso T. The roles of perineuronal nets and the perinodal extracellular matrix in neuronal function. *Nat Rev Neurosci* 2019;**20**(8):451—65.

43. McKeon RJ, et al. Reduction of neurite outgrowth in a model of glial scarring following CNS injury is correlated with the expression of inhibitory molecules on reactive astrocytes. *J Neurosci* 1991;**11**(11):3398—411.

44. Morgenstern DA, Asher RA, Fawcett JW. Chondroitin sulphate proteoglycans in the CNS injury response. *Prog Brain Res* 2002;**137**:313—32.

45. Tran AP, Warren PM, Silver J. The biology of regeneration failure and success after spinal cord injury. *Physiol Rev* 2018;**98**(2):881—917.

46. Monnier PP, et al. The Rho/ROCK pathway mediates neurite growth-inhibitory activity associated with the chondroitin sulfate proteoglycans of the CNS glial scar. *Mol Cell Neurosci* 2003;**22**(3):319—30.

47. Siebert JR, Osterhout DJ. The inhibitory effects of chondroitin sulfate proteoglycans on oligodendrocytes. *J Neurochem* 2011;**119**(1):176—88.

48. Buss A, et al. NG2 and phosphacan are present in the astroglial scar after human traumatic spinal cord injury. *BMC Neurol* 2009;**9**:32.

49. Fitch MT, et al. Cellular and molecular mechanisms of glial scarring and progressive cavitation: in vivo and in vitro analysis of inflammation-induced secondary injury after CNS trauma. *J Neurosci* 1999;**19**(19): 8182—98.

50. Rolls A, et al. Two faces of chondroitin sulfate proteoglycan in spinal cord repair: a role in microglia/macrophage activation. *PLoS Med* 2008;**5**(8):e171.

51. Raspa A, et al. Feasible stabilization of chondroitinase abc enables reduced astrogliosis in a chronic model of spinal cord injury. *CNS Neurosci Ther* 2019;**25**(1): 86—100.

52. Moon LD, et al. Regeneration of CNS axons back to their target following treatment of adult rat brain with chondroitinase ABC. *Nat Neurosci* 2001;**4**(5): 465—6.

53. Fletcher EJR, Moon LDF, Duty S. Chondroitinase ABC reduces dopaminergic nigral cell death and striatal terminal loss in a 6-hydroxydopamine partial lesion mouse model of Parkinson's disease. *BMC Neurosci* 2019;**20**(1).

54. Bradbury EJ, et al. Chondroitinase ABC promotes functional recovery after spinal cord injury. *Nature* 2002;**416**(6881):636—40.

55. Tester NJ, Howland DR. Chondroitinase ABC improves basic and skilled locomotion in spinal cord injured cats. *Exp Neurol* 2008;**209**(2):483—96.

56. Curinga GM, et al. Mammalian-produced chondroitinase AC mitigates axon inhibition by chondroitin sulfate proteoglycans. *J Neurochem* 2007;**102**(1):275—88.

57. Rosenzweig ES, et al. Chondroitinase improves anatomical and functional outcomes after primate spinal cord injury. *Nat Neurosci* 2019;**22**(8):1269–75.

58. Hu HZ, et al. Therapeutic efficacy of microtube-embedded chondroitinase ABC in a canine clinical model of spinal cord injury. *Brain* 2018;**141**(4):1017–27.

59. Lee H, McKeon RJ, Bellamkonda RV. Sustained delivery of thermostabilized chABC enhances axonal sprouting and functional recovery after spinal cord injury. *Proc Natl Acad Sci USA* 2010;**107**(8):3340–5.

60. Hyatt AJ, et al. Controlled release of chondroitinase ABC from fibrin gel reduces the level of inhibitory glycosaminoglycan chains in lesioned spinal cord. *J Control Rel* 2010;**147**(1):24–9.

61. Rossi F, et al. Sustained delivery of chondroitinase ABC from hydrogel system. *J Funct Biomater* 2012;**3**(1):199–208.

62. Kanno H, et al. Combination of engineered Schwann cell grafts to secrete neurotrophin and chondroitinase promotes axonal regeneration and locomotion after spinal cord injury. *J Neurosci* 2014;**34**(5):1838–55.

63. Bartus K, et al. Large-scale chondroitin sulfate proteoglycan digestion with chondroitinase gene therapy leads to reduced pathology and modulates macrophage phenotype following spinal cord contusion injury. *J Neurosci* 2014;**34**(14):4822–36.

64. Muir EM, et al. Modification of N-glycosylation sites allows secretion of bacterial chondroitinase ABC from mammalian cells. *J Biotechnol* 2010;**145**(2):103–10.

65. Cafferty WB, et al. Functional axonal regeneration through astrocytic scar genetically modified to digest chondroitin sulfate proteoglycans. *J Neurosci* 2007;**27**(9):2176–85.

66. Caggiano AO, et al. Chondroitinase ABCI improves locomotion and bladder function following contusion injury of the rat spinal cord. *J Neurotrauma* 2005;**22**(2):226–39.

67. Shields LB, et al. Benefit of chondroitinase ABC on sensory axon regeneration in a laceration model of spinal cord injury in the rat. *Surg Neurol* 2008;**69**(6):568–77. discussion 577.

68. Yick LW, et al. Axonal regeneration of Clarke's neurons beyond the spinal cord injury scar after treatment with chondroitinase ABC. *Exp Neurol* 2003;**182**(1):160–8.

69. Starkey ML, et al. Chondroitinase ABC promotes compensatory sprouting of the intact corticospinal tract and recovery of forelimb function following unilateral pyramidotomy in adult mice. *Eur J Neurosci* 2012;**36**(12):3665–78.

70. Massey JM, et al. Chondroitinase ABC digestion of the perineuronal net promotes functional collateral sprouting in the cuneate nucleus after cervical spinal cord injury. *J Neurosci* 2006;**26**(16):4406–14.

71. Barritt AW, et al. Chondroitinase ABC promotes sprouting of intact and injured spinal systems after spinal cord injury. *J Neurosci* 2006;**26**(42):10856–67.

72. Carter LM, et al. The yellow fluorescent protein (YFP-H) mouse reveals neuroprotection as a novel mechanism underlying chondroitinase ABC-mediated repair after spinal cord injury. *J Neurosci* 2008;**28**(52):14107–20.

73. Carter LM, McMahon SB, Bradbury EJ. Delayed treatment with chondroitinase ABC reverses chronic atrophy of rubrospinal neurons following spinal cord injury. *Exp Neurol* 2011;**228**(1):149–56.

74. Shen Y, et al. PTPsigma is a receptor for chondroitin sulfate proteoglycan, an inhibitor of neural regeneration. *Science* 2009;**326**(5952):592–6.

75. Fisher D, et al. Leukocyte common antigen-related phosphatase is a functional receptor for chondroitin sulfate proteoglycan axon growth inhibitors. *J Neurosci* 2011;**31**(40):14051–66.

76. Dickendesher TL, et al. NgR1 and NgR3 are receptors for chondroitin sulfate proteoglycans. *Nat Neurosci* 2012;**15**(5):703–12.

77. Duan Y, Giger RJ. A new role for RPTPσ in spinal cord injury: signaling chondroitin sulfate proteoglycan inhibition. *Sci Signal* 2010;**3**(110). p. pe6-pe6.

78. Fry EJ, et al. Corticospinal tract regeneration after spinal cord injury in receptor protein tyrosine phosphatase sigma deficient mice. *Glia* 2010;**58**(4):423–33.

79. Lang BT, et al. Modulation of the proteoglycan receptor PTPsigma promotes recovery after spinal cord injury. *Nature* 2015;**518**(7539):404–8.

80. Tran AP, et al. Modulation of receptor protein tyrosine phosphatase sigma increases chondroitin sulfate proteoglycan degradation through Cathepsin B secretion to enhance axon outgrowth. *J Neurosci* 2018;**38**(23):5399–414.

81. Dyck S, et al. Perturbing chondroitin sulfate proteoglycan signaling through LAR and PTPsigma receptors promotes a beneficial inflammatory response following spinal cord injury. *J Neuroinflammation* 2018;**15**(1):90.

82. Dyck S, et al. LAR and PTPsigma receptors are negative regulators of oligodendrogenesis and oligodendrocyte integrity in spinal cord injury. *Glia* 2019;**67**(1):125–45.

83. Dergham P, et al. Rho signaling pathway targeted to promote spinal cord repair. *J Neurosci* 2002;**22**(15):6570–7.

84. Lord-Fontaine S, et al. Local inhibition of Rho signaling by cell-permeable recombinant protein BA-210 prevents secondary damage and promotes functional recovery following acute spinal cord injury. *J Neurotrauma* 2008;**25**(11):1309—22.

85. Fehlings MG, et al. A phase I/IIa clinical trial of a recombinant Rho protein antagonist in acute spinal cord injury. *J Neurotrauma* 2011;**28**(5):787—96.

86. Kirshblum S, et al. Characterizing natural recovery after traumatic spinal cord injury. *J Neurotrauma* 2021;**38**(9):1267—84.

87. Fehlings MG, et al. A randomized controlled trial of local delivery of a Rho inhibitor (VX-210) in patients with acute traumatic cervical spinal cord injury. *J Neurotrauma* 2021;**38**(15):2065—72.

88. Klingseisen A, Lyons DA. Axonal regulation of central nervous system myelination: structure and function. *Neuroscientist* 2018;**24**(1):7—21.

89. Fields RD. A new mechanism of nervous system plasticity: activity-dependent myelination. *Nat Rev Neurosci* 2015;**16**(12):756—67.

90. McKenzie IA, et al. Motor skill learning requires active central myelination. *Science* 2014;**346**(6207):318—22.

91. Keirstead HS, et al. Suppression of the onset of myelination extends the permissive period for the functional repair of embryonic spinal-cord. *Proc Natl Acad Sci USA* 1992;**89**(24):11664—8.

92. Varga ZM, et al. The critical period for repair of CNS of neonatal opossum (*Monodelphis domestica*) in culture: correlation with development of glial cells, myelin and growth-inhibitory molecules. *Eur J Neurosci* 1995;**7**(10):2119—29.

93. Tang S, et al. Soluble myelin-associated glycoprotein released from damaged white matter inhibits axonal regeneration. *Mol Cell Neurosci* 2001;**18**(3):259—69.

94. Caroni P, Schwab ME. Antibody against myelin-associated inhibitor of neurite growth neutralizes nonpermissive substrate properties of CNS white matter. *Neuron* 1988;**1**(1):85—96.

95. George R, Griffin JW. Delayed macrophage responses and myelin clearance during Wallerian degeneration in the central nervous system: the dorsal radiculotomy model. *Exp Neurol* 1994;**129**(2):225—36.

96. Vanselow J, Schwab ME, Thanos S. Responses of regenerating rat retinal ganglion cell axons to contacts with central nervous myelin in vitro. *Eur J Neurosci* 1990;**2**(2):121—5.

97. Schnell L, Schwab ME. Axonal regeneration in the rat spinal cord produced by an antibody against myelin-associated neurite growth inhibitors. *Nature* 1990;**343**(6255):269—72.

98. Schnell L, Schwab ME. Sprouting and regeneration of lesioned corticospinal tract fibres in the adult rat spinal cord. *Eur J Neurosci* 1993;**5**(9):1156—71.

99. Chen MS, et al. Nogo-A is a myelin-associated neurite outgrowth inhibitor and an antigen for monoclonal antibody IN-1. *Nature* 2000;**403**(6768):434—9.

100. GrandPre T, Li S, Strittmatter SM. Nogo-66 receptor antagonist peptide promotes axonal regeneration. *Nature* 2002;**417**(6888):547—51.

101. McGee AW, Strittmatter SM. The Nogo-66 receptor: focusing myelin inhibition of axon regeneration. *Trends Neurosci* 2003;**26**(4):193—8.

102. Fournier AE, et al. Truncated soluble Nogo receptor binds Nogo-66 and blocks inhibition of axon growth by myelin. *J Neurosci* 2002;**22**(20):8876—83.

103. Fournier AE, et al. Nogo and the Nogo-66 receptor. In: *Progress in brain research*. Elsevier; 2002. p. 361—9.

104. Ng CEL, Tang BL. Nogos and the Nogo-66 receptor: factors inhibiting CNS neuron regeneration. *J Neurosci Res* 2002;**67**(5):559—65.

105. Freund P, et al. Anti-Nogo-A antibody treatment enhances sprouting of corticospinal axons rostral to a unilateral cervical spinal cord lesion in adult macaque monkey. *J Comp Neurol* 2007;**502**(4):644—59.

106. Kim J-E, et al. Axon regeneration in young adult mice lacking Nogo-A/B. *Neuron* 2003;**38**(2):187—99.

107. Zheng B, et al. Lack of enhanced spinal regeneration in Nogo-deficient mice. *Neuron* 2003;**38**(2):213—24.

108. Teng FYH, Tang BL. Why do Nogo/Nogo-66 receptor gene knockouts result in inferior regeneration compared to treatment with neutralizing agents? *J Neurochem* 2005;**94**(4):865—74.

109. Dimou L, et al. Nogo-A-deficient mice reveal strain-dependent differences in axonal regeneration. *J Neurosci* 2006;**26**(21):5591—603.

110. Lee JK, et al. Assessing spinal axon regeneration and sprouting in Nogo-, MAG-, and OMgp-deficient mice. *Neuron* 2010;**66**(5):663—70.

111. Van Norman GA. Limitations of animal studies for predicting toxicity in clinical trials is it time to rethink our current approach? *JACC-Basic Trans Sci* 2019;**4**(7):845—54.

112. Li M, et al. Myelin-associated glycoprotein inhibits neurite/axon growth and causes growth cone collapse. *J Neurosci Res* 1996;**46**(4):404—14.

113. Cai D, et al. Neuronal cyclic AMP controls the developmental loss in ability of axons to regenerate. *J Neurosci* 2001;**21**(13):4731—9.

114. McKerracher LA, et al. Identification of myelin-associated glycoprotein as a major myelin-derived inhibitor of neurite growth. *Neuron* 1994;**13**(4):805—11.

115. Mukhopadhyay G, et al. A novel role for myelin-associated glycoprotein as an inhibitor of axonal regeneration. *Neuron* 1994;**13**(3):757−67.

116. Bartsch U, et al. Lack of evidence that myelin-associated glycoprotein is a major inhibitor of axonal regeneration in the CNS. *Neuron* 1995;**15**(6):1375−81.

117. Schafer M, et al. Disruption of the gene for the myelin-associated glycoprotein improves axonal regrowth along myelin in C57BL/Wlds mice. *Neuron* 1996;**16**(6):1107−13.

118. Wang KC, et al. Oligodendrocyte-myelin glycoprotein is a Nogo receptor ligand that inhibits neurite outgrowth. *Nature* 2002;**417**(6892):941−4.

119. Gerhards R, et al. Oligodendrocyte myelin glycoprotein as a novel target for pathogenic autoimmunity in the CNS. *Acta Neuropathol Commun* 2020;**8**(1):207.

120. Barton WA, et al. Structure and axon outgrowth inhibitor binding of the Nogo-66 receptor and related proteins. *EMBO J* 2003;**22**(13):3291−302.

121. De Bellard M, Filbin MT. Myelin-associated glycoprotein, MAG, selectively binds several neuronal proteins. *J Neurosci Res* 1999;**56**(2):213−8.

122. DeBellard M-E, et al. Myelin-associated glycoprotein inhibits axonal regeneration from a variety of neurons via interaction with a sialoglycoprotein. *Mol Cell Neurosci* 1996;**7**(2):89−101.

123. Hu F, Strittmatter SM. Regulating axon growth within the postnatal central nervous system. In: *Seminars in perinatology*. Elsevier; 2004.

124. Hunt D, Coffin R, Anderson PN. The Nogo receptor, its ligands and axonal regeneration in the spinal cord; a review. *J Neurocytol* 2002;**31**(2):93−120.

125. Laurén J, et al. Characterization of myelin ligand complexes with neuronal Nogo-66 receptor family members. *J Biol Chem* 2007;**282**(8):5715−25.

126. Li S, et al. Blockade of Nogo-66, myelin-associated glycoprotein, and oligodendrocyte myelin glycoprotein by soluble Nogo-66 receptor promotes axonal sprouting and recovery after spinal injury. *J Neurosci* 2004;**24**(46):10511−20.

127. Schwab ME. Increasing plasticity and functional recovery of the lesioned spinal cord. In: *Progress in brain research*. Elsevier; 2002. p. 351−9.

128. Sicotte M, et al. Immunization with myelin or recombinant Nogo-66/MAG in alum promotes axon regeneration and sprouting after corticospinal tract lesions in the spinal cord. *Mol Cell Neurosci* 2003;**23**(2):251−63.

129. Zhou Y, et al. Expression of PirB in normal and injured spinal cord of rats. *J Huazhong Univ Sci Technol Med Sci* 2010;**30**(4):482−5.

130. Cafferty WB, et al. MAG and OMgp synergize with Nogo-A to restrict axonal growth and neurological recovery after spinal cord trauma. *J Neurosci* 2010;**30**(20):6825−37.

131. Cao Z, et al. Receptors for myelin inhibitors: structures and therapeutic opportunities. *Mol Cell Neurosci* 2010;**43**(1):1−14.

132. Gou Z, et al. PirB is a novel potential therapeutic target for enhancing axonal regeneration and synaptic plasticity following CNS injury in mammals. *J Drug Target* 2014;**22**(5):365−71.

133. Akbik F, Cafferty WB, Strittmatter SM. Myelin associated inhibitors: a link between injury-induced and experience-dependent plasticity. *Exp Neurol* 2012;**235**(1):43−52.

134. De Vries M, Cooper HM. Emerging roles for neogenin and its ligands in CNS development. *J Neurochem* 2008;**106**(4):1483−92.

135. Schwab JM, et al. Spinal cord injury-induced lesional expression of the repulsive guidance molecule (RGM). *Eur J Neurosci* 2005;**21**(6):1569−76.

136. Nash M, et al. Central nervous system regeneration inhibitors and their intracellular substrates. *Mol Neurobiol* 2009;**40**(3):224−35.

137. Hata K, et al. RGMa inhibition promotes axonal growth and recovery after spinal cord injury. *J Cell Biol* 2006;**173**(1):47−58.

138. Mothe AJ, et al. Delayed administration of the human anti-RGMa monoclonal antibody elezanumab promotes functional recovery including spontaneous voiding after spinal cord injury in rats. *Neurobiol Dis* 2020;**143**:104995.

139. Ziemann A, et al. A phase 1, multiple-dose study of elezanumab (ABT-555) in patients with relapsing forms of multiple sclerosis (S56.001). *Neurology* 2019;**92**(Suppl. 15):S56.001.

140. Zorner B, Schwab ME. Anti-Nogo on the go: from animal models to a clinical trial. *Ann N Y Acad Sci* 2010;**1198**(Suppl. 1):E22−34.

141. Kucher K, et al. First-in-Man intrathecal application of neurite growth-promoting anti-Nogo-A antibodies in acute spinal cord injury. *Neurorehabil Neural Rep* 2018;**32**(6−7):578−89.

142. Wang X, et al. Human NgR-Fc decoy protein via lumbar intrathecal bolus administration enhances recovery from rat spinal cord contusion. *J Neurotrauma* 2014;**31**(24):1955−66.

143. Wang X, et al. Nogo receptor decoy promotes recovery and corticospinal growth in non-human primate spinal cord injury. *Brain* 2020;**143**(6):1697−713.

144. Erickson KI, et al. Brain-derived neurotrophic factor is associated with age-related decline in hippocampal volume. *J Neurosci* 2010;**30**(15):5368−75.

145. Levi-Montalcini R, Angeletti PU. Essential role of the nerve growth factor in the survival and maintenance of dissociated sensory and sympathetic embryonic nerve cells in vitro. *Dev Biol* 1963;**6**:653−9.

146. Halegoua S, Patrick J. Nerve growth factor mediates phosphorylation of specific proteins. *Cell* 1980;**22**(2 Pt 2):571−81.

147. Kaplan DR, Martin-Zanca D, Parada LF. Tyrosine phosphorylation and tyrosine kinase activity of the trk proto-oncogene product induced by NGF. *Nature* 1991;**350**(6314):158−60.

148. Kwon BK, et al. Survival and regeneration of rubrospinal neurons 1 year after spinal cord injury. *Proc Natl Acad Sci USA* 2002;**99**(5):3246−51.

149. Barde YA, Edgar D, Thoenen H. Purification of a new neurotrophic factor from mammalian brain. *EMBO J* 1982;**1**(5):549−53.

150. Maisonpierre PC, et al. Neurotrophin-3: a neurotrophic factor related to NGF and BDNF. *Science* 1990;**247**(4949 Pt 1):1446−51.

151. Olson L. NGF and the treatment of Alzheimer's disease. *Exp Neurol* 1993;**124**(1):5−15.

152. Eriksdotter Jonhagen M, et al. Intracerebroventricular infusion of nerve growth factor in three patients with Alzheimer's disease. *Dement Geriatr Cogn Disord* 1998;**9**(5):246−57.

153. Romero MI, et al. Extensive sprouting of sensory afferents and hyperalgesia induced by conditional expression of nerve growth factor in the adult spinal cord. *J Neurosci* 2000;**20**(12):4435−45.

154. Tuszynski MH, et al. A phase 1 clinical trial of nerve growth factor gene therapy for Alzheimer disease. *Nat Med* 2005;**11**(5):551−5.

155. Rafii MS, et al. Adeno-associated viral vector (serotype 2)-nerve growth factor for patients with Alzheimer disease: a randomized clinical trial. *JAMA Neurol* 2018;**75**(7):834−41.

156. Eyjolfsdottir H, et al. Targeted delivery of nerve growth factor to the cholinergic basal forebrain of Alzheimer's disease patients: application of a second-generation encapsulated cell biodelivery device. *Alzheimer Res Ther* 2016;**8**(1):30.

157. Liebl DJ, et al. Regulation of Trk receptors following contusion of the rat spinal cord. *Exp Neurol* 2001;**167**(1):15−26.

158. Hollis 2nd ER, Tuszynski MH. Neurotrophins: potential therapeutic tools for the treatment of spinal cord injury. *Neurotherapeutics* 2011;**8**(4):694−703.

159. King VR, et al. Changes in truncated trkB and p75 receptor expression in the rat spinal cord following spinal cord hemisection and spinal cord hemisection plus neurotrophin treatment. *Exp Neurol* 2000;**165**(2):327−41.

160. Harvey AR, et al. Neurotrophic factors for spinal cord repair: which, where, how and when to apply, and for what period of time? *Brain Res* 2015;**1619**:36−71.

161. Deng LX, et al. A novel growth-promoting pathway formed by GDNF-overexpressing Schwann cells promotes propriospinal axonal regeneration, synapse formation, and partial recovery of function after spinal cord injury. *J Neurosci* 2013;**33**(13):5655−67.

162. Chen Q, Zhou L, Shine HD. Expression of neurotrophin-3 promotes axonal plasticity in the acute but not chronic injured spinal cord. *J Neurotrauma* 2006;**23**(8):1254−60.

163. Alto LT, et al. Chemotropic guidance facilitates axonal regeneration and synapse formation after spinal cord injury. *Nat Neurosci* 2009;**12**(9):1106−13.

164. Grill RJ, Blesch A, Tuszynski MH. Robust growth of chronically injured spinal cord axons induced by grafts of genetically modified NGF-secreting cells. *Exp Neurol* 1997;**148**(2):444−52.

165. Kobayashi NR, et al. BDNF and NT-4/5 prevent atrophy of rat rubrospinal neurons after cervical axotomy, stimulate GAP-43 and Talpha1-tubulin mRNA expression, and promote axonal regeneration. *J Neurosci* 1997;**17**(24):9583−95.

166. Bradbury EJ, et al. NT-3 promotes growth of lesioned adult rat sensory axons ascending in the dorsal columns of the spinal cord. *Eur J Neurosci* 1999;**11**(11):3873−83.

167. Hiebert GW, et al. Brain-derived neurotrophic factor applied to the motor cortex promotes sprouting of corticospinal fibers but not regeneration into a peripheral nerve transplant. *J Neurosci Res* 2002;**69**(2):160−8.

168. Cao Q, et al. Functional recovery in traumatic spinal cord injury after transplantation of multineurotrophin-expressing glial-restricted precursor cells. *J Neurosci* 2005;**25**(30):6947−57.

169. Frank L, et al. BDNF down-regulates neurotrophin responsiveness, TrkB protein and TrkB mRNA levels in cultured rat hippocampal neurons. *Eur J Neurosci* 1996;**8**(6):1220−30.

170. Sommerfeld MT, et al. Down-regulation of the neurotrophin receptor TrkB following ligand binding. Evidence for an involvement of the proteasome and differential regulation of TrkA and TrkB. *J Biol Chem* 2000;**275**(12):8982−90.

171. Fawcett JW. The struggle to make CNS axons regenerate: why has it been so difficult? *Neurochem Res* 2020;**45**(1):144−58.

172. McQuarrie IG, Grafstein B. Axon outgrowth enhanced by a previous nerve injury. *Arch Neurol* 1973;**29**(1):53−5.

173. Richardson PM, Issa VM. Peripheral injury enhances central regeneration of primary sensory neurones. *Nature* 1984;**309**(5971):791−3.

174. Smith DS, Skene JH. A transcription-dependent switch controls competence of adult neurons for distinct modes of axon growth. *J Neurosci* 1997;**17**(2):646–58.

175. Neumann S, Woolf CJ. Regeneration of dorsal column fibers into and beyond the lesion site following adult spinal cord injury. *Neuron* 1999;**23**(1):83–91.

176. Costigan M, et al. Replicate high-density rat genome oligonucleotide microarrays reveal hundreds of regulated genes in the dorsal root ganglion after peripheral nerve injury. *BMC Neurosci* 2002;**3**:16.

177. Michaelevski I, et al. Signaling to transcription networks in the neuronal retrograde injury response. *Sci Signal* 2010;**3**(130):ra53.

178. Blesch A, et al. Conditioning lesions before or after spinal cord injury recruit broad genetic mechanisms that sustain axonal regeneration: superiority to camp-mediated effects. *Exp Neurol* 2012;**235**(1):162–73.

179. Smith TP, et al. Intra-axonal mechanisms driving axon regeneration. *Brain Res* 2020;**1740**:146864.

180. Hunter T. Why nature chose phosphate to modify proteins. *Philos Trans R Soc Lond B Biol Sci* 2012;**367**(1602):2513–6.

181. Wayman GA, et al. Synergistic activation of the type I adenylyl cyclase by Ca^{2+} and Gs-coupled receptors in vivo. *J Biol Chem* 1994;**269**(41):25400–5.

182. Knott EP, Assi M, Pearse DD. Cyclic AMP signaling: a molecular determinant of peripheral nerve regeneration. *BioMed Res Int* 2014;**2014**:651625.

183. Hannila SS, Filbin MT. The role of cyclic AMP signaling in promoting axonal regeneration after spinal cord injury. *Exp Neurol* 2008;**209**(2):321–32.

184. Ghosh-Roy A, et al. Calcium and cyclic AMP promote axonal regeneration in *Caenorhabditis elegans* and require DLK-1 kinase. *J Neurosci* 2010;**30**(9):3175–83.

185. Neumann S, et al. Regeneration of sensory axons within the injured spinal cord induced by intraganglionic cAMP elevation. *Neuron* 2002;**34**(6):885–93.

186. Lu P, et al. Combinatorial therapy with neurotrophins and cAMP promotes axonal regeneration beyond sites of spinal cord injury. *J Neurosci* 2004;**24**(28):6402–9.

187. Qiu J, et al. Spinal axon regeneration induced by elevation of cyclic AMP. *Neuron* 2002;**34**(6):895–903.

188. Nikulina E, et al. The phosphodiesterase inhibitor rolipram delivered after a spinal cord lesion promotes axonal regeneration and functional recovery. *Proc Natl Acad Sci USA* 2004;**101**(23):8786.

189. Pearse DD, et al. cAMP and Schwann cells promote axonal growth and functional recovery after spinal cord injury. *Nat Med* 2004;**10**(6):610–6.

190. Bretzner F, et al. Combination of olfactory ensheathing cells with local versus systemic cAMP treatment after a cervical rubrospinal tract injury. *J Neuro Sci* 2010;**88**(13):2833–46.

191. Sluka KA. Activation of the cAMP transduction cascade contributes to the mechanical hyperalgesia and allodynia induced by intradermal injection of capsaicin 1997;**122**(6):1165–73.

192. Rock EM, et al. Potential of the rat model of conditioned gaping to detect nausea produced by rolipram, a phosphodiesterase-4 (PDE4) inhibitor. *Pharmacol Biochem Behav* 2009;**91**(4):537–41.

193. Wahane S, et al. Epigenetic regulation of axon regeneration and glial activation in injury responses. *Front Genet* 2019;**10**:640.

194. Kouzarides T. Chromatin modifications and their function. *Cell* 2007;**128**(4):693–705.

195. Palmisano I, et al. Epigenomic signatures underpin the axonal regenerative ability of dorsal root ganglia sensory neurons. *Nat Neurosci* 2019;**22**(11):1913–24.

196. Lin S, Smith GM. Acetylation as a mechanism that regulates axonal regeneration. *Neural Regen Res* 2015;**10**(7):1034–6.

197. Mann BS, et al. FDA approval summary: vorinostat for treatment of advanced primary cutaneous T-cell lymphoma. *Oncol* 2007;**12**(10):1247–52.

198. Finelli MJ, Wong JK, Zou HY. Epigenetic regulation of sensory axon regeneration after spinal cord injury. *J Neurosci* 2013;**33**(50):19664–76.

199. Cho Y, et al. Injury-induced HDAC5 nuclear export is essential for axon regeneration. *Cell* 2013;**155**(4):894–908.

200. Penas C, Navarro X. Epigenetic modifications associated to neuroinflammation and neuropathic pain after neural trauma. *Front Cell Neurosci* 2018;**12**:13.

201. Hutson TH, et al. Cbp-dependent histone acetylation mediates axon regeneration induced by environmental enrichment in rodent spinal cord injury models. *Sci Transl Med* 2019;**11**(487):13.

202. Chatterjee S, et al. A novel activator of CBP/p300 acetyltransferases promotes neurogenesis and extends memory duration in adult mice. *J Neurosci* 2013;**33**(26):10698–712.

203. Gaub P, et al. HDAC inhibition promotes neuronal outgrowth and counteracts growth cone collapse through CBP/p300 and P/CAF-dependent p53 acetylation. *Cell Death Differ* 2010;**17**(9):1392–408.

204. Gaub P, et al. The histone acetyltransferase p300 promotes intrinsic axonal regeneration. *Brain* 2011;**134**(Pt 7):2134–48.

205. Venkatesh I, et al. Epigenetic profiling reveals a developmental decrease in promoter accessibility during cortical maturation in vivo. *Neuroepigenetics* 2016;**8**:19–26.

206. Cho Y, Cavalli V. HDAC signaling in neuronal development and axon regeneration. *Curr Opin Neurobiol* 2014;**27**:118–26.

207. Ertürk A, et al. Disorganized microtubules underlie the formation of retraction bulbs and the failure of axonal regeneration. *J Neurosci* 2007;**27**(34):9169−80.

208. Cho Y, Cavalli V. HDAC5 is a novel injury-regulated tubulin deacetylase controlling axon regeneration. *EMBO J* 2012;**31**(14):3063−78.

209. Smith SS, et al. Mechanism of human methyl-directed DNA methyltransferase and the fidelity of cytosine methylation. *Proc Natl Acad Sci USA* 1992;**89**(10):4744−8.

210. Loh YE, et al. Comprehensive mapping of 5-hydroxymethylcytosine epigenetic dynamics in axon regeneration. *Epigenetics* 2017;**12**(2):77−92.

211. Weng Y-L, et al. An intrinsic epigenetic barrier for functional axon regeneration. *Neuron* 2017;**94**(2):337−346. e6.

212. Oh YM, et al. Epigenetic regulator UHRF1 inactivates REST and growth suppressor gene expression via DNA methylation to promote axon regeneration. *PNAS* 2018;**115**(52):E12417−26.

213. Bartel DP. MicroRNAs: genomics, biogenesis, mechanism, and function. *Cell* 2004;**116**(2):281−97.

214. Fasanaro P, et al. An integrated approach for experimental target identification of hypoxia-induced miR-210. *J Biol Chem* 2009;**284**(50):35134−43.

215. Gutierrez DV, et al. Intermittent hypoxia training after C2 hemisection modifies the expression of PTEN and mTOR. *Exp Neurol* 2013;**248**:45−52.

216. Strickland IT, et al. Axotomy-induced miR-21 promotes axon growth in adult dorsal root ganglion neurons. *PLoS One* 2011;**6**(8):e23423.

217. Cong M, et al. Improvement of sensory neuron growth and survival via negatively regulating PTEN by miR-21-5p-contained small extracellular vesicles from skin precursor-derived Schwann cells. *Stem Cell Res Ther* 2021;**12**(1):80.

218. Liu CM, et al. MicroRNA-138 and SIRT1 form a mutual negative feedback loop to regulate mammalian axon regeneration. *Genes Dev* 2013;**27**(13):1473−83.

219. Weng YL, et al. Epigenetic regulation of axonal regenerative capacity. *Epigenomics* 2016;**8**(10):1429−42.

220. Tamura M, et al. PTEN interactions with focal adhesion kinase and suppression of the extracellular matrix-dependent phosphatidylinositol 3-kinase/Akt cell survival pathway. *J Biol Chem* 1999;**274**(29):20693−703.

221. Li B, Sun HJMMR. MiR-26a promotes neurite outgrowth by repressing PTEN expression. *Mol Med Rep* 2013;**8**(2):676−80.

222. Zhou S, et al. microRNA-222 targeting PTEN promotes neurite outgrowth from adult dorsal root ganglion neurons following sciatic nerve transection. *PLoS One* 2012;**7**(9):e44768.

223. Liu NK, et al. Altered microRNA expression following traumatic spinal cord injury. *Exp Neurol* 2009;**219**(2):424−9.

224. Yang L, et al. Severity-dependent expression of pro-inflammatory cytokines in traumatic spinal cord injury in the rat. *J Clin Neurosci* 2005;**12**(3):276−84.

225. Yu YM, et al. MicroRNA miR-133b is essential for functional recovery after spinal cord injury in adult zebrafish. *Eur J Neurosci* 2011;**33**(9):1587−97.

226. Conrad S, et al. Prolonged lesional expression of RhoA and RhoB following spinal cord injury. *J Comp Neurol* 2005;**487**(2):166−75.

227. Erschbamer MK, Hofstetter CP, Olson L. RhoA, RhoB, RhoC, Rac1, Cdc42, and Tc10 mRNA levels in spinal cord, sensory ganglia, and corticospinal tract neurons and long-lasting specific changes following spinal cord injury. *J Comp Neurol* 2005;**484**(2):224−33.

228. Wullschleger S, Loewith R, Hall MN. TOR signaling in growth and metabolism. *Cell* 2006;**124**(3):471−84.

229. Ravikumar B, et al. Inhibition of mTOR induces autophagy and reduces toxicity of polyglutamine expansions in fly and mouse models of Huntington disease. *Nat Genet* 2004;**36**(6):585−95.

230. Santini E, et al. Inhibition of mTOR signaling in Parkinson's disease prevents L-DOPA-induced dyskinesia. *Sci Signal* 2009;**2**(80):ra36.

231. Ma T, et al. Dysregulation of the mTOR pathway mediates impairment of synaptic plasticity in a mouse model of Alzheimer's disease. *PLoS One* 2010;**5**(9).

232. Spilman P, et al. Inhibition of mTOR by rapamycin abolishes cognitive deficits and reduces amyloid-beta levels in a mouse model of Alzheimer's disease. *PLoS One* 2010;**5**(4):e9979.

233. Sekiguchi A, et al. Rapamycin promotes autophagy and reduces neural tissue damage and locomotor impairment after spinal cord injury in mice. *J Neurotrauma* 2012;**29**(5):946−56.

234. Park KK, et al. Promoting axon regeneration in the adult CNS by modulation of the PTEN/mTOR pathway. *Science* 2008;**322**(5903):963−6.

235. Liu K, et al. PTEN deletion enhances the regenerative ability of adult corticospinal neurons. *Nat Neurosci* 2010;**13**(9). 1075-U64.

236. Smith PD, et al. SOCS3 deletion promotes optic nerve regeneration in vivo. *Neuron* 2009;**64**(5):617−23.

237. Sun F, et al. Sustained axon regeneration induced by co-deletion of PTEN and SOCS3. *Nature* 2011;**480**(7377):372−5.

238. Moore DL, et al. KLF family members regulate intrinsic axon regeneration ability. *Science* 2009;**326**(5950):298−301.

239. Bonett RM, et al. Stressor and glucocorticoid-dependent induction of the immediate early gene

kruppel-like factor 9: implications for neural development and plasticity. *Endocrinology* 2009;**150**(4): 1757—65.

240. Lebrun C, et al. Klf9 is necessary and sufficient for Purkinje cell survival in organotypic culture. *Mol Cell Neurosci* 2013;**54**:9—21.

241. Cayrou C, Denver RJ, Puymirat J. Suppression of the basic transcription element-binding protein in brain neuronal cultures inhibits thyroid hormone-induced neurite branching. *Endocrinology* 2002;**143**(6):2242—9.

242. Apara A, et al. KLF9 and JNK3 interact to suppress axon regeneration in the adult CNS. *J Neurosci* 2017; **37**(40):9632—44.

243. Burks JD, et al. Imaging characteristics of chronic spinal cord injury identified during screening for a cell transplantation clinical trial. *Neurosurg Focus* 2019; **46**(3):E8.

244. Yang H, et al. Endogenous neurogenesis replaces oligodendrocytes and astrocytes after primate spinal cord injury. *J Neurosci* 2006;**26**(8):2157—66.

245. Guest JD, Hiester ED, Bunge RP. Demyelination and Schwann cell responses adjacent to injury epicenter cavities following chronic human spinal cord injury. *Exp Neurol* 2005;**192**(2):384—93.

246. Chen CZ, et al. Schwann cell remyelination of the central nervous system: why does it happen and what are the benefits? *Open Biol* 2021;**11**(1):200352.

247. Freeman TB, et al. Bilateral fetal nigral transplantation into the postcommissural putamen in Parkinson's disease. *Ann Neurol* 1995;**38**(3):379—88.

248. Wirth 3rd ED, et al. Feasibility and safety of neural tissue transplantation in patients with syringomyelia. *J Neurotrauma* 2001;**18**(9):911—29.

249. Giovanini MA, et al. Characteristics of human fetal spinal cord grafts in the adult rat spinal cord: influences of lesion and grafting conditions. *Exp Neurol* 1997;**148**(2): 523—43.

250. Santamaria AJ, et al. Intraspinal delivery of Schwann cells for spinal cord injury. *Methods Mol Biol* 2018; **1739**:467—84.

251. Federici T, et al. A stereotactic device for intraparenchymal spinal cord injections: Latest developments for practical clinical use. *Stereotact Funct Neurosurg* 2021:1—7.

252. Kutikov AB, et al. Method and apparatus for the automated delivery of continuous neural stem cell trails into the spinal cord of small and large animals. *Neurosurgery* 2019;**85**(4):560—73.

253. Xiong M, et al. Human stem cell-derived neurons repair circuits and restore neural function. *Cell Stem Cell* 2021;**28**(1):112—+.

254. Fleming JC, et al. The cellular inflammatory response in human spinal cords after injury. *Brain* 2006;**129**(Pt 12):3249—69.

255. Milich LM, Ryan CB, Lee JK. The origin, fate, and contribution of macrophages to spinal cord injury pathology. *Acta Neuropathol* 2019;**137**(5):785—97.

256. Stout RD, Suttles J. Functional plasticity of macrophages: reversible adaptation to changing microenvironments. *J Leukoc Biol* 2004;**76**(3):509—13.

257. Stout RD, et al. Macrophages sequentially change their functional phenotype in response to changes in microenvironmental influences. *J Immunol* 2005;**175**(1): 342—9.

258. Griffin JW, et al. Macrophage responses and myelin clearance during Wallerian degeneration: relevance to immune-mediated demyelination. *J Neuroimmunol* 1992;**40**(2—3):153—65.

259. Stoll G, Jander S. The role of microglia and macrophages in the pathophysiology of the CNS. *Prog Neurobiol* 1999;**58**(3):233—47.

260. Perry VH, Brown MC, Gordon S. The macrophage response to central and peripheral nerve injury. A possible role for macrophages in regeneration. *J Exp Med* 1987;**165**(4):1218—23.

261. Zeev-Brann AB, et al. Differential effects of central and peripheral nerves on macrophages and microglia. *Glia* 1998;**23**(3):181—90.

262. Lazarov-Spiegler O, et al. Transplantation of activated macrophages overcomes central nervous system regrowth failure. *FASEB J* 1996;**10**(11):1296—302.

263. Rapalino O, et al. Implantation of stimulated homologous macrophages results in partial recovery of paraplegic rats. *Nat Med* 1998;**4**(7):814—21.

264. Bomstein Y, et al. Features of skin-coincubated macrophages that promote recovery from spinal cord injury. *J Neuroimmunol* 2003;**142**(1—2):10—6.

265. Schwartz M, Yoles E. Immune-based therapy for spinal cord repair: autologous macrophages and beyond. *J Neurotrauma* 2006;**23**(3—4):360—70.

266. Knoller N, et al. Clinical experience using incubated autologous macrophages as a treatment for complete spinal cord injury: phase I study results. *J Neurosurg Spine* 2005;**3**(3):173.

267. Lammertse DP, et al. Autologous incubated macrophage therapy in acute, complete spinal cord injury: results of the phase 2 randomized controlled multicenter trial. *Spinal Cord* 2012;**50**(9):661—71.

268. Kigerl KA, et al. Identification of two distinct macrophage subsets with divergent effects causing either neurotoxicity or regeneration in the injured mouse spinal cord. *J Neurosci* 2009;**29**(43):13435—44.

269. Herrera LP, et al. Ultrastructural study of the primary olfactory pathway in *Macaca fascicularis*. *J Comp Neurol* 2005;**488**(4):427−41.

270. Graziadei PP, Monti Graziadei GA. Neurogenesis and plasticity of the olfactory sensory neurons. *Ann N Y Acad Sci* 1985;**457**:127−42.

271. Barber PC. Neurogenesis and regeneration in the primary olfactory pathway of mammals. *Bibl Anat* 1982;(23):12−25.

272. Tennent R, Chuah MI. Ultrastructural study of ensheathing cells in early development of olfactory axons. *Brain Res Dev Brain Res* 1996;**95**(1):135−9.

273. Raisman G. Specialized neuroglial arrangement may explain the capacity of vomeronasal axons to reinnervate central neurons. *Neuroscience* 1985;**14**(1):237−54.

274. Li Y, Field PM, Raisman G. Repair of adult rat corticospinal tract by transplants of olfactory ensheathing cells. *Science* 1997;**277**(5334):2000−2.

275. Li Y, Field PM, Raisman G. Regeneration of adult rat corticospinal axons induced by transplanted olfactory ensheathing cells. *J Neurosci* 1998;**18**(24):10514−24.

276. Doucette R. PNS-CNS transitional zone of the first cranial nerve. *J Comp Neurol* 1991;**312**(3):451−66.

277. Ramon-Cueto A, Nieto-Sampedro M. Regeneration into the spinal cord of transected dorsal root axons is promoted by ensheathing glia transplants. *Exp Neurol* 1994;**127**(2):232−44.

278. Ramer LM, et al. Peripheral olfactory ensheathing cells reduce scar and cavity formation and promote regeneration after spinal cord injury. *J Comp Neurol* 2004;**473**(1):1−15.

279. Richter MW, et al. Lamina propria and olfactory bulb ensheathing cells exhibit differential integration and migration and promote differential axon sprouting in the lesioned spinal cord. *J Neurosci* 2005;**25**(46):10700−11.

280. Plant GW, et al. Delayed transplantation of olfactory ensheathing glia promotes sparing/regeneration of supraspinal axons in the contused adult rat spinal cord. *J Neurotrauma* 2003;**20**(1):1−16.

281. Nazareth L, et al. Olfactory ensheathing cells are the main phagocytic cells that remove axon debris during early development of the olfactory system. *J Comp Neurol* 2015;**523**(3):479−94.

282. Su Z, et al. Olfactory ensheathing cells: the primary innate immunocytes in the olfactory pathway to engulf apoptotic olfactory nerve debris. *Glia* 2013;**61**(4):490−503.

283. Toft A, et al. Electrophysiological evidence that olfactory cell transplants improve function after spinal cord injury. *Brain* 2007;**130**(Pt 4):970−84.

284. Ramon-Cueto A, et al. Functional recovery of paraplegic rats and motor axon regeneration in their spinal cords by olfactory ensheathing glia. *Neuron* 2000;**25**(2):425−35.

285. Féron F, et al. Autologous olfactory ensheathing cell transplantation in human spinal cord injury. *Brain* 2005;**128**(12):2951−60.

286. Mackay-Sim A, et al. Autologous olfactory ensheathing cell transplantation in human paraplegia: a 3-year clinical trial. *Brain* 2008;**131**(9):2376−86.

287. Lima C, et al. Olfactory mucosa autografts in human spinal cord injury: a pilot clinical study. *J Spinal Cord Med* 2006;**29**(3):191−203. discussion 204-6.

288. Huang H, et al. Influence factors for functional improvement after olfactory ensheathing cell transplantation for chronic spinal cord injury. *Zhongguo Xiu Fu Chong Jian Wai Ke Za Zhi* 2006;**20**(4):434−8.

289. Guest J, Herrera LP, Qian T. Rapid recovery of segmental neurological function in a tetraplegic patient following transplantation of fetal olfactory bulb-derived cells. *Spinal Cord* 2006;**44**(3):135−42.

290. Dobkin BH, Curt A, Guest J. Cellular transplants in China: observational study from the largest human experiment in chronic spinal cord injury. *Neurorehabil Neural Rep* 2006;**20**(1):5−13.

291. Watzlawick R, et al. Olfactory ensheathing cell transplantation in experimental spinal cord injury: effect size and reporting bias of 62 experimental treatments: a systematic review and meta-analysis. *PLoS Biol* 2016;**14**(5):e1002468.

292. Granger N, et al. Autologous olfactory mucosal cell transplants in clinical spinal cord injury: a randomized double-blinded trial in a canine translational model. *Brain* 2012;**135**(Pt 11):3227−37.

293. Czyz M, et al. Obtaining the olfactory bulb as a source of olfactory ensheathing cells with the use of minimally invasive neuroendoscopy-assisted supraorbital keyhole approach–cadaveric feasibility study. *Br J Neurosurg* 2015;**29**(3):362−70.

294. Tabakow P, et al. Transplantation of autologous olfactory ensheathing cells in complete human spinal cord injury. *Cell Transplant* 2013;**22**(9):1591−612.

295. Woodworth CF, et al. Intramedullary cervical spinal mass after stem cell transplantation using an olfactory mucosal cell autograft. *Can Med Assoc J* 2019;**191**(27):E761−4.

296. Bruce JH, et al. Schwannosis: role of gliosis and proteoglycan in human spinal cord injury. *J Neurotrauma* 2000;**17**(9):781−8.

297. Xu XM, et al. Bridging Schwann cell transplants promote axonal regeneration from both the rostral and caudal stumps of transected adult rat spinal cord. *J Neurocytol* 1997;**26**(1):1−16.

298. Barakat DJ, et al. Survival, integration, and axon growth support of glia transplanted into the chronically contused spinal cord. *Cell Transplant* 2005;**14**(4):225−40.

299. Taylor JS, Bampton ET. Factors secreted by Schwann cells stimulate the regeneration of neonatal retinal ganglion cells. *J Anat* 2004;**204**(1):25−31.

300. Toyota B, Carbonetto S, David S. A dual laminin/collagen receptor acts in peripheral nerve regeneration. *Proc Natl Acad Sci USA* 1990;**87**(4):1319−22.

301. Xu XM, et al. Axonal regeneration into Schwann cell-seeded guidance channels grafted into transected adult rat spinal cord. *J Comp Neurol* 1995;**351**(1):145−60.

302. Xu XM, et al. Regrowth of axons into the distal spinal cord through a Schwann-cell-seeded mini-channel implanted into hemisected adult rat spinal cord. *Eur J Neurosci* 1999;**11**(5):1723−40.

303. Bamber NI, et al. Neurotrophins BDNF and NT-3 promote axonal re-entry into the distal host spinal cord through Schwann cell-seeded mini-channels. *Eur J Neurosci* 2001;**13**(2):257−68.

304. Williams RR, et al. Permissive Schwann cell graft/spinal cord interfaces for axon regeneration. *Cell Transplant* 2015;**24**(1):115−31.

305. Takami T, et al. Schwann cell but not olfactory ensheathing glia transplants improve hindlimb locomotor performance in the moderately contused adult rat thoracic spinal cord. *J Neurosci* 2002;**22**(15):6670−81.

306. Schaal SM, et al. Schwann cell transplantation improves reticulospinal axon growth and forelimb strength after severe cervical spinal cord contusion. *Cell Transplant* 2007;**16**(3):207−28.

307. Anderson KD, et al. Safety of autologous human Schwann cell transplantation in subacute thoracic spinal cord injury. *J Neurotrauma* 2017;**34**(21):2950−63.

308. Santamaria AJ, et al. Neurophysiological changes in the first year after cell transplantation in sub-acute complete paraplegia. *Front Neurol* 2020;**11**:514181.

309. Gant KL, et al. Phase 1 safety trial of autologous human Schwann cell transplantation in chronic spinal cord injury. *J Neurotrauma* 2021. https://doi.org/10.1089/neu.2020.7590.

310. Reynolds BA, Weiss S. Generation of neurons and astrocytes from isolated cells of the adult mammalian central nervous system. *Science* 1992;**255**(5052):1707−10.

311. Thomson JA, et al. Embryonic stem cell lines derived from human blastocysts. *Science* 1998;**282**(5391):1145−7.

312. Keirstead HS, et al. Human embryonic stem cell-derived oligodendrocyte progenitor cell transplants remyelinate and restore locomotion after spinal cord injury. *J Neurosci* 2005;**25**(19):4694−705.

313. Strauss S. Geron trial resumes, but standards for stem cell trials remain elusive. *Nat Biotechnol* 2010;**28**(10):989−90.

314. Scott CT, Magnus D. Wrongful termination: lessons from the Geron clinical trial. *Stem Cells Trans Med* 2014;**3**(12):1398−401.

315. Biotherapeutics, A., Asterias provides top line 12 Month data update for its OPC1 phase 1/2a clinical trial in severe spinal cord injury. https://www.globe newswire.com/news-release/2019/01/24/1704757/0/en/Asterias-Provides-Top-Line-12-Month-Data-Update-for-its-OPC1-Phase-1-2a-Clinical-Trial-in-Severe-Spinal-Cord-Injury.html.

316. Therapeutics LC. OPCI. https://lineagecell.com/products-pipeline/opc1/.

317. Glass JD, et al. Lumbar intraspinal injection of neural stem cells in patients with amyotrophic lateral sclerosis: results of a phase I trial in 12 patients. *Stem Cell* 2012;**30**(6):1144−51.

318. Lu P, et al. Long-distance growth and connectivity of neural stem cells after severe spinal cord injury. *Cell* 2012;**150**(6):1264−73.

319. Usvald D, et al. Analysis of dosing regimen and reproducibility of intraspinal grafting of human spinal stem cells in immunosuppressed minipigs. *Cell Transplant* 2010;**19**(9):1103−22.

320. van Gorp S, et al. Amelioration of motor/sensory dysfunction and spasticity in a rat model of acute lumbar spinal cord injury by human neural stem cell transplantation. *Stem Cell Res Ther* 2013;**4**(3):57.

321. Curtis E, et al. A first-in-human, phase I study of neural stem cell transplantation for chronic spinal cord injury. *Cell Stem Cell* 2018;**22**(6):941−950 e6.

322. Levi AD, et al. Emerging safety of intramedullary transplantation of human neural stem cells in chronic cervical and thoracic spinal cord injury. *Neurosurgery* 2018;**82**(4):562−75.

323. Levi AD, et al. Clinical outcomes from a multi-center study of human neural stem cell transplantation in chronic cervical spinal cord injury. *J Neurotrauma* 2019;**36**(6):891−902.

324. Takahashi K, Yamanaka S. Induction of pluripotent stem cells from mouse embryonic and adult fibroblast cultures by defined factors. *Cell* 2006;**126**(4):663−76.

325. Nagoshi N, Okano H. iPSC-derived neural precursor cells: potential for cell transplantation therapy in spinal cord injury. *Cell Mol Life Sci* 2018;**75**(6):989−1000.

326. Warren L, et al. Highly efficient reprogramming to pluripotency and directed differentiation of human cells with synthetic modified mRNA. *Cell Stem Cell* 2010;**7**(5):618−30.

327. Takahashi J. Clinical trial for Parkinson's disease gets a green light in the US. *Cell Stem Cell* 2021;**28**(2):182−3.

328. Meirelles LDS, Chagastelles PC, Nardi NB. Mesenchymal stem cells reside in virtually all post-natal organs and tissues. *J Cell Sci* 2006;**119**(11):2204−13.

329. Chamberlain G, et al. Concise review: mesenchymal stem cells: their phenotype, differentiation capacity, immunological features, and potential for homing. *Stem Cell* 2007;**25**(11):2739—49.

330. Turner L, Knoepfler P. Selling stem cells in the USA: assessing the direct-to-consumer industry. *Cell Stem Cell* 2016;**19**(2):154—7.

331. Berger I, et al. Global distribution of businesses marketing stem cell-based interventions. *Cell Stem Cell* 2016;**19**(2):158—62.

332. Taylor-Weiner H, Graff Zivin J. Medicine's wild west—unlicensed stem-cell clinics in the United States. *N Engl J Med* 2015;**373**(11):985—7.

333. Ryan KA, et al. Tracking the rise of stem cell tourism. *Regen Med* 2010;**5**(1):27—33.

334. Petersen A, et al. Stem cell miracles or Russian roulette?: patients' use of digital media to campaign for access to clinically unproven treatments. *Health Risk Soc* 2016;**17**(7—8):592—604.

335. McLean AK, Stewart C, Kerridge I. Untested, unproven, and unethical: the promotion and provision of autologous stem cell therapies in Australia. *Stem Cell Res Ther* 2015;**6**:33.

336. Dominici M, et al. Minimal criteria for defining multipotent mesenchymal stromal cells. The International Society for Cellular Therapy position statement. *Cytotherapy* 2006;**8**(4):315—7.

337. Friedenstein A, Chailakhjan R, Lalykina KJCP. The development of fibroblast colonies in monolayer cultures of Guinea-pig bone marrow and spleen cells. *Cell Tissue Kinet* 1970;**3**(4):393—403.

338. Owen M, Friedenstein A. Stromal stem cells: marrow-derived osteogenic precursors. In: *Ciba found symp*. Wiley Online Library; 1988.

339. Haynesworth S, et al. Characterization of cells with osteogenic potential from human marrow. *Bone* 1992;**13**(1):81—8.

340. Lazarus H, et al. Ex vivo expansion and subsequent infusion of human bone marrow-derived stromal progenitor cells (mesenchymal progenitor cells): implications for therapeutic use. *Bone Marrow Transplant* 1995;**16**(4):557—64.

341. Halvorsen Y, Wilkison W, Gimble JM. Adipose-derived stromal cells—their utility and potential in bone formation. *Int J Obs* 2000;**24**(S4):S41.

342. Zuk PA, et al. Multilineage cells from human adipose tissue: implications for cell-based therapies. *Tissue Eng* 2001;**7**(2):211—28.

343. Romanov YA, Svintsitskaya VA, Smirnov VN. Searching for alternative sources of postnatal human mesenchymal stem cells: candidate MSC-like cells from umbilical cord. *Stem Cell* 2003;**21**(1):105—10.

344. Weimann JM, et al. Contribution of transplanted bone marrow cells to Purkinje neurons in human adult brains. *Proc Natl Acad Sci USA* 2003;**100**(4):2088—93.

345. Johansson CB, et al. Extensive fusion of haematopoietic cells with Purkinje neurons in response to chronic inflammation. *Nat Cell Biol* 2008;**10**(5):575—83.

346. Rustad KC, Gurtner GC. Mesenchymal stem cells home to sites of injury and inflammation. *Adv Wound Care* 2012;**1**(4):147—52.

347. Bakshi A, et al. Lumbar puncture delivery of bone marrow stromal cells in spinal cord contusion: a novel method for minimally invasive cell transplantation. *J Neurotrauma* 2006;**23**(1):55—65.

348. Nakano N, et al. Characterization of conditioned medium of cultured bone marrow stromal cells. *Neurosci Lett* 2010;**483**(1):57—61.

349. Satake K, Lou J, Lenke LG. Migration of mesenchymal stem cells through cerebrospinal fluid into injured spinal cord tissue. *Spine* 2004;**29**(18):1971—9.

350. Chang Q, et al. Bone marrow mesenchymal stem cell-derived exosomal microRNA-125a promotes M2 macrophage polarization in spinal cord injury by downregulating IRF5. *Brain Res Bull* 2021;**170**:199—210.

351. Li HYZ, et al. Mitochondrial transfer from bone marrow mesenchymal stem cells to motor neurons in spinal cord injury rats via gap junction. *Theranostics* 2019;**9**(7):2017—35.

352. Sipp D. Conditional approval: japan lowers the bar for regenerative medicine products. *Cell Stem Cell* 2015;**16**(4):353—6.

353. Cyranoski D. Japan's approval of stem-cell treatment for spinal-cord injury concerns scientists. *Nature* 2019;**565**(7741):544—5.

354. Bydon M, et al. CELLTOP clinical trial: first report from a phase 1 trial of autologous adipose tissue-derived mesenchymal stem cells in the treatment of paralysis due to traumatic spinal cord injury. *Mayo Clin Proc* 2020;**95**(2):406—14.

355. Kishk NA, et al. Case control series of intrathecal autologous bone marrow mesenchymal stem cell therapy for chronic spinal cord injury. *Neurorehabil Neural Rep* 2010;**24**(8):702—8.

356. Mendonca MVP, et al. Safety and neurological assessments after autologous transplantation of bone marrow mesenchymal stem cells in subjects with chronic spinal cord injury. *Stem Cell Res Ther* 2014;**5**:11.

357. Liu J, et al. Clinical analysis of the treatment of spinal cord injury with umbilical cord mesenchymal stem cells. *Cytotherapy* 2013;**15**(2):185—91.

358. Albu S, et al. Clinical effects of intrathecal administration of expanded Wharton jelly mesenchymal stromal cells in patients with chronic complete spinal cord injury: a randomized controlled study. *Cytotherapy* 2021;**23**(2):146—56.

359. Aspinall P, et al. A systematic review of safety reporting in acute spinal cord injury clinical trials: challenges and recommendations. *J Neurotrauma* 2021;**38**(15):2047—54.

360. Santamaria AJ, et al. Clinical and neurophysiological changes after targeted intrathecal injections of bone marrow stem cells in a C3 tetraplegic subject. *J Neurotrauma* 2019;**36**(3):500—16.

361. Lo B, Parham L. Ethical issues in stem cell research. *Endocr Rev* 2009;**30**(3):204—13.

362. Thirabanjasak D, Tantiwongse K, Thorner PS. Angio-myeloproliferative lesions following autologous stem cell therapy. *J Am Soc Nephrol* 2010;**21**(7):1218—22.

363. Kuriyan AE, et al. Vision loss after intravitreal injection of autologous "stem cells" for AMD. *N Engl J Med* 2017;**376**(11):1047—53.

364. Alderazi YJ, Coons SW, Chapman K. Catastrophic demyelinating encephalomyelitis after intrathecal and intravenous stem cell transplantation in a patient with multiple sclerosis. *J Child Neurol* 2012;**27**(5):632—5.

365. Xin HQ, et al. Exosome-mediated transfer of miR-133b from multipotent mesenchymal stromal cells to neural cells contributes to neurite outgrowth. *Stem Cell* 2012;**30**(7):1556—64.

366. Soares AR, et al. Gap junctional protein Cx43 is involved in the communication between extracellular vesicles and mammalian cells. *Sci Rep* 2015;**5**:13243.

367. Liu W, et al. Exosome-shuttled miR-216a-5p from hypoxic preconditioned mesenchymal stem cells repair traumatic spinal cord injury by shifting microglial M1/M2 polarization. *J Neuroinflam* 2020;**17**(1):22.

368. Menei P, et al. Schwann cells genetically modified to secrete human BDNF promote enhanced axonal regrowth across transected adult rat spinal cord. *Eur J Neurosci* 1998;**10**(2):607—21.

369. Golden KL, et al. Transduced Schwann cells promote axon growth and myelination after spinal cord injury. *Exp Neurol* 2007;**207**(2):203—17.

370. Nakahara Y, Gage FH, Tuszynski MH. Grafts of fibroblasts genetically modified to secrete NGF, BDNF, NT-3, or basic FGF elicit differential responses in the adult spinal cord. *Cell Transplant* 1996;**5**(2):191—204.

371. Mendell JR, et al. Five-year extension results of the phase 1 START trial of onasemnogene abeparvovec in spinal muscular atrophy. *Jama Neurology* 2021:8.

372. Al-Zaidy SA, Mendell JR. From clinical trials to clinical practice: practical considerations for gene replacement therapy in SMA type 1. *Pediatr Neurol* 2019;**100**:3—11.

373. van Zundert B, Brown Jr RH. Silencing strategies for therapy of SOD1-mediated ALS. *Neurosci Lett* 2017;**636**:32—9.

374. Burnside ER, et al. Immune-evasive gene switch enables regulated delivery of chondroitinase after spinal cord injury. *Brain* 2018;**141**:2362—81.

375. Metcalfe AD, Ferguson MWJ. Tissue engineering of replacement skin: the crossroads of biomaterials, wound healing, embryonic development, stem cells and regeneration. *J R Soc Interface* 2007;**4**(14):413—37.

376. Paino CL, Bunge MB. Induction of axon growth into Schwann cell implants grafted into lesioned adult rat spinal cord. *Exp Neurol* 1991;**114**(2):254—7.

377. Park J, et al. Reducing inflammation through delivery of lentivirus encoding for anti-inflammatory cytokines attenuates neuropathic pain after spinal cord injury. *J Control Release* 2018;**290**:88—101.

378. Thomas AM, et al. Sonic hedgehog and neurotrophin-3 increase oligodendrocyte numbers and myelination after spinal cord injury. *Integr Biol* 2014;**6**(7):694—705.

379. He L, et al. Manufacture of PLGA multiple-channel conduits with precise hierarchical pore architectures and in vitro/vivo evaluation for spinal cord injury. *Tissue Eng Part C Methods* 2009;**15**(2):243—55.

380. Tsai EC, et al. Matrix inclusion within synthetic hydrogel guidance channels improves specific supraspinal and local axonal regeneration after complete spinal cord transection. *Biomaterials* 2006;**27**(3):519—33.

381. Yang Y, et al. Multiple channel bridges for spinal cord injury: cellular characterization of host response. *Tissue Eng Part A* 2009;**15**(11):3283—95.

382. Olson HE, et al. Neural stem cell- and Schwann cell-loaded biodegradable polymer scaffolds support axonal regeneration in the transected spinal cord. *Tissue Eng Part A* 2009;**15**(7):1797—805.

383. Dumont CM, et al. Aligned hydrogel tubes guide regeneration following spinal cord injury. *Acta Biomater* 2019;**86**:312—22.

384. Kim H, Tator CH, Shoichet MS. Chitosan implants in the rat spinal cord: biocompatibility and biodegradation. *J Biomed Mater Res A* 2011;**97**(4):395—404.

385. Guest JD, et al. Internal decompression of the acutely contused spinal cord: differential effects of irrigation only versus biodegradable scaffold implantation. *Biomaterials* 2018;**185**:284—300.

386. Tysseling-Mattiace VM, et al. Self-assembling nanofibers inhibit glial scar formation and promote axon elongation after spinal cord injury. *J Neurosci* 2008;**28**(14):3814—23.

387. Li X, et al. A therapeutic strategy for spinal cord defect: human dental follicle cells combined with aligned PCL/PLGA electrospun material. *BioMed Res Int* 2015;**2015**:197183.

388. Macaya D, Spector M. Injectable hydrogel materials for spinal cord regeneration: a review. *Biomed Mater* 2012;**7**:012001.

389. Perale G, et al. Hydrogels in spinal cord injury repair strategies. *ACS Chem Neurosci* 2011;**2**(7):336—45.

390. Woerly S, et al. Neural tissue formation within porous hydrogels implanted in brain and spinal cord lesions: ultrastructural, immunohistochemical, and diffusion studies. *Tissue Eng* 1999;**5**(5):467—88.

391. Hejcl A, et al. Macroporous hydrogels based on 2-hydroxyethyl methacrylate. Part 6: 3D hydrogels with positive and negative surface charges and polyelectrolyte complexes in spinal cord injury repair. *J Mater Sci Mater Med* 2009;**20**(7):1571—7.

392. Kataoka K, et al. Alginate enhances elongation of early regenerating axons in spinal cord of young rats. *Tissue Eng* 2004;**10**(3—4):493—504.

393. Lin H, et al. Novel nerve guidance material prepared from bovine aponeurosis. *J Biomed Mater Res A* 2006;**79**(3):591—8.

394. Xiao Z, et al. One-year clinical study of NeuroRegen scaffold implantation following scar resection in complete chronic spinal cord injury patients. *Sci China Life Sci* 2016;**59**(7):647—55.

395. Chen WG, et al. NeuroRegen scaffolds combined with autologous bone marrow mononuclear cells for the repair of acute complete spinal cord injury: a 3-year clinical study. *Cell Transplant* 2020;**29**:11.

396. Oudega M, et al. Validation study of neurotrophin-3-releasing chitosan facilitation of neural tissue generation in the severely injured adult rat spinal cord. *Exp Neurol* 2019;**312**:51—62.

397. Slotkin JR, et al. Biodegradable scaffolds promote tissue remodeling and functional improvement in non-human primates with acute spinal cord injury. *Biomaterials* 2017;**123**:63—76.

398. Teng YD, et al. Functional recovery following traumatic spinal cord injury mediated by a unique polymer scaffold seeded with neural stem cells. *Proc Natl Acad Sci USA* 2002;**99**(5):3024—9.

399. Pritchard CD, et al. Establishing a model spinal cord injury in the African green monkey for the preclinical evaluation of biodegradable polymer scaffolds seeded with human neural stem cells. *J Neurosci Methods* 2010;**188**(2):258—69.

400. Theodore N, et al. First human implantation of a bioresorbable polymer scaffold for acute traumatic spinal cord injury: a clinical pilot study for safety and feasibility. *Neurosurgery* 2016;**79**(2):E305—12.

401. Kim KD, et al. A study of probable benefit of a bioresorbable polymer scaffold for safety and neurological recovery in patients with complete thoracic spinal cord injury: 6-month results from the INSPIRE study. *J Neurosurg Spine* 2021:1—10.

402. Mesbah S, et al. Predictors of volitional motor recovery with epidural stimulation in individuals with chronic spinal cord injury. *Brain* 2021;**144**(2):420—33.

403. McCaig CD, Song B, Rajnicek AM. Electrical dimensions in cell science. *J Cell Sci* 2009;**122**(23):4267—76.

404. Al-Majed AA, Tam SL, Gordon T. Electrical stimulation accelerates and enhances expression of regeneration-associated genes in regenerating rat femoral motoneurons. *Cell Mol Neurobiol* 2004;**24**(3):379—402.

405. Wang WJ, et al. Electrical stimulation promotes BDNF expression in spinal cord neurons through Ca^{2+}- and Erk-dependent signaling pathways. *Cell Mol Neurobiol* 2011;**31**(3):459—67.

406. Liu Y, Grumbles RM, Thomas CK. Electrical stimulation of embryonic neurons for 1 hour improves axon regeneration and the number of reinnervated muscles that function. *J Neuropathol Exp Neurol* 2013;**72**(7):697—707.

407. Pale T, Frisch EB, McClellan AD. Cyclic amp stimulates neurite outgrowth of lamprey reticulospinal neurons without substantially altering their biophysical properties. *Neuroscience* 2013;**245**:74—89.

408. Li Q, et al. Electrical stimulation of the medullary pyramid promotes proliferation and differentiation of oligodendrocyte progenitor cells in the corticospinal tract of the adult rat. *Neurosci Lett* 2010;**479**(2):128—33.

409. Carmel JB, Martin JH. Motor cortex electrical stimulation augments sprouting of the corticospinal tract and promotes recovery of motor function. *Front Integr Neurosci* 2014;**8**:51.

410. Angeli CA, et al. Recovery of over-ground walking after chronic motor complete spinal cord injury. *N Engl J Med* 2018;**379**(13):1244—50.

411. Gerasimenko Y, et al. Transcutaneous electrical spinal-cord stimulation in humans. *Ann Phys Rehabil Med* 2015;**58**(4):225—31.

412. Raineteau O, Schwab ME. Plasticity of motor systems after incomplete spinal cord injury. *Nat Rev Neurosci* 2001;**2**(4):263—73.

413. Flynn JR, et al. The role of propriospinal interneurons in recovery from spinal cord injury. *Neuropharmacology* 2011;**60**(5):809—22.

414. Campos L, et al. Engineering novel spinal circuits to promote recovery after spinal injury. *J Neurosci* 2004;**24**(9):2090—101.

415. Rosenzweig ES, et al. Extensive spontaneous plasticity of corticospinal projections after primate spinal cord injury. *Nat Neurosci* 2010;**13**(12):1505—10.

416. Kumamaru H, et al. Regenerating corticospinal axons innervate phenotypically appropriate neurons within neural stem cell grafts. *Cell Rep* 2019;**26**(9):2329—2339 e4.

417. Cooper DJ, et al. Phenotypic screening with primary neurons to identify drug targets for regeneration and degeneration. *Mol Cell Neurosci* 2017;**80**:161—9.

418. Kim HS, et al. A microchip for high-throughput axon growth drug screening. *Micromachines* 2016;**7**(7).

419. Callahan A, et al. Developing a data sharing community for spinal cord injury research. *Exp Neurol* 2017;**295**:135—43.

Emerging concepts in the clinical management of SCI for the future

Laureen D. Hachem[1], Jetan H. Badhiwala[1], Fan Jiang[2], Brian K. Kwon[3], Mark R.N. Kotter[4], Jefferson R. Wilson[1,5], Alexander R. Vaccaro[6], F. Cumhur Oner[7], Michael G. Fehlings[1,2]

[1]Division of Neurosurgery, Department of Surgery, University of Toronto, Toronto, ON, Canada; [2]Division of Neurosurgery, Toronto Western Hospital, University Health Network, Toronto, ON, Canada; [3]Vancouver Spine Surgery Institute, Vancouver General Hospital, Department of Orthopaedics, UBC, Vancouver, BC, Canada; [4]Division of Neurosurgery, Department of Clinical Neurosciences, University of Cambridge, Cambridge, United Kingdom; [5]St. Michael's Hospital, Toronto, ON, Canada; [6]Department of Orthopedic Surgery, Rothman Orthopedic Institute, Thomas Jefferson University, Philadelphia, PA, United States; [7]Department of Orthopaedics, University Medical Center Utrecht, Utrecht, the Netherlands

Introduction

Advances across the spectrum of clinical neurosurgery and neuroscience research have transformed the management of spinal cord injury (SCI). The integration of high-resolution imaging techniques along with electrophysiological measures into clinical practice has provided novel methods to quantify injury severity and temporal plasticity. The emergence of molecular and genomic technologies has shed light on the underlying heterogeneity seen across patients with SCI and revealed candidate biomarkers with both predictive and prognostic relevance. Furthermore, novel methods to interrogate the central nervous system (CNS) at both a cellular and systems level have uncovered important circuitry that underlies basic spinal cord functions and serves as the substrate for understanding disease mechanisms and developing targeted therapies.[1]

In this chapter, we identify a number of priority themes in the field of SCI that will continue to redefine the landscape of clinical practice in the next decade. Specifically, we highlight the importance of early interventions in the treatment of SCI and describe how measures of SCI physiology, novel predictive and prognostic models along with noninvasive treatment modalities can be utilized to improve patient

Neural Repair and Regeneration after Spinal Cord Injury and Spine Trauma
https://doi.org/10.1016/B978-0-12-819835-3.00029-0

575

© 2022 Elsevier Inc. All rights reserved.

outcomes. Moreover, we outline methods to better characterize the heterogeneity seen across SCI patients at a genetic, molecular, and phenotypic level in order to develop a personalized approach to SCI care. Lastly, we describe how novel insights obtained from research into regenerative medicine, neural networks, and brain—computer interfaces (BCIs) have reshaped our understanding of SCI pathophysiology and expanded the boundaries of treatment potential for the future.

Time is spine

Pivotal work in the last decade has demonstrated the importance of early interventions after acute traumatic SCI. This concept, referred to as "Time is Spine," emphasizes the need to mitigate early events of the secondary injury cascade of SCI in order to limit further tissue damage.[2] Early transfer to a specialized trauma/acute SCI unit for appropriate cardiorespiratory management and stabilization is critical in achieving improved recovery. Furthermore, surgical decompression within 24h of acute SCI has been shown to be associated with improved outcomes, with the first 36 h representing an important window to achieve optimal recovery.[3] Current clinical practice guidelines thus recommend early surgical decompression in the setting of ongoing compression after SCI.[4,5] In addition, the efficacy of neuroprotective agents that have been trialed in SCI is highly dependent on early administration. As such, the importance of expeditious and timely management of SCI serves as the backbone of developing systematized approaches to facilitate early interventions for SCI.

Measures of spinal cord physiology

For decades, measures of brain physiology have been an integral part of clinical practice guidelines in managing traumatic brain injury patients. Only recently have similar parameters of spinal cord physiology been incorporated into the management of SCI. Integration of these measures into clinical practice will likely become an important factor in optimizing patient outcomes and measuring treatment response in both the clinical and research setting.

Spinal cord blood flow and perfusion pressure

SCI leads to hypoxic-ischemic injuries due to direct injury to blood vessels compromising perfusion, hypotension from autoregulatory failure, prostaglandin-induced vasoconstriction, and the formation of free radicals. Maintenance of spinal cord blood flow is of utmost importance during this critical time and, as a result, current guidelines recommend maintaining an MAP between 85 and 90 mmHg for 7 days following acute SCI.[6] However, recent work has suggested that spinal cord perfusion pressure (SCPP), as measured by lumbar intrathecal catheter, may be a more reliable measure that correlates with outcomes. A multicenter prospective trial of 92 individuals with acute SCI found that SCPP is a predictor of neurological outcome postinjury and that low SCPP increases the risk of poor motor recovery.[7] As such, it is recommended that SCPP be maintained above 50 mm Hg by augmentation of MAP (>70 mm Hg) and maintenance of CSFP (<29). Incorporation of this parameter into future clinical practice may therefore provide a novel means to better optimize neurological outcome after SCI.

Electrophysiology

Within the field of SCI there is a significant need to objectively measure spinal cord function for both clinical prognostication and as an outcome measure in clinical trial design. Diagnosis and grading of SCI is currently limited to clinical examination, with the American Spinal Injury Association Impairment Scale (AIS) being the most widely used system. However, sole reliance on this method of assessment presents a number of limitations in both clinical practice and trial design due to a lack of sensitivity and objectivity.

Electrophysiological measures of spinal cord function are an important adjunct to the clinical examination.[8] Application of electrophysiological studies can establish the topography of SCI and the severity of injury within various regions (Table 30.1). The use of somatosensory evoked potentials (SSEPs) can determine posterior cord dysfunction, while contact heat evoked potentials (CHEPs) are more sensitive in assessing central and anterior cord damage. motor evoked potentials (MEPs) may be utilized to assess the degree of corticospinal tract (CST) dysfunction and have demonstrated reliability in predicting lower extremity motor score and walking function after SCI in patients.[9,10] Moreover, assessment of MEPs in SCI AIS grade A patients uncovered a subgroup of patients with preservation of evoked potentials.[11] This suggests that the electrophysiological profile of patients within a given AIS grade can vary and may be a factor contributing to differential conversion rates. Furthermore, these tools may detect subclinical changes in motor or sensory function that could be an early predictor of recovery or treatment response. Electrophysiological parameters can also facilitate the detection of concomitant disorders such as polyneuropathies or plexus injuries that may confound clinical

examination. Moreover, they provide a valuable aid in determining function in patients with altered mental status or extremity fractures. Nevertheless, the application of electrophysiological parameters to predict outcomes after SCI must be used with caution and should not supplant standard clinical assessments as electrophysiological changes may not always correlate with a meaningful clinical effect.

Integration of imaging data into predictive and prognostic models

Advanced imaging techniques have facilitated better quantification of injury burden and offered potential in monitoring treatment response. The use of magnetic resonance (MR) diffusion tensor imaging in experimental models of SCI has been shown to correlate with structural changes induced by stem cell transplantation.[12] Imaging may also be of use in detecting microstructural changes that occur on a subclinical level.

In addition, the application of machine learning techniques to various imaging platforms has enabled promising predictive outcome models. Using fractional anisotropy (a measure of directional water diffusion in tissue) obtained from diffusion tensor images of the spinal cord, machine learning algorithms have been developed to identify injured tissue with both high specificity and sensitivity.[13] Convolutional neural networks have also been used to determine contusion volume which correlated with motor scores at both admission and discharge.[14]

Novel imaging classification scores have also added to our armamentarium of tools to predict outcomes after SCI. A 5-point scoring system based on the transverse extent of cord signal change on T2-weighted magnetic resonance imaging (MRI) at the lesion epicenter (termed the BASIC score) strongly correlated with both admission and discharge AIS grade. Furthermore, this score distinguished a subset of AIS grade A patients that had recovery at discharge.[15,16]

TABLE 30.1 Electrophysiological measures of spinal cord function.

Pathway	Electrophysiological measure
Corticospinal tract	TMS, MEP
Dorsal columns	EPT, SSEP, dSSEP
Sympathetic	SSR
Spinothalamic	CHEP, LEP
Peripheral, spinal	NCS, Reflex EMG

Adapted from Hubli et al.[8]

Minimally invasive treatment modalities

The last decade has witnessed rapid growth in minimally invasive technologies across all disciplines of medicine. Application of these novel approaches to the field of SCI will offer important strategies for drug delivery, gene therapy, and neuromodulation.

Focused ultrasound

Despite the disruption in blood–spinal cord barrier (BSCB) after acute SCI, administration of therapeutic agents to the lesion site remains a challenge. There is significant variability in the extent of BSCB permeability between patients and over time thus leading to unpredictable drug bioavailability within the CNS. The use of MR-guided focused ultrasound (MRgFUS) has emerged as a promising tool to overcome the blood–brain barrier (BBB) or BSCB. MRgFUS combined with injected lipid microbubbles has been shown to transiently disrupt the BBB and enhance the delivery of therapeutic agents in patients with ALS,[17] Alzheimer's disease,[18] and primary brain tumors.[19] Preclinical studies in animal models of SCI have shown promise in this technology for use to deliver targeted gene therapy or growth factors to the injured cord.[20,21] MRgFUS could therefore offer a noninvasive method to deliver therapeutics to degrade the glial scar, neuroprotectants to improve cell survival, or stem cell therapies to regenerate the injured spinal cord.[20]

Neuromodulation

A number of noninvasive and minimally invasive neuromodulatory techniques have shown promising potential to improve functional recovery after SCI.[22] Transcranial direct current stimulation applies low-intensity currents by way of electrodes placed on the scalp and in combination with exercise training may facilitate activity-dependent plasticity to enhance limb function. Transcranial magnetic stimulation (TMS) has also been applied to SCI patients where magnetic fields are delivered through the scalp in order to stimulate cortical neurons. Evidence for both techniques in the context of SCI are limited to small patient cohorts, however, have shown promising results in improving hand function and locomotion.[23-26] Deep brain stimulation (DBS) has been used to stimulate subcortical locomotor regions in animal models of SCI where stimulation of the mesencephalic locomotor region improved hind limb function in a rodent model of chronic SCI.[27] A Phase I/II open-label multicenter clinical trial is currently underway to examine the safety and preliminary efficacy of mesencephalic locomotor region DBS in patients with incomplete SCI (NCT03053791). Direct spinal cord modulation techniques by way of transcutaneous stimulation and closed-loop epidural stimulation are also under investigation for SCI. Further development and application of neuromodulatory technology to SCI patients hold promise in restoring functional circuitry and combined with task-specific training will redefine the modern-day approach to neurorehabilitation.

Redefining clinical heterogeneity

It is well recognized that SCI is a heterogeneous condition both in terms of injury pattern and severity. Traditional classification systems and grading scales, while useful in defining patient subgroups, lack the sensitivity to predict patient outcomes and response to treatment. As such, the coming years will see a focus on redefining SCI patient subgroups with clinical relevance to better incorporate into both the clinical and research realms.

Severity of injury

Currently, assessment of injury severity after SCI is based on thorough clinical examination. While AIS is the most widely used grading system for injury severity, it has a number of limitations. There is significant amount of AIS conversion following SCI, many of which due to changes only within sacral segments. This suggests that conversions may be an artifact of the assessment tool rather than a meaningful change in neurological function.[28] Moreover, AIS D patients show low rates of conversion, possibly due to a ceiling effect. There is also significant patient heterogeneity within each AIS grade with variable rates of conversion seen within the same group. This is borne out by findings from a large observational study incorporating 836 SCI patients whereby there was significant variability in motor recovery within each AIS grade depending on the level of injury.[29] Multiple patient- and injury-related factors must therefore be incorporated into grading scales of injury severity to more reliably predict motor recovery. Utilization of clinical, imaging, and electrophysiological data will facilitate the identification of patient severity subgroups with prognostic and therapeutic value.[15]

Pattern of injury

Incomplete SCI can be classified on the basis of various injury patterns that affect specific sensory and motor tracts. These subgroups include Brown-Sequard syndrome, anterior cord syndrome, posterior cord syndrome, and central cord syndrome (CCS). However, each type demonstrates significant heterogeneity in patient outcomes. This is best illustrated in the case of CCS which has traditionally been viewed as a single disease entity. Clinically, patients with CCS present with bilateral motor weakness worse in the upper extremities compared to lower extremities. Recent studies have shed light on the variability across patients with CCS redefining it as a heterogeneous condition with diverging recovery trajectories that depend on a multitude of patient- and injury-related factors.[30,31] Identifying subgroups of SCI patients with both prognostic and therapeutic relevance will help guide future management and inform clinical trial design.[32]

Even at a physiological level, our understanding of the molecular underpinnings of CCS has recently been challenged. Classic teaching posited that sudden injury to the central region of the cord disproportionately affects medial CST fibers controlling upper limbs, with relative preservation of lateral CST fibers controlling lower limbs.[33] To date there has yet to be strong evidence suggesting a somatotopic organization of the CST in humans but rather preservation of extrapyramidal tracts may underlie sparring of lower limb function seen in CCS.[2] This presents an important consideration when determining variation in relative sparring across patients. Incorporation of electrophysiological techniques may therefore lend important insight in better subtyping these patients.

Nontraumatic SCI

While traumatic mechanisms of SCI have traditionally been the focus of research efforts in the field, it is recognized that nontraumatic forms of SCI, particularly degenerative cervical myelopathy (DCM), make up a large portion of injuries. As such, there is a need to broaden the scope of SCI to include these nontraumatic forms of injury. Among these patients, there is considerable heterogeneity in clinical presentations and response to surgical intervention. Determining distinct clinical phenotypes of DCM patients that are more responsive to therapy[32] and identifying systemic biomarkers[34] or underlying genetic contributors to treatment response[35] will aid in more accurate prognostication and optimization of treatment plans.

Aging and frailty

With an aging population, SCI and spine trauma are increasingly involving the management of elderly patients. This shift in patient demographics imparts distinct considerations in the clinical management of SCI. Age-related changes to the spine result in specific patterns of injury and unique fracture healing patterns which must be factored into treatment decisions.[36,37] Furthermore, the decision to operate must be carefully weighed against surgical complication profile, which is often complicated by the presence of comorbidities and lower functional reserve among elderly patients. While increasing age has traditionally been associated with poorer surgical outcomes, it is now recognized that age alone is not a sole marker of treatment response. Rather, factors such as frailty index and performance status will likely be increasingly incorporated into treatment decision-making paradigms in the coming years.[38]

A personalized approach to SCI management

Emerging research has begun to demonstrate that both molecular and genetic variability may contribute to the clinical heterogeneity seen across SCI patients. A number of promising biomarkers have been identified after SCI that may predict prognosis and treatment response (Table 30.2). Furthermore, genetic and epigenetic differences between patients may further form the substrate of different injury trajectories.

Harnessing the potential of molecular and genomic technology will facilitate the integration of novel biomarkers into the clinical management of SCI to personalize treatment care plans.

Molecular biomarkers

Measurement of CSF and serum molecular markers holds promise in defining distinct "fingerprints" that correlate with clinical outcomes and treatment response. A clinical trial obtaining CSF samples from 27 SCI patients found distinct patterns of inflammatory cytokine and structural protein expression across different severities of SCI after 24h of injury.[39] Moreover distinct patterns were also found between individuals with AIS grade conversion compared to those that remained the same. Specifically, IL-6, IL-8, MCP-1 along with tau, S100beta, and GFAP could reliably predict failure to convert AIS grade. Importantly, many of the inflammatory mediators elevated in the CSF of SCI patients were also elevated in the serum, albeit at much lower concentrations. In the future, determining molecular patterns across patients may supplement clinical grading scores and add to our armamentarium of tools to predict recovery trajectories.

TABLE 30.2 Identified molecular/genetic signature and association with SCI outcomes.

Marker	Type	Association with SCI
Inflammatory cytokines (IL-6, IL-8, MCP-1)	CSF marker	Elevations correlate with injury severity and could reliably predict AIS conversion
Structural proteins (tau, S100beta, GFAP)	CSF marker	Elevations correlate with injury severity and could reliably predict AIS conversion
Val66Met SNP on BDNF gene	Single-nucleotide polymorphism	Associated with decreased exercise-induced BDNF release in patients with incomplete SCI
APOE ε4 allele	Single-nucleotide polymorphism	Associated with worse motor recovery in patients with SCI
TT homozygous rs13255063 variant of TRPA1 channel	Single-nucleotide polymorphism	More prevalent in SCI patients with neuropathic pain

Genetic heterogeneity

The human genome serves as the fundamental substrate of biology. Single-nucleotide polymorphisms (SNPs) are alterations in a single nucleotide at a specific point within the DNA seen across individuals. SNPs are the most common form of genetic variation and can underlie differences in disease susceptibility or severity along with drug response.

A specific polymorphism in the apolipoprotein E (APOE) gene (APOE ε4 allele) that encodes the corresponding protein which is important in lipid transport and tissue repair was associated with worse motor recovery in patients with SCI.[40] Genetic heterogeneity may also underlie differential responses to treatment or rehabilitation regimens. Activity-dependent neurotrophin release is an important mechanism of neuroplasticity after injury. A Val66Met SNP on the BDNF gene has recently been shown to decrease exercise-induced BDNF release in patients with incomplete SCI.[41]

Genetic variation may also explain differences in maladaptive plasticity after SCI. The TRPA1 channel (encoded by the corresponding TRPA1 gene) is important in pain transduction serving as a chemosensor of nociception. It is also postulated that TRPA1 expressed on astrocytes within the spinal cord may be associated with central pain hypersensitivity.[42] A study of 24 patients with SCI (12 with neuropathic pain and 12 without) found that rs11988795 GG homozygous variant in the TRPA1 gene was more commonly found in patients without pain while TT homozygous rs13255063 variant was prevalent in patients with neuropathic pain.[43] A number of other SNPs in genes important for various mechanisms of SCI have been identified as potential candidate targets to examine heterogeneity in disease course and treatment response.[44]

Epigenetic changes

Epigenetic mechanisms can regulate gene expression profiles through DNA methylation, histone modifications, and gene imprinting. These changes can lead to considerable differences in phenotypes across individuals. Epigenetic changes are important for CNS development and plasticity.[45] Emerging evidence has identified the role of epigenetics in recovery after injury in both the central and peripheral nervous system. Indeed, environmental enrichment prior to injury enhanced proprioceptive dorsal root ganglion neuron regenerative potential which was dependent on Creb-binding protein (Cbp)-mediated epigenetic histone acetylation.[46] There may be significant heterogeneity between individuals at an epigenetic level that can alter the clinical course of SCI. Identification of these changes will help better understand the pathogenesis of this condition and tailor treatment plans particularity in regard to motor recovery and rehabilitation.

Postinjury plasticity: a connectomics approach

Subacute and chronic SCI is a period of both adaptive and maladaptive plasticity. Alterations in brain and spinal cord neural connectivity and synaptic plasticity form the substrate of locomotor recovery, neuropathic pain, and spasticity. The emergence of connectomic profiling has facilitated identification of important networks that are altered in various disease states. Recently, this approach has been applied to traumatic brain injury to identify alterations in structural and functional connectivity that underlie brain dysfunction.[47] The use of this approach will likely lead to the identification of patient subgroups with different "connectomes" of clinical relevance.

While still in its infancy, application of connectomic-based analyses to the field of SCI will be important to determine changes in brain circuitry that may underlie maladaptive and adaptive plasticity. Indeed, functional MRI studies in rodents with SCI revealed decreased strength of functional connections between primary motor and primary sensory regions of the brain. In contrast, functional connectivity between primary sensory cortex and pain-related areas such as caudoputamen and anterior cingulate was strengthened in the chronic phase of injury likely serving as a substrate for neuropathic pain.[48] Similar alterations have been observed in studies on SCI patients,[49] suggesting that quantification of altered connections could conceivably allow for the generation of "connectivity fingerprints" across individuals to measure response to treatment.[50]

Regenerative medicine

Decades of research into the cellular and molecular mechanisms of SCI pathophysiology have led to the discovery of numerous candidate strategies to enhance regeneration after SCI. Stem cell—based therapies to replace lost cells have shown promise in ameliorating damage and enhancing functional plasticity. Engineering of cell transplants to target barriers to regeneration is under active investigation and is likely to provide important strategies to ameliorate the inhibitory environment of the injured spinal cord. Genetic modification of transplanted stem cells to express proneuronal factors has been shown to enhance neuronal differentiation and increase functional recovery.[51] Furthermore, cells modified to express growth factors such as bFGF, NT3, or BDNF may also increase the survival of cell transplants, improve axon outgrowth, and promote plasticity.[52–54] In addition to direct modification of cell therapies, small molecule approaches to target growth inhibitory factors in the CNS that limit regeneration are undergoing active clinical trial investigation for translation to patients.[2,55] Combinatorial therapies integrating these various approaches to target deleterious components of the SCI microenvironment are anticipated to take off in the next decade.

Interrogating neural circuitry to understand disease states

The CNS is a highly organized network of complex connections between numerous specialized cells. Early physiological studies laid the foundation of our knowledge on information conduction and relay pathways within the CNS. However, recent advances in pharmacogenetics, optogenetics, and electrophysiological techniques have allowed us to interrogate the CNS to discover novel circuits and pathways that underlie basic functions such as locomotion and respiration in both normal and disease states. Indeed, emerging work has suggested a direct corticospinal pathway from the somatosensory cortex, independent of the motor cortex, that directly influences locomotion.[56] Moreover, recent work has discovered the descending motor system that controls left—right locomotor asymmetries in animals,[57] and identified important pathways involved in the neural control of respiration.[58,59] Further study into these novel pathways will allow the development of targeted treatments to specifically restore abnormal circuitry.

Application of brain—computer interfaces

Given the hostile environment of the injured spinal cord, emerging therapies have attempted to bypass the lesion epicenter through the application of BCIs. BCI relies on the premise of acquiring neurophysiological signals from the cortex by way of either invasive or noninvasive recordings which are then translated and relayed to an output device that performs the intended action. This technology has shown promise in regaining hand function and improving gait in SCI patients.[60,61] Recently, the development of a brain—spinal cord interface by way of connecting leg motor cortex activity with epidural electrical stimulation of the lumbar spinal cord improved locomotion following SCI in monkeys.[62]

While still far from clinical implementation, the application of BCIs and ultimately a brain—spinal cord interface to the field of SCI will have significant implications for the future. This will likely redefine our approach to early rehabilitation. Moreover, this technology may have long-lasting implications in the chronic setting where maintaining functional cortical connectivity may avoid maladaptive plasticity that underlies complications such as spasticity and neuropathic pain.[63]

Conclusion

The intersection of clinical neurosurgery with advances in imaging, electrophysiological measures, and systems neuroscience along with molecular and genomic technologies has redefined our approach to SCI management. Emerging insight into spinal cord circuitry and plasticity will continue to inform targeted therapies and novel rehabilitation strategies, while greater understanding of disease heterogeneity will usher in an era of personalized SCI care.

References

1. Krishna V, Andrews H, Varma A, Mintzer J, Kindy MS, Guest J. Spinal cord injury: how can we improve the classification and quantification of its severity and prognosis? *J Neurotrauma* 2014;**31**(3):215—27.
2. Badhiwala JH, Ahuja CS, Fehlings MG. Time is spine: a review of translational advances in spinal cord injury. *J Neurosurg Spine* 2018;**30**(1):1—18.
3. Badhiwala JH, Wilson JR, Witiw CD, et al. The influence of timing of surgical decompression for acute spinal cord injury: a pooled analysis of individual patient data. *Lancet Neurol* 2021;**20**(2):117—26.
4. Fehlings MG, Vaccaro A, Wilson JR, et al. Early versus delayed decompression for traumatic cervical spinal cord injury: results of the Surgical Timing in Acute Spinal Cord Injury Study (STASCIS). *PLoS One* 2012;**7**(2):e32037.
5. Fehlings MG, Tetreault LA, Wilson JR, et al. A clinical practice guideline for the management of patients with acute spinal cord injury and central cord syndrome: recommendations on the timing (≤24 hours versus >24 hours) of decompressive surgery. *Global Spine J* 2017;**7**(3 Suppl. 1):195s—202s.
6. Ryken TC, Hurlbert RJ, Hadley MN, et al. The acute cardiopulmonary management of patients with cervical spinal cord injuries. *Neurosurgery* 2013;**72**(Suppl. 2):84—92.
7. Squair JW, Bélanger LM, Tsang A, et al. Spinal cord perfusion pressure predicts neurologic recovery in acute spinal cord injury. *Neurology* 2017;**89**(16):1660—7.
8. Hubli M, Kramer JLK, Jutzeler CR, et al. Application of electrophysiological measures in spinal cord injury clinical trials: a narrative review. *Spinal Cord* 2019;**57**(11):909—23.
9. Petersen JA, Spiess M, Curt A, Dietz V, Schubert M. Spinal cord injury: one-year evolution of motor-evoked potentials and recovery of leg motor function in 255 patients. *Neurorehabilitation Neural Repair* 2012;**26**(8):939—48.
10. Dhall SS, Haefeli J, Talbott JF, et al. Motor evoked potentials correlate with magnetic resonance imaging and early recovery after acute spinal cord injury. *Neurosurgery* 2018;**82**(6):870—6.
11. Petersen JA, Spiess M, Curt A, et al. Upper limb recovery in spinal cord injury: involvement of central and peripheral motor pathways. *Neurorehabilitation Neural Repair* 2017;**31**(5):432—41.
12. Jirjis MB, Valdez C, Vedantam A, Schmit BD, Kurpad SN. Diffusion tensor imaging as a biomarker for assessing neuronal stem cell treatments affecting areas distal to the site of spinal cord injury. *J Neurosurg Spine* 2017;**26**(2):243—51.
13. Tay B, Hyun JK, Oh S. A machine learning approach for specification of spinal cord injuries using fractional anisotropy values obtained from diffusion tensor images. *Comput Math Methods Med* 2014;**2014**:276589.
14. McCoy DB, Dupont SM, Gros C, et al. Convolutional neural network-based automated segmentation of the spinal cord and contusion injury: deep learning biomarker correlates of motor impairment in acute spinal cord injury. *AJNR Am J Neuroradiol* 2019;**40**(4):737—44.
15. Talbott JF, Whetstone WD, Readdy WJ, et al. The Brain and Spinal Injury Center score: a novel, simple, and reproducible method for assessing the severity of acute cervical spinal cord injury with axial T2-weighted MRI findings. *J Neurosurg Spine* 2015;**23**(4):495—504.

16. Mabray MC, Talbott JF, Whetstone WD, et al. Multidimensional analysis of magnetic resonance imaging predicts early impairment in thoracic and thoracolumbar spinal cord injury. *J Neurotrauma* 2016;**33**(10): 954—62.

17. Abrahao A, Meng Y, Llinas M, et al. First-in-human trial of blood-brain barrier opening in amyotrophic lateral sclerosis using MR-guided focused ultrasound. *Nat Commun* 2019;**10**(1):4373.

18. Lipsman N, Meng Y, Bethune AJ, et al. Blood-brain barrier opening in Alzheimer's disease using MR-guided focused ultrasound. *Nat Commun* 2018;**9**(1):2336.

19. Mainprize T, Lipsman N, Huang Y, et al. Blood-brain barrier opening in primary brain tumors with non-invasive MR-guided focused ultrasound: a clinical safety and feasibility study. *Sci Rep* 2019;**9**(1):321.

20. Weber-Adrian D, Thévenot E, O'Reilly MA, et al. Gene delivery to the spinal cord using MRI-guided focused ultrasound. *Gene Therapy* 2015;**22**(7):568—77.

21. Song Z, Wang Z, Shen J, Xu S, Hu Z. Nerve growth factor delivery by ultrasound-mediated nanobubble destruction as a treatment for acute spinal cord injury in rats. *Int J Nanomed* 2017;**12**:1717—29.

22. James ND, McMahon SB, Field-Fote EC, Bradbury EJ. Neuromodulation in the restoration of function after spinal cord injury. *Lancet Neurol* 2018;**17**(10):905—17.

23. Kumru H, Murillo N, Benito-Penalva J, Tormos JM, Vidal J. Transcranial direct current stimulation is not effective in the motor strength and gait recovery following motor incomplete spinal cord injury during Lokomat(®) gait training. *Neurosci Lett* 2016;**620**:143—7.

24. Cortes M, Medeiros AH, Gandhi A, et al. Improved grasp function with transcranial direct current stimulation in chronic spinal cord injury. *NeuroRehabilitation* 2017;**41**(1):51—9.

25. Kumru H, Benito-Penalva J, Valls-Sole J, et al. Placebo-controlled study of rTMS combined with Lokomat(®) gait training for treatment in subjects with motor incomplete spinal cord injury. *Exp Brain Res* 2016;**234**(12): 3447—55.

26. Gomes-Osman J, Field-Fote EC. Improvements in hand function in adults with chronic tetraplegia following a multiday 10-Hz repetitive transcranial magnetic stimulation intervention combined with repetitive task practice. *J Neurol Phys Ther J Neurol Phys Ther* 2015; **39**(1):23—30.

27. Bachmann LC, Matis A, Lindau NT, Felder P, Gullo M, Schwab ME. Deep brain stimulation of the midbrain locomotor region improves paretic hindlimb function after spinal cord injury in rats. *Sci Transl Med* 2013; **5**(208). 208ra146.

28. Spiess MR, Müller RM, Rupp R, Schuld C, van Hedel HJ. Conversion in ASIA impairment scale during the first year after traumatic spinal cord injury. *J Neurotrauma* 2009;**26**(11):2027—36.

29. Dvorak MF, Noonan VK, Fallah N, et al. Minimizing errors in acute traumatic spinal cord injury trials by acknowledging the heterogeneity of spinal cord anatomy and injury severity: an observational Canadian cohort analysis. *J Neurotrauma* 2014;**31**(18):1540—7.

30. Badhiwala JH, Wilson JR, Fehlings MG. The case for revisiting central cord syndrome. *Spinal Cord* 2020; **58**(1):125—7.

31. Dvorak MF, Fisher CG, Hoekema J, et al. Factors predicting motor recovery and functional outcome after traumatic central cord syndrome: a long-term follow-up. *Spine* 2005;**30**(20):2303—11.

32. Badhiwala JH, Hachem LD, Merali Z, et al. Predicting outcomes after surgical decompression for mild degenerative cervical myelopathy: moving beyond the mJOA to identify surgical candidates. *Neurosurgery* 2020;**86**(4): 565—73.

33. Harrop JS, Sharan A, Ratliff J. Central cord injury: pathophysiology, management, and outcomes. *Spine J* 2006; **6**(6 Suppl. l):198s—206s.

34. Laliberte AM, Karadimas SK, Vidal PM, Satkunendrarajah K, Fehlings MG. Mir21 modulates inflammation and sensorimotor deficits in cervical myelopathy: data from humans and animal models. *Brain Commun* 2021;**3**(1):fcaa234.

35. Setzer M, Vrionis FD, Hermann EJ, Seifert V, Marquardt G. Effect of apolipoprotein E genotype on the outcome after anterior cervical decompression and fusion in patients with cervical spondylotic myelopathy. *J Neurosurg Spine* 2009;**11**(6):659—66.

36. Lomoschitz FM, Blackmore CC, Mirza SK, Mann FA. Cervical spine injuries in patients 65 years old and older: epidemiologic analysis regarding the effects of age and injury mechanism on distribution, type, and stability of injuries. *AJR Am J Roentgenol* 2002;**178**(3):573—7.

37. Raudenbush B, Molinari R. Longer-Term outcomes of geriatric odontoid fracture nonunion. *Geriatr Orthop Surg Rehabil* 2015;**6**(4):251—7.

38. Wilson JRF, Badhiwala JH, Moghaddamjou A, Yee A, Wilson JR, Fehlings MG. Frailty is a better predictor than age of mortality and perioperative complications after surgery for degenerative cervical myelopathy: an analysis of 41,369 patients from the NSQIP database 2010-2018. *J Clin Med* 2020;**9**(11).

39. Kwon BK, Stammers AM, Belanger LM, et al. Cerebrospinal fluid inflammatory cytokines and biomarkers of injury severity in acute human spinal cord injury. *J Neurotrauma* 2010;**27**(4):669—82.

40. Jha A, Lammertse DP, Coll JR, et al. Apolipoprotein E epsilon4 allele and outcomes of traumatic spinal cord injury. *J Spinal Cord Med* 2008;**31**(2):171—6.

41. Leech KA, Hornby TG. High-intensity locomotor exercise increases brain-derived neurotrophic factor in individuals with incomplete spinal cord injury. *J Neurotrauma* 2017;**34**(6):1240—8.

42. Wei H, Koivisto A, Saarnilehto M, et al. Spinal transient receptor potential ankyrin 1 channel contributes to central pain hypersensitivity in various pathophysiological conditions in the rat. *Pain* 2011;**152**(3):582—91.

43. Vidal Rodriguez S, Castillo Aguilar I, Cuesta Villa L, Serrano Saenz de Tejada F. TRPA1 polymorphisms in chronic and complete spinal cord injury patients with neuropathic pain: a pilot study. *Spinal Cord Series Cases* 2017;**3**:17089.

44. Guimarães PE, Fridman C, Gregório SP, et al. DNA polymorphisms as tools for spinal cord injury research. *Spinal Cord* 2009;**47**(2):171—5.

45. York EM, Petit A, Roskams AJ. Epigenetics of neural repair following spinal cord injury. *Neurotherapeutics* 2013;**10**(4):757—70.

46. Hutson TH, Kathe C, Palmisano I, et al. Cbp-dependent histone acetylation mediates axon regeneration induced by environmental enrichment in rodent spinal cord injury models. *Sci Transl Med* 2019;**11**(487).

47. Iraji A, Chen H, Wiseman N, et al. Connectome-scale assessment of structural and functional connectivity in mild traumatic brain injury at the acute stage. *NeuroImage Clin* 2016;**12**:100—15.

48. Matsubayashi K, Nagoshi N, Komaki Y, et al. Assessing cortical plasticity after spinal cord injury by using resting-state functional magnetic resonance imaging in awake adult mice. *Sci Rep* 2018;**8**(1):14406.

49. Oni-Orisan A, Kaushal M, Li W, et al. Alterations in cortical sensorimotor connectivity following complete cervical spinal cord injury: a prospective resting-state fMRI study. *PLoS One* 2016;**11**(3):e0150351.

50. Voets NL, Parker Jones O, Mars RB, et al. Characterising neural plasticity at the single patient level using connectivity fingerprints. *NeuroImage Clin* 2019;**24**:101952.

51. Khazaei M, Ahuja CS, Nakashima H, et al. GDNF rescues the fate of neural progenitor grafts by attenuating Notch signals in the injured spinal cord in rodents. *Sci Transl Med* 2020;**12**(525).

52. Liu WG, Wang ZY, Huang ZS. Bone marrow-derived mesenchymal stem cells expressing the bFGF transgene promote axon regeneration and functional recovery after spinal cord injury in rats. *Neurol Res* 2011;**33**(7):686—93.

53. Gransee HM, Zhan WZ, Sieck GC, Mantilla CB. Localized delivery of brain-derived neurotrophic factor-expressing mesenchymal stem cells enhances functional recovery following cervical spinal cord injury. *J Neurotrauma* 2015;**32**(3):185—93.

54. Rooney GE, McMahon SS, Ritter T, et al. Neurotrophic factor-expressing mesenchymal stem cells survive transplantation into the contused spinal cord without differentiating into neural cells. *Tiss Eng Part A* 2009;**15**(10):3049—59.

55. Ahuja CS, Mothe A, Khazaei M, et al. The leading edge: emerging neuroprotective and neuroregenerative cell-based therapies for spinal cord injury. *Stem Cells Transl Med* 2020;**9**(12):1509—30.

56. Karadimas SK, Satkunendrarajah K, Laliberte AM, et al. Sensory cortical control of movement. *Nat Neurosci* 2020;**23**(1):75—84.

57. Cregg JM, Leiras R, Montalant A, Wanken P, Wickersham IR, Kiehn O. Brainstem neurons that command mammalian locomotor asymmetries. *Nat Neurosci* 2020;**23**(6):730—40.

58. Satkunendrarajah K, Karadimas SK, Laliberte AM, Montandon G, Fehlings MG. Cervical excitatory neurons sustain breathing after spinal cord injury. *Nature* 2018;**562**(7727):419—22.

59. Cregg JM, Chu KA, Hager LE, et al. A latent propriospinal network can restore diaphragm function after high cervical spinal cord injury. *Cell Rep* 2017;**21**(3):654—65.

60. Collinger JL, Wodlinger B, Downey JE, et al. High-performance neuroprosthetic control by an individual with tetraplegia. *Lancet* 2013;**381**(9866):557—64.

61. King CE, Wang PT, McCrimmon CM, Chou CC, Do AH, Nenadic Z. Brain-computer interface driven functional electrical stimulation system for overground walking in spinal cord injury participant. *Conf Proc Annu Int Conf IEEE Eng Med Biol Soc IEEE Eng Med Biol Soc Annu Conf* 2014:1238—42. 2014.

62. Capogrosso M, Milekovic T, Borton D, et al. A brain-spine interface alleviating gait deficits after spinal cord injury in primates. *Nature* 2016;**539**(7628):284—8.

63. Rupp R. Challenges in clinical applications of brain computer interfaces in individuals with spinal cord injury. *Front Neuroeng* 2014;**7**:38.

Translational research in spinal cord injury — What is in the future?

Nader Hejrati[1], William Brett McIntyre[6], Katarzyna Pieczonka[6], Sophie Ostmeier[1], Christopher S. Ahuja[5], Brian K. Kwon[4], Alexander R. Vaccaro[2], F. Cumhur Oner[3], Michael G. Fehlings[5]

[1]Division of Genetics and Development, Krembil Brain Institute, University Health Network, Toronto, ON, Canada; [2]Department of Orthopaedic Surgery, Rothman Orthopaedic Institute, Thomas Jefferson University, Philadelphia, PA, United States; [3]Department of Orthopaedics, University Medical Center Utrecht, Utrecht, the Netherlands; [4]International Collaboration on Repair Discoveries, University of British Columbia, Vancouver, BC, Canada; [5]Division of Neurosurgery, Department of Surgery, University of Toronto, Toronto, ON, Canada; [6]Institute of Medical Science, University of Toronto, Toronto, ON, Canada

Introduction

Survival rates after spinal cord injury (SCI) have significantly improved with advanced trauma protocols, specialized hospitals, and modern ICU care.[1] As patients survived longer, new strategies to improve long-term functional outcomes emerged. Pivotal basic science studies during the last three decades have demonstrated the importance of the secondary injury cascade,[2] which facilitated the translation of successful neuroprotective interventions such as early surgical decompression,[3] blood pressure augmentation, and the use of steroids in cervical SCI.[4] Enhancing outcomes further will require new classes of treatment in both the early and late injury period. These

discoveries will require a deep understanding of the pathophysiologic mechanisms underlying SCI, including causes of heterogeneity in outcomes among patients.[5] With varied neuroprotective and neuroregenerative therapeutics on the rise, future clinical trial designs and translational research strategies will need to consider several key factors: (1) the epidemiologic shift toward a geriatric patient population, (2) the development of outcome assessment tools capable of capturing meaningful clinical improvements, (3) identification of genetic factors and biomarkers capable of predicting neurologic outcomes and personalizing therapeutic strategies, (4) investigation of the utility of real-time patient monitoring techniques, and (5) the development of standards that define *optimal*

Neural Repair and Regeneration after Spinal Cord Injury and Spine Trauma
https://doi.org/10.1016/B978-0-12-819835-3.00013-7

587

© 2022 Elsevier Inc. All rights reserved.

surgical decompression. This chapter provides an overview of the key concepts that need future consideration when translating from preclinical models to clinical trials and discusses promising future directions in translational SCI research, including neuroprotective, neuroregenerative, and neuromodulative methods. Finally, we will discuss the potential role of artificial intelligence in the treatment of SCIs.

Epidemiologic changes

SCI prevalence was historically linked to a younger population, where the majority of injuries were reported from sport-related trauma or violence.[6,7] As the world's population progressively ages, the prevalence of SCI has begun to shift toward an older, geriatric population. Regardless of the socioeconomic class, falls appear to be the main contemporary cause of SCI,[8] which more prominently affects the older population.[9] Moreover, the increased prevalence of degenerative changes at the cervical spine puts the older generation at particular risk of incomplete SCI, even after minor trauma. These epidemiologic trends are also reflected in increased mean ages of study populations over time,[10–12] which is prominent in both sexes.[13] The shifts in SCI patient demographics may impact how we evaluate relative risk of injury for an older population, as well as how we test for new therapeutics in preclinical studies.

Implications for preclinical investigations

The extent of trauma, as well as the level of injury in recent epidemiological studies, should also be reflected in current preclinical modeling of SCI. Within the clinical population, the most common level of injury occurs at the cervical spine ($\sim$60%).[1] However, trends in preclinical injury modeling do not reflect what we see in the clinic. This may be because modeling cervical SCI is particularly cumbersome due to high mortality rates from respiratory complications in animal models. Corroborating this, a recent (2017)

systematic review evaluating 2209 animal SCI studies revealed that the majority (81%) of injury models are at the thoracic level.[14] Unfortunately, these studies cannot be translated between thoracic and cervical contexts, as there is a myriad of variable motor,[15] autonomic,[16] and anatomical outcomes between models. The extent of injury also requires additional allocation of incomplete models within a preclinical context. Notably, there are slightly more reported complete transection models (35%; 2017) as compared to incomplete transection models (32%; 2017) in preclinical studies.[14] This inaccurately portrays what is observed in the clinic, as the most common form of injury is an incomplete injury.[17] The most troubling aspect of these studies is the amount of preclinical studies that do not report injury extent (33%; 2017),[14] which reflects a lack of consideration regarding clinical translation. With the continuous prevalence of incomplete cervical SCI in the clinic, there is a need for more experimental studies evaluating pathophysiology and potential therapeutics using a more representative preclinical population.

Implications for clinical practice

The prevalence of SCI in the elderly is particularly troubling, as previous economic trends have stressed the allocation of educational and therapeutic resources toward a younger demographic. Recognizing these updated trends can both educate health-care professionals regarding appropriate risk assessment, as well as help accommodate treatment interventions for a more representative postinjury population. As such, current systematic practices can be improved for older adults at a higher risk for falls and subsequent trauma. This may involve quick assessments regarding history of falls, or any issues regarding balance and mobility during routine assessments with health-care professionals. Depending on the level of risk determined by the health-care provider, modifiable risk factors can then be identified and adjusted to help prevent possible injuries.[18] Modifiable factors that may increase a patient's risk of injury at an older age are commonly referred to as reduced physical mobility, cognitive function, as well as risk of adverse events associated with medication use. In order to accurately identify and accommodate these factors, the frequency of screening should be increased.[18] Ultimately, demographic changes in SCI patient age and etiology demand a shift in prevention measures to reduce the occurrence of traumatic injury in older individuals.

In the context of prevention, the next steps in further improving SCI care depend on geography. In the developing world, injury rates remain staggeringly high and improved prevention efforts would have a significant impact in these areas.[19]

Clinical considerations

The clinical management of traumatic SCI has experienced an intense evolution over the past decades, leading to measurable improvements in patient outcomes. One such example is the timing of surgical decompression, where recent efforts culminated in the finding that early surgical decompression within 24 hours of injury improves neurologic function.[20] And yet, other aspects of the clinical management require further consideration and may potentially benefit from additional optimization. Optimizing patient management by providing evidence-based guidelines will eventually facilitate the measurability of translationally relevant neuroprotective and/or neuroregenerative strategies.

Clinical outcome prediction

Predicting functional outcomes of patients with SCIs has garnered increasing significance as outcome prediction tools not only provide valuable prognostic information to patients and their families, but also guide patient management in the setting of future emerging therapies and limited resources. The initial severity of

SCIs, as indicated by the AIS grading scale, has been found to be one such prognosticator, which correlates with the rate of neurologic recovery. However, despite being one of the most widely used tools to predict functional recovery, it fails to account for the heterogeneous SCI patient population. Of particular note are patients with motor and sensory complete SCIs — the group with the highest functional compromise and therefore the greatest potential for neurologic improvement. Historical data have ascribed patients suffering from AIS A SCIs a very limited neurologic recovery, which has been typically quoted at 15%–20%. However, recent studies have shown higher-than-expected recovery rates of up to 28.1%.[21] Notably, a prospective cohort study termed Surgical Timing in Acute Spinal Cord Injury Study (STASCIS) reported that 37% and 40% of patients with a preoperative AIS grade A experienced at least a 1 grade improvement, depending on whether they were operated early (<24 hours) or late (≥24 hours), respectively.[22] These findings suggest a previously unrecognized potential for neurologic recovery, which may have been unveiled by recent advancements in SCI management.

Moreover, while AIS conversion rates are being widely used in clinical trials as surrogate markers to assess the efficacy of novel investigative therapies, minimal clinically important differences (MCIDs) remain elusive.[23,24] For example, regaining sensory function, which indicates a conversion from ASIA A to B, might not constitute a MCID, whereas regaining hand function, as indicated by an improvement of the neurologic level of injury, is clearly of benefit for the patient's quality of life.[25]

Clinical trials in SCI research are usually cost and labor intensive. Therefore, positive efficacy results are warranted at an early stage for clinical trials to be carried forward. To gauge the effectiveness of novel therapies and avoid the risk of premature abortion of cost-intensive clinical trials due to undetected therapeutic response, it is therefore essential to develop assessment/prediction tools that further stratify patients with the greatest potential benefit from novel therapeutic strategies.

Hemodynamic monitoring

As discussed earlier, ischemia due to compromised spinal cord perfusion has been identified as one of the major components of the secondary injury cascade. Supporting the MAP and avoidance of systemic hypotension has therefore become a central dogma in the early phases of SCI treatment. Yet, while the currently valid guidelines regarding blood pressure management were developed in 2002,[26] a paucity of new data did not result in a change to these recommendations in the 2013 guidelines update.[27] These class III recommendation guidelines, stating that systemic hypotension should be avoided by maintaining a MAP of 85—90 during the first 5—7 days after SCI, have given rise to several concerns: First, depending on several patient-related factors, such as age and comorbidities, or concomitant injuries, the side effects of using intravenous (IV) fluids or vasopressors to maintain the MAP might outweigh their benefits of preserving spinal cord perfusion; second, the application of vasopressors or IV fluids and the invasive monitoring setting hinder early mobilization, thereby increasing the risk for immobilization-associated complications; and third, the current body of evidence is sparse, as to whether patients with different ASIA scores or AIS grades necessitate the same degree of monitoring and blood pressure management.[110]

In analogy to the field of traumatic brain injuries, invasive technologies, such as intrathecal pressure monitors and epidural near-infrared spectroscopy (NIRS) sensors,[28] hold the potential to provide us with real-time information that can help us tailor patient-specific therapies. Recent studies have demonstrated the benefits of adhering to spinal cord perfusion pressure (SCPP) target ranges rather than solely focusing on MAP targets.[29] This can be achieved

by the placement of intrathecal cerebrospinal fluid (CSF) catheters. Of important note, CSF drainage, which can be accomplished through the same catheter, reduces intrathecal CSF pressure thereby indirectly reducing the need for artificial blood pressure augmentation in order to meet SCPP targets.[30] Eventually, intrathecal pressure monitors or NIRS sensors will need further evaluation to assess their utility and risks during everyday clinical practice. Future trials are required to provide customized evidence-based guidelines for the hemodynamic management of patients with acute SCIs thereby accounting for the heterogeneity of patient presentation and the complex, varied clinical features.

Surgical considerations

The importance of early surgical decompression to relieve pressure on the spinal cord vasculature, restore spinal cord blood flow as well as mitigate the harmful biochemical effects of the secondary injury cascade, has been underscored with a recent publication.[20] However, several questions related to the surgical procedure itself currently remain poorly understood: First, what is considered adequate surgical decompression? And second, what is the role of duroplasty?

Surgical decompression constitutes a heterogeneous definition encompassing a variety of surgical techniques, all of which aim to relieve pressure from the spinal cord. Importantly, in accordance with the abovementioned pathophysiological process, the extent of spinal cord decompression has been shown to correlate with AIS grade conversions at 6-month follow-up.[31] And yet, the current body of evidence does not provide precise guidance regarding surgical management. A recent study of 184 motor complete SCI patients (AIS A = 119, AIS B = 65) demonstrated that the addition of a laminectomy to an anterior cervical discectomy and fusion (ACDF) or an anterior cervical corpectomy and fusion (ACCF) increased the rate of decompression, as defined by the presence of a patent subarachnoid space around a swollen spinal cord, from 46.8% (ACDF) and 58.6% (ACCF) to 72% and 73.1%, respectively.[32] However, the importance of a patent subarachnoid space and its correlation to the intrathecal pressure as well as spinal cord perfusion warrant further corroboration. Hence, the uncertainty about what is considered adequate surgical decompression and if a supplemental or stand-alone laminectomy might be necessary to adequately relieve the pressure from the spinal cord require further investigation.

A concept that closely refers to the pathophysiological principles of diffuse traumatic brain injury is the role of duroplasty in spinal cord edema due to traumatic SCI. The hypothesis of the Monro-Kellie Doctrine states that the volumes of brain, CSF, and intracranial blood are constant. Therefore, an increase in one compartment will eventually result in a decrease in one or both of the remaining compartments.[33] This concept has been suggested to be similar in the setting of traumatic SCI, where an increase in intraspinal pressure will eventually result in reduced SCPP, as well as changes in the vascular pressure reactivity index (sPRx) and therefore spinal cord autoregulation.[32,34] This has resulted in ongoing debates as to whether expansion of the limiting compartment, i.e., the dura, by means of an expansile duroplasty is necessary to enhance spinal cord perfusion. In a series of 21 patients, Phang et al. have shown that performing a laminectomy and duroplasty was significantly associated with improvements to intraspinal pressures, SCPPs, and sPRxs as compared to laminectomy alone.[34] These results support the hypothesis that laminectomy alone might not be sufficient in certain SCI patient populations to enhance spinal cord perfusion. Further trials are clearly needed to shed light on the concept of duroplasty in traumatic SCI. As such, a Phase III multicenter randomized

clinical trial termed DISCUS (NIHR130048) is currently underway investigating the role of duroplasty in patients with cervical SCIs.[35] Estimated study completion is December 2026.

Personalized approaches

Genetic variability

Despite advances in SCI treatment and care, patients with similar pathologies demonstrate heterogenous extents of recovery, thus demonstrating an increasing need to consider patient variability. Factors such as genetic background vary across patient populations and may have a significant impact on clinical outcomes. A handful of gene variants have been identified as factors that may make patients more vulnerable to various complications or reduce the response to certain treatment strategies (Table 31.1). In this regard, genetic screening may become an effective prediction tool that can help determine which patients are suitable candidates for specific treatments.

TABLE 31.1 Overview of gene variants that may influence patients' susceptibility to complications and their treatment efficacy.

Gene variant	Potential implications
CHRFAM7A[36,37]	• Inflammatory complications • Neuropathic pain • Pressure ulcers
TRPA1[38]	• Neuropathic pain
HLA-B27[39]	• Heterotopic ossification
ApoE4[40,41]	• Reduced motor recovery following rehabilitation
BDNF[42]	• Not evaluated in SCI; may reduce recovery following spinal stimulation
Prothrombin[43]	• Increased risk to developing deep vein thrombosis

Genetic variability and susceptibility to SCI complications

The *CHRFAM7A* gene has been found to be one such relevant gene. It encodes the alpha 7 nicotinic acetylcholine receptor that plays an anti-inflammatory role by reducing tumor necrosis factor alpha (TNFα). However, a subpopulation of individuals have a functional polymorphism in the gene, which prevents these antiinflammatory effects. Importantly, individuals that have this polymorphism tend to demonstrate higher inflammatory responses following SCI, as marked by higher levels of circulating inflammatory mediators compared to individuals that do not carry this polymorphism. Functionally, these inflammatory responses translate to an increased risk of developing neuropathic pain and pressure ulcers.[36,37] As such, this genetic variant may provide useful prognostic information that can help allocate this subpopulation of patients to specific treatment plans that tackle inflammation-mediated neuropathic pain. Another relevant gene that has been identified is the *TRPA1* gene, which encodes a nonselective cation channel that is involved in pain processes. Notably, a single-nucleotide polymorphism in this gene has been found to make SCI patients more susceptible to developing neuropathic pain.[38] The gene encoding the histocompatibility antigen HLA-B27 is another genetic risk factor that may be useful in allocating patients to specific therapeutic interventions: SCI patients with the HLA-B27 antigen have been found to be more vulnerable to developing heterotopic ossification.[39] Therefore, early HLA-B27 screening after SCI could help allocate these individuals to treatment plans that can prevent the onset of these complications. Future studies are warranted in order to further elucidate the association between genetic variability and susceptibility to poorer outcomes in SCI patients.

Genetic variability and treatment efficacy

In addition to serving as predictors of SCI complications, genetic factors can also help predict responses to treatments and to determine which patients may not benefit from specific types of treatments. Of note, the apolipoprotein E gene has demonstrated its significance as an important genetic factor in a number of diseases and injuries, including SCI. It is a polymorphic gene that encodes a protein which is involved in cholesterol trafficking in the central nervous system (CNS).[44] Importantly, the apolipoprotein E4 (ApoE4) gene variant has been associated with reduced motor recovery in SCI patients during rehabilitation, as well as an increased length of stay in rehabilitation.[40,41] Transgenic mouse models that express the human ApoE4 gene have demonstrated reduced extents of spontaneous and treatment-induced neuroplasticity following injury,[45,46] which may explain the resistance to rehabilitative interventions. In a clinical setting, this may suggest that SCI patients carrying the ApoE4 variant may require more intensive rehabilitation regimens in an effort to harness greater neuroplasticity and therapeutic effects. Neuroplasticity is also related to the *BDNF* gene. A variant in this gene results in a Val66Met substitution, which causes healthy individuals with this genotype to experience less transcutaneous spinal direct current stimulation-induced neuroplasticity.[42] Although it has not been tested in the context of traumatic SCI, future studies investigating whether this polymorphism affects responses to different spinal cord stimulation modalities are warranted. Treatment responses in venous thromboembolism therapy may also be related to genetic predispositions. Patients carrying a G20210A mutation in the prothrombin gene have been found to have a greater risk of being resistant to heparin treatment, ultimately making them more prone to developing thrombosis post-SCI in cases where heparin is administered as a prophylactic treatment.[43] This becomes all the more important, as patients suffering from traumatic SCI are particularly prone to immobilization-associated complications, including deep vein thrombosis and pulmonary embolism. Therefore, clinicians may install alternative preventive measures in order to avoid these complications in patients carrying this mutation.

Further investigation of genetic variants that are associated with specific treatment outcomes is warranted, with the ultimate goal of establishing genetic screening algorithms that may allow clinicians to tailor personalized treatments that target specific complications and responsiveness to therapeutics.

Future translational perspectives for neuroprotective therapies

Neuroprotection in SCI aims to mitigate the harmful effects of the secondary injury cascade, which can cause protracted neuroglial cell death and aggravate neurological deficits and outcomes.[47,48] Considering severely injured SCI patients (such as motor complete SCIs), we need to be cognizant that the vast majority of axonal damage, loss of resident neuroglial cells, impairment of the vascular supply, and disruption of networks required for sensorimotor function are caused by the primary mechanical trauma.[1] Hence, excessive expectations for the role of neuroprotection in providing meaningful clinical improvements in these patient populations need to be dampened.

Instead, patients with incomplete SCIs, and therefore more salvageable neurologic function, need a different approach: more weight needs to be added to the identification of patients with the greatest potential to benefit from the varied neuroprotective therapeutic strategies currently under investigation (Table 31.2).

Pharmacotherapies

The secondary injury cascade encompasses consequences of blood—brain barrier disruption, aberrant pro-inflammatory responses, free radical

TABLE 31.2 Overview of prospective therapeutic strategies for SCI.

Therapeutic	Mechanism
Neuroprotective strategies	
Minocycline	Antiinflammatory
G-CSF	Antiinflammatory
IVIG	Antiinflammatory
Riluzole	Prevents excitotoxicity (sodium modulation)
Gacyclidine	Prevents excitotoxicity (calcium modulation)
Neuroregenerative strategies	
Chondroitinase ABC (ChABC)	Promotes neuroplasticity, removal of physical barriers to axonal outgrowth
NOGO A inhibition	Reduces inhibitory environment via inhibition of myelin damage—associated proteins, promotes neurite outgrowth
Rho/ROCK inhibition	Modulates inhibitory environment, promotes neurite growth
RGMa inhibition	Modulates inhibitory environment, promotes neurite growth
Neural stem cells	Promotes neuroplasticity, trophic support, neural cell replacement
In vivo Reprogramming	Trophic support, neural cell replacement
Neuromodulation	Promotes plasticity and neurite outgrowth

production, and eventually programmed neural cell death. As such, the multifaceted pathophysiology of SCI requires an equally versatile therapeutic response.

The excessive pro-inflammatory response in SCI has been successfully neutralized in experimental studies using immunomodulatory agents such as minocycline, granulocyte colony-stimulating factor (G-CSF), and intravenous immunoglobulin G (IVIG).[49–51] Of particular interest, these agents are regulatory-body approved, thereby providing a strong starting position for their use in SCI. Although regulated by different therapeutic mechanisms, these pharmacotherapies can broadly suppress synthesis of pro-inflammatory cytokines in reactive glia, such as INF-γ[51] and TNFα.[49,52]

Apart from reactive pro-inflammation, excessive proapoptotic factors released from the primary physical insult are another pathophysiological consequence of injury that can be targeted by regulating excitotoxicity. Excessive glutamate release and receptor activation can be mediated by blocking either sodium or calcium influx with riluzole or gacyclidine, respectively.[53,54] As a result, regulated glutamate release can prevent the formation of excessive free radical production, mitochondrial dysfunction, and subsequently neural cell death.

As several different facets can potentially exacerbate the physical injuries to the spinal cord, it is important to investigate various pathophysiological targets concurrently that can mitigate the extent of the secondary injury. Ideally, either one or several protective agents could synergistically optimize therapeutic efficacy. Their application in the clinic is of particular interest, as limiting the extent of injury enhances the cellular environment that can be ultimately targeted by regenerative strategies.

The identification of appropriate biomarkers may help clinicians in stratifying which patients would benefit from certain types of neuroprotective treatments by allowing them to choose therapeutic strategies that align with the mechanisms and pathophysiological processes associated with the biomarkers seen in these individuals. As previously mentioned, Riluzole blocks presynaptic voltage-driven sodium channels and glutamatergic transmission, ultimately making it effective in attenuating excitotoxicity.[55] Therefore, early screening for excitotoxicity biomarkers may help to discriminate between patients that would benefit from Riluzole. Although no distinct excitotoxicity biomarkers exist to date, IGF-1 and S100β have both been suggested as potential candidates which may warrant further investigation.[56,57] Similarly, inflammatory biomarkers may allow for the identification of patients with a greater risk of developing more excessive secondary, inflammation-mediated injuries. For example, patients with higher levels of IL-1 and IL-6 during screening may benefit from minocycline, which has been shown to exert neuroprotective effects in part by suppressing the microglial synthesis of these two cytokines,[49] or IVIG, which has also been shown to reduce the levels of these pro-inflammatory cytokines.[58] However, it is important to acknowledge that biomarker discovery is difficult given that many neurochemical biomarkers are inherently known to fluctuate. Nonetheless, future investigations may contribute to the personalization of neuroprotective treatment administration.

Future translational perspectives for neuroregenerative therapies

Overcoming the inhibitory microenvironment

Failure of spontaneous regeneration after SCI may be attributable to a loss of multipotent progenitor populations, ongoing neuroglial degeneration, and the inhibitory properties of the postinjury microenvironment.[1] Therefore, enhancing neuronal regeneration of both endogenous and grafted cell populations will likely benefit from combinatorial approaches in which the harsh microenvironment is addressed as well. As such, promising results have been observed when counterbalancing the inhibitory effects of Nogo-A, a protein found in central nervous system myelin that potently inhibits neurite outgrowth in the acute post-SCI microenvironment,[59,60] while in preclinical models of chronic SCIs, unblocking the chondroitin sulfate proteoglycan barrier, a major inhibitory component of the astroglial scar, has been shown to improve neurobehavioral outcomes.[61] Finally, targeting the respective downstream Rho/Rho-associated kinase (ROCK) signaling pathway of Nogo-A and CSPGs evolved to another area of increasing interest that will warrant further attention.[62,63]

Another promising target of the inhibitory microenvironment is the glycoprotein repulsive guidance molecule A (RGMa), which has been found to be significantly upregulated by neurons, astrocytes, oligodendrocytes, as well as macrophages and microglia around the SCI lesion site, where it acts as a strong inhibitor of neurite outgrowth.[64,65] Administration of RGMa antibodies has demonstrated promising results in preclinical studies, where it has enhanced neuronal survival, plasticity, and corticospinal tract regeneration (Table 31.2).[65–67] These exciting results have paved the way for a clinical trial, which is currently recruiting patients.

Optimizing neural stem/progenitor cell−based therapies

Neural stem/progenitor cells (NSCs) are a well-suited therapeutic for SCI because they can replace the diverse cellular niche that is present in the healthy spinal cord. Notably, exogenous NSC therapies can replace metabolically supportive and homeostatic astrocytes, myelinating oligodendrocytes, and motor/sensory-promoting neuronal functions. However, it is likely that traditional NSCs are not optimally suited to treat all contexts of SCI, without modifications. As such, research efforts, derived from neurodevelopmental concepts, present solutions that can uniquely regenerate a variety of SCI contexts.

Promoting optimal graft: host integration

Quite notably, the functional recovery of experimental rodent SCI is dependent on the integration of grafted neuronal cells with spared host circuitry.[68] However, the extent of integration in contemporary research efforts to date has been modest if the cell source does not match the region to which the graft was transplanted.[69–71] Of note, the use of fetal-derived spinal cord tissue can overcome this barrier by supporting a greater extent of functional excitatory synapses, and in turn functional motor recovery through corticospinal tract regeneration.[70] By approaching this concept from a neurodevelopmental perspective, the concept of cellular identity has been applied to develop identity-specific induced pluripotent stem cell (iPSC) therapies that are uniquely suited to regenerate the injured cord.[72] These caudal, or spinal-identifying, iPSC grafts were able to promote functional recovery, whereas rostralized, or forebrain identifying iPSC therapies were unable to do so. This presents a promising strategy to personalize stem cell therapies for SCI. For example, in certain SCI cases, a gradient of caudalized cell identity could be utilized to target level specific injuries. Gradients of neurodevelopmental ques, involving retinoic acid, fibroblast growth factor, and bone morphogenic protein (BMP)-signaling, can work harmoniously to promote cell grafts with cervical, thoracic, and lumbar identity in the developing spinal cord.[73] Fine-tuning the concept of regionalized cell grafts could in turn represent a promising regenerative strategy to combat level-specific SCIs.

Neuronal subtypes and task-specific recovery

Disruption of neuronal circuits in SCI impairs a variety of motor, sensory, nociceptive, and autonomic functions. Of importance, in vitro modeling of developmental morphogen gradients can be spatiotemporally coordinated to produce specific progenitor cells that are primed to replace cells along the rostral—caudal and dorsal—ventral axes. Therefore, matching appropriate cells to the respective transplant environment can specifically address compromised neuronal circuits.[73] In more detail, these extracellular ques dictate intrinsic cellular properties that define morphology, migration, as well as neurotransmitter activity in their terminally differentiated state. This developmental knowledge can be applied to the generation of motor neurons,[74] GABAergic,[75] glutamatergic,[76] and cholinergic progenitors.[77] These primed or "biased" cells can be applied to patients exhibiting function-specific deficits. Differences in clinical presentation as seen in incomplete SCI syndromes, such as central cord, anterior, or posterior cord syndromes, may benefit from additional corroboration of the neuroanatomical location of injury as evaluated through axial magnetic resonance imaging techniques.[78] In combination, information retrieved from meticulous clinical assessments and imaging studies can aid in guiding patient treatment. As an example of applying biased cell types as a personalized therapy, central cord syndrome presenting patients may benefit the most from a motor neuron progenitor-biased replacement therapy. Likewise, patients presenting with a posterior cord injury may benefit from V2a interneuron cell replacement to regulate proprioception and fine motor control.[79]

Patient-specific benefits of remyelination

Demyelination is a rapid onset pathophysiological process that extends well into the chronic phase of SCI.[80] Although it is controversial whether endogenous remyelination is essential to promote functional recovery,[81] promoting additive myelination through exogenous stem cell support exhibits significant therapeutic benefits.[82,83] Improved functional recovery through increased remyelination can be accomplished by promoting an "oligodendrogenic" cell fate in transplanted cells, which also can improve tissue sparing compared to conventional NSC sources. This additive therapeutic mechanism can both maintain the integrity of endogenous and newly transplanted exogenous circuitry, as well as attenuate the extent of axonal loss in harsh injury microenvironments. By modifying neurodevelopmental morphogen signaling in vitro, particularly through noggin, Shh, and the inhibition of BMP signaling,[83,84] exogenous cell therapies can be optimized to target patients that present with a greater extent of demyelination. Therefore, future investigational efforts should aim to identify either (1) biomarkers that indicate increased rates of demyelination via myelin damage—associated molecules (e.g., myelin-associated glycoprotein (MAG), Nogo-A, etc.[85]) or (2) the quantity or extent of demyelination using reproducible myelin MRI methods.[86] Thus, either screening method could identify patients with the greatest potential to benefit from oligodendrogenically biased NSC treatments.

In vivo direct reprogramming

As the injury progresses, the maladaptive reactive astroglial response provides a population of cells which can theoretically be used for direct reprogramming into adaptive cells that can promote neuroregeneration. *In vivo* direct reprogramming strategies utilize viral delivery of genetic constructs in order to drive the conversion of one cell type into another. Importantly, several reports demonstrate the in vivo conversion of either GFAP or NG^{2+} glial cells into neurons through the delivery of genes such as *Sox2*, *Asc11*, or *NeuroD1* following CNS injuries,[87–90] including SCI.[90] These efforts may have synergistic benefits by both reducing the maladaptive effects of reactive astrogliosis and producing new neurons that can help facilitate neuroregeneration. Nonetheless, these strategies undoubtedly present several hurdles that would need to be optimized in order to translate these findings to the clinic, such as the safety of the viral delivery and the specificity of the promoters used in order to ensure optimal targeting of the genes to the glial cells of interest. Moreover, an important consideration is the timing of the treatment, which relates to the phase of injury. This treatment strategy will presumably target chronic SCI patients, rather than patients during their acute phase of SCI. This is due to the fact that the glial response is an adaptive process that allows for the formation of a glial scar which restricts the damage from spreading during the earlier stages of injury.[91] In this regard, the conversion of adaptive glial cells during the acute phase of injury would likely be counterproductive, whereas it would be more appropriate in chronic SCI, at which point the glial scar becomes an impediment to regeneration.

Future perspectives in neuromodulation

With further improvements in outcomes, the importance of out-of-hospital care is now being emphasized. Intensive inpatient rehabilitation programs are becoming standard of care with many adjuvant therapies coming to the forefront such as electrical stimulation, exoskeleton-based locomotion retraining, and programs aimed at teaching skills for independence in the community.[92]

The innate regeneration in the injured cord is limited due to the inability of supraspinal axons to reform circuitry with neurons caudal to the lesion.[93] To bridge these severed circuits, spared neuronal networks must induce plasticity and sprout past the lesion site. However, a myriad of obstacles, including the physical glial scar barrier, as well as excessive axonal dieback due to prolonged cytotoxic signaling, reduces innate plasticity in the injured cord.

Epidural spinal cord stimulation

To promote neuroplasticity, a promising therapeutic option is neuromodulation via epidural electrical stimulation (EES). EES can be applied either to supraspinal or infraspinal networks, where either application promotes plasticity and neurite outgrowth on the opposing lesion side. Notably, EES in combination with intense rehabilitation has been shown to hold the

potential to restore locomotion.[94,95] However, the most notable difference compared to current treatment methodologies is that the chronic phase of complete injuries can be targeted to improve previously paralyzed muscles.[81,94−98] Further technical refinements of spatiotemporally well-defined stimulations through EES have been shown to reestablish adaptive control of paralyzed muscles during overground walking as early as 1 week after beginning stimulation.[96] While there has been a growing body of literature in a small series of patients demonstrating exciting results of EES for SCI, future trials with larger cohorts are needed to establish efficacy.

Brain—machine interfaces

Corticospinal tract activation in conjunction with somatosensory feedback is crucial for recovering optimal functional control. However, the lesion site physically impedes the circuitry between the brain and spinal cord. Bypassing the disrupted communication between the brain and spinal circuits can be achieved by so-called brain—machine interfaces (BMIs). Hereby, electro-cortical activities, often related to "motor intentions," are recorded, decoded, and eventually translated into functional outputs.[99] These functional outputs encompass movement-eliciting peripheral muscle stimulations,[100] epidural spinal cord stimulations,[101] or extracorporal devices, such as exoskeletons.[102] Preclinical studies in rats and rhesus monkeys using brain—spine interfaces have shown restoration of weight-bearing locomotion.[101,103] Notably, in a small clinical series of eight patients with chronic complete paraplegia, long-term training with a BMI combined with electroencephalography-controlled exoskeleton devices improved somatic sensation and voluntary motor control. This finding suggests that BMIs combined with training potentially exceed the provision of assistive functions and could enhance neurologic recovery by triggering brain and spinal cord neuroplasticity.[102] Challenges necessitating further attention when translating BMIs into everyday clinical practice include (1) the expertise of and accessibility to the specialized technology, (2) affordability of the cost-intensive equipment, (3) longevity and stability of implanted electrodes, (4) patient tethering to external computers complicating independent use, and (5) establishing optimal stimulation parameters for optimal outcomes.[99]

Exoskeleton

SCI patients relying on manual wheelchair propulsion as their primary means of locomotion encounter several challenges during everyday life. These may include architectural and environmental restrictions as well as insufficient eye-to-eye social interactions with able-bodied peers.[104] Exoskeletons are technical devices that enable overground walking for SCI patients, thereby overcoming the abovementioned hurdles.[104,105] In particular, exoskeletons hold the potential to improve rehabilitation programs by enhancing the performance of functional tasks, as the loss of strength and coordination considerably limit a patient's capacity for overground or treadmill ambulation training.[106]

Future perspectives related to artificial intelligence

We have previously mentioned how factors such as epidemiology, genetic factors, individualized biomarkers including inflammatory cytokines, and injury phase may influence individual patients' outcomes following SCI and their responses to different treatments. However, as large amounts of data are continuously being curated, it is becoming increasingly clear that a holistic approach that incorporates a multitude of these factors will be needed in order to determine the optimal standard of care for any individual patient. In this regard, artificial intelligence represents a relevant tool that is capable of tackling these enormous datasets. Indeed, it has been reported that a series of factors, such as age, can be used to predict walking ability following SCI by engaging statistical machine learning—based methods.[107] Therefore, these methods are capable of providing valuable prognostic information. As this field continues to progress, machine learning approaches can theoretically filter through large amounts of patient information in order to predict alternative outcomes such as patients' risks for specific complications, thereby allowing physicians to intervene earlier. In addition, machine learning strategies also involve the development of algorithms which are capable of assessing imaging data. These strategies can be used not only to automatically assess imaging data (thereby improving and accelerating diagnoses) but also for the development of novel prognostic biomarkers that may be related to functional outcomes. For example, the development of an automated machine learning—based segmentation tool for 3T MR images led to the finding that the lesion volume was an effective predictor of motor impairment.[108] In another study, an algorithm was created in order to identify regions of intraparenchymal hypointensity on T2-weight MR images from dogs. Importantly, these regions of hypointensity correlated with lower functional recovery post-SCI.[109] Therefore, this type of automated imaging analysis may present a useful predictive tool. Ultimately, further development of machine learning techniques will provide physicians with useful prognostic and diagnostic information in order to guide accurate decision-making in the clinic in the future.

Conclusion

Targeting the multifaceted and complex pathophysiology of SCI will likely necessitate novel approaches to enhance neural repair and regeneration. While cell-based treatments continue to be an attractive strategy, future treatments will most likely benefit from combinatorial approaches, where genetic engineering, biomaterials, and scar degrading enzymes meet to enhance clinical outcomes. Identification of markers to stratify patients with the greatest potential to benefit from combinatorial neuroprotective and/or neuroregenerative strategies will further aid in maximizing treatment outcomes.

References

1. Ahuja CS, et al. Traumatic spinal cord injury. *Nat Rev Dis Primers* 2017;**3**:1−21.
2. Tator CH, Fehlings MG. Review of the secondary injury theory of acute spinal cord trauma with emphasis on vascular mechanisms. *J Neurosurg* 1991;**75**:15−26.
3. Ahuja CS, Badhiwala JH, Fehlings MG. 'Time is spine': the importance of early intervention for traumatic spinal cord injury. *Spinal Cord* 2020;**58**:1037−9.
4. Fehlings MG, Ahuja CS, Mroz T, Hsu W, Harrop J. Future advances in spine surgery: the AOSpine North America perspective. *Neurosurgery* 2017;**80**:S1−8.
5. Ahuja CS, Schroeder GD, Vaccaro AR, Fehlings MG. Spinal cord injury-what are the controversies? *J Orthop Trauma* 2017;**31**(Suppl. 4):S7−13.
6. Bellucci CHS, et al. Contemporary trends in the epidemiology of traumatic spinal cord injury: changes in age and etiology. *NED* 2015;**44**:85−90.
7. Shin JC, Kim DH, Yu SJ, Yang HE, Yoon SY. Epidemiologic change of patients with spinal cord injury. *Ann Rehabil Med* 2013;**37**:50−6.
8. Kang Y, et al. Epidemiology of worldwide spinal cord injury: a literature review. *JN J Nephrol* 2017;**6**:1−9.
9. Chen Y, He Y, DeVivo MJ. Changing demographics and injury profile of new traumatic spinal cord injuries in the United States, 1972−2014. *Arch Phys Med Rehabil* 2016;**97**:1610−9.
10. Bárbara-Bataller E, Méndez-Suárez JL, Alemán-Sánchez C, Sánchez-Enríquez J, Sosa-Henríquez M. Change in the profile of traumatic spinal cord injury over 15 years in Spain. *Scand J Trauma Resuscitation Emerg Med* 2018;**26**:27.
11. Li H-L, et al. Epidemiology of traumatic spinal cord injury in Tianjin, China: an 18-year retrospective study of 735 cases. *J Spinal Cord Med* 2019;**42**:778−85.
12. Mitchell J, et al. Epidemiology of traumatic spinal cord injury in New Zealand (2007−2016). *N Z Med J* 2020;**133**:11.
13. Liu H, et al. The changing demographics of traumatic spinal cord injury in Beijing, China: a single-centre report of 2448 cases over 7 years. *Spinal Cord* 2021;**59**:298−305.
14. Sharif-Alhoseini M, et al. Animal models of spinal cord injury: a systematic review. *Spinal Cord* 2017;**55**:714−21.
15. Tracy LF, Kwak PE, Bayan SL, Van Stan JH, Burns JA. Vocal fold motion recovery in patients with iatrogenic unilateral immobility: cervical versus thoracic injury. *Ann Otol Rhinol Laryngol* 2019;**128**:44−9.
16. Lujan HL, DiCarlo SE. Direct comparison of cervical and high thoracic spinal cord injury reveals distinct autonomic and cardiovascular consequences. *J Appl Physiol* 2020;**128**:554−64.
17. Hamid R, et al. Epidemiology and pathophysiology of neurogenic bladder after spinal cord injury. *World J Urol* 2018;**36**:1517−27.
18. FALL_PREVENTION_WEB_1207-17.pdf.
19. Burns AS, O'Connell C. The challenge of spinal cord injury care in the developing world. *J Spinal Cord Med* 2012;**35**:3−8.
20. Badhiwala JH, et al. The influence of timing of surgical decompression for acute spinal cord injury: a pooled analysis of individual patient data. *Lancet Neurol* 2021;**20**:117−26.
21. El Tecle NE, et al. The natural history of complete spinal cord injury: a pooled analysis of 1162 patients and a meta-analysis of modern data. *J Neurosurg Spine* 2018;**28**:436−43.
22. Fehlings MG, et al. Early versus delayed decompression for traumatic cervical spinal cord injury: results of the Surgical Timing in Acute Spinal Cord Injury Study (STASCIS). *PLoS One* 2012;**7**:e32037.
23. van Middendorp JJ, Hosman AJF, Pouw MH, Van de Meent H. ASIA impairment scale conversion in traumatic SCI: is it related with the ability to walk? A descriptive comparison with functional ambulation outcome measures in 273 patients. *Spinal Cord* 2009;**47**:555−60.
24. Wu X, et al. Challenges for defining minimal clinically important difference (MCID) after spinal cord injury. *Spinal Cord* 2015;**53**:84−91.
25. Anderson KD. Targeting recovery: priorities of the spinal cord-injured population. *J Neurotrauma* 2004;**21**:1371−83.
26. Hadley MN, et al. Guidelines for the management of acute cervical spine and spinal cord injuries. *Clin Neurosurg* 2002;**49**:407−98.
27. Walters BC, et al. Guidelines for the management of acute cervical spine and spinal cord injuries: 2013 update. *Neurosurgery* 2013;**60**:82−91.
28. Shadgan B, et al. Optical assessment of spinal cord tissue oxygenation using a miniaturized near infrared spectroscopy sensor. *J Neurotrauma* 2019;**36**:3034−43.
29. Squair JW, et al. Empirical targets for acute hemodynamic management of individuals with spinal cord injury. *Neurology* 2019;**93**:e1205−11.

30. Martirosyan NL, et al. Cerebrospinal fluid drainage and induced hypertension improve spinal cord perfusion after acute spinal cord injury in pigs. *Neurosurgery* 2015;**76**:461—8. discussion 468—469.

31. Aarabi B, et al. Intramedullary lesion length on postoperative magnetic resonance imaging is a strong predictor of ASIA impairment scale grade conversion following decompressive surgery in cervical spinal cord injury. *Neurosurgery* 2017;**80**:610—20.

32. Aarabi B, et al. Extent of spinal cord decompression in motor complete (American spinal injury association impairment scale grades A and B) traumatic spinal cord injury patients: post-operative magnetic resonance imaging analysis of standard operative approaches. *J Neurotrauma* 2019;**36**:862—76.

33. Mokri B. The Monro—Kellie hypothesis: applications in CSF volume depletion. *Neurology* 2001;**56**:1746—8.

34. Phang I, et al. Expansion duroplasty improves intraspinal pressure, spinal cord perfusion pressure, and vascular pressure reactivity index in patients with traumatic spinal cord injury: injured spinal cord pressure evaluation study. *J Neurotrauma* 2015;**32**:865—74.

35. NIHR funding and awards search website. https://fundingawards.nihr.ac.uk/award/NIHR130048.

36. Huang W, et al. Association of a functional polymorphism in the CHRFAM7A gene with inflammatory response mediators and neuropathic pain after spinal cord injury. *J Neurotrauma* 2019;**36**:3026—33.

37. Lin M, et al. Effect of CHRFAM7A Δ2bp gene variant on secondary inflammation after spinal cord injury. *PLoS One* 2021;**16**:e0251110.

38. Vidal Rodriguez S, Castillo Aguilar I, Cuesta Villa L, Serrano Saenz de Tejada F. TRPA1 polymorphisms in chronic and complete spinal cord injury patients with neuropathic pain: a pilot study. *Spinal Cord Ser Cases* 2017;**3**:17089.

39. Larson JM, et al. Increased prevalence of HLA-B27 in patients with ectopic ossification following traumatic spinal cord injury. *Rheumatol Rehabil* 1981;**20**:193—7.

40. Jha A, et al. Apolipoprotein E epsilon4 allele and outcomes of traumatic spinal cord injury. *J Spinal Cord Med* 2008;**31**:171—6.

41. Sun C, Ji G, Liu Q, Yao M. Apolipoprotein E epsilon 4 allele and outcomes of traumatic spinal cord injury in a Chinese Han population. *Mol Biol Rep* 2011;**38**:4793—6.

42. Lamy J-C, Boakye M. BDNF Val66Met polymorphism alters spinal DC stimulation-induced plasticity in humans. *J Neurophysiol* 2013;**110**:109—16.

43. Rubin-Asher D, et al. Risk factors for failure of heparin thromboprophylaxis in patients with acute traumatic spinal cord injury. *Thromb Res* 2010;**125**:501—4.

44. Toro CA, Das DK, Cai D, Cardozo CP. Elucidating the role of apolipoprotein E isoforms in spinal cord injury-associated neuropathology. *J Neurotrauma* 2019;**36**:3317—22.

45. Toro CA, et al. The human ApoE4 variant reduces functional recovery and neuronal sprouting after incomplete spinal cord injury in male mice. *Front Cell Neurosci* 2021;**15**.

46. Strattan LE, et al. Novel influences of sex and APOE genotype on spinal plasticity and recovery of function after spinal cord injury. *eNeuro* 2021;**8**.

47. Norenberg MD, Smith J, Marcillo A. The pathology of human spinal cord injury: defining the problems. *J Neurotrauma* 2004;**21**:429—40.

48. Yip PK, Malaspina A. Spinal cord trauma and the molecular point of no return. *Mol Neurodegener* 2012;**7**:6.

49. Seabrook TJ, Jiang L, Maier M, Lemere CA. Minocycline affects microglia activation, Abeta deposition, and behavior in APP-tg mice. *Glia* 2006;**53**:776—82.

50. Brennan FH, et al. IVIg attenuates complement and improves spinal cord injury outcomes in mice. *Ann Clin Transl Neurol* 2016;**3**:495—511.

51. Malashchenko VV, et al. Direct anti-inflammatory effects of granulocyte colony-stimulating factor (G-CSF) on activation and functional properties of human T cell subpopulations in vitro. *Cell Immunol* 2018;**325**:23—32.

52. Chio JCT, et al. Delayed administration of high dose human immunoglobulin G enhances recovery after traumatic cervical spinal cord injury by modulation of neuroinflammation and protection of the blood spinal cord barrier. *Neurobiol Dis* 2021;**148**:105187.

53. Srinivas S, Wali AR, Pham MH. Efficacy of riluzole in the treatment of spinal cord injury: a systematic review of the literature. *Neurosurg Focus* 2019;**46**:E6.

54. Gerber YN, Privat A, Perrin FE. Gacyclidine improves the survival and reduces motor deficits in a mouse model of amyotrophic lateral sclerosis. *Front Cell Neurosci* 2013;**7**.

55. Nagoshi N, Nakashima H, Fehlings MG. Riluzole as a neuroprotective drug for spinal cord injury: from bench to bedside. *Molecules* 2015;**20**:7775—89.

56. Albayar AA, et al. Biomarkers in spinal cord injury: prognostic insights and future potentials. *Front Neurol* 2019;**10**.

57. Mazzone GL, Nistri A. S100β as an early biomarker of excitotoxic damage in spinal cord organotypic cultures. *J Neurochem* 2014;**130**:598—604.

58. Nguyen DH, et al. Immunoglobulin G (IgG) attenuates neuroinflammation and improves neurobehavioral recovery after cervical spinal cord injury. *J Neuroinflamm* 2012;**9**:224.

59. Bregman BS, et al. Recovery from spinal cord injury mediated by antibodies to neurite growth inhibitors. *Nature* 1995;**378**:498—501.

60. Freund P, et al. Anti-Nogo-A antibody treatment enhances sprouting of corticospinal axons rostral to a unilateral cervical spinal cord lesion in adult macaque monkey. *J Comp Neurol* 2007;**502**:644–59.

61. Bradbury EJ, et al. Chondroitinase ABC promotes functional recovery after spinal cord injury. *Nature* 2002;**416**:636–40.

62. Forgione N, Fehlings MG. Rho-ROCK inhibition in the treatment of spinal cord injury. *World Neurosurg* 2014;**82**:e535–9.

63. Lord-Fontaine S, et al. Local inhibition of Rho signaling by cell-permeable recombinant protein BA-210 prevents secondary damage and promotes functional recovery following acute spinal cord injury. *J Neurotrauma* 2008;**25**:1309–22.

64. Schwab JM, et al. Spinal cord injury-induced lesional expression of the repulsive guidance molecule (RGM). *Eur J Neurosci* 2005;**21**:1569–76.

65. Mothe AJ, et al. RGMa inhibition with human monoclonal antibodies promotes regeneration, plasticity and repair, and attenuates neuropathic pain after spinal cord injury. *Sci Rep* 2017;**7**.

66. Hata K, et al. RGMa inhibition promotes axonal growth and recovery after spinal cord injury. *J Cell Biol* 2006;**173**:47–58.

67. Mothe AJ, et al. Delayed administration of the human anti-RGMa monoclonal antibody elezanumab promotes functional recovery including spontaneous voiding after spinal cord injury in rats. *Neurobiol Dis* 2020;**143**:104995.

68. Ceto S, Sekiguchi KJ, Takashima Y, Nimmerjahn A, Tuszynski MH. Neural stem cell grafts form extensive synaptic networks that integrate with host circuits after spinal cord injury. *Cell Stem Cell* 2020;**27**. 430.e5–440.e5.

69. Dulin JN, et al. Injured adult motor and sensory axons regenerate into appropriate organotypic domains of neural progenitor grafts. *Nat Commun* 2018;**9**:84.

70. Kadoya K, et al. Spinal cord reconstitution with homologous neural grafts enables robust corticospinal regeneration. *Nat Med* 2016;**22**:479–87.

71. Dell'Anno MT, et al. Human neuroepithelial stem cell regional specificity enables spinal cord repair through a relay circuit. *Nat Commun* 2018;**9**:3419.

72. Kajikawa K, et al. Cell therapy for spinal cord injury by using human iPSC-derived region-specific neural progenitor cells. *Mol Brain* 2020;**13**:120.

73. Tao Y, Zhang S-C. Neural subtype specification from human pluripotent stem cells. *Cell Stem Cell* 2016;**19**:573–86.

74. Du Z-W, et al. Generation and expansion of highly pure motor neuron progenitors from human pluripotent stem cells. *Nat Commun* 2015;**6**:6626.

75. Abeysinghe HCS, et al. Pre-differentiation of human neural stem cells into GABAergic neurons prior to transplant results in greater repopulation of the damaged brain and accelerates functional recovery after transient ischemic stroke. *Stem Cell Res Ther* 2015;**6**:186.

76. Cao S-Y, et al. Enhanced derivation of human pluripotent stem cell-derived cortical glutamatergic neurons by a small molecule. *Sci Rep* 2017;**7**:3282.

77. Wu P, et al. Region-specific generation of cholinergic neurons from fetal human neural stem cells grafted in adult rat. *Nat Neurosci* 2002;**5**:1271–8.

78. Kunam VK, et al. Incomplete cord syndromes: clinical and imaging review. *Radiographics* 2018;**38**:1201–22.

79. Zholudeva LV, et al. Transplantation of neural progenitors and V2a interneurons after spinal cord injury. *J Neurotrauma* 2018;**35**:2883–903.

80. Totoiu MO, Keirstead HS. Spinal cord injury is accompanied by chronic progressive demyelination. *J Comp Neurol* 2005;**486**:373–83.

81. Duncan GJ, et al. Locomotor recovery following contusive spinal cord injury does not require oligodendrocyte remyelination. *Nat Commun* 2018;**9**:3066.

82. Nori S, et al. Human oligodendrogenic neural progenitor cells delivered with chondroitinase ABC facilitate functional repair of chronic spinal cord injury. *Stem Cell Rep* 2018;**11**:1433–48.

83. Nagoshi N, et al. Human spinal oligodendrogenic neural progenitor cells promote functional recovery after spinal cord injury by axonal remyelination and tissue sparing. *Stem Cells Transl Med* 2018;**7**:806–18.

84. Tang BL, Low CB. Genetic manipulation of neural stem cells for transplantation into the injured spinal cord. *Cell Mol Neurobiol* 2007;**27**:75–85.

85. Hulme CH, et al. The developing landscape of diagnostic and prognostic biomarkers for spinal cord injury in cerebrospinal fluid and blood. *Spinal Cord* 2017;**55**:114–25.

86. van der Weijden CWJ, et al. Myelin quantification with MRI: a systematic review of accuracy and reproducibility. *Neuroimage* 2021;**226**:117561.

87. Chen Y-C, et al. A NeuroD1 AAV-based gene therapy for functional brain repair after ischemic injury through in vivo astrocyte-to-neuron conversion. *Mol Ther* 2020;**28**:217–34.

88. Guo Z, et al. In vivo direct reprogramming of reactive glial cells into functional neurons after brain injury and in an Alzheimer's disease model. *Cell Stem Cell* 2014;**14**:188–202.

89. Heinrich C, et al. Sox2-mediated conversion of NG2 glia into induced neurons in the injured adult cerebral cortex. *Stem Cell Rep* 2014;**3**:1000–14.

90. Su Z, Niu W, Liu M-L, Zou Y, Zhang C-L. In vivo conversion of astrocytes to neurons in the injured adult spinal cord. *Nat Commun* 2014;**5**:3338.

91. Herrmann JE, et al. STAT3 is a critical regulator of astrogliosis and scar formation after spinal cord injury. *J Neurosci* 2008;**28**:7231—43.

92. Nori S, Ahuja CS, Fehlings MG. Translational advances in the management of acute spinal cord injury: what is new? What is hot? *Neurosurgery* 2017;**64**:119—28.

93. Eisdorfer JT, et al. Epidural electrical stimulation: a review of plasticity mechanisms that are hypothesized to underlie enhanced recovery from spinal cord injury with stimulation. *Front Mol Neurosci* 2020;**13**.

94. Angeli CA, et al. Recovery of over-ground walking after chronic motor complete spinal cord injury. *N Engl J Med* 2018;**379**(13):1244—50. https://doi.org/10.1056/NEJMoa1803588. Epub 2018 Sep 24. PMID: 30247091.

95. Gill ML, et al. Neuromodulation of lumbosacral spinal networks enables independent stepping after complete paraplegia. *Nat Med* 2018;**24**:1677—82.

96. Wagner FB, et al. Targeted neurotechnology restores walking in humans with spinal cord injury. *Nature* 2018;**563**:65—71.

97. Peña Pino I, et al. Long-term spinal cord stimulation after chronic complete spinal cord injury enables volitional movement in the absence of stimulation. *Front Syst Neurosci* 2020;**14**.

98. Rejc E, Angeli CA, Atkinson D, Harkema SJ. Motor recovery after activity-based training with spinal cord epidural stimulation in a chronic motor complete paraplegic. *Sci Rep* 2017;**7**:13476.

99. James ND, McMahon SB, Field-Fote EC, Bradbury EJ. Neuromodulation in the restoration of function after spinal cord injury. *Lancet Neurol* 2018;**17**:905—17.

100. Ajiboye AB, et al. Restoration of reaching and grasping movements through brain-controlled muscle stimulation in a person with tetraplegia: a proof-of-concept demonstration. *Lancet* 2017;**389**:1821—30.

101. Capogrosso M, et al. A brain—spinal interface alleviating gait deficits after spinal cord injury in primates. *Nature* 2016;**539**:284—8.

102. Donati ARC, et al. Long-term training with a brain-machine interface-based gait protocol induces partial neurological recovery in paraplegic patients. *Sci Rep* 2016;**6**:30383.

103. Bonizzato M, et al. Brain-controlled modulation of spinal circuits improves recovery from spinal cord injury. *Nat Commun* 2018;**9**:3015.

104. Esquenazi A, Talaty M, Packel A, Saulino M. The ReWalk powered exoskeleton to restore ambulatory function to individuals with thoracic-level motor-complete spinal cord injury. *Am J Phys Med Rehabil* 2012;**91**:911—21.

105. del-Ama AJ, et al. Review of hybrid exoskeletons to restore gait following spinal cord injury. *JRRD (J Rehabil Res Dev)* 2012;**49**:497.

106. Aach M, et al. Voluntary driven exoskeleton as a new tool for rehabilitation in chronic spinal cord injury: a pilot study. *Spine J* 2014;**14**:2847—53.

107. DeVries Z, et al. Development of an unsupervised machine learning algorithm for the prognostication of walking ability in spinal cord injury patients. *Spine J* 2020;**20**:213—24.

108. McCoy DB, et al. Convolutional neural network—based automated segmentation of the spinal cord and contusion injury: deep learning biomarker correlates of motor impairment in acute spinal cord injury. *AJNR Am J Neuroradiol* 2019;**40**:737—44.

109. Boudreau E, et al. Relationship between machine-learning image classification of T2-weighted intramedullary hypointensity on 3 Tesla magnetic resonance imaging and clinical outcome in dogs with severe spinal cord injury. *J Neurotrauma* 2021;**38**:725—33.

110. Hejrati N, Fehlings MG. Commentary on Hemodynamic Management of Acute Spinal Cord Injury. *Neurospine.* 2021;**18**(1):15—6. https://doi.org/10.14245/ns.2121077.075. Epub 2021 Mar 31. PMID: 33819931; PMCID: PMC8021840.

Index

Note: 'Page numbers followed by 'f ' indicate figures those followed by 't' indicate tables and 'b' indicate boxes.'

Printed and bound by CPI Group (UK) Ltd, Croydon, CR0 4YY

21/03/2025

01835501-0002